# 15th International Conference on Strangeness in Quark Matter (SQM 2015)



Journal of Physics: Conference Series Volume 668

Dubna, Russia
6 – 11 July 2015

**Editors:**

**David Alvarez-Castillo**
**David Blaschke**
**Vladimir Kekelize**
**Victor Mateev**
**Alexander Sorin**

ISBN: 978-1-5108-1959-7
ISSN: 1742-6588

# 15th International Conference on Strangeness in Quark Matter (SQM 2015)

Journal of Physics: Conference Series Volume 668

Dubna, Russia
6 – 11 July 2015

IOP
Institute of Physics

**Printed from e-media with permission by:**

Curran Associates, Inc.
57 Morehouse Lane
Red Hook, NY 12571

**Some format issues inherent in the e-media version may also appear in this print version.**



Printed by Curran Associates, Inc. (2016)

For permission requests, please contact the Institute of Physics
at the address below.

Institute of Physics
Dirac House, Temple Back
Bristol BS1 6BE UK

Phone: 44 1 17 929 7481
Fax: 44 1 17 920 0979

techtracking@iop.org

**Additional copies of this publication are available from:**

Curran Associates, Inc.
57 Morehouse Lane
Red Hook, NY 12571 USA
Phone: 845-758-0400
Fax: 845-758-2634
Email: curran@proceedings.com
Web: www.proceedings.com

# Table of contents

**Volume 668**

15th International Conference on Strangeness in Quark Matter (SQM2015)
6–11 July 2015, Dubna, Russia

**Accepted papers received: 19 November 2016**
**Published online: 18 January 2016**

**Preface**



## Papers

**General Lectures**

**Plenary Talks**

**Parallel Talks**

**Poster Contributions**

**Summary Talks**



**Additional Paper**



## Additional Papers

012130

Possible effect of mixed phase and deconfinement upon spin correlations in the ΛΛ pairs generated in relativistic heavy-ion collisions OPEN ACCESS Valery V Lyuboshitz and Vladimir L Lyuboshitz pg. 605

# Preface

The 15th International Conference *Strangeness in Quark Matter (SQM)* took place at the Veksler and Baldin Laboratory of High Energy Physics (VBLHEP) of the Joint Institute for Nuclear Research (JINR) in Dubna in the period July 6 -11, with a record participation of 244 people from 31 countries! The previous meeting of the series in Birmingham 2013 had collected 158 physicists from 25 countries [J. Phys. Conf. Ser. 509, 011001 (2014)]. At SQM-2015, there was also a record participation of young scientist; every 4$^{th}$ conference attendee did not yet hold a PhD degree! There was a special program of 4 general lectures, a devoted session of parallel talks for *Young Talents* and the Helmholtz International Summer School (HISS) with 16 lecturers on the topics regarding *Dense Matter* (29.06.-11.07.) as a satellite event at the Bogoliubov Laboratory of Theoretical Physics (BLTP) and at VBLHEP. Another satellite event was the Round Table Workshop on *Physics at NICA*, jointly organized by JINR and the Republic of South Africa on July 5, 2015.

The selection of Dubna as the place for SQM-2015 conference by the International Advisory Committee (IAC) demonstrates the broad interest of the community in the progress of the Russian Megascience Project on the *Nuclotron-based Ion Collider Facility (NICA)* hosted at JINR Dubna. In a few years from now the experiments planned at NICA will produce data that provide new information of unprecedented accuracy which will help to answer some of the key questions which are topical at this conference.

The SQM-2015 conference had an ambitious scientific program with 38 plenary talks, 97 parallel talks in 7 topical directions and 39 posters reporting the state of the research and the future directions in the fields of *strangeness, heavy flavors and bulk physics*, suggested by the IAC to be the subtitle of the conference from 2016 onwards. Most of the contributions are represented in these Proceedings which we recommend to the community!

We gratefully acknowledge support from the JINR Dubna, the Russian Foundation for Basic Research, the Bundesministerium für Bildung und Forschung via the Heisenberg-Landau program, the Ministerstwo Nauki i Szkolnistwa Wyższego via the Bogoliubov-Infeld program, the LOEWE program via HIC for FAIR, the Helmholtz Association with their centres DESY, FZ Jülich, GSI Darmstadt, HZ Dresden-Rossendorf, Karlsruhe Institute of Technology and the Helmholtz Institutes in Mainz and Jena via the HISS programme. We thank the IAC for their help and advice in planning the conference, and we are grateful to the members of the Local Organisation Committee for their help in during the conference as well as to Niels-Uwe Bastian, Alexandra Friesen, Mark Alexander Kaltenborn and Irina Pirozhenko for their assistance in preparing these Proceedings.

**David Alvarez-Castillo** *BLTP, JINR Dubna*
**David Blaschke** *BLTP, JINR Dubna & University of Wroclaw*
**Vladimir Kekelidze** *VBLHEP, JINR Dubna*
**Victor Matveev** *JINR Dubna & Institute for Nuclear Research RAS, Moscow*
**Alexander Sorin** *VBLHEP & BLTP, JINR Dubna*

*Editors*

Journal of Physics: Conference Series **668** (2016) 011001 doi:10.1088/1742-6596/668/1/011001

**Conference Photo**

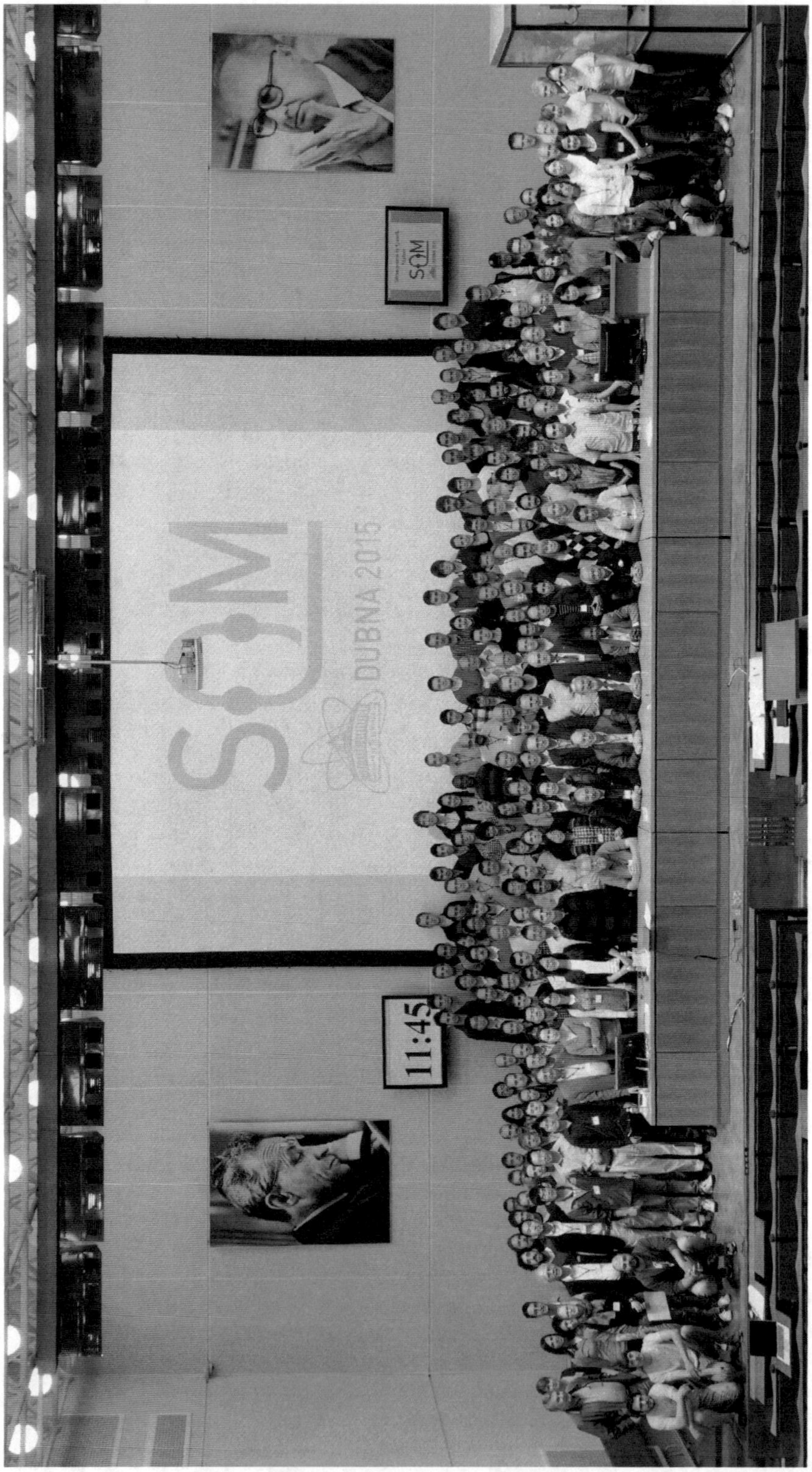

# Committees

### Interational advisory committee

J. Aichelin (Nantes)
F. Antinori (Padova)
F. Becattini (Firenze)
G. Barnaföldi (Budapest)
M. Bleicher (Frankfurt)
P. Braun-Munzinger (GSI)
J. Cleymans (Cape Town)
L. Csernai (Bergen)
B. Cole (Columbia)
A. Dainese (Padova)
W. Florkowski (Krakow)
M. Gaździcki (Frankfurt)
P. Giubellino (Torino)
R. Granier de Cassagnac (Paris)
V. Kekelidze (Dubna)
B. Mohanty (Niser)
B. Müller (Duke)
S. Nagamiya (Tsukuba)
J. Nagle (Colorado)
G. Odyniec (Berkeley)
J. Rafelski (Arizona)
H. Ritter (Berkeley)
G. Roland (Cambridge)
K. Šafařík (Geneva)
L. Šandor (Kosice)
H. Satz (Bielefeld)
G. Stephans (Cambridge)
H. Stöcker (Darmstadt)
J. Stroth (Frankfurt)
J. Takahashi (Campinas)
D. Villalobos Baillie (Birmingham)
F. Wang (Purdue)
F. Weber (San Diego)
N. Xu (Berkeley)
P. Zhuang (Tsinghua)
G. Zinovyev (Kiev)

### Local organizing committee

D. Alvarez-Castillo (Dubna)
Yu. Anisimov (Dubna)
D. Blaschke (Dubna & Wroclaw) (co-chair)
A. Friesen (Dubna)
D. Kamanin (Dubna)
V. Kekelidze (Dubna) (co-chair)
A. Khvorostukhin (Dubna)
E. Kolganova (Dubna) (coordinator)
A. Korotkova (Dubna)
S. Krivonos (Dubna)
R. Lednicky (Dubna)
V. Matveev (Dubna & Moscow) (chair)
O. Matyukhina (Dubna) (secretary)
D. Peshekhonov (Dubna)
I. Pirozhenko (Dubna) (scientific secretary)
Yu. Potrebenikov (Dubna)
A. Sorin (Dubna) (co-chair)
O. Teryaev (Dubna)
G. Trubnikov (Dubna)
V. Zhuravlev (Dubna)

# Peer review statement

All papers published in this volume of *Journal of Physics: Conference Series* have been peer reviewed through processes administered by the proceedings Editors. Reviews were conducted by expert referees to the professional and scientific standards expected of a proceedings journal published by IOP Publishing.

Journal of Physics: Conference Series **668** (2016) 012001 doi:10.1088/1742-6596/668/1/012001

# Transport Theory from the Nambu-Jona-Lasinio Lagrangian

**R. Marty**[a,b,c], **J. M. Torres-Rincon**[a], **E. Bratkovskaya**[b,c] **and J. Aichelin**[a 1]

[a] SUBATECH, Laboratoire de Physique Subatomique et des Technologies Associées,
Université de Nantes - IN2P3/CNRS - Ecole des Mines de Nantes
4 rue Alfred Kastler, F-44072 Nantes, Cedex 03, France
[b] Institute for Theoretical Physics, JWG Universität, Frankfurt/M, Germany
[c] Frankfurt Institute for Advanced Studies, JWG Universität, Frankfurt/M, Germany

E-mail: `aichelin@in2p3.fr`

**Abstract.** Starting from the (Polyakov-) Nambu-Jona-Lasinio Lagrangian, (P)NJL, we formulate a transport theory which allows for describing the expansion of a quark-antiquark plasma and the subsequent transition to the hadronic world —without adding any new parameter to the standard (P)NJL approach, whose parameters are fixed to vacuum physics. This transport theory can be used to describe ultrarelativistic heavy-ion reaction data as well as to study the (first-order) phase transition during the expansion of the plasma. (P)NJL predicts such a phase transition for finite chemical potentials. In this contribution we give an outline of the necessary steps to obtain such a transport theory and present first results.

## 1. Introduction

The Nambu-Jona-Lasinio (NJL) Lagrangian shares the symmetries of the Lagrangian of Quantum Chromo Dynamics (QCD). Since symmetries are nowadays considered as the most important properties of a Lagrangian, it has been used for many studies of ground-state properties of strongly interaction matter at zero temperature, as well as at finite temperatures and finite chemical potentials. This includes the studies of the hadron-parton transition and of the color-flavor locking. It has also been used very successfully in the past to study meson and baryon properties as functions of the temperature and of the baryon density like hadron masses, widths and decay constants. The NJL Lagrangian contains quark fields but no gluon fields. The quarks interact by a four-point interaction. Recently the standard NJL Lagrangian has been extended to the Polyakov NJL (PNJL) which includes gluons on a mean-field level [1].

Having a four-fermion interaction the (P)NJL Lagrangian is not renormalizable. It does also not confine the quarks into colorless objects. Therefore it is a challenge to describe the transition of a plasma of quarks and antiquarks to hadrons with such an approach. Usually the calculations using the (P)NJL Lagrangian have been performed at the mean-field level. Going calculationally beyond the mean field by including systematically the fluctuations, the meson contribution to the equation of state could be identified [2, 3, 4].

In these short proceedings we cannot cover all these developments and therefore, we refer to the recent reviews for details [5, 6, 7, 8]. Here we concentrate on the steps which are necessary to develop a transport theory from the (P)NJL Lagrangian. After having introduced the Lagrangian in section 2, discuss in section 3 how this approach is extended to finite temperature and density and present the calculation of the masses of quarks, mesons and baryons at finite temperature. Section 4 contains the outline of the covariant transport theory which can be built with the help of the NJL Lagrangian. In section 5 we present first results of this transport theory in comparison with experimental data of ultrarelativistic heavy-ion collisions. Section 6, finally, concludes this lecture with a summary.

[1] invited lecturer

## 2. The starting point: The (P)NJL Lagrangian

The thermal properties of QCD, the theory of strong interactions, can presently only be calculated computationally ("lattice QCD"), and this can be done exclusively for the case of grand canonical systems with zero chemical potential. This allows for the determination of the equation of state and other thermal properties, but already the thermal properties of the constituents of the theory are not precisely known. In addition, only few of the transport coefficients which can shed light on the evolution of a system out of (but close to) equilibrium are yet calculated on lattice-QCD with still considerable error bars.

If one wants to study the dynamics of strongly interacting matter, one has therefore to rely on effective theories. One of these effective theories is the NJL model based on the Lagrangian (in the SU(3) version)

$$\begin{aligned}\mathcal{L}_{\mathrm{NJL}} &= \sum_i \bar{\psi}_i(i\gamma_\mu\partial^\mu - \hat{m}_i)\psi_i - G_c^2 \sum_{ijkl} [\bar{\psi}_i\gamma^\mu T^{a'}\delta_{ij}\psi_j]\ [\bar{\psi}_k\gamma_\mu T^{a'}\delta_{kl}\psi_l] \\ &+ H \det_{ij}[\bar{\psi}_i(1-\gamma_5)\psi_j] - H\det_{ij}[\bar{\psi}_i(1+\gamma_5)\psi_j] \qquad (1)\end{aligned}$$

where $\psi_i$ are spinors of quarks of flavor $i$. The second term is a four-point interaction in Dirac (green), color (red) and flavor (blue) spaces. This term can be related to perturbative QCD by realizing that it describes in Born approximation the interaction of two quark currents mediated by a gluon whose mass is large as compared to the momentum transfer [7]. This term is followed by the 't Hooft determinant which gives rise to the $U_A(1)$ symmetry breaking. $T^{a'}$ are the color SU(3) Gell-Mann matrices with tr$_c$ $(T^{a'}T^{b'}) = 2\delta^{a'b'}$. The mass of the strange quarks is different from that of up and down quarks which are considered as degenerated. Being an effective model of QCD the NJL Lagrangian respects the symmetries of the underlying theory, in particular the $U_V(1)\otimes SU_V(N_f)\otimes SU_A(N_f)$ global symmetries of the massless QCD Lagrangian. The $U_V(1)$ symmetry leads to the baryon number conservation, while the chiral symmetry $SU_V(N_f)\otimes SU_A(N_f)$ is spontaneously broken down to $SU_V(N_f)$ at low temperatures. The $U_A(1)$ symmetry is broken by the axial anomaly.

After performing a Fierz transformation [5, 8, 9] this Lagrangian can be reexpressed in a convenient way to describe the $qq$, $\bar{q}\bar{q}$ and $\bar{q}q$ interactions. The $\bar{q}q$ sector of the effective theory will allow us to describe mesons. For instance, the pseudoscalar sector of the interacting Lagrangian (1) reads (all repeated indices are to be summed)

$$\mathcal{L}_{\bar{q}q} = G\,(\bar{\psi}_i\, i\gamma_5\, \tau^a_{ij}\, \psi_j)(\bar{\psi}_k\, i\gamma_5\, \tau^a_{kl}\, \psi_l)\,, \qquad (2)$$

where $a = 1, ..., N_f^2 - 1$ and $G$ is a coupling constant. We will take $G$ as a free parameter to be fixed by comparing our results with the experimental hadron spectrum. The flavor Gell-Mann matrices $\tau^a$ follow the normalization

$$\mathrm{tr}\,(\tau^a\tau^b) = 2\delta^{ab}\,. \qquad (3)$$

The Fierz transformation produces diquarks as well. We detail here the Lagrangian describing the scalar diquark sector

$$\mathcal{L}_{qq} = G_{DIQ}\,(\bar{\psi} i\gamma_5 C\tau^A T^{A'}\bar{\psi}^T)(\psi^T C i\gamma_5 \tau^A T^{A'}\psi)\,, \qquad (4)$$

and the one for the axial diquark sector

$$\mathcal{L}_{qq} = G_{DIQ,V}\,(\bar{\psi}\gamma^\mu C\tau^S T^{A'}\bar{\psi}^T)(\psi^T C\gamma_\mu\tau^S T^{A'}\psi)\,, \qquad (5)$$

where $G_{DIQ}$ and $G_{DIQ,V}$ are coupling constants (related to the original $G_c^2$ but taken here as free parameters) and $C = i\gamma_0\gamma_2$ represents the charge-conjugation operator. Finally, we have denoted by $\tau^A$ and $\tau^S$ the antisymmetric and symmetric SU(3) flavor matrices, respectively; and by $T^{A'}$ the antisymmetric color matrices. In particular, the presence of the latter reflects that the diquarks cannot be color singlets.

The (P)NJL Lagrangian has several parameters which have to be fixed. In our calculation we take those of Table 1.

$\Lambda$ is the momentum cut-off which has to be applied in the loop integrations. Using the parameter set in Table 1 we obtain at $T = 0$: the light-quark condensate $\langle\bar{\psi}_u\psi_u\rangle = -(241.3\ \mathrm{MeV})^3$, the pion decay constant $f_\pi = 92.2$ MeV, the pion mass $m_\pi = 134.8$ MeV, the kaon mass $m_K = 492.1$ MeV, the $\eta - \eta'$ mass splitting of 475.5 MeV, the proton mass 932.0 MeV and the $\Delta$ baryon mass 1221.4 MeV.

This means that all our parameters are fixed by vacuum hadron masses and decay constants. We would like to stress, and this is the great advantage of the (P)NJL model, that no other parameters are needed to describe hadron masses at finite temperature and density, the pressure, trace anomaly and entropy density as a function of $T$ and $\mu$, the cross sections for scattering among quarks and antiquarks and the hadronization cross sections. Finally the whole transport approach and its prediction for the observables do not need more than these parameters.

| Parameter | $m_{q0}$ | $m_{s0}$ | $\Lambda$ | $G$ | $H$ | $G_{DIQ}$ | $G_{DIQ,V}$ |
|---|---|---|---|---|---|---|---|
| Value | 5.5 MeV | 134 MeV | 569 MeV | $2.3/\Lambda^2$ | $11/\Lambda^5$ | $1.56\,G$ | $-0.639\,G_{DIQ}$ |
| Parameter | $a_0$ | $a_1$ | $a_2$ | $a_3$ | $b_3$ | $b_4$ | $T_0$ |
| Value | 6.75 | -1.95 | 2.625 | -7.44 | 0.75 | 7.5 | 190 MeV |

**Table 1.** Parameters of the NJL and PNJL model used in this study. In the isospin limit we have $m_{q0} = m_{u0} = m_{d0}$.

### 3. Finite-temperature physics

To calculate the hadron properties at finite temperature, we use the imaginary time formalism with the prescription

$$\int \frac{d^4k}{(2\pi)^4} \to iT \sum_{n\in\mathbb{Z}} \int \frac{d^3k}{(2\pi)^3} , \tag{6}$$

with $T$ the temperature and $k^0 \to i\omega_n$ the fermionic Matsubara frequencies $i\omega_n = i\pi T(2n+1)$.

To account for the finite baryonic density we can introduce a quark chemical potential by adding to the Lagrangian the term

$$\mathcal{L}_\mu = \sum_{ij} \bar{\psi}_i\, \mu_{ij}\gamma_0\, \psi_j , \tag{7}$$

where $\mu_{ij} =$ diag $(\mu_u, \mu_d, \mu_s)$ contains the quark chemical potentials (which can be alternatively expressed in terms of the baryon, charge, and strangeness chemical potentials, $\mu_B, \mu_Q, \mu_S$). In this work we will restrict ourselves to a vanishing chemical potential $\mu_u = \mu_d = \mu_s = 0$.

In the PNJL model an effective potential $\mathcal{U}(\Phi, \bar{\Phi}, T)$, is added to the effective NJL Lagrangian $\mathcal{L} \to \mathcal{L} - \mathcal{U}$. $\mathcal{U}$ is a function of the Polyakov loop $\Phi$

$$\Phi = \frac{1}{N_c}\,\mathrm{tr}_c\langle L\rangle \quad ; \quad L(\mathbf{x}) = \mathcal{P}\exp\left(i\int_0^\beta d\tau A_4(\tau,\mathbf{x})\right) , \tag{8}$$

and its complex conjugate, which are taken to be homogeneous fields. The form of the effective potential is inspired by the $\mathbb{Z}_3$ center symmetry [1]

$$\frac{\mathcal{U}(T,\Phi,\bar{\Phi})}{T^4} = -\frac{b_2(T)}{2}\bar{\Phi}\Phi - \frac{b_3}{6}\left(\Phi^3 + \bar{\Phi}^3\right) + \frac{b_4}{4}\left(\bar{\Phi}\Phi\right)^2 , \tag{9}$$

with

$$b_2(T) = a_0 + a_1\frac{T_0}{T} + a_2\left(\frac{T_0}{T}\right)^2 + a_3\left(\frac{T_0}{T}\right)^3 . \tag{10}$$

The parameters $a_0, a_1, a_2, a_3, b_3, b_4$ and $T_0$ are fitted from the pure-gauge lattice-QCD equation of state (with a reduction of $T_0$ following the arguments in [10]) and can be found in Table 1.

A detailed calculation shows [11, 12] that including the Polyakov loop leads to the replacement of the Fermi distribution function, which appears in the finite temperature calculation using the NJL Lagrangian, by

$$f_\Phi^+(E_i) = \frac{(\Phi + 2\bar{\Phi}e^{-E_i/T})e^{-E_i/T} + e^{-3E_i/T}}{1 + 3(\Phi + \bar{\Phi}e^{-E_i/T})e^{-E_i/T} + e^{-3E_i/T}} , \tag{11}$$

$$f_\Phi^-(E_i) = \frac{(\bar{\Phi} + 2\Phi e^{-E_i/T})e^{-E_i/T} + e^{-3E_i/T}}{1 + 3(\bar{\Phi} + \Phi e^{-E_i/T})e^{-E_i/T} + e^{-3E_i/T}} . \tag{12}$$

where + and - refer to a negative (positive) chemical potential $(E+\mu)((E-\mu))$ and $\Phi(\bar{\Phi})$ is obtained by minimizing the PNJL grand-canonical potential,

$$\begin{aligned}\Omega_{PNJL}(\Phi,\bar{\Phi},m_i,T) &= \mathcal{U}(T,\Phi,\bar{\Phi}) + 2G\sum_i\langle\bar{\psi}_i\psi_i\rangle^2 - 4H\prod_i\langle\bar{\psi}_i\psi_i\rangle - 2N_c\sum_i\int\frac{d^3k}{(2\pi)^3}E_i \\ &- 2T\sum_i\int\frac{d^3k}{(2\pi)^3}\left[\mathrm{tr}_c\log\left(1+Le^{-E_i/T}\right) + \mathrm{tr}_c\log\left(1+L^\dagger e^{-E_i/T}\right)\right] , \end{aligned} \tag{13}$$

(with $E_i = \sqrt{k^2 + m_i^2}$) with respect to $\Phi(\bar{\Phi})$,

$$\frac{\partial \Omega_{PNJL}}{\partial \Phi} = 0 \,, \quad \frac{\partial \Omega_{PNJL}}{\partial \bar{\Phi}} = 0. \tag{14}$$

The minimization of the PNJL grand-canonical potential with respect to $\langle \bar{\psi}_i \psi_i \rangle$ gives the PNJL quark condensate

$$\langle \bar{\psi}_i \psi_i \rangle = -2N_c \int \frac{d^3k}{(2\pi)^3} \frac{m_i}{E_i} \left[ 1 - f_\Phi^+(E_i) - f_\Phi^-(E_i) \right] . \tag{15}$$

### 3.1. *Quark and meson masses*

In this model [9] chiral symmetry restoration —expressed by the decreasing quark condensate (Eq. 15) with temperature— lowers the quark masses with increasing temperature

$$M_i = m_{i0} - 4G\langle \bar{\psi}_i \psi_i \rangle - 2H \langle \bar{\psi}_j \psi_j \rangle \langle \bar{\psi}_k \psi_k \rangle \tag{16}$$

with $i, j, k = u, d, s$ (to be fixed in cyclic order). Mesons are obtained as poles of the quark-antiquark scattering amplitude in the complex-energy plane. The scattering amplitude is the solution of the Bethe-Salpeter equation, formally displayed in Fig. 1 $q_i + \bar{q}_j \to q_k + \bar{q}_l$ (where the indices refer to the quark flavor) and which is solved

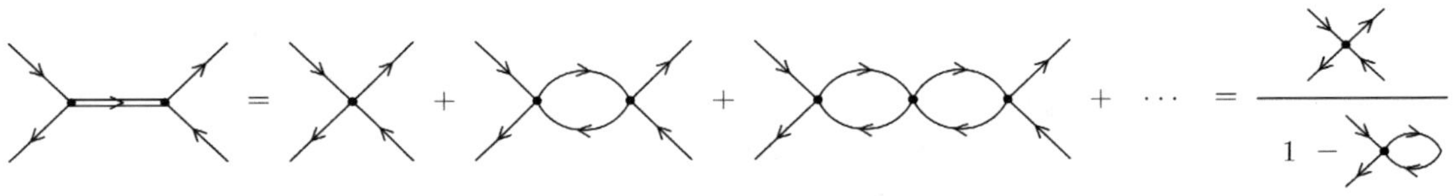

**Figure 1.** Effective interaction between a quark and an antiquark in the Random Phase Approximation.

in the random-phase or Brueckner approximation (which are the same for local four-point interactions)

$$T^{ab}_{i\bar{j},m\bar{n}}(p^2) = \mathcal{K}^{ab}_{i\bar{j},m\bar{n}} + i \int \frac{d^4k}{(2\pi)^4} \mathcal{K}^{ac}_{i\bar{j},p\bar{q}} \, S_p\left(k + \frac{p}{2}\right) S_{\bar{q}}\left(k - \frac{p}{2}\right) T^{cb}_{p\bar{q},m\bar{n}}(p^2) \,, \tag{17}$$

where $a, b$ denotes the meson flavor channel. The kernel $\mathcal{K}$ reads

$$\mathcal{K}^{ab}_{i\bar{j},m\bar{n}} = \Omega^a_{i\bar{j}} \, 2K^{ab} \, \bar{\Omega}^b_{\bar{n}m} \,, \tag{18}$$

with the vertex factors containing color, flavor and spin matrices

$$\Omega^a_{i\bar{j}} = \left( \mathbb{I}_{\text{color}} \otimes \tau^a_{i\bar{j}} \otimes \Gamma \right) , \tag{19}$$

as well as a combinatorial factor of 2. The Dirac structure —whose indices we have omitted in the BS equation— can be chosen to be $\Gamma = \{1, i\gamma_5, \gamma^\mu, \gamma_5\gamma^\mu\}$ for scalar, pseudoscalar, vector, and axial-vector mesons, respectively. Figure 2 displays the masses of mesons and their constituent quarks as a function of the temperature for $\mu_i = 0$. We see that the quark masses (Eq. 16) are about the constituent quark masses ($\approx M_{proton}/3$.) at low temperature and decrease as a function of the temperature due to the melting of the quark condensate. At very large temperature they are very close to the bare mass. The pseudoscalar mesons are the Goldstone bosons that result from the spontaneous breakdown of the chiral-flavor symmetries of QCD, and are therefore little influenced by the temperature. At low temperature their masses are lower than that of the two constituent quarks, and therefore they are the thermodynamically relevant degrees of freedom. Beyond the Mott temperature the sum of the constituent quark masses is lower than the mass of the meson which develops a width because it may decay into its constituents. The Mott temperatures of $K$'s and $\pi$'s are not identical but close by. Beyond the Mott temperature the quarks are the thermodynamically relevant quantities and a plasma of quarks and antiquarks is formed.

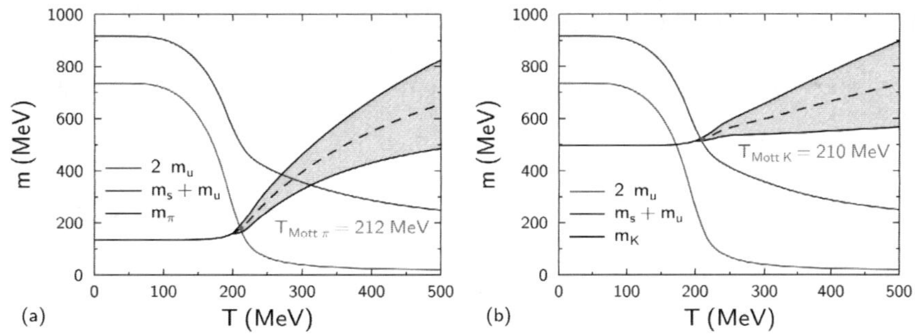


**Figure 2.** Left panel: Mass of $\pi$ meson and of light quarks, right panel: Mass of K meson and light and heavy quark as a function of the temperature.

*3.2. Baryons*

Diquarks can form — together with a third quark— color neutral baryons [9]. For $N_f = 3$ we will consider both the octet and decuplet flavor representations of baryons. Scalar diquarks (those belonging to flavor $\bar{\mathbf{3}}$ representation) and axial diquarks (**6**) will be used to build up the baryon octet and decuplet states, respectively, according to the decomposition,

$$\mathbf{3} \otimes (\bar{\mathbf{3}} \oplus \mathbf{6}) = (\mathbf{1} \oplus \mathbf{8}) \oplus (\mathbf{8} \oplus \mathbf{10}) . \tag{20}$$

The starting point to describe baryons is the Fadeev equation [13, 14]:

$$X_j^{\bar{j},\alpha}(P^2,q) - \int \frac{d^4k}{(2\pi)^4} L_{jk}^{\bar{j}k,\alpha\beta}(P^2,q,k) X_k^{\bar{k},\beta}(P^2,k)\bigg|_{P^2=M_B^2} = 0 , \tag{21}$$

where the baryon wave function is denoted by $X_j^{\bar{j},\alpha}$ and it carries a quark index ($j$), diquark index ($\bar{j}$), and a possible spin index $\alpha$.

The kernel reads [13]

$$L_{jk}^{\bar{j}k,\alpha\beta}(P^2,q,k) = \mathcal{G}_{k\bar{k}}^{\gamma\beta}(P^2,q) Z_{jk}^{\bar{k}\bar{j},\alpha\gamma}(q,k) , \tag{22}$$

with a first term which accounts for the free quark and diquark propagators (see right panel of Fig. 3)

$$\mathcal{G}_{k\bar{k}}^{\gamma\beta}(P^2,q) = S_k(P/2+q)\; it_{\bar{k}}^{\gamma\beta}(P/2-q) \tag{23}$$

and a second term

$$Z_{jk}^{\bar{k}\bar{j},\alpha\gamma}(q,k) = \Omega_{jl}^{\bar{k},\gamma}\; S_l(-q-k)\; \Omega_{lk}^{\bar{j},\alpha} , \tag{24}$$

which represents an interaction with an exchanged quark (displayed in the left panel of Fig. 3). We do not

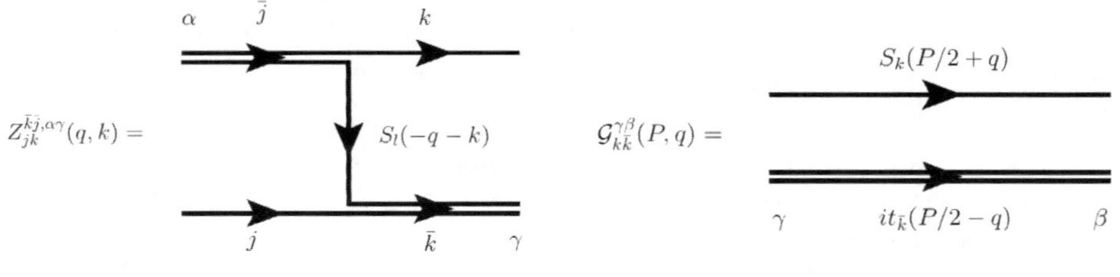


**Figure 3.** Left panel: Effective coupling in the Fadeev equation which contains a quark exchange. Right panel: Two-particle (quark+diquark) propagator in the Fadeev kernel.

attempt here to justify the form of the Fadeev equation (21) and we refer the reader to the original papers [15, 13] to learn the rigorous derivation and to learn more details.

Nevertheless, we can provide a simple motivation for Eq. (21): If we denote by $\mathbf{G}(P^2)$ the full baryon propagator, one can form a Dyson equation by taking $\mathcal{G}$ as the leading order approximation (free propagation), and then considering $Z$ as the elementary interaction (see Fig. 4). The Dyson equation can be symbolically

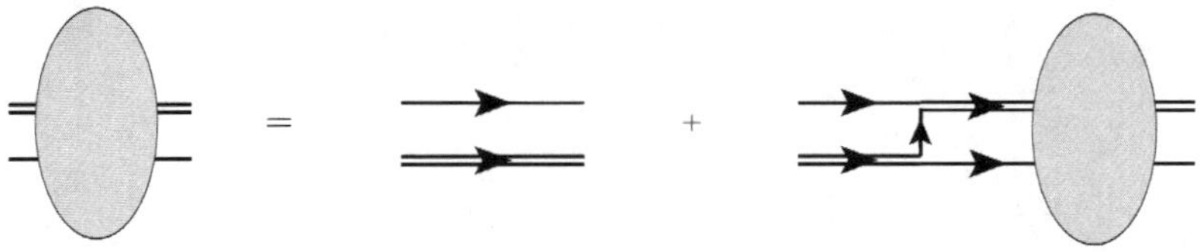


**Figure 4.** Dyson equation for the baryon propagator $\mathbf{G} = \mathcal{G} + \mathcal{G}Z\mathbf{G}$.

written as

$$\mathbf{G} = \mathcal{G} + \mathcal{G}Z\mathbf{G}\,, \tag{25}$$

whose solution reads

$$\mathbf{G} = \frac{\mathcal{G}}{1 - \mathcal{G}Z}\,. \tag{26}$$

The baryon masses are now extracted as the poles of the baryon propagator, so one needs to solve $\mathbf{G}^{-1}X(P^2 = M_B^2) = 0$, where $X$ is the baryon wave function. Explicitly,

$$(1 - \mathcal{G}Z)X(P^2 = M_B^2) = 0\,, \tag{27}$$

at $P^2 = M_B^2$, which is a simplified version of the more complete Eq. (21). Our results for the baryon masses at finite temperature for vanishing chemical potential in both, the octet and decuplet, representations are shown in the PNJL model in Fig. 5. We observe first of all the the ground state masses are quite well reproduced in

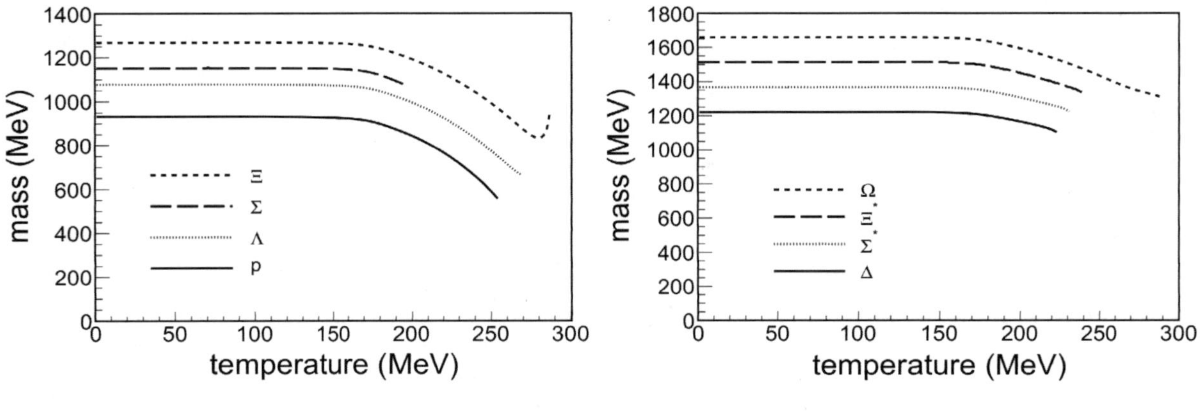


**Figure 5.** Baryon masses as a function of temperature for vanishing chemical potential in the PNJL model.

the PNJL approach. Deviations from the experimental values are smaller than 5%. This is astonishing in view of the fact that besides the parameters fixed already by the meson sector only one new parameter enters, the diquark coupling constant. We observe as well that with increasing temperature the baryon mass decreases and become finally unstable at the temperature where the lines end.

## 4. Relativistic Transport Theory

We have seen that at high temperatures quarks and antiquarks are the essential degrees of freedom whereas at low energies the formation of mesons are thermodynamically preferred because the quark condensate becomes large at low temperature and hence the quarks become heavy. Mesons, on the contrary, become unstable at high temperature and develop a width due to their decay into quarks and antiquarks. The question we address in this section is how a finite size system develops, when we have brought it to a high temperature where the system is characterized by the quark degrees of freedom. In order to answer this question we need a transport theory. We could start out with a hydrodynamical approach but since we are interested not in the energy and particle density but in the processes by which quarks and antiquark pairs form mesons during the expansion and how the observables fluctuate in the vicinity of the phase transition, we search for a transport theory in which the time evolution of the quarks is addressed i.e. a molecular dynamics approach. The smallness of the quark mass at high temperature requires that this approach is covariant.

### *4.1. Relativistic molecular dynamics*

The starting point for the nonrelativistic equations of motion is the time evolution equation for a function $A$ defined on the $6N + 1$-dimensional phase space:

$$\frac{dA}{dt} = \frac{\partial A}{\partial t} + \{A, \mathcal{H}\}, \tag{28}$$

where $\mathcal{H}(\mathbf{q}, \mathbf{p})$ is the Hamiltonian of the system and $\{A, B\}$ is the Poisson bracket of $A$ and $B$, defined as

$$\{A, B\} = \sum_k^N \frac{\partial A}{\partial \mathbf{q}_k}\frac{\partial B}{\partial \mathbf{p}_k} - \frac{\partial A}{\partial \mathbf{p}_k}\frac{\partial B}{\partial \mathbf{q}_k} . \tag{29}$$

If we replace $A$ by either $\mathbf{q}$ or $\mathbf{p}$ (which do not depend explicitly on time) we recover the Hamilton-Jacobi equations

$$\frac{d\mathbf{q}}{dt} = \{\mathbf{q}, \mathcal{H}\} = \frac{\partial \mathcal{H}}{\partial \mathbf{p}} , \quad \frac{d\mathbf{p}}{dt} = \{\mathbf{p}, \mathcal{H}\} = -\frac{\partial \mathcal{H}}{\partial \mathbf{q}} . \tag{30}$$

which give, for a initial condition $(\mathbf{q}_0, \mathbf{p}_0)$, the desired trajectory of the particle in the phase space. For the later discussion it is important to note that Eqs. (30) are the differential equations for the trajectory in the $6N + 1$ dimensional phase space on which the energy $\mathcal{H}$ is conserved.

Equation (30) cannot directly be made covariant containing the fourth component of the coordinate and of the energy-momentum 4-vector. Therefore, for the relativistic equations of motions we have to go back to the Poisson equations for 4-vectors

$$\{A, B\} = \sum_{k=1}^N \frac{\partial A}{\partial q_k^\mu}\frac{\partial B}{\partial p_{k\mu}} - \frac{\partial A}{\partial p_k^\mu}\frac{\partial B}{\partial q_{k\mu}} , \tag{31}$$

because in a dynamical system $q^\mu$ and $p^\mu$ depend on the time evolution parameter $\tau$, these quantities have to be taken at equal time $\tau$.

If the system is composed of several particles we find for the generators for the translation group

$$P^\mu = \sum_k^N p_k^\mu , \tag{32}$$

and for the Lorentz group $[SL(n = 2, \mathbb{C}) \to \dim = 2(n^2 - 1) = 6]$

$$M^{\mu\nu} = \sum_{k=1}^N q_k^\mu p_k^\nu - q_k^\nu p_k^\mu . \tag{33}$$

These 10 generators define the Poincaré algebra:

$$[P_\mu, P_\nu] = 0, \quad ; \quad [M_{\mu\nu}, P_\rho] = g_{\mu\rho}P_\nu - g_{\nu\rho}P_\mu, \tag{34}$$

$$[M_{\mu\nu}, M_{\rho\sigma}] = g_{\mu\rho}M_{\nu\sigma} - g_{\mu\sigma}M_{\nu\rho} - g_{\nu\rho}M_{\mu\sigma} + g_{\nu\sigma}M_{\mu\rho}. \tag{35}$$

The generator of a Poincaré transformation is given by

$$G = \frac{1}{2}\omega^{\mu\nu}M_{\mu\nu} - a^\mu P_\mu. \tag{36}$$

The space-time coordinates of the same event in two inertial frames $O$ and $O'$ are related by

$$q'^{\mu} = q^{\mu} + \{q, G\} = q^{\mu} + \omega^{\mu}_{\nu} q^{\nu} + a^{\mu} = \Lambda^{\mu}_{\nu} q^{\nu} + a^{\mu}. \tag{37}$$

World lines of particles are given by $(\mathbf{q}_i(\tau), \mathbf{p}_i(\tau))$ and therefore physical trajectories (position and momentum of the particles as functions of the time $\tau$) have $6N + 1$ dimensions. Thus we need *constraints* to reduce the number of degrees of freedom in the relativistic phase space from $8N$ to $6N + 1$. These constraints $K$ have to be Poincaré-invariant quantities.

$$\{K, M_{\mu\nu}\} = 0, \quad \{K, P_{\mu}\} = 0. \tag{38}$$

In this lecture we demonstrate the idea and the consequence of this reduction for the case of one free particle. For the general case we have to refer to [16]. The trajectory in phase space on which a constraint $K$ is satisfied, is given by the solution of

$$\begin{aligned} \frac{dq^{\mu}(\tau)}{d\tau} &= \lambda\{q^{\mu}(\tau), K\}, \\ \frac{dp^{\mu}(\tau)}{d\tau} &= \lambda\{p^{\mu}(\tau), K\}, \end{aligned} \tag{39}$$

with the initial condition $q(0) = q_0$ and $p(0) = p_0$. $\lambda$ is a Lagrange multiplier. In order to associate to each value of $\tau$ *one* point in phase space $(q(\tau), p(\tau))$ or, in other words, in order to create a world line a second constraint, $\chi(q^{\mu}, p^{\mu}, \tau) = 0$, has to be employed to fix $\lambda$. It relates the time $q^0$ of the particle with a Lorentz-invariant system time $\tau$. The subspace we are interested in is determined by conserved $\chi$ and $K$ constraints. This is expressed by

$$\frac{d\chi}{d\tau} = \frac{\partial\chi}{\partial\tau} + \lambda\{\chi(\tau), K\} = 0 \rightarrow \lambda = -\frac{\partial\chi}{\partial\tau}\{\chi, K\}^{-1}. \tag{40}$$

To obtain the desired time evolution equations we replace $\chi$ in Eq. (40) by $q^{\mu}$ and $p^{\mu}$. To discuss the trajectory it is better to take a concrete example. Let be $K$ the mass shell constraint $K = p^{\mu}p_{\mu} - m^2 = 0$. The time constraint $\chi$ is much less self-evident. We investigate here the equations of motion for two choices of $\chi$: $\chi = q^0 - \tau = 0$ or $\chi = x_{\mu}p^{\mu} - m\tau = 0$. In the first case we obtain

$$\frac{dq^{\mu}}{d\tau} = \lambda\{q^{\mu}, K\} = \frac{p^{\mu}}{p^0}, \tag{41}$$

whereas in the second

$$\frac{dq^{\mu}}{d\tau} = \lambda\{q^{\mu}, K\} = \frac{p^{\mu}}{m}. \tag{42}$$

In both cases we have $dp^{\mu}/d\tau = 0$. These examples show the essential property of relativistic kinematics: In order to determine a trajectory in the 6N+1 dimensional phase space we have to impose constraints but these constraints are not unique. Different choices of the constraints yield a different time evolution of the system. Thus the time evolution in the 6N+1 dimensional phase space is not uniquely defined and other criteria like cluster separability have to be addressed.

This method can be extended to a system of $N$ interacting particles [16]. In the (P)NJL approach the interaction is expressed by the change of the mass of the quarks, Eq. (16). Assuming that the system is locally not far from equilibrium we calculate the local temperature of the environment in which the quark is localized. Knowing the temperature we can determine the masses. After having determined the masses of all particles we use the energy constraints to calculate the trajectories. For a given initial condition and using the temperature and density dependent masses of the (P)NJL approach we can then follow the trajectories of all particles using Poincaré invariant kinematics.

### *4.2. Collisional interactions*

If created in heavy-ion collisions, the QGP will expand rapidly. Therefore, the cross sections between the constituents become dominant over the mean-field properties of the theory. In the NJL model these cross sections can be calculated via a $1/N_c$ expansion of the Lagrangian [17]. All the details can be found in [17, 18, 19] and we refer to these references for details. There are two kinds of cross sections: cross sections with quarks or antiquarks in the entrance and exit channels: $q(\bar{q}) + q(\bar{q}) \rightarrow q(\bar{q}) + q(\bar{q})$ and hadronization

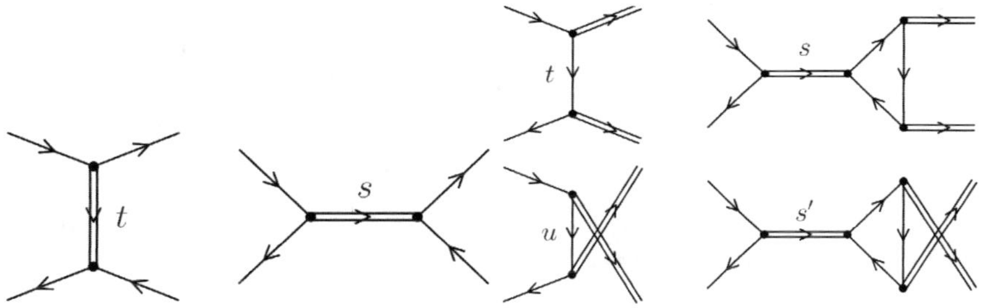


**Figure 6.** Feynman diagrams for elastic $q\bar{q}$ scattering (left) and for hadronization (right).

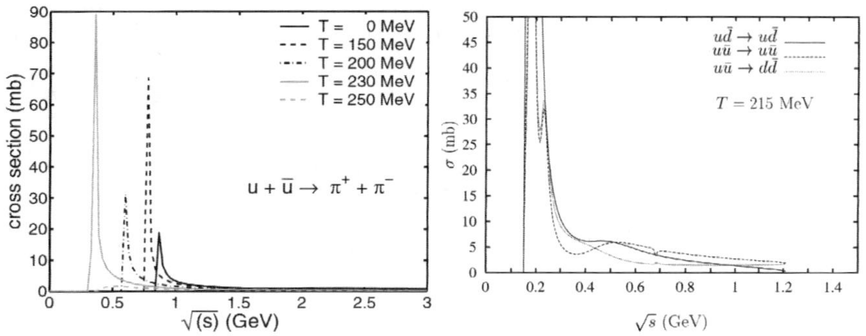


**Figure 7.** The cross section for the hadronization channel $u + \bar{u} \to \pi^+ + \pi^-$ (left) and the $q + \bar{q} \to q + \bar{q}$ cross section (right) as a function of the center of mass energy of the colliding quarks .

cross sections $q + \bar{q} \to M + M$ where $M$ represents a meson. The Feynman diagrams for $q + \bar{q} \to q + \bar{q}$ collisions are given in the left side of Fig. 4.2, whereas in the right side of this figure we display the Feynman diagrams for hadronization cross sections. $s$ and $s'$ denote two $s$-channel diagrams with different intermediate mesons. The calculation of these diagrams requires the knowledge of the quark and meson masses and the vertex of the $q - \bar{q} - M$ interaction which we have already obtained when calculating the meson masses. Hence no new parameters are needed for calculating the cross sections. The (P)NJL cross sections share a common property which is essential if we want to calculate with the help of the potential and of the cross sections the time evolution of a system of quarks and anti-quarks. These cross sections have the typical size of perturbative-QCD cross sections, a couple of milibarns, if we are away from the Mott temperature. Close to the Mott temperature, due to a resonant $s$-channel, the cross sections become huge and arrive at values of more than 100 mb. This is shown in Fig. 7. For an expanding plasma this means that early during the expansion —when the density is high— a lot of collisions take place, but during the further expansion the rate decreases due to a decreasing density. If the system arrives at the Mott temperature the rate increases suddenly and elastic cross sections thermalize the system locally whereas the still larger hadronization cross sections convert the quarks into mesons. Finally, at temperatures below the Mott temperature we find a gas of mesons. The large hadronization cross sections provide an effective confinement in a theory which does not contain confinement.

## 5. Expansion of a quark-antiquark plasma and comparison with other theories and data

In the preceding sections we have developed all the theoretical ingredients for a covariant transport theory based in the (P)NJL Lagrangian. In this section we present the results of the numerical realization of such a transport approach - dubbed RSP model (relativistic quantum molecular dynamics for strongly interacting matter with phase transition or cross-over) - applied to the expansion of a quark+anti-quark plasma created in relativistic heavy-ion collisions [16, 20]. The mechanism which brings the colliding nuclei in the time scale of 1 fm close to local equilibrium is not known yet and therefore we start our calculations at the time at which the highest density is reached (see Ref. [20] for details). We use two initial conditions for the system at this time point: **a)** either we assume a completely equilibrated system to perform model studies or **b)** we use the initial condition of the PHSD model [21, 20] which has an anisotropic momentum distribution with a smaller component in transverse direction than in longitudinal direction and large energy density fluctuations. This initial energy

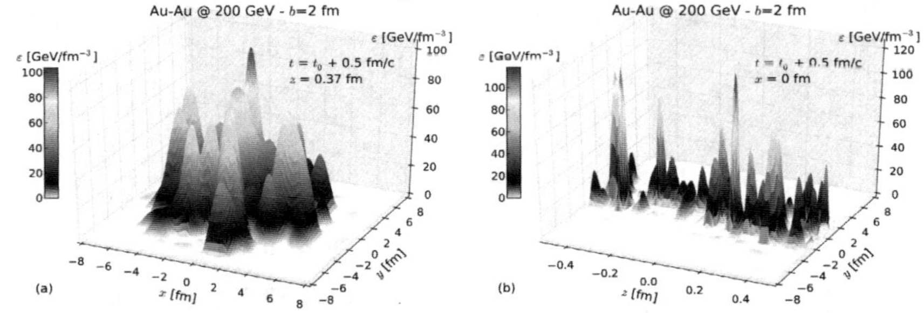


**Figure 8.** Initial energy density $\epsilon$ of the expanding plasma as given by the PHSD initial condition. On the left panel the energy density in the $xy$ plane, on the right panel in the $zy$ plane. $z$ is the direction of the beam and $x$ the direction of the impact parameter.

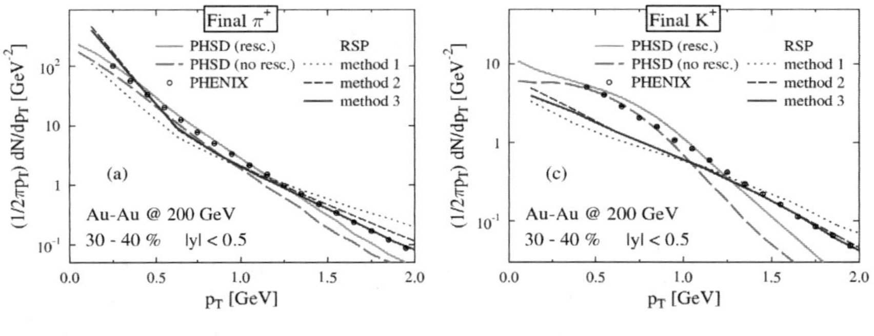


**Figure 9.** The final $\pi^+$ (left) and $K^+$ (right) transverse momentum distribution of the NJL transport theory for AuAu with $\sqrt{s}$= 200 AGeV and a centrality of 30-40% using the PHSD initial energy density distribution. Method 1-3 describe 3 different methods to convert this initial energy density distribution into the NJL quarks. Our results are compared with that of PHSD and with the experimental data.

density distribution is shown in Fig. 8. On the left panel we display the energy density distribution in the plane perpendicular to the beam direction, on the right panel in the $yz$ plane, where $z$ is the beam direction and $x$ the direction of the impact parameter. In Fig. 9 we display the final distribution of $\pi^+$ and $K^+$ at the end of the expansion in a simulation of Au+Au reactions at $\sqrt{s}$ = 200$A$GeV and a centrality of 30-40% with the PHSD initial condition. These results are compared with the results of PHSD calculations and the experimental data. The PHSD approach has been proven to describe a multitude of experimental data. Methods 1-3 characterize three methods to translate the PHSD initial energy density distribution into the quark degrees of freedom of the NJL approach. For details we refer to [20]. We see that the $\pi^+$ distribution is quite reasonable described, whereas for the $K^+$ we reproduce the high transverse momentum part but miss the yield at low momentum. The origin is that in our approach the $s + \bar{s} \to u(d) + \bar{u}(\bar{d})$ cross section is higher than in PHSD, so for the same initial condition we form less $K^+$.

Another observable of interest is the elliptic flow. In a hydrodynamical approach to describe the expanding plasma a elliptic flow is expected because the almond shape form of the overlap region between projectile and target is transformed into a larger pressure in the reaction plane that out of the reaction plane. In ideal hydrodynamics we expect that $v_2/\epsilon$, where $\epsilon$ is the eccentricity of the overlap region and $v_2$ the second coefficient in the expansion $\frac{dN}{d\phi} = C(1 + 2v_1 \cos\phi + 2v_2 \cos 2\phi + ..)$ of the azimuthal distribution of particles, is independent of the impact parameter. Since none of the two transport approaches which we discuss here is of hydrodynamical nature the $v_2$ they produce cannot be deduced from general considerations. Fig. 10 displays the results as compared to the experimental data. We see that PHSD agrees with the data. The NJL transport approach does not contain hadronic rescattering yet and therefore we have to compare its result with PHSD without rescattering, which is presented by the red line. Thus without rescattering we find good agreement between the NJL transport approach and PHSD. In order to study a bit more the expansion of the plasma we perform calculations in which we use a simplified initial condition, an expanding fireball whose geometry is given by the overlap of projectile and target. Fig. 11, left, shows that for such an initial condition the results

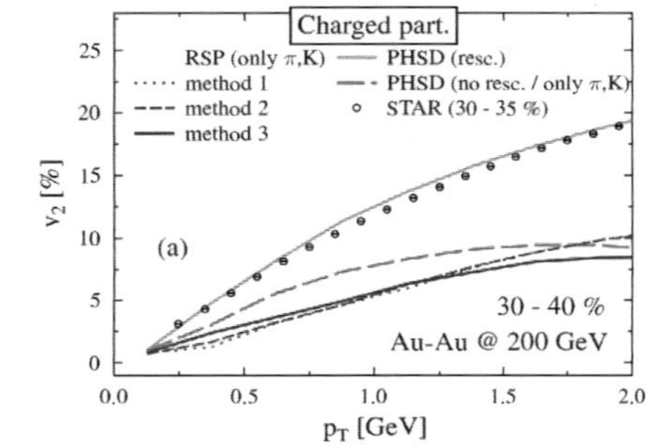


**Figure 10.** Elliptic flow, $v_2$ produced in the NJL based transport approach (RSP) and in PHSD in comparison with the experimental data for AuAu, $\sqrt{s}$ = 200 AGeV for a centrality of 30-40%.

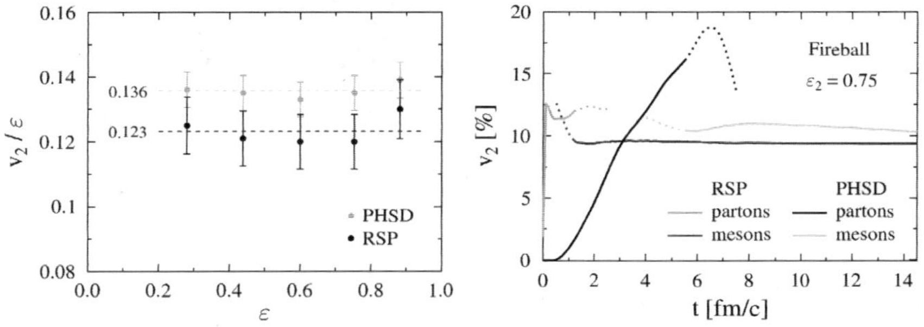


**Figure 11.** Elliptic flow $v_2$ for a fireball initial condition (see text). On the right panel we display the centrality dependence of $v_2/\epsilon$, for PHSD and the NJL based transport approach (RSP), on the right panel the time evolution of $v_2$ for the expanding fireball with an eccentricity of $\epsilon = 0.75$.

for both transport approaches are compatible with hydrodynamical calculations, means $v_2/\epsilon$ is independent of the centrality of the interaction and that the $v_2$ value of both approaches differ by only 10%. The dynamical creation of this $v_2$ value is, however, very different in the approaches. This is shown on the right hand side. The almost massless partons in the NJL approach produce the $v_2$ very fast and the hadronization lowers the $v_2$ value only little whereas in the PHSD approach where the partons are heavy the expansion as well as the creation of $v_2$ is much slower. The hadronization lowers the $v_2$ by 30%. So quite different scenarios for the expansion of the plasma give rather similar values for the final $v_2$.

It is interesting to study when the finally observed mesons are produced. As discussed the mesons get stable at temperatures below the Mott temperature. In an expanding plasma this transition from quarks to hadrons is not that sudden. We observe a broad distribution as seen in Fig.12. Mesons created above the Mott temperature can survive the time in which the local temperature passes the Mott temperature and mesons can be created well below the Mott temperature if only then the $q$ and $\bar{q}$ come sufficiently close to form mesons and the reaction becomes exothermic and hence favoured.

## 6. Summary

In summary, we have presented in this lecture the whole way from the definition of the (P)NJL Lagrangian to a transport approach based on this Lagrangian which can be compared with experimental data and other theories. The transport approach does not need any additional input as compared to the (P)NJL Lagrangian whose parameters are fixed by vacuum values of meson masses and decay constants. For a given initial condition for the expansion of a quark-antiquark plasma, which we borrow from the PHSD approach because it can presently not be calculated, in this transport approach the predictions for the observables are close to the experimental results (and close to that obtained in the PHSD approach). Future steps will be to include baryons in the transport approach and to include the interaction between gluons, presented by the Polyakov loop and the quarks, which has recently been calculated in a Dyson-Schwinger approach. This interaction will bring the

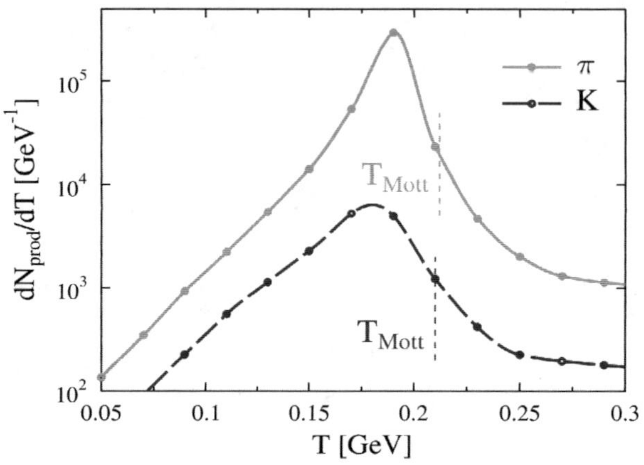


**Figure 12.** Local temperature distribution around the $q\bar{q}$ pair at the moment in which it creates pions which are finally observed.

equation of state of the P(NJL) approach close to that of recent lattice gauge calculations.

This work was supported by the LOEWE center "HIC for FAIR". JMTR is supported by the Program TOGETHER from Region Pays de la Loire and the European I3-Hadron Physics program. The computational resources have been provided by the LOEWE-CSC.

[1] C. Ratti, M. A. Thaler and W. Weise, Phys. Rev. D **73**, 014019 (2006).
[2] P. Zhuang, J. Hüfner and S. P. Klevansky, Nucl. Phys. A **576**, 525 (1994).
[3] J. Hüfner, S. P. Klevansky, P. Zhuang and H. Voss, Annals Phys. **234**, 225 (1994).
[4] D. Blaschke, M. Buballa, A. Dubinin, G. Röpke and D. Zablocki, Annals Phys. **348**, 228 (2014).
[5] S. P. Klevansky, Rev. Mod. Phys. **64**, 649 (1992).
[6] U. Vogl and W. Weise, Prog. Part. Nucl. Phys. **27**, 195 (1991).
[7] D. Ebert, H. Reinhardt and M. K. Volkov, Prog. Part. Nucl. Phys. **33**, 1 (1994).
[8] M. Buballa, Phys. Rept. **407**, 205 (2005).
[9] J. M. Torres-Rincon, B. Sintes and J. Aichelin, Phys. Rev. C **91**, 065206 (2015).
[10] B. J. Schaefer, J. M. Pawlowski and J. Wambach, Phys. Rev. D **76**, 074023 (2007).
[11] H. Hansen, W. M. Alberico, A. Beraudo, A. Molinari, M. Nardi and C. Ratti, Phys. Rev. D **75**, 065004 (2007).
[12] P. Costa, M. C. Ruivo, C. A. de Sousa, H. Hansen and W. M. Alberico, Phys. Rev. D **79**, 116003 (2009).
[13] A. Buck, R. Alkofer and H. Reinhardt, Phys. Lett. B **286**, 29 (1992).
[14] N. Ishii, W. Bentz and K. Yazaki, Nucl. Phys. A **587**, 617 (1995).
[15] H. Reinhardt, "Effective Hadron Theory of QCD," Egyptian-German Springschool Conference "Particle and Nuclear Physics", Forschungzentrum Juelich GmbH (Germany), (1992)
[16] R. Marty and J. Aichelin, Phys. Rev. C **87**, 034912 (2013).
[17] E. Quack and S. P. Klevansky, Phys. Rev. C **49**, 3283 (1994).
[18] P. Rehberg, S. P. Klevansky and J. Hüfner, Phys. Rev. C **53**, 410 (1996).
[19] J. Hüfner, S. P. Klevansky, E. Quack and P. Zhuang, Phys. Lett. B **337**, 30 (1994).
[20] R. Marty, E. Bratkovskaya, W. Cassing and J. Aichelin, Phys. Rev. C **92**, 015201 (2015).
[21] W. Cassing and E.L. Bratkovskaya, Nucl. Phys. A **831**, 2 (2009).

Journal of Physics: Conference Series **668** (2016) 012002 doi:10.1088/1742-6596/668/1/012002

# Freeze-out, Hadronization and Statistical Model

**Paolo Castorina**

Dipartimento di Fisica ed Astronomia, Universita' di Catania and INFN Sezione di Catania, Italy

E-mail: paolo.castorina@ct.infn.it

**Abstract.** The comparison of the statistical hadronization model with experimental data and lattice QCD results is not always straightforward. Indeed, the interpretation of the $\phi$ meson production, of the proton to pion multiplicity ratio at LHC and the agreement of the freeze-out curve with the lattice critical line in the $T - \mu_B$ plane, require further analyses. Moreover the dynamics of the hadronization has to be compatible with: 1) the statitical behavior also observed in elementary high energy collisions; 2) a universal hadronization temperature for all high energy collisions; 3) the freeze-out criteria. In these lecture notes the SHM is recalled and some explanations of the puzzling aspects of its comparison with data are discussed.

## 1. Introduction

The statistical hadronization model (SHM) [1, 2, 3, 4, 5] is a useful tool to describe the hadron formation at a scale where QCD is no longer perturbative and it has to be interpreted as a coarse grained approach which, with a small number of parameters, successfully describes the experimental data of the particle yields, of the low transverse momentum spectra and gives the dependence on the baryochemical potential, $\mu_B$, of the chemical freeze-out temperature, $T_{ch}$, i.e. the temperature where subsequent inelastic collisions between hadrons cease.

However, the comparison of the SHM with experimental and lattice results is not always straightforward and, for example, the interpretation of the $\phi$ meson production, of the proton to pion multiplicity ratio at LHC and the comparison of the freeze-out curve with the lattice critical line in the $T - \mu_B$ plane [6, 7] require further analyses. In these lecture notes the SHM and, in particular, some puzzling aspects of its comparison with data are reviewed.

In Sec. 2 the SHM is briefly summarized and Sec. 3 contains its comparison with specific experimental data and lattice results which opens interesting questions on the underlying dynamics. Sec.4 discusses the possible answers to the previous questions within the SHM or by introducing some modifications, completely consistent with the statistical approach. A possible understanding of other puzzling aspects, concerning the application of the SHM to elementary ($e^+e^-$ and hadron-hadron) collisions, the universality of the freeze-out temperature in all high energy collisions and the theoretical basis of the freeze-out criteria, are recalled in Sec. 5, where hadronization is interpreted as the Hawking-Unruh radiation in QCD. Comments and conclusions are given in Sec. 6.

## 2. Statistical hadronization model

The statistical hadronization model assumes that hadronization in high energy collisions is a universal process proceeding through the formation of multiple colorless massive clusters (or fireballs) of finite spatial extension. These clusters are taken to decay into hadrons according to a purely statistical law: every multi-hadron state of the cluster phase space defined by its mass, volume and charges is equally probable. The mass distribution and the distribution of charges (electric, baryonic and strange) among the clusters and their (fluctuating) number are determined in the prior dynamical stage of the process. Once these distributions are known, each cluster can be hadronized on the basis of statistical equilibrium, leading to the calculation of averages in the *microcanonical ensemble*, enforcing the exact conservation of energy and charges of each cluster.

Hence, in principle, one would need the mentioned dynamical information in order to make definite quantitative predictions to be compared with data. Nevertheless, for Lorentz-invariant quantities such as multiplicities, one can introduce a simplifying assumption and thereby obtain a simple analytical expression in terms of a temperature. The key point is to assume that the distribution of masses and charges among clusters is again purely statistical [5]. Therefore, as far as the calculation of multiplicities is concerned, the set of clusters becomes equivalent, on average, to a large cluster (*equivalent global cluster*) whose volume is the sum of proper cluster volumina and whose charge is the sum of cluster charges (and thus the conserved charge of the initial colliding system). In such a global averaging process, the equivalent cluster generally turns out to be large enough in mass and volume so that the canonical ensemble becomes a good approximation. In other words, a temperature can be introduced which replaces the *a priori* more fundamental description in terms of an energy density.

To obtain a simple expression for our further discussion, we neglect for the moment an aspect which is important in any actual analysis. Although in elementary collisions the conservation of the various discrete Abelian charges (electric charge, baryon number, strangeness, heavy flavour) has to be taken into account *exactly* [8, 9], as we shall discuss in detail in Sec. 4, we here consider for the moment a grand canonical picture. In general the average multiplicity of a given hadronic species $i$ then becomes

$$\langle n_i \rangle^{\text{primary}} = d_i \frac{VT}{(2\pi)^3} \int d^3p \frac{1}{\gamma_s^{-s_i} \exp[(E_i - \vec{\mu}\vec{q}_i)/T_{ch}] \pm 1}, \tag{1}$$

where $d_i$ is the degeneracy factor, $E_i = \sqrt{p^2 + m_i^2}$ is the particle energy, $s_i$ is the number of strange quarks/antiquarks it contains, $\vec{\mu}_i$ are the chemical potentials associated with the conserved charges $\vec{q}_i$, $V$ denotes the overall equivalent cluster volume, $\gamma_s$ is the parameter describing the strangeness suppression.

Here primary indicates that it gives the average number at the hadronization point, prior to all subsequent resonance decays. As a second step, all resonances in the gas which are unstable against strong decays are allowed to decay in lighter stable hadrons, using appropriate branching ratios for the decay published by the Particle Data Group [10].

A typical fit of the experimental data in the SHM contains four parameters: $T_{ch}, V, \mu_B$ and $\gamma_s$ and the strangeness suppression implies $\gamma_s < 1$ in elementary collisions.

In conclusion, the dynamics of the SHM corresponds to a gas of free hadronic resonances (HRG) at fixed temperature and chemical potentials which includes the known resonances up to a large mass (in general $2 - 2.5$ GeV).

In the next section we shall discuss some aspects of the comparison of the SHM with data.

## 3. Statistical hadronization model versus experimental data

Intuitively one expects a statistical behavior when there is a large number of partons in the initial state and/or of hadrons in the final state, but an interesting aspect of the SHM is that it is in good agreement with data for elementary collisions also [3, 4].

For example, the comparison of the SHM fit to the p+p data for the abundances of about 20 different species for collisions at $\sqrt{s} = 200$ GeV, reported in table 1 of Ref. [11], gives a $\chi^2/$ d.o.f. $= 15.6/14$, with $\gamma_s \simeq 0.6$ and $T_{ch} \simeq 170$ MeV. Surprising enough, also for $e^+e^-$ there is a good agreement with data, although the implementation of the SHM, the effective meaning of the $\chi^2$/d.o.f and the role of heavy quarks produced in the initial state by the virtual photon need some specifications [11, 12, 13, 14].

The strangeness suppression in elementary collisions is reported in Fig. 1 [15] and the enhancement in nucleus-nucleus scattering [16] is depicted in Fig. 2 ( $\gamma_s \to 1$) [15].

The compilation of the freeze-out temperatures for high energy ($\sqrt{s} \geq 20$ GeV) elementary and nucleus-nucleus collisions up to RHIC energy is reported in Fig. 3 and the independence of the hadronization temperature on the initial setting of the scattering (elementary particles or nuclei) clearly calls for a universal temperature for all high energy collisions.

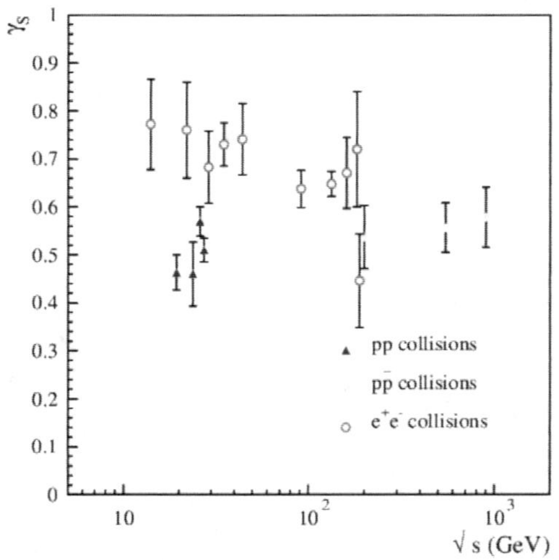


**Figure 1.** Strangeness suppression, i.e. $\gamma_s < 1$, in elementary collisions, from Ref. [15].

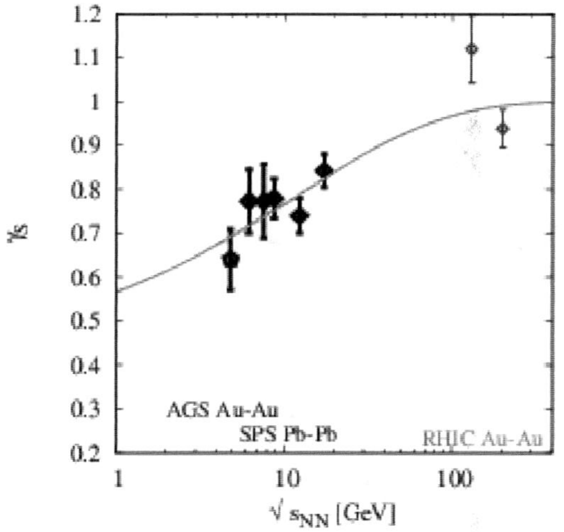


**Figure 2.** Strangeness enhancement in heavy ion collisions, i.e. $\gamma_s \to 1$, in the grand canonical description, Eq. (2). From Ref. [15].

The fits to the relativistic heavy ion data require the further parameter $\mu_B$, describing the baryon density, which turns out to be directly correlatd with the collision energy. The baryon-chemical potential sharply decreases with increasing energy ($\mu_B$ is only a few MeV at RHIC energy and $\mu_B \simeq 0$ at LHC). The freeze-out temperature strongly depends on $\sqrt{s}$ at low energy, i.e. on the baryon density at large $\mu_B$, and one obtains the freeze-out curve in fig.4 where is also reported the result of the recent fit to LHC ALICE data [17] (see later) and depicted are the curves obtained by imposing specific, intriguing, criteria. Indeed, the phenomenological freeze-out curve can be described by requiring that the average energy per particle , $< E > / < N >$ is about 1.08 GeV and/or that the ratio between the entropy density, $s$ and $T^3_{ch}$ is about 7 [18, 19, 20, 21, 22, 23, 24]. The latter result is completely consistent with lattice data on $s/T^3$ at the critical temperature, for $\mu_B = 0$ [25].

Finally, the recent LHC data are in good agreement with the SHM, however the proton to pion multiplicity ratio shows a discrepancy of about 3 standard deviations (see Fig. 6) [26, 27].

The previous brief summary of the comparison of the SHM with data calls for an understanding of the hadronization process which answers the following questions:

- Why is strangeness production universally suppressed in elementary collisions?

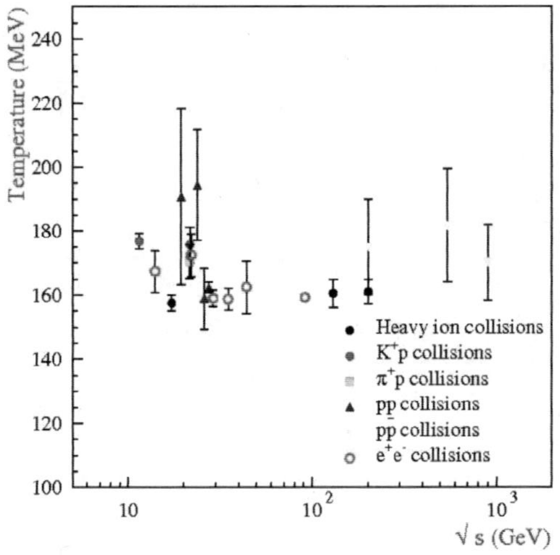


**Figure 3.** Universal behavior of freeze-out temperature for elementary and heavy ion collisions at high energy.

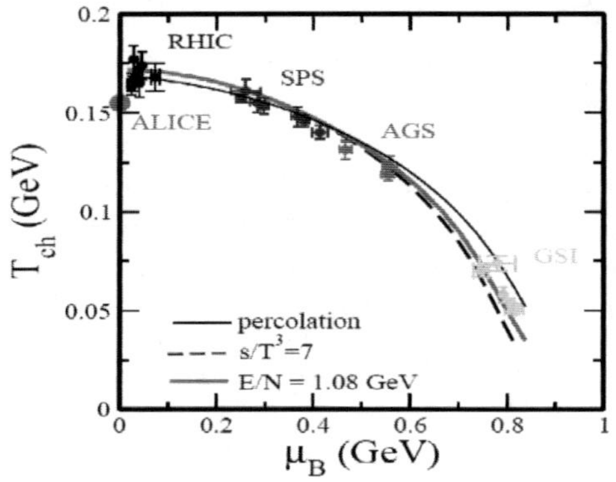


**Figure 4.** Freeze-out curve from SHM and comparison with the freeze-out criteria, see text.

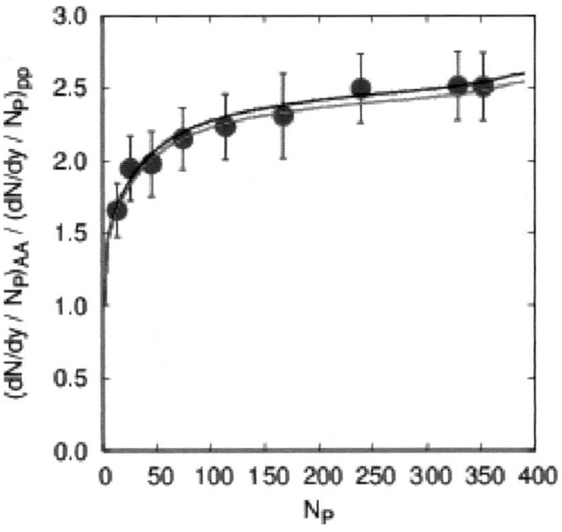


**Figure 5.** Suppression of $\phi$ meson production with centrality for Au-Au collision. The red curve is the fit based on the core-corona model [34].

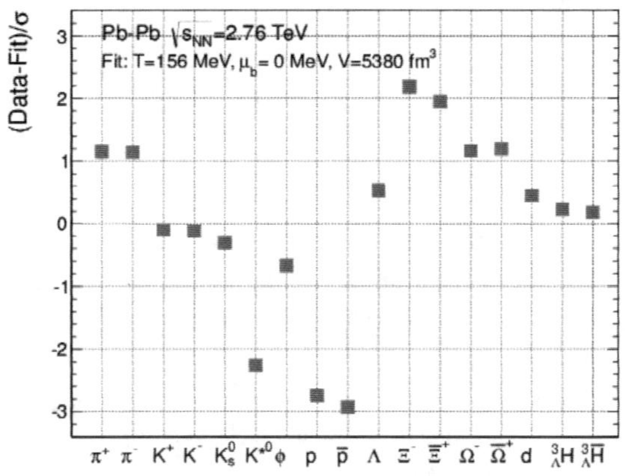


**Figure 6.** SHM fit to LHC data. Difference (data-fit) in standard deviation units, from Ref. [26].

- Why is there (almost) no strangeness suppression in nuclear collisions?
- Why is the proton/pion ratio at LHC in disagreement with the SHM fit?
- Is the freeze-out curve consistent with the lattice deconfinement critical line?

and, at a more fundamental level,

- Why do elementary high energy collisions show a statistical behavior?

Journal of Physics: Conference Series **668** (2016) 012002 doi:10.1088/1742-6596/668/1/012002

- Why is there a universal hadronization temperature for all high energy collisions?
- Why does hadron freeze-out correspond to $s/T^3 \simeq 7$ or $< E > / < N > \simeq 1.08$ GeV?

In the next sections we shall discuss some possible answers and the separation of the questions in two different groups will become more clear.

## 4. Statistical hadronization: conservative answers to open questions

Let us now consider the first set of questions for which some answers can be obtained within the SHM or consistently with it.

### *4.1. Canonical suppression, strangeness enhancement and the $\phi$ meson*

The average particle number in the grand canonical ensemble is given by Eq. (1), which by using the Boltzmann approximation can be written as

$$\langle n_i \rangle^{\text{primary}} = \frac{d_i V T m_i^2}{2\pi^2} \gamma_s^{s_i} \mathrm{K}_2 \left( \frac{m_j}{T} \right) \tag{2}$$

where $K_2(x)$ is the Hankel function ( $K(x) \sim \exp\{-x\}$ for large $x$). Notice that the ratio of particle yields is $V$ independent, however, as we shall clarify later, the nature of $V$ in elementary collisions is quite different from that in nuclear collisions and this can effectively lead to different behavior in the two cases.

The local conservation of charges, and in particular of strangeness, has been proposed for quite some time as the mechanism responsible for strangeness suppression [8, 9, 2]. The local strangeness suppression is based on two features. First, one imposes exact strangeness conservation, which leads to a volume-dependent strangeness reduction. However, in elementary collisions with the corresponding overall equivalent cluster volume, the resulting reduction is not sufficient to account for the observed strange particle rates. Hence it was argued that if in a given collision only one pair of strange hadrons is produced, these should appear close to each other spatially, the more so if the medium is relatively short-lived. This approach thus introduces somewhat *ad hoc* a strangeness correlation volume $V_c < V$, within which strangeness has to be conserved exactly. The corresponding model thus now has $T$, $V$ and $V_c$ as the parameters to be specified by the data, and fits based on such a model provide as good an account for the data as the earlier $\gamma_s$ scheme [28, 29], with the exception of the $\phi$, to which we return later. In Fig. 7 is shown the effect of the strangeness correlation radius, $R_c$, in the particle yields and the comparison with the grand canonical results. For $R_c \geq 2-3$ fm the canonical suppression almost disappears. In Fig. 8 the values of $R_c$ to fit the data at SPS and RHIC energies is given and extrapolated at LHC energy [30].

*A priori* little is known about $V_c$, and in particular, what happens to it in nuclear collisions. The physical origin of $V_c$ has been clarified in Ref. [31] and is due to the causality constraint in the hadronization mechanism.

The evolution of elementary high energy collisions is generally described in terms of an inside-outside cascade [32]. It specifies a boost-invariant proper time $\tau_q$, at which local volume elements experience the transition from an initial state of frozen virtual partons ("color glass") to the almost on-shell partons which will eventually form hadrons. This partonisation time can be estimated most easily in $e^+e^-$ annihilation where the initial quark-antiquark pair is bound by a string of tension $\sigma$. When the separation distance $R$ of the initial pair exceeds the energy $2\omega_q$ of an additional $\bar{q}q$ pair, the string breaks and the virtual pair is brought on-shell. For quarks of mass $m_q$, this energy is determined by $\sigma R = 2\sqrt{m_q^2 + k_T^2}$, where $k_T$ is the transverse momentum of each quark in the newly formed pair. Through lattice simulation [33] $k_T = \sqrt{\pi\sigma/2}$, leading to $R \simeq \sqrt{2\pi/\sigma} \simeq 1$ fm, using $\sigma \simeq 0.2$ GeV$^2$ and $m_q \ll \sigma$. From this,

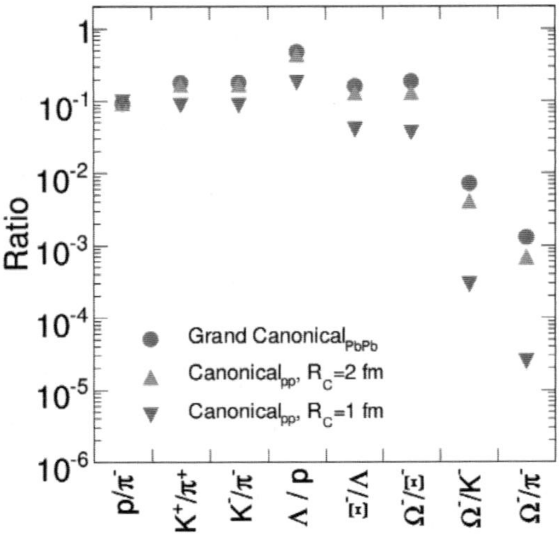


**Figure 7.** Comparison of grand canonical calculations with the canonical suppression ones, including a correlation volume with radius $R_c$ , from Ref. [30].

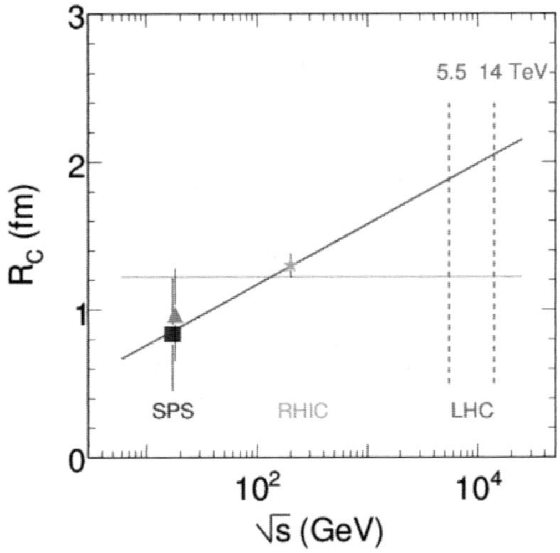


**Figure 8.** Extrapolation of $R_c$ at LHC energy [30].

we estimate $\tau_q \simeq \sqrt{\sigma/2\pi} \simeq 1$ fm. This process is subsequently iterated, leading to a cascade of emitted $\bar{q}q$ pairs; while the first pair appears at rest in the center of mass of the annihilation process, the subsequent pairs are produced at increasing rapidities. The different pairs will eventually bind to form free-streaming hadrons; for a boost-invariant evolution, this defines a second time threshold, the hadronization time $\tau_h > \tau_q$. The generalization to $pp$ collisions is straightforward: again there is a finite time $\tau_q$ needed to bring the partons on-shell, and after a larger time $\tau_h$, these combine to form hadrons. We denote the bubbles of medium for proper time $\tau$, with $\tau_q < \tau < \tau_h$, as "fireballs". Hadronization thus occurs through the formation of partonic fireballs in a cascade of increasing rapidities. It turns out from simple calculations and it is clear from Fig. 9 that there are hadronizing clusters with different spatial rapidity $\eta$ that are causally disconnected. This remains valid also including the spatial extension of the fireball, defined an evaluated in Ref. [31] as depicted in Fig 10.

The strangeness correlation volume $V_c$ should be identified in fact with that one of a causally connected cluster; causal connectivity thus provides the fundamental reason for local strangeness conservation and hence for the strangeness suppression observed in elementary interactions. It is moreover clear that in nucleus-nucleus interactions, the overlapping fireballs produced at fixed rapidity by the different nucleon-nucleon collisions will give rise to a much larger causally connected volume and thus effectively remove the locality constraints. Moreover, if *very* high energy $pp$ interactions lead to multiple jet production, this could eventually lead to a similar effect, with overlapping clusters from the different jet directions. This is indeed experimentally observed and the strangeness suppression decreases by increasing the energy of the pp collision (in the description of the suppression with $\gamma_s$ , one has $\gamma_s \simeq 0.6$ for $\sqrt{s} = 200$ MeV and $\gamma_s \simeq 0.8$ at $\sqrt{s} = 900$ MeV).

One can show [31] that the size of the causally connected cluster volumes varies with the fireball life-time,$\tau_h$. An obvious question therefore is whether the fits to production data lead to reasonable cluster sizes. It is found [28, 29] that good fits to data at $\sqrt{s} = 17.3$ and 200 GeV require a strangeness correlation radius of about 1 fm, while leading to the same

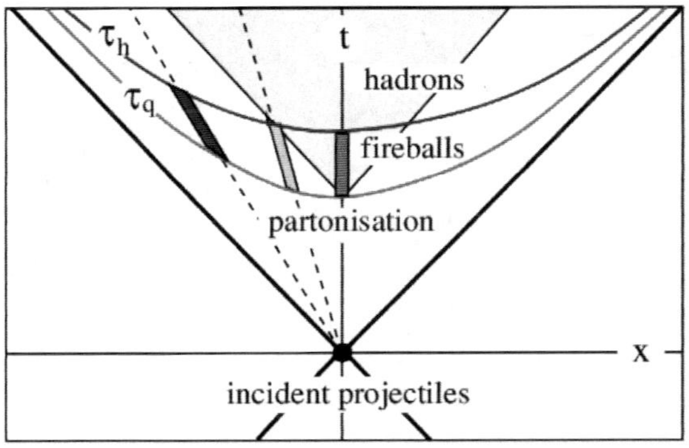


**Figure 9.** Evolution of the space-time diagram of fireball with no spatial extension. The yellow area indicates the space-time events causally connected with the fireball at rest in the center of mass system.

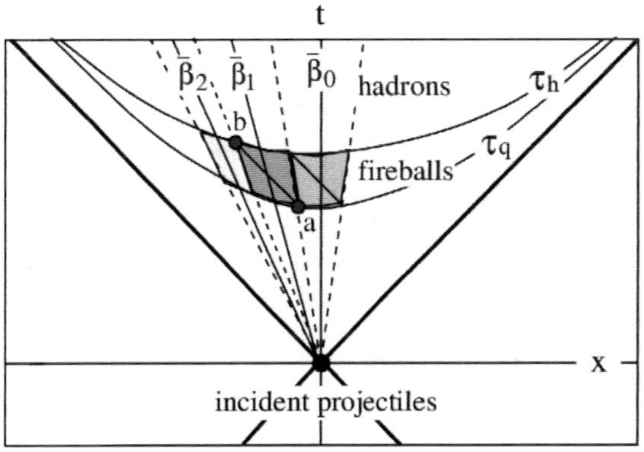


**Figure 10.** Fireballs with finite spatial extension and their causally connected regions. See Ref. [31] for details.

universal hadronization temperature of about 160 MeV. This is seen [31] to be in accord with a hadronization time $\tau_h$ of about 2 - 3 fm. For an evolution of the kind shown in Fig 10 that makes good sense: it takes about 1 fm to form the first $\bar{q}q$ pair, and another to have it hadronize. The causality constraints in elementary high energy collisions thus appears to provide the reason for the observed strangeness suppression, thereby justifying the strangeness correlation model.

However, it should be stressed that the canonical suppression is implemented at hadronic level and therefore there is no suppression for hidden strangeness, i.e. the $\phi$ meson. Indeed, It is found experimentally that the $\phi$ meson, consisting of an $s\bar{s}$ pair, is also suppressed, although from a hadronic point of view, it is of zero strangeness. Fig. 5 shows the data on the $\phi$ suppression for peripheral collisions (similar to elementary pp-collisions) with respect to central ones [34]. In the conventional statistical model with a strangeness suppression factor, the power of $\gamma_s$ is determined by the number of $s$ plus $\bar{s}$ quarks a given hadron contains. Hence the $\phi$ gets a factor $\gamma_s^2$, which leads to rough agreement with the data. In contrast, in a canonical formulation on a hadronic level, the $\phi$ does not present any quantum number to be conserved exactly and is not subject to any suppression.

The disagreement of the $\phi$ abundance may be a signal that strangeness correlation really occurs already on a pre-hadronic level. Requiring exact strangeness conservation for the quark system in the fireball prior to hadronization would in fact result in canonical strangeness suppression of both open and hidden strangeness. A more conservative answer is the core-corona model [34, 35], which is a minimal modification of the SHM and describe the heavy ion scattering as a superposition of a fully equilibrated core (where $\gamma_s = 1$) and single nucleon-nucleon collisions ($\gamma_s < 1$). The corona is indeed defined by the number of nucleons colliding only once , evaluated by Glauber Monte Carlo method. The model introduces a new parameter, essentially describing the non geometrical aspects of the collisions, which is fixed by data in one centrality bin. The red curve in Fig. 5 is the core-corona fit to the $\phi$ production and the model is successfully applied to many other particle species.

### *4.2. The proton/pion puzzle at LHC*

The unsatisfactory SHM fit to proton production at LHC energy, reported in Fig. 6, is a surprising result since the SHM works very well for nucleus-nucleus scattering up to RHIC energy for particle multiplicities which differ of many orders of magnitude.

Whithin the SHM a solution could be a modification of the resonance spectrum, crucial ingredient in the calculation of the yields, due to low mass resonances (still undetected but predicted by lattice QCD) or to high mass Hagedorn states (HS), with mass $m$ and spectrum $\rho(m) \simeq exp(m/T)$ (see Ref. [36] for a review of the Hagedorn states). At a temperature about 160 MeV, one can populate the hadrons using multihadronic decay reactions, $n\pi \rightarrow HS \rightarrow n^{'}\pi + \bar{X}X$ where $\bar{X}X = \bar{p}p, \bar{K}K, \bar{\Lambda}\Lambda$ and the results are reported in Fig. 11 [37] where the solid black dots represent the situation where there are no initial protons, kaons, and lambdas in the system (while the pions and Hagedorn states begin in chemical equilibrium) whereas the outlined circles represent the scenario when all hadrons begin in chemical equilibrium,

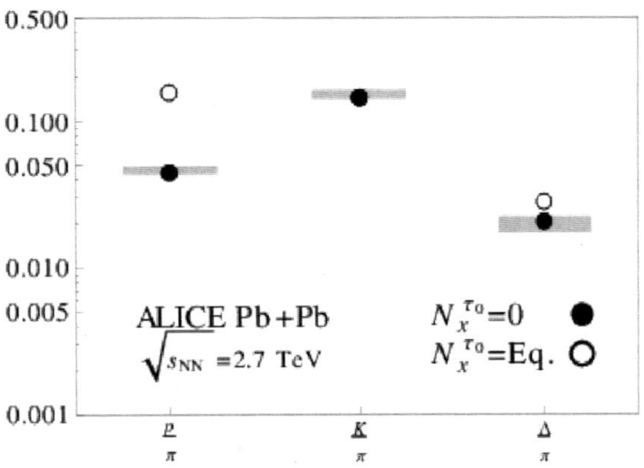


**Figure 11.** Calculations vs. experimental data points of ALICE at Pb+Pb at $\sqrt{s_{NN}} =$ 2.76 TeV from Ref. [37].

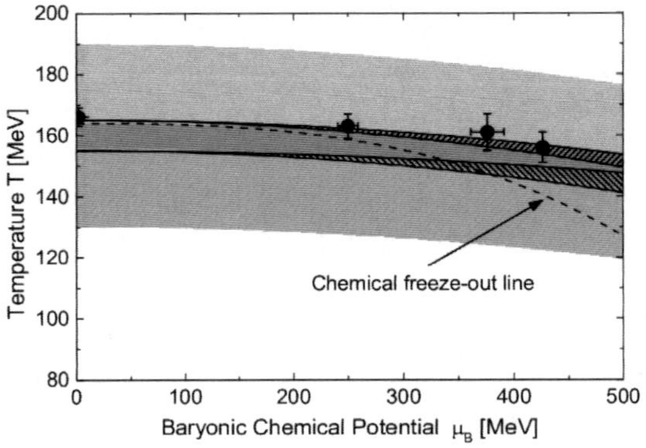


**Figure 12.** Phase diagram of strongly interacting matter: the coloured areas represent the widths of the crossover transitions. The dashed line is the chemical freeze-out line whilst the reconstructed original chemical equilibrium points in this work are shown as closed circles, see ref.[38].

While the previous proposal depends on a somehow arbitrary modification of the spectrum and a different understanding of the proton/pion puzzle could be the effect of the inelastic reactions between hadrons in the final state which induce changes in the baryon yields and in the chemical freeze-out curve. More precisely, all previous determinations of the points in the $(T, \mu_B)$ plane, in the framework of SHM, have implicitly assumed that the primordial hadro-chemical equilibrium (an intrinsic feature of the hadronization process as observed in elementary collisions) remains frozen-in throughout the final expansion phase in heavy ion collisions. On the other hand, the baryon-antibaryon annihilation and regeneration processes do not fall away with the onset of expansive cooling (Ref. [38] and references therein). Their final effect consists of a distortion of the initial, post-hadronization equilibrium yield distribution, in the antibaryon and baryon sector. The dynamical implemenation of this inelastic collisions in the final hadron/resonance expansion phase requires a specific code and the UrQMD hybrid model has been used in Ref. [38, 39]. The differences between the initial and the final hadronic multiplicities after the rescattering stage resemble the pattern of data deviation from the statistical equilibrium calculations, that is the "afterburner" corrections considerably reduces the $\chi^2$ per d.o.f. in Fig. 6 and the freeze-out temperature [38, 39].

The previous "correction" to the standard SHM seems to help to solve another problem. The freeze-out curve converges towards the lattice QCD phase critical line, at small $\mu_B$, but it appears to fall well below the line toward higher values of $\mu_B$. By taking these corrections into account in the statistical model analysis of the data, one is able to reconstruct the original hadrochemical equilibrium points in the $(T, \mu_B)$ plane [38] which closely follow the parton-hadron phase boundary recently predicted by lattice QCD at finite density.

However, the UrQMD hybrid model does not take into account the baryon and antibaryon regeneration. In Ref. [40] it has been clarified that the regeneration process reduces the "afterburner" correction of about $10 - 20\%$ and moreover one has to understand if the same mechanism is at work for hyperons since the resulting suppression is not observed in LHC data.

In conclusions, within the SHM or by introducing some modifications , not in contradiction with the statistical approach, one can answer the first set of questions of Sec. 3. However, the other questions (universal freeze-out temperature at high energy; success of the SHM in elementary collisions; freeze-out criteria) require a more detailed analysis of the dynamics underlying the hadronization process. The recent proposal of considering the hadron production as the Hawking-Unruh in QCD [41, 42] is recalled in the next section.

## 5. Confinement, string breaking and freeze-out criteria

In high energy heavy ion collisions multiple parton scattering could lead to kinetic thermalization, but $e^+e^-$ or elementary hadron interactions do not readily allow such a description. Moreover, the universality of the observed temperature suggests a common behavior in all high energy collisions or, in other terms, another non-kinetic mechanism providing a common origin of the statistical features.

Indeed a similar mechanism is well known in general relativity: for a black-hole the Hawking radiation [43] has a thermal spectrum (with the limits discussed in Ref. [44]), with the Hawking temperature $T_{haw} = k/2\pi = 1/8\pi GM$, where $k$ is the surface gravity (roughly, the acceleration at the event horizon) and $G$ is the Newton constant. As shown in Ref. [45, 46] this result can be understood in terms of tunneling through the event horizon.

More generally, Unruh [47] discovered that for an observer in uniform acceleration $a$ the Minkowski vacuum corresponds to a thermal bath with a temperature $T_u = a/2\pi$. For the uniformly accelerated observer (Rindler observer) there is an event horizon and the analogue of the Hawking radiation can be evaluated by particle tunneling through the event horizon [48]. It has a thermal spectrum with temperature $T_u$ [47, 49, 41, 48].

It is rather interesting to see if some of the previous considerations apply to QCD dynamics. Indeed lattice simulations and phenomenological analyses indicate that the large distances behavior of the potential $V$ between two static color charges increases linearly with the separation $r$, $V = \sigma r$. Therefore, QCD at large distances has a typical Rindler force, i.e. a constant acceleration, usually associated with a flux tube (a string) between a quark and an antiquark.

The phenomenological consequences of a constant acceleration and of the breaking of the flux tube in the high energy hadronization process have been analyzed in Ref. [41, 12, 50] with the conclusion that: 1) there is a universal Unruh temperature associated with the hadronization, $T_h \simeq 160 \pm 10$ MeV for massless quarks; 2) a small difference in the acceleration due to quark masses ($m_s \neq m_u = m_d$) explains the strangeness suppression in elementary collisions; 3) this suppression is almost removed in heavy ion collisions.

In this framework a possible understanding [51] of the unexplained freeze-out condition $\langle E\rangle/\langle N\rangle \simeq 1.09$ GeV , for $\mu_B \simeq 0$, seems rather natural. Indeed, the fundamental mechanism of hadronization is quark acceleration or deceleration, leading to string breaking with the resulting pair production and to the universal Unruh temperature. The maximum $\bar{q}q$ separation distance $R$, previously evaluated for $m_q = 0$, is given by $R = \sqrt{2\pi/\sigma}$.

The Unruh phenomenon, in high energy collisions, is responsible for the production of *newly formed* hadrons, therefore, it cannot address the role of the nucleons *already present* in the initial state of heavy ion collisions. As such, it captures the whole of the freeze-out process only as long as there are no significant baryon-density effects, i.e., only for $\mu_B \simeq 0$, corresponding to high energy collisions where the large part of produced hadrons are $q\bar{q}$ mesons. Therefore, the energy of the pair produced by string breaking, i.e. of the newly formed hadron is from previous equation, $E_h = \sigma R = \sqrt{2\pi\sigma}$.

In the central rapidity region of high energy collisions, one has $\mu_B \simeq 0$ so that $E_h$ is in fact the average energy $< E >$ per hadron, with an average number $< N >$ of newly produced hadrons. Hence one obtains

$$\frac{< E >}{< N >} = \sqrt{2\pi\sigma} = 1.09 \pm 0.08\ GeV \tag{3}$$

for $\sigma = 0.19 \pm 0.03$ GeV$^2$.

The possible explanation of the other freeze-out condition, $s/T_{ch}^3 \simeq 7$, is much more subtle. Indeed, in the hadronization process one deals with quantum particles, relativistically accelerated by the strong Rindler force, which experience a spacetime with an event horizon. From this point of view, since there is no gravitational interaction, the corresponding entropy has to be considered as an entanglement entropy due to causally disconnected regions. On the other hand, a quantum field near an event horizon has modes belonging to both sides, inside and outside the horizon, whose entanglement entropy can be computed to give $S_{ent} = \alpha A/r^2$, see, e.g., [52, 53, 54]. Here $A$ is the area of the event horizon, $r$ is the scale of characteristic quantum fluctuations, $\alpha$ is an undetermined numerical constant.

For a black-hole $\alpha = 1/4$ and $r^2 = G$ (in natural units). However, the metric of a uniformly accelerated observer corresponds to the near horizon approximation of a black-hole metric if the acceleration $a$ is equal to the surface gravity $k$ and therefore, by taking the phenomenological point of view of the analogy with the corresponding gravitational horizon, in Ref. [51] $\alpha = 1/4$ has been chosen. The scale of the characteristic quantum fluctuations for the quark-antiquark string is given by $r_T = 1/k_T$.

If for the entropy associated with hadron production at the string breaking the formula

$$S_h = \frac{1}{4}\frac{A_h}{r_T^2} = \frac{1}{4}\frac{4\pi R^2}{r_T^2} \tag{4}$$

is used with $R = \sqrt{2\pi/\sigma}$ and $T \simeq \sqrt{\sigma/2\pi}$ [41], one gets $S_h = \pi^3$. The entropy *density* divided by $T_h^3$ at freeze out (for $\mu_B \simeq 0$) turns out to be

$$\frac{s}{T_h^3} = \frac{S_h}{(4\pi/3)R^3 T_h^3} = \frac{3\pi^2}{4} \simeq 7.4. \tag{5}$$

compatible with the phenomenological value.

## 6. Comments and Conclusions

We have seen that all abundances in high energy collisions, including those of strange hadrons, are indeed given by an ideal resonance gas with some improvements/modifications compatible with a statistical approach. The underlying hadronization mechanism of the string breaking can answer some aspects of the comparison of the SHM with data.

We thus find that high energy heavy ion collisions produce essentially a medium which can be considered as hadronic matter in equilibrium, formed at the pseudocritical hadronization temperature predicted by lattice QCD.

However, we want to find in the collision data some sign of the QCD transition, of something like a critical behavior ([55] and reference therein). For example, sufficiently close to a continuous transition, correlations appear at all scales because the correlation length diverges, and in QCD this must produce strong deviations from the ideal hadron gas behavior.

Journal of Physics: Conference Series **668** (2016) 012002 doi:10.1088/1742-6596/668/1/012002

An interesting ansatz to understand hadronization and freeze-out parameter is provided in Refs. [56, 57] which reverse the mechanism of Mott dissociation of hadrons, occuring under compression and heating of hadronic matter as a delocalization of bound states of quarks, and picture the chemical freeze-out as a localization of the quark wave functions in the expanding and cooling fireball. This leads to a description of the chemical freeze-out line in the $(T, \mu_B)$ plane.

The study the fluctuations of conserved quantum numbers in an ideal hadron gas, compared to both lattice results and heavy ion data, has been proposed as an signature of the critical behavior (see, for example, Ref. [58]). However, the difference between the HRG and lattice simulations is effectively visible only for quantities which involve high derivatives of $P/T^4$ with respect to the chemical potentials (i.e. higher cumulants [58, 59]), and therefore the experimental detection requires a large number of events or physical conditions which enhance the fluctuations. Therefore, since a first order phase transition is expected at large $\mu_B$, the future accelerators for heavy ion collisions will have the chance to detect the deconfinement transition by analyzing the strong fluctuations of the conserved charges.

## References


[1] J. Cleymans et al., Phys. Lett. **B 242**, 111 (1990); J. Cleymans and H. Satz, Z. Phys. **C57**, 135 (1993); K. Redlich et al., Nucl. Phys. **A566**, 391 (1994); P. Braun-Munzinger et al., Phys. Lett. **B344**, 43 (1995); F. Becattini, M. Gazdzicki and J. Sollfrank, Eur. Phys. J. C **5**, 143 (1998); F. Becattini et al., Phys. Rev. C **64**, 024901 (2001);

[2] P. Braun-Munzinger, K. Redlich and J. Stachel, in *Quark-Gluon Plasma 3*, R. C. Hwa and X.-N Wang (Eds.), World Scientific, Singapore 2004.

[3] F. Becattini, *Universality of thermal hadron production in pp, p$\bar{p}$ and $e^+e^-$ collisions*, in *Universality features in multihadron production and the leading effect*, Singapore, World Scientific (1996), p. 74-104 [arXiv:hep-ph/9701275].

[4] F. Becattini, Z. Phys. **C69**, 485 (1996).

[5] F. Becattini and G. Passaleva, Eur. Phys. J. C**23**, 551 (2002).

[6] S. Borsanyi et al., Phys. Lett. **B370**, 99 (2014).

[7] A. Bazavov (HotQCD collaboration) Phys. Rev. D **90**, 094503 (2014). for a recent review see H-T. Ding, F. Karsch and S. Mukherjee, Thermodynamics of strong-interaction matter from Lattice QCD, arXiv:1504.05274.

[8] R. Hagedorn and K. Redlich, Z. Phys. C27, 541 (1985).

[9] S. Hamieh. K. Redlich, A. Tounsi, Phys. Lett. **B486**, 61 (2000).

[10] K. A. Olive et al. (Particle Data Group), Chin. Phys. **C38**, 090001 (2014).

[11] F. Becattini, P. Castorina, A. Milov and H. Satz, Eur. Phys. J. C **66** 377 (2010).

[12] F. Becattini, P. Castorina, J. Manninen, H. Satz, Eur. Phys. J. C **56** 493 (2008).

[13] A. Andronic, F. Beutler, P. Braun-Munzinger, K. Redlich and J. Stachel, Phys. Lett. **B675**, 312 (2009).

[14] K. Redlich, A. Andronic, F. Beutler, P. Braun-Munzinger and J. Stachel, J. Phys. G **36**, 064021 (2009).

[15] F. Becattini, J. Manninen, M. Gazdzicki, Phys. Rev. C **73**, 044905 (2006).

[16] J. Rafelski and B. Müller, Phys. Rev. Lett. **48**, 1066 (1982) [Erratum-ibid. 56 (1986) 2334]; for a recent review see J. Rafelski and M. Petran, Acta Phys. Polon. Supp. **7**, 35 (2014).

[17] M. Floris, Nucl. Phys. **A931**, 103 (2014).

[18] J. Cleymans and K. Redlich, Phys. Rev. Lett. **81**, 5284 (1998).

[19] J. Cleymans and K. Redlich, Phys. Rev. C **60**, 054908 (1999).

[20] J. Cleymans et al., arXiv:hep-ph/0511094.

[21] P. Braun-Munzinger and J. Stachel, J. Phys. G **28** , 1991 (2002).

[22] V. Magas and H. Satz, Eur. Phys. J. C **32**, 115 (2003).

[23] J. Cleymans et al., Phys. Lett. **B615**, 50 (2005).

[24] A. Tawfik, J. Phys. G **31**, S1105 (2005); Nucl. Phys. **A764**, 387 (2006).

[25] A. Bazazov et al. (HotQCD Collaboration), arXiv:1407.6387.

[26] J. Stachel, A. Andronic, P. Braun-Munzinger and K. Redlich, J. Phys. Conf. Ser. **509**, 012019 (2014).

[27] F. Becattini, P. Castorina, A. Milov and H. Satz, J. Phys. G **38**, 025002 (2011).

[28] I. Kraus et al., Phys. Rev. C **76**, 064903 (2007).

[29] I. Kraus et al., Phys. Rev. C **79**, 014901 (2009).

[30] I. Kraus, J. Cleymans, H. Oeschler and K. Redlich, J. Phys. G **37**, 094021 (2010).

[31] P.Castorina and H.Satz, Int. J. Mod. Phys. **E23** 4, 1450019 (2014).
[32] J. D. Bjorken, Lecture Notes in Physics (Springer) 56, 93 (1976).
[33] M. Luscher, G. Münster and P. Weisz, Nucl. Phys. **B180**, 1 (1981).
[34] F.Becattini and J.Manninen Phys. Lett. **B673**, 19 (2009).
[35] J. Aichelin and K. Werner, Phys. Rev. C **79**, 064907 (2009).
[36] K. Redlich and H. Satz, The Legacy of Rolf Hagedorn: Statistical Bootstrap and Ultimate Temperature, arxiv: 1501.07523.
[37] J. Noronha-Hostler and C. Greiner, Nucl. Phys. **A931**, 1108 (2014).
[38] R. Stock et al., PoS **CPOD 2013**, 011 (2013).
[39] J. Steinheimer, J. Aichelin and M. Bleicher, Phys. Rev. Lett. **110**, 042501 (2013).
[40] Yinghua Pan and S. Pratt, Phys. Rev. C **89**, 044911 (2014).
[41] P. Castorina, D. Kharzeev and H. Satz, Eur. Phys. J. C**52**, 187 (2007).
[42] D. Kharzeev and K. Tuchin, Nucl. Phys. **A753**, 316 (2005)
[43] S. W. Hawking, Comm. Math. Phys. **43**, 199 (1975).
[44] M. Visser , "Thermality of the Hawking flux", arXiv. 1409.7754.
[45] M.K. Parikh and F. Wilczek, Phys. Rev. Lett. **85**, 5042 (2000).
[46] M.K. Parikh, Gen. Rel. Grav. **36**, 2419 (2004).
[47] W. G. Unruh, Phys. Rev. **D14**, 870 (1976).
[48] See for example, A. de Gill, D. Singleton, V. Akhmedova and T. Pilling, Am.J.Phys. **78**, 685 (2010).
[49] W. G. Unruh and N. Weiss, Phys. Rev. **D29**, 165 (1984).
[50] P. Castorina and H.Satz, Adv. High Energy Phys. **2014**, 376982 (2014).
[51] P. Castorina, A. Iorio and H. Satz, Int. J. Mod. Phys. **E24**, 1550056 (2015).
[52] M. Srenidcky, Phys. Rev. Lett. **71**, 666 (1993).
[53] H. Terashima, Phys. Rev. D **61**, 104016 (2000).
[54] A. Iorio, G. Lambiase, G. Vitiello, Ann. Phys. **309**, 151 (2004).
[55] H. Satz, Int. J. Mod. Phys. **A28**, 1330043 (2013).
[56] D. B. Blaschke, J. Berdermann, J. Cleymans and K. Redlich, Phys. Part. Nucl. Lett. **8**, 811 (2011).
[57] D. Blaschke, J. Berdermann, J. Cleymans and K. Redlich, Few Body Syst. **53**, 99 (2012).
[58] See for example, F. Karsch and K. Redlich, Phys. Lett. **B695**, 136 (2011).
[59] C. Schmidt, Nucl. Phys. **A904-905**, 865c (2013).

Journal of Physics: Conference Series **668** (2016) 012003 doi:10.1088/1742-6596/668/1/012003

# Flavoured aspects of the QCD thermodynamics

**O. Kaczmarek**

Fakultät für Physik, Universität Bielefeld, D-33615 Bielefeld, Germany

E-mail: okacz@physik.uni-bielefeld.de

**Abstract.** We discuss recent progress in lattice QCD studies on various aspects involving strange and heavy quarks. Appropriate combinations of conserved net strange and net charm fluctuations and their correlations with other conserved charges provide evidence that in the hadronic phase so far unobserved hadrons contribute to the thermodynamics and need to be included in hadron resonance gas models. In the strange sector this leads to significant reductions of the chemical freeze-out temperature of strange hadrons. We have found that a description of the thermodynamics of open strange and open charm degrees of freedom in terms of an uncorrelated hadron gas is valid only up to temperatures close to the chiral crossover temperature. This suggests that in addition to light and strange hadrons also open charm hadrons start to dissolve already close to the chiral crossover. Further indications that open charm mesons start to melt in the vicinity of $T_c$ is obtained from an analysis of screening masses, while in the charmonium sector these screening masses show a behavior compatible with a sequential melting pattern. In the last section we discuss resent progress in extracting a transport coefficient, the heavy quark momentum diffusion coefficient, based on continuum extrapolated color-electric field correlation functions and estimate the time scale associated with the kinetic equilibration of heavy quarks. Close to $T_c$ this suggests that charm quark kinetic equilibration appears almost as fast as that of light partons in agreement with qualitative discoveries at RHIC and LHC that charm quarks appear to flow about as effectively as light quarks.

## 1. Additional strange hadrons and strangeness freeze-out in Heavy Ion Collisions

Fluctuations and correlations of conserved quantum numbers like baryon number B, electric charge Q, strangeness S and charm C are important quantities that can be used to analyze different contributions to thermodynamic quantities. Especially in the strange and charm sector they can serve as probes for the liberation of degrees of freedom and the appearance of fractional charge quantum numbers, e.g. by comparing to the limiting cases, i.e. Hadron Resonance Gas (HRG) in the hadronic phase and the non-interaction Stefan Boltzmann limit at high temperatures. Furthermore, by comparing to hadron resonance gas models information on the abundance of open strange and open charm hadrons can be gained.

The mesonic (M) and baryonic (B) pressure contributions of open strange hadrons in the hadron resonance gas model, using the classical Boltzmann approximation, can be evaluated separately by summing over all open strange mesons and baryons respectively,

$$P_{M/B}^{S,X}(T,\vec{\mu}) = \frac{T^4}{2\pi^2}\sum_{i\in X} g_i \left(\frac{m_i}{T}\right)^2 K_2(m_i/T)\cdot\cosh\left(B_i\hat{\mu}_B + Q_i\hat{\mu}_Q + S_i\hat{\mu}_S\right) \ , \tag{1}$$

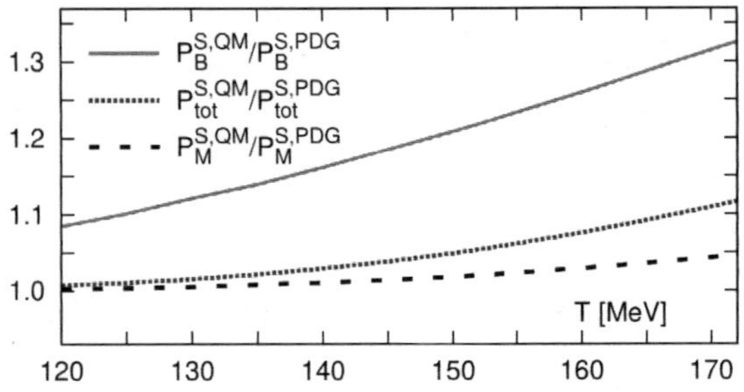


**Figure 1.** Ratios of partial pressures of open strange hadrons ($P_{\rm tot}^{S,X}$), mesons ($P_M^{S,X}$) and baryons ($P_B^{S,X}$) calculated in HRG with particle spectra from the particle data table, $X = PDG$, and with additional resonances predicted by the relativistic quark model, $X = QM$, respectively, from [1].

and both sum up to the total partial pressure of open strange hadrons,

$$P_{\rm tot}^{S,X} = P_M^{S,X} + P_B^{S,X} \ . \tag{2}$$

Here the chemical potentials related to the conserved quantum numbers are denoted by $\hat{\mu}_q \equiv \mu_q/T$ with $q = B,\ Q,\ S$ and $g_i$ is the degeneracy factor for the hadrons of mass $m_i$. The quark model (X=QM) predicts a larger number of open strange mesons [2] and open strange baryons [3] compared to the ones listed in the particle data tables (X=PDG). While the effect of the additional mesonic contribution on the partial pressure in the QM-HRG compared to the PDG-HRG is small, reflecting a smaller number of experimentally still-unobserved open strange mesons, the partial baryonic pressure is largely enhanced as can be seen in Fig. 1.

Suitable observables to analyze different contributions to the pressure $P$, that are related to the underlying effective degrees of freedom in the system, are generalized susceptibilities defined by derivatives with respect to the corresponding chemical potentials,

$$\chi_{klm}^{BQS} = \left. \frac{\partial^{(k+l+m)} P(T, \hat{\mu}_B, \hat{\mu}_Q, \hat{\mu}_S)/T^4}{\partial \hat{\mu}_B^k \partial \hat{\mu}_Q^l \partial \hat{\mu}_S^m} \right|_{\vec{\mu}=0} \tag{3}$$

In Fig. 2(top) the ratio of two such suitable, strangeness sensitive, cumulatants are shown, the correlation of net strangeness with net baryon number fluctuations, $\chi_{11}^{BS}$, normalized by the second cumulant of net strangeness fluctuations, $\chi_2^S$. While in the validity range of HRG models, the former only contains partial pressure contributions of strange baryons, the latter is dominated by strange mesons. Therefore this ratio is especially sensitive to the enhancement of partial baryonic pressure of strange baryons in the QM-HRG compared to the PDG-HRG. The results for two different lattice sizes together with an estimate for the continuum result shown in Fig. 2 are obtained in (2+1)-flavor QCD using the highly improved staggered quark (HISQ) action with a physical strange quark mass and light quark masses, $m_l = m_s/20$, corresponding to a pion mass of around 160 MeV [1].

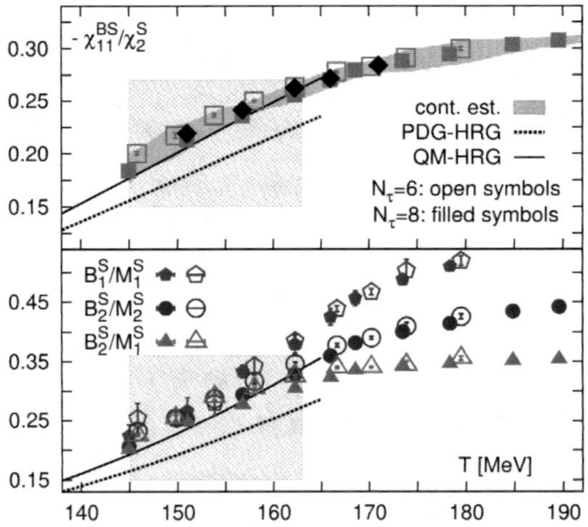


**Figure 2.** Top: $BS$ correlations normalized to the second cumulant of net strangeness fluctuations. The band shows an improved estimate for the continuum result. Bottom: Ratios of baryonic ($B_i^S$) to mesonic ($M_i^S$) pressure observables. The dotted and solid lines show results from PDG-HRG and QM-HRG model calculations, respectively. The shaded region denotes the continuum extrapolated chiral crossover temperature, from [1].

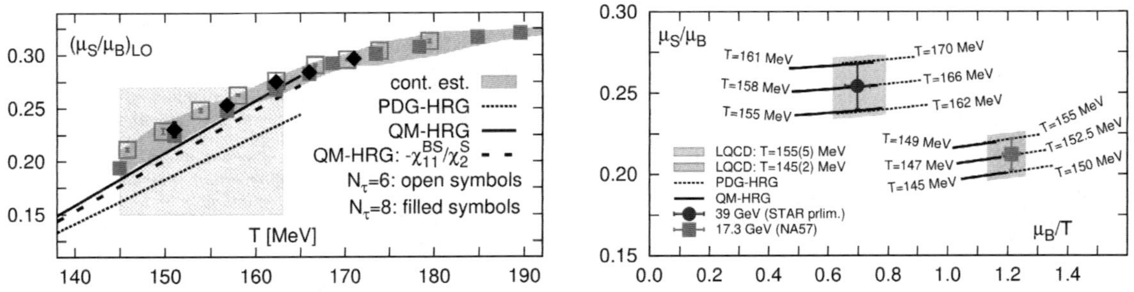


**Figure 3.** Left: Leading-order result for the ratio of strangeness and baryon chemical potentials versus temperature. Data and HRG model results are for a strangeness neutral system having a ratio of net electric charge to net baryon number $N_Q/N_B = 0.4$. The dashed line shows the QM-HRG result for vanishing electric charge chemical potential [1].
Right: Experimental values of $(\mu_S^f/\mu_B^f, \mu_B^f/T^f)$ extracted from fits to multiple strange hadrons yields are compared to $\mu_S/\mu_B$ predictions, obtained by imposing strangeness neutrality, from lattice QCD calculations (shaded bands) as well as from QM-HRG (solid lines) and PDG-HRG (dotted lines) models [1].

From the comparison of lattice QCD data with the two HRG models we find a better description of the QCD thermodynamics in the hadronic phase using the QM-HRG model. This shows the importance of the contributions of additional strange baryons to the thermodynamics below the QCD crossover and provides evidence for the existence of additional strange baryons.

This is further corroborated when comparing other combinations of second- and fourth-order cumulants of strangeness and net baryon number fluctuations [1], which are identical for a gas of uncorrelated hadrons, shown in Fig. 2(bottom). All three combinations agree up to the chiral crossover temperature, indicating again that a hadron resonance gas model is a valid description in the hadronic phase when additional states are included, before they start to deviate indicating the liberation of strangeness degrees of freedom in the QGP.

In a net strangeness neutral system, like in heavy ion collisions, the ratio $\chi_{11}^{BS}/\chi_2^S$ can be related to the strangeness chemical potential in terms of a Taylor expansion in $\mu_B/T$ [4, 5], where the leading order only receives a small correction from the nonzero electric charge potential,

$$\left(\frac{\mu_S}{\mu_B}\right)_{\rm LO} \equiv s_1(T) = -\frac{\chi_{11}^{BS}}{\chi_2^S} - \frac{\chi_{11}^{QS}}{\chi_2^S}\frac{\mu_Q}{\mu_B}\,, \tag{4}$$

and the next-to-leading order correction is small in the relevant temperature region. Up to the chiral crossover temperature, the leading order results of the QM-HRG model agree with the lattice QCD results in Fig. 3(left), while the PDG-HRG model gives substantially lower values.

This has implications for the determination of the chemical freeze-out temperature, illustrated in Fig. 3(right) where the experimental freeze-out chemical potentials were obtained from the relative yields of strange anti-baryons to baryons for the NA57 experiment [6] and the STAR preliminary result [7]. Comparing the freeze-out chemical potentials to the leading order results for $\mu_S/\mu_B$ in 3(left) determines the freeze-out temperature, $T^f$. While the QM-HRG and lattice QCD results are in good agreement and give consistent values for $T^f$, the freeze-out temperatures obtained from the PDG-HRG model are larger by about 8 MeV for the STAR data and 5 MeV for the NA57 data.

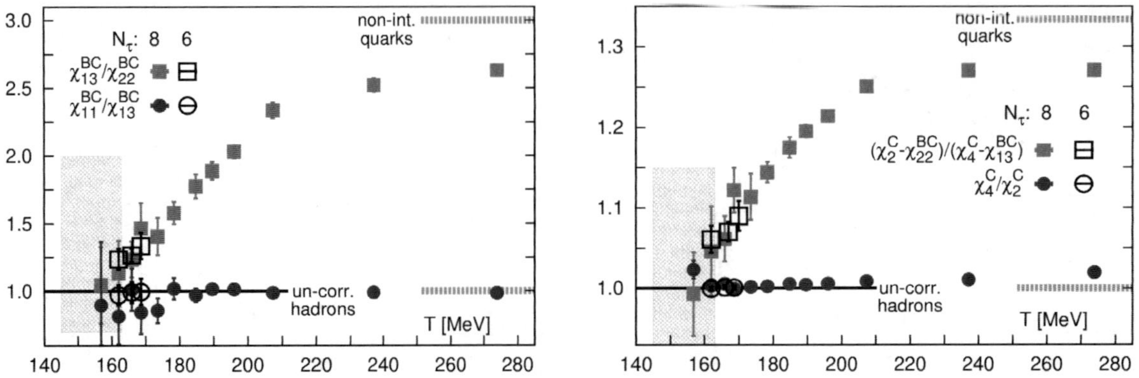


**Figure 4.** Left:Ratios of fourth order baryon-charm ($BC$) correlations. In an uncorrelated hadron gas both ratios receive contributions only from charmed baryons. The horizontal lines on the right hand side show the infinite temperature non-interacting charm quark gas limits. Right: Ratio $\chi_4^C/\chi_2^C$, which is dominated by open charm meson contributions and $(\chi_2^C - \chi_{22}^{BC})/(\chi_4^C - \chi_{13}^{BC})$ which only receives contributions from open charm mesons. The horizontal lines on the right hand side show the infinite temperature non-interacting charm quark gas limits. Both taken from [8].

## 2. The melting of open charm hadrons

Due to their relation to chiral symmetry, deconfinement of light quark bound states happens at or close to the chiral crossover temperature, leading to sudden changes in bulk thermodynamic observables apparent in the behavior of fluctuations of conserved charges. For open strange hadrons, studying results of charge fluctuation observables containing strangeness, a description in terms of uncorrelated hadronic degrees of freedom breaks down in the chiral crossover region [9]. This is an indication that also strangeness gets dissolved at or close to $T_c$.

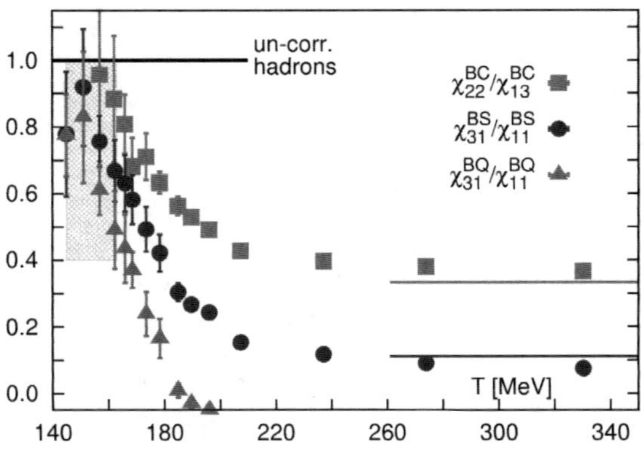


**Figure 5.** Ratios of baryon-electric charge ($BQ$), baryon-strangeness ($BS$) and baryon-charm ($BC$) correlations. The shaded region shows the chiral crossover region. Horizontal lines on the right side show corresponding results for an uncorrelated quark gas. This limiting value is not defined for $\chi_{31}^{BQ}/\chi_{11}^{BQ}$ since the denominator and numerator vanishes in perturbation theory up to $\mathcal{O}(g^4)$. From [8].

The situation of heavy quarks becomes more interesting as they are more related to the deconfinement aspects and less affected by chiral symmetry and at least some charmonium and bottomonium bound states are expected to dissolve at temperatures in the deconfined medium well above the chiral crossover region. In this section we will discuss the open charm sector which can be analyzed by charge fluctuation observable introduced in the previous section, adding a charm quark chemical potential.

In our lattice QCD results [8] the charm quark sector is treated within the quenched approximation, which is well justified in the temperature region analyzed here [10, 11]. We utilize the same gauge field configurations as in the previous section, generated with the HISQ

Journal of Physics: Conference Series **668** (2016) 012003 doi:10.1088/1742-6596/668/1/012003

action and (2+1)-flavor with physical strange quark masses and almost physical light quark masses. To reduce discretization errors in the charm sector the so called $\epsilon$-term [12] is used for the charm contribution in the generalized susceptibilities of conserved charges,

$$\chi_{klmn}^{BQSC} = \left. \frac{\partial^{(k+l+m+n)}[P(\hat{\mu}_B, \hat{\mu}_Q, \hat{\mu}_S, \hat{\mu}_C)/T^4]}{\partial \hat{\mu}_B^k \partial \hat{\mu}_Q^l \hat{\mu}_S^m \partial \hat{\mu}_C^n} \right|_{\vec{\mu}=0} . \tag{5}$$

Using suitable combinations of cumulants of net charm fluctuations, $\chi_n^C$, and of correlations between net charm and net baryon number fluctuations, $\chi_{mn}^{BC}$, the open charm baryonic and mesonic sectors can be analyzed separately.

The ratio $\chi_{13}^{BC}/\chi_{22}^{BC}$ shown in Fig. 4(left) is unity in a hadron resonance gas and approaches a value of 3 in the infinite temperature non-interacting charm quark gas limit. The results suggest that above the crossover region, an uncorrelated gas of charmed baryons no longer provides an appropriate description of the net charm to net baryon correlations. The second ratio shown in Fig. 4(left), $\chi_{11}^{BC}/\chi_{13}^{BC}$, is unity in both limits and also the lattice QCD results are consistent with unity for all temperatures.

Fig. 4(right) shows ratios which are sensitive to the open charm meson sector. Also here it is obvious that a hadron resonance gas model description is only justified up to the crossover region.

The behavior of the correlations between net charm fluctuations and net baryon number fluctuations is similar to the behavior of corresponding correlations in the strangeness [9] as well as in the light quark sector [13]. Fig. 5 shows results for the three sectors. All the ratios should be unity in a gas of uncorrelated hadrons and approach their non-interacting quark gas limit at infinite temperatures. It is apparent that a hadron resonance gas model description breaks down just above the chiral crossover temperature not only for the light and open strange hadrons but also for the open charm hadrons.

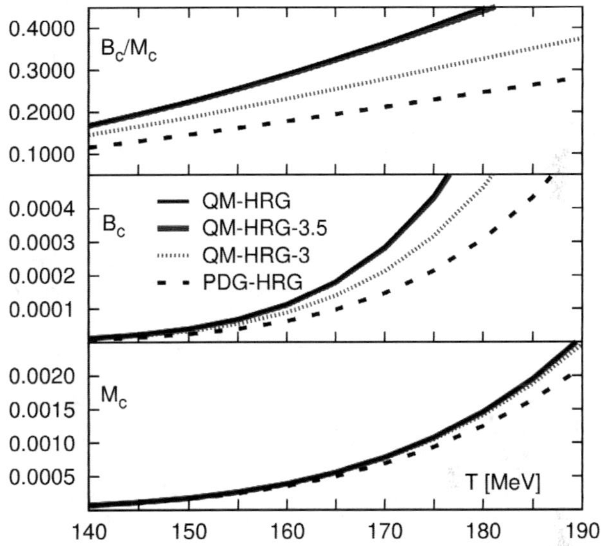


**Figure 6.** Partial pressure of open charm mesons ($M_c$, bottom), baryons ($B_c$, middle) and the ratio $B_C/M_C$ (top) in a gas of uncorrelated hadrons, using all open charm resonances listed in the particle data table (PDG-HRG) and using additional charm resonances calculated in a relativistic quark model (QM-HRG, solid lines) Also shown are results from HRG model calculations where the open charm resonance spectrum is cut off at mass 3 GeV (QM-HRG-3) and 3.5 GeV (QM-HRG-3.5). From [8].

## 3. The abundance of open charm hadrons

The results from suitable cumulants and correlations from conserved quantum numbers in the strangeness sector, as discussed in the previous sections, open the question about the importance of the inclusion of additional yet unobserved open charm mesons and charmed baryons in hadron resonance gas models and their contribution to the thermodynamics.

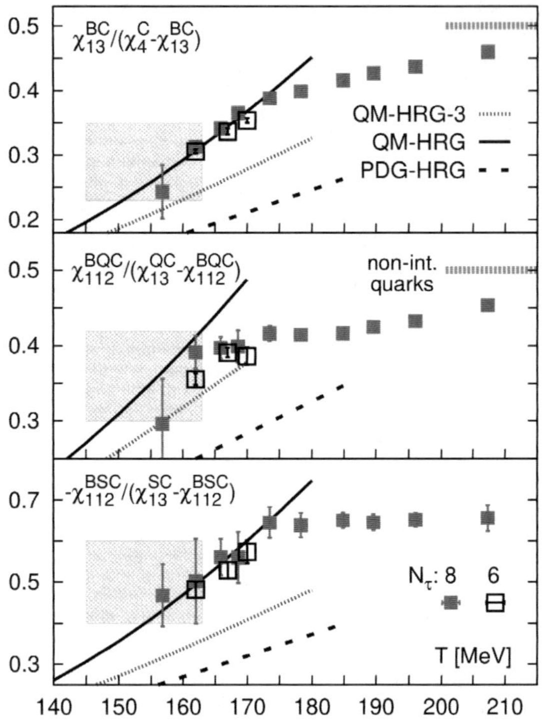


**Figure 7.** Thermodynamic contributions of all charmed baryons, $R_{13}^{BC}$ (top), all charged charmed baryons, $R_{13}^{QC}$ (middle) and all strange charmed baryons, $R_{13}^{SC}$ (bottom) relative to that of corresponding charmed mesons. The dashed lines (PDG-HRG) are predictions for an uncorrelated hadron gas using only the PDG states. The solid lines (QM-HRG) are similar HRG predictions including also the states predicted by the quark model. The dotted lines (QM-HRG-3) are the same QM predictions, but only including states having masses $< 3$ GeV. The shaded region shows the QCD crossover region. The horizontal lines on the right hand side denote the infinite temperature non-interacting charm quark gas limits for the respective quantities. The lattice QCD data have been obtained on lattices of size $32^3 \cdot 8$ (filled symbols) and $24^3 \cdot 6$ (open symbols). From [8].

In Fig. 6 the partial pressures of open charm mesons and charmed baryons are shown for different HRG models, i.e. using all open charm resonances listed in the particle data table (PDG-HRG), including additional charm resonances calculated in a relativistic quark model (QM-HRG), the latter cut-off at mass 3 GeV (QM-HRG-3) and 3.5 GeV (QM-HRG-3.5). Due to the experimentally better known meson sector, the differences between the PDG-HRG and QM-HRG models are negligible up to the chiral crossover temperature, while in the baryonic sector the influence of predicted additional states becomes as large as 40% and the largest contributions come from states up to 3 GeV.

The HRG model predictions for the different open charmed baryon contents are shown in Fig. 7 for three ratios of charmed charge correlations and fluctuations that are sensitive to all charmed baryons (top), all charged charmed baryons (middle) and all strange charmed baryons (bottom). The lattice QCD results for those ratios show good agreement with a hadron resonance gas model that contains all open charm meson and baryon spectra calculated in a relativistic quark model [14, 15]. In contrast, the PDG-HRG model results are far below the lattice QCD data in the temperature range of the QCD crossover region. This observation provides evidence for a substantial contribution of experimentally so far unobserved charmed baryons to the pressure of a hadron resonance gas. Nevertheless the charm contribution to the total pressure is of course strongly suppressed relative to the contribution of the non-charmed sector.

## 4. Screening masses of quarkonia

The bulk thermodynamic charge fluctuations discussed in the previous sections provided information about the melting and abundance of open strange and open charm baryons. Due to their relation to conserved quantum numbers, these quantities are not sensitive to strange mesons ($s\bar{s}$) and charmonia ($c\bar{c}$). In this sector hadronic current-current correlation functions are the appropriate tools to obtain spectral properties on the thermal modification of bound states in the medium.

In contrast to temporal correlation functions which are limited by the finite temporal extent, $1/T$, of the lattice, spatial correlation functions can be studied at larger separations and can directly be compared to the corresponding vacuum correlation functions to quantify modifications of the in-medium spectral functions. Although the relation to the spectral function, $\rho(\omega, p_z, T)$, is even more involved than in the temporal case as the spatial correlation functions are sensitive to the momentum dependence of spectral functions,

$$G(z,T) = \int_{-\infty}^{\infty} \frac{dp_s}{2\pi} e^{ip_z z} \int_0^{\infty} \frac{d\omega}{2\pi} \frac{\rho(\omega, p_z, T)}{\omega}, \quad (6)$$

the large distance behavior of correlation functions, $G(z,T) \sim \exp(-M(T)z)$, is dominated by an exponential decay characterized by the screening mass $M$. While at low temperatures, in the case of a bound state in the spectrum, the screening mass is equal to the pole mass of the ground state, at high temperatures when the mesonic excitations are completely melted, the screening mass becomes strongly temperature dependent. In the non-interacting limit it is related to the lowest non-zero Matsubara frequency [17],

$$M_{\text{free}} = \sqrt{m_{q_1}^2 + (\pi T)^2} + \sqrt{m_{q_2}^2 + (\pi T)^2}, \quad (7)$$

with the masses $m_{q_1}$ and $m_{q_2}$ of the quark and anti-quark of the corresponding meson.

Therefore a change of the temperature dependence of screening masses between those two limiting cases can be used as an indicator for thermal modifications and dissociation of charmonia and open charm mesons.

Fig. 8 shows results, obtained using the HISQ action [16], for $s\bar{s}$ mesons (top), $s\bar{c}$ (middle) and charmonium (bottom) for various states compared to the corresponding zero-temperature masses (solid lines) and non-interaction limit Eq. 7 (dashed line).

The screening masses for open charm mesons start to deviate from the vacuum masses already in the chiral crossover region and their temperature dependence is comparable to that in the strange meson sector. This is a further indication that open charm mesons start to melt in the vicinity of $T_c$ in agreement with the results of open quantum number fluctuations of the previous sections.

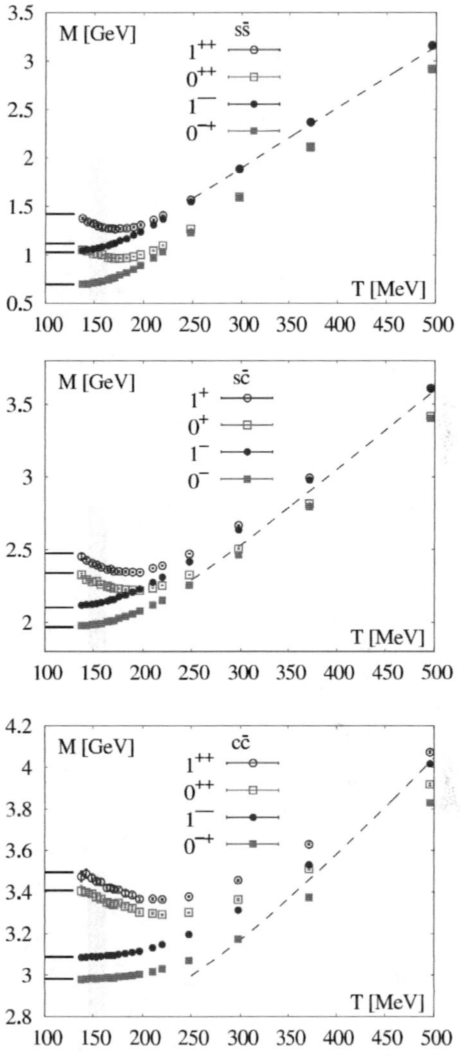


**Figure 8.** Screening masses for different channels in $s\bar{s}$ (top), $s\bar{c}$ (middle) and $c\bar{c}$ (bottom) sectors as functions of the temperature. The solid horizontal lines on the left depict the corresponding zero temperature meson masses. The shaded regions indicate the chiral crossover temperature $T_c = (154 \pm 9)$ MeV. The dashed lines are the corresponding free field theory results. Taken from [16].

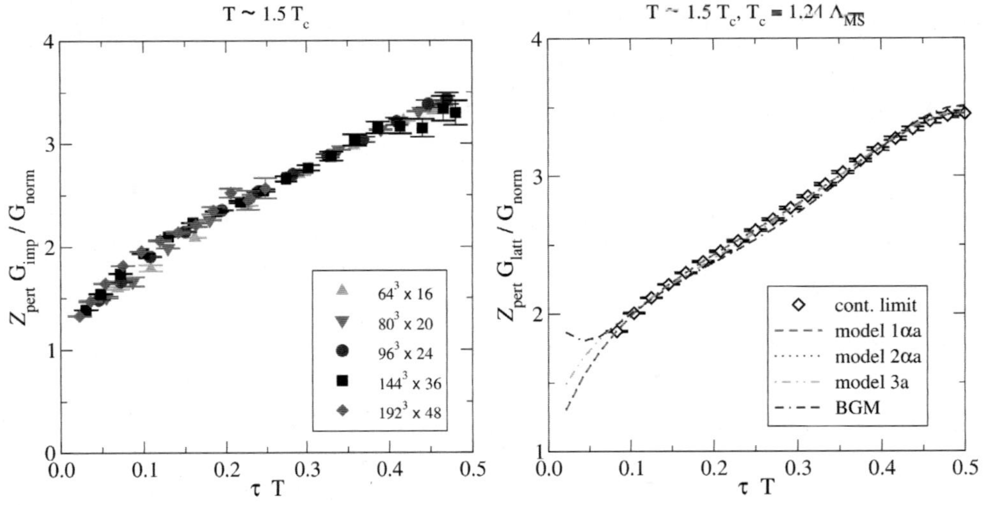


**Figure 9.** Left: Lattice data of the color-electric correlator after perturbative renormalization and the use of tree-level improved distances. Right: Continuum-extrapolated lattice data and examples of different interpolation results, as described in the text. The results have been normalized to the leading order correlator, $G_{\rm norm}$. Taken from [18].

The results in the charmonium sector show a behavior compatible with a sequential melting pattern. While the $J/\Psi$ $(1^{--})$ and $\eta_c$ $(0^{-+})$ screening masses remain almost identical to their vacuum pole masses up to around 1.1 $T_c$, the screening masses for $\chi_{c0}$ and $\chi_{c1}$ show thermal modifications already below $T_c$. At temperatures above two times $T_c$ all channels show a linear rise of the screening masses with temperature.

## 5. Heavy quark diffusion coefficients

Besides information on the medium modifications of bound states, also transport coefficients, in the case of heavy quarks these are heavy quark diffusion coefficients, are encoded in vector meson correlation and spectral functions. However due to the different contributions and different scales entering, it is more convenient to use an operator that is more sensitive to the transport contribution and does not contain any bound state contribution.

In the limit of heavy quarks, using Heavy Quark Effective Theory, a purely gluonic operator, the color-electric correlator can be derived [19, 20],

$$G_{\rm E}(\tau) \equiv -\frac{1}{3}\sum_{i=1}^{3} \frac{\Big\langle {\rm ReTr}\Big[U(\beta;\tau)\, gE_i(\tau,\vec{0})\, U(\tau;0)\, gE_i(0,\vec{0})\Big]\Big\rangle}{\Big\langle {\rm ReTr}[U(\beta;0)]\Big\rangle}\,, \quad \beta \equiv \frac{1}{T}\,, \tag{8}$$

where $gE_i$ denotes the color-electric field, $T$ the temperature, and $U(\tau_2;\tau_1)$ a Wilson line in the Euclidean time direction.

In a recent study we have calculated the color-electric correlator (8) on large quenched lattices [18] and performed the continuum extrapolation for $G_{\rm E}(\tau)$ at a temperature of about 1.5 $T_c$. Fig. 9 shows the results on the different lattices (left) and the continuum extrapolated correlator (right).

The correlator is related to the spectral function by

$$G_{\rm E}(\tau) = \int_0^\infty \frac{{\rm d}\omega}{\pi}\, \rho_{\rm E}(\omega)\, \frac{\cosh[\omega(\frac{\beta}{2}-\tau)]}{\sinh[\frac{\omega\beta}{2}]}\,. \tag{9}$$

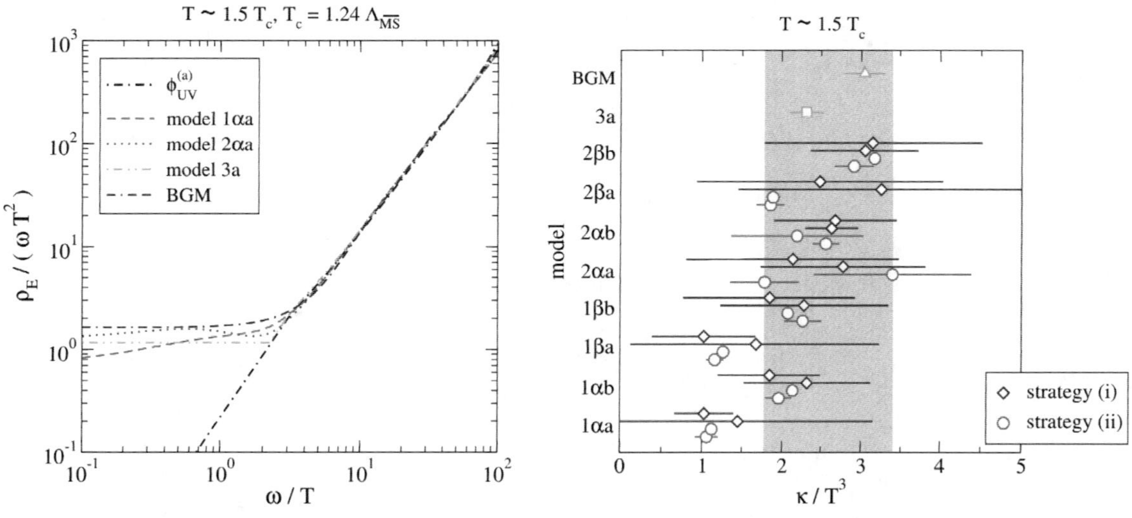


**Figure 10.** Left: Examples of model spectral functions, compared with the UV perturbative spectral function $\phi^{(a)}_{\rm UV}$ and the result a Backus-Gilbert method (BGM) study, from [18]. Right: Results for the heavy quark momentum diffusion coefficient $\kappa$ based on different models and different fitting strategies. The grey band illustrates the final estimate based on models 2, $3a$ and BGM [18].

In the infrared regime ($\omega \ll T$), the heavy quark momentum diffusion coefficient can be defined as [19]

$$\kappa \equiv \lim_{\omega\to 0} \frac{2T\rho_{\rm E}(\omega)}{\omega} \ . \tag{10}$$

In contrast to the case of correlation and spectral functions of conserved currents where a transport peak is expected, the approach to this limit appears to be smooth and $\rho_{\rm E}$ is rather a monotonically increasing function for small $\omega$,

$$\rho_{\rm E}^{\rm IR} = \frac{\kappa\omega}{2T}. \tag{11}$$

In the ultraviolet regime ($\omega \gg T$) due to asymptotic freedom, the spectral function can be computed in perturbation theory. We used various interpolations between the two limiting cases to perform a detailed analysis of the systematic uncertainties in extracting the heavy quark momentum diffusion $\kappa$ based on fits to the continuum extrapolated correlation function. In addition to these fits, we have also made use of a variant of the Backus-Gilbert method (BGM) [21, 22]. Some of the resulting correlation and spectral functions are shown in Fig. 9(right) and Fig.10(left) respectively.

The results of our full systematic analysis on the transport coefficient $\kappa$ are shown in Fig. 10(right). The grey band illustrates the central value,

$$\kappa/T^3 = 1.8 \ldots 3.4 \ . \tag{12}$$

In the non-relativistic limit, i.e. for a heavy quark mass $M \gg \pi T$ the heavy quark momentum diffusion $\kappa$ is related to the diffusion coefficient, $D = 2T^2/\kappa$, for which we obtain

$$DT = 0.59 \ldots 1.1 \ . \tag{13}$$

In this limit $\kappa$ is also related to the drag coefficient, $\eta_D = \kappa/(2MT)$. The latter can be interpreted as the kinetic equilibration time scale associated with heavy quarks, $\tau_{\text{kin}} = \eta_D^{-1}$. With our result for $\kappa$ (12) we obtain

$$\tau_{\text{kin}} = \frac{1}{\eta_{\text{D}}} = (1.8 \ldots 3.4) \left(\frac{T_c}{T}\right)^2 \left(\frac{M}{1.5\ \text{GeV}}\right) \text{fm/c}\ . \tag{14}$$

Close to $T_c$, charm quark kinetic equilibration appears therefore to be almost as fast as that of light partons, for which a time scale $\sim 1$ fm/c is generally considered.

## Conclusions

Studying various observables of net strangeness and net charm fluctuations in comparison with predictions from hadron resonance gas models, we have found evidence that additional, experimentally so far unobserved strange and charm hadrons become thermodynamically relevant in the vicinity of the QCD crossover region. In the strange sector this leads to significant reductions of the chemical freeze-out temperature of strange hadrons. When including these additional states in hadron resonance gas models, the thermodynamics of open strange and open charm degrees of freedom in terms of an uncorrelated hadron gas gives a good description in the hadronic phase, but only up to the chiral crossover region. Deviations at higher temperatures suggest the liberation of strangeness degrees of freedom in the crossover region and that open charm hadrons start to dissolve already close to $T_c$. The analysis of screening masses are consistent with these findings for open charm mesons ($s\bar{c}$), but suggest a sequential melting pattern for charmonium ($c\bar{c}$) states in the QGP. We have determined a continuum estimate of the heavy quark momentum diffusion coefficient. It's value can be used to serve as an indication that close to $T_c$, charm quark kinetic equilibration appears almost as fast as that for light partons in agreement with experimental results on the flow of charm quarks.

## References

[1] A. Bazavov *et al. Phys. Rev. Lett.* **113** no. 7, (2014) 072001.
[2] D. Ebert, R. N. Faustov, and V. O. Galkin *Phys. Rev.* **D79** (2009) 114029.
[3] S. Capstick and N. Isgur *Phys. Rev.* **D34** (1986) 2809.
[4] A. Bazavov *et al. Phys. Rev. Lett.* **109** (2012) 192302.
[5] S. Borsanyi *et al. Phys. Rev. Lett.* **111** (2013) 062005.
[6] F. Antinori *et al. Phys. Lett.* **B595** (2004) 68–74.
[7] F. Zhao *PoS* **CPOD2013** (2013) 036.
[8] A. Bazavov *et al. Phys. Lett.* **B737** (2014) 210–215.
[9] A. Bazavov *et al. Phys. Rev. Lett.* **111** (2013) 082301.
[10] A. Bazavov *et al. PoS* **LATTICE2013** (2014) 154.
[11] S. Borsanyi *et al. PoS* **LATTICE2011** (2011) 201.
[12] E. Follana *et al. Phys. Rev.* **D75** (2007) 054502.
[13] A. Bazavov *et al. Phys. Rev.* **D86** (2012) 034509.
[14] D. Ebert, R. N. Faustov, and V. O. Galkin *Eur. Phys. J.* **C66** (2010) 197–206.
[15] D. Ebert, R. N. Faustov, and V. O. Galkin *Phys. Rev.* **D84** (2011) 014025.
[16] A. Bazavov, F. Karsch, Y. Maezawa, S. Mukherjee, and P. Petreczky *Phys. Rev.* **D91** no. 5, (2015) 054503.
[17] W. Florkowski and B. L. Friman *Z. Phys.* **A347** (1994) 271–276.
[18] A. Francis, O. Kaczmarek, M. Laine, T. Neuhaus, and H. Ohno `arXiv:1508.04543`.
[19] S. Caron-Huot, M. Laine, and G. D. Moore *JHEP* **04** (2009) 053.
[20] J. Casalderrey-Solana and D. Teaney *Phys. Rev.* **D74** (2006) 085012.
[21] H. B. Meyer *Phys. Rev. Lett.* **100** (2008) 162001.
[22] B. B. Brandt, A. Francis, H. B. Meyer, and D. Robaina `arXiv:1506.05732`.

Journal of Physics: Conference Series **668** (2016) 012004 doi:10.1088/1742-6596/668/1/012004

# Approaches to QCD phase diagram; effective models, strong-coupling lattice QCD, and compact stars

**Akira Ohnishi**

Yukawa Institute for Theoretical Physics, Kyoto University, Kyoto 606-8502, Japan

E-mail: ohnishi@yukawa.kyoto-u.ac.jp

**Abstract.** The outline of the two lectures given in "Dense Matter School 2015" is presented. After an overview on the relevance of the phase diagram to heavy-ion collisions and compact star phenomena, I show some basic formulae to discuss the QCD phase diagram in the mean field treatment of the Nambu-Jona-Lasinio model. Next, I introduce the strong-coupling lattice QCD, which is one of the promising methods to access the QCD phase diagram including the first order phase boundary. In the last part, I discuss the QCD phase diagram in asymmetric matter, which should be formed in compact star phenomena.

## 1. Introduction

QCD phase diagram is closely related to the history of our universe and can be probed by using laboratory experiments including heavy-ion collisions. Recent developments in heavy-ion collision and compact star physics have been providing hints to understand of the QCD phase diagram at finite temperature ($T$) and/or finite baryon density ($\rho_B$) as schematically shown in the left panel of Fig. 1. Heavy-ion collision data obtained at RHIC top energy ($\sqrt{s_{NN}} = 200$ GeV) and LHC energy (run 1, $\sqrt{s_{NN}} = 2.76$ TeV) imply the formation of strongly interacting matter consisting of quarks and gluons at high $T$ and almost zero $\rho_B$ [1]. The phase transition at zero $\rho_B$ is known to be a crossover from lattice QCD Monte-Carlo (LQCD-MC) simulations [2]. Crossover nature of the transition justifies the standard scenario of the homogeneous big bang nucleosynthesis. In heavy-ion collisions at lower energies, the rapidity gap between projectile and target nuclei becomes smaller. Nucleons in the projectile and target lose their energies and can get into the mid-rapidity region. Thus the nuclear stopping gives rise to the formation of hot matter at finite $\rho_B$. In the right panel of Fig. 1, we plot the density and temperature reachable in Au+Au collisions at 25$A$GeV ($\sqrt{s_{NN}} = 7.1$ GeV) calculated in a hadron transport model, JAM [3]. In a small central box of 1 fm$^3$, the event averaged density may reach $9\rho_0$. Unfortunately, the LQCD-MC simulations suffer from the notorious sign problem, which prevents us from obtaining precise phase boundary at finite $\rho_B$ from the first principles non-perturbative calculation. Effective model calculations suggest the existence of the first order phase transition boundary at finite density [6]. It is also possible that we have the region with inhomogeneous chiral condensate instead of a sharp phase boudary [7]. If the phase transition density is not very high, we may have quark matter in the neutron star core, whose density could reach $(5-10)\rho_0$. Between the first order boundary (if exists) and the crossover transition, there should be a critical point (CP) [6], where the transition is the second order. Around CP, we expect large fluctuations of the order parameter and other observables which couple with the

Journal of Physics: Conference Series **668** (2016) 012004 doi:10.1088/1742-6596/668/1/012004

order parameter [8]. The location of CP determines the shape of the phase diagram, and CP hunting is in progress in the current beam energy scan (BES) program at RHIC [9] and will be performed in the forthcoming facsilities such as FAIR, NICA, and heavy-ion beam facsilities at J-PARC.

In this proceedings contribution, I give the outline of the two lectures given in "Dense Matter School 2015". In the first lecture (before SQM), I explained basic idea to describe the chiral phase transition based on a mean-field treatment of a chiral effective model. In the second lecture (during SQM), I discussed how we can describe the QCD phase diagram in the strong-coupling lattice QCD. I also discussed the effects of isospin chemical potential on the phase boundary, and their implication to the compact star phenomena.

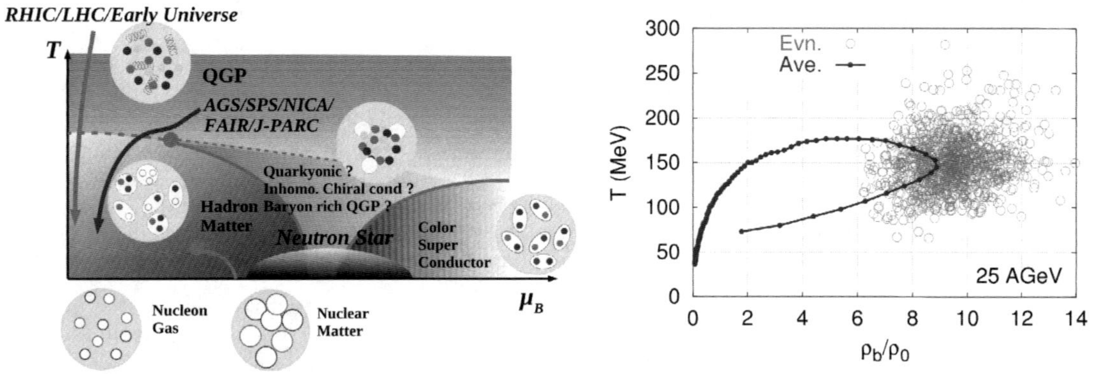


**Figure 1.** Left: Schematic QCD phase diagram [4]. Right: Temperature and density in 25 AGeV Au+Au collisions [5].

## 2. QCD phase diagram in chiral effective models

The QCD phase transition involves chiral and deconfinement transitions, and LQCD-MC results show that two transitions takes place at similar temperatures. I here concentrate on the chiral transition.

Chiral symmetry is the symmetry of QCD with massless quarks. The QCD Lagrangian is given as

$$\begin{aligned}\mathcal{L}_{\rm QCD} =&\bar{q}(i\gamma^\mu D_\mu - m)q - \frac{1}{2}{\rm tr}F^{\mu\nu}F_{\mu\nu} \\ =&\bar{q}_L(i\gamma^\mu D_\mu)q_L + \bar{q}_R(i\gamma^\mu D_\mu)q_R - (\bar{q}_L m q_R + \bar{q}_R m q_L) - \frac{1}{2}{\rm tr}F^{\mu\nu}F_{\mu\nu}\ ,\end{aligned} \tag{1}$$

where $q_{L,R} = (1 \mp \gamma_5)q/2$ are left- and right-handed quarks and $F_{\mu\nu}$ is the field strength tensor of gluon. When the quarks are massless ($m = 0$), $\mathcal{L}_{\rm QCD}$ is invariant under independent rotation of $q_L$ and $q_R$ in the flavor space, $q_{L,R} \to q'_{L,R} = U_{L,R}q_{L,R}$, where $U_{L,R}$ are SU($N_f$) matrices. In terms of hadrons, the chiral transformation mixes $\sigma = \bar{q}q$ and $\pi^a = \bar{q}i\gamma_5\tau^a q$. We should have a scalar meson having the same mass as pions if the vacuum is also invariant under the chiral transformation. In our world, the chiral condensate has a finite value in vacuum, $\langle\bar{q}q\rangle \neq 0$, and the vacuum is not invariant under the chiral transformation. Thus the symmetry of the QCD action with massless quarks is spontaneously broken[1], $\mathrm{SU}(N_f)_L \otimes \mathrm{SU}(N_f)_R \otimes \mathrm{U}(1)_B \otimes \mathrm{SU}(N_c) \to$

[1] $\mathrm{U}(1)_A$ is broken by the anomaly.

$\mathrm{SU}(N_f)_V \otimes \mathrm{U}(1)_B \otimes \mathrm{SU}(N_c)$. This spontaneous breaking of chiral symmetry in addition to the explicit breaking (finite bare quark mass) gives rise to the mass difference of $\sigma$ and $\pi$ modes. At high $T$ and/or $\rho_B$, spontaneously broken chiral symmetry is restored, at least partially, *i.e.* the chiral transition takes place.

The chiral transition has been studied extensively by using chiral effective models such as the Nambu-Jona-Lasinio (NJL) model [10], the Quark Meson (QM) model [11], and Polyakov loop extended versions of them such as PNJL [12] and PQM [13] models. These models have chiral symmetry, and we expect that they could be obtained from QCD by integrating out high-momentum (short-range) degrees of freedom. Let us consider the NJL model Lagrangian[2],

$$\mathcal{L}_{\mathrm{NJL}} = \bar{q}(i\gamma^\mu\partial_\mu - m_0 + \gamma_0\mu)q + \frac{G^2}{2\Lambda^2}\left[(\bar{q}q)^2 + (\bar{q}i\gamma_5\boldsymbol{\tau}q)^2\right] , \tag{2}$$

$$\mathcal{L}^{(E)}_{\mathrm{NJL}} = \bar{q}(-i\gamma_\mu\partial_\mu + m_0 + i\gamma_4\mu)q - \frac{G^2}{2\Lambda^2}\left[(\bar{q}q)^2 + (\bar{q}i\gamma_5\boldsymbol{\tau}q)^2\right] . \tag{3}$$

In the second line, we introduce the Euclidean action, $\mathcal{L}^{(E)} = -\mathcal{L}(t = -i\tau, \gamma^0 = -i\gamma_4, A^0 = -iA_4)$, where Euclidean coordinate and $\gamma$ matrices are $(x_\mu)_E = (\tau = it, \mathbf{x})$ and $(\gamma_\mu)_E = (i\gamma^0, \boldsymbol{\gamma})$, respectively. The partition function in NJL is given as

$$\begin{aligned}
\mathcal{Z}_{\mathrm{NJL}} &= \int \mathcal{D}[q,\bar{q}] e^{-\int_0^{1/T} d\tau \int d\mathbf{x} \mathcal{L}^{(E)}_{\mathrm{NJL}}} = \int \mathcal{D}[q,\bar{q},\sigma,\boldsymbol{\pi}] e^{-\int d^4x \left\{\bar{q}Dq + \frac{\Lambda^2}{2}(\sigma^2+\boldsymbol{\pi}^2)\right\}} \\
&= \int \mathcal{D}[\sigma,\boldsymbol{\pi}] \exp\left[-S_{\mathrm{eff}}(\sigma,\boldsymbol{\pi};T)\right] , && (4) \\
D &= -i\gamma\partial + m_0 + i\gamma_4\mu + G(\sigma + i\gamma_5\boldsymbol{\tau}\cdot\boldsymbol{\pi}) , \\
S_{\mathrm{eff}} &= -\log\det D + \int d^4x \frac{\Lambda^2}{2}\left[\sigma^2(x) + \boldsymbol{\pi}^2(x)\right] . && (5)
\end{aligned}$$

In the second line, we have converted the four-Fermi interaction term in $\mathcal{L}^{(E)}_{\mathrm{NJL}}$ by using the Hubbard-Stratonovich transformation.

In order to calculate the partition function (or the free energy), we need to obtain the determinant of the fermion matrix $D$ under the anti-periodic boundary condition in the temporal coordinate, $q(1/T) = -q(0)$, for a given configuration of the auxiliary fields, $(\sigma(x), \boldsymbol{\pi}(x))$, and to integrate over the auxiliary fields. We here work in the mean field approximation in order to illustrate how the chiral transition occurs in a simple manner. In the mean field approximation, we replace the auxiliary fields with constant numbers, $(\sigma(x), \boldsymbol{\pi}(x)) \to (\sigma, 0)$, and $\sigma$ is chosen to minimize $S_{\mathrm{eff}}$. Then the Fermion matrix becomes a Dirac operator with modified mass, $D = -i\gamma\partial + i\gamma_4\mu + m(m = m_0 + G\sigma)$. $D$ is diagonal in plane wave basis states, $D = -\gamma_0(i\omega + \mu) + \boldsymbol{\gamma}\cdot\mathbf{k} + m$, and the determinant is found to be[3] $\det D = \prod_{n,\mathbf{k}}((\omega_n - i\mu)^2 + E_k^2)^{d_f/2}$ ($E_k = \sqrt{\mathbf{k}^2 + m^2}$), where $d_f = 4N_cN_f$ is the Fermion degrees of freedom and $\omega_n = \pi T(2n-1)$ is the Matsubara frequency. Now the free energy

[2] We follow the notation in Ref. [14]

[3] The determinant of the usual (Minkowski) Dirac operator $D^{(M)}$ becomes zero for $E = \pm E_k$, then we get $\det D^{(M)} = \prod(E^2 - E_k^2)^{d_f/2}$. By replacing $E \to i\omega + \mu$, we get $\det D$.

density is obtained as

$$\begin{aligned}
\mathcal{F}_{\text{eff}} =& -\frac{T}{V}\log\mathcal{Z} = \frac{\Lambda^2}{2}\sigma^2 - \frac{T}{V}\sum_{n,\mathbf{k}}\log\left((\omega_n - i\mu)^2 + E_k^2\right)^{d/2} \\
=& \frac{\Lambda^2}{2}\sigma^2 - d_f\int\frac{d^3k}{(2\pi)^3}\left[\frac{E_k}{2} + \frac{k^2}{3E_k}\frac{1}{2}\left(\frac{1}{e^{(E_k-\mu)/T}+1} + \frac{1}{e^{(E_k+\mu)/T}+1}\right)\right] \\
=& \frac{\Lambda^2}{2}\sigma^2 - \frac{d_f}{2}\Lambda^4 I\left(\frac{m}{\Lambda}\right) - P^{(F)} , \qquad (6)
\end{aligned}$$

$$I(x) = \frac{1}{16\pi^2}\left[\sqrt{1+x^2}(2+x^2) - x^4\log\frac{1+\sqrt{1+x^2}}{x}\right] \simeq \frac{1 + x^2 + \frac{x^4}{8}\left(1 + 4\log\frac{x}{2}\right) + \mathcal{O}(x^6)}{8\pi^2} \qquad (7)$$

From the first to the second line in Eq. (6), Matsubara frequency sum is performed. The first term in Eq. (6) comes from bosonization, and the second and third terms come from the Fermion determinant and show the zero point energy and thermal pressure, respectively.

The spontaneous symmetry breaking is understood from the shape of $\mathcal{F}_{\text{eff}}$ as a function of $\sigma$. Let us consider the case in the chiral limit in vacuum ($m = 0$, $T = 0$ and $\mu = 0$). We can ignore the thermal contribution $P^{(F)}$, and $\mathcal{F}_{\text{eff}}$ can be written as

$$\mathcal{F}_{\text{eff}}/\Lambda^4 = -\frac{d_f}{16\pi^2} + \frac{x^2}{2}\left[\frac{1}{G^2} - \frac{1}{G_c^2}\right] + \mathcal{O}(x^4\log x) \quad (x = m/\Lambda,\ G_c^2 = 8\pi^2/d_f) . \qquad (8)$$

In the case $G > G_c$, the coefficient of $x^2$ is negative, the $\mathcal{F}_{\text{eff}}$ minimum appears at $x \neq 0$, and the chiral condensate is non-zero in vacuum as shown schematically in the left panel of Fig. 2. Thus if the interaction is strong enough, $\sigma(= -\langle\bar{q}q\rangle)$ condensates and constituent quark mass $m = G\sigma$ is generated.

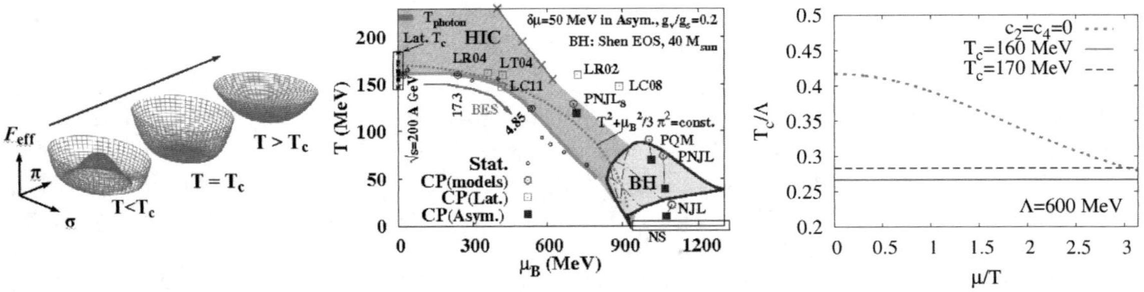


**Figure 2.** Left: Schematic view of the chiral phase transition. Middle: Predicted critical points and observed chemical freeze-out points [4]. Right: Tricritical point condition.

Now let us discuss the chiral transition at finite $T$ and $\mu$, where we need the mass dependence of the thermal pressure.

$$\begin{aligned}
P^{(F)}/d_f =& \frac{7}{8}\frac{\pi^2}{90}T^4 + \frac{\mu^2}{24}T^2 + \frac{\mu^4}{48\pi^2} - \frac{m^2}{16\pi^2}\left[\frac{\pi^2}{3}T^2 + \mu^2\right] \\
&- \frac{m^4}{32\pi^2}\left[\log\left(\frac{m}{\pi T}\right) - \frac{3}{4} + \gamma_E - H^\nu\left(\frac{\mu}{T}\right)\right] + \mathcal{O}(m^6) , \qquad (9)
\end{aligned}$$

$$H^\nu(\nu) = \left(\frac{\nu}{\pi}\right)^2\sum_{l=1}^{\infty}\frac{2}{(2l-1)[(2l-1)^2 + (\nu/\pi)^2]} = \sum_{k=1}^{\infty}(-1)^{k-1}\left(\frac{\nu}{\pi}\right)^{2k}\left(2 - \frac{1}{2^{2k}}\right)\zeta(2k+1) , \qquad (10)$$

where $\gamma_E = 0.5772156649\cdots$ is the Euler's constant. The first three terms in Eq. (9) are the massless quark contributions (Stefan-Boltzmann law), and other terms show mass dependence of the pressure [15]. We again consider the chiral limit ($m_0 = 0$), then we can expand $\mathcal{F}_{\text{eff}}$ in the Taylor series of $m$ as

$$\mathcal{F}_{\text{eff}}(m;T,\mu) = \mathcal{F}_{\text{eff}}(0;T,\mu) + \frac{c_2(T,\mu)}{2}m^2 + \frac{c_4(T,\mu)}{24}m^4 + \mathcal{O}(m^6)\ , \tag{11}$$

$$c_2(T,\mu) = \frac{d_f}{24}\left[T^2 + \frac{3}{\pi^2}\mu^2 - T_c^2\right] \quad \left(T_c^2 = \frac{3\Lambda^2}{\pi^2}\left[1 - \frac{G_c^2}{G^2}\right]\right)\ , \tag{12}$$

$$c_4(T,\mu) = \frac{3d_f}{4}\left[\gamma_{\text{E}} - 1 - \log\left(\frac{\pi T}{2\Lambda}\right) - H^\nu\left(\frac{\mu}{T}\right)\right]\ . \tag{13}$$

We assume that the interaction is strong enough and chiral symmetry is spontaneously broken in vacuum, then $T_c^2 > 0$. At $\mu = 0$, $c_2$ changes its sign at $T = T_c$, then the chiral restored state ($m = G\sigma = 0$) is realized in equilibrium at $T > T_c$. This chiral transition should occur at $T < \sqrt{3}\Lambda/\pi$, which is less than the cutoff.

The second order phase boundary is given by the condition $c_2(T,\mu) = 0$ provided that $c_4 > 0$, and is found to be $T^2 + 3\mu^2/\pi^2 = T_c^2$. If this boundary reaches the $\mu$ axis, the critical baryon chemical potential is evaluated to be $N_c\mu_c = \pi N_c T_c/\sqrt{3}$, which amounts to be around the nucleon mass $N_c\mu_c \simeq 925$ MeV for $T_c \sim 170$ MeV. This elliptically shaped phase boundary roughly explains the chemical freeze-out points probed in heavy-ion collisions, as shown in the middle panel of Fig. 2.

The second order boundary may be terminated by the tricritical point (TCP), where $c_2 = 0$ and $c_4 = 0$ is simultaneously satisfied. For the free energy density in Eq. (9), the TCP condition is expressed as

$$\frac{T_c}{\Lambda} = \frac{2\sqrt{1 + 3\nu^2/\pi^2}}{\pi}e^{\gamma_E - 1 - H^\nu(\nu)} \quad (\nu = \mu/T)\ . \tag{14}$$

In the right panel of Fig. 2, dotted line shows the right hand side of Eq. (14), and solid and dashed lines show $T_c/\Lambda$ values with $\Lambda = 600$ MeV for $T_c = 160$ MeV and 170 MeV, respectively. For $T_c = 170$ MeV, the TCP condition is satisfied around the conversion radius of the high-temperature expansion, $|\mu/T| = \pi$.

When the quark mass is finite, the transition at small $\mu$ becomes the crossover, and TCP becomes CP if the first order boundary exists. The location of CP is sensitive to the details of the model. The NJL model in the present simple treatment predicts a large $\mu/T$ value of TCP, $\mu/T \gtrsim \pi$, as discussed above. When the confinement and/or the 8-Fermi interaction effects are taken into account [16], CP temperature increases. Some lattice QCD calculations predict $T_{\text{CP}}$ similar to $T_c(\mu = 0)$. We need experimental data on CP and the first order phase transition, and/or some breakthrough in lattice QCD at finite density in order to pin down the location of CP and to understand the structure of the QCD phase diagram.

## 3. QCD phase diagram in strong-coupling lattice QCD

The lattice QCD Monte-Carlo (LQCD-MC) simulations is the non-perturbative and first-principles method of QCD, and have been applied to various observables such as hadron masses, hadron-hadron interactions, and QCD thermodynamics at $\mu = 0$. At finite density, however, the fermion determinant becomes complex and it becomes difficult to perform precise calculation at large $\mu$. There are many methods proposed so far in order to avoid the sign problem. The strong-coupling lattice QCD is one of the promising methods.

In this section, I briefly introduce the basic idea of the lattice QCD and the sign problem. Next, I explain the strong-coupling lattice QCD.

### 3.1. Lattice QCD

We consider here a lattice QCD action for color SU($N_c$) with one species of unrooted staggered fermion in the $d(=3)+1$ dimensional Euclidean spacetime with $N_\tau$ temporal and $L$ spatial lattice sizes.

$$\begin{aligned} S_{\rm LQCD} =& \bar{\chi} D \chi + S_G \qquad (15)\\ =& \frac{1}{2}\sum_x \left[\bar{\chi}_x U_0(x)\, e^{\mu} \chi_{x+\hat{0}} - \bar{\chi}_{x+\hat{0}} U_0^\dagger(x)\, e^{-\mu}\chi_x\right] \\ &+\frac{1}{2}\sum_{x,j} \eta_j(x) \left[\bar{\chi}_x U_j(x)\, e^{\mu} \chi_{x+\hat{j}} - \bar{\chi}_{x+\hat{j}} U_j^\dagger(x)\, e^{-\mu}\chi_x\right] + m_0 \sum_x \bar{\chi}_x \chi_x \\ &+\frac{2N_c}{g^2}\sum_{\rm plaq.}\left[1-\frac{1}{N_c}{\rm Re\,tr}\, U_{\mu\nu}(x)\right] . \qquad (16)\end{aligned}$$

Spacetime is discritized on the lattice, $(\tau, x, y, z) = (ai_\tau, ai_x, ai_y, ai_z)$, where $a$ is the lattice spacing and $i_\mu$ is an integer, $1 \le i_\tau \le N_\tau$ and $1 \le i_j \le L$ $(j = 1, 2, 3)$. In this proceedings, we adopt the lattice unit, $a = 1$. Quarks are represented as anti-commuting Grassmann numbers on the lattice sites, $\chi_x^a$ (color index $a$ is suppressed in Eq. (16)). Since the sqare of a Grassmann number is zero, we just need to define $\int d\chi = 0$ and $\int \chi d\chi = 1$ to evaluate the path integral; the former is *an anti-commuting constant* which should be zero, and the latter is a commuting constant which can be defined as unity. Using these definitions, we find $\int \exp(\bar{\chi}A\chi) d\bar{\chi} d\chi = \det A$. The link variable $U_\mu(x) = P\exp\left[ig\int_x^{x+\hat{\mu}} dx A_\mu(x)\right]$ is an $N_c \times N_c$ matrix function of the gluon field and defined on a link $(x, x+\hat{\mu})$. The gauge transformation of the quark and link variables are given as $\chi_x \to V_x \chi_x$, $\bar{\chi}_x \to \bar{\chi}_x V_x$, and $U_\mu(x) \to V_x U_\mu(x) V_{x+\hat{\mu}}$. Then the combination $\bar{\chi}_x U_\mu(x) \chi_{x+\hat{\mu}}$ is gauge invariant. The staggered factor $\eta_j(x) = (-1)^{x_0+\cdots+x_{j-1}}$ represents $\gamma_j$ for staggered fermions, then the first two lines in Eq. (16) leads to the quark action in the continuum QCD. The trace of the plaquette $U_{\mu\nu}(x) = U_\mu(x)U_\nu(x+\hat{\mu})U_\mu^\dagger(x+\hat{\mu}+\hat{\nu})U_\nu^\dagger(x+\hat{\nu})$ is also gauge invariant. The loop integral $\oint dx A$ appears in the exponent in $U_{\mu\nu}$ and generates the *rotation* $F_{\mu\nu}$, as deduced from the Stokes theorem in the U(1) gauge case.

The partition function of lattice QCD is given as

$$\mathcal{Z}_{\rm LQCD} = \int \mathcal{D}[\chi, \bar{\chi}, U_0, U_j]\, e^{-\bar{\chi}D\chi - S_G} = \int \mathcal{D}[U_0, U_j]\, \det D\, e^{-S_G} . \qquad (17)$$

The integrand $\det U\, \exp(-S_G)$ is regarded as a statistical weight for a configuration of link variables. The Fermion matrix $D$ has $\gamma_5$ Hermiticity, and its determinant is real at $\mu = 0$,

$$[\gamma_5 D(\mu)\gamma_5]^\dagger = D(-\mu^*) , \quad [\det D(\mu)]^* = \det[D(-\mu^*)] . \qquad (18)$$

Then it is possible to perform the Monte-Carlo integral by regarding $\det D\, \exp(-S_G)$ as a statistical weight for a configuration of $U$ *at zero* $\mu$. At finite $\mu$, $\det D$ becomes a complex number, and a naïve probability interpretation fails down. There are many attempts to avoid the sign problem. Many of these methods are useful for $\mu/T < 1$, while it is difficult to perform the Monte-Carlo simulation in the larger $\mu/T$ region, where we may expect the first order phase transition.

### 3.2. Strong-coupling lattice QCD

In the strong coupling region, it is possible to explore the phase diagram including the first phase transition boundary. The strong-coupling lattice QCD is a method based on the $1/g^2$ expansion of the partition function, and has a long history of study. In 1974, Wilson showed in the first work on the lattice gauge field theory that the Wilson loop follows the area law in the strong coupling limit [17],

$$\langle W(C = L \times N_\tau)\rangle = \frac{1}{\mathcal{Z}} \int \mathcal{D}[U] W(C)\, e^{-S_G} \to N_c \left(\frac{1}{g^2 N_c}\right)^{LN_\tau} \quad (g^2 \to \infty)\ , \tag{19}$$

where $LN_\tau$ is the area of the Wilson loop as shown in the left panel of Fig. 3. We need at least $LN_\tau$ plaquettes in order to kill all unpaired link variables. By using the one-link integral formula shown later in Eq. (20), each plaquette generates a factor $1/N_c g^2$ to the Wilson loop. The Wilson loop $W(C) = \mathrm{tr}\, P \exp(i \oint_C A_\mu dx_\mu)$ is related with the potential between heavy quarks, $\langle W(C)\rangle = \exp(-V(L)N_\tau)$. The area law tell us that $V(L)$ contains a linear confining potential, $V(L) = L \log(g^2 N_c)$. Thus color is found to be confined in the strong coupling region. This confining feature was confirmed by Creutz in the first lattice QCD Monte-Carlo simulation in 1980 [18]. The observed string tension is found to connect the strong coupling region and the weak coupling region smoothly. Strong-coupling expansion with higher order terms was performed by Münster [19], and the MC results are well reproduced.

Strong-coupling lattice QCD with quarks was first explored by Kawamoto and Smit [20]. It was shown that chiral symmetry is spontaneously broken in the strong-couling limit [20, 21]. This idea is extended to finite $T$ and $\mu$ [22, 23, 24, 26, 27, 28, 30, 29]. The phase transition in the strong-coupling and chiral limit is found to be the second order at $\mu = 0$ and the first order at $T = 0$, in the mean field treatment [22, 23] and with fluctuation effects [27, 28]. Effects of plaquette terms at finite $1/g^2$ on the phase diagram are also studied [24, 30, 29].

### *3.3. Phase diagram at strong coupling*

The plaquette term $S_G$ in the lattice QCD action Eq. (16) is proportional to $1/g^2$, and can be treated as perturbation in the strong coupling region. Especially, we can neglect $S_G$ in the strong coupling limit, then it is possible to integrate out the link integral independently. We can perform the one-link integral by using the formulae,

$$\int dU U_{ab} = 0\ , \int dU U_{ab} U^\dagger_{cd} = \frac{1}{N_c}\, \delta_{ad}\, \delta_{bc}\ , \int dU U_{ab} U_{cd} U_{ef} = \frac{1}{N_c!}\, \varepsilon_{ace}\, \varepsilon_{bdf}\ , \ldots \tag{20}$$

The second formula gives an effecitve action terms after integrating out spatial link variables,

$$S_{\mathrm{eff}}^{(\mathrm{SCL})} = S_F^{(t)} - \frac{1}{4N_c} \sum_{x,j} M_x M_{x+\hat{j}} + m_0 \sum_x M_x \quad (M_x = \bar{\chi}_x \chi_x)\ , \tag{21}$$

where $S_F^{(t)}$ is the temporal hopping term of quarks shown in the first term in Eq. (16). This effective action contains the nearest neighbor four-Fermi interaction. The third one-link integral formula generates six-Fermi interaction (baryon hopping term), which is in the sub-leading order in the large dimensional $(1/d)$ expansion [25]. Let us consider the case with large spatial dimension, $d \gg 1$. In order to keep the four-Fermi interaction term finite at $d \to \infty$, quark field should scale as $\chi \propto d^{-1/4}$ to compensate a factor $d$ from the sum over $j$. The six-quark terms are $\mathcal{O}(d^{-1/2})$ for $N_c = 3$, and we ignore them in the later discussion.

Once an action with four-Fermi interaction is given, there are several standard methods to solve the quantum many-body problem. One of these methods is the bosonization of the action followed by the mean field approximation and the Grassmann integral. The phase diagram in the strong coupling limit has been studied in the mean field treatment of the effective action given in Eq. (21) [23, 24]. The obtained phase diagram has similar features of that prediced in chiral effective models; the second order (crossover) transition at low $\mu$, the first order transition at large $\mu$, and the existence of TCP (CP) in (off) the chiral limit, as shown in the left panel

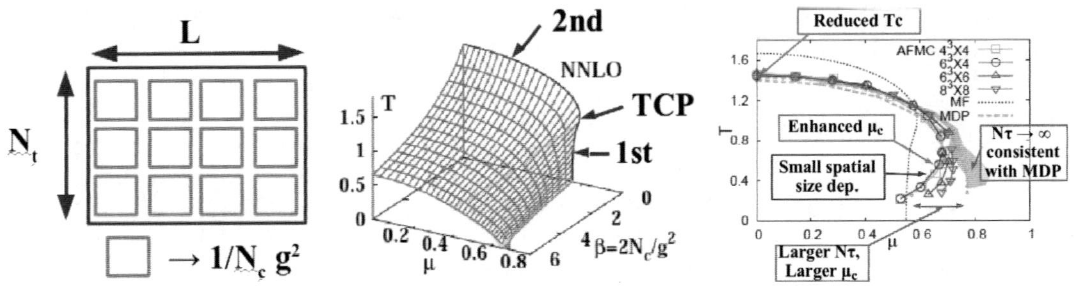


**Figure 3.** Left: Schematic figure of the Wilson loop. Middle: Phase diagram as a function of inverse coupling, $\beta_g = 2N_c/g^2$ [24]. Right: Phase diagram in the strong coupling limit [28].

of Fig. 3 (the $\beta = 2N_c/g^2 = 0$ surface). This similarity of the phase diagrams owes to the symmetry of the theory. $S_{\rm eff}^{\rm (SCL)}$ in Eq. (21) [23, 24] in the chiral limit ($m_0 = 0$) is invariant under the O(2) chiral transformation ($\chi_x \to e^{i\varepsilon_x\alpha/2}\chi_x$, $\bar{\chi}_x \to e^{i\varepsilon_x\alpha/2}\bar{\chi}_x$), and anomaly is not realized in the strong coupling region. From the universality argument, the system with O(2) symmetry without anomaly shows the second order phase transition. This can be proxy for the system with O(4) symmetry with anomaly, as in the $N_f = 2+1$ QCD in the chiral limit for $u$ and $d$ quarks.

The above QCD phase diagram is not satisfactory in several points; mean field approximation (no fluctuation effects), the strong coupling limit ($1/g^2 = 0$), only the leading order of the $1/d$ expansion, and one species of unrooted staggered Fermion. There are some recent progress on the fluctuation effects. One of the methods to exactly evaluate the partition function is the Monomer-Dimer-Polymer (MDP) simulation [26, 27]. We integrate out all link variables in the strong coupling limit, and we find that the partition function is represented as a sum of weights for monomer-dimer-polymer configurations. Some of the MDP configurations have negative weights, but the weight cancellation is weak. In MDP, we can also include the baryon hopping term in the sub-leading order of the $1/d$ expansion. Another method proposed so far is the auxiliary field Monte-Carlo (AFMC) method [28], which is a straightforward extension of the mean field treatment. After bosonizing the effective action, we keep the auxiliary fields as they are without putting them as constant mean fields, and integrate over the auxiliary fields using the Monte-Carlo method. We show the phase diagram obtained in these two methods in the strong coupling and chiral limit in the right panel of Fig. 3. Anisotropic lattice ($a_\tau \neq a_s$) is adopted in these calculations. Green dashed lines show the phase boundaries in MDP with $N_\tau = 4$ and $\infty$. Solid lines with symbols show the phase boundaries in AFMC for $4^4, 6^3 \times 4, 6^4$ and $8^4$ lattices. Compared with the mean field results, fluctuation effects reduces the transition temperature at $\mu = 0$, and enhance the hadron phase at large $\mu/T$. For larger $N_\tau$, the transition $\mu$ becomes larger. These features are common in the two methods, and the phase boundaries are found to agree with each other.

Plaquette term effects (finite $1/g^2$ effects) are also investigated in the mean field treatment [24]. In the middle panel of Fig. 3, we show the phase diagram as a function of $1/g^2$. The transition temperature decreases with increasing $1/g^2$ and is found to be consistent with the standard MC calculation results in the strong coupling region ($\beta_g = 2N_c/g^2 \lesssim 4$), when we take account of the $1/g^2$ terms and the Polyakov loop terms ($(1/g^2)^{N_\tau}$). Calculations including both plaquette and fluctuation effects have been performed recently.

An interesting application of the strong-coupling lattice QCD is the net-baryon number cumulants across the phase boundary [31]. Around the phase transition, we expect large fluctuations of order parameters. The order parameter at CP is a linear combination of the chiral

Journal of Physics: Conference Series **668** (2016) 012004 doi:10.1088/1742-6596/668/1/012004

condensate and the baryon density [32], then larger baryon number fluctuations are expected in heavy-ion collisions at $\sqrt{s_{NN}} = (5 - 20)$ GeV, where the formed matter is considered to go around CP. Net-baryon number cumulants

$$\chi^{(n)} = \partial^n (P/T^4)/\partial(N_c\mu/T)^n = \langle(\delta B)^n\rangle_c \tag{22}$$

show the baryon number fluctuations, higher-order cumulants are known to be more sensitive to the criticality [33]. Recently observed data in the beam energy scan show that the cumulant ratio $\kappa\sigma^2 = \chi^{(4)}/\chi^{(2)}$ shows a non-monotonic behavior as a funciton of $\sqrt{s_{NN}}$, which might signal the existence of CP. Since the susceptibility ($\chi^{(2)}$) diverges at CP, it is dangerous to rely on the Taylor expansion in $\mu/T$ from $\mu = 0$.

We have recently obtained the net-baryon number cumulants in the strong coupling and chiral limit. The cumulant ratio $\kappa\sigma^2$ shows oscillatory behavior as a function of $T$ for a given $\mu/T$, and the negative $\kappa\sigma^2$ region appears along the phase boundary at $\mu/T \gtrsim 0.2$. The lattice size dependence shows diverging behavior in agreement with the scaling function analysis [34]. It is important to study the finite mass effects in order to understand the observed non-monotonic behavior of $\kappa\sigma^2$.

## 4. QCD phase diagram of isospin-asymmetric matter and compact stars

Now let us turn to the compact star phenomena. In the previous sections, we have discussed the phase diagram of symmetric matter, where chemical potential is the same for all quarks. Dense matter probed in compact star phenomena, however, is not symmetric. Before the core collapse, the electron-to-baryon ratio[4] in the iron core is around $Y_e \sim 0.46$, and it decreases to around $Y_e \sim 0.3$ as the electron capture proceeds. In neutron stars, it is further small, $Y_e \lesssim 0.1$. In binary neutron star mergers, $Y_e$ at high density is similar to that in neutron stars, while $Y_e$ has variety in the low density region leading to the r-process nucleosynthesis [35]. If we ignore hyperons, charge neutrality requires the same number of protons and electrons, then the proton fraction (proton-to-baryon ratio) $Y_e = Y_p = Z/(Z + N) = (0 - 0.46)$ in compact star phenomena. It is well known that symmetric ($N \sim Z$) and neutron-rich ($N > Z$) nuclei have different properties such as the binding energy, nuclear radius, and excitation spectra. Then we should ask ourselves. How does the QCD phase diagram evolve as a function of the proton fraction ?

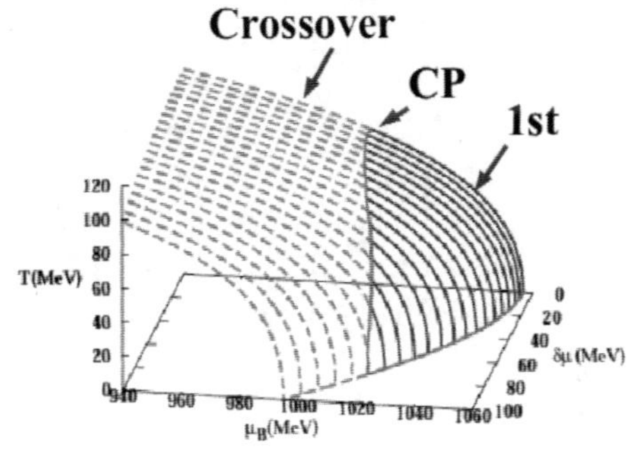


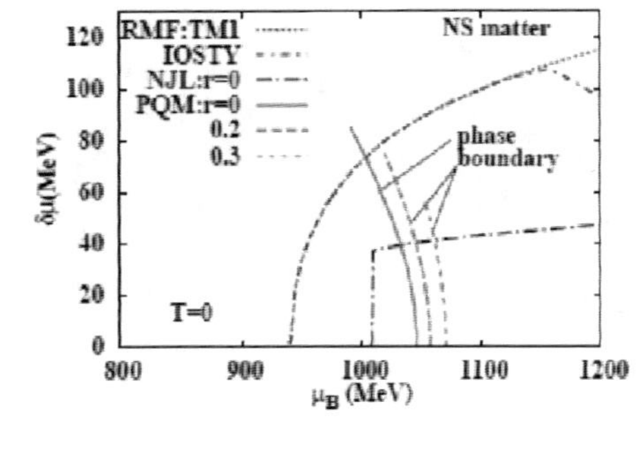


**Figure 4.** Left: Phase diagram as a function of isospin chemical potential [39]. Right: First order phase boundary in comparison with isospin chemical potential in neutron star matter [39].

[4] $Y_e$ in this proceedings should be interpreted as the ratio of electric charge of hadrons and baryon number ($Y_Q$) when muons and hyperons are taken into account.

Journal of Physics: Conference Series **668** (2016) 012004 doi:10.1088/1742-6596/668/1/012004

Chiral effective model approach is again useful to explore the QCD phase diagram in asymmetric matter. We consider the Polyakov-loop extended Quark Meson (PQM) model with vector interaction,

$$\begin{aligned}\mathcal{L}_{\rm PQM} =&\bar{q}\left[i\gamma^\mu D_\mu - g_\sigma(\sigma + i\gamma_5\boldsymbol{\tau}\cdot\boldsymbol{\pi})\right]q + \frac{1}{2}\partial^\mu\sigma\partial_\mu\sigma + \frac{1}{2}\partial^\mu\boldsymbol{\pi}\partial_\mu\boldsymbol{\pi} - U_\sigma(\sigma,\boldsymbol{\pi}) - U_\Phi(\Phi,\bar{\Phi})\\ &-g_v\bar{q}\gamma^\mu(\omega_\mu + \boldsymbol{\tau}\cdot\boldsymbol{R}_\mu)q - \frac{1}{4}\omega_{\mu\nu}\omega^{\mu\nu} - \frac{1}{4}\boldsymbol{R}_{\mu\nu}\cdot\boldsymbol{R}^{\mu\nu} + \frac{1}{2}m_\omega^2\omega_\mu\omega^\mu + \frac{1}{2}m_\rho^2\boldsymbol{R}_\mu\cdot\boldsymbol{R}^\mu\ . \end{aligned} \tag{23}$$

In addition to quarks, and $\sigma$ and $\pi$ mesons, we consider $\omega$ and $\rho$ (represented by $\boldsymbol{R}$) vector mesons. $\Phi$ and $\bar{\Phi}$ are the Polyakov and anti-Polyakov loops. The meson potential is chosen to be the sum of the zero-point energy of quarks, polynomial of $\sigma^2+\boldsymbol{\pi}^2$, and explicit breaking term $-h\sigma$. The Polyakov-loop potential is tuned to fit to the lattice results. The vector coupling of quarks are not well fixed, then we regard it is a free parameter.

In the left panel of Fig. 4, we show the QCD phase diagram in $(T, \mu_B, \delta\mu)$ space, where $\delta\mu = (\mu_d - \mu_u)/2 = (\mu_n - \mu_p)/2$ is the isospin chemical potential. We have applied the mean field approximation with an assumption that the pions do not condensate, according to the $s$-wave $\pi N$ repulsion argument [38]. Compared with the phase diagram of symmetric matter ($\delta\mu = 0$), the transition temperature decreases with increasing $\delta\mu$. CP temperature also decreases with $\delta\mu$, and eventually CP disappears at $\delta\mu = (70 - 80)$ MeV. The reduction of the transition temperature can be understood in part by using the high-temperature expansion formula in Eq. (9) and the curvature $c_2$ in Eq. (12) in the chiral limit. Since $u$- and $d$-quark chemical potentials are given as $\mu_u = \mu - \delta\mu$ and $\mu_d = \mu + \delta\mu$, we find $c_2 = d_f/24\left[T^2 + 3(\mu^2 + (\delta\mu)^2)/\pi^2 - T_c^2\right]$ for the $N_f = 2$ NJL model in the chiral limit, and the second order phase transition temperature decreases with increasing $\delta\mu$ for given $(T, \mu = \mu_B/3)$. We expect that the effects of the vector coupling, the Polyakov loop, and finite quark mass are less sensitive to $\delta\mu$. This 3D phase diagram is similar to that obtained in the functional renomalization group (FRG) method [37], while it contradict to the effective model results with pion condensation [36].

Shrinkage of the hadron phase and the first order phase transition boundary could be relevant to compact star phenomena. First, the disappearance of the first order phase boundary at large $\delta\mu$ suggest the possibility of the crossover transition in the neutron star core. In the middle panel of Fig. 4, we show isospin chemical potential of neutron star matter in $\beta$ equilibrium as a function of baryon chemical potential calculated by using a relativistic mean field (RMF) model, TM1, in comparison with the first order phase boundary at $T = 0$ in PQM with the vector-scalar coupling ratio $r = g_v/g_\sigma = 0, 0.2$ and $0.3$. In the cold neutron star core, $\delta\mu$ is calculated to reach around 100 MeV. When the vector coupling is not very small ($r > 0.2$), the first order phase boundary disappears below $\delta\mu = 80$ MeV and the chiral phase transition could be the crossover. If this is the case, the softening of the equation of state from the first order phase transition is weakened. This supports the crossover scenario [43] to support the $2M_\odot$ neutron stars [44].

Another interesting possibility is the critical point sweep during the dynamical black hole formation [40]. We consider here the numerical results in [41]; Gravitational collapse of a 40 $M_\odot$ star is simulated using the $\nu$ radiation 1D (spherical) hydrodynamics with the Shen EOS [42]. In failed supernovae, core collapse of massive stars directly form black holes. Temperature can reach $T \sim 90$ MeV at off-center by the shock heat, which is above the CP temperature of asymmetric matter in many of effective models. In the center, temperature is relatively low ($T < 40$ MeV) while the density is high ($\mu_B \sim 1300$ MeV just before the black hole formation). Thus there is a possibility that matter at the center goes through the first order phase transition boundary below CP, the off-center goes above CP, and some of the mass-shell hit CP. We expect large baryon number fluctuations of that mass-shell, but we have not yet found possible observational signals, unfortunately.

## 5. Summary

In this proceedings, I gave the outline of my two lectures in the Dense Matter 2015. We expect the existence of QCD phase transition and the critical point from chiral effective model studies. This point is discussed based on the Nambu-Jona-Lasinio model. When interaction is strong enough, chiral symmetry is spontaneously broken in vacuum. Chiral symmetry should be restored at high temperature. Density effect reduces the 4-th order coefficient in constituent quark mass which is proportional to the chiral condensate in the chiral limit. Thus we can expect the existence of the critical point and the first order transition at high density. In the lecture I also explained some technical part, such as the Matsubara sum, Hubbard-Stratonovich transformation, and high-temperature expansion. Since the first principle calculation of QCD has difficulties at finite densities, we need studies using effective models, approximate treatment of QCD, and of course, experiments.

In the second lecture, I explained other two approaches to the QCD phase diagram. The first one is the strong-coupling lattice QCD. While we have the sign problem in lattice QCD at finite $\mu$, the phase diagram study is on going using various ideas. I have shown recent results based on the strong-coupling lattice QCD. Smaller weight cancellation allow us to study phase transition at high density. Phase diagram in the strong coupling limit has been confirmed, through the agreement of MDP and AFMC results. While the strong-coupling limit is the opposite limit of the continuum limit, cumulant ratio calculation would be valuable to understand the appearance of the criticality in finite volume.

The second one is the phase diagram of asymmetric matter and compact stars. Compact stars are also good laboratories of dense matter. In neutron stars, supernovae, black hole formation, and binary neutron star mergers, we expect the formation of dense and isospin asymmetric matter. With the first order boundary (and CP) and isospin chemical potential, there are many ways of realizing phase transition in compact star phenomena.

Finally, I would like to emphasize that dense matter is "terra incognita", where there are many unsolved problems. In heavy-ion collisions at $\sqrt{s_{NN}} = (5-10)$GeV, we expect formation of highest baryon density matter, whose density may exceed $5\rho_0$. In equilibrium, this would be above the transition density. In compact star phenomena, hydro simulations with hadronic matter EOS suggest the formation of dense matter ($(4-5)\rho_0, \mu_B \sim 1300$ MeV), which is above the transition density in many effective models. We need more experimental, observational, and theoretical works to explore dense matter.


## Acknowledgments

The author would like to thank David Blaschke and other members in JINR for their hospitality. The author also thanks Kenta Kiuchi and Yuichiro Sekiguchi for preparing the figure for the lecture. This work is supported in part by the Grants-in-Aid for Scientific Research from JSPS (Nos. 23340067 24340054 24540271 and 15K05079 ), the Grants-in-Aid for Scientific Research on Innovative Areas from MEXT (No. 2404: 24105001, 24105008), and by the Yukawa International Program for Quark-Hadron Sciences.


## References


[1] Braun-Munzinger P, Magestro D, Redlich K and Stachel J 2001 *Phys. Lett.* B **518** 41
[2] Aoki Y et al. 2006 *Nature* **443** 675
[3] Nara Y, Otuka N, Ohnishi A, Niita K, Chiba S 2000 *Phys. Rev.* C **61** 024901
[4] Ohnishi A 2012 *Prog. Thoer. Phys. Suppl.* **193** 1
[5] Ohnishi A 2002 *Proc. of the 2nd theory workshop on JHF nuclear physics* KEK-PROC–2002-13
[6] Asakawa M and Yazaki K 1989 *Nucl. Phys.* A **504** 668
[7] Nakano E and Tatsumi T 2005 *Phys. Rev.* D **71** 114006; Nickel D 2009 *Phys. Rev.* D**80** 074025
[8] Stephanov MA, Rajagopal K and Shuryak EV 1998 *Phys. Rev. Lett.* **81** 4816; Asakawa M, Heinz UW and Müller B, *Phys. Rev. Lett.* **85** 2072; Jeon S and Koch V 2000 *Phys. Rev. Lett.* **85** 2076

[9] Adamczyk L et al. (STAR Collaboration) 2014 *Phys. Rcv. Lett.* **112** 032302
[10] Nambu Y and Jona-Lasinio G 1961 *Phys. Rev.* **122** 345; *ibid.* **124** 246; Vogl U and Weise W 1991 *Prog. Part. Nucl. Phys.* **27** 195; Klevansky SP 1992 *Rev. Mod. Phys.* **64** 649; Hatsuda T and Kunihiro T 1994 *Phys. Rept.* **247** 221; Buballa M 2005 *Phys. Rept.* **407** 205
[11] Jungnickel DU and Wetterich C 1996 *Phys. Rev.* D **53** 5142
[12] Fukushima K 2004 *Phys. Lett.* B **591** 277
[13] Schaefer BJ, Pawlowski JM and Wambach J 2007 *Phys. Rev.* D**76** 074023; Skokov V, Friman B, Nakano E, Redlich K and Schaefer BJ 2010 *Phys. Rev.* D **82** 034029
[14] Yagi K, Hatsuda T, Miake Y 2005 *Quark-Gluon Plasma: From Big Bang to Little Bang* (Cambridge, Cambridge University Press)
[15] Kapusta JI and Gale C 2006 *Finite-Temperature Field Theory: Principles and Applications* (Cambridge, Cambridge University Press)
[16] Sasaki T, Sakai Y, Kouno H and Yahiro M 2010 *Phys. Rev.* D **82** 116004
[17] Wilson KG 1974 *Phys. Rev.* D **10** 2445
[18] Creutz M 1980 *Phys. Rev.* D **21** 2308
[19] Munster G 1981 *Nucl. Phys.* B **180** 23
[20] Kawamoto N 1981 *Nucl. Phys.* B**190** **[FS3]** 617; Kawamoto N and Smit J 1981 *Nucl. Phys.* B **192** 100
[21] Aoki S 1984 *Phys.Rev.* D**30** 2653
[22] Damgaard PH, Kawamoto N and Shigemoto K 1984 *Phys. Lett.* B**114** 152; Faldt G and Petersson B 1986 *Nucl. Phys.* B**265** 197;
[23] Ilgenfritz EM and Kripfganz J 1985 *Z. Phys.* C**29** 79; Bilic N, Karsch F and Redlich K 1992 *Phys. Rev.* D**45** 3228; Fukushima K 2004 *Prog. Theor. Phys. Suppl.* **153** 204; Nishida Y 2004 *Phys. Rev.* D **69** 094501; Kawamoto N, Miura K, Ohnishi A and Ohnuma T 2007 *Phys. Rev.* D**75** 014502
[24] Miura K, Nakano TZ and Ohnishi A 2009 *Prog. Theor. Phys.* **122** 1045; Miura K, Nakano TZ, Ohnishi A and Kawamoto N 2009 *Phys. Rev.* D **80** 074034; Nakano TZ, Miura K and Ohnishi A 2010 *Prog. Theor. Phys.* **123** 825; Nakano TZ, Miura K and Ohnishi A 2011 *Phys. Rev.* D **83** 016014; Ohnishi A, Miura K, Nakano TZ and Kawamoto K 2009 *PoS* **LATTICE 2009** 160
[25] Kluberg-Stern H, Morel A and Petersson B 1983 *Nucl. Phys.* B **215** **[FS7]** 527
[26] Karsch F and Mutter KH 1989 *Nucl. Phys.* B **313** 541
[27] de Forcrand P and Fromm M 2010 *Phys. Rev. Lett.* **104** 112005
[28] Ichihara T, Ohnishi A and Nakano TZ 2014 *Prog. Theor. Exp. Phys.* **2014** 123D02
[29] Ichihara T and Ohnishi A 2014 *PoS* **LATTICE2014** 188 (*Preprint* arXiv:1503.07049 [hep-lat])
[30] de Forcrand P, Langelage J, Philipsen O and Unger W 2014 *Phys. Rev. Lett.* **113** 152002
[31] Ichihara T, Morita K and Ohnishi 2015 *Prog. Theor. Exp. Phys.* **2015** in press (*Preprint* arXiv::1507.04527 [hep-lat])
[32] Fujii H and Ohtani M 2004 *Phys. Rev.* D **70** 014016
[33] Stephanov MA 2009 *Phys. Rev. Lett.* **102** 032301
[34] Friman B, Karsch F, Redlich K and Skokov V 2011 *Eur. Phys. J.* C **71** 1694
[35] Sekiguchi Y, Kiuchi K, Kyotoku K, Shibata M 2015 *Phys. Rev.* D **91** 064059
[36] Sasaki T, Sakai Y, Kouno H and Yahiro M 2010 *Phys. Rev.* D **82** 116004; Andersen JO and Kyllingstad L 2010 *J. Phys.* G **37** 015003
[37] Kamikado K, Strodthoff N, von Smekal L and Wambach J 2013 *Phys. Lett.* B **718** 1044
[38] Ohnishi A, Jido D, Sekihara T and Tsubakihara K 2009 *Phys. Rev.* C **80** 038202
[39] Ueda H, Nakano TZ, Ohnishi A, Ruggieri M and Sumiyoshi K 2013 *Phys. Rev.* D **88** 074006
[40] Ohnishi A, Ueda H, Nakano TZ, Ruggieri M, Sumiyoshi K 2011 *Phys. Lett.* B **704** 284
[41] Sumiyoshi K, Yamada S, Suzuki H and Chiba S 2006 *Phys. Rev. Lett.* **97** 091101
[42] Shen H, Toki H, Oyamatsu K and Sumiyoshi K 1998 *Nucl. Phys.* A **637** 436; *Prog. Theor. Phys.* **100** 1013
[43] Masuda K, Hatsuda T and Takatsuka T 2013 *Astrophys. J.* **764** 12; *Prog. Theor. Exp. Phys.* **2013** 073D01
[44] Demorest P et al. 2010 *Nature* **467** 1081; Antoniadis J et al. 2013 *Science* **340** 448

Journal of Physics: Conference Series **668** (2016) 012005 doi:10.1088/1742-6596/668/1/012005

# Open heavy-flavor measurements with ALICE at the LHC

**R. Bailhache on behalf of the ALICE Collaboration**

Goethe-University Frankfurt am Main, Germany

E-mail: `rbailhache@ikf.uni-frankfurt.de`

**Abstract.** ALICE is well equipped to reconstruct heavy-flavor particles down to low transverse momentum $p_{\mathrm{T}}$ at mid and forward rapidity. An overview of the ALICE results obtained with the Run 1 data in pp ($\sqrt{s}$=2.76 TeV and $\sqrt{s}$=7 TeV), Pb-Pb ($\sqrt{s_{\mathrm{NN}}}$=2.76 TeV) and p-Pb ($\sqrt{s_{\mathrm{NN}}}$=5.02 TeV) collisions is presented. In pp collisions the measured cross sections are well described by pQCD calculations. The charged-particle multiplicity dependence of heavy-flavor yields indicates that Multi-Parton Interactions contribute to the high-multiplicity pp collisions and affect charm and beauty production in a similar way. In p-Pb and Pb-Pb collisions the measured nuclear modification factors indicate a final-state energy loss of heavy-quarks in central Pb-Pb collisions. Furthermore, the observed positive heavy-flavor elliptic flow in semi-central Pb-Pb collisions gives a hint that charm quarks participate in the collective expansion of the medium at low $p_{\mathrm{T}}$. In high-multiplicity p-Pb collisions, a double-ridge structure is observed in the heavy-flavor decay electron-hadron azimuthal correlations at low $p_{\mathrm{T}}$ similar to what is measured in the light-flavor sector. Such long-range correlations in $\eta$ could originate from a collective expansion of the system, as well as from gluon saturation in the initial state (color-glass condensate) or other mechanisms.

## 1. Introduction

Ultra-relativistic nuclear collisions allow the study of strongly-interacting matter under extreme conditions. A deconfined state of quarks and gluons, the Quark-Gluon Plasma (QGP), is expected to be formed in such collisions. The study of heavy-quark production is of particular interest, since charm and beauty quarks are produced via hard partonic scattering processes at the beginning of the collision and interact strongly with the surrounding medium. Therefore, the measurements of open heavy-flavor yields and elliptic flow allow one to test models of in-medium parton energy loss and probe the level of thermalization of heavy quarks in the medium at low transverse momentum $p_{\mathrm{T}}$. For this purpose, open heavy-flavor production in pp and p-Pb collisions has to be measured as well to provide the necessary baselines for heavy-ion studies. In pp collisions, the measurements of cross sections allow the testing of perturbative QCD calculations, whereas more differential measurements give deeper insight into the charm-production mechanisms. In p-Pb collisions, cold nuclear matter effects can be studied such that these can be better disentangled from hot and dense matter effects in Pb-Pb collisions. Possible cold nuclear matter effects are modification of Parton Distribution Functions (PDFs) in nuclei (shadowing [1]/gluon saturation [2] at small Bjorken-x), $k_{\mathrm{T}}$-broadening from multiple soft scatterings [3] and possible energy loss.

ALICE (A Large Ion Collider Experiment) collected data during the pp, Pb-Pb and p-Pb running periods at center-of-mass energies $\sqrt{s}$=7 and 2.76 TeV, $\sqrt{s_{\rm NN}}$=5.02 TeV, and $\sqrt{s_{\rm NN}}$=2.76 TeV, respectively, at the LHC (Large Hadron Collider) at CERN. A detailed description of the experimental setup can be found in [4]. At mid rapidity ($|\eta|$<0.9), open heavy-flavor production is measured in the central barrel via the hadronic decay of D mesons ($D^0$, $D^+$, $D^{*+}$ and $D_s^+$), the semileptonic decays to electrons of charm and beauty hadrons and the non-prompt $J/\psi$ from beauty-hadron decays. At forward rapidity (-4<$\eta$<-2.5), muons from heavy-flavor hadron decays are reconstructed in the muon spectrometer. The data readout was triggered by a minimum-bias interaction trigger based on the trigger signals from two forward scintillator hodoscopes (VZERO-A and VZERO-C). The summed amplitudes in the VZERO scintillator tiles were used to determine the centrality of Pb-Pb collisions, employing a model of particle production based on a Glauber description of the nuclear collision geometry. In addition to the minimum-bias trigger, the muon spectrometer provided a single muon trigger at foward rapidity for the heavy-flavor decay muon analysis. At mid rapidity, the ElectroMagnetic Calorimeter (EMCal) allowed us to trigger on single electrons for the heavy-flavor decay electron analysis.

## 2. Results in pp collisions at $\sqrt{s}$= 2.76 and 7 TeV

The differential cross sections of prompt D mesons ($D^0$, $D^+$, $D^{*+}$ and $D_s^+$), heavy-flavor decay electrons, electrons from beauty-hadron decays and heavy-flavor decay muons were measured in pp collisions at $\sqrt{s}$=7 [5, 6, 7, 8, 9] and 2.76 TeV[10, 11, 12, 13]. They are well described by pQCD calculations[14, 15, 16].

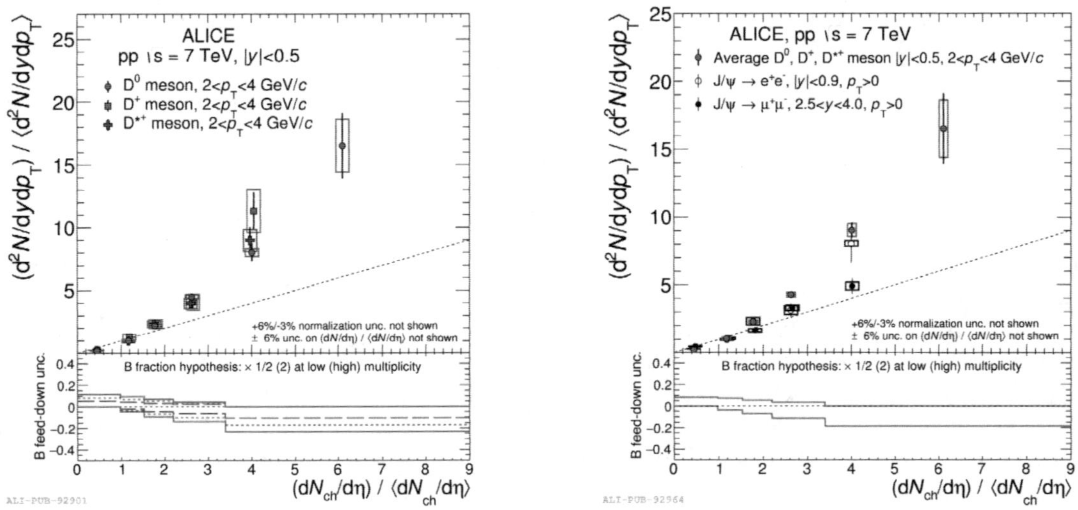


**Figure 1.** Prompt $D^0$, $D^+$, $D^{*+}$ meson self-normalized yields (left panel) and average prompt D meson and inclusive $J/\psi$ self-normalized yields (right panel) as a function of charged-particle multiplicity at mid rapidity. Vertical bars and boxes are statistical and systematic errors, respectively. The feed-down fraction uncertainty for the D mesons is drawn separately in the bottom panel [17].

More differential measurements, e.g. concerning the multiplicity dependence of D-meson yields, were also performed in pp collisions at $\sqrt{s}$=7 TeV [17]. The self-normalized D-meson yields, i.e. the corrected per-event yield normalized to the multiplicity-integrated value, increase faster than linear with the self-normalized charged-particle multiplicity in pp collisions, as shown in the left panel of Fig.1. The results for $D^0$, $D^+$, $D^{*+}$ are compatible with each other within uncertainties. A similar trend is observed for inclusive $J/\psi$ at mid rapidity, see right panel of Fig.1, as well as for non-prompt $J/\psi$ [17]. This suggests that Multi-Parton Interactions (MPI) play a role in high-multiplicity pp collisions and affect charm (open and hidden) and beauty in a similar way. More precise measurements are needed to discriminate among the possible origins of the effect.

## 3. Results in Pb-Pb collisions at $\sqrt{s_{\rm NN}}$= 2.76 TeV

One of the main topics being studied in heavy-ion collisions is the parton energy loss in the medium, where both inelastic (radiative energy loss via medium-induced gluon radiation) and elastic (collisional energy loss) processes play a role. The energy loss for quarks is expected to be smaller than for gluons, because of the smaller color coupling factor of quarks with respect to gluons. In addition, the dead-cone effect should reduce small-angle gluon radiation for quarks that have moderate energy-over-mass values, i.e. for charm and beauty quarks with momenta up to about 10 and 30 GeV/$c$, respectively [18]. Both effects together lead to an expected ordering of the energy loss in the QGP at the parton level ($\Delta E_{\rm g}>\Delta E_{\rm u,d,s}\geq\Delta E_{\rm c}>\Delta E_{\rm b}$). Experimentally, the modification of the measured spectra in Pb-Pb collisions compared to pp collisions at the same energy is quantified by the nuclear modification factor $R_{\rm AA}=1/\langle T_{\rm AA}\rangle$x$({\rm d}N_{\rm AA}/{\rm d}p_{\rm T})/({\rm d}\sigma_{\rm pp}/{\rm d}p_{\rm T})$ where $\langle T_{\rm AA}\rangle$ is the average nuclear overlap function for a given centrality class, ${\rm d}N_{\rm AA}/{\rm d}p_{\rm T}$ is the measured yield for Pb-Pb collisions in this centrality class and ${\rm d}\sigma_{\rm pp}/{\rm d}p_{\rm T}$ is the corresponding cross section in pp collisions. In the absence of any medium effect, the nuclear modification factor is equal to unity.

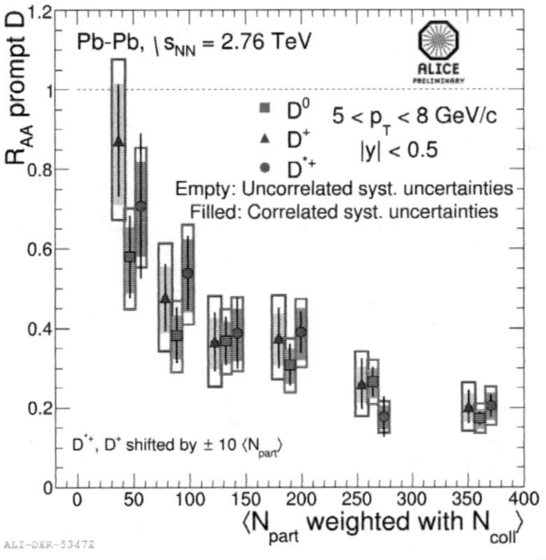


**Figure 2.** $R_{\rm AA}$ as a function of the centrality of Pb-Pb collisions (see text) of $D^0$, $D^+$ and $D^{*+}$. The bars represent the statistical uncertainty, while the filled (empty) boxes represent the systematic uncertainties that are correlated (uncorrelated) among centrality intervals [19].

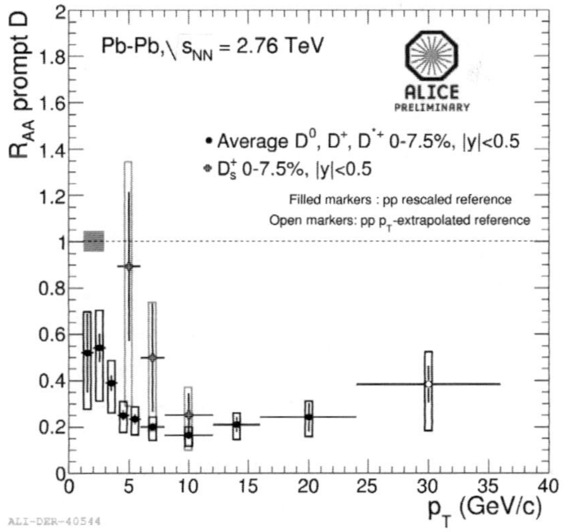


**Figure 3.** Average $D^0$, $D^+$ and $D^{*+}$-meson $R_{\rm AA}$ as a function of $p_{\rm T}$ compared to $D_s^+$ $R_{\rm AA}$ in the most central Pb-Pb collisions. The bars represent the statistical uncertainty, while the boxes represent the systematic uncertainties.

Fig.2 shows $R_{\rm AA}$ of high $p_{\rm T}$ $D^0$, $D^+$ and $D^{*+}$ mesons as function of the average number of participant nucleons in the Pb-Pb collision [19]. A large suppression similar for the three

D-meson species is observed in central Pb-Pb collisions. The $p_{\mathrm{T}}$ dependence of $R_{\mathrm{AA}}$ is presented in Fig.3 (see also [20]). Low-$p_{\mathrm{T}}$ measurements in all systems are crucial to study the expected binary scaling of the total charm yield in Pb-Pb collisions. In addition, the comparison of $R_{\mathrm{AA}}$ of $\mathrm{D}_s^+$ in the 0-7.5 % centrality class with the average nuclear modification factor of $\mathrm{D}^0$, $\mathrm{D}^+$ and $\mathrm{D}^{*+}$ mesons measured in the same centrality class is presented in Fig.3. An enhancement of the strange $\mathrm{D}_s$-meson yield is expected compared to non-strange D mesons because of the in-medium strangeness enhancement observed in the light-flavor sector and hadronization via in-medium coalescence at intermediate and low momentum [21, 22, 23]. The results for the strange and non-strange D mesons are compatible within uncertainties with a hint for an enhancement of $\mathrm{D}_s^+$ at low $p_{\mathrm{T}}$. Nevertheless, the uncertainties need to be reduced to make any quantitative statement.

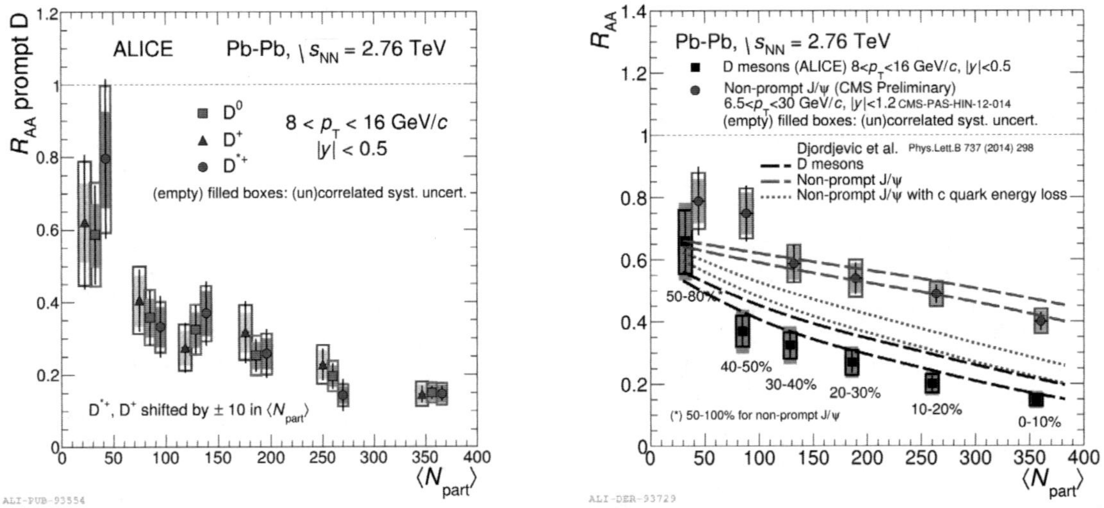


**Figure 4.** Comparison of the D-meson $R_{\mathrm{AA}}$ and of the charged-pion $R_{\mathrm{AA}}$ (left panel) and of the D-meson $R_{\mathrm{AA}}$ and of non-prompt J/$\psi$ $R_{\mathrm{AA}}$ measured by the CMS Collaboration [25, 26] (right panel). The vertical bars represent the statistical uncertainties and the boxes show the systematic uncertainties [19].

The color-charge dependence of the parton energy loss can a priori be studied by comparing $R_{\mathrm{AA}}$ of D mesons and charged pions, as shown in the left panel of Fig.4, as a function of the centrality of the Pb-Pb collision [19]. A similar suppression is observed for both species. The measurements are reproduced by calculations including a color-charge dependence of the parton energy loss [24]. The latter is compensated by the softer fragmentation and $p_{\mathrm{T}}$ spectrum of gluons with respect to those of charm quarks leading to a very similar $R_{\mathrm{AA}}$ for D mesons and charged pions. The mass dependence of parton energy loss was investigated by comparing $R_{\mathrm{AA}}$ of prompt D mesons and non-prompt J/$\psi$ from beauty-hadron decays measured by the CMS Collaboration [25, 26] in a similar kinematic region ($p_{\mathrm{T}}^{\mathrm{D}} \approx p_{\mathrm{T}}^{\mathrm{B}}$). A larger suppression is observed for D mesons in central collisions than for non-prompt J/$\psi$, as shown in right panel of Fig.4. In this case, the quark mass used in the model calculation [24] is crucial to reproduce the data, as demonstrated by the non-prompt J/$\psi$ $R_{AA}$ calculated considering the charm-quark mass for the energy loss of beauty quarks. The ALICE collaboration measured at mid rapidity non-prompt J/$\psi$ at lower $p_{\mathrm{T}}$ ($1.5<p_T<10$ GeV/$c$) compared to the CMS collaboration [27]. The measured $R_{\mathrm{AA}}$ seems to be overestimated by some models in the range $4.5<p_{\mathrm{T}}<10$ GeV/$c$. More precise data is nevertheless needed to discrimate among models.

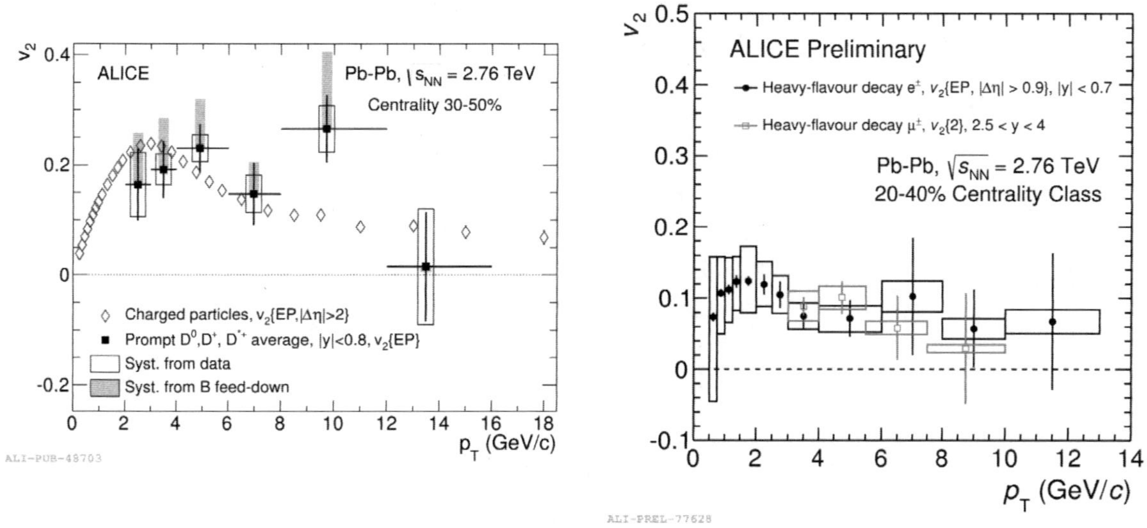


**Figure 5.** Left panel: Average of $D^0$, $D^+$ and $D^{*+}$ $v_2$ at mid rapidity as a function of $p_T$, compared to the charged-particle $v_2$. [28, 29]; Right panel: Heavy-flavor decay muon $v_2$ at forward rapidity as a function of $p_T$, compared to the heavy-flavor decay electron $v_2$ at mid rapidity [30].

Collectivity phenomena can be studied in Pb-Pb collisions via the azimuthal distribution of heavy-flavor hadrons, which reflects the initial spatial anisotropy of the heavy-ion collision in case of sufficient re-scattering of the heavy quarks in the hot and dense matter. The heavy-flavor elliptic flow, the second harmonic in the Fourier expansion of the particle azimuthal distribution, is an observable sensitive to the degree of thermalization of charm and beauty quarks in the medium at low $p_T$, as well as to the path-length dependence of the energy loss of heavy quarks at high $p_T$. The left panel of Fig.5 shows that the $v_2$ of D mesons in Pb-Pb collisions, in the 30-50% centrality class [18, 19], is similar to the charged-particle $v_2$ [28, 29]. A positive $v_2$ of D mesons (5.7$\sigma$ effect in the interval $2<p_T<6$ GeV/$c$), heavy-flavor decay electrons (3$\sigma$ effect in the interval $2<p_T<5$ GeV/$c$) and heavy-flavor decay muons [30] (3$\sigma$ effect in the interval $3<p_T<6$ GeV/$c$) is measured in semi-central Pb-Pb collisions, see Fig.5. This indicates that charm quarks participate in the collective expansion of the medium.

## 4. Results in p-Pb collisions at $\sqrt{s_{NN}}$=5.02 TeV

Fig.6 shows $R_{pPb}$ of D mesons at mid rapidity in minimum-bias p-Pb collisions at $\sqrt{s_{NN}}$=5.02 TeV [31]. Calculations that include cold nuclear matter effects [32, 3, 2] can describe the measurement. Smaller uncertainties are nevertheless needed to discrimate among models. Cold nuclear matter effects do not lead to a strong suppression in the measured $p_T$ range. $R_{pPb}$ is compatible with unity for $p_T>2$ GeV/$c$, within uncertainties.

Fig.7 shows $R_{pPb}$ of heavy-flavor decay muons at forward ($2.5<y_{cms}<3.54$, proton-beam direction) and at backward rapidity ($-4<y_{cms}<-2.96$, Pb-beam direction), respectively. $R_{pPb}$ at forward rapidity is consistent with unity, as well as $R_{pPb}$ backward rapidity for $p_T>4$ GeV/$c$. Both observables are well described by calculations [32, 3, 33], within uncertainties. From the results of D mesons, heavy-flavor decay electrons and heavy-flavor decay muons in p-Pb collisions, it can be concluded that the strong suppression of heavy-flavor yields at intermediate and high $p_T$ measured in central Pb-Pb collisions is due to the interaction of heavy quarks with

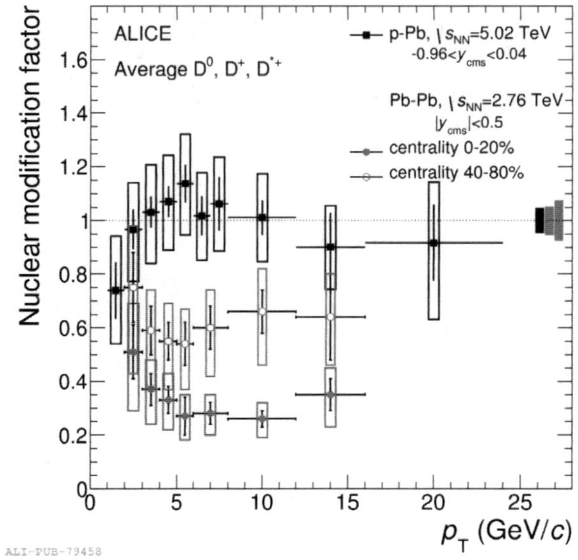


**Figure 6.** Average $R_{\rm pPb}$ of prompt $D^0$, $D^+$ and $D^{*+}$ mesons as a function of $p_T$ compared to the D-meson $R_{AA}$ in the 20 % most central and in 40-80 % central Pb-Pb collisions. Statistical (bars), systematic (empty boxes) and normalization (full boxes) uncertainties are shown [31].

the hot and dense medium.

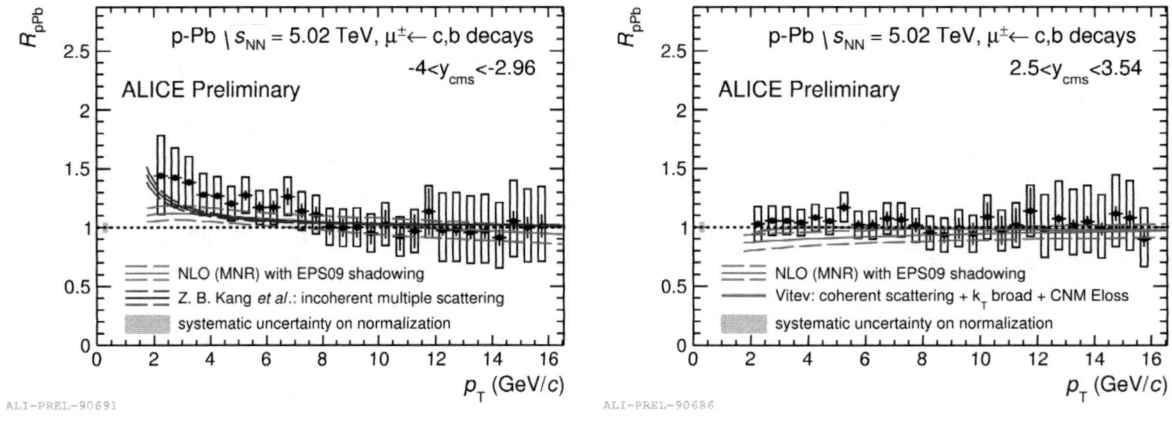


**Figure 7.** Nuclear modification factors of heavy-flavor decay muons in p-Pb collisions at -4<$y_{\rm cms}$<-2.96 (left) and 2.5<$y_{\rm cms}$<3.54 (right) are compared with model calculations [32, 3, 33].

Further insight into the charm quark production mechanisms and quark fragmentation in pp collisions, as well as information about additional possible cold-nuclear matter effects in p-Pb collisions and hot and dense matter effects in Pb-Pb collisions can be obtained with measurements of angular correlations between heavy-flavor hadrons (or their decay electrons) and charged particles. Fig.8 shows the azimuthal-correlation distribution between D mesons and charged particles in the interval $|\Delta\eta|<1$ for $8<p_T^{D}<16\,{\rm GeV}/c$ and $p_T^{\rm assoc}>1\,{\rm GeV}/c$. The azimuthal-correlation distributions measured in pp and p-Pb collisions agree within uncertainties after subtracting the baseline because of the uncorrelated background. Various PYTHIA tunes can describe the data within the relatively large statistical uncertainties. Smaller statistical errors are expected with data from Run 2. The angular-correlation distribution between heavy-flavor decay electrons and charged particles has been measured in p-Pb collisions. After subtracting the correlation distribution of the 60-100% multiplicity class from the one of the 0-20% multiplicity class to remove jet-like contributions, a long-range double-ridge structure is observed as shown in Fig.9. This distribution is qualitatively similar to the one observed in the hadron-hadron correlation analysis [34, 35]. This suggests that the mechanism that generates these long-range correlation structures (by gluon saturation in the initial state as predicted by the Color Glass Condensate (CGC) calculations [36] and/or by a collectively expanding system in the final state as it is assumed in hydrodynamical model calculations [37] or other mechanims) may also affect heavy quarks.

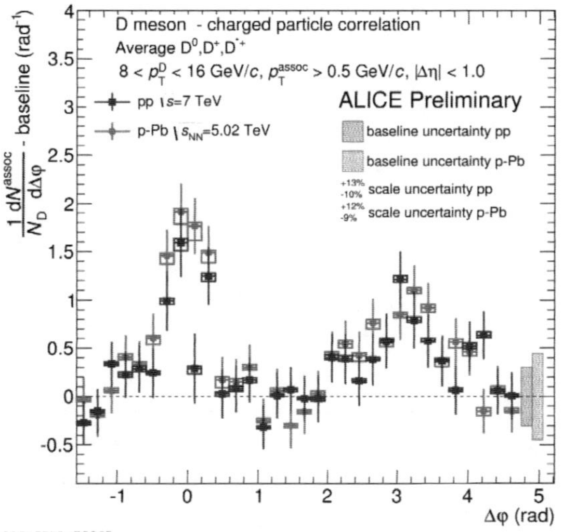


**Figure 8.** D meson-charged particle $\Delta\varphi$ correlations in minimum bias pp collisions at $\sqrt{s}$=7 TeV and p-Pb collisions at $\sqrt{s}_{NN}$=5.02 TeV.

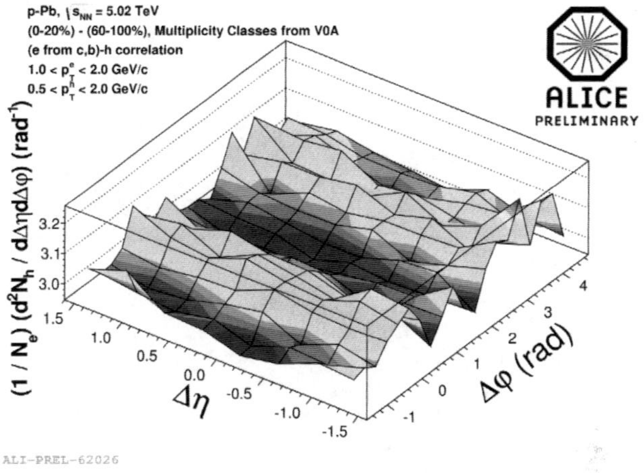


**Figure 9.** Angular correlations in azimuth and pseudorapidity between heavy-flavor decay electrons and charged particles in the 20 % highest multiplicity p-Pb collisions after subtracting the correlations in the multiplicity class 60-100 %.

## 5. Conclusion and Outlook

In Pb-Pb collisions at $\sqrt{s_{\rm NN}}$=2.76 TeV, a large suppression of the high-$p_{\rm T}$ heavy-flavor yields is observed in the most-central collisions. The measured nuclear modification factor of prompt D mesons is smaller than the one of non-prompt J/$\psi$ measured by the CMS Collaboration. The quark mass used in model calculations is crucial to reproduce the data. The measured positive heavy-flavor elliptic flow in semi-central Pb-Pb collisions indicates that the interactions of charm quarks with the expanding medium transfer information on the azimuthal anisotropy of the system to the heavy quarks. In pp collisions at $\sqrt{s}$= 2.76 and 7 TeV, the measured cross sections are described by perturbative calculations. Measurements of the multiplicity dependence of heavy-flavor yields indicate that Multi-Parton Interactions contribute to the high-multiplicity pp collisions and affect charm and beauty in a similar way. In p-Pb collisions at

Journal of Physics: Conference Series **668** (2016) 012005 doi:10.1088/1742-6596/668/1/012005

$\sqrt{s_{\rm NN}}$=5.02 TeV, cold nuclear matter effects do not lead to a strong suppression in the measured $p_{\rm T}$ range. Therefore, the strong suppression observed in Pb-Pb collisions is due to the presence of a hot and dense medium. A double-ridge structure is observed in the heavy-flavor decay electron-hadron azimuthal correlations at low $p_{\rm T}$ in p-Pb collisions, which indicates that heavy quarks are possibly affected by the processes leading to long-range correlations in $\eta$ of light-flavor hadrons.

The ongoing Run 2 data-taking period at the LHC will allow for the extention of the $p_{\rm T}$ range of the measurements, for more precise measurements of the azimuthal correlations, beauty and heavy flavor in jets, and for better study of the hadronization of heavy quarks by measuring the heavy baryons and $D_s$ mesons. The major ALICE upgrade between Run 2 and Run 3, including a new Inner Tracking System and a Muon Forward Tracker, opens the possibility to measure beauty production via displaced D mesons or directly via hadronic decays of beauty hadrons, as well as to separate the charm from the beauty contribution at forward rapidity [38].

## References

[1] K. J. Eskola, H. Paukkunen and C. A. Salgado, JHEP 0904 (2009) 065.
[2] H. Fujii and K. Watanabe, NPA 920 (2013) 78.
[3] I. Vitev, PRC 75 (2007) 064906.
[4] K. Aamodt *et al.*, ALICE Collaboration, JINST, 3 (2008) S08002
[5] B. Abelev *et al.*, ALICE Collaboration, JHEP 1201 (2012) 128.
[6] B. Abelev *et al.*, ALICE Collaboration, PLB 718 (2012) 279.
[7] B. Abelev *et al.*, ALICE Collaboration, PRD 86 (2012) 112007.
[8] B. Abelev *et al.*, ALICE Collaboration, PLB 721 (2013) 13.
[9] B. Abelev *et al.*, ALICE Collaboration, PLB 708 (2012) 265.
[10] B. Abelev *et al.*, ALICE Collaboration, JHEP 1207 (2012) 191.
[11] B. Abelev *et al.*, ALICE Collaboration, PRD 91 (2015) 012001.
[12] B. Abelev *et al.*, ALICE Collaboration, PLB 738 (2014) 97.
[13] B. Abelev *et al.*, ALICE Collaboration, PRL 109 (2012) 112301.
[14] M. Cacciari *et al.*, JHEP 1210 (2012) 137.
[15] B. Kniehl, G. Kramer, I. Schienbein and H. Spiesberger, EPJ C72 (2012) 2082.
[16] R. Maciula and A. Szczurek, PRD 87 (2013) 094022.
[17] B. Abelev *et al.*, ALICE Collaboration, arXiv: 1505.00664 (submitted to JHEP).
[18] Y. L. Dokshitzer, D. E. Kharzeev, PLB 519 (2001) 199
[19] B. Abelev *et al.*, ALICE Collaboration, arXiv: 1506.06604 (submitted to JHEP).
[20] B. Abelev *et al.*, ALICE Collaboration, JHEP 09 (2012) 112.
[21] M. He, R. J. Fries and R. Rapp, PRL 110 (2013) 112301.
[22] A. Andronic *et al.*, PLB 659 (2008) 149.
[23] I. Kuznetsova and J. Rafelski, EPJ C51 (2007) 113.
[24] M. Djordjevic and M. Djordjevic, PLB 734 (2014) 286.
[25] S. Chatrchyan *et al.*, CMS Collaboration, JHEP 05 (2012) 063.
[26] S. Chatrchyan *et al.*, CMS Collaboration, CMS-PAS-HIN-12-014.
[27] B. Abelev *et al.*, ALICE Collaboration, JHEP 07 (2015) 051
[28] B. Abelev *et al.*, ALICE Collaboration, PRC 90 (2014) 034904.
[29] B. Abelev *et al.*, ALICE Collaboration, PRL 111 (2013) 102301.
[30] B. Abelev *et al.*, ALICE Collaboration, arXiv:1507.03134v1 (submitted to PLB).
[31] B. Abelev *et al.*, ALICE Collaboration, PRL 113 (2014) 232301.
[32] M. Mangano *et al.*, NPB 373 (1992) 295.
[33] Z. B. Kang *et al.*, PLB 740 (2015) 23.
[34] B. Abelev *et al.*, ALICE Collaboration, PLB 719 (2013) 29.
[35] B. Abelev *et al.*, ALICE Collaboration, PLB 726 (2013) 164.
[36] K. Dusling and R. Venugopalan, PRD 87 (2013) 094034.
[37] P. Bozek and W. Broniowski, PLB 718 (2013) 1557.
[38] A. Dainese on behalf of the ALICE Collaboration, sQM 2015 proceeding.

Journal of Physics: Conference Series **668** (2016) 012006 doi:10.1088/1742-6596/668/1/012006

# Heavy-flavor transport

**Andrea Beraudo**
INFN, Sezione di Torino, Via Pietro Giuria 1, I-10125 Torino
E-mail: beraudo@to.infn.it

**Abstract.** The formation of a hot deconfined medium (Quark-Gluon Plasma) in high-energy nuclear collisions affects heavy-flavor observables. In the low/moderate-$p_T$ range transport calculations allow one to simulate the propagation of heavy quarks in the plasma and to evaluate the effect of the medium on the final hadronic spectra: results obtained with transport coefficients arising from different theoretical approaches can be compared to experimental data. Finally, a discussion of possible effects on heavy-flavor observables due to the possible formation of a hot-medium in small systems (like in p-A collisions) is presented.

## 1. Introduction

Heavy quarks play an important and peculiar role in characterizing the medium produced in heavy-ion collisions. The description of soft observables (low-$p_T$ hadrons) is based on hydrodynamics, assuming that one deals with a system (to which the measured particles belonged before decoupling) close to local thermal equilibrium. Jet-quenching studies address the energy-degradation of external probes (high-$p_T$ partons) crossing the medium. The description of heavy-flavor observables on the other hand requires the employment/development of a setup (transport theory) allowing one to face a more general situation and in particular to describe how particles would (asymptotically) approach kinetic equilibrium with the medium in which they are embedded. At high $p_T$ the interest in heavy flavor is no longer related to their possible thermalization, but rather to the study of the mass and color-charge dependence of jet-quenching: this last aspect however will not be addressed by our discussion.

Why are charm and beauty quarks considered *heavy*? This is due to the large value of their mass compared to other physical scales entering into the problem. First of all, $M \gg \Lambda_{\rm QCD}$, so that their initial hard production is under theoretical control, well described by pQCD. Secondly $M \gg T$, so that their thermal abundance is negligible and their final total multiplicity in the plasma (rapidly expanding and with a lifetime of at most $\sim$10 fm/c) created in the heavy-ion collisions is set by the initial hard production (possibly affected by nuclear corrections to the proton PDFs). Finally, in the context of a weakly-coupled relativistic plasma, their masses are also much larger than the typical momentum exchange with the medium particles, $M \gg gT$ ($g$ being the QCD coupling), such that many independent collisions are necessary to significantly change the trajectory/momentum of a heavy quark; this last observation will be of importance for the development of numerical schemes to simulate the propagation of heavy-flavor particles in the plasma.

The present contribution is organized as follows. In Sec. 2 we discuss how heavy-flavor production in elementary pp collisions can be described and nicely reproduced by pQCD. In Sec. 3 we address the case of nucleus-nucleus (A-A) collisions, where a hot Quark-Gluon Plasma

Journal of Physics: Conference Series **668** (2016) 012006 doi:10.1088/1742-6596/668/1/012006

(QGP) is expected to be produced and to modify heavy-flavor observables. Finally, in Sec. 4, we move to small systems, like the ones produced in p-Pb or d-Au collisions, where, looking at soft observables, signatures of the formation of a medium displaying a collective behavior were found. Can this also affect heavy-flavor spectra? Some first estimates are provided.

## 2. The benchmark: heavy flavor in pp collisions

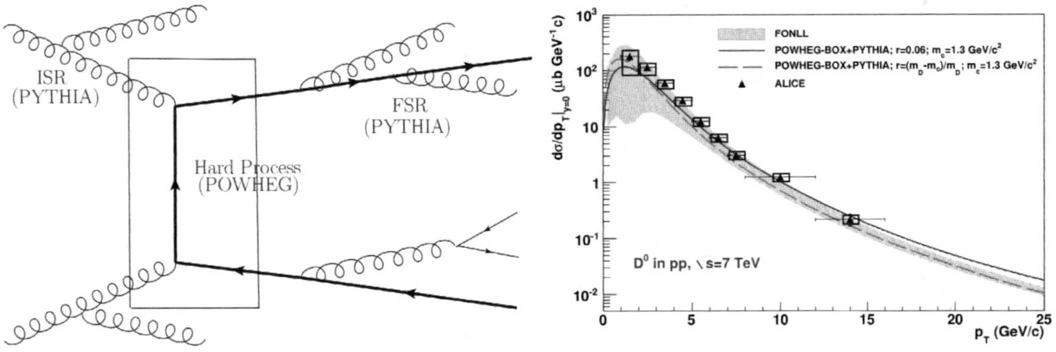


**Figure 1.** A cartoon of heavy-quark production within the POWHEG-BOX setup [1], with the hard process interfaced with a parton-shower stage. Results for the final D-meson spectra in pp collisions at $\sqrt{s}=7$ TeV are displayed in the right panel [2], compared to FONLL predictions [3] and ALICE data [4].

Because of their large mass, heavy ($c$ and $b$) quarks are produced in hard processes occurring on a very short time-scale. Hence, in A-A collisions, their initial production is not expected to suffer any modification due to the presence of a medium, which does not have time to form and in case would not be able to affect physics at very short distances. The only initial-state effect is thought to arise from the nuclear modification of the Parton Distribution Functions (PDFs) and from the larger transverse momentum broadening of the incoming partons, which have to cross a slice of nuclear matter before colliding. Hence, the initial hard $Q\overline{Q}$ production can be thought to occur in independent nucleon-nucleon collisions and, in order to quantify later medium effects present in A-A collisions, one has to check that the pp case is under control. The state of the art in the description of the initial $Q\overline{Q}$ production is represented by NLO pQCD calculations (POWHEG, MC@NLO) for the hard process interfaced to some event generator (PYTHIA, HERWIG) to simulate the Initial and Final State Radiation and other non perturbative processes (intrinsic $k_T$, Underlying Event and hadronization). Automated tools, such as the POWHEG-BOX package [1], are available to perform such calculations producing fully-exclusive information of the final state. As displayed in the right panel of Fig. 2, a satisfactory agreement with the experimental data can be obtained, both for the inclusive single-particle spectra and (not shown) for more differential observables sensitive to the initial $Q\overline{Q}$ correlations. Unfortunately, current measurements are rather indirect (e-h or D-h) and, in the A-A case, medium effects are difficult to model; actually $Q\overline{Q}$ production can occur in processes (gluon-splitting and flavor excitation) characterized by the presence of a third hard parton on the away-side, which will affect the final observables and whose propagation in the medium should be also described.

## 3. Heavy flavor in nucleus-nucleus collisions

Heavy quarks, whose initial hard production is quite well under control, can be used as calibrated probes of the medium formed in heavy-ion collisions, where, on average, they have

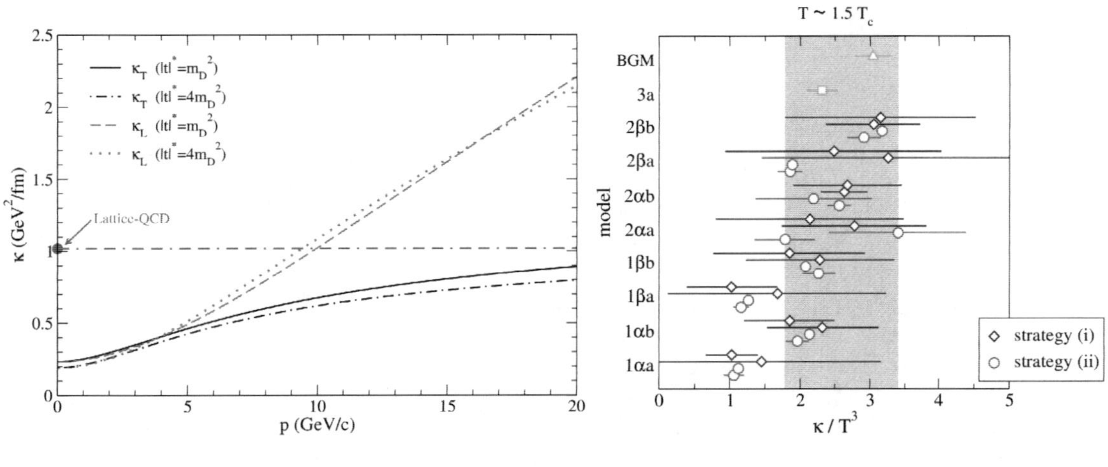


**Figure 2.** Heavy-flavor transport coefficients. Weak coupling (HTL+pQCD) results for $b$ quarks (left panel, [2]) and the systematic uncertainty band of lattice-QCD estimates (right panel, [5]) are displayed.

to propagate for several fm/c's before hadronizing and finally decoupling. Experimental heavy-flavor measurements (D mesons, $J/\psi$'s from B decays, heavy-flavor electrons...) show that the $p_T$-spectra of charm and beauty are suppressed with respect to pp collisions (indicating an important in-medium energy loss) and provide evidence of a non-vanishing elliptic flow $v_2$ in non-central collisions: charm seems to follow the collective flow of the medium. This last observation raises the question to what extent charm quarks approach kinetic equilibrium with the rest of the medium during the limited lifetime of the latter, which furthermore undergoes a hydrodynamic expansion. Transport calculations represent the tool to study heavy-quark propagation in the QGP. As long as the medium admits a particle description the starting point of all transport calculation is the Boltzmann equation for the evolution of the heavy quark phase-space distribution:

$$\frac{d}{dt}f_Q(t,\boldsymbol{x},\boldsymbol{p}) = C[f_Q] \quad \text{with} \quad C[f_Q] = \int d\boldsymbol{k}[w(\boldsymbol{p}+\boldsymbol{k},\boldsymbol{k})f_Q(t,\boldsymbol{x},\boldsymbol{p}+\boldsymbol{k}) - w(\boldsymbol{p},\boldsymbol{k})f_Q(t,\boldsymbol{x},\boldsymbol{p})], \tag{1}$$

where the collision integral $C[f_Q]$ is expressed in terms of the $\boldsymbol{p}\to\boldsymbol{p}-\boldsymbol{k}$ transition rate $w(\boldsymbol{p},\boldsymbol{k})$ The direct solution of the Boltzmann equation in the evolving inhomogeneous fireball produced in heavy-ion collisions is challenging; results can be found in [6, 7]. However as long as $k \ll p$ ($k$ being typically of order $gT$) one can expand the collision integral in powers of the momentum exchange. Relying on the above mentioned hypothesis of dominance of soft scatterings one can approximate the exact Boltzmann equation with a form more suited to an easy numerical implementation: the relativistic Langevin equation. The latter allows one to study the evolution of the momentum $\vec{p}$ of each heavy quark in the medium, through the combined effect of a friction and a noise term, both arising from the collisions suffered in the plasma:

$$\frac{\Delta\vec{p}}{\Delta t} = -\eta_D(p)\vec{p} + \vec{\xi}(t) \quad \text{with} \quad \langle\xi^i(t)\xi^j(t')\rangle = b^{ij}(\vec{p})\delta_{tt'}/\Delta t \equiv \left[\kappa_L(p)\hat{p}^i\hat{p}^j + \kappa_T(p)(\delta^{ij}-\hat{p}^i\hat{p}^j)\right]\delta_{tt'}/\Delta t. \tag{2}$$

The strength of the noise is set by the momentum-diffusion coefficients $\kappa_{L/T}$, reflecting the average longitudinal/transverse squared momentum exchanged with the plasma. The friction coefficient $\eta_D(p)$ must be fixed in order to ensure the approach to thermal equilibrium by the

Einstein fluctuation-dissipation relation

$$\eta_D(p) \equiv \frac{\kappa_L(p)}{2TE_p} + \text{corrections}, \tag{3}$$

where the corrections, subleading by a factor $\mathcal{O}(T/E_p)$, depend on the discretization scheme and are fixed in order to reproduce the correct continuum result in the $\Delta t \to 0$ limit. The advantage of the Langevin approach is that, independently of the nature of the medium (even when the latter does not admit a quasi-particle description), it allows one to summarize the interaction of the heavy quarks with the plasma in a few transport coefficients with a clean physical interpretation. Results for the momentum-diffusion coefficients, obtained with very different approaches (resummed weak-coupling calculations and lattice-QCD simulations) are displayed in Fig. 2. One finds that, at zero momentum, the lattice result is much larger than the perturbative value; however weak-coupling calculations provide evidence, in particular for the case (not shown) of charm, of a quite strong momentum dependence of $\kappa_{L/T}$ (the latter is found to be even larger in other approaches, like AdS/CFT calculations [8]), which unfortunately at present cannot be accessed on the lattice. This represents a limit on the information one can extract from experimental charm measurements if one cannot access the low-$p_T$ region. On the other hand, one can see how, for beauty, $\kappa_{L/T}$ values stay close to each other and quite constant for an extended momentum range, opening the possibility with the ALICE upgrade to perform measurements able to tightly constrain the value of the QGP transport coefficients.

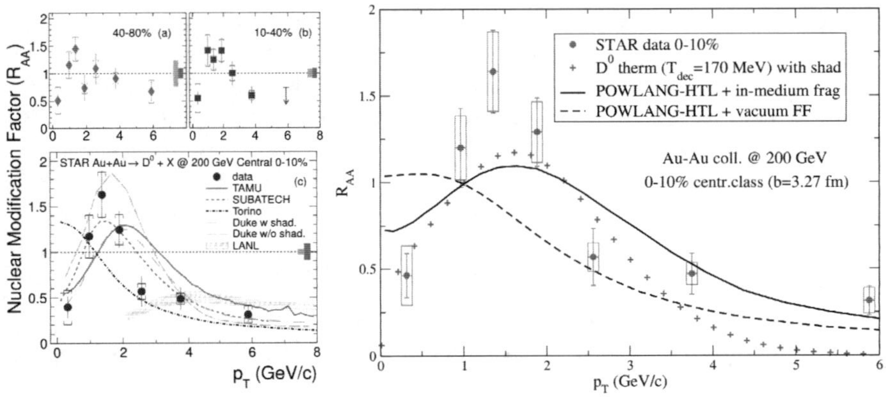


**Figure 3.** The D-meson $R_{\mathrm{AA}}$ in central Au-Au collisions at $\sqrt{s_{\mathrm{NN}}} = 200$ GeV. Left panel: STAR results [9] are compared to various model calculations. Right panel: POWLANG results (black curves) with and without in-medium hadronization, together with the limit of full kinetic equilibrium (green crosses), are displayed [10].

The presence of a deconfined medium, besides modifying the heavy quark propagation, can also affect their hadronization, changing both the relative particle abundances and their momentum spectra. While in the vacuum hadrons come from the fragmentation of partons produced in the hard event, in A-A collisions recombination with light thermal partons may be at work. Since, in particular, the environment is very rich in strange quarks, changes in the heavy-flavor hadrochemistry may be expected, leading for instance to an enhanced production of $D_s$ mesons [11]. The effect of heavy-quark coalescence with light partons on the final particle spectra was studied in detail in [12], where it turned out to allow a better description of the $R_{\mathrm{AA}}$ and $v_2$ of heavy-flavor electrons at RHIC. In the left panel of Fig. 3 it is shown how models including coalescence display a better agreement with STAR data [9] for the D-meson $R_{\mathrm{AA}}$.

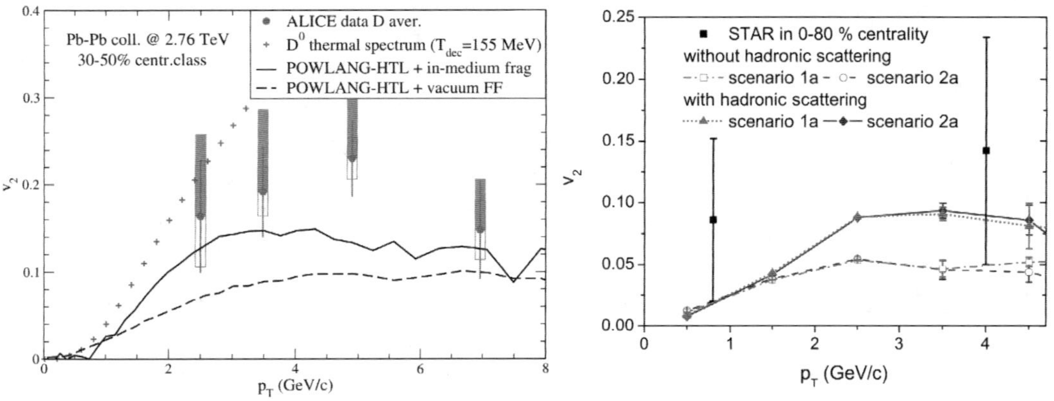


**Figure 4.** D-meson elliptic flow. Left panel: POWLANG results (black curves) with and without in-medium hadronization, together with the limit of full kinetic equilibrium (green crosses), are shown [10] and compared to ALICE data in non-central Pb-Pb collisions at the LHC [13]. Right panel: the relevance of the possible D-meson rescattering in the hadronic phase is displayed [14]. Results are compared to STAR data in minimum-bias Au-Au collisions [16].

In the coalescence picture, the recombination probability is large when the wave-function of the final D meson (if one considers charm) and the wave-packets of the two partons (the $c$ quark and a light antiquark from the medium) display a sizable overlap; this occurs when the initial partons are sufficiently close in space and have comparable velocities. An alternative picture is the one recently implemented in the POWLANG setup [10]. Here, heavy quarks, once they reach a cell below a decoupling temperature, independently of their kinematics, are combined with light thermal partons from the medium, giving rise to $Q\overline{q}$ strings eventually fragmented according to the PYTHIA [17] implementation of the Lund model into the final charmed (or beauty) hadrons. Although the mechanism is slightly different from coalescence, the result is similar: heavy-flavor hadrons tend to inherit part of the collective flow of the medium from the light thermal partons, leading to a better agreement with the data. This can be appreciated in the right panel of Fig. 3, where STAR data for the D-meson $R_{\rm AA}$ are compared to different POWLANG predictions. As it can be seen, with the same transport coefficients, results including in-medium hadronization better reproduce the qualitative trend of the data, with the peak at moderate $p_T$ arising from radial flow. The same occurs for the elliptic flow, as displayed in the left panel of Fig. 4 where ALICE data [13] in non-central Pb-Pb collisions are plotted together with POWLANG predictions: also in this case the additional flow acquired via recombination with light partons from the medium moves theory outcomes closer to the experimental findings.

Finally, although very often neglected, heavy-flavor hadrons can go on scattering also in the hadronic phase. Effective chiral lagrangians can be employed to describe the interactions between D mesons and the other light particles like pions or kaons [14, 15]. Of course the hadronic phase is characterized by a lower value of the the temperature and hence of the transport coefficients compared to the QGP, but it is also the stage in which the system has acquired the largest flow: a non-negligible $v_2$ can be acquired also in the hadron gas, as show in the right panel of Fig. 4.

## 4. Heavy flavor in small systems

One of the surprises in the experimental search for QGP formation came from the observation of collective effects in small systems: the ridge in long-range rapidity correlation in high-multiplicity pp and p-Pb events, the non-vanishing flow harmonics measured in p-Pb and d-Au collisions, the

Journal of Physics: Conference Series **668** (2016) 012006 doi:10.1088/1742-6596/668/1/012006

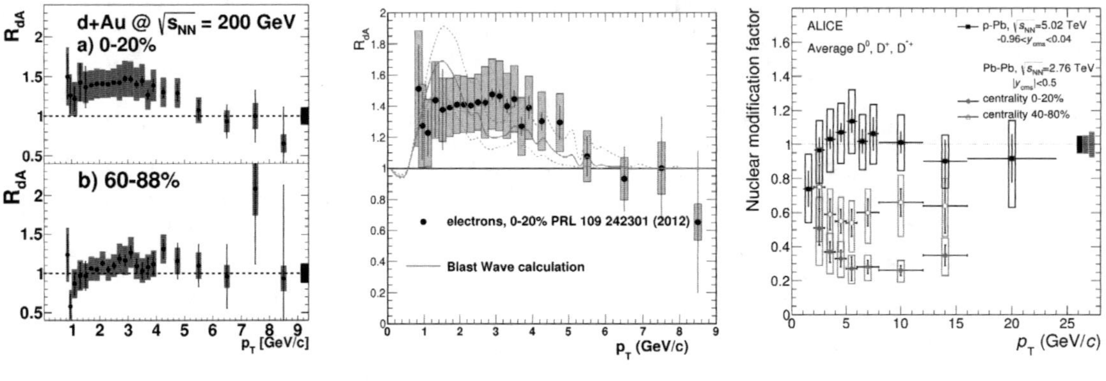


**Figure 5.** Heavy-flavor nuclear modification in small systems. Left and middle panels: PHENIX results for heavy-flavor electrons in central d-Au collisions at RHIC [18] compared to peripheral events and to a blast-wave fit [19], assuming a common collective flow with the medium. Right panel: ALICE results for D-mesons in minimum-bias p-Pb collisions at the LHC [20], compared to central and peripheral Pb-Pb events [21].

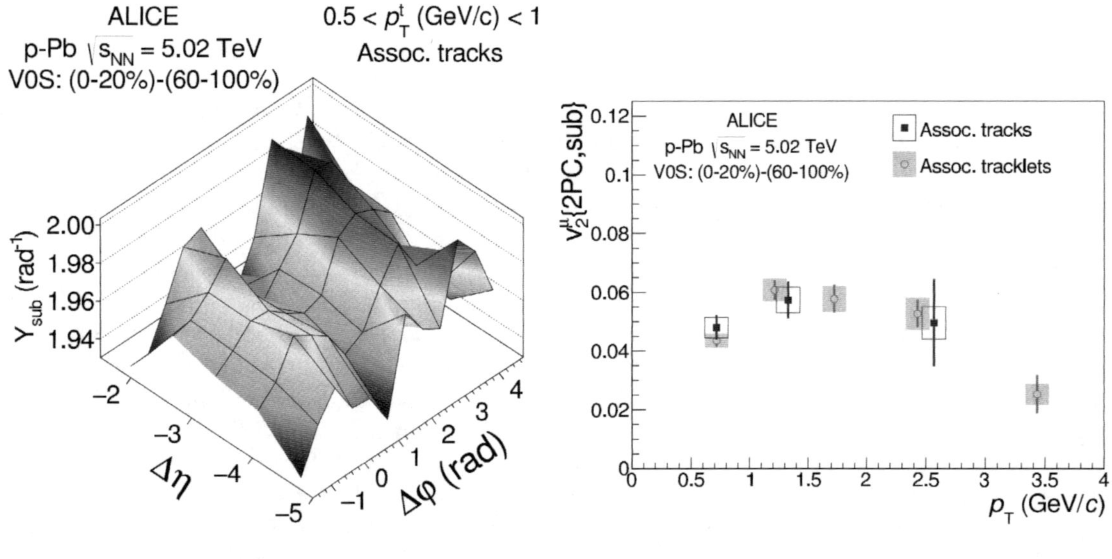


**Figure 6.** ALICE results for muon-hadron correlations in p-Pb collisions, after subtraction of the results in peripheral events [22]. The corresponding estimate of the muon elliptic flow is displayed in the left panel.

$p/\pi$ ratio of the $p_T$-spectra, etc. All the above effects involve soft light hadrons and suggest the formation of a hot medium with a hydrodynamic behavior even in these collisions. On the other hand, no significant effect was observed in the high-$p_T$ domain; neither jets nor particle spectra turn out to be suppressed and minor modifications can be interpreted as an initial-state effect due to the nuclear PDFs. Actually, the two observations are not necessarily in contradiction. In spite of the possible strongly interacting nature of the medium (responsible for the hydrodynamic expansion driven by pressure gradients), the energy loss of hard partons depends in any case on the system size ($\langle \Delta E \rangle \sim L^2$ in the presence of coherence effects).

One may wonder what happens to heavy-flavor production in these small systems. The situation is not completely settled. At RHIC (see Fig. 5) the nuclear modification factor of heavy-flavor electrons in central d-Au collisions was found to slightly exceed 1 [18]. Although error bars are large this looks puzzling; a possible explanation was proposed in Ref. [19], where a blast-wave spectrum for D and B mesons was employed, assuming they share the same collective flow of the medium. At the LHC the D-meson $R_{\mathrm{pA}}$ in minimum-bias p-Pb collisions turns out to be compatible with 1 within the large systematic error-bars [20], with a depletion at low $p_T$ to attribute to the nPDFs. On the other hand, other observations suggest signatures of collective effects also in the heavy-flavor sector, such as (see Fig. 6) the appearance of a double ridge in e-h and $\mu$-h correlations [22] measured in central p-Pb events, where electrons and (part of the) muons come from heavy-flavor decays, and from which it is possible to extract an estimate of the elliptic-flow of the trigger particle.

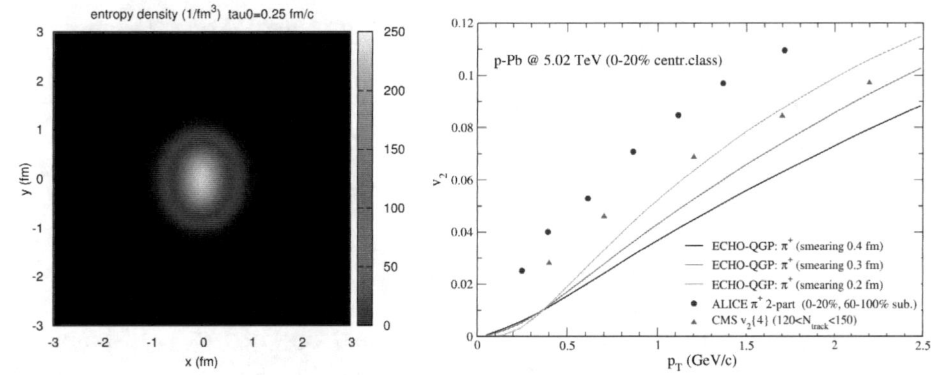


**Figure 7.** Average initial condition for 0-20% more central p-Pb events at $\sqrt{s_{\mathrm{NN}}} = 5.02$ TeV and the corresponding pion elliptic flow (for various smearing parameters) arising from the hydrodynamic evolution of the medium compared to ALICE and CMS data.

Can the propagation and subsequent hadronization of heavy quarks in the presence of a small-size hot deconfined medium provide results compatible with the present data? We attempt to model the situation as follows (see Ref. [23] for details). A smooth hydrodynamic background is constructed summing (depending on the centrality selection) hundreds/thousands Glauber-MC fluctuating initial conditions, each one rotated by the event-plane angle $\psi_2$ so as to get a realistic estimate of the average initial eccentricity $\epsilon_2$; the resulting profile is then used as an initial condition for the numerical solution of viscous hydrodynamic equations performed with the ECHO-QGP code [24]. As shown in Fig. 7, the procedure allows one to roughly reproduce the size of the light-hadron elliptic flow, so that one can be confident to have a quite realistic description of the medium. The solution of hydrodynamic equations can then be used as a background for the heavy quark propagation, simulated through the POWLANG setup. As done in the A-A case, below a certain decoupling temperature $c$ quarks are then combined with thermal partons to form strings whose fragmentation will give rise to the final D-mesons. Results are shown in Fig. 8. As the crossed medium is very thin, the elliptic flow of $c$ quarks at decoupling is very small; however, recombination with light thermal partons may enhance the $v_2$ up to $\sim$5% because of the additional flow inherited from the medium, producing a result in qualitative agreement with the kinetic equilibrium scenario. Also for what concerns the $p_T$-spectrum, the recombination with the light background quarks leaves its fingerprints, giving rise to a bumb because of radial flow. With the present experimental error bars such an interpretation is not in contradiction with data.

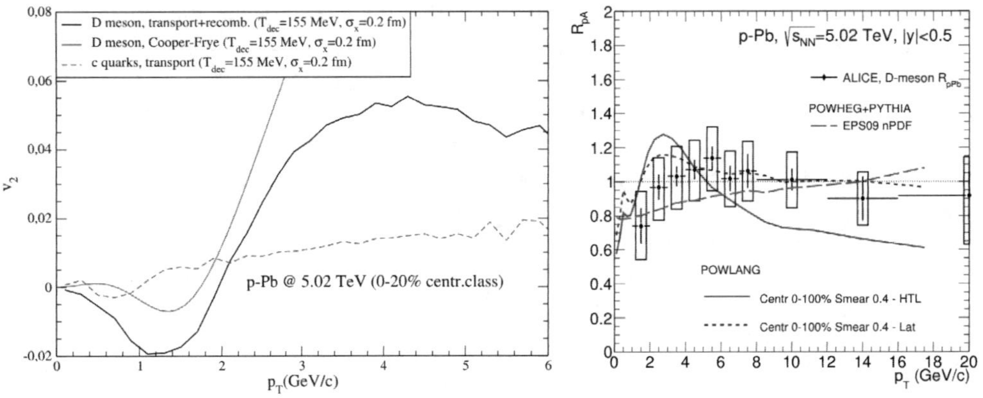


**Figure 8.** Left panel: POWLANG results for charm elliptic flow in 0-20% more central p-Pb events. The $v_2$ of $c$ quarks, D-meson (after in-medium hadronization) as well as the limit of full kinetic equilibrium are compared. Right panel: POWLANG predictions for the D-meson nuclear modification factor in minimum-bias p-Pb collisions.

## 5. Conclusions and perspectives

Heavy-flavor observables allow one to get rich information on the medium produced in heavy-ion collisions. In order to describe the data one needs to properly model both the dynamics and the hadronization of heavy quarks in the presence of a QGP. Recently, signatures of final-state medium effects were observed also in small systems (like in p-A collisions), previously considered just a benchmark to estimate cold nuclear-matter effects. This surely deserves deeper investigation in the future, together with the study of heavy-flavor correlations and of beauty at low $p_T$, allowing a more solid extraction of the heavy-flavor transport coefficients.

## References

[1] Alioli S, Nason P, Oleari C and Re E 2010 *JHEP* **06** 043 (*Preprint* 1002.2581)
[2] Alberico W M *et al.* 2013 *Eur. Phys. J.* **C73** 2481 (*Preprint* 1305.7421)
[3] Cacciari M, Greco M and Nason P 1998 *JHEP* **05** 007 (*Preprint* hep-ph/9803400)
[4] Abelev B *et al.* (ALICE) 2012 *JHEP* **01** 128 (*Preprint* 1111.1553)
[5] Francis A *et al.* 2015 (*Preprint* 1508.04543)
[6] Gossiaux P B and Aichelin J 2008 *Phys. Rev.* **C78** 014904 (*Preprint* 0802.2525)
[7] Uphoff J, Fochler O, Xu Z and Greiner C 2011 *Phys. Rev.* **C84** 024908 (*Preprint* 1104.2295)
[8] Gubser S S 2008 *Nucl. Phys.* **B790** 175–199 (*Preprint* hep-th/0612143)
[9] Adamczyk L *et al.* (STAR) 2014 *Phys. Rev. Lett.* **113** 142301 (*Preprint* 1404.6185)
[10] Beraudo A *et al.* 2015 *Eur. Phys. J.* **C75** 121 (*Preprint* 1410.6082)
[11] Innocenti G M (ALICE) 2013 *Nucl. Phys.* **A904-905** 433c–436c (*Preprint* 1210.6388)
[12] van Hees H, Mannarelli M, Greco V and Rapp R 2008 *Phys. Rev. Lett.* **100** 192301 (*Preprint* 0709.2884)
[13] Abelev B *et al.* (ALICE) 2013 *Phys. Rev. Lett.* **111** 102301 (*Preprint* 1305.2707)
[14] Song T *et al.* 2015 *Phys. Rev.* **C92** 014910 (*Preprint* 1503.03039)
[15] He M *et al.* 2014 *Phys. Lett.* **B701** 445-450 (*Preprint* 1103.6279)
[16] Tlusty D (STAR) 2013 *Nucl. Phys.* **A904-905** 639c–642c (*Preprint* 1211.5995)
[17] Sjostrand T, Mrenna S and Skands P Z 2006 *JHEP* **05** 026 (*Preprint* hep-ph/0603175)
[18] Adare A *et al.* (PHENIX) 2012 *Phys. Rev. Lett.* **109** 242301 (*Preprint* 1208.1293)
[19] Sickles A M 2014 *Phys. Lett.* **B731** 51–56 (*Preprint* 1309.6924)
[20] Abelev B B *et al.* (ALICE) 2014 *Phys. Rev. Lett.* **113** 232301 (*Preprint* 1405.3452)
[21] Abelev B *et al.* (ALICE) 2012 *JHEP* **09** 112 (*Preprint* 1203.2160)
[22] Adam J *et al.* (ALICE) 2015 (*Preprint* 1506.08032)
[23] Nardi M, Beraudo A, De Pace A, Monteno M and Prino F 2015 (*Preprint* 1509.04083)
[24] Del Zanna L *et al.* 2013 *Eur. Phys. J.* **C73** 2524 (*Preprint* 1305.7052)

Journal of Physics: Conference Series **668** (2016) 012007 doi:10.1088/1742-6596/668/1/012007

# Recent results from strangeness in transport models

**J. Steinheimer[1], A.S. Botvina[1,3] and M. Bleicher[1,2]**

[1]Frankfurt Institute for Advanced Studies, Ruth-Moufang-Str. 1, 60438 Frankfurt am Main, Germany

[2]Institut für Theoretische Physik, Goethe Universität Frankfurt, Max-von-Laue-Strasse 1, D-60438 Frankfurt am Main, Germany

[3]Institute for Nuclear Research, Russian Academy of Sciences, 117312 Moscow, Russia

E-mail: bleicher@fias.uni-frankfurt.de

**Abstract.** In these proceedings we discuss recent developments in the microscopic description of strange particle production in nuclear collisions. We put a special emphasis on the production of hypernuclei at the upcoming FAIR and NICA facilities as well as the deep sub threshold, $\phi$ and $\Xi^-$ production yields measured with the HADES experiment. Employing new resonance decay channels we obtain a satisfactory description of $\phi$ and $\Xi^-$ production in deep sub threshold Ar+KCl reactions. Our results implicate that no new medium effects are required to describe the rare strange particle production data from low energy nuclear collisions.

## 1. Introduction

Strange hadrons have long since been considered to be a good probe for the properties of dense hadronic matter [1, 2] created in nuclear collisions. The properties of such hot and dense systems are subject of investigation of running and planned experimental programs at the GSI/FAIR [3], NICA [4] and RHIC facilities. To understand the dynamics in such collisions one usually employs microscopic transport models.

For example, by modifying the medium properties of strange hadrons in a transport study, it was found that the production rates and properties of Kaons are a promising probe to extract their medium interactions in low energy nuclear collisions [2, 5, 6, 7, 8, 9, 10, 11, 12, 13, 14, 15, 16].

Furthermore recent progress has been made in the understanding of strange hadron dynamics at the LHC [17] and RHIC [18] with special attention payed to the role of strange particle dynamics in the hadronic, non-equilibrium, phase of ultra relativistic nuclear collisions [19, 20, 21, 22, 23].

Similarly, the investigation of hypernuclei is a rapidly progressing field of nuclear physics, since these nuclei provide methods to study nuclear interactions in particle physics and nuclear astrophysics (see, e.g., [24, 25, 26, 27, 28] and references therein). Presently, hypernuclear physics is still focused on spectroscopic information and is dominated by a set of reactions induced by high-energy hadrons and leptons leading to the production of only few particles. Many experimental collaborations have started or plan to investigate hypernuclei and their properties in hadron and heavy ion induced reactions.

In the following we will highlight recent results on the formation of hypernuclei in nuclear collisions and explore new mechanisms for the production of the $\phi$ meson and $\Xi$ baryons at sub-

Journal of Physics: Conference Series **668** (2016) 012007 doi:10.1088/1742-6596/668/1/012007

threshold energies, in the microscopic transport model UrQMD. These proceedings are based on [47, 34]

## 2. Particle production in UrQMD

Strange hadron production in the UrQMD transport model [29, 30] proceeds through different channels: the excitation and de-excitation (decay) of hadronic resonances, the excitation and de-excitation of a string and the annihilation of a particle with its anti-particle. The probabilities of the different processes are governed by their reaction cross sections. These cross sections serve as input for the model and are, whenever possible, taken from experimental measurements of elementary (binary) collisions. For example the total and inelastic resonance cross sections of binary proton+proton collisions has been measured in many experiments over a wide range of beam energies [31].

Another important channel for the description of strange particle production in nuclear collisions is the strangeness exchange reaction which can change the flavor content of a hadron. This includes reactions of the type $N+\overline{K} \leftrightarrow Y+\pi$ as well as $Y+\overline{K} \leftrightarrow \Xi+\pi$. Such reactions are included in the UrQMD transport approach however it was shown that they are not sufficient to explain the large $\Xi/\Lambda$ ratio in Ar+KCl and p+Nb reactions measured with the HADES experiment (see [32] for a detailed discussion).

## 3. Hypernuclei

To study hyperon production in nuclear collisions we apply the Ultra-relativistic Quantum Molecular Dynamics model (UrQMD), described above, and a recently developed, alternative, formulation of the coalescence model, the coalescence of baryons (CB), which is suitable for event by event simulations [33]. Baryons (nucleons and hyperons) can produce a cluster with mass number $A$ if their velocities relative to the center-of-mass velocity of the cluster is less than $v_c$. Accordingly we require $|\vec{v}_i - \vec{v}_{cm}| < v_c$ for all $i = 1, ..., A$, where $\vec{v}_{cm} = \frac{1}{E_A}\sum_{i=1}^{A}\vec{p}_i$ ($\vec{p}_i$ are momenta and $E_A$ is the sum energy of the baryons in the cluster). This is performed by sequential comparison of the velocities of all baryons.

The results on the total mass yields of the normal fragments and hyper-fragments (with one bound $\Lambda$) are shown in Fig. 1. The coalescence of baryons (the CB model) was applied on output from UrQMD, for the reactions presented in Fig. 1. The yields are normalized per one inelastic event (More detailed information on the model can be found in [34]). However, one should take into account that only events with at least one hyperon produced are analysed in this case. For this reason there is no characteristic increase of the yield of normal fragments with masses around the projectile/target mass, which are caused by very peripheral collisions. The explanation of this behaviour was already suggested in Ref. [35]; production of hyperons usually requires many particle collisions leading to a considerable emission of fast nucleons from the residues.

One can see that the production of fragments of all sizes is possible. As expected the yield of conventional fragments is by few orders of magnitude higher than the yield of hyper-fragments. Nevertheless, the production of hyper-fragments is sufficient to be experimentally measured (see also [36, 37]). It is a natural result of the coalescence that the yield of the lightest hyper-fragments is dominating. However, the capture of hyperons by residues saturates the yield for large masses and leads to abundant production of heavy hyper-fragments. Within this approach one can clearly see nearly all the normal fragments and hyper-fragments with $A > 3-4$ in the carbon collisions and with $A > 10$ in the gold collisions originate from the capture of $\Lambda$ hyperons by spectator residues (dotted lines). As was mentioned, we believe that these hyper-fragments represent excited pieces of hyper-matter whose evolution can be calculated with statistical models [38, 28]. The excitation energy of such primary fragments can also be evaluated from the analysis of experimental data [39, 40, 41, 42]. This demonstrates that big

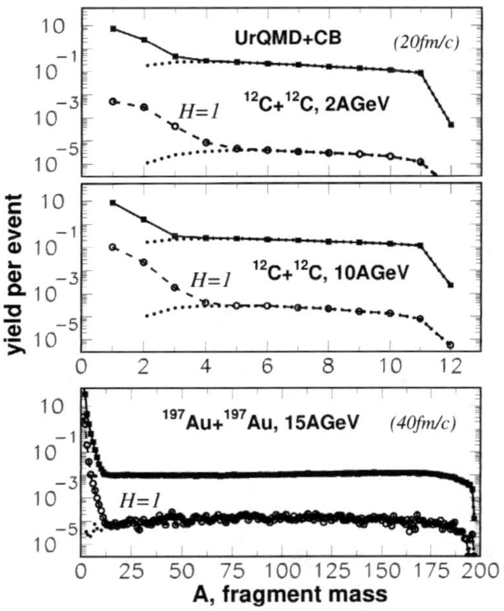


**Figure 1.** Yields (per one inelastic event) of normal fragments (solid lines with squares) and hyper-fragments with one captured Λ (notation H=1, dashed lines with circles) versus their mass number (A) in reactions induced by carbon and gold collisions. The dotted lines present the corresponding fragments originated from the spectator residues. The calculations are performed within the hybrid UrQMD plus CB model, with the coalescence parameter $v_c = 0.22c$, and integration over all impact parameters. The projectile lab energies and the transition times from UrQMD to CB are shown in panels.

hyper-fragments are mostly produced from the spectator residues, while the light ones can be formed at all rapidities. We expect that some large species of hypermatter will be excited and decay as in usual fragmentation and multifragmentation reactions. Such a mechanism should allow the investigation of possible phase transitions in hypermatter with statistical models describing the secondary disintegration. The significant production yields at beam energies higher than 5–10 GeV per nucleon open up the possibility to study hypernuclei at GSI/FAIR (Darmstadt), Nuclotron/NICA (Dubna), RHIC (Brookhaven), HIAF (Lanzhou) and other heavy-ion accelerators of moderate relativistic energies.

## 4. Deep sub threshold $\phi$ and $\Xi$ production

When discussing sub-threshold production of $\phi$'s and $\Xi$'s in nuclear collisions, one should note that, apart from in-medium modifications of hadron properties (see e.g. recent results in [43] and references therein), there are two distinct mechanisms, which allow for the production of hadrons with masses higher than what would be energetically forbidden in elementary reactions:

(i) One is the fact, that in a nucleus, the nucleons acquire a Fermi momentum due to their bound state. This allows for collisions of nucleons at energies higher than the actual beam energy.

(ii) Furthermore, energy can be accumulated due to secondary interactions of already excited

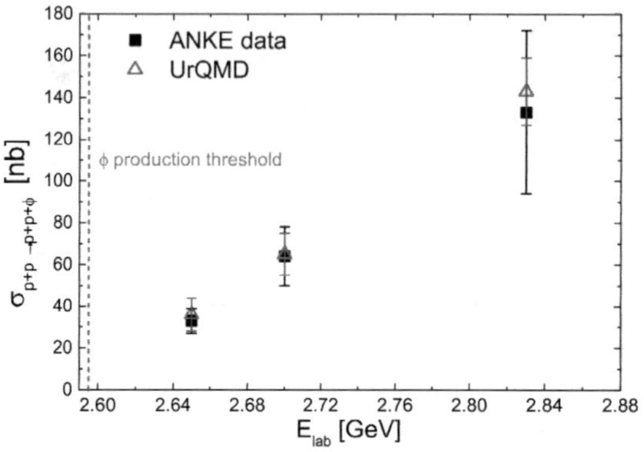


**Figure 2.** Total production cross section of the $\phi$ meson in p+p reactions near the $\phi$ production threshold. We compare experimental data from [46] with UrQMD results including the $\phi$ decay of the $N^*$ resonances.

states, produced earlier in the collision [44, 45].

From this one can conclude that a moderate amount of excited states with sufficiently high mass are available that may produce $\phi$ mesons as well as $\Xi$ baryons. In the following, we will show how decays of the most massive $N^*$ resonances implemented in UrQMD, namely the $N^* \to N + \phi$ and $N^* \to \Xi + K + K$ channels, can be used to describe the production of $\phi$ and $\Xi$ particles near and below their elementary threshold energies (more details on the method and results is given in [47]).

To determine the probability that a heavy baryonic resonance state decays into the specific final states introduced above, one can use recently published ANKE data on the cross section of single $\phi$ production in near threshold $p$+$p$ collisions [46]. These cross sections are shown in figure 2 as black squares. Using these, we find that a (constant) branching fraction of $\frac{\Gamma_{N+\phi}}{\Gamma_{\rm tot}} = 0.2\%$, for all the above mentioned $N^*$ resonances, provides a very good description (red triangles) of the measured $\phi$ production cross section.

Determining a similar branching fraction of $N^* \to \Xi + K + K$ is not as straight forward as it is with the $\phi$ decay. There exists no experimental data on $\Xi$ production in elementary collisions near its production threshold. Using the new HADES data on $\Xi$ production in p+Nb reactions as a proxy for the unavailable elementary collision data to fix the $N^* \to \Xi + K + K$ branching fraction, one obtains a branching $\frac{\Gamma_{\Xi+K+K}}{\Gamma_{\rm tot}} = 10\%$ for all $N^*$ states mentioned above that have a sufficiently high mass for this decay. A branching fraction of 10% appears to be large; however, one should keep in mind that this branching fraction applies only in the high mass tails of the resonances and the integrated fraction is less than one percent. Table 1 summarizes the results for $\Xi^-$ production in $p + Nb$ at $E_{\rm lab} = 3.5$ GeV, measured with the HADES experiment, as well as results from the simulations.

A ratio which has shown an interesting beam energy dependence, especially below the $\phi$ production threshold, is the $\phi/K^-$ ratio, which is depicted in figure 3 for nuclear collisions at different beam energies, measured by several experiments [49, 50, 51, 52, 53, 54]. Results from simulations, including the new branching gractions, for most central ($b < 3.4$ fm) Au+Au collisions are shown as the red line.

From the comparison in figure 3, it is clearly visible that the qualitative behavior of the data, a rapid increase of the $\phi/K^-$ ratio for sub-threshold energies, is nicely reproduced by the new process. Also the value of the ratio is in agreement for beam energies at and above the HADES Ar+KCl data with $E_{\rm lab} = 1.76$ A GeV as well as the FOPI results for Ni+Ni collisions at 1.91

| HADES data | |
|---|---|
| $\langle \Xi^- \rangle$ | $\Xi^-/\Lambda$ |
| $(2.0 \pm 0.3 \pm 0.4) \times 10^{-4}$ | $(1.2 \pm 0.3 \pm 0.4) \times 10^{-2}$ |
| UrQMD | |
| $\langle \Xi^- \rangle$ | $\Xi^-/\Lambda$ |
| $(1.44 \pm 0.05) \times 10^{-4}$ | $(0.71 \pm 0.03) \times 10^{-2}$ |

**Table 1.** $\Xi^-$ production yield and $\Xi^-/\Lambda$ ratio for minimum bias $p + Nb$ collision at a beam energy of $E_{\text{lab}} = 3.5$ GeV, compared with recent HADES results [48]

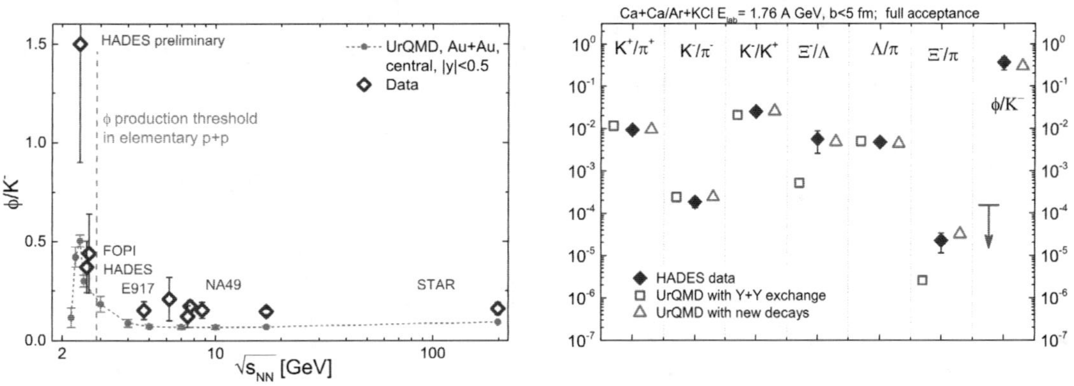


**Figure 3.** Left: Excitation function of the $\phi/K$ ratio for central ($b < 5$ fm) Au+Au collisions, calculated with the UrQMD model including the new $N^*$ decays (red line). Experimental results from different beam energies and systems are shown as symbols [49, 50, 51, 52, 53, 54]. Right: Different strange particle ratios from the UrQMD model in its default settings (green squares), compared to our results including the new $N^*$ decays (red triangles). We compare the simulations for Ca+Ca at $E_{\text{lab}} = 1.76$ A GeV and $b < 5$ fm with published HADES data [55, 56, 49, 57] for Ar+KCl collisions at the same beam energy (blue diamonds).

A GeV. However, one also observes that above the low SPS energy regime the present model underpredicts the $\phi/K^-$ ratio. This can be understood as a result of the high threshold for $\phi$ production in the string break-up. Because string excitation dominate the particle production at beam energies above $\sqrt{s_{NN}} > 5$ GeV, the $\phi$ must always be produced together with a Kaon-Antikaon pair, which strongly suppresses the $\phi$ production.

Finally, one can compare the multitude of strange particles produced in UrQMD, including the new $N^*$ decays, with sub-threshold nuclear collision data. Figure 3 shows results on strange particle ratios, in Ca+Ca collisions at $E_{\text{lab}} = 1.76$ A GeV from the UrQMD model. The default calculation with the previously released UrQMD version (v3.4) is shown as as green squares. Compared to the default calculation the new results, including the $\phi$ and $\Xi$ decay channels of the $N^*$ are depicted as red triangles. A considerable increase in the $\phi$ and $\Xi^-$ production is visible. More importantly, when compared to data from the HADES experiment (blue diamonds), one observes a very good description of all measured ratios, including the $\phi$ and $\Xi$. Such a good description of the full set of data has not been achieved in any previous study. Hence, we conclude that strange particle production in Ar+KCl collisions at the HADES experiment can be explained, and is in fully consistent, with production cross sections obtained in elementary reactions.

Journal of Physics: Conference Series **668** (2016) 012007 doi:10.1088/1742-6596/668/1/012007

## 5. Nuclear absorption of the $\phi$

An important aspect of $\phi$ production in nuclear collisions is the possibility of medium modifications of the $\phi$ meson. An interesting observable in this context is the so-called nuclear transparency, which measures the suppression (absorption) of $\phi$ meson production on nuclear targets of increasing mass number. It is thought to be sensitive on the $\phi + N$ absorption cross section in a nuclear medium and should give constraints on in-medium models of hadronic interactions. First experiments measured the $\phi$ absorption in photo-production reactions at the CLAS and LEPS-SPring8 experiments [58, 59] to be of the order of 35 mb. A rather large $\phi + N$ absorption cross section of 14-21 mb was also inferred from hadronic production at the ANKE experiment [60]. These numbers were extracted using certain assumptions. For example the LEPS-SPring8 results used a Glauber model analysis and assumed that the $\phi$ is essentially produced instantaneously, while the ANKE analysis relied on comparisons to different models which may or may not be valid descriptions.

Using the same parameters for the $\phi$ production by resonance as used in the previous section one can calculate the momentum dependent $\phi$ production cross section in different $p+A$ collisions at a beam energy of $E_{\text{lab}} = 2.83$ GeV and within the ANKE acceptance.
The transparency ratio $R$,

$$R = \frac{12}{A} \frac{\sigma_\phi^A}{\sigma_\phi^C}, \tag{1}$$

obtained with UrQMD, which is essentially the scaled ratio of the $\phi$ production cross section on different size nuclear targets, is compared to ANKE data in figure 4. Again the ratio $R$ is well described by the model, which does not include any explicit medium modification of the $\phi$. The explanation of the strong suppression of the $\phi$ meson production in the resonance approach cannot be a large inelastic $\phi + N$ cross section (which is of the order of $< 1$ mb while the elastic cross section is set to 5 mb). Because the $\phi$ is hardly absorbed (in our model) in the nuclear medium, it is in fact the heavy mother resonance of the $\phi$ which rescatters with another nucleon before it can decay into $N + \phi$. During this inelastic rescattering, the resonance is likely to create one or two other resonances, however with smaller masses, which then are unable to decay into a $\phi$, effectively suppressing the $\phi$ production probability in a nuclear environment. The same processes can also be responsible for the apparently large $\phi$ cross section measured in the LEPS-SPring8 experiment.

## 6. Summary

Important progress has been made in the description of strange hadron dynamics in nuclear collisions. It was found that relativistic heavy-ion reactions are a very promising source of hypermatter and hyper-fragments. A large amount of hypernuclei of all masses can be produced. Their properties can be investigated taking into account the advantages of relativistic velocities, e.g., for the life-time and correlation measurements.

Furthermore, a new mechanism for $\phi$ and $\Xi$ production in elementary and nuclear collisions, namely the decay of heavy resonances, has been proposed. For the $\phi$ production, the unknown branching ratios of the baryon resonances were extracted from $p+p \rightarrow p+p+\phi$ data measured by ANKE. The branching fraction necessary to describe the data is of the order of 0.2%.

The same approach to $\phi$ production is able to describe the nuclear transparency ratio measured in proton induced reactions at the ANKE experiment, without the inclusion of any additional in-medium effects and only a very small inelastic $\phi + p$ cross section. The large $\phi + p$ cross sections, extracted from the LEPS-SPring8 and ANKE experiments, might therefore just be an artifact of not taking into account the $\phi$ production processes.

For the $\Xi^-$ production the branching ratios of the heavy baryon resonances were extracted

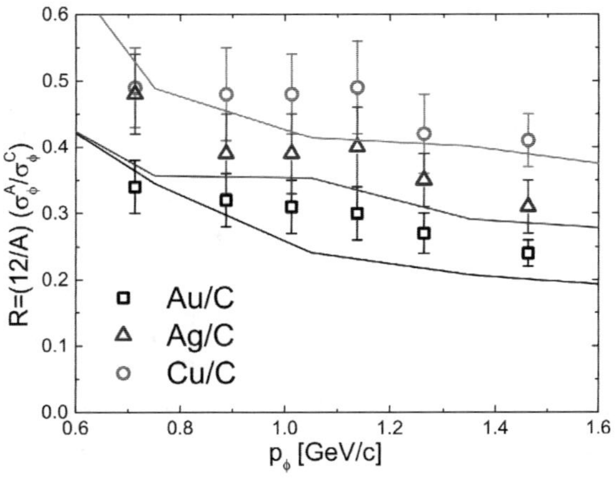


**Figure 4.** The transparency ratio $R$ calculated within our approach (lines) compared to ANKE data from [60] (symbols).

from p+Nb data of the HADES collaboration. Here, a larger branching fraction of 10% for $R^* \to \Xi + K + K$ is required. With this input from elementary reactions, a good description of the HADES Ar+KCl data is achieved.

Consequently, these results highlight the importance of resonance physics and dynamics in elementary and nuclear collisions in the energy regime of the GSI/FAIR, NICA and BESII accelerator programms. Rare probes, like the multi-strange hadrons and hypernuclei, discussed in this paper can be very sensitive to unknown resonance states and their properties. Therefore, if any conclusions on new physics are to be drawn from measuring such rare probes, it is necessary to have a detailed understanding of the hadronic resonances and their dynamics in nuclear collisions.

## 7. Acknowledgments
This work was supported by GSI and the Hessian initiative for excellence (LOEWE) through the Helmholtz International Center for FAIR (HIC for FAIR). The computational resources were provided by the LOEWE Frankfurt Center for Scientific Computing (LOEWE-CSC).

[1] P. Koch, B. Muller and J. Rafelski, Phys. Rept. **142**, 167 (1986).
[2] J. Aichelin and C. M. Ko, Phys. Rev. Lett. **55**, 2661 (1985).
[3] B. Friman, C. Hohne, J. Knoll, S. Leupold, J. Randrup, R. Rapp and P. Senger, Lect. Notes Phys. **814**, 1 (2011).
[4] 2014. NICA White Paper 10.01.
[5] A. Shor *et al.*, Phys. Rev. Lett. **63**, 2192 (1989).
[6] G. Hartnack, L. Sehn, J. Jaenicke, H. Stoecker and J. Aichelin, Nucl. Phys. A **580**, 643 (1994).
[7] X. S. Fang, C. M. Ko, G. Q. Li and Y. M. Zheng, Nucl. Phys. A **575**, 766 (1994)
[8] G. Q. Li, C. M. Ko and X. S. Fang, Phys. Lett. B **329**, 149 (1994).
[9] U. Mosel, Ann. Rev. Nucl. Part. Sci. **41**, 29 (1991).
[10] D. Miskowiec *et al.*, Phys. Rev. Lett. **72**, 3650 (1994).
[11] W. Cassing, E. L. Bratkovskaya, U. Mosel, S. Teis and A. Sibirtsev, Nucl. Phys. A **614**, 415 (1997)
[12] E. L. Bratkovskaya, W. Cassing and U. Mosel, Nucl. Phys. A **622**, 593 (1997)
[13] C. Hartnack, H. Oeschler and J. Aichelin, Phys. Rev. Lett. **90**, 102302 (2003)
[14] C. Hartnack, H. Oeschler and J. Aichelin, Phys. Rev. Lett. **96**, 012302 (2006)
[15] C. Hartnack, H. Oeschler, Y. Leifels, E. L. Bratkovskaya and J. Aichelin, Phys. Rept. **510**, 119 (2012)
[16] S. A. Bass *et al.*, Phys. Rev. Lett. **81**, 4092 (1998) [nucl-th/9711032].
[17] K. Werner, M. Bleicher, B. Guiot, I. Karpenko and T. Pierog, Phys. Rev. Lett. **112**, no. 23, 232301 (2014)
[18] A. Ilner, D. Cabrera, P. Srisawad and E. Bratkovskaya, Nucl. Phys. A **927**, 249 (2014)

[19] S. Takeuchi, K. Murase, T. Hirano, P. Huovinen and Y. Nara, arXiv:1505.05961 [nucl-th].
[20] J. Steinheimer and M. Bleicher, EPJ Web Conf. **97**, 00026 (2015).
[21] J. Steinheimer, J. Aichelin and M. Bleicher, Phys. Rev. Lett. **110**, no. 4, 042501 (2013)
[22] F. Becattini, E. Grossi, M. Bleicher, J. Steinheimer and R. Stock, Phys. Rev. C **90**, no. 5, 054907 (2014)
[23] A. G. Knospe, C. Markert, K. Werner, J. Steinheimer and M. Bleicher, arXiv:1509.07895 [nucl-th].
[24] H. Bando, T. Motoba and J. Zofka, Int. J. Mod. Phys. A **5**, 4021 (1990).
[25] J. Schaffner, C. B. Dover, A. Gal, C. Greiner and H. Stoecker, Phys. Rev. Lett. **71**, 1328 (1993).
[26] O. Hashimoto and H. Tamura, Prog. Part. Nucl. Phys. **57**, 564 (2006).
[27] J. Schaffner-Bielich, Nucl. Phys. A **804**, 309 (2008)
[28] N. Buyukcizmeci, A. S. Botvina, J. Pochodzalla and M. Bleicher, Phys. Rev. C **88**, no. 1, 014611 (2013)
[29] S. A. Bass *et al.*, Prog. Part. Nucl. Phys. **41**, 255 (1998) [Prog. Part. Nucl. Phys. **41**, 225 (1998)]
[30] M. Bleicher *et al.*, J. Phys. **G25**, 1859 (1999)
[31] K. A. Olive *et al.* [Particle Data Group Collaboration], Chin. Phys. C **38**, 090001 (2014).
[32] G. Graef, J. Steinheimer, F. Li and M. Bleicher, Phys. Rev. C **90**, 064909 (2014)
[33] W. Neubert and A. S. Botvina, Eur. Phys. J. A **7**, 101 (2000)
[34] A. S. Botvina, J. Steinheimer, E. Bratkovskaya, M. Bleicher and J. Pochodzalla, Phys. Lett. B **742**, 7 (2015)
[35] A. S. Botvina, K. K. Gudima, J. Steinheimer, M. Bleicher and I. N. Mishustin, Phys. Rev. C **84**, 064904 (2011)
[36] A. S. Botvina, K. K. Gudima and J. Pochodzalla, Phys. Rev. C **88**, no. 5, 054605 (2013)
[37] J. Steinheimer, K. Gudima, A. Botvina, I. Mishustin, M. Bleicher and H. Stocker, Phys. Lett. B **714**, 85 (2012)
[38] A. S. Botvina and J. Pochodzalla, Phys. Rev. C **76**, 024909 (2007)
[39] J. P. Bondorf, A. S. Botvina, A. S. Ilinov, I. N. Mishustin and K. Sneppen, Phys. Rept. **257**, 133 (1995).
[40] H. f. Xi *et al.* [ALADIN Collaboration], Z. Phys. A **359**, 397 (1997) [Eur. Phys. J. A **1**, 235 (1998)]
[41] R. P. Scharenberg *et al.* [EOS Collaboration], Phys. Rev. C **64**, 054602 (2001)
[42] R. Ogul *et al.*, Phys. Rev. C **83**, 024608 (2011) [Erratum-ibid. C **85**, 019903 (2012)]
[43] D. Cabrera, L. Tolos, J. Aichelin and E. Bratkovskaya, Phys. Rev. C **90**, no. 5, 055207 (2014)
[44] C. Spieles, A. Jahns, H. Sorge, H. Stoecker and W. Greiner, Mod. Phys. Lett. A **8**, 2547 (1993).
[45] G. Zeeb, M. Reiter and M. Bleicher, Phys. Lett. B **586**, 297 (2004)
[46] Y. Maeda *et al.* [ANKE Collaboration], Phys. Rev. C **77**, 015204 (2008)
[47] J. Steinheimer and M. Bleicher, arXiv:1503.07305 [nucl-th].
[48] G. Agakishiev *et al.*, Phys. Rev. Lett. **114**, no. 21, 212301 (2015)
[49] G. Agakishiev *et al.* [HADES Collaboration], Phys. Rev. C **80**, 025209 (2009)
[50] M. Lorenz [HADES Collaboration], Nucl. Phys. A **931**, 785 (2014).
[51] K. Piasecki *et al.* [FOPI Collaboration], Phys. Rev. C **91**, no. 5, 054904 (2015)
[52] B. Holzman *et al.* [E917 Collaboration], Nucl. Phys. A **698**, 643 (2002)
[53] S. V. Afanasev *et al.* [NA49 Collaboration], Phys. Lett. B **491**, 59 (2000).
[54] J. Adams *et al.* [STAR Collaboration], Phys. Lett. B **612**, 181 (2005)
[55] G. Agakishiev *et al.* [HADES Collaboration], Eur. Phys. J. A **47**, 21 (2011)
[56] P. Tlusty *et al.* [HADES Collaboration], arXiv:0906.2309 [nucl-ex].
[57] G. Agakishiev *et al.* [HADES collaboration], Phys. Rev. Lett. **103**, 132301 (2009)
[58] T. Ishikawa *et al.*, Phys. Lett. B **608**, 215 (2005)
[59] M. H. Wood *et al.* [CLAS Collaboration], Phys. Rev. Lett. **105**, 112301 (2010)
[60] M. Hartmann *et al.*, Phys. Rev. C **85**, 035206 (2012)

Journal of Physics: Conference Series **668** (2016) 012008 doi:10.1088/1742-6596/668/1/012008

# Heavy flavor in relativistic heavy-ion collisions

**E. L. Bratkovskaya[1,2], T. Song[1,2], H. Berrehrah[1,2], D. Cabrera[1,2], J. M. Torres-Rincon[3], L. Tolos[2,4], W. Cassing[5]**

[1] Institute for Theoretical Physics, JWG Universität, Frankfurt/M, Germany
[2] Frankfurt Institute for Advanced Studies, JWG Universität, Frankfurt/M, Germany
[3] Subatech, UMR 6457, IN2P3/CNRS, Université de Nantes, École des Mines de Nantes
[4] Institut de Ciencies de l'Espai (IEEC/CSIC), Campus Univ. Autonoma de Barcelona, Spain
[5]Institut für Theoretische Physik, Universität Gießen, Germany

E-mail: Elena.Bratkovskaya@th.physik.uni-frankfurt.de

**Abstract.** We study charm production in ultra-relativistic heavy-ion collisions by using the Parton-Hadron-String Dynamics (PHSD) transport approach. The initial charm quarks are produced by the PYTHIA event generator tuned to fit the transverse momentum spectrum and rapidity distribution of charm quarks from Fixed-Order Next-to-Leading Logarithm (FONLL) calculations. The produced charm quarks scatter in the quark-gluon plasma (QGP) with the off-shell partons whose masses and widths are given by the Dynamical Quasi-Particle Model (DQPM), which reproduces the lattice QCD equation-of-state in thermal equilibrium. The relevant cross sections are calculated in a consistent way by employing the effective propagators and couplings from the DQPM. Close to the critical energy density of the phase transition, the charm quarks are hadronized into $D$ mesons through coalescence and/or fragmentation. The hadronized $D$ mesons then interact with the various hadrons in the hadronic phase with cross sections calculated in an effective lagrangian approach with heavy-quark spin symmetry. The nuclear modification factor $R_{AA}$ and the elliptic flow $v_2$ of $D^0$ mesons from PHSD are compared with the experimental data from the STAR Collaboration for Au+Au collisions at $\sqrt{s_{NN}}$ =200 GeV and to the ALICE data for Pb+Pb collisions at $\sqrt{s_{NN}}$ =2.76 TeV. We find that in the PHSD the energy loss of $D$ mesons at high $p_T$ can be dominantly attributed to partonic scattering while the actual shape of $R_{AA}$ versus $p_T$ reflects the heavy-quark hadronization scenario, i.e. coalescence versus fragmentation. Also the hadronic rescattering is important for the $R_{AA}$ at low $p_T$ and enhances the $D$-meson elliptic flow $v_2$.

## 1. Introduction

The Quantum Chromo Dynamics (QCD) [1, 2, 3, 4] predicts that matter changes its phase at high temperature and density and bound (colorless) hadrons dissolve to interacting (colored) quarks and gluons, i.e. to the Quark-Gluon-Plasma (QGP). Such extreme conditions have existed in the early expansion of the universe and now can be realized in the laboratory by collisions of heavy-ions at ultra-relativistic energies. The study of the phase boundary and the properties of the QGP are the main goal of several present and future heavy-ion experiments at SPS (Super Proton Synchrotron), RHIC (Relativistic Heavy-Ion Collider), LHC (Large Hadron Collider) and the future FAIR (Facility for Antiproton and Ion Research) and NICA (Nuclotron-based Ion Collider fAcility) [5]. Since the QGP is created only for a short time (a couple of fm/c) it is quite challenging to study its properties and to find the most sensible probes. The first advantage of 'hard probes' such as mesons containing heavy quarks (charm and beauty) is that as a result of their heavy masses they are dominantly produced in the very early stages of the reactions with large energy-momentum transfer, contrary to the light hadrons and electromagnetic probes. Secondly, they are not in an equilibrium with the surrounding matter due to smaller interaction cross sections relative to the light quarks and, thus, may provide information on their creation mechanisms. Moreover, due to the hard scale, perturbative QCD (pQCD) should be applicable for the calculation of heavy quark production. As shown in Ref. [6], the FONLL calculations are in agreement with the experimental observables on charm meson spectra in p+p collisions. This provides a solid reference frame for studying the heavy-meson production and their flow pattern in heavy-ion collisions.

The first charm measurements at RHIC energies by the PHENIX[7] and STAR [8] collaborations were related to the single non-photonic electrons emitted from the decay of charm mesons. However, recently the STAR Collaboration directly measured the nuclear modification factor and the elliptic flow of $D^0$ mesons in Au+Au collisions at $\sqrt{s_{\rm NN}}$ = 200 GeV [9, 10] which allows for a straightforward comparison with the theoretical model calculations. It has been observed that the $R_{\rm AA}$ and $v_2$ of charm mesons show a similar behavior as light hadrons contrary to expectations from pQCD. Similar observations have been made at LHC energies [11].

It still remains a challenge for the theory to reproduce the experimental data and to simultaneously explain the large energy loss of charm quarks ($R_{\rm AA}$) and the strong collectivity ($v_2$) (cf. e.g. [12, 13, 14, 16]). Commonly, the interactions of charm quarks with the partonic medium are based on pQCD with massless light quarks and a fixed or running coupling. However, lattice QCD and quasiparticle approaches lead to the notion of massive degrees-of-freedom with finite spectral width [15]. In this case the pQCD cross sections are no longer meaningful. Moreover, the conclusions on the amount of suppression due to collisional energy loss by means of the elastic interactions of charm quarks with the QGP partons versus the radiative energy loss due to the emission of soft gluons (i.e. gluon bremsstrahlung) are still far from being robust. Also the influence of hadronization and especially hadronic rescattering is not yet settled.

Our goal here is to study the charm dynamics based on a consistent microscopic transport approach for the charm production, hadronization and rescattering with the partonic and hadronic medium. In this study we will compare our calculations within the Parton-Hadron-String Dynamics (PHSD) approach to the experimental data on charm at RHIC and LHC energies and discuss the perspectives/problems of using the charm quarks for the tomography of the QGP. To achieve this goal we embed the heavy-quark physics in the existing PHSD transport approach [17] which incorporates explicit partonic degrees-of-freedom in terms of strongly interacting quasiparticles (quarks and gluons) in line with an equation-of-state from lattice QCD (lQCD) in addition to dynamical hadronization and hadronic elastic and inelastic collisions in the final reaction phase. Since PHSD has been successfully applied to describe the final distribution of mesons (with light quark content) from lower SPS up to LHC energies [17, 18, 19, 20], it provides a solid framework for the description of the creation, expansion and hadronization of the QGP as well as the hadronic expansion with which the heavy quarks interact either as quarks or as bound states (e.g., $D$-mesons). We note in passing that in the hadronic phase PHSD merges with the familiar Hadron-String-Dynamics (HSD) approach [21, 22].

## 2. Description of charm production and dynamics in PHSD

### *2.1. Charm production in p+p collisions*

Before studying the charm production in relativistic heavy-ion collisions, we discuss the charm production in p+p collisions at the top RHIC energy of $\sqrt{s_{\rm NN}} = 200$ GeV. The charm production in p+p collisions also plays a role as a reference for the nuclear modification factor $R_{\rm AA}$ in heavy-ion collisions.

We use the Pythia event generator to produce charm and anticharm quarks in p+p collisions

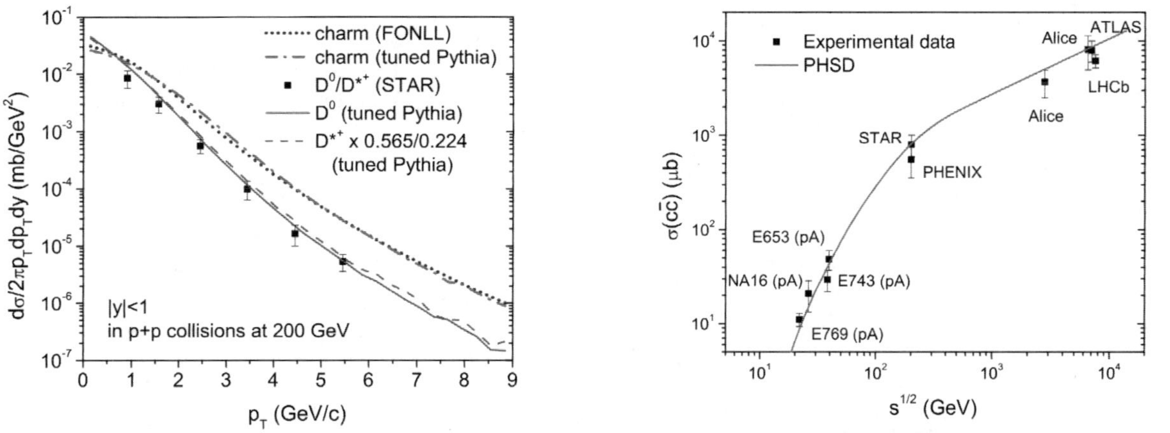


**Figure 1.** (l.h.s.) Transverse momentum distributions (upper lines) of charm quarks in p+p collisions at $\sqrt{s_{\rm NN}} = 200$ GeV from FONLL (dotted lines) and the tuned Pythia event generator (dot-dashed lines); in the lower lines the transverse momentum spectrum of $D^0$ mesons – which are fragmented from charm quarks with the contribution from $D^*$ decay included (solid line) – and that of $D^{*+}$ after scaling are compared with the experimental data from the STAR Collaboration [24]. (r.h.s.) The total cross section for charm production in p+p collisions (as parameterized in the PHSD) in comparison with the experimental data at various collision energies [24, 25].

with the parameters PARP(91)=1.0 GeV/c and PARP(67)=1.0 as in Ref. [23]. The former parameter is the Gaussian width of the primordial transverse momentum of a parton which initiates a shower in hadrons and the latter the parton shower level parameter. We note that by an additional suppression of the transverse momenta of charm and anticharm quarks by 10 % and a suppression of rapidities by 16 % the transverse momentum spectrum as well as rapidity distribution of charm and anticharm quarks from the Pythia event generator are very similar to those from the FONLL calculations in p+p collisions at $\sqrt{s_{\rm NN}}$ =200 GeV as shown in Fig. 1 (l.h.s.). Here the red dot-dashed lines are from the Pythia event generator (after tuning) and the blue dotted lines from the FONLL calculations, respectively.

The produced charm and anticharm quarks hadronize by emitting soft gluons. The probabilities for a charm quark to hadronize into $D^0$, $D^+$, $D^{*0}$, $D^{*+}$, $D_s^+$, and $\Lambda_c$ are, respectively, specified in Ref. [23] together with the fragmentation function. The solid line in Fig. 1 (l.h.s.) shows the transverse momentum spectrum of $D^0$ mesons after charm quark fragmentation including the contribution from the decay of $D^{*0}$ and $D^{*+}$. We can see that our results reproduce the experimental data from the STAR Collaboration [24] reasonably well.

Since charm-quark production requires a high energy-momentum transfer, the number of produced charm quark pairs in relativistic heavy-ion collisions is proportional to the number of binary nucleon-nucleon collisions $N_{bin}$. Since the probability to produce a charm quark pair depends on invariant energy and is less than that for primary hard collision in the PHSD, the binary nucleon-nucleon collisions producing charm quark pairs in PHSD are chosen by Monte Carlo from the ratio of the cross section for charm production in nucleon-nucleon collisions, $\sigma_{c\bar{c}}^{pp}(\sqrt{s})$, to the inelastic nucleon-nucleon cross section. The actual charm cross section $\sigma_{c\bar{c}}^{pp}(\sqrt{s})$ (red solid line) is shown in Fig. 1 (r.h.s.) in comparison to the experimental data from [24, 25].

Journal of Physics: Conference Series **668** (2016) 012008 doi:10.1088/1742-6596/668/1/012008

### *2.2. Heavy quark scattering in the QGP*

Quarks, antiquarks, and gluons are dressed in the QGP and have temperature-dependent effective masses and widths. In the DQPM, the mass and width of the light partons are given by thermal quantum-field theory assuming leading order diagrams but the strong coupling $g^2(T)$ is fitted to lattice data on energy and entropy densities, etc. [15]. Note that a nonzero width of a parton reflects the off-shell nature as well as the strong interaction of the quasi-particle and finite life-time.

The charm and anticharm quarks produced in early hard collisions interact with the dressed off-shell partons in the QGP. The cross sections for the charm quark scattering with massive off-shell partons are calculated considering the mass spectra of final state particles [26, 27]. In the current study the charm quark mass is taken to be 1.5 GeV and its mass spectrum is neglected for simplicity.

In the present study we consider only elastic scattering of charm quarks by light quarks and gluons. We do not yet consider the radiative processes which generate radiative energy loss because we expect that, because of the large gluon mass in the DQPM, the radiative processes are sub-dominant as compared to the collisional ones, especially for low charm-quark momenta ($p_T$). We expect the radiative energy loss to contribute at very high $p_T$ as accessible experimentally at the LHC [28].

We emphasize that the transport coefficient $D_s$ for charm quark diffusion – extracted from our microscopic calculations – and its agreement with the lQCD results (within errors) and the corresponding $D$ meson $D_s$ in hadronic medium validate our description for the coupling of charm with the QGP matter [29].

### *2.3. Hadronization of charm quarks*

Since the hot and dense matter created by a relativistic heavy-ion collision expands with time, the energy density of the matter decreases and the deconfined degrees-of-freedom hadronize to color neutral hadronic states. Once the local energy density in PHSD becomes lower than 0.5 GeV/fm$^3$, the partons are hadronized. First we look for all combinations of a charm quark and light antiquarks or of an anticharm quark and a light quark and then calculate the probability for each combination to form a $D$ or $D^*$ (or $D_s, D_s^*$) meson. The probability for a quark and an anti-quark to form a meson is given by

$$f(\vec{\rho}, \mathbf{k}_\rho) = \frac{8g_M}{6^2} \exp\left[-\frac{\vec{\rho}^2}{\delta^2} - \mathbf{k}_\rho^2\delta^2\right], \tag{1}$$

where $g_M$ is the degeneracy of the meson $M$, and

$$\vec{\rho} = \frac{1}{\sqrt{2}}(\mathbf{r}_1 - \mathbf{r}_2), \quad \mathbf{k}_\rho = \sqrt{2}\,\frac{m_2\mathbf{k}_1 - m_1\mathbf{k}_2}{m_1 + m_2}, \tag{2}$$

with $m_i$, $\mathbf{r}_i$ and $\mathbf{k}_i$ being the mass, position and momentum of the quark or antiquark $i$, respectively. The width parameter $\delta$ is related to the root-mean-square radius of the meson produced through $\langle r^2\rangle = 3(m_1^2 + m_2^2)/(2(m_1 + m_2)^2)\ \delta^2$ and thus determined by experiment (if available). However, we will use this radius as a free parameter and evaluate the coalescence for a $D$-meson radius of 0.5 and 0.9 fm (see below). For further details we refer the reader to [23].

Collecting all possible combinations of a charm or an anticharm quark with light quarks or antiquarks and calculating the coalescence probability for each combination from Eq. (1), we obtain the probability for the charm or the anticharm quark to hadronize by coalescence in the actual space-time volume $\Delta t\Delta x\Delta y\Delta z$. Whether the charm or the anticharm quark is actually hadronized by coalescence is decided by Monte Carlo. Once the charm or the anticharm quark is decided to be hadronized by coalescence, then we find its partner again by Monte Carlo on the basis on the probability of each combination in the selected local ensemble. In case the charm or anticharm quark is decided not to hadronize by coalescence, it is hadronized by the fragmentation method as in p+p collisions (cf. Section 2.1). Since the hadronization by coalescence is absent in p+p collisions, it can be interpreted as a nuclear matter effect on the hadronization of charm and anticharm quarks.

### *2.4. D meson scattering in the hadron gas*

The $D$ and $D^*$ mesons produced through coalescence or fragmentation interact with the surrounding hadrons in PHSD. The presence of several resonances close to threshold energies with dominant decay modes involving open-charm mesons and light hadrons suggests that the scattering cross sections of a $D/D^*$ off a meson or baryon, highly abundant in the post-hadronization medium, could manifest a non-trivial energy, isospin and flavor dependence. An example of these states is the broad scalar resonance $D_0(2400)$, which decays into the pseudoscalar ground state $D$ by emitting a pion in the $s$-wave (similarly to the heavy-quark spin partner $D_1(2420)$, decaying into $D^*\pi$). Moreover, the similarity between the $\Lambda(1405)$ and the $\Lambda_c(2595)$ has driven attention to the fact that the latter could be playing the role of a sub-threshold resonance in the $DN$ system, connected to the latter by coupled-channel dynamics.

All these features have been addressed within several recent approaches based on hadronic effective models which incorporate chiral symmetry breaking in the light sector. The additional freedom stemming from the coupling to heavy-flavored mesons is constrained by imposing heavy-quark spin symmetry (HQSS). Whereas chiral symmetry fully determines the scattering amplitudes of Goldstone bosons with other hadrons at leading order in a model independent way, by means of HQSS the dynamics of the pseudoscalar and the vector mesons containing heavy quarks can be connected, since all kinds of spin interactions are suppressed in the limit of infinite quark masses [30]. For further details we refer the reader to Ref. [23] and the original literature cited therein.

## 3. Comparison to experiment

We now turn to actual nucleus-nucleus collisions at RHIC and LHC energies. Here the nuclear modification of $D$ mesons is expressed in term of the ratio $\mathrm{R_{AA}}$ which is defined as

$$\mathrm{R_{AA}}(p_T) \equiv \frac{dN_D^{\mathrm{Au+Au}}/dp_T}{N_{\mathrm{binary}}^{\mathrm{Au+Au}} \times dN_D^{\mathrm{p+p}}/dp_T}, \tag{3}$$

where $N_D^{\mathrm{Au+Au}}$ and $N_D^{\mathrm{p+p}}$ are, respectively, the number of $D$ mesons produced in Au+Au collisions and that in p+p collisions, and $N_{\mathrm{binary}}^{\mathrm{Au+Au}}$ is the number of binary nucleon-nucleon collisions in Au+Au (Pb+Pb) collisions for the centrality class considered. If the matter produced in relativistic heavy-ion collisions does not modify the $D$ meson production and propagation, the numerator of Eq. (3) should be equal to the denominator. Therefore, an $\mathrm{R_{AA}}$ smaller or larger than one implies that the nuclear matter suppresses or enhances $D$ mesons, respectively. The elliptic anisotropy in the azimuthal angle $\psi$ is characterized by

$$v_2 = \langle \cos(2\psi - 2\Psi_{RP}) \rangle = \left\langle \frac{p_x^2 - p_y^2}{p_x^2 + p_y^2} \right\rangle , \tag{4}$$

where $p_x$ and $p_y$ are the $x$ and $y$ components of the particle momenta and $\Psi_{RP}$ is the azimuth of the reaction plane. The brackets denote averaging over particles and events. The $v_2$ coefficient can be considered as a function of centrality, rapidity $y$ and/or transverse momentum $p_T$. We note that the reaction plane in PHSD is given by the $(x - z)$ plane with the $z$-axis in the beam direction.

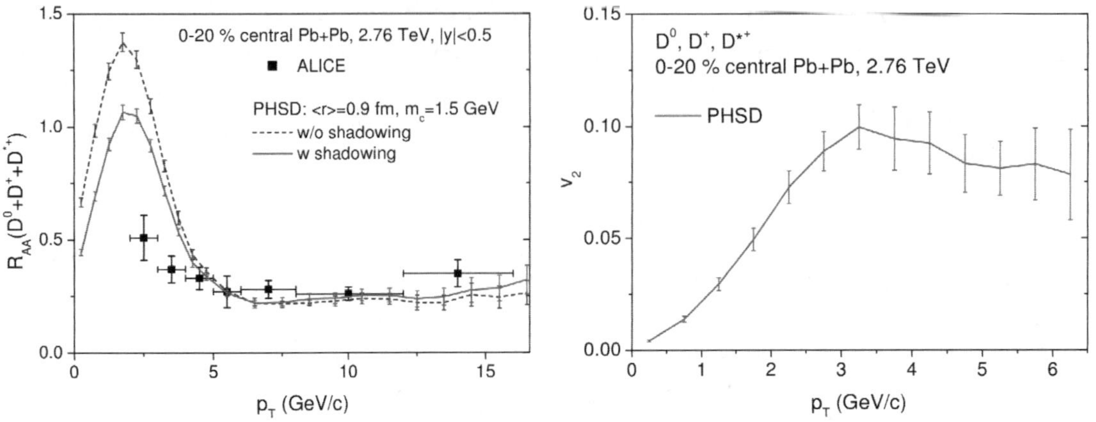


**Figure 3.** (l.h.s.) The PHSD results for the $R_{AA}$ of $D^0 + D^+ + D^{*+}$ mesons at midrapidity for 0-20% central Pb+Pb collisions at $\sqrt{s_{NN}}$ =2.6 TeV calculated without (dashed blue line) and with (solid red line) shadowing in comparison to the experimental data from the ALICE Collaboration [32]. (r.h.s.) The PHSD results for the $v_2$ versus $p_T$ of $D^0 + D^+ + D^{*+}$ mesons for 20% central Pb+Pb collisions at $\sqrt{s_{NN}}$ =2.6 TeV including the shadowing effect.

## 4. Summary

We have studied charm production in relativistic heavy-ion collisions by using the Parton-Hadron-String Dynamics (PHSD) approach [18] where the initial charm quark pairs are produced in binary nucleon-nucleon collisions from the PYTHIA event generator taking into account the smearing of the collision energy due to the Fermi motion of nucleons in the initial nuclei. The produced charm and anticharm quarks interact with the dressed quarks and gluons in the QGP which are described by the Dynamical Quasi-Particle Model [15] in PHSD. The interactions of the charm quarks with the QGP partons have been evaluated with the DQPM propagators and couplings consistently [26]. We mention that when extracting the spatial diffusion coefficient $D_s$ from our cross sections as a function of the temperature we observe a minimum of $D_s$ close to $T_c$ which is in line with lattice data above $T_c$ and hadronic many-body calculations below $T_c$ (cf. Ref. [29]).

We have found that the interaction of the charm quarks with the dynamical partons of the QGP softens the $p_T$ spectrum of charm and anticharm quarks but does not lead to a full thermalization for transverse momenta $p_T > 2$ GeV/c. The charm and anticharm quarks, furthermore, are hadronized to $D$ mesons either through the coalescence with a light quark or antiquark or through the fragmentation by emitting soft 'perturbative' gluons. Since the hadronization through coalescence is absent in p+p collisions, it can be interpreted as a nuclear matter effect on the $D$ meson production in relativistic heavy-ion collisions. In the coalescence mechanism the charm or anticharm quark gains momentum by fusing with a light quark or antiquark while it looses momentum in the fragmentation process (as in p+p reactions). This partly contributes to the large $R_{AA}$ of $D$ mesons between 1 and 2 GeV/c of transverse momentum. Finally, the formed $D$ mesons interact with hadrons by using the cross sections calculated in an effective lagrangian approach with heavy quark spin-symmetry [30], which is state-of-the art. We have found that the contribution from hadronic scattering both to the $R_{AA}$ and to the elliptic flow of $D$ mesons is appreciable, especially in central collisions, and produces additional elliptic flow $v_2$ [23].

Since the PHSD results reproduce the experimental data from the STAR and ALICE Collaborations without radiative energy loss in the $p_T$ range considered, we conclude that collisional energy loss is dominant at least up to $p_T$ = 6 (15) GeV/c in relativistic heavy-ion collisions at RHIC (LHC). In our approach this is essentially due to the infrared enhanced coupling $\alpha_s(T)$ in the DQPM leading to large scattering cross sections of charm quarks with partons at temperatures close to $T_c$.

## Acknowledgements

The authors acknowledge inspiring discussions with J. Aichelin, P. B. Gossiaux, C. M. Ko, O. Linnyk, R. Marty, V. Ozvenchuk, and R. Vogt. This work was supported by DFG under contract BR 4000/3-1, and by the LOEWE center "HIC for FAIR". JMTR is supported by the Program TOGETHER from Region Pays de la Loire and the European I3-Hadron Physics program. LT acknowledges support from the Ramon y Cajal Research Programme and contracts FPA2010-16963 and FPA2013-43425-P from Ministerio de Ciencia e Innovación, as well as from FP7-PEOPLE-2011-CIG under Contract No. PCIG09-GA-2011-291679. The computational resources have been provided by the LOEWE-CSC.

## References

[1] Aoki Y *et al.* 2006 *Phys. Lett.* B **643** 46
[2] Borsanyi S *et al.* 2010 *JHEP* **1009** 073; 2010 *JHEP* **1011** 077; 2012 *JHEP* **1208** 126; 2014 *Phys. Lett.* B **730** 99; 2015 *Phys. Rev.* D **92** 014505
[3] Petreczky P [HotQCD Collaboration] 2012 *PoS LATTICE* **2012** 069
[4] Ding H-T Karsch F Mukherjee S 2015 arXiv:1504.05274
[5] Proceedings of Quark Matter-2014 2014 *Nucl. Phys.* A **931** 1
[6] Cacciari M Nason P Vogt R 2005 *Phys. Rev. Lett.* **95** 122001
[7] Adare A *et al.* (PHENIX Collaboration) 2007 *Phys. Rev. Lett.* **98** 172301
[8] Abelev B I *et al.* [STAR Collaboration] 2007 *Phys. Rev. Lett.* **98** 192301
[9] Adamczyk L *et al.* [STAR Collaboration] 2014 *Phys. Rev. Lett.* **113** 142301
[10] Tlusty D [STAR Collaboration] 2013 *Nucl. Phys.* A **904-905** 639c
[11] Sakai S *et al.* 2013 *Nucl. Phys.* A **904-905** 661c
[12] van Hees H Greco V and Rapp R 2006 *Phys. Rev.* C **73** 034913
[13] Uphoff J *et al.* 2013 *Nucl. Phys.* A **910-911** 401; *ibid* 2014 **931** 535.
[14] Das S K Scardina F Plumari S Greco V 2014 *Phys. Rev.* C **90** 044901
[15] Cassing W 2009 *Eur. Phys. J. ST* **168** 3; 2007 *Nucl. Phys.* A **795** 70
[16] Ozvenchuk V *et al.* 2014 *Phys. Rev.* C **90** 054909
[17] Cassing W Bratkovskaya E L 2009 *Nucl. Phys.* A **831** 215
[18] Bratkovskaya E L Cassing W Konchakovski V P Linnyk O 2011 *Nucl. Phys.* A **856** 162
[19] Konchakovski V P *et al.* 2015 *J. Phys.* G **42** 055106; 2014 *J. Phys.* G **41** 105004; 2014 *Phys. Rev.* C **90** 014903; 2012 *Phys. Rev.* C **85** 044922; 2012 *Phys. Rev.* C **85** 011902
[20] Linnyk O *et al.* 2014 *Phys. Rev.* C **89** 034908; 2013 *Phys. Rev.* C **88** 034904; 2013 *Phys. Rev.* C **87** 014905; 2012 *Phys. Rev.* C **85** 024910; 2011 *Phys. Rev.* C **84** 054917; 2011 *Nucl. Phys.* A **855** 273
[21] Cassing W Bratkovskaya E L 1999 *Phys. Rep.* **308** 65
[22] Cassing W Bratkovskaya E L Juchem S 2000 *Nucl. Phys.* A **674** 249
[23] Song T Berrehrah H Cabrera D Torres-Rincon J M Tolos L Cassing W Bratkovskaya E L 2015 *Phys. Rev.* C **92** 014910
[24] Adamczyk L *et al.* [STAR Collaboration] 2012 *Phys. Rev.* D **86** 072013
[25] Conesa del Valle Z [ALICE Collaboration] 2012 *AIP Conf. Proc.* **1441** 886
[26] Berrehrah H *et al.* 2014 *Phys. Rev.* C **89** 054901
[27] Berrehrah H *et al.* 2014 *Phys. Rev.* C **90** 064906
[28] Younus M Coleman-Smith C E Bass S A Srivastava D K 2015 *Phys. Rev.* C **91** 024912
[29] Berrehrah H *et al.* 2014 *Phys. Rev.* C **90** 051901
[30] Tolos L Torres-Rincon J M 2013 *Phys. Rev.* D **88** 074019
[31] Eskola K J Paukkunen H Salgado C A 2009 *JHEP* **0904** 065
[32] Abelev B *et al.* [ALICE Collaboration] 2012 *JHEP* **1209** 112
[33] Abelev B *et al.* [ALICE Collaboration] 2013 *Phys. Rev. Lett.* **111** 102301

# Measurements of $\Lambda\Lambda$ correlation function in heavy-ion collisions at RHIC

**J.H. Chen for the STAR Collaboration**
Shanghai Institute of Applied Physics, Chinese Academy of Science, Shanghai-201800, China
E-mail: chenjinhui@sinap.ac.cn

**Abstract.** The information on the hyperon-hyperon interaction is crucial for our understanding of baryon-baryon interactions in hot and dense medium from high-energy nuclear collisions. In this proceeding we discuss the measurement of the $\Lambda\Lambda$ correlation function from the STAR experiment and the status of an H-dibaryon search based on this measurement.

## 1. Introduction

Based on a bag-model calculation, R. Jaffe [1] predicted a deeply bound state, $H$-dibaryon, with isospin ($I$) = 0 and strangeness ($S$) = -2. Lattice QCD calculations from the HAL [2] and NPLQCD [3] collaborations indicate existence of a bound $H$-dibaryon for pion masses larger than the physical mass, while chiral extrapolation to a physical quark mass leads to an unbound $H$-dibaryon [4]. Since its prediction, many experimental searches and theoretical calculations have been performed. Various experiments have explored the $\Lambda\Lambda$, $\Sigma p$ and $\Lambda p\pi$ invariant mass spectrum to look for an $H$-dibaryon signal. However the results from those experiments did not have enough statistical significance to establish any conclusion about its existence [5, 6, 7, 8]. At present, the constraint on the binding energy for the $H$-dibaryon comes from the so-called Nagara event: measurement of ${}^{6}_{\Lambda\Lambda}$He candidate with very small binding energy [9]. The observed enhancement in $\Lambda\Lambda$ invariant mass spectrum from the KEK-E522 experiment [5] showed that it is important to understand the interaction between two $\Lambda$s to extract an $H$-dibaryon signal.

At the densities in the interior of neutron stars, the neutron chemical potential exceeds the mass of various members of the baryon octet. Therefore, in addition to nucleons and electrons, abundant hyperons are expected to exist in neutron stars [10]. If the $\Lambda\Lambda$ interaction is attractive enough, a significant fraction of $\Lambda$s in a neutron star may combine to form $H$-dibaryons. If the densities are high enough, this could lead to the formation of $H$-matter in the core of moderately dense neutron stars [11]. The knowledge of the strength of the $\Lambda\Lambda$ interaction will improve our understanding of the equation of state for neutron stars.

Measurements of the correlation function for a pair of particles with small relative momenta have been used to get insight into the geometry and lifetime of the particle-emitting source in relativistic heavy-ion collisions [12]. These measurements are sensitive to the separation distribution of the source, as well as effects from quantum statistics (QS) and interactions in the final state (FSI). This allows one to use the measurement of a two particle correlation function to get information about the final state interactions between two particles and the H-dibaryon signal. In 2015, the STAR experiment published the first high statistics measurement of the $\Lambda\Lambda$ correlation function [13], which provided a better constraint on the $\Lambda\Lambda$ interaction parameters.

## 2. $\Lambda$ reconstruction

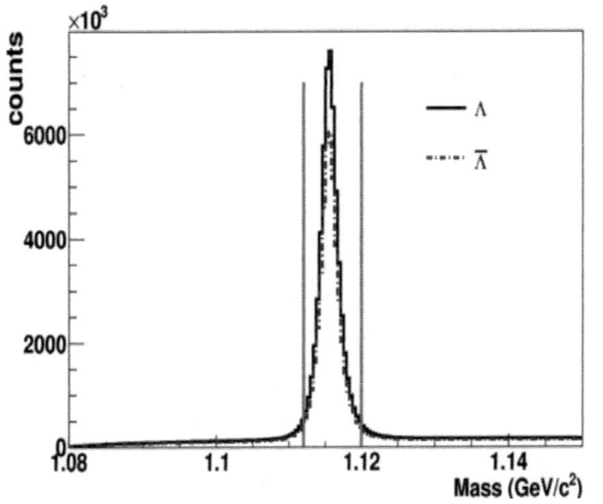


**Figure 1.** (color online). The invariant mass distribution for $\Lambda$ and $\bar{\Lambda}$ in Au+Au collisions at $\sqrt{s_{NN}} = 200$ GeV from 0-80% centrality [13]. The solid red vertical lines show a mass window from 1.112 to 1.120 GeV/$c^2$, which were selected for the correlation function measurement.

Identification and momentum determination for the $\Lambda$ particle was carried out with the Time Projection Chamber. The decay channel $\Lambda \to p\pi$ with a branching ratio of 63% was used for the reconstruction of $\Lambda$. The $\Lambda$ candidates were formed out of a pair of positive ($p$) and negative ($\pi$) tracks whose trajectories point to a common secondary decay vertex which was well separated from the primary vertex. The decay length of $\Lambda$ candidate was asked to be more than 5 cm and the distance of closest approach from the primary vertex was taken to be less than 0.4 cm. Approximately $2.87 \times 10^8$ events from the year 2010 run and $5.0 \times 10^8$ events from the year 2011 run were analyzed. The invariant mass distribution of the $\Lambda$ and $\bar{\Lambda}$ candidates at 0-80% centrality under these conditions are shown in Fig. 1. It has an excellent signal (S) to background (B) ratio of $S/(S+B) \approx 0.97$. All candidates with invariant mass between 1.112 and 1.120 GeV/$c^2$ were used in the correlation analysis.

## 3. $\Lambda\Lambda$ interactions

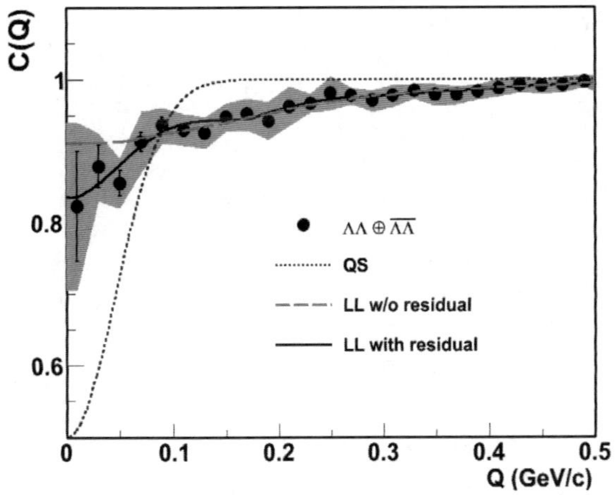


**Figure 2.** (color online). The combined $\Lambda\Lambda$ and $\bar{\Lambda}\bar{\Lambda}$ correlation function for 0-80% centrality Au+Au collisions at $\sqrt{s_{NN}} = 200$ GeV [13]. Curves correspond to fits using the Lednický-Lyuboshitz (LL) analytical model with and without a residual correlation term [14]. The dotted line corresponds to quantum statistics with a source size of 3.13 fm. The shaded band corresponds to the systematic error.

The combined $\Lambda\Lambda$ and $\bar{\Lambda}\bar{\Lambda}$ correlation function as a function of relative momentum (Q) for 0-80% centrality is shown in Fig. 2. The systematic error is shown separately as a shaded band in Fig. 2 and the dotted line corresponds to QS with a source size of 3.13 fm. The following function from Lednický and Lyuboshitz analytical model [14] was used to fit the correlation function data

$$\begin{aligned} C(Q) &= N[1+\lambda(-\frac{1}{2}\exp(-r_0^2Q^2)+ \\ &\frac{1}{4}\frac{|f(k)|^2}{r_0^2}(1-\frac{1}{2\sqrt{\pi}}\frac{d_0}{r_0})+ \\ &\frac{\mathrm{Re}\, f(k)}{\sqrt{\pi}r_0}F_1(Qr_0)-\frac{\mathrm{Im}\, f(k)}{2r_0}F_2(Qr_0))], \end{aligned} \tag{1}$$

where $k = Q/2$, $F_1(z) = \int_0^1 e^{x^2-z^2}/z\,dx$ and $F_2(z) = (1-e^{-z^2})/z$ in Eq. (1). The scattering amplitude is given by

$$f(k) = (\frac{1}{f_0}+\frac{1}{2}d_0k^2-ik)^{-1}, \tag{2}$$

where $f_0 = a_0$ is the scattering length and $d_0 = r_{\mathrm{eff}}$ is the effective range. Note that a universal sign convention is used rather than the traditional sign convention for the s-wave scattering length, $a_0 = -f_0$ for baryon-baryon systems. More details about the model can be found in Ref. [14]. The free parameters of the LL model are the normalization ($N$), a suppression parameter ($\lambda$), an emission radius ($r_0$), scattering length ($a_0$) and effective radius ($r_{\mathrm{eff}}$). In the absence of FSI, $\lambda$ equals unity for a fully chaotic Gaussian source. The impurity in the sample used and finite momentum resolution can suppress the value of $\lambda$-parameter. In addition to this the non-Gaussian form of the correlation function and the FSI between particles can affect (suppress or enhance) its value.

A fit to data using the above equation gives a larger $\chi^2$/NDF (dashed line in Fig. 2). Also, the obtained $r_0$ is much smaller than the expected $r_0$ from previous measurements [15, 16, 17], which suggests that the measured correlation is wider than that indicated by the fit in this scenario.

One possible explanation for this is the presence of a negative residual correlation in the data, which is expected to be wider than the correlation from the parent particles like $\Sigma^0$ and $\Xi$. Therefore a Gaussian term $a_{\mathrm{res}}\exp(-Q^2r_{\mathrm{res}}^2)$ is added in Eq. (1), shown as the solid line in Fig. 2, where $a_{\mathrm{res}}$ is the residual amplitude and $r_{\mathrm{res}}$ is the width of the Gaussian. A negative residual correlation contribution is required with $a_{\mathrm{res}} = -0.044 \pm 0.004^{+0.048}_{-0.009}$ and $r_{\mathrm{res}} = 0.43 \pm 0.04^{+0.43}_{-0.03}$ fm. The fit parameters obtained with the residual correlation term are $N = 1.006 \pm 0.001$, $\lambda = 0.18 \pm 0.05^{+0.12}_{-0.06}$, $a_0 = -1.10 \pm 0.37^{+0.68}_{-0.08}$ fm, $r_{\mathrm{eff}} = 8.52 \pm 2.56^{+2.09}_{-0.74}$ fm and $r_0 = 2.96 \pm 0.38^{+0.96}_{-0.02}$ fm with $\chi^2/\mathrm{NDF} = 0.56$.

From the scattering length and the effective radius obtained from the model fit we conclude that $|a_{\Lambda\Lambda}| < |a_{p\Lambda}| < |a_{NN}|$, as shown in Fig. 3. This means that a weaker interaction exists between $\Lambda\Lambda$ pairs compared to nucleon-nucleon and proton-hyperon. In case of a resonance state that decays to two $\Lambda$ particles, one would expect to observe a peak structure in the two-particle correlations [22]. However, our data does not show such a peak in the correlation function for the relative momentum above 0.3 GeV/c. Numerical analysis of the final-state interaction effect using an $s$-wave scattering amplitude suggests the non-existence of a $\Lambda\Lambda$ resonance saturating the $s$-wave below the $N\Xi$ and $\Sigma\Sigma$ thresholds (for more details please refer to ref [13]).

Assuming that $H$-dibaryons are produced through coalescence of $\Lambda\Lambda$ pairs, the yield for the $H$-dibaryon can be related to the $\Lambda$ yield by $d^2N_H/(2\pi p_T dp_T dy) = 16B[d^2N_\Lambda/(2\pi p_T dp_T dy)]^2$, where $B$ is a constant known as the coalescence coefficient. We obtain the upper limit for the $p_T$-integrated $dN_H/dy = (1.23 \pm 0.47_{stat} \pm 0.61_{sys}) \times 10^{-4}$ [13].

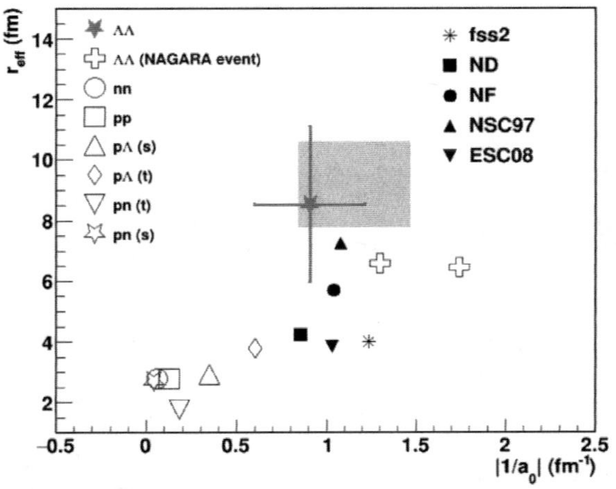


**Figure 3.** (color online). The $\Lambda\Lambda$ interaction parameters from this experiment (solid star), where the shaded band represents the systematic error. The interaction parameters from $pp$, $pn$ singlet ($s$) and triplet ($t$) states, and from $nn$, $p\Lambda$ ($s$) and $p\Lambda$ ($t$) states are shown as open markers [18]. Also, the $\Lambda\Lambda$ interaction parameters which reproduce the NAGARA event are shown as open cross [19, 20]. The filled points are fit to STAR data from K. Morita, T. Furumoto and A. Ohnishi [21].

## 4. Summary

We report $\Lambda\Lambda$ correlation function from Au+Au collisions at $\sqrt{s_{NN}} = 200$ GeV at RHIC. Weakly attractive interaction between $\Lambda$ is observed. The measured interaction parameter suggests the non-existence of $\Lambda\Lambda$ resonance below the N$\Xi$ and $\Sigma\Sigma$ thresholds. We also obtained an upper limit on the $H$-dibaryon cross section in such collisions [13].

## 5. Acknowledgments

We thank the organizer for the invitation to give this talk in the conference. Valuable discussions with Dr. Ohnishi and Dr. Rappold are appreciated. This work was supported in part by the NNSF of China under Grant No. 11322547 and 11275250.

## References

[1] R. Jaffe, Phys. Rev. Lett. **38** (1977) 195.
[2] T. Inoue et al. [HAL QCD Collaboration], Phys. Rev. Lett. **106** (2011) 162002.
[3] S. Beane et al. [NPLQCD Collaboration], Phys. Rev. Lett. **106** (2011) 162001.
[4] P.E. Shanahan, A.W. Thomas, and R.D. Young, Phys. Rev. Lett. **107** (2011) 092004.
[5] C. J. Yoon et al.[KEK-PS E522 Collaboration], Phys. Rev. C **75** (2007) 022201(R).
[6] B. A. Cole *et al.* (E910 Collaboration), Nucl. Phys. A **639** (1998) 407c.
[7] A. Alavi-Harati *et al.* (KTeV Collaboration), Phys. Rev. Lett. **84** (2000) 2593.
[8] H. Caines *et al.* (E896 Collaboration), J. Phys. G **27** (2001) 311.
[9] H. Takahashi et al., Phys. Rev. Lett. **87** (2001) 212502.
[10] N. K. Glendenning, Astrophys. J. **293** (1985) 470.
[11] R. Tamagaki, Prog. Theor. Phys. **85** (1991) 321.
[12] M. A. Lisa, S. Pratt, R. Soltz, and U. Wiedemann, Annu. Rev. Nucl. Part. Sci. **55** (2005) 357.
[13] L. Adamczyk et al. (STAR Collaboration), Phys. Rev. Lett. **114** (2015) 022301.
[14] R. Lednický, V. L. Lyuboshitz, Sov. J. Nucl. Phys. **35** (1982) 770.
[15] J. Adams *et al.* (STAR Collaboration), Phys. Rev. C **71** (2005) 044906.
[16] B. I. Abelev *et al.* (STAR Collaboration), Phys. Rev. C **74** (2006) 054902.
[17] J. Adams *et al.* (STAR Collaboration), Phys. Rev. C **74** (2006) 064906.
[18] L. Heller, Rev. Mod. Phys. **39** (1967) 584, W. von Witsch, B. Gomez Moreno, W. Rosenstock, K. Ettling and J. Bruinsma, Phys. Lett. B **80** (1979) 187, J. Deng, A. Siepe and W. von Witsch, Phys. Rev. C **66** (2002) 047001 , and F. Wang and S. Pratt, Phys. Rev. Lett. **83** (1999) 3138.
[19] E. Hiyama, M. Kamimura, T. Motoba, T. Yamada and Y. Yamamoto, Phys. Rev. C **66** (2002) 024007.
[20] I. N. Filikhin and A. Gal, Nucl. Phys. A **707** (2002) 491.
[21] K. Morita, T. Furumoto and A. Ohnishi, Phys. Rev. C **91** (2015) 024916.
[22] C. Greiner and B. Müller, Phys. Lett. B **219** (1989) 199.

# Suppression and energy loss in Quark-Gluon Plasma

**M Djordjevic**

Institute of Physics Belgrade, Pregrevica 118, 11080 Zemun, Serbia

E-mail: magda@ipb.ac.rs

**Abstract.** High momentum suppression of light and heavy flavor observables is considered to be an excellent probe of jet-medium interactions in QCD matter created at RHIC and LHC. Utilizing this tool requires accurate suppression predictions for different experiments, probes and experimental conditions, and their unbiased comparison with experimental data. We developed the dynamical energy loss formalism which takes into account both radiative and collision energy loss computed within the same theoretical framework, dynamical (as opposed to static) scattering centers, finite magnetic mass, running coupling and uses no free parameters in comparison with experimental data. Within this formalism, we provide predictions, and a systematic comparison with the experimental data, for a diverse set of probes, various centrality ranges and various collision energies at RHIC and LHC. We also provide clear qualitative and quantitative predictions for the upcoming LHC experiments. A comprehensive agreement between our predictions and experimental results suggests that our dynamical energy loss formalism can well explain the jet-medium interactions in QGP, which will be further tested by the obtained predictions for the upcoming data.

## 1. Introduction

Traditionally, light and heavy flavour suppressions are considered to be excellent probes of QCD matter [1–3]. Since suppression for a number of observables has been measured at RHIC and LHC, their comparison with theoretical predictions allows for testing our understanding of QCD matter. Due to this, we will mainly concentrate on theoretical energy loss and suppression predictions, and how well these predictions can explain the experimental data. Note that in these proceedings only the main results will be reviewed; for the full account, the reader should refer to [4–8].

Figure 1 shows a scheme, which represents the main steps in the suppression calculations. These steps involve jet production, energy loss, fragmentation and decay. Therefore, to have reliable calculation of jet suppression, reliable calculations of all the underlying processes are needed. It is generally considered that the critical step in the jet suppression calculations is the jet energy loss [9], so we will first concentrate on this step. We will then go back to discuss the entire numerical procedure, and when discussing the experimental data, we will argue that other steps can be very important as well.

## 2. Energy loss overview

Initially, most of the energy loss calculations were based on the assumption of static scattering centers [10], and only radiative energy loss was included (see e.g. [11–15]). These calculations, however, lead to obvious disagreements with the experimental data (for overview, see [16]),

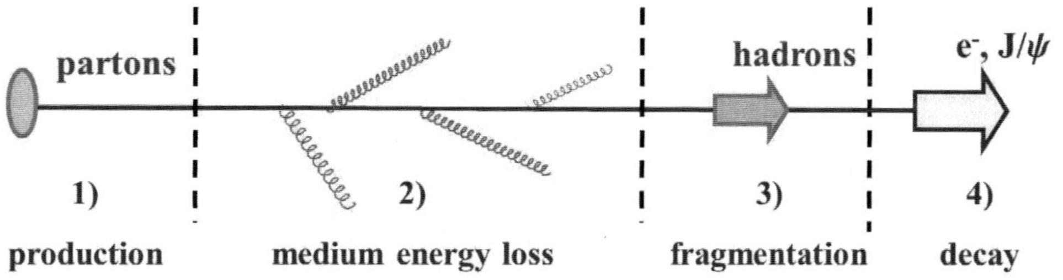


**Figure 1. Suppression scheme.** The scheme represents important steps in the suppression calculations.

which opened a question whether radiative energy loss really controls the energy loss in QGP, or collision energy loss is also important. Several calculations of the collision energy loss were then performed[17–20], showing that the collision and radiative energy losses are actually comparable, and concluding that collision energy loss is important and should indeed be taken into account in the calculations of jet suppression.

However, the fact that collision energy loss is non-zero opened a fundamental problem with the radiative energy loss formalisms built on the static QCD medium approximation. One consequence of such an approximation is that in a static medium, collision energy loss has to be exactly equal to zero [21]! However, contrary to this expectation, collision energy loss computations[17–20] showed that collision and radiative energy losses are comparable. This leads to the conclusion that collision energy loss results are inconsistent with static approximation, since static medium approximation necessarily leads to zero collision energy loss. So this inconsistency leads to the second conclusion that QCD medium cannot be modeled by static scattering centers, and that dynamical effects have to be included in the radiative energy loss calculations [21].

To address this important issue, we developed the radiative jet energy loss formalism in a finite size dynamical QCD medium [19, 22, 23]. The formalism takes into account that the medium constituents are in reality dynamical, i.e. moving particles, and that the medium has finite size. To calculate the energy loss we used the generalized two hard thermal loop approach, which removes assumption of static scattering centers. The calculations are based on the finite temperature field theory, take into account both radiative and collision energy losses, and are applicable to both light and heavy flavor. The formalism has also been recently generalized to the case of finite magnetic mass [24] and running coupling [4]. Regarding the formalism, note that all the dynamical energy loss ingredients are included based on the theoretical grounds that we discussed above. Furthermore, in [25] and the contribution by B. Blagojevic in these proceedings, it is discussed how these model ingredients are important in obtaining accurate numerical predictions.

## 3. Suppression predictions

### *3.1. Comparison with the existing experimental data at central collisions*

To generate suppression predictions, we incorporated this formalism into a numerical procedure [4], which also includes light and heavy flavor production [26], path-length [27, 28] and multi-gluon [29] fluctuations, fragmentation [30] for light and heavy flavor and, in the case of heavy flavor, decays to single electrons and $J/\Psi$ [31]. As a starting point in our calculations, for LHC we used effective temperature of 304 MeV as extracted by ALICE [32], and for RHIC we used 221 MeV as extracted by PHENIX [33].

We will first concentrate on the existing RHIC and LHC suppression data and our goal is to see how our energy loss formalism can help us in understanding this data. Within this goal, we will first generate a comprehensive set of joint predictions for all the available light and heavy flavor $R_{AA}$ central collision data. Here, we will concentrate on some puzzling data (i.e. data which seems counterintuitive). We will then test how our model works for different centrality regions. All the predictions will be generated by the same formalism, with the same numerical procedure, and with no free parameters used in model testing. Actually, all used parameters correspond to the standard literature values, as stated in [7].

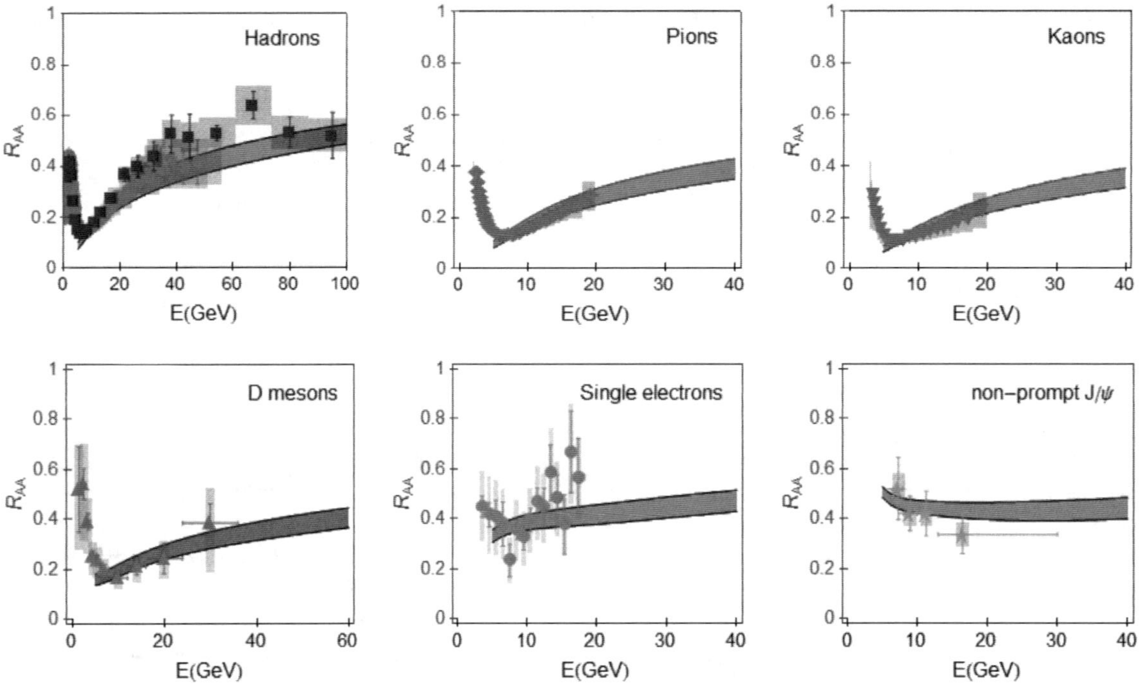


**Figure 2. Comparison of suppression predictions with $R_{AA}$ experimental data in central 2.76 TeV Pb+Pb collisions at the LHC.** Our suppression predictions (shown as gray bands) are compared with the available experimental $R_{AA}$ data for charged hadrons, pions, kaons, D mesons, non-photonic single electrons and $J/\psi$. The figure is adapted from [4]. The bands comes from the uncertainty in the magnetic mass value, see Ref. [4].

In Fig. 2 we show the predictions for central collisions at LHC, which are, as we said, generated for a diverse set of probes, for which LHC measurements are available. Specifically, we compare predictions with experimental data [34–39] for charged hadrons, pions, kaons, D mesons, single electrons and non-prompt $J/\Psi$. We see that we obtain excellent agreement for light flavor and D mesons. We see that the single electron data is quite noisy, but we still obtain good agreement with the predictions. There is also good agreement for non-prompt $J/\psi$, except for the last data point, which comes with large error bars. Now, if we look at the D mesons and charged hadrons, we can see that they reveal a puzzle, which we call the heavy flavor puzzle at LHC, and which we will briefly overview below (for more details, please refer to [6]).

### *3.2. Heavy flavor puzzles at RHIC and LHC*

While D meson suppression is a clear charm probe, charged hadrons are composed of both light quarks and gluons. Moreover, the gluons have significant (even dominant in the lower momentum region) contribution to charged hadron production [6]. Also, from the left panel in Fig. 3 we see that charm and light quarks have about the same suppression, while the suppression of gluons is much higher. Based on this and on the common expectation that energy loss is the most important step in the jet suppression, we clearly expect that $R_{AA}(D) > R_{AA}(h)$.

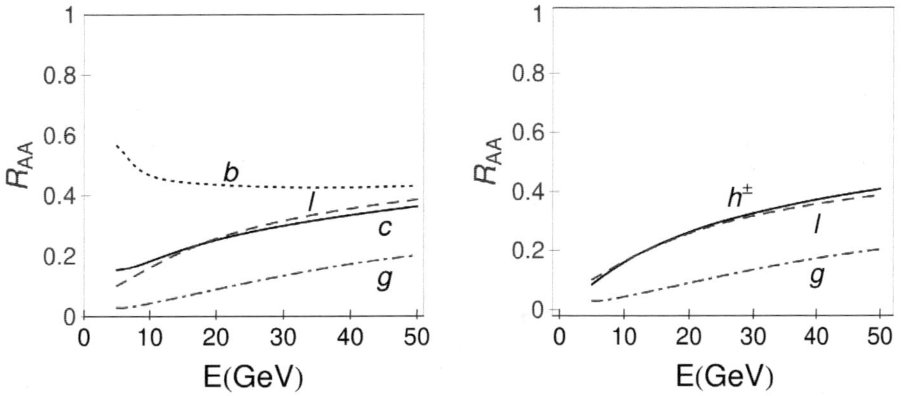


**Figure 3. Comparison of the light and heavy flavor suppression predictions.** The left panel shows the momentum dependence of the jet suppression for bottom quarks (the dotted curve), charm quarks (the full curve), light quarks (the dashed curve) and gluons (the dot-dashed curve). The right panel shows the comparison of charged hadron suppression predictions (the full curve) with light quark (the dashed curve) and gluon (the dot-dashed curve) suppression predictions, as a function of momentum. The figure is adapted from [6].

However, contrary to this expectation, LHC experimental results actually show the same $R_{AA}$ for charged hadrons and D mesons [34, 37]. Having in mind the above, a naive conclusion would be that quarks and gluons lose the same amount of energy, which is however not in accordance with pQCD. Even more surprisingly, as shown in Fig. 2 (see also [6]), we see that our theoretical predictions, which are based on pQCD, are actually in excellent agreement with the data. We therefore ask why both the experimental data and the theoretical predictions are not in agreement with the qualitative expectations that we summarized in the previous slide.

Therefore, for explaining the puzzling data, steps in the jet suppression shown in Fig. 1, other than the energy loss, must also be important. In the case of the heavy flavor puzzle, we expect that the other important step is jet fragmentation, since it defines a transfer from the parton to hadron level. With regard to the D mesons, it is well known that fragmentation does not modify bare charm quark suppression (see e.g. [6]). Consequently, the D meson suppression is indeed a genuine probe of the charm quark suppression in QCD medium. However, things are much more complicated for charged hadrons, that is, fragmentation functions significantly modify the suppression patterns of light quarks and gluons, as demonstrated in the right panel of Fig. 3 (see also [6]). In particular, due to fragmentation functions, charged hadron/pion suppression unexpectedly becomes almost the same as bare light quark suppression.

Therefore, as a summary of the puzzle, the main result that we obtained is that charged hadron suppression is almost identical to the light quark suppression which is despite the dominant gluon contribution in the charged hadron production. Since we also have that D meson $R_{AA}$ is the same as the charm $R_{AA}$ and that the light and charm quarks have almost the same $R_{AA}$s, we obtain the solution of the puzzle, i.e. that the charged hadrons and D mesons indeed should have the same suppression, as observed in the experiment and confirmed by our theoretical predictions. Consequently, it is not only the energy loss, but an unexpected interplay between the energy loss and the fragmentation functions, which is responsible for explaining the heavy flavor puzzle at the LHC.

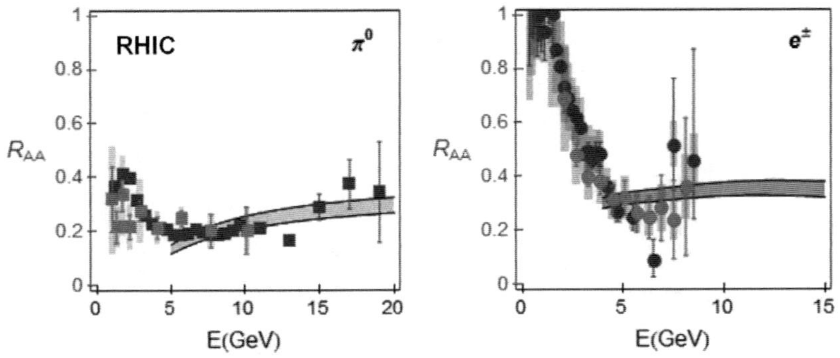


**Figure 4. Comparison of suppression predictions with $R_{AA}$ experimental data in central 200 MeV Au+Au collisions at RHIC.** Our suppression predictions (shown as gray bands) are compared with the available experimental $R_{AA}$ data for neutral pions and non-photonic single electrons. The figure is adapted from [5]. The bands comes from the uncertainty in the magnetic mass value, see Ref. [5].

We now test how our model works for RHIC central-collision data, since we know that the suppression data shown here leads to a well-known heavy flavor puzzle at RHIC; that is, the previous static energy loss models were not able to explain this data. However, from Fig. 4 we see that the dynamical energy loss formalism outlined in the previous slides can well explain the data (see also [5]). Note that no free parameters are used in model testing.

### *3.3. Comparison with the data for non-central collisions*

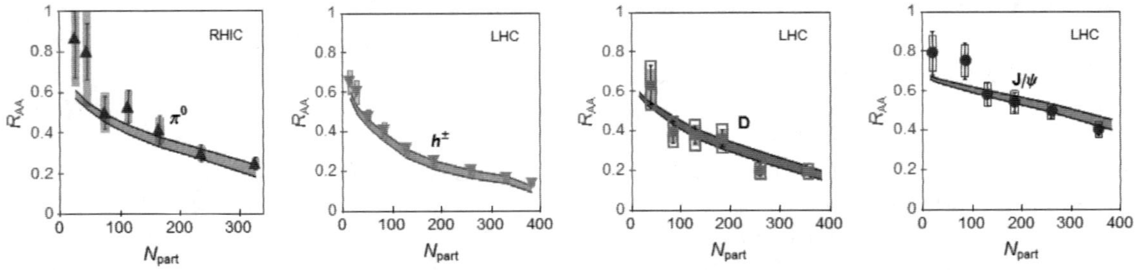


**Figure 5. Comparison of suppression predictions with $R_{AA}$ experimental data in non-central collisions at RHIC and the LHC.** Our suppression predictions (shown as gray bands), as a function of a number of participants, are compared with the available experimental $R_{AA}$ data for neutral pions at RHIC and charged hadrons, D mesons and non-prompt $J/\Psi$ at LHC. The figure is adapted from [7].

We next generate predictions for non-central collisions and compare them with available LHC [41–43] and RHIC [40] data. In these proceedings, we concentrate on the case when the suppression is measured for fixed momentum range and the changing centrality (for additional comparisons, see [7]), for different probes at RHIC and LHC. We see RHIC data for neutral pions, and LHC data for charged hadrons, D mesons, and non-photonic $J/\Psi$. We also obtain a very good agreement with the experiments. Therefore, our model is able to explain the non-central measurements at the RHIC and LHC very well.

### *3.4. Predictions for the upcoming 5.1 TeV Pb+Pb at LHC*

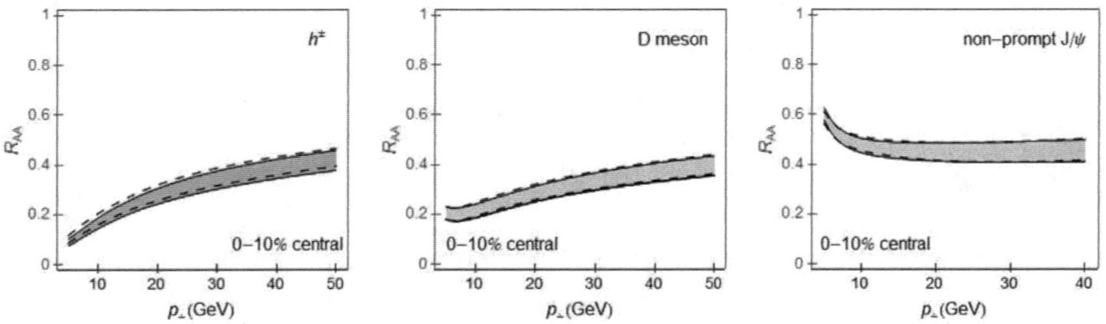


**Figure 6. Suppression predictions for central 5.1 TeV Pb+Pb collisions the LHC.** Charged hadron, D meson and non-prompt $J/\psi$ suppression predictions, as a function of transverse momentum, are shown on the left, central and the right panel, respectively. Full (dashed) curves correspond to $R_{AA}$ predictions at 5.1 TeV (2.76 TeV) Pb+Pb collisions at the LHC. The figure is adapted from [8]. The bands comes from the uncertainty in the magnetic mass value, see Ref. [8].

We will now concentrate on suppression predictions for the upcoming 5.1 TeV Pb+Pb collision energies at the LHC. Within this, we will concentrate on most central collisions and provide suppression predictions for charged hadrons, D mesons and J/Psi, as a function of momentum, which we also compare with analogous predictions for 2.76 TeV beam energy, see Fig. 6. We see that we obtain the same suppressions predictions between these two collision energies, for all types of probes. Moreover, this agreement is in line with BES energy scan, which show similar suppressions between RHIC and LHC. Furthermore, we predict that this similar suppression behavior will continue to higher beam energies as well.

We next ask why we observe the same suppressions between 2.76 and 5.1 TeV, which is addressed in Fig. 7. We see that the reason for the same suppression is an interplay between initial distribution and energy loss effects. That is, the initial distributions become flatter with an increase in the beam energy, which has a tendency to lower the suppression from 2.76 to 5.1 TeV, as seen in the left panel of Fig. 7. On the other hand, the energy loss increases with an increase in beam energy, which has a tendency to increase the suppression from 2.76 to 5.1 TeV, as shown in the right panel of Fig. 7. These two effects approximately cancel, leading to the same suppression patterns between the two collision energies. Therefore, contrary to the heavy flavor puzzle, where the interplay between energy loss and fragmentation functions was crucial for explaining the puzzle, this case provides a clear example on the importance of initial distributions; that is, similarly predicted (and measured) suppressions for different collision energies is a consequence of an interplay between initial production and energy loss.

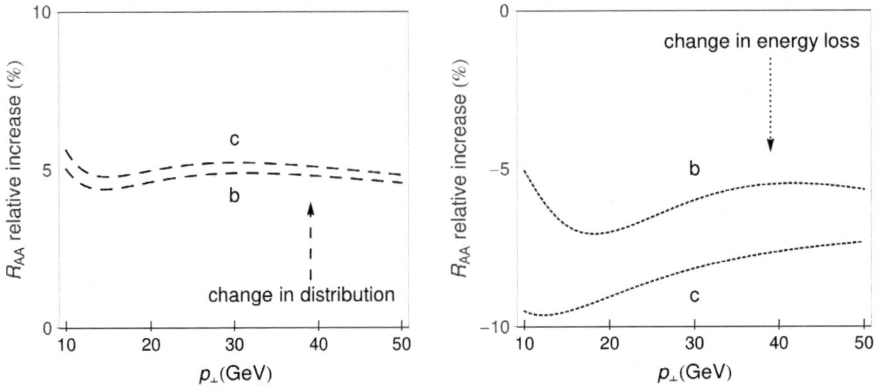


**Figure 7. Relative increase in $R_{AA}$ between 2.76 and 5.1 TeV.** The left panel shows momentum dependence of the relative increase in $R_{AA}$ between 2.76 and 5.1 TeV Pb+Pb collisions at the LHC due to differences in the initial momentum distributions; to calculate the relative increase in $R_{AA}$ due to different initial momentum distributions, the energy loss is kept fixed and calculated for 2.76 TeV case, while the distributions are varied between 2.76 and 5.1 TeV. The right panel shows momentum dependence of the relative increase in $R_{AA}$ between 2.76 and 5.1 TeV collisions at the LHC due to differences in the energy loss; to calculate the relative increase in $R_{AA}$ due to different energy losses, the initial momentum distribution is kept fixed and calculated for 2.76 TeV case, while the energy loss is varied between 2.76 and 5.1 TeV. The figure is adapted from [8].

## 4. Summary

We developed the dynamical energy loss formalism, which includes different important effects. The predictions of the model were then comprehensively tested against angular averaged $R_{AA}$ data, which is considered to be largely insensitive to the medium evolution [44,45]. We obtained robust agreement for wide range of probes, centralities and beam energies. We have seen that the model can also obtain an understanding of the intuitively puzzling data. Finally, we obtained clear predictions for future experiments. Therefore, these results provide us with a good level of confidence that our dynamical energy loss formalism can explain the jet-medium interactions in QGP well.

As an outlook, since we have the jet-medium interactions under control, and since the model can explain the wide range of data without free parameters, we can combine this model with the sophisticated simulations of the bulk medium evolution [46]. We can than generate predictions of high $p_t$ angular differential suppression variables, which are, in distinction to the angular average suppression variables, considered to be highly sensitive to the medium evolution.

## Acknowledgments
This work is supported by Marie Curie International Reintegration Grant within the $7^{th}$ European Community Framework Programme PIRG08-GA-2010-276913 and by the Ministry of Science and Technological Development of the Republic of Serbia, under project No. ON171004.

## References
[1] Bjorken J D 1982 FERMILAB-PUB-82-059-THY pp 287-92
[2] Brambilla N et al 2014 *Eur. Phys. J.* C **74** 2981
[3] Gyulassy M 2002 *Lect. Notes Phys.* **583** 37

[4] Djordjevic M and Djordjevic M, Phys. Lett. B **734** 286 (2014).
[5] Djordjevic M and Djordjevic M, Phys. Rev. C **90** 034910 (2014).
[6] Djordjevic M, Phys. Rev. Lett. **112** 4, 042302 (2014).
[7] Djordjevic M, Djordjevic M and Blagojevic B, Phys. Lett. B **737** 298 (2014).
[8] Djordjevic M and Djordjevic M, Phys. Rev. C **92** 024918 (2015).
[9] Gyulassy M and McLerran L, Nucl. Phys. A 750, 30 (2005).
[10] Gyulassy M, Wang XN, Nucl. Phys. B **420** 583 (1994).
[11] Baier R, Dokshitzer Y, Mueller A, Peigne S, and Schiff D, Nucl. Phys. B **484**, 265 (1997).
[12] Zakharov B, JETP Lett. 63, **952** (1996); ibid **65**, 615 (1997).
[13] Armesto N, Salgado C A, and Wiedemann U A, Phys. Rev. D **69**, 114003 (2004).
[14] Gyulassy M, Levai P, and Vitev I, Nucl. Phys. B 594, **371** (2001).
[15] Djordjevic M and Gyulassy M, Nucl. Phys. A **733** (2004) 265-298.
[16] Djordjevic M, J. Phys. G **32** (2006) S333.
[17] M. G. Mustafa, Phys. Rev. C **72**, 014905 (2005); M. G. Mustafa and M. H. Thoma, Acta Phys. Hung. A **22**, 93 (2005).
[18] A. K. Dutt-Mazumder, Jan-e Alam, P. Roy and B. Sinha, Phys. Rev. D **71**, 094016 (2005).
[19] M. Djordjevic, Phys. Rev. C **74**, 064907 (2006).
[20] A. Adil, M. Gyulassy, W. A. Horowitz and S. Wicks, arXiv:nucl-th/0606010.
[21] Djordjevic M and Heinz U, Phys. Rev. C **77** (2008) 024905.
[22] Djordjevic M, Phys. Rev. C **80** 064909 (2009).
[23] Djordjevic M and Heinz U, Phys. Rev. Lett. **101** 022302 (2008).
[24] Djordjevic M and Djordjevic M 2012 *Phys. Lett.* B **709** 229
[25] Blagojevic B and Djordjevic M, J. Phys. G **42** (2015) 075105.
[26] Kang Z B, Vitev I and Xing H, Phys. Lett. B **718**, 482 (2012), Sharma R, Vitev I and Zhang B W, Phys. Rev. C **80**, 054902 (2009)
[27] Dainese A, Eur. Phys. J. C **33** 495 (2004).
[28] Wicks S, Horowitz W, Djordjevic M and Gyulassy M, Nucl. Phys. A **784**, 426 (2007).
[29] Gyulassy M, Levai P and Vitev I, Phys. Lett. B **538**, 282 (2002).
[30] de Florian D, Sassot R and Stratmann M, Phys. Rev. D **75**, 114010 (2007).
[31] Cacciari M, Frixione S, Houdeau N, Mangano M L, Nason P and Ridolfi G, JHEP **1210**, 137 (2012)
[32] Wilde M 2013 *Nucl. Phys.* A **904-905** 573c
[33] Adare A *et al* 2010 *Phys. Rev. Lett.* **104** 132301
[34] Aamodt K *et al*: (ALICE collaboration), Phys. Lett. B **720**, 52 (2013).
[35] Velasquez A O (for the ALICE collaboration), Nucl. Phys. A **904905** 763c (2013)
[36] Chatrchyan S *et al*: (CMS collaboration), Eur. Phys. J. C **72**, 1945 (2012).
[37] A. Grelli (for the ALICE collaboration), Nucl. Phys. A **904905** 635c (2013).
[38] S. Sakai (for the ALICE collaboration), Nucl. Phys. A **904-905**, 661c (2013).
[39] J. Mihee (for the CMS collaboration) Nucl. Phys. A **904-905**, 657c (2013).
[40] A. Adare *et al.*: (PHENIX Collaboration), Phys. Rev. C **87**, 034911 (2013); arXiv:1204.1526
[41] K. Aamodt *et al.* (ALICE Collaboration), Phys. Lett. B **720**, 52 (2013)
[42] B. Abelev *et al.* (ALICE Collaboration), JHEP **9** 112 (2012).
[43] S. Chatrchyan *et al*: (CMS Collaboration), JHEP **5** 063 (2012).
[44] Molnar D and Sun D, Nucl. Phys. A **932** (2014) 140; **910-911** 486 (2013).
[45] Renk T, Phys. Rev. C **85** 044903 (2012).
[46] Molnar E, Holopainen H, Huovinen P, and Niemi H, Phys. Rev. C 90, 044904 (2014).

15th International Conference on Strangeness in Quark Matter (SQM2015) IOP Publishing
Journal of Physics: Conference Series **668** (2016) 012011 doi:10.1088/1742-6596/668/1/012011

# ATLAS results on heavy flavour production and its relation to quark matter

**Marisilvia Donadelli on behalf of the ATLAS Collaboration**
University of Sao Paulo, Brazil
E-mail: marisilvia.donadelli@cern.ch

**Abstract.** We present measurements of $J/\psi$ and $\psi$(2S) production in proton-lead collisions at a centre of mass energy per nucleon-nucleon interaction of $\sqrt{s_{NN}} = 5.02$ TeV and in proton-proton collisions at $\sqrt{s} = 2.76$ TeV, with integrated luminosities of 28.1 $nb^{-1}$ and 4.0 $pb^{-1}$, respectively . Both measurements are based on data recorded by the ATLAS detector in 2013 and are performed in the dimuon decay channel. Prompt components of $J/\psi$ and $\psi$(2S) are separated from the nonprompt components resulting from $b$-hadron decays through an analysis of the distance between the $J/\psi$ and $\psi$(2S) decay vertexes and the event primary vertex. The measurements of nuclear modification factors and self-normalized yields for the $J/\psi$ and $\psi$(2S) are performed in the following ranges: $8.5 < p_{\mathrm{T}}^{\mu\mu} < 30$ GeV in transverse momentum, $-1.5 < y^*_{\mu\mu} < 1.5$ in centre-of-mass rapidity and $0 - 90\%$ in centrality. The differential cross section for production of nonprompt $J/\psi$ is compared to a FONLL calculation which does not include nuclear effects for the ranges of $8 < p_{\mathrm{T}}^{\mu\mu} < 30$ GeV in transverse momentum, and for $|y^*|_{\mu\mu} < 1.94$ in centre-of-mass rapidity. Forward-backward production ratios are presented and compared to theoretical predictions.

## 1. Introduction

Intense theoretical and experimental efforts have been deployed in the ultra-relativistic heavy ion physics field [1, 2] since Matsui and Satz first suggested that quarkonium ($Q\bar{Q}$) suppression could be a signal of the quark gluon plasma (QGP) formation in nucleus-nucleus (A+A) collisions [3]. With the idea that color screening would prevent the formation of $Q\bar{Q}$ states when the screening length becomes shorter than the $Q\bar{Q}$ size and given the fact that this length is directly related to the temperature, a measurement of a suppressed quarkonium yield would thus provide direct experimental sensitivity to the temperature of the medium created in high energy nuclear collisions. However, before using the suppression of quarkonium yields in A+A collisions relative to $pp$ collisions as a probe to study QGP, it is important to measure the quarkonium production also in $p(d)$+A collision where no QGP effect is expected. Quarkonium suppression has also been observed in $p(d)$+A interactions [4, 5, 6, 7], where the so-called cold nuclear matter (CNM) effects take place, with several phenomenological interpretations. The CNM effects not only affect quarkonia production but can also affect $b$-quark production. For instance, the effects of gluon saturation and shadowing are expected to be similar to those for charmonium production however, nuclear absorption and parton energy loss are expected to be less pronounced. Therefore, additional constraints can be obtained by measuring $b$-quark production, which can be accomplished by measuring the cross sections for $J/\psi$ and $\psi$(2S) production in the decay chains of $b$-hadrons.

## 2. The ATLAS detector

The ATLAS detector [8] at the LHC covers nearly the entire solid angle around the collision point, and it is designed to measure the properties of a wide range of physics processes in $pp$, $p$+Pb and Pb+Pb interactions. It consists of an inner tracking detector surrounded by a thin superconducting solenoid, electromagnetic and hadronic calorimeters, and a muon spectrometer incorporating three superconducting toroid magnet systems. The inner-detector system (ID) is immersed in a 2 T axial magnetic field and provides charged particle tracking in the pseudorapidity range $|\eta| < 2.5$. The high-granularity silicon pixel detector covers the vertex region and is surrounded by the silicon microstrip tracker and transition radiation tracker. The calorimeter system covers the range $|\eta| < 4.9$. Within the region $|\eta| < 3.2$, electromagnetic calorimetry is provided by barrel and endcap high-granularity lead liquid-argon (LAr) calorimeters, with an additional thin LAr presampler covering $|\eta| < 1.8$. Forward calorimeters (FCal) are located in the range $3.1 < |\eta| < 4.9$. The muon spectrometer comprises separate trigger and high-precision tracking chambers that measure the deflection of muons in a magnetic field generated by superconducting air-core toroids. The precision tracking chamber system covers the region $|\eta| < 2.7$, with trigger coverage in the range $|\eta| < 2.4$.

## 3. $J/\psi$ and $\psi(2S)$ production in $\sqrt{s_{NN}} = 5.02$ TeV $p$+Pb and $\sqrt{s} = 2.76$ TeV $pp$ collisions

### 3.1. Signal extraction

$J/\psi$ and $\psi(2S)$ production measurements were performed in the dimuon decay channel in $p$+Pb collisions at $\sqrt{s_{NN}} = 5.02$ TeV with an integrated luminosity of 28.1 $nb^{-1}$, and in $pp$ collisions at $\sqrt{s} = 2.76$ TeV with an integrated luminosity of 4.0 $pb^{-1}$ [9]. The $p$+Pb collisions were produced from a 4 TeV proton beam and a 1.57 TeV per nucleon lead beam. The asymmetric energy of the beams resulted in a shift of the centre of mass by 0.465 units of rapidity relative to the laboratory frame. The centre-of-mass rapidity, $y^* = y^{\text{lab}} - 0.465$, is defined to be positive in the proton-going side. The number of produced $J/\psi$ and $\psi(2S)$ mesons and the relative fraction of nonprompt[1] $J/\psi$ and $\psi(2S)$ with respect to inclusive production are determined using a two-dimensional likelihood fit of the dimuon invariant mass and pseudoproper time[2] ($m_{\mu\mu}$,$\tau$) spectrum of weighted $J/\psi$ and $\psi(2S)$ candidates. The probability density function for the fit follows the method described in [10]. From vertices reconstruction, it is possible to obtain the pseudoproper time to disentangle prompt and nonprompt components. Figure 1 shows the distributions of dimuon invariant mass (upper panels) and pseudoproper time (bottom panels) of weighted $J/\psi$ and $\psi(2S)$ candidates in the range of $8.5 < p_{\text{T}} < 30$ GeV and $-1.5 < y^* < 1.5$ [9].

### 3.2. Self-normalized yields and nuclear modification factors

In order to characterize the $p$+Pb collision impact parameter, each event is assigned to a centrality class based on the total transverse energy measured in the FCal on the Pb-going side ($-4.9 < \eta < -3.2$ with the negative $\eta$ being defined as Pb-going side), $E_{\text{T}}^{\text{FCal}}$ [11]. Each centrality class corresponds to a fixed percentile in the $E_{\text{T}}^{\text{FCal}}$ distribution. The standard Glauber model [12] is used to calculate the mean number of participant nucleons, $\langle N_{\text{part}}\rangle$, and the mean nuclear thickness function, $\langle T_{p\text{Pb}}\rangle$, for each centrality class. Based on the observed centrality dependence of charged particle multiplicity [11], the Glauber-Gribov Color Fluctuation (GGCF) model [13, 14], an extension to the Glauber model which allows event-by-event fluctuations of the nucleon-nucleon cross section, is also considered. Recently, calculations of a centrality bias have

[1] Nonprompt component originates from decays of $b$-hadrons.

[2] The pseudopoper time is defined as $\tau = L_{xy}\frac{m_{\mu\mu}}{p_{\text{T}}}$, where $m_{\mu\mu}$ is the invariant mass of the dimuon, $p_{\text{T}}$ is its transverse momentum and $L_{xy}$ is the signed transverse distance between the primary interaction vertex and the $J/\psi \to \mu^+\mu^-$, or the $\psi(2S) \to \mu^+\mu^-$ vertex.

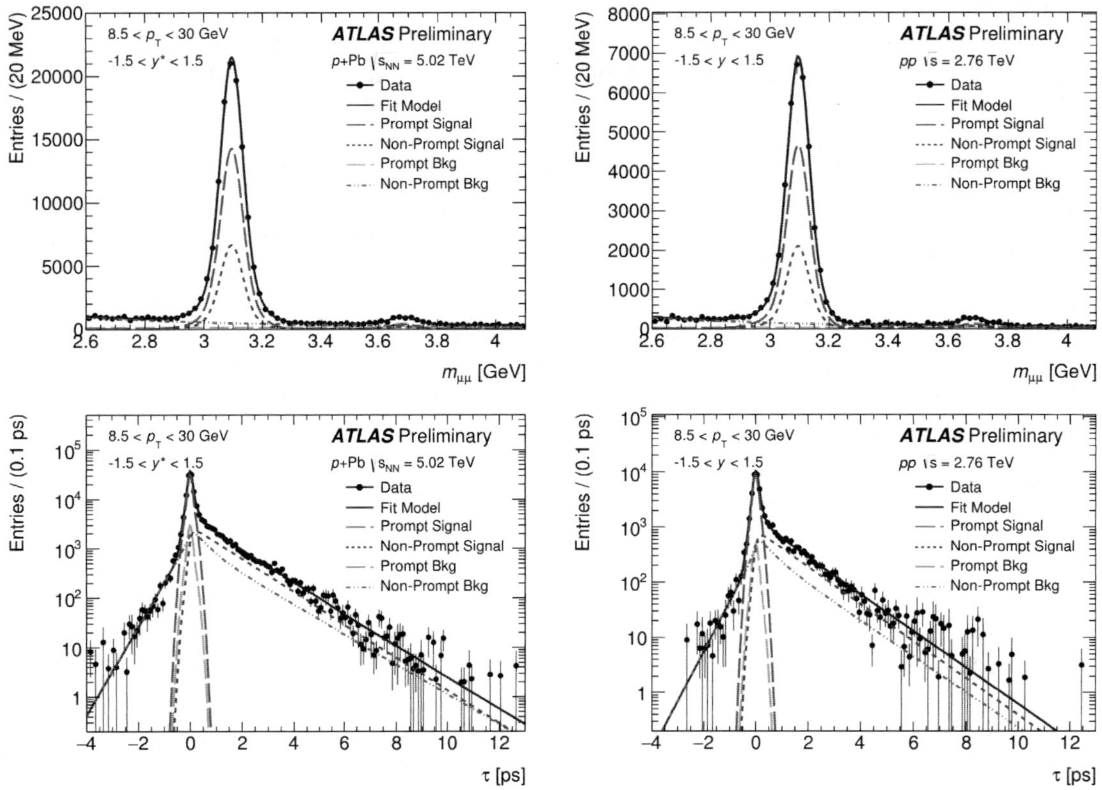


**Figure 1.** Data are projected onto the dimuon invariant mass range of $2.6 - 4.1$ GeV (top), and dimuon pseudoproper lifetime (bottom), respectively. The left column is for $p$+Pb collisions at 5.02 TeV, and the right column is for $pp$ collisions at 2.76 TeV. The dimuon candidates are selected in the range of $8.5 < p_{\mathrm{T}} < 30$ GeV and $-1.5 < y^* < 1.5$. The blue solid lines superimposed on the data are the projections of the total fit PDF [9].

been made in which the average yield from hard scattering processes in each nucleon-nucleon collision is taken to be proportional to the contribution from that collision to the $E_{\mathrm{T}}$, which determines the centrality ($E_{\mathrm{T}}^{\mathrm{FCal}}$) [15]. This model is applied to the ATLAS $p$+Pb centrality classification for both the standard Glauber model and the GGCF model.

The yields of $J/\psi$ and $\psi(2\mathrm{S})$ per event in each centrality class can be normalized to the yield per event measured in all centrality classes (0-90% centrality). These quantities are called self-normalized yields, $\psi/\langle\psi\rangle$. The correlation of charmonium production with the size of the underlying event activity is done by comparing the self-normalized charmonium yields with the similar self-normalized transverse energy, $E_{\mathrm{T}}^{\mathrm{FCal}}/\langle E_{\mathrm{T}}^{\mathrm{FCal}}\rangle$, measured in the Pb-going side of the FCal. $\psi/\langle\psi\rangle$ and $E_{\mathrm{T}}^{\mathrm{FCal}}/\langle E_{\mathrm{T}}^{\mathrm{FCal}}\rangle$ are defined as:

$$\frac{\psi}{\langle\psi\rangle} = \frac{N_{\psi}/N_{\mathrm{evt}}\mid_{\mathrm{cent}}}{N_{\psi}^{0-90\%}/N_{\mathrm{evt}}^{0-90\%}}, \tag{1}$$

$$\frac{E_{\mathrm{T}}^{\mathrm{FCal}}}{\langle E_{\mathrm{T}}^{\mathrm{FCal}}\rangle} = \frac{\langle E_{\mathrm{T}}^{\mathrm{FCal}}\rangle\mid_{\mathrm{cent}}}{\langle E_{\mathrm{T}}^{\mathrm{FCal}}\rangle\mid_{0-90\%}}, \tag{2}$$

where $N_{\psi}$ is the yield extracted from the fit for one centrality class. The yield is divided by

$\langle T_{p\mathrm{Pb}} \rangle$ in the case of bias-corrected self-normalized ratios.

The prompt $J/\psi$, nonprompt $J/\psi$ and prompt $\psi(2\mathrm{S})$ self-normalized yields, $\psi/\langle\psi\rangle$, in bins of self-normalized FCal $E_{\mathrm{T}}$, $E_{\mathrm{T}}^{\mathrm{FCal}}/\langle E_{\mathrm{T}}^{\mathrm{FCal}}\rangle$, are shown in Figure 2 [9]. The self-normalized ratios without centrality bias corrections are compared with CMS results [16]. The uncorrected results are consistent with unity in the slope of the line as well as CMS results at small $E_{\mathrm{T}}^{\mathrm{FCal}}/\langle E_{\mathrm{T}}^{\mathrm{FCal}}\rangle$. However, at large $E_{\mathrm{T}}^{\mathrm{FCal}}$ region $E_{\mathrm{T}}^{\mathrm{FCal}}/\langle E_{\mathrm{T}}^{\mathrm{FCal}}\rangle > 3$), significant deviations from the linear scaling were observed for all three resonances. The deviations become more significant when the centrality bias correction is applied. This deviation at large $E_{\mathrm{T}}^{\mathrm{FCal}}$ is also observed for $Z$ and thus it has probably no relation with quarkonium suppression but it could rather be due to a kinematic effect.

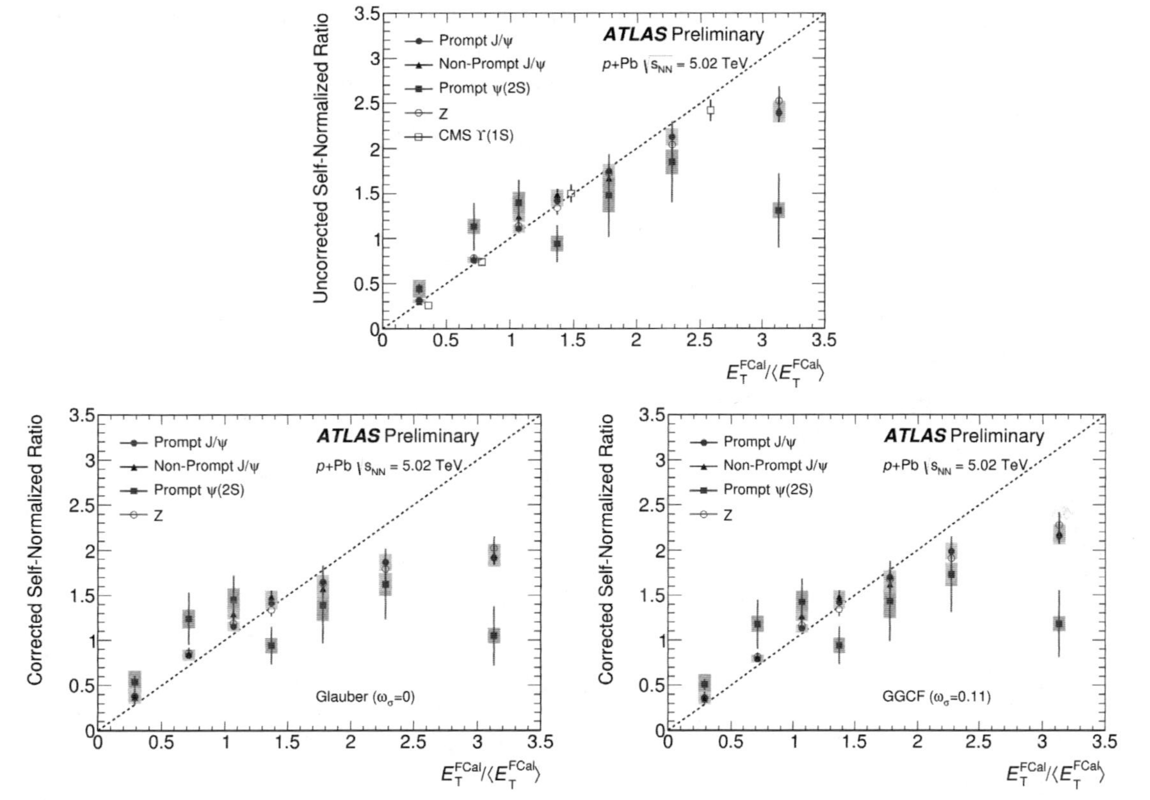


**Figure 2.** Self-normalized ratio for prompt $J/\psi$, nonprompt $J/\psi$ and prompt $\psi(2\mathrm{S})$ with no centrality bias corrections(top) and with bias corrections for the standard Glauber (bottom left) and GGCF (bottom right) configurations. The x-axis variables are normalized by their corresponding $E_{\mathrm{T}}^{\mathrm{FCal}}$ integrated values. For all points, the abscissae are at the mean value in each $E_{\mathrm{T}}^{\mathrm{FCal}}$ bin. The dotted line is a linear function with a slope equal to unity. The error bars indicate only the statistical uncertainties, and the boxes represent the systematic uncertainties. The $E_{\mathrm{T}}/\langle E_{\mathrm{T}}\rangle$ value of CMS results shown in the top plot was measured at $4 < |\eta| < 5.2$ [9].

The nuclear modification factor, $R_{p\mathrm{Pb}}$, is used to compare the production of $J/\psi$ and $\psi(2\mathrm{S})$ in $p$+Pb collisions to the same processes in $pp$ collisions:

$$R_{p\mathrm{Pb}} = \frac{1}{A^{\mathrm{Pb}}} \frac{d^2\sigma_{\psi}^{p+\mathrm{Pb}}/dy^* dp_{\mathrm{T}}}{d^2\sigma_{\psi}^{pp}/dy dp_{\mathrm{T}}} \tag{3}$$

where $\psi$ refers to either prompt or nonprompt $J/\psi$ or prompt $\psi$(2S) and $A^{\text{Pb}} = 208$. The nuclear modification factor may also be defined for a given centrality class as:

$$R_{p\text{Pb}} = \frac{1}{\langle T_{p\text{Pb}} \rangle_{\text{cent}}} \frac{1/N_{\text{evt}} \; d^2 N_{\psi}^{p+\text{Pb}} / dy^* dp_{\text{T}} \big|_{\text{cent}}}{d^2 \sigma_{\psi}^{pp} / dy dp_{\text{T}}} \tag{4}$$

where $N_{\text{evt}}$ is the total number of minimum bias $p$+Pb events, $\langle T_{p\text{Pb}} \rangle_{\text{cent}}$ is the mean nuclear thickness function in the centrality class, and $N_{\psi}^{p+Pb}$ is the number of charmonium signal events in the same centrality class. In the absence of a measured $pp$ reference cross section at $\sqrt{s} =$ 5.02 TeV, an interpolation of the cross sections measured at nearby energies is used. The interpolation is based on the cross section measured at $\sqrt{s} = 2.76$ TeV [9] and the cross sections measured at $\sqrt{s} = 7$ and 8 TeV [10]. For that, three functional forms in $\sqrt{s}$ (linear, power law and exponential) were used for the interpolation, with the central value of the cross section at $\sqrt{s} = 5.02$ TeV derived from the power law form. The other functions are used in the calculation of the uncertainty associated with the interpolation [9]. The prompt $\psi$(2S) to $J/\psi$ production ratio in $p$+Pb collisions is shown in Figure 3 in bins of $E_{\text{T}}^{\text{FCal}}$ along with the same ratio in $pp$ collisions at 2.76 TeV.

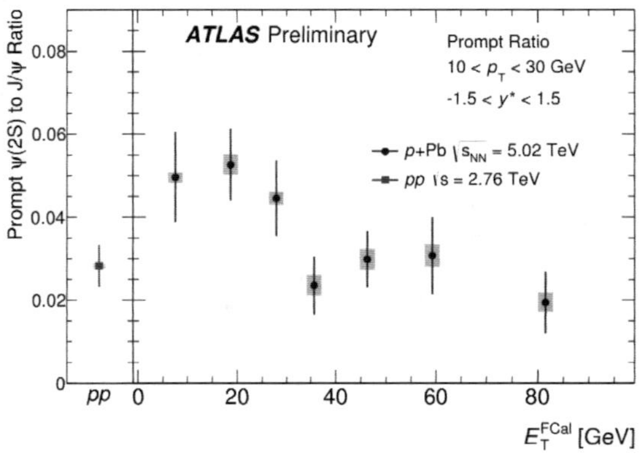


**Figure 3.** Prompt $\psi$(2S) to $J/\psi$ production ratio as a function of total FCal $E_{\text{T}}$ for $p$+Pb collisions. The error bars indicate only the statistical uncertainties, and the boxes represent the systematic uncertainties. As a reference, the ratio of $pp$ collisions at 2.76 TeV is also shown in the square green point [9].

Prompt $\psi$(2S) to $J/\psi$ double ratios $R_{p\text{Pb}}(\psi(2S))/R_{p\text{Pb}}(J/\psi)$ in bins of $y^*$ and in bins of $E_{\text{T}}^{\text{FCal}}$ are shown in Figure 4. The prompt production ratios at 2.76 TeV $pp$ collisions are used as the reference, since the single production ratios in $pp$ collisions do not depend significantly on $\sqrt{s}$. The right panel of Figure 4 is the single prompt excited-to-ground state ratio measured in $p$+Pb collisions in Figure 3 divided by the $pp$ point for $\sqrt{s} = 2.76$ TeV with the errors convolved. The prompt $\psi$(2S) to $J/\psi$ double ratios show no obvious rapidity dependence but a decreasing trend as the event activity increases.

## 4. Differential $J/\psi$ production cross section and forward-backward ratio in $\sqrt{s_{NN}} =$ 5.02 TeV $p$+Pb collisions

### *4.1. Forward-backward ratio*

Using the same data set of $p$+Pb collisions with an integrated luminosity of 28.1 $nb^{-1}$, $J/\psi$ mesons were reconstructed in the dimuon decay channel over the transverse momentum range

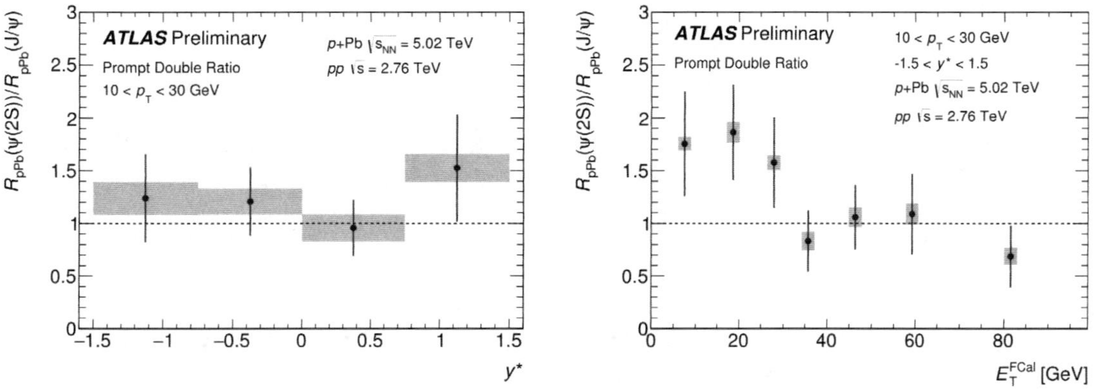


**Figure 4.** Prompt $\psi(2S)$ to $J/\psi$ double ratio in bins of center-of-mass rapidity, $y^*$ (left) and total FCal $E_T$ (right). The error bars indicate only the statistical uncertainties, and the boxes represent the systematic uncertainties. The horizontal size of box represents bin width. Prompt production ratios at 2.76 TeV $pp$ collisions are used as the reference when calculating the double ratio [9].

of $8 < p_T < 30$ GeV and $-2.87 < y^* < 1.94$ [17]. The number of produced $J/\psi$ mesons and the nonprompt fraction are determined through similar procedure as already described in Section 3.1. Considering that previous measurements in p+Pb collisions at the LHC [18, 19] show that the differential cross section for $J/\psi$ production at forward $y^*$ (proton direction) are significantly smaller than at backward $y^*$ (heavy-ion direction), an observable to quantify this asymmetry is proposed, namely, the forward-backward ratio, $R_{FB}$:

$$R_{FB}(p_T, y^*) \equiv \frac{d^2\sigma(p_T, y^* > 0)/dp_T/dy^*}{d^2\sigma(p_T, y^* < 0)/dp_T/dy^*} \tag{5}$$

which has the advantage of not depending on the knowledge of the $J/\psi$ production cross section in $pp$ collisions, not to mention that the experimental and theoretical uncertainties partially cancel in the ratio. Figure 5 shows $R_{FB}$ as a function of transverse momentum in the range $8 < p_T < 30$ GeV for prompt $J/\psi$ (upper panel) and for nonprompt $J/\psi$ (bottom panel). Figure 5 shows $R_{FB}$ as a function of $y^*$ in the range $|y^*| < 1.94$ for prompt $J/\psi$ (upper panel) and for nonprompt $J/\psi$ (bottom panel). These results are consistent with unity within experimental uncertainties, with no significant $p_T$ or $y^*$ dependence, for both prompt and nonprompt $J/\psi$ [17]. These results differ from measurements at more forward $y^*$ and lower $p_T$ performed by ALICE and LHCb Collaborations, suggesting a strong kinematic dependence of the CNM effects on both charmonium and $b$-quark production. The $R_{FB}$ ratio for prompt $J/\psi$ agrees with theoretical predictions [20, 21] that include shadowing effects based on the EPS09 nuclear parton distribution functions [22].

### *4.2. Differential cross sections and comparison with FONLL calculation*

The differential cross sections of nonprompt $J/\psi$ production are compared to a scaled $pp$ reference based on FONLL calculations [23]. They are multiplied by a factor of 208 to account for the number of nucleons in the Pb ion. The FONLL calculations are performed using CTEQ6.6 [24] parton distribution functions that do not include any nuclear modification. As can be seen in Figure 7, the measured cross sections are found to be consistent with FONLL calculation within uncertainties.

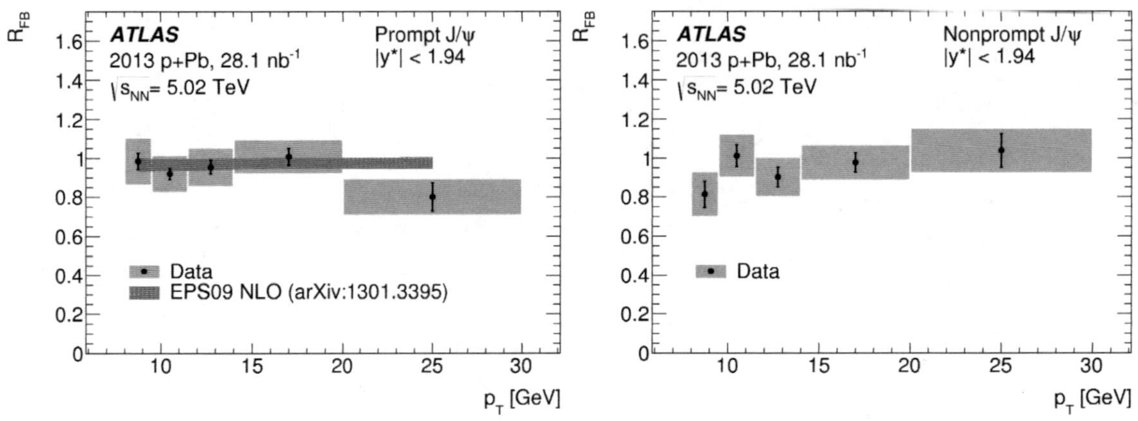


**Figure 5.** Forward-backward production ratio $R_{\rm FB}$ measured in the centre-of-mass rapidity range $|y^*| < 1.94$ as a function of $J/\psi$ transverse momentum, for prompt $J/\psi$ (left panel) and nonprompt $J/\psi$ (right panel) [17].

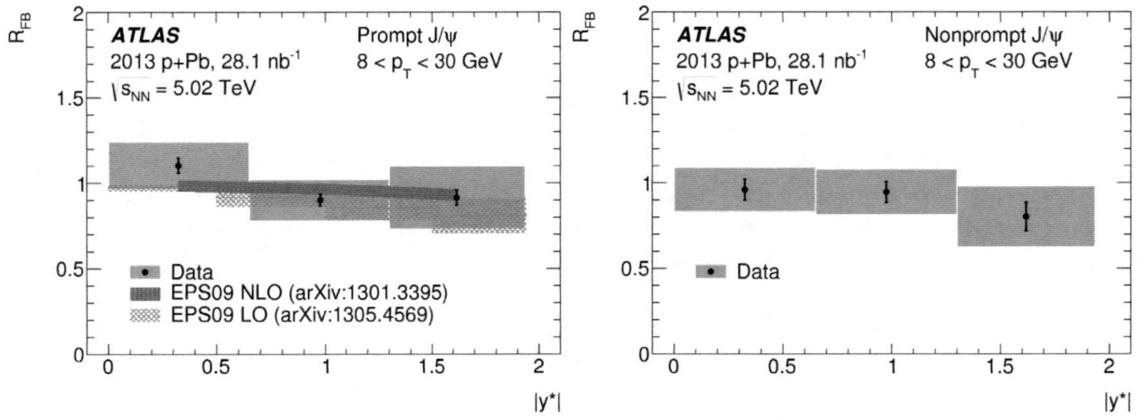


**Figure 6.** Forward-backward production ratio $R_{\rm FB}$ as a function of centre-of-mass rapidity $y^*$ for prompt $J/\psi$ (left panel) and nonprompt $J/\psi$ (right panel) [17].

## 5. Conclusions

ATLAS has performed measurements of prompt and nonprompt components of $J/\psi$ and $\psi(2S)$ production in $p$+Pb collisions at 5.02 TeV and in $pp$ collisions at 2.76 TeV. Results were presented for differential cross sections, forward-backward ratio for $J/\psi$, $R_{p\rm Pb}$ and self-normalized yields for $J/\psi$ and $\psi(2S)$ via $pp$ interpolation. CNM effects were seen in a number of observables.

## References

[1] Lansberg J, Rakotozafindrabe A, Artoisenet P, Blaschke D, Cugnon J *et al.* 2008 *AIP Conf. Proc.* **1038** 15–44 (*Preprint* `arXiv:0807.3666`)
[2] Brambilla N, Eidelman S, Heltsley B, Vogt R, Bodwin G *et al.* 2011 *Eur. Phys. J. C* **71** 1534 (*Preprint* `arXiv:1010.5827`)
[3] Matsui T and Satz H 1986 *Phys. Lett. B* **178** 416–422
[4] B Allessandro *et al*, (NA50 Collaboration) 2004 *Eur. Phys. J. C* **33** 31–40
[5] S S Adler et al, (PHENIX Collaboration) 2006 *Phys. Rev. Lett.* **96** 012304 (*Preprint* `arXiv:0507032`)
[6] A Adare et al, (PHENIX Collaboration) 2008 *Phys. Rev. C* **77**(2) 024912

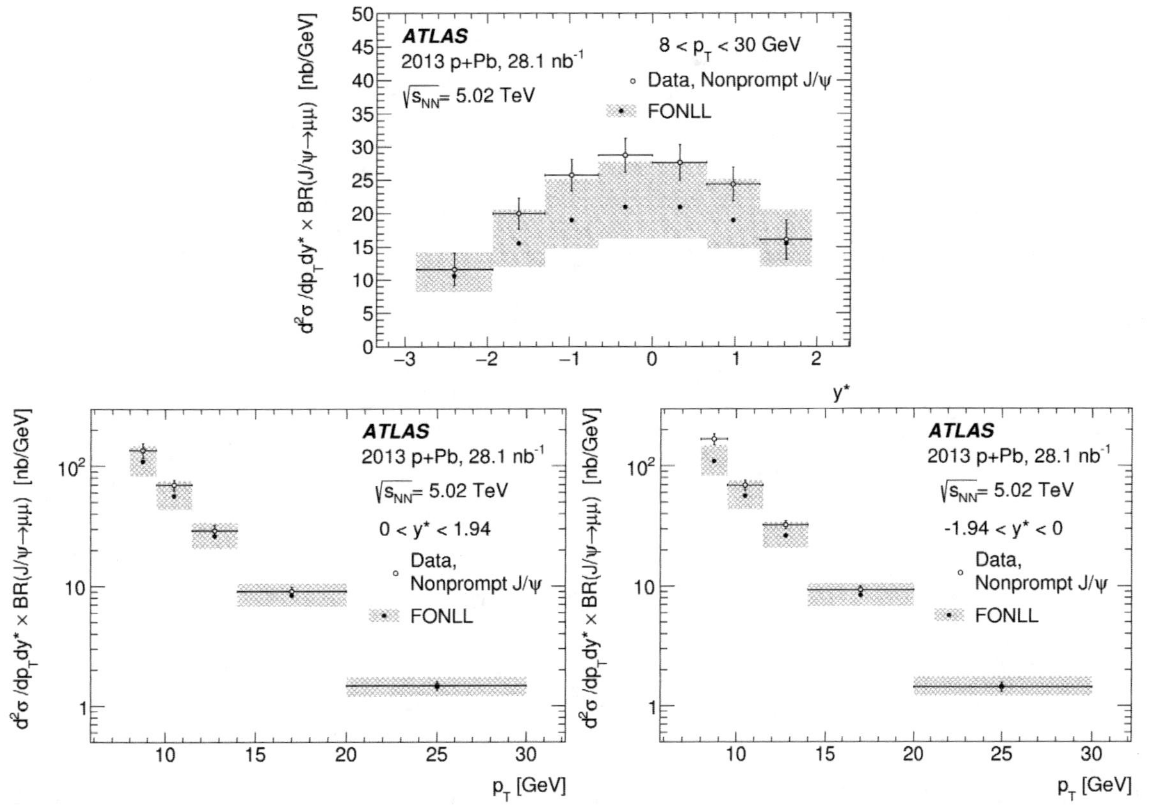


**Figure 7.** Differential cross section for production of nonprompt $J/\psi$ as a function of $J/\psi$ centre-of-mass rapidity (upper panel) and transverse momentum (bottom panel) compared with a FONLL calculation for *pp* collisions scaled by the number of nucleons in the Pb ion [17].

[7] A Adare et al, (PHENIX Collaboration) 2011 *Phys. Rev. Lett.* **107**(14) 142301
[8] ATLAS Collaboration 2008 *JINST* **3** S08003
[9] ATLAS Collaboration *ATLAS-CONF-2015-023* URL `http://cdsweb.cern.ch/record/2029577`
[10] ATLAS Collaboration *ATLAS-BPHY-2012-02* URL `http://cdsweb.cern.ch/record/2023931`
[11] ATLAS Collaboration *ATLAS-CONF-2013-096* URL `http://cdsweb.cern.ch/record/1599773`
[12] Miller M L, Reygers K, Sanders S J and Steinberg P 2007 *Ann.Rev.Nucl.Part.Sci.* **57** 205–243 (*Preprint* `arXiv:0701025`)
[13] Guzey V and Strikman M 2006 *Phys. Lett. B* **633** 245–252 (*Preprint* `arXiv:0505088`)
[14] Alvioli M and Strikman M 2013 *Phys. Lett. B* **722** 347–354 (*Preprint* `arXiv:1301.0728`)
[15] Perepelitsa D V and Steinberg P A 2014 (*Preprint* `arXiv:1412.0976`)
[16] CMS Collaboration 2014 *JHEP* **1404** 103 (*Preprint* `arXiv:1312.6300`)
[17] ATLAS Collaboration 2015 (*Preprint* `arXiv:1505.08141`)
[18] B Abelev et al, (ALICE Collaboration) 2014 *JHEP* **1402** 073 (*Preprint* `arXiv:1308.6726`)
[19] R AAij et al, (LHCb) 2014 *JHEP* **02** 072 (*Preprint* `arXiv:1308.6729`)
[20] Albacete J L *et al.* 2013 *Int. J. Mod. Phys.* **E22** 1330007 (*Preprint* `arXiv:1301.3395`)
[21] Ferreiro E G, Fleuret F, Lansberg J P and Rakotozafindrabe A 2013 *Phys. Rev.* **C88** 047901 (*Preprint* `arXiv:1305.4569`)
[22] Eskola K J, Paukkunen H and Salgado C A 2009 *JHEP* **04** 065 (*Preprint* `arXiv:0902.4154`)
[23] Cacciari M, Frixione S, Houdeau N Mangano M, P N and Ridolfi G 2012 *JHEP* **137** 1210 (*Preprint* `arXiv:1205.6344`)
[24] Nadolsky P M *et al.* 2008 *Phys. Rev. D* **78**(1) 013004 URL `http://link.aps.org/doi/10.1103/PhysRevD.78.013004`

# NA61/SHINE vertex detector for open charm measurements

**Grigory Feofilov (for NA61/SHINE Collaboration)**
St.Petersburg State University, RF
E-mail: grigory-feofilov@yandex.ru

**Abstract.**
In this report we present the motivation and general design of a new vertex detector (VD) for NA61/SHINE. Several practical solutions to meet the requirements for this VD are proposed and demonstrated.

## 1. Introduction

The SPS Heavy Ion and Neutrino Experiment (NA61/SHINE) [1] at CERN was designed to study the properties of the onset of deconfinement and search for the critical point of strongly interacting matter. These goals are being pursued by investigating proton-proton, proton-nucleus and nucleus-nucleus collisions at different collision energies from 13A to 158A GeV. Precise hadron production measurements have also been done by NA61/SHINE in order to improve calculations of the initial neutrino beam flux in long-baseline neutrino oscillation experiments as well as for more reliable simulations of cosmic-ray air showers.

Recently it was proposed [2, 3, 4] that measurements of hadrons containing charm quarks could be used in investigations of the initial stage of nucleus-nucleus collision at relativistic energies. Charm is considered an important probe in a deconfined medium formed in high energy A+A collisions and of the dynamic processes of thermalization of the QGP that could be used to test the color screening effects on the binding of charm quarks to color neutral $J/\psi$ mesons [2]. A schematic view illustrating $J/\psi$ production in p+p collisions is shown in Fig. 1 ([3]). According to the standard scenario of charmonium production and suppression in AA collisions [2]:

- Charmonia are produced before QGP formation
- Suppression takes place in the QGP
- Some charmonia may survive beyond some critical temperature $T_c$.

Matter induced changes in the yield of quarkonia were discussed in a recent paper [4] where the following contributing mechanisms are considered: color screening (see Fig. 2 upper left), ionization by thermal gluons (see Fig. 2 lower left) and recombination (see Fig. 2 right). Experimental studies in the charm sector provide important information for discriminating among theoretical predictions.

Figure 3 shows the results of a comparison [9] of meson abundances from central Au+Au reactions from SIS to RHIC energies predicted by Hadron String Dynamics (HSD) transport

Journal of Physics: Conference Series **668** (2016) 012012 doi:10.1088/1742-6596/668/1/012012

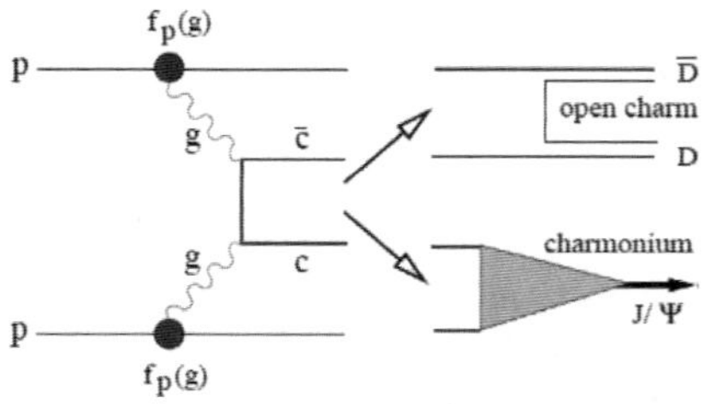


**Figure 1.** Schematic view of $J/\psi$ production in p+p collisions [3]

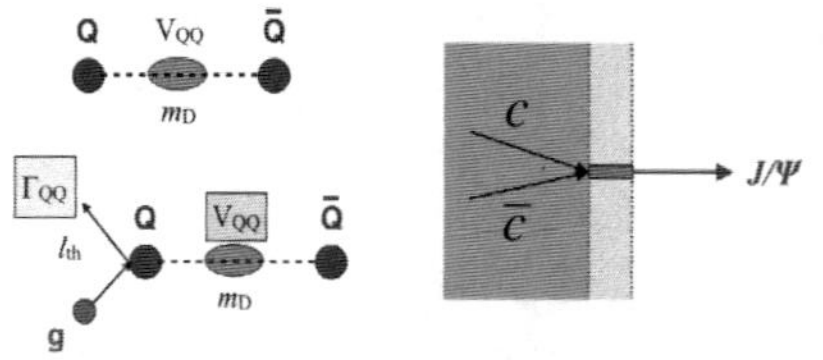

**Figure 2.** Mechanisms contributing to matter induced changes in the yield of quarkonia: upper left – color screening, lower left – ionization by thermal gluons, right – recombination [4].

calculations [8] and experimental data from central nucleus-nucleus collisions Ref.[9]. Calculated meson multiplicities show a monotonic increase with bombarding energy that is in approximate agreement with data. This allows for the use of the HSD predictions for $D(\bar{c})$, $D(c)$ and $J/\psi$ multiplicities as a basis for estimating the required efficiency of registration and beam intensity at the SPS. As one may see from these predictions, the measurements of $D(\bar{c})$, $D(c)$ and $J/\psi$ yields are rather challenging because the multiplicity is lower by orders of magnitude when compared to the other mesons.

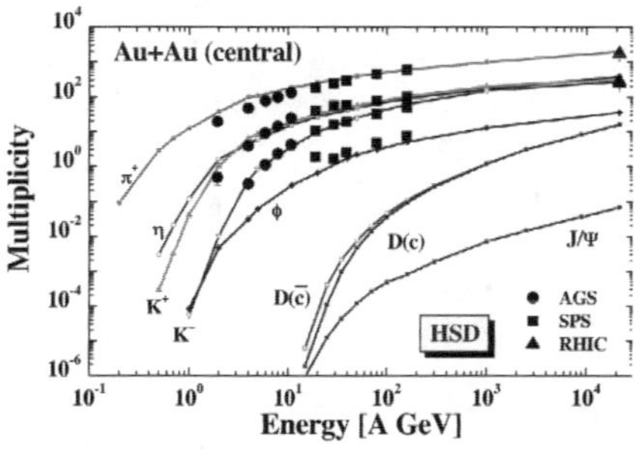


**Figure 3.** Overview of the experimental meson abundances and HSD model predictions [11] for the multiplicities of $\pi^+$, $\eta$ , $K^+, K^-, \phi, D, \overline{D}$ and $J/\psi$ - mesons in central Au+Au collisions as a function of bombarding energy from SIS to RHIC energies. (Figure of Ref. [8] is modified in Ref.[9].)

Below, we present the motivations for open charm measurements in nucleus-nucleus collisions at SPS energies and the challenges and requirements to be met by the experiment and in the design of the new Vertex Detector for the NA61/SHINE apparatus. Some practical solutions are proposed and demonstrated based on previous experience.

## 2. Open charm measurements at SPS energies

The first experimental look at the open charm production at SPS energies was done by the NA60 experiment in In+In collisions [14].

A feasibility study for possible $D^o$ meson (Open charm) measurements via its decay into two charged daughter particles, pion and kaon, in central Pb+Pb collisions at the SPS energies was presented earlier [5]. A simulation framework was developed for these studies using such MC models as AMPT (A Multi-Phase Transport model [10]), HSD (Hadron String Dynamics model [11]) and the event generator PYTHIA[12]. It was shown [7] that 12 hours of beam time at the SPS should be sufficient to record about 200K events. and to provide reconstruction for $D^o \to K^-\pi^+$ in 0-10% central Pb+Pb collisions at 158 A GeV. Model calculations and GEANT4 simulation of particle transport through the experimental setup [5, 6, 7] proved that

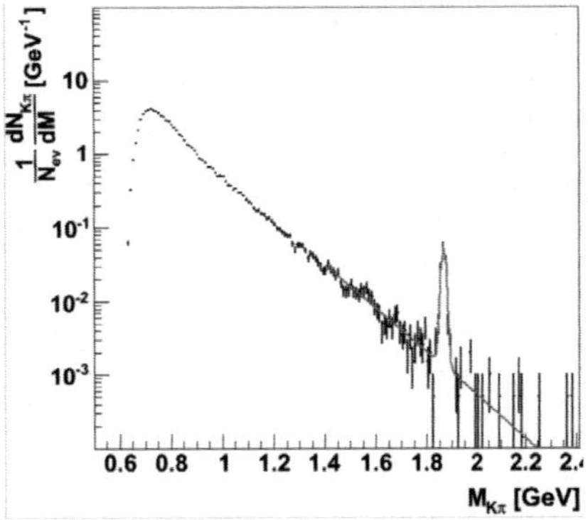


**Figure 4.** Reconstruction of simulated events: yield for $D^o- >K^-\pi^+$ in 200k events (12 hours of beam) for the most central 0-10% Pb+Pb collisions at 158A GeV ( perfect PID is assumed) [7].

| Meson | Decay Channel | $c\tau$ | branching ratio |
|---|---|---|---|
| $D^0$ | $D^0 \to K^- + \pi^+$ | 122.9 $\mu$m | $(3.91 \pm 0.05)\%$ |
| $D^0$ | $D^0 \to K^- + \pi^+ + \pi^+ + \pi^-$ | 122.9 $\mu$m | $(8.14 \pm 0.20)\%$ |
| $D^+$ | $D^+ \to K^- + \pi^+ + \pi^+$ | 311.8 $\mu$m | $(9.2 \pm 0.25)\%$ |
| $D_S^+$ | $D_S^+ \to K^+ + K^- + \pi^+$ | 149.9 $\mu$m | $(5.50 \pm 0.28)\%$ |
| $D^{*+}$ | $D^{*+} \to D^0 + \pi^+$ | — | $(61.9 \pm 2.9)\%$ |

**Table 1.** Decay channels, branching ratios and values of $c\tau$ for various D mesons.

such experiments are possible. The result of reconstruction of $D^o \to K^-\pi^+$ in these 0-10% most central Pb+Pb collisions at 158A GeV shows a clear $D^o$ peak (see Fig. 4).

A number of challenges and requirements for the design of the VD for the NA61/SHINE were established. The first challenge is the low yields of $D(\bar{c})$, $D(c)$ and $J/\psi$ mesons which require sufficiently fast detectors capable to provide high time resolution and high efficiency of track registration.

The second challenge comes from the short mean life-time of D mesons (see values of $c\tau$ in Table.1). Therefore, one expects a rather small distance between the decay vertices of D mesons and the primary production vertex.

The following general requirements for the VD were obtained [5, 6, 7] as a result of the simulations:

- Precise vertexing and tracking accuracy (at the level of 10 - 20 $\mu$m in the transverse plane)
- The lowest possible material budget in the tracking region in order to increase the efficiency of open charm measurements
- High granularity of vertex tracking detectors capable to register the multiple tracks in nucleus-nucleus collisions
- Large acceptance.

CMOS technologies and the use of MIMOSA-26 chips were proposed [5] to meet these requirements and to ensure high accuracy of tracking of charged particles and of secondary vertex reconstruction. A high precision VD needs to be developed that is capable of providing an accuracy of charged particle tracking to the vertices on the level of tens of microns. It is also required to have an extremely low material budget (of the order of below 0.3 % $X_0$) in order to mimimise multiple scattering of low-momentum particles, which would reduce the efficiency of $D^o$ reconstruction.

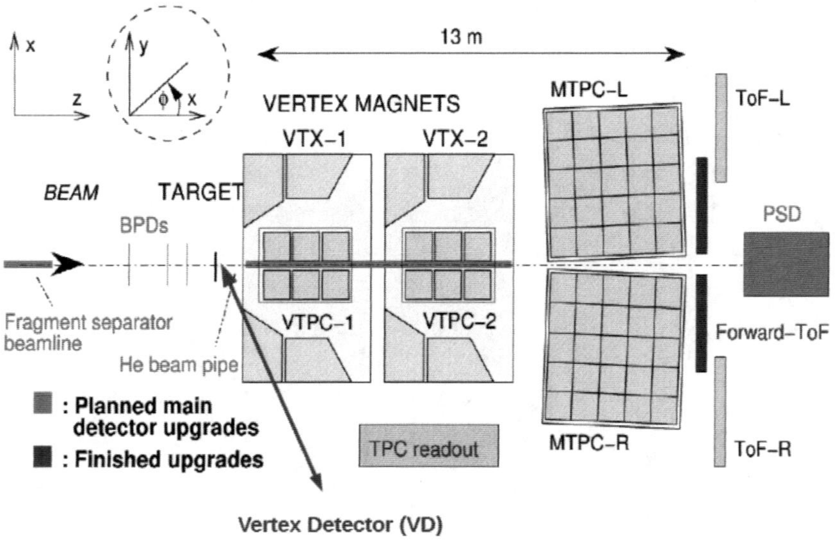


**Figure 5.** Layout of the NA61/SHINE apparatus with the new VD planned to be positioned downstream of the target.

## 3. The NA61/SHINE Vertex Detector project based on CMOS pixel detectors

The Vertex Detector will be positioned between the primary vertex and the first VTPC of the NA61/SHINE apparatus (see Fig. 5). Four planes of coordinate-sensitive detectors will provide the first four hits on charged particle tracks coming from the vertices. Detectors and the relevant supporting elements will form stations, called VDS1-VDS4, located at 5, 10, 15 and 20 cm distance from the target. The tracks registered by the VD will be matched with the tracks of the time-projection chambers (VTPCs) of NA61/SHINE in which the PID will be performed.

High coordinate resolution MIMOSA-26 sensors [15] whose architecture is based on the Monolithic Active Pixel Sensor (MAPS) with fast binary readout have been chosen [7] as the basic detection element of these VD stations.

### *3.1. MIMOSA-26 sensor*

The sensitive area of the MIMOSA-26 chip is 1.06x2.12 $\mathrm{cm}^2$ with a pixel pitch equal to 18.4 $\mu$m (= 663.5k pixels/chip). This dimension of the pixels of 18.4 $\mu$m were found to be sufficient for the required charged particle tracking accuracy. The MIMOSA-26 Si-chips are thinned down to 50 $\mu$m, providing a significant reduction in material budget. Therefore, the multiple scattering of low momentum charged particles in the detector volume is negligible ($X/X_0 = 0.064\%$).

MIMOSA-26 is suited for detecting charged particles with density up to $10^6$ hits/$\mathrm{cm}^2$/s. The selection of this chip was mostly driven by the very high hit occupancy expected in the stations. Estimates show that the hit occupancy generated by a single central Pb+Pb collision at the top SPS energy of 158A GeV reaches value of about 5 hits/$\mathrm{mm}^2$/event in the innermost part of the VDS1 station.

MIMOSA-26 pixel columns are read out in parallel, row by row. The chip readout time is 115.2 $\mu$s. Each pixel includes amplification and Correlated Double Sampling (CDS) units and each end of column is equipped with a discriminator. After the analogue to digital conversion, digital signals pass through zero suppression circuits. The readout electronics boards for MIMOSA-26 chips are being developed by NA61/SHINE in synergy with the CBM collaboration

| Material | Thickness ($\mu$m) | $/X_0$ $(cm)$ | $X/X_0$ (% ) |
|---|---|---|---|
| Polyimide cooling pipe wall | 25 | 28.4 | 0.003 |
| Carbon fleece | 40 | 106.8 | 0.004 |
| Water | 1000 | 35.76 | 0.032 |
| Carbon fiber plate K13D2U | 70 | 26.08 | 0.027 |
| Graphite foil | 30 | 26.56 | 0.011 |
| Thermal greese (glue) | 100 | 44.37 | 0.023 |
| Si–sensor | 50 | 9.36 | 0.064 |
| Total | | | 0.154 |

**Table 2.** Material budget for the carbon fiber cooling and support structure with the MIMOSA-26 Si-sensor according to the design developed in synergy with ALICE [16]

at FAIR. The major components include the Front-end Board connected to the Si-sensors by the flex polyimide cable and the Converter Board.

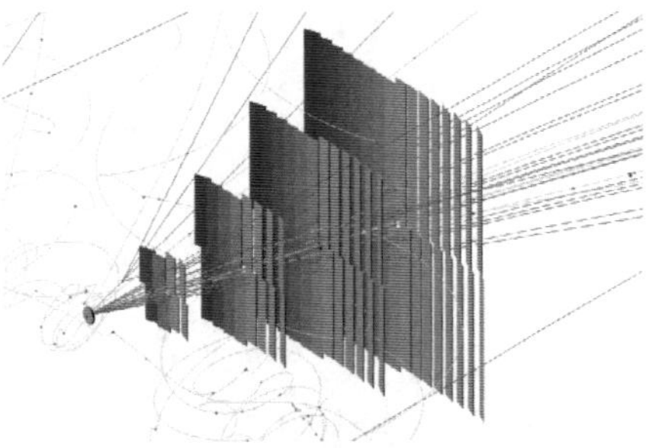

**Figure 6.** The full acceptance VD in GEANT4 simulations.

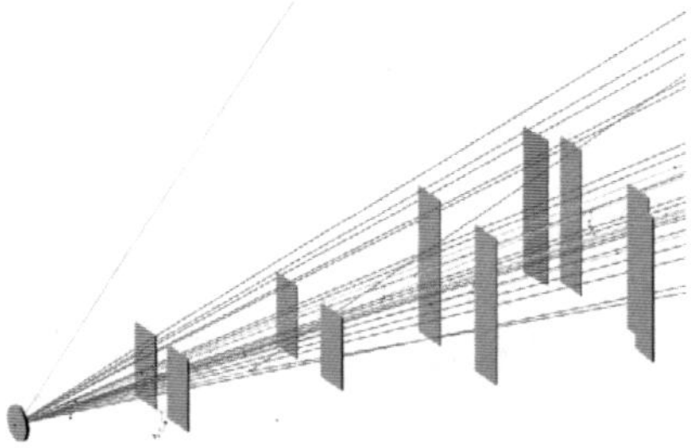

**Figure 7.** The limited acceptance VD.

*3.2. General design of the Vertex Detector*

The geometry of the full acceptance VD used in the GEANT4 simulations is shown in Fig. 6. As the first stage of the VD development it is planned to use the limited acceptance VD pictured in Fig. 7. Straight track reconstruction was performed for these simulations and provided the following results for the accuracy of secondary vertex determination: $\sigma_x$, $\sigma_y$, $\sim 7\mu m$, $\sigma_z$, $\sim 70\mu m$, where a realistic material budget (see Table 2) of the stations was taken into account in the GEANT4 transport of charged particles. This design of the NA61/SHINE VD ensures $X/X_0 \leq 0.2\%$ per sensitive layer of Si-detectors (including the chip and services).

The conceptual layout of the VD for NA61/SHINE is shown in Fig. 8. The design includes four VD stations – four planes of coordinate sensitive detectors – that are formed by MIMOSA-26 chips mounted on extra-lightweight vertical carbon fiber panels.

These extra-light support, cooling and precise positioning carbon fiber structures with MIMOSA-26 chips are precisely mounted on the C-shape support frames depicted in Fig. 9. The number of MIMOSA-26 chips, forming each sensitive plane, will vary depending on the distance to the target.

The C-shape support frames are mounted on two movable platforms. Thus the MIMOSA-26 chips can be moved in and out of the measuring position by using the high accuracy ($5\mu$m) linear translation stage step motor. The C-shape frames also provide cooling water manifolds (see Fig. 10) for cooling panels with the detectors mounted.

**Figure 8.** Conceptual view of the Vertex Detector for NA61/SHINE. The four sensor planes are formed by MIMOSA-26 chips mounted with support and cooling carbon fiber panels on the C-shape frames and placed inside the He-filled box.

**Figure 9.** The C-shape supporting frames mounted on two movable platforms (called "left" and "right"). Only one carbon fiber structure with MIMOSA-26 chips is shown per each C-shape frame. The flat polyimide cables (FPCs) and the readout boards will also be fixed to the additional holders of these C-shape supporting frames (not shown here).

**Figure 10.** Conceptual design of cooling liquid distribution manifolds mounted on the C-shape frames.

All VD components are housed inside a hermetic box structure (see Fig. 8 filled with He (at atmospheric pressure). This He gas inside the box helps to further minimise the multiple scattering in the sensitive region of tracking of low-momentum charged particles. The movable target holder of the VD (not shown) will be positioned on the beam line in front of the VDS1 station.

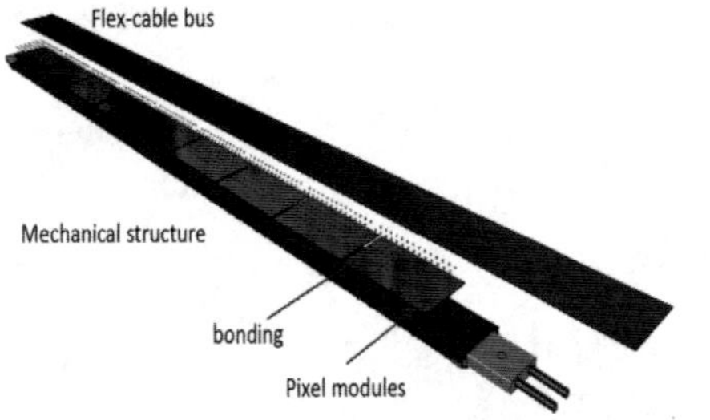


**Figure 11.** The ALICE type [16] design of the extra-lightweight (1.5 g) support and cooling carbon fiber structure of 30 cm length used for precise positioning of MIMOSA-26 chips.

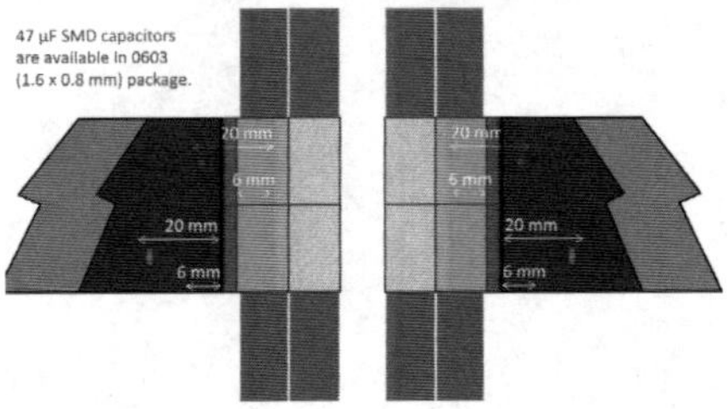


**Figure 12.** Eight MIMOSA-26 chips of the 1st station VDS1 mounted on vertical extra-lightweight support and cooling carbon fiber structures. Flat readout cables, going left and right from the center of the VDS1 station, are shown schematically.

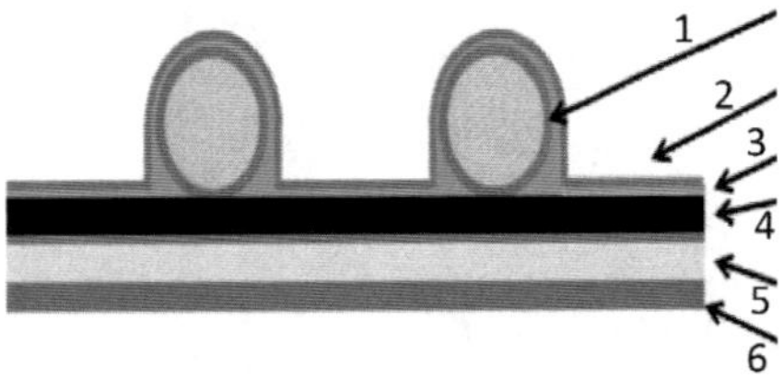


**Figure 13.** The cross-section of the ALICE type [16] design of the cooling and support carbon fiber panel for MIMOSA-26 chips: 1 – polyimide cooling pipe (D=1mm); 2 – carbon fleece (20 $\mu$m); 3 – graphite foil (30 $\mu$m); 4 – carbon fiber plate K13D2U (70 $\mu$m); 5 — thermal grease; 6— Si-sensor (60 $\mu$m).

The ALICE design of the extra-light support, cooling and precise positioning carbon fiber structures (Fig. 11) is used as a baseline. The proven ALICE technology is currently being improved for the upgrade of the ALICE Inner Tracking System [16]. The extra lightweight carbon fiber cooling panel of 30 cm length was designed and tested for support and efficient drain of heat from the MIMOSA-26 chips. It is equipped with precise positioning elements and allows high accuracy alignment and mounting of chips.

The layout of eight MIMOSA-26 chips, that are forming the first detector plane VDS1, are shown in Fig. 12 together with the readout flat polyimide cables (FPCs). In the current NA61/SHINE application, contrary to ALICE, the FPCs used for readout of the MIMOSA-26 chips are oriented horizontally (see Fig. 12). This design ensures the required minimisation of multiple scattering in the sensitive area.

The cross section layout of the thermoconductive carbon fiber cooling panel with embedded miniature cooling arteries is shown in Fig. 13. The panel ensures heat drain capability up to 0.5 W/cm$^2$ and will allow the operation of MIMOSA-26 chips at ambient temperature.

The first carbon fiber panels for detector support and cooling were successfully produced (see Fig.14) and are being used for mounting of detector chips. This will allow to instrument the limited acceptance VD and start with the first measuments of $D^o$ production. Subsequently the large acceptance vertex detector will be completed which is planned to be ready after the Long Shutdown-2 of the CERN accelerator chain in 2020.

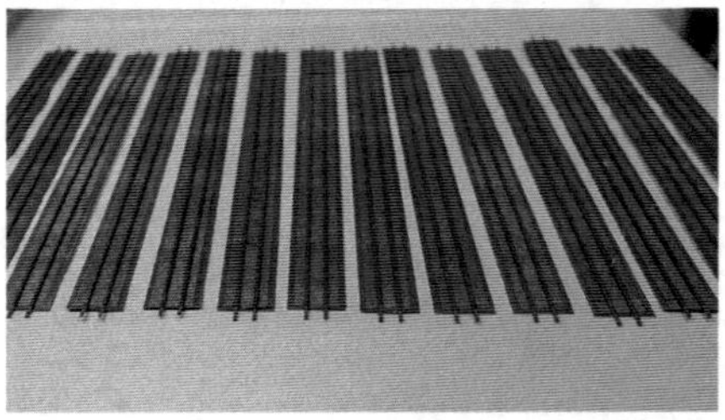

**Figure 14.** The first batch of extra-lightweight 30 cm long detector support and carbon fiber panels with the embedded cooling arteries produced for the NA61/SHINE VD.

## 4. Summary

Open charm measurements in A+A collisions at top SPS energy are possible with NA61/SHINE after construction of the Vertex Detector. The proposed design is based on proven technologies and includes:

- Sensors — MIMOSA-26 CMOS chips (IPHC-Strasbourg).
- Readout electronics boards – in synergy with CBM.
- Extra-light support, cooling and precise positioning – in synergy with ALICE.
- Mechanical layout of the VD by NA61/SHINE.

### *4.1. Acknowledgments*

This work is supported for G.F. by the Saint-Petersburg State University research grant 11.38.242.2015.

[1] N.Abgrall et al. (NA61 Collaboration), 2014, JINST 9, P06005.
[2] T. Matsui and H. Satz, 1986, Phys. Lett. **B 178**, 416.
[3] Helmut Satz, 2013, arXiv 1303.3493
[4] Berndt Müller, 2013, arXiv:1309.7616.
[5] Yasir Ali and Pawel Staszel for the NA61/SHINE collaboration, 2014, in Proceedings of 14th International Conference on Strangeness in Quark Matter (SQM2013), Journal of Physics: Conference Series **509** 012083.
[6] Yasir Ali, Pawel Staszel, 2014, EPJ Web of Conferences, **71**, 00004.
[7] Yasir Ali, Pawel Staszel, 2013, Acta Physica Polonica B Proceedings Supplement **6**, No 4, 1081.
[8] W. Cassing, E. L. Bratkovskaya, and A. Sibirtsev, 2001, Nucl. Phys. **A691**, 753.
[9] O. Linnyk et al., Int. J. Mod. Phys., 2008, E17, 1367; [arXiv:0808:1504 [nucl-th]].
[10] Zi-Wei Lin, Che Ming Ko, Bao-An Li, Bin Zhang, and Subrata Pal, 2005, Phys.Rev.C72, 064901.
[11] W. Cassing, E.L. Bratkovskaya, arXiv:0907.5331; The open Hadron String Dynamics source code and papers can be found at http://www.th.physik.uni- frankfurt.de/ brat/hsd.html.
[12] PYTHIA, 2001, Comput.Phys.Commun.**135**, 238.
[13] NA60 Collaboration, A. Foerster et al., 2006, J. Phys. bf G32, S51.
[14] NA60 Collaboration, R. Shahoian et al., 2005, Phys.J. **C43**, 209-213.
[15] MIMOSA26 User Manual, Institut Pluridisciplinaire Hubert Curien IN2P3-CNRS.
[16] The ALICE Collaboration, 2013, Technical Design Report for the Upgrade of the ALICE Inner Tracking System ALICE-TDR-017, CERN-LHCC-2013-024.

# Overview of ALICE Results

**Michele Floris on behalf of the ALICE collaboration**
CERN, Geneva, Switzerland
E-mail: michele.floris@cern.ch

**Abstract.** Recent results from the ALICE Collaboration are presented and discussed. Special emphasis is given to measurements relevant for strangeness and open and hidden heavy flavor physics.

## 1. Introduction

The main goal of ultra-relativistic heavy-ion collisions is the study of the deconfined phase of QCD matter, known as the Quark-Gluon Plasma (QGP). Current research aims at precisely determining properties of the QGP. In particular, a significant effort is being dedicated to the determination of its transport properties (such as shear and bulk viscosity), to the understanding of its microscopic structure (what are the relevant degrees of freedom?) and the mechanism of hadronization. Flavor plays a crucial role in this program, as extensively discussed during this conference.

The recent observations in pp and pA collisions of features traditionally associated with the formation of a deconfined phase, highlights the importance of studying the energy and system-size dependence of the relevant observables, rather than simply focusing on the largest system at the highest center of mass energy.

In this work, recent results from the ALICE collaboration are reviewed, focusing on the measurements presented at this conference, with particular emphasis on strangeness and heavy flavor measurements.

The ALICE apparatus is composed of a central barrel, covering $|\eta| < 0.8$ for full-length tracks and providing tracking and particle identification in the transverse momentum region $p_{\mathrm{T}} \gtrsim 0.1$ GeV/$c$. The first two layers of the central barrel are made of silicon pixel detectors (SPD) and can be used to reconstruct short track segments pointing to a common origin (tracklets). The central barrel is complemented by a forward muon arm, covering $-4.0 < \eta < -2.5$. A detailed description of the ALICE apparatus can be found in [1].

## 2. Flow and bulk properties

One of the main surprises from the first pA data taken at the LHC was the observation of long-range structures in two-particle correlation measurements (the so-called ridges) [2, 3, 4, 5, 6]. These structures, in fact, are typically associated with collective flow effects, and indeed hydrodynamic models could give a satisfactory description of the data (see e.g. [7]). At the same time, alternative explanations based in particular on saturation physics (e.g. color-glass condensate [8]) could qualitatively account for the effects seen in the data.

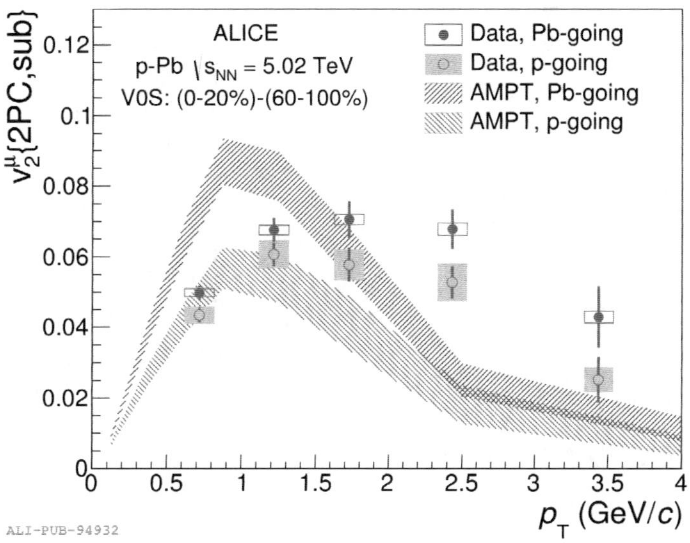


**Figure 1.** $v_2^{\mu}\{2\text{PC}, \text{sub}\}$ as a function of $p_T$ in the Pb-going and in the p-going directions.

These measurements have recently been extended significantly in $\Delta\eta$, with the measurement of correlations between muons observed at large rapidities in the muon arm and tracklets reconstructed at mid-rapidity in p–Pb collisions at $\sqrt{s_{NN}} = 5.02$ TeV [9, 10]. The LHC provided p–Pb collisions in two configurations, with Pb nuclei circulating either clockwise or counter-clockwise. This allowed the study of muons both in the Pb-going and p-going directions, with the correlation measurements covering the range $1.5 < |\Delta\eta| < 5$. Figure 1 shows the elliptic flow coefficient $v_2$, obtained subtracting the correlation function measured in low multiplicity collisions from the one in high multiplicity collisions, to reduce the contribution of jet fragments correlations ($v_2^{\mu}\{2\text{PC}, \text{sub}\}$) [10]. The $v_2^{\mu}\{2\text{PC}, \text{sub}\}$ is about 16% higher in the Pb-going direction, as one would qualitatively expect in case of hydrodynamic flow [11]. The measurement is also compared to a corresponding AMPT [12] calculation. The $v_2^{\mu}\{2\text{PC}, \text{sub}\}$ in the data is significantly higher than in in the simulation for $p_T > 2$ GeV/$c$. In this $p_T$ range, the production of muons is dominated by heavy flavor decays, which have zero $v_2$ in AMPT. This result could indicate a finite $v_2$ of heavy flavor particles or significantly different parent distributions in data and AMPT.

Other observables traditionally associated with radial flow are the ratios between baryon and mesons, such as p/$\pi$ and $\Lambda/K_s^0$, which show a strong enhancement (up to a factor $\sim 5$) in central Pb–Pb collisions. A similar enhancement was also observed in high multiplicity pA collisions [13, 14]. The ALICE collaboration has shown that, in the 0-5% most central Pb–Pb collisions, the main factor driving these ratios is the different mass of the particles: the ratio $\phi/p$ was indeed found to be flat out to $p_T \sim 5$ GeV/$c$ [15, 16]. In a new measurement presented at this conference [17], it was shown that the ratio $\Lambda/K_s^0$ is consistent with the one measured in pp collisions for particles associated with a reconstructed jet (Fig. 2). This poses a challenge for models which describe the enhancement via the recombination of semi-hard quarks from jet fragmentation with thermal partons from the medium.

## 3. Energy loss

The detailed study of high-$p_T$ partons energy loss and energy redistribution provides a tool to identify the microscopic nature of the QGP.

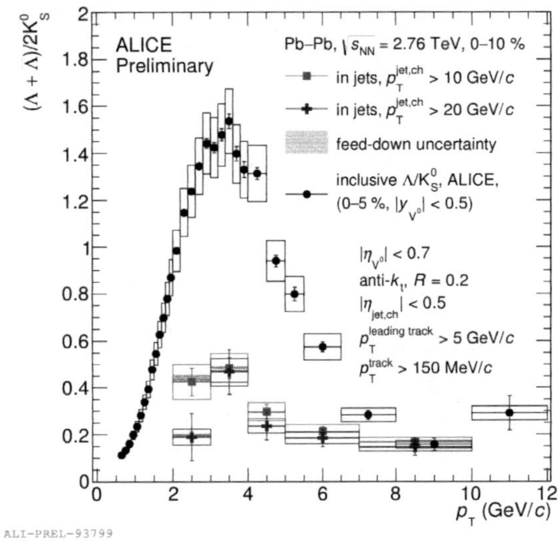


**Figure 2.** $\Lambda/K^0_S$ ratio inside jets.

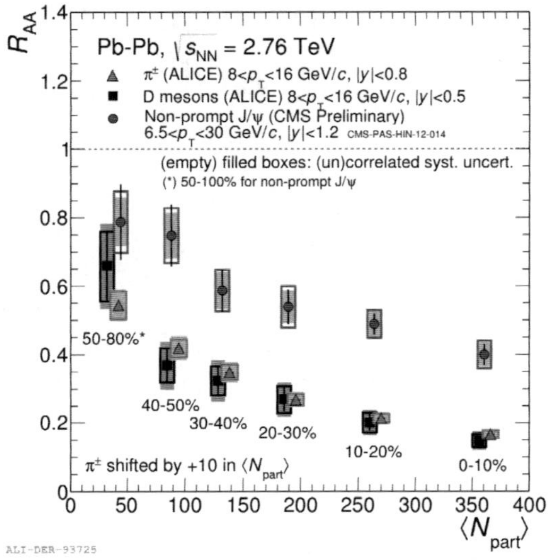


**Figure 3.** $R_{\mathrm{AA}}$ of D mesons, pions and non-prompt $J/\psi$.

The ALICE experiment has recently published results on a new observable $\Delta_{\mathrm{recoil}}$, built from hadron-jet correlations subtracting the contributions of random associations on a statistical basis, see [17, 18] for a detailed discussion. This allows the study of jet modifications out to larger radii (up to 0.5 in the resolution parameter $R = \sqrt{\Delta\eta^2 + \Delta\varphi^2}$) and down to lower jet momenta ($(p_{\mathrm{T}})^{\mathrm{ch}}_{\mathrm{T,jet}} = 20$ GeV) than previous measurements. This variable was also used to study also the angular distribution of the recoil jets. The results do not show any evidence for intra-jet broadening, no evidence of medium-induced acoplanarity and no signal for large-angle (Moliere) parton-medium scattering: they are consistent with a largely homogeneus medium.

The study of the particle-species dependence of the energy loss is a powerful tool to differentiate different energy loss models. Species-dependent effects could be caused by in-medium modifications of jet fragmentation and are expected to arise as a consequence of the mass dependence of energy loss. In general, due to color factors, gluons are expected to lose more energy than light quarks and light quarks, due to the "dead cone" effect, are expected to lose more energy than heavy quarks. Quarks and gluons, however, are not directly accessible experimentally and the simplest way to study these effects is the "nuclear modification factor" $R_{\mathrm{AA}}$ for different particle species.

The ALICE experiment has performed an exhaustive study of identified-particles $R_{\mathrm{AA}}$, both in the light and in the heavy quark sector. In the light flavor sector, no difference between different particle species has been observed above $p_{\mathrm{T}} > 10$ GeV/$c$, establishing the baseline for heavy flavor studies [19]. Below this $p_{\mathrm{T}}$ threshold, mass-dependent bulk effects become important.

The $R_{\mathrm{AA}}$ for D mesons is found to be in agreement with that of the pions in Fig. 3 [20, 21]. At the momenta depicted in the figure ($p_{\mathrm{T}} > 8$ GeV/$c$), the charm quark behaves essentially as a light quark, while pions are expected to originate mostly from gluons at LHC energies. Considering hadrons $R_{\mathrm{AA}}$ as a proxy for the corresponding parton (quark or gluon) $R_{\mathrm{AA}}$, one would naively expect pions to be more suppressed than D mesons. The agreement observed in Fig. 3 can however be understood as being due to an interplay of parton energy loss and the effect of fragmentation functions (see e.g. [22]).

The same figure also shows the CMS measurement of the $R_{\rm AA}$ of non-prompt $J/\psi$ [23], which originate from the decay of B mesons. The difference between non-prompt $J/\psi$ and D mesons is the first clear indication of mass-dependent energy loss. These measurements have also been recently extended, with the ALICE study of low $p_{\rm T}$ non prompt $J/\psi$ [24] and of the $\rm D_s^+$ meson [25]. The latter is of particular interest because of the interplay with strangeness enhancement and could be sensitive to recombination effects.

## 4. Study of possible nuclear modifications at High-$p_{\rm T}$ in pA

The observation of collective-like effects in pA collisions (suggestive of the formation of a thermalized medium) and the ample evidence for energy loss and high-$p_{\rm T}$ suppression in Pb–Pb collisions lead to the natural question of whether high-$p_{\rm T}$ energy loss can also be observed in small systems.

A number of early publications did not find any indication for such effects in minimum-bias p–Pb collisions. The $R_{\rm pPb}$ was found to be consistent with unity for charged hadrons [26, 27], jets [28], D mesons [29] and heavy flavor decay leptons [30]. Other correlations and jet measurements [31] also indicate result consistent with proton-proton collisions.

A natural extension is the study of high multiplicity p-Pb collisions, which is however complex due to the difficulty in defining a reliable measurement of "centrality". While centrality has a clear and non-ambiguous geometrical interpretation in Pb–Pb collisions, multiplicity dependent studies in small systems are sensitive to local multiplicity fluctuations and the relatively low multiplicity leads to large non-trivial biases for the centrality selection. The nuclear modification factor is called in this case $Q_{\rm pPb}$ to highlight that it is potentially affected by dynamical biases. These become stronger when the regions used for the measurements and to define the centrality classes are overlapping or close in phase space. They can be quantitatively understood in a simulation in which $n$ Pythia collisions are superimposed, based on the number of binary collisions sampled from a Glauber Monte Carlo [32, 33]. To minimize their effect, the ALICE collaboration employed a hybrid approach [32, 33], in which the multiplicity classes are selected using the neutron Zero Degree Calorimeter in the Pb-going direction (ZN, the ALICE detector farther away from mid-rapidity). The number of binary collisions is then calculated with several assumptions for scaling properties of particle production: *i)* assuming that $\mathrm{d}N_{\rm ch}/\mathrm{d}\eta$ at mid-rapidity scales with the number of participants, *ii)* assuming that the multiplicity in the Pb-going direction scales with the number of wounded target nucleons or *iii)* assuming that particle production at high $p_{\rm T}$ scales with the number of binary collisions. The three assumptions give consistent results. As an example, the $Q_{\rm pPb}$ of charged particles, computed with the first two assumptions is shown in Fig. 4. It is consistent with unity for $p_{\rm T} \gtrsim 10$ GeV/$c$.

## 5. Cold Nuclear Matter effects

The traditional motivation for the study of pA collision is to provide a reference for Pb–Pb measurements and assess the so-called "Cold Nuclear Matter" (CNM) effects, that is effects related to the presence of the nucleus, but not associated to the formation of a deconfined medium. Examples of CNM effects include, for instance, nuclear modifications of the parton distribution functions or the effect of hadronic co-movers.

While recent results indicate phenomena beyond CNM in pA collisions (Sec. 2), it remains of central importance for the interpretation of heavy-ion collisions results to understand CNM effects. Figure 5 shows the $R_{\rm pPb}$ of the $J/\psi$ and of the $\Psi(2S)$ as a function of rapidity, $y$ [34, 35]. The $R_{\rm pPb}$ of the $J/\psi$ is consistent with unity at backward rapidity (the Pb-going direction), and suppressed at mid- and forward rapidity (the p-going direction). This trend can be explained by models which include nuclear modifications of the parton distribution functions (shadowing), with a contribution from energy loss. This is studied in more details in Fig. 6, which shows

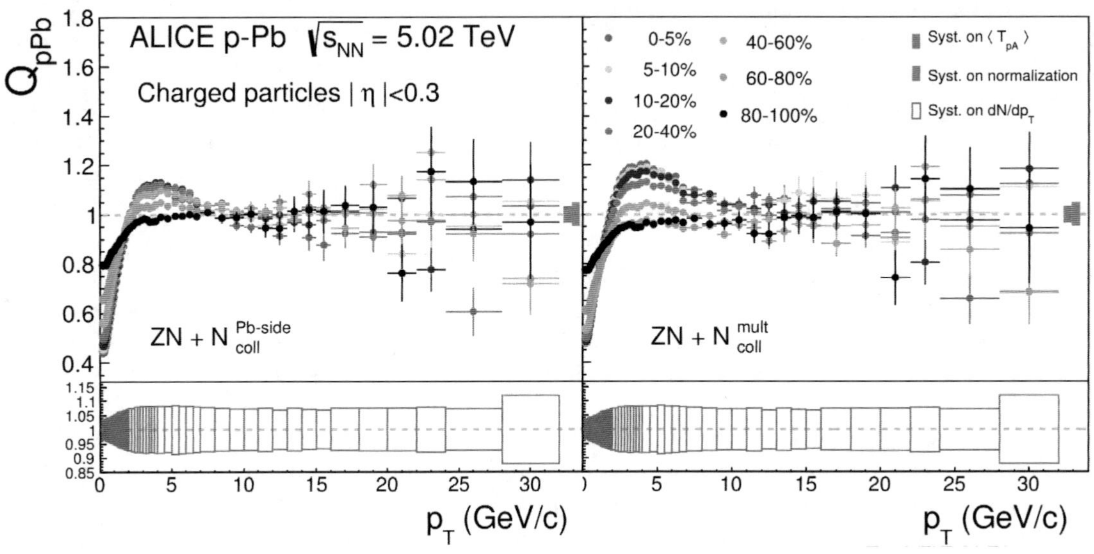


**Figure 4.** $Q_{\rm pPb}$ for charged particles, as a function of centrality, obtained in bins of ZN energy, with $N_{\rm coll}$ computed assuming a that the $dN_{\rm ch}/d\eta$ at mid-rapidity scales with the number of participant (left) or assuming that the multiplicity in the Pb-going direction scales with the number of wounded target nucleons (right).

the $Q_{\rm pPb}$ of the $J/\psi$ as a function of the number of binary collisions (obtained with the hybrid method described in Sec. 4) [36]. Shadowing is necessary to describe the multiplicity dependence of the data, especially at backward rapidity.

Figure 6 also depicts the $R_{\rm pPb}$ of the $\Psi(2S)$, which shows a different trend as compared to that of $J/\psi$, being similarly suppressed at backward and forward rapidities. This trend is somewhat surprising, as initial state effects should not differentiate between the two charmonium states. It can be understood in models which incorporate the effect of hadronic co-movers. While the effect of co-movers is expected to be stronger in the Pb-going direction, the effect of shadowing is stronger in the p-going direction. The two contributions seem to balance each other. At the same time, it should be noted that the co-movers model would predict a flatter-than-measured multiplicity dependence of the $J/\psi$ $Q_{\rm pPb}$ (Fig. 6).

An alternative way of studying shadowing effects and the initial-state nuclear wave function is through "ultra peripheral collisions." With this term, we refer to inelastic collisions where the impact parameter is larger than twice the nuclear radius. In these collisions, the scattering process is initiated by the photons produced in the strong electric field of the nuclei ("photo-production"). Models including moderate shadowing give a good description of the data on photo-production of vector mesons. Recent results are discussed in [37]. The ALICE collaboration has also recently reported an excess of $J/\psi$ production at low $p_{\rm T}$ in peripheral Pb–Pb events: this could be the first indication of $J/\psi$ photo-production in collisions with impact parameter smaller than twice the nuclear radius [38] (i.e. with a finite nuclear overlap).

## 6. Hadrons and the hadronic phase

The ALICE experiment recently reported new results on the enhancement of strange particles at the LHC. Figure 7 shows the ratios $\Xi/\pi$ (left) and $\Omega/\pi$ (right) measured in pp, p–Pb and Pb–Pb collisions [14]. The two ratios are seen to increase as a function of multiplicity in p–Pb reaching, for the highest measured multiplicities, values comparable to central (peripheral)

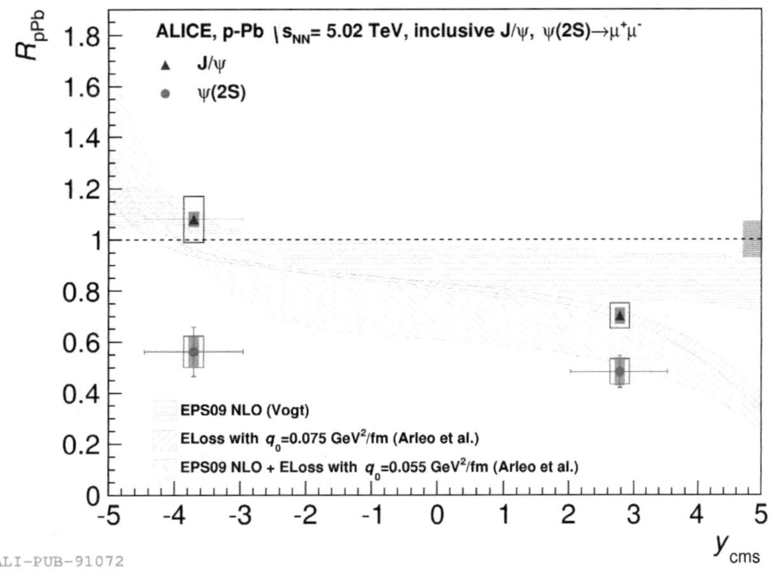


**Figure 5.** $R_{\mathrm{pPb}}$ of the $J/\psi$ and of the $\Psi(2S)$.

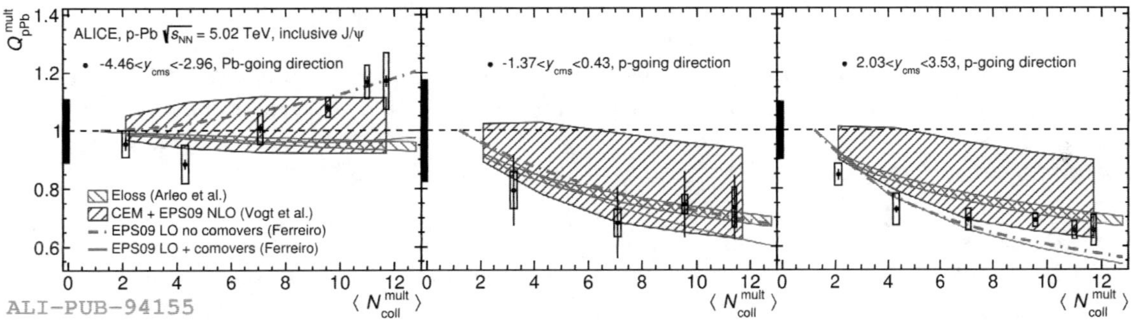


**Figure 6.** $Q_{\mathrm{pPb}}$of the $J/\psi$ as a function of the number of binary collisions.

Pb–Pb collisions in the case of the $\Xi$ ($\Omega$). This could be interpreted as a release of canonical suppression in p–Pb collisions, even if the quality of a grand canonical fit of high multiplicity p–Pb collisions is rather poor ($\chi^2/ndf \sim 5$) [14].

On the other hand, a grand canonical fit in central Pb–Pb collisions yields fair agreement with the data, although some tensions are still observed ($\chi^2/ndf \sim 2$) [39]. In particular the measured proton yield is lower than predicted by the model (protons would prefer a lower temperature), while strange baryons are in general higher (would point to a larger temperature). These tensions have been extensively debated in the literature, and several explanations have been proposed, including an incomplete hadron spectrum [40, 41], sequential freeze-out [42, 43, 44] and interactions in the hadronic phase [45] (for a review, see [46, 47]).

Other measurements reported at this conference will allow to better constrain the effect of the hadronic phase, including one-dimensional femptoscopic radii of identified particles [48] and resonance production as a function of centrality [15]. These result suggests that at least elastic scattering in the hadronic phase is playing a relevant role. The same hadronic-cascade model

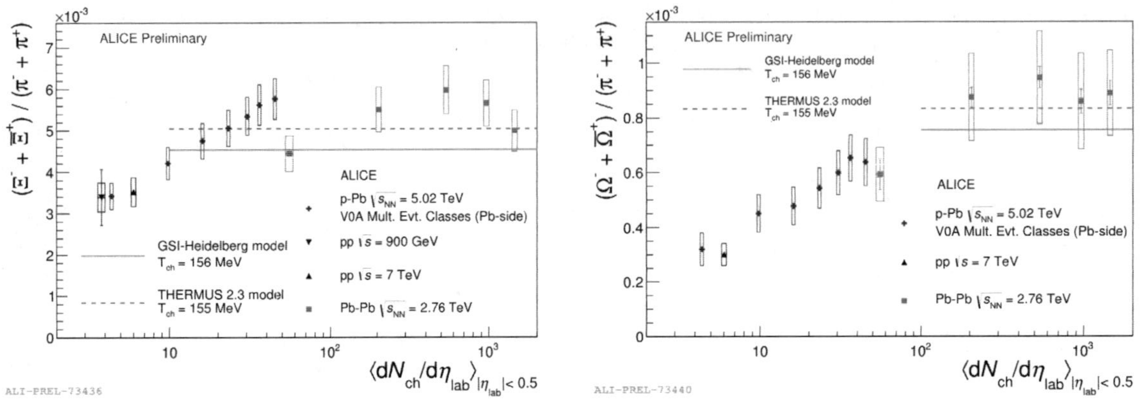


**Figure 7.** Ratios $\Xi/\pi$ (left) and $\Omega/\pi$ (right) as a function of multiplicity.

used to describe these data (UrQMD) predicts a significant suppression of baryons, because of baryon-antibaryon annihilation. However, this point is not yet established, for instance because of the lack of some inverse reactions in the model. Moreover, it is not yet clear how the presence of significant inelastic processes in the hadronic phase can be reconciled with the fact that loosely bound objects such as light (anti-)nuclei and hypertritons survive the hadronic phase, being produced with a rate consistent with the equilibrium thermal model calculations [49].

These studies of particle abundances also provide a baseline for exotica searches. In particular, the ALICE experiment searched for $\Lambda\Lambda$ and $\Lambda n$ bound states in Pb–Pb collisions at $\sqrt{s_{\rm NN}} = 2.76$ TeV, placing upper limits $\gtrsim$4 times smaller than the yield expected from the thermal models [49], when lifetimes consistent with theoretical expectations are assumed for the dibaryons, namely not exceeding $10^{-8}$ s.

## 7. Outlook

The ALICE collaboration is completing the campaign of analysis based on the data collected during the first run of the LHC. Additional results presented at this conference, but not covered in this summary include dilepton production [50, 51], W meson production [52] and D meson production as a function of multiplicity in pp and p–Pb collisions [53]. These new results are posing stringent constraints on the modelling of the hydrodynamic evolution of the system created in heavy-ion collisions, of its microscopic structure, of the initial state configuration and of the late stage of evolution. At the same time, the study of small systems has been extended with more differential measurements, in order to investigate the possible formation of a collective medium.

The LHC resumed the operations for the second data taking run recently, with the first pp collisions at $\sqrt{s} = 13$ TeV taking place at the beginning of June. A Pb–Pb run at $\sqrt{s_{\rm NN}} = 5$ TeV will follow, pushing the energy frontier of heavy-ion collisions. At the same time, the ALICE experiment is preparing for a significant upgrade for the high luminosity measurements, foreseen in run four [54, 55].

## References

[1] Aamodt K *et al.* (ALICE) 2008 *JINST* **3** S08002
[2] Chatrchyan S *et al.* (CMS) 2012 (*Preprint* 1210.5482)
[3] Khachatryan V *et al.* (CMS) 2010 *JHEP* **09** 091 (*Preprint* 1009.4122)
[4] Aad G *et al.* (ATLAS) 2013 *Phys.Rev.Lett.* **110** 182302 (*Preprint* 1212.5198)
[5] Abelev B B *et al.* (ALICE) 2013 *Phys.Lett.* **B719** 29–41 (*Preprint* 1212.2001)

[6] Abelev B B *et al.* (ALICE) 2013 *Phys.Lett.* **B726** 164–177 (*Preprint* 1307.3237)
[7] Bozek P, Broniowski W and Torrieri G 2013 *Phys.Rev.Lett.* **111** 172303 (*Preprint* 1307.5060)
[8] Dusling K and Venugopalan R 2014 *Nucl. Phys.* **A931** 283–287
[9] Ohlson A (for the ALICE Collaboration) 2015 *these proceedings*
[10] Adam J *et al.* (ALICE) 2015 (*Preprint* 1506.08032)
[11] Bozek P, Bzdak A and Ma G L 2015 *Phys. Lett.* **B748** 301–305 (*Preprint* 1503.03655)
[12] Lin Z W, Ko C M, Li B A, Zhang B and Pal S 2005 *Phys.Rev.* **C72** 064901 (*Preprint* nucl-th/0411110)
[13] Chinellato D (for the ALICE Collaboration) 2015 *these proceedings*
[14] Nicassio M (for the ALICE Collaboration) 2015 *these proceedings*
[15] Riabov V (for the ALICE Collaboration) 2015 *these proceedings*
[16] Abelev B B *et al.* (ALICE) 2015 *Phys. Rev.* **C91** 024609 (*Preprint* 1404.0495)
[17] Krizek F (for the ALICE Collaboration) 2015 *these proceedings*
[18] Adam J *et al.* (ALICE) 2015 (*Preprint* 1506.03984)
[19] Volpe G (for the ALICE Collaboration) 2015 *these proceedings*
[20] Bailhache R (for the ALICE Collaboration) 2015 *these proceedings*
[21] Festanti A (for the ALICE Collaboration) 2015 *these proceedings*
[22] Djordjevic M 2015 *these proceedings*
[23] CMS Collaboration (CMS) 2012 CMS-PAS-HIN-12-014
[24] Adam J *et al.* (ALICE) 2015 *JHEP* **07** 051 (*Preprint* 1504.07151)
[25] Barbano A (for the ALICE Collaboration) 2015 *these proceedings*
[26] Abelev B B *et al.* (ALICE) 2014 *Eur. Phys. J.* **C74** 3054 (*Preprint* 1405.2737)
[27] Abelev B *et al.* (ALICE) 2013 *Phys. Rev. Lett.* **110** 082302 (*Preprint* 1210.4520)
[28] Adam J *et al.* (ALICE) 2015 *Phys. Lett.* **B749** 68–81 (*Preprint* 1503.00681)
[29] Abelev B B *et al.* (ALICE) 2014 *Phys. Rev. Lett.* **113** 232301 (*Preprint* 1405.3452)
[30] Li S (ALICE) 2014 *Nucl. Phys.* **A931** 546–551 (*Preprint* 1408.1915)
[31] Adam J *et al.* (ALICE) 2015 *Phys. Lett.* **B746** 385–395 (*Preprint* 1503.03050)
[32] Toia A (for the ALICE Collaboration) 2015 *these proceedings*
[33] Adam J *et al.* (ALICE) 2015 *Phys. Rev.* **C91** 064905 (*Preprint* 1412.6828)
[34] Arsene I 2015 *these proceedings*
[35] Weber S (for the ALICE Collaboration) 2015 *these proceedings*
[36] Adam J *et al.* (ALICE) 2015 (*Preprint* 1506.08808)
[37] Villalobos Baillie O (for the ALICE Collaboration) 2015 *these proceedings*
[38] Adam J *et al.* (ALICE) 2015 (*Preprint* 1509.08802)
[39] Guerzoni B (for the ALICE Collaboration) 2015 *these proceedings*
[40] Stachel J, Andronic A, Braun-Munzinger P and Redlich K 2014 *J. Phys. Conf. Ser.* **509** 012019 (*Preprint* 1311.4662)
[41] Noronha-Hostler J and Greiner C 2014 (*Preprint* 1405.7298)
[42] Chatterjee S, Godbole R and Gupta S 2013 *Phys.Lett.* **B727** 554–557 (*Preprint* 1306.2006)
[43] Alba P, Alberico W, Bellwied R, Bluhm M, Mantovani Sarti V, Nahrgang M and Ratti C 2014 *Phys. Lett.* **B738** 305–310 (*Preprint* 1403.4903)
[44] Borsanyi S, Fodor Z, Katz S D, Krieg S, Ratti C and Szabo K K 2014 *Phys. Rev. Lett.* **113** 052301 (*Preprint* 1403.4576)
[45] Becattini F, Grossi E, Bleicher M, Steinheimer J and Stock R 2014 *Phys. Rev.* **C90** 054907 (*Preprint* 1405.0710)
[46] Floris M 2014 *Nucl. Phys.* **A931** 103–112 (*Preprint* 1408.6403)
[47] Castorina P 2015 *these proceedings*
[48] Mikhaylov K (for the ALICE Collaboration) 2015 *these proceedings*
[49] Barile F (for the ALICE Collaboration) 2015 *these proceedings*
[50] Uras A (for the ALICE Collaboration) 2015 *these proceedings*
[51] Özdemir M (for the ALICE Collaboration) 2015 *these proceedings*
[52] Bossù F (for the ALICE Collaboration) 2015 *these proceedings*
[53] Terrevoli C (for the ALICE Collaboration) 2015 *these proceedings*
[54] Fionda F (for the ALICE Collaboration) 2015 *these proceedings*
[55] Dainese A (for the ALICE Collaboration) 2015 *these proceedings*

# Strangeness Prospects with the CBM Experiment

**Volker Friese for the CBM Collaboration**

Helmholtz-Gesellschaft für Schwerionenforschung mbH, Planckstraße 1, 64291 Darmstadt, Germany

E-mail: v.friese@gsi.de

**Abstract.** The CBM experiment will study strongly interacting matter at high net-baryon densities with nuclear collisions up to 45$A$ GeV beam energy at the future FAIR facility. With interaction rates unprecedented in heavy-ion collisions, CBM will give access also to extremely rare probes and thus to the early stage of the collisions, in search for the first-order phase transition from confined to deconfined matter and the QCD critical point. The CBM physics programme will be started with beams delivered by the SIS-100 synchrotron, providing energies from 2 to 11 GeV/nucleon for heavy nuclei, up to 14 GeV/nucleon for light nuclei, and 30 GeV for protons. The highest net baryon densities will be explored with ion beams up to 45 GeV/nucleon energy delivered by SIS-300 in a later stage of the FAIR project.

After several years of preparation, the CBM experiment now enters the realisation phase. In this article, we report on the current status of the system developments and the expected physics performance for strange and charmed observables, as well as on the roadmap towards the first data taking.

## 1. Introduction

In parallel to the investigation of strongly interacting matter at the highest energy densities achievable in the terrestrial laboratory as currently performed with the heavy-ion programme at the CERN-LHC, an increasing number of experimental projects address heavy-ion collisions at moderate energies, corresponding roughly to the beam energy range of BNL-AGS and CERN-SPS. Such collisions are expected to create matter at extreme net-baryon density but moderate temperature, as opposed to collisions at the LHC, where the temperature is highest, but the net-baryon density is almost vanishing. According to model predictions, the maximum net-baryon density at freeze-out is expected from nuclear collisions at about 30 GeV per nucleon [1], corresponding to a centre-of-momentum energy of $\sqrt{s_{NN}} \approx 8$ GeV.

Heavy-ion reactions in this energy range allow us to explore a region of the phase diagram of strongly interacting matter which has recently attracted renewed interest since it may bear a rich structure. Potential landmarks are the first-order phase transition from confined to deconfined matter, the restoration of chiral symmetry at high baryon densities, the critical point of QCD separating the region of first-order phase transition from the cross-over transition, and the transition to Quarkyonic matter. The physics of such "compressed baryonic matter" was repeatedly discussed in this conference series and elsewhere (see e.g., [2]); an extensive survey can be found in [3].

## 2. The CBM Experiment

The CBM ("Compressed Baryonic Matter") experiment will conduct a systematic research programme to study strongly interacting matter with external ion beams delivered from the FAIR accelerator complex currently under construction in Darmstadt. It will be operated in fixed-target mode with beam energies from $2A$ to $45A$ GeV and with a variety of projectile species ranging from protons up to heavy nuclei. CBM aims at a close-to-complete characterisation of the particles emitted from the final state of the collisions by measuring hadrons, electrons and muons in a large part of the forward rapidity hemisphere.

A key feature of the CBM design considerations is the capability to operate at very high interaction rates of up to $10^7$ collisions per second – several orders of magnitude higher than any other existing or currently planned heavy-ion experiment. This high-rate capability will give CBM a unique discovery potential for extremely rare probes like multi-strange anti-hyperons, exotic hadronic states, or charm production near the kinematic threshold. It obviously puts severe requirements on detector design in terms of radiation hardness and read-out speed, as well as on data acquisition and online data processing.

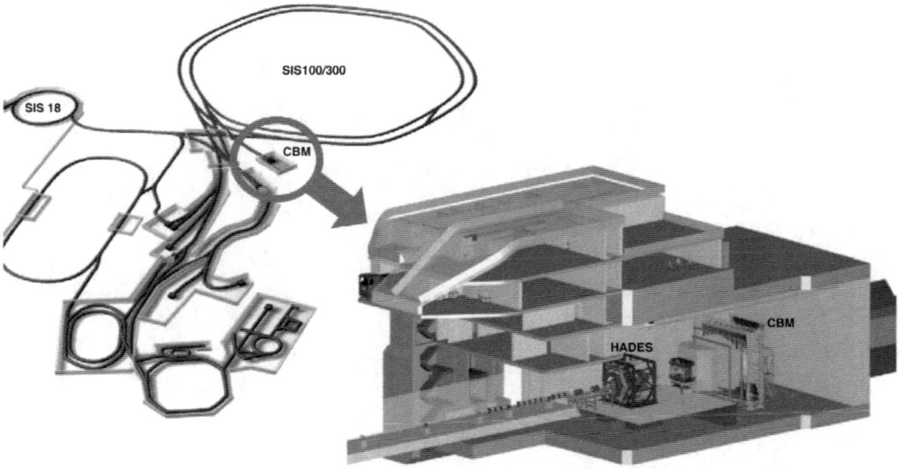


**Figure 1.** Left: The FAIR accelerator complex and the location of the CBM cave. Beams are extracted from the two synchrotron rings at the top. Right: layout of the CBM building and its underground hall.

CBM will be set up in an underground hall of the dimensions 37 m x 27 m x 17 m to which beams from the SIS-100 and SIS-300 synchrotrons will be extracted (Fig. 1). The experimental set-up is versatile in order to cope with the specific requirements of hadron, electron or muon measurements as well as with the different ranges of beam energy delivered by SIS-100 and SIS-300, respectively. Figure 2 shows a configuration for hadron and electron measurements, comprising the dipole magnet and the main tracker (STS), the RICH and TRD detectors for electron identification, the TOF detector for hadron identification, and the PSD calorimeter for event characterisation. The muon detection system, being installed on rails, will be interchangeable with the RICH detector.

The FAIR facility will be realised in different stages. The so-called Modular Start Version comprises the SIS-100 synchrotron with Au beams up to $11A$ GeV and proton beams up to 30 GeV, as well as the CBM experimental hall and the extraction beam line. In a later stage, the SIS-300 accelerator will enlarge the energy range to $35A$ GeV for Au, $45A$ GeV for symmetric nuclei, and 90 GeV for protons. The time line for this extension is not yet fixed. Thus, the first

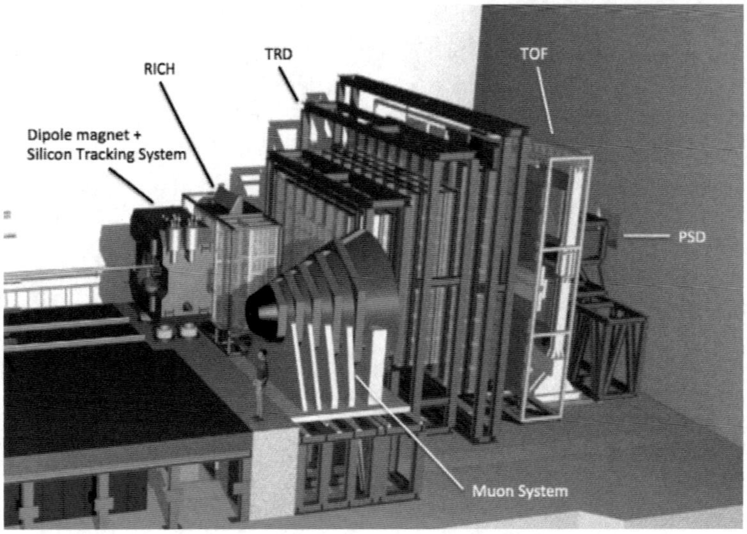


**Figure 2.** CBM experimental set-up for hadron and electron measurements. The beam enters the cave from the left side. The Silicon Tracking System is placed in the dipole magnet. Downstream are the RICH and TRD detectors, the TOF detector and the PSD calorimeter. The muon system is in its parking position; it can be moved into the acceptance on rails when the RICH detector is removed.

years of CBM operations will be restricted to beams from SIS-100, with which already a rich and unique research programme can be pursued. Some of the detectors systems (TRD, muon detector) come with a reduced start version for SIS-100, being optimised for the conditions in this beam energy range; others (STS, RICH, TOF, PSD) can be operated with beams from both accelerators.

## 3. Instrumentation

In the following, we give a brief overview for the detector systems relevant for strangeness and charm measurements. A recent review of all sub-systems can be found as shown in [4].

### *3.1. Silicon Tracking System*

The core of the CBM setup is the silicon tracking system (STS) intended for the reconstruction of the trajectories of charged particles produced in the interaction and the measurement of their momenta. It is installed between the yokes of a superconducting dipole magnet [5], providing a field integral of 1 Tm and an aperture of 140 cm in vertical and 280 cm in horizontal direction.

The STS is realised as an array of eight tracking stations constructed from thin double-sided silicon micro-strip detectors. The roughly 1,200 sensors will be mounted on a light-weight carbon structure, the signals from the inner part being routed to the front-end electronics at the periphery of the station by ultra-thin cables. The aim is to keep the material budget below 1% radiation length per station. Figure 3 shows the conceptual design of the detector system and the current engineering model [6]. The development of the detector is advanced; production readiness is targeted for 2016.

Algorithms for track reconstruction is this system are already developed. They show a track finding efficiency of above 95% for tracks above 1 GeV momentum and a momentum resolution below 2% for the same track class.

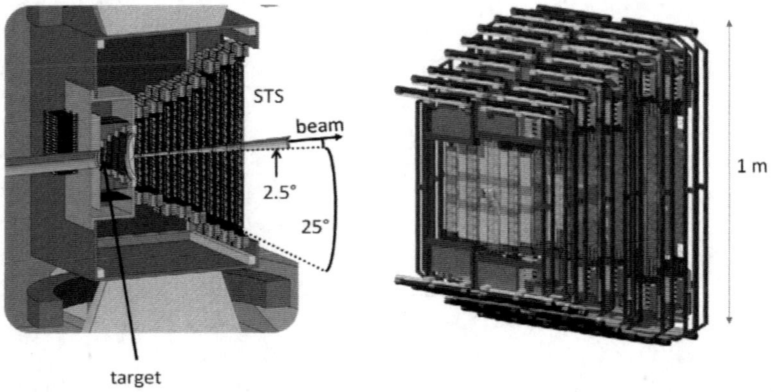


**Figure 3.** Left: conceptual design of the Silicon Tracking System, showing a cut through the detector in the dipole magnet. Right: engineering model of the STS, with the eight tracking stations to be mounted on the main support frame.

*3.2. Micro-Vertex Detector*

The MVD adds very high spatial resolution to the tracking in order to detect the decay products of short-lived particles like D mesons. The system will make use of Monolithic Active Pixel Sensors (MAPS), which combine excellent position resolution ($< 3\,\mu$m) with low material budget. However, the read-out speed of the current prototype MIMOSA-26 limits the interaction rate to about $10^5$/s in runs where the MVD is operated. The sensors will be arranged on four stations in the target vacuum chamber [7]. The system under development arranges the sensors on both sides of a diamond carrier equipped with flex-print circuits. The material budget of such an assembly is below $0.5\%\,X_0$.

*3.3. Time-of-Flight Detector*

The identification of hadrons in CBM is achieved through the measurements of their time-of-flight. The TOF detector [8] is based on the technology of resistive plate chambers, providing a time resolution of about 60 ps. RPC modules will be assembled in a wall with an active area of about 120 m$^2$, the granularity being adapted to cope with the variation of hit rates, which range from 2 kHz/cm$^2$ at the periphery to 25 kHz/cm$^2$ in the inner sections. Prototypes were tested in various in-beam experiments, and production readiness will be achieved in 2016.

*3.4. Projectile Spectator Detector*

The Projectile Spectator (PSD) serves the determination of collision centrality and orientation of the event plane by the detection of the projectile spectator nucleons and fragments [9]. It is a fully compensating, modular lead-scintillator calorimeter with high and uniform energy resolution. The detector will be built from 44 individual modules, with 60 lead/scintillator layers, facing to the beam a surface of $20\,\text{cm} \times 20\,\text{cm}$. The read-out is realized by multi-anode photo detectors via wavelength shifting fibers. A similar device using the same technology is currently being installed at the NA61 experiment at the CERN-SPS.

*3.5. Data acquisition and online data processing*

At the maximum design interaction rate ($10^7$ events per second), the expected raw data flow from the front-end electronics (FEE) is about 1 TB/s, which has to be reduced by more than two orders of magnitude before archiving by online data selection. The high rates and complicated trigger patterns (e.g., open charm decays) forbid the use of conventional, latency-limited triggers; instead, the FEE will be self-triggered, i.e., autonomously streaming hit information with time stamps through the data acquisition chain (DAQ) to a first-level event selector, a dedicated computer farm where the association of hits to physical events, partial event reconstruction and event selection is performed. Several free-running read-out ASICs as well as DAQ components are under development, prototypes of which were already successfully operated in beam tests.

The online selection of signal events requires efficient and fast reconstruction algorithms [10]. The required speed is realised by applying massive parallelization to reconstruction algorithms and analysis tasks [11]. With the current software performance, we estimate the required size of the online computing farm to be of the order of 60,000 - 100,000 cores. This cluster will be hosted by the FAIR Tier-0 data centre currently under construction on the GSI campus. Prototypes of the computing architecture are being operated in Frankfurt (LOEWE-CSC), at GSI and at the FAIR-Russia Research Center.

## 4. Physics Reach

The physics capabilities of the CBM experiment are assessed by detailed simulations down to the raw data level, employing realistic detector geometries and response models which are constantly adapted to the developments in detector design and to the results of prototype tests in the laboratory and in beam. The performance figures shown in the following are the result of the full reconstruction software chain being applied to these simulated data.

Particles containing strange quarks continue to be important probes for hot and dense matter. Of particular interest for CBM are multi-strange anti-hyperons, since their production is strongly suppressed in a dense baryonic medium - a constraint which would be relaxed should deconfined matter be formed. This is demonstrated by calculations using the hadronic model HSD and its successor pHSD which includes both hadronic and partonic degrees of freedom. Figure 4 shows the predicted multiplicities for $\Omega^-$ and $\overline{\Omega}^+$ as function of beam energy for both models. While only small differences are seen for $\Omega^-$, its anti-particle is enhanced by more than an order of magnitude in the CBM energy range if the creation of a QGP in the centre of the fireball is allowed. Still, the predicted multiplicities are extremely low; CBM with its high-rate capability will be the only experiment in this energy regime able to measure such particles.

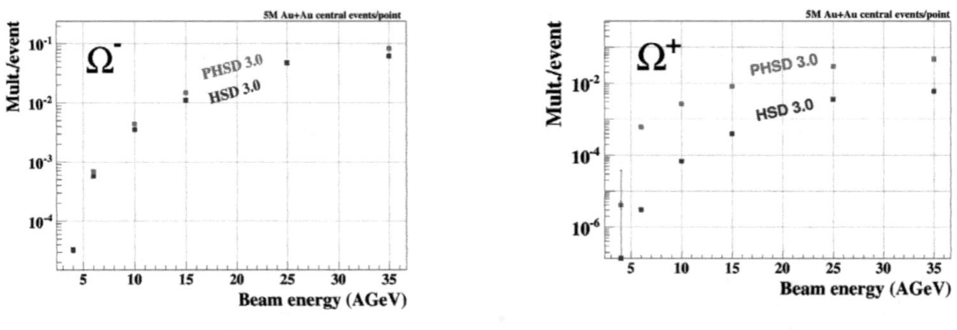


**Figure 4.** Multiplicity of (left) $\Omega^-$ and (right) $\overline{\Omega}^+$ in central Au+Au collisions as function of beam energy as predicted from the HSD and the pHSD model.

CBM will reconstruct anti-hyperons by the decay topology in the STS, using TOF identification of the daughter particles to further reduce the combinatorial background. The good position resolution of the STS detector results in an excellent signal-to-background ratio

Journal of Physics: Conference Series **668** (2016) 012014 doi:10.1088/1742-6596/668/1/012014

as demonstrated in Fig. 5. At 10$A$ GeV and an interaction rate of $10^6$/s, we expect to reconstruct about $10^5$ $\overline{\Omega}^+$ per week of runtime in the 10% most central collisions, assuming the (lower) HSD multiplicity.

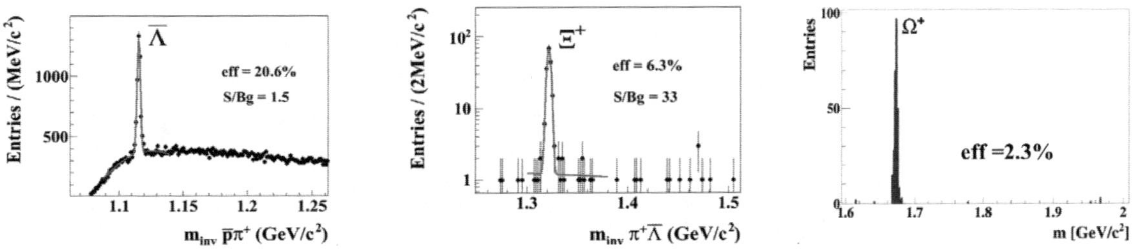


**Figure 5.** Reconstructed invariant-mass spectra for (left) $\overline{\Lambda}$, (centre) $\overline{\Xi}^+$ and (right) $\overline{\Omega}^+$ in central Au+Au collisions at 10$A$ GeV. The input to the simulations was generated with UrQMD for $\overline{\Lambda}$ and $\overline{\Xi}^+$ and with pHSD for $\overline{\Omega}^+$.

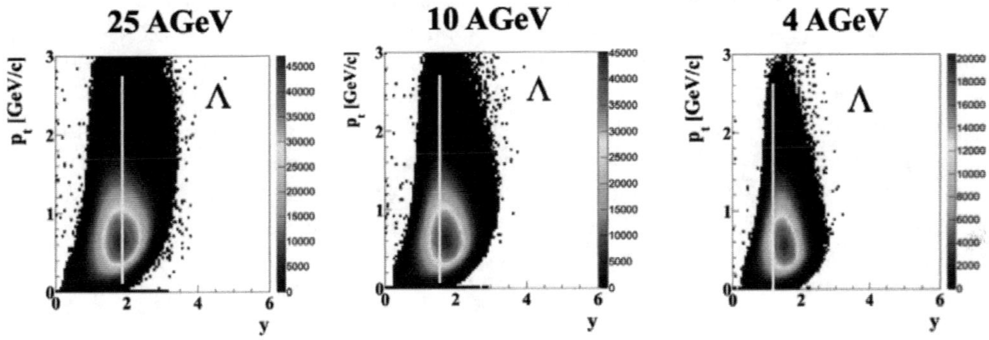


**Figure 6.** Phase-space coverage of CBM for $\Lambda$ at three different energies in terms of rapidity and transverse momentum. Mid-rapidity is indicated by the vertical lines.

The phase-space coverage of the CBM setup for the reconstruction of hyperons is demonstrated in Fig. 6 for the $\Lambda$ baryon at three different beam energies. At all of them, the acceptance covers the bulk production near mid-rapidity as well as a large part of the forward rapidity hemisphere. The acceptance figures for $\Xi$ and $\Omega$ look similar.

The energy range of CBM is also particularly well suited for the study of hyper-nuclei and the search for exotic, meta-stable strange objects like di-baryons. In heavy-ion collisions, hyper-nuclei can be formed by the coalescence of $\Lambda$ with nucleons or light nuclei in the final state of the reaction. Model calculations suggest that the maximum yield of hyper-nuclei is found in the SIS-100 energy domain [12]. The multiplicity of the $(\Xi^0\Lambda)_b$ bound state was calculated within a hybrid (microscopic transport + hydrodynamics) model to be about $10^{-2}$ in central collisions at 30$A$ GeV [13] – well within reach for a high-rate experiment like CBM.

The detection principle - the weak-decay topology - is similar to that for hyperons. As examples for our performance studies, Fig. 7 shows the reconstructed invariant-mass spectra for $He^3_\Lambda$ and for $(\Xi^0\Lambda)_b$ in central Au+Au collisions at 10$A$ GeV. We again expect very good signal-to-background ratios and conclude that the CBM detector is capable to detect even exotic multi-strange objects, should they exist.

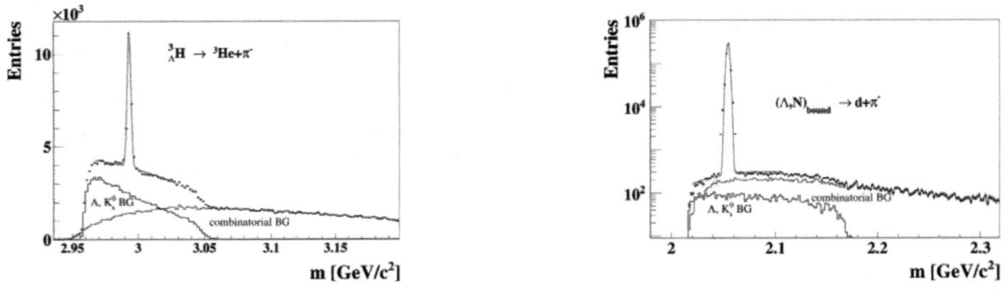


**Figure 7.** Reconstructed invariant-mass spectra for (left) $He^3_\Lambda$ and (right) $(\Xi^0\Lambda)_b$ in central Au+Au collisions at 10$A$ GeV. The blue line shows the correlated, the green line the uncorrelated background.

The study of charm production in heavy-ion collisions at moderate beam energies is both particularly interesting and challenging. Because of the very low charm cross section at these energies, CBM is again the only experiment which is sensitive to probes containing charm quarks. A systematic measurement of both open and hidden charm in nuclear collisions will only be possible with the SIS-300 accelerator. Already at SIS-100, however, precursor measurements are within reach; symmetric nuclei (e.g., Ni) can be accelerated up to 14$A$ GeV, i.e., above the kinematic threshold for both D mesons and charmonium. At the top SIS-100 energy for Au+Au, a sub-threshold measurement of J/$\psi$ might be feasible. Furthermore, the production and propagation of charm in cold nuclear matter can be investigated with p+A collisions up to 30 GeV. As an example of a feasibility study in this regards, Fig. 8 shows the sensitivity of CBM to charged D mesons in p+C collisions at 30 GeV. The crucial device here is the Micro-Vertex Detector, delivering a resolution for secondary vertices of about 60 $\mu$m along the beam, which is well sufficient to detect the open charm decay above the combinatorial background.

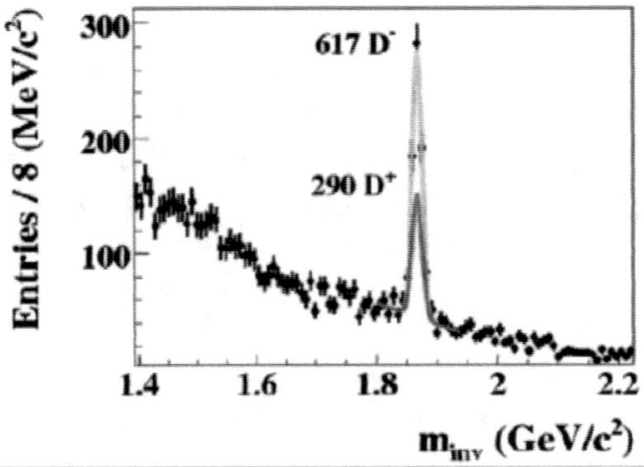


**Figure 8.** Reconstructed invariant-mass spectra for $D^+$ and $D^-$ in p+C collisions at 30 GeV. The multiplicity of the D mesons was taken from the HSD model.

## 5. Status of preparations

The development of detector and technical systems for CBM is in an advanced stage [14] and will be largely completed by the end of 2016. For most of the systems, Technical Design Reports are already approved by FAIR or are under evaluation. Serial component production and the construction of the systems are planned to be completed by the end of 2019. Installation in the CBM cave and commissioning without beam will then follow in 2020.

Compared to the original planning, the current FAIR timeline is delayed by 2 – 3 years. First beams from SIS-100 into the CBM cave are not expected before 2022. In view of this delay, the CBM collaboration is investigating possibilities to install and commission parts of detector systems at other experimental facilities. Examples are the installation of a part of the TOF wall at the STAR experiment at RHIC or the co-operation of CBM with BM@N at NICA in the construction of the silicon detectors. Furthermore, CBM intends to install a reduced set-up at a beam line at the SIS-18 accelerator at GSI in order to test the interplay of systems and the full data chain including data processing in real-time. These activities have the objective to gain operational experience in order to minimize the start-up phase once beams from SIS-100 will become available. The first measurements of CBM will then comprise systematic and multi-differential measurements of multi-strange (anti-)hyperons and hyper-nuclei, a search for exotic strange objects and (not covered in this article) the measurement of di-lepton spectra in the energy range up to 11$A$ GeV.

## References

[1] Randrup J and Cleymans J 2006 *Phys. Rev. C* **74** 047901
[2] Friese V 2015 *EPJ Web Conf.* **95** 03010
[3] Friman B, Höhne C, Knoll J, Leupold S, Randrup J, Rapp R and Senger P (eds.) 2011 The CBM Physics Book - Compressed Baryonic Matter in Laboratory Experiments, *Lect. Notes Phys.* **814**
[4] Balog T 2014 *J. Phys. Conf. Ser.* **503** 012019
[5] The CBM Collaboration 2013 *Technical Design Report for the CBM Superconducting Dipole Magnet* (GSI-2015-02000, `https://repository.gsi.de/record/109025`)
[6] The CBM Collaboration 2013 *Technical Design Report for the CBM Silicon Tracking System* (GSI-2013-05499, `https://repository.gsi.de/record/54798`)
[7] Deveaux M and Heuser J 2013 *PoS(Vertex2013)* 009
[8] The CBM Collaboration 2014 *Technical Design Report for the CBM Time-of-Flight System* (GSI-2015-01999, `https://repository.gsi.de/record/109024`)
[9] The CBM Collaboration 2015 *Technical Design Report for the CBM Projectile Spectator Detector* (GSI-2015-02020, `https://repository.gsi.de/record/109059`)
[10] Friese V 2015 *Lect. Notes Comp. Sc.* **7125** 17
[11] Kisel I 2015 *EPJ Web Conf.* **95** 01007
[12] Andronic A *et al.* 2011 *Phys. Lett. B* **697** 203
[13] Stöcker H *et al.* 2009 *Nucl. Phys. A* **827** 624c
[14] Friese V and Sturm C (eds.) 2015 *CBM Progress Report 2014* (GSI-2015-01521, `https://repository.gsi.de/record/97909`)

Journal of Physics: Conference Series **668** (2016) 012015 doi:10.1088/1742-6596/668/1/012015

# Prospects for the dense baryonic matter research at NICA

**V. Golovatyuk, M. Kapishin, V. Kekelidze, V. Kolesnikov, O. Rogachevky and A. Sorin**

Russia, 141980 Moscow region, Dubna, JINR

E-mail: kekelidze@jinr.ru

**Abstract.** The project NICA (Nuclotron-based Ion Collider fAcility) is aimed to study hot and dense baryonic matter in heavy ion collisions in the energy range up to $\sqrt{s_{NN}}$=11 GeV at the Nuclotron extracted beams and at the NICA collider with average luminosity of $L = 10^{27}$ cm$^{-2}$s$^{-1}$ (for $^{197}$Au$^{79}$). This study will be performed with two experiments, BM@N (Baryonic Matter at Nuclotron) and MPD (MultiPurpose Detector) at the NICA collider.

## 1. Introduction

The main goal of the NICA physics program is the study of hot and dense nuclear matter in the energy region of the maximum baryonic density [1–3]. NICA is a new accelerator complex being constructed in JINR (Dubna, Russia) to provide beams of ions over a wide range of atomic masses (from $p$ to Au) at the average luminosity of $L = 10^{27}$ cm$^{-2}$s$^{-1}$ and $10^{32}$ cm$^{-2}$s$^{-1}$ for gold-gold and proton-proton collisions, respectively. The center-of-mass energy will be $\sqrt{s}_{NN} = 4-11$ GeV for Au+Au and up to 27 GeV for $p + p$. This energy range will be extended to cover the lower region providing the experiment with the beam extracted from the Nuclotron.

In this paper, we overview the main physics objectives of the NICA program and current status of the project realization.

## 2. Physics objectives of the NICA program

Lattice QCD calculations predict both the deconfinement phase transition and chiral symmetry restoration to occur at high enough energy densities, and there is strong experimental evidence that the deconfined phase of nuclear matter (Quark-Gluon Plasma - QGP) can be created in ultra-high-energy nuclear collisions [4, 5]. As the collision energy is decreased down to several GeV per nucleon-nucleon pair, there is a certain transition region of collision energy below which it is no longer possible to access the plasma phase in the course of the collision. Existing data on single-particle spectra and mean multiplicities [6] suggest that this transition occurs within the NICA energy range ($4 < \sqrt{s} < 11$ GeV). Moreover, the energy range of NICA is sufficiently large to encompass both collisions in which the plasma phase is well developed and collisions in which the matter remains purely hadronic throughout. However, the theoretical understanding of this transition, and how it manifests itself in nuclear collision observables, is still rather poor, and quantitative predictions cannot yet be made with confidence. In particular, lattice gauge calculations can presently not be employed at the finite baryon densities, which will be studied by NICA. Furthermore, in addition to determining the existence and location of the transition region, where the new plasma phase is first being created, it is of fundamental interest to establish the character of the associated phase transformation, whether it remains a smooth crossover or becomes a first-order transition, as many models predict. In the latter case, the phase diagram of strongly interacting matter must contain a critical point, and its experimental identification forms a focal point for this research field. Therefore, it has been suggested [7] that the first round of NICA experiments should concentrate on a variety of diagnostic observables that have already been employed in experimental programs at RHIC and the SPS. The experimental setup at the NICA collider has to be optimized for the study of fluctuations and correlations of bulk event properties with a primary goal to measure the excitation functions and the dependence of fluctuations and correlations on centrality and system size.

Theory predicts that the deconfinement phase transition can be accomplished by partial restoration of the chiral symmetry in heavy-ion collisions [8,9] leading to possible modifications of hadronic spectral functions in dense hadronic matter. The correlated $e^+e^-$ pairs (*dileptons*) from decays of vector mesons $(\rho, \omega, \phi)$ are the best candidates to study such in-medium modifications, since they escape the interaction region unaffected by subsequent strong interactions. Until now, no dilepton measurements have been performed at center-of-mass energies of several GeV and the dilepton program at NICA is aimed to close this gap. The range of collision energies from $\sqrt{s}$=4 to 11A GeV is very promising for dilepton studies, since the effect of modifications is expected to be sensitive to the baryon density, while the latter happens to reach the maximum in central Au+Au collisions at NICA [10].

Study of strangeness, in particular, hyperon production is of interest due to several reasons. First, theory predicts that the strangeness enhancement in heavy-ion induced interactions (relative to elementary p+p collisions) might be a signature for the deconfinement phase transition [11]. Secondly, since the hadronic cross-sections of multi-strange hyperons are small, additional re-scattering effects of hyperons in the dense nuclear matter are not as strong as for other hadrons. Thus, measured phase-space distributions of strange hyperons reveal important characteristics of the fireball at the early stages of the system evolution. Moreover, under such conditions nuclear objects with strangeness, hypernuclei [12], can be formed, and, since the energy range of the NICA covers the region of the maximal baryon density, the production rates of nuclear clusters with strangeness are predicted to be enhanced considerably [13]. The outcome of the NICA program, in particular, new experimental data on (anti)hyperon and hypernuclei production, which will be taken with the MPD detector, will provide valuable insight into the reaction dynamics and properties of the QCD matter.

The studies in the collider mode will be complemented by a fixed target experiment at the Nuclotron at energies up to 6 GeV per nucleon. The Baryonic Matter at Nuclotron (BM@N) research program on heavy-ion collisions includes [20] the following topics: investigation of the reaction dynamics and nuclear EOS, study of the in-medium properties of hadrons, production of (multi)-strange hyperons at the threshold and search for hypernuclei. Particle yields, ratios, transverse momentum spectra, rapidity and angular distributions, as well as fluctuations and correlations, will be studied at BM@N as a function of the collision energy and centrality.

## 3. MPD detector at NICA

The Multi Purpose Detector (MPD) is designed as a $4\pi$ spectrometer capable of detecting charged hadrons, electrons and photons in heavy-ion collisions at high luminosity in the energy range of the NICA collider [14]. The experimental setup (see Fig. 1) will comprise a precise 3D tracking system and a high-performance particle identification (PID) system based on the time-of-flight measurements and calorimetry. At the design NICA luminosity, the event rate of minimum bias interactions is of about 7 kHz, and the total charge particle multiplicity exceeds 1000 in the most central Au+Au collisions at $\sqrt{s_{NN}}$=11 GeV. As the average transverse

Journal of Physics: Conference Series **668** (2016) 012015 doi:10.1088/1742-6596/668/1/012015

momentum of the particles produced in a collision at NICA energies is below 500 MeV/c, the detector design requires a very low material budget. The whole detector setup includes Central Detector (CD) covering pseudorapidity interval $|\eta|<2$ and two forward spectrometers (FS-A,B) for $2 < |\eta| < 3$ (optional). The cross-sectional view of the MPD Central Detector (CD) is shown

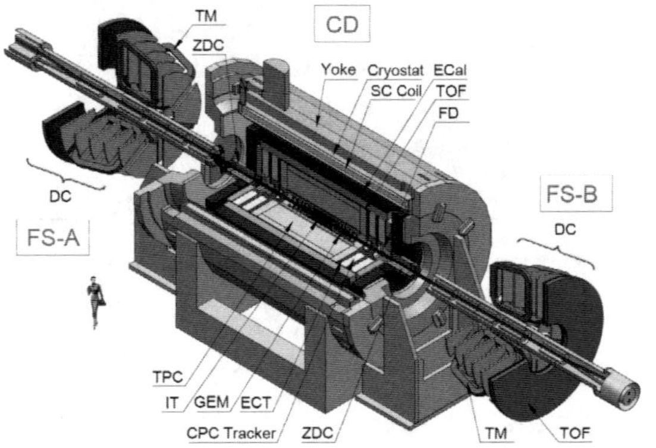


**Figure 1.** MPD setup.

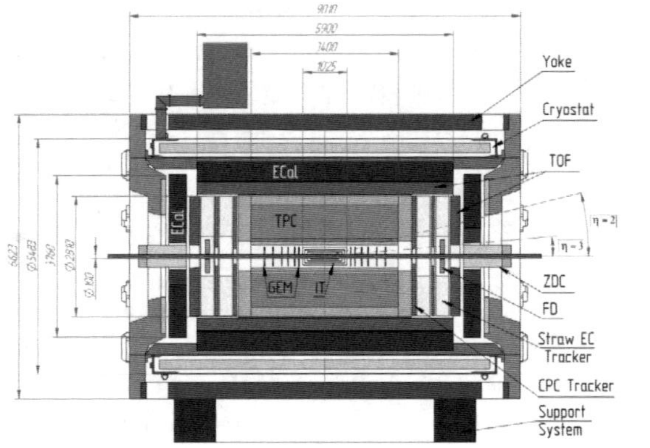


**Figure 2.** MPD detector cross-sectional view.

in Fig. 2. The MPD CD is 9 meters long and 6.6 meters in diameter and will be constructed in two stages. At the first stage (at the end of 2019), MPD will consist of the superconducting solenoid with a flux return yoke, Time-Projection Chamber (TPC), barrel Time-Of-Flight system (TOF), Electromagnetic Calorimeter (ECal), Zero-Degree Calorimeter (ZDC) and Fast Forward Detector (FFD). A detailed description of the detector components can be found in Ref. [15].

The Time-Projection Chamber is the main tracking detector of the MPD and will provide tracking and particle ID (dE/dx measurements with a resolution better than 8%) in the pseudorapidity range $|\eta| < 1.2$, momentum resolution for charge particles better than 3% in the transverse momentum range $0.1 < p_t < 1$ GeV/c and two-track resolution of about 1 cm. The TPC has an inner diameter of 54 cm, an outer diameter of 280 cm, and an overall length along the beam direction of 340 cm. The uniform electric field in the active volume is created by a thin central HV electrode together with a voltage dividing network at the surface of the outer and inner cylinders and at the readout end-caps. Since the amount of material traversed by particles in the MPD has to be kept as low as possible, the TPC field cage will be made of composite materials (kevlar and tedlar), thus we estimate the TPC overall thickness to be less than 5% X0. It has been shown that for precise tracking the relative radial magnetic and electric field components in MPD have not to exceed $\approx 5.2 \cdot 10^{-4}$. Hence, the mechanical structure and electric field defining network have to be designed in a way to keep radial field non-uniformity below $10^{-4}$ . The TPC readout system is based on the Multi-Wire Proportional Chambers (MWPC) with cathode pad readout. The gas mixture of 90% argon and 10% methane (P10), or 90%Ar and 10% CO2, is supposed to be used. In Fig. 3, a relative momentum resolution $(\Delta P/P)$ for charged tracks in the MPD TPC from Monte-Carlo simulations is shown.

Identification of charged hadrons in MPD in the range of momentum from 0.1 to 2 GeV/c will be achieved by the time-of-flight (TOF) measurements, which are complemented by the energy loss (dE/dx) information from the TPC (see Fig. 4). The cylindrical part of the TOF MPD has a full azimuthal and $|\eta|$ <1.2 coverage. The TOF outer diameter is 3.4 m, and the total surface is of about 53 $m^2$. The basic element of the TOF system will be a Multigap Resistive Plate Chamber (MRPC). This technology, which provides the intrinsic time resolution in the range 60÷80 pc, is widely used in heavy-ion experiments at RHIC and LHC. MRPCs are easily

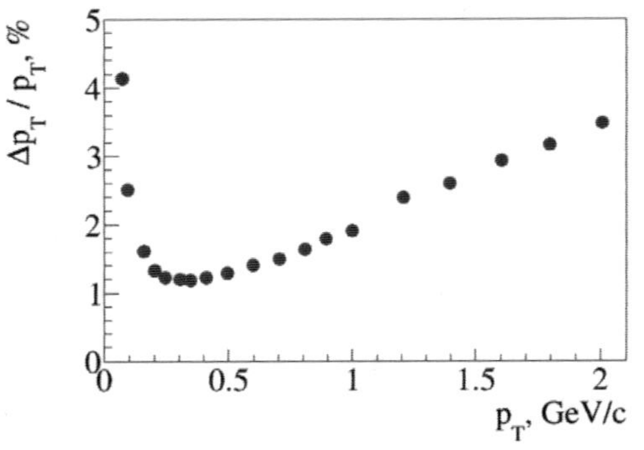


**Figure 3.** $\Delta P_t/P_t$ as a function of transverse momentum in MPD.

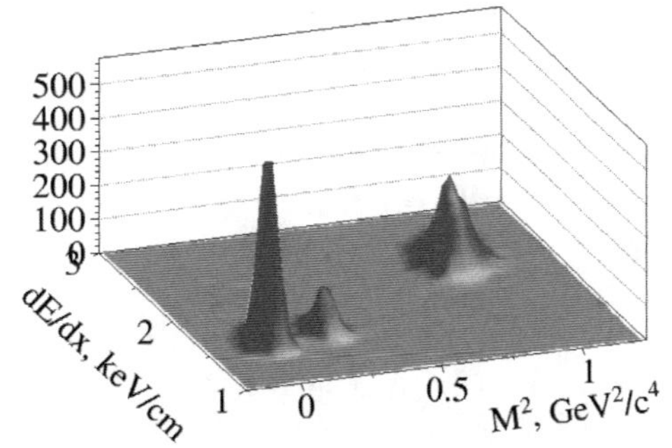


**Figure 4.** A typical distribution of the specific energy loss dE/dx versus mass-squared for $\pi^+, K^+$, and protons (from left to right) at p = 1 GeV/c.

manufactured and are relatively inexpensive. The TOF electronics will be based on the NINO specific integrated circuit [16], and TDC based on a multihit High Performance ASIC (HPTDC) chip developed at CERN [17]. The detector is segmented in azimuth into 12 sectors of 5.7 m length. Each sector consists of 4 modules, and each module comprises 6 MRPCs detectors arranged in 2 layers. The overall TOF geometrical efficiency is above 95%, and time resolution is better than 100 pc. It allows us to separate pions from kaons up to $p_t$ < 1.5 GeV/c and identify (anti)protons up to $p_t \approx 3$ GeV/c, thus more than 95% of the transverse momentum spectra of hadrons will be measured. As can be seen from Fig. 4, the basic detector parameters, namely, dE/dx resolution of $\sigma_{dE/dx}$ ≈8% and TOF resolution of $\sigma_{TOF}$ ≈100 ps, will provide a high degree of selectivity for hadrons at momenta below 2 GeV/c.

The primary role of the electromagnetic calorimeter is to measure the spatial position and energy of electrons and photons produced in heavy ion collisions. The MPD electromagnetic calorimeter is proposed to be built of towers as basic building elements of about 3 $cm^2$ cross section. Each lead-scintillator sampling tower contains 250 alternating tiles of Pb (0.275 mm) and plastic scintillator (1.5 mm). A module of 18 radiation length thickness will be ≈40 cm long. The cells of each tower are optically combined by 9 longitudinally penetrating wavelength shifting fibers for light collection. The light collected by these fibers will be read out by avalanche photodiodes (MAPD) units of 33 $\text{mm}^2$ sensitive area. The towers, mechanically grouped together, make 139 trapezium-shape modules which are combined into 48 detector sectors. The heat-producing electronics will be thus separated from the modules and mounted on the upper parts of sectors.

The goal of the FFD detector, located at ≈75 cm on both sides of the interaction point, is to provide a fast interaction trigger and timing signal for TOF. A FFD module is made of 1.5 cm thick quartz Cherenkov radiator, optically coupled to a multianode MCP-PMT Planacon capable of registering high energy photons and relativistic charged particles of velocity $\beta > 0.69$ with a timing resolution better than 50 ps.

The MPD Zero Degree Calorimeter (ZDC) located in the both sides of MPD at a distance of 3.2 m from the MPD center will be used for classification events by centrality by measuring the energy of spectators. Each ZDC semi-part consists of identical modules in parallelepiped-shaped with a transverse dimension of 10x10 $\text{cm}^2$ and a longitudinal size of 90 cm. A distinctive feature of the proposed design is the existence of square hole in the center of the calorimeter for the passage of the primary beam. Each ZDC module of consists of 60 lead-scintillator tile sandwiches with thickness of 16 and 4 mm for Pb and scintillator, respectively. The sampling ratio of 4÷1 will provide the compensation condition. Light readouts will be provided by WLS-fibers embedded in the round grooves in the scintillator plates, which will ensure a high efficiency and uniformity of light collection over the scintillator tiles within a few percent. WLS-fibers from each 6 consecutive scintillator tiles are collected together and viewed by an avalanche photodiode at the end of each module.

## 4. MPD feasibility study

All the Monte-Carlo simulation and feasibility study tasks for MPD were performed in the framework of a dedicated program tool - MPDRoot comprising interfaces to many event generators and Geant as well as detector response simulation and reconstruction algorithms.

*4.0.1. Study of hyperon production at MPD: feasibility* The performance of the MPD detector for hyperon measurements was studied for a combination of Time Projection Chamber (TPC) and Time-Of-Flight system (TOF) covering the mid-rapidity region ($|\eta| < 1.3$). A small material budget (not exceeding 10% of the radiation length in the region of interest) and powerful particle ID based on ionization loss (dE/dx) measurements in the TPC and time-of-flight information from the TOF allow precise trajectory reconstruction and reliable separation of particle species.

Hyperons and cascades were reconstructed using the secondary vertex finding technique with an optimized set of topological and track quality cuts as described in [19]. In Fig. 5 invariant mass distributions of $(p, \pi^-)$, $(\Lambda,\pi^-)$ and $(\Lambda,K^-)$ pairs from central Au+Au collisions at center-of-mass energy $\sqrt{s} = 9$ GeV are shown. The estimated efficiencies, Signal-to-Background (S/B) ratio and significance are also printed. Based on the results of this study, we have estimated the expected yields of particle species under interest for 10 weeks of data taking at the nominal NICA collider luminosity of $\approx 10^7$ and $\approx 10^6$ for $\Xi^-$ and $\Omega^-$, respectively. The accumulated statistics will allow us to get unique experimental information on cascade production over a broad region of the QCD phase diagram and a large phase-space of the reaction.

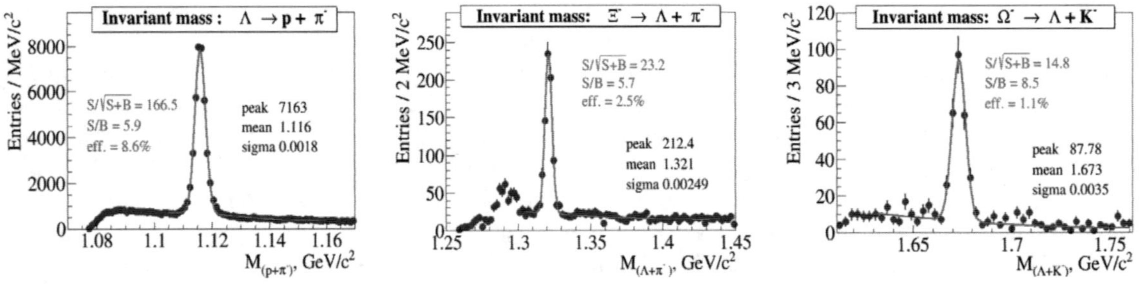


**Figure 5.** Invariant mass of $p$ and $\pi^-$ (left), $\Lambda$ and $\pi^-$ (center), $\Lambda$ and $K^-$ (right) candidates.

*4.0.2. MPD prospects for dileptons* We investigated MPD performance for dileptons, in particular focusing on determination of the hadron suppression factor of the MPD particle ID system, and achieved signal-to-background ratio and invariant mass resolution in the region of masses of vector mesons. A detailed description of the analysis procedure and results can be found in Ref. [18]. The charged hadrons and dileptons from event generators (UrQMD+Pluto) were transported through the detector setup, which includes the Time Projection Chamber (TPC), Time-Of-Flight system (TOF) and Electromagnetic Calorimeter (EMC) covering $|\eta|<1.2$. A realistic detector response was simulated followed by track reconstruction, TPC-TOF matching and particle identification procedures, which were performed within the MPDRoot framework. Electron identification was achieved by using combined information about the specific energy loss dE/dx from TPC, time-of- flight from TOF and $E/p$ information from EMC. After applying several quality and PID cuts, the achieved overall hadron rejection factor was of about 3200. The background from conversion pairs was eliminated by an extra set of topological and kinematical cuts. Fig. 6 shows the spectrum of reconstructed low-mass dielectrons after background rejection (dots) and true dielectrons from the event generator (line). $S/B$ ratio in mass region $0.2 < M_{e^+e^-} < 1.2$ GeV/$c^2$ was estimated to be of about 10%. This number is shown in Fig. 7 along with the published data from other experiments; the expected parameters of the MPD setup are among the best over the world.

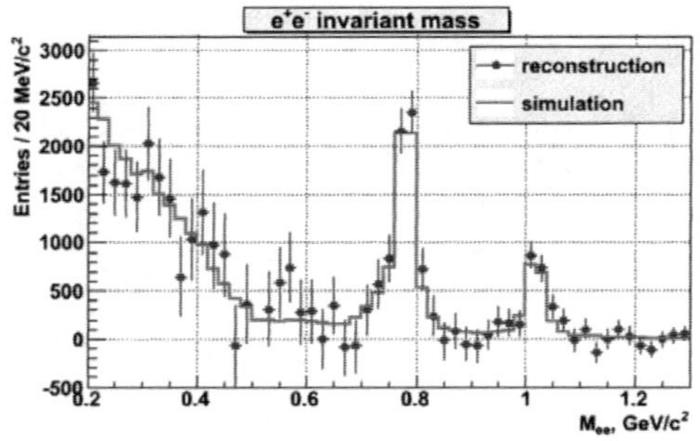


**Figure 6.** Invariant mass of dileptons after background subtraction.

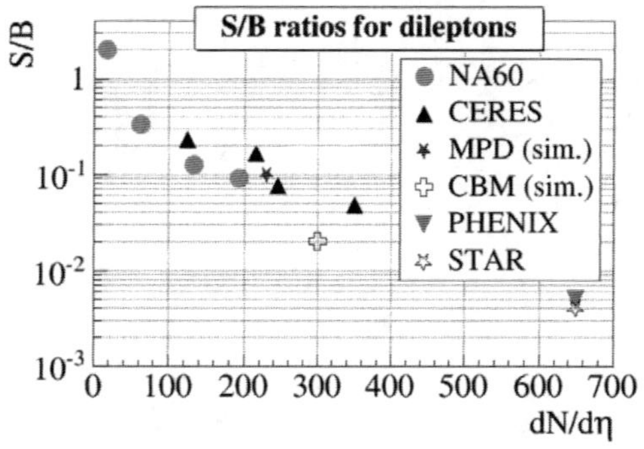


**Figure 7.** Overall Signal-to-Backgrond (S/B) ratio from heavy ion experiments ($0.2<M_{ee}<1.2$ GeV/$c^2$).

*4.0.3. Observation of hypernuclei at MPD: feasibility* In this section, we present the results of a study of MPD capability for reconstruction of hypertritons. The statistics of events from the DCM-QGSM event generator ($5 \cdot 10^5$ events were used in this study) corresponds to about 30 minutes of data taking time at NICA. The full event reconstruction, realistic particle ID and secondary vertex finding are described in [19]. The obtained invariant mass spectra of helium-3 and $\pi^-$ candidates is shown in Fig. 8. These results demonstrate good sensitivity of the MPD

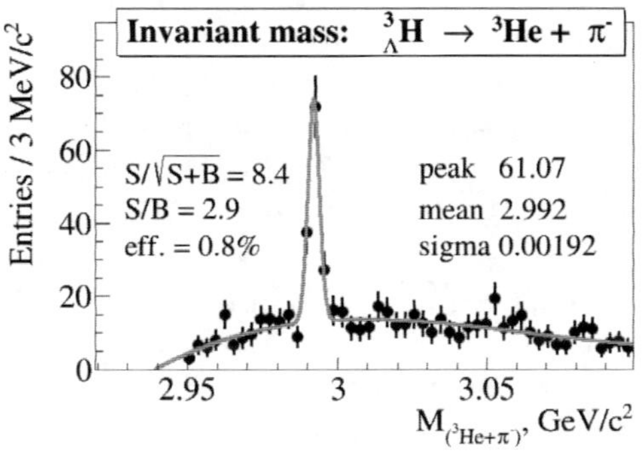


**Figure 8.** Invariant mass of $^3$He and $\pi^-$ candidates (MPD).

setup for hypernuclei; with a typical event rate of 7 kHz for the design NICA luminosity of $10^{27}$cm$^{-2}$c$^{-1}$, we are able to register of about $10^5$ hypertritons in a week of data taking. Thus, a detailed study of the production mechanism of single hypernuclei, as well as an observation of double hypernuclei at NICA, look feasible.

Journal of Physics: Conference Series **668** (2016) 012015 doi:10.1088/1742-6596/668/1/012015

## 5. BM@N experiment

Baryonic Matter at Nuclotron (BM@N) is an experimental set-up at the Nuclotron extracted beam to accomplish the NICA physics program at low energy range. A sketch of the BM@N

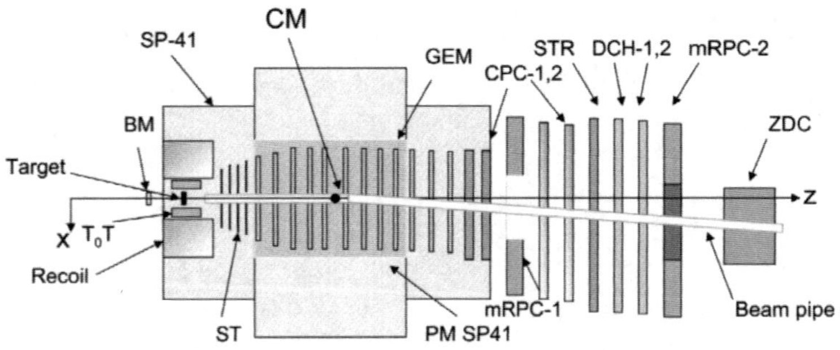


**Figure 9.** The BM@N experimental set-up.

proposed experimental set-up is shown in Fig. 9. It combines high precision track measurements with time-of-flight information for particle identification and total energy measurements for the analysis of the collision centrality. The charged track momentum and multiplicity will be measured with the set of 12 two-coordinate planes of GEM (Gaseous Electron Multipliers) detectors located downstream of the target inside the analyzing magnet complemented by drift and straw tube chambers situated outside the magnetic field. The GEM detectors sustain high rates of particles and are operational in strong magnetic fields. At the second stage of the BM@N experiment, at least 4 planes of two-coordinate silicon strip detectors will be installed between the GEM tracker and the target. The design parameters of the time-of-flight detectors based on multi-gap Resistive Plate Chambers (mRPC-1,2) with a strip read-out allow us to discriminate between hadrons $(\pi, K, p)$, as well as light nuclei with the momentum up to a few GeV/c. The Zero Degree Calorimeter (ZDC) is designed for the analysis of the collision centrality by measuring the energy of forward going particles. The T0 detector, partially covering the backward hemisphere around the target, is planned to trigger on central heavy ion collisions and provide a start signal for the mRPC-1,2 detectors. The Nuclotron will provide variety of beams from protons to gold ions with the kinetic energy from 1 to 6 GeV per nucleon, and the length of the beam line between the Nuclotron and the BM@N is around 160 meters. In order to provide beams of high quality, it comprises 26 elements of magnetic optics including 8 dipole magnets and 18 quadruple lenses. The planned intensity of the gold ion beam for the BM@N experiment (by the end of 2018) is expected to be of $10^7$ ions per second, and beams of xenon are expected by the end of 2017. The first technical run of the BM@N detectors was performed with deuteron and carbon beams already in spring 2015.

At present, the activities on the detector and beam line construction are complemented with intensive Monte Carlo simulation studies to optimize the detector set-up. A focus is made on the efficiency of the measurement of strange hyperons and hyper-nuclei in central Au+Au collisions. In Figures 10 and 11 invariant mass of $(\Lambda, \pi^-)$ and $(^3He, \pi^-)$ candidates, respectively, reconstructed with the GEM tracker from central Au+Au collisions at the beam kinetic energy of 4.5A GeV are shown. The obtained results indicate that the proposed set-up has a reasonable reconstruction capability for strange hyperons produced in high multiplicity central Au+Au collisions. The reconstructed signals of $\Xi^-$- hyperon and hyper-triton correspond to 0.9M and 2M of central collisions, respectively. Taking into account the signal reconstruction efficiency, data acquisition capacity and the duty factor of the Nuclotron beam, the expected statistics of $\Xi^-$-hyperons and hypertritons for a month of the BM@N operation are expected to be of 2.7M

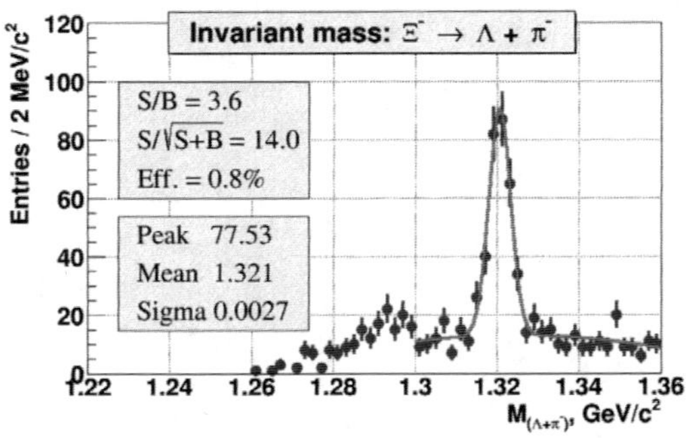


**Figure 10.** Invariant mass of $\Lambda$ and $\pi^-$ (BM@N)

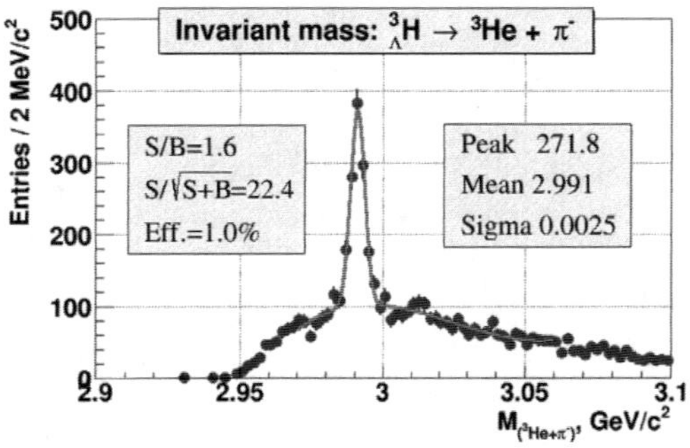


**Figure 11.** Invariant mass of $^3$He and $\pi^-$ candidates (BM@N)

and 4M, respectively. These statistics are sufficient to perform studies of strange hyperon and hyper-nuclei production by measuring their transverse momentum spectra, rapidity and angular distributions.

## 6. Summary and Conclusions

The NICA project realization plan foresees a staged construction and commissioning with the goal to start the BM@N experiment with the Nuclotron extracted beams in 2017. The research program will be continued at higher energies after putting the startup configuration of the NICA collider into operation in 2019. The commissioning of the design configuration of the NICA accelerator complex is foreseen in 2023.

## References

[1] G. Trubnikov, A. Kovalenko, V. Kekelidze, I. Meshkov, R. Lednicky, A. Sissakian, A. Sorin, PoS (ICHEP 2010) 523.
[2] G. Trubnikov, N. Agapov, V.Kekelidze, A. Kovalenko, V. Matveev, I. Meshkov, R. Lednicky, A. Sorin, PoS 36th International Conference of High Energy Physics (ICHEP2012), July 4-11,Melbourne, Australia.
[3] V.Kekelidze, A. Kovalenko, R. Lednicky, V. Matveev, I. Meshkov, A. Sorin, G. Trubnikov, 37th International Conference on High Energy Physics, ICHEP 2014, Valencia, Spain, 2 Jul 2014 - 9 Jul 2014 Nuclear physics B Proceedings supplements (2014).
[4] J. Adams et al, *Nucl. Phys.* A **757**, 102-183 (2005).
[5] K. Adcox et al, *Nucl. Phys.* A **757**, 184-283 (2005).
[6] C. Alt et al, *Phys. Rev.* C **77**, 024903 (2008).
[7] NICA White Paper, http://nica.jinr.ru, free access
[8] R. Rapp and J. Wambach, *Adv. Nucl. Phys.* **25**, 1 (2000).
[9] G. E. Brown and M. Rho, *Phys. Rev. Lett.* **66**, 2720 (1991).
[10] J. Randrup and J. Cleymans, *Phys. Rev.* C **74**, 047901 (2006).
[11] J. Rafelski and B. Müller, *Phys. Rev. Lett.* **48**, 1066 (1982).
[12] A. K. Kermann and M. S. Weiss, *Phys. Rev.* C **8**, 408 (1973).
[13] J. Steinheimer et al., *Phys. Lett.* B **714**, 85 (2012).
[14] K. U. Abraamyan et al., *Nucl. Instrum. Meth.* A **628**, 99 (2011).
[15] MPD Conceptual Design Report. http://nica.jinr.ru/files/CDR_MPD/MPD_CDR_en.pdf.
[16] F. Anghinolfi et al., *Nucl. Instr. Meth.* A **533**, 183-187 (2004).
[17] J. Christiansen, HPTDC V 1.3, March 2004, http://tdc.web.cern.ch/tdc/hptdc/docs/hptdc_manual_ver2.2.pdf.
[18] see paper of A. Zinchenko et al. from this Proceedings.
[19] see paper of M. Ilieva et al. from this Proceedings.
[20] BMN Conceptual Design Report. http://nica.jinr.ru/files/BMN/BMN_CDR.pdf.

Journal of Physics: Conference Series **668** (2016) 012016 doi:10.1088/1742-6596/668/1/012016

# CMS Heavy Ion Results on Heavy Flavour and Quarkonia

**Yongsun Kim on behalf of the CMS collaboration**
Korea University, 145 Anam-ro, Seonbuk-gu, Seoul, Republic of Korea
E-mail: kimy@cern.ch

**Abstract.** The CMS collaboration has recorded 150 $\mu b^{-1}$ and 35 $nb^{-1}$ of PbPb and pPb collisions at 2.76 TeV and 5.02 TeV respectively. This presentation reports the results of the heavy quark measurements including quarkonia and $B$ mesons. A particular emphasis is given to the most recent results regarding the suppression of J/$\psi$, $\psi$(2S) and $\Upsilon$ 's in PbPb collisions, and the study of cold nuclear matter effects based on pPb collisions.

## 1. Introduction

Understanding the production mechanisms of quarkonium bound states, from first principle calculations based on quantum chromodynamics (QCD), remains an open question in proton-proton (pp) and heavy ion collisions. In pp collisions, despite the extensive progress, none of the existing models can reproduce simultaneously the cross sections and polarization results measured experimentally. In heavy ion collisions, it was predicted that the Debye screening of the heavy quark potential would result in a sequential suppression of the quarkonium states reflecting their binding energies [1]. The quarkonium production can be affected also by several other processes apart from the existence of QGP, such as the nuclear parton distribution functions (nPDFs), parton energy loss, and nuclear absorption [2]. For this reason, the study of quarkonium production in pA collisions, besides PbPb and pp, is critical in order to understand and disentangle such Cold Nuclear Matter (CNM) effects.

This conference proceeding reports the measurement of $J/\psi$, $\psi(2S)$, $\Upsilon$ in pp, pPb and PbPb data collected during Run I period of CMS experiment. In addition, $B$ meson results in pPb collisions are also discussed to study the CNM effects on open heavy flavor hadrons.

## 2. Experimental setup and analysis

### 2.1. CMS detector

A detailed description of the CMS detector can be found in Ref. [3]. Its central feature is a superconducting solenoid with an internal diameter of 6 meter, providing a magnetic field of 3.8 T. Within the field volume are the silicon pixel and strip tracker, the crystal electromagnetic calorimeter, and the brass/scintillator hadronic calorimeter. The silicon pixel and strip tracker measure charged-particle trajectories in the range $|\eta| < 2.5$. It consists of 66 million pixel and 10 million strip sensor elements. Muons are detected in the range $|\eta| < 2.4$, with detection planes based on three technologies: drift tubes, cathode strip chambers, and resistive plate chambers. Because of the strong magnetic field and the fine granularity of the tracker, the muon

Journal of Physics: Conference Series **668** (2016) 012016 doi:10.1088/1742-6596/668/1/012016

$p_T$ measurement based on information from the tracker alone has a resolution between 1 and 2% for typical muons in this analysis.

The CMS apparatus also has extensive forward calorimetry, including two steel/quartz-fiber Čerenkov hadron forward calorimeters (HF), which cover $2.9 < |\eta| < 5.2$. These detectors are used for online event selection and the impact parameter-like characterization of the events such as centrality in PbPb.

### *2.2. Reconstruction of quarkonia and B mesons*

Analyses in this report commonly use muons as the basic object and quarkonia are identified via their di-muon decay channels. To extract the quarkonia yields, unbinned maximum likelihood fits to the $\mu^+\mu^-$ invariant mass spectra are performed. In general, the $\Upsilon$ and $J/\psi$ resonances are modelled by the sum of two Crystal Ball (CB) functions, that is, Gaussian resolution functions with the low side tail replaced with a power law describing final state radiation. Given the relatively large statistical uncertainties of the 2.76 TeV samples , most of the parameters are fixed to values provided by simulations, then allowed to vary to compute the associated systematic uncertainties. The background is modelled by an exponential function multiplied by an error function describing the low-mass turn-on.

$J/\psi$ and $\psi(2S)$ peaks are fitted simultaneously using the same invariant mass distribution. At mid-rapidity, a CB function describes well the $J/\psi$ signal. At forward rapidity, due to rapid changes in muon resolution with $|\eta^\mu|$, the addition of a Gaussian function, with same mean $m_0$ but different width than the CB function, is needed to obtain a good fit. Various other fits are also performed to compute the systematic uncertainty.

The $B^+$ meson is reconstructed by combining a $J/\psi$ with an extra track, while $B^0$, $B^0_s$ candidates are reconstructed by combining a $J/\psi$ with two tracks. For four-prong channels ($B^0 \to J/\psi\ K^*(892)$ and $B^0_s \to J/\psi\ \phi$), two selected tracks are required to originate from a common vertex to form a $K^*(892)$ or a $\phi$ candidate. Finally, the two muons together with the tracks are required to form a B-meson candidate with the constraint that the di-muon invariant mass is equal to the nominal mass value of $J/\psi$. B-meson candidates with rapidity in the center-of-mass frame $|y_{CM}| < 1.93$ and $p_T > 10\text{GeV/c}$ are stored for further analysis.

Measured yields are corrected for acceptance and efficiency obtained from Monte Carlo (MC) simulation and further corrected by a data-driven technique called tag-and-probe [4].

## 3. Analysis methods and Results

### *3.1. $\Upsilon(nS)$ in PbPb*

The yields of $\Upsilon(1S)$, $\Upsilon(2S)$ and $\Upsilon(3S)$ mesons are reported, from PbPb and pp data sets at the same center-of-mass energy of $\sqrt{s_{\mathrm{NN}}} = 2.76$ TeV, corresponding to 166 $\mu\mathrm{b}^{-1}$ and 5.4 $\mathrm{pb}^{-1}$, respectively [5]. With respect to the previous result [6], the PbPb data reconstruction was improved yielding an increase of $\approx 30\%$ in the $\Upsilon(1S)$ yields. The pp dataset, recorded in 2013, is new and allows for further differential studies as functions of the $\Upsilon$ rapidity ($y$) and transverse momentum. $\mathrm{R_{AA}}$ are derived from the PbPb and pp yields

Figure 1 shows $\mathrm{R_{AA}}$ as a function of centrality, displayed as the number of participating nucleons, $\mathrm{N_{Part}}$. The global, fully-correlated uncertainties come from the uncertainty in the pp cross sections (which is different for the two $\Upsilon$ states) and the PbPb tracking efficiency. The strong centrality dependence, already observed in Ref. [6], is mapped out with more precision. The dependence of $\mathrm{R_{AA}}$ on kinematic variables is found to be weak as shown in figure 2.

### *3.2. $J/\psi$ and $\psi(2S)$ in PbPb*

The cross-section results of $J/\psi$ and $\psi(2S)$ are reported in the form of double ratio (Fig. 3), instead of $\mathrm{R_{AA}}$ so that efficiency and acceptance corrections cancel to a large extent, reducing the systematic uncertainties.

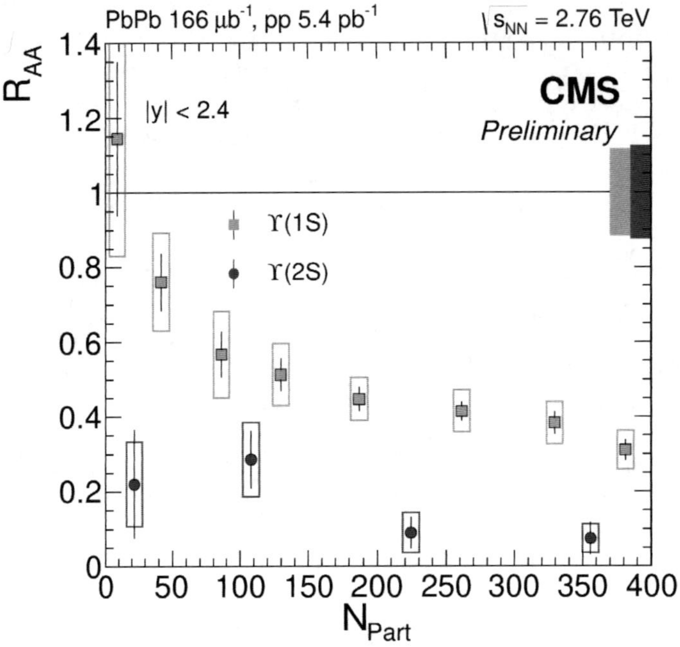


**Figure 1.** Nuclear modification factor of $\Upsilon(1S)$ and $\Upsilon(2S)$ in PbPb as a function of centrality, displayed as the number of participating nucleons. Statistical (systematic) uncertainties are displayed as error bars (boxes), while the global fully-correlated uncertainty is displayed as a box at unity.

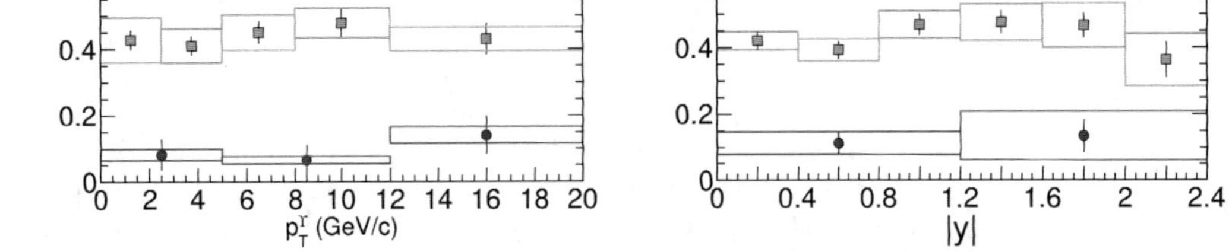


**Figure 2.** Nuclear modification factor of $\Upsilon(1S)$ and $\Upsilon(2S)$ in PbPb collisions as a function of transverse momentum (left) and rapidity (right). Statistical (systematic) uncertainties are displayed as error bars (boxes), while the global fully-correlated uncertainty is displayed as a box at unity.

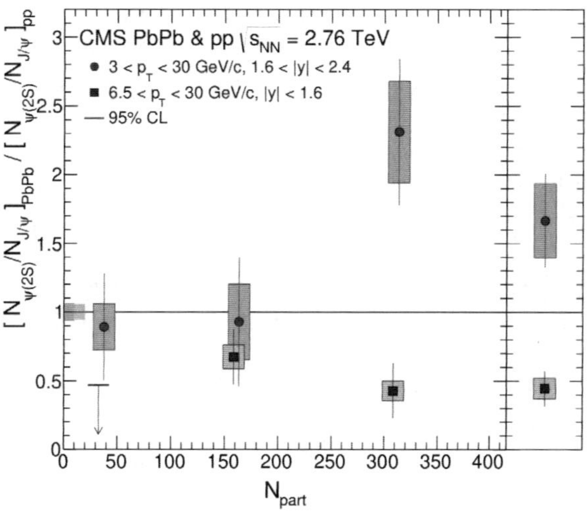


**Figure 3.** Measured yield double ratio as a function of centrality. The $p_{\mathrm{T}}$ and rapidity bins are $6.5 < \mathrm{p_T} < 30$ GeV/c and $|y| < 1.6$ (left), and $3.0 < p_{\mathrm{T}} < 30$GeV/c and $1.6 < |y| < 2.4$ (right). The error bars and boxes stand for the PbPb statistical and systematic uncertainties, respectively. The shaded band is the uncertainty on the pp measurement, common to all double-ratio points.

For mid-rapidity, $|y| < 1.6$ the double ratio is always less than unity, meaning that $\psi(2S)$ are more suppressed than $J/\psi$. Within uncertainties, no centrality dependence is observed. However, going to the forward rapidity range, the data show an increase of the double ratio with centrality, though with large uncertainties. In the most central collisions the double ratio is about 1.7 which means that more $\psi(2S)$ are produced compared to $J/\psi$ than in pp collisions, again with large uncertainties.

These results exhibit a clear $\psi(2S)$ suppression in the midrapidity and higher $p_{\mathrm{T}}$ region, while the pp uncertainty is too large to draw a firm conclusion in the forward rapidity lower $p_{\mathrm{T}}$ region.

### *3.3. Polarization of $\Upsilon(nS)$ in pp collisions*

Polarization is a significant input to the non-relativistic QCD phenomenology of quarkonia production. In this model, the nonperturbative phenomenon of bound state formation involves the interaction with surrounding hadrons in the event. Therefore, the pre-resonant $Q\overline{Q}$ must neutralize its colour by absorbing or emitting soft gluons. It is natural to wonder whether the $Q\overline{Q}$ bound-state formation process could be different between pp and nucleus-nucleus collisions, or between "peripheral" and "central" pp collisions. So, polarization measurements can provide the cleanest indications on how the QCD medium surrounding the production of the $Q\overline{Q}$ pair affects its binding into the final-state quarkonium. The polarizations of the $\Upsilon(1S)$, $\Upsilon(2S)$, and $\Upsilon(3S)$ mesons produced in pp collisions at 7 TeV have been determined as a function of the charged particle multiplicity of the event, in two $p_{\mathrm{T}}$ranges. [7]

To maximize the physical information and avoid interpretation ambiguities in comparisons with calculations or other experimental measurements [8], the three frame-dependent angular anisotropy parameters, $\vec{\lambda} = (\lambda_\vartheta, \lambda_\varphi, \lambda_{\vartheta\varphi})$, as well as the frame-invariant parameter [9] $\tilde{\lambda} = (\lambda_\vartheta + 3\,\lambda_\varphi)/(1 - \lambda_\varphi)$, are measured in three frames, with different directions of the quantization axis: the center-of-mass helicity (HX) frame, where the polar axis coincides with the direction

Journal of Physics: Conference Series **668** (2016) 012016 doi:10.1088/1742-6596/668/1/012016

of the $\Upsilon$ momentum; the Collins–Soper (CS) frame, whose axis is the average of the two beam directions in the $\Upsilon$ rest frame; and the perpendicular helicity (PX) frame, orthogonal to the CS frame.

Figure 4 displays the corresponding results for the frame-invariant parameter $\tilde{\lambda}$ . The results obtained in the CS, HX, and PX frames also showed good agreement. No significant changes of the polarizations are seen as a function of $N_{\rm ch}$, for any of the three states.

### 3.4. $J/\psi$ in pPb

The rapidity dependence of the $J/\psi$ production can provide key information on the nPDFs and other CNM effects. The effect was quantified by investigating the forward-to-backward production ratio $\rm R_{FB}$ defined by

$$R_{\rm FB}(p_{\rm T}, y) = \frac{d^2\sigma(p_{\rm T}, y > 0)/dp_{\rm T}dy}{d^2\sigma(p_{\rm T}, y < 0)/dp_{\rm T}dy}, \tag{1}$$

where the $d^2\sigma/dp_{\rm T}dy$ is the J/$\psi$ production cross section decayed into muon pairs in a given $p_{\rm T}$ and $y$ bin, and forward regions ($y > 0$) are defined by the proton-going direction.

Fig. 5 shows the $p_{\rm T}$ dependence of $\rm R_{FB}$ in three different rapidity ranges for prompt J/$\psi$. The data points are plotted at the bin-averaged values. The $\rm R_{FB}$ is closer to unity at high $p_{\rm T}$, and decreases at low $p_{\rm T}$. No strong rapidity dependence has been observed.

The $\rm R_{FB}$ of prompt J/$\psi$ is further analyzed in terms of an event activity variable $E_T^{HF|\eta|>4}$, the transverse energy deposited in the forward hadronic calorimeter in $4 < |\eta| < 5.2$. The centers of the bin abscissae are given by their bin-averaged values, but slightly shifted for higher $p_{\rm T}$ points (left) or higher rapidity points (right), so that they do not overlap with each other. The clear decreasing tendency of the $\rm R_{FB}$ with increasing $E_T^{HF|\eta|>4}$ has been observed over the whole kinematic ranges, and it is more pronounced at low $p_{\rm T}$ in the forward rapidity region.

### 3.5. $\rm B^+$, $\rm B^0_s$, $\rm B^0$ in pPb

The study of CNM effects is extended to heavy-quark production by performing the first measurement of exclusive $B$ meson decays in pPb collisions. The $\rm R_{AA}$ and $\rm R_{FB}$ of $B$ mesons are measured above transverse momentum of 10GeV/c.

The $p_{\rm T}$-differential production cross sections of all three $B$ mesons in the interval $|y_{\rm lab}| < 2.4$ are measured [10], with data points placed at the center of each bin. They are compared to the pp cross sections obtained from fixed-order plus next-to-leading-logarithm (FONLL) calculations [11] The FONLL predictions are scaled by A(=208), the atomic mass of the Pb nucleus, to account for the number of binary NN collisions. The FONLL uncertainties, larger than the ones from data, represent the quadratic sum of several variations made to the calculation: of the factorization and renormalisation scales, of the c quark mass, and of the uncertainty associated with PDFs (providing the largest contribution) [11, 12, 13].

The theoretical uncertainties represented by the open blue boxes in Fig. 6 are computed by recalculating $R_{\rm pA}{}^{\rm FONLL}(p_{\rm T})$ with the upper and lower values of the FONLL predictions. The nuclear modification factors of the three $B$ mesons do not show evidence for modification of pPb data compared to the FONLL reference, in the considered $p_{\rm T}$ range within the quoted uncertainties. No significant differences are observed between the three $B$ meson species.

The production cross section of $B^+$ is also studied as a function of its rapidity in the center-of-mass frame ($y_{\rm CM}$). The $y_{\rm CM}$-differential cross section of $B^+$ in the interval $10 < p_{\rm T} < 60$GeV/c is shown in Fig. 7 (left). In Fig. 7 (right), the rapidity dependence of the nuclear modification factor of $B^+$ is shown. No strong evidence of rapidity dependence of $R_{\rm pA}{}^{\rm FONLL}$ is observed within the uncertainties.

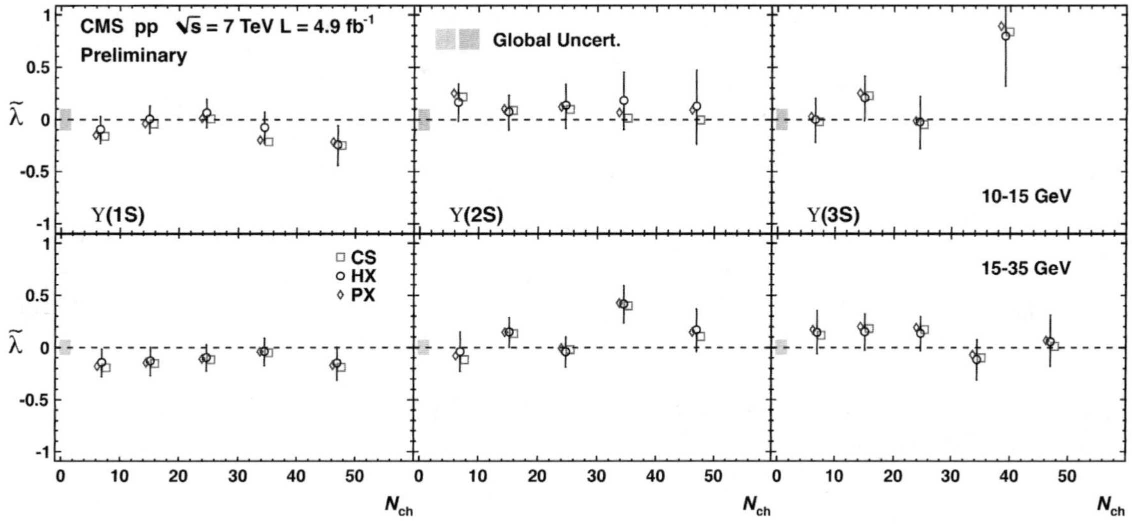


**Figure 4.** Values of $\tilde{\lambda}$ for the $\Upsilon(1S)$, $\Upsilon(2S)$, and $\Upsilon(3S)$states (left to right), in the HX, CS, and PX frames. The vertical bars represent the 68.3% CL intervals of the total uncertainties, while the global uncertainties are represented by boxes at the zero horizontal line. The points have been slightly offset for easier viewing.

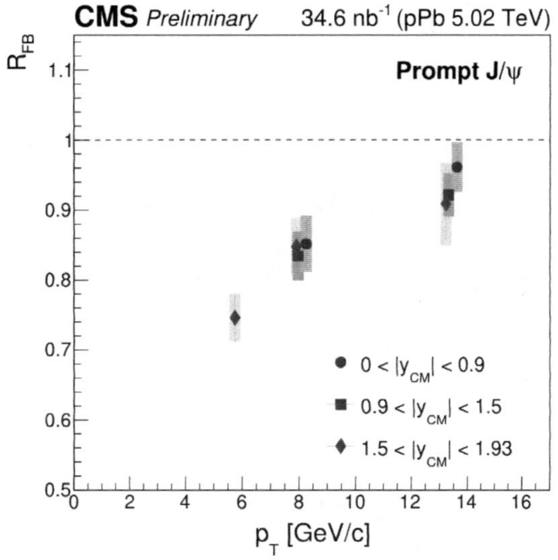


**Figure 5.** $p_{\mathrm{T}}$ dependence of $\mathrm{R_{FB}}$ for prompt $\mathrm{J}/\psi$ in three rapidity ranges. The error bars show the statistical uncertainties, and the shaded boxes represent the quadratic sum of systematic uncertainties.

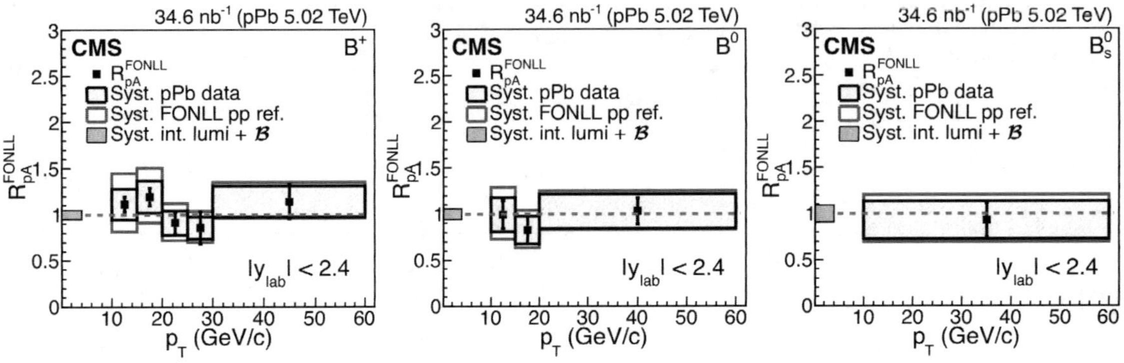


**Figure 6.** The nuclear modification factors $R_{\mathrm{pA}}{}^{\mathrm{FONLL}}(p_{\mathrm{T}})$ of $B^+$ (left), $B^0$ (center), $B_s^0$ (right) measured in pPb collisions at 5.02 TeV. The statistical and systematic uncertainties on the pPb data are shown as bars and yellow boxes around the data points, respectively.

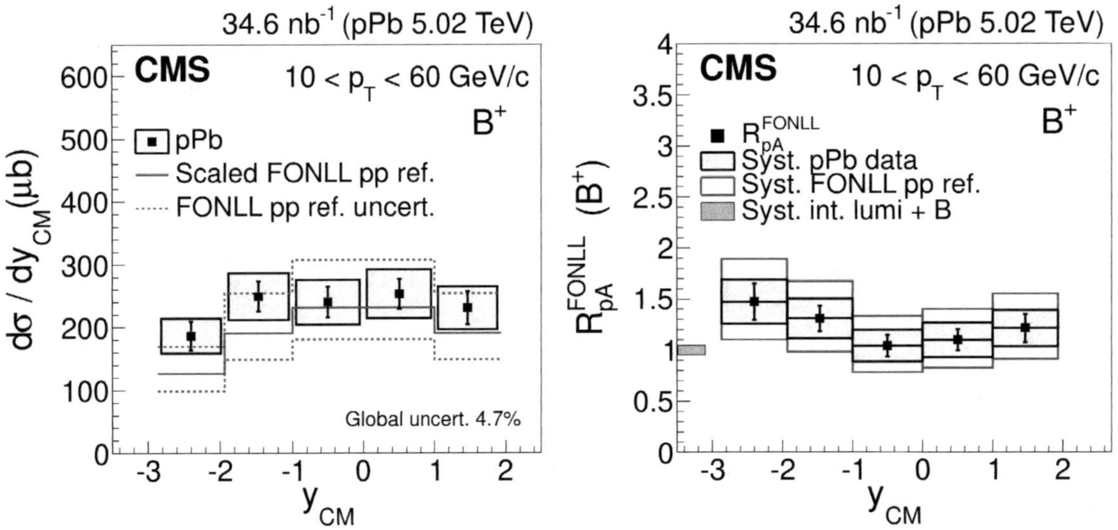


**Figure 7.** (left) The $y_{\mathrm{CM}}$-differential production cross section of $B^+$ measured in pPb collisions at 5.02 TeV. Vertical bars (boxes) correspond to statistical (systematic) uncertainties. The listed global systematic uncertainty is not included in the data points. The result is compared to a FONLL calculation [12, 13, 11] represented by a continuous histogram. The dashed histograms represent the theoretical uncertainties for the FONLL reference. (right) The nuclear modification factor $R_{\mathrm{pA}}{}^{\mathrm{FONLL}}$ of $B^+$ as a function of $y_{\mathrm{CM}}$. The statistical and systematic uncertainties on the pPb data are shown as bars and yellow boxes around the data points, respectively. The systematic uncertainty from the FONLL reference is plotted separately as open blue boxes. The global systematic uncertainty is shown as a full grey box at unity, and is not included in the data points.

## 4. Summary

Recent results of quarkonia and $B$ mesons from CMS heavy ion experiment were reviewed. $\Upsilon(1S)$, $\Upsilon(2S)$ and $\Upsilon(3S)$ meson yields have been measured in PbPb and pp collisions at the same energy per nucleon pair, $\sqrt{s_{NN}} = 2.76$ TeV. The $\Upsilon(1S)$ and $\Upsilon(2S)$ are suppressed by a factor of $\approx 2$ and 10, respectively, while the unobserved $\Upsilon(3S)$ corresponds to a suppression by a factor larger than 7, at 95% confidence level. Though a strong centrality dependence of the suppression is observed for the $\Upsilon(1S)$ and $\Upsilon(2S)$, no noticeable dependence is observed as a function of transverse momentum or rapidity.

The modification of the $\psi(2S)$ meson production and its ratio to $J/\psi$ in PbPb was also discussed. At lower $\mathrm{p_T}$, the data suggest an enhancement of the $\psi(2S)$ to $J/\psi$ ratio, with respect to the corresponding pp measurement, that increases with PbPb collision centrality. Both the centrality dependence and the double ratio being larger than unity are opposite to the expected behavior in the sequential melting scenario.

The polarization of $\Upsilon(1S)$, $\Upsilon(2S)$ and $\Upsilon(3S)$ in pp collisions and their dependence on the particle multiplicity was investigated to observe the surrounding materials's effect to the angular states of quarkonia. There was not evidence of interference in high multiplicity pp collisions, however this study opens the way for analogous measurements in pPb and PbPb, which can provide the cleanest indications on how quark-antiquark bound-state formation is influenced by the surrounding medium, an essential input for the interpretation of quarkonium nuclear suppression patterns.

In the end, the measurement of prompt and non-prompt $J/\psi$ and $B$ mesons in pPb collisions at $\sqrt{s_{\mathrm{NN}}} = 5.02$ TeV has been reported to discuss the cold nuclear matter effects. For $J/\psi$, the forward-to-backward production ratio $\mathrm{R_{FB}}$ is observed to be smaller than unity at low $\mathrm{p_T}$ and become consistent with unity for $p_{\mathrm{T}}$ above 10GeV/c. The $\mathrm{R_{FB}}$ and $\mathrm{R_{AA}}$ of $B$ mesons were found to be consistent to unity in the overall measurement ranges, $p_{\mathrm{T}}$ above 10GeV/c, which is consistent to $J/\psi$ result.

## 5. References

[1] T. Matsui, H. Satz, $\mathrm{J}/\psi$ suppression by quark-gluon plasma formation, Phys. Lett. B 178 (1986) 416. doi:10.1016/0370-2693(86)91404-8.

[2] A. Andronic, et al., Heavy-flavour and quarkonium production in the LHC era: from proton-proton to heavy-ion collisions, arXiv:1506.03981.

[3] S. Chatrchyan, et al., The CMS experiment at the CERN LHC, JINST 3 (2008) S08004. doi:10.1088/1748-0221/3/08/S08004.

[4] V. Khachatryan, et al., Measurements of Inclusive $W$ and $Z$ Cross Sections in $pp$ Collisions at $\sqrt{s} = 7$ TeV, JHEP 1101 (2011) 080. arXiv:1012.2466, doi:10.1007/JHEP01(2011)080.

[5] CMS Collaboration, Nuclear modification of Y states in PbPb, Tech. Rep. CMS-PAS-HIN-15-001, CERN, Geneva (2015).
URL `https://cds.cern.ch/record/2030083`

[6] CMS Collaboration, Observation of sequential Upsilon suppression in PbPb collisions, Phys. Rev. Lett. 109 (2012) 222301. arXiv:1208.2826, doi:10.1103/PhysRevLett.109.222301.

[7] CMS Collaboration, Measurement of the Y(nS) polarizations in pp collisions as a function of charged particle multiplicity, Tech. Rep. CMS-PAS-HIN-15-003, CERN, Geneva (2015).
URL `http://cds.cern.ch/record/2030082`

[8] P. Faccioli, C. Lourenço, J. Seixas, H. Wöhri, Towards the experimental clarification of quarkonium polarization, Eur. Phys. J. C 69 (2010) 657. arXiv:1006.2738, doi:10.1140/epjc/s10052-010-1420-5.

[9] P. Faccioli, C. Lourenço, J. Seixas, New approach to quarkonium polarization studies, Phys. Rev. D 81 (2010) 111502(R). arXiv:1005.2855, doi:10.1103/PhysRevD.81.111502.

[10] CMS Collaboration, Study of B meson production in pPb collisions at $\sqrt{s_{\mathrm{NN}}} = 5.02$ TeV, arXiv:1508.06678.

[11] M. Cacciari, S. Frixione, N. Houdeau, M. L. Mangano, P. Nason, G. Ridolfi, Theoretical predictions for charm and bottom production at the LHC, JHEP 10 (2012) 137. arXiv:1205.6344, doi:10.1007/JHEP10(2012)137.

[12] M. Cacciari, M. Greco, P. Nason, The $p_T$ spectrum in heavy-flavour hadroproduction, JHEP 05 (1998) 007. arXiv:hep-ph/9803400, doi:10.1088/1126-6708/1998/05/007.

[13] M. Cacciari, P. Nason, Charm cross sections for the Tevatron Run II, JHEP 09 (2003) 006. arXiv:hep-ph/0306212, doi:10.1088/1126-6708/2003/09/006.

Journal of Physics: Conference Series **668** (2016) 012017 doi:10.1088/1742-6596/668/1/012017

# Strangeness production in PHENIX experiment

**D O Kotov[1,2] (for the PHENIX collaboration)**

[1] Peter the Great St. Petersburg Polytechnic University, Polytechnicheskaya st. 29, St. Petersburg, 195251, Russia
[2] National Research Center "Kurchatov Institute", Petersburg Nuclear Physics Institute, Orlova roscha, Gatchina, Leningrad district, 188300, Russia

E-mail: dm_kotov@phmf.spbstu.ru

**Abstract.** The PHENIX experiment at RHIC has measured production of $K^{\pm}$, $K_s$, $K^*$ and $\phi$-mesons in p+p, $d$+Au, Cu+Cu and Au+Au collisions at $\sqrt{s_{NN}}$ = 62.4 and 200 GeV. While p+p collisions provide a baseline and are used for precision tests of pQCD calculations, for heavier colliding systems such as $d$+Au, Cu+Cu and Au+Au nuclear modification factors are studied at different centralities. These systematic studies enrich current understanding of the strange meson production and its difference from light quark hadrons. The role of radial flow and coalescence in particle production is discussed.

## 1. Introduction

A strongly coupled quark gluon plasma (QGP) was discovered in heavy ion collisions at the Relativistic Heavy Ion Collider (RHIC) [1]. Unlike up and down quarks, strange quarks are not present in colliding nuclei and are formed in collisions between constituents of the QGP. Therefore, the measurement of particles, which contain strange quarks, is an effective way to compare with hadrons that contain only light quarks and investigate the properties of the hot and dense matter formed in heavy ion collisions. Particles with strangeness content cover a wide range of masses and include mesons and baryons, which makes them a perfect tool to study features of hadron production such as radial flow and recombination at intermediate $p_T$ and energy loss flavor dependence at high $p_T$.

RHIC [2] at Brookhaven National Lab is one of the biggest operating particle accelerators designed to study heavy ion collisions at high energies. Two big experimental setups operate at RHIC nowadays: STAR [3] and PHENIX [4]. RHIC is a flexible and reliable accelerator complex with an extensive experimental program. A lot of operational time is devoted to beam energy scanning and switching between colliding nuclei. Up to now RHIC has provided 9 combinations of nuclei and 11 energies, while beam luminosity has been permanently increased from to run to run.

The PHENIX experimental setup consists of central and forward arms. Rapidity and azimuthal coverage are $|\eta| < 0.35$ and 90 degrees for each central arm (east and west), respectively. For forward arms, rapidity coverage expands from $1.2 < |\eta| < 2.2$ and azimuthal coverage is 360 degrees for each forward arm (north and south).

Charged particle track reconstruction in the PHENIX central arms is performed with drift and pad chambers [5], which have high momentum and spatial resolution. The main purpose of the electromagnetic calorimeter [6] is to measure the energy and coordinates of photons and electrons originating from the interaction region. Two time-of-flight systems [7, 8] in the east and west central arms of the PHENIX spectrometer are used for particle identification in low and intermediate transverse

Journal of Physics: Conference Series **668** (2016) 012017 doi:10.1088/1742-6596/668/1/012017

momentum regions. PHENIX also provides excellent capabilities for muon measurements at forward rapidity. The main detectors responsible for muon registration are Muon Trackers and MuID [9].

PHENIX central tracking starts to work from an approximately two meter radius outside of the central magnetic field. In order to measure the momenta of charged particles, the assumption that all of the charged particles originate from primary vertex of the collision is made. Therefore, momenta values of all particles originating from the decay of long lived hadrons, like $\Lambda$, $K_s$ etc, are not measured accurately, and production measurements of such particles become problematic. The PHENIX experiment is mostly focused on short lived particle production measurements. Due to small central arms acceptance, it is also difficult to measure multi-particle decays in low $p_T$ region. This paper is devoted to measurements of short lived hadron production in intermediate and high $p_T$ region except for $K^{\pm}$ and $\phi$-mesons in $e^+e^-$ decay mode also measured at low $p_T$.

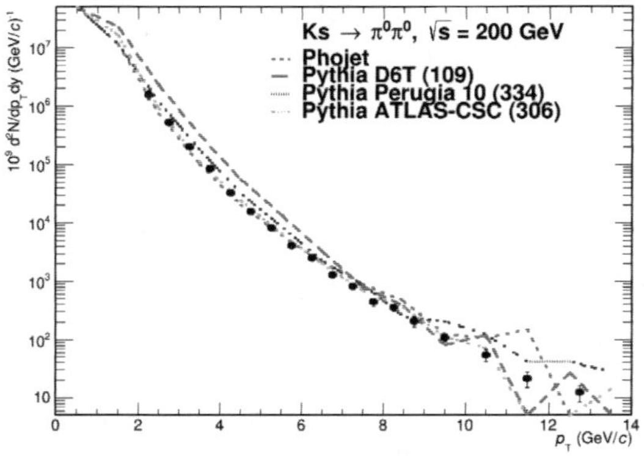


**Figure 1.** Invariant transverse momentum spectrum of $K_s$-mesons in p+p collisions at $\sqrt{s_{NN}}$ = 200 GeV compared to predictions of different Pythia tunes and Phojet.

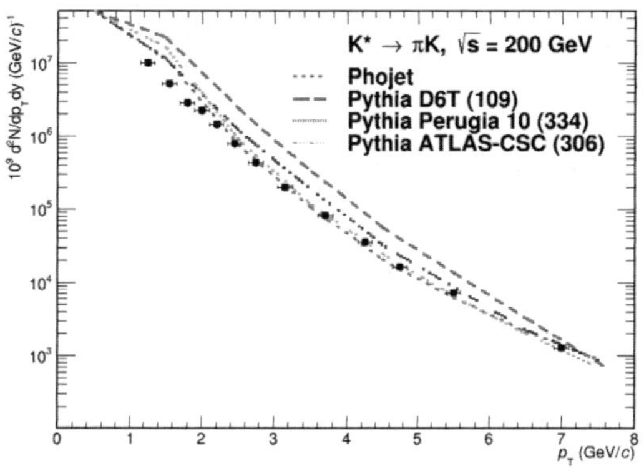


**Figure 2.** Invariant transverse momentum spectrum of $K^*$-meson in p+p collisions at $\sqrt{s_{NN}}$ = 200 GeV compared to predictions of different Pythia tunes and Phojet.

## 2. Measurements in p+p collisions

The invariant transverse momentum spectra for $K_s$ and $K^*$-mesons measured by the PHENIX experiment in p+p collisions at 200 GeV in central arms are shown in figure 1 and figure 2 respectively. Results for $K^{\pm}$ and $\phi$-mesons were obtained as well and are shown in [10]. These spectra are used as a baseline to compare with more complex and heavy colliding systems, such as p+$A$ and $A$+$A$ interactions. Moreover, these spectra are needed for event generator tuning, pQCD checks and available parameterizations of fragmentation functions. As seen from the figures available, the tunes of Pythia and Phojet cannot describe the measured spectra; some of the tunes manage to partially describe spectra, while others fail.

The PHENIX experiment has also measured $\phi$-meson production in p+p collisions at $\sqrt{s_{NN}}$ = 200 GeV at forward rapidity in $\mu^+\mu^-$ decay channel. The invariant transverse momentum spectrum of $\phi$-meson obtained in p+p collisions at $\sqrt{s_{NN}}$ = 200 GeV at forward rapidity is presented in figure 3. Rapidity dependence of $\phi$-meson production is presented in figure 4. These dependencies are used for pQCD checks, event generator tuning and to study rapidity dependence of nuclear modification factors of $\phi$-meson. As seen from the figures different tunes of Pythia and Phojet are not able to fully describe presented results.

The Pythia ATLAS-CSC and default tunes describe forward rapidity data except for the $\phi$-rapidity distribution. For the midrapidity $K_s$ and $K^*$-mesons data the PYTHIA ATLAS-CSC tune and Phojet seem to work reasonably well in $p_T$ region above 2 GeV/$c$.

Journal of Physics: Conference Series **668** (2016) 012017 doi:10.1088/1742-6596/668/1/012017

## 3. $R_{AB}$ measurements in *d*+Au, Cu+Cu and Au+Au collisions

In the high $p_T$ region particle production is governed by fragmentation of hard-scattered partons. Hard processes are characterized by small cross section and large value of transferred momentum. This fact allows to describe hard processes in heavy ion collisions as superposition of nucleon nucleon interactions. Such processes can be compared with p+p collisions when scaled by the number of binary collisions – $N_{coll}$.

The difference between heavy ion collisions and simple superposition of nucleon nucleon interactions lies in the presence of collective effects. These effects are often being studied with nuclear modification factor $R_{AB}$, which is calculated as the ratio of particle yield in heavy ion collision to yield of same particles in proton-proton collisions scaled by number of binary collisions - $N_{coll}$.

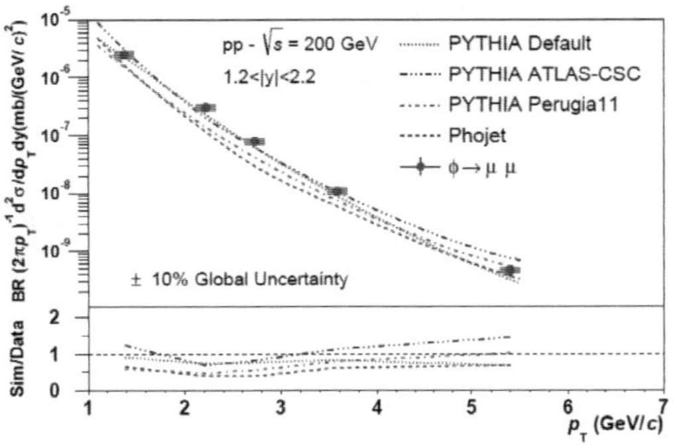


**Figure 3.** Invariant transverse momentum spectrum of $\phi$-mesons in p+p collisions at $\sqrt{s_{NN}}$ = 200 GeV compared to predictions of different Pythia tunes and Phojet.

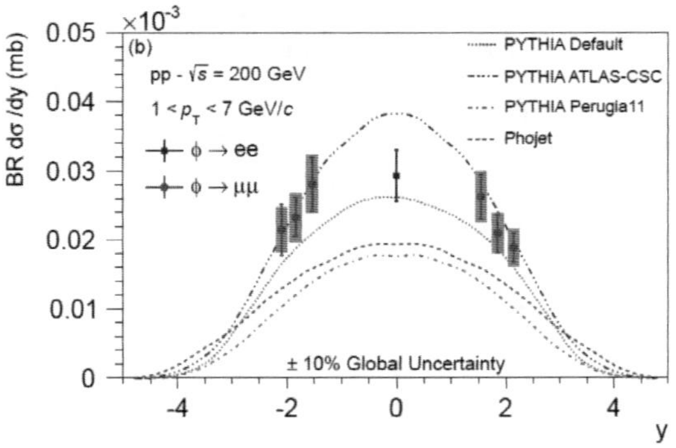


**Figure 4.** Rapidity dependence of $\phi$-mesons production in p+p collisions at $\sqrt{s_{NN}}$ = 200 GeV compared to predictions of different Pythia tunes and Phojet.

### *3.1. $R_{dA}$ measurements*

Since the number of nucleons participating in *d*+Au interaction is rather small, such collisions are used as a control experiment where QGP is not formed. Figure 5 shows nuclear modification factors obtained in *d*+Au collisions at $\sqrt{s_{NN}}$ = 200 GeV in PHENIX central arms for most central (a) and peripheral (b) collisions.

Nuclear modification factors in peripheral collisions are equal to unity, which means that peripheral *d*+Au collisions can be described as a sequence of non-correlated binary collisions. In central *d*+Au collisions, cold nuclear matter effects can be seen: non-zero enhancement in particle production at intermediate $p_T$ region, which is significantly different between baryons and mesons and a hint of hadron suppression in high $p_T$ region. It is important to note that the behavior of mesons with strange quarks is similar to other mesons even though particle masses and quark contents are different.

Particle enhancement at intermediate $p_T$ is a characteristic of Cronin effect which is often explained by multiple parton scatterings in the initial state; although this explanation cannot describe baryon/meson differences seen in the central *d*+Au collisions.

Figure 6 presents nuclear modification factors of $\phi$-mesons obtained in *d*+Au collisions at 200 GeV at forward and backward rapidities from most central to peripheral collisions. In the Au-going direction, clear enhancement can be seen, which is a characteristic of Cronin effect, while in *d*-going direction $\phi$-meson yields are suppressed, which may suggest influence of shadowing. This effect was previously seen by PHOBOS in charged hadron density [11]. The enhancement and suppression gradually decrease from central to peripheral collisions.

At the present moment, there is no numerical description of rapidity dependence observed for nuclear modification factors of $\phi$-mesons in $d$+Au collisions at $\sqrt{s_{NN}}$ = 200 GeV. Therefore, the explanation of the Cronin effect as soft multiple parton scatterings in the initial state may not be fully justified.

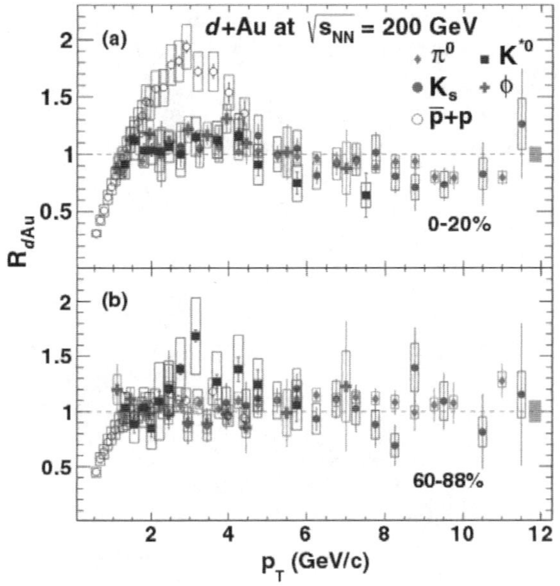


**Figure 5.** Nuclear modification factors obtained for $\pi^0$, $K^*$, $K_s$, $\phi$-mesons and protons in PHENIX central arms in most central (a) and peripheral (b) $d$+Au collisions at $\sqrt{s_{NN}}$ = 200 GeV. Lines and boxes represent statistic and systematic uncertainties respectively.

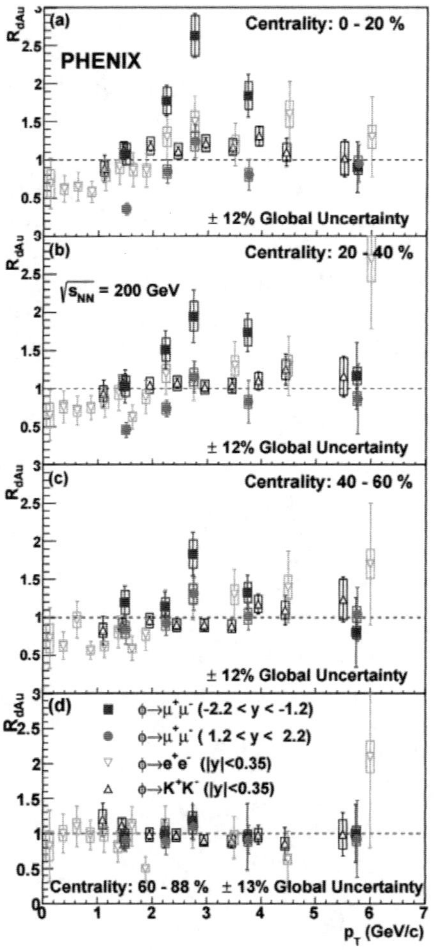


**Figure 6.** Nuclear modification factors obtained for $\phi$-mesons in $d$+Au collisions at $\sqrt{s_{NN}}$ = 200 GeV in forward and backward rapidities from most central to peripheral collisions. Lines and boxes represent statistic and systematic uncertainties respectively.

### *3.2. $R_{AA}$ measurements*

Figure 7 shows the comparison of nuclear modification factors obtained in Cu+Cu collisions at $\sqrt{s_{NN}}$ = 200 GeV for $K_s$, $K^*$, $\phi$ and $\pi^0$-mesons at different centralities. In peripheral collisions, $R_{CuCu}$ factors are equal to unity, which suggests sequential non-correlated nucleon-nucleon interactions. In central collisions, all particle yields are equally suppressed at high $p_T$ by a factor of 2. In the intermediate $p_T$

region, the suppression of particles containing strange quarks is smaller than that of pions. Despite the mass difference, all particles with strange quarks have the same suppression pattern.

Figure 8 presents $R_{AA}$ comparison between Cu+Cu and Au+Au results at $\sqrt{s_{NN}}$ = 200 GeV for the similar number of nucleons participating in the interaction – $N_{part}$. In peripheral collisions, nuclear modification factors are equal to unity except for protons, which show non-zero enhancement. In central collisions, a clear suppression hierarchy can be seen: the pion's yield is suppressed, the proton's yield is enhanced, while particles containing strange quarks lie in between.

Intermediate $p_T$ results obtained in Cu+Cu and Au+Au collisions at $\sqrt{s_{NN}}$ = 200 GeV can be qualitatively described with recombination models [12]. After comparison with data, these models indicate that for $\phi$-mesons recombination of thermal partons dominates over hard processes in a wider $p_T$ range than for pions, which can qualitatively explain the difference in $R_{AA}$ values between particles with strange quarks and pions. The difference between baryons and mesons originates from baryons obtaining larger kick in momentum than mesons, because of the presence of 3 quarks instead of 2.

Recombination models assume that there should be a source of thermal partons, which can be associated with quark gluon plasma. It is important to note that for RHIC recombination models do not give numerical predictions of particle production spectra, which can be compared with data.

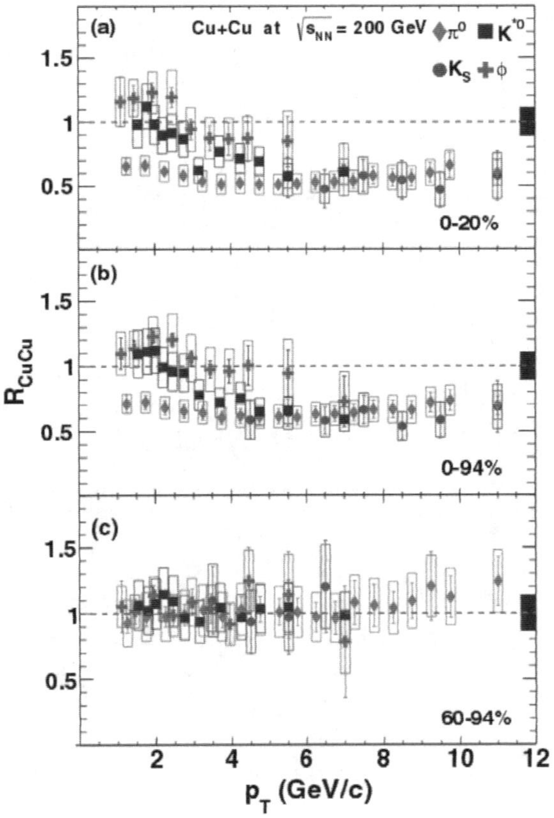


**Figure 7.** Comparison of nuclear modification factors obtained in Cu+Cu collisions at $\sqrt{s_{NN}}$ = 200 GeV for $K_s$, $K^*$, $\phi$ and $\pi^0$-mesons at different centralities. Lines and boxes represent statistic and systematic uncertainties respectively.

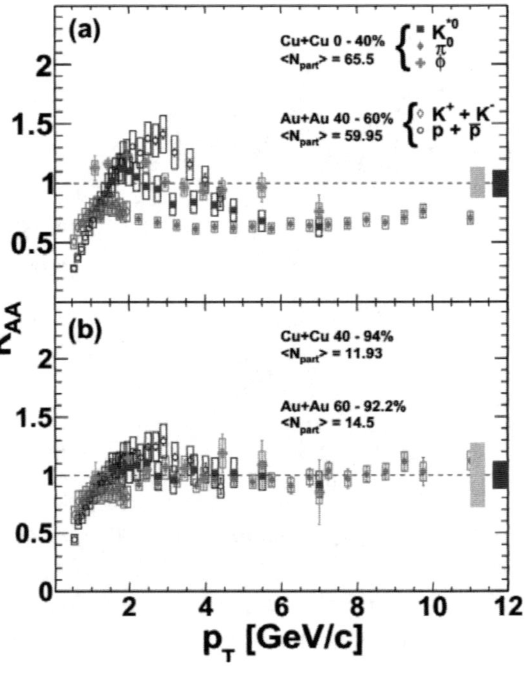


**Figure 8.** $R_{AA}$ comparison between Cu+Cu and Au+Au results at $\sqrt{s_{NN}}$ = 200 GeV for the similar number of nucleons participating in the interaction. Lines and boxes represent statistic and systematic uncertainties respectively.

Another approach to describe intermediate $p_T$ difference in $R_{AA}$ is radial flow. High multiplicity of particles produced in heavy ion collisions leads to intense interaction between hadrons. The evolution of heavy ion collisions suggests that a phase of a quickly expanding strongly interacting system exists. Each hadron gets increases in velocity equal to velocity of the wave front. The heavier the particle the more momentum it gets with the same velocity increase. This approach can qualitatively describe the difference in $R_{AA}$ between mesons containing light quarks and baryons.

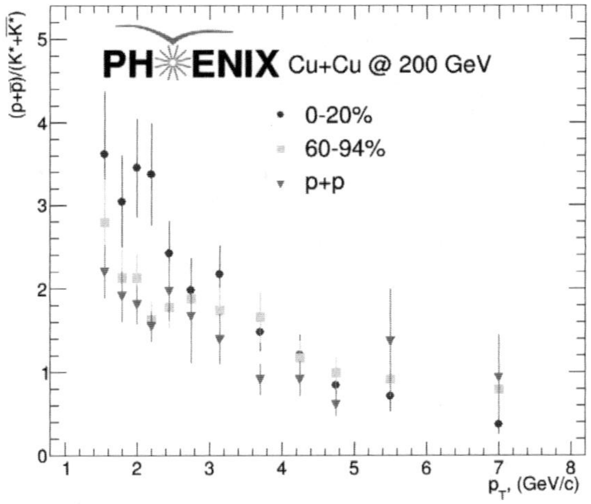


**Figure 9.** Proton to $K^*$-meson yields ratio in p+p (Cu+Cu) collisions at $\sqrt{s_{NN}}$ = 200 GeV. Lines represent combined statistic and systematic uncertainties.

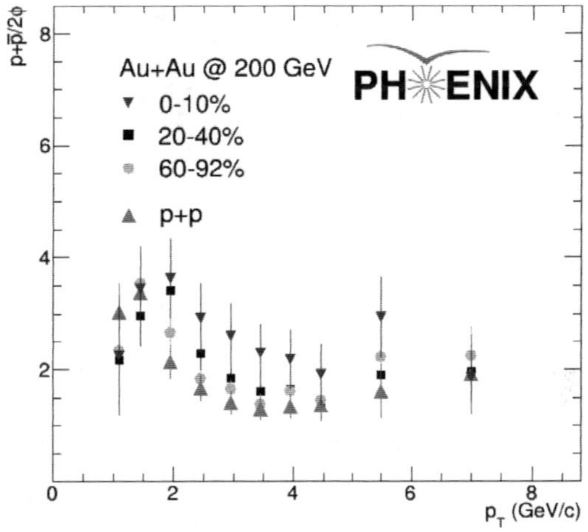


**Figure 10.** Proton to $\phi$-meson yields ratio in p+p (Au+Au) collisions at $\sqrt{s_{NN}}$ = 200 GeV. Lines represent combined statistic and systematic uncertainties.

Figure 9 and figure 10 show particle yield ratios: baryons to mesons with strange quarks. Particles for these ratios were chosen to have similar masses but different quark content. Both ratios (protons to $K^*$-mesons in p+p (Cu+Cu) collisions at $\sqrt{s_{NN}}$ = 200 GeV and protons to $\phi$-mesons in p+p (Au+Au) collisions at $\sqrt{s_{NN}}$ = 200 GeV) show a hint of flattening in a $p_T$ region up to 2.5 GeV/$c$, which may suggest that spectra shapes are determined by the mass of the particle in this $p_T$ region. This fact may be the evidence of radial flow development.

PHENIX has also measured the production of particles with strange quarks in $\sqrt{s_{NN}}$ = 62.4 GeV collisions. High $p_T$ particle suppression is explained by parton energy loss in a hot and dense matter formed in central heavy ion collisions. The change in collision energy changes parton energy loss and system size, and therefore allows for the study of particle production in different conditions.

Figure 11 shows the comparison of nuclear modification factors obtained for $\phi$-mesons, pions and protons in central Cu+Cu and Au+Au collisions at $\sqrt{s_{NN}}$ = 62.4 GeV. Hadron yields are less suppressed; although the suppression pattern is similar to the one observed in heavy ion collisions at $\sqrt{s_{NN}}$ = 200 GeV. Pion yields are suppressed, and proton yields are enhanced while $\phi$-mesons, particles with strange quarks, lie in between. This fact may suggest that recombination works here as well and the source of thermal partons can be associated with quark gluon plasma.

## 4. Conclusions

Particles with strangeness content are a perfect tool to study hadron production mechanisms and properties of hot and dense matter formed in heavy ion collisions. Peripheral $d$+Au and heavy-ion collisions at $\sqrt{s_{NN}}$ = 200 GeV follow binary scaling. Central $d$+Au collisions at $\sqrt{s_{NN}}$ = 200 GeV show non-zero Cronin enhancement at intermediate $p_T$ and a hint of suppression at high $p_T$. Rapidity dependence of nuclear modification factors for $\phi$-mesons has been observed in $d$+Au collisions at $\sqrt{s_{NN}}$ = 200 GeV.

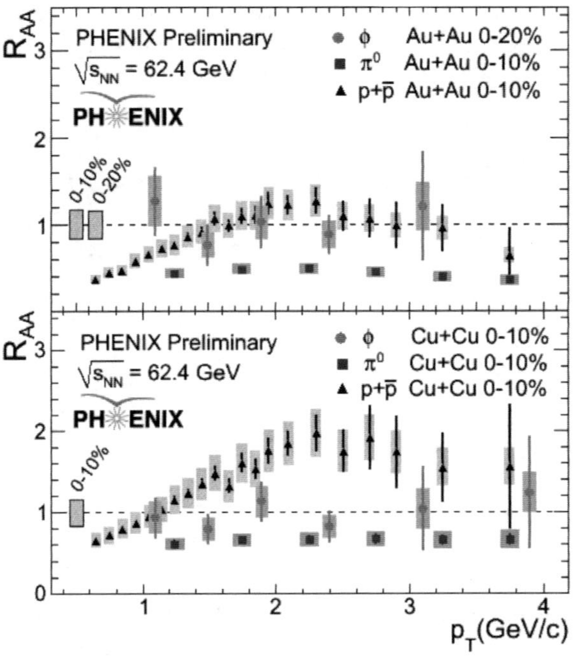


**Figure 11.** Nuclear modification factors for $\phi$, $\pi^0$-mesons and protons in central Cu+Cu and Au+Au collisions at $\sqrt{s_{NN}}$ = 62.4 GeV. Lines and boxes represent statistic and systematic uncertainties respectively.

In central heavy ion collisions at $\sqrt{s_{NN}}$ = 200 GeV at high $p_T$ all hadrons are equally suppressed while at intermediate $p_T$ suppression of strange mesons lie between pions and protons. A similar suppression pattern is observed in heavy ion collisions at $\sqrt{s_{NN}}$ = 62.4 GeV. Recombination and radial flow are two alternative explanations of experimental results.

## References

[1] Adcox K, Adler S S, Afanasiev S, Aidala C, Ajitanand N N *et al.* 2005 *Nucl. Phys. A* **757** 184–283
[2] Baym G 2002 *Nucl. Phys. A.* **698** 23–32
[3] Ackermann K H, Adams N, Adler C, Ahammed Z, Ahmad S *et al.* 2003 *NIM A* **499** 624–632
[4] Adcox K, Adler S S, Aizama M, Ajitanand N N, Akiba Y *et al.* 2003 *NIM A* **499** 469–479
[5] Adcox K, Ajitanand N N, Alexander J, Autrey D, Averbeck R *et al.* 2003 *NIM A* **499** 489–507
[6] Aphetceche L, Awes T C, Banning J, Bathe S, Bazilevsky A *et al.* 2003 *NIM A* **499** 521–536
[7] Aizawa M, Akiba Y, Begay R, Burward-Hoy J, Chappell R B *et al.* 2003 *NIM A* **499** 508–520
[8] Llope W J, Nussbaum T, Eppley G, Velkovska J, Chujo T *et al.* 2008 *NIM A* **596** 430–433
[9] Akikawa H, Al-Jamel A, Archuleta J B, Archuleta J R, Armendariz R *et al.* 2003 *NIM A* **499** 537–548
[10] Adare A, Afanasiev S, Aidala C, Ajitanand N N, Akiba Y *et al.* 2011 *Phys. Rev. C* **83** 024909
[11] Back B B, Baker M D, Ballintijn M, Barton D S, Becker B *et al.* 2005 *Phys. Rev. C* **72** 031901
[12] Hwa R and Yang C B 2007 Phys. Rev. C 75 054904

# Correlations and flavors in jets in ALICE

**Filip Krizek for the ALICE collaboration**
Nuclear Physics Institute, Hlavní 130, Řež, Czech Republic
E-mail: filip.krizek@cern.ch

**Abstract.** We report on the measurement of hadron composition in charged jets in pp at $\sqrt{s}$ = 7 TeV and show the first data on particle type dependent jet fragmentation at the LHC. Further, we present $(\Lambda + \bar{\Lambda})/2K^0_S$ ratios measured in charged jets in Pb–Pb collisions at $\sqrt{s_{NN}} = 2.76$ TeV and in p–Pb collisions at $\sqrt{s_{NN}} = 5.02$ TeV. While the ratio of the inclusive $p_T$ spectra of $\Lambda$ and $K^0_S$ exhibits centrality dependent enhancement both in the Pb–Pb and p–Pb system, the $(\Lambda + \bar{\Lambda})/2K^0_S$ ratio measured in charged jets reveals that jet fragmentation does not contribute to the observed baryon anomaly.

Finally, we discuss the measurement of semi-inclusive $p_T$ spectra of charged jets that recoil from a high-$p_T$ hadron trigger in Pb–Pb and pp collisions at $\sqrt{s_{NN}} = 2.76$ TeV and $\sqrt{s} = 7$ TeV, respectively. The jet yield uncorrelated with the trigger hadron is removed at the event-ensemble level without introducing a bias on the jet population, which is therefore infrared and collinear safe. The recoil jet yield in central Pb–Pb is found to be suppressed w.r.t. that from pp PYTHIA reference. On the other hand, there is no sign of intra-jet broadening even for anti-$k_T$ jets with a resolution parameter as large as $R = 0.5$.

## 1. Introduction

A jet is a collimated spray of particles that is produced in the process of parton showering, when a highly virtual parton fragments. Since it is ambiguous to associate final state hadrons with initial partons, the assignment of hadrons to jets is based on the decision of a reconstruction algorithm. Jet reconstruction algorithms are designed to recover the four-momentum of the initial parton by summing up momenta of particles in the final state. In order to have a clear correspondence between jets in theory and experiment, jet algorithms have to be infrared and collinear safe [1].

Jets have several useful properties, which predestine them as convenient probes to study the medium created in a collision of two relativistic nuclei. First of all, we have a good understanding of jet production on perturbative QCD level in elementary reactions. Further, it is known that processes with large four-momentum transfer $Q^2$ occur in the initial stage of the nucleus-nucleus collision, before the quark gluon plasma is formed. Hard-scattered partons then interact with the medium of deconfined quarks and gluons and lose energy. This is manifested by the so-called jet quenching phenomenon, when a jet pair with a high transverse momentum imbalance is observed [2, 3]. The principal goal, why the jet-medium interaction is studied, is to understand the nature of medium-induced parton energy loss mechanism and its possible connection to the strongly coupled limit of QCD.

Journal of Physics: Conference Series **668** (2016) 012018 doi:10.1088/1742-6596/668/1/012018

## 2. Jets in ALICE

A detailed description of the ALICE detector can be found in [4]. Here let us just briefly mention that precise tracking is provided by a six-layered silicon vertex tracker in combination with a large Time Projection Chamber (TPC). Both detectors are placed in the core of ALICE central barrel, where the solenoidal magnetic field reaches 0.5 T. ALICE has nearly uniform acceptance and efficiency coverage for tracks in the pseudorapidity range $|\eta| < 0.9$ in the full azimuth throughout a wide transverse momentum range from $\approx 150$ MeV/$c$ to 100 GeV/$c$. ALICE has also partial coverage by the electromagnetic calorimeter EMCal, which covers 107 deg in azimuth and $|\eta| < 0.7$ in pseudorapidity.

Jets are reconstructed either from tracks only (*charged jets*) or by combining tracks and calorimeter clusters (*full jets*). Whereas charged jets can be reconstructed in the full azimuth; acceptance of full jets is restricted by the incomplete coverage of the electromagnetic calorimeter. In order to remove partially reconstructed jets that appear at the boarder of the ALICE acceptance, the jet pseudorapidity range is limited by means of a fiducial cut, namely for charged jets we use $|\eta_{\rm jet}| < 0.9 - R$, where $R$ is the resolution parameter of the given jet algorithm. Jets are reconstructed using the infrared and collinear safe anti-$k_{\rm T}$ and $k_{\rm T}$ algorithms [5].

Soft bulk hadrons created in a heavy-ion collision constitute a significant background for jet reconstruction and cause a shift in the jet energy scale. The transverse momentum of reconstructed jets $p_{\rm T,jet}^{\rm raw}$ is therefore corrected for the average contribution of the bulk hadrons using the formula,

$$p_{\rm T,jet} = p_{\rm T,jet}^{\rm raw} - \rho A_{\rm jet} \ ,$$

where $A_{\rm jet}$ denotes jet area and $\rho$ is the mean background density. The mean background density is estimated on an event-by-event basis employing the standard area based method [6]. The finite width of the jet energy resolution is given by two sources. The first source of jet energy smearing are the local background fluctuations that occur in each event. The second contributor to the jet energy resolution is detector effects. The response matrix that relates the $p_{\rm T}$ of reconstructed and true jets is assumed to factorize into a product of matrices that describe momentum smearing by local background fluctuations and momentum smearing by detector effects. The matrix is inverted by means of common regularized unfolding techniques based on the SVD decomposition [7] or Bayes' theorem [8] implemented in the RooUnfold package [9].

## 3. Hadron structure of charged jets in pp collisions at $\sqrt{s} = 7$ TeV

Data on hadron composition of jets provides an important benchmark for the theory and is also needed for the fine tuning of the commonly used event generators such as PYTHIA [10]. ALICE has measured the inclusive $p_{\rm T}$ spectra of hadrons in charged jets in pp collisions at $\sqrt{s} = 7$ TeV [11]. To extract the yields of identified particles in a given track $p_{\rm T}$ bin, we use the ionization energy loss d$E$/d$x$ measured by the ALICE TPC. The fractions of the most abundant particle species, π, K, p, and e, are extracted using two methods. The first one parameterizes the d$E$/d$x$ distribution corresponding to individual particle species in a given track $p_{\rm T}$ bin with Gaussian functions that have the width and mean constrained by an analytic model [12, 13]. The second approach substitutes the Gaussian functions with data driven templates of particle d$E$/d$x$ [14]. Both models provide consistent results.

In Fig. 1, we present the fully corrected $p_{\rm T}$ spectra of pions, kaons, and protons in charged anti-$k_{\rm T}$ jets with $R = 0.4$. This data is the first measurement of particle type dependent jet fragmentation at the LHC. Let us just briefly mention that the relative abundance of K w.r.t. π grows with the increasing fraction of $p_{\rm T}$ carried by the particle in the jet. On the other hand, with the increasing fraction of $p_{\rm T}$ carried by the jet constituent, protons are getting suppressed w.r.t. pions, which signals the suppression of the leading baryon in the jet.

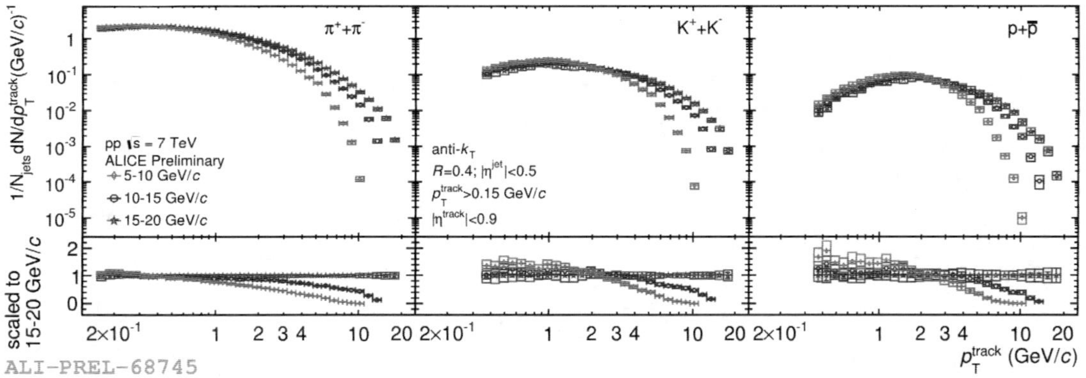


**Figure 1.** $p_T$ spectra of π, K, and p in anti-$k_T$ jets with $R = 0.4$ in pp collisions at $\sqrt{s} = 7$ TeV. The spectra are shown for three jet transverse momentum bins: 5–10 GeV/$c$, 10–15 GeV/$c$, and 15–20 GeV/$c$.

## 4. $\Lambda/K^0_S$ ratio in charged jets in Pb–Pb and p–Pb

In [15], PHENIX reported that the ratio of the inclusive $p_T$ spectra of p to π exhibits a strong centrality dependence in Au–Au collisions at $\sqrt{s_{NN}} = 200$ GeV. A similar behavior of the ratio was observed also for other baryon to meson ratios, e.g., $\Lambda/K^0_S$ [16]. Hence, this effect is sometimes referred to as the baryon anomaly.

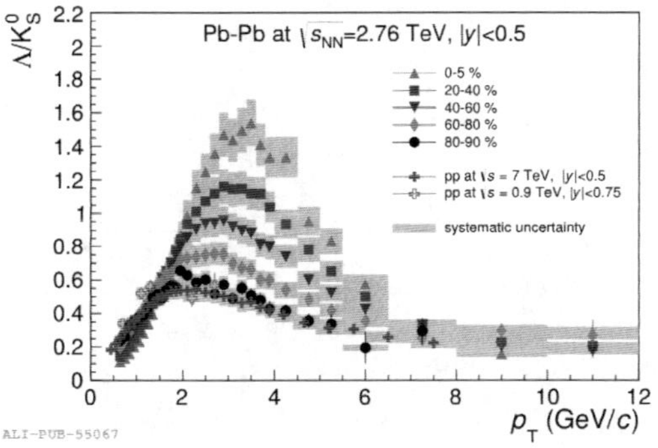


**Figure 2.** Ratio of inclusive $p_T$ spectra of Λ and $K^0_S$ measured in Pb–Pb and pp collisions. Centrality bins used in Pb–Pb are quoted in legend. For more details see [17].

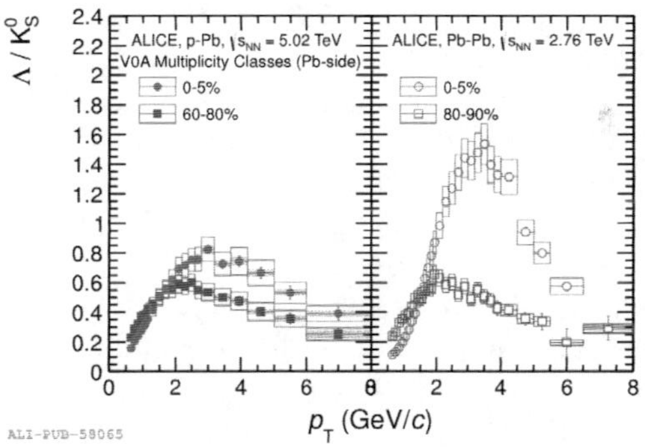


**Figure 3.** Ratio of inclusive $p_T$ spectra of Λ and $K^0_S$ measured in p–Pb (left) and Pb–Pb (right) collisions. The corresponding centrality selections are quoted in the legend. See [25] for more details.

Using Λ and $K^0_S$ particles allows to extend particle identification to higher transverse momentum, when compared to the p/π case. Figure 2 shows the ratio of inclusive spectra of Λ and $K^0_S$ as measured by the ALICE experiment in Pb–Pb collisions at $\sqrt{s_{NN}} = 2.76$ TeV and in pp at $\sqrt{s} = 0.9$ TeV and 7 TeV [17]. While pp and peripheral Pb–Pb collisions have compatible $\Lambda/K^0_S$ ratios, in more central Pb–Pb collisions, a gradually increasing enhancement in the range 2–7 GeV/$c$ is observed. Above 7 GeV/$c$, all data sets tend to pp, which indicates that the dominant mechanism for particle production in this region is the jet fragmentation. On the other hand, the evolution of the ratio below 2 GeV/$c$ with $p_T$ and centrality is well reproduced by a hydro model calculation [17, 18, 19, 20]. The underlying mechanism of particle

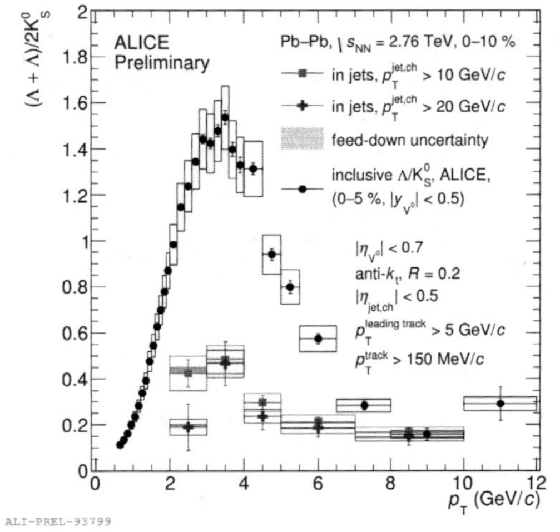


**Figure 4.** Ratio of $p_T$ spectra of $\Lambda + \bar{\Lambda}$ and $K^0_S$ measured in 0–10% most central Pb–Pb collisions at $\sqrt{s_{NN}} = 2.76$ TeV. Black circles show the ratio obtained from inclusive spectra, red squares, and blue crosses correspond to the ratios measured in charged jets with transverse momentum larger than 10 GeV/$c$ and 20 GeV/$c$, respectively.

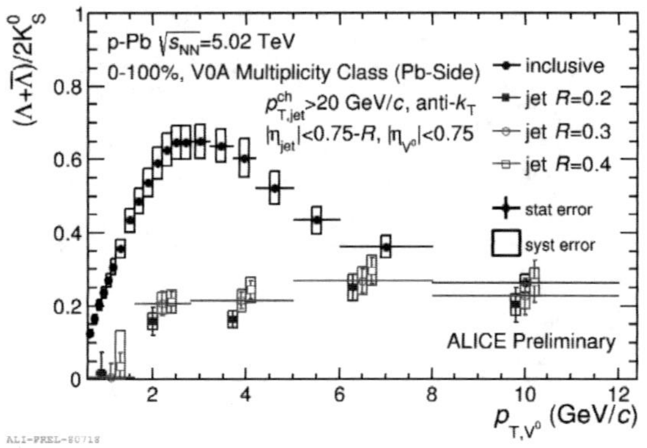


**Figure 5.** Ratio of $p_T$ spectra of $\Lambda + \bar{\Lambda}$ and $K^0_S$ measured in minimum bias p–Pb collisions at $\sqrt{s_{NN}} = 5.02$ TeV. Black circles show the ratio obtained from inclusive spectra. Color markers show the ratio measured in charged anti-$k_T$ jets for three sizes of the resolution parameter $R$. Charged jets have transverse momentum larger than 20 GeV/$c$.

production in the range 2–7 GeV/$c$ is still a matter of discussion [21, 23, 22, 24], but note that the signature of baryon anomaly is seen also in the p–Pb at $\sqrt{s_{NN}} = 5.02$ TeV system [25] (see Fig. 3). In these small systems, flow-like patterns might emerge because of the color reconnection as suggested by [26].

A natural question to be asked is whether the baryon anomaly is produced only in the bulk or whether some in-medium modified jet fragmentation also contributes. An answer to this question was searched for in the Pb–Pb and p–Pb data in parallel. The analysis required to divide $V^0$ candidates into two groups: i) $V^0$s that can be associated with a jet and ii) $V^0$ that come from the underlying event. Charged jets were reconstructed using the anti-$k_T$ algorithm with resolution parameter $R = 0.2$, 0.3, and 0.4. In order to have in the Pb–Pb system a better separation between the combinatorial background jets that are composed of underlying event particles only and the real jets that contain particles produced by parton fragmentation, jets were required to contain a leading track with $p_T$ larger than 5 GeV/$c$. Note that such condition makes the selected sample of jets more biased as will be discussed in the next section.

$V^0$ reconstruction was based on charged decay channels $\Lambda \to p + \pi^-$ and $K^0_S \to \pi^+ + \pi^-$ and employed topological cuts. As the decay daughters of $V^0$ particles are secondary tracks, they were excluded from the jet reconstruction procedure. The decision to associate $\Lambda$ or $K^0_S$ candidate with a given jet was done based on their mutual angular distance,

$$\sqrt{\Delta\varphi^2_{\mathrm{jet},V^0} + \Delta\eta^2_{\mathrm{jet},V^0}} < R .$$

Here $\Delta\varphi_{\mathrm{jet},V^0}$ ($\Delta\eta_{\mathrm{jet},V^0}$) denote the azimuthal angle (pseudorapidity) distance between the $V^0$ candidate momentum and the jet axis, and $R$ is the jet resolution parameter. The extracted yield of $V^0$ candidates in jets was corrected for the expected contribution of $V^0$ from the underlying event and for the reconstruction efficiency. In addition, the yield of $\Lambda$ particles was corrected for a contribution coming from the feed down of $\Xi$ cascades.

Figures 4 and 5 compare the $(\Lambda + \bar{\Lambda})/2K^0_S$ ratio obtained from the inclusive $V^0$s and from $V^0$s associated with jets with $p^{ch}_{T,jet} > 20$ GeV/$c$ in central Pb–Pb collisions at $\sqrt{s_{NN}} = 2.76$ TeV and in two centrality bins of p–Pb collisions at $\sqrt{s_{NN}} = 5.02$ TeV. In both systems, the ratio measured in jets is significantly lower than the inclusive one. This observation thus supports the picture that fragmentation of hard partons producing charged jets with $p^{ch}_{T,jet} > 20$ GeV/$c$ does not contribute to the observed baryon anomaly. A similar conclusion was reported earlier for p/$\pi$ ratio using two particle correlations [27]. The source of the baryon anomaly thus seems to be intimately connected with partons having smaller transverse momenta [21].

## 5. h-jet correlation measurement

The multiplicity of hadrons in a central heavy-ion collision is large. In such an environment, the jet reconstruction algorithm often clusters together just soft bulk particles. As a result an artificial jet is created. One possibility how to suppress the number of these artificial jets is to require that the reconstructed jet contains at least one constituent with $p_T$ above some preset threshold. This condition, however, imposes a bias on the jet fragmentation, which can be unwanted especially in the situation, when we look for quenched jets.

Hadron-jet coincidence measurements offer an elegant way how to overcome this problem. The approach, which we will discuss, allows to remove the contribution of combinatorial background jets including multi-parton interaction without imposing fragmentation bias on the reconstructed jet. The method is data driven and can be used also for jets with large $R$ and low $p_T$. The detailed description can be found in [28, 29].

The basic steps of the analysis can be summarized as follows. We select events that contain a high-$p_T$ hadron (trigger track). The presence of a high-$p_T$ hadron identifies collisions, where a hard scattering occurred. In these events, we analyze jets that recoil nearly back to back in the azimuth w.r.t. to the trigger track; e.g., in this analysis, it was requested that the opening angle between the trigger track and the recoiling jet is larger than $\pi - 0.6$ rad in the azimuth. Figure 6 shows a comparison of two per trigger normalized background density corrected semi-inclusive $p_T$ distributions of recoil jets associated to exclusive trigger $p_T$ bins, $p_{T,trig} \in \{20, 50\}$ GeV/$c$ and $p_{T,trig} \in \{8, 9\}$ GeV/$c$. For brevity from now on $p_{T,trig}$ will be labeled as TT. In general, a trigger track with larger $p_T$ comes on average from a hard scattering process with larger $Q^2$; therefore, the corresponding recoil jet $p_T$ spectrum is harder. Nevertheless, in the region, where the mean background density corrected jet $p_T$ is around or below zero, both spectra turn out to be nearly identical. This part of the distribution is dominated by accidental combinations of the trigger track with uncorrelated combinatorial background jets. The data suggests that the number of such combinations is largely independent of TT bin.

Assuming that the number of combinatorial background jets associated with a trigger track is independent of trigger track $p_T$, it is possible to introduce the following observable

$$\Delta_{recoil} = \left.\frac{1}{N_{trig}} \frac{d^2 N_{jet}}{dp^{ch}_{T,jet} d\eta_{jet}}\right|_{TT\{20,50\}} - \left.\frac{1}{N_{trig}} \frac{d^2 N_{jet}}{dp^{ch}_{T,jet} d\eta_{jet}}\right|_{TT\{8,9\}}, \quad (1)$$

where $N_{trig}$ is the number of trigger tracks in a given TT bin. In this observable the contribution of combinatorial background jets is removed. Further, let us point out that $\Delta_{recoil}$ has a direct link to theory, since the per trigger yield of recoil jets can be expressed in terms of the cross-section to produce a high-$p_T$ hadron and the cross section to produce a high-$p_T$ hadron together with a jet, i.e.,

$$\left.\frac{1}{N^{AA}_{trig}} \frac{d^2 N^{AA}_{jet}}{dp^{ch}_{T,jet} d\eta_{jet}}\right|_{p_{T,trig} \in TT} = \left.\left(\frac{1}{\sigma^{AA \to h+X}} \cdot \frac{d^2 \sigma^{AA \to h+jet+X}}{dp^{ch}_{T,jet} d\eta_{jet}}\right)\right|_{p_{T,h} \in TT}. \quad (2)$$

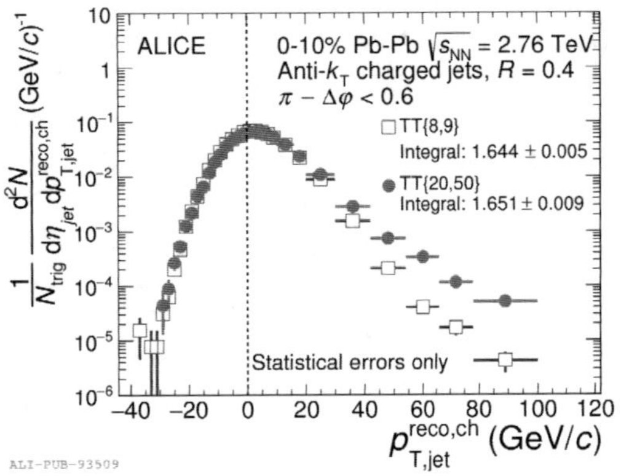


**Figure 6.** Semi-inclusive distributions of jets recoiling from a hadron trigger with $p_{\mathrm{T,trig}} \in \{20, 50\}$ GeV/$c$ (red circles) and $p_{\mathrm{T,trig}} \in \{8, 9\}$ GeV/$c$ (blue squares) for 0–10% central Pb–Pb collisions at $\sqrt{s_{\mathrm{NN}}} = 2.76$ TeV.

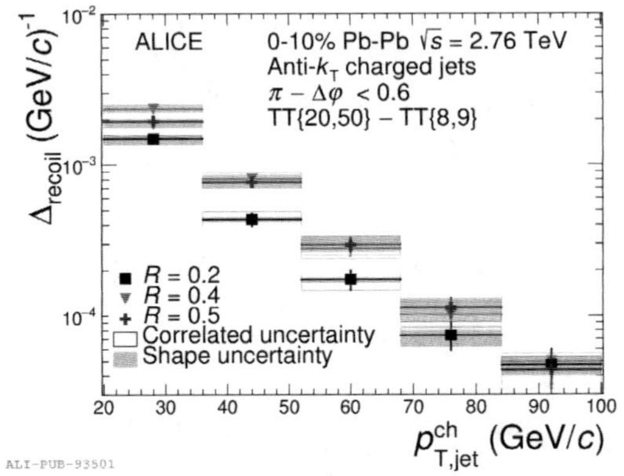


**Figure 7.** Fully corrected $\Delta_{\mathrm{recoil}}$ distributions measured for 0–10% central Pb–Pb collisions at $\sqrt{s_{\mathrm{NN}}} = 2.76$ TeV for anti-$k_{\mathrm{T}}$ jets with $R = 0.2$, 0.4, and 0.5. The boxes indicate correlated and shape systematic uncertainties, see [29] for more details.

The requirement to have a high-$p_{\mathrm{T}}$ trigger in event introduces some interesting biases on the selected sample of events and recoil jets. E.g., the parton that produces the trigger particle is more biased to be close to the surface of the collision zone [30]; therefore, the recoiling jet will have on average longer path length through the medium. Events containing a hard particle are also more biased to have larger number of participants or in other words to be more central. In both TT bins that are used, this bias is the same.

In Figure 7, we present $\Delta_{\mathrm{recoil}}$ distributions for several values of resolution parameter. The distributions are corrected for background fluctuations and detector effects. The medium-induced modification of jet fragmentation can be then searched for by means of the ratio

$$\Delta I_{\mathrm{AA}} = \frac{\Delta^{\mathrm{PbPb}}_{\mathrm{recoil}}}{\Delta^{\mathrm{pp}}_{\mathrm{recoil}}}, \tag{3}$$

where the $\Delta_{\mathrm{recoil}}$ distribution measured in Pb–Pb is divided by the reference $\Delta_{\mathrm{recoil}}$ distribution from pp collisions at the same center of mass energy per nucleon-nucleon pair. As ALICE data from pp at $\sqrt{s} = 2.76$ TeV have poor statistics, we use a reference spectrum generated by the PYTHIA Perugia 2010 tune [10]. To test the reliability of the PYTHIA prediction, we have cross-checked the PYTHIA calculation with the measured $\Delta_{\mathrm{recoil}}$ spectrum obtained from pp at $\sqrt{s} = 7$ TeV data, see Fig. 8. The pp analysis closely followed what was done for Pb–Pb. In general, we can say that PYTHIA Perugia tunes provide predictions that are compatible with the measured data within the quoted statistical and systematic uncertainties.

Figure 9 shows the $\Delta I_{AA}$ obtained with PYTHIA Perugia 10 reference for anti-$k_{\mathrm{T}}$ jets with $R = 0.5$. The ratio exhibits a deficit in the yield of recoil jets. The magnitude of this suppression for jets with $R = 0.2$ and 0.4 is similar.

The $\Delta_{\mathrm{recoil}}$ observable allows also to look for possible medium-induced modification of the jet shape. Figure 10 presents the ratio of $\Delta_{\mathrm{recoil}}$ spectra of jets obtained for different $R$. The measured data are compared with vacuum pp trend predicted by PYTHIA. Within the quoted errors there is no evidence for significant energy redistribution w.r.t. PYTHIA pp data. This also means that there is no evidence for intra-jet broadening up to $R = 0.5$.

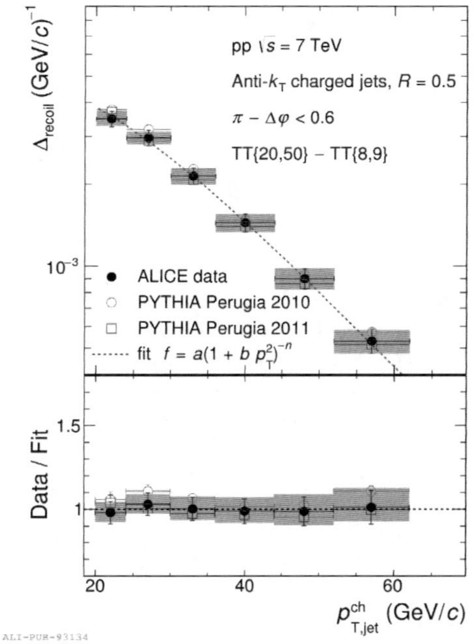


**Figure 8.** Upper panel: $\Delta_{\mathrm{recoil}}$ distributions corresponding to anti-$k_{\mathrm{T}}$ jets with $R = 0.5$ in pp collisions at $\sqrt{s} = 7$ TeV. Comparison of measured data and PYTHIA Perugia 2010 and 2011 calculation. Systematic uncertainties on the measured data points are shown as gray boxes. The measured data are fit with a smooth function. Bottom panel: ratios of the data sets to the fit.

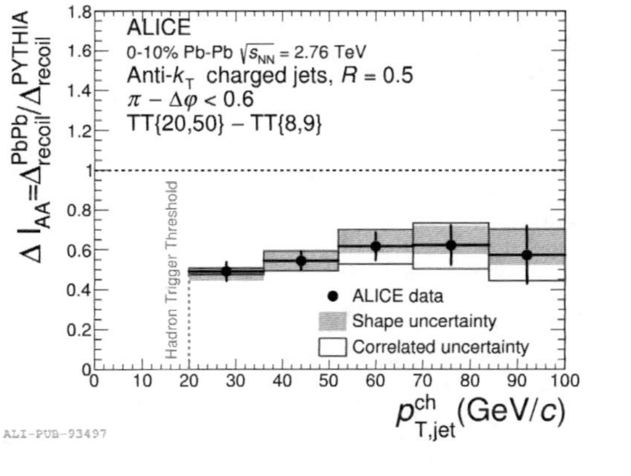


**Figure 9.** $\Delta I_{\mathrm{AA}}$, the ratio of $\Delta_{\mathrm{recoil}}$ measured in central Pb–Pb and pp collisions at $\sqrt{s_{\mathrm{NN}}} = 2.76$ TeV. $\Delta_{\mathrm{recoil}}$ for pp collision was calculated with PYTHIA Perugia 10. The data corresponds to charged anti-$k_{\mathrm{T}}$ jets with $R = 0.5$.

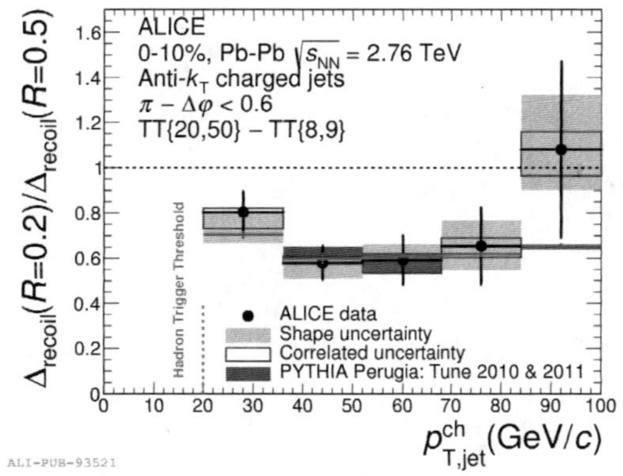


**Figure 10.** Ratio of $\Delta_{\mathrm{recoil}}$ distributions corresponding to charged anti-$k_{\mathrm{T}}$ jets with $R = 0.5$ and $R = 0.2$. The black circles show the data measured in central Pb–Pb at $\sqrt{s_{\mathrm{NN}}} = 2.76$ TeV. An estimate of the ratio in pp was obtained using PYTHIA Perugia tunes and is represented by the red bend. The width of the red bend is given by the difference between the Perugia 2010 and 2011 tune.

## 6. Summary

Jets as well as other hard probes bring us information about the early states of matter produced in a collision of two nuclei. The picture of jet-medium interaction would, however, be incomplete without the precise understanding of jet production and properties in elementary reactions such as pp. The measurement of particle type dependent jet fragmentation by ALICE in pp collisions at $\sqrt{s} = 7$ TeV thus helps to constraint available fragmentation models and event generators that are on the market.

Complex final state interaction gives rise to the baryon anomaly, which is observed both in Pb–Pb and p–Pb systems at LHC energies. Measurements of $\Lambda/\mathrm{K}^0_\mathrm{S}$ ratio done with inclusive particles and jet fragments indicate that fragmentation of partons producing charged jets with $p_{\mathrm{T,jet}}^{\mathrm{ch}} > 20$ GeV/$c$ is not a relevant source for the observed anomaly, which thus seems to emerge from processes where the transferred $Q^2$ was lower.

Hadron-jet correlation measurements offer a new variety of observables well suited to study also low-$p_\mathrm{T}$ jets with large resolution parameter without inducing a fragmentation bias. PYTHIA Perugia tunes seem to provide reliable prediction for the measured h-jet observables in pp collisions, such as $\Delta_{\mathrm{recoil}}$. In Pb–Pb system, we see a suppression of the recoil jet yield but without a sign of intra-jet broadening for charged anti-$k_\mathrm{T}$ jets up to resolution parameter $R = 0.5$.

## Acknowledgments

The work has been supported by the MEYS grant CZ.1.07/2.3.00/20.0207 of the European Social Fund (ESF) in the Czech Republic: Education for Competitiveness Operational Programme (ECOP) and by the grant LG 13031 of the Ministry of Education of the Czech Republic.

## References

[1] Salam G 2010 *Eur. Phys. J.* C **67** 637
[2] Aad G *et al.* (ATLAS collaboration) 2010 *Phys. Rev. Lett.* **105** 252303
[3] Chatrchyan S *et al.* (CMS collaboration) 2011 *Phys. Rev.* C **84** 024906
[4] Aamodt K *et al.* (ALICE collaboration) 2008 *J. Instrum.* **3** S08002
[5] Cacciari M, Salam G P and Soyez G 2012 *Eur. Phys. J.* C **72** 1896
[6] Cacciari M and Salam G P 2008 *Phys. Lett.* B **659** 119
[7] Höcker A and Kartvelishvili V 1996 *Nucl. Instrum. Meth. A* **372** 469
[8] D'Agostini G 1995 *Nucl. Instrum. Meth. A* **362** 487
[9] Adye T *CERN–2011–006* 313
[10] Skands P 2010 *Phys. Rev.* D **82** 074018
[11] Abelev B *et al.* (ALICE collaboration) 2014 nucl-ex/1411.4969
[12] Lu X *et al.* (ALICE collaboration) 2014 *Nucl. Phys.* A **931** 428-432
[13] Lu X *CERN-THESIS–2013–179*
[14] Hess B *et al.* (ALICE collaboration) 2014 hep-ex/1408.5723
[15] Adler S S *et al.* (PHENIX collaboration) 2003 *Phys. Rev. Lett.* **91** 172301
[16] Adams J *et al.* (STAR collaboration) 2012 *Phys. Rev. Lett.* **108** 072301
[17] Abelev B *et al.* (ALICE collaboration) 2013 *Phys. Rev. Lett.* **111** 222301
[18] Song H and Heinz U W 2008 *Phys. Lett.* B **658** 279
[19] Song H and Heinz U W 2008 *Phys. Rev.* C **77** 064901
[20] Song H and Heinz U W 2008 *Phys. Rev.* C **78** 024902
[21] Cuautle E, Jimenez R, Maldonado I, Ortiz A, Paic G and Perez E 2014 hep-ph/1404.2372
[22] Topor Pop V, Gyulassy M, Barrette J and Gale C 2011 *Phys. Rev.* C **84** 044909
[23] Brodsky S J and Sickles A 2008 *Phys. Lett.* B **668** 111-115
[24] Werner K 2012 *Phys. Rev. Lett.* **109** 102301
[25] Abelev B *et al.* (ALICE collaboration) 2014 *Phys. Lett.* B **728** 25
[26] Ortiz A, Christiansen P, Cuautle E, Maldonado I and Paic G 2012 *Phys. Rev. Lett.* **111** 042001
[27] Veldhoen M (ALICE collaboration) 2013 *Nucl. Phys.* A **910-911** 306-309
[28] de Barros G O V *et al.* 2012 hep-ex/1208.1518
[29] Abelev B *et al.* (ALICE collaboration) 2015 hep-ex/1506.03984
[30] Renk T 2013 *Phys. Rev.* C **87** 2

Journal of Physics: Conference Series **668** (2016) 012019 doi:10.1088/1742-6596/668/1/012019

# Theory status of quarkonium production in proton-nucleus collisions

**J.P. Lansberg**

Institut de Physique Nucléaire, CNRS-IN2P3, Univ. Paris-Sud, Université Paris-Saclay, 91406 Orsay Cedex, France

E-mail: Jean-Philippe.Lansberg@in2p3.fr

**Abstract.** I give a brief overview of the recent theoretical progress in the study of quarkonium production in proton-nucleus collisions in view of the recent LHC and RHIC results. A special emphasis is put on the excited states such as the $\psi'$, $\Upsilon(2S)$ and $\Upsilon(3S)$.

## 1. On the importance of understanding conventional nuclear effects: the CMS $\Upsilon(nS)$ sequential suppression example.

One of the highlights of the Run1-LHC heavy-ion results is admittedly the observation made by the CMS collaboration of a relative suppression of the $n = 2, 3$ $\Upsilon$ states with respect to the $n = 1$ $\Upsilon$ state in PbPb collisions at $\sqrt{s_{NN}} = 2.76$ TeV. Such an observation was expected and even awaited for by the supporters of the phenomenon of "sequential quarkonium suppression" according to which the excited states with larger radii should suffer more from the colour screening in the deconfined matter [1].

Actually, CMS published a set of two papers [2, 3] describing this observation, the second reinforcing the first by quoting measurements of individual nuclear modification factor[1]. In discussing their results, they had in mind that other nuclear-matter effects than the creation of a quark-gluon plasma (QGP) could suppress the $\Upsilon(nS)$, yet noting :

> *[...] such effects should have a small impact on the double ratios reported here. Initial-state nuclear effects are expected to affect similarly each of the three $\Upsilon$ states, thereby canceling out in the ratio. Final-state nuclear absorption becomes weaker with increasing energy and is expected to be negligible at the LHC.*

At this point, it is fair to say that nobody claimed the contrary and that this assertion was supported by a number of theoretical studies, e.g. [4, 5, 6], where the excited states were expected to be suppressed as much as the $n = 1$ state. In particular, such expectations were in line with the idea that any phenomenon taking place into the very fast moving lead nucleus (the nuclear matter in question) and which would impact differently the $\Upsilon$ states would *de facto* violate the Heisenberg principle. Indeed, the time they take to escape this nuclear matter is way smaller, given the huge boost between their rest frame and the nucleus rest frame, than the time they

[1] For minimum bias $AB$ collisions, such factors are defined such that $R_{AB} = \sigma_{AB}/(AB\sigma_{pp})$; they equal unity in absence of nuclear effects, that is when the nucleus-nucleus collision is like an incoherent superposition of nucleon-nucleon collisions.

**Table 1.** Double ratios expressing the relative nuclear suppression of the excited $\Upsilon(2S, 3S)$ states with respect to the $\Upsilon(1S)$ state as observed by CMS [3, 7] in PbPb collisions at $\sqrt{s_{NN}} = 2.76$ TeV and in $p$Pb collisions at $\sqrt{s_{NN}} = 5$ TeV.

| $\frac{[\Upsilon(\text{nS})/\Upsilon(1\text{S})]_{ij}}{[\Upsilon(\text{nS})/\Upsilon(1\text{S})]_{pp}}$ | $2S$ | $3S$ |
|---|---|---|
| PbPb | $0.21 \pm 0.07$ (stat.) $\pm 0.02$ (syst.) | $0.06 \pm 0.06$ (stat.) $\pm 0.06$ (syst.) |
| $p$Pb | $0.83 \pm 0.05$ (stat.) $\pm 0.05$ (syst.) | $0.71 \pm 0.08$ (stat.) $\pm 0.09$ (syst.) |

need to form and to be distinguished. The situation is obviously different if one considers effects resulting from a nucleus-nucleus collision, such as that of a QGP which is produced at small velocities/rapidities in the center-of-mass frame of the nucleus-nucleus reaction. This is precisely the region covered the LHC detectors which are used to detect quarkonia.

However, in 2013, the CMS collaboration carried out the corresponding analysis in proton-nucleus collisions at 5 TeV [7]. At the great surprise of many, they uncovered a significant relative suppression of the $2S$ and $3S$ states with respect to the $1S$ state. Quantitatively speaking, if the effects responsible for the relative $nS/1S$ suppression in $p$Pb collisions quoted in Table 1 factorise (and is thus squared in the PbPb case), they could be responsible for half of the PbPb relative suppression strongly contradicting all previous expectations! This illustrates how important state-of-the-art and dedicated studies in $p$Pb collisions are.

In addition to be of crucial importance to correctly interpret nucleus-nucleus observations – as just illustrated, such proton-nucleus reactions involve many physics effects, which are of high interest on their own, and which can certainly tell us much about QCD at the interface between its pertubative and nonpertubative domain. They indeed provide means to study :

- the nuclear modification of the parton distributions of nucleons inside nuclei, also referred as to shadowing, antishadowing, EMC and Fermi-motion effects;
- the phenomenon of gluon saturation and the high-energy limit of QCD, the so-called low-$x$ physics;
- the time-evolution of a $Q\overline{Q}$ pair and the dynamics of its hadronisation;
- the propagation of quarks and gluons in a dense medium, the process of energy loss and the Cronin effect;
- the quarkonium-production mechanisms, in particular the colour octet vs. singlet contributions;
- the nonperturbative charm content of the proton;
- the collinear-factorisation framework in the nuclear medium;
- the quarkonium-hadron interactions;
- the mechanisms underlying single-spin asymetries, ...

In this proceedings contribution, I briefly review the current knowledge of quarkonium production in proton-nucleus collisions at ultra-relativistic energies, with a specific focus on quarkonium radially-excited states.

## 2. A baseline to understand the basics

Let us start by discussing a simple picture where only two effects are accounted for, namely (i) the inelastic scattering of the quarkonia with some nucleons in the nuclear matter, sometimes also called the nuclear absorption and (ii) the modification of the partonic densities in nucleons embedded in nuclei, historically referred to as the EMC effect.

The differential cross section as a function, for instance, of $y$, $P_T$ and $\vec{b}$ for the production of a quarkonium in a proton-nucleus collision can be assumed to be obtained from the partonic one via the following convolution:

$$\frac{d\sigma_{pA\to \mathcal{Q}X}}{dy\, dP_T\, d\vec{b}} = \int dx_1 dx_2\, g(x_1,\mu_F) \int dz_A \mathcal{F}_g^A(x_2,\vec{b},z_B,\mu_F) \\ \times 2\hat{s}P_T \frac{d\sigma_{gg\to \mathcal{Q}+g}}{d\hat{t}} \delta(\hat{s}-\hat{t}-\hat{u}-M^2) S_A(\vec{b},z_A) \quad (1)$$

The partonic differential cross section $\frac{d\sigma_{gg\to \mathcal{Q}+g}}{d\hat{t}}$ can in principle be evaluated from any model (Colour Singlet, Colour Octet or Colour Evaporation Model for instance. See [8, 9, 10] for reviews.) provided that it satisfactorily describes the spectra under scrutiny, *i.e.* $y$ or $P_T$. To do so, we have developped a probabilistic Glauber Monte Carlo code, JIN, which can handle any expression for $\frac{d\sigma_{gg\to \mathcal{Q}+g}}{d\hat{t}}$ [11, 12, 13], yet accounting for the impact-parameter $\vec{b}$ dependence of any nuclear effect.

Since theoretical predictions are usually compared to experimental measurements in the form of $R_{pA}$, the absolute normalisation of $\frac{d\sigma_{gg\to \mathcal{Q}+g}}{d\hat{t}}$ cancels out. In general, its normalisation uncertainty (from the heavy-quark mass, the scales and nonpertubative inputs), which can easily be on the order of a factor of 2 or 3, is significantly larger than the expected nuclear effects, up to 50% at best. It makes sense to keep looking at $R_{pA}$, rather than at $\sigma_{pA}$ not to be misguided by normalisation issues unrelated to nuclear effects.

### *2.1. The so-called nuclear break-up*

The first basic effect and ingredient entering this simple approach is the survival probability for a $Q\overline{Q}$ produced at the point $(\vec{r}_A, z_A)$ to escape the nuclear medium unscathed. It is usually parametrised as

$$S_A(\vec{r}_A,\, z_A) = \exp\left(-A\,\sigma_{\text{break-up}} \int_{z_A}^{\infty} d\tilde{z}\; \rho_A(\vec{r}_A,\tilde{z})\right), \quad (2)$$

by introducing $\sigma_{\text{break-up}}$ as the cross section for inelastic collisions between a nucleon in the nuclear matter–the nucleus–and the quarkonium ($\rho_A(\vec{r}_A, z)$ is the nuclear density). Although it is sometimes meant to account for any effect beyond those of the nuclear PDF (see below), $\sigma_{\text{break-up}}$ is nevertheless considered to be somehow connected to the size of the propagating object. In particular, *if* the meson is formed when it traverses the nuclear matter, one expects $\sigma_{\text{break-up}} \propto r^2_{\text{meson}}$. In particular, $2S$ (and $3S$) states should significantly be more suppressed owing to their significantly larger radius.

In principle, $\sigma_{\text{break-up}}$ is also connected to the meson-photoproduction cross section on nuclear target. This connection is however not always trivial [14]. Both charmonia and bottomonia typically need $0.3 \div 0.4$ fm/$c$ to form – in other words for one to be able to distinguish the $1S-2S$ energy levels in the meson rest frame. If the quarkonium momentum in the nucleus rest frame increases, this time gets boosted and the states form way outside the nuclear matter. In such a case, they are less suppressed and such a break-up mechanism should act on the same way on any state of a family. What propagates in the nuclear matter is a sort of pre-resonant state whose quantum numbers are not yet determined.

To fix the idea, in the case of a quarkonium produced at $y = 0$ in a $d$Au collision at RHIC at $\sqrt{s_{NN}} = 200$ GeV, the boost between the quarkonium rest frame and that of the gold nucleus[2] is $\gamma = E_{\text{beam,cms}}/m_N \simeq 107$. It thus takes 30 fm/$c$ for a quarkonium to form and to

[2] $\gamma(y=0) = \cosh(y_{\text{beam}} = 5.36)$.

become distinguishable from its excited states. At the LHC (5 TeV), still for a particle with $y = 0$, $\gamma \simeq 2660$ ($y_{\text{beam}} = 8.58$). The quarkonium-formation time seen from the nuclear-matter viewpoint is now $800 \div 1000$ fm/$c$. No nuclear-matter effect taking place over the size of the nucleus, *i.e.* less than 15 fm, can in principle act differently on a $J/\psi$ or $\Upsilon$ and its excited states. The boost at $y = 0$ increases when one increases the collision energies, and obviously does so when one increases $y$ in the colliding nucleon direction. This formation-time effect has for instance been invoked to explain the disappearance of the different $\psi'$ vs $J/\psi$ suppression measured by E866 [15] at Fermilab when moving from $y \simeq 0$ ($x_F \simeq 0$) to forward rapidities ($x_F \to 1$).

For extremely large values of $\gamma$, what propagates in the nucleus is probably a mere $Q\bar{Q}$ pair–whose colour state is however not known–and, in such a case, one could assume a naive high energy limit, $\sigma_{\text{break-up}} \propto \pi/m_Q^2$, that is on the order of 0.5 mb for the charmonia. The situation is however much more complex, not only because of the suppression of the gluon densities, which I discuss below, but also because of non-linear coherent effects [16] arising at high energies.

In general, one admits that the "conventional" nuclear break-up is not among the dominant effects at LHC energies–except perhaps in the backward edge of the LHCb acceptance–and can even be neglected. According to the above discussion about the formation time, what should matter to determine if $\sigma_{\text{break-up}}$ is large or not is $\sqrt{s_{\psi N}}$, or the energy of the quarkonium in the nucleus rest frame, rather than the collision energy, $\sqrt{s_{NN}}$. A global survey of data at $x_F \simeq 0$ however showed [17] that $\sigma_{\text{break-up}}$ seemed to scale and to decrease with $\sqrt{s_{NN}}$. However, I stress that a scaling and a decrease with $\sqrt{s_{\psi N}}$ was not ruled out in presence of a strong gluon shadowing (see Fig. 11 (d) of [17]) as I discuss now.

### *2.2. Implementing nuclear PDFs and their impact-parameter dependences*

In addition to final-state interactions–once the meson is produced–in the nuclear matter, one should indeed account for the nuclear modification of the initial parton densities. To do so, the nuclear gluon PDF and its spatial dependence, $\mathcal{F}_g^A(x_1, \vec{r}_A, z_A, \mu_F)$, are assumed to be factorisable [18] in terms of the nucleon gluon PDFs :

$$\mathcal{F}_g^A(x_1, \vec{r}_A, z_A; \mu_F) = \rho_A(\vec{r}_A, z_A) \times g(x_1; \mu_F) \times \left(1 + \left[R_g^A(x, \mu_F) - 1\right] N_{\rho_A} \frac{\int dz\, \rho_A(\vec{r}_A, z)}{\int dz\, \rho_A(0, z)}\right) \quad (3)$$

where[3] $R_g^A(x, \mu_F)$ is the ratio of the gluon density per nucleon in a nucleus $A$ by that in a free proton at a momentum fraction $x$ and a factorisation scale $\mu_F$. When discussing $R_g^A(x, \mu_F)$, four regions are distinguished depending on the value of $x$, namely the (i) Fermi-motion region ($x > 0.7$), (ii) the EMC region ($0.3 < x < 0.7$), (iii) the anti-shadowing region ($0.05 < x < 0.3$), (iv) the shadowing region ($x < 0.05$). Regarding the gluons, I stress that only the shadowing depletion is established, but its magnitude is still discussed. The gluon antishadowing has not yet been observed although assumed in many studies. For instance, it is absent in some nPDF fits [19]. The gluon EMC effect is even less known. We have claimed in [5] that the backward $\Upsilon$ data at RHIC may hint at a significant gluon EMC effect, perhaps stronger than for quarks.

### *2.3. Overall*

Typical comparisons at LHC energies when the break-up is neglected are shown on Figure 1. A few comments are in order: (i) as discussed in [20, 5], such evaluations do depend on the factorisation scale as well as on the nPDF set chosen[4]; The corresponding uncertainties are not

[3] $N_{\rho_A}$ is fixed such that $A^{-1}\int d^2\vec{r}_A \int dz_A \rho_A(\vec{r}_A, z_A)\left(1 + \left[R_g^A(x, \mu_F) - 1\right] N_{\rho_A} \frac{\int dz\, \rho_A(\vec{r}_A, z)}{\int dz\, \rho_A(0, z)}\right) = R_g^A(x, \mu_F)$.

[4] In addition to EPS09 [21] and nDSg [19] used here, other common nPDFs sets which could have been used are nCTEQ15[22], HKN [23], DSSZ[24], FSG[25], ...

**Figure 1.** $R_{p\text{Pb}}$ at $\sqrt{s_{NN}} = 5$ TeV: typical comparisons between the ALICE and LHCb data and the effect induced by the gluon shadowing only (see text).

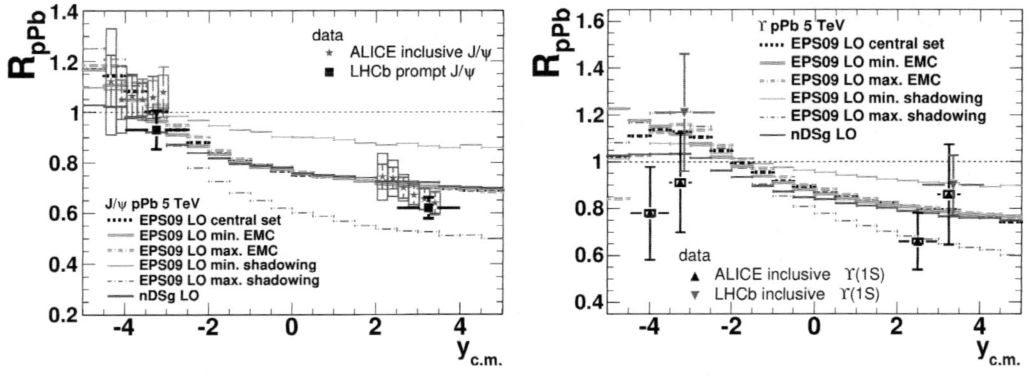


(a) $J/\psi$ (b) $\Upsilon(1S)$

shown and can be as large as than of a given nPDF set; (ii) the global normalisation uncertainties of the data due to the absence of *pp* data at 5 TeV are also not shown. These however cancel in $R_{\text{FB}}$ (see [20]). (iii) such a computation accounts for a $2 \to 2$ parton scattering. Evaluations with a simplified $2 \to 1$ kinematics and based on the colour evaporation model give similar results [4, 26] (See [13] for a detailed discussion).

RHIC results have been discussed at length within this simple picture in e.g. [12, 13, 27, 28]. $\sigma_{\text{break-up}}$ was fitted to the PHENIX data. In general, this method works except in some parts of the phase space: a counter-intuitive increase of $\sigma_{\text{break-up}}$ at forward rapidities[5], a slight $p_T$ enhancement difficult to explain at backward rapidites, a hint at a different centrality dependence as compared to nPDFs proportional to the local nuclear density, etc. For more details, I refer the interested reader to the recent review [8] and references therein.

## 3. Going further by adding some nuclear effects

Whereas the simple model discussed above probably accounts reasonably well for initial state effects via nuclear PDFs, the treatment of the final state effects is indeed rather empirical for $\sigma_{\text{break-up}}$ is fitted to the data and its value *a posteriori* confronted to qualitative expectations. As just said, it works, but some "anomalies" have lately been observed.

This has motivated more advanced theoretical studies. In particular, it was argued [29] that a coherent energy loss, scaling like the projectile energy, could explain well the data from fixed-target to LHC energies. In presence of such an energy loss, arising from interferences between initial and final state radiations, no additional break-up is needed. In a sense, it accounts for it. However, it is not clear if gluon shadowing is needed when such an energy loss is "switched on."

The treatment of small-$x$ effects like the gluon saturation, which were initially predicted [30] to be stronger than the expected gluon-shadowing suppression from global nPDF fits, have been improved. Current postdictions [31] are in better agreement with data[6]. They are thus on the same order as nPDF predictions, and it is legitimate to wonder if different physics is really at work.

Finally, it was recently noted in [8] that the nuclear suppression seen in the data seems not to show an increase from RHIC to LHC energies as most of the models had predicted it: higher

[5] except in presence of a strong gluon shadowing [13].

[6] It has also recently been claimed that the suppression may essentially depend on the quantum state of propagating pair [32], emphasising that such a small-$x$ suppression is not merely an initial state effect.

**Figure 2.** Ratios expressing the relative nuclear suppression of the $\psi'$ over the $J/\psi$ at $\sqrt{s_{NN}} = 200$ GeV in $d$Au collisions and at $\sqrt{s_{NN}} = 5$ TeV in $p$Pb collisions: comparisons between the PHENIX [33] and ALICE [35] data and the results of the comover interaction model [36]. [A 28% systematical normalisation uncertainty on the PHENIX data is not shown]. Plot from [8].

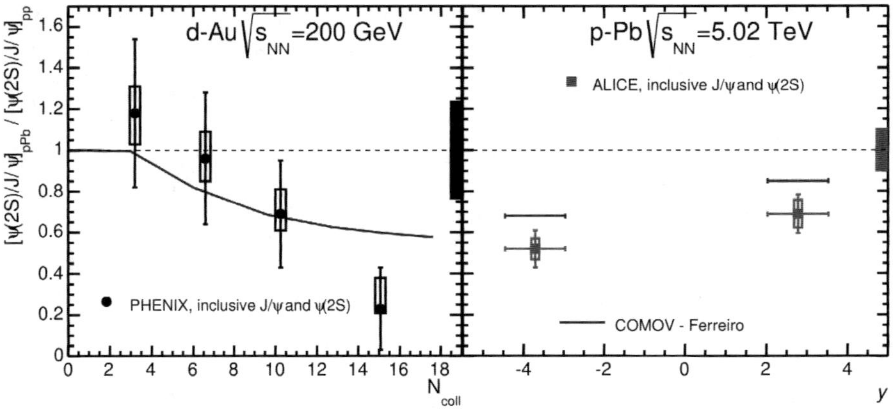


energy, smaller $x$ and stronger shadowing. Yet, it seems that the nuclear suppression decreases when progressively going to heavier systems, *i.e.* from $J/\psi$ to $\Upsilon$, passing by $b(\to J/\psi)$.

## 4. The excited state puzzle(s)

As I have reported in the introduction, CMS uncovered an unexpected relative suppression of the $\Upsilon(2S, 3S)$ state to the $\Upsilon(1S)$. As it can be read on Table 1, it is on the order 20% for the $\Upsilon(2S)$ and 30% for the $\Upsilon(3S)$. Along the lines of the discussion of section 2.1, it is extremely counter-intuitive because of the very large boost between the quarkonia and the nucleus at LHC energies.

In fact, a similar observation was also made by the PHENIX experiment at 200 GeV. They reported [33] a relative suppression of the $\psi'$ to the $J/\psi$ increasing with the centrality (but with a large ($\simeq$ 25%) global systematical uncertainty). Integrating over centrality, the ratio $[\psi'/J/\psi]_{d\mathrm{Au}}/[\psi'/J/\psi]_{pp} = 0.68 \pm 0.14^{+0.13}_{-0.08} \pm 0.16$, *i.e.* smaller than unity but with significant uncertainties. Following this analysis, we confirmed [34] that the theoretical uncertainties in the nPDF impact evaluation (mass in the kinematics, scale) is at most 2 %, *i.e.*much smaller. ALICE also performed an analysis [35] in $p$Pb collisions at 5 TeV (see Figure 2 (right panel)) and also uncovered a relative suppression of larger magnitude in the backward region than that in the forward region.

As for now, the only approach, which can describe these results without tuning parameters[7], is the comover interaction model (CIM) (see [38] and references therein). Such an approach accounts for final-state scatterings with comoving hadrons along the quarkonium trajectory. These can occur over times long enough for the quarkonium to be formed; differences between the $1S$ and $2S$ suppression can thus occur. As for the break-up cross section, one introduces an empirical comover-quarkonium inelastic cross section, which is however fixed to reproduce $AA$ data where the comover density is very large. Such a picture also naturally explains a larger relative suppression in the nucleus rapidity region (negative $y$ at the LHC) where the comover density–connected to the charged particle multiplicity–is larger. Comparisons between the CIM results and data from PHENIX and ALICE are shown on Figure 2.

[7] In [37], Du and Rapp studied the impact of some of their model parameters on the dAu RHIC relative suppression, which is then used to study the PbPb LHC relative suppression.

## 5. Conclusions and outlooks

As I argued, the study of quarkonium production in proton-nucleus collisions is of fundamental importance, if one wishes to properly interpret–quantitatively but even also qualitatively–the physics underlying the quarkonium suppression in nucleus-nucleus collisions. The case of the sequential suppression of $\Upsilon$ measured by CMS is a prime example. Just measuring quarkonium production in nucleus-nucleus collisions is not enough!

Moreover, such studies are of highest relevance for themselves. It is very important to identify the mechanisms responsible for the relative suppression of quarkonium excited states seen at RHIC and the LHC. Indeed, at high energies, final-state interactions within the nucleus–as the ones included in the conventional models discussed here–occur too early to impact differently the different energy levels. One of the mechanisms, which can naturally explain such a relative suppression is the possibility of scatterings over “long” times with comoving hadrons. This is included in the comover interaction model, which was introduced long ago to explain part of the anomalous quarkonium suppression in $AA$ collisions.

As outlooks, let us stress that LHCb recently reported preliminary results on $\psi'$ production in $p$Pb collisions [39]. Their measurement confirms the ALICE one reported above. It also has the potential to extend it with the extraction of the (relative) suppression of non-prompt $\psi'$. On general grounds, $\psi'$ from $b$ come from a process at a higher scale where some nuclear effect may be smaller but, more importantly, such $\psi'$ are produced a fraction of a millimeter–as opposed to femtometers–after the collisions. As such, it is hard to believe that they would suffer from any nuclear final-state effect. As for now, the LHCb uncertainty is somewhat too large to draw any strong conclusion, the non-prompt $\psi'$ suppression is both compatible with that of non-prompt $J/\psi$–which would be compatible with the absence of final-state interactions–but also with that of prompt $\psi'$, which would be another stunning result. As stunning as it could be, such a possibility should, however, not be ruled out given the previous surprising observations reported here which clearly challenged the established ideas.

In the longer run, let us add that ATLAS has also recently reported $J/\psi$ measurements in $p$Pb collisions [40] at finite $P_T$ and could also probably add some information on the excited state sector. Along these lines, the $P_T$ dependence of the relative $\psi'$ over $J/\psi$ suppression has still to be addressed theoretically and experimental uncertainties can certainly be reduced. Complementary measurements by LHCb in $p$Pb collisions bearing on Drell-Yan and open charm production may also shed some light on some nuclear-matter effects at work in the same kinematical region as for quarkonia. The same also holds for measurements in the fixed-target mode at the LHC, in the energy range of RHIC, but at negative $x_F$ and with unprecedented luminosities [41, 42, 43]. For an overview, I guide the reader to [44, 45].

To complete the picture, let us mention the importance of looking at quarkonium-associated production in $p$Pb collisions which could uncover a new geometric scaling regime $A^{3/2}$ vs $A^1$ [46, 47]. Recent theoretical [48, 49, 50, 51, 52, 53, 54] and experimental [55, 56, 57, 58, 59] studies in $pp$ collisions paved the way for such studies.

**Acknowledgments**
I would like to thank A. Rakotazafindrabe for the plots of Figure 1 as well as the SQM organisers, in particular D. Blaschke and J. Aichelin for the kind invitation to present this overview. This work is supported in part by the CNRS (Défi Inphyniti – Théorie LHC France).

[1] Matsui T and Satz H 1986 *Phys. Lett.* **B178** 416
[2] Chatrchyan S *et al.* (CMS) 2011 *Phys. Rev. Lett.* **107** 052302 (*Preprint* 1105.4894)
[3] Chatrchyan S *et al.* (CMS) 2012 *Phys. Rev. Lett.* **109** 222301 (*Preprint* 1208.2826)
[4] Vogt R 2010 *Phys. Rev.* **C81** 044903 (*Preprint* 1003.3497)
[5] Ferreiro E G, Fleuret F, Lansberg J P, Matagne N and Rakotozafindrabe A 2013 *Eur. Phys. J.* **C73** 2427 (*Preprint* 1110.5047)
[6] Arleo F and Peigne S 2013 *JHEP* **03** 122 (*Preprint* 1212.0434)
[7] Chatrchyan S *et al.* (CMS) 2014 *JHEP* **04** 103 (*Preprint* 1312.6300)

[8] Andronic A *et al.* 2015 (*Preprint* 1506.03981)
[9] Brambilla N *et al.* 2011 *Eur. Phys. J.* **C71** 1534 (*Preprint* 1010.5827)
[10] Lansberg J P 2006 *Int. J. Mod. Phys.* **A21** 3857–3916 (*Preprint* hep-ph/0602091)
[11] Ferreiro E G, Fleuret F and Rakotozafindrabe A 2009 *Eur. Phys. J.* **C61** 859–864 (*Preprint* 0801.4949)
[12] Ferreiro E G, Fleuret F, Lansberg J P and Rakotozafindrabe A 2009 *Phys. Lett.* **B680** 50–55 (*Preprint* 0809.4684)
[13] Ferreiro E G, Fleuret F, Lansberg J P and Rakotozafindrabe A 2010 *Phys. Rev.* **C81** 064911 (*Preprint* 0912.4498)
[14] Benhar O, Kopeliovich B Z, Mariotti C, Nicolaev N N and Zakharov B G 1992 *Phys. Rev. Lett.* **69** 1156–1159
[15] Leitch M J *et al.* (NuSea) 2000 *Phys. Rev. Lett.* **84** 3256–3260 (*Preprint* nucl-ex/9909007)
[16] Kopeliovich B, Tarasov A and Hufner J 2001 *Nucl. Phys.* **A696** 669–714 (*Preprint* hep-ph/0104256)
[17] Lourenco C, Vogt R and Woehri H K 2009 *JHEP* **02** 014 (*Preprint* 0901.3054)
[18] Klein S R and Vogt R 2003 *Phys. Rev. Lett.* **91** 142301 (*Preprint* nucl-th/0305046)
[19] de Florian D and Sassot R 2004 *Phys. Rev.* **D69** 074028 (*Preprint* hep-ph/0311227)
[20] Ferreiro E G, Fleuret F, Lansberg J P and Rakotozafindrabe A 2013 *Phys. Rev.* **C88** 047901 (*Preprint* 1305.4569)
[21] Eskola K J, Paukkunen H and Salgado C A 2009 *JHEP* **04** 065 (*Preprint* 0902.4154)
[22] Kovarik K *et al.* 2015 (*Preprint* 1509.00792)
[23] Hirai M, Kumano S and Nagai T H 2007 *Phys. Rev.* **C76** 065207 (*Preprint* 0709.3038)
[24] de Florian D, Sassot R, Zurita P and Stratmann M 2012 *Phys. Rev.* **D85** 074028 (*Preprint* 1112.6324)
[25] Frankfurt L, Guzey V and Strikman M 2012 *Phys. Rept.* **512** 255–393 (*Preprint* 1106.2091)
[26] Vogt R 2015 *Phys. Rev.* **C92** 034909 (*Preprint* 1507.04418)
[27] Ferreiro E G, Fleuret F, Lansberg J P, Matagne N and Rakotozafindrabe A 2012 *Few Body Syst.* **53** 27–36 (*Preprint* 1201.5574)
[28] McGlinchey D C, Frawley A D and Vogt R 2013 *Phys. Rev.* **C87** 054910 (*Preprint* 1208.2667)
[29] Arleo F, Peigne S and Sami T 2011 *Phys. Rev.* **D83** 114036 (*Preprint* 1006.0818)
[30] Fujii H and Watanabe K 2013 *Nucl. Phys.* **A915** 1–23 (*Preprint* 1304.2221)
[31] Ducloué B, Lappi T and Mantysaari H 2015 *Phys. Rev.* **D91** 114005 (*Preprint* 1503.02789)
[32] Ma Y Q, Venugopalan R and Zhang H F 2015 *Phys. Rev.* **D92** 071901 (*Preprint* 1503.07772)
[33] Adare A *et al.* (PHENIX) 2013 *Phys. Rev. Lett.* **111** 202301 (*Preprint* 1305.5516)
[34] Ferreiro E G, Fleuret F, Lansberg J P and Rakotozafindrabe A 2012 [J. Phys. Conf. Ser.422,012018(2013)] (*Preprint* 1211.4749)
[35] Abelev B B *et al.* (ALICE) 2014 *JHEP* **12** 073 (*Preprint* 1405.3796)
[36] Ferreiro E G 2015 *Phys. Lett.* **B749** 98–103 (*Preprint* 1411.0549)
[37] Du X and Rapp R 2015 *Nucl. Phys.* **A943** 147–158 (*Preprint* 1504.00670)
[38] Ferreiro E G 2014 *Phys. Lett.* **B731** 57–63 (*Preprint* 1210.3209)
[39] Yang Z LHCb-CONF-2015-005
[40] Aad G *et al.* (ATLAS) 2015 *Phys. Rev.* **C92** 034904 (*Preprint* 1505.08141)
[41] Massacrier L, Trzeciak B, Fleuret F, Hadjidakis C, Kikola D, Lansberg J P and Shao H S 2015 *Adv. High Energy Phys.* **2015** 986348 (*Preprint* 1504.05145)
[42] Arleo F and Peigné S 2015 *Adv. High Energy Phys.* **2015** 961951 (*Preprint* 1504.07428)
[43] Vogt R 2015 *Adv. High Energy Phys.* **2015** 492302
[44] Brodsky S J, Fleuret F, Hadjidakis C and Lansberg J P 2013 *Phys. Rept.* **522** 239–255 (*Preprint* 1202.6585)
[45] Lansberg J P, Brodsky S J, Fleuret F and Hadjidakis C 2012 *Few Body Syst.* **53** 11–25 (*Preprint* 1204.5793)
[46] Strikman M and Treleani D 2002 *Phys. Rev. Lett.* **88** 031801 (*Preprint* hep-ph/0111468)
[47] d'Enterria D and Snigirev A M 2014 *Nucl. Phys. A* **932** 296301 (*Preprint* 1403.2335)
[48] Artoisenet P, Lansberg J P and Maltoni F 2007 *Phys. Lett.* **B653** 60–66 (*Preprint* hep-ph/0703129)
[49] Lansberg J P 2009 *Phys. Lett.* **B679** 340–346 (*Preprint* 0901.4777)
[50] Gong B, Lansberg J P, Lorce C and Wang J 2013 *JHEP* **03** 115 (*Preprint* 1210.2430)
[51] Lansberg J P and Lorce C 2013 *Phys. Lett.* **B726** 218–222 [Erratum: Phys. Lett.B738,529(2014)] (*Preprint* 1303.5327)
[52] Lansberg J P and Shao H S 2013 *Phys. Rev. Lett.* **111** 122001 (*Preprint* 1308.0474)
[53] Lansberg J P and Shao H S 2014 (*Preprint* 1410.8822)
[54] Lansberg J P and Shao H S 2015 *Nucl. Phys. B* (*Preprint* 1504.06531)
[55] Aaij R *et al.* (LHCb) 2012 *Phys. Lett.* **B707** 52–59 (*Preprint* 1109.0963)
[56] Aaij R *et al.* (LHCb) 2012 *JHEP* **06** 141 [Addendum: JHEP03,108(2014)] (*Preprint* 1205.0975)
[57] Aad G *et al.* (ATLAS) 2015 *Eur. Phys. J.* **C75** 229 (*Preprint* 1412.6428)
[58] Aad G *et al.* (ATLAS) 2014 *JHEP* **04** 172 (*Preprint* 1401.2831)
[59] Khachatryan V *et al.* (CMS) 2014 *JHEP* **09** 094 (*Preprint* 1406.0484)

# Energy dependence of hadron spectra and yields in p+p and $^7$Be+$^9$Be collisions from the NA61/SHINE experiment at the CERN SPS

**D.T. Larsen for the NA61/SHINE Collaboration**
Uniwersytet Jagielloński, Łojasiewicza 11, 30-348 Kraków, Poland
E-mail: dlarsen@cern.ch

**Abstract.** The NA61/SHINE programme on strong interactions covers the study of the onset of deconfinement and aims to discover the critical point of strongly interacting matter by performing an energy and system-size scan over the full CERN SPS momentum range. So far the scans of p+p, $^7$Be+$^9$Be, as well as Ar+Sc have been completed. Results from p+p and Be+Be collisions are now emerging, in particular the energy dependence of hadron spectra and yields. Status and preliminary results from this effort will be presented.

## 1. Introduction
NA61/SHINE [1] is a fixed target experiment at the CERN SPS. Tracking is provided by five time projection chambers. Additionally, particle identification is aided by time-of-flight detectors. A modular calorimeter is used to determine the collision centrality for nucleus-nucleus collisions. Both primary and secondary beams are available to the experiment, allowing data taking with projectile sizes ranging from proton to lead, as well as with pions and kaons. Besides studying strong interactions, the experiment also performs precise hadron production reference measurements for neutrino (Fermilab and T2K) and cosmic-ray (KASCADE and Pierre Auger Observatory) physics.

## 2. Analysis of identified hadron spectra
More than 90% of primary negative particles produced in inelastic interactions at SPS energies are $\pi^-$. Thus, the $\pi^-$ spectra may be obtained by subtracting the small non-pion contribution from the overall spectra for negatively charged hadrons. This non-pion contribution is taken from the EPOS model [2]. Since particle identification is not required, a very large phase-space acceptance is obtained. In addition to the $h^-$ method [3], $dE/dx$ and time-of-flight may be used for proper identification. The $dE/dx$ method is based on the particle energy loss in the TPC gas, while the particle mass can be identified using the time-of-flight information. The acceptances for these methods are shown in Figure 1 for p+p interactions at 158 GeV/c. The results are corrected for detector inefficiencies, feed-down from weak decays and secondary interactions, contribution from non-target interactions, as well as trigger and event selection biases.

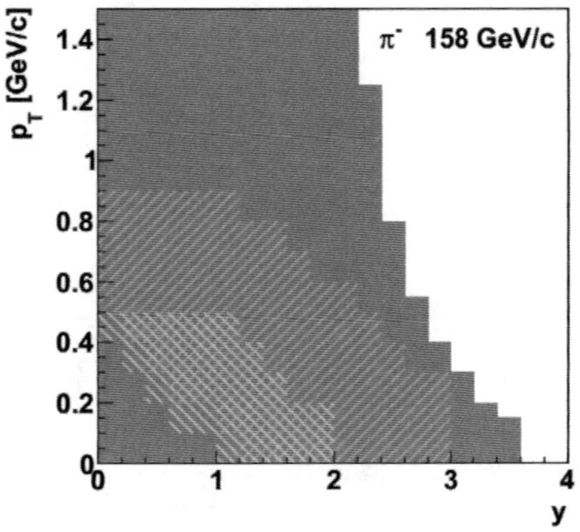


**Figure 1.** Acceptance for various data analysis methods. Blue, solid area: h$^-$; magenta lines: $dE/dx$; yellow lines: time-of-flight.

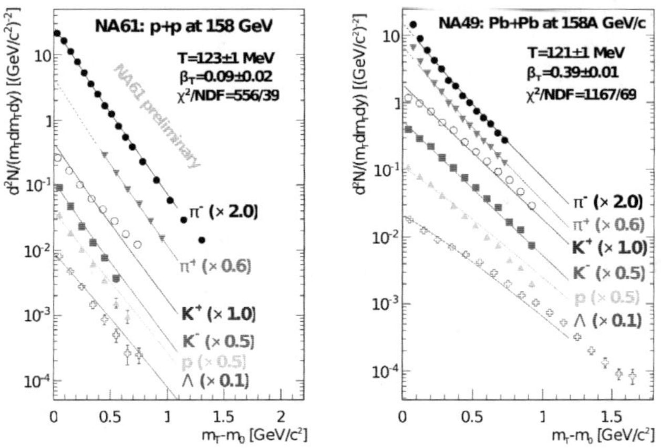


**Figure 2.** Transverse mass spectra at mid-rapidity. Left: NA61/SHINE inelastic p+p interactions at 158 GeV/c; right: NA49 central Pb+Pb collisions at 158$A$ GeV/c.

## 3. p+p results

The results presented below were obtained for identified hadrons produced in inelastic p+p interactions at 20, 31, 40, 80 and 158 GeV/c [4].

Figure 2 shows spectra of transverse mass of negatively and positively charged π, K, p and Λ produced in inelastic p+p interactions at mid-rapidity. Corresponding NA49 measurements [5, 6, 7] for central Pb+Pb collisions are also shown. The data was fitted using a blast wave model parametrisation [8] $\frac{dN_i}{m_T dm_T dy} = A_i m_T K_1\left(\frac{m_T \cosh\rho}{T}\right) I_0\left(\frac{p_T \sinh\rho}{T}\right)$. The parameter $\rho$ is related to the transverse flow velocity $\beta_T$ by $\rho = \tanh^{-1}\beta_T$. One finds that $\beta_T$ is significantly smaller in p+p than Pb+Pb collisions. While the spectra are approximately exponential in p+p reactions, this exponential dependence is modified in Pb+Pb interactions by the transverse flow.

Figure 3 shows rapidity spectra of $\pi^-$ produced in inelastic p+p interactions (left), the energy dependence of the rms width $\sigma$ of those spectra divided by beam rapidity (centre) and divided by predictions ($\sigma_{LS}$) from the hydro-dynamical model [9, 10] (right). Although the shapes of the rapidity spectra are approximately Gaussian, the best fit was obtained using a sum of two Gaussians. It can be seen in Figure 3 (centre) that $\sigma/y_{beam}$ for $\pi^-$ decreases with increasing collision energy. While $\sigma/\sigma_{LS}$ and $\sigma/y_{beam}$ are smaller in p+p than Pb+Pb collisions, no qualitative difference is observed for their energy dependence.

The "kink", "horn", and "step" [6, 14] are important signals for the onset of deconfinement observed in central Pb+Pb interactions. Figure 4 shows that the energy dependence of mean π multiplicity increases slower in p+p than in Pb+Pb collisions ("kink"). Hence, the two dependences cross each other at around 40$A$ GeV/c.

The new measurements of NA61/SHINE significantly improve the world data for the inverse-slope parameter $T$ of $m_T$ spectra of Kaons [16] and for yields of $\pi^+$ mesons at mid-rapidity [17, 18]. The $m_T$ spectra were fitted to the exponential function $\frac{d^2n}{dp_T dy} = \frac{S p_T}{T^2 + m_K T}\exp\left(-\frac{\sqrt{p_T^2+m_K^2}-m_K}{T}\right)$. $S$ is the yield integral and $m_K$ is the K mass. Figure 5 shows the result of the fit to the spectra of K$^+$ and K$^-$ at mid-rapidity.

Figure 6 shows the energy dependence of $T$ for K$^+$ and K$^-$. Surprisingly, the results from inelastic p+p interactions exhibit rapid changes similar to those observed in central Pb+Pb

Journal of Physics: Conference Series **668** (2016) 012020 doi:10.1088/1742-6596/668/1/012020

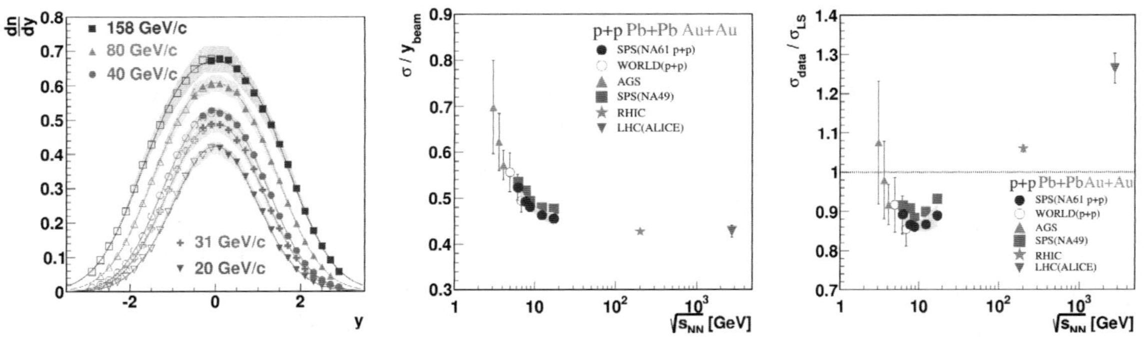


**Figure 3.** Left: $\pi^-$ rapidity spectra in inelastic p+p interactions. Centre: energy dependence of the scaled width of rapidity spectra. Right: energy dependence of the ratio of the measured width to that predicted by the hydro-dynamical model [9, 10]. World data from Refs. [11, 12, 13]. Not corrected for isospin effects.

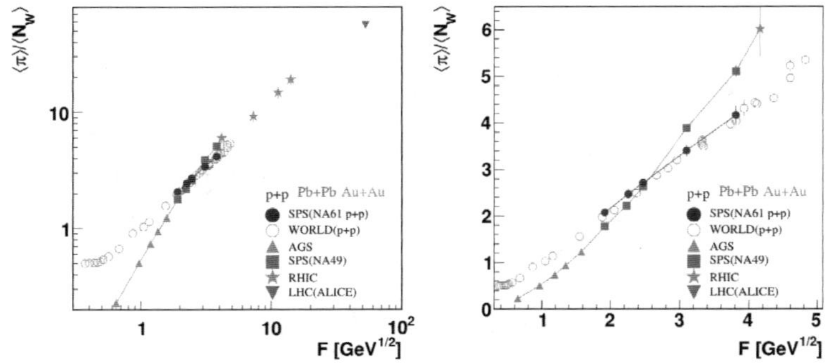


**Figure 4.** Energy dependence of mean $\pi$ multiplicity per wounded nucleon in inelastic p+p interactions and central Pb+Pb (Au+Au) collisions. World data from Refs. [12, 15].

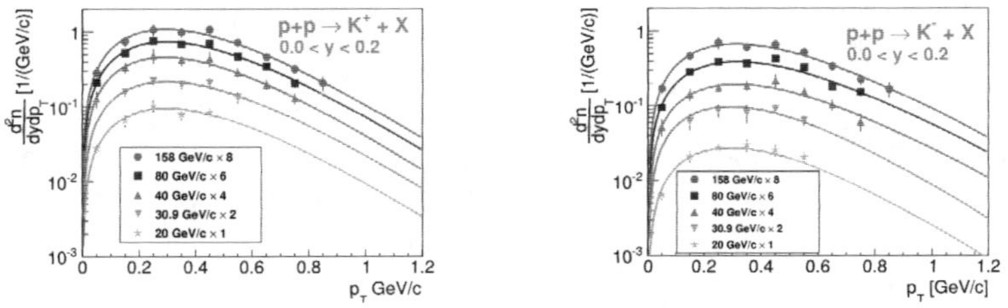


**Figure 5.** Transverse momentum spectra of K mesons at mid-rapidity in inelastic p+p interactions at CERN SPS energies.

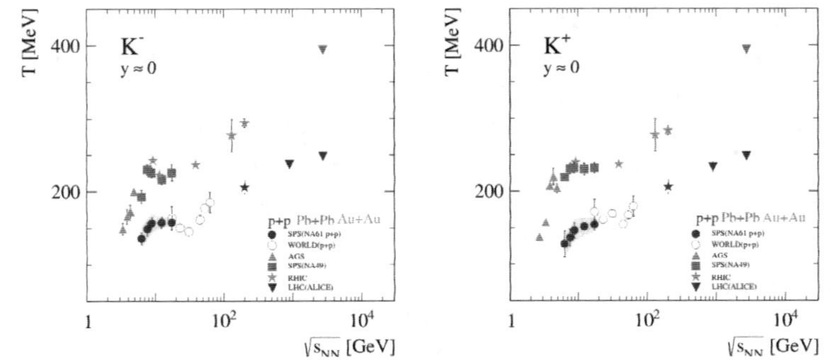


**Figure 6.** Energy dependence of the inverse slope parameter T of transverse mass spectra of $K^+$ and $K^-$ in inelastic p+p and central Pb+Pb/Au+Au interactions. World data from Refs. [16, 19, 20, 21].

collisions ("step").

While the $h^-$ method [3] can be used to measure spectra for $\pi^-$, a similar approach cannot be used for $\pi^+$ because of the large K and p contamination. Instead, the ratio of the measured $\pi^+$

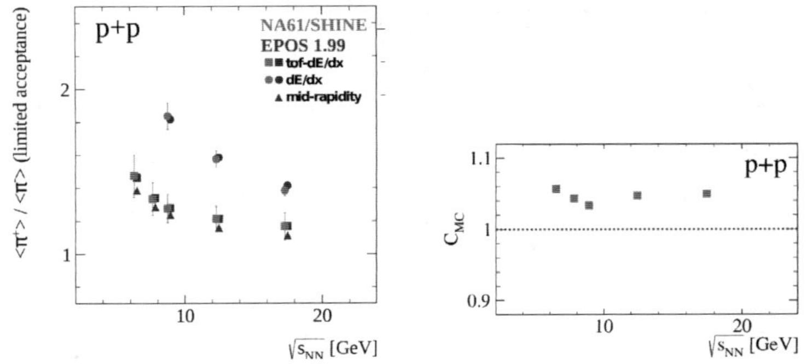


**Figure 7.** Left: energy dependency of $\pi^+/\pi^-$ ratio in inelastic p+p interactions compared with EPOS predictions. Right: correction factors used to obtain $\pi^+$ yields from $\pi^-$ yields.

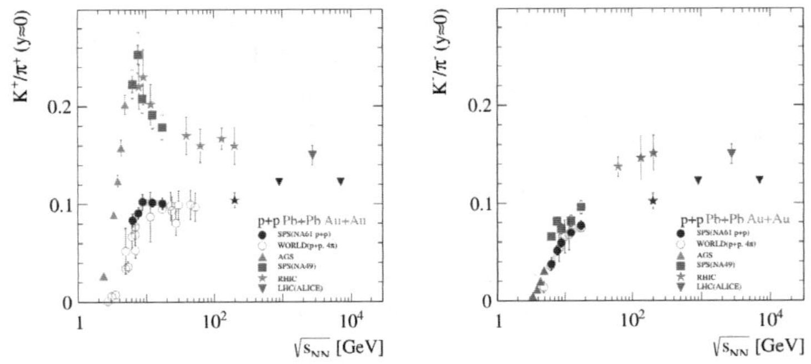


**Figure 8.** Energy dependence of the $K^+/\pi^+$ and $K^-/\pi^-$ ratios in inelastic p+p and central Pb+Pb and Au+Au interactions. World data from Refs. [15, 17, 18, 21, 22].

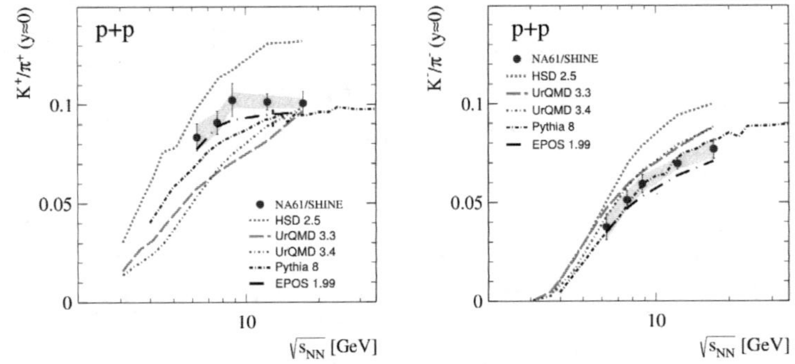


**Figure 9.** Energy dependence of the $K^+/\pi^+$ and $K^-/\pi^-$ ratios in inelastic p+p interactions compared [23] with theoretical models EPOS [2], UrQMD [24, 25], Pythia 8 [26] and HSD [27].

and $\pi^-$ yields was calculated within the acceptance of the time-of-flight and/or $dE/dx$ methods, and compared with model predictions. As can be seen in Figure 7 (left), the agreement between the measurements and the EPOS model [2] is better than 0.1%. Finally, the $\pi^+$ mid-rapidity yield was calculated as the product of the measured $\pi^-$ yield at $y = 0$, the measured $\pi^+/\pi^-$ ratio within the time-of-flight and $dE/dx$ acceptance, and the EPOS correction factor $C_{MC}$. This is shown in Figure 7 (right): $\pi^+(y = 0) = \pi^-(y = 0)\pi^+/\pi^-(tof - dE/dx)C_{MC}$, where $C_{MC} = \left[\frac{\pi^+/\pi^-(y=0)}{\pi^+/\pi^-(tof)}\right]_{MC} \approx 5\%$.

Figure 8 shows the energy dependence of the $K^+/\pi^+$ and $K^-/\pi^-$ ratios observed in inelastic p+p interactions, as well as world data for central Pb+Pb and Au+Au interactions as reference. Again surprisingly, the data suggests that even inelastic p+p interactions exhibit a step structure in the energy dependence of the $K^+/\pi^+$ ratio, which may be considered a precursor of the "horn" found in central Pb+Pb collisions.

Furthermore, the measured ratios were compared to theoretical models, as shown in Figure 9. It is evident that none of the models adequately describes the data.

Figure 10 shows for protons from inelastic p+p interactions the spectra of $p_T$ (left) and $\langle m_T \rangle$

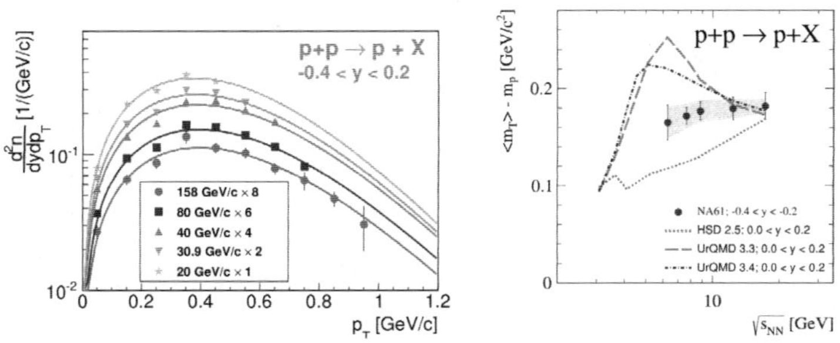


**Figure 10.** Left: transverse momentum spectra of protons at mid-rapidity in inelastic p+p interactions. Right: comparison of $\langle m_T \rangle - m_p$ calculated using the spectra from the left plot with predictions of models [23].

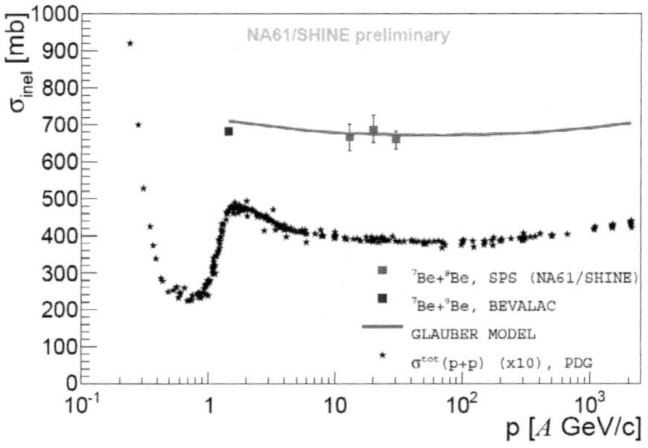


**Figure 11.** Inelastic cross section of $^{7}$Be+$^{9}$Be collisions as well as total cross section for p+p interactions as a function of beam momentum. The curve shows the prediction of the Glauber model.

(right) as a function of collision energy. Neither the magnitude nor the energy dependence of $\langle m_T \rangle$ is reproduced by the UrQMD or the HSD model.

## 4. Be+Be results

Previously, the inelastic cross section for $^{7}$Be+$^{9}$Be was only measured at 1.45$A$ GeV/c [28]. The new measurements of NA61/SHINE now extend this to 13$A$, 20$A$ and 30$A$ GeV/c [29]. A scintillator counter placed behind the target measures the square of the charge of a particle passing through it. Inelastic interactions were selected by requiring a signal below that expected for an intact beam nucleus. Data was also taken with the target removed to be able to subtract the background caused by beam interactions with detector material. The interaction probability is given by $P_{int} = \frac{P_I - P_R}{1 - P_R}$, where $P_I$ and $P_R$ are the interaction probabilities when the target is inserted and removed, respectively. Using $P_{int}$, the cross section can be calculated from $\sigma_{inel} = \frac{1}{\rho L_{eff} N_A / A} P_{int}$, $L_{eff} = \lambda_{abs}(1 - e^{-L/\lambda_{abs}})$ and $\lambda_{abs} = \frac{A}{\rho N_A \sigma_{inel}}$. $N_A$, $\rho$, $L$ and $A$ are the Avogadro constant, the target density, length and atomic number, respectively. Figure 11 shows that the new measurements are in agreement with the previous measurement as well as the Glauber model [30] prediction.

Preliminary results on spectra were obtained for $^{7}$Be+$^{9}$Be interactions at beam momenta of 20$A$, 30$A$, 40$A$, 75$A$ and 150$A$ GeV/c in the centrality classes 0–5%, 5–10%, 10–15% and 15–20% [31]. Unless otherwise stated, only statistical errors are shown. The centrality was derived from the energy deposited in the forward calorimeter PSD. Figure 12 shows the relationship between deposited energy and the centrality classes. Due to the modularity of this detector, it is possible to change the acceptance during off-line analysis. A smaller acceptance will cause some of the projectile spectators to be lost, but at the same time reduces the contribution from particles produced in the collision. Thus, the result will depend on the selected acceptance. This effect is largest for low beam momenta. For example for the total $\pi^-$ multiplicity at 20$A$ GeV/c, the spread for different acceptances is up to 5%.

Journal of Physics: Conference Series **668** (2016) 012020 doi:10.1088/1742-6596/668/1/012020

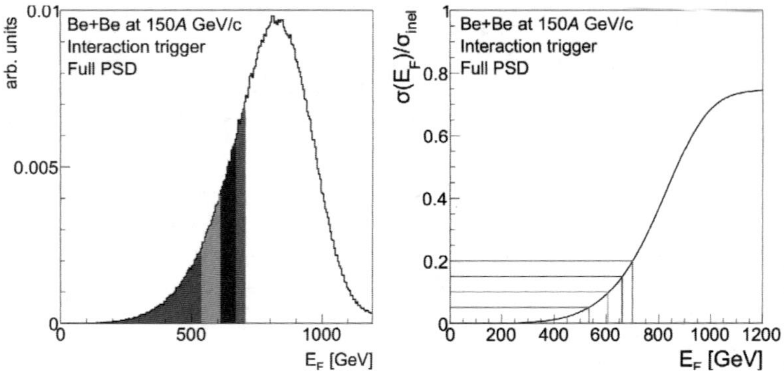


**Figure 12.** Left: distribution of forward energy $E_F$ measured in the PSD calorimeter in $^7$Be+$^9$Be collisions at $150A$ GeV/c. Right: definition of forward energy event classes.

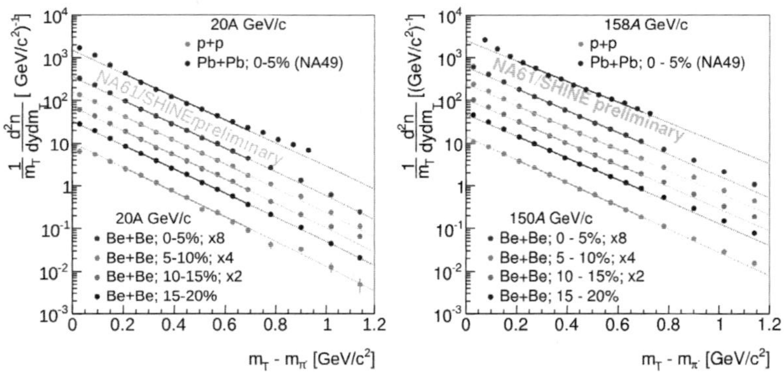


**Figure 13.** Transverse mass spectra of $\pi^-$ mesons for $^7$Be+$^9$Be, p+p and Pb+Pb interactions. Left: $20A$ GeV/c; right: $150A$ GeV/c beam momentum.

Figure 13 shows for $^7$Be+$^9$Be $\pi^-$ collisions the $m_T$ spectra for the different centrality classes for the beam momenta $20A$ and $150A$ GeV/c, as well as for inelastic p+p and central Pb+Pb interactions. The p+p data is described very well by an exponential function, while both $^7$Be+$^9$Be and Pb+Pb spectra deviate from this function at low and high values of $m_T$.

Calculating normalised $\pi^-$ $m_T$ spectra as the ratios ($^7$Be+$^9$Be)/(p+p) and (Pb+Pb)/(p+p) facilitates the comparison of the shapes of the spectra produced in these different size systems. Figure 15 shows qualitatively similar behaviour for both $^7$Be+$^9$Be and Pb+Pb reactions. The high-$m_T$ regions of both $^7$Be+$^9$Be and Pb+Pb exhibit an increase of the ratio, while for the intermediate regions a decrease is seen. This effect is stronger for central Pb+Pb collisions. Usually, this is attributed to collective flow. Also, for $^7$Be+$^9$Be reactions the increase of the ratio at high values of $m_T$ appears to be larger at higher beam momenta. This may suggest an increase of the magnitude of collective effects in $^7$Be+$^9$Be collisions with increasing beam momentum. The low-$m_T$ regions of both $^7$Be+$^9$Be and Pb+Pb reactions exhibit an increase of the ratio. Possible explanations include isospin asymmetry of p+p data or electromagnetic effects.

Figure 15 shows the $\pi^-$ rapidity spectra from $^7$Be+$^9$Be collisions in the different centrality classes for beam momenta $20A$ and $150A$ GeV/c, as well as for inelastic p+p reactions. The data were fitted to two Gaussians symmetrically displaced with respect to mid-rapidity. Although their widths are the same, the amplitudes are different due to the asymmetry of $^7$Be+$^9$Be collisions. By extending the fit range into the backward rapidity region, it was possible to obtain a stable fit.

The rms width $\sigma_y$ of the spectra can be obtained from the fitted function. Figure 16 (left) shows the $\sigma_y/y_{beam}$ for $^7$Be+$^9$Be, as well as inelastic p+p and central Pb+Pb interactions. The widths of the spectra for all systems decrease monotonically with respect to collision energy. However, the widths of the spectra do not behave monotonically with respect to the system

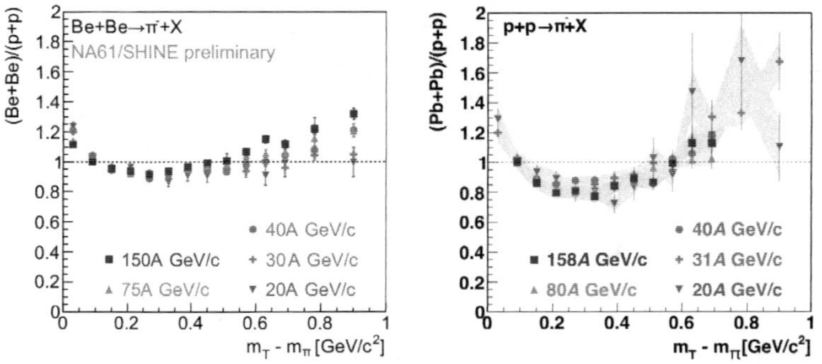


**Figure 14.** Ratio of normalised transverse mass spectra of $\pi^-$ mesons for different system sizes. Left: $(^7Be+^9Be)/(p+p)$; right: (Pb+Pb)/(p+p).

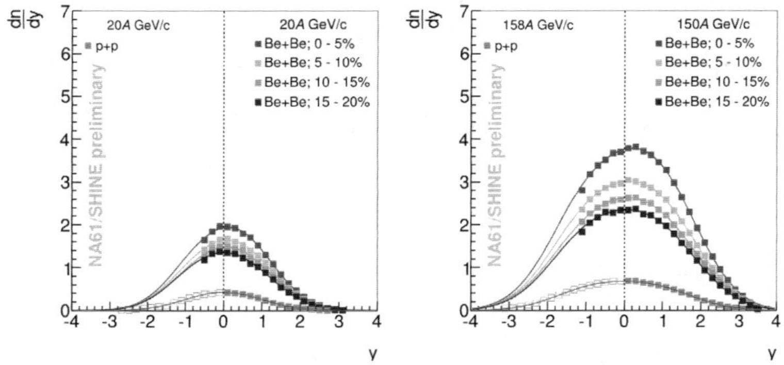


**Figure 15.** Rapidity spectra of $\pi^-$ mesons for $^7Be+^9Be$ and inelastic p+p interactions. Left: $20A$ GeV/c; right: $150A$ GeV/c.

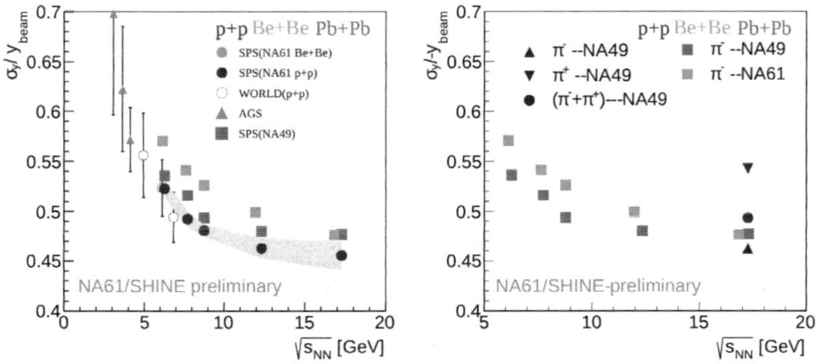


**Figure 16.** Left: energy dependence of scaled width of $^7Be+^9Be$ $\pi^-$ rapidity spectra; right: effect of isospin asymmetry in p+p interactions. World data from Refs. [7].

size for a given collision energy. The width of the spectra from Pb+Pb lies between those of p+p and $^7Be+^9Be$ collisions. When comparing the rapidity spectra from different systems, one must consider that p+p has larger isospin asymmetry than $^7Be+^9Be$ and Pb+Pb. This isospin asymmetry can be accounted for by calculating the average spectra for $\pi^-$ and $\pi^+$. The only large-acceptance data available for $\pi^+$ production in p+p in the relevant energy domain is from NA49 at 158 GeV/c [32]. Unfortunately, this is also the energy with the smallest differences for the rapidity widths of the different system sizes. Still, Figure 16 (right) shows for 158 GeV/c beam momentum that by taking into account the different isospin asymmetries, the dependence of the rapidity width with respect to system size becomes monotonic.

## 5. Conclusion

The analysis of inelastic p+p and $^7Be+^9Be$ interactions at CERN SPS energies showed many interesting effects. In particular, the p+p data exhibited step-like structures in the energy region where the onset of deconfinement was found in central Pb+Pb collisions. Also, many of the measurements could not be explained well by theoretical models. For $^7Be+^9Be$ reactions new results on the inelastic cross section were obtained at several energies. Transverse mass and rapidity spectra were measured in the SPS energy range for three centrality classes. An indication of a transverse flow effect was found at the highest beam momenta.

## Acknowledgements

Supported by the National Science Centre of Poland (grants UMO-2014/13/N/ST2/02565, UMO-2014/12/T/ST2/00692, UMO-2013/11/N/ST2/03879 and UMO-2012/04/M/ST2/00816).

## References

[1] N. Abgrall *et al.*, [NA61/SHINE Collab.] *JINST* **9** (2014) P06005, arXiv:1401.4699 [physics.ins-det].
[2] T. Pierog and K. Werner *Nucl.Phys.Proc.Suppl.* **196** (2009) 102–105, arXiv:0905.1198 [hep-ph].
[3] N. Abgrall *et al.*, [NA61/SHINE Collab.] *Eur.Phys.J.* **C74** (2014) 2794, arXiv:1310.2417 [hep-ex].
[4] S. Puławski, [NA61 Collab.] arXiv:1502.07916 [nucl-ex].
[5] C. Alt *et al.*, [NA49 Collab.] *Phys.Rev.* **C73** (2006) 044910.
[6] C. Alt *et al.*, [NA49 Collab.] *Phys. Rev.* **C77** (2008) 024903.
[7] S. Afanasiev *et al.*, [NA49 Collab.] *Phys. Rev.* **C66** (2002) 054902.
[8] E. Schnedermann, J. Sollfrank, and U. W. Heinz *Phys.Rev.* **C48** (1993) 2462–2475, arXiv:nucl-th/9307020 [nucl-th].
[9] L. Landau *Izv.Akad.Nauk Ser.Fiz.* **17** (1953) 51–64.
[10] E. V. Shuryak *IYF-75-4* (1974) .
[11] J. Klay *et al.*, [E895 Collab.] *Phys. Rev.* **C68** (2003) 054905.
[12] E. Abbas *et al.*, [ALICE Collab.] *Phys.Lett.* **B726** (2013) 610–622, arXiv:1304.0347 [nucl-ex].
[13] S. Adler *et al.*, [PHENIX Collab.] *Phys.Rev.* **C69** (2004) 034909.
[14] M. Gazdzicki, M. Gorenstein, and P. Seyboth *Acta Phys.Polon.* **B42** (2011) 307–351, arXiv:1006.1765 [hep-ph].
[15] B. Abelev *et al.*, [ALICE Collab.] *Phys.Rev.Lett.* **109** (2012) 252301, arXiv:1208.1974 [hep-ex].
[16] M. Kliemant, B. Lungwitz, and M. Gazdzicki *Phys.Rev.* **C69** (2004) 044903, arXiv:hep-ex/0308002 [hep-ex].
[17] M. Gazdzicki and D. Roehrich *Z.Phys.* **C65** (1995) 215.
[18] M. Gazdzicki and D. Rohrich *Z.Phys.* **C71** (1996) 55–64, arXiv:hep-ex/9607004 [hep-ex].
[19] B. Abelev *et al.*, [STAR Collab.] *Phys.Rev.* **C79** (2009) 034909, arXiv:0808.2041 [nucl-ex].
[20] B. B. Abelev *et al.*, [ALICE Collab.] *Phys.Lett.* **B736** (2014) 196–207, arXiv:1401.1250 [nucl-ex].
[21] K. Aamodt *et al.*, [ALICE Collab.] *Eur.Phys.J.* **C71** (2011) 1655, arXiv:1101.4110 [hep-ex].
[22] I. Arsene *et al.*, [BRAHMS Collab.] *Phys.Rev.* **C72** (2005) 014908, arXiv:nucl-ex/0503010 [nucl-ex].
[23] V. Y. Vovchenko, D. Anchishkin, and M. Gorenstein *Phys.Rev.* **C90** no. 2, (2014) 024916, arXiv:1407.0629 [nucl-th].
[24] M. Bleicher *et al. J.Phys.* **G25** (1999) 1859–1896, arXiv:hep-ph/9909407 [hep-ph].
[25] S. Bass *et al. Prog.Part.Nucl.Phys.* **41** (1998) 255–369, arXiv:nucl-th/9803035 [nucl-th].
[26] T. Sjostrand, S. Ask, J. R. Christiansen, R. Corke, N. Desai, *et al. Comput.Phys.Commun.* **191** (2015) 159–177, arXiv:1410.3012 [hep-ph].
[27] W. Ehehalt and W. Cassing *Nucl.Phys.* **A602** (1996) 449–486.
[28] I. Tanihata *et al. Phys. Rev. Lett.* **55** (1985) 2676.
[29] I. Weimer https://edms.cern.ch/file/1308546/1. Bachelor Thesis, Ludwig-Maximilian University Munich.
[30] W. Broniowski, M. Rybczynski, and P. Bozek *Comput.Phys.Commun.* **180** (2009) 69.
[31] E. Kaptur, [NA61/SHINE Collab.] *PoS* **CPOD** (2014) , http://pos.sissa.it/archive/conferences/217/053/CPOD2014_053.pdf.
[32] C. Alt *et al.*, [NA49 Collab.] *Eur. Phys. J.* **C45** (2006) 343.

# FRIGA, A New Approach To Identify Isotopes and Hypernuclei In N-Body Transport Models.

**A. Le Fèvre[1], Y. Leifels[1], J. Aichelin[2], Ch. Hartnack[2], V. Kireyev[3] and E. Bratkovskaya[4]**

[1]GSI Helmholtzzentrum für Schwerionenforschung, Darmstadt, Germany;
[2]Subatech Nantes, France; [3]JINR Dubna, Russia; [4]FIAS and ITP Frankfurt, Germany.

E-mail: A.LeFevre@gsi.de

**Abstract.** We present a new algorithm to identify fragments in computer simulations of relativistic heavy ion collisions. It is based on the simulated annealing technique and can be applied to n-body transport models like the Quantum Molecular Dynamics. This new approach is able to predict isotope yields as well as hyper-nucleus production. In order to illustrate its predicting power, we confront this new method to experimental data, and show the sensitivity on the parameters which govern the cluster formation.

## Introduction

In heavy ion reactions at energies between 20 A.MeV and several A.GeV, many clusters are formed. This cluster formation presents a big challenge for transport models in which nucleons are the degrees of freedom which are propagated. Identifying clusters in a transport code which transports nucleons is all but simple and therefore in many approaches the fragment formation is simply omitted. This invalidates the prediction of single particle observables as well, because the cluster formation – and therefore the modification of the single particle spectra due to the fragment formation – depends on the phase space region and, as a consequence, cannot be approximated by a momentum independent scaling factor.

The simplest way to identify clusters is by employing coalescence or a minimum spanning tree procedure. The first needs a multitude of free parameters, whereas the second allows only for an identification at the end of the reaction which excludes any study on the physical origin [1]. In addition, quantum effects, like additional binding energies due to closed shells or pairing energies, are not supplied by the underlying transport theory which is semi-classical.

## The principles of the fragment recognition.

If one wants to identify fragments early, while the reaction is still going on, one has to use the momentum as well as the coordinate space informations. An idea how to do this has been launched by Dorso et al. [2]. It has been further developed into the Simulated Annealing Clusterisation Algorithm (SACA) [3] in the late 1990's and has been successfully applied to understand the measured fragment charge distribution and spectra as well as bimodality [4, 5]. Starting from the positions and momenta of the nucleons at a given time during the reaction, nucleons are combined in all possible ways into fragments or single nucleons applying a simulated annealing technique. Neglecting the interaction among nucleons in different clusters, but taking into account the interaction among the nucleons in the same fragment, this algorithm identifies that combination of fragments and free nucleons which has the *highest binding energy.* If applied after the time when the energetic initial collisions are over, this most bound configuration has been proven to be close to the final distribution of fragments identified by the minimum spanning tree method at the end of the reaction[3]. The reason for this is the fact that fragments are not a random collection of nucleons at the end, but an initial-final state correlation. SACA can be applied at any moment during the reaction and allows therefore for a detailed study of the fragment production mechanism.

In SACA, for accounting the interaction in-between nucleons, only the bulk Skyrme interaction supplemented by a Yukawa potential is used – which is also the potential used for the propagation of the nucleons in the QMD transport model. To obtain more realistic fragment observables and to be able to predict observables for isotopes and hyper-nuclei, we employ in our new approach a more realistic interaction and add the secondary decay because the fragments, when identified, have a (moderate) excitation energy. This new approach is dubbed FRIGA ("Fragment Recognition In General Application").

**The new features of FRIGA.**
In order to predict the absolute multiplicity of the isotope yields, we have added new features to the SACA cluster identification. They include the asymmetry energy, pairing and quantum effects.

For the asymmetry energy, we adopt the parametrisation from IQMD [8], a transport code which we use in the present article for the transport of nucleons. For a proton the single particle energy thus reads:

$$B_{asy} = E_0\left(\frac{<\rho_B>}{\rho_0}\right)^{\gamma-1}\frac{\rho_n-\rho_p}{\rho_B}$$

where $E_0$=23.3 MeV, and $\rho_n$, $\rho_p$, $\rho_B$, $\rho_0$ are the neutron, proton, baryonic and saturation densities, respectively. In the present work, we take $\gamma$=1 ("stiff" asymmetry potential).

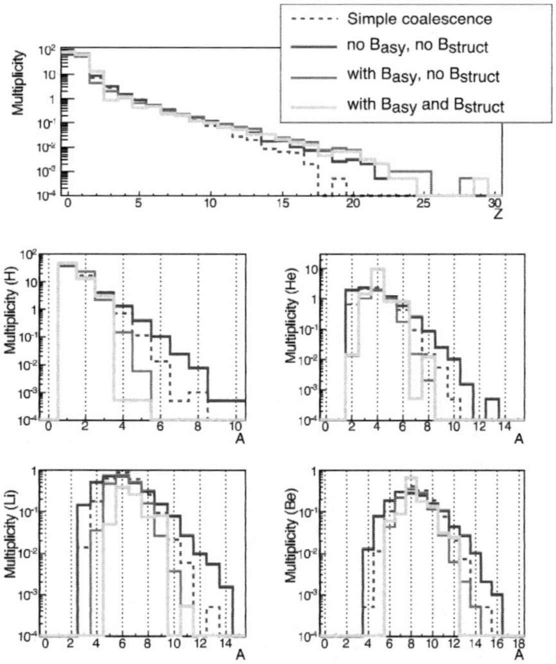


**Figure 1.** IQMD predictions for the central ($b < 0.2b_{max}$) collisions of $^{124}Xe + ^{112}Sn$ at 100 A.MeV incident energy. Dashed line for the MST (coalescence) algorithm alone (performed at the late time 200 fm/c), blue line for the initial SACA model, which has been extended into FRIGA with an asymmetry term (red) and additional nuclear structure contribution (green). The top panel shows the mean multiplicity distribution of fragments as a function of their charge. The four others depict the yields of H, He, Be and Li isotopes.

Another significant part of the binding energy of light isotopes are the shell structure and odd-even effects (pairing). In the conditions of high pressure and temperature where FRIGA is used to determine the pre-fragments, these structure effects are not well known. E. Khan et al. [9] showed that there are some indications that they affect the primary fragments. The authors demonstrate that the pairing vanishes above a nuclear temperature $T_V \approx 0.5\Delta_{pairing}$ (pairing energy). At normal density the pairing energy tends to be negligible for heavy nuclei, with $\Delta_{pairing} = \frac{12}{\sqrt{A}}MeV$, whereas it is strong for light isotopes, like $^4He$ and $^3He$ with 12 MeV and 6.9

MeV, respectively. In FRIGA, the primary fragments are usually produced slightly below the saturation density (typically around half of it) and quite cold, with $T < 1 - 2MeV$, and hence below $T_V$. Therefore, one cannot neglect the pairing energy. The same is true for shell effects which produce experimentally a visible enhancement of the fragment yield for closed shell nuclei.

In order to determine the contribution of all structure effects to the binding energy of clusters, we make two hypotheses independent of the density and the average kinetic energy of the fragment environment.

First, the relative ratio of this nuclear structure contribution to the overall binding energy remains unchanged at the moderate temperatures and at the density at which clusters are formed which is not far away from the saturation density.

Introducing among the nucleons, initialised with the right root-mean-square radius , two body interactions, which corresponds in infinite matter to the Skyrme equation of state, the total fragment energy

$$\begin{aligned} E_B(N,Z) = \langle H \rangle &= \langle T \rangle + \langle V \rangle \\ &= \sum_i \frac{p_i^2}{2m_i} + \sum_i \sum_{j>i} \int f_i(\mathbf{r},\mathbf{p},t)\, V(\mathbf{r},\mathbf{r}',\mathbf{p},\mathbf{p}') f_j(\mathbf{r}',\mathbf{p}',t)\, d\mathbf{r}\, d\mathbf{r}' d\mathbf{p}\, d\mathbf{p}' \quad (1) \end{aligned}$$

where $f_i$ is the single-particle Wigner density

$$f_i(\mathbf{r},\mathbf{p},t) = \frac{1}{\pi^3 \hbar^3} e^{-\frac{2}{L}(\mathbf{r}-\mathbf{r_i}(t))^2} e^{-\frac{L}{2\hbar^2}(\mathbf{p}-\mathbf{p_i}(t))^2} \quad (2)$$

reproduces very well the nuclear binding energy given by the Weizsäcker mass formula for ground state nuclei, $B_0$ [6]. (Fig.12).

Our second hypothesis is that eq. 2 remains the right description of the binding energy if the nuclei are deformed or excited when the fragments are identified by the FRIGA algorithm.

Taken both assumptions together, we can express the nuclear structure contribution to the binding energy of a deformed cluster with Z protons and N neutrons in the following way:

$$B_{struct} = E_B(N,Z)\frac{B_{exp}(Z,N) - B_{BW}(Z,N)}{B_{BW}(Z,N)}$$

where $B_{exp}$ and $B_{BW}$ are the experimentally measured binding energy (which contains the structure contribution and the sum of the volume and surface terms of the Bethe-Weizsäcker formula). (Hyper-)Isotopes which are not stable at all in nature, are discarded in FRIGA by assigning to them a very repulsive $B_{struct}$. The total binding energy of a cluster with N and Z, which is used in the annealing algorithm, will then be:

$$B = E_B(N,Z) + B_{asy} + B_{struct}.$$

in contradistinction to SACA in which only the first term is used.

Fig. 1 shows the influence of the asymmetry energy and of the structure energy on the isotope yield in the reaction $^{124}Xe+^{112}Sn$ at 100 A.MeV. We display here the results for central collisions ($b < 0.2\ b_{max}$). This figure illustrates as well how the various ingredients influence the fragments yield obtained in FRIGA, assuming an early clusterisation at t=60 fm/c. From that time on, the size of the pre-fragments does not change anymore. We see that the charge distributions are not strongly modified for the different options, whereas details of the isotopic yield are strongly influenced: the asymmetry energy tends to narrow the distributions towards the valley of stability, whereas the structure effects contribute to restore the natural abundances, particularly strong for the $^4He$ clusters.

Journal of Physics: Conference Series **668** (2016) 012021 doi:10.1088/1742-6596/668/1/012021

**Excitation energy and density of the primary fragments.**
The pre-fragments, called also "primary" fragments, created in FRIGA, are often produced non relaxed in shape and density. When turning to their ground state, the shape surface energy is converted into excitation energy. Using QMD simulations, for beam energies between 50 A.MeV - 1 A.GeV, FRIGA obtains for central heavy ion collisions a mean excitation energy of the intermediate mass fragments between 0 and 3 A.MeV, depending on the fragment size and very similarly to the experimental measurements of [12]. This excitation energy is sufficiently large that the secondary decay of the pre-fragments causes a significant contribution to the yield of small clusters. For this reason, we optionally allow in FRIGA the excited cluster to undergo sequential secondary decays, using the GEMINI algorithm [13].

Another interesting feature of the primary clusters in FRIGA is their internal density. Although the medium is close to $\rho_0$, at the stage of the collision when the primary cluster formation is stabilised, just after the colliding system begins to separate, the fragments predicted by FRIGA are produced quite dilute, typically around $\rho = \rho_0/2$ for intermediate mass fragments, and around $\rho = \rho_0/5$ for the light $Z < 3$ isotopes. This is explained by the fact that the dense clusters are disfavoured, because they would contain nucleons which are moving against each other. In this case the nucleons have a too high relative momenta to form a cluster. Therefore, in the FRIGA approach, fragment formation tests only the low density behaviour of the potentials, which are contributing to the binding energy.

**Another application of FRIGA: the hypernucleus formation.**
An hyper-nucleus is a nucleus which contains at least one hyperon ($\Lambda(uds)$, ...) in addition to nucleons. Extending FRIGA to the strange sector requires the knowledge of the $\Lambda N$ potential. In this first study, we consider the strange quark as inert and use $V_{\Lambda N} = \frac{2}{3}V_{nN}$ for protons as well as for neutrons. Similarly, we consider the case of multiple strange nuclei as well, in which more than one hyperon is part of the fragment. There, the coupling of 2 $\Lambda's$ contributes with the potential $V_{\Lambda\Lambda} = (\frac{2}{3})^2 V_{nN}$. In the present approach we neglect $B_{asy}$ for the hyperons, and take the contribution of the core nucleus (partner of the hyperons) as if it were decoupled from the hyperon. Since the pairing and shell contributions in the binding energy are not yet well known for hyper-nuclei, we neglect the $B_{struct}$ contribution.

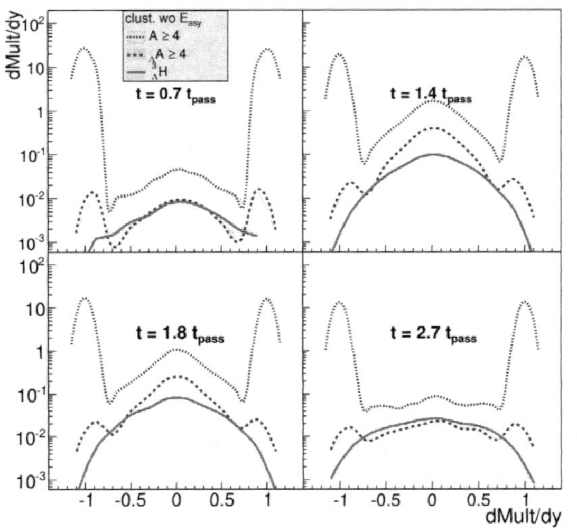


**Figure 2.** Predictions of FRIGA from PHSD [7] simulations of Au+Au collisions at 11.45 A.GeV incident energy, b=6 fm. it shows the multiplicity per event as a function of the rapidity in the centre of the collision scaled to the projectile rapidity, for various (hyper-)clusters ($A_\Lambda$ are hyper-clusters with $A \geq 4$) and clusterisation times. $t_{pass}$=7.5 fm/c is the passing time of the projectile and target in central collisions. Like in the following figures, the shaded areas depict the statistical uncertainties.

Using these modifications of the potentials, FRIGA produces hyper-nuclei with the same procedures as non strange fragments. In the underlying QMD-like programs, which propagate the hadrons, $\Lambda$'s are produced in different reactions: $\bar{K} + N \to \Lambda + \pi$, $\pi + n \to \Lambda + K^+$, $\pi^- + p \to \Lambda + K_0$, $p + p \to \Lambda + X$. Their abundance, position and momentum distributions are strongly influenced by the reaction kinematics, the nuclear equation of state and the in-medium properties of the $K^+$ (kaon potential, etc.) which are implemented in the transport model [14].

Journal of Physics: Conference Series **668** (2016) 012021 doi:10.1088/1742-6596/668/1/012021

Due to their composition, the yields of hyper-nuclei are produced when a cluster in coordinate and momentum space absorbs a hyperons. In heavy ions collisions at relativistic energies, the hyperon distributions are strongly peaked around the mid-rapidity region whereas the large fragments have rapidities close to the beam or target rapidity. The closer the rapidity of the hyperon approaches – by production or by subsequent collisions – the target/beam rapidity, the larger is the probability that it can be absorbed by one of this larger clusters. Heavy hyper-nuclei are therefore observed not far away from beam/target rapidity. At the same time hyperons can also form with other nucleons light clusters at mid-rapidity. There, the probability decreases with the cluster size because it is increasingly difficult to form large cluster out of a gas of nucleons. Whereas the large clusters in the beam/target rapidity regime can be identified quite early, the light clusters at mid-rapidty are formed later and many of them decay due to the interactions with the surrounding nucleons which form a gas of a large temperature as compared to the cluster binding energy. This is illustrated in Fig. 2.

As seen previously, in Fig. 1, the ingredients of the cluster binding energy influence the light isotope yields in FRIGA. The same is observed for hypernuclei. Adopting the factor $\frac{2}{3}$ in $V_{\Lambda N}$ has a strong effect, decreasing on the average the hypernucleus yields by around 20 percent. The asymmetry energy in the cluster can have a similar effect, depending on the isotope (Z,N) asymmetry.

In order to illustrate the predicting power of the FRIGA algorithm, we confront it to experimental observations of light hypernuclei produced in the spectator region in collisions of the light system $^6Li + ^{12}C$ at 2 AGeV incident energy, measured by the HyPHI Collaboration at the SIS synchrotron of GSI Darmstadt. The data presented here are taken from [15]. Fig. 3 compares the IQMD-FRIGA predictions for the rapidity distributions of $^3_\Lambda H$ and $^4_\Lambda H$ with the experiment. The best agreement in the experimentally resolved rapidity region (close to the projectile spectator, $y/y_{beam} > 0.7$) has been obtained while excluding the most central collisions (taking b>3 fm). This procedure is a very basic approach to the simulate the effect of

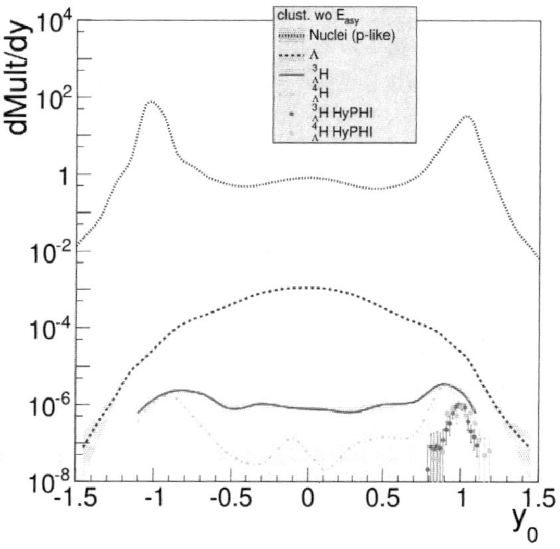


**Figure 3.** Predictions of FRIGA (clustering at $2t_{pass}$, binding energy excluding $B_{asy}$) from IQMD simulations of $^6Li + ^{12}C$ collisions at 2 A.GeV incident energy, $b > 3fm$ compared to the HyPHI experimental data. The results of the model calculations are not filtered for the experimental acceptance. It shows the multiplicity per event per unit of rapidity, as a function of the rapidity in the centre of the collision scaled to the projectile rapidity, for all clusters (in proton-like weighting), $\Lambda's$, $^3_\Lambda H$ and $^4_\Lambda H$. Markers are experimental data.

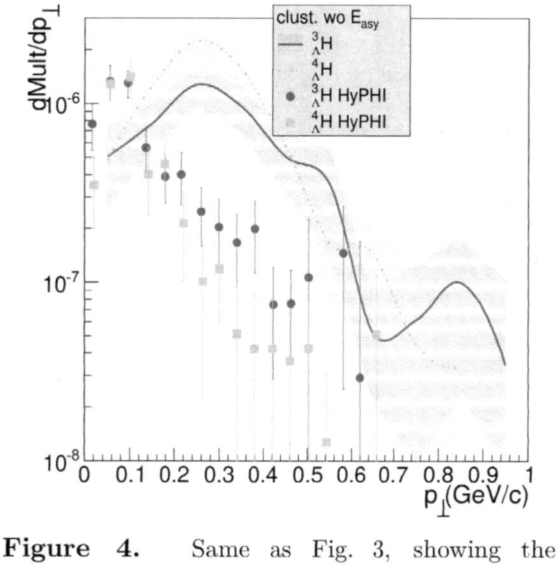


**Figure 4.** Same as Fig. 3, showing the transverse momentum multiplicity distributions in the projectile spectator region ($y_0 > 0.9$).

Journal of Physics: Conference Series **668** (2016) 012021 doi:10.1088/1742-6596/668/1/012021

the complex experimental trigger. The chosen rapidity region has the highest hadronic yield and contains still the tail of the $\Lambda$ distribution, as predicted by IQMD-FRIGA. At these rapidities, the experiment has measured a yield ratio $Y(^3_\Lambda H)/Y(^4_\Lambda H) = 1.4 \pm 0.8$, with which IQMD-FRIGA agrees within the experimental uncertainty with $1.3 \pm 0.2$. Including the asymmetry contribution $B_{asy}$ in the cluster binding energy in FRIGA, we obtain a yield ratio of $1.9 \pm 0.4$ which is still within this uncertainty. Therefore, at this level, the role of the asymmetry energy is difficult to judge. Fig. 4 shows that the transverse momentum distributions in the spectator region agree as well, in the slopes and the absolute yield. Just a slight discrepancy has to be noticed with a shift of 0.1 GeV/c in $p_\perp$. Here, we present the results without $B_{asy}$ contribution, but the conclusion is similar when including it.

## Conclusion

We present here the first step towards an understanding of the production of isotopic yields and hypernuclei in heavy ion reactions. Our clusterisation algorithm FRIGA, an improved version of the SACA approach, which includes pairing and asymmetry energies as well as other structure effects is able to describe more precisely the nuclear binding energy and allows for realistic predictions of absolute (hyper-)isotope yields. We have seen that the asymmetry and pairing potentials can have a strong influence on both, the yields and momentum anisotropies for the (hyper-)isotopes. According to this model, the nucleons which form fragments have initially a fairly low density. They contract and form finally slightly excited fragments which may undergo secondary decays. Therefore, the fragment formation is sensitive to the density dependence of the asymmetry energy and the pairing energy. However, fragments test this dependence only for densities below the saturation density.


[1] P.B. Gossiaux, R. Puri, Ch. Hartnack, J. Aichelin, Nucl. Phys. A **619** (1997) 379-390.
[2] C. O. Dorso and J. Randrup, Phys. Lett. B **301**, 328 (1993).
[3] R. K. Puri and J. Aichelin, J. Comput. Phys. **162**, 245 (2000).
[4] K. Zbiri, A. Le Fèvre, J. Aichelin et al., Phys. Rev. C **75** (2007) 034612.
[5] A. Le Fèvre et al., Phys. Rev. C **80** (2009) 044615.
[6] J. Aichelin. Phys. Reports **202**, 233 (1991).
[7] W. Cassing, E.L. Bratkovskaya, Nucl. Phys. A **831** (2009) 2.
[8] Ch. Hartnack et al., Eur. Phys. J. A **1** (1998) 151.
[9] E. Khan, Nguyen Van Giai, N. Sandulescu, Nucl. Phys. A **789** (2007) 94.
[10] E. Khan, M. Grasso and J. Margueron, Phys. Rev. C **80** (2009) 044328.
[11] A. Le Fèvre et al., Nucl. Phys. A **735** (2004) 219.
[12] S. Hudan et al. (INDRA collaboration), Phys. Rev. C **67** (2003) 064613.
[13] R.J. Charity et al., Nucl. Phys. A **476** (1988) 516. [14]
[14] C. Hartnack, H. Oeschler, Y. Leifels, E. L. Bratkovskaya and J. Aichelin, Phys. Rept. **510** (2012) 119 [arXiv:1106.2083 [nucl-th]].
[15] Ch. Rappold et al., Phys. Lett. B **747** (2015) 129.

# Strange hadron production at SIS energies: an update from HADES

**M. Lorenz[8,f], J. Adamczewski-Musch[4], O. Arnold[10,9], E.T. Atomssa[15], C. Behnke[8], J.C. Berger-Chen[10,9], J. Biernat[3], A. Blanco[2], C. Blume[8], M. Böhmer[10], P. Bordalo[2], S. Chernenko[7], C. Deveaux[11], A. Dybczak[3], E. Epple[10,9], L. Fabbietti[10,9], O. Fateev[7], P. Fonte[2,a], C. Franco[2], J. Friese[10], I. Fröhlich[8], T. Galatyuk[5,b], J. A. Garzón[17], K. Gill[8], M. Golubeva[12], F. Guber[12], M. Gumberidze[5,b], S. Harabasz[5,3], T. Hennino[15], S. Hlavac[1], C. Höhne[11], R. Holzmann[4], A. Ierusalimov[7], A. Ivashkin[12], M. Jurkovic[10], B. Kämpfer[6,c], T. Karavicheva[12], B. Kardan[8], I. Koenig[4], W. Koenig[4], B. W. Kolb[4], G. Korcyl[3], G. Kornakov[5], R. Kotte[6], A. Krása[16], E. Krebs[8], H. Kuc[3,15], A. Kugler[16], T. Kunz[10], A. Kurepin[12], A. Kurilkin[7], P. Kurilkin[7], V. Ladygin[7], R. Lalik[10,9], K. Lapidus[10,9], A. Lebedev[13], L. Lopes[2], T. Mahmoud[11], L. Maier[10], A. Mangiarotti[2], J. Markert[8], V. Metag[11], J. Michel[8], C. Müntz[8], R. Münzer[10,9], L. Naumann[6], M. Palka[3], Y. Parpottas[14,d], V. Pechenov[4], O. Pechenova[8], V. Petousis[14], J. Pietraszko[4], W. Przygoda[3], B. Ramstein[15], L. Rehnisch[8], A. Reshetin[12], A. Rost[5], A. Rustamov[8], A. Sadovsky[12], P. Salabura[3], T. Scheib[8], K. Schmidt-Sommerfeld[10], H. Schuldes[8], P. Sellheim[8], J. Siebenson[10], L. Silva[2], Yu.G. Sobolev[16], S. Spataro[e], H. Ströbele[8], J. Stroth[8,4], P. Strzempek[3], C. Sturm[4], O. Svoboda[16], A. Tarantola[8], K. Teilab[8], P. Tlusty[16], M. Traxler[4], H. Tsertos[14], T. Vasiliev[7], V. Wagner[16], C. Wendisch[4], J. Wirth[10,9], J. Wüstenfeld[6], Y. Zanevsky[7], P. Zumbruch[4]**

(HADES collaboration)

[1]Institute of Physics, Slovak Academy of Sciences, 84228 Bratislava, Slovakia
[2]LIP-Laboratório de Instrumentação e Física Experimental de Partículas , Coimbra, Portugal
[3]Smoluchowski Institute of Physics, Jagiellonian University of Cracow, 30-059 Kraków, Poland
[4]GSI Helmholtzzentrum für Schwerionenforschung GmbH, 64291 Darmstadt, Germany
[5]Technische Universität Darmstadt, 64289 Darmstadt, Germany
[6]Institut für Strahlenphysik, Helmholtz-Zentrum Dresden-Rossendorf, 01314 Dresden, Germany
[7]Joint Institute of Nuclear Research, 141980 Dubna, Russia
[8]Institut für Kernphysik, Goethe-Universität, 60438 Frankfurt, Germany
[9]Excellence Cluster 'Origin and Structure of the Universe' , 85748 Garching, Germany
[10]Physik Department E12, Technische Universität München, 85748 Garching, Germany
[11]II.Physikalisches Institut, Justus Liebig Universität Giessen, 35392 Giessen, Germany
[12]Institute for Nuclear Research, Russian Academy of Science, 117312 Moscow, Russia
[13]Institute of Theoretical and Experimental Physics, 117218 Moscow, Russia
[14]Department of Physics, University of Cyprus, 1678 Nicosia, Cyprus
[15]Institut de Physique Nucléaire (UMR 8608), F-91406 Orsay Cedex, France
[16]Nuclear Physics Institute, Academy of Sciences of Czech Republic, 25068 Rez, Czech Republic
[17]LabCAF. F. Física, Univ. de Santiago de Compostela, 15706 Santiago de Compostela, Spain

[a] also at ISEC Coimbra, Coimbra, Portugal
[b] also at ExtreMe Matter Institute EMMI, 64291 Darmstadt, Germany
[c] also at Technische Universität Dresden, 01062 Dresden, Germany
[d] also at Frederick University, 1036 Nicosia, Cyprus
[e] also at Dipartimento di Fisica and INFN, Università di Torino, 10125 Torino, Italy
[f] also at Utrecht University, 3584 CC Utrecht, The Netherlands

E-mail: `m.lorenz@gsi.de`

**Abstract.** We present and discuss recent experimental activities of the HADES collaboration on open and hidden strangeness production close or below the elementary NN threshold. Special emphasis is put on the feed-down from $\phi$ mesons to antikaons, the presence of the $\Xi^-$ excess in cold nuclear matter and the comparison of statistical model rates to elementary p+p data. The implications for the interpretation of heavy-ion data are discussed as well.

## 1. Introduction

In the last two decades strange hadrons have been considered to represent particularly suitable probes of the high density phase of nuclear matter produced in relativistic heavy-ion collisions. For example, the relation of subthreshold-produced $K^+$ in light and heavy systems has been connected to the nuclear equation of state (EOS) at up to three times nuclear saturation [1, 2, 3], while phase space distributions and flow patterns are considered to be sensitive to the in-medium kaon-nucleon potential [4, 5, 6, 7].

Various aspects of strangeness production at energies available the Schwerionen Synchrotron (SIS) at GSI Darmstadt (up to kinetic beam energies of 2 A GeV) have been investigated by the FOPI and KaoS experiments and lately also by HADES (for reviews see [8, 9]). In contrast to strange particle production in elementary NN collisions, in heavy ion reactions multi-step processes as well as strangeness-exchange reactions like $\pi\Lambda \rightarrow NK^-$ are possible. Hence for any detailed understanding of strangeness production one has also to understand strangeness dynamics in the hot and dense medium. It is therefore important to measure as many particles with open or hidden strangeness as possible. Only recently data from FOPI [10] and HADES [11] has come close to fulfilling this requirement, and several still puzzling aspects have been revealed, e.g.:

- The strength of the kaon-nucleon potential, as derived from the comparison of data to models, yields values varying by about 50% for different data samples [12, 13, 14].
- The discovery of neutron stars with masses larger than 1.5 solar masses [15, 16] challenges the previously extracted constraints on the softness of the EOS [17].
- The large $\phi/K^-$ ratio reported in [18, 19] and the resulting feed-down which affects significantly the slope of the $K^-$ questions the common interpretation of a later freeze-out of the $K^-$ compared to $K^+$ [20]. In addition, new mechanisms seem to be required to reproduce microscopically the enhanced $\phi$ production [21, 22].
- Although transport codes do not reach chemical equilibrium within the lifetime of such collisions [9], the yields of produced hadrons can be well described within the framework of statistical models [11, 23], except for the $\Xi^-$ (see below).
- The deep-subthreshold production of the $\Xi^-$ which exceeds thermal model predictions by an order of magnitude [24, 25], can up to now, only be reproduced by one model [26].

In this paper we will review and investigate the status of the last three aspects.

## 2. Antikaons: slopes, yields and freeze-out

Systematic investigations of the KaoS collaboration in the nineties revealed a similar rise of kaon and antikaon yields with increasing centrality of the collision. In addition, the inverse slope parameters of the antikaons obtained from transverse spectra are systematically lower

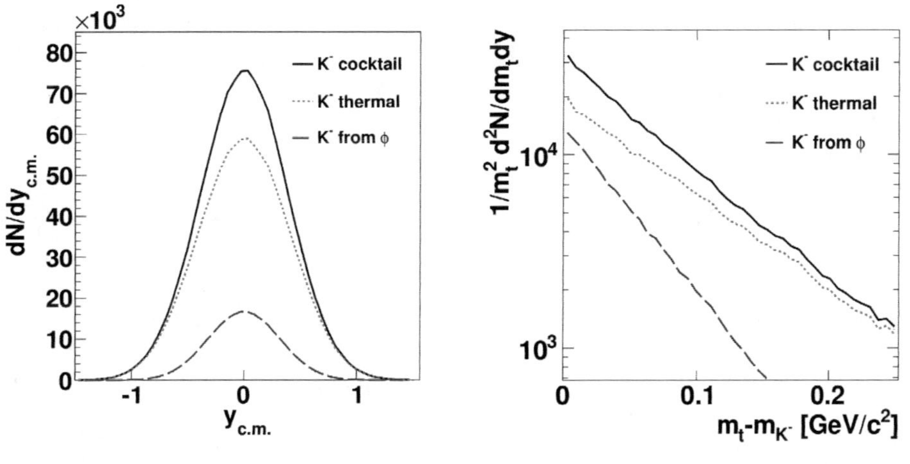


**Figure 1.** Left panel: Normalized rapidity density distributions of the thermal $K^-$ and the ones resulting of a $\phi$ according to the measured yields in [19] . The black solid line shows the sum of the two distributions. Right panel: Simulated transverse mass spectra for $K^-$ coming from a thermal source with a temperature of 89 MeV and those which stem from a $\phi$ decay. The resulting cocktail spectrum is shown as a black solid line.

than those observed for the kaons [8]. These findings have been interpreted with the help of transport models and the following conclusion has been drawn: the production of kaons and antikaons is coupled via strangeness exchange reactions, e.g. $\pi^- + \Lambda \rightarrow p + K^-$. As a consequence, the antikaons experience a later freeze-out than the kaons and hence have steeper transverse spectra. Furthermore people concluded that strangeness production close or below the elementary nucleon-nucleon (NN) threshold is very different to the one in elementary reactions [9]. The antikaon abundance has also been connected to the reduction of the effective antikaon mass via the strength of the strangeness exchange channels with an antikaon in the final state in [27].

Newer data from FOPI and HADES showed that for energies around $\sqrt{s_{NN}}$= 2.6 GeV the fraction of antikaons originating from a decay of the $\phi$ meson is about 20% and is hence not negligible [18, 19]. The resulting effect of the $\phi$ meson feed-down on the antikaon slopes has first been investigated in our previous work [20]. Simulating a cocktail of thermal and antikaons resulting from $\phi$ meson feed-down, according to the measured contributions in the Ar+KCl system, we found that within errors the inverse slope of the resulting $K^-$ spectra agrees with the experimentally observed one, see Fig. 1. Hence the different slopes of the kaon and antikaon spectra do not necessarily call for a sequential freeze-out of the two kaon species but can be explained solely by $\phi$ feed-down. Meanwhile this issue has been investigated in greater detail by the FOPI collaboration [28].

Preliminary HADES Au+Au data taken at $\sqrt{s_{NN}}$= 2.4 GeV, where the complete set of particles carrying open and hidden strangeness is produced below the free nucleon-nucleon threshold, indicate a strong rise of the $\phi/K^-$ ratio towards lower energies [29], as predicted in a statistical model calculation [19]. For minimization of systematic errors due to efficiency corrections and extrapolations in rapidity, ratios of the corrected yields at mid-rapidity are used in this analysis. The resulting $K^-/K^+$ ratio can be directly compared to the previously obtained systematics at similar energy. The measured ratio $K^-/K^+$ fits into the trend observed at higher energies and extrapolated down to the beam energy of 1.23 A GeV, see left side of Fig.2: The $\phi/K^-$

ratio is displayed on the right side of Fig.2; it shows a flat trend at high energies, and a rise towards lower energies. This rise can be reproduced in the framework of a statistical model, if the suppression of strangeness is handled by introducing a strangeness correlation radius $R_c$ within which strangeness has to be exactly conserved [30]. In this context it is important to realize that, as the $\phi$ conserves strangeness by definition, it is not suppressed by the strangeness correlation parameter in contrast to the other particles containing strange quarks.
Recently, feed down from higher lying baryonic resonances have been included in UrQMD in order to be able to describe the energy dependence of the $\phi/K^-$ ratio also in a transport code [31]. The authors tuned the mass depending branching ratio of high lying baryon resonances, namely the $N^*(1990)$, $N^*(2080)$, $N^*(2190)$, $N^*(2220)$ and $N^*(2250)$, in a transport code to match elementary data on $\phi$ meson production. As a result, the $\phi/K^-$ in Ar+KCl is successfully reproduced, as well as the shape of the excitation function. At higher energies the model undershoots the data and the preliminary Au+Au data point is also not fully reproduced.
It will be very interesting, as soon as the final data becomes available, to compare besides the pure yields, the shape of the transverse particle spectra, as they are strongly dependent on the contribution from resonances as investigated in [32] for kaons with respect to the sensitivity of the spectral shape to the kaon nucleon potential.

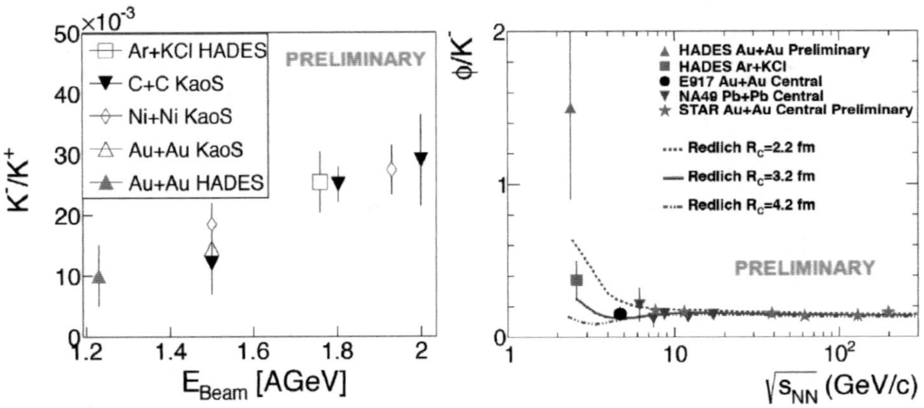


**Figure 2.** Left: Kinetic beam energy excitation function of the $K^-/K^+$ ratio for various colliding systems. Data is taken from [8, 19]. The preliminary ratio at midrapidity extracted in this work, fits nicely to the trend. Right: The $\phi/K^-$ ratio shows a flat trend at high energies and a sharp rise towards lower energies, which can be explained within the statistical model framework using a proper strangeness correlation radius $R_c$. The data is taken from [19].

## 3. Statistical equilibrium rates

In the last 30 years statistical hadronization models have been established as a successful tool to fit particle yields or yield ratios from relativistic and ultrarelativistic heavy-ion collisions [33, 34, 35] with only a few parameters. The extracted freeze-out parameters show a striking regularity, lining up on a curve in the temperature-baryochemical potential plane, connecting smoothly data from the GeV to the TeV regime [36] . These findings have been widely interpreted as a hint that (local) chemical equilibrium is achieved in such collisions.
Since the days of Hagedorn [37], statistical methods have also been used to predict particle production in elementary reactions. Recently two groups applied the same model, which successfully describes hadron yields in heavy-ion collisions, also for elementary reactions but reached very different conclusions. While the authors of [38, 39] found good agreement for yields and even transverse momentum spectra obtained in elementary $e^+ + e^-$, the authors of [40] found significantly worse agreement between model and experimental yields.

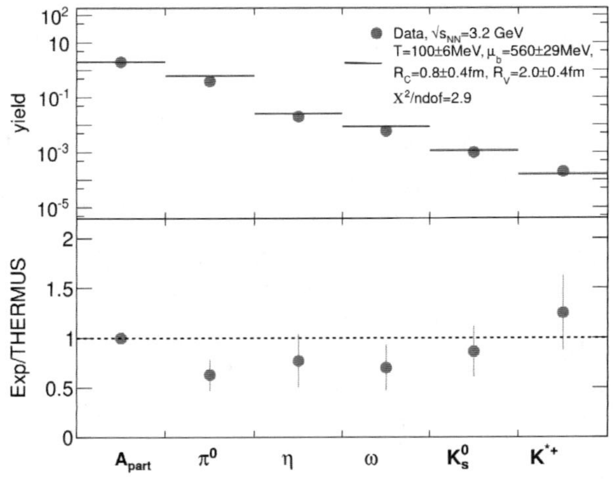


**Figure 3.** The upper plot shows the yields of secondary hadrons in p+p reactions (filled red circles) and the corresponding THERMUS fit values (blue bars) as well as the obtained fit parameters. The lower plot shows the ratio of the experimental value and the statistical model value.

In order to investigate this issue, we use the yields of hadrons produced in elementary p+p collisions at $\sqrt{s_{NN}}$= 3.2 GeV measured with HADES, which have recently become available [41, 42, 43]. We apply a similar fit as in our previous work [11] but use an updated version of THERMUS (v3.0) [44]. We use the mixed canonical ensemble where strangeness is exactly conserved while all other quantum numbers are calculated grand canonically and constrain the charge chemical potential $\mu_Q$ using the ratio of the baryon and charge numbers of the collision system.

We state that the yield of the $\phi$ meson is of particular interest, because of its different sensitivity to the strangeness suppression parameters $\gamma_s$ and $R_c$. As the $\phi$ conserves strangeness by definition as a $s\bar{s}$ state its yield is not suppressed in the $R_c$ formalism, while strongly suppressed when $\gamma_s$ is used. We found in [11] that the $\phi$ yield is well described using $R_c$ and therefore stick to this way of suppressing strange particle yields in our statistical model calculations.

We simultaneously fit the yields of the neutral pions, the $\eta$, the $\omega$ and the kaons, as well as the mean number of participants $\langle A_{part}\rangle$ and constrain the charge chemical potential $\mu_Q$ by the initial charge to baryon ratio. We find the following values for chemical freeze-out parameters: $T_{chem}$ = (100 ± 6) MeV, $\mu_b$ = (560 ± 29) MeV, the strangeness correlation radius results as $R_c = (0.8 \pm 0.4)$ fm and the radius of the whole fireball $R = (2.0 \pm 0.4)$ fm with $\chi^2$/d.o.f.=2.9. A detailed comparison of the data with our the statistical model fit is shown in the upper part of Fig. 3, while the lower part of this figure depicts the ratio of data. The overall agreement between model and data is comparable to the outcome of our previous work where we fitted hadron yields obtained from Ar+KCl collisions at $\sqrt{s_{NN}}$= 2.6 GeV. This is rather surprising as one naively expects a larger amount of thermalization in the larger Ar+KCl system and hence less deviation from statistical equilibrium values. Furthermore, the p+p freeze-out point fits quite well to the previously observed regularity of freeze-out points in the $T - \mu_b$ plane, displayed in Fig. 4, where the extracted point of this work are displayed together with similar points extracted in [45] and [36].

This brings us back to the introduction; while the success of the statistical model in describing particle rates from heavy-ion collision is often implicitly connected to a thermalization of the created system, the success of the model in describing the p+p data at $\sqrt{s_{NN}}$= 3.2 GeV strongly questions this connection.

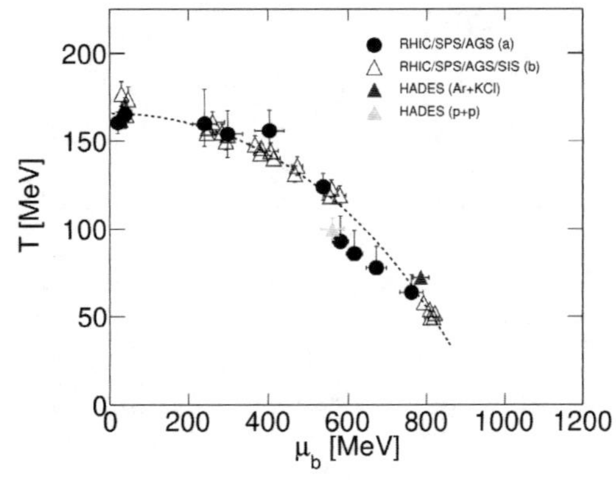


**Figure 4.** Freeze-out points in the $T - \mu_b$ plane obtained from statistical model fits to hadron yields or yield ratios [45] (filled symbols) and [36] (open symbols). In addition, the HADES points for Ar+KCl [11] (blue triangle) and the p+p point (cyan triangle) from this work are displayed. The dotted line corresponds to a common freeze-out condition of 1 GeV fixed energy per baryon.

## 4. The $\Xi^-$ excess

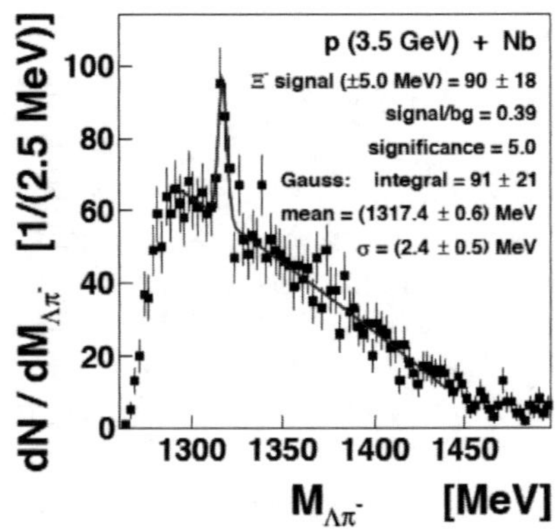


**Figure 5.** Invariant mass spectrum of $\Lambda$ hyperons and negative pions obtained in p+Nb collisions at $\sqrt{s_{NN}}$= 3.2 GeV. A clear peak corresponding to the $\Xi^-$ signal is visible.

Recent data of HADES revealed that the excess of the $\Xi^-$ over several state of the art models shows up also in p+Nb reactions at $\sqrt{s_{NN}}$= 3.2 GeV [46]. The extracted experimental signal in the invariant mass spectrum is displayed in Fig. 5. In addition, the energy excitation function of the $\Xi^-$ Lambda ratio for various colliding systems is displayed in Fig. 6. While the open symbols correspond to heavy-ion data, the filled symbols represent proton induced collisions on nuclei. The black line corresponds to a simple parameterization fitted to the proton induced data of the form

$$F(\sqrt{s_{NN}}) = C \times \left(1 - \left[\frac{D}{\sqrt{s_{NN}}}\right]^G\right)^H . \qquad (1)$$

The inlay shows a zoom to the area of the HADES p+Nb data point and the comparison to three different models. The upper star symbol corresponds to the yield obtained from a THERMUS (v2.3) fit to p+Nb data similar to the one discussed in this work. While the lower star symbol corresponds to the yield obtained via the UrQMD (v3.4) transport code, the intermediate point results from a GiBUU transport calculation.
The presence of the $\Xi^-$ excess in cold nuclear matter has several interesting implications for the interpretation of the heavy-ion data as its origin seems to be already in the elementary channels without the involvement of many body effects in the medium [26, 47]. Therefore, the increased cross sections of strangeness exchange reactions, which were found to be sufficient to explain the high yield in, seem to be questionable as they are highly unlikely to play an important role in

p+Nb reactions. Also, the invoked [23] catalytic strangeness production by secondary processes like $\pi + Y \rightarrow \Xi + K$ are strongly suppressed in cold nuclear matter.

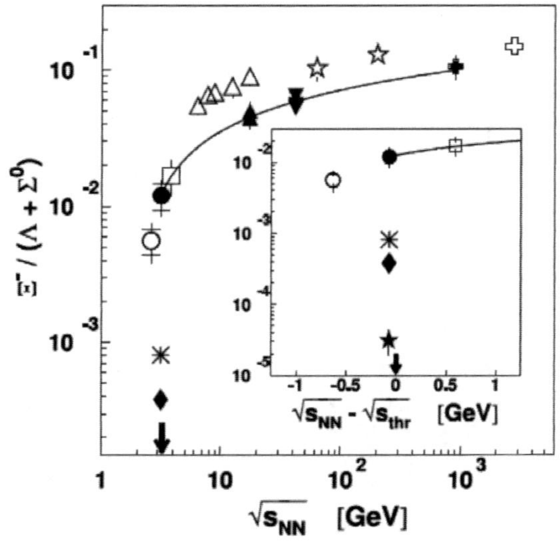


**Figure 6.** Excitation function of the $\Xi^-/\Lambda$ ratio for various colliding systems. The open symbols correspond to heavy-ion data, the filled symbols represent data from p+A collisions. The black line corresponds to a simple parameterization fitted to the p+A data, see text for details. The inlay shows a zoom to the area of the HADES p+Nb data point and the comparison to three different models. The upper star symbol corresponds to a the yield obtained from a THERMUS (v2.3) fit to p+Nb data. The lower star symbol corresponds to the yield obtained via the UrQMD (v3.4) transport code and the intermediate point a GiBUU transport calculation.

Recently the same mechanism of feed-down of the high invariant mass tails of baryonic resonances used to explain the $\phi$ meson yield by the UrQMD group was also used for the $\Xi^-$ hyperon. Due to the lack of elementary data the model is tuned to match our p+Nb data [32]. After this tuning it is not too surprising that the model is able to describe the excess observed in Ar+KCl without further tuning.
The technique of mass dependent branching ratios for broad resonances has been successfully applied in order to describe the dilepton spectra at low energies as pointed out in several publications [48, 49]. Note, how even that tuned branching ratios are still consistent with the OZI rule there is yet no experimental evidence for the decay of the $N^*$ resonance to final states containing a $\phi$ meson or a $\Xi$ hyperon.

## 5. Summary

We showed that for all three topics discussed here elementary and cold nuclear matter data are needed for the interpretation of heavy-ion data. In particular:

- The observed large $\phi/K^-$ ratio and the resulting feed-down lowers significantly the slope of the $K^-$. Hence a later freeze-out of antikaons compared to kaons is not necessarily needed to explain the different slopes of both kaon species. Furthermore, the rise towards lower energies can also be reproduced, at least to some extent, in transport codes after tuning of the $\phi$ cross sections on elementary data.
- We find that a statistical model fit to elementary p+p data is able to describe the yields with the same parameters as used for Ar+KCl with comparable accuracy and this questions the often implicitly used connection to a thermalized system.
- The presence of the $\Xi^-$ excess also in cold nuclear matter rules out more exotic explanations involving multistep processes in heavy-ion collisions. Feed-down from high mass tails of baryonic resonances remains a possible explanation but at the moment it is not very well constrained by both the experimental data and the theoretical treatment.

The collaboration gratefully acknowledges support by PTDC/FIS/113339/2009 (LIP Coimbra), NCN grant 2013/10/M/ST2/00042 (SIP JUC Cracow), Helmholtz Alliance

HA216/EMMI (GSI Darmstadt), VH-NG-823, Helmholtz Alliance HA216/EMMI (TU Darmstadt), 283286, 05P12CRGHE (HZDR), Helmholtz Alliance HA216/EMMI, HIC for FAIR (LOEWE), (GSI F&E Goethe-University Frankfurt) VH-NG-330, BMBF 06MT7180 (TU München), Garching BMBF:05P12RGGHM (JLU Giessen, Giessen) UCY/3411-23100 (University Cyprus) CNRS/IN2P3, (IPN Orsay), Orsay MSMT LG 12007, AS CR M100481202, GACR 13-06759S NPI AS CR, Rez, EU Contract No. HP3-283286. One of us (M.L.) acknowledges the support of the Humboldt Foundation.

[1] C. Sturm *et al.* [KaoS Collaboration], Phys. Rev. Lett. **86**, 39 (2001).
[2] C. Fuchs, A. Faessler, E. Zabrodin, Phys. Rev. Lett. **86**, 1974 (2001).
[3] C. Hartnack, H. Oeschler, J. Aichelin, Phys. Rev. Lett. **96**, 012302 (2006).
[4] J. Schaffner-Bielich, J. Bondorf, A. Mishustin, Nucl. Phys. A **625**, 325 (1997).
[5] W. Cassing, E. L. Bratkovskaya, U. Mosel, S. Teis, A. Sibirtsev, Nucl . Phys. A **614** 415 (1997).
[6] L. Adamczyk *et al.* [STAR Collaboration], Phys. Rev. Lett. **110** 142301 (2013).
[7] V. Zinyuk *et al.* [FOPI Collaboration], arXiv:1403.1504 [nucl-ex].
[8] A. Forster, F. Uhlig, I. Bottcher, D. Brill, M. Debowski *et al.*, Phys. Rev. C **75** 024906 (2007).
[9] C. Hartnack, H. Oeschler, Y. Leifels, E. L. Bratkovskaya, J. Aichelin, Phys. Rept. **510** 119 (2012).
[10] N. Bastid *et al.* [FOPI Collaboration], Phys. Rev. C **76** 024906 (2007).
[11] G. Agakishiev *et al.* [HADES Collaboration], Eur. Phys. J. A **47** 21 (2011).
[12] M. L. Benabderrahmane *et al.* [FOPI Collaboration], Phys. Rev. Lett. **102** 182501 (2009).
[13] G. Agakishiev *et al.* [HADES Collaboration], Phys. Rev. C **82** 044907 (2010).
[14] G. Agakishiev *et al.* [HADES Collaboration], arXiv:1404.7011 [nucl-ex].
[15] P. Demorest *et al.*, Nature **476**, 1081 (2010).
[16] J. Antoniadis *et al.*, Science **340**, 6131 (2013).
[17] A. W. Thomas *et al.*, EPJ Web of Conferences **63**, 03004 (2013).
[18] A. Mangiarotti *et al.* [FOPI Collaboration], Nucl. Phys. A **714** 89 (2003).
[19] G. Agakishiev *et al.* [HADES Collaboration], Phys. Rev. C **80** 025209 (2009).
[20] M. Lorenz *et al.* [HADES Collaboration], PoS BORMIO **2010** 038 (2010).
[21] B. Tomasik and E. E. Kolomeitsev, Acta Phys. Polon. Supp. **5** 201 (2012).
[22] B. Kampfer, R. Kotte, C. Hartnack, J. Aichelin, J. Phys. G **28** 2035 (2002).
[23] E. E. Kolomeitsev, B. Tomasik, D. N. Voskresensky, Phys. Rev. C **86** 054909 (2012).
[24] G. Agakishiev *et al.* [HADES Collaboration], Phys. Rev. Lett. **103** 132301 (2009).
[25] J. Steinheimer, T. Lang, H. van Hees, A. S. Botvina et al., J. Phys. Conf. Ser. **509** 012002 (2014).
[26] F. Li, L. -W. Chen, C. M. Ko and S. H. Lee, Phys. Rev. C **85** 064902 (2012).
[27] W. Cassing, E. L. Bratkovskaya, U. Mosel, S. Teis, A. Sibirtsev, Nucl. Phys. A **614** (1997) 415.
[28] K. Piasecki *et al.* [FOPI Collaboration], Phys. Rev. C **91** 5, 054904, (2015).
[29] M. Lorenz [HADES Collaboration], Nucl. Phys. A **931**, 785, (2014).
[30] J. Cleymans *et al.*, Phys. Rev. C 59, 1663 (1999).
[31] J. Steinheimer, M. Bleicher, arXiv:1503.07305 [nucl-th].
[32] J. Steinheimer, M. Bleicher, EPJ Web Conf. **97**, 00026 (2015).
[33] P. Braun-Munzinger, K. Redlich and J. Stachel, arXiv:nucl-th/0304013 (2003).
[34] A. N. Tawfik, Int. J. Mod. Phys. A **29** 17, 1430021 (2014).
[35] M. Floris, Nucl. Phys. A **931** 103 (2014).
[36] J. Cleymans, H. Oeschler, K. Redlich and S. Wheaton, Phys. Rev. C **73**, 034905 (2006).
[37] R. Hagedorn, Nuovo Cim. Suppl. **3** 147 (1965).
[38] F. Becattini, G. Passaleva, Eur. Phys. J. C **23** 551 (2002).
[39] F. Becattini, P. Castorina, J. Manninen, H. Satz, Eur. Phys. J. C **56** 493 (2008).
[40] A. Andronic, F. Beutler, P. Braun-Munzinger, K. Redlich, J. Stachel, Phys. Lett. B **675**, 312, (2009).
[41] G. Agakishiev *et al.* [HADES Collaboration], Eur. Phys. J. A **48**, 64, (2012).
[42] G. Agakishiev *et al.* [HADES Collaboration], Phys. Rev. C **90**, 015202, (2014).
[43] G. Agakishiev *et al.* [HADES Collaboration], Phys. Rev. C **92**, 024903, (2015).
[44] S. Wheaton, J. Cleymans, Comput. Phys. Commun. **180** 84 (2009).
[45] A. Andronic, P. Braun-Munzinger and J. Stachel, Nucl. Phys. A **772**, 167 (2006).
[46] G. Agakishiev *et al.* [HADES Collaboration], Phys. Rev. Lett. **114** 21, 212301, (2015).
[47] G. Graef, J. Steinheimer, F. Li, M. Bleicher, Phys. Rev. C **90**, 064909, (2014).
[48] E. L. Bratkovskaya, J. Aichelin, M. Thomere, S. Vogel, M. Bleicher, Phys. Rev. C **87** 064907 (2013).
[49] J. Weil, H. van Hees, U. Mosel, Eur. Phys. J. A **48** 111 (2012).

Journal of Physics: Conference Series **668** (2016) 012023 doi:10.1088/1742-6596/668/1/012023

# Three Stages of the NICA Accelerator Complex

**V D Kekelidze, R Lednicky, V A Matveev, I N Meshkov, A S Sorin, G V Trubnikov**

Joint Institute for Nuclear Research, Dubna 141980, Russia

E-mail: meshkov@jinr.ru

**Abstract.** The Nuclotron-based Ion Collider fAcility (NICA) project is under development at JINR (Dubna). The general goals of the project are to provide colliding beams for experimental studies of both hot and dense strongly interacting baryonic matter and spin physics (in collisions of polarized protons and deuterons). The first program will require the running of heavy ion mode in the energy range of $\sqrt{s_{NN}} = 4\div11$ GeV at an average luminosity of $L = 1\cdot10^{27}$ cm$^{-2}\cdot$s$^{-1}$ for $^{197}$Au$^{79}$ nuclei. This stage of the project will be preceded by fixed target experiments with the heavy ion beam to be extracted from the Nuclotron at kinetic energies up to 4.5 GeV/u. The polarized beam mode is proposed to be used in the energy range of $\sqrt{s_{NN}} = 12 \div 27$ GeV (protons) at luminosities up to $1 \cdot 10^{32}$ cm$^{-2}\cdot$s$^{-1}$ . This report contains a brief description of the facility scheme and characteristics in the heavy ion operation mode, the description of the MultiPurpose Detector (MPD), and characteristics of the reactions of the colliding ions, which will allow us to detect the mixed phase formation. The plans and status of the project development are presented.

## 1. Introduction: The NICA Project at JINR

The NICA project is aimed to be developed, constructed, and commissioned at the Joint Institute for Nuclear Research (Dubna, Russia), a modern accelerator complex Nuclotron-based Ion Collider fAcility (NICA) equipped with two detectors MultiPurpose Detector (MPD) & Spin Physics Detector (SPD). Experiments shall be performed in search of the mixed phase of baryonic matter and the nature of nucleon/particle spin.

A study of hot and dense baryonic matter should shed light on in-medium properties of hadrons and the nuclear matter equation of state; onset of deconfinement and/or chiral symmetry restoration; phase transition, mixed phase and the critical end-point; and possible local parity violation in strong interactions [1]. It has been indicated in series of theoretical works, in particular, in [2], that heavy ion collisions at the nucleon-nucleon center-of-mass energy $\sqrt{s_{NN}} \sim 10$ GeV allow one to reach the highest possible net baryon density.

The NICA project is under development as a flagship JINR project in high energy physics. In addition to the beams extracted from the Nuclotron, the project foresees the construction of collider facility providing ion collisions in collider mode at the energy range of $\sqrt{s_{NN}} = 4 \div 11$ GeV for Au$^{79+}$ with the luminosities up to $L = 10^{27}$cm$^{-2}\cdot$s$^{-1}$.

NICA will also provide the polarized proton and deuteron beams up to the c.m.s. energy of 27 GeV for pp collisions with luminosity up to $L = 10^{32}$cm$^{-2}\cdot$s$^{-1}$.

The high intensity and high polarization ($> 50\%$) of the colliding beams will present a unique possibility for spin physics research, which is of crucial importance for the solution of the nucleon spin problem ("spin puzzle")–one of the main tasks of the modern hadron physics.

## 2. Accelerators and Colliders in NICA Energy range

The comparison of the parameters of the NICA accelerator complex with existing and currently developing machines of heavy ions and polarized beams (see Figures 1, 2) shows that NICA is perfectly situated to approach the research goals formulated above.

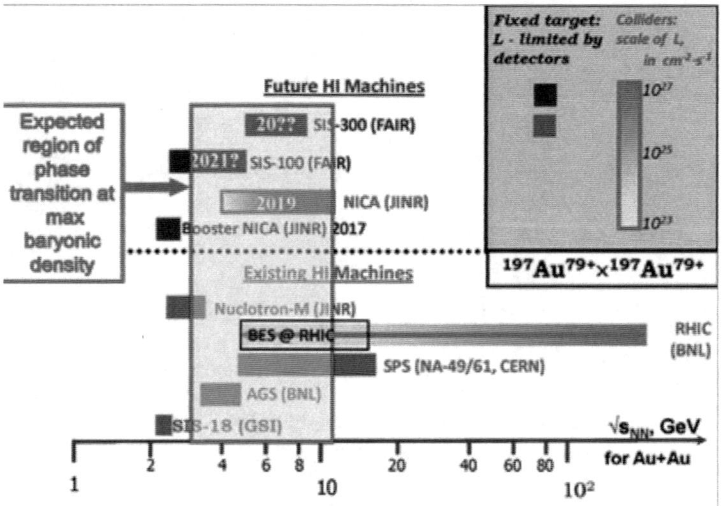


**Figure 1.** The energy ranges of heavy ion machines in operation (lower part in green) and under construction (upper part in blue and red); the rectangular frame indicates the region of maximum net baryonic density.

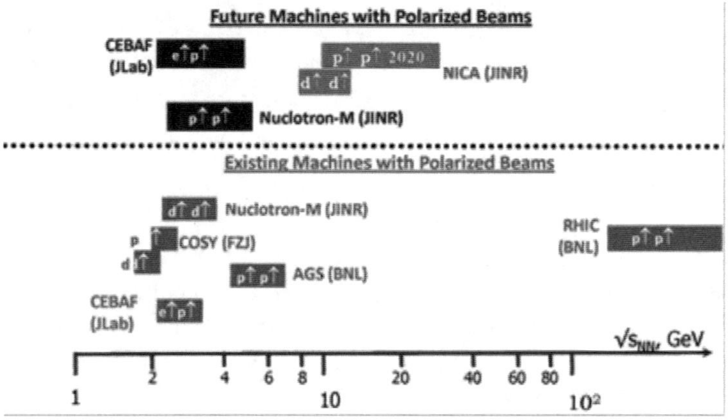


**Figure 2.** The existing machines of polarized beams (lower part in green) and ones under construction (upper part in blue and red).

## 3. NICA - Stage I

The Nuclotron facility currently consists of the "Old injector" and the Nuclotron. The "Old injector" contains set of light ion sources including a source of polarized protons and deuterons and an Alvarez-type linac LU-20 (Figure 3, pos.1).

The Nuclotron itself is a SC proton synchrotron (pos. 5) that has a maximum magnetic rigidity of 45 $T \cdot m$ and the circumference of 251.52 $m$. It can provide the acceleration of completely stripped $^{197}Au^{79+}$ ions, up to the kinetic energy in the range of $1 \div 4.5$ GeV/u, and protons, up to maximum kinetic energy of 12.6 GeV. It is used presently for fixed target experiments with extracted beams and experiments with an internal target. The program includes experimental studies on relativistic nuclear physics, spin physics in few-body nuclear systems (with polarized deuterons), and physics of flavours. This part of the program, "The Baryonic Matter at Nuclotron" (BM@N), is under active development.

The development of Stage I of the NICA project will be completed after construction of the "New injector" and the Booster-synchrotron and after the commissioning of the BM@N detector.

The "New injector" (pos. 2) is under construction. It contains the ESIS-type ion source, which will provide $^{197}Au^{32+}$ ions of intensity of $2 \cdot 10^9$ ions per pulse of about 7 $\mu s$ duration at a repetition rate of 10 Hz, and the heavy ion linear accelerator (HILaC), consisting of RFQ and RFQ Drift Tube Linac sections. The linac accelerates the ions at $A/Z \leq 8$ up to the energy of 3 MeV/u, at efficiency no less than 80% (A, Z are ion mass and charge numbers). It was delivered by the BEVATECH Company (Germany) in 2014-2015 and is presently being commissioned. This stage is intended to be complementary to the stage to be realized with Collider in heavy ion beam mode.

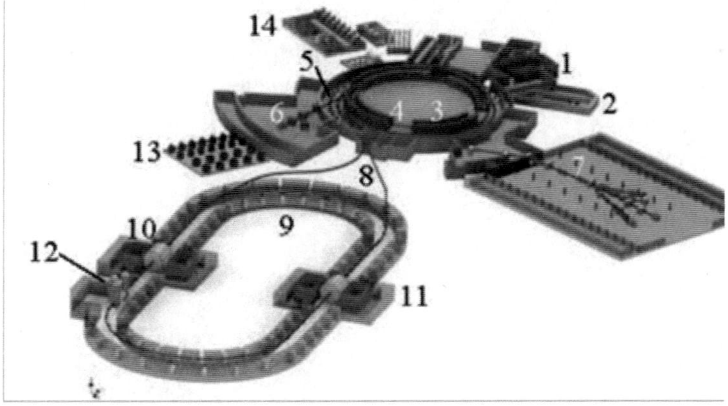


**Figure 3.** Scheme of the NICA facility. The description is given in the text.

Housed inside the Synchrophasotron yoke (pos. 3), the Booster-synchrotron (pos. 4) has a superconducting (SC) magnetic system that provides maximum magnetic rigidity of 25 T m at the ring circumference of 215 m. It is equipped with an electron cooling system, constructed by Budker INP, which will provide cooling for the ion beam in the energy range from the injection energy up to 100 MeV/u. The maximum energy of $^{197}Au^{31+}$ ions, accelerated in the Booster, is of 600 MeV/u. Stripping foil placed in the ion transfer line from the Booster to the Nuclotron provides the stripping efficiency at the maximum Booster energy, no less than 80%. The Booster elements are being manufactured, and the machine is planned to be commissioned in 2017.

The Nuclotron beams also find use in radiobiology research and other applied research. Moreover, the Nuclotron is used for testing the collider equipment and operational regimes, elements and prototypes for the MPD using extracted beams ($C^{6+}$ ions at 3.5 GeV/u and deuterons at 4 GeV/u presently). Particularly, in the run #45 (Feb. 2012), the circulation of a 3.5 GeV/u deuteron beam for 1000 seconds was demonstrated. During 2011- 2013, the first version of the stochastic cooling system was designed, constructed, and tested at the Nuclotron at an ion kinetic energy of 3.5 GeV/u with deuteron and carbon ($^{12}C^{+6}$) ion beams. This work was performed in close collaboration with the Forschungszentrum Jülich. The results will also be used for the design of the stochastic cooling system for the High-Energy Storage Ring (HESR, FAIR).

Two transfer lines, in order to transport particle beams extracted from the Booster (pos. 6), and the Nuclotron (pos. 7), in order to research areas for both basic and applied research using fixed target experiments, will be constructed.

## 4. NICA - Stage II

Stage II of the NICA project includes construction of the Collider, the beam transfer line from the Nuclotron to the Collider, and the MultiPurpose Detector (MPD).

The transfer line (pos. 8) will transport the particles from Nuclotron to the Collider rings. The line is currently in the design stage.

Two SC collider rings (pos. 9) of racetrack shape have maximum magnetic rigidity of 45 T m and a circumference of 503 m. The maximum field of SC dipole magnets is of 1.8 T. For luminosity preservation, the electron and stochastic cooling systems will be constructed. The collider design is in progress; the prototypes of its magnets were fabricated and tested in 2013; the mass production is scheduled for 2016-2018.

Two detectors – the MultiPurpose Detector (MPD, pos. 10) and the Spin Physics Detector (SPD, pos. 11) are located in opposite straight sections of the racetrack rings. The MPD is being designed presently; prototypes of the sub-detectors are under construction and testing. The SPD is under conceptual design and is planned to be constructed during Stage III.

The electron cooler with the electron kinetic energy of $0.5 \div 2.5$ MeV will be placed in a special building (pos. 12). Cryogenics and the auxiliary equipment supply facility (pos. 13, 14) provide LHe, LN2, electric power and cooling water to feed the accelerator complex and detectors. The NICA parameters allow us to reach the project luminosity at the ion kinetic energy higher than 3 GeV/u (Figure 4).

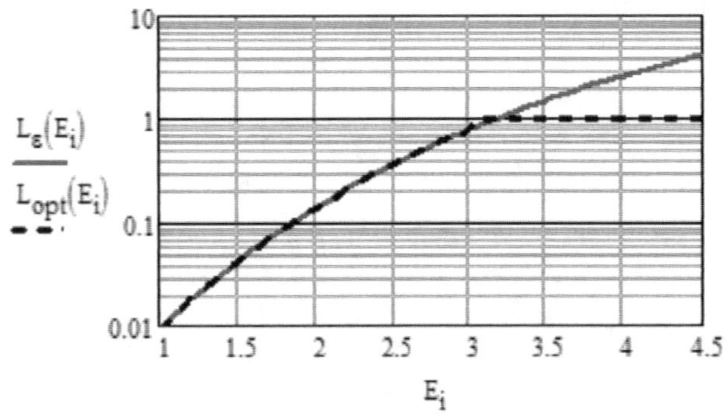


**Figure 4.** Project luminosity ($10^{27}$ cm$^{-2}$·s$^{-1}$) of the NICA collider versus ion kinetic energy per nucleon (GeV/u); two modes: ion number per bunch is limited by the ion bunch space charge ($L_\varepsilon(E_i)$, solid curve) and is optimized (dashed line, $L_{opt}(E_i)$).

## 5. The MPD

The MPD [4] is a typical collider detector based on the solenoidal superconducting magnet (Figure 5), with a magnetic field of 0.66 T (6.623 m in diameter and 9.010 m in length). The major sub-detectors of the MPD are the time projection chamber (TPC), the inner tracker (IT), the time-of-flight (TOF) system, the electromagnetic calorimeter (ECal), the end cap tracker (ECT), and two forward spectrometers based on toroidal magnets (optional). Three stages are foreseen in order to bring MPD into operation. The first stage of operation involves the magnet, TPC, TOF, ECal, and IT (partially).

The MPD experiment should be competitive and, at the same time, complementary to experiments carried out at RHIC [5] and those planned within the FAIR project [6].

There are several MPD detection tasks that should be fulfilled first [1]. A study of the elliptic flow of the secondary particles in the momentum space, in order to quantify the collective behavior of the central fireball matter.

A detailed measurement of the well-known "Horn effect" can give information about peculiarity of the heavy ion collisions. The effect has been observed in experiments where energy dependence of the multiplicity ratio $R = \langle K^+\rangle/\langle\pi^+\rangle$ was measured at the pseudorapidity $y^* \approx 0$ (i.e, at the scattering angle $\theta \approx \pi/2$). The non-monotonic dependence of the $R(0)$ ratio on energy can be regarded as an indication of the onset of deconfinement and/or the onset of a meson dominated production regime.

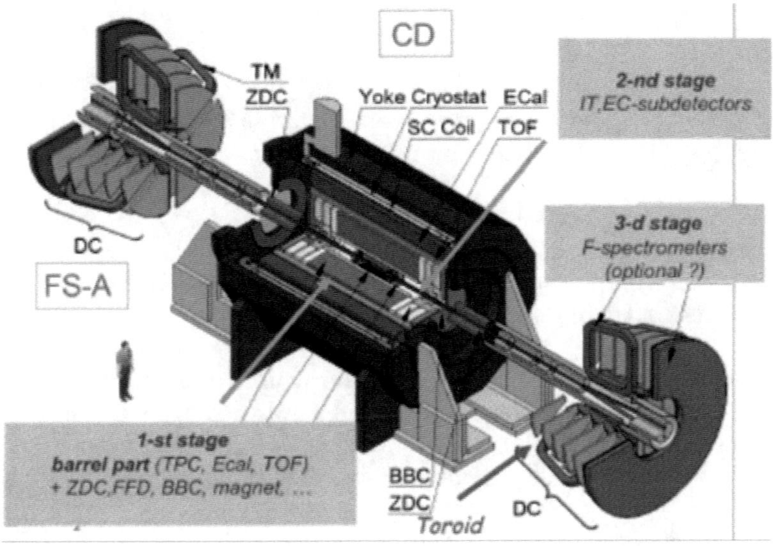


**Figure 5.** General view of the MPD and sets of sub-detectors to be put in operation at different stages.

A lot of information can be obtained from the detection of leptons and photons. Leptons result from decays of mesons, such as the $\pi$, $\rho$, $\omega$, $\phi$, $J/\Psi$, etc, which give rise to $e^+e^-$, $\mu^+\mu^-$, $\nu_e, \nu_\mu$ (the latter are undetectable in the MPD). They provide information about the QGP-phase structure. Detection of photons allows one to estimate the QGP temperature.

The fluctuation studies of the collision products parameters are very informative. Large fluctuations can be considered as a signature of the mixed phase formation. Indeed, the system becomes unstable at the two-phase stage (as in a classic process of boiling water–a flow of bubbles fluctuates tremendously). The idea is to locate the critical point using correlation/fluctuation of particle characteristics, e.g, dispersion and higher moments of $R(0)$: $D_R = \langle(R - \langle R\rangle)^2\rangle$, $M_{3R} = \langle(R - \langle R\rangle)^3\rangle$. An attempt to detect the fluctuations in the Beam Energy Scan at RHIC likely failed because of a lack of statistics, i.e, a low luminosity of the RHIC at lower energies.

An important signal for the onset of deconfinement, $1^{st}$ order phase transition, is a substantial increase of the evolution time of the excited matter. This signal can be revealed with the help of femtoscopic momentum correlations as a tail in the reconstructed space-time separation of particle emitters.

The measurement of the charge asymmetry with respect to the reaction plane characterizes the electric dipole moment of QCD matter and is a possible signature of local parity violation.

Processes studied with MPD were simulated using the dedicated software framework (MpdRoot). The evaluated rate in $Au + Au$ collisions, at $\sqrt{s_{NN}} = 7$ GeV/u (10% central interactions) for the luminosity of $10^{27}$ cm$^{-2}$s$^{-1}$, is of the order of 7 kHz. The MPD performance in general meets the required parameters for the proposed experimental program.

## 6. The Stage III - polarized ions and SPD

The polarized beam mode of NICA is being implemented in two steps. The first is the acceleration of polarized deuterons at the Nuclotron, which has been performed since the 1990s. This beam is used in fixed target experiments, and the beam intensity will be increased with commissioning of the new Source of Polarized Particles by the end of 2015. The second step is the development of the collider lattice for storing polarized beams and keeping them circulating in the collider mode. This work is in the design stage.

The SPD will be constructed in the second IP. Its elaboration is also postponed to the second phase of the NICA project. Nevertheless, the SPD concept has been formulated and creation of motivated international collaboration has begun.

## 7. Start-up version of the NICA project

The very important and difficult task of the NICA project development is to begin its commissioning by the end of 2019. It is planned to be done in a reduced version of the facility and element parameters. Nevertheless, this will allow us to start experiments in the colliding beams' mode, with the test and tuning of the MPD detector and the majority of the accelerators' elements.

The start-up version of the NICA assumes the following:

1. An increased length of colliding beams' bunches equal to $\sigma_{bunch} = 0.6$ m has been chosen to provide the "concentration" of the luminosity at the inner tracker area of the MPD.

2. Maximum ion number per bunch is limited by the value of the betatron tune shift $\Delta Q \leq 0.05$.

3. Maximum emittance of the colliding bunches is less than $1.1\pi\cdot$ mm $\cdot$ mrad; the ratio of the horizontal emittance to the vertical one and the momentum spread of the ions is defined by the equilibrium state of the bunches in the presence of the intrabeam scattering (IBS).

4. The bunch number per ring is limited by the requirement of avoiding the parasitic collisions and is equal to $n_{nbunch} - 22$.

5. RF systems consist of the barrier voltage system ("RF-1) and the RF system of the $3^{rd}$ harmonics of the revolution frequency (RF-2). RF-1 is used for storage in the collider rings of the injected ions; RF-2 is used for formation of the bunched ion beams, where each third separatrix is filled with the ions. The square of the separatrix for the RF-2 is 25 times larger than the longitudinal r.m.s. emittance of the bunch.

6. For suppression of the IBS one, a stochastic cooling system, for longitudinal degrees of freedom (the "filter method"), will be constructed.

As a result, the maximum peak luminosity can be provided at the level of $5 \cdot 10^{25}$ $\text{cm}^{-2}\cdot$ $\text{s}^{-1}$ at the kinetic energy of the $^{197}\text{Au}^{79+}$ ions in the range of $3 - 4.5$ GeV/u.

## 8. Conclusion

The main characteristics of the NICA project, its status, and the principle problems related to the NICA creation have been considered in this report. The NICA project as a whole has passed the phase of design and is currently in the stage of accelerator element manufacturing and construction.

The project realization plan foresees a staged construction and commissioning of the facility accelerators.

## References

[1] NICA White Paper: http://theor.jinr.ru/twiki-cgi/view/NICA/WebHome
[2] Cleymans J and Randrup J 2006 *Phys. Rev.* C **74** 047901
[3] Meshkov I 2012 *Phys. of At. Nucl.* **75** 594
[4] Kekelidze V D et al. 2012 *Proc. of ICHEP2012* p. 411
[5] Stephans G S 2006 *J. Phys.* G **32** S447 nucl-ex/0607030
[6] Senger P 2004 *J. Phys.* G **30** S1087

# Open heavy-flavor observables at the LHC and the importance of higher-order flow harmonics

**Marlene Nahrgang**
Department of Physics, Duke University, Durham, NC 27708, USA
E-mail: marlene.nahrgang@phy.duke.edu

**Jörg Aichelin, Pol Bernard Gossiaux, Klaus Werner**
SUBATECH, UMR 6457, Université de Nantes, Ecole des Mines de Nantes, IN2P3/CNRS. 4 rue Alfred Kastler, 44307 Nantes cedex 3, France

**Abstract.** Heavy-quark dynamics in the quark-gluon plasma provide information about the medium properties and the details of the heavy-quark-medium interaction. Traditional observables like the nuclear modification factor and the elliptic flow of $D$ mesons have shown suppression at intermediate and high momenta and collective flow at low momenta. Thanks to the improving accuracy of the experimental data a combined analysis of these two observables starts to have discriminating power between different heavy-quark transport coefficients or features of the energy-loss models. In this overview, we summarize the modern strategy of describing heavy-quark dynamics and show how recent advances to include more differential observables such as azimuthal correlations and higher-order flow harmonics can help us understand the properties of heavy-flavor transport.

## 1. Heavy-flavor observables in heavy-ion collisions

Understanding thermodynamic and transport properties of strongly interacting many-body systems is the main motivation of perfoming (ultra)relativistic heavy-ion collisions and the discovery of the quark-gluon plasma (QGP) as an almost perfect fluid [1] has been one of the highlights of these endeavors. Key observables are linked to probes which do not fully thermalize with the medium and thus do not lose memory of their interactions with the medium constituents. An extreme example are electromagnetic probes, like photons and dileptons, which do not reinteract strongly with the produced QCD matter and thus give us information about the circumstances of their production. High-$p_T$ jets are another important probe. They are produced in the initial hard scatterings of partons inside the nucleons of the incoming nuclei. Upon scattering with the medium constituents they lose some fraction of their energy and their final spectra are thus expected to be modified compared to elementary reactions.

The mass of heavy quarks is another scale in the system, which allows us to probe both the regime of low heavy-quark momenta, where $p_T \ll m_{\rm HQ}$, and the high-momentum regime, where $p_T \gg m_{\rm HQ}$. The dynamics of heavy quarks at low momenta is mostly governed by diffusive processes and one may ask about the degree of thermalization inside the medium. At high momenta, the heavy quarks behave similarily to light partons and the contributions from different energy loss mechanisms are the main subject of interest. The intermediate-momentum regime is especially interesting, because most experimental data exist in this kinematic region.

Partial thermalization can be studied in terms of collective flow pattern of heavy flavor compared to the flow of the bulk medium consisting of light partons. Most models give relaxation times for charm quarks which are much longer than the evolution of the QGP [2]. Surprisingly, the measured light-flavor and heavy-flavor elliptic flow coefficients are quite similar within the large experimental error bars on the latter [3].

While nowadays the nuclear modification factor $R_{\mathrm{AA}}$ is mostly used to calibrate parameters in the energy loss models - explicitly or implicitly - in connection with the model of the medium evolution, the $v_2$ as well as higher-order flow harmonics for the low-$p_T$ sector and more differential observables like azimuthal correlations provide further opportunities to distinguish between energy loss models and exclude some of them.

In this overview, we will first present the current strategies to model heavy-flavor dynamics in heavy-ion collisions in section 2 and then discuss recent theoretical advances in heavy-flavor observables in particular higher-order flow harmonics in section 3.

## 2. Heavy-flavor dynamics

One of the main questions is how to get access to fundamental properties of the QGP from a comparison of theoretical models to experimental data and which observables can serve which purpose. It needs to be distinguished between the purpose of tomography, i.e. learning about medium properties, such as temperature, viscosity, but also about initial state fluctuations, expansion velocities, density distributions, etc., and studying the heavy-quark-medium interaction itself to learn about the individual processes of the underlying theory. In [4] it has been pointed out that jet observables can to some extent be classified in one or the other category, but both aspect remain very much entangled. Neither tomography of the medium nor studies of the energy loss model can be achieved alone but one needs to well calibrating the other.

pQCD and pQCD inspired models of collisional [5, 6, 7] and radiative processes [8], as well as non-perturbative resonance scattering [9, 10] and strong-coupling approaches [11, 12] have been proposed and studied to great detail in the past decades. Most of the mentioned models are derived in some limit or kinematic region. Usually, however, experimental observables are not sensitive to only one asymptotic limit and span over various kinematic regions including several scales. Most models are therefore not applicable over the whole kinematic range mostly in the intermediate $p_T$ regime, which is most interesting. However, even if the underlying theory for heavy-quark-medium interaction could be fully solved, there are many other sources of uncertainties involved, as will become clear by outlining the complete strategy of modeling heavy-flavor dynamics in heavy-ion collisions in the following:

In the initial hard scatterings the heavy quarks are produced according to partonic cross sections. The most elaborate model for inclusive spectra is provided by the FONLL (first-order next-to-leading log) [13] pQCD framework. One has to remember, however, that the comparison with data is only possible, when the pQCD production cross sections are folded with the parton distribution functions and a fragmentation function. For the FONLL framework the fragmentation functions in [14] work best to reproduce the data. This should be kept in mind and done consistently for the heavy-quark fragmentation after the evolution stage in the medium as described below. For more exclusive spectra, like $Q\bar{Q}$ azimuthal correlations, NLO pQCD matrix elements are coupled to parton shower event generators [15, 16]. Thanks to the heavy-quark mass the virtuality is significantly reduced compared to light partons and the charm and bottom quarks can be assumed onshell at the equilibration time of the QGP medium. The stage of preequilibirum dynamics remains the least understood phase during a heavy-ion collision and this is no different for the heavy-flavor sector.

The interaction of the heavy-quarks with the QGP constituents can occur via elastic or inelastic scattering processes. The elastic processes, also called collisional energy loss processes, in the perturbative regime need to be regularized in the $t$-channel. This is usually done schematically by the Debye mass and can be extended to replacing the bare gluon propagator with the full hard-thermal loop (HTL) resummed propagator at small momentum exchange. In the nonperturbative QCD regime, however, it turns out that the energy loss will not be independent of the intermediate scale between bare and HTL-propagators. Here, it is necessary to resort to modeling in order to extend the validity to the full kinematic regime, e.g. like in [28], or use nonperturbative approaches, such as resonance scattering effective theory [9, 10] or the strong-coupling AdS/CFT correspondance [11, 12]. For very energetic partons the radiative gluon emission resulting from the scattering with the medium constituents is considered the dominant contribution to energy loss. The gluon formation in this regime is long and thus several scatterings contribute coherently to the emission spectra. For heavy quarks at intermediate energies, the gluon formation time gets reduced by the mass and the radiative energy loss can be described by several incoherent scatterings [17, 18]. Such a prescription can be readily applied in Monte-Carlo simulations for heavy-quark dynamics. Several theoretical descriptions have addressed the radiative energy loss of light and heavy quarks in various kinematic regimes [8] resulting in averaged formulas assuming a medium at constant temperatures. As pointed out above, however, it is crucially important to assume a most realistic background evolution of the medium in order to compare to experimental data in a meaningful way. The QGP medium provides the scattering partners of the heavy-quarks, which can either be described by a Boltzmann parton cascade [19], a fluid dynamical evolution or a microscopic off-shell transport model [20]. Most importantly the medium description needs to be well tested for and constraint by the light and soft bulk observables. Factor-of-2 differences in the $R_{\rm AA}$ at intermediate and high $p_T$ can solely stem from using a simplistic versus realistic medium. Another source of inconsistency in the coupling between light- and heavy-flavor sectors can be the choice of the active degrees of freedom [21]. In a fluid dynamical evolution at LHC and top-RHIC energies the equation of state comes from lattice QCD [22, 23]. In the temperature regime of interest a gas of massless, noninteracting partons, as it is often sampled from via thermal distributions, does not reproduce the lattice equation of state. A representation of the medium quasi-particles which matches the lattice equation of state [24] can affect the scattering cross sections and the extracted transport coefficients [25]. Further aspects of the medium evolution, including initial state fluctuations [26] and viscous corrections are also of importance.

Once the medium has expanded and cooled down the heavy quarks hadronize via either recombination [30] at low momenta or fragmentation [14] at higher momenta. Both descriptions as well as their respective probabilities have a couple of parameters, which ideally one would like to obtain from other experiments. This is to some degree possible for the fragmentation process, as it is constraint by the initial production mechanisms and comparison to data from elementary reactions. After hadronization the final-state hadronic interaction are expected to affect the heavy-flavor observables on the order of 10-20% [27].

## 3. Recent advances in heavy-flavor observables

Traditional heavy-flavor observables are the nuclear modification factor, $R_{AA}$, and the elliptic flow $v_2$. As mentioned in the introduction the $R_{\rm AA}$ with its rather generic $p_T$ dependence, has become a good observable to fix the overall normalization of the energy loss model in conjunction with the medium evolution model. Various approaches with different ingredients and largely different heavy-quark transport coefficients are so able to describe the available $R_{\rm AA}$ data remarkably well. For a review and collection of data-to-model comparisons, see [33].

In the light-flavor sector, important contributions have become possible by looking into correlations and higher-order flow harmonics. For example, the better understanding of the

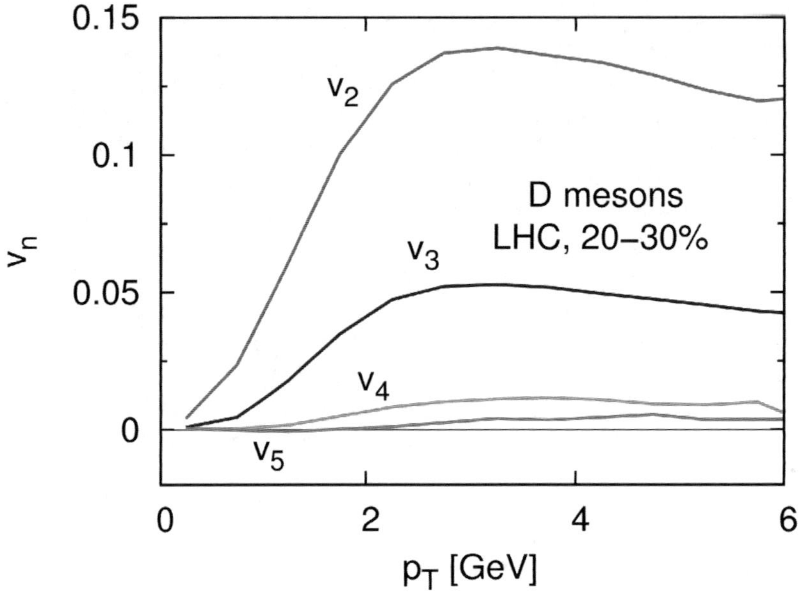


**Figure 1.** Higher-order flow harmonics of D mesons for the 20 − 30% most central Pb+Pb collisions at $\sqrt{s}$ = 2.76 TeV. Calculations are done in the MC@sHQ+EPOS2 model for a collisional+radiative(+LPM) energy loss mechanism

initial conditions for fluid dynamical modeling in correspondance with the extracted values for the ratio of shear viscosity over entropy density $\eta/s$ was made possible by the exact measurement of $v_n$ [34, 35, 36, 37, 38].

In recent works [39, 40] it has been pointed out that azimuthal correlations of $Q\bar{Q}$ pairs are promising observables for disentangling purely elastic and inelastic contributions to the energy loss mechanisms. There are, however, a couple of challenges, that need to be overcome, before a meaningful comparison between theory and experiment can be achieved. From the theory side already the $c\bar{c}$ proton-proton baseline is not well understood, as can be seen in the differences between NLO pQCD + parton shower calculations and available $D\bar{D}$ measurements in proton-proton collisions [41]. Since experimentally the feasibility of measuring $D\bar{D}$ correlations directly will only become possible with data from future runs, theoretical tools need to couple the light- and the heavy-flavor sectors more consistently in order to get reliable results for correlations between $D$ mesons and hadrons or heavy-flavor decay electrons and hadrons. This would in particular concerns the initial production of heavy quarks and the soft bulk matter, the fragmentation process, as well as the final state hadronic interactions.

In [42] we have predicted a finite triangular flow of $D$ mesons for RHIC and LHC energies using the MC@sHQ+EPOS2 model [28]. It couples a Boltzmann propagation of heavy quarks to a fluid dynamical evolution with the event-by-event EPOS initial conditions [43, 44]. This bulk description yields an excellent agreement with the available light hadron sector and can thus be considered a realistic modeling of the QGP medium. This is in particular true for LHC energies. With the upcoming coupling of MC@sHQ to EPOS3 the reliability will be extended to lower energies and different collision systems. The energy loss mechanism can describe either be a purely elastic or inelastic scatterings. The elastic scattering cross sections are obtained from the the Born approximation of pQCD and includes a running strong coupling coupling $\alpha_s$ [45]. The $t$-channel infrared divergence is regularized by the Debye screening mass, which is calculated self-consistently [46], albeit as the reduced semi-hard gluon propagator tuned

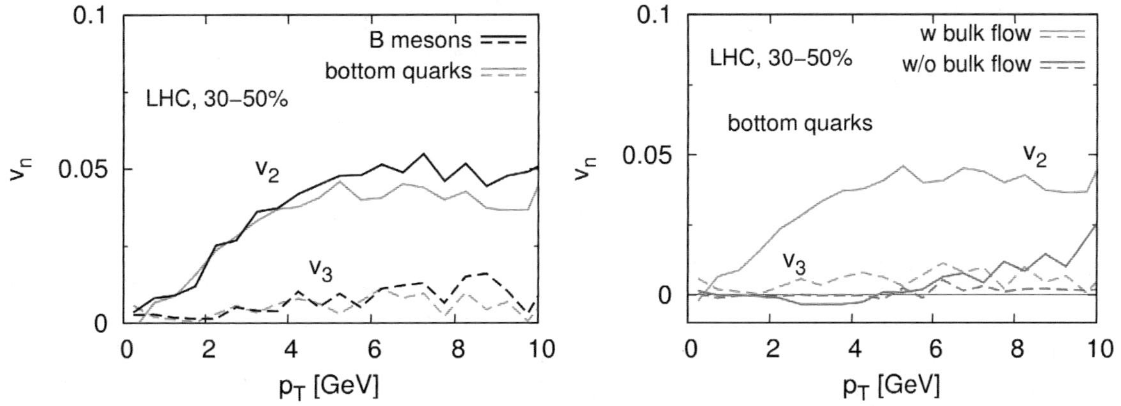


**Figure 2.** The contribution of hadronization (recombination) to the $B$ meson elliptic (solid) and triangular (dashed) flow (left plot) and the contribution of the bulk flow to the bottom quark elliptic (solid) and triangular (dashed) flow (right plot) for the 30 − 50% most central Pb+Pb collisions at $\sqrt{s} = 2.76$ TeV. Calculations are done in the MC@sHQ+EPOS2 model for a collisional+radiative(+LPM) energy loss mechanism

to reproduce the HTL+semi-hard energy loss [28]. The radiative processes are included as stemming from incoherent scatterings. The emission spectra are an extention of the calculation in [29] to finite quark masses [18]. Coherent emission is accounted for in form of a suppression of the power spectra at high $p_T$ [47]. This model reproduces well the $R_{\mathrm{AA}}$ at RHIC and LHC after rescaling the scattering rates by a global factor $K$, which is $K_{\mathrm{coll}} = 1.5$ in the case of purely elastic scatterings and $K_{\mathrm{coll+rad}} = 0.8$ in the case of both elastic and inelastic scatterings, which is close to unity. While the collective flow observables of the light charged hadrons are determined by the initial state geometry and fluctuations, the flow of the charm quarks comes from the interaction with the medium constituents. The stronger the charm quarks couple to the medium the more they thermalize inside the flowing medium and pick up the momentum anisotropy from the bulk. We could show that the $v_3$ of $D$ mesons is more sensitive to the decoupling of the charm quarks from the medium evolution than the $v_2$.

In Fig. 1 we can now present the higher-order coefficients $v_4$ and $v_5$, which require to run many more events for sufficient statistics. Calculations are performed for the collisional plus radiative(+LPM) energy loss mechanism. The expected D meson $v_4$ has a maximal value of $\sim 1 - 2\%$, while the values of charged hadron $v_4$ measured by the ALICE and ATLAS collaborations [38] show a maximal value of $\sim 8\%$. The decoupling of heavy quarks from the bulk is thus clearly seen.

Our calculations show that the elliptic flow of the much heavier bottom quarks is already significantly lower than that for charm quarks or the light charged hadrons. In the left plot of Fig. 2 we can further demonstrate that the $v_3$ of bottom quarks is vanishingly small. Furthermore, the $v_2$ of $B$ mesons is almost entirely due to the flow of the bottom quarks. The hadronization mechanism of recombination adds only a small contribution to the overall $v_2$. Due to the path length difference in an spatially anisotropic medium heavy quarks are expected to be more strongly deflected from their original azimuthal direction along the long axis than along the short axis, which leads to a $v_2$ signal independent of the flow of the medium. Theoretically one can mimick this situation by artifically neglecting the local fluid velocity in the scattering processes. At low momenta this contribution to the $v_2$ signal is negligible for the D mesons, but is the dominant source of $v_2$ at higher momenta. We repeat the same analysis for the bottom quarks and see in the right plot of Fig. 2, that the energy loss along the path length anisotropy does not lead to a substantial contribution to the bottom quark $v_2$.

Both the $D$ meson triangular flow and the $B$ meson elliptic flow would be valuable experimental data to enhance our understanding of heavy-flavor dynamics beyond the current level.

## 4. Conclusions

Heavy quarks probe partial thermalization with the QGP at low $p_T$ and in-medium energy loss at high $p_T$. In the most interesting region around intermediate $p_T$, many aspects play an important role, in particular the onset of coherent gluon emission, gluon thermal mass and width [48], finite path length and the importance of nonperturbative scatterings.

Modern models propagate heavy quarks by either the Langevin or the Boltzmann equation, where the former is a good approximation for bottom quarks but its reliability for charm quarks is questionable [49]. These approaches couple the heavy-flavor dynamics to dynamical evolutions of the QGP, which are well tested in the light-flavor sector and are able to reproduce $p_T$ spectra and flow coefficients of charged hadrons.

Yet, the traditional observables of heavy flavor, the D meson $R_{AA}$ and $v_2$ are described well by many models with to some extent very different ingredients and transport coefficients. It is therefore important to go beyond these observables in AA collisions. Currently, effort goes into the study of systems like proton-lead collisions, where initial shadowing effects at low transverse momenta play a role and first experimental data do not see a suppression at higher $p_T$.

More differential observables like $Q\bar{Q}$ correlations and higher-order flow coefficients of $D$ mesons as well as the elliptic flow of $B$ mesons indicate to have a larger discriminating power and are good candidates for reaching another level of understanding of the heavy-flavor dynamics and its relation to the bulk medium evoluion. They can thus help us to identify most dominant features of the heavy-quark-medium interaction in order to connect experimental data to fundamental properties of QCD.

## Acknowledgments

M.N. acknowledges support from a fellowship within the Postdoc-Program of the German Academic Exchange Service (DAAD). This work was supported by the U.S. department of Energy under grant DE-FG02-05ER41367.

## References

[1] U. W. Heinz, nucl-th/0512051.
[2] Moore G D and Teaney D, *Phys. Rev. C* **71** (2005) 064904;
[3] B. B. Abelev *et al.* [ALICE Collaboration], Phys. Rev. C **90** (2014) 3, 034904.
[4] T. Renk, Phys. Rev. C **85** (2012) 044903 [arXiv:1112.2503 [hep-ph]].
[5] Bjorken J D, Fermilab preprint Pub-82/59-THY (1982).
[6] Peshier A, *Phys. Rev. Lett.* **97** (2006) 212301;
[7] Peigne S and Peshier A, *Phys. Rev. D* **77** (2008) 114017.
[8] Gyulassy M and Wang X N, Nucl. Phys. B **420** (1994) 583; Wang X N, Gyulassy M, and Plümer M, *Phys. Rev. D* **51** (1995) 3436; Baier R, Dokshitzer Y L, Peigné S, and Schiff D, *Phys. Lett. B* **345** (1995) 277; Baier R, Dokshitzer Y L, Müller A H, Peigné S, and Schiff D, *Nucl. Phys. B* **483** (1997) 291; *Nucl. Phys. B* **484** (1997) 265; Zakharov B G, *JETP Lett.* **63** (1996) 952; *JETP Lett.* **64** (1996) 781; *JETP Lett.* **65** (1997) 615; *JETP Lett.* **73** (2001) 49; *JETP Lett.* **78** (2003) 759; *JETP Lett.* **80** (2004) 617; Gyulassy M, Levai P, and Vitev I, *Phys. Rev. Lett.* **85** (2000) 5535; *Nucl. Phys. B* **571** (2000) 197; *Nucl. Phys. B* **594** (2001) 371; Dokshitzer Y L and Kharzeev D E, *Phys. Lett. B* **519** (2001) 199; Arnold P B, Moore G D, and Yaffe L G, *JHEP* 0011 (2000) 001; *JHEP* 0305 (2003) 051; Armesto N, Salgado C A, and Wiedemann U A, *Phys. Rev. D* **69** (2004) 114003; *Phys. Rev. C* **72** (2005) 064910; Zhang B W, Wang E and Wang X N, *Phys. Rev. Lett.* **93** (2004) 072301; M. Djordjevic and M. Gyulassy, Nucl.Phys. A **733** (2004) 265-298; M. Djordjevic, M. Gyulassy and S. Wicks, Phys. Rev. Lett. **94** (2005) 112301; M. Djordjevic, Phys. Rev. C **80** (2009) 064909.
[9] H. van Hees and R. Rapp, Phys. Rev. C **71**, 034907 (2005);

[10] H. van Hees, V. Greco and R. Rapp, Phys. Rev. C **73**, 034913 (2006);
[11] S. S. Gubser, Nucl. Phys. B **790** (2008) 175 [hep-th/0612143].
[12] W. A. Horowitz, AIP Conf. Proc. **1441** (2012) 889 [arXiv:1108.5876 [hep-ph]].
[13] M. Cacciari, M. Greco and P. Nason, JHEP **9805** (1998) 007; M. Cacciari, S. Frixione and P. Nason, JHEP **0103** (2001) 006; M. Cacciari, S. Frixione, N. Houdeau, M. L. Mangano, P. Nason and G. Ridolfi, JHEP **1210** (2012) 137.
[14] M. Cacciari and P. Nason, JHEP **0309** (2003) 006; [hep-ph/0306212]. E. Braaten, K. m. Cheung, S. Fleming and T. C. Yuan, Phys. Rev. D **51** (1995) 4819; [hep-ph/9409316]. V. G. Kartvelishvili, A. K. Likhoded and V. A. Petrov, Phys. Lett. B **78** (1978) 615.
[15] S. Frixione, P. Nason and G. Ridolfi, arXiv:0707.3081 [hep-ph]; S. Frixione, P. Nason and G. Ridolfi, JHEP **0709** (2007) 126.
[16] S. Frixione and B. R. Webber, JHEP **0206** (2002) 029; S. Frixione, P. Nason and B. R. Webber, JHEP **0308** (2003) 007.
[17] O. Fochler, J. Uphoff, Z. Xu and C. Greiner, Phys. Rev. D **88** (2013) 1, 014018.
[18] J. Aichelin, P. B. Gossiaux and T. Gousset, Phys. Rev. D **89** (2014) 7, 074018.
[19] Uphoff J, Fochler O, Xu Z and Greiner C, *Phys. Rev. C* **82** (2010) 044906; Uphoff J, Fochler O, Xu Z and Greiner C, *Phys. Rev. C* **84** (2011) 024908.
[20] T. Song, H. Berrehrah, D. Cabrera, J. M. Torres-Rincon, L. Tolos, W. Cassing and E. Bratkovskaya, Phys. Rev. C **92** (2015) 1, 014910.
[21] M. Nahrgang, J. Aichelin, P. B. Gossiaux and K. Werner, Phys. Rev. C **89** (2014) 014905.
[22] S. Borsanyi, Z. Fodor, C. Hoelbling, S. D. Katz, S. Krieg and K. K. Szabo, Phys. Lett. B **730** (2014) 99.
[23] A. Bazavov *et al.* [HotQCD Collaboration], Phys. Rev. D **90** (2014) 9, 094503.
[24] M. Bluhm, B. Kampfer and G. Soff, Phys. Lett. B **620** (2005) 131.
[25] M. Nahrgang, J. Aichelin, P. B. Gossiaux and K. Werner, to be published soon.
[26] M. Nahrgang, J. Aichelin, P. B. Gossiaux and K. Werner, arXiv:1405.0938 [hep-ph].
[27] V. Ozvenchuk, J. M. Torres-Rincon, P. B. Gossiaux, L. Tolos and J. Aichelin, Phys. Rev. C **90** (2014) 054909; S. Cao, G. Y. Qin and S. A. Bass, Phys. Rev. C **92** (2015) 2, 024907.
[28] Gossiaux P B and Aichelin J, *Phys. Rev. C* **78** (2008) 014904.
[29] J. F. Gunion and G. Bertsch, Phys. Rev. D **25** (1982) 746.
[30] C. B. Dover, U. W. Heinz, E. Schnedermann and J. Zimanyi, Phys. Rev. C **44** (1991) 1636; V. Greco, C. M. Ko and R. Rapp, Phys. Lett. B **595** (2004) 202; P. B. Gossiaux, R. Bierkandt and J. Aichelin, Phys. Rev. C **79** (2009) 044906.
[31] Conesa del Valle Z [ALICE Collaboration], *Nucl. Phys. A*904-905 **2013** (2013) 178c; B. B. Abelev *et al.* [ALICE Collaboration], Phys. Rev. C **90** (2014) 3, 034904.
[32] K. J. Eskola, H. Paukkunen and C. A. Salgado, JHEP **0904** (2009) 065.
[33] A. Andronic *et al.*, arXiv:1506.03981 [nucl-ex].
[34] K. Aamodt *et al.* [ALICE Collaboration], Phys. Rev. Lett. **107** (2011) 032301 [arXiv:1105.3865 [nucl-ex]].
[35] A. Adare *et al.* [PHENIX Collaboration], Phys. Rev. Lett. **107** (2011) 252301 [arXiv:1105.3928 [nucl-ex]].
[36] S. Chatrchyan *et al.* [CMS Collaboration], Eur. Phys. J. C **72** (2012) 2012 [arXiv:1201.3158 [nucl-ex]].
[37] G. Aad *et al.* [ATLAS Collaboration], Phys. Rev. C **86** (2012) 014907 [arXiv:1203.3087 [hep-ex]].
[38] B. Abelev *et al.* [ALICE Collaboration], Phys. Lett. B **719** (2013) 18 [arXiv:1205.5761 [nucl-ex]].
[39] M. Nahrgang, J. Aichelin, P. B. Gossiaux and K. Werner, Phys. Rev. C **90** (2014) 2, 024907.
[40] S. Cao, G. Y. Qin and S. A. Bass, arXiv:1505.01869 [nucl-th].
[41] R. Aaij *et al.* [LHCb Collaboration], JHEP **1206** (2012) 141 [JHEP **1403** (2014) 108] [arXiv:1205.0975 [hep-ex]].
[42] M. Nahrgang, J. Aichelin, S. Bass, P. B. Gossiaux and K. Werner, Phys. Rev. C **91** (2015) 1, 014904
[43] K. Werner, I. Karpenko, T. Pierog, M. Bleicher and K. Mikhailov, Phys. Rev. C **82** (2010) 044904;
[44] K. Werner, I. Karpenko, M. Bleicher, T. Pierog and S. Porteboeuf-Houssais, Phys. Rev. C **85** (2012) 064907
[45] Y. L. Dokshitzer, G. Marchesini and B. R. Webber, Nucl. Phys. B **469** (1996) 93 [hep-ph/9512336].
[46] A. Peshier, hep-ph/0601119.
[47] P. B. Gossiaux, Nucl. Phys. A **910-911** (2013) 301.
[48] M. Bluhm, P. B. Gossiaux and J. Aichelin, Phys. Rev. Lett. **107** (2011) 265004 [Phys. Rev. Lett. **109** (2012) 169902].
[49] S. K. Das, F. Scardina, S. Plumari and V. Greco, Phys. Rev. C **90** (2014) 044901.

Journal of Physics: Conference Series **668** (2016) 012025 doi:10.1088/1742-6596/668/1/012025

# Overview of the hypernuclear production in heavy-ion collision experiments

**Christophe Rappold**

GSI Helmholtz Centre for Heavy Ion Research - Darmstadt, Germany & Universidad de Castilla-La Mancha, Institute of Mathematics applied to Science and Engineering - Ciudad Real, Spain

E-mail: c.rappold@gsi.de

**Abstract.**
In the last decade, heavy-ion collision experiments have brought new insight to the study of hypernucleus. Experiments using ion induced reactions for hypernuclear research focus on two distinct aspects: the spectroscopy and probing the nuclear reaction. In the case of the experimental spectroscopy, the internal structure of hypernuclei is investigated in order to determine the baryon-baryon interaction in the strangeness sector for the hyper-matter equation of state. The dynamical aspect of the nucleus-nucleus reaction can also be explored by studying the production of hypernuclei. The experimental observations of the production mechanisms responsible for the formation of the hypernuclei in ion collisions will be presented. Depending of the center-of-mass energy and the type of experiment, fixed target or collider, hypernuclei can be produced in the mid-rapidity and/or in the spectator regions. The experimental results from both cases will be presented and discussed.

## 1. Introduction

The hypernucleus, a bound state of ordinary nucleons and hyperons, has been a subject of research for the understanding of the baryon-baryon interaction in an unified way. The main purpose is to tackle the problem of the neutron stars or the equation of state of baryonic matter up to the strangeness sector. The baryon-baryon interaction could be easily extracted from the hyperon-nucleon and hyperon-hyperon scattering data; however the scarce data sets from the hyperon-nucleon scattering experiments and the impossibility to have hyperon-hyperon scattering experiments have made the direct determination of the baryon-baryon interaction strength within $SU(3)_f$ extremely difficult. By performing precise spectroscopy of Λ-hypernuclear bound states, interaction strength of Λ-N has been deduced indirectly by means of the experimental data and theoretical models [1, 2, 3, 4, 5, 6]. Those spectroscopy experiments have been performed mainly by means of secondary meson beams via ($\pi$, K), (K, $\pi$), or ($K^-$,$K^+$) elementary reactions or by means of $e^-$ primary beam via ($e^-$,$e'^-$ $K^+$) photoelectric production on fixed target [7]. Additionally to those experimental approaches, hypernuclear production by heavy-ion induced reaction has been theoretically investigated and predicted in 1973 [8]. This first theoretical prospect has led to new experimental approaches for the spectroscopic study of hypernuclei. Indeed, heavy-ion induced reaction on fixed targets and heavy-ion collision in colliders have been preformed in the most recent years. It has renewed interest in the production of hypernuclei in central collisions in which those hypernuclear bound states could be used as a probe to the possible phase transition or the presence of quark-gluon plasma.

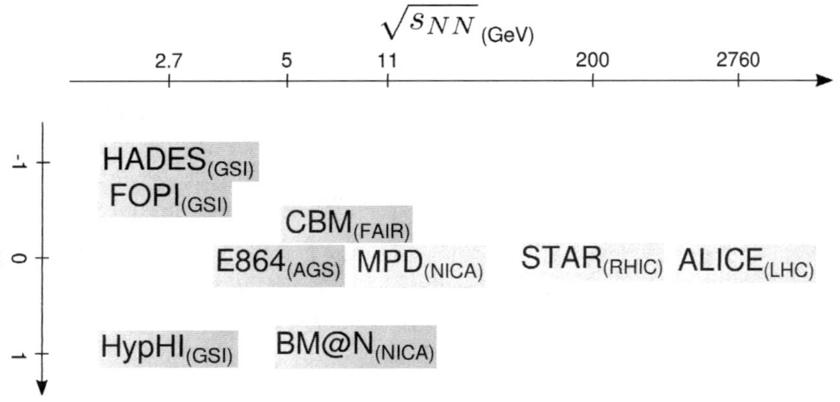


**Figure 1.** heavy-ion experiments with their respective facility where hypernuclear production have been studied during the last decades.

Figure 1 shows the past, current and future experiments that have published results on the hypernuclear production or dedicates parts of their research plans to the hypernuclear study. First of all, those experiments are sorted by their experimental apparatus: fixed target (in blue rectangle) or collider mode (in green rectangle). Then those experiments are distributed as a function of $y_0$ and $\sqrt{s_{NN}}$, the rapidity in center-of-mass system region of hypernuclear observations and the center-of-mass energy of the reactions. The discrete repartition between the observed rapidity regions (central or peripheral collisions) and the collision energy has concrete consequences on the physical phenomenon of the hypernuclear production that will be presented in the following sections. Besides experiments that have heretofore been published or reported study of hypernuclei, namely ALICE [9], STAR [10], E864 [11], HADES [12], FOPI [13, 14] and HypHI [15, 16], the NICA and FAIR facilities will host novel experiments (CBM [17], BM@N [18], and MPD [18]) that have a dedicated hypernuclear physics program.

The experimental observables for describing the hypernuclear production mechanism in heavy-ion experiments are mainly the yields, cross sections and multiplicities of the observed hypernuclei. Most of them are reported either as an absolute value or within the experimental acceptance. Those observables can also be determined as a function of the rapidity $y_0$ or/and the transverse momentum $P_t$. Those primal observables are essential when compared to theoretical models of the dynamics of the reaction, such as transport models. Additionally, yield ratios to the hyperon yield or to the nuclear yield, such as ${}^{3}_{\Lambda}\mathrm{H}/\Lambda$ or ${}^{3}_{\Lambda}\mathrm{H}/{}^{3}\mathrm{He}$ are useful to describe sensitive aspect of the dynamics, and can be more easily extracted from the experimental data, where the efficiency corrections can be canceled out. A double ratio, $\mathrm{S}_3 = {}^{3}_{\Lambda}\mathrm{H}/{}^{3}\mathrm{He} \cdot \mathrm{p}/\Lambda$ or $\mathrm{S}_4 = {}^{4}_{\Lambda}\mathrm{H}/{}^{4}\mathrm{He} \cdot \mathrm{p}/\Lambda$, has been introduced by the E864 collaboration [11]; afterwards it was used by the other STAR and ALICE collider experiments to describe possible phase transition and the QGP formation [19]. Those experimental observables will be the presented in the following sections.

## 2. Ultra-relativistic energy regime

The STAR experiment has been the first to show evidences of the observation of the hypertriton, ${}^{3}_{\Lambda}\mathrm{H}$, and the anti-hypertriton hypernuclei in Au-Au collisions at $\sqrt{s_{NN}}$=200 GeV in the ultra-relativistic energy regime [10]. Those hypernuclear bound states are considered to be formed at the freeze-out of the quark gluon plasma. The published results presented some

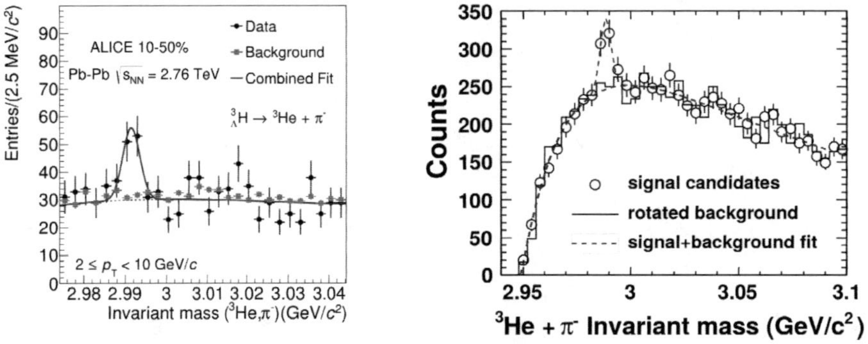


**Figure 2.** Invariant mass spectrum of the two-body decay reconstruction of ${}^3_\Lambda$H hypernuclei from the ALICE [9] and STAR experiments [10] in the left and right panel, respectively. Similar invariant mass spectra of anti-hypertriton can be found in the respective references [10, 9].

spectroscopy aspects, such as invariant mass spectrum shown in Figure 2 or its extracted lifetime besides several yield ratios and $S_3$ factor [10]. It was first claimed that the enhancement of the production of hypertriton at the mid-rapidity would be a signature of the phase transition and formation of the QGP [10] based on theoretical prediction [19]. After the beam scan program of the STAR experiment, additional results on the $S_3$ factor as a function of the energy in the center-of-mass system have been presented and reported [20]. As well, the ALICE experiment has reported and published the evidence of the hypertriton and anti-hypertriton observation in Pb-Pb collisions at 2.76 TeV [9], shown in Figure 2, with the extracted lifetime and yield ratios [9].

The production mechanism in such collision has been theoretically investigated following the STAR experiment publication and ALICE experiment reports. On one hand, coalescence models through hadronic transport models describe the hypernuclear formation of hyper-cluster at the chemical freeze-out of the QGP phase [19, 21]. On the other hand, thermal models propose a statistical approach to describe the production yield of particle, nucleus, and hypernucleus with their associated anti-particle and bound state at the chemical freeze-out equilibrium, where the entropy per baryon is defined and consequently conserved [22].

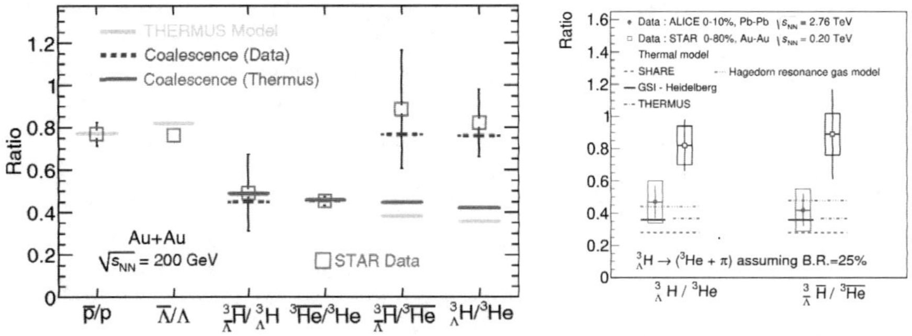


**Figure 3.** Yield ratios extracted by STAR [23] and ALICE experiment [9] compared to the expected yield estimated by several models.

Several coalescence calculation approaches have been developed [19, 21] and showed similar trends. It is proposed that the $S_3$ factor could carry the information of the local correlation between the baryon number and the strangeness, qualitatively similar to the $C_{BS}$ factor [19]. Some theoretical results were compared directly to the published STAR experiment results, as shown in the left panel of Figure 3. In this case, the coalescence model [23] reproduced the yield ratio ${}^{3}_{\Lambda}\mathrm{H}/{}^{3}\mathrm{He}$, while the THERMUS model [24, 25] used for this comparison could not. Additionally, the theoretical estimation of $0.35 \pm 0.003$ of this yield ratio by the thermal model [22] could not reproduce the STAR experimental results. Considering the ALICE experimental results on the same yield ratio ${}^{3}_{\Lambda}\mathrm{H}/{}^{3}\mathrm{He}$, shown in right panel of Figure 3, the situation seems to favor a good description from the thermal models [22, 21]. Besides those yield ratios, it should be noted the thermal model fit of the production yields [22, 26] reproduces the experimental data fairly well either for STAR or ALICE experimental data within several orders of magnitude difference, and especially the hypertriton and anti-hypertriton yield [22, 26].

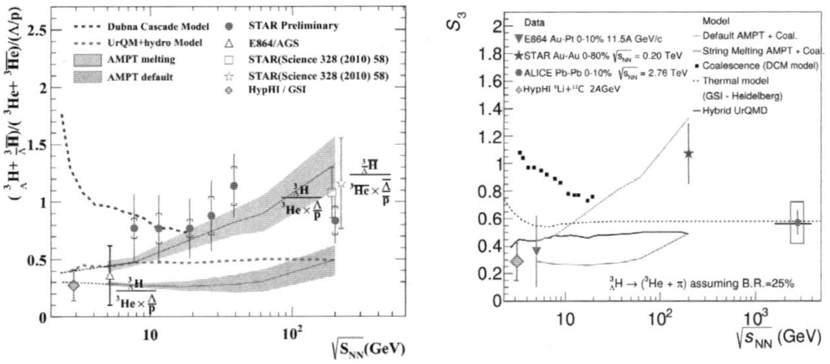


**Figure 4.** $S_3$ factor extracted by STAR and ALICE experiments with different model comparisons [20, 9].

The case of the double ratio $S_3$ is also interesting, since the distinct theoretical approaches for the hypernuclear production in the QGP describe differently the evolution of the $S_3$ factor. The thermal models predict generally that this double ratio is mostly independent from the energy of the collision [22, 21], while the coalescence models predict different energy dependencies, decreasing or increasing as the beam energy is increased [21, 19]. Figure 4 shows the results from the reported STAR beam energy scan program, the published STAR and ALICE results [10, 20, 9]. The recent $S_3$ estimation from the HypHI experiment at $\sqrt{s_{NN}} = 2.7$ GeV has been added to those published figures. Several theoretical models are shown and compared to the experimental data available. With the current situation from the preliminary results of STAR beam scan program, an increasing trend is noticeable from the left panel of Figure 4, which is not confirm by the ALICE experimental results shown in the right panel of Figure 4. The estimation from thermal model [22] reproduces the ALICE results when a feed-down from the contribution of the strong decay of resonances is included, increasing the $\Lambda/p$ ratio contribution to the $S_3$. A similar tendency can be noticed from the second compared thermal model [21]; however, the theoretical calculation within this framework will need to extended to the beam energy of ALICE experiment to assure a proper comparison. Contrarily coalescence model calculations reported an higher $S_3$ factor and energy dependency. The model [19] predicted a value close to unity which could corresponds to the manifestation of the quark deconfinement. The published STAR results have been considered as a first hint as expressed in [10]; however, the recent ALICE results [9] bring new contradictory light onto this possible interpretation of the "AMPT+melting string" model [19].

## 3. Relativistic energy regime

At a lower center-of-mass energy, several other experiments have already been performed in the past and more recently. With a lower beam energy than in ultra-relativistic regime, experiments have focused on different rapidity regions for the hypernuclear study. The E864 experiment at AGS facility published results of the ${}^{3}_{\Lambda}$H production yield and upper limit of ${}^{4}_{\Lambda}$H yield in the central collision of Au beam at 11.5 GeV/c per nucleon on a fixed Pt target [11]. Figure 5 shows the invariant mass spectrum of ${}^{3}_{\Lambda}$H published by the E864 experiment with a reported ${}^{3}_{\Lambda}$H signal significance of $2\sigma$ [11]. The $S_3$ double ratio was for the first time extracted and published by the E864 experiment [11]; a value of 0.36±0.26 was reported. This double ratio was introduced initially in order to compare the production of hypernuclei in central collision with the production of ordinary nuclei with a same baryon number A. This comparison would lead to estimate the additional penalty factor required, when a strange baryon is included in the coalescence process. It was notified that there was an extra penalty factor of 3 to produce ${}^{3}_{\Lambda}$H compared to ${}^{3}$He; however, consideration about the smaller binding energy of ${}^{3}_{\Lambda}$H (around 2.2 MeV) than the one of ${}^{3}$He (8.1 MeV) and the larger size of ${}^{3}_{\Lambda}$H than ${}^{3}$He may explain the lower production yield of ${}^{3}_{\Lambda}$H in the central collision. Besides ${}^{3}_{\Lambda}$H, the study of ${}^{4}_{\Lambda}$H could not be performed completely since the ${}^{4}_{\Lambda}$H signal was not significant enough. An upper limit at 90% confidence level was then estimated for the ${}^{4}_{\Lambda}$H production yield and the factor $S_4 = {}^{4}_{\Lambda}\mathrm{H}/{}^{4}\mathrm{He}\cdot\mathrm{p}/\Lambda$ [11].

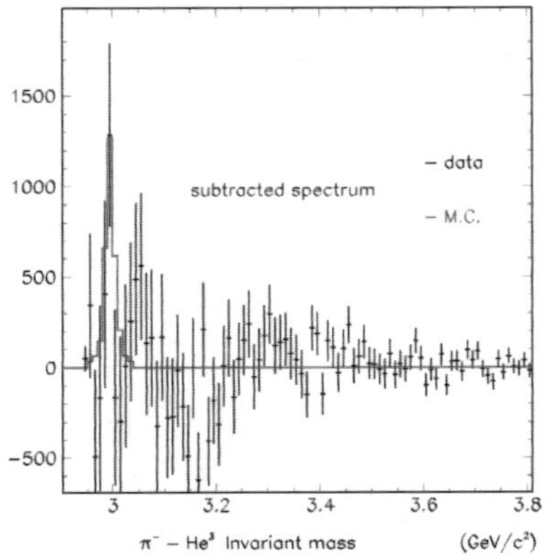


**Figure 5.** Invariant mass spectrum of the two-body decay reconstruction of hypertriton from E864 experiment [11].

At a lower beam energy, the HADES and FOPI experiments from the GSI facility have conducted hypernuclear study in the target rapidity region. HADES experiment analyzing the reaction Ar+KCl at 1.76 *A*GeV within the range of rapidity $y_0$, [-0.7, 0.1], did not observed any ${}^{3}_{\Lambda}$H signal [12], and reported a upper limit at 99.7% confidence level for the hypertriton production yield and for the ${}^{3}_{\Lambda}$H/$\Lambda$ yield ratio [12]. The yield ratio ${}^{3}_{\Lambda}$H/$\Lambda$ was estimated to be inferior to $2.5 \pm 0.3 \times 10^{-2}$ [12]. FOPI experiment has already reported preliminary results of evidence of the observation of ${}^{3}_{\Lambda}$H and ${}^{4}_{\Lambda}$H in specific rapidity range $y_0$ and transverse momentum interval $P_t/m$, [-0.56, -0.11]× [0.2, 0.4] and [-0.67, -0.11]×[0.15, 0.35], respectively [13, 14]. While the preliminary observation of ${}^{3}_{\Lambda}$H and ${}^{4}_{\Lambda}$H appeared clear, the comparison between the published upper limit of the yield ratio ${}^{3}_{\Lambda}$H/$\Lambda$ from HADES experiment within a similar rapidity interval and the estimated yield ratio ${}^{3}_{\Lambda}$H/$\Lambda$ of 0.52± 0.04 casts some incompatibility with the HADES upper limit. Additional considerations for the final FOPI experimental results are pending.

In the projectile fragment rapidity region, hypernuclear study has also been carried out, firstly at LBNL Bevatron with ${}^{16}$O beams at 2.1 *A* GeV on a polyethylene target [27] and then at JINR

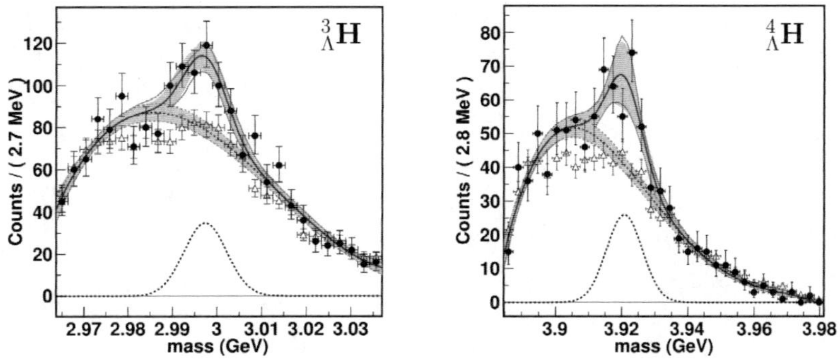


**Figure 6.** Invariant mass spectrum of two body decay reconstruction of $^3_\Lambda$H and $^4_\Lambda$H of the HypHI experiment in left and right panel respectively [15].

Dubna in the reaction of a beam of $^4$He at 3.7 *A* GeV and $^7$Li at 3.0 *A* GeV on a polyethylene target [28, 29]. The results on hypernuclear production from those first experiments were not clear enough because of the low statistics acquired and analyzed. More recently, the results from the HypHI experiment on the hypernuclear production have been published [15, 16]. $^3_\Lambda$H and $^4_\Lambda$H hypernuclei were observed in the reaction of $^6$Li beam at 2 *A*GeV on a $^{12}$C fixed target. Figure 6 shows the invariant mass spectrum of observed $^3_\Lambda$H and $^4_\Lambda$H by the HypHI experiment [15]. The $^3_\Lambda$H and $^4_\Lambda$H production cross section of 3.9±1.4μb and 3.1±1.0μb, and several yield ratios in the projectile rapidity region were extracted and published [16]. For example the $^3_\Lambda$H/Λ yield ratio of $2.6 \pm 1.4 \times 10^{-3}$ is compatible with one of the HADES experiment, if we mirror observed rapidity ranges. Additionally, the differential multiplicity as a function of the rapidity and transverse moment for $^3_\Lambda$H and $^4_\Lambda$H were extracted for the first time and shown in Figure 7. Those published results from HypHI experiment has been compared with theoretical transport model and shows fair agreement with the theoretical calculations [30]. Besides, the $S_3$ and $S_4$ double ratio have been estimated to be 0.28±0.14 and 0.08±0.04. The $S_3$ value from the HypHI experiment has been added to two published figures of Figure 4 and will help to constrain models in the hadronic phase at lower beam energy.

## 4. Conclusion

In the ultra-relativistic regime, STAR and ALICE experiments have provided new insights on the hypertriton and anti-hypertriton study [10, 20, 9]. The production aspects have been investigated in details by the ALICE experiment in [9]. Combining results from both experiments shows theoretical models draw different conclusion on the hypertriton formation in the QGP phase. Improvement on the experimental estimations will help to draw clearer conclusion on the model predictions especially on the yield ratios like $^3_\Lambda$H/$^3$He and the $S_3$ factor. At lower beam energy, in the relativistic regime, recent results on the projectile rapidity region by the HypHI experiment [15, 16] will help transport models and coalescence models to obtain a more precise description of the reaction mechanism and of the dynamics between the participant and spectator region [21, 31, 32, 33, 34]. Additional experiments, such as the FOPI experiment, are also focusing on the hypernuclear study and will provide further information on the hypernuclear production in the target rapidity region with their final experimental results. In the future, more results from the HypHI experiment are also expected with the analysis of the second experiment $^{20}$Ne+$^{12}$C reaction at 2*A*GeV. Moreover, the STAR and ALICE experiments are still accumulating new experimental data and will provide more precise estimation in future, and for instance, the ALICE experiment might access the production of $^4_\Lambda$H. Besides those running

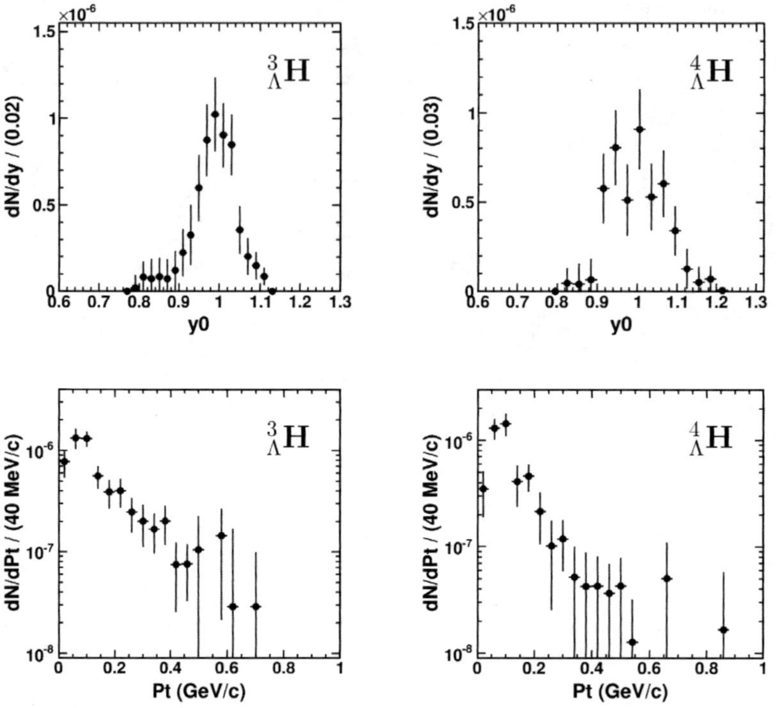


**Figure 7.** The differential multiplicity as a function of the rapidity $y_0$ and $Pt$ of extracted $^{3}_{\Lambda}$H and $^{4}_{\Lambda}$H signal [16].

experiments, new experiments are in preparation within the new FAIR and NICA facilities, which will contribute to the study of the hypernuclear production in heavy-ion reaction.

## Acknowledgements

This work has been supported by the HypHI project funded by the Helmholtz association as Helmholtz-University Young Investigators Group VH-NG-239 at GSI, and the German Research Foundation (DFG) under contract number SA 1696/1-1 and EU FP7 Hadron-Physics-2 SPHERE. This work has also been supported by the co-funded program of the University of Castilla-La Mancha "Ayudas para estancias de investigadores invitados en la UCLM para el ao 2015" and FEDER 2014-2020.

## References

[1] Nagels M.M. *et al.* 1977 *Phys. Rev.* D **15** 2547
[2] Nagels M.M. *et al.* 1979 *Phys. Rev.* D **20** 1633
[3] Maessen P.M. *et al.* 1989 *Phys. Rev.* C **40** 2226
[4] Rijken Th.A. and Stokes V.G.J. 1999 *Phys. Rev.* C **59** 21
[5] Dalitz R.H. and A. Gal A. 1978 *Ann. Phys.* **116** 167
[6] Millener D.J., Gal A., Dover C.B. and Dalitz R.H. 1985 *Phys. Rev.* C **31** 499
[7] Hashimoto O. and Tamura H. 2006 *Prog. Part. Nucl. Phys.* **57** 564
[8] Kerman A. K. and Weiss M. S. 1973 *Phys. Rev.* C **8** 408
[9] ALICE Collaboration, 2015 *Preprint* arXiv:1506.08453
[10] STAR Collaboration 2010 *Science* **328** 58
[11] Armstrong T. A. *et al.* 2004 *Phys. Rev.* C **70** 024902
[12] HADES Collaboration 2013 *Eur. Phys. J.* A **49** 146
[13] Y. Zhang, PhD Thesis 2013

[14] Leifels Y. 2013 *J. Phys.: Conf. Series* **420** 012019
[15] Rappold C. *et al.* 2013 *Nucl. Phys.* A **913** 170
[16] Rappold C. *et al.* 2015 *Phys. Lett.* B **747** 129
[17] Friman B., Hohne C., Knoll J., Leupold S., Randrup J., Rapp R. and Senger P. 2011 *Lect.Notes Phys.* **814** 980
[18] Kovalenko A.D. *et al.* 2015 *PoS BaldinISHEPPXXII* 002
[19] Zhang S., Chen J. H., Crawford H., Keane D., Ma Y.G. and Xu Z.B. 2010 *Phys. Lett.* B **684** 224
[20] Ma Y.G. for STAR collaboration 2014 *EPJ Conf.* **66** 04020
[21] Steinheimer J., Gudima K. , Botvina A. , Mishustin I., Bleicher M. and Stocker H. 2012 *Phys. Lett.* B **714** 85
[22] Andronic A., Braun-Munzinger P., Stachel J., and Stocker H. 2011 *Phys. Lett.* B **697** 203
[23] Cleymans J. *et al.* 2011 *Phys. Rev.* C **84** 054916
[24] Wheaton S. and Cleymans J. 2005 *J. Phys.* G **31** S1069
[25] Wheaton S., Cleymans J. and Hauer M. 2009 *Comput. Phys.Commun.* **180** 84
[26] Elia D. for ALICE Collaboration 2015 *EPJ Conf.* **95** 03008
[27] Nield K. *et al.*, 1976 *Phys. Rev.* C **13** 1263
[28] Abdurakhimov A.U. *et al.* 1989 *Nuovo Cimento* A **102** 645
[29] Avramenko S. *et al.* 1992 *Nucl. Phys.* A **547** 95c
[30] Le Fevre A. 2015 *these proceedings*
[31] Gaitanos T., Lenske H. and Mosel U. 2009 *Phys. Lett.* B **675** 297
[32] Botvina A. S., Mishustin I. N. and Pochodzalla J. 2012 *Phys. Rev.* C **86** 011601
[33] Topor Pop V. and Das Gupta S. 2010 *Phys. Rev.* C **81** 054911
[34] Botvina A. S. and Pochodzalla J. 2007 *Phys. Rev.* C **76** 024909

Journal of Physics: Conference Series **668** (2016) 012026 doi:10.1088/1742-6596/668/1/012026

# In-medium heavy-quarkonium from lattice QCD spectral functions

**Alexander Rothkopf**

Institute for Theoretical Physics, Heidelberg University, Philosophenweg 16, 69120 Heidelberg, Germany

E-mail: rothkopf@thphys.uni-heidelberg.de

**Abstract.** We discuss recent progress in the study of in-medium heavy quarkonium using first-principles lattice QCD calculations and effective field theory. In particular our focus lies on real-time information carried by QCD spectral functions and we report on a novel strategy for their extraction from Euclidean-time simulations. Combining an effective description for heavy quarks with finite temperature lattice QCD we compute in-medium spectra providing insight in heavy quarkonium melting in a static thermal medium. On the other hand we use spectral information to extract the complex in-medium potential which provides the basis for a dynamical description of quarkonium real-time evolution via the Schrödinger equation.

## 1. Introduction

Heavy quarkonium, the bound state of a b or c-quark with its anti-partner, is a valuable probe of the strong interactions both in vacuum, as well as at high energy densities, such as in the droplet of quark-gluon-plasma created in relativistic heavy-ion collisions at RHIC and LHC [1].

Recent experimental studies of vector channel Bottomonium ($\Upsilon(nS)$) and Charmonium ($J/\Psi, \Psi(nS)$) by the CMS and ALICE collaborations have revealed two distinct faces of quarkonium, represented by each different. For $b\bar{b}$, a comparison of di-muon spectra in $p+p$ and $Pb+Pb$ collisions provided evidence for a pattern of sequential suppression, i.e. less deeply bound excited states are dissolved more easily than the ground state [2]. These findings are in line with the concept of Bottomonium forming early on in a collision and traversing the plasma as well defined probe. On the other hand ALICE observed [3] at mid-rapidity that the suppression of Charmonium reduces with increasing centrality. This supports the picture of $J/\Psi$ formation from fully thermalized $c\bar{c}$ at the phase boundary, as advocated for in the statistical model of hadronization [4]. Understanding the intricate in-medium physics leading to each of the two scenarios from first principles has to be the central goal for theory.

Heavy quarkonium is characterized by the well pronounced separation of scales between the constituent rest mass ($m_c = 1.275(25)$GeV, $m_b = 4.66(3)$GeV) [5] and all other scales present in current experiments. This distinguishes them as well defined observables, measurable with high precision, and opens up the possibility for theory to employ effective descriptions that allow a more intuitive understanding of the bound state physics in the presence of a thermal medium.

Already at zero temperature it is the intrinsic scale of quantum fluctuations of QCD, often referred to as $\Lambda_{\rm QCD} \simeq 200$MeV that lies well below $m_Q$, which in turn tells us that quantum fluctuations cannot spontaneously create a $Q\bar{Q}$ pair. Such a system, where creation and annihilation do not play a central role can be described by a non-relativistic effective field-theory (EFT) called NRQCD, in fact one can show that it can even be described to a good degree by a non-relativistic potential entering a Schrödinger equation [6]. At finite temperature the additional scale $T$ enters and one might ask how its presence changes the properties of the bound state, which eventually will melt if thermal fluctuations overcome its typical binding energy. NRQCD when used at T=0 only requires a separation of $m_Q$ from the next lower scale and thus in principle is also valid in-medium. We report on its application to lattice QCD studies of heavy quarkonium at $T > 0$ in more detail in sec.3.

On the other hand, the question whether an in-medium potential $V(r)$ can be defined and what its values are, has only recently been answered in a satisfactory way. For more than two decades, in the absence of the means to derive a potential from QCD, theorists have resorted to model potentials, among which two candidates have gained popularity. The purely real color singlet free energies in Coulomb gauge $e^{F^{(1)}(\mathbf{x})} = \langle \mathrm{Tr}\big[\Omega(\mathbf{x})\Omega^\dagger(0)\big]\rangle_{\rm CG}$ from Polyakov-loop $\Omega(\mathbf{x}) = \exp\Big[ig\int_0^\beta d\tau A^4(\tau,\mathbf{x})\Big]$ correlators [7] and the internal energies $U^{(1)} = F^{(1)} - TS$ derived from it [8], as well as linear combinations [9] have been employed.

It is important to note that while these quantities are readily accessible in lattice QCD [10] they have not been linked to a potential in QCD. In fact progress in the formulation of EFTs [11, 12] made it possible to actually derive an expression for the potential between two static quarks in a thermal medium starting from the QCD Lagrangian, revealing that neither $F^{(1)}$ nor $U^{(1)}$ alone can play the role of $V(r)$.

The reason is that the evaluation of the potential both in resummed hard-thermal loop perturbation theory [13, 14] and on the lattice [15, 16] showed that $V(r)$ is actually a complex quantity. While at $T > T_C$, Re[$V$] is closely related to the phenomenon of Debye screening; Im[$V$] is intimately linked to the scattering of the medium partons with the color string spanning in between the $Q\bar{Q}$ [14, 12]. Using a purely real model potential neglects these effects and necessarily misses an important aspect of the in-medium physics of heavy quarkonium. Recent lattice QCD results on the extraction of the complex in-medium potential are presented in sec.4.

While the non-relativistic character of the heavy quarks allows a simplification of their description, the confinement aspects of the QCD vacuum as well as the physics of the strongly coupled QGP at $T \simeq T_C$ require fully relativistic and non-perturbative methods. One truly first principles approach is lattice QCD, where spacetime is discretized on a hyper-cubic grid with finite lattice spacing $a$ and box size $N_x a$. Gauge fields denoted in continuum by $A^a_\mu$ are represented by SU(3) valued link variables $U_\mu(x) = \exp\big[igaA^a_\mu T^a\big]$, where $T^a$ represent the generators of the gauge group. These object reside on the links connecting the nodes of the spacetime grid. The light fermions of the QCD medium on the other hand populate the nodes.

The strength of the lattice approach lies in the fact that the Feynman path integral of this theory can be evaluated non-perturbatively via stochastic Monte-Carlo methods, after rotating the Minkowski-time axis onto imaginary time. Since the simulation is not carried out in physical time a form of analytic continuation is required if dynamical information is to be extracted. This represents an inherently ill-defined problem. One strategy, Bayesian inference, proven successful in many areas of physics to overcome such a challenge is discussed in sec.2.

At $T = 0$ the combination of lattice QCD and NRQCD, have been highly successful in reproducing the quarkonium spectrum (see e.g. [17, 18]). There is no modeling input necessary in these computations, thus they truly link the QCD Lagrangian to physically observable quantities, such as particle masses or matrix elements. Note that since all simulations are carried out using dimensionless numbers, at some point experimental input is required to set the overall scale. Once the scale is fixed however ab-initio predictions are possible; one example being the existence of the $\eta_b(2S)$ state with mass $m^{\rm LNRQCD}_{\eta_b(2S)} = 9.988(3)$GeV predicted in 2011 [18], which was confirmed by the Belle collaboration [19] at $m^{\rm Belle}_{\eta_b(2S)} = 9.999(3.5)(2.8)$ in the following year.

Journal of Physics: Conference Series **668** (2016) 012026 doi:10.1088/1742-6596/668/1/012026

At finite temperature lattice QCD can also help us to elucidate the static properties of the QCD medium, now that consistent continuum extrapolated results for thermodynamic quantities, such as the energy density and pressure have been obtained [21, 23, 22]. While the transition between the confined and deconfined phase is a mere crossover at physical quark masses [20], a pseudo-critical temperature of $T_C = 155(9)$MeV has been established that characterizes the change from the hadronic to QGP regime. Evaluation of the trace anomaly [23, 22, 24], a non-perturbative measure of the interaction strength revealed a region of strong coupling exists in the regime just above $T_C$. This reinforces the need for non-perturbative methods when approaching the physics of heavy quarkonium in heavy-ion collisions.

## 2. Spectral functions from lattice QCD

Physical information relevant for an understanding of in-medium heavy quarkonium can be found in spectral functions $\rho(\omega)$. They inform us about the presence and strength of bound states, which appear as peaked features of skewed Breit-Wigner type. The position encodes the mass of the state, the width its transition probability to other bound states or into the continuum, which is manifested as a featureless structure at large $\omega$. What makes spectral functions interesting from a technical point of view is that they constitute a bridge between correlation functions in real-time and those analytically continued to Euclidean time, which will play a crucial role in the extraction of the heavy quark potential from lattice QCD simulations in sec.4.

In sec.3 we will calculate heavy quarkonium correlation functions $D(\tau)$ on the lattice. These are related to spectra via a Kernel $K(\tau,\omega)$, which is determined by QCD and exactly known

$$D(\tau,\mathbf{x}) = \int d\omega\, K(\tau,\omega)\,\rho(\omega,\mathbf{x}), \quad D(\tau_i) = \sum_{l=1}^{N_\omega} \Delta\omega_l\, K_{il}\,\rho_l. \tag{1}$$

In both studies presented in sec.3 and sec.4 we will encounter the $T$ independent Kernel $K(\tau,\omega) = e^{-\omega\tau}$, with which the above expressions amount to Laplace transforms.

Writing the convolution on the right of Eq.(1) with $N_\tau$ steps along imaginary time and $N_\omega$ frequency points, reveals the challenge we face in inverting this relation. I.e. a function $\rho$ with an intricate set of features has to be computed from a finite set of noisy datapoints obtained from a Monte Carlo simulation. This is an ill-defined problem, since for any discretization $\tau_i = \beta/N_\tau$ and in the presence of inevitable noise on $D$, an infinite number of choices for the $\rho_l$'s exist that will reproduce the data, even in the presence of the unitarity constraint $\rho_l > 0$.

One possibility to give meaning to the problem is through so called Bayesian inference, which is based on Bayes theorem $P[\rho|D,I] \propto P[D|\rho,I]P[\rho|I]$. It tells us that the probability for a test function $\rho$ to be the correct spectral function, given the simulated lattice data $D$ and any further prior information $I$ on the spectrum, is essentially the product of two terms. The likelihood probability $P[D|\rho,I] = e^{-L}$ denotes the probability of the lattice data if $\rho$ were the correct spectrum and $L$ amounts to nothing but the usual $\chi^2$ quadratic distance functional. The second term $P[\rho|I] = e^S$ denotes the prior probability, which encodes prior information on the spectrum and acts as regulator of the degenerate $\chi^2$ fit. On the one hand the functional form of $S$ encodes part of $I$ and on the other hand $S[\rho,m]$ depends on a function $m(\omega)$, the default model, which defines the extremum of $S$ and thus represents the correct spectrum in the absence of data. We have to carry out a numerical optimization to find the most probable $\rho$ given $D$ and $I$, with the competing $L$ and $S$ leading to a unique answer.

One popular implementation of the Bayesian strategy is the Maximum Entropy Method (MEM) [25, 26, 27]. In it the Shannon-Jaynes entropy is used as regulator term $S = S_{\rm SJ}$. One difficulty lies in the fact that $S_{\rm SJ}$ contains flat directions, which can adversely affect the convergence properties of the method and in practice necessitates the use of a priori restrictions of the functional space from which the spectra can be chosen. To avoid this resolution impediment we instead use a recently developed Bayesian approach [28] that relies on a prior $S$ derived from the following three physical consideration: positivity; smoothness of the spectrum, where data has not introduced sharp features and the independence of the result from the units chosen. As discussed in [28] these lead us to a prior functional $S = \alpha \sum_{l=1}^{N_\omega} \Delta\omega_l \Big(1 - \frac{\rho_l}{m_l} + \log\Big[\frac{\rho_l}{m_l}\Big]\Big)$, which is devoid of the flat directions of $S_{\rm SJ}$. Note that the weighting factor $\alpha$ here in contrast to the MEM can actually be integrated out explicitly in a Bayesian fashion [28]. With $S$ set, we only need to specify $m(\omega)$, for which we will use a constant, the most neutral choice.

## 3. Bottomonium spectra from lattice NRQCD

Let me paint in broad strokes the ingredients to investigating heavy-quarkonium in the non-relativistic field theory NRQCD on the lattice at finite temperature using Bottomonium as example. Readers interested in more technical details can consult [17]. Due to the separation of scales between the heavy-quark rest mass and the two other characteristic scales $\Lambda_{\rm QCD}$ and $T$, the creation and annihilation of heavy quark pairs is highly suppressed. This allows us to go over from a description of the quark fields in terms of four-component Dirac spinors $Q(x)$ to a set of two two-component Pauli spinors $\psi(t,\mathbf{x})$ and $\chi(t,\mathbf{x})$, representing the quark and anti-quark contribution respectively. Formally this separation of the upper and lower components of the four-spinor is obtained from a systematic expansion of the QCD Lagrangian in terms of increasing powers of the inverse rest mass (or more precisely the heavy quark velocity $v$) via the Foldy-Tani-Wouthuysen transformation. This is the same approach used to derive the Hydrogen energy levels from the Dirac equation.

If we place the non-relativistic theory on a Euclidean spacetime lattice, the Pauli spinors obey the corresponding non-relativistic Hamiltonian,

$$H = H_0 + \delta H = -\frac{\Delta^{(2)}}{2M_b} - \frac{g}{2M_b}\boldsymbol{\sigma}\cdot\mathbf{B} + \frac{ig}{8M_b^2}(\boldsymbol{\Delta}^{\pm}\cdot\mathbf{E} - \mathbf{E}\cdot\boldsymbol{\Delta}^{\pm}) + \ldots \qquad (2)$$

Here and $\Delta^{\pm}$ and $\Delta^{(2)}$ denote the lattice derivative and Laplace operator respectively. If all factors of the lattice spacing $a$ were made explicit it turns out that the expansion parameter amounts to $aM_b$ on the lattice. For the UV behavior to be well-defined its value must be larger than 1.5 [29], which reflects the fact that NRQCD works better on coarser lattices.

The fact that quark and antiquark modes are separately propagating in the medium background leads to an initial-value problem for the heavy quark propagator $G(\mathbf{x},\tau_i) = \langle\psi(\mathbf{x},\tau_i)\psi(0,0)\rangle$ in the background of medium fields. Written in terms of the link variables provided as external fields by conventional lattice simulations we have

$$G(\mathbf{x},\tau_i) = \left(1 - \frac{H_0}{2n}\right)^n U_4^\dagger(\mathbf{x},\tau)\left(1 - \frac{H_0}{2n}\right)^n \times (1-\delta H)\, G(\mathbf{x},\tau_{i-1})$$

The parameter $n$ reflects a freedom in the choice of the time discretization and is set to $n = 2$ here. From $G(\mathbf{x},\tau_i)$ we can construct heavy quarkonium propagators $D(\mathbf{x},\tau) = \sum_{\mathbf{x}_0}\langle O(\mathbf{x},\tau)G(\mathbf{x},\tau)O^\dagger(\mathbf{x}_0,\tau_0)G^\dagger(\mathbf{x},\tau)\rangle_{\rm med}$ , selecting individual channels by appropriate vertex operators $O(^3S_1;\mathbf{x},\tau) = \sigma_i$ for the S-wave and $O(^3P_1;\mathbf{x},\tau) = \overset{\leftrightarrow_s}{\Delta}_i\sigma_j - \overset{\leftrightarrow_s}{\Delta}_j\sigma_i$ for the P-wave.

The medium expectation value is obtained by measuring and averaging these correlators on realistic gauge field configurations containing light u,d, and s (HISQ) quarks, generated by the HotQCD collaboration [21]. These $48^3 \times 12$ lattices with $T_C = 159MeV$ feature an almost physical pion mass of $m_\pi = 161$MeV and allow a fine temperature scan between $T = 140 - 249$MeV, since the lattice spacing is used to regulate the extend in Euclidean time $\beta = 6.664 - 7.280$. The expansion parameter lies between $aM_b = 2.76 - 1.56 > 1.5$ allowing for the application of NRQCD. For a series of studies using a complementary fixed scale approach see [30]. Experimental input, e.g. the $\Upsilon(1S)$ mass is required to set the relative and absolute frequency scale, for which we use low temperature ensembles with $48^3 \times 48, 64$.

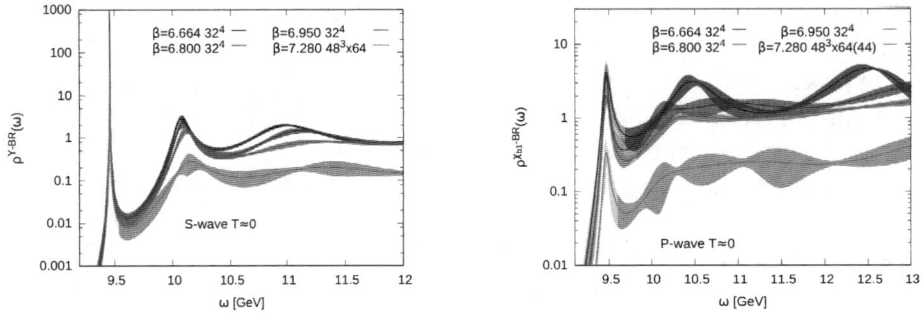


**Figure 1.** S-wave (left) and P-wave (right) Bottomonium spectra at $T \approx 0$.

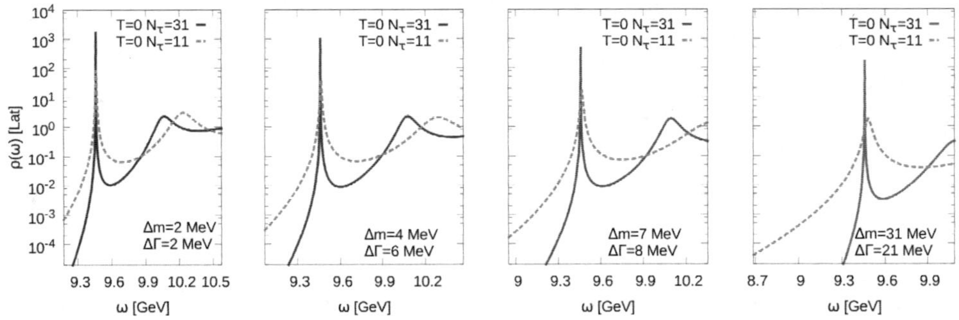


**Figure 2.** Spectra of the full (solid) and truncated (dashed) $T = 0$ correlator sets.

At $T \approx 0$ we have extracted [32] (left) the S- and (right) P-wave spectra shown in Fig.1. The S-wave ground state peak is very well resolved, around an order of magnitude more clearly than what is possible with the MEM. The second bump contains mostly contributions from $\Upsilon(2S)$, the higher lying wiggles are numerical ringing. After fixing the scale by the S-wave, we can estimate the systematic error in the computation from our postdiction for the $\chi_b(1S)$ mass. We obtain a value of $m^{\rm NRQCD}_{\chi_b(1P)} = 9.917(3)$GeV slightly larger than the PDG value of $m^{\rm PDG}_{\chi_b(1P)} = 9.89278(26)(31)$GeV. The P-wave ground state appears more washed out than the S-wave, not only because of a lower signal to noise ratio in the underlying correlators. Even in continuum the ratio of P-wave peak height to background is smaller in comparison [31].

To investigate in-medium modification we have to ascertain how well our spectral reconstruction fares at $T > 0$. The difference to vacuum lattices $N_\tau = 48$ is the physical extend in Euclidean time, reflected in a smaller number of $N_\tau = 12$ correlator points at $T > 0$. Hence we take the $T = 0$ correlator data set and truncate it after twelve points ($K(\omega, \tau)$ is $T$ independent). Supplying it to the reconstruction yields the spectra shown as gray dashed curves in Fig.2. On the coarsest lattices ($\beta = 6.664$) where NRQCD works best, the ground state peak reconstruction remains unaffected ($\Delta m < 2$MeV,$\Delta\Gamma < 2$MeV), while on the finest lattices ($\beta = 7.280$) corresponding to the highest temperature in the following, we find ($\Delta m < 40$MeV,$\Delta\Gamma < 21$MeV). As expected for the P-wave the accuracy limits are even larger [32]. Note that the changes are on the level of individual MeV, which could not have been resolved in the same manner with the MEM. If at finite temperature we were to find changes larger than the stringent limits set here, we can attribute them to in-medium modification.

In Fig.3 we show the $T > 0$ results for (left) the S-wave and (right) the P-wave spectra at the fourteen different temperatures available. In the S-wave we find at all $T$ a well-defined ground state peak, much larger than any further numerical ringing at higher $\omega$. If zoomed in, the peak is seen to slightly shift and broaden with $T$; however the changes are smaller than the

Journal of Physics: Conference Series **668** (2016) 012026 doi:10.1088/1742-6596/668/1/012026

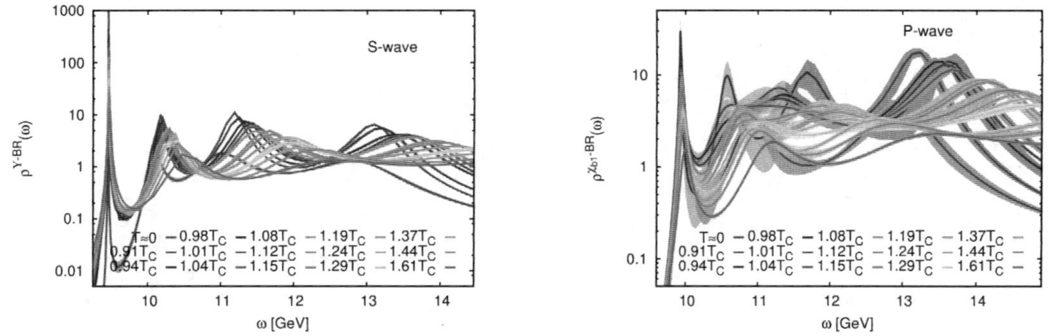

**Figure 3.** S-wave (left) and P-wave (right) Bottomonium spectra at $T > 0$.

accuracy limits established before. Hence we refrain from attributing them to medium effects. Nevertheless by a simple inspection by eye it is evident that $\Upsilon(1S)$ survives up to $T = 249$MeV. This conclusion agrees with the results from recent a MEM based study [30].

The P-wave case is more subtle, because of the larger errors due to the lower signal to noise ratio in the correlators. While we find a well defined lowest lying peak at all $T$ its height is comparable to that of higher lying numerical ringing at the highest $T$. A naive inspection by eye fails to give conclusive insight. Therefore we propose to put the decision of survival or melting on a more systematic footing via a comparison with non-interacting spectral functions.

The $Q\bar{Q}$ spectrum in the absence of interactions is analytically known not to show any bound state peaks. Such spectra can also be extracted numerically from lattices with all links set to unity and subsequent Bayesian reconstruction. Just as with Gibbs ringing in the Fourier transform we expect wiggly structures to appear, even if absent in the actual free spectrum. This can be seen in the gray curve in Fig.4 where the free spectrum has been shifted to match the $\omega$ range of the full spectrum (red). We see that while numerical ringing dominates the second bump, the first peak is a factor of three larger and thus constitutes a signal for bound state survival. Hence we conclude that $\chi_b(1P)$ survives up to $T = 249$MeV ($1.61T_C$). This conclusion differs from a recent MEM based study, which found melting already at $T = 1.27T_C$ [30]. One explanation might lie in the restricted search space of the MEM, which shrinks with the number of available datapoints and thus entangles physical effects and numerical limitations.

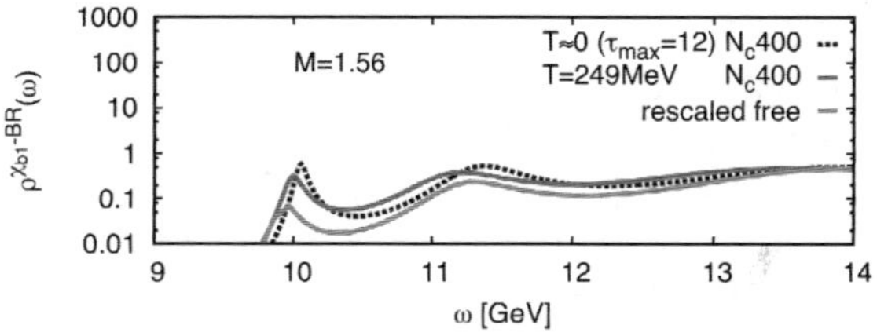

**Figure 4.** Comparison of free (gray), interacting (red) and the spectral function from the truncated $T = 0$ dataset (blue dashed)

## 4. The complex in-medium heavy quark potential from lattice QCD

We now wish to address the question of how to determine the potential $V(r)$ acting between two static quarks at finite temperature non-perturbatively. One starts from an EFT description in terms of $Q\bar{Q}$ wave-functions called pNRQCD [6, 12]. The values of $V(r)$ are determined by the procedure of matching, in which a suitable QCD correlation function is identified, which carries the same physics content as a correlation functions in pNRQCD and both quantities are set equal at a certain scale. At $m_Q \to \infty$ the correlator of two color-singlet wavefunctions is identified with the real-time Wilson loop $W_\square(t,r)$. The potential arises from its late Minkowski-time limit

$$W_\square(t,r) = \langle \mathrm{Tr}\Big(\exp\big[ig\int_\square dz^\mu A_\mu(z)\big]\Big)\rangle, \quad V(r) = \lim_{t\to\infty} \frac{i\partial_t W_\square(t,r)}{W_\square(t,r)} \tag{3}$$

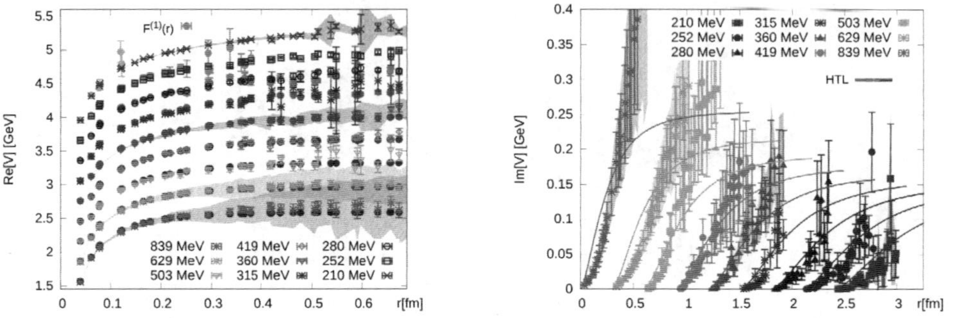


**Figure 5.** (left) Real and (left) imaginary part of the complex in-medium potential extracted from quenched QCD simulations.

The limit reflects the fact that replacing an interaction mediated by gluons with an instantaneous potential is only valid at timescales much larger than those of the medium gluons.

The challenge to a non-perturbatively evaluation lies in the real-time character of the definition, not directly amenable to a lattice QCD simulation. Here the concept of spectral function comes to our rescue [33, 34] since the Wilson loop can be rewritten as a Fourier transform of a positive definite $\rho_\square(\omega, r)$. In fact the Euclidean Wilson loop is related to the same $\rho_\square$ via the Laplace transform, which can be inverted using the Bayesian method of sec.2. Inserting the spectral decomposition in the potential definition (3) leads to $V(r) = \lim_{t\to\infty} \int d\omega\, \omega e^{-i\omega t}\rho(\omega, r)/\int d\omega\, e^{-i\omega t}\rho(\omega, r)$. The late time limit tells us that only the lowest lying spectral structure contributes to $V(r)$, which has been shown to take on the form of a skewed Lorentzian [35]. Its position and width correspond to the real and imaginary part of the potential respectively. Hence we fit the lowest lying peak and read off Re$[V]$ and Im$[V]$.

In Fig.5 we show (colored points) the results [15, 16] for (left) the real and (right) imaginary part obtained from anisotropic $32^3 \times N_\tau$ lattices with $\beta = 7$ and bare anisotropy $\xi = 3.5$ in the quenched approximation ($T_C = 270$MeV). Changing $N_\tau = 24 - 96$ we scan a range of $T = 210 - 839$MeV. Note that instead of Wilson loops we here use Wilson line correlators in Coulomb gauge since they do not posses cusp divergences.

Re$[V]$ shows a smooth transition from the well-known Cornell type behavior at low $T$ to a Debye screened form at $T > T_C$, as expected from perturbation theory. Interestingly the values of $F^{(1)}(r)$ (gray points) lie very close to the values of the real part. Hence this first principles calculation supports $F^{(1)}(r)$ as an approximation of Re$[V]$. Im$[V]$ is non-zero on the lattice at $T > T_C$ and hence cannot be neglected in an in-medium potential description. Its shooting off at larger $r$ is an artifact of the reconstruction because of a diminishing signal to noise ratio in the correlators.

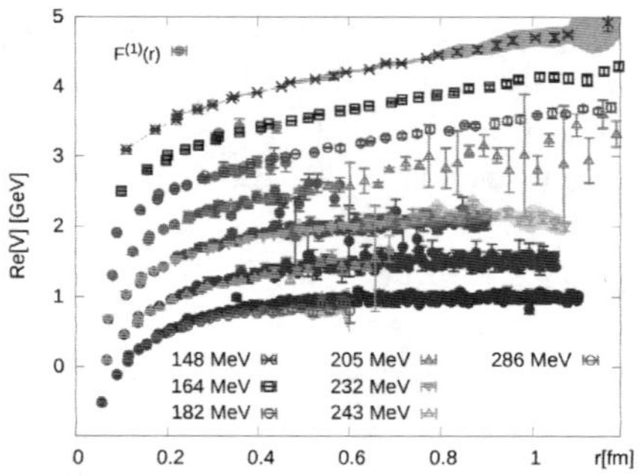


**Figure 6.** Re$[V]$ in full QCD with dynamical u,d and s quarks.

We present in Fig.6 the values of Re$[V]$ in a true QGP with u,d, and s quarks. It is extracted on isotropic $48^3 \times 12$ lattices by the HotQCD collaboration [15, 16], spanning a temperature range of $T = 148 - 286$MeV corresponding to a lattice parameter range of $\beta = 6.8 - 7.48$. Based on the asqtad action with $m_\pi \simeq 300$MeV the transition temperature here is $T = 172.5$MeV. The behavior we find is qualitatively similar to the quenched case with a smooth transition from Cornell to Debye screened behavior, and again $F^{(1)}(r)$ appears to be a good approximation of Re[$V$]. We refrain from showing the tentative values of Im[$V$], since spectral widths extracted from $N_\tau = 12$ points are generally considered not reliable.

## 5. Conclusion

Heavy quarkonium presents a formidable testing ground for our capability to connect a first principles understanding of the strong interactions to phenomenologically relevant observables. The combination of non-relativistic EFTs with non-perturbative simulations of the QCD medium, already highly successful at $T \simeq 0$, are nowadays providing valuable insight also at $T \simeq T_C$, reached in relativistic heavy ion-collisions. We reported on recent progress made in the determination of dynamical information from such simulations through a novel Bayesian spectral reconstruction approach and two studies based on its application were discussed. The first work extracted the in-medium spectra of bottomonium directly from lattice NRQCD and found survival of both S-wave and P-wave ground states up to $T = 249$MeV. The second work used spectral functions to extract the values of the complex in-medium heavy-quark potential. Besides a smooth transition from a confining to Debye screened behavior it confirmed the existence of an imaginary part in the potential in the quark-gluon-plasma.

The author thanks Y. Burnier, O. Kaczmarek, S. Kim and P. Petreczky for the fruitful collaboration that underlies the lattice QCD spectra results presented in this proceeding.

## References

[1] N. Brambilla *et al.*, Eur. Phys. J. C **71**, 1534 (2011)
[2] S. Chatrchyan *et al.* [CMS], Phys. Rev. Lett. **107** (2011) 052302,Phys.Rev.Lett. **109** (2012) 222301 .
[3] B. Abelev *et al.* [ALICE], Phys. Rev. Lett. **109**, 072301 (2012)
[4] A. Andronic, F. Beutler, P. Braun-Munzinger, K. Redlich and J. Stachel, Phys. Lett. B **678**, 350 (2009)
[5] K. A. Olive *et al.* [Particle Data Group], Chin. Phys. C **38**, 090001 (2014).
[6] N. Brambilla, A. Pineda, J. Soto and A. Vairo, Rev. Mod. Phys. **77**, 1423 (2005)
[7] S. Nadkarni, Phys. Rev. D **34**, 3904 (1986);
[8] H. Satz, J. Phys. G **36**, 064011 (2009).
[9] C. Y. Wong, Phys. Rev. C **72**, 034906 (2005);
[10] O. Kaczmarek, PoS CPOD **07**, 043 (2007);
[11] A. Barchielli, E. Montaldi and G. M. Prosperi, Nucl. Phys. B **296**, 625 (1988),
[12] N. Brambilla, J. Ghiglieri, A. Vairo, P. Petreczky, Phys. Rev. D **78**, 014017 (2008)
[13] M. Laine, O. Philipsen, P. Romatschke and M. Tassler, JHEP **0703**, 054 (2007)
[14] A. Beraudo, J. -P. Blaizot, C. Ratti, Nucl. Phys. A **806**, 312 (2008)
[15] Y. Burnier, O. Kaczmarek and A. Rothkopf, Phys. Rev. Lett. **114**, no. 8, 082001 (2015)
[16] Y. Burnier, O. Kaczmarek and A. Rothkopf, PoS LATTICE **2014**, 220 (2015)
[17] B. A. Thacker and G. P. Lepage, Phys. Rev. D **43**, 196 (1991).
[18] R. J. Dowdall *et al.* [HPQCD], Phys. Rev. D **85**, 054509 (2012)
[19] R. Mizuk *et al.* [Belle], Phys. Rev. Lett. **109**, 232002 (2012)
[20] Y. Aoki, G. Endrodi, Z. Fodor, S. D. Katz and K. K. Szabo, Nature **443**, 675 (2006)
[21] A. Bazavov *et al.*, Phys. Rev. D **85**, 054503 (2012)
[22] A. Bazavov *et al.* [HotQCD], Phys. Rev. D **90**, no. 9, 094503 (2014)
[23] S. Borsanyi, Z. Fodor, C. Hoelbling, S. D. Katz, S. Krieg and K. K. Szabo, Phys. Lett. B **730**, 99 (2014)
[24] F. Burger *et al.* [tmfT], Phys. Rev. D **91**, no. 7, 074504 (2015)
[25] M. Asakawa, T. Hatsuda and Y. Nakahara, Prog. Part. Nucl. Phys. **46** (2001) 459
[26] A. Jakovac, P. Petreczky, K. Petrov and A. Velytsky, Phys. Rev. D **75**, 014506 (2007)
[27] A. Rothkopf, J. Comput. Phys. **238**, 106 (2013)
[28] Y. Burnier and A. Rothkopf, Phys. Rev. Lett. **111**, 182003 (2013),PoS LATTICE **2013**, 490 (2014)
[29] C. T. H. Davies and B. A. Thacker, Phys. Rev. D **45**, 915 (1992).
[30] G. Aarts, *et al.* [FASTSUM] JHEP **1111**, 103 (2011),JHEP **1312**, 064 (2013) ,JHEP **1407**, 097 (2014)
[31] Y. Burnier, M. Laine and M. Vepsalainen, JHEP **0801**, 043 (2008)
[32] S. Kim, P. Petreczky and A. Rothkopf, Phys. Rev. D **91**, 054511 (2015),PoS LATTICE **2014**, 208 (2014)
[33] A. Rothkopf, T. Hatsuda and S. Sasaki, PoS LAT **2009**, 162 (2009)
[34] A. Rothkopf, T. Hatsuda and S. Sasaki, Phys. Rev. Lett. **108**, 162001 (2012)
[35] Y. Burnier and A. Rothkopf, Phys. Rev. D **86**, 051503 (2012)

15th International Conference on Strangeness in Quark Matter (SQM2015) IOP Publishing
Journal of Physics: Conference Series **668** (2016) 012027 doi:10.1088/1742-6596/668/1/012027

# Search for the Critical Point of Strongly Interacting Matter in Heavy-Ion Collisions

**Peter Seyboth**
Max-Planck-Institut für Physik, Munich, Germany
and
Jan Kochanowski University, Kielce, Poland

E-mail: pxs@mppmu.mpg.de

**Abstract.** Results obtained from the search for the critical point of strongly interacting matter in relativistic heavy-ion collisions at the CERN-SPS (experiments NA49 and NA61/SHINE) and the RHIC beam energy scan program (experiments STAR and PHENIX) are discussed. Although some intriguing signals were found, establishing firm evidence for the critical point remains a challenging enterprise.

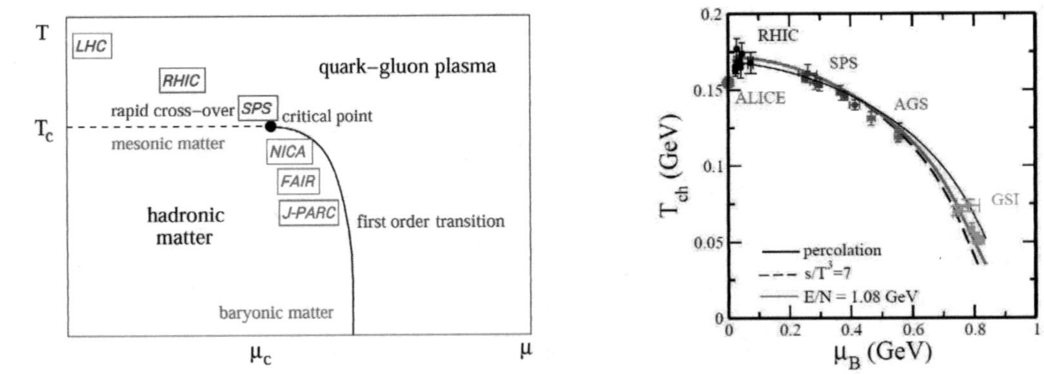


**Figure 1.** *Left:* Sketch of the phase diagram of strongly interacting matter. Italic labels added to the sketch show regions probed by the early stage of heavy-ion collisions studied in current (in red: CERN SPS, RHIC, LHC) and future (in green: NICA, FAIR, J-PARC) experimental programmes. *Right:* Compilation of chemical freeze-out points of central Pb+Pb (Au+Au) collisions [8].

## 1. Introduction

The structure of the phase diagram of strongly interacting matter is an important aspect of QCD in the non-perturbative sector. The most popular suggestion is shown schematically in Fig. 1 *left* (see e.g. Ref. [1]). The deconfinement transition at low temperature $T$ and high baryochemical potential $\mu_B$ is believed to be of the first order and happen along a line which ends with decreasing $\mu_B$ in a critical point CP (of the second order) and then turns into a crossover region. Present lattice QCD calculations can neither provide consistent estimates of the location of the CP [2, 3, 4] nor a definitive answer to the question of its existence [5, 6] due to the so-called sign problem encountered in calculations for finite $\mu_B$ values. Experimentally the structure of the phase diagram is explored by studying the final states produced in relativistic nucleus-nucleus collisions. By varying collision energy and size of colliding nuclei one changes temperature $T$ and chemical potentials $\mu_B$ of the matter at the freeze-out stage. The values of $T$ and $\mu_B$ for the system are obtained from fits of the statistical model of hadronisation [7, 8] to the composition of the observed hadrons (see Fig. 1 *right*). Model calculations suggest [9] that at high collision energies these values are close to those of the deconfinement transition (hadronisation) of the system in the phase diagram. Of course, effects of the phase transition and the CP can only be seen if the matter droplet initially reaches deconfinement. Rapid changes found in the freeze-out properties indicate that this happens for collision energies above about $E_{LAB} \approx 30A$ GeV ($\sqrt{s_{NN}} = 8$ GeV) [14].

At the CP the correlation length $\xi$ diverges resulting in a maximum of fluctuations of the multiplicity of produced particles. This leads to scale invariance, the effect of intermittency [10] as well as to a large increase of the moments of the distributions of event-by-event multiplicities, net charge or transverse momentum [11]. However, the growth of $\xi$ is limited by the small size and short lifetime of the system [12] to estimated values of about 3 fm in collisions of heavy nuclei. Moreover, the CP signal might be weakened or erased during the evolution of the hadron phase.

This contribution discusses results obtained from the search for the CP from experiments NA49 and NA61/SHINE at the CERN-SPS [13] and experiments STAR and PHENIX in the RHIC beam energy scan program. Emphasis is put on showing the experimental results. For more details on the experiments and the analysis procedure the given references should be consulted.

## 2. Intermittency studies

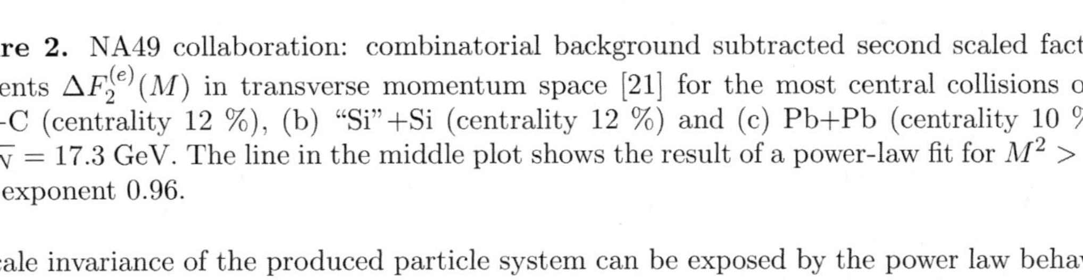


**Figure 2.** NA49 collaboration: combinatorial background subtracted second scaled factorial moments $\Delta F_2^{(e)}(M)$ in transverse momentum space [21] for the most central collisions of (a) "C"+C (centrality 12 %), (b) "Si"+Si (centrality 12 %) and (c) Pb+Pb (centrality 10 %) at $\sqrt{s_{NN}} = 17.3$ GeV. The line in the middle plot shows the result of a power-law fit for $M^2 > 6000$ with exponent 0.96.

Scale invariance of the produced particle system can be exposed by the power law behaviour of scaled factorial moments $F_r(\delta)$ of the particle multiplicity $N$ on the selected phase-space bin size $\delta$:

$$F_r(\delta) = \langle \frac{1}{M} \sum_{i=1}^{M} N_i(N_i - 1)...(N_i - r + 1) \rangle / \langle \frac{1}{M} \sum_{i=1}^{M} N_i \rangle^r \ , \tag{1}$$

Journal of Physics: Conference Series **668** (2016) 012027 doi:10.1088/1742-6596/668/1/012027

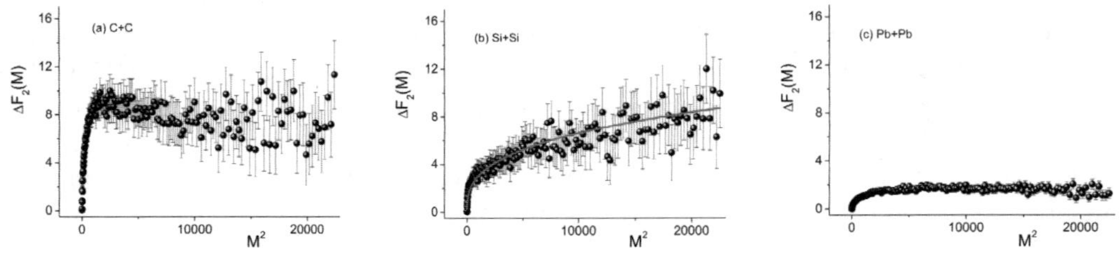


**Figure 3.** NA49 collaboration: combinatorial background subtracted second scaled factorial moments $\Delta F_2$ in transverse momentum space [22] for: (a) C+C, (b) Si+Si and (c) Pb+Pb systems. The line in (b) shows the result of a power-law fit for $M^2 > 2000$ with exponent 0.33.

where $r$ is the rank of the moment, $\delta$ the size of the subdivision intervals of the momentum phase space region $\Delta$ and $M = \Delta/\delta$ the number of intervals. $N_i$ refers to particle multiplicity in the interval $i$ and $\langle ... \rangle$ indicates averaging over the analyzed collisions. Scale invariance implies the relation:

$$F_r(\delta) = F_r(\Delta) \cdot (\Delta/\delta)^{\phi_r}, \quad \phi_r = (r-1) \cdot d \;, \tag{2}$$

with a rank independent anomalous fractal dimension $d$.

Early searches for critical fluctuations in nucleus-nucleus collisions investigated scaled factorial moments from charged particle production [10]. However, the observed effects were small and inconclusive [17, 18, 19]. More recent QCD considerations suggested [1] that the order parameter of the phase transition is the chiral condensate $\langle \bar{q}q \rangle$ ($q$ is the quark field), which at the hadronisation stage transforms into $\sigma$ mesons. The critical fluctuations at the CP were predicted to be observable via the intermittent behaviour of $\pi^+\pi^-$ pairs from the decay of the $\sigma$ mesons with invariant mass just above twice the pion mass [15] or of protons which also couple to the chiral condensate [16]. The corresponding power law indices were predicted to have the values $\phi_2 = 2/3$ [15] for $\pi^+\pi^-$ pairs and $\phi_2 = 5/6$ [20] for protons respectively.

The NA49 collaboration searched for intermittency in central nucleus-nucleus collisions at the highest SPS energy of 17.3 GeV. The resulting combinatorial background subtracted second scaled factorial moments $\Delta F_2(M)$ are plotted versus $M^2$ in Figs. 2 and 3 for protons and for $\pi^+\pi^-$ pairs near mass threshold, respectively. One observes power law behavior at large $M^2$ for central Si+Si collisions (freeze-out at $T \approx 162$ MeV, $\mu_B \approx 260$ MeV [7]) with exponents $\phi_2 = 0.96^{+0.38}_{-0.25}(stat) \pm 0.16(syst)$ and $\phi_2 = 0.33 \pm 0.04$. The result for protons agrees with the CP prediction whereas that for $\pi^+\pi^-$ pairs is smaller, probably as a consequence of the difficulty of isolating the pairs from $\sigma$ decay. Although the freeze-out states of C+C and Pb+Pb are close to that of Si+Si the critical fluctuation pattern probably cannot develop in C+C due to the small size of the system and may be erased in Pb+Pb during the longer evolution of the hadron phase.

## 3. Finite size scaling in collisions of heavy nuclei

An interesting non-monotonic behaviour was observed by R. Lacey [23] for the difference of matter droplet radius parameters $R^2_{out} - R^2_{side}$ as obtained from pion Bose-Einstein correlation analysis of Au+Au and Pb+Pb collisions (see Fig.4 *left*). A maximum appears at RHIC energies which might be related to a minimum of the compressibility as expected at the CP. Finite size scaling is then predicted [24] when changing the radius $\bar{R}$ of the system by selecting different collision centralities. As seen in Fig.4 *right* this is indeed observed with suitable scaling parameters and suggests the existence of the CP at $\mu_B \approx 95$ MeV and $T \approx 165$ MeV.

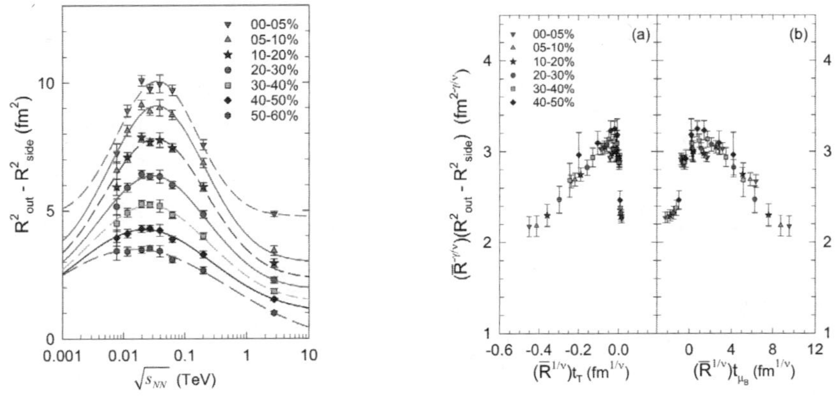


**Figure 4.** Figures from Ref. [23]. *Left*: difference $R^2_{out} - R^2_{side}$ of radius parameters in Au+Au and Pb+Pb collisions versus collision energy for various centralities. *Right*: demonstration of finite size scaling of the radius parameter difference with system radius $\bar{R}$.

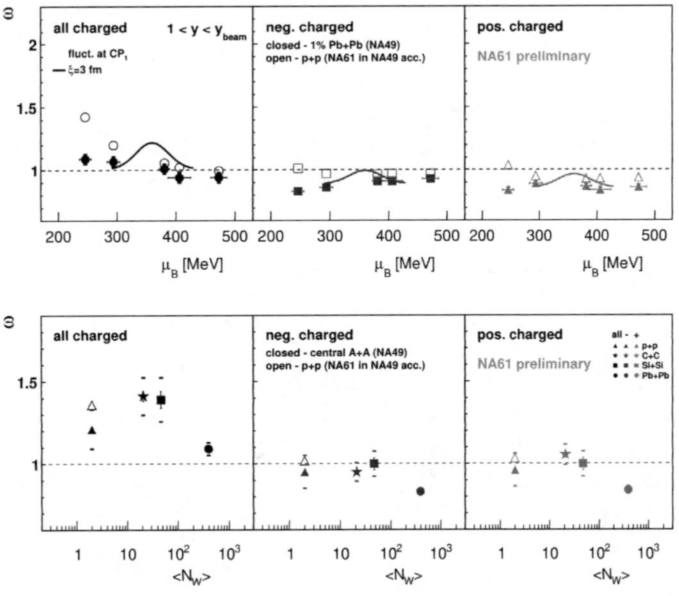


**Figure 5.** NA49 and NA61/SHINE collaborations: scaled variance $\omega$ of the multiplicity distribution of charged particles for the 1 % most central Pb+Pb collisions and inelastic p+p reactions for $1.0 < y < y_{beam}$ (assuming the pion mass). *Top:* versus $\mu_B$. *Bottom:* versus the number of wounded nucleons $N_W$. Full symbols show results of NA49 [25], open symbols preliminary measurements of NA61/SHINE [26]. Curves illustrate the effect of a critical point [27].

## 4. Multiplicity and transverse momentum fluctuations

Multiplicity and net-charge fluctuations in nucleus-nucleus collisions have most often been studied via the moments of their distributions. Cumulants of the distribution $P(N)$ (here $N$

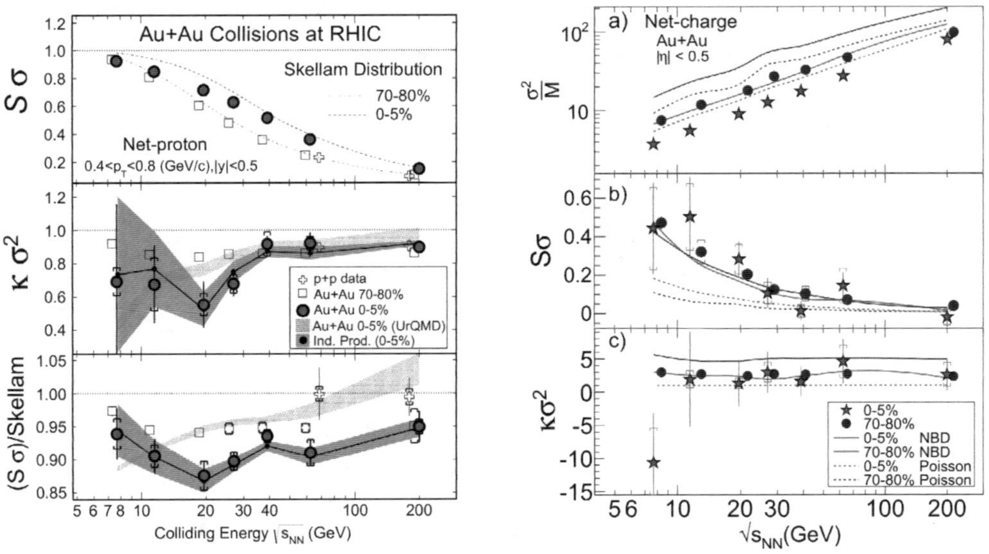


**Figure 6.** STAR collaboration: fluctuations of net-proton multiplicity (*left*) [29] and net-charge (*right*) [30] in central Au+Au collisions measured by the width $\sigma$, mean M, skewness $S$ and kurtosis $\kappa$ of the distributions. (Rapidity range $|y| < 0.5$ and $|\eta| < 0.5$ respectively)

can be e.g. the multiplicity or the net charge in the event) are defined as follows:

$$\kappa_2 = \langle(\delta N)^2\rangle \ , \quad \kappa_3 = \langle(\delta N)^3\rangle \ , \quad \kappa_4 = \langle(\delta N)^4\rangle - 3\kappa_2^2 \ , \ and \ \ \delta N := N - \langle N\rangle \qquad (3)$$

Cumulants are extensive quantities, proportional to the size of the system as given e.g. by the number of participant nucleons in the collision. Since the latter cannot be tightly controlled or measured in nucleus-nucleus collisions one prefers to use size independent normalised cumulants $\omega_k = \kappa_k/\langle N\rangle$. For independent particle production one expects $\omega_2 = 1$ and $\omega_{k>2} = 0$. Moreover the cumulant ratios $S\sigma = \kappa_3/\kappa_2$ and $\kappa\sigma^2 = \kappa_4/\kappa_2$ ($\sigma$, $S$ and $\kappa$ denote the variance, skewness and kurtosis of the distribution, respectively) are of special interest since they are related to susceptibilities of conserved charges calculable in lattice QCD.

The simplest non-trivial normalised cumulant is the widely used scaled variance $\omega$, an intensive quantity, i.e. independent of the size of the system, but not of the inevitable fluctuations thereof. In order to minimise their effect NA49 [25] selected the 1 % most central collisions. The obtained results on $\omega$ for charged particles are plotted in Fig. 5 *top* for the energy dependence in Pb+Pb reactions and in Fig. 5 *bottom* for the nuclear size dependence at highest SPS energy. Clearly no significant maximum is observed in the energy scan, while there may be a bump for medium-size nuclei (Si+Si).

Higher moments have been shown to be more divergent near the CP. This applies in particular to the net proton multiplicity [28] which was investigated by the STAR collaboration [29] in central Au+Au collisions at RHIC. The energy dependence of combinations of moments and cumulants (width $\sigma$, skewness $S$ and kurtosis $\kappa$) is plotted in Fig. 6 *left*. The results are close to expectations for independent particle production (Skellam distribution for a difference of

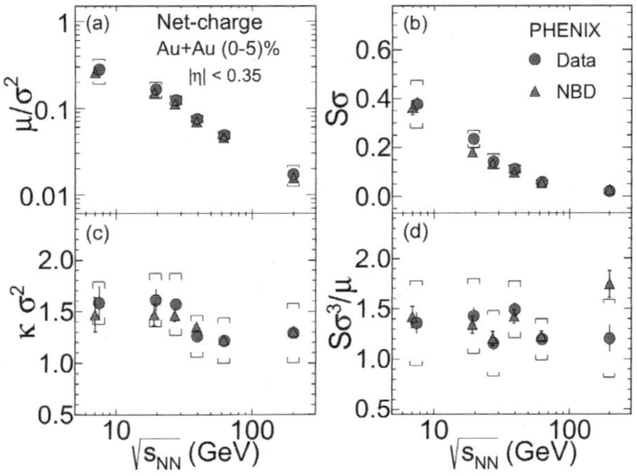


**Figure 7.** PHENIX collaboration: fluctuations of net-charge [31] in central Au+Au collisions at RHIC measured by the width $\sigma$, mean $\mu$, skewness $S$ and kurtosis $\kappa$ of the distribution. (Pseudo-rapidity range $|\eta| < 0.35$)

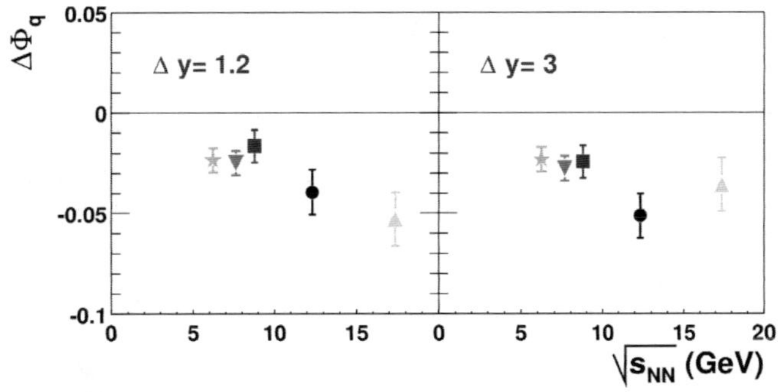


**Figure 8.** NA49 collaboration: energy dependence of $\Delta\Phi_q$ measured in central Pb+Pb collisions at 20$A$, 30$A$, 40$A$, 80$A$ and 158$A$ GeV [33] for a narrow rapidity interval $\Delta y = 1.2$ (*left*) and a broad rapidity interval $\Delta y = 3$ (*right*).

multilicities) and the trend with collision energy is similar in the microscopic model UrQMD. Thus no structure was found which might be attributable to the CP.

Charge fluctuations are also expected to show a non-monotonic dependence in a collision energy scan when the system passes near the CP in the phase diagram. The results from STAR [30] for the energy dependence of combinations of moments and cumulants of the net-charge multiplicity distribution in central Au+Au collisions is presented in Fig. 6 *right*. The observed smooth behaviour does not provide evidence for the CP. A similar analysis was performed by the PHENIX collaboration [31]. The measurements shown in Fig. 7 likewise do not find indications for the CP.

Some years back NA49 studied charge fluctuations in central Pb+Pb collisions at the SPS [33] and employed the strongly intensive measure $\Phi_x$ [32, 34]:

$$\Phi_x = \sqrt{\langle Z^2 \rangle / \langle N \rangle} - \sqrt{\overline{(x - \overline{x})^2}} \quad , \qquad Z = \sum_{i=1}^{N} (x_i - \overline{x}) \, , \tag{4}$$

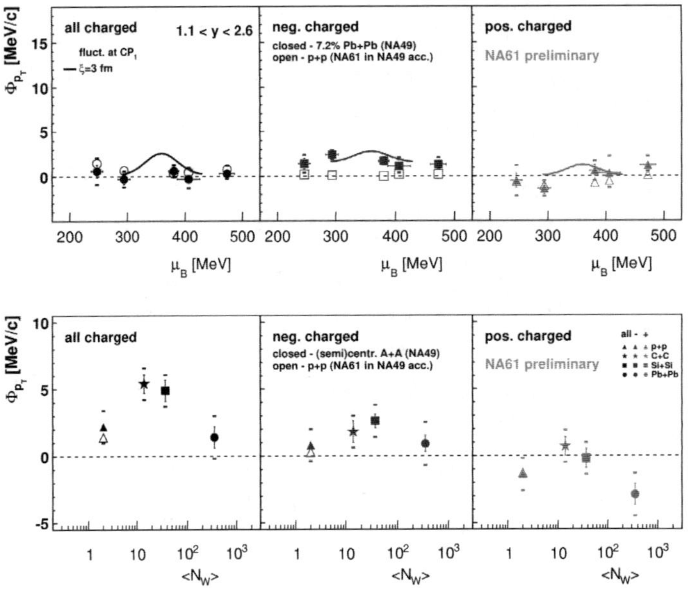


**Figure 9.** NA49 and NA61 collaborations: measure $\Phi_{p_T}$ of transverse momentum - multiplicity fluctuations of charged particles. *Top:* versus $\mu_B$ for the 7.2 % most central Pb+Pb collisions (full symbols, NA49 [35]) and inelastic p+p reactions (open symbols, NA61 preliminary). *Bottom:* versus the number of wounded nucleons $N_W$ in central C+C, Si+Si and Pb+Pb collisions at 158*A* GeV (NA49 [36]) and inelastic p+p reactions (NA61/SHINE preliminary [26]). Results are for cms rapidity $1.1 < y < 2.6$ assuming the pion mass. Curves illustrate the effect of a critical point [27].

where $x$ is a particle observable and the bar and brackets denote the inclusive average and averaging over events, respectively. $\Phi_x$ is independent of both the system volume and its fluctuations. For the study of charge fluctuations the observable $x$ is taken to be the charge $q$ of the particle. It was found that there were large contributions from global charge conservation which were calculated and subtracted. The resulting values of $\Delta\Phi_q$ are plotted in Fig. 8 as function of collision energy. Again no structure due to a CP is seen.

A maximum of fluctuations of the transverse momentum $p_T$ of produced particles was also proposed as a signature of the CP [11]. Such fluctuations were measured for nucleus-nucleus and p+p collisions at the SPS by experiments NA49 and NA61, respectively, again using the quantity $\Phi_x$ now with $x = p_T$. The results for charged particles are shown in Fig. 9 as function of energy (*top*) and size of the interacting nuclei (*bottom*). Similar to the measurements of charged particle multiplicity fluctuations (see Fig. 5) one finds no peak in the energy dependence but perhaps a weak indication of a maximum in the system size study.

## 5. Summary and conclusion

A search for evidence of the critical point of strongly interacting matter has been performed by pilot scans of the phase diagram with meaurements of fluctuations in nucleus-nucleus collisions in the most promising energy region, i.e at the CERN SPS and in the RHIC beam energy scan program. While some intriguing effects were found in studies of intermittency and multiplicity fluctuations in the medium size Si+Si system as well as of finite size scaling as function of centrality in Au+Au collisions no definitive evidence of the critical point has yet emerged.

Further more systematic and precise studies are ongoing by the NA61/SHINE collaboration, are scheduled for the second beam energy scan at RHIC and are planned by future experiments at NICA, FAIR and J-PARC. Exploration of the phase diagram and discovering the critical point remains a challenging and exciting enterprise.

## Acknowledgments

The author wishes to thank the organisers of SQM2015 for the invitation to present this talk. Also numerous helpful discussions with M. Gazdzicki are gratefully acknowledged.

## References

[1] M. A. Stephanov, Prog. Theor. Phys. Suppl. **153**, 139 (2004) [Int. J. Mod. Phys. A **20**, 4387 (2005)] [hep-ph/0402115].
[2] Z. Fodor and S. D. Katz, JHEP **0404**, 050 (2004) [hep-lat/0402006].
[3] A. Li, A. Alexandru and K. F. Liu, Phys. Rev. D **84**, 071503 (2011) [arXiv:1103.3045 [hep-ph]].
[4] S. Datta, R. V. Gavai and S. Gupta, Nucl. Phys. A **904-905**, 883c (2013) [arXiv:1210.6784 [hep-lat]].
[5] P. de Forcrand and O. Philipsen, JHEP **0811**, 012 (2008) [arXiv:0808.1096 [hep-lat]].
[6] G. Endrodi, Z. Fodor, S. D. Katz and K. K. Szabo, JHEP **1104**, 001 (2011) [arXiv:1102.1356 [hep-lat]].
[7] F. Becattini, J. Manninen and M. Gazdzicki, Phys. Rev. C **73**, 044905 (2006) [arXiv:hep-ph/0511092].
[8] J. Cleymans, EPJ Web Conf. **95** (2015) 03004 [arXiv:1412.7045 [hep-ph]].
[9] R. Stock, F. Becattini, M. Bleicher, T. Kollegger, T. Schuster and J. Steinheimer, PoS CPOD **2013**, 011 (2013) [arXiv:1306.4201 [nucl-th]].
[10] A. Bialas and R. C. Hwa, Phys. Lett. B **253** (1991) 436.
[11] M. A. Stephanov, K. Rajagopal and E. V. Shuryak, Phys. Rev. D **60**, 114028 (1999) [hep-ph/9903292].
[12] B. Berdnikov and K. Rajagopal, Phys. Rev. D **61**, 105017 (2000) [hep-ph/9912274].
[13] for recent review see: M. Gazdzicki and P. Seyboth, arXiv:1506.08141 [nucl-ex].
[14] C. Alt *et al.* [NA49 Collaboration], Phys. Rev. C **77**, 024903 (2008) [arXiv:0710.0118 [nucl-ex]].
[15] N. G. Antoniou, Y. F. Contoyiannis, F. K. Diakonos and G. Mavromanolakis, Nucl. Phys. A **761**, 149 (2005) [hep-ph/0505185].
[16] Y. Hatta and M. A. Stephanov, Phys. Rev. Lett. **91**, 102003 (2003) [Erratum-ibid. **91**, 129901 (2003)] [hep-ph/0302002].
[17] R. Holynski *et al.*, Phys. Rev. C **40**, 2449 (1989).
[18] R. Albrecht *et al.* [WA80 Collaboration], Phys. Rev. C **50**, 1048 (1994).
[19] J. Bachler *et al.* [NA35 Collaboration], Z. Phys. C **61**, 551 (1994).
[20] N. G. Antoniou, F. K. Diakonos, A. S. Kapoyannis and K. S. Kousouris, Phys. Rev. Lett. **97**, 032002 (2006) [hep-ph/0602051].
[21] T. Anticic *et al.* [NA49 Collaboration], arXiv:1208.5292 [nucl-ex].
[22] T. Anticic *et al.* [NA49 Collaboration], Phys. Rev. C **81**, 064907 (2010) [arXiv:0912.4198 [nucl-ex]].
[23] R. A. Lacey, Phys. Rev. Lett. **114**, no. 14, 142301 (2015) [arXiv:1411.7931 [nucl-ex]].
[24] M. Ladrem and A. Ait-El-Djoudi, Eur. Phys. J. C **44**, 257 (2005) [hep-ph/0412407].
[25] C. Alt *et al.* [NA49 Collaboration], Phys. Rev. C **78**, 034914 (2008) [arXiv:0712.3216 [nucl-ex]].
[26] T. Czopowicz [NA61/SHINE Collaboration], arXiv:1503.01619 [nucl-ex].
[27] K. Grebieszkow [NA49 Collaboration], Nucl. Phys. A **830**, 547C (2009) [arXiv:0907.4101 [nucl-ex]]; M. Stephanov, private communication.
[28] C. Athanasiou, K. Rajagopal and M. Stephanov, Phys. Rev. D **82**, 074008 (2010) [arXiv:1006.4636 [hep-ph]].
[29] L. Adamczyk *et al.* [STAR Collaboration], Phys. Rev. Lett. **112**, 032302 (2014) [arXiv:1309.5681 [nucl-ex]].
[30] L. Adamczyk *et al.* [STAR Collaboration], Phys. Rev. Lett. **113**, 092301 (2014) [arXiv:1402.1558 [nucl-ex]].
[31] A. Adare *et al.* [PHENIX Collaboration], arXiv:1506.07834 [nucl-ex]. [32]
[32] M. Gazdzicki and S. Mrowczynski, Z. Phys. C **54**, 127 (1992).
[33] C. Alt *et al.* [NA49 Collaboration], Phys. Rev. C **70**, 064903 (2004) [nucl-ex/0406013].
[34] for a general discussion of strongly intensive measures see: M. I. Gorenstein and M. Gazdzicki, Phys. Rev. C **84**, 014904 (2011) [arXiv:1101.4865 [nucl-th]].
[35] T. Anticic *et al.* [NA49 Collaboration], Phys. Rev. C **79**, 044904 (2009) [arXiv:0810.5580 [nucl-ex]].
[36] T. Anticic *et al.* [NA49 Collaboration], Phys. Rev. C **70**, 034902 (2004) [hep-ex/0311009].

# Cold dark matter and the cosmic phase transition

**Bikash Sinha**

Variable Energy Cyclotron Centre, 1/AF Bidhannagar
Kolkata 700064, India

E-mail: bikash@vecc.gov.in; bsinha1945@gmail.com

**Abstract.** It is entirely plausible that during the primordial quark- hadron phase transition in the universe, microseconds after the Big Bang, supercooling takes place, accompanied by mini-inflation. With $\mu/T \sim 1$ ($\mu$ is chemical potential), leading to a first order phase transition from quarks to hadrons; there will be relics in the form of quark nuggets, and, that they consist of Strange Quark Matter. The possibility that these SQM nuggets may well be the candidates of cold dark matter is critically examined. A cursory comparison with the neutron star is presented at the end.

## 1. Introduction

Over the years abundant evidence has been accumulated indicating the presence of large quantities of unseen matter surrounding normal galaxies including our own [1,2,3]. The nature of this dark matter, however, remains unknown except for that it cannot be normal, such as the matter in stars, gas or dust.

The evidence of the existence of dark matter is primarily gravitational. The discrepancy between the luminous mass and the gravitational mass clearly indicates the presence of a large amount of unseen mass of the universe, usually referred to as dark matter. By now, it is also established that dark matter accounts for approximately 27% of the energy budget of the universe.

All kinds of speculations are in circulation. Axions, massive neutrinos, other kind of yet unknown exotic particles, and even weakly interacting massive particles, referred to usually as WIMPs have been proposed as candidates of dark matter, but none of which are experimentally observed so far. Alcock et al. [4] and Aubourg et al. [5] (using gravitational microlensing method suggested by Paczynski [6]) discovered the existence of this dark matter. Alcock et al. [4] suggested that the dark matter can be explained by normal matter, known collectively as Massive Compact Halo Objects (MACHOs) [4]. They went on to speculate that MACHOs might be brown dwarfs, Jupiter like objects, neutron stars, or even black holes. It has been argued by us [8,9,10,11] quite exhaustively, that a natural explanation will be that the MACHOs are the relics from the putative cosmic phase transition from quark to hadrons about a microsecond after the Big Bang. MACHOs, it is argued, can be the quark nuggets, which have survived from that primordial epoch. It is entirely plausible that these relic quark nuggets are made of strange matter, the true ground state of QCD, originally suggested by Witten [12].

This assertion has acquired further credibility from the recent experimental observations of the Bose Institute group [13,14] engaged in studying cosmic strangelets at a mountain altitude. Although heavy- ion collision is unlikely to produce strangelets, strange quark nuggets in the form of strangelets hurtling down through the cosmos will tend to pick up mass [13] from the atmospheric atoms, as they fall to the earth. The analysis of Banerjee et al. [13] indicates that a strangelet with an initial mass $\sim$ 64 amu and charge $\sim$ 2 (typically for strangelet Z/A << 1) acquires mass as it passes through the atmosphere, and goes on to, eventually, evolve to a mass $\sim$ of 340 amu or so, at the end of its cosmic journey. This is at an altitude of $\sim$ 3.6 km above sea level, typically the Himalayan mountain region in India such as Darjeeling. The collected data, just mentioned, suggest the interpretation of exotic cosmic ray events of very small Z/A, arising from the Strange Quark Matter droplets.

Recently, Harvey et al. [15] have suggested that collisions between galaxy clusters provide a test of the non-gravitational forces acting on dark matter. Using the Chandra and Hubble Space Telescopes, they observe 72 collisions, including both "major" and "minor" mergers. Combining these measurements statistically, they find the existence of dark mass of 7.6 $\sigma$ significance. The position of the dark mass has remained closely aligned, implying a self-interaction cross section $\sigma_{DM}/m < 0.47$ cm$^2$/g and disfavoring some proposed extensions to the standard model.

This observation [15] is in general agreement, it seems with the identification of dark matter with strange quark nuggets, relics of the quark-hadron phase transition. It is interesting [10] to observe that going over from a radiation dominated universe to gravity dominated universe these strange quark nuggets begin to clump around the temperature $\sim$ 1MeV of the universe. It was found by us [10] that after clumping is over, these objects feel only the influence of gravity as has been observed by Harvey et al. [15]. **Thus, SQNs are in some sense the WIMPs, massive but weakly interacting, by gravity only.**

Armed with the recent observations [13,14] of strangelets at mountain top and the very first observations of MACHO collaboration at Mount Stromlo by Alcock et al. [4] and by Aubourg et al. [5] of EROS collaboration at La Silla, Chile, it is proposed to go through a critical analysis of the origin of Strange Quark Matter (SQM) and it's survival through the cosmological time scale, and, finally identifying SQMs as the candidates for cold dark matter (CDM) which may indeed close the universe [10].

There are two central issues: one the formation of quark nuggets immediately after the quark-hadron phase transition and second, their survival from that primordial epoch to now.

Since the chemical potential of the universe at that point of time was close to zero and the temperature was around 200 MeV, the wisdom of lattice will lead the universe to cross over to the hadrons with no relic of QCD phase transition. It should be noted however that reliability of lattice calculation with bare quarks in an expanding universe is a non trivial issue.

In the next section, I analyze the circumstances precipitated essentially by some degree of supercooling accompanied by a "mini inflation." With $\mu/T \sim 1$, the entry point of the universe of quarks to the world of hadrons is shifted along the phase boundary, and a first order phase transition takes place. In the following, we analyse this scenario critically.

## 2. Cosmic Phase Transition from Quark to Hadrons

It is conventionally assumed that the baryon asymmetry $\eta = (n_B - n_{\bar{B}})/\gamma$ at that primordial epoch of phase transition is the same as that of today's universe $\eta \sim 10^{-10}$. There are however reasonably straightforward arguments [12,16,17] that $\eta$ at that epoch is much higher and indeed of the order of $\eta \sim \mathcal{O}(1)$ unity. However, it is seen that after the phase transition $\eta$ goes back to $10^{-10}$, as it is today. The consequence of such a possibility is discussed in the following.

Witten [12] and others [8,11,16,17,18] have argued that a first order phase transition is plausible with a "small" supercooling. In a recent private communication Witten [19] further asserted that if $\eta_B \approx n_B/\gamma$ remains $\sim 10^{-10}$ at the point of q-h phase transition, as it is in the current universe, then supercooling is implausible. However, he also points out [19] that if the baryon to photon ratio is not small during the QCD phase transition and becomes small because of some phenomena at later times, then supercooling is plausible in principle. In the following,

Journal of Physics: Conference Series **668** (2016) 012028 doi:10.1088/1742-6596/668/1/012028

we demonstrate that this is entirely possible.

This is the central issue, the relevance of baryon asymmetry at that primordial epoch.

One of the more compelling scenarios of baryogenesis is based on its generation from leptogenesis through topological sphaleron transitions occurring around the electroweak transition temperature. Leptogenesis occurs through out-of-equilibrium decays of heavy right-handed neutrinos which occur naturally within a seesaw mechanism, leading to Majorana masses for neutrinos, (as well as neutrino oscillation parameters) within observable ranges. Fermions with only Majorana masses and no Dirac masses ( Majorana fermions) are charge self-conjugate spin- 1/2 particles for any global U(1) charge. If this U(1) charge is associated with lepton number, then the charge self-conjugate property automatically implies that Majorana mass terms violate lepton number. Thus, it is this supposedly Majorana nature of neutrinos (even if they have a Dirac component as well) which lies at the heart of the incipient lepton number violation. The positive aspect of this mechanism of leptogenesis-induced baryogenesis is that one obtains a numerical result close to the observed baryon-photon ratio of $\mathcal{O}(10^{-10} - 10^{-9})$ without any fine tuning.

The resolution of the issue as to whether neutrinos are predominantly Majorana fermions, as happens to be the common prejudice currently, is to be decided by the currently ongoing experiments on neutrinoless double beta decay. If, contrary to extant belief, such experiments happen to yield null results, and neutrinos are confirmed to be Dirac fermions, this scenario of baryogenesis loses its prime attraction, entailing unsavoury fine tuning.

Given such a volatile situation, alternative scenarios of baryogenesis cannot be ruled out. Prominent among these is the nonthermal Affleck-Dine mechanism [20].

The Affleck-Dine mechanism [20] has the potential to produce a baryon asymmetry of $\mathcal{O}(1)$ without requiring superhigh temperatures. However, the observed baryon asymmetry of $\mathcal{O}(10^{-10})$ at CMB temperatures needs to emerge naturally from such a scenario. This is what is achieved through a "little inflation" of about 7 e-folding occurring at a lower temperature, which may be identified with the QCD first order phase transition [16]. Such an inflation naturally dilutes the baryon photon ratio to the observed range, even though the baryon potential before the first order phase transition may have been high (of $\mathcal{O}(1)$ in photon units). Comparing this "little inflation" with the more standard Guth's inflationary model [21], one finds that the patterns of entropy variation in the two cases are very different. In the standard inflationary model [21] the entropy is conserved during exponential expansion, and increases, due to reheating when bubbles collide, at the end of the transition. However in Guth's scenario, supercooling is there though very large; In the little inflation scenario for the case of quark-hadron phase transition, on the other hand, the entropy is constantly increasing during the quark-hadron phase transition.

The possibility and the criterion of a mini-inflationary epoch can be demonstrated in a simple way within the Friedman model of a spatially flat universe, which is homogeneous and isotropic along with an appropriate equation of state (EOS). Let the scale factor be R with an energy density $\epsilon$, and then the Friedman equation reads

$$\dot{R} - CR\sqrt{\epsilon} = 0 \tag{1}$$

$$\dot{\epsilon} - 3(\dot{R}/R)(\epsilon + P) = 0 \tag{2}$$

with $C = (8\pi/3)^{1/2}/M_p$ the Planck mass $M_p = 1.2 \times 10^9$ GeV. The corresponding equation of state, relating energy density $\epsilon$ and the pressure $p$ using the bag model reads for QGP

$$\epsilon_{qg} = (37\pi^2/90)T^4 + B \tag{3}$$

$$p_{qg} = (\epsilon_{qg} - 4B)/3 \tag{4}$$

$$p = p_{qg} + p_{bg}; \epsilon = \epsilon_{qg} + \epsilon_{bg}; \epsilon_{bg} = 3p_{bg} \tag{5}$$

$$p_{bg} = 14.25\pi^2 T^4/90 \tag{6}$$

The cosmic evolution will be an inflationary one if the expansion is accelerated, $\ddot{R} \geq 0$ which leads to using equations (1) and (2)

$$\ddot{R} = -C^2 R(\epsilon + p)/2 \geq 0 \tag{7}$$

and

$$3p + \epsilon < 0, \tag{8}$$

with the solution

$$\epsilon = B\coth^2[2C\sqrt{B}(t - t_c) + \text{arcth}(\sqrt{\epsilon_c/B})], \tag{9}$$

$$R = \sinh^{1/2}[2C\sqrt{B}(t - t_c) + \text{arcth}(\sqrt{\epsilon_c/B})\sinh^{-1/2}(\text{arcth}\sqrt{\epsilon_c/B})] \tag{10}$$

at $t >> t_{\text{exp}} = (2C\sqrt{B})^{-1}$; the expansion proceeds exponentially as $R \propto \exp(C\sqrt{B}t)$, with $\epsilon_c = 5.5B$. It is worthwhile to note that in the present case, $R \propto \exp(C\sqrt{B}t)$, where $C = (8\pi/3)^{1/2}/M_p$; for Guth inflation [21] $R \propto \exp(\chi t)$ and $\chi = \sqrt{(8\pi/3)G\rho_0}$, $\rho_0$ is the initial energy density of the universe.

As can be seen Guth's inflation involves the gravitational constant, whereas mini-inflation involves the Bag constant; supercooling in the standard inflation model of the universe is by 28 or more orders of magnitude, whereas here in the quark hadron phase transition its only 7 e-folds.

Thus in the cosmological context, phase transition seems to be intimately connected with supercooling.

Equations (3) to (8) are satisfied for $T < T_i$ with $T_i \cong 0.5B^{14}$.

For temperatures below $T_0 = 0.65B^{1/4}$, the pressure becomes negative leading to acceleration of the universe. This is exactly what is achieved by the mini-inflation Fig. 1.

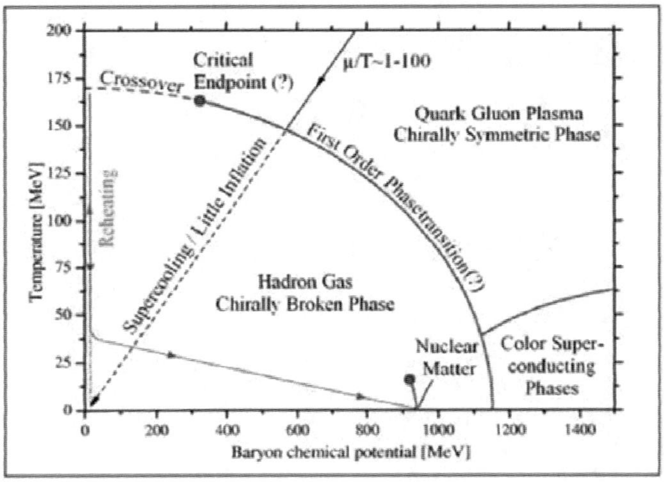


**Figure 1.** Sketch of a possible QCD phase diagram with the evolution path of the universe in the little inflation scenario, see Refs. [16], [18].

The temperature drops primarily as per the standard $t^{-1/2}$ law. Then it increases only slightly with the release of latent heat as quarks go to hadrons and the degrees of freedom are quenched. Finally $T$ again decreases as $t^{-1/2}$ again.

The scale factor $R$ evidently does not follow the standard $t^{1/2}$ law of the standard model. As shown in Fig. 2, the guiding equation for mini-inflation reads as $R \propto \exp[C\sqrt{B}t]$; here the scale factor grows exponentially. In more general terms, the scale factor $R$ gets multiplied by $\sim 10^4$, where $T$ only decreases by only $\sim$ 70 MeV and the entropy increases by $3 \times 10^9$ times (see Fig. 2). Thus in the "little inflation" scenario, the entropy is constantly increasing during the quark-hadron phase transition and the product $R(t)T(t)$ does not remain constant where $R$ is the scaling factor.

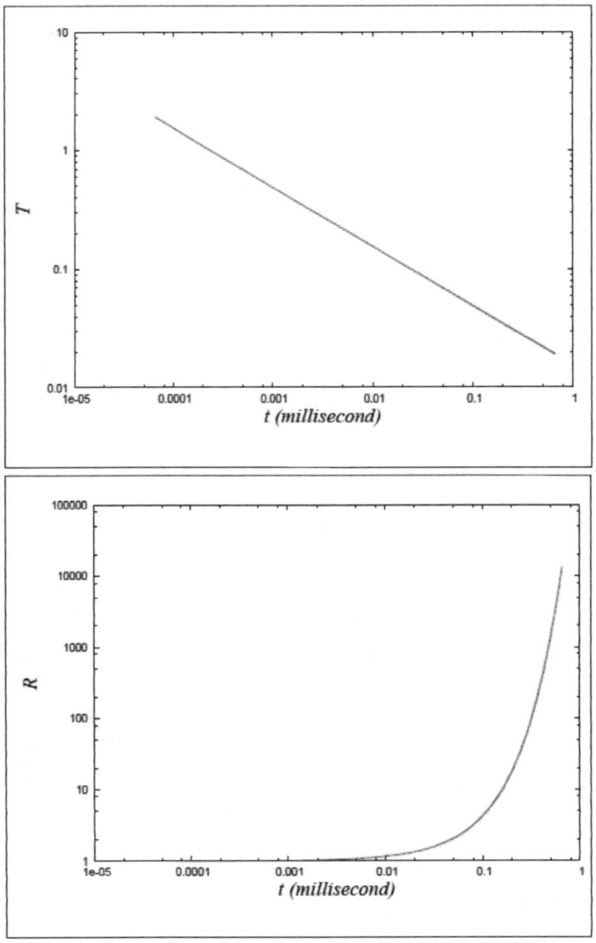


**Figure 2.** (Upper panel) The temperature as a function of time, Standard Cosmology. (Lower panel) The Scale factor as a function of time at the time of mini inflation: $R \propto \exp(C\sqrt{B}t)$, $C = \sqrt{8\pi/3}/M_p$.

The increase in entropy changes $n_q/n_\gamma$ radically with dramatic consequences. The quark number decreases as $R^{-3}$, and $n_\gamma \propto T^3$, and thus the ratio $n_q/n_\gamma \propto (RT)^{-3}$ is proportional to the inverse of the entropy. More specifically, the scale factor $R$ is multiplied by $\sim 10^4$ where $T$ only decreases from (say) 90 to 18 MeV. The product $RT$ is multiplied by more than $1.4 \times 10^3$, and thus the entropy increases by $3 \times 10^9$.

In conclusion for this part of argument, the entropy is multiplied by $10^9$ during the quark-hadron phase transition, while the baryon asymmetry is divided by the same factor. Thus we get back $n_q/n_\gamma \sim 10^{-9}$ at the end of the phase transition, as required for correct prediction of primordial nucleosynthesis, it also means, for $T > T_c$, the ratio $n_q/n_\gamma \sim 1$, and we get back our first order phase transition. This result contrasts sharply with normal adiabatic expansion in which baryon asymmetry does not change and remains as $n_q/n_\gamma \approx 10^{-9}$.

To recapitulate, the universe is assumed to begin with a large baryon chemical potential acquired through an Affleck-Dine [20] type of mechanism. It then undergoes a period of inflation, Fig. 1, crossing the QCD first order phase transition line, while remaining in a deconfined and in a chirally symmetric phase. The universe is then trapped in a false metastable QCD vacuum state.

The delayed phase transition then releases the latent heat and produces concomitantly a large entropy density, which effectively reduces the baryon asymmetry to currently observable values. It then enters a reheating phase all the way up to the usual reheating temperature with no significant change in the baryon potential, and then the universe follows the standard path to lower temperatures.

Experimental observations of Alcock et al. [4], Aubourg et al. [5] and more recently by the Bose Institute group [13,14] are the clinching proof of such a scenario.

Finally, as suggested by Witten [19] stable quark lumps (nuggets) are extremely optimistic. It would be very lucky to be true. The relics as pointed out in this paper are the acid test of that luck. The recent observations [13,14] along with the old MACHO observations [4,5] seem to survive the acid test. Thus Cold Dark Matter is the relic of the cosmic phase transition.

## 3. Survivability of Cosmological Strange Quark Nuggets (SQN)

In Ref. [11] the present author presents a detailed discussion on the nature of the phase transition and about the survivability of SQNs. Tracing the history from Alcock and Farhi (see Ref. [11]) onwards to Madsen, Heiselberg and Riisager [22], and then to the detailed work of Bhattacharya et al. [8] using Chromo Electric Flux Tube (CEFT), we come to the conclusion that QNs with baryon number $\geq 10^{39} - 10^{40}$ will indeed be cosmologically stable. It is thus very relevant to ask what fraction of the dark matter could be accounted for by the surviving QNs. To put it yet another way what we wish to address in this paper is whether the cosmological dark matter, accounting for 90% or more of all the dark matter in the universe, can be made up entirely of QNs?

As per ref. [9,10,11], the universe is closed by the baryonic dark matter trapped in QNs, and we should have

$$N_B^H(t_p) = N_B^{QN} n_{QN} V^H(t_p), \tag{11}$$

where $N_B^H(t_p)$ is the total number of baryons required to close the universe ($\Omega_B = 1$) at $t_p$, $N_B^{QN}$ is the total number of baryons contained in a single QN, and $V^H(t_p = (4\pi/3)(ct_p)^3$ is the horizon volume. With $v/c \approx 1/\sqrt{3}$

$$N_B^{QN} \leq 10^{-4.7} N_B^H(t_p) \tag{12}$$

As per standard Big Bang Nucleosynthesis (SBBN) $\eta \equiv n_B/n_\gamma(10^{-9}, 10^{-10})$ for convenience (as an estimate), the baryon number is well within the horizon limit at the QCD epoch $\approx 10^{49}$.

These usual baryons constitute only $\sim 10\%$ of the closure density ($\Omega_B \sim 0.1$ from SBBN), a total number of $10^{50}$ within the horizon at a temperature of $\sim 100$ MeV would close the universe baryonically, provided these baryons do not take part in SBBN,and a criterion fulfilled by QNs. This would require $N_B^{QN} \leq 10^{45.3}$, clearly above the survivability limit of QNs.

## 4. Epilogue

The mini-bang between two nuclei, although mimicing the big bang of the universe, is somewhat different that this phenomenon. This is discussed in detail elsewhere [11].

It is clear however, that [23] the role played by the Newtonian constant of Gravity, G, in the big bang is somewhat analogous to the role played by the vacuum energy density often referred to as the bag constant, B. The Big Bang is a display of gravity, space, and time, whereas the little bang is essentially dealing with confinement and subsequently deconfinement in extreme conditions.

On the other extreme end of the phase diagram lies a domain of very high baryon density at rather low temperature, the scenario of neutron star matter, of compressed baryonic matter (CBM) at zero temperature. It will be of some interest to compare and contrast the "perfect fluid" property of the quark matter in the microsecond universe with the "perfect fluid" of the core of the neutron star.

It is interesting to note that for the early universe, we have depleting quark matter as hadronisation progresses and the universe expands in space and time. From the canonical value of $\eta/s \leq 1/4\pi$, with hadronisation, $\eta/s$ will go on increasing as pointed at Roy et al. [24].

Eventually, the SQNs will be floating in a dilute hadronic fluid, which is not so perfect, facing more viscous drag than its quark matter counter part.

In the case of the neutron star, however, the scenario is opposite; more hadrons will be transformed to quarks, so $\eta/s$ will decrease towards the canonical value $\eta/s \leq 1/4\pi$ . For the neutron star , an approximate estimate of $\eta/s \sim T\lambda_F C_s$ will indicate that (with very low value of $\lambda_F \equiv (\rho\sigma)^{-1}$, a very high $\rho$ and an extremely low temperature $\eta/s$) for the quark core of the star with $c_p \sim 1/\sqrt{3}$ (say) may well go down below the generic value $1/4\pi$, close to zero, making the core, a perfect fluid splashing on the membrane [12] of hybrid hadronic matter and quark core. It will be of great interest to explore this in future.

Thus, 'G' ensures gravitational attraction and 'B' ensures confinement [23]. Similarly, the Hawking radiation from a black hole, with its celebrated connection to entropy [25], has an analogous scenario to that of the quark nuggets that survive the cosmic phase transition. The strange quark nuggets radiate neutrons but remain dark and cold (non-relativistic). These so-called Strange Quark Nuggets are somewhat analogous to black holes, in the sense that they tend to absorb matter as they hurtle through the cosmos [14]. An attempt is being made here to find out a modified entropy "entropy equivalence" between black hole and SQNs.

For SQNs, the deciding role is played by the Bag Pressure, B, making the SQNs dark, veiled to an exterior observer. The natural length scale for SQNs, heuristically, can be argued as $L_B = \frac{M}{M_\odot}(B^{1/4})^{-1}$, so that the entropy of SQNs is $S_{QN} = \frac{A}{4L_B^2} = \pi R^2 B^{1/2}(M_N/M_\odot)^{-2}$, $M_N$ being the mass of the quark nugget and $M_\odot$ is the solar mass, whereas, for the black hole, $S_{BH} \equiv \frac{A}{4L_p^2} = \frac{c^3 A}{4G\hbar} : A = 16\pi(GM/C^2)^2$, where M is the mass of the black hole. At this preliminary stage, the above ansatz of SQNs' entropy seem to broadly agree with other result (more will be reported later [26]).

## 5. Conclusion

Driven by the familiar standard model and introducing a "mini- inflation" at the cosmic quark-hadron phase transition, one can precipitate a first order phase transition from quarks to hadrons.

It has been argued in this paper that it does not introduce any non-standard cosmological scenario and indeed this mini-inflation comes in quite naturally, raising the value of $\eta = n_B/n_\gamma$ substantially. It is argued that the MACHOs will survive the cosmological time scale and beyond a critical baryon number window of $\sim 10^{40}$, are the candidates of cold dark matter observed some years ago [4,5] and rather more recently [13,14].

The interestingly satisfying thought lingers that we can accommodate all this in the framework of the standard model without invoking yet unobserved exotic physics. Clearly, we should look for SQM nuggets, as candidates for cold dark matter, more vigorously.

### *5.1. Acknowledgment*

First of all I like to thank Edward Witten for his insightful comments and Jürgen Schaffner-Bielich for explaining to me the purpose of mini inflation some time ago.

I like to thank my colleagues Partha Mazumdar, Debasis Mazumdar, Pijush Bhattacharya, and Sibaji Raha; Chiranjib Barman's help is gratefully acknowledged. Finally, I would like to thank Victor Matveev for inviting me to the Dubna Conference on Strange Quark Matter, and the Indian National Science Academy for my emeritus position.

## 6. References

[1] Trimble V A 1987 *Rev. Astr. Astrophys.* **25** 425
[2] Fich M, Tremaine S A 1991 *Rev. Astr. Astrophys.* **29** 409
[3] Griest K 1991 *Astrophys. J.* **366** 412
[4] Alcock C et al. 1993 *Nature* **365** 621
[5] Aubourg E et al. 1993 *Nature* **365** 623
[6] Paczynski B 1986 *Astrophys. J.* **304** 1
[7] Berrett D et al. 1993 *Ann. N. Y. Acad. Sci.* **688** 619
[8] Bhattacharya P et al. 1993 *Phys. Rev. D* **48** 4630
[9] Bhattacharye A et al. 2000 *Phys. Rev. D* **61** 083509
[10] Banerjee S et al. 2003 *Mon. Not. R. Astron. Soc.* **340** 284
[11] Sinha B 2014 *Int. JMP A* **29** 1432004
[12] Witten E 1984 *Phys. Rev. D* **30** 272
[13] Banerjee S et al. 1999 *Phys. Rev. Lett.* **85** 2000
[14] Basu B et al. 2015 *Astro. Par. Phys.* **61** 88
[15] Harvey D et al. 2015 arXiv:1503.07675v2 [astro-ph.co]
[16] Boeckel T and Schaffner- Bielich J 2010 *Phys. Rev. Lett.* **105** 041301
[17] Borghini N et al. 2000 *J. Phys. G* **26** 771
[18] Boeckel T and Schaffner- Bielich J 2012 *Phys. Rev. D* **85** 103506
[19] Witten E 2014 Private Communication
[20] Affleck I and Dine M 1985 *Nucl. Phys. B* **249** 361
[21] Guth A H 1981 *Phys. Rev. D* **23** 347
[22] Madsen J, Heiselberg H and Riisager K 1986 *Phys. Rev. D* **34** 2947
[23] Shuryak E, Private Communication
[24] Lacey R A et al. 2007 *Phys. Rev. Lett.* **98** 092301
[25] Hawking S W 1974 *Nature* **24 B** 30; Hawking S W 1976 *Phys. Rev. D* **13** 191
[26] Sinha B to be published

15th International Conference on Strangeness in Quark Matter (SQM2015) IOP Publishing
Journal of Physics: Conference Series **668** (2016) 012029 doi:10.1088/1742-6596/668/1/012029

# PHENIX Measurements of Correlations at RHIC

**Arkadiy Taranenko (for the PHENIX Collaboration)**
National Research Nuclear University "MEPhI" (Moscow Engineering Physics Institute), Kashirskoe Shosse 31, 115409 Moscow, Russia

E-mail: AVTaranenko@mephi.ru

**Abstract.** Relativistic heavy-ion collisions provide a unique opportunity to study the expansion dynamics and the transport properties of the produced strongly interacting quark gluon plasma (QGP). This article reviews the recent soft physics results obtained via correlation measurements from the PHENIX experiment at RHIC: space-time extent of the pion emission source and azimuthal anisotropy of the particle production.

## 1. Introduction

Since the discovery of the strongly interacting matter, quark gluon plasma (QGP) produced in relativistic Au+Au collisions at RHIC in 2005 [1, 2], the central goal of the experimental heavy-ion programs at both the RHIC and the LHC is the full characterization of the transport properties of the novel state of matter. The experimental observation of the large flow anisotropies at both RHIC and LHC indicates that the QGP has very low specific shear viscosity $\eta/s$ close to its quantum limit of $\eta/s = 1/4\pi$ [3, 4]. A crucial open question is whether a fundamental change occurs in the reaction dynamics and the particle production mechanism, when the collision system-size is reduced from the values produced in central and mid-central heavy-ion collisions, to those obtained in p+p, p+A, d+A, and peripheral heavy-ion collisions [4].

At top energies of the RHIC and the LHC, the produced QGP is essentially baryon free and the quark-hadron transition is thus a smooth crossover according to the results from the Lattice QCD calculations. For larger values of baryon chemical potential $\mu_B$ or lower $\sqrt{s_{NN}}$, several model calculations have indicated a first order transition [5, 6] and hence, the possible existence of a critical end point (CEP). The next crucial open question is how to locate the possible CEP point. Thus, a current strategy for experimental mapping of the phase diagram is centered on the RHIC beam energy scans, which sample reaction trajectories with the broadest possible range of $\mu_B$ and temperatures.

This article reviews the recent soft physics results obtained via correlation measurements from the PHENIX experiment at RHIC: anisotropic flow of produced charged particles and space-time extent of the pion emission source.

Track and momentum reconstruction for charged particles were performed by combining hits from the drift chambers (DC) and pad chambers in the PHENIX central spectrometers ($|\eta| < 0.35$). Charged particles were identified by combining time-of-flight from the two time-of-flight detectors, as well as the electromagnetic calorimeters [7, 8], with momentum reconstructed from the DC and pad-chamber hits in the magnetic field.

The magnitude of anisotropic flow of produced charged particles was quantified by the Fourier

coefficients: $v_n\{\Psi_n\}=\left\langle\cos n(\phi_p - \Psi_n^{\rm obs})\right\rangle/{\rm Res}\{n,\Psi_n\}$, where n is the order of the harmonics, $\phi_p$ is the azimuthal angle of the charged track reconstructed in the PHENIX central arms ($|\eta| \leq 0.35$), $\Psi_n^{\rm obs}$ is the azimuth of the estimated n-th order event planes and ${\rm Res}\{n,\Psi_n\}$ is the event plane resolution. The azimuth of the estimated n-th order event planes $\Psi_n^{\rm obs}$, determined via hits in the two BBCs, and the two inner (i), outer (o) and combined (io) rings of the Reaction-Plane Detectors (RXN). The respective $\eta$ coverage for these event-plane detector pairs are $3.1 < |\eta_{\rm BBC}| < 3.9$, $1.5 < |\eta_{\rm RXN_i}| < 2.8$ and $1.0 < |\eta_{\rm RXN_o}| < 1.5$. This references to us the Event Plane (EP) method of $v_n$ measurements [7, 8, 9]. In addition we used the the long-range two-particle correlation (2PC) method, which pairs the hadrons with deposited charges in the RXN segments [9]. The relative azimuthal angle of particle hits in separate $\eta_p$ ranges $A$ and $B$, $\Delta\phi \equiv \phi^A - \phi^B$, reflects the products of the $v_n$ via $dN/d\Delta\phi \propto 1 + \sum_{n=1} 2v_n^A v_n^B \cos(n\Delta\phi)$ [9]. We analyze the $\Delta\phi$ correlations using event mixing technique for two pair combinations; $(A,B)$=(HAD,RXN) and $(A,B)$=(RXN-N,RXN-S). These correlations then fix the event-averaged products $\left\langle v_n^{\rm HAD} v_n^{\rm RXN}\right\rangle$ and $\left\langle v_n^{\rm RXN} v_n^{\rm RXN}\right\rangle$, and allow us to obtain the $v_n$ for hadrons as $v_n^{HAD} = \left\langle v_n^{HAD} v_n^{\rm RXN}\right\rangle/\sqrt{\langle v_n^{\rm RXN} v_n^{\rm RXN}\rangle}$. Results in a wide centrality range are obtained by integrating centrality differential measurements with their multiplicity weight [9].
The interferometry technique of Hanbury Brown and Twiss (HBT) [10] was used to perform detailed differential measurements of two-pion correlation functions. The two-pion correlation function is defined as the ratio $C_2(\mathbf{q}) = A(\mathbf{q})/B(\mathbf{q})$, where $A(\mathbf{q})$ is the measured distribution of the relative momentum difference $\mathbf{q} = \mathbf{p}_2 - \mathbf{p}_1$ between particle pairs with momenta $\mathbf{p}_1$ and $\mathbf{p}_2$; $B(\mathbf{q})$ is the uncorrelated distribution, obtained from particle pairs in which each particle is selected from a different event but with similar event centralities, vertex positions, and charge sign. The relative momentum $\mathbf{q}$ is calculated in the longitudinally co-moving system, where the longitudinal pair momentum (along the beam direction) is zero. It is also decomposed into its three components, $q_{\rm out}$, $q_{\rm side}$, and $q_{\rm long}$, following the Bertsch–Pratt convention [10, 11, 13], *i.e.* the "out" axis points along the pair transverse momentum, the "side" axis is perpendicular to the out axis, and the "long" axis points along the beam. Correlation functions were studied as a function of collision centrality, as well as for different pion-pair transverse momenta $k_T = |\mathbf{p}_{T,1} + \mathbf{p}_{T,2}|/2$ or transverse mass $m_T = \sqrt{(k_T^2 + m_\pi^2)}$, where $m_\pi$ is the pion mass. The correlation functions were fitted with the following expression (in which cross-terms are assumed to be negligible) which accounts for the Bose–Einstein enhancement and the Coulomb interaction between pion pairs [11, 13]:

$$C_2(\mathbf{q}) = N[(\lambda(1 + G(\mathbf{q})))F_c + (1-\lambda)],$$
$$G(\mathbf{q}) \cong \exp(-R_{\rm side}^2 q_{\rm side}^2 - R_{\rm out}^2 q_{\rm out}^2 - R_{\rm long}^2 q_{\rm long}^2), \quad (1)$$

where $N$ is a normalization factor, $\lambda$ is the correlation strength, $F_c$ is the Coulomb correction factor [11, 13] evaluated with the Coulomb wave function, and $R_{\rm out}$, $R_{\rm side}$ and $R_{\rm long}$ are the Gaussian HBT radii which characterize the emission source. $R_{\rm long}$ is related to medium lifetime and $(R_{\rm out}^2 - R_{\rm side}^2)$ is sensitive to emission duration $\Delta\tau$ [10]. Similarly, $(R_{\rm side} - \sqrt{2}\bar{R})$ gives an estimate for the expansion radius for small values of $m_T$.
Collision centrality was determined from the charge distribution measured in the beam-beam counters. In a Monte Carlo Glauber (MC-Glauber) calculation [13] a subset of the nucleons become participants ($N_{\rm part}$) in each collision by undergoing an initial inelastic N+N interaction. The transverse distribution of these participants in the X-Y plane has RMS widths $\sigma_x$ and $\sigma_y$ along its principal axes. We define $\bar{R}$, the characteristic initial transverse size, as $1/\bar{R} = \sqrt{\left(1/\sigma_x^2 + 1/\sigma_y^2\right)}$ [12, 13]. The $\bar{R}$, $N_{\rm part}$ and participant eccentricity $\varepsilon$ were computed as a function of collision centrality. This simulation includes modeling of the BBC response.

## 2. System size and beam energy dependence of $v_2$ and HBT radii

The $N_{part}$ dependence of the elliptic flow signal $v_2$ for charged hadrons from Au+Au and Cu+Cu collisions at $\sqrt{s_{NN}}$= 62.4 - 200 GeV is presented in the left part of Fig. 1 for three bins in transverse momentum $p_T$ [8]. As seen in Fig. 1, the measured $v_2$ in Cu+Cu collisions is smaller than that of Au+Au at a comparable $N_{part}$. The lower panel of the part of Fig. 1 shows the same $v_2$ data normalized by participant eccentricity $\varepsilon$. $v_2/\varepsilon$ plots show strong centrality dependence. For ideal fluids one should observe the eccentricity scaling of $v_2$: that at a given energy $v_2/\varepsilon$ should be independent on the transverse size of the system $\bar{R}$. If equilibration is incomplete, then eccentricity scaling is broken and $v_2/\varepsilon$ also depends on the Knudsen number $K = \lambda/\bar{R}$, where $\lambda$ is the mean free path [14]. Several works based on the assumption of the acoustic nature of the anisotropic flow suggest that viscous corrections to $v_n/\varepsilon_n$ grow exponentially as $n^2$ and $1/\bar{R}$ [15, 16];

$$\ln\left(\frac{v_n(cent)}{\varepsilon_n(cent)}\right) \propto \frac{-\beta''}{\bar{R}}, \quad \beta'' = \frac{4}{3}\frac{n^2\eta}{Ts}, \tag{2}$$

or a given n, Eq. 2 indicates a characteristic linear dependence of $\ln(v_n/\varepsilon_n)$ on $1/\bar{R}$, with slope $\beta'' \propto \eta/s$. The validation of the linear dependence of $\ln(v_n/\varepsilon_n)$ on $1/\bar{R}$ for the viscous hydrodynamical calculations for $v_n$ of charged hadrons from Au+Au collisions at $\sqrt{s_{NN}}$= 200 GeV was performed in [16]. The right part of Fig. 1 shows $\ln(v_n/\varepsilon_n)$ vs. $1/\bar{R}$ for the measured $v_2$ values of charged hadrons with $p_T$ = 1-2 GeV. The dashed and dot-dashed curves are linear fits.

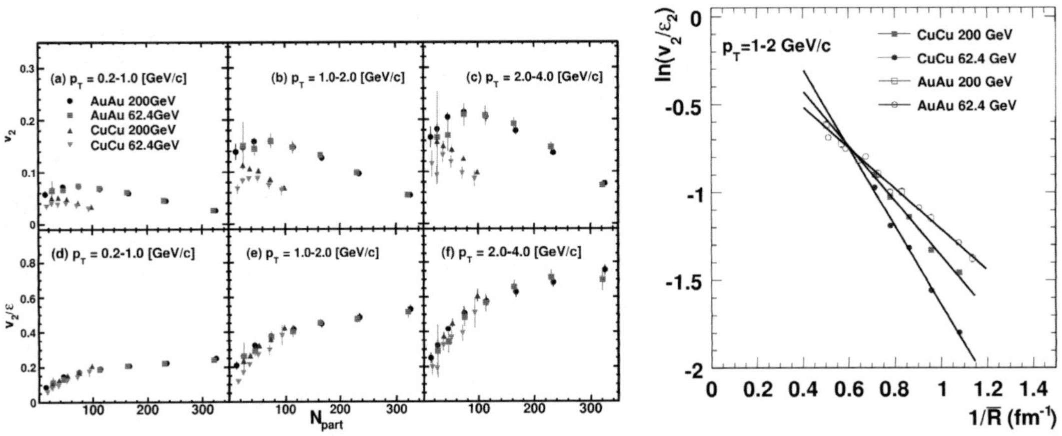


**Figure 1.** (Color online) (left) $N_{part}$ dependence of $v_2$ and $v_2/\varepsilon_2$ for charged hadrons with $p_T$ = 1-2 GeV/c emerged from Au+Au and Cu+Cu collisions at $\sqrt{s_{NN}}$= 62.4 GeV and 200 GeV. (right) $\ln(v_2/\varepsilon_2)$ vs $1/\bar{R}$ for the same set of data points. The solid lines are linear fits. The data points are from [8].

The data confirm the linear dependence of $\ln(v_n/\varepsilon_n)$ on $1/\bar{R}$. The slopes $\beta''$ are found to be the same withing errors for Au+Au collisions at $\sqrt{s_{NN}}$= 62.4 and 200 GeV. However, they are different for Au+Au and Cu+Cu at $\sqrt{s_{NN}}$= 200 GeV, which may indicate the larger viscous damping for smaller colliding systems. The slope $\beta''$ is larger for Cu+Cu at 62.4 GeV than for 200 GeV. The later observation may suggest the beam energy scan performed for several colliding systems will provide more information which will help to constrain the temperature dependence of $\eta/s$.

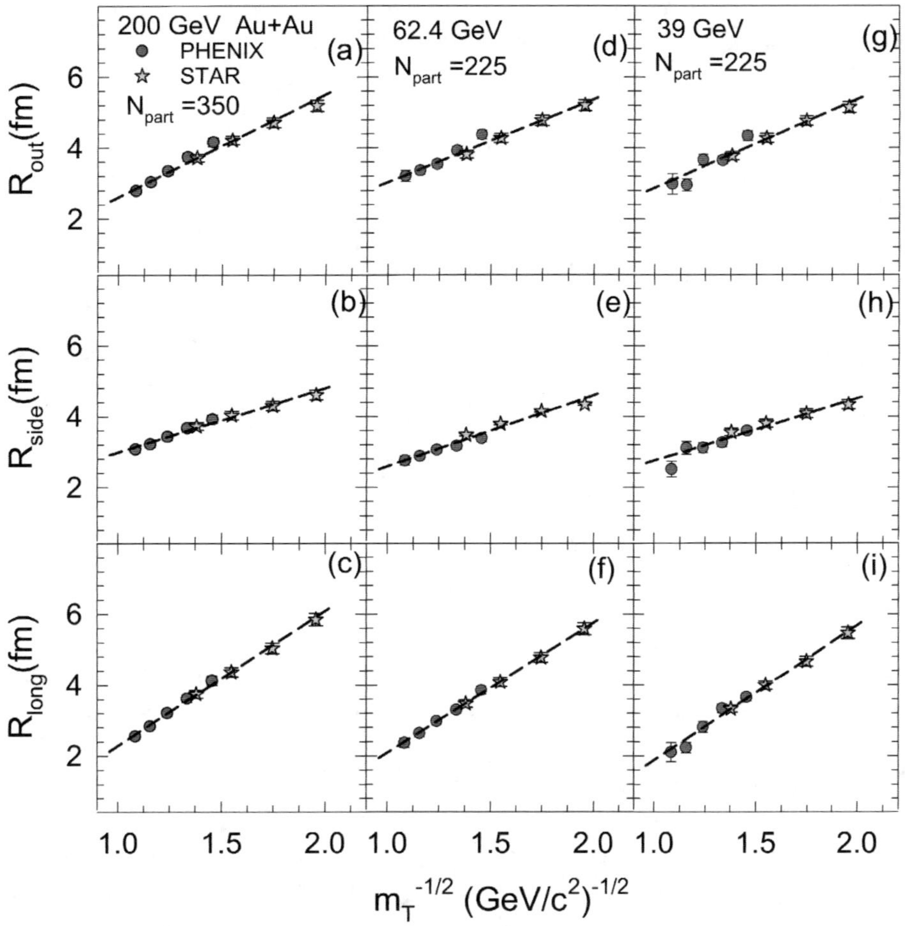


**Figure 2.** (Color online) Comparison of PHENIX and STAR HBT radii for Au+Au collisions at $\sqrt{s_{NN}}$= 39.0, 62.4 and 200 GeV as indicated. The STAR data are taken from Ref. [17]. The dashed curves are linear fits to the combined data sets.

PHENIX Collaboration has also measured HBT radii ($R_{\rm out}$, $R_{\rm side}$, and $R_{\rm long}$) for charged pion pairs in Au+Au collisions at $\sqrt{s_{NN}}$= 200, 62.4 and 39 GeV. Figure 2 shows the resulting $m_T$ dependence of the HBT radii and comparison with published results from the STAR collaboration. The results for two centrality or $N_{\rm part}$ selections indicate the characteristic $1/\sqrt{m_T}$ dependence of $R_{\rm out}$, $R_{\rm side}$, and $R_{\rm long}$ for each beam energy presented. It is consistent with hydrodynamic expansion and also indicates good agreement between the PHENIX and STAR data sets [13, 17].

Figure 3 shows the detailed centrality and $m_T$ dependence of the extracted radii for Cu+Cu and Au+Au collisions at $\sqrt{s_{NN}}$= 200 GeV. Figures 3(a), (b) and (c) show that the $R_{\rm side}$, $R_{\rm out}$ and $R_{\rm long}$ scale linearly in transverse size of the system $\bar{R}$ and that the magnitudes of the radii for each system, are comparable at similar values of $\bar{R}$ and $m_T$. The space-time extent of the emission source at freeze-out, as measured by the HBT radii, reflects the initial size of the system, the expansion time, as well as transverse expansion rate, visible via the $1/\sqrt{m_T}$ dependence of the HBT radii. Since the expansion time is expected to scale with $\bar{R}$ [16, 15], the HBT radii are also expected to scale with $\bar{R}$ in a given $m_T$ range.

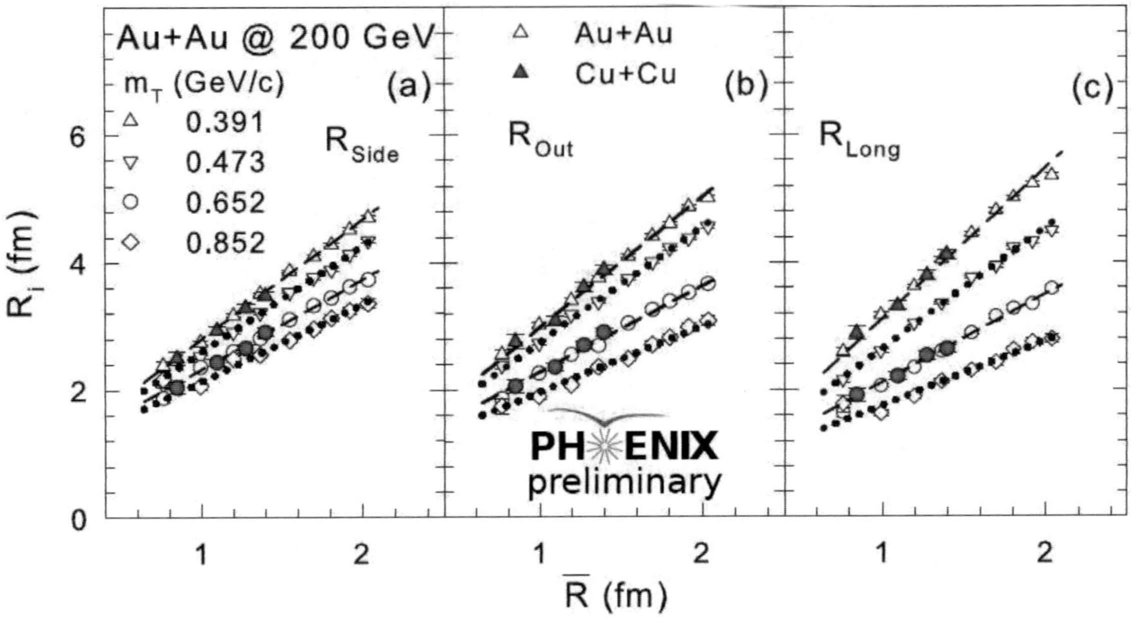


**Figure 3.** (Color online) HBT radii vs. $\bar{R}$ for several $m_T$ cuts (as indicated) for (a) $R_{\rm side}$, (b) $R_{\rm out}$ and (c) $R_{\rm long}$ for 0%–10%, 10%–20%, 20%–30% and 30%–40% Cu+Cu collisions, and 0%–5%, 5-10%, 10%–15%, 15-20%, 20%–30%, 30%–40%, 40%–50%, 50%–60% and 60%–70% Au+Au collisions at $\sqrt{s_{NN}}$=200 GeV.

The quantities $(R^2_{\rm out} - R^2_{\rm side})$ and $[(R_{\rm side} - \sqrt{2}\bar{R})/R_{\rm long}]$ are related to the emission duration and expansion velocity respectively, were investigated as a function of $\sqrt{s_{NN}}$. Figures 4(a) and (b) show these dependencies obtained at $m_T$ = 0.26 GeV/$c^2$ for the 5% most central Au+Au collisions of the combined PHENIX and STAR data sets [13, 17]. The observed non-monotonic pattern near $\sqrt{s_{NN}}$=40 GeV could be an indication of the softening of the equation of state near the deconfinement transition. Further detailed studies are required to make a more precise mapping, as well as to confirm that the observed patterns are linked to the critical end point in the QCD phase diagram [13].

## 3. Scaling properties of $v_n$ for identified hadrons

After 15 years of experiments at RHIC, the most extensive set of elliptic flow $v_2$ measurements for various hadrons with different masses, charges, quark content and hadronic cross-sections became available for the first time in the history of heavy-ion collisions. They show that, for a given centrality, elliptic flow for all observed hadrons at the top RHIC energy scale to a single curve when plotted as $v_2/n_q$ versus $KE_T/n_q$, where $n_q$ is the number of constituent quarks in a given hadron species ($n_q = 2$ for mesons, 3 for baryons) and $KE_T$ is the transverse kinetic energy for these hadrons [18, 19]. The observed $v_2$ scaling suggests that the bulk of the elliptic flow at the top RHIC energy is partonic, rather than hadronic [18, 19]. PHENIX recently measured the anisotropic flow coefficients $v_n$ for identified charged pions, kaons and protons, relative to the n-th order event planes, in Au+Au collisions at $\sqrt{s_{NN}}$= 200 GeV [9].

Figures 5(a) - (c) show a comparison of $v_2(p_T)$, $v_3(p_T)$, and $v_4(p_T)$ for $\pi^\pm$, $K^\pm$, and $p+\bar{p}$ for the EP (solid points) and 2PC (open points) methods in 0%–50% central Au+Au collisions; they indicate good agreement between the two methods of analysis. $v_3(p_T)$ and $v_4(p_T)$ results show exactly the same patterns, which have been observed previously in $v_2$ measurements for identified

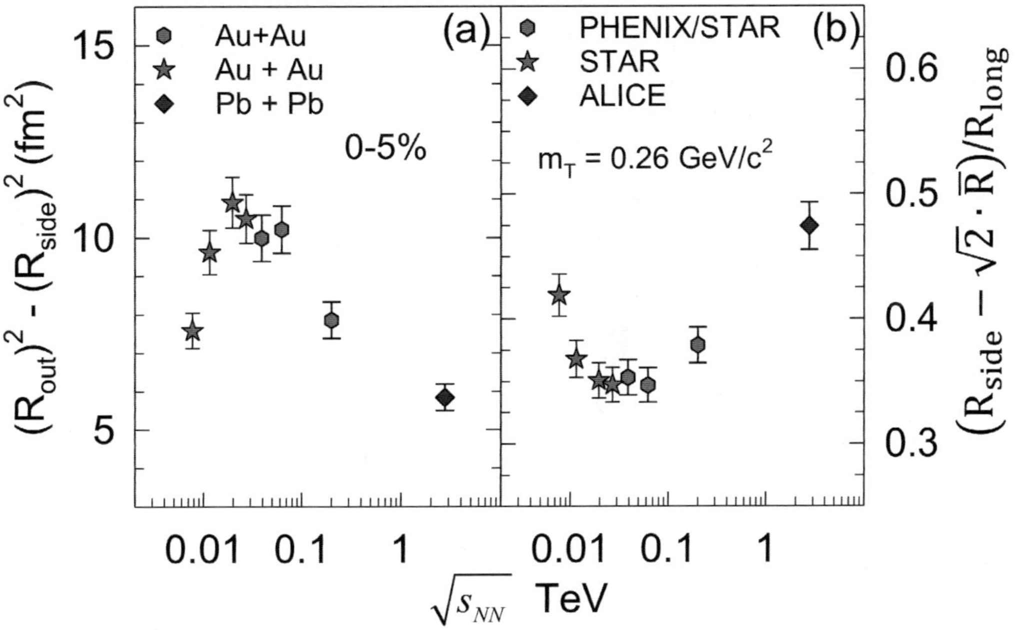


**Figure 4.** (Color online) The $\sqrt{s_{NN}}$ dependence of (a) $(R_{\rm out}^2 - R_{\rm side}^2)$, (b) $[(R_{\rm side} - \sqrt{2}\bar{R})/R_{\rm long}]$. The HBT radii are taken from the present work and Refs. [13, 17]. The PHENIX and STAR data points represent the results from fits to the $m_T$ dependence of the combined data sets.

particles in Au+Au collisions at RHIC. In the low-$p_T$ region the azimuthal anisotropy $v_n$ for all harmonics appears largest for the lightest hadron and smallest for the heaviest hadron, reflecting the mass ordering effect originating from the common velocity field, which is a generic feature of hydrodynamical models. In the intermediate-$p_T$ $(2.5 - 4$ GeV$/c)$ region the anisotropy $v_n$ is greater for the (anti-)baryons $(n_q = 3)$ than for the mesons $(n_q = 2)$ at the same p$_T$. This may suggest that all the measured n-orders of azimuthal anisotropy $v_n$ are driven by similar physics. Figure 6 shows that the present data obey a generalization of the constituent quark number scaling, observed for $v_2$: for each harmonic order $n$, the values of $v_n/(n_q)^{n/2}$ vs. $KE_T/n_q$ lie on a single curve for all the measured species within a $\pm 10 - 15\%$ range. The general constituent quark number scaling may reflect the combination of the quark number scaling for $v_2$ and the empirical observation of $v_n(p_T) \propto (v_2)^{n/2}$ [18, 19, 16] for charged hadrons, which may stem from the acoustic nature of anisotropic flow [16].

## 4. Conclusions

In summary, the recent soft physics results obtained via correlation measurements from the PHENIX experiment at RHIC were presented and discussed. At top RHIC energy $\sqrt{s_{NN}}$= 200 GeV the higher-order harmonics $v_n$ of azimuthal anisotropy show mass ordering at low $p_T$ and and baryon-meson difference at intermediate $p_T$ very similar to what has been seen already for $v_2$. The anisotropies obey a generalization of quark number scaling where $v_n/(n_q)^{n/2}$ falls on a common trend against $KE_T/n_q$ for each $n$, which can be taken as an indication of hydrodynamic expansion, a quark coalescence process, and/or an acoustic nature of the azimuthal anisotropy. The observation of the linear dependence of $\ln(v_n/\varepsilon_n)$ on $1/\bar{R}$ for charged hadrons emerged from Au+Au and Cu+Cu collisions at $\sqrt{s_{NN}}$= 62.4 GeV and 200 GeV provides another confirmation of an acoustic nature of the azimuthal anisotropy at RHIC. The extracted HBT radii for charged pions, which are compared to recent STAR and ALICE data allow an investigation of the

$\sqrt{s_{NN}}$dependence of the quantities $R^2_{\rm out} - R^2_{\rm side}$ and $R_{\rm side} - \sqrt{2}\bar{R}/R_{\rm long}$ which are sensitive to the emission duration and expansion velocity, respectively. Non-monotonic dependencies observed in these variables close to $\sqrt{s_{NN}}$= 40 GeV may be linked to trajectories that spend a significant fraction of time near the softest point in the equation of state. Further detailed studies are required from both experimental and theory side before any quantitative conclusions can be drawn from these results.

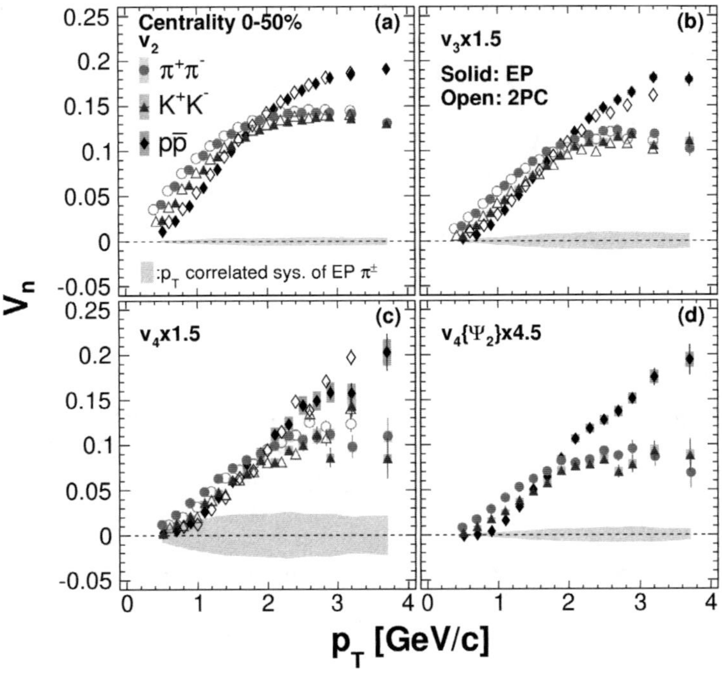


**Figure 5.** (Color online) Fourier coefficients for charge combined $\pi^{\pm}$, $K^{\pm}$, and $p+\bar{p}$ at midrapidity for 0%–50% central Au+Au collisions at $\sqrt{s_{NN}}$ = 200 GeV. The green bands indicate the $p_T$ correlated systematic uncertainties of the $\pi^{\pm}$ results from the EP method. The shaded boxes around the data points are $p_T$-uncorrelated systematic uncertainties.

## References

[1] K. Adcox *et al.* [PHENIX Collaboration], Nucl. Phys. A **757**, 184 (2005).
[2] J. Adams *et al.* [STAR Collaboration], Nucl. Phys. A **757**, 102 (2005)
[3] U. Heinz and R. Snellings, Ann. Rev. Nucl. Part. Sci. **63**, 123 (2013)
[4] Y. Akiba *et al.*, arXiv:1502.02730 [nucl-ex].
[5] M. A. Stephanov, Prog. Theor. Phys. Suppl. **153** (2004) 139 [Int. J. Mod. Phys. A **20** (2005) 4387]
[6] S. Ejiri, Phys. Rev. D **78**, 074507(2008)
[7] A. Adare *et al.* [PHENIX Collaboration], Phys. Rev. Lett. **107** (2011) 252301
[8] A. Adare *et al.* [PHENIX Collaboration], Phys. Rev. C **92** 034913 (2015)
[9] A. Adare *et al.* [PHENIX Collaboration], arXiv:1412.1038 [nucl-ex].
[10] M. A. Lisa, S. Pratt, R. Soltz and U. Wiedemann, Ann. Rev. Nucl. Part. Sci. **55** (2005) 357
[11] S. S. Adler *et al.* [PHENIX Collaboration], Phys. Rev. Lett. **93** (2004) 152302
[12] M. L. Miller, K. Reygers, S. J. Sanders and P. Steinberg, Ann. Rev. Nucl. Part. Sci. **57** (2007) 205
[13] A. Adare *et al.* [PHENIX Collaboration], arXiv:1410.2559 [nucl-ex].
[14] R. S. Bhalerao, J. P. Blaizot, N. Borghini and J. Y. Ollitrault, Phys. Lett. B **627** (2005) 49
[15] P. Staig and E. Shuryak, Phys. Rev. C **84** (2011) 034908
[16] R. A. Lacey, A. Taranenko, J. Jia, D. Reynolds, N. N. Ajitanand, J. M. Alexander, Y. Gu and A. Mwai, Phys. Rev. Lett. **112** (2014) 8, 082302
[17] L. Adamczyk *et al.* [STAR Collaboration], Phys. Rev. C **92** (2015) 1, 014904
[18] A. Adare *et al.* [PHENIX Collaboration], Phys. Rev. Lett. **98** (2007) 162301
[19] R. A. Lacey and A. Taranenko, PoS C **FRNC2006** (2006) 021

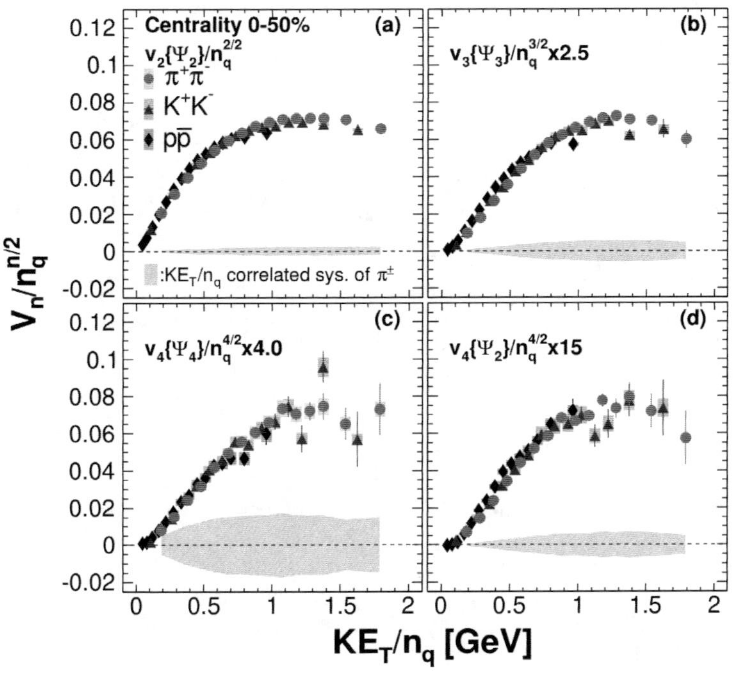


**Figure 6.** (Color online) Quark-number ($n_q$) scaling for 0%–50% central Au+Au collisions at $\sqrt{s_{NN}} = 200$ GeV, where $n_q$ is the constituent valence quark number of each hadron. Systematic uncertainties are shown as in Fig. 5.

# Flow of strange and charged particles in pPb and PbPb collisions at LHC energies

**Zhoudunming Tu, on behalf of the CMS Collaboration**
Physics & Astronomy MS-61, Rice University, Houston, US, 77005
E-mail: zt7@rice.edu

**Abstract.** Observation of a long-range near-side two-particle correlation (known as the "Ridge") in high-multiplicity pPb and pp collisions opened up new opportunities of exploring novel QCD dynamics in small collision systems. Latest CMS results in pPb and PbPb collisions will be shown: (1) The multi-particle correlation in pPb collisions will be presented for the high multiplicity events, indicating the collective behavior in small collision systems. (2) Identified $p_T$ spectra of $\pi^+/\pi^-$, $K^+/K^-$, and $p/\bar{p}$ in pPb collisions show a strong multiplicity dependence, which indicates radial flow at high multiplicity events. (3) The second-order anisotropy harmonics ($v_2$) of strange particle $K_s^0$ and $\Lambda/\bar{\Lambda}$ are extracted from long-range correlations as a function of particle multiplicity and $p_T$. The mass ordering effect of $v_n$ at low $p_T$ as predicted by hydrodynamics also points to the strong collective nature of expanding medium in small collision systems. Finally, the possible constituent quark number of scaling of $v_2$ between mesons and baryons may indicate the deconfinement in small systems.

## 1. Introduction

The study of strange and charged particle production and anisotropy harmonics ($v_n$) in high-energy collisions provide deep insights into the nature of the produced systems as well as become an important tool to investigate the properties of soft QCD matter. Earlier studies of the second-order anisotropy harmonics (known as "$v_2$") in AA collisions indicate that the produced systems would undergo a hydrodynamic expansion, and the medium that was created behaves like a fluid [1, 2].

However, in recent years, CMS has observed the long-range near side two-particle correlation, known as "ridge" structure, in high multiplicity pPb [3] and pp [4]collisions, which is not expected in small systems. The origin of such structure in smaller collision system is still under intense debate. Even though it is natural to think that the collective flow in hydrodynamics can explain the phenomenon [5–8], other models propose to attribute to the initial gluon correlations [9–11].

From the previous studies of two-particle correlation, one may wonder whether the observed behavior is due to correlation from particle pair or that can be also true for multi-particle, which latter case could indicate the collective nature of the systems. Using the multi-particle cumulant and Lee-Yang zeros (LYZ) method, the $v_2$ from 4, 6, 8 and all particles can be extracted. One of the advantage of using multi-particle correlation is to highly suppress the non-flow contribution to the $v_2$ signal, such as back-to-back jet correlations. In small collision systems, such as pPb, the multi-particle correlation could be an essential study to see whether there is collectivity in small systems. In this paper, the presented results come from the previous published CMS results, which could be found in Ref. [12].

In order to further understand the underlying mechanism of elliptic flow or triangular flow in pA, directly measuring the radial flow in small system could be another important study to theoretical calculations. The identified particle transverse momentum spectra in different multiplicity classes can clearly show the modification of the spectra at different multiplicities. The mass ordering of the radial flow can be directly compared by extracting the slope of the spectra, where the particles, $\pi^+$ (includes $\pi^-$) , $K^+$ (includes $K^-$) and $p$ (includes $\bar{p}$), are the most abundant species in high-energy collisions. With their mass differences, the modification of the spectra can be easily seen in pPb collisions, if there is radial flow in such small systems. In this paper, the results that are presented are taken from previous CMS study, which could be found in Ref. [13].

The studies of strange particles, the most abundant produced heavy quark generations in heavy-ion collisions, can provide information of formation of strongly interacting medium, where earlier studies have shown the strangeness enhancement, with respect to pp reference, as an indication [14]. The strange hadrons, $K_S^0$ and $\Lambda$ (includes $\bar{\Lambda}$), can be useful candidates to calculate the $v_n$ in small systems. In hydrodynamic models, the collective flow would result in a final-state particles distribution with an azimuthal anisotropy and its magnitude would depend on the mass of the particles. The heavier particles, due to the radial boost from the medium, would be more likely to show up at a higher momentum. This mass ordering effect is well studied in AA collisions as one of the evidences of hydrodynamic flow [1,2]. In this paper, the $v_n$ study of $K_S^0$ and $\Lambda$ in pPb and PbPb collisions is taken from previous CMS published result, which can be found in Ref. [15].

## 2. Reconstruction of $\pi^+$ , $K^+$, $p$, $K_S^0$ and $\Lambda$ candidates

The particles $\pi^+$ , $K^+$ and $p$ reconstruction are based on the special relationship between energy loss rate and total momentum which is shown in Fig. 1. The details of reconstruction could be found in Ref. [13]. The reconstruction of $K_S^0$ and $\Lambda$ candidates can be found in Ref. [15], where the invariant mass distribution can be clearly reconstructed as shown in Fig. 2. The $K_S^0$ and $\Lambda$ candidates (generally referred to as $V^0$s) are reconstructed first, by combining oppositely charged tracks with at least 4 hits and both transverse and longitudinal impact parameter significances greater than 1 (with respect to the primary vertex) to form a secondary vertex. The distance of closest approach of the pair of tracks is required to be less than 0.5cm. The fitted vertex in $x$, $y$, $z$ of each pair of tracks is required to have a $\chi^2$ value normalized by the number of degrees of freedom less than 7. The pair of tracks is assumed to be $\pi^+$ $\pi^-$ in $K_S^0$ reconstruction, while the assumption of $\pi^-$ $p$ is used in $\Lambda$ reconstruction. For $\Lambda$ ($\bar{\Lambda}$), the lower-momentum track is assumed to be the $\pi^-$ ($\pi^+$).

Due to the assumptions of $\pi$ ($p$) tracks when reconstructing the $K_S^0$ ($\Lambda$) invariant mass, there are non-negligible probabilities of mis-identifying $K_S^0$ or $\Lambda$ candidates. Therefore, using the daughter tracks of $K_S^0$ ($\Lambda$), and assuming it has $p$ ($\pi$) mass, to see whether it could be reconstructed within the mass window of $\Lambda$ ($K_S^0$). Finally, those candidates would be removed.

In addition to the mis-identifications, the $\pi^+$, $p$ and $\Lambda$ candidates could come from weak decay instead of primary collisions(known as "feed-down" effect), where $\pi^+$ could be decayed from $K_S^0$, $p$ could come from $\Lambda$ and $\Lambda$ could come from $\Xi^+$. The contributions from weak decay could be up to 10–15% depending on the transverse momentum. However, only the ratios of $\Lambda/p$, $\Lambda/\bar{p}$, $K_S^0/\pi^+$, $K_S^0/\pi^-$ and $\Xi^+/\Lambda$ matter in the study, since the feed-down is given as a fraction of primary particles. If these ratios are similar between the Monte Carlo simulation and measurement, this feed-down effect would be corrected when applying efficiency corrections. The probability of having

$K^+$ or $K_S^0$ come from weak decay is negligible. The detail descriptions and studies could be found in Ref. [13].

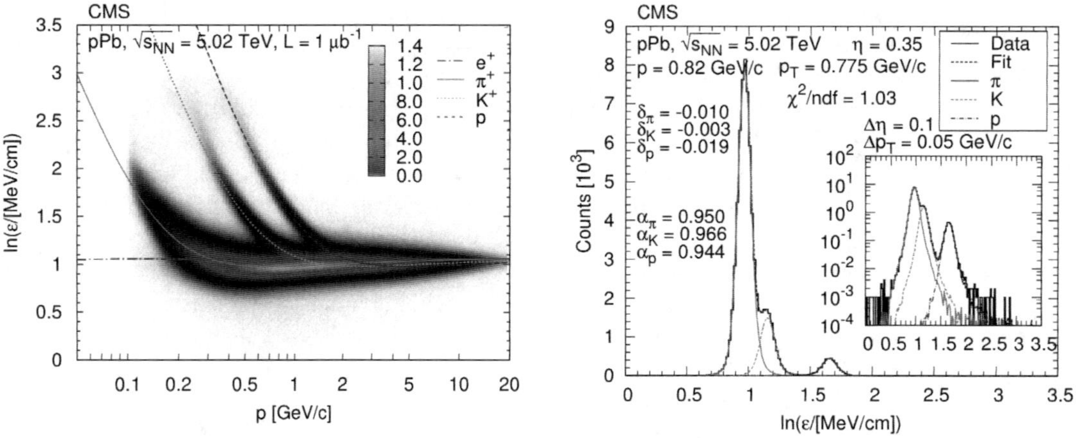


**Figure 1.** The distribution of $ln\epsilon$ as a function of total momentum for different particles are presented, where $\epsilon$ is the most probable energy loss rate at a reference path-length $l_0 = 450\mu m$.

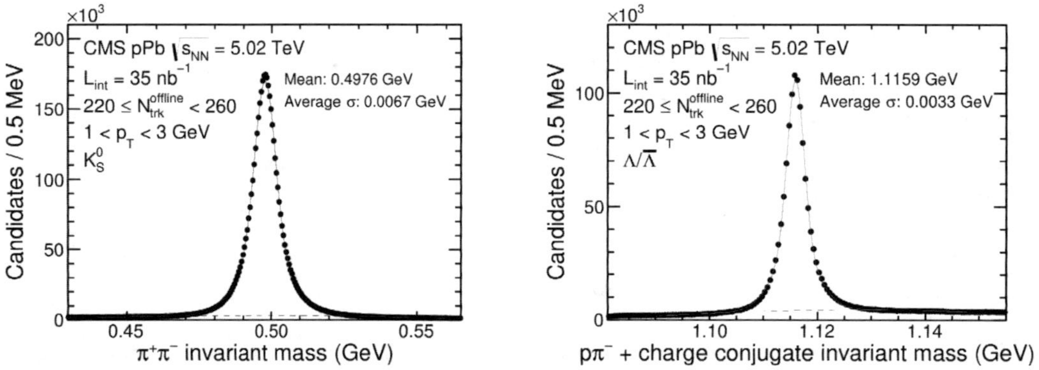


**Figure 2.** The invariant mass distribution of $K_S^0$ and $\Lambda/\bar{\Lambda}$ are presented at multiplicity range of $220 \leq N_{\rm trk}^{\rm offline} < 260$ and transverse momentum 1-3 GeV in pPb collisions. The signal peak is fitted by a double Gaussian function (with a common mean) and the background is described by a 4th-order polynomials.

## 3. Results

The second-order anisotropy Fourier harmonics, $v_2$, as a function of multiplicity ($N_{\rm trk}^{\rm offline}$) is shown in Fig. 3 based on six- and eight-particle cumulants for 2.76 TeV PbPb (left) and 5.02 TeV pPb (right) collisions, averaged over transverse momentum 0.3-3 GeV. The data of $v2\{2\}$ and $v2\{4\}$ are taken from the previous measurement from CMS [16], with $\eta$ gap greater than 2 for $v2\{2\}$ to suppress the non-flow effects. The $v_2$ values from LYZ method that involves all particles correlations are also shown. For each multiplicity range, all the $v_2$ values from different number of particles (except $v2\{2\}$) agree with each other within 10% for pPb collisions and better agreement as within 2% for PbPb collisions. This result is strongly indicating the collective nature of the produced medium in high multiplicity pPb collisions, where all particles are involved in the correlations.This behavior is not consistent with jet correlation. In the context of hydrodynamics, ratios of $v_2$ values from different orders could follow the universal behavior from fluctuation-driven initial-state conditions [17,18]. The $v_2$ values from 2 particles are consistently higher than other $v_2$ values from multi-particles, which is under expected as event-by-event participant geometry fluctuations of the $v_2$ values could be affected differently by two- and multi-particle cumulants [19,20].

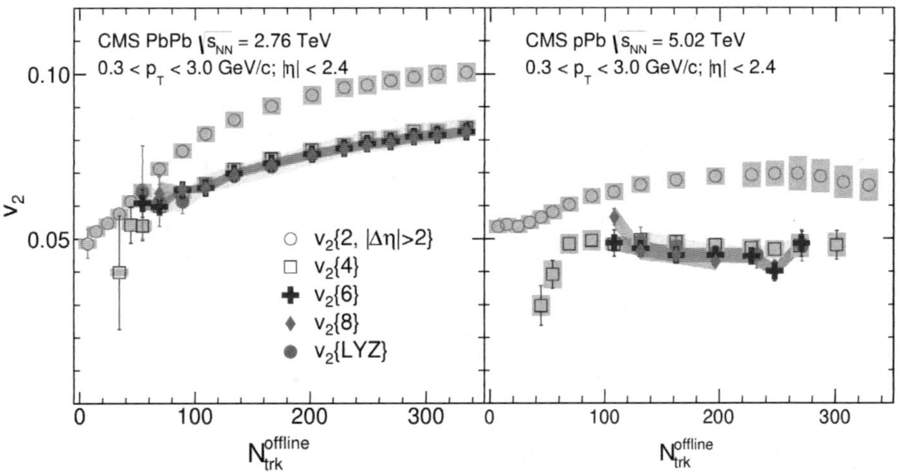


**Figure 3.** The second-order anisotropy harmonics $v_2$ from different number of particle cumulant are presented in pPb and PbPb collisions. The transverse momentum range is chosen to be 0.3-3 GeV/c and the $v2\{2\}$ and $v2\{4\}$ are taken from previous CMS published results.

The transverse momentum spectra of $\pi^+$ , $K^+$ and $p$ can be well fitted from a function such as $m_T e^{-m_T/T'}$, motivated by the success of Boltzmann-type distributions in nucleus-nucleus collisions [21], which is shown in Fig. 4. The $m_T = \sqrt{p_T^2 + m^2}$ and $T'$ is the inverse slope parameter of the fit function. As particle multiplicity increases, the spectra becomes "harder" as the slope of the spectra becomes more flatten so that $T'$ increases with multiplicity for all particles. The relation between $T'$ and particle mass can be interpreted by using radial flow phenomenon, supported by hydrodynamics. As the system goes to higher multiplicity, the $T'$ becomes more linear with the particle mass. This can be explained if the system has radial flow effect at high multiplicity, the system would provide a common radial flow velocity and kinetic freeze-out temperature to all particles. Therefore, the inverse slope parameter $T'$ would be linear proportional to the particle mass if $T' \propto T_{kin} + m\langle\beta_{\mathrm{T}}\rangle$. A similar trend has been observed in PbPb collisions, interpreted as radial flow velocity boost as shown in Ref. [22]. For the lower multiplicity range, this relation is clearly broken.

The second-order anisotropy harmonics $v_2$ is shown as a function of $p_T$ for $K_S^0$ (filled squares), $\Lambda$ (filled circles) and previously published result of inclusive charge particles(open crosses) in Fig. 5. The four different panels are corresponding to the four high multiplicity ranges in pPb collisions. In the range of $p_T < 2$ GeV, the $v_2$ of the heavier particle $\Lambda$, is lower than that of $K_S^0$ particles, while

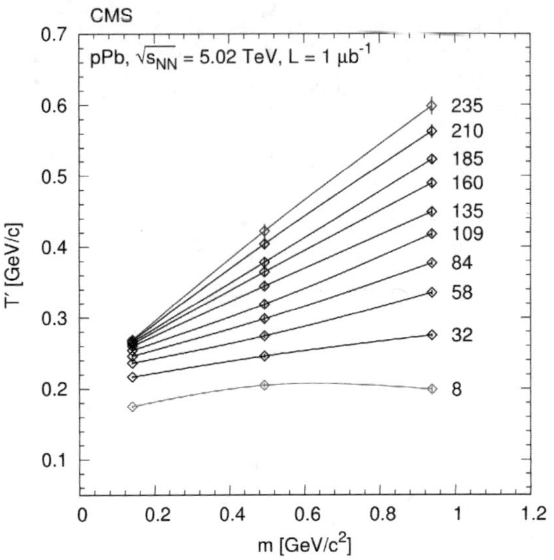


**Figure 4.** The inverse slope parameter $T'$ is shown as a function of particle mass at different multiplicities in pPb collisions, where $T'$ is extracted from fit function $m_T e^{-m_T/T'}$ of the $m_T$ spectra of $\pi^+$ , $K^+$ and $p$. The lines that connect the data point is used as guiding lines to see the linear relation between $T'$ and particle mass.

both $K_S^0$ and $\Lambda$ are below the inclusive charged hadron. In hydrodynamics, this mass ordering effect can be an indication of collective radial flow, which pushes the heavier particles to a higher momentum if the system has a common radial flow velocity. Interestingly, if we compare Fig. 6, which is also $v_2$ as function of $p_T$ of $K_S^0$ and $\Lambda$ in PbPb collisions, the mass ordering effect at low $p_T$ is smaller in PbPb than that in pPb at the same multiplicities. This might indicate the strength of the flow effect in small systems is larger than that of large systems has been predicted by hydrodynamic models such as Ref. [23].

The scaling behavior of $v_2$ value divided by number of quarks, as a function of kinetic energy of each particle divided by the number of quarks in their bound states, is also investigated in high multiplicity pPb events. As shown in the middle row of Fig. 5, the $K_S^0$ and $\Lambda$ are in a good agreement except $KE_T/n_q < 0.2$ GeV, which has a deviation of 20%. This could be understood as the impact of radial flow at very low $p_T$ [15]. In PbPb collisions, this approximate scaling behavior can be related to the quark recombination models, which states that the collective flow is developed among constituent quarks before hadronizations [24–26].

The third-order anisotropy harmonics $v_3$, known as the triangular flow, is also shown in Fig. 7 as a function of $p_T$ for $K_S^0$ (filled squares), $\Lambda$ (filled circles), and previously published result of inclusive charge particles(open crosses) [15]. Similar scaling trend can be seen in $v_3$ results.

## 4. Summary

From the latest CMS results, the multi-particle correlation of charged hadron in high multiplicity pPb collisions indicates the collective nature of the produced medium in small systems. The identified particle spectra, $\pi^+/\pi^-$, $K^+/K^-$ and $p/\bar{p}$, show a strong dependence of particle multiplicity in pPb collisions, indicating radial flow phenomenon at high multiplicity. The strange

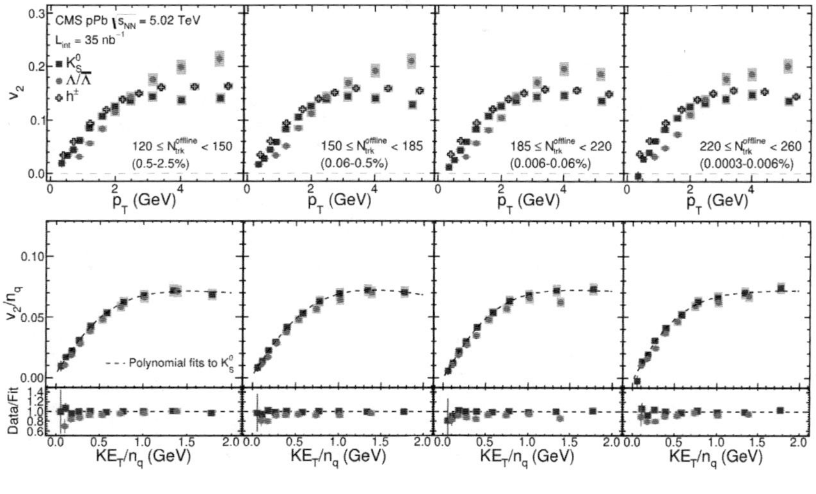


**Figure 5.** Top row: The $v_2$ as a function of $p_T$ for four different high multiplicity ranges for $K_S^0$ (filled squares), $\Lambda$ (filled circles) and inclusive charge particles(open crosses) are shown in pPb collisions. Middle row: $v_2/n_q$ is plotted as a function $KE_T/n_q$ for the same multiplicity ranges as the top row, along with a polynomial fit function of $K_S^0$. Bottom row: the fit to data ratio of $K_S^0$ and $\Lambda$ are presented as function of $KE_T/n_q$. The error bar corresponds to statistical uncertainties and the boxes denote systematic uncertainties.

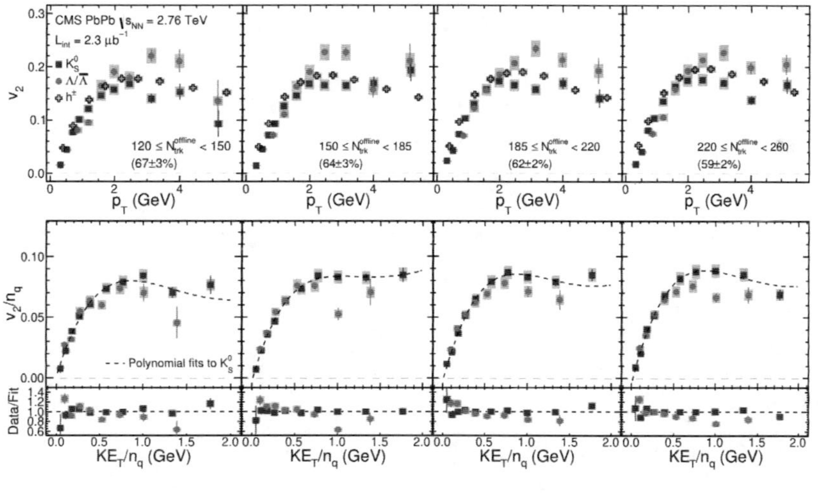


**Figure 6.** Top row: The $v_2$ as a function of $p_T$ for four different high multiplicity ranges for $K_S^0$ (filled squares), $\Lambda$ (filled circles) and inclusive charge particles(open crosses) are shown in PbPb collisions. Middle row: $v_2/n_q$ is plotted as a function $KE_T/n_q$ for the same multiplicity ranges as the top row, along with a polynomial fit function of $K_S^0$. Bottom row: the fit to data ratio of $K_S^0$ and $\Lambda$ are presented as function of $KE_T/n_q$. The error bar corresponds to statistical uncertainties and the boxes denote systematic uncertainties.

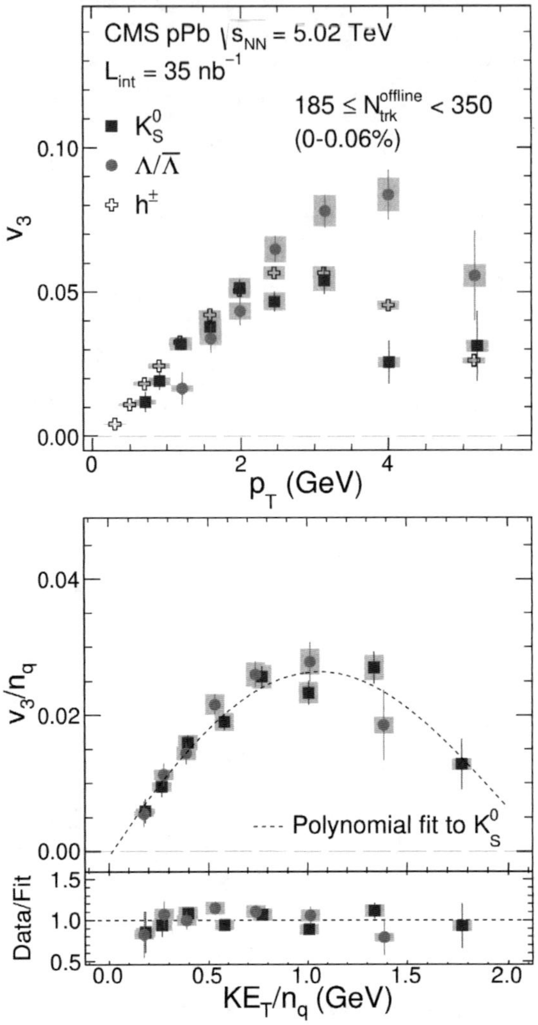


**Figure 7.** Top row: The $v_3$ as a function of $p_T$ for four different high multiplicity ranges for $K_S^0$ (filled squares), $\Lambda$ (filled circles) and inclusive charge particles(open crosses) are shown in pPb collisions. Middle row: $v_3/n_q$ is plotted as a function $KE_T/n_q$ for the same multiplicity ranges as the top row, along with a polynomial fit function of $K_S^0$. Bottom row: the fit to data ratio of $K_S^0$ and $\Lambda$ are presented as function of $KE_T/n_q$. The error bar corresponds to statistical uncertainties and the boxes denote systematic uncertainties.

hadron $K_S^0$ and $\Lambda/\bar{\Lambda}$ have shown comparable values of $v_2$ and $v_3$ in pPb and PbPb collisions, while in the low $p_T$ regime, the mass ordering effect is larger in pPb than that in PbPb collisions. In addition, the number of consitutent quark scaling have been observed in pPb and PbPb collisions.

## References

[1] J. Adams *et al.*, "Experimental and theoretical challenges in the search for the quark gluon plasma: The STAR Collaboration's critical assessment of the evidence from RHIC collisions," *Nucl. Phys. A*, vol. 757, p. 102, 2005.
[2] K. Adcox *et al.*, "Formation of dense partonic matter in relativistic nucleus-nucleus collisions at RHIC: Experimental evaluation by the PHENIX collaboration," *Nucl. Phys. A*, vol. 757, p. 184, 2005.
[3] K. Chatrchyan *et al.*, "Observation of long-range, near-side angular correlations in pPb collisions at the LHC," *Phys.Lett.*, vol. B718, p. 795, 2013.
[4] V. Khachatryan *et al.*, "Observation of Long-Range Near-Side Angular Correlations in Proton-Proton Collisions at the LHC," *JHEP*, vol. 1009, p. 091, 2010.
[5] P. Bozek, "Collective flow in p-Pb and d-Pd collisions at TeV energies," *Phys. Rev. C*, vol. 85, p. 014911, 2012.
[6] P. Bożek and W. Broniowski, "Correlations from hydrodynamic flow in pPb collisions," *Phys. Lett. B*, vol. 718, p. 1557, 2013.
[7] A. Bzdak, B. Schenke, P. Tribedy, and R. Venugopalan, "Initial state geometry and the role of hydrodynamics in proton-proton, proton-nucleus and deuteron-nucleus collisions," *Phys. Rev. C*, vol. 87, no. 6, p. 064906, 2013.
[8] K. Werner, I. Karpenko, and T. Pierog, "The 'Ridge' in Proton-Proton Scattering at 7 TeV," *Phys. Rev. Lett.*, vol. 106, p. 122004, 2011.
[9] A. Dumitru, L. McLerran, and V. Skokov, "Azimuthal asymmetries and the emergence of collectivity from multi-particle correlations in high-energy pA collisions," *Phys. Lett. B*, vol. 743, p. 134, 2015.
[10] M. Gyulassy, P. Levai, I. Vitev, and T. Biro, "Non-Abelian Bremsstrahlung and Azimuthal Asymmetries in High Energy $p + A$ Reactions," *Phys. Rev. D*, vol. 90, p. 054025, 2014.
[11] W. Li, "Observation of a 'Ridge' correlation structure in high multiplicity proton-proton collisions: A brief review," *Mod.Phys.Lett.*, vol. A27, p. 1230018, 2012.
[12] V. Khachatryan *et al.*, "Evidence for Collective Multiparticle Correlations in p-Pb Collisions," *Phys. Rev. Lett.*, vol. 115, no. 1, p. 012301, 2015.
[13] S. Chatrchyan *et al.*, "Study of the production of charged pions, kaons, and protons in pPb collisions at $\sqrt{s_{NN}} =$ 5.02 TeV," *Eur. Phys. J. C*, vol. 74, p. 2847, 2014.
[14] E. Andersen, F. Antinori, N. Armenise, H. Bakke, J. Ban, *et al.*, "Enhancement of central Lambda, Xi and Omega yields in Pb - Pb collisions at 158 A-GeV/c," *Phys. Lett. B*, vol. 433, p. 209, 1998.
[15] V. Khachatryan *et al.*, "Long-range two-particle correlations of strange hadrons with charged particles in pPb and PbPb collisions at LHC energies," *Phys. Lett. B*, vol. 742, p. 200, 2015.
[16] S. Chatrchyan *et al.*, "Multiplicity and transverse momentum dependence of two- and four-particle correlations in pPb and PbPb collisions," *Phys. Lett. B*, vol. 724, p. 213, 2013.
[17] L. Yan and J.-Y. Ollitrault, "Universal fluctuation-driven eccentricities in proton-proton, proton-nucleus and nucleus-nucleus collisions," *Phys. Rev. Lett.*, vol. 112, p. 082301, 2014.
[18] A. Bzdak, P. Bozek, and L. McLerran, "Fluctuation induced equality of multi-particle eccentricities for four or more particles," *Nucl. Phys.*, vol. A927, pp. 15–23, 2014.
[19] J.-Y. Ollitrault, A. M. Poskanzer, and S. A. Voloshin, "Effect of flow fluctuations and nonflow on elliptic flow methods," *Phys. Rev.*, vol. C80, p. 014904, 2009.
[20] J.-Y. Ollitrault, A. M. Poskanzer, and S. A. Voloshin, "Effect of nonflow and flow fluctuations on elliptic flow methods," *Nucl. Phys.*, vol. A830, pp. 279C–282C, 2009.
[21] P. Braun-Munzinger, D. Magestro, K. Redlich, and J. Stachel, "Hadron production in Au - Au collisions at RHIC," *Phys. Lett.*, vol. B518, pp. 41–46, 2001.
[22] E. Schnedermann, J. Sollfrank, and U. W. Heinz, "Thermal phenomenology of hadrons from 200-A/GeV S+S collisions," *Phys. Rev.*, vol. C48, pp. 2462–2475, 1993.
[23] E. Shuryak and I. Zahed, "High-multiplicity pp and pA collisions: Hydrodynamics at its edge," *Phys. Rev. C*, vol. 88, p. 044915, 2013.
[24] D. Molnar and S. A. Voloshin, "Elliptic flow at large transverse momenta from quark coalescence," *Phys. Rev. Lett.*, vol. 91, p. 092301, 2003.
[25] V. Greco, C. M. Ko, and P. Levai, "Parton coalescence and anti-proton / pion anomaly at RHIC," *Phys. Rev. Lett.*, vol. 90, p. 202302, 2003.
[26] R. J. Fries, B. Muller, C. Nonaka, and S. A. Bass, "Hadronization in heavy ion collisions: Recombination and fragmentation of partons," *Phys. Rev. Lett.*, vol. 90, p. 202303, 2003.

# Hyperons in Neutron Stars

**Isaac Vidaña**

Centro de Física da Universidade de Coimbra (CFisUC), Department of Physics, University of Coimbra, PT-3004-516 Coimbra, Portugal

E-mail: ividana@fis.uc.pt

**Abstract.** In this work I briefly review some of the effects of hyperons on the properties of neutron and proto-neutron stars. In particular, I revise the problem of the strong softening of the EoS, and the consequent reduction of the maximum mass, induced by the presence of hyperons, a puzzle which has become more intringuing and difficult to solve because of the recent measurements of the unusually high masses of the millisecond pulsars PSR J1903+0327 ($1.667 \pm 0.021 M_\odot$), PSR J1614-2230 ($1.97 \pm 0.04 M_\odot$), and PSR J0348+0432 ($2.01 \pm 0.04 M_\odot$). Some of the solutions proposed to tackle this problem are discussed. Finally, I re-examine also the role of hyperons on the cooling properties of newly born neutron stars and on the so-called r-mode instability.

## 1. Introduction

Neutron stars are the remmants of the gravitational collapse of massive stars during a Type-II, Ib or Ic supernova event. Their masses and radii are typically of the order of $1 - 2M_\odot$ ($M_\odot \simeq 2 \times 10^{33}$g being the mass of the Sun) and $10 - 12$ km, respectively. With central densities in the range of $4 - 8$ times the normal nuclear matter saturation density, $\epsilon_0 \sim 2.7 \times 10^{14}$ g/cm$^3$ ($\rho_0 \sim 0.16$ fm$^{-3}$), neutron stars are most likely among the densest objects in the Universe [1]. These objects are an excellent observatory to test our present understanding of the theory of strong interacting matter at extreme conditions, and they offer an interesting interplay between nuclear processes and astrophysical observables.

Conditions of matter inside neutron stars are very different from those one can find in the Earth; therefore, a good knowledge of the Equation of State (EoS) of dense matter is required to understand the properties of neutron stars. Nowadays, it is still an open question which is the true nature of neutron stars. Traditionally the core of neutron stars has been modeled as a uniform fluid of neutron-rich nuclear matter in equilibrium with respect to the weak interaction ($\beta$-stable matter). Nevertheless, due to the large value of the density, new hadronic degrees of freedom are expected to appear in addition to nucleons. Hyperons, baryons with a strangeness content, are an example of these new degrees of freedom. Contrary to terrestial conditions, where hyperons are unstable and decay into nucleons through the weak interaction, the equilibrium conditions in neutron stars can make the inverse process happen. Hyperons may appear in the inner core of neutron stars at densities of about $2 - 3\rho_0$. Their presence on the neutron star interior leads to a softening of the EoS and cosequently to a reduction of the maximum mass.

Other neutron star properties, such as their thermal and structural evolution, can also be very sensitive to the composition, and therefore to the hyperonic content of neutron star interiors. In particular, the cooling of neutron stars may be affected by the presence of hyperons, since they

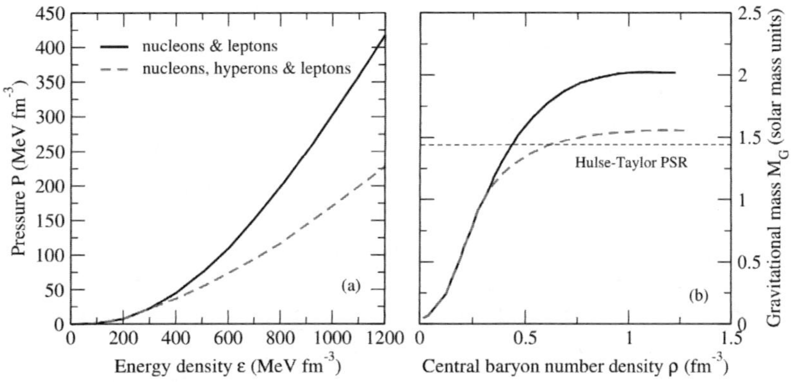


**Figure 1.** (Color online) Illustration of the effect of the presence of hyperons on the EoS (panel (a)) and mass of a neutron star (panel (b)). A generic model with (black solid line) and without (red dashed line) hyperons has been considered. The horizontal line shows the observational mass of the Hulse–Taylor [14] pulsar.

can modify neutrino emissivities and can allow for fast cooling mechanisms. Furthermore, the emission of gravitational waves in hot and rapidly rotating neutron stars due to the so-called r-mode instability can also be afffected by the presence of hyperons in neutron stars, because the bulk viscosity of neutron star matter is dominated by the contribution of hyperons as soon as they appear in the neutron star interior.

In the following I review briefly the so-called hyperon puzzle and other implications of hyperons on the properties of neutron stars.

## 2. The hyperon puzzle

The presence of hyperons in neutron stars was considered for the first time in the pioneering work of Ambartsumyan and Saakyan in 1960 [2]. Since then, their effects on the properties of these objects have been studied by many authors using either phenomenological [3, 4] or microscopic [5, 6, 7, 8, 9] approaches for the neutron star matter EoS with hyperons. Phenomenological approaches, either relativistic or non-relativistic, are based on effective density-dependent interactions which typically contain a certain number of parameters adjusted to reproduce nuclear and hypernuclear observables, and neutron star properties. Relativistic mean field (RMF) models [3] and Skyrme-type interactions [4] are among the most commonly used ones within this type of approach. Microscopic approaches, on the other hand, are based on realistic two-body baryon-baryon interactions that describe the scattering data in free space. These realistic interactions have been mainly constructed within the framework of a meson-exchange theory [10, 11]. Recently a new approach based on chiral perturbation theory has emerged as a powerful tool [12]. In order to obtain the EoS one has to solve then the many-body problem [13]. However, the repulsive core of the baryon-baryon interaction, which dominates its short-range behavior, is very difficult to treat. Although different microscopic many-body methods have been extensively used to the study of nuclear matter, up to my knowledge, only the Brueckner–Hartree–Fock (BHF) approximation [5] of the Brueckner–Bethe–Goldstone theory, the $V_{low\,k}$ approach [6], the Dirac–Brueckner–Hartree–Fock theory [7, 8], and very recently the Auxiliary Field Diffusion Monte Carlo method [9], have been extended to the hyperonic sector.

All these approaches agree that hyperons may appear in the inner core of neutron stars at densities of $\sim 2 - 3\rho_0$ as it has been said. At such densities, the nucleon chemical potential is large enough to make the conversion of nucleons into hyperons energetically favorable. This conversion relieves the Fermi pressure exerted by the baryons and makes the EoS softer, as it is illustrated in panel (a) of Fig. 1 for a generic model with (black solid line) and without (red dashed line) hyperons. As a consequence (see panel (b)) the mass of the star, and in particular the maximum one, is substantially reduced. In microscopic calculations (see *e.g.,* Refs. [5, 6]), the reduction of the maximum mass can be even below the "canonical" one of $1.4 - 1.5M_\odot$ [14]. This is not the case, however, of phenomenological calculations for which the maximum mass obtained is still compatible with the canonical value. In fact, most relativistic models including hyperons obtain maximum masses in the range $1.4 - 1.8M_\odot$ [3].

Although the presence of hyperons in neutron stars seems to be energetically unavoidable, however, their strong softening of the EoS leads (mainly in microscopic models) to maximum masses not compatible with observation. The solution of this so-called "hyperon puzzle" is not easy, and it is presently a subject of very active research, specially in view of the recent measurements of unusually high masses of the millisecond pulsars PSR J1903+0327 ($1.667\pm0.021$) [15], PSR J1614-2230 ($1.97\pm0.04M_\odot$) [16], and PSR J0348+0432 ($2.01\pm0.04M_\odot$) [17] which rule out almost all currently proposed EoS with hyperons (both microscopic and phenomenological). To solve this problem it is necessary a mechanism that could eventually provide the additional repulsion needed to make the EoS stiffer and, therefore the maximum mass compatible with the current observational limits. Three different mechanisms that could provide such additional repulsion that have been proposed are: (i) the inclusion of a repulsive hyperon-hyperon interaction through the exchange of vector mesons [18, 19, 20, 21], (ii) the inclusion of repulsive hyperonic three-body forces [22, 23, 24, 25], or (iii) the possibility of a phase transition to deconfined quark matter at densities below the hyperon threshold [27, 28, 29, 30, 31]. In the following I briefly revise these three possible solutions.

### *2.1. Hyperon-hyperon repulsion*

This solution has been mainly explored in the context of RMF models (see *e.g.,* Refs. [18, 19, 20, 21]) and it is based on the well-known fact that, in a meson-exchange model of nuclear forces, vector mesons generate repulsion at short distances. If the interaction of hyperons with vector mesons is repulsive enough then it could provide the required stiffness to explain the current pulsar mass observations. However, hypernuclear data indicates that, at least, the $\Lambda$N interaction is attractive [32]. Therefore, in order to be consistent with experimental data of hypernuclei, the repulsion in the hyperonic sector is included only in the hyperon-hyperon interaction through the exchange of the hidden strangeness $\phi$ vector meson coupled only to the hyperons. In this way, the onset of hyperons is shifted to higher densities and neutron stars with maximum masses larger than $2M_\odot$ and a significant hyperon fraction can be successfully obtained. For further information the interested reader is referred to any of the works that have explored this solution in the last years.

### *2.2. Hyperonic three-body forces*

It is well known that the inclusion of three-nucleon forces in the nuclear Hamiltonian is fundamental to reproduce properly the properties of few-nucleon systems as well as the empirical saturation point of symmetric nuclear matter in calculations based on non-relativistic many-body approaches. Therefore, it seems natural to think that three-body forces involving one or more hyperons (*i.e.,* NNY, NYY and YYY) could also play an important role in the determination of the neutron star matter EoS, and contribute to the solution of the hyperon puzzle. These forces could eventually provide, as in the case of the three-nucleon ones, the additional repulsion needed to make the EoS stiffer at high densities and, therefore, make the maximum mass of the star compatible with the recent observations. This idea was suggested even before the observation of neutron stars with $\sim 2M_\odot$ (see *e.g.,* Ref. [22]), and it has been explored by some authors in the last years [23, 24, 25]. However, the results of these works show that there is not yet a general

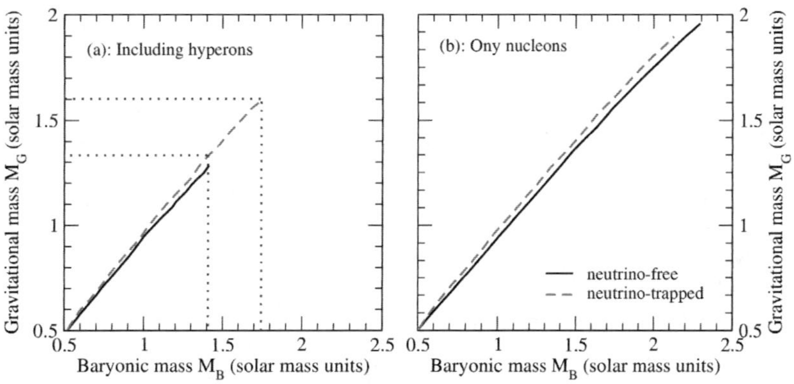


**Figure 2.** (Color online) Gravitational mass as a function of the baryonic mass for neutrino-free (solid lines) and neutrino-trapped (dashed lines) matter. Panel (a) shows the results for matter containing nucleons and hyperons, whereas the results for pure nucleonic mater are shown in panel (b). Dotted horizontal and vertical lines show the window of metastability in the gravitational and baryonic masses. Figure adapted from Ref. [34].

consensus regarding the role of hyperonic three-body forces on the hyperon puzzle. Whereas in Refs. [22, 24] these forces allow to obtain hyperon stars with $2M_\odot$, in Ref. [23] the larger maximum mass that they can support is $1.6M_\odot$, and the results of Ref. [25] are not conclusive enough because of their strong dependence on the $\Lambda$NN force employed. Therefore, it seems that hyperonic three-body forces are not the full solution to the hyperon puzzle, although, most probably they can contribute to it in a very important way. The interested reader is referred to these works for the specific details of the calculations.

### *2.3. Quarks in neutron stars*

Several authors have suggested that an early phase transition from hadronic matter to deconfined quark matter at densities below the hyperon threshold could provide a solution to the hyperon puzzle. Therefore, massive neutron stars could actually be hybrid stars with a stiff quark matter core. The question that arises in this case is then whether quarks can provide the sufficient repulsion required to produce a $2M_\odot$ neutron star. To yield maximum masses larger than $2M_\odot$, quark matter should have two important and necessary features: (i) a significant overall quark repulsion resulting in a stiff EoS, and (ii) a strong attraction in a particular channel resulting in a strong color superconductivity, needed to make the deconfined quark matter phase energetically favorable over the hadronic one [26]. Several models of hybrid stars with the necessary properties to generate $2M_\odot$ neutron stars have been proposed in the recent years [27, 28, 29, 30, 31]. Conversely, the observation of $2M_\odot$ neutron stars may also helped to impose important constraints on the models of hybrid and strange stars with a quark matter core, and improve our present understanding of the hadron-quark phase transition. Here the interested reader is also referred to the original works for detail information on this possible solution.

## 3. Hyperon stars at birth and neutron star cooling

Neutron stars are formed after a successful supernova explosion. Properties of newly born neutron stars are affected by thermal effects and neutrino trapping. These two effects have a strong influence on the overall stiffness of the EoS and the composition of the star. In particular (see *e.g.*, [33, 34, 35]) matter becomes more proton rich, the number of muons is significantly reduced, and the onset of hyperons is shifted to higher densities. In addition, the number of strange particles is on average smaller, and the EoS is stiffer in comparison with the cold and neutrino-free case.

A very important implication of neutrino trapping in dense matter is the possibility of having metastable neutron stars and a delayed formation of a "low-mass" ($M = 1 - 2M_\odot$) black hole. This is illustrated in Fig. 2 for the case of a BHF calculation of Ref. [34]. The figure shows the gravitational mass $M_G$ of the star as a function of its baryonic mass $M_B$. If hyperons are present (panel (a)), then deleptonization lowers the range of gravitational masses that can be supported by the EoS from about $1.59M_\odot$ to about $1.28M_\odot$ (see dotted horizontal lines in the figure). Since most of the matter accretion on the forming neutron star happens in a very early stages after birth ($t < 1$ s), with a good approximation, the neutron star baryonic mass stays constant during the evolution from the initial proto-neutron star configuration to the final neutrino-free one. Then, for this particular model, proto-neutron stars which at birth have a gravitational mass between $1.28 - 1.59M_\odot$ (a baryonic mass between $1.40 - 1.72M_\odot$) will be stabilized by neutrino trapping effects long enough to carry out nucleosynthesis accompanying a Type-II supernova explosion. After neutrinos leave the star, the EoS is softened and it cannot support anymore the star against its own gravity. The newborn star collapses then to a black hole [33]. On the other hand, if only nucleons are considered to be the relevant baryonic degrees of freedom (panel (b)), no metastability occurs and a black hole is unlikely to be formed during the deleptonization since the gravitational mass increases during this stage which happens at (almost) constant baryonic mass. If a black hole were to form from a star with only nucleons, it is much more likely to form during the post-bounce accretion stage.

The cooling of the newly born hot neutron stars is driven first by the neutrino emission from the interior, and then by the emission of photons at the surface. Neutrino emission processes can be divided into slow and fast processes depending on whether one or two baryons participate. The simplest possible neutrino emission process is the so-called direct Urca process ($n \to p+l+\bar{\nu}_l$, $p+l \to n+\nu_l$). This is a fast mechanism which however, due to momentum conservation, is only possible when the proton fraction exceeds a critical value $x_{DURCA} \sim 11\%$ to $15\%$ [36]. Other neutrino processes which lead to medium or slow cooling scenarios but are operative at any density and proton fraction, are the so-called modified Urca processes ($N+n \to N+p+l+\bar{\nu}_l$, $N+p+l \to N+n+\nu_l$), the bremsstrahlung ($N+N \to N+N+\nu+\bar{\nu}$), or the Cooper pair formation ($n+n \to [nn]+\nu+\bar{\nu}$, $p+p \to [pp]+\nu+\bar{\nu}$), this last operating only when the temperature of the star drops below the critical temperature for neutron superfluidity or proton superconductivity. If hyperons are present in the neutron star interior new neutrino emission processes, like *e.g.*, $Y \to B+l+\bar{\nu}_l$, may occur providing additional fast cooling mechanisms. Such additional rapid cooling mechanisms, however, can lead to surface temperatures much lower than that observed, unless they are suppressed by hyperon pairing gaps. Therefore, the study of hyperon superfluidity becomes of particular interest since it could play a key role in the thermal history of neutron stars. Nevertheless, whereas the presence of superfluid neutrons in the inner crust of neutron stars, and superfluid neutrons together with superconducting protons in their quantum fluid interior is well established and has been the subject of many studies, a quantitative estimation of the hyperon pairing has not received so much attention, and just few calculations exists in the literature [37].

## 4. Hyperons and the r-mode instability of neutron stars

It is well known that the upper limit on the rotational frequency of a neutron star is set by its Kepler frequency $\Omega_{Kepler}$, above which matter is ejected from the star's equator [38]. However, a neutron star may be unstable against some perturbations which prevent it from reaching rotational frequencies as high as $\Omega_{Kepler}$, setting, therefore, a more stringent limit on its rotation [39]. Many different instabilities can operate in a neutron star. Among them, the so called r-

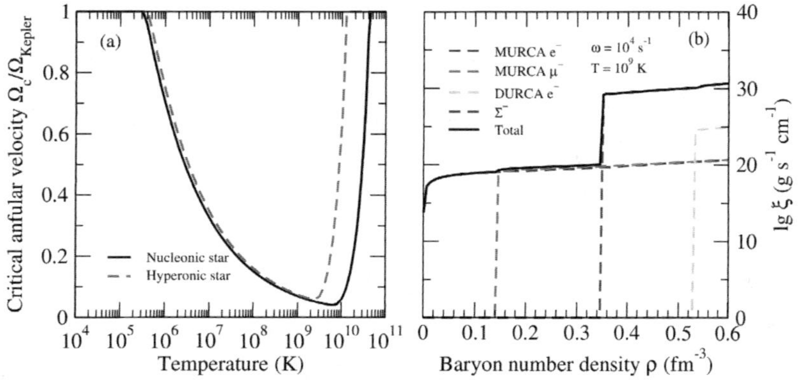


**Figure 3.** (Color online) Panel (a): r-mode instability region for a pure nucleonic and a hyperonic star wth $1.27M_{\odot}$. The frequency of the mode is taken as $\omega = 10^4$ s$^{-1}$. Panel (b): Bulk viscosity as a function of the density for $T = 10^9$ K and $\omega = 10^4$ s$^{-1}$. Contributions direct and modified nucleonic Urca processes as well as from the weak non-leptonic process $n + n \leftrightarrow p + \Sigma^-$ are included. Figure adapted from Ref. [43].

mode instability [40], a toroidal mode of oscillation whose restoring force is the Coriolis force, is particularly interesting. This oscillation mode leads to the emission of gravitational waves in hot and rapidly rotating neutron stars though the Chandrasekhar–Friedman–Schutz mechanism [41]. Gravitational radiation makes an r-mode grow, whereas viscosity stabilizes it. Therefore, an r-mode is unstable if the gravitational radiation driving time is shorter than the damping time due to viscous processes. In this case, a rapidly rotating neutron star could transfer a significant fraction of its rotational energy and angular momentum to the emitted gravitational waves. These waves, potentially detectable, could provide invaluable information on the internal structure of the star and constraints on the EoS.

Bulk ($\xi$) and shear ($\eta$) viscosities are usually considered the main dissipation mechanism of r- and other pulsation modes in neutron stars. Bulk viscosity is the dominant one at high temperatures ($T > 10^9$ K) and, therefore, it is important for hot young neutron stars. It is produced when the pulsation modes induce variations in pressure and density that drive the star away from $\beta$-equilibrium. As a result, energy is dissipated as the weak interaction tries to reestablish the equilibrium. In the absence of hyperons or other exotic components, the bulk viscority of neutron star matter is mainly determined by the reactions of direct and modified Urca processes. However, has soon as hyperons appear new mechanisms such as weak non-leptonic hyperon reactions ($N + N \leftrightarrow N + Y$, $N + Y \leftrightarrow Y + Y$), direct ($Y \to B + l + \bar{\nu}_l$, $B + l \to Y + \nu_l$) and modified hyperonic Urca ($B' + Y \to B' + B + l + \bar{\nu}_l$, $B' + B + l \to B' + Y + \nu_l$), or strong interactions ($Y + Y \leftrightarrow N + Y$, $N + \Xi \leftrightarrow Y + Y$, $Y + Y \leftrightarrow Y + Y$) contribute to the bulk viscosity and dominate it for $\rho > 2 - 3\rho_0$. Several works have been devoted to the study of the hyperon bulk viscosity [42]. The interested reader is referred to these works for detailed studies.

The time dependence of an r-mode oscillation is given by $e^{i\omega t - t/\tau}$, where $\omega$ is the frequency of the mode, and $\tau$ is an overall time scale of the mode which describes both its exponential growth due to gravitational wave emission as well as its decay due to viscous damping. It can be written as $1/\tau(\Omega, T) = -1/\tau_{GW} + 1/\tau_\xi + 1/\tau_\eta$. If $\tau_{GW}$ is shorter than both $\tau_\xi$ and $\tau_\eta$ the mode will exponentially grow, whereas in the opposite case it will be quickly damped away. For each star at a given temperature T one can define a critical angular velocity $\Omega_c$ as the smallest root of the equation $1/\tau(\Omega_c, T) = 0$. This equation defines the boundary of the so-called r-mode instability region. A star will be stable against the r-mode instability if its angular velocity is smaller than its corresponding $\Omega_c$. On the contrary, a star with $\Omega > \Omega_c$ will develope an instability that will cause a rapid loss of angular momentum through gravitational radiation until its angular velocity falls below the critical value. On panel (a) of Fig. 3 it is presented, as example, the r-mode instability region for a pure nucleonic (black solid line) and a hyperonic (red dashed line) star with $1.27M_\odot$ [43]. The contributions to the bulk viscosity from direct and modified nucleonic Urca processes as well as from the weak non-leptonic process $n+n \leftrightarrow p+\Sigma^-$ included in the calculation are shown in the panel (b) of the figure. Clearly the r-mode instability is smaller for the hyperonic star. The reason being simply the increase of the bulk viscosity due to the presence of hyperons which makes the damping of the mode more efficient.

## 5. Summary

I have briefly reviewed the main effects of hyperons on the properties of neutron. In particular, I have revised the problem of the strong softening of the EoS, and the consequent reduction of the maximum mass, due to the presence of hyperons. I have discussed three different solutions proposed to tackle this problem: (i) the inclusion of a repulsive hyperon-hyperon interaction through the exchange of vector mesons (ii) the inclusion of repulsive hyperonic three-body forces, and (iii) the possibility of a phase transition to deconfined quark matter at densities below the hyperon threshold. The role of hyperons on the properties of proto-neutron stars, as well as on the cooling properties of newly born neutron stars, and on the so-called r-mode instability has been also re-examined.

## Acknowledgments

The author is very grateful to the organizers of SQM 2015 for their invitation and to D. Unkel for interesting discussions. This work is partly supported by the project PEst-OE/FIS/UI0405/2014 developed under the inititative QREN financed by the UE/FEDER through the program COMPETE-Programa Operacional Factores de Competitividade, and by NewCompstar, COST Action MP1304.

## References

[1] Shapiro S L and Teukolsky S A, *Black Holes, White Dwarf and Neutron Stars* (Wiley, New York) 1983; Glendenning N K , *Compact Stars: Nuclear Physics, Particle Physics and General Relativity*, 2nd edition (Springer, Berlin) 2000; Haensel P, Potekhin A Y and Yakovlev D G, *Neutron Stars 1: Equation of State and Structure, Astrophysics and Space Science Library Series*, Vol. **326** (Springer, New York) 2007.

[2] Ambartsumyan V A and Saakyan G S 1960 Sov. Astron. **4** 187.

[3] Glendenning N K 1982 Phys. Lett. B **114**, 392; Glendenning N K 1985 Astrophys. J. **293**, 470; Glendenning N K 1987 Z. Phys. A **326**, 57; Glendenning N K and Moszkowski S A 1991 Phys. Rev. Lett. **67**, 2414; Weber F and Weigel M K 1989 Nucl. Phys. A **505**, 779. Knorren R, Prakash M and Ellis P J 1995 Phys. Rev. C **52**, 3470; Schaffner J and Mishustin I 1996 Phys. Rev. C **53**, 1416; Huber H, Weber F, Weigel M K and Schaab Ch 1998 Int. J. Mod. Phys. E **7**, 310.

[4] Balberg S and Gal A 1997, Nucl. Phys. A **625**, 435; Balberg S, Lichtenstadt I and Cook G B 1999 Astrophys. J. Suppl. Ser. **121**, 515; Lanskoy D E and Yamamoto Y 1997 Phys. Rev. C **55**, 2330; Tretyakova T Y and Lanskoy D E 1999 Eur. Phys. J. A **5**, 391; Cugnon J, Lejeune A and Schulze H-J 2000 Phys. Rev. C **62**, 064308; Vidaña I, Polls A, Ramos A and Schulze H-J 2001 Phys. Rev. C **64**, 044301; Zhou X-R , Schulze H-J, Sagawa H, Wu C-Xand Zhao E-G 2007 Phys. Rev. C **76**, 034312; Zhou X-R, Polls A, Schulze H-J and Vidaña I 2008, Phys. Rev. C **78**, 054306.

[5] Schulze H-J, Baldo M, Lombardo U, Cugnon J and Lejeune A 1995 Phys. Lett. B **355** 21; Schulze H-J, Baldo M, Lombardo U, Cugnon J and Lejeune A 1998 Phys. Rev. C **57** 704; Baldo M, Burgio G F and Schulze H-J 1998 Phys. Rev. C **58** 3688; Baldo M, Burgio G F and Schulze H-J 2000 Phys. Rev. C **61** 055801; Vidaña I, Polls A, Ramos A, Hjorth-Jensen M and Stoks V G J 2000 Phys. Rev. C **61** 025802; Vidaña I, Polls A, Ramos A, Engvik L and Hjorth-Jensen M 2000 Phys. Rev. C **62** 035801; Schulze H-J, Polls A, Ramos A and Vidaña I 2006 Phys. Rev. C **73** 058801; Schulze H-J and Rijken T 2011 Phys. Rev. C **84** 035801.

[6] Djapo H, Schaefer B-J and Wambach J 2010 Phys. Rev. C **81** 035803.

[7] Sammarruca F 2009 Phys. Rev. C **79** 034301.
[8] Katayama T and Saito K 2014 arXiv:1410.7166; 2015 arXiv:1501.05419.
[9] Lonardoni D, Pederiva F and Gandolfi S 2014 Phys. Rev. C **89** 014314.
[10] Maessen P M M , Rijken T A and de Swart J J 1989 Phys. Rev. C **40**, 2226; Rijken Th A, Stoks V G J and Yamamoto Y 1999 Phys. Rev. C **59**, 21; Stoks V G J and Rijken Th A 1999 Phys. Rev C **59**, 3009; Rijken Th A 2006 Phys. Rev. C **73**, 044007; Rijken Th A and Yamamoto Y 2006 Phys. Rev. C **73**, 044008.
[11] Holzenkamp B, Holinde K and Speth J 1989 Nucl. Phys. A **500**, 485; Haidenbauer J and Meissner U-G 2005 Phys. Rev. C **72**, 044005.
[12] Polinder H, Haidenbauer J and Meissner U-G 2006 Nucl. Phys. A **779**, 244; Haidenbauer J. Petschauer S, Kaiser N, Meissner U-G, Nogga A and Weise W 2013 Nucl. Phys. A **915** 24
[13] *Nuclear Methods and Nuclear Equation of State* (International Review of Nuclear Physics, Vol 8. Edited by Marcello Baldo. World Scientific Publishing Company 1999).
[14] Hulse R A and Taylor J H 1975 Astrophys. J. Lett. **195** L51.
[15] Champion D J *et al.,* 2008 Science **320** 1309; Freire P C C 2009 arXiv:0907.32199; arXiv:1011.5809.
[16] Demorest P *et al.,* 2010 Nature **467** 1081.
[17] Antoniadis J *et al.,* 2013 Science **340** 6131.
[18] Bednarek I, Haensel P, Zdunik L, Bejger M and Mańka R 2012 Astron. and Astrophys. **157** 543.
[19] Weissenborn S, Chatterjee D and Schaffner-Bielich J 2012 Phys. Rev. C **85**, 065802; 2014 Phys. Rev. C **90** (E) 019904.
[20] Oertel M, Providência C, Gulminelli F and Raduta Ad R 2015 J. Phys. G. **42** 075202.
[21] Maslov K A, Kolomeitsev E E and Voskresensky D N 2015 Phys. Lett. B **748** 369.
[22] Takatsuka T *et al.,* 2002 Eur. Phys. J. A **13** 213; 2008 Prog. Theor. Phys. Suppl. **174** 80.
[23] Vidaña I, Logoteta D, Providência C, Polls A and Bombaci I 2011 Eur. Phys. Lett. **94**11002.
[24] Yamamoto Y, Furumoto T, Yasutake B and Rijken Th A 2013 Phys. Rev. C **88** 022801; 2014, Phys. Rev. C **90** 045805.
[25] Lonardoni D, Lovato A, Gandolfi S and Pederiva F 2015 Phys. Rev. Lett. **114** 092301.
[26] Zdunik J L and Haensel P 2013 Astron. and Astrophys. **551** A61.
[27] Özel F Psaltis D, Ransom S, Demorest P and Alford M 2010 Astrophys. J. Lett. **724** L199.
[28] Weissenborn S, Sagert I, Pagliara G, Hempel M, Schaeffner-Bielich J 2011 Astophys. J. Lett. **740** L14.
[29] Klähn T, B. Blaschke D and Lastowiecki D 2013 Phys. Rev. D **88** 085001.
[30] Bonanno L and Sedrakian A 2012 Astron. and Astrophys. **539** 416.
[31] Lastowiecki R, Blaschke D, Grigorian H and Typel S 2012 Acta Phys. Polon. Suppl., **5** 535.
[32] Hashimoto O and Tamura H 2006, Prog. Part. Nucl. Phys. **57** 564.
[33] Keil W and Janka H-Th 1996 Astron. and Astrophys. **296** 145; Bombaci I 1996 Astron. and Astrophys. **305** 871; Prakash M, Bombaci I, Prakash M, Ellis P J, Knorren R and Lattimer J M 1997 Phys. Rep. **280** 1.
[34] Vidaña I, Bombaci I, Polls I and Ramos A 2003 Astron. and Astrophys. **399** 687.
[35] Burgio G F, Schulze H-J and Li A 2011 Phys. Rev. C **83** 025804.
[36] Lattimer J M, Pethick C J , Prakash M and Haensel P 1991 Phys. Rev. Lett. **66** 2701.
[37] Balberg S and Barnea N 1998 Phys. Rev. C **57** 409; Takatsuka T and Tamagaki R 1999 Prog. Theor. Phys. **102** 1043; Takatsuka T, Nishizaki S, Yamamoto Y and Tamagaki R 2000 Prog. Theor. Phys. **105** 179; Takatsuka T, Nishizaki S, Yamamoto Y and Tamagaki R 2002 Prog. Theor. Phys. Suppl. **146** 279; Vidaña I and Tolós 2004 Phys. Rev. C **70** 028802; Zhou X-R, Schulze H-J, Pan F and Drayer J P 2005 Phys. Rev. Lett. **95** 051101; Wag Y N and Shen H 2010 Phys. Rev. C **81** 025801.
[38] Lindblom L 1986 Astrophys. J. **303** 146; Friedman J L, Ipser J R and Parker L 1986 Astrophys. J. **304** 115.
[39] Lindblom L 1995 Astrophys. J. **438** 265.
[40] Anderson N 1998 Astrophys. J. **502** 708; Friedman J L and Morsink S M 1998 Astrophys. J. **502** 714.
[41] Chandrasekhar S 1970 Phys. Rev. Lett. **24** 611; Friedman J L and Schutz B F 1978 Astrophys. J. **221** 937; Friedman J L and Schutz B F 1978 Astrophys. J. **222** 281; Friedman J L 1978 Commun. Math. Phys. **62** 247.
[42] Langer W D and Cameron A G W 1969 Astrophys. Space Sci. **5** 213; Jones P B Proc. R. Soc. A 1971 **323** 111; Levin Y 1999 Astrophys. J. **517** 328; Jones P B 2001 Phys. Rev. Lett. **86** 1384; Jones P B 2001 Phys. Rev. D **64** 084003; Lindblom L and Owen B J 2002 Phys. Rev. D **65** 0653006; Haensel P, Levenfish K P and Yakovlev D G 2002 Astron. and Astrophys. **381** 1080; van Dalen E N E and Dieperink A E 2004 Phys. Rev. C **69** 025802; Chatterjee D and Bandyopadhyay D 2006 Phys. Rev. D **74** 023003; Bondarescu R, Teukolsky S A and Wasserman I 2007 Phys. Rev. D **76** 064019; Chatterjee D and Bandyopadhyay D 2008 Astrophys. J. **680** 686; Gusakov M E and Kantor E M 2008 Phys. Rev. D **78** 083006; Sinha M and Bandyopadhyay D 2009 Phys. Rev. D **79** 123001; Patruno A 2010 Astrophys. J. **722** 909; Jha T K, Mishra H and Sreekanth V 2010 Phys. Rev. C **82** 025803.
[43] Vidaña I and Albertus C 2015 in preparation.

Journal of Physics: Conference Series **668** (2016) 012032 doi:10.1088/1742-6596/668/1/012032

# Heavy flavor production and the properties of sQGP at top energies

**Wei Xie for the STAR collaboration**

Department of Physics and Astronomy, PURDUE University, West Lafayette, IN

E-mail: wxie@purdue.edu

**Abstract.** We present the recent heavy flavor measurements in the STAR experiment and briefly discuss the STAR future heavy flavor physics program with the newly installed Heavy Flavor Tracker and Muon Telescope Detector.

## 1. Introduction

Measurements of open heavy flavor and heavy quarkonia may provide further insight into understanding the properties of the strongly coupled Quark-Gluon plasma (sQGP). The interaction between heavy quarks and the medium involves different mechanisms from that of light quarks and can be studied through measuring open heavy flavor production. In addition to the radiative energy loss mechanism which describes very well the charged hadron $R_{AA}$ measurements [1], other mechanisms such as collisional energy loss [1, 2], collisional dissociation [3] and models based on the AdS/CFT correspondence [4] have been proposed to describe the heavy quark energy loss in the medium. The quarkonium production is sensitive to the medium's temperature and the state of deconfinement due to different quarkonium binding energies. Besides the large suppression of production rate due to mechanisms like the Debye screening [5], the coalescence process was found to be important for quarkonium production because of the large number of heavy quarks produced at RHIC and LHC [6]. Furthermore, cold nuclear matter effects [7], can also lead to enhancement or suppression in heavy-ion collisions. To disentangle all these effects, we need to study a wide variety of species of heavy flavor hadrons in a broad $p_T$ range in $p+p$, $d$+A and A+A collisions.

The STAR experiment is a large acceptance, multipurpose experiment. STAR measures heavy quarkonium production through dilepton decay channels [8,9,11]. The production of open heavy flavor hadrons are measured both indirectly through semi-leptonic decay channels (non-photonic electron) [12] and directly through hadronic decay channels [13]. With the completion of the Heavy Flavor Tracker (HFT) [14] and Muon Telescope Detector (MTD) [15], STAR is able to provide more powerful constraints on heavy quark and medium interaction. In this article, we first present the STAR heavy quarkonia measurements including $J/\psi$ and $\Upsilon$ production and correlation in Au+Au and $p+p$ collisions. Then we present the STAR measurements of open heavy flavor production in heavy-ion collisions, including the $R_{AA}$ as a function of $p_T$ for $D^0$ from direct reconstruction and non-photonic electrons in different centrality bins as well as the non-photonic electron $v_2$. At the end, we provide a summary and a brief overview of the STAR ongoing heavy flavor physics program with the newly installed HFT and MTD.

## 2. Heavy Quarkonium Measurements

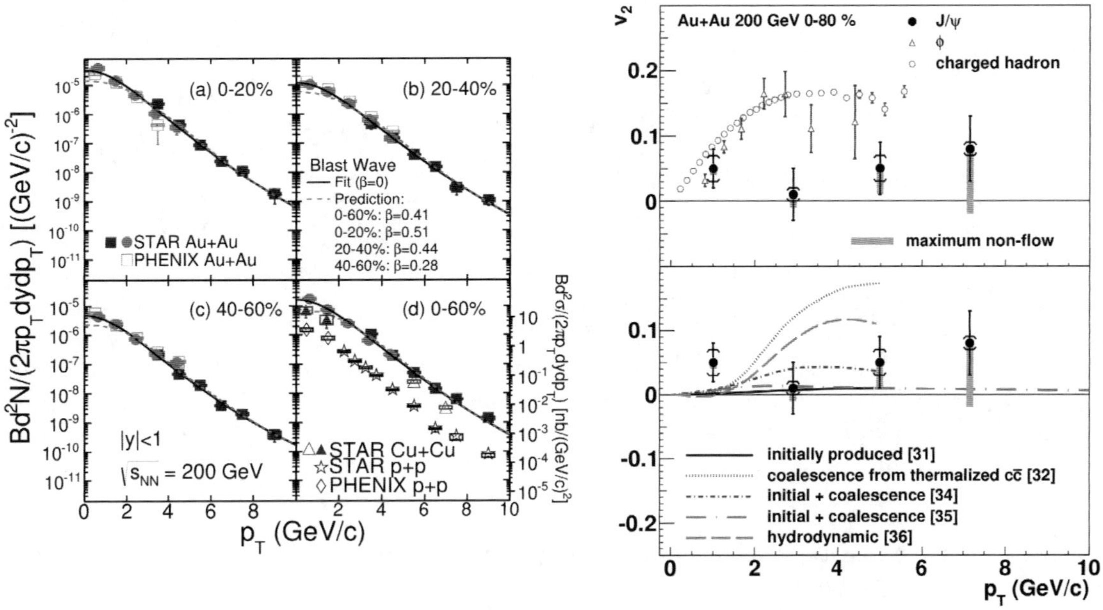


**Figure 1.** (Left) The invariant yield of $J/\psi$ production as a function of $p_T$ in 200 GeV Au+Au collisions. The predictions from Tsallis Blast-wave model for different $J/\psi$ radial flow are represented by the dashed and solid lines. (Right) $J/\psi$ $v_2$ as a function of $p_T$ in 0-80% Au+Au collisions at $\sqrt{s_{NN}} = 200$ GeV. The open circles and open triangles represent $v_2$ of charged particles and $\phi$ meson, respectively. The various lines represent predictions from different models.

The left panel of Fig. 1 shows $J/\psi$ invariant yield as a function of $p_T$ in 200 GeV Au+Au collisions in different centralities [8]. The solid and dashed curves are the predictions of Tsallis Blast-wave (TBW) model [16] assuming the radial flow velocity of $J/\psi$ is zero or the same as that of light flavor hadrons, respectively. Although the model describes the shape of the specra at high $p_T$ assuming non-zero radial flow velocity, it predicts much harder slopes at low $p_T$. In contrast, with the assumption of zero velocity the model can describe the data very well in the whole $p_T$ region, which may suggest that the radial flow of $J/\psi$ is smaller than that of light flavor hadrons or the coalescence production plays an important role at low $p_T$ . The right panel shows the $J/\psi$ $v_2$ as a function of $p_T$ [9]. At low $p_T$, we can not rule out the possibility of non-zero $J/\psi$ flow for lack of statistics. If coalescence production is significant in this region, the $J/\psi$ flow can be significant since we observed large charm $v_2$ from non-photonic electron measurements (See Fig. 7). For $p_T > 2$ GeV/$c$ the result is consistent with zero disfavoring the scenario that coalescence dominates $J/\psi$ production.

Figure 2 shows the measurements of $J/\psi$ $R_{AA}$ as a function of $N_{part}$ in 200 GeV, 62.4 GeV and 39 GeV Au+Au collisions [10]. At $\sqrt{s_{NN}} = 200$ GeV, $J/\psi$ production is more suppressed in more central collisions because of larger medium effects. Within the same centrality, $J/\psi$ production become less suppressed at higher $p_T$ possibly because of the Cronin effect or longer $J/\psi$ formation time. Two model predictions [17] are compared with the data. Both models require significant coalescence production at low $p_T$ and can describe the general features of the data. This might indicate that the abovementioned differences between the $J/\psi$ spectra and TBW model predictions assuming non-zero $J/\psi$ velocity are due to the missing coalescence

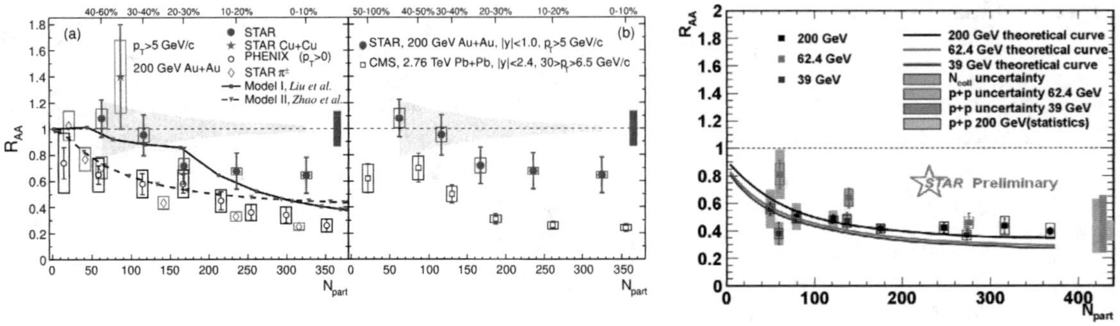


**Figure 2.** (left) The $J/\psi$ $R_{AA}$ as a function of $N_{part}$ in 200 GeV Au+Au and Cu+Cu collisions. The STAR (closed circles) and PHENIX (open circles) are for $p_T > 5$ GeV/$c$ and 0 GeV/$c$, respectively. The solid and dashed lines are the predictions from two different models. (middle) STAR and CMS measurements of high $p_T$ $J/\psi$ $R_{AA}$ as a function of $N_{part}$. (right) $J/\psi$ $R_{AA}$ as a function of $N_{part}$ in 200 GeV (black square), 62.4 GeV (red square) and 39 GeV (blue square) Au+Au collisions. The lines of different colors are model predictions.

production component in the model. Furthermore, the comparison between STAR and CMS results [18] at high $p_T$ shows that $J/\psi$ production is more suppressed at 2.76 TeV possibly because of higher energy density or larger shadowing effect. However, the production at low $p_T$ is found to be less suppressed at LHC [19] indicating the effect of larger coalescence production due to larger charm quark production cross section. When comparing the 200 GeV result with the 62.4 GeV and 39 GeV measurements, the $R_{AA}$ as a function of $N_{part}$ are found to be surprisingly similar. This similarity may come from the interplay between suppression and coalescence production and can be described by models taking these two effects into account [20].

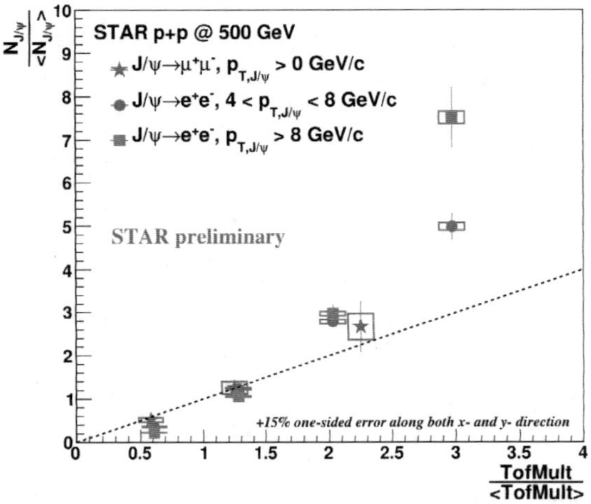


**Figure 3.** Relative $J/\psi$ yield as a function of event activity for $p_T > 0$ GeV/$c$ (solid stars), 4 GeV/$c$ $< p_T <$ 8 GeV/$c$ (solid circles) and $p_T > 8$ GeV/$c$ (solid squares).

Recently, LHC experiments observed the nonlinear dependence of open heavy flavor and quarkonium production on event activity in small collision systems [21] and found the dependence is very different in $p+p$ , $p$+Pb and Pb+Pb collisions. STAR is able to perform a similar measurement at $\sqrt{s_{NN}} = 500$ GeV in $p+p$ collisions. Figure 3 shows the relative $J/\psi$ yield as a function of event activity defined as the ratio of TOF-matched track multiplicity to the its average. The relative yield is defined as the ratio of $J/\psi$ yield in a certain event activity bin to the average $J/\psi$ yield. The nonlinearity of the dependence seems to be larger in higher $p_T$ region and is in quantitative agreement with the PYTHIA prediction involving multiparton interactions.

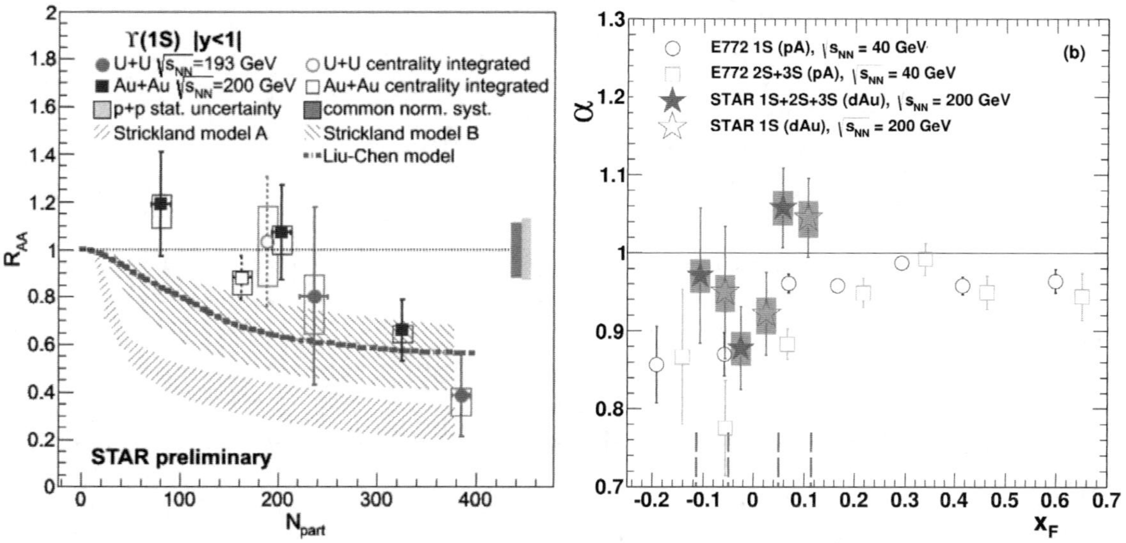


**Figure 4.** (Left) $\Upsilon(1S)$ $R_{AA}$ as a function of $N_{part}$ in 200 GeV Au+Au (squares) and 193 GeV U+U (circles) collisions. The bands and dashed lines are model predictions. (right) $\alpha$ as a function of $X_F$ for $\Upsilon(1S)$ (open stars) and $\Upsilon(1S+2S+3S)$ (closed stars) in 200 GeV $d$+Au collisions.

Unlike $J/\psi$, $\Upsilon$ production is much less affected by the coalescence process at RHIC because of the small number of bottom quarks produced in heavy-ion collisions. Furthermore, the $\Upsilon$ dissociation from the interaction with final state co-movers is also small. It is thus considered to be a cleaner probe to study the sQGP medium properties. The left panel of Fig. 4 shows the $\Upsilon(1S)$ $R_{AA}$ as a function of $N_{part}$ in 200 GeV Au+Au collisions from Run2010 [11]. The results are consistent with the Internal Energy Potential model [22], which predicts a strong 2S and complete 3S suppression. However, this model does not include cold nuclear matter effect. The right panel of the figure shows the STAR measurement of the exponent $\alpha$ as a function of Feynman's x variable ($X_F$). The $\alpha$ is defined as $\sigma_{pA} = A^{\alpha}\sigma_{pp}$. The result indicates a suppression of $\Upsilon(1S+2S+3S)$ production in $d$+Au collisions in $|X_F| < 0.05$ and is consistent with the E772 measurements. More precise measurements are needed to quantify the cold nuclear matter effects which are important in understanding the $\Upsilon$ production in the medium.

## 3. Open Heavy Flavor Production

### *3.1. $D^0$ Production*

The left panel of Fig 5 shows the $D_0$ $R_{AA}$ as a function of $p_T$ in 0-10%, 10-40% and 40-80% Au+Au collisions at $\sqrt{s_{NN}} = 200$ GeV [13]. The results are obtained from around 800 M minimum-bias events and 240 M 0-10% most central events combining Run2010 and Run2011 datasets. At high $p_T$ , the $D^0$ $R_{AA}$ is consistent with unity in the 40-80% centrality bin, while

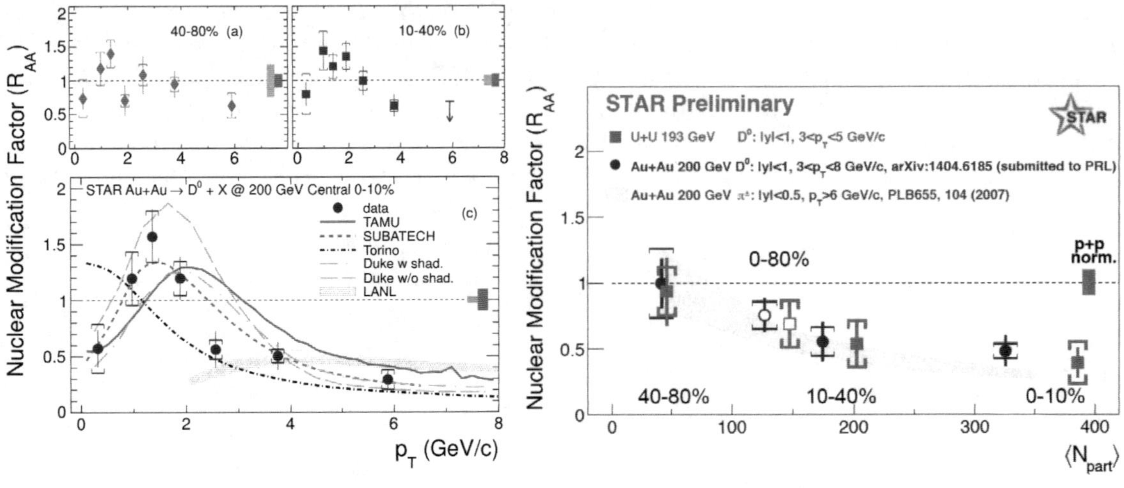


**Figure 5.** (Left) $D_0$ $R_{AA}$ as a function of $p_T$ in 200 GeV Au+Au collisions of different centralities. Various lines and band represent model predictions. (right) $D_0$ $R_{AA}$ as a function of $N_{part}$ in 200 GeV Au+Au and 193 U+U collisions. The band represents STAR charged $\pi$ measurements.

in 0-10% central Au+Au collisions, $D^0$ production is significantly suppressed and the level of suppression is comparable to that of charged $\pi$ production [23]. A clear maximum can be seen at $p_T \sim 2$ GeV/$c$ in 10-40% and 0-10% centrality bins and the shape is in qualitative agreement with the predictions from models including charm quark and light quark coalescence which boost the radial flow of charm hadrons leading to an enhancement at low $p_T$ [24]. The model without this effect [25] can not describe the data. The model including effects of cold nuclear matter and meson dissociation in the hot and dense medium is consistent with the data for $p_T > 2$ GeV/$c$ where it is applicable [26]. The right panel of the figure shows the $D_0$ $R_{AA}$ as a function of $N_{part}$ in 200 GeV Au+Au collisions and 193 GeV U+U collisions. The higher energy density of U+U collisions allows the measurements in $N_{part}$ to be extended beyond the most central Au+Au collisions. The two measurements are consistent with each other in their overlapping region in $N_{part}$ . For $N_{part} > 300$, the data seem to indicate larger suppression in U+U data point but more precise measurement is needed to draw a conclusion.

### *3.2. Non-photonic Electron Production*

Figure 6 left panel shows the non-photonic electron $R_{AA}$ as a function of $p_T$ for 0-10% Au+Au collisions. The level of suppression for non-photonic electron production is comparable to light flavor hadrons represented as gray bands in the figure at high $p_T$. The brackets in the figure represent systematic uncertainty from Au+Au data. The box around each data point stands for systematic uncertainty from the $p+p$ measurement. Although the Au+Au results are precise, the $p+p$ uncertainty is large and leads to large uncertainty in $R_{AA}$ . Predictions from some models with different energy loss mechanisms are also shown [1–4]. Models involving only radiative energy loss mechanism can not describe the data. Higher statistics, especially for the $p+p$ reference result are needed to have a better understanding of the partonic energy loss mechanism. A more precise $p+p$ measurement is ongoing using the $p+p$ data collected in STAR during Run2012. The right panel of the figure shows the non-photonic electron invariant yield as a function of $p_T$ in 62.4 GeV Au+Au collisions [27]. The FONLL mean and upper limit scaled with $N_{coll}$ are also shown to compare with the data. Unlike the measurements at 200 GeV, the 62.4 GeV invariant yields are about two times higher than the FONLL upper limit

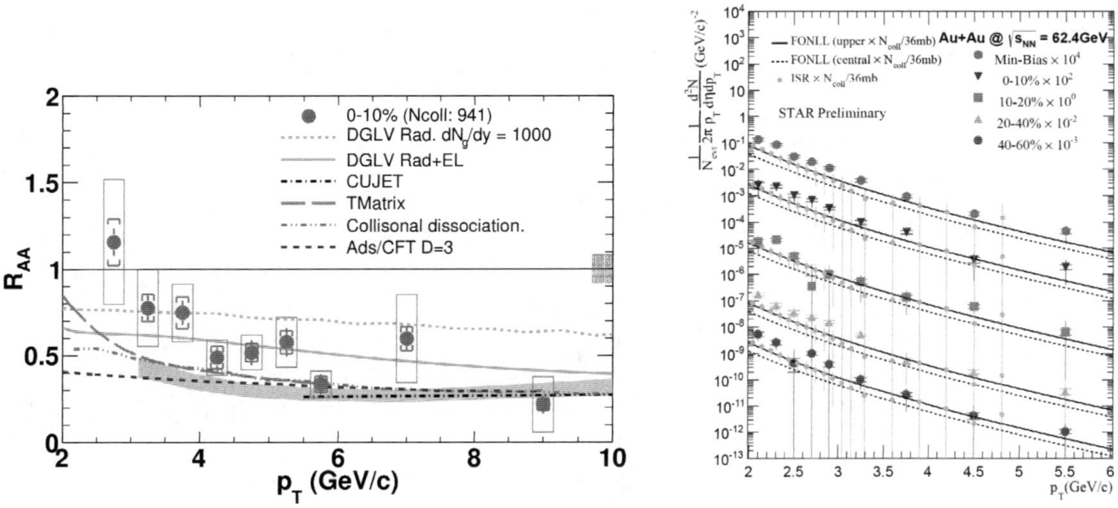


**Figure 6.** (Left) Non-photonic electron $R_{AA}$ as a function of $p_T$ in 0-10% Au+Au collisions at $\sqrt{s_{NN}} = 200$ GeV. The band represented STAR charged $\pi$ measurements. The lines represent model predictions. (right) Non-photonic electron invariant yield as a function of $p_T$ in 62.4 GeV Au+Au collisions of different centralities. The FONLL mean (dashed lines) and upper limit (solid lines) are also plotted for comparison.

indicating an enhancement of non-photonic electron production. These enhancement may come from a combination of Cronin effect and radial flow because of coalescence with light quarks [28].

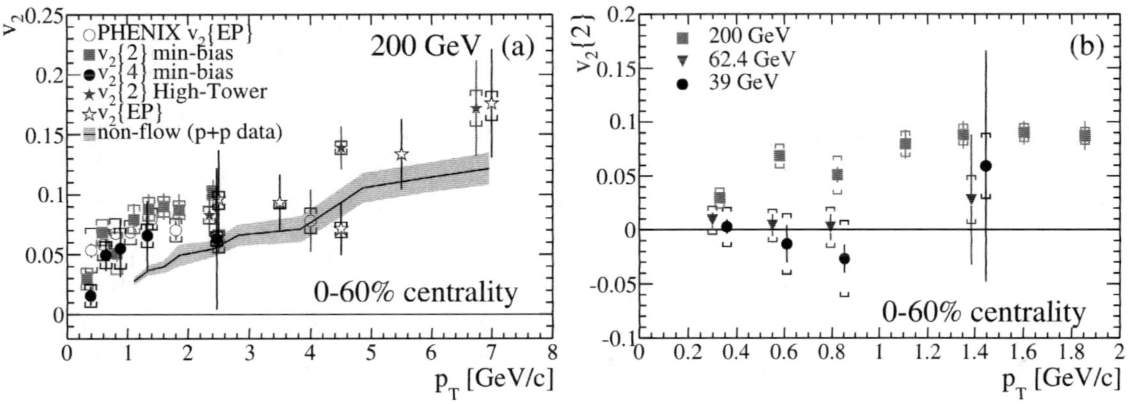


**Figure 7.** (Left) Non-photonic electron $v_2\{2\}$ (closed squares), $v_2\{4\}$ (closed circles) and $v_2$ from event plane method (open stars) as a function a $p_T$ in 200 GeV Au+Au collisions in 0-60% centrality bin. The band is the non-flow estimated from PYTHIA. (Right) Non-photonic electron $v_2\{2\}$ in 200 GeV (closed squares), 62.4 GeV (closed triangle up) and 39 GeV (closed triangle down) Au+Au collisions in 0-60% centrality bin.

In addition to measuring medium modification of heavy quark production, quantifying heavy flavor anisotropic flow is very important because it is complementary to $R_{AA}$ measurements and applies further constraints in understanding energy loss mechanism. It also carries information on the degree of thermalization of the sQGP medium. The left panel of Figure 7 shows the non-photonic electron $v_2$ measurements in 0-80% Au+Au collisions at $\sqrt{s_{NN}} = 200$ GeV. The non-photonic electron $v_2$ are measured through both the two-particle ($v_2\{2\}$) and four-particle ($v_2\{4\}$) correlation methods [29]. $v_2\{4\}$ is less affected by non-flow correlations such as jet correlations. Flow fluctuation component is positive for $v_2\{2\}$ and negative for $v_2\{4\}$ [30]. Therefore $v_2\{4\}$ would provide the lower limit on non-photonic electron anisotropic flow. The consistency between measured $v_2\{2\}$ and $v_2\{4\}$ at $p_T < 3$ GeV/$c$ indicates strong interaction between charm quark and sQGP medium in 200 GeV Au+Au collisions since low $p_T$ electrons are dominated by charm hadron decays [31]. The enhancement for $p_T > 3$ GeV/$c$ is mostly due to the correlations between electrons and hadrons from the same $c\bar{c}$ or $b\bar{b}$ pair. We studied this effect using PYTHIA represented as the gray band in the figure and can reproduce the increasing trend toward high $p_T$ . The right panel of Fig. 7 shows the $v_2\{2\}$ in 200 GeV, 62.4 GeV and 39 GeV Au+Au collisions in 0-60% centrality bin. Although the uncertainty is large, the low energy results are systematically lower than 200 GeV measurements. This may indicate that charm-medium interaction is stronger at 200 GeV.

## 4. Summary and Outlook

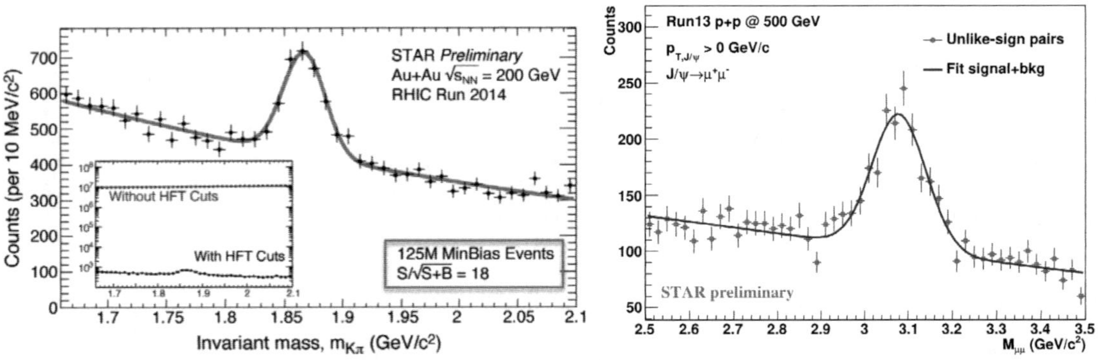


**Figure 8.** (Left) $D_0$ signals observed from a sub sample of Run2014 200 GeV Au+Au collisions with HFT. (right) $J/\psi$ signals observed from a sub sample of Run2013 500 GeV $p+p$ collisions with MTD.

The STAR $J/\psi$ $p_T$ spectra can be well described by Tsallis Blast-wave model with zero radial flow velocity indicating other source of production like coalescence or the smaller $J/\psi$ radial flow than that of light hadrons. The $J/\psi$ $v_2$ results disfavor the scenario that coalescence dominates $J/\psi$ production at $p_T > 2$ GeV/$c$. We observe no significant energy dependence of $J/\psi$ $R_{AA}$ , which may be due to the interplay between suppression and coalescence. Similar to LHC measurements, a nonlinear dependence of relative $J/\psi$ production on event activity in $p+p$ collision at $\sqrt{s_{NN}} = 500$ GeV is observed. The $\Upsilon$ $R_{AA}$ measurements are consistent with a model predicting strong 2S and complete 3S state suppression. The measurements in $d$+Au collisions indicate a $\Upsilon$ suppression due to cold nuclear effects and is consistent with the E772 results. We observe large suppression of heavy quark production at high $p_T$ in central Au+Au collisions through measuring the production of non-photonic electron and directly reconstructed $D^0$ meson. A maximum is observed for the $D^0$ $R_{AA}$ at $p_T \sim 2$ GeV/$c$, which might be induced by the transverse flow from charm quark and light quark coalescence. Large non-photonic electron $v_2$ is observed in 200 GeV Au+Au collisions and is systematically higher than the measurements in 62.4 GeV and 39 GeV. This may indicate a stronger interaction between heavy quarks and sQGP medium in higher energy Au+Au collisions.

In Run2014, STAR successfully completed HFT and MTD upgrades. HFT can determine the secondary vertices precisely and allows precision measurement of directly reconstructed charm hadrons. Figure 8 left panel shows the $D_0$ signal observed using a small data sample in Run2014. The significance is already much better than the $D_0$ signal observed with the combined Run2010 and Run2011 data. With MTD STAR can study low $p_T$ quarkonia through dimuon decay channels and disentangle $\Upsilon(1S)$, $\Upsilon(2S)$ and $\Upsilon(3S)$. The right panel of the figure shows the low $p_T$ $J/\psi$ signal observed in 500 GeV $p+p$ collisions with MTD dimuon trigger. The analyses using Run2014 data with the two new detectors are ongoing. STAR heavy flavor physics program has entered a new era of precision and comprehensive measurements.

## References


[1] M. Djordjevic *et al.*, Phys. Lett. **B 632**, 81 (2006).
[2] H Van Hees *et al.*, Phys. Rev. Lett. **100**, 192301 (2008); P.B. Gossiaux *et al.*, Nucl. Phys. **A 830**, 203 (2009); X. Zhao *et al.*, Phys. Rev. **C 82**, 064905 (2010); A. Buzzatti *et al.*, arXiv:1207.6020.
[3] R. Sharma *et al.*, Phys. Rev. **C 80**, 054902 (2009).
[4] W.A. Horowitz *et al.*, J. Phys. **G 35**, 104152 (2008).
[5] T. Matsui, and H. Satz, Phys. Lett. **B 178**, 416 (1986).
[6] P. Braun-Munzinger and J. Stachel, Phys. Lett. **B 490**, 196 (2000); R.L. Thews *et al.*, Phys. Rev. **C 63**, 054905 (2001).
[7] K. Tuchin, Nucl. Phys. **A 904-905**, 627c-630c (2013).
[8] L. Adamczyk *et al.* (STAR Collaboration) Phys. Rev. **C 90**, 024906 (2014).
[9] L. Adamczyk *et al.* (STAR Collaboration) Phys. Rev. Lett. **111**, 052301 (2013).
[10] L. Adamczyk *et al.* (STAR Collaboration) Phys. Lett. **B 722**, 55 (2013).
[11] L. Adamczyk *et al.* (STAR Collaboration) , Phys. Lett. **B 735**, 127 (2014).
[12] H. Agakishiev *et al.* (STAR Collaboration), Phys. Rev. **D 83**, 052006 (2011).
[13] L. Adamczyk *et al.* (STAR Collaboration) , Phys. Rev. Lett. 113, 142301 (2014).
[14] J. Bouchet for the STAR Collaboration, Nucl. Phys. **A 830**, 636c (2009); J. Kapitan for the STAR Collaboration, Eur. Phys. J. **C 62**, 217 (2009).
[15] L. Ruan *et al.*, J. Phys **G 36**, 095001 (2009).
[16] Z. Tang *et al.*, Chin. Phys. Lett. **30**, 031201 (2013).
[17] Y. Liu *et al.*, Phys. Lett. **B 678**, 72 (2009) and private communications; X. Zhao *et al.*, Phys. Rev. **C 82**, 064905 (2010) and private communications.
[18] The CMS Collaboration, CMS PAS HIN-12-014.
[19] B. Abelev *et al.* (ALICE Collaboration) , Phys. Lett. **B 734**, 314 (2014).
[20] X. Zhao, R. Rapp, Phys. Rev. **C 82**, 064905 (2010). R. Rapp, provide communications.
[21] S. Chatrchyan *et al.* (CMS Collaboration) JHEP **04**, 103 (2014); B. Abelev *et al.* (ALICE Collaboration) Phys. Lett **B 712**, 65-175 (2012).
[22] M. Strickland *et al.*, arXiv:1112.2761v4.
[23] G. Agakishiev *et al.* (STAR Collaboration), Phys. Rev. Lett. **108**, 072302 (2012).
[24] M. He *et al.*, Phys. Rev. Lett. **110**, 112301 (2013); P.Gossiaux *et al.*, Nucl. Phys. **A 904-905**, 992c (2013); S. S. Cao *et al.*, Phys. Rev. **C 88**, 044907 (2013).
[25] W. M. Alberico *et al.*, Eur. Phys. J. **C 73**, 2481 (2013).
[26] R. Sharma *et al.*, Phys. Rev. **C 80**, 054902 (2009).
[27] M. Mustafa (STAR Collaboration), Nucl. Phys. **A 904-905**, 665c (2013).
[28] M. He, R. Fries, R. Rapp, Phys. Rev. **C 91**, 024904 (2015).
[29] L. Adamczyk *et al.* (STAR Collaboration), arXiv:1405.6348.
[30] L. Yi *et al.*, arXiv:1101.4646; S. A. Voloshin *et al.*, Phys. Lett. **B 659**, 537 (2008).
[31] M. M. Aggarwal *et al.* (STAR Collaboration), Phys. Rev. Lett. **105**, 202301 (2010).

Journal of Physics: Conference Series **668** (2016) 012033 doi:10.1088/1742-6596/668/1/012033

# STAR results on strangeness production in beam energy scan program

**Xiaoping Zhang for the STAR Collaboration**

Lunar and Planetary Science Laboratory, Macau University of Science and Technology, Macau, China

Department of Engineering Physics, Tsinghua University, Beijing 100084, China

E-mail: xpzhangnju@gmail.com

**Abstract.** We present the recent STAR results on the production of strange hadrons ($K^0_S$, $\phi$, $\Lambda$, $\Xi$ and $\Omega$) in $\sqrt{s_{\rm NN}} = 7.7-39$ GeV Au+Au collisions in the RHIC beam energy scan program. We investigate the strangeness enhancement and strangeness equilibration as a function of beam energy and system size at RHIC. Nuclear modification factors and particle ratios will be highlighted. Implications on partonic vs. hadronic dynamics as a function of the beam energy will be discussed.

## 1. Introduction

Lattice quantum chromodynamics (QCD) calculations suggest that, at high temperature and low baryon chemical potential ($\mu_{\rm B}$), the transition from the Quark Gluon Plasma (QGP) to the state of a hadron gas is smooth and continuous (cross-over transition) [1]. At lower temperatures and high $\mu_{\rm B}$, theoretical calculations predict a first order phase transition [2] which may end at a critical point [3]. The mapping of the QCD phase diagram is the main motivation of RHIC heavy ion program.

Strange quark production is expected to be sensitive to the phase transition, because its mass is presumably much higher than system temperature in hadronic phase, but lower than system temperature in partonic phase [4]. This suggests the enhancement of strange hadron yields compared to light hadrons in the partonic phase, which has been discovered in central Pb-Pb collisions at Super Proton Synchrotron (SPS) at $\sqrt{s_{\rm NN}} = 7.7 - 17.3$ GeV [5–8]. In addition, the strange quark production is specially important because of its small hadronic cross-section and sensitivity to the partonic phase of collision dynamics. Strange hadrons, especially the multistrange hadrons like $\Omega$ and $\phi$ are excellent probes to study the QGP properties and the QCD phase boundary [9]. Differential measurements are performed at top RHIC energy of $\sqrt{s_{\rm NN}} = 200$ GeV to understand the dynamics of partons in the QGP phase. Baryon to meson ratios have been found to be large compared to those from elementary collisions [10–14]. The measured elliptic flow $v_2$ has been found to scale with the number of constituent quarks (NCQ) for both baryons and mesons [15] in Au+Au collisions at the top RHIC energy. In order to explain these observations, theoretical model calculations require the development of collectivity among constituent quarks during the partonic phase [16–20]. This partonic collectivity has been considered as an important evidence for the formation of deconfined QCD matter with partonic degrees of freedom in Au+Au collisions at the highest RHIC energy [10–20].

Journal of Physics: Conference Series **668** (2016) 012033 doi:10.1088/1742-6596/668/1/012033

To fully map out the phase diagram of the QCD matter, a Beam Energy Scan (BES) program has been initiated at RHIC with Au+Au collisions at $\sqrt{s_{NN}} = 7.7$, 11.5, 19.6, 27 and 39 GeV [21] in year 2010 and 2011, and $\sqrt{s_{NN}} = 14.5$ in year 2014. These collisions allow us to reach a broad range of temperature and $\mu_B$ in the QCD phase diagram [22] and search for a possible beam energy region where the underlying dynamics are different from those of partonic matter observed in Au+Au collisions at the top RHIC energy. Here we present the STAR results on strange hadrons production ($K^0_S$, $\phi$, $\Lambda$, $\Xi$ and $\Omega$) in above collisions. We investigate the strangeness enhancement and strangeness equilibration as a function of beam energy and system size at RHIC. Nuclear modification factors and particle ratios will be highlighted. The implications on partonic vs. hadronic dynamics as a function of the beam energy will be discussed.

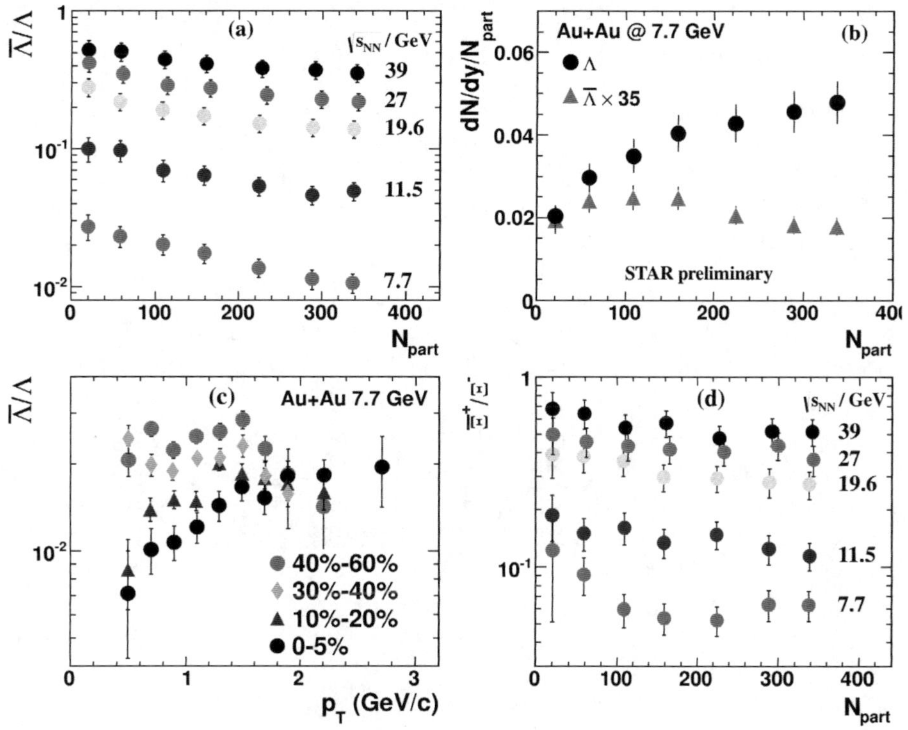


**Figure 1.** (a) Yield ratios of $dN/dy(\overline{\Lambda})/dN/dy(\Lambda)$ in mid-rapidity ($|y| < 0.5$) as a function of number of participant nucleons ($N_{\text{part}}$) in $\sqrt{s_{NN}} = 7.7 - 39$ GeV Au+Au collisions. (b) Mid-rapidity $dN/dy/N_{\text{part}}$ as a function of $N_{\text{part}}$ for $\Lambda$ and $\overline{\Lambda}$ in Au+Au collisions at 7.7 GeV. (c) $p_T$ dependence of invariant yield ratios of $\overline{\Lambda}/\Lambda$ in mid-rapidity Au+Au at 7.7 GeV. (d) Similar as Figure 1(a), but for $dN/dy(\Xi^+)/dN/dy(\Xi^-)$ ratios.

## 2. Results

Figure 1(a) shows the yield ratios of $dN/dy(\overline{\Lambda})/dN/dy(\Lambda)$ in mid-rapidity ($|y| < 0.5$) as a function of $N_{\text{part}}$ in $\sqrt{s_{NN}} = 7.7 - 39$ GeV Au+Au collisions. The ratios decrease with $N_{\text{part}}$ in each collision energy, suggesting higher $\overline{\Lambda}/\Lambda$ ratios in peripheral collisions compared to central collisions. The peripheral-central differences are more prominent at lower collisional energies such as 7.7 GeV. From Figure 1(b) one can see that the $N_{\text{part}}$-scaled $\Lambda$ yields show a generally increasing trend with increasing $N_{\text{part}}$, suggesting a stronger baryon transport to mid-rapidity

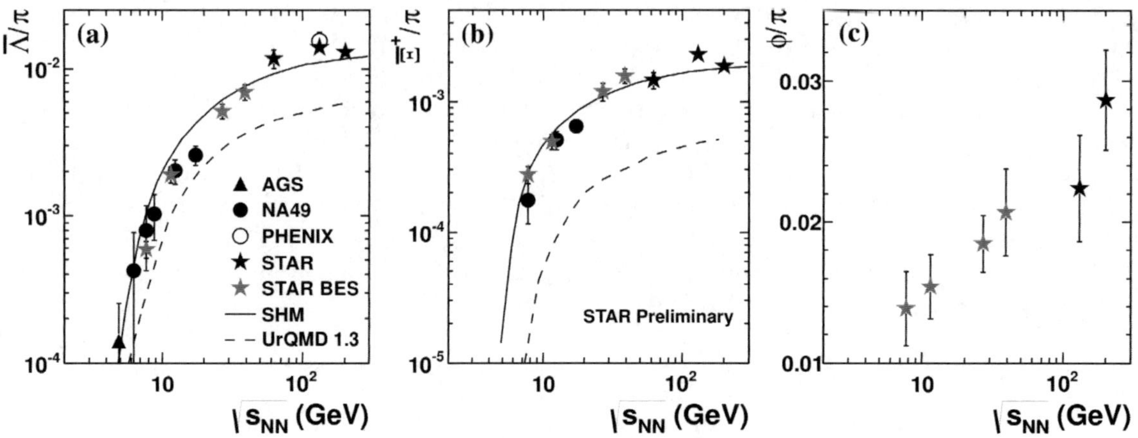


**Figure 2.** Energy dependence of $\overline{\Lambda}$, $\overline{\Xi}$ and $\phi$ to pions $(1.5(\pi^++\pi^-))$ ratios at mid-rapidity in central Au+Au and Pb+Pb collisions. The pions yield in STAR BES is taken from [25]. Also shown are AGS [26–29], NA49 [5], PHENIX [30, 31] and STAR [32–37] data, as well as calculations from the hadronic transport models: UrQMD (v1.3) [38] and a statistical hadron gas model (SHM) [39]. Errors are the quadratic sum of statistical and systematic errors.

due to more multiple scatterings in central collisions compared to peripheral collisions. For $\overline{\Lambda}$, the $N_{\text{part}}$-scaled yields increase with $N_{\text{part}}$ at small $N_{\text{part}}$. It turns down in central collisions with large $N_{\text{part}}$. There may be two effects here: one is the strangeness enhancement with increasing $N_{\text{part}}$; the other is more $\overline{\Lambda}$ absorption in the central collisions where the baryon density is higher. Experimental data [23] and model calculations [24] suggest a maximum net baryon density around $\sqrt{s_{\text{NN}}} = 7.7$ GeV. This means a significant anti-baryon absorption at 7.7 GeV, which is consistent with our data. The $p_T$ dependence of $\overline{\Lambda}/\Lambda$ invariant yield ratios in 7.7 GeV (Fig. 1(c)) shows the loss of low $p_T$ $\overline{\Lambda}$ in central collisions. This may be caused by larger absorption cross sections of $\overline{\Lambda}$ hyperons at low $p_T$. The $dN/dy(\overline{\Xi}^+)/dN/dy(\Xi^-)$ ratios show a similar pattern as $dN/dy(\overline{\Lambda})/dN/dy(\Lambda)$ ratios, which might be understood with similar mechanism of $\Lambda(\overline{\Lambda})$ production.

Figure 2 shows the ratios of strange particles to pions. $\overline{\Lambda}/\pi$ and $\overline{\Xi}^+/\pi$ ratios agree with SHM model predictions, but they are higher than the UrQMD predictions. The deviation from UrQMD predictions is larger for multi-strange baryons $\overline{\Xi}^+$. From these ratios, we can see clearly $\overline{\Lambda}$, and $\overline{\Xi}^+$ yield enhancement compared to pions with increasing collision energy. Similar behavior has been found for hidden strangeness ($\phi$ meson). The enhancement factor for $\phi/\pi$ ratio is smaller than that of $\overline{\Lambda}/\pi$ and $\overline{\Xi}^+/\pi$ ratios, partly because of anti-baryon absorption at low energy at which the new baryon density is high.

In Figure 3, we present the $R_{\text{CP}}$ of $K^0_S$, $\phi$, $\Lambda$, $\Xi^-$, and $\Omega^-$ in Au+Au $\sqrt{s_{\text{NN}}} = 7.7 - 39$ GeV collisions. The $R_{\text{CP}}$ is defined as the ratios of particle yields in central collisions over those in peripheral ones scaled by the number of inelastic binary collisions $N_{\text{bin}}$. Here, $N_{\text{bin}}$ is determined from Monte Carlo Glauber model calculations [40]. The $R_{\text{CP}}$ will be unity if nucleus-nucleus collisions are just simple superpositions of nucleon-nucleon collisions. Deviation of these ratios from unity would imply contributions from nuclear or medium effects. For $p_T$ above 2 GeV/$c$, one can see from Figure 3 that $K^0_S$ $R_{\text{CP}}$ increases with decreasing beam energies, indicating that the partonic energy loss effect becomes less important. While the cold nuclear matter effect (Cronin effect) [41] starts to take over at $\sqrt{s_{\text{NN}}} = 11.5$ and 7.7 GeV, and enhances the hadron yields at intermediate $p_T$. The energy evolution of strange hadron $R_{\text{CP}}$ reflects the decreasing

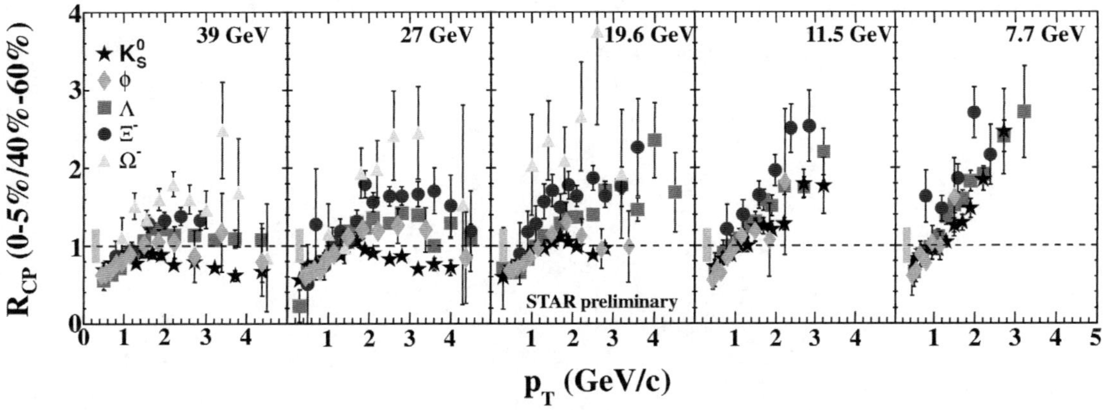


**Figure 3.** $K^0_S$, $\Lambda$, $\Xi^-$, and $\Omega^-$ $R_{\rm CP}$ (0-5%/40%-60%) at mid-rapidity ($|y| < 0.5$) in Au+Au $\sqrt{s_{NN}} = 7.7 - 39$ GeV collisions. $\phi$-meson $R_{\rm CP}$ is from 0-10% and 40%-60% centralities. Errors are statistical only. The left grey band is the normalization error from $N_{\rm bin}$.

partonic effects with decreasing beam energies. In addition, the particle $R_{\rm CP}$ differences are apparent at $\sqrt{s_{\rm NN}} \geq 19.6$ GeV. However, the differences becomes smaller at $\sqrt{s_{NN}} = 11.5$ and 7.7 GeV, which may suggest different properties of the system created in $\sqrt{s_{\rm NN}} = 11.5$ and 7.7 GeV Au+Au collisions compared to those in $\sqrt{s_{\rm NN}} \geq 19.6$ GeV.

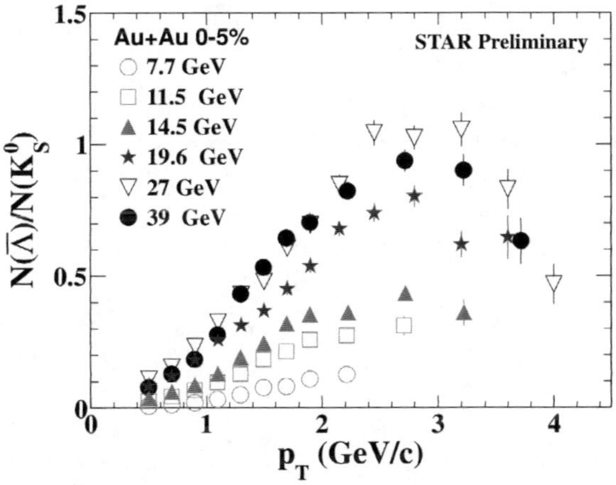


**Figure 4.** The baryon-to-meson ratios, $N(\overline{\Lambda})/N(K^0_S)$, as a function of $p_T$ in mid-rapidity ($|y| < 0.5$) from Au+Au collisions at $\sqrt{s_{\rm NN}} = 7.7 - 39$ GeV.

The enhancement of baryon-to-meson ratios at intermediate-$p_T$ compared to elementary collisions is interpreted as a consequence of hadron formation from parton recombination and parton collectivity [16–20]. Therefore, the baryon-to-meson ratios are expected to be sensitive to parton dynamics of the collision system. Figure 4 shows the baryon-to-meson ratios, $N(\overline{\Lambda})/N(K^0_S)$, as a function of $p_T$ in mid-rapidity ($|y| < 0.5$) from Au+Au collisions at $\sqrt{s_{\rm NN}} = 7.7 - 39$ GeV. The intermediate-$p_T$ $\overline{\Lambda}/K^0_S$ ratios are close to each other at 27 and 39 GeV, and show a slight decrease at 19.6 GeV. There is a sudden decrease of intermediate-$p_T$ $\overline{\Lambda}/K^0_S$ ratios between 19.6 and 14.5 GeV, suggesting a possible change of underlying hadron formation mechanism and/or parton dynamics between these two energies. For a detailed study on the centrality dependence of $\overline{\Lambda}/K^0_S$ ratios, one can refer to Ref. [42].

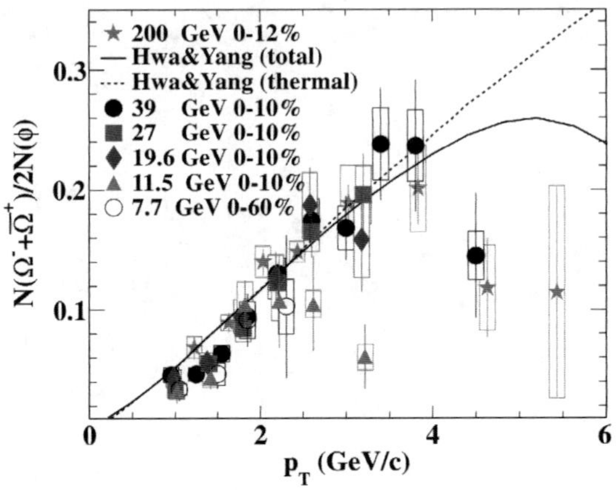


**Figure 5.** The baryon-to-meson ratio, $N(\Omega^- + \overline{\Omega}^+)/(2N(\phi))$, as a function of $p_T$ in mid-rapidity ($|y| < 0.5$) from Au+Au collisions at $\sqrt{s_{\rm NN}} = 7.7 - 200$ GeV. The boxes denote systematical errors. The solid and dashed lines represent recombination model calculations for central collisions at $\sqrt{s_{\rm NN}} = 200$ GeV [17] with total and thermal strange quark contributions, respectively.

We use multi-strange hadrons, $N(\Omega^- + \overline{\Omega}^+)/(2N(\phi))$ ratios, to further study the energy dependence of baryon-to-meson ratios. The $\Omega$ hyperons and $\phi$ mesons are expected to have relatively small hadronic interaction cross sections [43, 44]. Therefore, they can carry the information directly from the chemical freeze-out stage with little or no distortion due to hadronic rescattering. In addition, the measured $\Omega$ and $\phi$ yields suffer minimal distortion from decay feed-down. We present baryon-to-meson ratios, $N(\Omega^- + \overline{\Omega}^+)/(2N(\phi))$, as a function of $p_T$ from Au+Au collisions for various beam energies from $\sqrt{s_{\rm NN}} = 7.7$ to 200 GeV in Figure 5 and for various collision centralities in Figure 6, respectively. Data from 200 GeV Au+Au collisions are from previously published STAR results [12]. Coalescence or recombination models [17–19] have been used to describe particle productions in nucleus-nucleus collisions at RHIC. In particular, a model calculation by Hwa and Yang for Au+Au collisions at $\sqrt{s_{\rm NN}} = 200$ GeV [17] predicted that most of the $\Omega$ and $\phi$ yields up to the intermediate $p_T$ region are from coalescence/recombination of thermal strange quarks. The straight dotted line assumed that these thermal strange quarks have exponential $p_T$ distributions. Deviations from the straight line at high $p_T$ were attributed to recombination with strange quarks from high $p_T$ showers. Deviations from the theory calculation at low $p_T$ could indicate that thermal strange quarks may not have an exponential distribution. Possibly, other particle production dynamics may also contribute deviations from coalescence model calculations.

In Figure 5 the measured $N(\Omega^- + \overline{\Omega}^+)/(2N(\phi))$ ratios from central Au+Au collisions at $\sqrt{s_{\rm NN}} = 19.6$, 27 and 39 GeV follow closely the ratio from 200 GeV and are consistent with a picture of coalescence/recombination dynamics over a broad $p_T$ range of $1-4$ GeV/$c$. The ratios at 11.5 GeV seem to deviate from the trend observed at higher beam energies. In particular, the ratios at 11.5 GeV appear to turn down around $p_T$ of 2 GeV/$c$ while those at higher beam energies such as 39 and 200 GeV peak at $p_T$ of 3 GeV/$c$ or above. The collision centrality dependence of the $N(\Omega^- + \overline{\Omega}^+)/(2N(\phi))$ ratios in Figure 6(a)-(d) shows significant differences between the 40% – 60% centrality bin and the other centrality bins for Au+Au collisions at 19.6 and 27 GeV. Furthermore, the ratios from the peripheral collisions of 40% – 60% at 27 GeV are similar in magnitude to the ratios from collisions at 11.5 GeV. Because the $\Omega$ and $\phi$ particles have small hadronic rescattering cross sections [45], the change in these ratios is likely

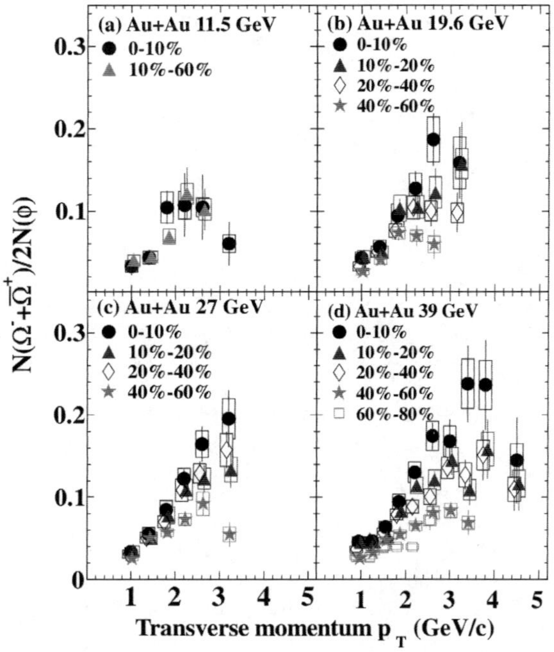


**Figure 6.** Centrality dependence of $N(\Omega^- + \overline{\Omega}^+)/(2N(\phi))$ ratios, as a function of $p_T$ in mid-rapidity ($|y| < 0.5$) from Au+Au collisions at $\sqrt{s_{\rm NN}} = 11.5$, 19.6, 27 and 39 GeV. The boxes denote systematical errors.

to originate from the partonic phase. The decrease in the $N(\Omega^- + \overline{\Omega}^+)/(2N(\phi))$ ratios from central collisions at 11.5 GeV compared to those at 19.6 GeV or above may indicate a significant change in the hadron formation dynamics and/or in strange quark $p_T$ distributions at the lower energy. Such a change may arise from a transition from hadronic to partonic dynamics with increasing beam energy. The turn-over in the ratios from Au+Au collisions below 11.5 GeV beam energy is unlikely due to contributions of high $p_T$ shower partons as suggested by model calculation from Hwa and Yang [17] because of relatively low $p_T$ particles involved.

## 3. Summary

In summary, we have presented the recent STAR results on the production of strange hadrons ($K^0_S$, $\phi$, $\Lambda$, $\Xi$ and $\Omega$) in $\sqrt{s_{\rm NN}} = 7.7 - 39$ GeV Au+Au collisions in the RHIC beam energy scan program. The comparisons between $\Lambda$ yields and $\overline{\Lambda}$ yields suggest that baryon transport and anti-baryon absorption processes are important in relatively low beam energies. There is a clear $\phi$, $\overline{\Lambda}$, $\overline{\Xi}^+$ yield enhancement compared to pions with increasing collision energy. Strange hadron $R_{\rm CP}$ increases with decreasing beam energies, indicating that the partonic energy loss effect becomes less important at low beam energies. At intermediate $p_T$, particle $R_{\rm CP}$ difference becomes smaller at $\sqrt{s_{\rm NN}} = 7.7$ and 11.5 GeV. The intermediate-$p_T$ $\overline{\Lambda}/K^0_S$ ratios between 19.6 and 14.5 GeV show a sudden decrease. The $N(\Omega^- + \overline{\Omega}^+)/(2N(\phi))$ ratios at intermediate $p_T$ in peripheral collisions are found to be lower than those in central collisions at 19.6, 27 and 39 GeV. The ratios from 11.5 GeV central collisions are systematically lower than those from collisions at 19.6 GeV or above for $p_T > 2.4$ GeV/$c$. These features suggest that there is likely a change in the underlying strange quark dynamics in the bulk QCD matter responsible for strange hadron production with the decreasing beam energy. Our measurements point to collision energies below 19.6 GeV for further investigation of a possible transition from partonic dominant matter

($\sqrt{s_{\mathrm{NN}}} > 19.6$ GeV) to hadronic dominant matter ($\sqrt{s_{\mathrm{NN}}} < 11.5$ GeV).

## Acknowledgments

X. Zhang thanks the support by the National Natural Science Foundation of China (Grant No. 11105079), and by the Science and Technology Development Fund of Macau (Grant Nos. 068/2011/A, 048/2012/A2, 039/2013/A2 and 020/2014/A1).

## References

[1] Y. Aoki *et al.*, Nature **443**, 675 (2006).
[2] S. Ejiri, Phys. Rev. D **78**,074507 (2008); E. S. Bowman and J. I. Kapusta, Phys. Rev. C **79**, 015202 (2009).
[3] M. A. Stephanov, Prog. Theor. Phys. Suppl. **153**, 139 (2004); Z. Fodor and S. D. Katz, JHEP **0404**, 50 (2004); R. V. Gavai and S. Gupta, Phys. Rev. D **78**, 114503 (2008).
[4] J. Rafelski, B. Müller, Phys. Rev. Lett. **48**, 1066 (1982).
[5] C. Alt *et al.* (NA49 Collaboration), Phys. Rev. C **78**, 034918 (2008)
[6] C. Alt *et al.* (NA49 Collaboration), Phys. Rev. Lett. **94**, 192301 (2005).
[7] D. Elia for the NA57 Collaboration, J. Phys. G **31**, S135 (2005).
[8] C. Alt *et al.* (NA49 Collaboration), Phys. Rev. C **78**, 044907 (2008).
[9] L. Adamczyk *et al.* (STAR Collaboration), arXiv:1506.07605 [nucl-ex]
[10] K. Adcox *et al.* (PHENIX Collaboration), Phys. Rev. C **69**, 024904 (2004).
[11] B. I. Abelev *et al.* (STAR Collaboration), Phys. Rev. Lett. **97**, 152301 (2006); Phys. Lett. B **655**, 104 (2007).
[12] B. I. Abelev *et al.* (STAR Collaboration), Phys. Rev. Lett. **99**, 112301 (2007); Phys. Rev. C **79**, 064903 (2009).
[13] G. Agakishiev *et al.* (STAR Collaboration), Phys. Rev. Lett. **108**, 072301 (2012).
[14] B. Abelev *et al.* (ALICE Collaboration), Phys. Rev. Lett. **111**, 222301 (2013).
[15] J. Adams *et al.* (STAR Collaboration), Phys. Rev. Lett. **92**, 052302 (2004).
[16] D. Molnár and S. A. Voloshin, Phys. Rev. Lett. **91**, 092301 (2003).
[17] R. C. Hwa and C. B. Yang, Phys. Rev. C **66**, 025205 (2002); Phys. Rev. C **75**, 054904 (2007).
[18] R. J. Fries, B. Müller, C. Nonaka, and S. A. Bass, Phys. Rev. Lett. **90**, 202303 (2003); Phys. Rev. C **68**, 044902 (2003).
[19] V. Greco, C. M. Ko, and P. Lévai, Phys. Rev. Lett. **90**, 202302 (2003); Phys. Rev. C **68**, 034904 (2003).
[20] J. H. Chen *et al.*, Phys. Rev. C **78**, 034907 (2008).
[21] M. M. Aggarwal *et al.* (STAR Collaboration), arXiv:1007.2613.
[22] J. Cleymans *et al.*, Phys. Rev. C **73**, 034905 (2006); F. Becattini, J. Manninen and M. Gaździcki, Phys. Rev. C **73**, 044905 (2006); A. Andronic, P. Braun-Munzinger and J. Stachel, Nucl. Phys. A **772**, 167 (2006).
[23] X. Zhu (for the STAR Collaboration), J. Phys.: Conf. Ser. **509**, 012004 (2014).
[24] J. Randrup and J. Cleymans, Phys. Rev. C **74**, 047901 (2006).
[25] L. Kumar (for the STAR Collaboration), J. Phys. G **38**, 124145 (2011); S. Das (for the STAR Collaboration), J. Phys.: Conf. Ser. **509**, 012066 (2014).
[26] S. Albergo *et al.*, Phys. Rev. Lett. **88**, 062301 (2002).
[27] L. Ahle *et al.*, Phys. Rev. C **57**, 466 (1998).
[28] B. B. Back *et al.*, Phys. Rev. Lett. **87**, 242301 (2001).
[29] S. Ahmad *et al.*, Phys. Lett. B **382**, 35 (1996); Nucl. Phys. A **636**, 507 (1998).
[30] K. Adcox *et al.* (PHENIX Collaboration)), Phys. Rev. Lett. **89**, 092302 (2002).
[31] K. Adcox *et al.* (PHENIX Collaboration), Phys. Rev. Lett. **88**, 242301 (2002).
[32] J. Adams *et al.* (STAR Collaboration), Phys. Rev. Lett. **98**, 062301 (2007).
[33] C. Adler *et al.* (STAR Collaboration), Phys. Rev. Lett. **89**, 092301 (2002).
[34] J. Adams *et al.* (STAR Collaboration), Phys. Rev. Lett. **92**, 182301 (2004).
[35] M. M. Aggarwal *et al.* (STAR Collaboration), Phys. Rev. C **83**, 024901 (2011).
[36] C. Adler *et al.* (STAR Collaboration), Phys. Lett. B **595**, 143 (2004).
[37] J. Adams *et al.* (STAR Collaboration), Phys. Rev. Lett. **92**, 112301 (2004).
[38] E. L. Bratkovskaya *et al.*, Phys. Rev. C **69**, 054907 (2004).
[39] A. Andronic *et al.*, Nucl. Phys. A **772**, 167 (2006).
[40] B. I. Abelev *et al.* (STAR Collaboration), Phys. Rev. C **79**, 034909 (2009).
[41] J. W. Cronin *et al.*, Phys. Rev. D **11**, 3105 (1975).
[42] M. U. Ashraf (for the STAR Collaboration), these proceedings.
[43] H. van Hecke, H. Sorge, and N. Xu, Phys. Rev. Lett. **81**, 5764 (1998).
[44] A. Shor, Phys. Rev. Lett. **54**, 1122 (1985).
[45] T. Hirano *et al.*, Phys. Rev. C **77**, 044909 (2008).

Journal of Physics: Conference Series **668** (2016) 012034 doi:10.1088/1742-6596/668/1/012034

# Forward-backward correlations with strange particles in PYTHIA

**I.G.Altsybeev**[1], **G.A.Feofilov**[1], **E. L. Gillies**[2]

[1] *Saint-Petersburg State University, RU*

[2] *University of Edinburgh, UK*

E-mail: Igor.Altsybeev@cern.ch

**Abstract.** We present studies of strange particle yields and correlations in *pp* collisions in the PYTHIA8 event generator by studying forward-backward correlations. Several key processes that give rise to these correlative effects are identified and manipulated to probe the fundamental properties of strange particle emitting sources. The sensitivity of strange particle production and correlations to PYTHIA's multiparton interaction, color reconnection, and explicit strangeness suppression are shown.

## 1. Introduction

Forward-backward (FB) correlations are a powerful tool for studying the initial stages of *pp* and AA collisions. FB correlations are studied in two intervals of pseudorapidity ($\eta$) which are selected one in the forward and another in the backward hemispheres in the center-of-mass system.

The FB correlation strength is characterized by the correlation coefficient $b_{\mathrm{corr}}$. This value is defined via a linear regression of the average value of a given quantity $B$ measured in the backward hemisphere ($\langle B\rangle_F$) as a function of value of another quantity $F$ measured in the forward hemisphere. Note that $F$ and $B$ can describe the kinematic quantity or distinct quantities.

$$\langle B\rangle_F = a + b_{\mathrm{corr}} \cdot F \; . \tag{1}$$

Taking $F$ and $B$ particle multiplicities, the relation (1) becomes $\langle n_{\mathrm{B}}\rangle_{n_{\mathrm{F}}} = a + b_{\mathrm{corr}} \cdot n_{\mathrm{F}}$, which was first experimentally observed in UA5 [1] and discussed in [2–4]. FB correlations between multiplicities have been recently studied in *pp* and Au-Au collisions by STAR [5] at RHIC, and in *pp* collisions by ATLAS [6] and ALICE [7] at LHC.

FB correlation studies are more informative when decoupled into short-range and long-range components [4,8]. Short-range correlations (SRC) are localized over a small range of $\eta$, typically up to one unit. They are induced by various short-range effects like decays of clusters or resonances, jet and mini-jet induced correlations. Long-range correlations (LRC) extend over a wider range in $\eta$ and originate from fluctuations in the number and properties of particle emitting sources, e.g. clusters, cut pomerons, strings, mini-jets etc. [4, 8–11]. In ALICE paper [7], the "classical" approach to the long-range correlation analysis in two pseudorapidity intervals was expanded using additional azimuthal ($\varphi$) sectors within these windows. This approach allows for a more thorough investigation of the SRC and LRC and their contributors, which can provide stronger constrains on phenomenological string models. Correlations with additional azimuthal segmentation of rapidity windows were also studied in PYTHIA6 [12].

## 2. Motivation for other variables in FB correlations

FB multiplicity correlations in *pp* collisions can be interpreted using the parametric string model [13] which implies event-by-event fluctuations in number of strings as *independent* particle emitters. However, independent emitters can not describe other types of correlations, such as a non-zero correlation between charged particle multiplicity and average transverse momentum ($\langle p_{\rm T}\rangle_{N_{\rm ch}}$ correlation) in a *single* $\eta$-window This was first established at ISR energies in [14].

The correlation of the mean $p_{\rm T}$ of charged particles and other observables can be explained via collective effects relevant to the formation of particle emitting sources. In *pp* and $p\bar{p}$ collisions, these collective effects were considered to be string fusion between quark-gluon strings [15, 16]. Specifically, the multi-Pomeron exchange model provided a description of the experimentally measured growth in $p_{\rm T}$ with event multiplicity over a wide energy range of collision energies (0.3-1.8 TeV). It was shown [17] that the use of color reconnection in string-based PYTHIA model can produce the positive $p_{\rm T}$-multiplicity correlation seen experimentally in *pp* collisions.

In string-based models, the Schwinger-like mechanism of string hadronization is used to obtain the production rate of $q\overline{q}$ pairs with opposite transverse momenta $p_{\rm T}$. The rate is proportional to $\exp\Big(-\frac{\pi}{\kappa}(m^2+p_{\rm T}^2)\Big)$, where $\kappa$ is a string tension and $m$ is a quark mass. This result can be used to estimate the relative production of different flavoured quarks and the $p_{\rm T}$ distribution. Collective effects could yield a higher effective string tension, as in the string fusion model [18–20] and the overlapping color ropes model in DIPSY event generator [21]. Larger string tension implies larger strangeness and baryon fractions as expected.

PYTHIA allows for *multiple parton interactions* (MPI) in *pp* events. This can cause non-negligible phase-space overlaps between final states from different MPI systems. The interaction between strings is implemented by color reconnection (CR), as proposed in [22]. In PYTHIA8 before reconnection, partons are connected in their respective MPI system. The color flow of two such systems can be fused such that the partons of the lower-$p_{\rm T}$ system are added to the strings defined by the higher-$p_{\rm T}$ system to give the smallest total string length. This is the default method in PYTHIA8. In the new CR model [23], junction structures are introduced in addition to the more common string-string reconnections. The new model has been tuned to reproduce measured ratios of kaons and hyperons such as $\Lambda/K^0_{\rm S}$ ratio. The use of junction structures introduces a slight enhancement of the strangeness and overall baryon production in this implementation.

## 3. Forward-backward correlations of strange particles in PYTHIA8

The effect of MPI and CR on FB correlations was studied in PYTHIA8. As shown in Fig. 1(a), the $\langle p_{\rm T}\rangle_{N_{\rm ch}}$ correlation between single $y$-window is preserved when considering two windows with large rapidity separation. When the color reconnection mechanism in PYTHIA8 is switched off the correlation drops to almost flat behavior as is shown in the Fig. 1(a) with open markers. Using the FB correlation approach, it is possible to examine string configurations and their interactions along $\eta$-range, accessible in an experiment, and also to get rid of short-range contributions coming from resonance decays, jets etc.

FB correlations involving strange particles can also be used to test string models. Fig. 1(b) shows that the $\langle\Lambda/\overline{\Lambda}\rangle_{N_{\rm ch}}$ correlation function is affected by the choice of the CR model. At $N^{\rm F}_{\rm ch}\approx 8$, a "knee" can be seen in the correlation function. This indicates some threshold behavior incorporated in PYTHIA8. Fig. 2 compares the FB strange particle multiplicity correlation in $\eta$-$\varphi$ windows, obtained in PYTHIA8. Specifically, it shows that the "plateau" level and the shapes of the near-side and away-side structures in this topology changes in the absence of CR.

The effective quark masses used in the Schwinger-mechanism are tuned the $s/u$ ratio seen in experimental data. In PYTHIA, this implies an explicit suppression of strange quark production, $u:d:s\approx 1:1:0.3$ [24]. In Fig. 2(c), the particle multiplicity correlation topology in $\eta$-$\varphi$ windows is shown for enhanced strangeness production with MPI turned off, revealing additional

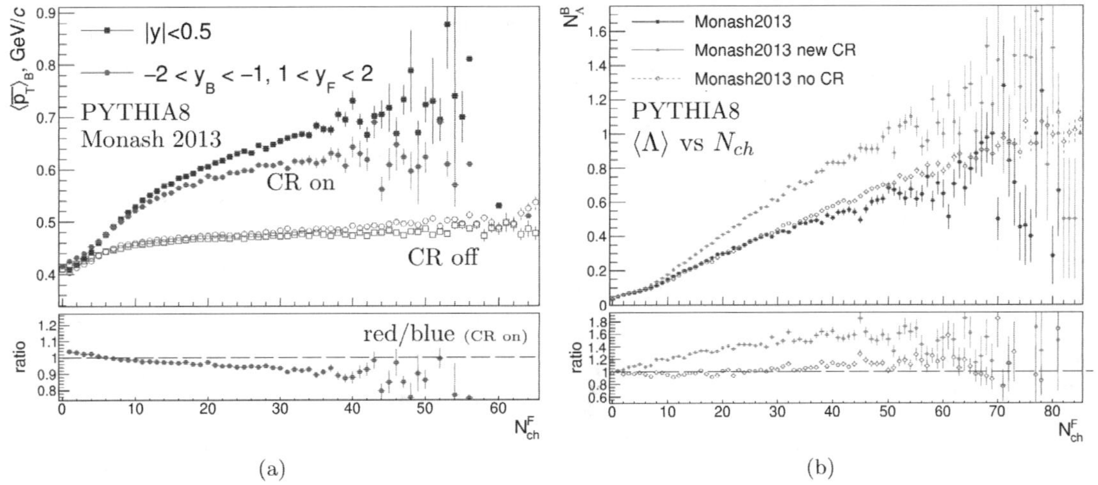


**Figure 1.** (a): $\langle p_{\mathrm{T}}\rangle_{N_{\mathrm{ch}}}$ correlations in PYTHIA8 with CR (filled markers) and without CR (open markers) for a single *eta*-window (blue markers) and the $-2.4 < \eta < -0.5$ and $0.5 < \eta < 2.4$ *eta*-windows (red markers). (b): FB correlation between $\Lambda/\overline{\Lambda}$ in backward window and $N_{\mathrm{ch}}$ in forward window, for three configurations of PYTHIA8.

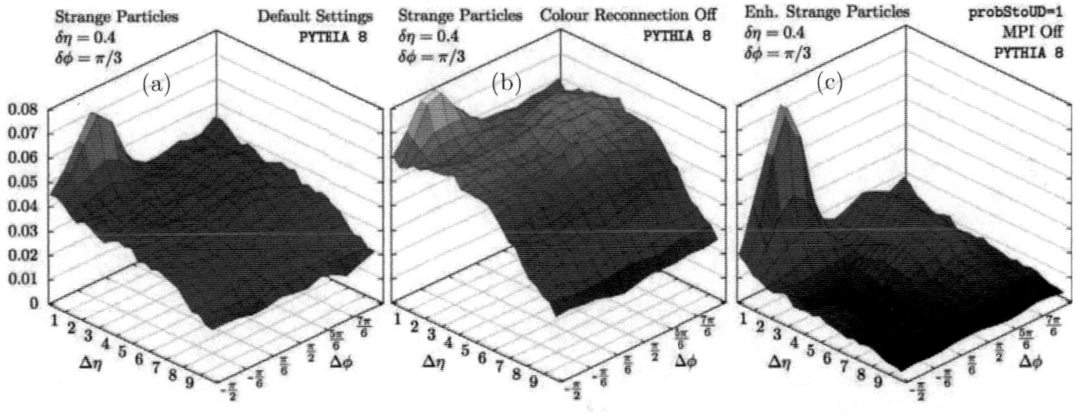


**Figure 2.** The strange particle multiplicity correlation topology in $\eta$-$\varphi$ windows in PYTHIA8 with CR on (a), CR off (b) and with enhanced strangeness production (c).

modifications of the correlation coefficient $b_{\mathrm{corr}}$. Additional figures can be found in attachments for these proceedings.

## 4. Conclusions

A number of observables in *pp* and AA collisions can not be described by independently hadronizing particle emitters, indicating the presence of collective effects. To study this collectivity, conventional analysis of forward-backward correlations can be extended from charged particle multiplicities to other observables chosen in the windows in phase-space. Usage of the "intensive" variables in FB correlations like $\langle p_{\mathrm{T}}\rangle$, $\Lambda/\pi$, $K/\pi$, as well as strange particle yields allows string interaction mechanisms in *pp* and AA collisions to be studied. To properly understand the underlying physics, the shape of the correlation function can be more informative than using the correlation coefficient $b_{\mathrm{corr}}$ alone.

It was shown that different color reconnection models in PYTHIA8 change the behavior of

FB correlations. The newest CR scheme in PYTHIA8 gives more baryons and demonstrates different slopes of FB correlation functions. The correlations are also affected by MPI and explicit strangeness suppression in this generator.

*Acknowledgements*
This work is supported for I.A. and G.F. by the Saint-Petersburg State University research grant 11.38.242.2015.

## References

[1] Alner G et al. 1987 UA5: A general study of proton-antiproton physics at $\sqrt{s} = 546$ GeV, Phys. Rep. **154**, 247.
[2] Capella A and Tran Thanh Van J 1984 Long Range Rapidity Correlations in Hadron - Nucleus Interactions, Phys.Rev. **D29**, 2512.
[3] Fowler G, Friedlander E, Pottag F, Weiner R, Wheeler J, et al. 1988 Rapidity Scaling of Multiplicity Distributions in a Quantum Statistical Approach, Phys.Rev. **D37**, 3127.
[4] Capella A and Krzywicki A 1978 Unitarity Corrections to short range order: Long range rapidity correlations, Phys.Rev. **D18**, 4120.
[5] STAR Collaboration, Abelev B et al. 2009 Growth of Long Range Forward-Backward Multiplicity Correlations with Centrality in Au+Au Collisions at $\sqrt{s_{NN}} = 200$ GeV, Phys.Rev.Lett. **103**, 172301, arXiv:0905.0237.
[6] ATLAS Collaboration, Aad G et al. 2012 Forward-backward correlations and charged-particle azimuthal distributions in *pp* interactions using the ATLAS detector, JHEP **07** 019, arXiv:1203.3100.
[7] Abelev 2009 et al. 2015 Forward-backward multiplicity correlations in *pp* collisions at $\sqrt{s} = 0.9$, 2.76 and 7 TeV, JHEP **05** 097.
[8] Capella A, Sukhatme U, C.-I. Tan and J. T. T. Van 1994 Dual parton model, Phys. Rep. **236** 225.
[9] Braun M, Pajares C and Vechernin V 2000 On the forward-backward correlations in a two stage scenario, Phys.Lett. **B493** 54.
[10] Amelin N, Armesto N, Braun M, Ferreiro E and Pajares C 1994 Long and short range correlations and the search of the quark gluon plasma, Phys.Rev.Lett. **73** 2813.
[11] Braun M, Kolevatov R, Pajares C and Vechernin V 2004 Correlations between multiplicities and average transverse momentum in the percolating color strings approach, Eur.Phys.J. **C32** (2004) 535.
[12] Wraight K and Skands P 2011 Forward-Backward Correlations and Event Shapes as probes of Minimum-Bias Event Properties. Eur.Phys.J., **C71** 1628.
[13] Vechernin V 2015 Forward-backward correlations between multiplicities in windows separated in azimuth and rapidity, Nucl.Phys. **A939** 21, arXiv:1210.7588.
[14] A. Breakstone et al. 1983 Multiplicity Dependence of Transverse Momentum Spectra at ISR Energies Phys.Lett. **B132** 463.
[15] Armesto N, Derkach D, Feofilov G 2008 Pt-Multiplicity Correlations in a Multi-pomeron Exchange Model With String Collectivity Effects, Physics of Atomic Nuclei **71** No.l2, 2087-2095(2008).
[16] Bodnya E, Derkach D, Feofilov G, Kovalenko V, Puchkov A 2013 Multi-pomeron exchange model for *pp* and $p\bar{p}$ collisions at ultra-high energy. PoS (QFTHEP 2013) 060, arXiv:1310.1627.
[17] Asryan A, Feofilov G, Ivanov A, Grebenyuk A, Naumenko P and Vechernin V 2008 Long-Range Multiplicity Correlation in pp-collisions in ALICE at the LHC, Proc. ISHEPP-XIX, vol.2, pp.208-214, JINR, Dubna.
[18] Braun M A, Dias de Deus J, Hirsch A S, Pajares C, Scharenberg R P, Srivastava B K 2015 De-confinement and clustering of color sources in nuclear collisions, Physics Reports, Available online 11 September 2015, http://dx.doi.org/10.1016/j.physrep.2015.09.003; arXiv:1501.01524 [nucl-th].
[19] Vechernin V and Kolevatov R 2007 On multiplicity and transverse-momentum correlations in collisions of ultrarelativistic ions, Phys.Atom.Nucl. **70** 1797.
[20] Kovalenko V N, 2013 Modelling of exclusive parton distributions and long-range rapidity correlations for *pp* collisions at the LHC energy, Phys. Atom. Nucl. **76**, 1189.
[21] Bierlich C, Gustafson G, Lonnblad L, Tarasov A, 2015 Effects of Overlapping Strings in *pp* Collisions, arXiv:1412.6259 [hep-ph].
[22] Lonnblad L 1996 Reconnecting coloured dipoles, Zeitschrift fur Physik C Particles and Fields, Volume **70**, Issue 1, *pp* 107-113.
[23] Christiansen J R, Skands P Z 2015 String Formation Beyond Leading Colour, JHEP08 003, arXiv:1505.01681 [hep-ph].
[24] Buckley A et. al. 2011 General-purpose event generators for LHC physics, Phys. Rept. 504 145, arXiv:1101.2599 [hep-ph].

Journal of Physics: Conference Series **668** (2016) 012035 doi:10.1088/1742-6596/668/1/012035

# Excluded volume effects in the hybrid star EoS

**David E. Alvarez-Castillo**[1,2], **Mark A. R. Kaltenborn**[3], **David Blaschke**[1,3,4]

Bogoliubov Laboratory for Theoretical Physics, JINR Dubna, 141980 Dubna, Russia
Universidad Autónoma de San Luis Potosí, S.L.P. 78290 México, Mexico
Instytut Fizyki Teoretycznej, Uniwersytet Wroclawski, 50-204 Wroclaw, Poland
National Research Nuclear University (MEPhI), 115409 Moscow, Russia

E-mail: alvarez@theor.jinr.ru

**Abstract.** In this contribution, we outline a new 2-phase description of the quark-nuclear matter hybrid equation of state that takes into account effects of phase space occupation (excluded volume) in both, the hadronic and the quark matter phases. For the nuclear matter phase, the reduction of the available volume at increasing density leads to a stiffening, while for the quark matter phase a reduction of the effective string tension in the confining density functional is obtained. The deconfinement phase transition in the resulting hybrid equation of state is sensitive to both excluded volume effects. As an application, we consider matter under compact star constraints of electric neutrality and $\beta$-equilibrium. We obtain mass-radius relations for hybrid stars that fulfill the $2M_\odot$ constraint and exhibit the high-mass twin phenomenon. Both features depend sensitively on the excluded volume.

## 1. Introduction

There is a growing interest in the physics of compact stars, since their masses and radii are observables that can constrain the dense matter equation of state (EoS) at zero temperature and thus provide benchmarks, which are not accessible in terrestrial experiments. The question arises: is there quark matter in compact star interiors, and how would it manifest itself in observations? In this context, the recent conjecture about the existence of high-mass twin stars [1, 2] is of special interest for the heavy-ion collision programmes searching for the critical point in the QCD phase diagram. If the pattern of a third family of compact stars, or just a quasi-horizontal branch in the population of high mass pulsars at $\sim 2M_\odot$ in their mass-radius ($M-R$) diagram could be identified by means of precise mass and radius measurements, then this would imply the existence of a strong 1$^{\rm st}$ order phase transition (PT) in compact star matter at $T=0$, which in turn proves the existence of at least one critical endpoint in the QCD phase diagram.

As it has been shown in [3], there are three necessary conditions on the hybrid star EoS in order to obtain high-mass twin configurations in the $M-R$ diagram: (i) a strong stiffening of nuclear matter at supersaturation densities, (ii) a density dependent stiffening of quark matter and (iii) a sufficiently soft behaviour of quark matter at the deconfinement PT in order to provide the necessary large jump in energy density in order to render the hybrid star branch disconnected from the hadronic one. For the EoS criteria on hybrid star branches being connected or disconnected to hadronic ones, see also [4].

In this contribution, we present an effective relativistic density functional approach to quark-nuclear matter at zero temperature that implements a mechanism of quark confinement based on the Cornell-type confining potential. To this end, we revive the ideas of the string-flip model (SFM) [5, 6] giving them a relativistic form as in [7]. In a simplified form such a model has recently already been applied to study massive hybrid stars [8]. The main new ingredient of the present work is a heuristic ansatz for the reduction and vanishing of the effective string tension in dense matter, which is assumed to scale with the available volume fraction. This will provide condition (ii) while (i) is realized by an excluded volume prescription [9] applied to the DD2 EoS [10, 11] and (iii) from multiquark interactions [12]. We will provide the two-flavor quark-nuclear matter EoS under neutron star conditions of charge neutrality and $\beta$-equilibrium and demonstrate the possibility to obtain high-mass twin stars with a third family branch that covers masses of $1.8 - 2.3\ M_\odot$ and radii of $11 - 14$ km that are well separated from the radii of the hadronic star branch of $15.0 - 15.5$ km in the same mass range.

## 2. Description of the quark matter model

The quark phase of matter in compact stars is considered by o be a combination of up, down quarks in $\beta$-equilibrium with electrons. We neglect the contribution from strange quarks since, due to the large threshold chemical potential of $\sim$ 1500MeV, they are insignificant. Here, we consider quark confinement to be achieved through a density dependence of the quark masses $m_{\rm I} = \Sigma_{\rm s}$, generated by the scalar quark selfenergy $\Sigma_{\rm s}$ that diverges for densities approaching zero. We notice that in the equation of motion for hadrons the diverging quark selfenergies are exactly compensated by those of the confining interaction [13].

As usual for fermions, there is also a vector selfenergy $\Sigma_{\rm v}$ that results in an energy correction contributing to the Hamiltonian functional,

$$H_{QCD} = H_{\rm k} + \sum_{q=u,d} m_{q0}\bar{q}q + H_{\rm I} \equiv \sum_{q=u,d} \bar{q}\left(\gamma_0\ \vec{\gamma}\cdot\vec{p} + m_q - \gamma_0\omega_0\right) q \ , \tag{1}$$

where $m_{q0}$ $(q = u, d)$ are the bare quark masses, which we neglect here in considering the chiral limit. $H_{\rm k}$ is the kinetic term, and $H_{\rm I}$ is the interaction term.

The effective mass, $m_q$, and the vector meanfield, $\omega_0$, embody all the medium effects of interacting quark matter. At $T = 0$, the density functional for the effective mass is taken from the string flip model (see also [8]),

$$m_q = m_{q0} + \Sigma_{\rm s} = m_{q0} + \underbrace{D(n_{\rm B})n_{\rm B}^{-\frac{1}{3}}}_{\text{confinement}} + \underbrace{Cn_{\rm B}^{\frac{1}{3}}}_{\text{Coulomb or OGE}} \ , \tag{2}$$

where OGE stands for one gluon exchange, and $\Sigma_{\rm s}$ is parameterized as a function of the baryon number density $n_{\rm B}$. The confinement interaction dominates at lower densities, while the perturbative OGE interaction becomes more important at higher densities and is neglected for now. The main new element of the present study is $D(n_B)$, the effective in-medium string tension resulting from multiplying the vacuum string tension $\sigma \sim D_0$ between quarks with the available volume fraction $\Phi(n_{\rm B})$,

$$D(n_{\rm B}) = \underbrace{D_0\ \Phi(n_{\rm B})}_{\text{effective string tension}} \ , \tag{3}$$

$$\Phi(n_{\rm B}) = \exp\left\{-\frac{\alpha}{2}(n_{\rm B} - n_0)\left(n_{\rm B} - n_0 + |n_{\rm B} - n_0|\right)\right\} \ , \quad \text{Gaussian} \tag{4}$$

$$\Phi(n_{\rm B}) = (1 + v_{\rm ex}n_{\rm B})^{-1} \ , \quad \text{inverse linear} \tag{5}$$

where $v_{\rm ex}$ is the excluded volume parameter and $\alpha = v_{\rm ex}^2$. The effect of the Gaussian versus the inverse linear scheme on the EoS are compared in Fig. 1. From [12], we have

$$\omega_0(n_{\rm B}) = \frac{a}{\Lambda^2}n_{\rm B} + \frac{b}{\Lambda^8}n_{\rm B}^3 \ , \tag{6}$$

where $\Lambda = 200$ MeV has been introduced to allow $a$ and $b$ to be dimensionless quantities. The quark quasiparticle dispersion relation in quark matter is then

$$E_q(p) \quad = \quad \sqrt{p^2 + m_q^2(n_{\rm B})} + \omega_0(n_{\rm B}) \ . \tag{7}$$

The EoS follows from the energy density $\varepsilon = \sum_{q=u,d} \varepsilon_q$, with the partial energy densities

$$\varepsilon_q \quad = \quad \frac{3}{\pi^2}\int_0^{p_q} dp\; p^2 E_q(p) = \underbrace{\frac{3}{\pi^2}\int_0^{p_q} \sqrt{p^2 + m_q^2(n_{\rm B})}\; p^2 dp}_{\varepsilon_{{\rm FG},q}} + \underbrace{n_q \omega_0(n_{\rm B})}_{\Delta\varepsilon_q} \ , \tag{8}$$

where $p_q$ is the quark Fermi momentum. Since the quark masses $m_q$ are density dependent, the quark chemical potentials $\mu_q$ have an additional term, $\Delta\mu_q$ with respect to the free Fermi gas

$$\mu_q \quad = \quad \frac{d\varepsilon}{dn_q} = \frac{\partial\varepsilon_{{\rm FG},q}}{\partial p_q}\frac{\partial p_q}{\partial n_q} + \sum_j\left(\frac{\partial\varepsilon}{\partial m_j}\frac{\partial m_j}{\partial n_q} + \frac{\partial\Delta\varepsilon_q}{\partial n_q}\right) = \sqrt{p_q^2 + m_q^2(n_{\rm B})} + \Delta\mu_q \ , \tag{9}$$

$$\Delta\mu_q \quad = \quad \sum_{i=u,d}\frac{3}{\pi^2}\int_0^{p_i} p^2 dp \frac{m_i}{\sqrt{p^2+m_i^2}}\frac{\partial m_i}{\partial n_q} + \omega_0(n_{\rm B}) + n_{\rm B}\frac{\partial\omega_0(n_{\rm B})}{\partial n_{\rm B}} . \tag{10}$$

The relevant chemical potentials satisfy the $\beta-$equilibrium condition $\mu_d = \mu_u + \mu_e$ , where we assume that neutrinos escape without interacting. The number densities are then constrained by the charge neutrality condition in compact stars $(2/3)n_u - (1/3)n_d - n_e = 0$. With these values we are able to calculate total pressure as the thermodynamic potential of the system,

$$P_{\rm tot} = -\sum_{i=u,d,e} \varepsilon_i + \sum_{i=u,d,e} \mu_i n_i \ . \tag{11}$$

## 3. Results and Conclusions

Our results for the compact star EoS are shown in the left panel of Fig. 1 and demonstrate that increasing the excluded volume parameter results in a stiffening of the nuclear matter and a softening of the quark matter branches so that at the intersection of both branches in the $P-\mu_B$ diagram the change in pressure slope increases. This corresponds to a larger jump in the energy density at the transition and thus to enforcing the instability induced by it for compact star configurations. The masses and radii for compact stars are obtained from solving the Tolman-Oppenheimer-Volkoff equations with the corresponding hybrid EoS as input. Results are shown in the right panel of Fig. 1 and illustrate the above statement that the increase of the excluded volume parameter $v_{\rm ex}$ leads to the occurrence and prominence of the high-mass twin star effect. For $v_{\rm ex} = 0$ the $2M_\odot$ mass constraint [14, 15] is not fulfilled.

We have presented an effective relativistic density functional approach to quark-nuclear matter at zero temperature that implements a mechanism of quark confinement based on the Cornell-type confining potential, thus reviving the ideas of the string-flip model in a relativistic fashion. We have demonstrated that the main new ingredient of the present work, the effective reduction of the string tension in dense matter by the available volume fraction $\Phi(n_B)$ results in a softening of quark matter at intermediate densities in the region of deconfinement transition. This provided a large jump in energy density at the transition as a necessary condition for obtaining high-mass twin star configurations and an extended branch of compact, massive hybrid stars with masses of 1.8 – 2.3 $M_\odot$ and radii of 11 – 14 km that are well separated from the hadronic star branch in the same mass range. It is a goal of future observational programmes to make sufficiently accurate neutron star radius measurements feasible.

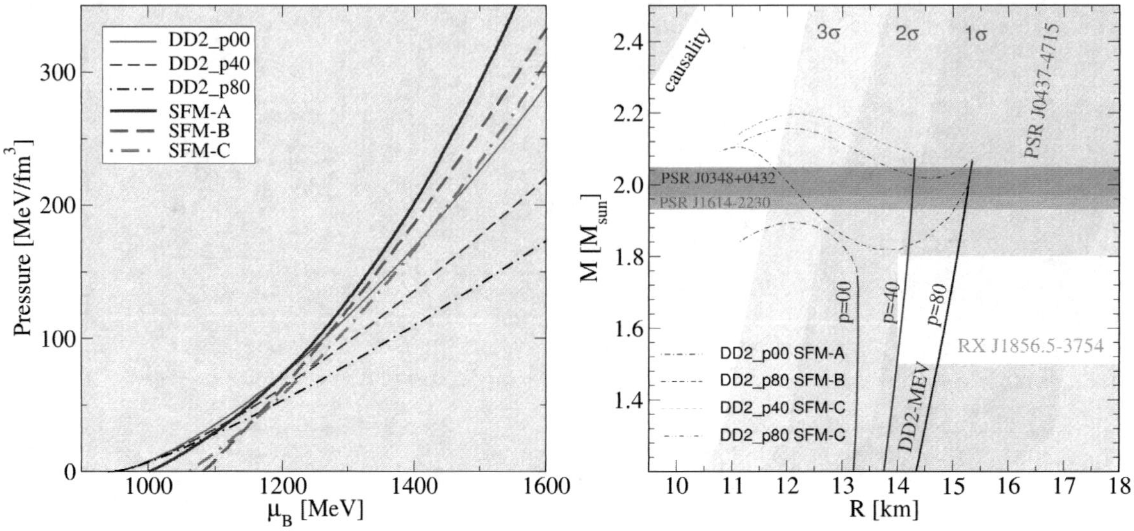


**Figure 1.** Left: EoS with deconfinement transition between DD2-MEV (thin lines) and SQM (bold lines) for different excluded volume parameters, ($v_{\rm ex}$[fm$^3$], $D_0$[MeV], a, b): SFM-A (0.0, 170, 0.0, 0.0), SFM-B Gaussian $\Phi(n_{\rm B})$ (2.0, 190, 0.29, 0.01) and SFM-C inverse linear $\Phi(n_{\rm B})$ (2.0, 220, 0.25, 0.04). The label "p80" stands for positive $v_{\rm ex} = 8.0$ fm$^3$, "p40" for $v_{\rm ex} = 4.0$ fm$^3$, and "p00" for $v_{\rm ex} = 0$. Right: Mass-Radius sequences for compact stars with the EoS of the left panel. The hadronic part of the sequences is shown by the solid lines, the broken lines correspond to hybrid stars whereby the branches with positive slope are formed by unstable configurations.

## Acknowledgements

The work of D.E.A.-C. and D.B. was supported by Narodowe Centrum Nauki (NCN) under contract number UMO-2014/13/B/ST9/02621 while M.A.R.K. received support from NCN under contract number UMO-2011/02/A/ST2/00306. D.E.A.-C. acknowledges the Heisenberg-Landau and Bogoliubov-Infeld programmes for supporting in part his collaboration visits at GSI Darmstadt (Germany) and at University of Wroclaw (Poland), respectively.

## References

[1] D. E. Alvarez-Castillo and D. Blaschke, *Proving the CEP with compact stars?*, arXiv:1304.7758 [astro-ph.HE].
[2] D. Blaschke, D. E. Alvarez-Castillo and S. Benic, PoS CPOD **2013**, 063 (2013) [arXiv:1310.3803 [nucl-th]].
[3] S. Benic, D. Blaschke, D. E. Alvarez-Castillo, T. Fischer and S. Typel, Astron. Astrophys. **577**, A40 (2015).
[4] M. G. Alford, S. Han and M. Prakash, Phys. Rev. D **88**, no. 8, 083013 (2013).
[5] C. J. Horowitz, E. J. Moniz and J. W. Negele, Phys. Rev. D **31**, 1689 (1985).
[6] G. Röpke, D. Blaschke and H. Schulz, Phys. Rev. D **34**, 3499 (1986).
[7] A. S. Khvorostukin, V. V. Skokov, V. D. Toneev and K. Redlich, Eur. Phys. J. C **48**, 531 (2006).
[8] A. Li, W. Zuo and G. X. Peng, Phys. Rev. C **91**, no. 3, 035803 (2015).
[9] S. Typel, private communication (2015).
[10] S. Typel, Phys. Rev. C **71**, 064301 (2005).
[11] S. Typel, G. Röpke, T. Klähn, D. Blaschke and H. H. Wolter, Phys. Rev. C **81**, 015803 (2010).
[12] S. Benic, Eur. Phys. J. A **50**, 111 (2014).
[13] L. Y. Glozman, Phys. Rev. D **79**, 037504 (2009).
[14] P. Demorest, T. Pennucci, S. Ransom, M. Roberts and J. Hessels, Nature **467**, 1081 (2010)
[15] J. Antoniadis *et al.*, Science **340**, 6131 (2013)

# Energy dependence of fluctuations in p+p and Be+Be collisions from NA61/SHINE

**Evgeny Andronov for the NA61/SHINE Collaboration**

Saint Petersburg State University, ul. Ulyanovskaya 1, 198504, Petrodvorets, Saint Petersburg, Russia

E-mail: evgeny.andronov@cern.ch

**Abstract.** The main aims of the NA61/SHINE experiment are to discover the critical point of strongly interacting matter and to study properties of the onset of deconfinement. For these goals a two dimensional scan of the phase diagram (T-$\mu_B$) is being performed at the SPS with measurements of hadron production in proton-proton, proton-nucleus and nucleus-nucleus interactions as a function of collision energy and system size. It is expected that fluctuations of various dynamical quantities will increase close to the critical point. In this contribution recent results of measurements of fluctuations in p+p and Be+Be collisions at SPS energies are presented as well as comparisons with model predictions and corresponding data of other experiments are discussed.

## 1. Introduction

The NA61/SHINE experiment [1] is a fixed target experiment at the Super Proton Synchrotron (SPS) of the European Organization for Nuclear Research (CERN). The strong interaction programme of NA61/SHINE consists of studies of the onset of deconfinement (OD) [2], [3] in heavy ion collisions and search for the critical point (CP) [4] of strongly interacting matter. A two-dimensional phase diagram scan - energy versus system size - is being performed by NA61/SHINE. It includes measurements of hadron production in collisions of protons and various nuclei (p+p, Be+Be, Ar+Sc, Xe+La) at a range of beam momenta (13$A$ - 158$A$ GeV/c). It is expected that there will be an enhancement of fluctuations of a number of observables due to the phase transition related to the CP and the OD [5]. Some hints of this enhancement were already observed by the NA49 experiment [6], predecessor of NA61/SHINE, at the top SPS energy [7].

## 2. Fluctuation measures

The size of the system created in collisions of two nuclei changes significantly from event to event influencing fluctuations of measured quantities. The following classification of observables was introduced: 1) extensive quantity - proportional to the system volume in the Grand Canonical Ensemble or the number of the wounded nucleons in the Wounded Nucleon Model [8] 2) intensive quantity - independent of the system volume 3) strongly intensive quantity [9] - independent of the system volume and fluctuations of this volume. Strongly intensive quantities are best suited to study fluctuations in nucleus-nucleus collisions because of the unavoidable event-to-event variations of the volume.

The most common way to characterize fluctuations of the quantity $A$ is to measure the variance of its distribution $Var(A)$ which is an extensive quantity. The scaled variance

$$\omega[A] = \frac{Var(A)}{\langle A \rangle} \tag{1}$$

is intensive. For the Poisson distribution $\omega = 1$. Section 2.1 is devoted to this quantity applied in the analysis of multiplicity fluctuations.

As was shown in [9], there are two ways to combine the first two moments of extensive observables $A$ and $B$ to get strongly intensive quantities:

$$\Delta[A, B] = \frac{1}{C_\Delta}\Big[\langle B \rangle \omega[A] - \langle A \rangle \omega[B]\Big] \tag{2}$$

$$\Sigma[A, B] = \frac{1}{C_\Sigma}\Big[\langle B \rangle \omega[A] + \langle A \rangle \omega[B] - 2(\langle AB \rangle - \langle A \rangle \langle B \rangle)\Big] \tag{3}$$

The normalization factors $C_\Delta$ and $C_\Sigma$ are usually chosen in such a way [10] that for independent particle emission $\Delta[A, B] = \Sigma[A, B] = 1$. In Section 2.2 results for electric charge fluctuations are presented. In this case:

$$A = N_+, \qquad B = N_-, \qquad C_\Delta = \langle N_- \rangle - \langle N_+ \rangle, \qquad C_\Sigma = \langle N_- \rangle + \langle N_+ \rangle.$$

### *2.1. NA61/SHINE multiplicity fluctuation studies*

The analysis of multiplicity fluctuations was performed for inelastic p+p interactions at $\sqrt{s_{NN}} =$ 7.6, 8.7, 12.3, 17.3 GeV and for centrality selected Be+Be interactions at $\sqrt{s_{NN}} = 6.27$, 8.73, 11.94, 16.83 GeV. Results are preliminary, include the statistical uncertainty and a first estimate of systematic uncertainty.

Figure 1 presents results for the scaled variance of the charged hadron multiplicity distribution from Be+Be collisions and inelastic p+p interactions [11]. $\omega$ increases with energy probably due to KNO scaling (variance of the multiplicity distribution increases faster than its mean). Larger values for Be+Be than for p+p collisions are probably due to volume fluctuations because $\omega$ is not a strongly intensive measure.

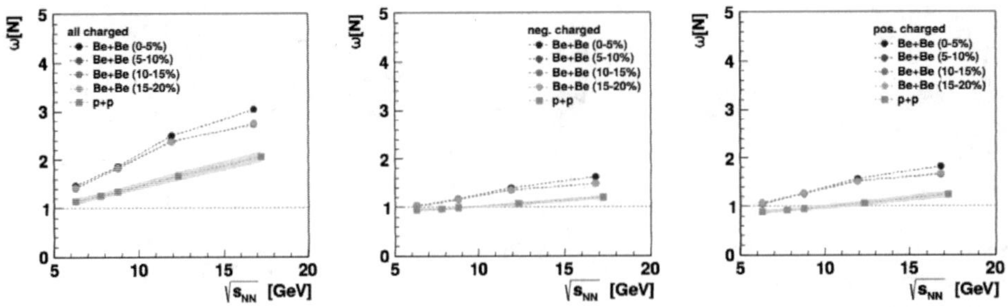


**Figure 1.** Scaled variance of the multiplicity distribution for Be+Be collisions of several collision centralities and energies for all charged (left), negatively charged (middle) and positively charged (right) particles compared with the corresponding results from the energy scan with inelastic p+p interactions [11].

In order to measure fluctuations of particles of specific types a new identification procedure (the identity method [12]) was developed. It allows to determine the second and third moments of the multiplicity distribution when the particle identification is not unique but can only be done on a statistical basis. Applying this method NA49 and NA61/SHINE determined the scaled variance of the multiplicity distribution of identified protons, kaons and pions in inelastic p+p and 3.5% most central Pb+Pb collisions [13]. For $\pi$ as well as for all charged particles we observe a linear increase of $\omega$ with the collision energy and higher values of $\omega$ for the larger system. For $K$ and p these properties are probably masked by effects of baryon number conservation and strangeness conservation.

Both for unidentified and identified charged particle fluctuations no indication of the CP is found.

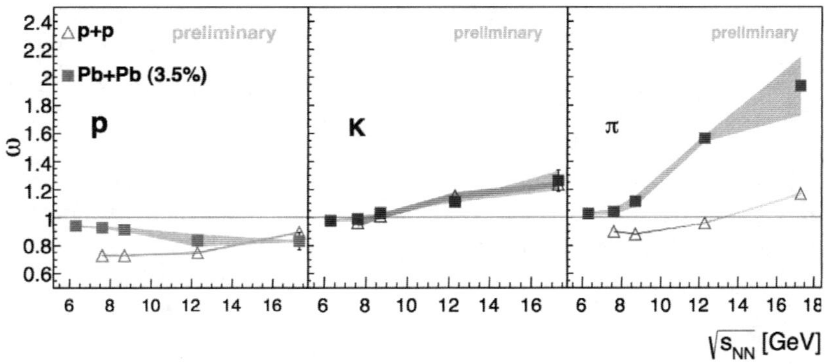


**Figure 2.** Scaled variance $\omega$ of the multiplicity distribution of protons, kaons, and pions versus nucleon-nucleon cms energy $\sqrt{s_{NN}}$ for the 3.5% most central Pb+Pb collisions (NA49, full squares) and inelastic p+p reactions (NA61, open triangles) [13].

### *2.2. NA61/SHINE electric charge fluctuation studies*

Fluctuations of conserved charges are expected to be sensitive to phase transitions [14]. An analysis of fluctuations of electric charge was performed for Be+Be interactions at $\sqrt{s_{NN}} = 16.83$ GeV for three centrality classes 0-5%, 0-10% and 0-20%. Centrality was determined from the energy deposited in the NA61/SHINE forward calorimeter - the Projectile Spectator Detector (PSD). $\Delta[N_+, N_-]$ and $\Sigma[N_+, N_-]$ were calculated for 9 pseudorapidity intervals centred at pseudorapidity $\eta_{lab} = 4.6$ in the laboratory reference frame. The width of the interval varies from 0.2 to 3.4 units of pseudorapidity. Results are preliminary and include the statistical uncertainty.

In Figure 3 results for $\Delta[N_+, N_-]$ and $\Sigma[N_+, N_-]$ are plotted as a function of the width of the pseudorapidity interval for three centrality classes. The weak dependence on centrality in Be+Be shows that the fluctuations do not depend on the collision volume. The suppressed (<1) values of $\Sigma[N_+, N_-]$ are probably due to global conservation of electric charge. This effect is less visible for $\Delta[N_+, N_-]$.

In Figure 4 a comparison of $\Delta[N_+, N_-]$ and $\Sigma[N_+, N_-]$ for the 0-20% centrality class with the prediction from the EPOS1.99 model [15] is shown. Theoretical predictions were restricted to the NA61/SHINE geometrical acceptance. One can see that the behaviour of the measured charge fluctuation quantities is well reproduced by the EPOS model.

## Acknowledgments

This work was supported by the Federal Agency of Education of the Ministry of Education and Science of the Russian Federation, SPbSU research grant 11.38.193.2014, the Hungarian

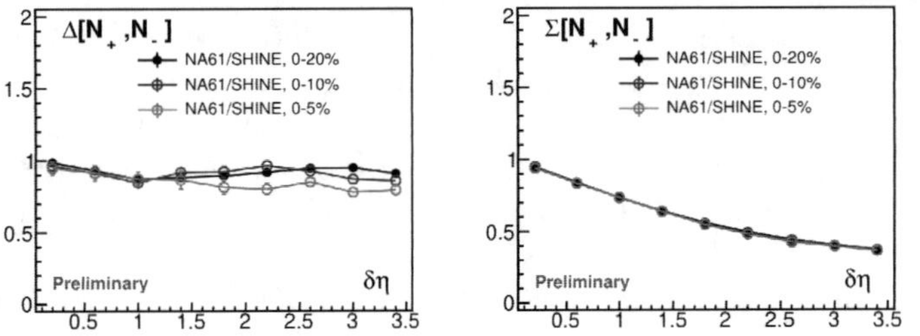


**Figure 3.** $\Delta[N_+, N_-]$ (left) and $\Sigma[N_+, N_-]$ (right) as a function of the width of pseudorapidity interval $\delta\eta$ for 0-5% (red), 0-10% (blue) and 0-20% (black) centrality classes for Be+Be interactions at $\sqrt{s_{NN}} =$ 16.83 GeV.

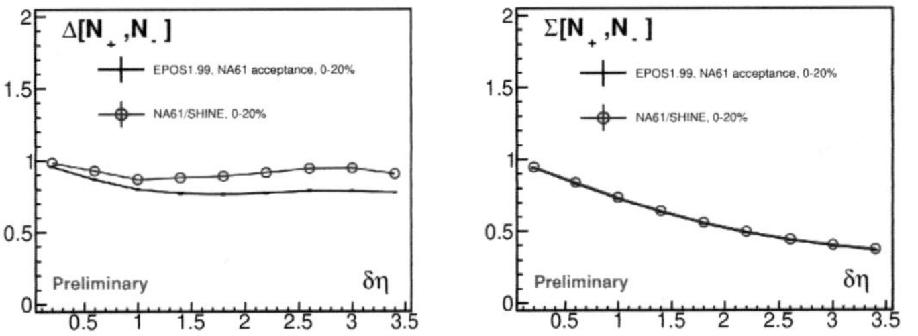


**Figure 4.** $\Delta[N_+, N_-]$ (left) and $\Sigma[N_+, N_-]$ (right) as a function of the width of pseudorapidity interval $\delta\eta$ for 0-20% centrality class (black) in comparison with predictions from the EPOS model (blue) for Be+Be interactions at $\sqrt{s_{NN}} = 16.83$ GeV.

Scientific Research Fund (grants OTKA 68506 and 71989), the János Bolyai Research Scholarship of the Hungarian Academy of Sciences, the Polish Ministry of Science and Higher Education (grants 667/N-CERN/2010/0, NN 202 48 4339 and NN 202 23 1837), the Polish National Center for Science (grants 2011/03/N/ST2/03691, 2012/04/M/ST2/00816 and 2013/11/N/ST2/03879), the Foundation for Polish Science — MPD program, co-financed by the European Union within the European Regional Development Fund, the Russian Academy of Science and the Russian Foundation for Basic Research (grants 08-02-00018, 09-02-00664 and 12-02-91503-CERN), the Ministry of Education, Culture, Sports, Science and Technology, Japan, Grant-in-Aid for Scientific Research (grants 18071005, 19034011, 19740162, 20740160 and 20039012), the German Research Foundation (grant GA 1480/2-2), the EU-funded Marie Curie Outgoing Fellowship, Grant PIOF-GA-2013-624803, the Bulgarian Nuclear Regulatory Agency and the Joint Institute for Nuclear Research, Dubna (bilateral contract No. 4418-1-15/17), Ministry of Education and Science of the Republic of Serbia (grant OI171002), Swiss Nationalfonds Foundation (grant 200020117913/1) and ETH Research Grant TH-01 07-3.

## References

[1] Abgrall N *et al.* (NA61 Coll.) 2014 *JINST* **9** P06005
[2] Gazdzicki M and Gorenstein M I 1999 *Acta Phys. Polon.* B **30** 2705
[3] Alt C *et al.* (NA49 Coll.) 2008 *Phys. Rev.* C **77** 024903
[4] Fodor Z and Katz S D 2004 *JHEP* **0404** 050
[5] Stephanov M A, Rajagopal K and Shuryak E V 1999 *Phys. Rev.* D **60** 114028
[6] Afanasiev S *et al.* (NA49 Coll.) 1999 *Nucl. Instrum. Meth.* A **430** 210
[7] Grebieszkow K for the NA49 Coll. 2009 *Nucl. Phys.* A **830** 546c50c
[8] Bialas A, Bleszynski M and Czyz W 1976 *Nucl. Phys.* B **111** 461
[9] Gorenstein M I and Gazdzicki M 2011 *Phys. Rev.* C **84** 014904
[10] Gazdzicki M, Gorenstein M I and Mackowiak-Pawlowska M 2013 *Phys. Rev.* C **88** 024907
[11] Czopowich T. for the NA61/SHINE Coll. 2014 Transverse momentum and multiplicity fluctuations in the Be+Be energy scan from NA61/SHINE *Proceedings of Science* PoS(CPOD2014)054
[12] Gazdzicki M, Grebieszkow K, Mackowiak M and Mrowczynski S 2011 *Phys. Rev.* C **83** 054907
[13] Mackowiak-Pawlowska M and Wilczek A for the NA61 Coll. 2014 *J. Phys.: Conf. Series* **509** 012044
[14] Jeon S and Koch V 2000 *Phys. Rev. Lett.* **85** 2076
[15] Werner K 2007 *Phys. Rev. Lett.* **98** 152301

# Two-particle correlation measurements in p+Nb reactions at $\sqrt{s_{\mathrm{NN}}} = 3.18$ GeV

**Oliver Arnold**[1,2] **for the HADES collaboration**

[1] Physik Department E12, Technische Universität München, James-Franck-Str. 1, 85748 Garching, Germany

[2] Excellence Cluster 'Origin and Structure of the Universe', Boltzmannstr. 2, 85748 Garching, Germany

E-mail: oliver.arnold@mytum.de

**Abstract.** We present a two-particle correlation measurement of proton- and of $\Lambda p$-pairs, measured with the HADES detector in p+Nb reactions at a kinetic beam energy of 3.5 GeV. The proton-proton correlation function is used to extract the size of the region of homogeneity. Using this information together with a UrQMD transport simulation opens the possibility to study the interaction of $\Lambda p$ pairs in terms of spin average scattering length and effective range.

## 1. Introduction

The collision of two heavy nuclei at high energies creates a complex system, which is believed to undergo several evolution steps in time including e.g. a hydrodynamic expansion controlled by the viscosity of the system. At the end of the time evolution, the formed hadrons freeze out and fly into the detector setup. To gain information about the underlying physics of the multi-particle production and the system dynamics, it is helpful to investigate the correlations between the produced particles. Measuring two-particle correlations between particles emitted from a source at freeze-out provides information about the spatio-temporal extension (radii) of the region of homogeneity on the femtometer scale. A dependence of these radii e.g. on the transverse mass $m_T$ of the pair is sensitive to various properties like the time scale of the evolution.

The general picture of two-particle correlations is that there is a probability that particles leaving the source after the freeze-out can still interact among each other. Knowing the final state interactions between the particles, one can solely concentrate on extracting the radii. However, it is also possible to turn the picture around: by knowing the radii of the particle emitting source one is able to investigate the final-state interaction especially between particle species where the interaction is not well determined. This is mainly the case for strange hadrons like pairs of $p\bar{\Lambda}, \Lambda\Lambda$ etc.. In most of these cases there is, if any, information from scattering experiments and therefore any data about the interaction is helpful in this respect. A couple of successful analyses using this technique were already performed with extractions of interaction parameters like cross sections, scattering length and effective ranges [1] [2].

## 2. The HADES Experiment

### *2.1. The individual components*

The High-Acceptance Di-Electron Spectrometer (HADES) is a fixed target experiment located in Darmstadt, Germany, at the GSI Helmholtzzentrum für Schwerionenforschung. Originally designed to measure rare electromagnetic decays of vector mesons like $\omega, \phi, \rho$ into dilepton pairs, the detector setup is also able to measure hadrons with an average momentum resolution of $\Delta p/p \approx 5\%$. Collision systems studied with HADES span a wide range, from elementary $p+p$ collisions up to more complex heavy-ion reactions of $Au + Au$, but also experiments with pion beams were performed quite recently to investigate e.g. excitations of $N^*$ resonances. HADES covers a nearly full azimuthal range of 85% and a polar angle interval between 18° and 85°.
HADES consists of several detector components which can be used for particle identification. We will present only these components which are used for the analysis. The superconducting magnets located between the Multiwire-Drift-chambers (MDCs), which information is also part of the tracking procedure, provide the momentum and polarity of the particles. The MDCs measure the energy loss of the traversing charged particles. This information together with theoretical Bethe-Bloch curves allows to identify the particles in a first step. In a second step we use additionally the energy loss measurement of the META system, which consists of two Time-Of-Flight walls named TOF and TOFINO, which distinguish themselves by different time resolutions and polar angle coverages.

### *2.2. The collision system under investigation*

The HADES collaboration measured in 2008 the collision of a proton with a niobium nucleus (p $+^{93}$ Nb), where the kinetic energy of the proton was $E_{\text{kin}} = 3.5$ GeV. This pA system offers the interesting environment to study particle productions and correlations between them in a rather low energy regime (compared to LHC or RHIC), which gives us the possibility to try to find similarities and trends which are already well established in heavy-ion reactions at large energies. A further advantage of the rather low energy is that the measured correlation function is free from correlation-disturbing effects induced for example by mini-jets, which can mimic a femtoscopy signal in the interesting region of low relative momenta and the absence of many high mass resonances which contribute via their feed down to residual correlations.

## 3. Data Analysis

As have been already mentioned, on the one hand the femtoscopy method can be used to study the size of the region of homogeneity, on the other hand by knowing the radii of the emitting source one can use this information to study final-state interactions. Because of this reason, the analysis was divided into two parts. In the first part we investigated the properties and size of the source by using proton-proton pairs, because the interaction is well established and a lot of pairs are produced in the pNb reaction. Additionally, it's a baryon pair like $\Lambda p$ in which we are interested to study its interaction, performed in the second part of the analysis.

### *3.1. Proton-Proton correlation function*

To identify protons, we use the energy loss information of the MDCs and the META system. With this method we obtain a proton sample which has a global purity of about 99%. The experimental correlation function is calculated by using pairs from the same event and divide them by a pair distribution obtained from mixed events as a function of the relative momentum of the pair in their rest frame $k = \frac{1}{2}|\mathbf{p}_1 - \mathbf{p}_2|$, $\mathbf{p}_1 + \mathbf{p}_2 = 0$. The idea is that pairs from event mixing reproduce the kinematics of the particle production but by construction are free from any other two-particle correlations. The experimental correlation function suffers from several detector efficiencies. First of all, correlated particles are usually emitted close together in space, which means the opening angle between the pair is small. This leads to the problem that at some point the opening angle is so small, that because of the finite granularity of the components, the detector is not able to distinguish between two pairs anymore, which leads to the well known track merging effect. Because the mixed event sample contains by construction always two distinct tracks, the track merging introduces a suppression of the correlation signal. We correct for this effect by looking at the ($\Delta\phi$, $\Delta\Theta$) plane, where a clear suppression for same events is visible. We exclude this angle region from the sample to get rid of the close track suppression. As a next step we explore the finite momentum resolution of the detector. A finite resolution usually leads to a broadening of the correlation signal which leads to an underestimation of the extracted source sizes. We simulate this effect by creating an ideal and a smeared mixed event distribution with the UrQMD event generator [3]. The smeared sample is obtained by simulating also the detector response of HADES. Both samples are weighted with a correlation weight, and the parameters of the weight are chosen in a way that the smeared correlation function matches the experimental measurement. Because we know also the ideal momenta we can correct for the momentum smearing. As a last step, we also have to deal with a more difficile effect that can be observed in the data. Usually the correlation signal is located at small relative momenta $k < 100$ MeV/c. Above this momentum region the particles are not correlated anymore in terms of femtoscopy effects and the correlation function should approach unity. Because we have to deal with a small pA system ($\langle A_{\mathrm{part}} \rangle = 2.7$) additional correlations are present and the baseline of the correlation function is still rising with increasing momenta. This effect can be explained by energy and momentum conservation which is not completely reproducible with the mixed event sample. To get rid of this so called long range correlations (LRC) we assume that the correlation signal and the LRC signal factorize because they stem from completely different sources $C_{\mathrm{meas}} = C_{\mathrm{femto}} \times C_{\mathrm{LRC}}$.

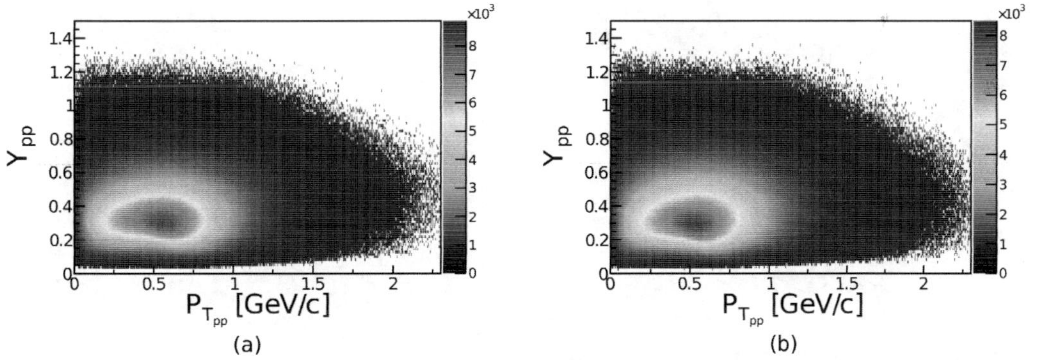


**Figure 1.** Two-particle phase-space of proton pairs as a function of the total transverse momentum and total rapidity inside of the HADES acceptance. The plot on the left side (a) is for experimental data and the right one (b) for simulations using the UrQMD model. The distributions look very similar, which means that UrQMD is able to describe the kinematics of the pairs.

To simulate the LRC, we use the UrQMD event generator, where the events are free from any femtoscopic correlations. The kinematics of the model is in good agreement with the experimental data which can be seen in Figure 1 illustrating the two-particle phase-space of protons as a function of the total transverse momentum and total rapidity of the pair in the HADES acceptance. The model is also able to describe the LRC of the correlation function and follows exactly the same trend. This motivates the introduction of a new correlation measurement defined as $C_{\text{LRC-free}} = C_{\text{meas}}/C_{\text{UrQMD}}$. After applying all the corrections we compare the measured correlation function to the theoretical prediction obtained by the Koonin-Pratt formalism [4]. The model takes the quantum statistics of the fermion pair into account and their Coulomb and strong interactions. For the source function we assume a Gaussian shape, which is very common in femtoscopy analyses and obtain a preliminary Gaussian source size of $R^{pp}_{G,meas} = 2.016 \pm 0.010^{+0.039}_{-0.027}$ fm including errors from statistics, momentum resolution and close track rejection.

### *3.2. $\Lambda p$ correlation function*

To obtain the $\Lambda p$ correlation function we use only the energy loss information from the MDCs for particle identification. This way we don't loose statistic with further PID cuts. The $\Lambda$-hyperon is reconstructed via the $p, \pi^-$ invariant mass ($\Lambda \rightarrow p + \pi^-$,BR $\sim 64\%$) technique. To reduce the combinatorial background of $p, \pi^-$-pairs, which do not originate from $\Lambda$ decays, we employ topological cuts on certain track parameters: A cut on the flight distance of the $\Lambda$, a cut on the distance between the daughter tracks, a cut on the distance of the daughter tracks to the primary vertex, and a cut on the pointing angle between the $\Lambda$ momentum vector and the vector pointing from the primary to the secondary vertex. To explore the systematics of these cuts on the correlation function we test three different cut combinations which give different $\Lambda$ purities of $P^{\Lambda}_{1,2,3} = 86.1, 89.6$ and $92.5\%$. An example plot showing the invariant mass distribution for the first cut combination is given in Figure 2. For all three topological cut combinations we

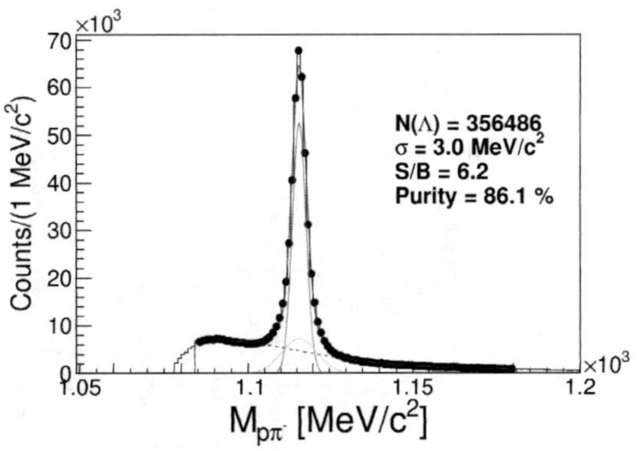


**Figure 2.** Invariant mass of $p, \pi^-$ pairs for one certain topological cut combination.

calculate the $\Lambda p$ correlation function and obtain the positive correlation signal, which is expected due to the attractive interaction between $\Lambda p$. All three correlation functions are corrected for close track efficiency, momentum resolution, purity and LRC. With help of UrQMD simulations, which include also the information of the freeze-out coordinates, we are able to determine the source functions and their sizes of the $pp$ and $\Lambda p$ system, where UrQMD predicts an about 20% smaller source size for $\Lambda p$, which results as a ratio of $\text{RF} = R^{pp}_{G,UrQMD}/R^{\Lambda p}_{G,UrQMD} = 1.23$. Together with the proton-proton measurement $R^{pp}_{G,meas}$ we are able to fix the source size of the $\Lambda p$ pair. By fitting the Lednicky-Lyuboshitz model [5], which depends on the scattering lengths and effective ranges, to the data we are able to extract spin averaged values for the scattering length and effective range.

[1] Adamczyk L *et al.* 2015 *Phys. Rev. Lett.* **114** 022301
[2] Kisiel A, Zbroszczyk H and Szymanski M 2014 *Phys. Rev.* C **89** 054916
[3] Bass S A *et al.* 1998 *Prog. Part. Nucl. Phys.* **41** 255
[4] Koonin S E 1977 *Phys. Lett.* B **70** 43
[5] Lednicky R and Lyuboshitz V L 1982 *Sov. J. Nucl. Phys.* **35** 770

# Mass-radius constraints for the neutron star EoS - Bayesian analysis

**A. Ayriyan**[1], **D.E. Alvarez-Castillo**[2,3], **D. Blaschke**[3,4,5], **H. Grigorian**[1,6]

[1]Laboratory for Information Technologies, JINR Dubna, 141980 Dubna, Russia
[2]Bogoliubov Laboratory for Theoretical Physics, JINR Dubna, 141980 Dubna, Russia
[3]Universidad Autónoma de San Luis Potosí, S.L.P. 78290 México, Mexico
[4]Institute of Theoretical Physics, University of Wrocław, 50-204 Wroclaw, Poland
[5]National Research Nuclear University (MEPhI), 115409 Moscow, Russia
[6]Department of Physics, Yerevan State University, Yerevan, Armenia

E-mail: `ayriyan@jinr.ru`

**Abstract.** We suggest a new Bayesian analysis (BA) using disjunct M-R constraints for extracting probability measures for cold, dense matter equations of state (EoS). One of the key issues of such an analysis is the question of a deconfinement transition in compact stars and whether it proceeds as a crossover or rather as a first order transition. We show by postulating results of not yet existing radius measurements for the known pulsars with a mass of $2M_\odot$ that a radius gap of about 3 km would clearly select an EoS with a strong first order phase transition as the most probably one. This would support the existence of a critical endpoint in the QCD phase diagram under scrutiny in present and upcoming heavy-ion collision experiments.

## 1. Introduction: EoS with a QCD phase transition and hybrid stars

There is a unique relationship between the EoS of cold dense matter and the sequence of compact star configurations in the mass-radius (M-R) diagram provided by the Tolman-Oppenheimer-Volkoff equations. It can be used to quantify the likelihood of EoS models by Bayesian analyses using a selection of mass and/or radius measurements as priors [1]. In this work we discuss a family of hybrid neutron stars described by an EoS with a first-order phase transition that is obtained from a Maxwell construction between hadronic and quark matter branches. Our approach to the hybrid EoS captures important physical features:

- excluded volume corrections in the hadronic EoS,
- multi-quark interactions for quark matter EoS.

The excluded volume corrections at the hadronic level take into account the inner structure of the nucleons which results in Pauli blocking due to quark exchange interactions. This can be seen as a precursor of the quark delocalization which is the expected mechanism of the deconfinement transition [2] that is still intensified by the simultaneous partial chiral restoration [3]. The excluded volume modification is implemented for baryon densities exceeding saturation $n_{\rm sat} = 0.16$ fm$^{-3}$ (the density in the interior of atomic nuclei) by considering the available volume $\Phi_N$ for the motion of nucleons at a given density $n$ as defined in [4]

$$\Phi_N = \begin{cases} 1 \,, & \text{if} \quad n \leq n_{\rm sat} \\ \exp[-v|v|(n - n_{\rm sat})^2/2] \,, & \text{if} \quad n > n_{\rm sat} \,, \end{cases} \tag{1}$$

where $v = 16\pi r_N^3/3$ is the van-der-Waals excluded volume that is related to the nucleon hard-core radius $r_N$. Note that in the form (1) assigning positive (negative) values for $v$ leads to a stiffening (softening) of the original EoS for which we adopt DD2 [5] with parameters from [6]. Here, we will consider only positive values of excluded volume and introduce for them the dimensionless parameter $p = 10 \times v[\mathrm{fm}^3]$, varying it in the range from 0 to 80.

We consider the multi-quark interactions as they were introduced by [7] and succesfully applied to massive hybrid star twins in [8]. It is expected that quark matter will still be strongly interacting at the energy densities present in neutron star interiors. Those fall in the range where in lattice QCD simulations at finite temperatures the transition from hadronic to quark matter is obtained. Therefore the introduction of the 8-quark vector channel interaction may be justified. In accordance with Refs. [7, 8], we denote the dimensionless coupling constant of this channel by $\eta_4$ and vary it in the range from 0 to 30. Increasing the value of this parameter has the effect of stiffening the EoS at higher densities, which allows to support hybrid stars as massive as 2 $M_\odot$ and thus fulfilling the constraint from recent observations [9, 10], see Fig. 1.

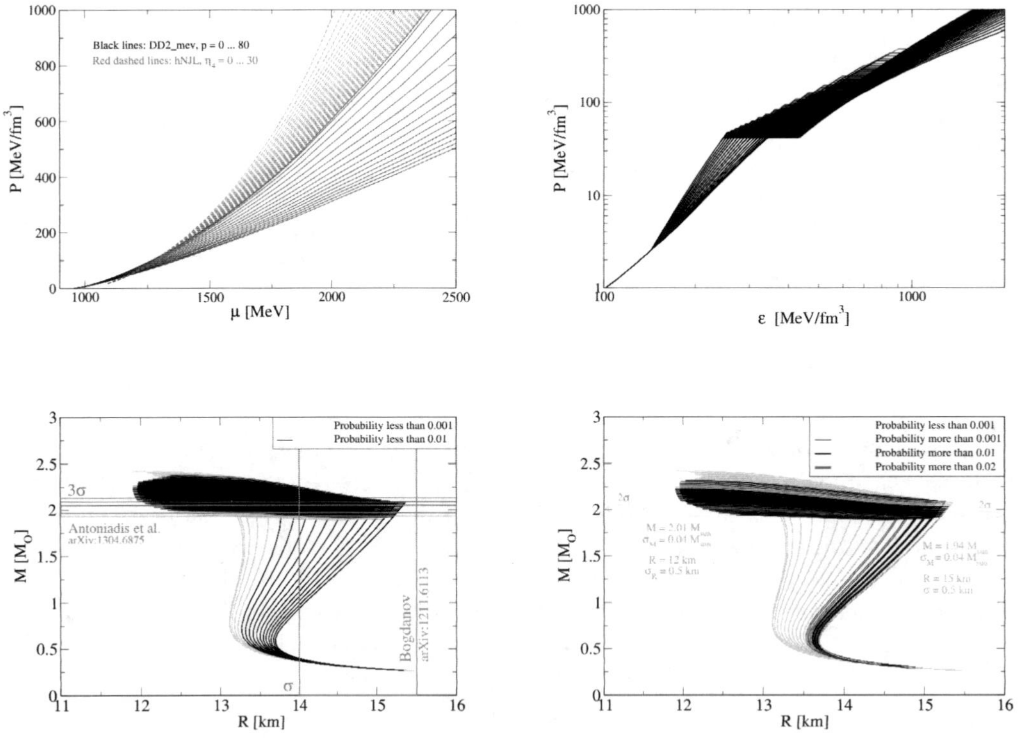


**Figure 1.** Hybrid EoS scheme for sets of two parameters $p$ and $\eta_4$ corresponding to hadronic and quark matter [11]. Upper panels: pressure vs. chemical potential (left) for hadronic (DD2_mev, black solid lines) and quark matter (hNJL, red dashed lines); pressure vs. energy density with Maxwell constructions of first order phase transitions (right). The two EoS parameters are varied in the range $(p, \eta_4) = (0 \ldots 80, 0 \ldots 30)$. Lower panels: The corresponding mass-radius relations are compared to existing mass and radius measurements (left) and to ficticious (right) priors for which it is assumed that for the two objects with measured high masses, PSR J0348+0432 [10] and PSR J1614-2230 [9] also their radii could be measured with a $1\sigma$ accuracy of $\sim 0.5$ km and turn out to be significantly different at the $3\sigma$ level with mean values of 15 km and 12 km, resp.

## 2. Observational constraints and Bayesian analysis

We want to compare the theoretical results for masses and radii of hybrid star sequences with suitable constraints from neutron star observations in order to conclude for the likeliness of the underlying EoS model. We consider two cases, a real (A) and a ficticious (B) set of measurements:

(i) case A
- a maximum mass constraint from PSR J0348+0432 [10],
- a radius constraint from the nearest millisecond pulsar and PSR J0437-4715 [12],

(ii) case B
- a radius measurement of $R = 12 \pm 0.5$ km for PSR J0348+0432 with its known mass,
- a radius measurement of $R = 15 \pm 0.5$ km for PSR J1614-2230 with its known mass,

which both are shown in the lower panels of Fig. 1.

For the BA of the most probable EoS for given prior from (real or ficticious) observations, we start by defining a vector of free parameters $\overrightarrow{\pi}(p, \eta_4)$, which correspond to all the possible models with phase transition from nuclear to quark matter using the EoS described above. The way we sample these parameters is

$$\pi_i = \overrightarrow{\pi}\left(p(k), \eta_4(l)\right), \tag{2}$$

where $i = 0, 1, \ldots, N-1$ with $N = N_1 \times N_2$ such that $i = N_2 \times k + l$ and $k = 0, 1, \ldots, N_1 - 1$, $l = 0, 1, \ldots, N_2 - 1$, with $N_1$ and $N_2$ as the total number of parameters $p_k$ and $\eta_{(4)l}$ respectively. The goal is to find the set $\pi_i$ corresponding to an EoS and thus a sequence of configurations which contains the most probable one based on the given constraints using BA. For initializing the BA we propose that *a priori* each vector of parameters $\pi_i$ has the same probability: $P(\pi_i) = 1/N$ for $\forall i$. We can calculate probability of $\pi_i$ using Bayes' theorem [13]

$$P(\pi_i \,|E) = \frac{P(E\,|\pi_i)\,P(\pi_i)}{\sum\limits_{j=0}^{N-1} P(E\,|\pi_j)\,P(\pi_j)}. \tag{3}$$

The results of our BA for the two cases defined above are shown in the two panels of Fig. 2. For the case A, due to the constraint of a large radius, the stiffest possible EoS among the accessible parameters was selected as the most probable hybrid star EoS. In particular, largest values of the parameters $p$ and $\eta_4$ that would correspond to hybrid star branches connected with hadronic ones, in the classification of [14]. For case B, the high-mass twin star solution is preferred which corresponds to a disconnected hybrid star branch ("third family") and disselects large values of the $\eta_4$ parameter. Herewith we could demonstrate the usefulness of the Bayesian analysis method. It allows to check the discriminating power of a certain experimental result before the actual measurement was made. In this sense the discovery potential of future programmes for compact star observations, like NICER or SKA could be quantified.

## 3. Conclusions

To test the interesting possibility that a strong first-order phase transition might occur in massive neutron stars of 2 $M_\odot$ [10] leading to the appearance of a "third family" of hybrid stars (mass twins) at this mass, one should perform radius measurements for these massive neutron stars [13]. This possibility offers bright prospects for future observational campaigns and bears the chance to "prove" the existence of a critical point [8, 11, 13, 15, 16, 17, 18] in the QCD phase diagram from astrophysics.

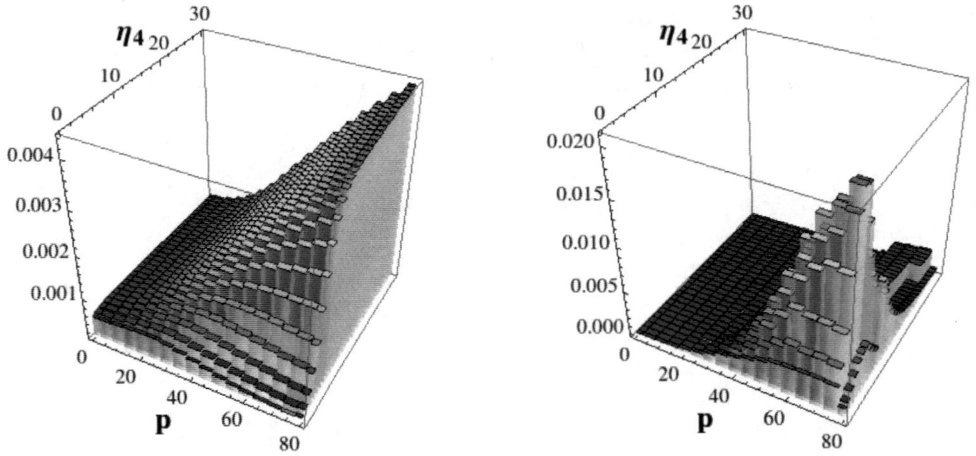


**Figure 2.** LEGO plots for Bayesian analysis in the plane of two parameters $(p, \eta_4)$; for their definition see text. The left panel uses the priors of case A: the mass measurement by Antoniadis et al. [10] and the radius measurement by Bogdanov [12] as given in the lower left panel of Fig. 1. The result in the right panel considers the priors of case B, with two ficticious radius measurement, as in the lower right panel of Fig. 1.

## 4. Acknowledgments

We acknowledge fruitful discussions with M. C. Miller and J. Trümper. We thank S. Typel and S. Benic for providing us with their EoS data. D.B., H.G. and D.E.A.-C. profited from support of networking activities by the COST Action MP1304 "NewCompStar". This work was supported by the NCN "Opus" programme under contract UMO-2014/13/B/ST9/02621. D.E.A-C., A.A. and H.G. received support from the Bogoliubov–Infeld and the Ter-Antonian–Smorodinsky programmes. A.A. acknowledges JINR grant No. 14-602-01.

## References

[1] A. W. Steiner, J. M. Lattimer and E. F. Brown, Astrophys. J. **722**, 33 (2010).
[2] G. Röpke, D. Blaschke and H. Schulz, Phys. Rev. D **34**, 3499 (1986).
[3] D. Blaschke, H. Grigorian and G. Röpke, in preparation (2015).
[4] S. Typel, to appear in the 2015 EPJA Topical Issue on "Exotic Matter in Neutron Stars".
[5] S. Typel and H. H. Wolter. Nucl. Phys. A **656**, 331 (1999).
[6] S. Typel, G. Röpke, T. Klähn, D. Blaschke and H. H. Wolter, Phys. Rev. C **81**, 015803 (2010).
[7] S. Benic. Eur. Phys. J. A **50**, 111 (2014).
[8] S. Benic, D. Blaschke, D. E. Alvarez-Castillo, T. Fischer and S. Typel, Astron. Astrophys. **577**, A40 (2015).
[9] P. Demorest, T. Pennucci, S. Ransom, M. Roberts and J. Hessels, Nature **467**, 1081 (2010).
[10] J. Antoniadis, P. C. C. Freire, N. Wex, T. M. Tauris, R. S. Lynch, M. H. van Kerkwijk, M. Kramer and C. Bassa, V. S. Dhillon, T. Driebe, J. W. T. Hessels, V. M. Kaspi, V. I. Kondratiev, N. Langer, T. R. Marsh, M. A. McLaughlin, T. T. Pennucci, S. M. Ransom, I. H. Stairs, J. Leeuwen, J. P. W. Verbiest, D. G. Whelan, Science **340**, 6131 (2013).
[11] D. E. Alvarez-Castillo, A. Ayriyan, D. Blaschke and H. Grigorian, arXiv:1506.07755
[12] S. Bogdanov, Astrophys. J. **762**, 96 (2013).
[13] D. Blaschke, H. Grigorian, D. E. Alvarez-Castillo and A. Ayriyan, J. Phys. Conf. Ser. **496**, 012002 (2014).
[14] M. G. Alford, S. Han and M. Prakash, Phys. Rev. D **88**, 083013 (2013).
[15] D. Blaschke, D. E. Alvarez-Castillo and S. Benic, PoS **CPOD 2013**, 063 (2013).
[16] D. E. Alvarez-Castillo and D. Blaschke, arXiv:1304.7758 [astro-ph.HE].
[17] A. Ayriyan, D. E. Alvarez-Castillo, D. Blaschke, H. Grigorian and M. Sokolowski, Phys. Part. Nucl. **46**, 854 (2015).
[18] D. E. Alvarez-Castillo and D. Blaschke, Phys. Part. Nucl. **46**, 846 (2015).

Journal of Physics: Conference Series **668** (2016) 012039 doi:10.1088/1742-6596/668/1/012039

# Azimuthal anisotropy of $\varphi$ meson in U+U and Au+Au collisions at RHIC

**Vipul Bairathi (for the STAR Collaboration)**

National Institute of Science Education and Research (NISER), Jatni-752050, India.

E-mail: vipul@rcf.rhic.bnl.gov/vipul.bairathi@gmail.com

**Abstract.** The measurements of the azimuthal anisotropy of $\varphi$ meson in the U+U and Au+Au collisions at the Relativistic Heavy-Ion Collider (RHIC) are reported. The centrality dependence of the Fourier coefficients $v_2$, $v_3$, $v_4$ and $v_5$ is presented for $\varphi$ meson at midrapidity ($|\eta| \leq 1.0$), in U+U and Au+Au collisions at $\sqrt{s_{NN}}$ = 193 and 200 GeV, respectively. The $\eta$-sub event plane method is used with a $\eta$ gap of 0.1 to suppress the non-flow effects. A strong centrality dependence is observed for the $\varphi$ meson elliptic flow ($v_2$), whereas no clear centrality dependence is observed for $v_3$, $v_4$ and $v_5$. Ratios of the Fourier coefficients, $v_3/v_2$ and $v_4/v_2^2$ as a function of transverse momentum ($p_T$) are also presented. A systematic comparison of the Fourier coefficients for the two systems U+U and Au+Au is discussed.

## 1. Introduction

Azimuthal anisotropy is an important tool for studying the properties of the hot and dense matter created in high energy heavy-ion collsions at the Relativistic Heavy Ion Collider (RHIC) [1]. It describes the momentum space anisotropy of produced particles in non-central heavy-ion collisions, which is expected to be caused by the pressure gradient developed in the early stages of collisions. The azimuthal distribution of the produced particles can be expanded as a Fourier series [2],

$$\frac{dN}{d\phi} \propto 1 + \sum_{n=1}^{\infty} 2v_n \cos\left[n(\phi - \psi_{RP})\right] \tag{1}$$

where $\phi$ is the azimuthal angle and $\psi_{RP}$ is the reaction plane angle. The reaction plane is defined by the beam direction and impact parameter vector between the two colliding nuclei. The reaction plane angle $\psi_{RP}$ is not measurable directly, so the Fourier coefficients are determined with respect to the estimated event plane angle ($\psi_n$) as $v_n = \langle \cos n\,(\phi - \psi_n) \rangle$. The observed $v_n$ coefficients are then corrected for event plane resolution.

The $\varphi$ meson is the lightest bound state of a strange ($s$) and an anti-strange ($\bar{s}$) quark, and is considered as a clean probe for the study of the properties of the matter created in heavy-ion collisions. The $\varphi$ meson has a life time of $\sim$ 42 fm/c [3]. $\varphi$ meson generally decays outside the fireball therefore its decay daughters do not undergo significant re-scattering in the hadronic phase. In addition, the hadronic interaction cross-section of the $\varphi$ meson is expected to be small compared to the other hadrons [4]. Although the generation of the azimuthal anisotropy is an early time phenomenon, its magnitude might still be affected by the late stage hadronic interactions. Due to the above properties of the $\varphi$ meson, its $v_n$ coefficients are relatively unaffected by late stage interactions [5, 6]. Therefore $\varphi$ meson $v_n$ could reflect the collective motion of the partonic phase. This makes the $\varphi$ meson a clean probe for the study of the properties of the matter created in heavy-ion collisions [7].

## 2. Data Analysis

The data analyzed were recorded with the STAR experiment at RHIC during 2011 and 2012 for Au+Au collisions at $\sqrt{s_{NN}}$ = 200 GeV and U+U collisions at $\sqrt{s_{NN}}$ = 193 GeV, respectively. The minimum bias triggered events with a primary collision vertex position along the longitudinal beam direction ($V_z$) within 30 cm from the center of the detector are selected for this analysis. An additional cut on the transverse position of the primary vertex radius ($V_r < 2.0$ cm) from the center of the beam pipe has also been used to minimize effects of beam pipe interactions. The total number of events analyzed are ~562 million and ~293 million for Au+Au and U+U collisions, respectively. The STAR's Time Projection Chamber (TPC) [8] and Time Of Flight (TOF) [9] detectors were used for particle identification. The TPC is capable of tracking charged particles within $|\eta| < 1.0$ and has full azimuthal coverage. The TPC detector measures the momentum of the charged tracks as well as the specific ionization energy loss. This energy loss information is then used to identify the individual tracks by comparing them with the theoretical predictions using Bichsel functions [10]. The TOF detector measures the time taken by a track to traverse the distance from the primary vertex to the TOF detector. The TOF also has full azimuthal coverage and works within $|\eta| < 0.9$. $\varphi$ mesons are identified using the invariant mass technique from their decay to $K^+$ and $K^-$ with branching ratio of 48.9%. The mixed event technique has been used for combinatorial background estimation [11]. The $\eta$-sub event plane method has been used for the flow analysis. In this method, one defines the event flow vector for each particle based on particles measured in the opposite hemisphere in pseudo-rapidity. An $\eta$ gap of 0.1 between positive and negative pseudo-rapidity sub-events has been introduced to suppress non-flow effects [1]. Invariant mass method has been used to extract the flow coefficients of the $\varphi$ meson [12].

## 3. Results

### *3.1. Flow Harmonics ($v_n$)*

The differential flow harmonics $v_n$ (n = 2-5) for $\varphi$ meson as a function of the transverse momentum ($p_T$) at mid-rapidity ($|y| < 1.0$) in Au+Au collisions at $\sqrt{s_{NN}}$ = 200 GeV and U+U collisions at $\sqrt{s_{NN}}$ = 193 GeV for minimum bias events are presented. Figure 1 shows the

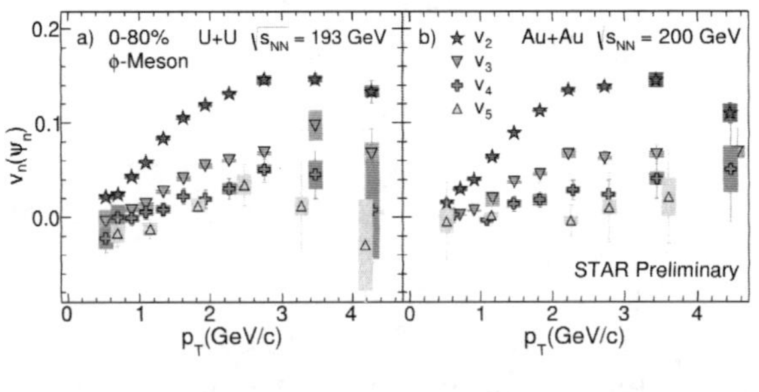


**Figure 1.** (Color online) $\varphi$ meson $v_n$ as function of $p_T$ in U+U collisions (a) and Au+Au collisions (b) for minimum bias (0 − 80% centrality). The different colored bands are the systematic uncertainties and vertical lines are the statistical uncertainties.

$\varphi$ meson flow coefficients $v_n$ for Au+Au and U+U. It is observed that the values are comparable between the two collision systems and $v_2 > v_3 > v_4 > v_5$. The $v_n$ coefficients as a function of

$p_T$ first increase for $p_T < 2.0$ GeV/c, saturate for intermediate $p_T$ and then decrease for higher $p_T > 3.5$ GeV/c. A similar type of $p_T$ dependence is observed for both systems.

### 3.2. Centrality Dependence

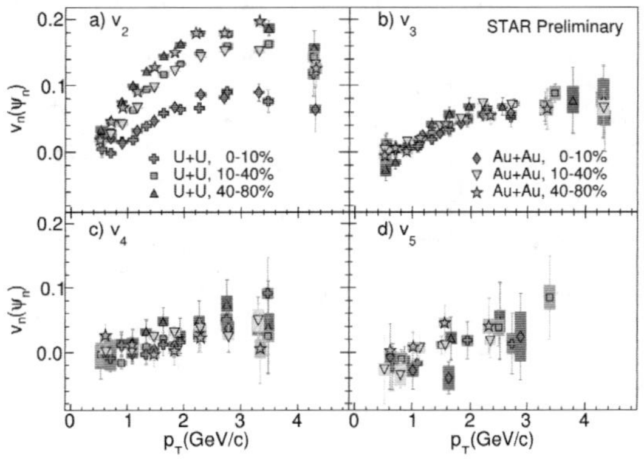


**Figure 2.** (Color online) $\varphi$ meson $v_2$, $v_3$, $v_4$ and $v_5$ as function of $p_T$ in U+U collisions and Au+Au collisions for three different centrality bins($0-10\%$, $10-40\%$ and $40-80\%$). The different colored bands are the systematic uncertainties and vertical lines are the statistical uncertainties.

Figure 2 shows the centrality dependence of $\varphi$ meson flow coefficients $v_n$ for Au+Au at $\sqrt{s_{NN}} = 200$ GeV and U+U collisions at $\sqrt{s_{NN}} = 193$ GeV. Strong centrality dependence is observed for the elliptic flow ($v_2$) in both Au+Au and U+U collisions. We found $v_2$ values higher in peripheral collisions ($40-80\%$ centrality) compare to central collisions ($0-10\%$ centrality). No clear centrality dependence is observed for other flow harmonics. This can be explained as the higher flow coefficients $v_3$, $v_4$ and $v_5$ are generated due to fluctuations corresponding to the initial states of the colliding nuclei.

### 3.3. Ratio ($v_3/v_2$)

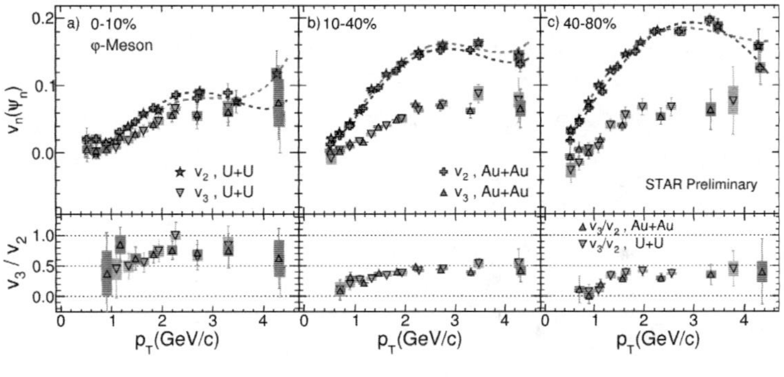


**Figure 3.** (Color online) $v_3/v_2$ ratio as a function of $p_T$ in U+U collisions and Au+Au collisions for three different centrality bins($0-10\%$, $10-40\%$ and $40-80\%$). Different colored bands are the systematic uncertainties and vertical lines are the statistical uncertainties. The dashed lines are the 4th order polynomial fit to the $v_2$ to get $v_2$ at the $p_T$ corresponding to $v_3$.

Figure 3 shows the ratio of $v_3$ to $v_2$ as a function of $p_T$ for Au+Au and U+U collisions at $\sqrt{s_{NN}} = 200$ GeV and 193 GeV, respectively. We observe the $v_3/v_2$ ratio for $p_T > 1.5$ GeV/c is constant for both U+U and Au+Au collisions and this is qualitatively consistent with the prediction from hydrodynamic calculations [13].

### *3.4. Ratio ($v_4/v_2^2$)*

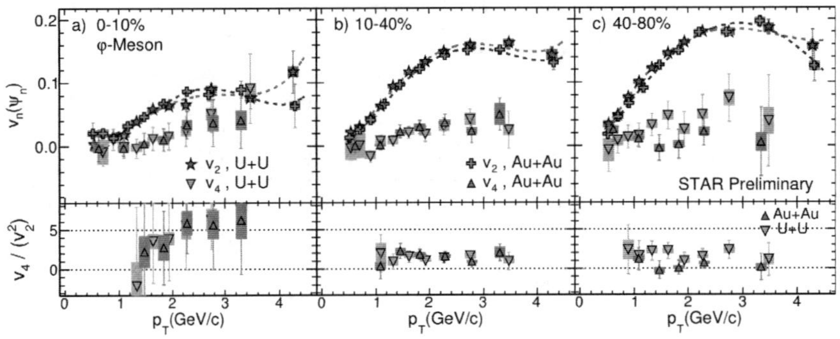


**Figure 4.** (Color online) $v_4/v_2^2$ ratio as a function of $p_T$ in U+U collisions and Au+Au collisions for three different centrality bins($0-10\%$, $10-40\%$ and $40-80\%$). The different colored bands are the systematic uncertainties and vertical lines are the statistical uncertainties.The dashed lines are the 4th order polynomial fit to the $v_2$ to get $v_2$ at the $p_T$ corresponding to $v_4$.

Figure 4 shows the ratio of $v_4$ to $v_2^2$ as a function of $p_T$ for Au+Au and U+U collisions at $\sqrt{s_{NN}}$ = 200 and 193 GeV, respectively. We observe the ratio $v_4/v_2^2$ of $\varphi$ meson is constant for U+U and Au+Au collisions. We found that the ratio has higher values for central collisions compared to peripheral collisions in both U+U and Au+Au collisions. We also found that the ratio $v_4/v_2^2$ has higher values than expected from a hydrodynamic model [13].

## 4. Summary

$\varphi$ meson azimuthal anisotropy coefficients $v_n$ (n = 2-5) has been measured in U+U and Au+Au collisions at $\sqrt{s_{NN}}$ = 193 GeV and 200 GeV, respectively. The $v_n$ results were compared as a function of the transverse momentum for minimum bias (0-80%) events and in the centrality ranges 0-10%, 10-40% and 40-80%. A strong centrality dependence is observed for elliptic flow ($v_2$), but no clear centrality dependence is observed for $v_3$, $v_4$ and $v_5$ for both the Au+Au and U+U systems. The ratios of flow coefficients has been measured as function of the transverse momentum. The ratio $v_3/v_2$ for $p_T > 1.5$ GeV/c is found to be constant for both the U+U and Au+Au systems, and is consistent with the expectations from a hydrodynamic model [13]. The ratio $v_4/v_2^2$ is found to be higher than expected from the hydrodynamic model [13].

## References

[1] A. M. Poskanzer and S. A. Voloshin, Phys. Rev. C **58**, 1671 (1998).
[2] S. A. Voloshin and Y. Zhang, Z. Phys. C **70**, 665-672 (1996).
[3] K. A. Olive *et al.* (Particle Data Group), Chin. Phys. C **38**, 090001 (2014).
[4] J. Adams *et al.* (STAR Collaboration), Nucl. Phys. A **757**, 102 (2005).
[5] A. Sibirtsev *et al.*, Eur. Phys. J. A **29**, 209-220 (2006).
[6] B. Mohanty and N. Xu, J. Phys. G **36**, 064022 (2009).
[7] M. M. Aggarwal *et al.* (STAR Collaboration), *Preprint*: nucl-ex/1007.2613
[8] M. Anderson *et al.*, Nucl. Instrum. Meth. A **499**, 659 (2003).
[9] W. J. Llope *et al.*, Nucl. Instrum. Meth. A **522**, 252-273 (2004).
[10] H. Bichsel, Nucl. Instrum. Meth. A **562**, 154 (2006).
[11] B. I. Abelev *et al.*, (STAR Collaboration), Phys. Lett. B **673**, 183 (2009).
[12] N. Borghini and J. Y. Ollitrault, Phys. Rev. C **70**, 064905 (2004).
[13] C. Lang and N. Borghini, Eur. Phys. J. C **74**, 2955 (2014).

Journal of Physics: Conference Series **668** (2016) 012040 doi:10.1088/1742-6596/668/1/012040

# Measurements of $D_s^+$ production in pp, p–Pb and Pb–Pb collisions with ALICE at the LHC

**Anastasia Barbano, for the ALICE Collaboration**
Università di Torino and INFN sez. di Torino, via Giuria 1, 10125 Torino, Italy
E-mail: barbano@to.infn.it

**Abstract.** ALICE has measured $D_s^+$ production at mid-rapidity in pp ($\sqrt{s}$ =7 TeV), p–Pb ($\sqrt{s_{\mathrm{NN}}}$ =5.02 TeV) and Pb–Pb ($\sqrt{s_{\mathrm{NN}}}$ =2.76 TeV) collisions, through the hadronic decay channel $D_s^+ \rightarrow \phi\pi^+ \rightarrow K^+ K^- \pi^+$. The study of the $D_s^+$-meson production could reveal important details about charm-quark hadronisation in the Quark-Gluon Plasma (QGP). The measured $D_s^+$ yields in pp and p–Pb collisions and the nuclear modification factors for central and semi-central Pb–Pb collisions will be shown here as a function of transverse momentum.

## 1. Introduction

Heavy quarks (charm and beauty) are excellent tools to study the properties of the hot and dense medium created in high-energy heavy-ion collisions. Charm and beauty quarks are produced in high-momentum transfer parton scatterings. Therefore, their production cross-section can be calculated using perturbative QCD (pQCD). Such processes occur, in Pb–Pb collisions, at the early stage of the reaction and on timescales shorter than the QGP formation time [1]. Heavy quarks hence experience the entire evolution of the fireball, interacting with its constituents via both elastic and inelastic processes, exchanging energy and momentum with the expanding medium [1, 2].

Evidence for heavy-quark energy loss is provided by the measurement of the nuclear modification factor $R_{\mathrm{AA}}$, defined as the ratio of the yield measured in nucleus–nucleus collisions and the cross-section in pp interactions scaled by the average nuclear overlap function for a given centrality class. If no medium effects are present, one expects $R_{\mathrm{AA}} = 1$. A value of $R_{\mathrm{AA}}$ that differs from unity implies modifications of the production of D(B) mesons relative to pp collisions, that can be induced by initial or final-state effects. In particular, the observed suppression of D mesons at high $p_{\mathrm{T}}$ with respect to the binary-scaled pp yield is described in terms of in-medium energy loss of charm quarks [3].

A significant fraction of charm quarks could undergo hadronisation via in-medium coalescence at intermediate and low momentum [4]. Furthermore, an enhancement of strangeness production with respect to pp collisions was long suggested as a possible signal of QGP formation [5]. Strange and multistrange hadron enhancement was observed from SPS to LHC energy, showing a dependence on the strangeness content of the particles and also on the collision centrality [6, 7, 8]. In this scenario, the measurement of $D_s^+$ production would provide knowledge on charm-quark hadronisation inside the medium. The possibility of coalescence of charm quarks with the medium constituents, together with the possible strangeness enhancement, should lead to a

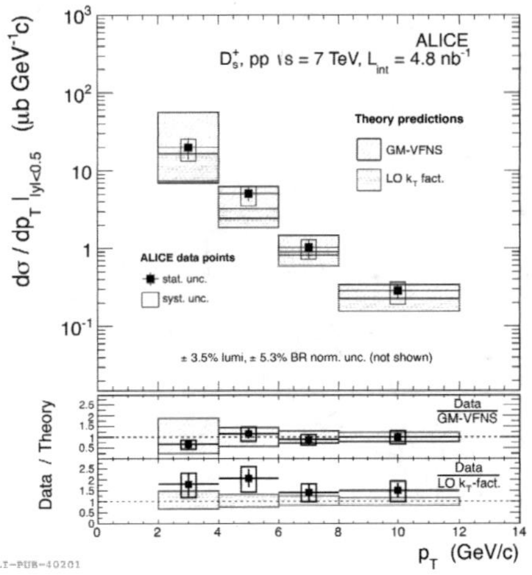


**Figure 1.** $p_{\rm T}$-differential cross section of prompt $D_s^+$-meson production in pp collisions at $\sqrt{s}$ = 7 TeV [13]. Theoretical predictions from GM-VFNS [14] and from $k_{\rm T}$-factorization at LO [15] are also shown.

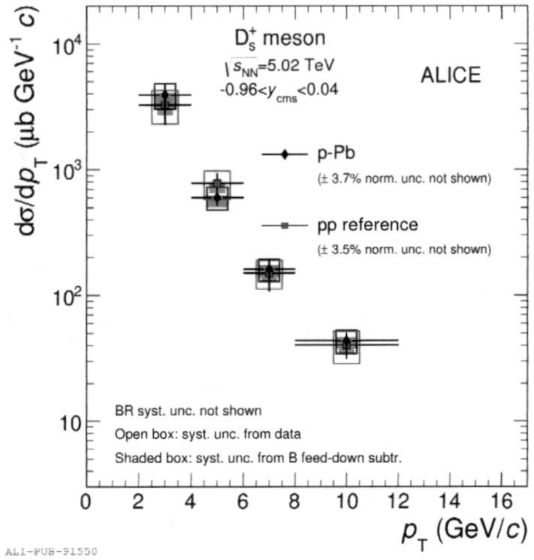


**Figure 2.** $p_{\rm T}$-differential cross section of prompt $D_s^+$-meson production in p–Pb collisions at $\sqrt{s}$ = 5.02 TeV compared to the pp rescaled reference cross section [16].

larger relative abundance of $D_s$ mesons compared to non-strange D mesons when going from pp to Pb–Pb collisions [10].

## 2. Reconstruction of $D_s^+$ mesons in ALICE

$D_s^+$ mesons and their charge conjugates are reconstructed at mid rapidity ($|y| < 0.8$) in their decay channel $D_s^+ \to \phi\pi^+ \to K^+ K^- \pi^+$, with a branching ratio of $(2.24 \pm 0.10)\%$ and a mean proper decay length of $(150 \pm 2)$ $\mu$m [11]. The analysis strategy is based on the reconstruction of three-prong decay topologies with a secondary vertex separated from the interaction vertex. The large combinatorial background is reduced by applying a selection based on the decay topology together with a selection on the invariant mass of the reconstructed $\phi$ meson.

To separate secondary and primary vertices, excellent vertex and impact parameter resolutions are mandatory. These are provided by the Inner Tracking System (ITS), a six-layer silicon detector covering the pseudorapidity interval $|\eta| < 0.9$ [12]. In the same pseudorapidity region, track reconstruction is performed using the ITS together with the Time Projection Chamber (TPC). Finally, particle identification is provided in $|\eta| < 0.9$ by the TPC via specific energy-loss measurements and by the Time-Of-Flight (TOF) detector. They provide K/$\pi$ separation up to a transverse momentum $p_{\rm T} \sim 2.5$ GeV/$c$. A track is considered compatible with the kaon or pion hypothesis if both its d$E$/d$x$ and time-of-flight are within 3$\sigma$ from the expected values. Tracks without a TOF signal (mostly at low momentum) are identified using only the TPC information and requiring a 2$\sigma$ compatibility with the expected d$E$/d$x$.

The raw $D_s^\pm$-meson yield was corrected to obtain the production yields of prompt mesons (i.e. not coming from weak decays of B mesons) [13]. A correction was also applied for the acceptance and selection efficiency of prompt $D_s^+$ mesons. The measurements presented here refer to prompt $D_s^+$ only, as they were computed as the average of particles and antiparticles under the assumption that the production cross section is the same for $D_s^+$ and $D_s^-$.

## 3. Results

### *3.1. pp collisions at $\sqrt{s} = 7$ TeV*

$D_s^+$ production in pp collisions was measured from a data sample of $3 \times 10^8$ events, corresponding to an integrated luminosity $L_{int} = 4.8$ nb$^{-1}$. The result in pp collisions provides a test for pQCD calculations at LHC energies, as well as a reference for p–Pb and Pb–Pb measurements. The production cross section of prompt $D_s^+$ mesons in four transverse momentum intervals in the range $2 < p_T < 12$ GeV/$c$ is shown in Fig. 1 [13]. The error bars represent the statistical uncertainties, the systematic uncertainties are shown as empty boxes. The measurement is well described within uncertainties by a pQCD next-to-leading order GM-VFNS calculation [14]. It is also compatible with the $k_T$-factorization model at LO [15], that uses unintegrated gluon distributions in the proton.

### *3.2. p–Pb collisions at $\sqrt{s_{NN}} = 5.02$ TeV*

The data sample of p–Pb collisions consists of $97.3 \times 10^6$ events ($L_{int} = 48.6$ $\mu$b$^{-1}$). Measurements with this colliding system are necessary to assess the role of possible cold nuclear matter effects in the modification of heavy-flavour production with respect to pp collisions. They are therefore crucial for the interpretation of the results in Pb–Pb collisions where a hot and dense medium is created. Figure 2 shows the $p_T$-differential cross section of prompt $D_s^+$-meson production in the range $2 < p_T < 12$ GeV/$c$ [16]. Data in p–Pb collisions (black points) are compared to the reference obtained by scaling the measured cross section in pp collisions at $\sqrt{s} = 7$ TeV to $\sqrt{s} = 5.02$ TeV (blue points) and multiplying it by the number of nucleons in the lead nucleus. The two cross sections are compatible within uncertainties. The $R_{pPb}$ obtained as the ratio of these cross sections is compatible with unity within uncertainties. This result indicates that cold nuclear matter effects are small in the measured $p_T$ range.

### *3.3. Pb–Pb collisions at $\sqrt{s_{NN}} = 2.76$ TeV*

Two different centrality classes were analysed, namely the 7.5% most central collisions ($16 \times 10^6$ events, $L_{int} = 21.5$ $\mu$b$^{-1}$) and the semi-central collisions in the 20–50% class ($13.5 \times 10^6$ events, $L_{int} = 5.9$ $\mu$b$^{-1}$). The $R_{AA}$ of $D_s^+$ mesons as a function of $p_T$ is shown in Fig. 3, in three $p_T$ intervals from 4 to 12 GeV/$c$ for the most central events, and in two $p_T$ intervals from 6 to 12 GeV/$c$ for semi-central events. A significant suppression of the $D_s^+$-meson yield by a factor about three in Pb–Pb collisions relative to the binary-scaled pp cross section is observed in the highest $p_T$ interval ($8 < p_T < 12$ GeV/$c$) for the most central events. This suppression is due to in-medium energy loss as $R_{pPb}$ in this $p_T$ range is compatible with unity. A hint for a smaller suppression is found for the more peripheral events at the same $p_T$, although the values are consistent within uncertainties. The comparison of $R_{AA}$ of $D_s^+$ in the 0–7.5% centrality class with the average nuclear modification factor of $D^0$, $D^+$ and $D^{*+}$ mesons measured in the same centrality class is presented in Fig. 4. The central values of the $D_s^+$ $R_{AA}$ are larger than those of non-strange D mesons in the intervals $4 < p_T < 6$ and $6 < p_T < 8$ GeV/$c$, but compatible within uncertainties. To conclude about the predicted enhancement of the $D_s^+$ yield relative to non-strange D mesons in heavy-ion collisions and about the centrality dependence of $D_s^+$ $R_{AA}$, uncertainties need to be reduced.

In Fig. 4 the results are also compared to predictions of the TAMU model [17], which currently is the only one that provides quantitative calculations for the $p_T$-differential $D_s^+$ nuclear modification factor. In this model the interactions of charm quarks with the medium constituents are described via elastic collisions. The hadronisation of charm quarks into non-strange D and $D_s$ mesons is performed via recombination with thermalized light and strange quarks, respectively. The $R_{AA}$ of $D_s$ is hence predicted to be higher relative to non-strange D mesons. The theoretical calculation results can describe the measurements within uncertainties.

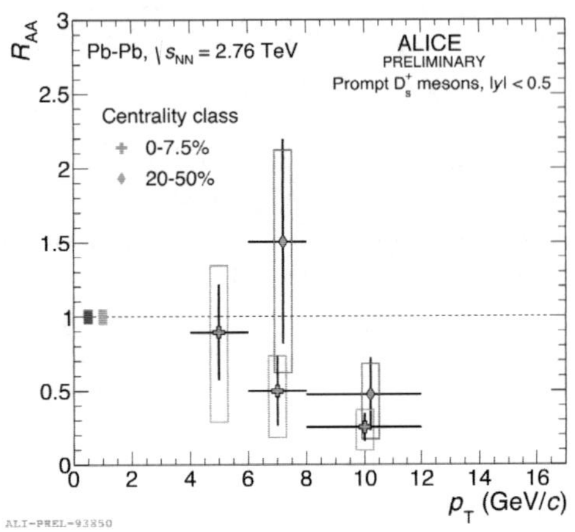


**Figure 3.** $R_{\mathrm{AA}}$ of prompt $\mathrm{D_s^+}$ mesons in the 0–7.5% and 20–50% centrality classes as a function of $p_{\mathrm{T}}$.

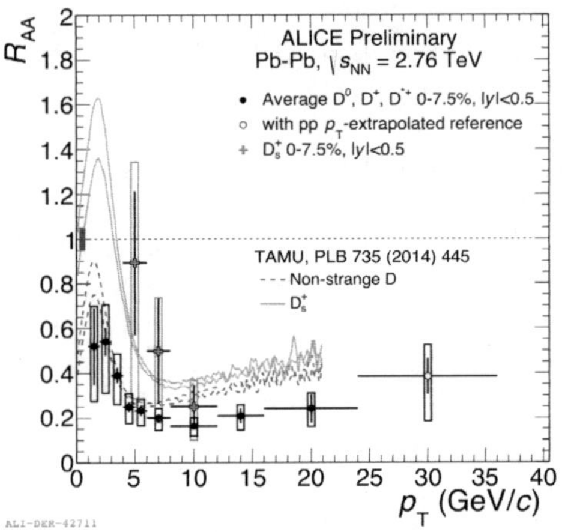


**Figure 4.** $R_{\mathrm{AA}}$ of prompt $\mathrm{D_s^+}$ and non-strange D mesons (average of $\mathrm{D^0}$, $\mathrm{D^+}$ and $\mathrm{D^{*+}}$ [3]) for the 7.5% most central events compared to predictions of the TAMU model [17].

## 4. Conclusions

We have measured $\mathrm{D_s^+}$ production in pp, p–Pb and Pb–Pb collisions.
The measurement of the $p_{\mathrm{T}}$-differential cross section in pp collisions is described by pQCD calculations at LHC energies. The results for the 7.5% most central events in Pb–Pb collisions indicate a suppression ($R_{\mathrm{AA}} \sim 0.3$) of the production of $\mathrm{D_s^+}$ mesons at high $p_{\mathrm{T}}$ ($8 < p_{\mathrm{T}} < 12$ GeV/$c$) with respect to the reference obtained from the binary-scaled pp yields. The observed suppression is compatible with that of non-strange D mesons and can be described by models including charm quark in-medium energy loss. At lower momenta ($4 < p_{\mathrm{T}} < 8$ GeV/$c$), where the relative abundance of strange and non-strange D mesons could be influenced by charm quark recombination in the medium, the values of the $\mathrm{D_s^+}$-meson nuclear modification factor are larger than those of non-strange D mesons, although compatible within uncertainties.
The larger data samples that will be collected during Run 2 at the LHC will allow us to substantially reduce the measurement uncertainties and to provide further constraints to models.

## References

[1] Rapp R and van Hees H, (2010) Hwa R C, Wang X N *Quark Gluon Plasma* 4, 111 (World Scientific)
[2] Armesto N, Dainese A, Salgado C A and Wiedemann U A, (2005) *Phys. Rev.* D **71** 054027
[3] Festanti A (ALICE) 2015 *These proceedings*
[4] Greco V, Ko C and Rapp R (2004) *Phys. Lett.* B **595** 202-208
[5] Rafelski J and Muller B (1982) *Phys. Rev. Lett.* **48** 1066
[6] NA49 Collaboration (1998) *Phys. Lett.* B **444** 523-530
[7] STAR Collaboration *Phys. Rev.* C **77**, 044908
[8] ALICE Collaboration (2014) *Phys. Lett.* B **728** 25-38
[9] Hamieh S, Redlich K and Tounsi A (2000) *Phys. Lett.* B **486** 61-66
[10] He M, Fries R and Rapp R (2013) *Phys. Rev. Lett.* **110**(11) 112301
[11] Particle Data Group Collaboration, Olive K *et al.* (2014) *Chin. Phys.* **C38** 090001
[12] ALICE Collaboration (2008) *JINST* **3** S08002
[13] ALICE Collaboration (2012) *Phys. Lett.* B **718** 279-294
[14] Kniehl B A, Kramer G, Schienbein I and Spiesberger H (2012) *Eur. Phys. J.* C **72** 2082
[15] Maciula R, Luszczak M and Szczurek A (2012) *EPJ Web Conf.* **37** 06008
[16] ALICE Collaboration (2014) *Phys. Rev. Lett.* **1130.2** 232301
[17] He M, Fries R and Rapp R (2014) *Phys. Lett.* B **735** 445-450

# Recent results for the (anti-)(hyper-)nuclei production and searches for exotica by ALICE at the LHC

**Francesco Barile for the ALICE Collaboration**
Università degli Studi di Bari and INFN Bari
E-mail: francesco.barile@ba.infn.it

**Abstract.** The excellent tracking and particle identification capabilities of the ALICE apparatus combined with the high particle production rates reached at the LHC in pp, p–Pb and in particular in Pb–Pb collisions allow for detailed study of the production of (anti-)(hyper-)nuclei and searches for exotica, like dibaryons. In this report, we present recent results on the production of the (anti-)deuteron and (anti-)helium. Further, we discuss recent results on the production and the lifetime of the (anti-)hypertriton. In addition, results from the searches for the weakly-decaying exotic nuclear bound states H-dibaryon and $\Lambda$n are shown. The results are compared with the expectations from statistical (thermal) particle production and coalescence models.

## 1. Introduction

Collisions of ultra-relativistic heavy ions are an efficient tool for producing (anti-)(hyper-)nuclei because a high energy density is reached over a large volume. The measurements presented here, have been performed in pp collisions at $\sqrt{s} = 7$ TeV, in Pb–Pb collisions at $\sqrt{s_{\rm NN}} = 2.76$ TeV as a function of collision centrality and in p–Pb collisions at $\sqrt{s_{\rm NN}} = 5.02$ TeV as a function of charged-particle multiplicity.

The excellent particle identification capabilities of the ALICE detector system [1–4] is suited to measure the production of (anti-)(hyper-)nuclei and to search for exotic bound states like $\Lambda$n and the H-dibaryon. The formation of (anti-)nuclei is very sensitive to the chemical freeze-out conditions, to the dynamics of the emitting source as well as to the final-state effects. The production mechanisms of these particles are typically discussed within two approaches: the thermal-statistical model and the coalescence model. In the thermal model [5–7] the chemical freeze-out temperature $T_{\rm chem}$ (predicted around 160 MeV) acts as the key parameter at LHC energies. The strong sensitivity of the nuclei production to the choice of $T_{\rm chem}$ is caused by the large mass $m$ and the exponential dependence of the yield given by the factor exp($-m/T_{\rm chem}$). In the coalescence model, nuclei are formed by protons and neutrons which are nearby in space and exhibit similar velocities [8, 9]. A quantitative description of this process, applied to many collision systems at various energies [10–17], is typically based on the coalescence parameter $B_{\rm A}$ (see Section 2). The two mechanisms give very similar predictions [18].

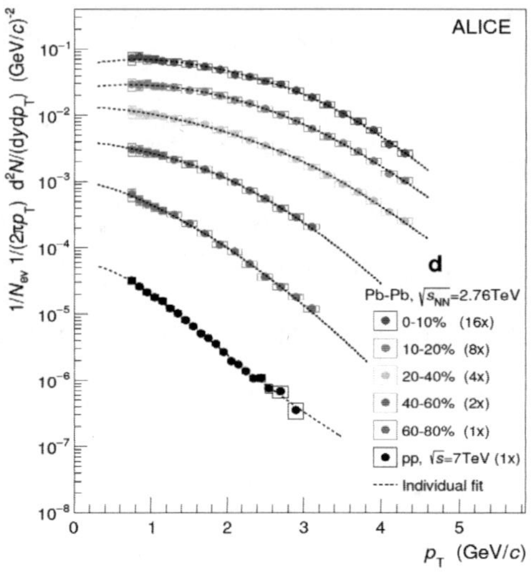


**Figure 1.** Efficiency and acceptance corrected deuteron spectra for Pb–Pb collisions at $\sqrt{s_{NN}}$ = 2.76 TeV in various centrality classes and for inelastic pp collisions at $\sqrt{s}$ = 7 TeV.

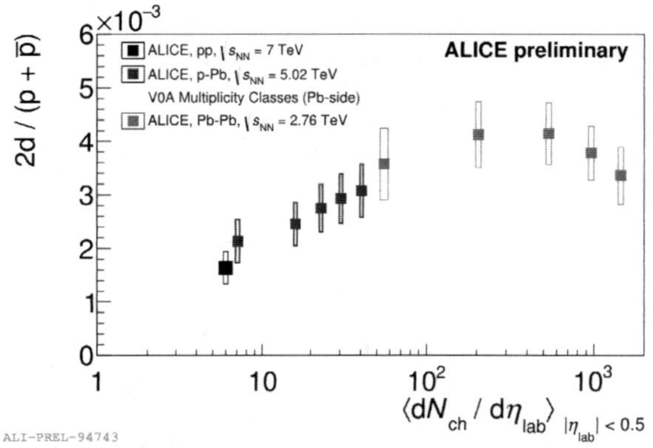


**Figure 2.** Deuteron-to-proton ratio as a function of charged-particle multiplicity at midrapidity.

## 2. Nuclei

Nuclei and anti-nuclei such as (anti-)deuterons, (anti-)tritons, and (anti-)$^3$He are identified using the specific energy loss measurement in the TPC. The final spectra of deuterons obtained in Pb–Pb and in pp collisions [19] are shown in Figure 1. The $p_T$-distributions in Pb–Pb show a clear evolution, becoming harder as the centrality increases, a behaviour similar to that of the protons which exhibit a significant radial flow [20]. The spectra obtained in Pb–Pb collisions are individually fitted with the blast wave model for the determination of $p_T$-integrated yields. Figure 2 shows the deuteron-to-proton ratio as a function of the multiplicity in pp, p–Pb and in Pb–Pb collisions. The p-Pb ratio rises with multiplicity (higher nucleon multiplicities would lead to an increased deuteron production) until a saturation (within the achieved experimental precision) is reached for Pb–Pb collisions: the increasing nucleon multiplicity is balanced by the increasing source volume, leading to a constant density. The observed value of about 3.6 $\times$ $10^{-3}$ for the d/p ratio in Pb–Pb collisions is in agreement with the expectation from the thermal-statistical models (with $T_{\rm chem}$ =156 MeV) [19].

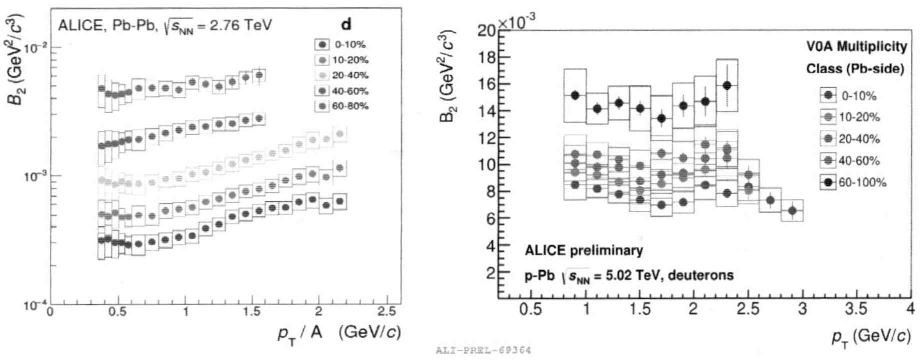


**Figure 3.** Coalescence parameter $B_2$ as a function of the transverse momentum in Pb–Pb collisions at $\sqrt{s_{\mathrm{NN}}} = 2.76$ TeV (left) and in p–Pb collisions at $\sqrt{s_{\mathrm{NN}}} = 5.02$ TeV (right).

In the coalescence approach, light nuclei are formed at kinetic freeze-out via coalescence of protons and neutrons which are close to each other in space and momentum phase space. In this mechanism, the spectral distribution of the composite nuclei is related to the one of the primordial nucleons via the following relationship:

$$E_i \frac{d^3 N_i}{(dp_i)^3} = B_A \left( E_{\mathrm{p}} \frac{d^3 N_{\mathrm{p}}}{(dp_{\mathrm{p}})^3} \right)^A \tag{1}$$

assuming that protons and neutrons have the same momentum distribution. $B_{\mathrm{A}}$ is the coalescence parameter of particle $i$ with mass number $A$ and a momentum of $p_{\mathrm{i}} = \mathrm{A}\ p_{\mathrm{p}}$.
Figure 3 shows the obtained $B_2$ values for deuterons for Pb–Pb and p–Pb (right) collisions. A clear decrease with increasing centrality is observed for Pb–Pb collisions. In the coalescence picture, this behaviour is explained by an increase in the source volume $V_{\mathrm{eff}}$. $B_2$ also shows an increasing trend with the transverse momentum for central collisions, in contrast with the most simple coalescence models. This behaviour can be qualitatively understood by position-momentum correlations which are caused by a radially expanding source. In p–Pb collisions the $B_2$ is slightly decreasing with multiplicity, again in contrast with the most simple coalescence model, that represents a solid ground for further studies.

## 3. (Anti-)hypertriton

The production of hypertriton ${}^{3}_{\Lambda}\mathrm{H}$ and ${}^{3}_{\overline{\Lambda}}\overline{\mathrm{H}}$ has been measured in Pb–Pb collisions via invariant mass reconstruction in the weak decay channel ${}^{3}_{\Lambda}\mathrm{H} \rightarrow {}^{3}\mathrm{He} + \pi^-$ and ${}^{3}_{\overline{\Lambda}}\overline{\mathrm{H}} \rightarrow {}^{3}\overline{\mathrm{He}} + \pi^+$, respectively [21]. The daughter tracks of the (anti-)hypertriton candidates are required to originate from a secondary vertex and identified as a ${}^{3}\mathrm{He}$ (${}^{3}\overline{\mathrm{He}}$) and a pion via TPC d$E$/d$x$. The total yield for the hypertriton, d$N$/d$y$ × B.R. = (3.86 ± 0.77 (stat) ± 0.68) × $10^{-5}$ in the 0–10% most central collisions, is consistent with the predictions from a statistical model using the same temperature as for the light hadrons ($T_{\mathrm{chem}} = 156 \pm 2$ MeV). Also the measured lifetime of the hypetriton ($\tau = 181^{+54}_{-39}$ (stat.) ± 33 (syst.) ps) is compatible with the world average value [21].

## 4. Exotica bound states

Searches for two hypothetical strange dibaryon states, i.e. the H-dibaryon and the $\overline{\Lambda \mathrm{n}}$ bound state is performed in 0–10% most central Pb–Pb collisions, by invariant mass analysis in the decay modes H-dibaryon → $\Lambda \mathrm{p} \pi^-$ and $\overline{\Lambda \mathrm{n}} \rightarrow \bar{d} \pi^+$ [22]. No signal in the invariant mass distributions is observed. The analysis provides stringent upper limits at 99% confidence level for the production of H-dibaryon and $\overline{\Lambda n}$ bound state, significantly below the thermal model prediction [22]. The upper limits are obtained for different lifetimes: the values are significantly below (about a factor 20) the model predictions when realistic branching ratios (64% for the H-dibaryon and 54% for $\overline{\Lambda n}$) and reasonable lifetimes are assumed.

## 5. Summary and conclusions

The most recent results of the ALICE experiment on the (anti-)deuteron production in pp, p–Pb and Pb–Pb collisions have been reported. The deuteron spectra in Pb–Pb collisions exhibit a significant hardening with increasing centrality. The d/p ratio, in Pb–Pb collisions, is found to be constant (within the achieved experimental precision) as a function of $dN_{ch}/d\eta$ which is in agreement with the thermal-statistical interpretation. The rising d/p ratio with the charged particle multiplicity in p-Pb is consistent with the coalescence approach: higher nucleon multiplicities would lead to an increased deuteron production. However, the d/p ratio for Pb–Pb collisions is constant with the increasing centrality: the increasing nucleon multiplicity is balanced by the increasing source volume, leading to a constant density. Extrapolations and model predictions based on the thermal-statistical or coalescence approach are, therefore, a solid ground for further studies. Measurements of hypertriton and anti-hypertriton in Pb–Pb collisions have been also presented. In addition, a search for the existence of loosely bound strange dibaryon has been discussed and no signals are observed.

## References

[1] F. Carminati et al., (ALICE Collaboration), Physics Performance Report, Volume I, J. Phys. G: Nucl. Part. Phys. 30 (2004) 1517
[2] B. Alessandro et al., (ALICE Collaboration), Physics Performance Report, Volume II, J. Phys. G. Nucl. Part. Phys. 32 (2006) 1295
[3] K. Aamodt et al. (ALICE Collaboration), J. Instrum. 3 (2008) SO8002
[4] B. Abelev et. al., (ALICE Collaboration), Int. J. Mod. Phys. A 29 (2014) 1430044
[5] A. Andronic, P. Braun-Munzinger, J. Stachel, and H. Stoecker, Phys. Lett. B 697 (2011) 203
[6] J. Cleymans, S. Kabana, I. Kraus, H. Oeschler, K. Redlich, et al., Phys. Rev. C 84 (2011) 054916
[7] A. Andronic, P. Braun-Munzinger, K. Redlich, and J. Stachel, J. Phys. G 38 (2011) 124081
[8] S. Butler and C. Pearson, Phys. Rev. 129 (1963) 836
[9] J. Kapusta, Phys. Rev. C 21 (1980) 1301
[10] S. Afanasev et al. (NA49 Collaboration), Phys. Lett. B 22 (2000) 486
[11] C. Adler et al. (STAR Collaboration), Phys. Rev. Lett. 87 (2001) 262301
[12] S. Schael et al. (ALEPH Collaboration), Phys. Lett. B 639 (2006) 192
[13] A. Aktas et al. (H1 Collaboration), Eur. Phys. J. C 36 (2004) 413
[14] V. Abramov, B. Y. Baldin, A. Buzulutskov, V. Y. Glebov, A. Dyshkant et al., Sov. J. Nucl. Phys. 45 (1987) 845
[15] T. Anticic et al. (NA49 Collaboration), Phys. Rev. C 69 (2004) 024902
[16] I. Bearden, H. Boggild, J. Boissevain, P. Christiansen, L. Conin et al., Phys. Rev. Lett. 85 (2000) 2681
[17] R. Scheibl and U.W. Heinz, Phys. Rev. C 59 (1999) 1585
[18] J. Steinheimer, K. Gudima, A. Botvina, I. Mishustin, M. Bleicher et al., Phys. Lett. B 85 (2012) 714
[19] L. Adam et al., (ALICE Collaboration), http://arxiv.org/abs/1506.08951
[20] B. Abelev et al. (ALICE Collaboration), Phys. Rev. Lett. 111 (2013) 222301
[21] L. Adam et al., (ALICE Collaboration), http://arxiv.org/abs/1506.08453
[22] L. Adam et al., (ALICE Collaboration), http://arxiv.org/abs/1506.07499

Journal of Physics: Conference Series **668** (2016) 012042 doi:10.1088/1742-6596/668/1/012042

# Towards a new quark-nuclear matter EoS for applications in astrophysics and heavy-ion collisions

**Niels-Uwe Bastian**[1], **David Blaschke**[1,2,3]

[1]Institute for Theoretical Physics, University of Wrocław, 50-204 Wrocław, Poland
[2]Bogoliubov Laboratory for Theoretical Physics, JINR Dubna, 141980 Dubna, Russia
[3]National Research Nuclear University (MEPhI), 115409 Moscow, Russia

E-mail: niels-uwe@bastian.science

**Abstract.** The aim of our work is to develop a unified equation of state (EoS) for nuclear and quark matter for a wide range in temperature, density and isospin so that it becomes applicable for heavy-ion collisions as well as for the astrophysics of neutron stars, their mergers and supernova explosions. As a first step, we use improved EoS for the hadronic and quark matter phases and join them via Maxwell construction. We discuss the limitations of a 2-phase description and outline steps beyond it, towards the formulation of a unified quark-nuclear matter EoS on a more fundamental level by a cluster virial expansion.

## 1. Introduction

The description of the hadron-to-quark matter phase transition (PT) in effective models for the equation of state (EoS) of strongly interacting matter is still a challenge, even half a century after the quark model has been developed. A modern description should consistently address the question how the breaking and restoration of the symmetries of the QCD Lagrangian take place as a function of the thermodynamic variables, temperature $T$ and baryon chemical potential $\mu_b$ (for asymmetric matter, there is a set of additional chemical potentials which eventually can be restricted by conditions like conservation laws). Particular emphasis is hereby on the dynamical chiral symmetry breaking and restoration, an aspect that can be very effectively addressed within models of the Nambu–Jona-Lasinio (NJL) type with the quark condensate or the dynamically generated quark mass serving as order parameters of the transition. A crucial role is played by quark and gluon confinement, i.e. the empirical principle that these QCD degrees of freedom carrying color charges have to be confined within color neutral bound states in the hadronic phase. Accordingly, within the deconfinement PT quarks and gluons appear as quasiparticles in the system and together with the chiral symmetry restoration in the quark sector the hadronic bound states get dissociated (Mott effect). However, as NJL models have no confinement, modifications are invoked to remedy this fact. Let us mention four possibilities: (i) a bag pressure $B$, (ii) a confining meanfield (density functional) [1, 2, 3], (iii) minimal coupling of the quark dynamics to the Polyakov-loop with an effective potential [4], and (iv) a nonlocal, covariant generalization of the current-current coupling [5, 6, 7]. A benchmark check for any effective model is the comparison with EoS data from lattice QCD simulations limited at present to $T$ for vanishing $\mu_b$ [8, 9].

The working horse of the QCD PT model studies is still the so-called 2-phase description, where separately developed EoS for the hadronic and the quark-gluon phases of matter are employed to obtain a hybrid quark-nuclear matter EoS by applying Gibbs conditions of phase equilibrium. See, e.g., Ref. [10] for a phase diagram.

In this contribution, we present a hybrid EoS obtained with a 2-phase construction using modern inputs in order to illustrate deficiencies that are inherent in such an approach when applied to the QCD transition.

Finally, we outline steps beyond a 2-phase description towards the formulation of a unified quark-nuclear matter EoS on the basis of a cluster virial expansion for quark matter. While being similar in spirit to developments for describing clusters in nuclear matter [11], for the present task one has to address additionally the questions of chiral symmetry restoration and deconfinement within a relativistic, field-theoretic approach. To this end, the $\Phi$-derivable approach [12] shall be generalized as described in [13].

## 2. Hybrid EoS from 2-phase construction with modern inputs

For the hadronic side we use the well known density-dependent relativistic meanfield theory parametrisation (DD2) by Stefan Typel [14, 15]. This EOS reproduces very well the known behavior of nuclear matter near the saturation density for symmetric matter and describes a liquid-gas PT. For applications at high temperatures ($T > 50\,\mathrm{MeV}$) the hadronic side should be improved by taking into account the appearance of additional hadrons (Pions, Kaons...).

For the quark side we use the higher order NJL (hNJL) model recently developed by Sanjin Benić for zero temperature [16] and subsequently applied to hybrid compact stars [17]. For extending it to finite temperature, we use the grand potential density (symmetric case)

$$\Omega(T,\mu) = U(\sigma,n) - 2N_cN_fT\int\frac{\mathrm{d}^3p}{(2\pi)^3}\left\{\frac{\tilde{E}}{T} + \ln\left\{\left[1+\mathrm{e}^{-\frac{(\tilde{E}-\tilde{\mu})}{T}}\right]\left[1+\mathrm{e}^{-\frac{(\tilde{E}+\tilde{\mu})}{T}}\right]\right\}\right\} + \Omega_0 \quad (1)$$

with an effective potential $U(\sigma,n)$ and a renormalized quark chemical potential $\tilde{\mu} = \mu - \Delta_\mu(n)$. The renormalized mass $M = m - \Delta_m(\sigma)$ in the dispersion relation $\tilde{E} = \sqrt{p^2+M^2}$ is obtained from solving the gap equation, i.e. minimizing $\Omega(T,\mu)$ [16, 17] and thus defining the quark matter pressure in equilibrium $P_{\mathrm{hNJL}}(T,\mu) = -\Omega(T,\mu)$. In Fig. 1 we show the solution of the mass gap equation (upper panel) and the corresponding quark pressure (middle panel) as a function of $\mu_b = 3\mu$ for selected isothermes which shows the chiral PT as a van-der-Waals wiggle similar to that for the nuclear liquid-gas transition in the DD2 EoS. Both EoS have a critical endpoint with a temperature above which the transition is a crossover, see Fig. 2. The crossing points with the hadronic pressure are denoted as *deconfinement PT* when for increasing $\mu_b$ the quark pressure lies above the hadronic one and is thus preferred. We notice that at finite $T$ unphysical second crossings occur and that already above $T = 55$ MeV no hadronic matter exists. This unphysical behaviour is due to the absence of confinement in the hNJL model. We cure this pathology in the simplest way, by including a bag pressure $P_{\mathrm{hNJLb}} = P_{\mathrm{hNJL}} - B$ , see Fig. 1 (lower panel) for two values of $B$ that result in PT lines in the phase diagram of Fig. 2 (left) that correspond to critical temperatures in the range $175 - 180$ MeV at $\mu_b = 0$.

The left panel of Figure 2 shows the phase transition between DD2 and hNJLb with $B = 160\,\mathrm{MeVfm}^{-3}$ for two temperatures $T = 0$ and $T = 150\,\mathrm{MeV}$ in the pressure over density plane compared to the EoS used by Yuri Ivanov in three-fluid hydrodynamic simulations [18, 19]. For our EoS, the deconfinement PT occurs at much lower densities, in particular for the temperatures expected in heavy-ion collisions for NICA and FAIR energies, what should result in interesting modifications of the baryon stopping signal of this transition predicted in [18].

## 3. Discussion and outlook

Applying the 2-phase construction as described above reveals a crucial problem of such an approach to the quark-nuclear transition. At lower chemical potentials a chiral transition occurs

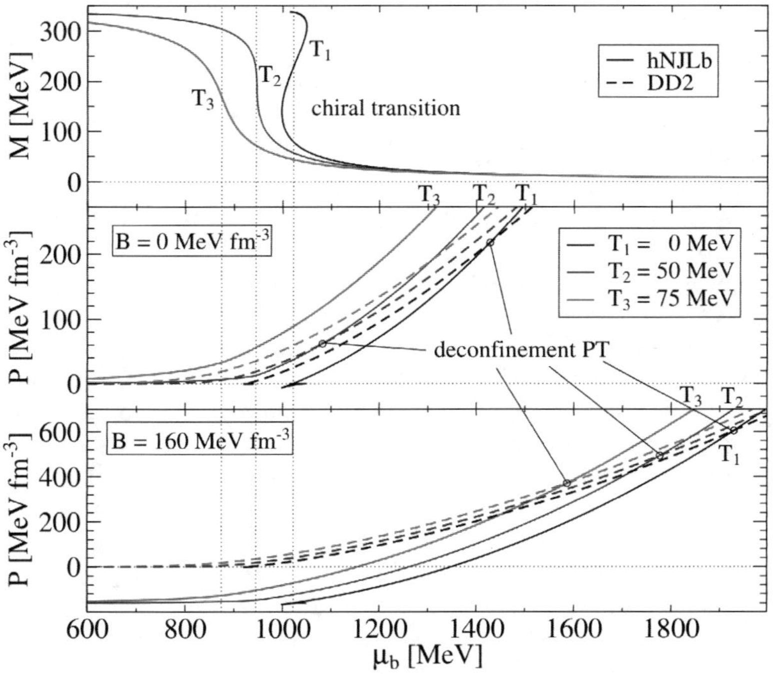


**Figure 1.** Effective mass $M$ (upper panel) and pressure $P$ (middle and lower panels) as functions of the baryochemical potential $\mu_b$ for the hNJLb EoS for $\eta_{02} = 0.08$ and $\eta_{04} = 5.0$ in the isospin symmetric case. The chiral transition is defined by the hNJLb EoS and deconfinement occurs at the crossing points with the hadronic (DD2) EoS for $B = 0$ (middle panel) and for $B = 160$ MeV/fm$^3$ (lower panel).

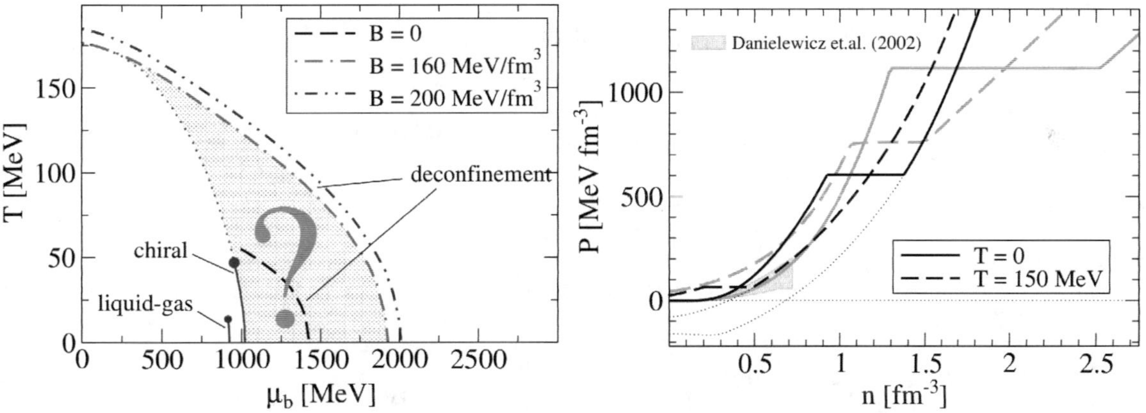


**Figure 2.** Left: Phase diagram for isospin symmetric quark-nuclear matter. The nuclear liquid-gas transition (blue line) and the chiral symmetry restoration (purple line) are first order PTs with endpoints. The deconfinement transition between DD2 and hNJLb is shown for different bag pressures: $B = 0$ (dashed line), $B = 160$ MeV/fm$^3$ (dash-dotted line) and $B = 200$ MeV/fm$^3$ (dash-double-dotted line). Right: EoS with deconfinement transition between DD2 and hNJLb for $\eta_{02} = 0.08$, $\eta_{04} = 5.0$ and bag constant $B = 160$ MeV/fm$^3$ compared to the EoS from Ref. [2] which is used in three-fluid hydro simulations [18, 19]. Shown are isothermes for $T = 0$ (solid lines) and $T = 150$ MeV (dashed lines) in symmetric matter and the flow constraint of Ref. [20]

within the NJL model, so we have chiral restoration of the quarks and they lose their masses. But in the construction at this point there are no quarks and only the hadrons exist, for which no chiral transition is considered. Therefore, in the resulting EOS the chiral transition is moved to the deconfinement transition, by the PT construction. On the other hand, if in the hatched region of Fig. 2 the quarks should already lose their masses while being still confined in hadrons, this should alter the hadron properties (e.g., the radii) and thus modify the behavior of their EoS in crucial way, so that the deconfinement transition probably should appear much earlier than shown in that figure, but this will probably still leave a window for a so-called *quarkyonic phase* [21] which may end in a triple point [22]. Such a phase might not exist and it can be avoided by construction, defining a temperature dependent bag constant so that the deconfinement phase transition coincides with the chiral phase transition [23]. On the other hand, one can explore this window between defined asymptotic EoS of hadronic and quark matter by interpolation [24, 25, 26], thus admitting the lack of fundamental theoretical concepts in that region.

To develop a unified quark-nuclear EoS we suggest to work out a cluster expansion [13] on the basis of the $\Phi$-derivable approach [12], where hadrons are considered as bound states of quarks. Their dissociation in hot, dense matter due to the Mott effect will be triggered by the lowering of the quark continuum thresholds that inevitably renders them unbound, but still allows their presence as strong correlations in the scattering continuum. This situation is accounted for within a generalized Beth-Uhlenbeck EoS that follows within this approach. As opposed to NJL models which have no confinement mechanism (and also their Polyakov-loop generalization which becomes ambigious at finite chemical potentials), we propose here the usage of relativistic density functionals which effectively account for confinement and can be motivated, e.g., within the concept of string-flip model [1]. Then, in the confined, hadronic phase before the Mott transition quark exchange effects among hadrons, due to their wave function overlap, will give rise to strongly repulsive forces. These Pauli blocking effects in dense hadronic matter justify the adoption of excluded volume prescriptions in the hadronic EoS in order to effectively account for them, see [17] for a recent application. In this way one should obtain a consistent unified description of quark-nuclear matter in the QCD phase diagram.

**Acknowledgement**
This work was supported by NCN under contract number UMO-2011/02/A/ST2/00306.

**References**
[1] Röpke G, Blaschke D and Schulz H 1986 *Phys. Rev.* **D34** 3499
[2] Khvorostukin A S, Skokov V V, Toneev V D and Redlich K 2006 *Eur. Phys. J.* **C48** 531
[3] Li A, Zuo W and Peng G X 2015 *Phys. Rev.* **C91** 035803
[4] Ratti C, Thaler M A and Weise W 2006 *Phys. Rev.* **D73** 014019
[5] Schmidt S M, Blaschke D and Kalinovsky Yu L 1994 *Phys. Rev.* **C50** 435
[6] Bowler R D and Birse M C 1995 *Nucl. Phys.* **A582** 655
[7] Gomez Dumm D, Blaschke D B, Grunfeld A G and Scoccola N N 2006 *Phys. Rev.* **D73** 114019
[8] Borsanyi S, Fodor Z, Hoelbling C, Katz S D, Krieg S and Szabo K K 2014 *Phys. Lett.* **B730** 99
[9] Bazavov A *et al.* (HotQCD) 2014 *Phys. Rev.* **D90** 094503
[10] Barz H W, Friman B L, Knoll J and Schulz H 1989 *Phys. Rev.* **D40** 157
[11] Röpke G, Bastian N U, Blaschke D, Klähn T, Typel S and Wolter H H 2013 *Nucl. Phys.* **A897** 70
[12] Baym G and Kadanoff L P 1961 *Phys. Rev.* **124** 287; Baym G 1962 *Phys. Rev.* **127** 1391
[13] Blaschke D 2015 *PoS* **BaldinISHEPPXXII** 113
[14] Typel S and Wolter H H 1999 *Nucl. Phys.* **A656** 331
[15] Typel S, Röpke G, Klähn T, Blaschke D and Wolter H H 2010 *Phys. Rev.* **C81** 015803
[16] Benic S 2014 *Eur. Phys. J.* **A50** 111
[17] Benic S, Blaschke D, Alvarez-Castillo D E, Fischer T and Typel S 2015 *Astron. Astrophys.* **577** A40
[18] Ivanov Yu B 2010 *Phys. Lett.* **B690** 358
[19] Ivanov Yu B and Blaschke D 2015 *Phys. Rev.* **C92** 024916
[20] Danielewicz P, Lacey R and Lynch W G 2002 *Science* **298** 1592
[21] McLerran L and Pisarski R D 2007 *Nucl. Phys.* **A796** 83
[22] Andronic A *et al.* 2010 *Nucl. Phys.* **A837** 65
[23] Klähn T and Fischer T 2015 *Astrophys. J.* **810** 134
[24] Masuda K, Hatsuda T and Takatsuka T 2013 *PTEP* **2013** 073D01
[25] Kojo T, Powell P D, Song Y and Baym G 2015 *Phys. Rev.* **D91** 045003
[26] Kojo T 2015 (*Preprint* 1508.04408)

# Beam Energy Scan of Specific Heat Through Temperature Fluctuations in Heavy Ion Collisions

**Sumit Basu[1], Basanta K Nandi[2], Sandeep Chatterjee[1], Rupa Chatterjee[1,3], Tapan Nayak[1]**

[1]Variable Energy Cyclotron Centre, 1/AF Bidhan Nagar, Kolkata - 700064, India
[2]Department of Physics, Indian Institute of Technology Bombay, Mumbai - 400067, India
[3]Department of Physics, P.O.Box 35, FI - 40014 University of Jyvāskylā, Finland

E-mail: sumit.basu@cern.ch

**Abstract.** Temperature fluctuations may have two distinct origins, first, quantum fluctuations that are initial state fluctuations, and second, thermodynamical fluctuations. We discuss a method of extracting the thermodynamic temperature from the mean transverse momentum of pions, by using controllable parameters such as centrality of the system, and range of the transverse momenta. Event-by-event fluctuations in global temperature over a large phase space provide the specific heat of the system. We present Beam Energy Scan of specific heat from data, AMPT and HRG model prediction. Experimental results from NA49, STAR, PHENIX, PHOBOS and ALICE are combined to obtain the specific heat as a function of beam energy. These results are compared to calculations from AMPT event generator, HRG model and lattice calculations, respectively.

## 1. Introduction

Experiments at RHIC and LHC are on the quest to unearth the nature of the QCD phase transition and to get a glimpse of how matter behaves at such extreme conditions. Phase transitions are governed by a set of thermodynamic parameters, like, temperature (T), pressure, entropy, and energy density (E), and can be further characterized by their response functions, like, specific heat, compressibility, and susceptibility. In thermodynamics, the heat capacity (C) is defined in terms of the ratio of the event-by-event fluctuations of the energy of a part of a finite system in thermal equilibrium to the energy $\langle \Delta E^2 \rangle = T^2 C(T)$ [1, 2, 3, 4]. This can be applied for a locally thermalized system produced during the evolution of heavy-ion collisions. But for a system at freeze-out, specific heat can expressed in terms of the event-by-event fluctuations in temperature of the system where volume is fixed: $1/C = (\langle T^2 \rangle - \langle T \rangle^2)/\langle T \rangle^2$. We define the specific heat as the heat capacity per pion multiplicity within the experimentally available phase space in rapidity and azimuth. For a system in equilibrium, the mean values of temperature and energy density are related by an equation of state. However, the fluctuations in energy and temperature have quite different behavior. Energy being an extensive quantity, its fluctuations have a component arising from the volume fluctuations, and not directly suited for obtaining the heat capacity. The heat capacity ($C_{\rm v}$), which is defined as the amount of heat required to change the temperature of the system by one unit. The specific heat is the heat capacity per unit mass or per particle in the system.

## 2. Methodology & Results

The temperature of the system can be obtained from the transverse momentum ($p_{\mathrm{T}}$) spectra of the emitted particles. An exponential Boltzmann-type fit to the $p_{\mathrm{T}}$ spectra gives a measure of the temperature:

$$F(p_{\mathrm{T}}) = \frac{1}{p_{\mathrm{T}}}\frac{dN}{dp_{\mathrm{T}}} \approx Ae^{-p_{\mathrm{T}}/T_{\mathrm{eff}}}, \tag{1}$$

where $A$ is a normalization factor and $T_{\mathrm{eff}}$ is the apparent or effective temperature of the system [5]. For obtaining the event-by-event fluctuation, the temperature needs to be estimated in every event either by fitting in a specified (thermal) region or by calculating mean transverse momentum ($\langle p_{\mathrm{T}}\rangle$) of particles. Since the calculation of the mean value is more stable, this method of temperature estimation can also be used for collisions at RHIC energies. As the $p_{\mathrm{T}}$ window is finite, For a range of $p_{\mathrm{T}}$ within $a$ to $b$, we obtain:

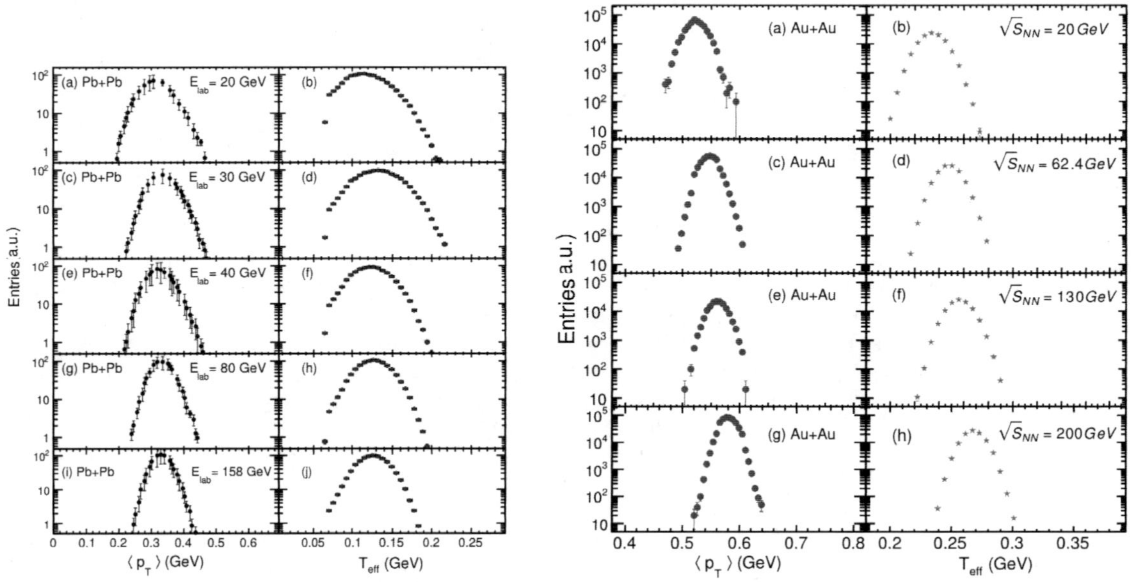


(a) NA49 (b) STAR Au-Au

Figure 1: $\langle p_{\mathrm{T}}\rangle$ and $T_{\mathrm{eff}}$ for both fixed-target and collider energy.

$$\langle p_{\mathrm{T}}\rangle = \frac{\int_a^b p_{\mathrm{T}}^2 F(p_{\mathrm{T}})dp_{\mathrm{T}}}{\int_a^b p_{\mathrm{T}} F(p_{\mathrm{T}})dp_{\mathrm{T}}} \tag{2}$$

$$= 2T_{\mathrm{eff}} + \frac{a^2e^{-a/T_{\mathrm{eff}}} - b^2e^{-b/T_{\mathrm{eff}}}}{(a+T_{\mathrm{eff}})e^{-a/T_{\mathrm{eff}}} - (b+T_{\mathrm{eff}})e^{-b/T_{\mathrm{eff}}}}. \tag{3}$$

Fig. 1 shows the published [6, 7]. $\langle p_{\mathrm{T}}\rangle$ distribution for central ($\sim$ top 5%) of (a) Fixed target experiment in NA49 by Pb+Pb at (1.1< y*< 2.6) with corresponding $T_{\mathrm{eff}}$ distribution extracted using 3 and (b) Collider experiment in STAR by Au+Au at mid-rapidity with the range $0.15 < p_{\mathrm{T}} < 2.0$ .

The event-by-event fluctuations of $\beta_{\mathrm{T}}$ need to be taken into account for calculating the fluctuation in kinetic temperature. Fluctuation in $\beta_{\mathrm{T}}$ has been discussed in the literature lately.

Experimental determination is only possible by a fitting event-by-event $p_{\mathrm{T}}$ distribution by a blast-wave fit. This has not yet been done. For the present work, we have assigned an error to the value of $\beta_{\mathrm{T}}$ from the available data and calculated its effect on heat capacity. We calculate $C_{\mathrm{v}}$ using the equation,

$$\frac{1}{C_v} = \frac{(\langle T_{\mathrm{kin}}^2\rangle - \langle T_{\mathrm{kin}}\rangle^2)}{\langle T_{\mathrm{kin}}\rangle^2} \approx \frac{(\langle T_{\mathrm{eff}}^2\rangle - \langle T_{\mathrm{eff}}\rangle^2)}{\langle T_{\mathrm{kin}}\rangle^2}, \tag{4}$$

and assign an error to $C_{\mathrm{v}}$ corresponding to the fluctuations in $\beta_{\mathrm{T}}$. The values of $\langle T_{\mathrm{kin}}\rangle$ can be obtained from the Blast Wave fits to the experimental data for large number events. Knowing the heat capacity, the specific heat ($c_{\mathrm{v}}$) is calculated as the heat capacity per particle [5] in the system and expressed in terms of $c_{\mathrm{v}}/T_{\mathrm{kin}}^3$ as a dimensionless quantity.

Fig. 2 shows the $\langle p_{\mathrm{T}}\rangle$ and $T_{\mathrm{eff}}$distribution for central ($\sim$ top 5%)(a) AMPT model with string melting (SM) mode ranging from 7GeV to 200GeV by Au+Au at mid-rapidity and by Pb+Pb at 2760 GeV [8, 9]. While (b) shows the same in a different system size at colliding mode via Cu+Cu [10]. Due to finite multiplicity, within a centrality window in the heavy-ion collisions contribute

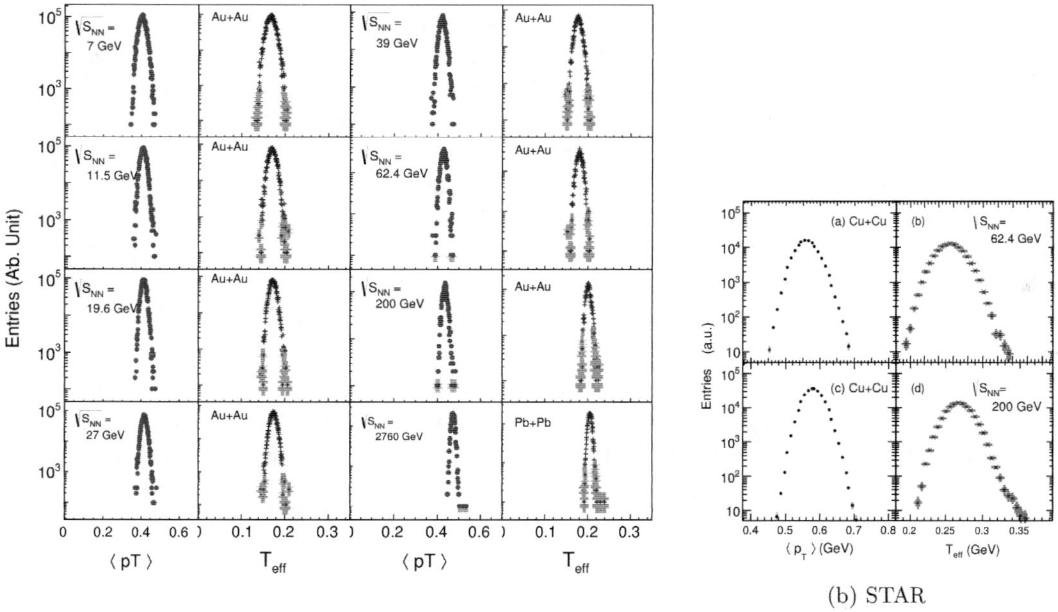


(a) AMPT

(b) STAR Cu+Cu

Figure 2: AMPT SM modes and system size dependence of $\langle p_{\mathrm{T}}\rangle$ and extraction of $T_{\mathrm{eff}}$.

to the statistical fluctuations in temperature fluctuations. The temperature fluctuations arising from statistical effects have been calculated in terms of a statistical model calculation. It can be either done via a synthetic event assuming similar slope of corresponding collision energy with zero fluctuations on $T_{\mathrm{eff}}$ and only fluctuation in its event multiplicity with a realistic approach or via mix event analysis. Then following the same mechanism of analysis one can estimate the statistical fluctuations in temperature ($\Delta T_{\mathrm{stat}}$ ). Now, $\Delta T_{\mathrm{eff}}^2 = \Delta T_{\mathrm{dyn}}^2 + \Delta T_{\mathrm{stat}}^2$. Knowing the $\Delta T_{\mathrm{dyn}}$ one can crudely assume that within a narrow centrality if the $\beta_{\mathrm{T}}$ varies too little event-to-event then $\Delta T_{\mathrm{dyn}}$ is solely arieses due to $\Delta T_{\mathrm{kin}}$. Even if $\beta_{\mathrm{T}}$ varies within the error one can calculate the modified contribution from $\Delta T_{\mathrm{kin}}$ and put it as a error. A similar estimation could be possible in a much simple way $\Delta T_{\mathrm{stat}}^2 = \frac{1}{M}$ where M is the particle number within specified pseudo-rapidity and $p_{\mathrm{T}}$ range [11].

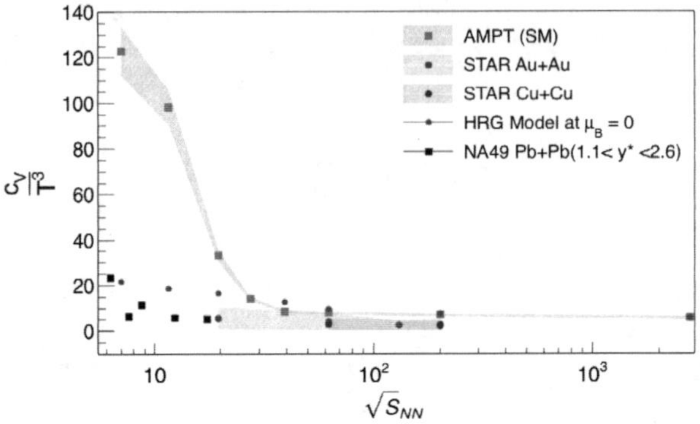


Figure 3: Beam energy dependence of $c_{\rm v}/T_{\rm kin}^3$ with data and model comparisons.

After subtracting out the statistical estimations, one can find out the heat capacity $C_{\rm v}$ via (4) and having the $c_{\rm v}$ by knowing the particle or pion numbers. Calculating the $c_{\rm v}$, we can measure the dimensionless quantity $c_{\rm v}/T_{\rm kin}^3$ and directly compare the values with theoretical prediction like thermal Hadron Gas (HRG) model at corresponding energy considering baryonic potential is zero ($\mu_{\rm B} = 0$ ). Fig. 3 shows that the variation of $c_{\rm v}/T_{\rm kin}^3$ with different beam energy and system size with one microscopic event generator (AMPT SM) and thermal model (HRG) prediction. This is the first result which can be experimentally measurable.

## 3. Discussions

This result is very important to understand the critical point and other transitions at phase boundary. This is also first time some result of specific heat form experimental result. Normally $\frac{c_{\rm v}}{T_{\rm kin}^3}$ quantity is widely used by the theoretical study assuming the chemical freeze-out. Also if the experimental results matches with the theory prediction, it would help to understand the QGP partonic phase and the equation of state more deeply. It should be observed that by increasing the energy then $\langle p_{\rm T}\rangle$ increases but the $\Delta\langle p_{\rm T}\rangle$ decreases, which is expected as chance of achieving thermalization is being higher. Thus specific heat from temperature fluctuation could play a key role in the order to fix the remaining mystery of the QGP and heavy ion collisions.

## References

[1] L.D. Landau.and E.M. Lifshitz, STATISTICAL PHYSICS, (Course of Theoretical Physics, v. 5, 3rd, rev. and enlarged ed.) Pergamon Press (1980).
[2] L. Stodolsky, Phys. Rev. Lett. **75**, 1044 (1995).
[3] E.V. Shuryak, Phys. Lett. B **423**, 9 (1998).
[4] M.A. Stephanov, K. Rajagopal and E. V. Shuryak, Phys. Rev. Lett. **81**, 4816 (1998).
[5] B-H Sa, X-M Li, S-Y Hu, S-P Li, J. Feng, and D-M Zhou, Phys. Rev. **75**, 054912 (2007).
[6] J. Adams *et al.* (STAR Collaboration) Phys. Rev. C **79** 044904 (2009).
[7] J. Adams *et al.* (STAR Collaboration) Phys. Rev. C**72** 044902 (2005).
[8] D. Solanki *et al.* Phys. Lett. **B720** 352-357 (2013).
[9] S. Basu *et al.* arXiv:1504.04502 [nucl-ex] (2015).
[10] B. I. Abelev *et al.* (STAR Collaboration) Phys. Rev. C**79** 034909 (2009).
[11] S. A. Voloshin *et al.* Phys. Rev. **C60** 024901 (1999).

Journal of Physics: Conference Series **668** (2016) 012044 doi:10.1088/1742-6596/668/1/012044

# Modeling jet-medium interactions at RHIC and LHC - which energy loss effect is crucial?

**B Blagojevic and M Djordjevic**

Institute of Physics Belgrade, Pregrevica 118, 11080 Zemun, Serbia

E-mail: bojanab@ipb.ac.rs

**Abstract.** High momentum hadrons suppression is considered to be excellent probe of QCD matter created in ultra-relativistic heavy ion collisions. Here we apply our recently developed dynamical energy loss formalism, which includes the following effects: dynamical scattering centers, QCD medium of a finite size, both radiative and collisional energy losses, running coupling and finite magnetic mass, and which we further incorporate into numerical procedure, to generate angular averaged $R_{AA}$ predictions and to compare them with experimental $R_{AA}$ data, by using no free parameters. A robust agreement of our predictions and experimentally measured $R_{AA}$ for different energies, probes and all available centrality regions, raised the question whether this agreement is consequence of a single effect or of a superposition of all these effects. We obtained that, although the inclusion of dynamical scattering center has the largest relative importance, all the other effects are also important, since they lead to fine improvements of the agreement. Therefore, the robust agreement is a cumulative effect of all these features, with dynamical effect being crucial for accurate $R_{AA}$ predictions.

## 1. Introduction

High momentum light and heavy flavor suppression [1] is considered to be excellent tool for studying QCD matter created in ultra-relativistic heavy ion collisions at RHIC and LHC. An abundance of suppression data, that has become available at RICH and LHC since recently, and its comparison with theoretical $R_{AA}$ predictions [2–4], allows testing our understanding of QGP matter. In order to generate these predictions, we recently developed dynamical energy loss formalism, which we further integrated into numerical procedure as described in [5]. This formalism includes the following energy loss effects: *i)* dynamical scattering centers, *ii)* QCD medium of a finite size [6, 7], *iii)* both radiative [6, 7] and collisional [8] energy losses, *iv)* running coupling [5] and *v)* finite magnetic mass [9]. Also, note that, accurate energy loss calculation is considered to be the main ingredient responsible for obtaining reliable $R_{AA}$ predictions.

In our previous papers [5, 10, 11], we demonstrated a robust agreement between our $R_{AA}$ predictions, obtained as explained in previous paragraph, and $R_{AA}$ data for both RHIC and LHC experiments, diverse set of probes and all available centrality ranges.

Here we address the relative importance of different energy loss effects in obtaining accurate angular averaged $R_{AA}$ predictions for D mesons (as the clearest energy loss probe), for which it was previously shown that fragmentation function does not alter bare charm quark $R_{AA}$ [10, 12]. High momentum D meson $R_{AA}$ data, obtained recently at LHC [13], serve as a baseline for testing the models. We concentrate on central 200 GeV Au+Au collisions at RHIC and 2.76 TeV Pb+Pb collisions at LHC. Our approach is to systematically include the effects in energy loss calculations [14], i.e. we first compare the relative importance of radiative and collisional contribution to $R_{AA}$ predictions, next we address the importance of including the dynamical scattering centers, then the running coupling and finally the finite magnetic mass. Note that only the main results are presented here; for the full account on the results, please see [14].

## 2. Theoretical and computational formalism

In this section, we concisely describe computational formalism, our dynamical energy loss formalism [5] and how each effect, when introduced, changed energy loss expressions.

For obtaining quenched spectra we apply generic pQCD convolution given by Eq.(2) from [14] ([15]). The initial charm quark spectrum is calculated in accordance with [16], while energy loss probability includes both radiative and collisional energy losses in a finite size dynamical QCD medium, multi-gluon [17] and path length fluctuations [15, 18].

The expression for radiative energy loss in a finite size dynamical QCD medium is given by Eq.(2.12) from [6], while the transition from static to dynamical scattering centers is explained in [7]. The collisional energy loss is calculated according to Eq.(14) from [8]. The running coupling is introduced in accordance with [5], while for constant coupling we use $\alpha_S = 0.3$ ($\alpha_S = 0.25$) in RHIC (LHC) case. Debye screening mass is $\mu_E = gT$ ($g = 2$). The finite magnetic mass is introduced as in [9], and its range ($0.4 < \mu_M/\mu_E < 0.6$) is set according to many non-perturbative approaches [19–23], otherwise $\mu_M = 0$ is used.

We model the medium by assuming an effective temperature of 221 MeV at RHIC [24] and 304 MeV at LHC [25]. No medium evolution is accounted. The validity of this assumption is discussed in [14]. For charm quark mass we use $M_c = 1.2$ GeV, and for the number of effective light quark flavors we use $n_f = 2.5$ ($n_f = 3$) in RHIC (LHC) case.

## 3. Results and discussion

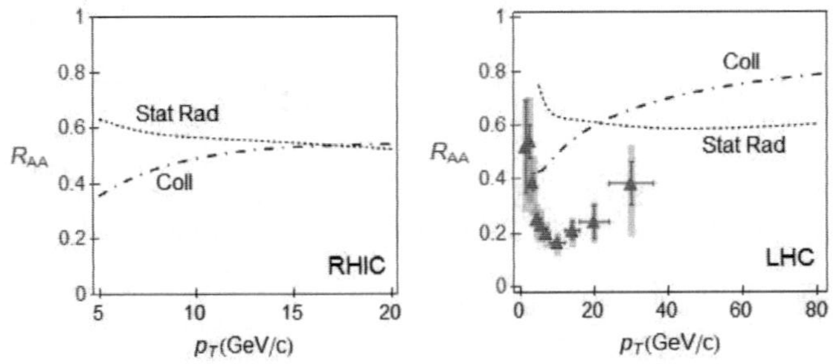


**Figure 1. Necessity of abolishing static approximation.** D meson $R_{AA}$ predictions, as a function of transverse momentum, are shown for only static radiative (dotted curve) and for only dynamical collisional (dot-dashed curve) contribution in a finite size QCD medium. Left (right) panel corresponds to RHIC (LHC) case. Right panel also displays D meson $R_{AA}$ data in 0 – 7.5% central 2.76 TeV Pb+Pb collisions at LHC (red triangles) [13]. Debye mass is $\mu_E = gT$, coupling constant is $\alpha_S = 0.3$ ($\alpha_S = 0.25$) for RHIC (LHC) and finite magnetic mass is not included ($\mu_M = 0$). Adapted from [14].

In this section we apply historically-driven approach, starting from static approximation [26, 27] and gradually adding energy loss effects. We display only the main results of our study carried out in [14]. Finite size QCD medium is assumed throughout the paper. The constant coupling and constant Debye mass (as mentioned above), and no finite magnetic mass are considered in Fig. 1 and Fig. 2. Static approximation, which assumes that the medium is composed of static scattering centers, was firstly commonly used. It entails also, that collisional energy loss can be neglected compared to radiative one. However, Fig. 1 clearly shows that static approximation has to be abolished in favor of dynamical scatering centers' approximation, since collisional $R_{AA}$ is comparable with radiative one. Further, we compute these both energy losses within the

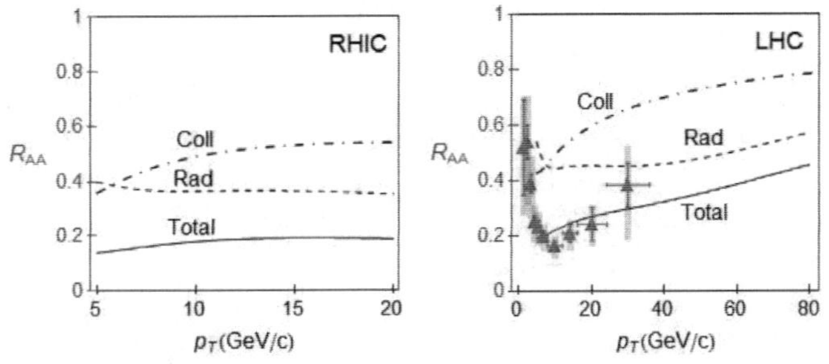


**Figure 2. Dynamical approximation as the main effect.** D meson $R_{AA}$ predictions, as a function of transverse momentum, are shown for radiative (dashed curve), collisional (dot-dashed curve) and radiative + collisional (solid curve) energy losses in a finite size dynamical QCD medium. Left (right) panel corresponds to RHIC (LHC) case. Right panel also displays D meson $R_{AA}$ data in $0-7.5\%$ central 2.76 TeV Pb+Pb collisions at LHC (red triangles) [13]. Debye mass is $\mu_E = gT$, coupling constant is $\alpha_S = 0.3$ ($\alpha_S = 0.25$) for RHIC (LHC) and finite magnetic mass is not included ($\mu_M = 0$). Adapted from [14].

same dynamical framework (Fig. 2) and we draw three conclusions: 1) dynamical radiative $R_{AA}$ alone is not su cient to explain qualitatively nor quantitatively the LHC experimental data; 2) radiative and collisional $R_{AA}$ are still both important; 3) the total $R_{AA}$ is in rough agreement with experimental data. Therefore, the inclusion of dynamical scattering centers is the main effect responsible for obtaining accurate $R_{AA}$ predictions. Finally, we address the importance of

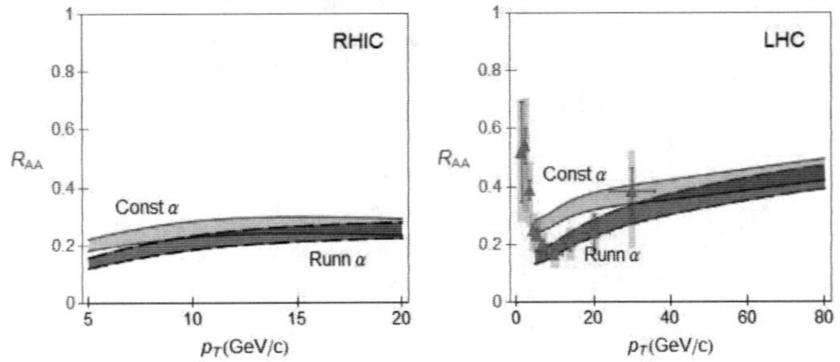


**Figure 3. Our dynamical energy loss formalism.** D meson $R_{AA}$ predictions, as a function of transverse momentum, are shown for the constant coupling $\alpha_S = 0.3$ ($\alpha_S = 0.25$) for RHIC (LHC) (light gray band) and for the running coupling (dark gray band). In both cases radiative + collisional contributions in a finite size dynamical QCD medium are accounted. Upper (lower) boundary of each band corresponds to $\mu_M/\mu_E = 0.6$ ($\mu_M/\mu_E = 0.4$). Left (right) panel corresponds to RHIC (LHC) case. Right panel also displays D meson $R_{AA}$ data in $0-7.5\%$ central 2.76 TeV Pb+Pb collisions at LHC (red triangles) [13]. Adapted from [14].

including the running coupling (leads to a significant $R_{AA}$ decrease at lower jet momenta) and finite magnetic mass (leads to a significant $R_{AA}$ increase) [14]. From Fig. 3 we see that these two effects (although taken alone worsens the agreement [14]) taken together lead to quantitatively and qualitatively better agreement with the LHC $R_{AA}$ data, compared to the case when these effects are omitted. This illustrates possible synergy in including these two effects.

## 4. Conclusions

A robust agreement of angular averaged $R_{AA}$ predictions, based on our dynamical energy loss formalism, with $R_{AA}$ data, for different energies, probes and centrality ranges, initiated the question: whether this agreement is a consequence of a one dominant energy loss effect or a joint effect of several smaller improvements [14]. With the LHC suppression data serving as a baseline, we here showed that (for the clearest energy loss probe: D meson $R_{AA}$), inclusion of dynamical scattering centers has the largest relative importance in obtaining accurate $R_{AA}$ predictions. Furthermore, we found that all other considered effects are also important and responsible for the finer agreement with the data. So the good agreement is a result of a superposition of all these effects. Therefore, detailed study of partons' energy loss, as well as, inclusion of all relevant medium effects is necessary to correctly model the jet-medium interactions.

## Acknowledgments

This work is supported by Marie Curie International Reintegration Grant within the $7^{th}$ European Community Framework Programme PIRG08-GA-2010-276913 and by the Ministry of Science and Technological Development of the Republic of Serbia, under project No. ON171004.

## References

[1] Bjorken J D 1982 FERMILAB-PUB-82-059-THY pp 287-92
[2] Brambilla N *et al* 2014 *Eur. Phys. J.* C **74** 2981
[3] Gyulassy M 2002 *Lect. Notes Phys.* **583** 37
[4] d Enterria D and Betz B 2010 *Lect. Notes Phys.* **785** 285
[5] Djordjevic M and Djordjevic M 2014 *Phys. Lett.* B **734** 286
[6] Djordjevic M 2009 *Phys. Rev.* C **80** 064909
[7] Djordjevic M and Heinz U 2008 *Phys. Rev. Lett.* **101** 022302
[8] Djordjevic M 2006 *Phys. Rev.* C **74** 064907
[9] Djordjevic M and Djordjevic M 2012 *Phys. Lett.* B **709** 229
[10] Djordjevic M and Djordjevic M 2014 *Phys. Rev.* C **90** 034910
[11] Djordjevic M, Djordjevic M and Blagojevic B 2014 *Phys. Lett.* B **737** 298
[12] Djordjevic M 2014 *Phys. Rev. Lett.* **112** 042302
[13] Grelli A 2013 *Nucl. Phys.* A **904-905** 635c
Abelev B *et al* 2012 *J. High Energy Phys.* JHEP1209(2012)112
[14] Blagojevic B and Djordjevic M 2015 *J. Phys.* G **42** 075105
[15] Wicks S, Horowitz W, Djordjevic M and Gyulassy M 2007 *Nucl. Phys.* A **784** 426
[16] Kang Z B, Vitev I and Xing H 2012 *Phys. Lett.* B **718** 482-7
[17] Gyulassy M, Levai P and Vitev I 2002 *Phys. Lett.* B **538** 282
[18] Dainese A 2004 *Eur. Phys. J.* C **33** 495
[19] Maezawa Yu, Aoki S, Ejiri S, Hatsuda T, Ishii N, Kanaya K, Ukita N and Umeda T 2010 *Phys. Rev.* D **81** 091501
[20] Maezawa Yu, Aoki S, Ejiri S, Hatsuda T, Ishii N, Kanaya K, Ukita N and Umeda T 2008 *Proc. of Science Lattice* 2008 (Williamsburg) p 194 (*Preprint* hep-lat/0811.0426)
[21] Nakamura A, Saito T and Sakai S 2004 *Phys. Rev.* D **69** 014506
[22] Hart A, Laine M and Philipsen O 2000 *Nucl. Phys.* B **586** 443
[23] Bak D, Karch A and Yaffe L G 2007 *J. High Energy Phys.* JHEP0708(2007)049
[24] Adare A *et al* 2010 *Phys. Rev. Lett.* **104** 132301
[25] Wilde M 2013 *Nucl. Phys.* A **904-905** 573c
[26] Djordjevic M and Gyulassy M 2004 *Nucl. Phys.* A **733** 265-98
[27] Gyulassy M and Wang X N 1994 *Nucl. Phys.* B **420** 583
Wang X N, Gyulassy M and Plumer M 1995 *Phys. Rev.* D **51** 3436

15th International Conference on Strangeness in Quark Matter (SQM2015) IOP Publishing
Journal of Physics: Conference Series **668** (2016) 012045 doi:10.1088/1742-6596/668/1/012045

# Scaling trends in proton-proton collisions from SPS to LHC in quark-gluon string model

**L V Bravina**[1] **and E E Zabrodin**[1,2]

[1] Department of Physics, University of Oslo, Norway
[2] Skobeltsyn Institute of Nuclear Physics, Moscow State University, Russian Federation

E-mail: larissa.bravina@fys.uio.no; eugen.zabrodin@fys.uio.no

**Abstract.** Fulfillment of scaling relations, such as Feynman scaling, extended longitudinal scaling and Koba-Nielsen-Olesen (KNO) scaling, in ultrarelativistic proton-proton collisions are checked within the microscopic quark-gluon string model (QGSM). The model is based on Reggeon Field theory accomplished by string phenomenology. Multiplicity, rapidity and transverse momentum spectra are described quite well in a wide span of the collision energy. Predictions are made for $pp$ collisions at $\sqrt{s} = 14\,\mathrm{TeV}$.

## 1. Introduction

The theory of multiparticle production in elementary proton-proton collisions at ultrarelativistic energies is not completed yet, because even at very high energies the main contribution to multiparticle production in hadronic interactions comes from the processes with small momentum transfer. The running coupling constant $\alpha_s(Q^2)$ is not small and other techniques based on non-perturbative methods should be utilized. The quark-gluon string model (QGSM) [1, 2] relying on the Reggeon Field Theory (RFT) [3] is one of the few possible approaches. Its basic principles are sketched below.

QGSM employs the $1/N$ series expansion of the amplitude of a process in QCD, where $N$ is either the number of colors [4] or the number of flavors [5]. This method is also called *topological expansion*, because of emergence of diagrams of various topologies. Although it is not possible to assign weights for the diagrams within the QCD, there is one-to-one mapping between the diagrams in $1/N$-expansion and the processes with exchange of Regge singularities in the $t$-channel. For instance, exchange of quantum numbers via Reggeons corresponds to *planar* diagrams, whereas the *cylinder* diagrams are represented by the reactions without the quantum number exchange. The latter proceed via the exchange of Pomerons. Therefore, the perturbative Reggeon Field Theory (RFT) [3] is directly linked to quantum chromodynamics.

The Monte Carlo version of the QGSM [6, 7, 8, 9] employs statistical weights, hadron structure functions and leading quark fragmentation functions obtained from the Regge approach in [1, 2] to choose subprocesses of string production, to compute mass and momentum of strings and to simulate string decays, respectively. The hadron-hadron collision part of the model includes single and double diffraction subprocesses, antibaryon-baryon annihilation and elastic scattering. The hadron inelastic interaction cross section $\sigma_{inel}(s) = \sigma_{tot}(s) - \sigma_{el}(s)$ is split further into the cross section for single diffractive interactions $\sigma_{SD}(s)$ and the cross section for non-diffractive

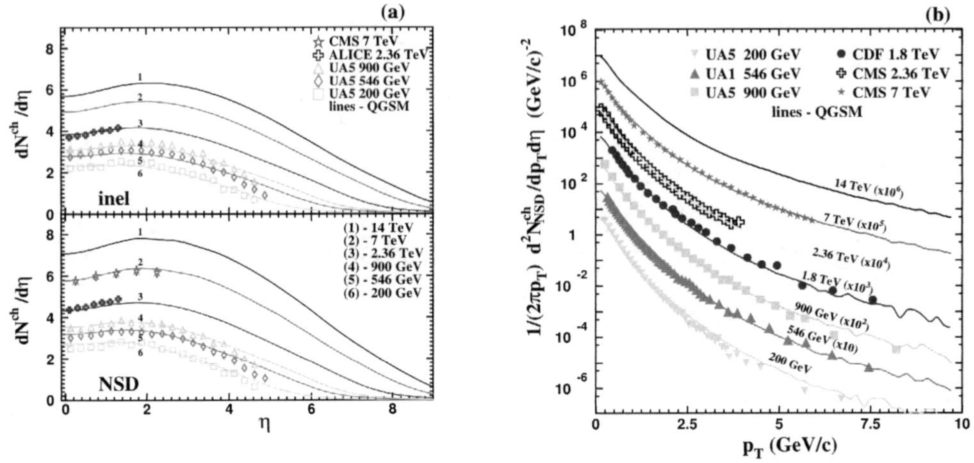


**Figure 1.** **(a)**: Pseudorapidity spectra for charged particles in inelastic and NSD $pp$ collisions at $200\text{GeV} \le \sqrt{s} \le 14\,\text{TeV}$. **(b)**: transverse momentum distribution of the invariant cross section in NSD $pp$ collisions in the same energy range.

reactions $\sigma_{ND}(s)$, similar to analysis of experimental data. By means of the Abramovskii-Gribov-Kancheli (AGK) cutting rules [10] the inelastic non-diffractive interaction cross section $\sigma_{ND}(s)$ can be expressed via the sum of the cross sections for the production of $n = 1, 2, \ldots$ pairs of quark-gluon strings, or cut Pomerons, and the cross section of double diffractive process $\sigma_{ND}(s) = \sum_{n=1}^{\infty} \sigma_n(s) + \sigma_{DD}(s)$. To find $\sigma_n(s)$ one can utilize the quasi-eikonal model [11, 12]. For the modeling of string fragmentation the Field-Feynman algorithm [13] is employed. It enables one to consider emission of hadrons from both ends of the string with equal probabilities. The break-up procedure invokes the energy-momentum conservation and the preservation of the quark numbers.

## 2. Results

*Feynman scaling.* Recall that at energies $\sqrt{s} \ge 50\,\text{GeV}$ the annihilation cross section is extremely small and, therefore, the main features of particle production in $pp$ interactions are similar to those in $\bar{p}p$ ones. Thus, for the comparison with the model results we utilized data obtained by the UA5 Collaboration for proton - antiproton collisions at c.m. energies $\sqrt{s} = 200\,\text{GeV}$, 546 GeV and 900 GeV [14], by the UA1 Collaboration for $\bar{p}p$ collisions at $\sqrt{s} = 546\,\text{GeV}$ [15, 16], by the CDF and the E735 Collaborations for $\bar{p}p$ collisions at $\sqrt{s} = 1800\,\text{GeV}$ [17, 18], and more recent CERN LHC data obtained in $pp$ interactions at $\sqrt{s} = 900\,\text{GeV}$, 2360 GeV, and 7 TeV by the ALICE Collaboration [19] and by the CMS Collaboration [20].

Pseudorapidity spectra of charged particles in elastic and non-single diffraction (NSD) proton-proton interactions at $200\,\text{GeV} \le \sqrt{s} \le 14\,\text{TeV}$ are shown in Fig. 1(a). Available experimental data are plotted onto the model calculations as well. The hypothesis of the so-called Feynman scaling [21] postulates that the density of produced charged particles at midrapidity $dN^{ch}/d\eta$ should be saturated somewhere at very high energies. This scaling regime is obviously not reached yet. Moreover, at LHC energies $dN^{ch}/d\eta\,|_{\eta=0}$ demonstrates a non-linear rise with $\ln s$, as suggested by the saturation of the Froissart bound. For $pp$ collisions at top LHC energy $\sqrt{s} = 14\,\text{TeV}$ the QGSM predicts $dN_{inel}/d\eta\,|_{\eta=0} = 6.1$, $dN_{NSD}/d\eta\,|_{\eta=0} = 7.0$, respectively. The transverse momentum distributions of charged hadrons in NSD collisions at energies in question

Journal of Physics: Conference Series **668** (2016) 012045 doi:10.1088/1742-6596/668/1/012045

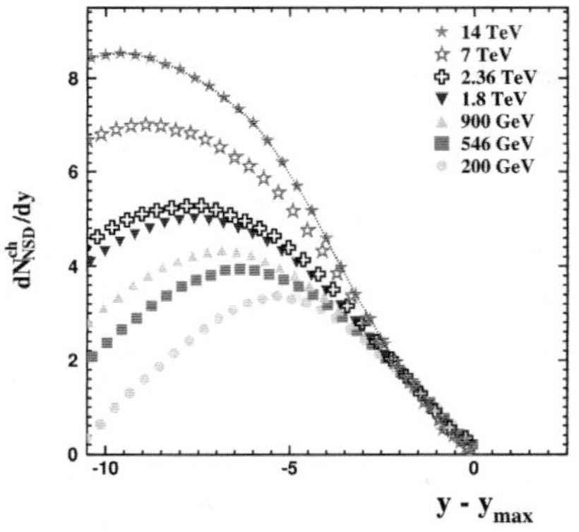


**Figure 2.** The distributions $\frac{1}{\sigma_{NSD}} \frac{d\sigma_{NSD}}{dy}$ of charged particles as functions of rapidity difference $y - y_{max}$ obtained in QGSM for $pp$ collisions.

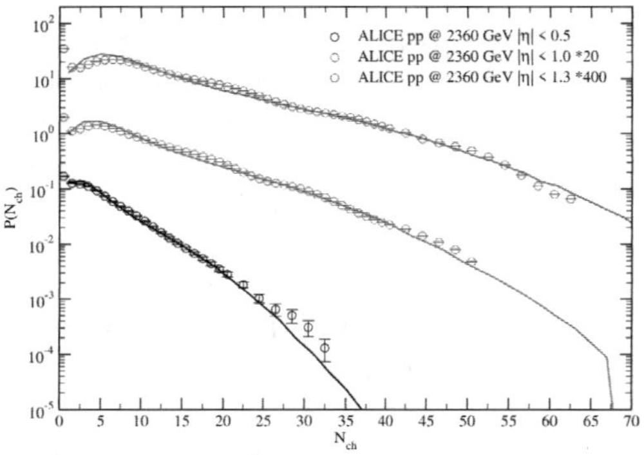


**Figure 3.** Charged-particle multiplicity distributions in $|\eta| < 0.5$, $|\eta| < 1.0$ and $|\eta| < 1.3$ intervals, obtained in QGSM for $pp$ collisions at $\sqrt{s} = 2360\,\text{GeV}$. Open symbols present the ALICE data [19].

are shown in Fig. 1(b) together with the experimental data. The agreement between the model results and the data is pretty good. Predictions for $\sqrt{s} = 14\,\text{TeV}$ are plotted as well.

*Extended longitudinal scaling and KNO scaling.* Another scaling relation is the extended longitudinal scaling (ELS) [22] exhibited by the slopes of (pseudo)rapidity spectra. In the QGSM these slopes are identical in the fragmentation region $y_{beam} \geq -2.5$ as shown in Fig. 2, where the distributions $\frac{1}{\sigma_{NSD}} \frac{d\sigma_{NSD}}{dy}$ are expressed as functions $y - y_{max}$. QGSM indicates that the ELS remains certainly valid at LHC. Obtained result contradicts to the prediction based on the statistical thermal model [23]. The latter fits the measured rapidity distributions to the Gaussian, extracts the widths of the Gaussians and implements the energy dependence of the obtained widths to simulate the rapidity spectra at LHC. The extrapolated distribution was found to be much narrower [23] compared to that presented in Fig. 2. Further LHC measurements of $pp$ collisions in the fragmentation regions are needed to resolve this obvious discrepancy. Note, that experimentally the extended longitudinal scaling was found to hold to 10% in a broad energy range from $\sqrt{s} = 30.8\,\text{GeV}$ to 900 GeV [14].

The extended longitudinal scaling in the QGSM emerges merely due to short range correlations in rapidity space. The correlation function of particle $i$ and particle $j$, produced as a result of a string fragmentation, drops exponentially with rising rapidity difference

$$C(y_i, y_j) = \frac{d^2\sigma}{\sigma_{inel}\, dy_i dy_j} - \frac{d\sigma}{\sigma_{inel}\, dy_i} \frac{d\sigma}{\sigma_{inel}\, dy_j} \propto \exp\left[-\lambda\left(y_i - y_j\right)\right] , \quad (1)$$

and, therefore, the particles with large rapidity difference are uncorrelated. Consider now the inclusive process $1 + 2 \to i + X$. Its single particle inclusive cross section $f_i = d^2\sigma(y_1 - y_i, y_i - y_2, p_{iT}^2)/dy_i d^2 p_{iT}$ becomes independent of $y_i - y_2$ at sufficiently high collision energy in the fragmentation region of particle *1*, provided $y_1 - y_i \approx 1$ and $y_i - y_2 \approx y_1 - y_2 \gg 1$. Thus, the inclusive densities $n_i \equiv f_i/\sigma_{inel}$ are determined by only two variables $n_i = \phi(y_1 - y_i,\, p_{iT}^2)$.

Next scaling dependence is known as Koba-Nielsen-Olesen (KNO) scaling [24]. It claims that at $\sqrt{s} \to \infty$ the normalized multiplicity distribution just scales up as $\ln s$ or, equivalently, that

$$\frac{\langle n \rangle\, \sigma_n}{\Sigma\, \sigma_n} = \Psi\left(\frac{n}{\langle n \rangle}\right) , \quad (2)$$

with $\sigma_n$ being the partial cross section for $n$-particle production, $\langle n \rangle$ - the average multiplicity and $\Psi(n/\langle n \rangle)$ - energy independent function. KNO-scaling was found to hold up to ISR energies, $\sqrt{s} \leq 62$ GeV. Violation of the KNO-scaling was predicted within the QGSM in [1, 2]. Later on the violation was observed experimentally by the UA5 and UA1 collaborations in $\bar{p}p$ collisions at $\sqrt{s} = 546$ GeV in the full phase space [14]. The rapidity range is crucial for this study. In very central pseudorapidity window $|\eta| < 0.5$ the KNO-scaling is still maintained at $\sqrt{s} = 2.36$ TeV [19], as seen in Fig. 3, whereas already the UA5 Collaboration observed progressive violation of the scaling with increasing $\eta$ intervals at much lower energies. For a bit broader midrapidity intervals at LHC a peak at low multiplicities seems to appear, see Fig. 3. The origin of this phenomenon in the model is the following. At ultrarelativistic energies the main contribution to particle multiplicity comes from the cut-Pomerons, and each cut results to formation of two strings. Short range correlations inside a single string lead to a Poisson-like multiplicity distribution of produced secondaries. At energies below 100 GeV the multi-string (or chain) processes are not very abundant and invariant masses of the strings are not very large. Therefore, different contributions to particle multiplicity overlap strongly, and the KNO-scaling is nearly fulfilled. With rising $\sqrt{s}$ the number of strings increases as $(s/s_0)^{\Delta}$ and their invariant masses increase as well. This leads to enhancement of high multiplicities, deviation of the multiplicity distribution from the Poisson-like behavior and violation of KNO-scaling [1, 2, 6].

## 3. Conclusions

Several scaling properties predicted or observed in particle production at relativistic energies have been examined. QGSM favors violation of Feynman scaling in the central rapidity region. Extended longitudinal scaling is shown to hold at LHC, whereas further violation of the KNO-scaling in multiplicity distributions is demonstrated. The origin of both conservation and violation of the scaling trends is traced to short range correlations of particles in the strings and interplay between the multi-Pomeron processes at ultra-relativistic energies.

## References

[1] Kaidalov A B 1982 *Phys. Lett. B* **116** 459
[2] Kaidalov A B and Ter-Martirosyan K A 1982 *Phys. Lett. B* **117** 247
[3] Gribov V N 1968 *Sov. Phys. JETP* **26** 414
[4] t'Hooft G 1974 *Nucl. Phys. B* **72** 461
[5] Veneziano G 1974 *Phys. Lett. B* **52** 220
[6] Amelin N S and Bravina L V 1990 *Sov. J. Nucl. Phys.* **51** 133
[7] Bravina L V et al. 1999 *Phys. Rev. C* **60** 044905
[8] Zabrodin E E et al. 2001 *Phys. Lett. B* **508** 184
[9] Burau G et al. 2005 *Phys. Rev. C* **71** 054905
[10] Abramovskii V, Gribov V and Kancheli O 1974 *Sov. J. Nucl. Phys.* **18** 308
[11] Kaidalov A B 1999 *Surveys in High Energy Phys.* **13** 265
[12] Baker M and Ter-Martirosyan K A 1976 *Phys. Rep.* **28C** 1
[13] Field R D and Feynman R P 1978 *Nucl. Phys. B* **136** 1
[14] Alner G J et al. (UA5 Collaboration) 1987 *Phys. Rep.* **154** 247
[15] Arnison G et al. (UA1 Collaboration) 1982 *Phys. Lett. B* **118** 167
[16] Albajar C et al. (UA1 Collaboration) 1990 *Nucl. Phys. B* **335** 261
[17] Abe F et al. (CDF Collaboration) 1988 *Phys. Rev. Lett.* **61** 1819; 1990 *Phys. Rev. D* **41** R2330
[18] Alexopoulos T et al. (E735 Collaboration) 1993 *Phys. Rev. D* **48** 984
[19] Aamodt K et al. (ALICE Collaboration) 2010 *Eur. Phys. J. C* **68** 89; 2010 *Eur. Phys. J. C* **68** 345; 2010 *Phys. Lett. B* **693** 53
[20] Khachatryan K et al. (CMS Collaboration) 2010 *JHEP* **1002** 041; 2010 *Phys. Rev. Lett.* **105** 022002
[21] Feynman R 1969 *Phys. Rev. Lett.* **23** 1415; Feynman R 1972 *Photon-hadron interactions* (Benjamin, 1972)
[22] Alner G J et al. (UA5 Collaboration) 1986 *Z. Phys. C* **33** 1
[23] Cleymans J, Strümpfer J and Turko L 2008 *Phys. Rev. C* **78** 017901
[24] Koba Z, Nielsen H B and Olesen P 1972 *Nucl. Phys. B* **40** 317

# Generation of Higher Flow Harmonics in Pb+Pb Collisions at LHC in HYDJET++ model

**B H Brusheim Johansson**

Department of Physics, P.b. 1048 Blindern, 0316 Oslo, Norge

E-mail: bhjohan@uio.no

**Abstract.** The observed particle distributions in Pb+Pb collisions at LHC are under investigation. Monte Carlo simulations of the azimuthal anisotropy i.e. flow, and transverse momentum spectra provide insights in hadron genesis and in medio scattering. Simulations are made using the HYDJET++ model, which is based on parameterization of soft processes and generation of hard physics. The second and third event planes are implemented, thus generating the second and third order flows which are believed to provide the largest part of the observed azimuthal distribution. The elliptic and triangular flows are simulated in good agreement with experimental data for ($p_T$ < 3.5) GeV/c and centralities 0 − 50%, along with the associated transverse momentum spectra, which also displays a high degree of precision. Also, flow projections are investigated, which further differentiates the observables. These simulations can be further elaborated upon, thus investigating multiparticle and multiplane (nonlinear) correlations.

## 1. Introduction

The particle distributions in relativistic heavy ion collisions are under investigation. Particle spectra provide information regarding fundamental processes such as hadronization, phase transitions; and *in medio* scattering [1–4]. In addition to the particle spectra; the azimuth is a direct observable which carries coherence information related to *in praegressus* processes. In present paper, the main topic is the yield anisotropy. The azimuthal part of the observed momentum distribution is commonly presented as the Fourier expansion [5, 6]

$$E\frac{d^3N}{d^3p} = \frac{1}{\pi}\frac{d^2N}{dp_T^2dy}\left[1 + 2\sum_{n=1}^{\infty} v_n \cos n\left(\phi - \Psi_n\right)\right]. \quad (1)$$

Here, $\phi$ and $\Psi_n$ are the particle azimuth and the n$^{\text{th}}$ event plane angle. The total momenta is p and the transverse momenta and rapidity are denoted by $p_T$ and y. The energy is denoted by E. From (1); the particle azimuth is characterized by the Fourier coe cients $v_n$, i.e. *flow*. The flow displays characteristics depending on order n and is to a large part determined by the initial conditions, which are characterized by *eccentricity* ($\epsilon$), with initial state event plane $\phi_n$ and transverse radius r,

$$\epsilon_n = -e^{-i\phi_n}\frac{\langle r^n e^{i\phi}\rangle}{\langle r^n\rangle}. \quad (2)$$

The eccentricity thus corresponds by the order n to the flow terms in (1).

Journal of Physics: Conference Series **668** (2016) 012046 doi:10.1088/1742-6596/668/1/012046

## 2. Method

The investigation of azimuthal anisotropy begins with the simulation of the first two modes: $v_2$ and $v_3$; which provides the foundation for the azimuthal part [7] of (1). In present model, i.e. HYDJET++ [8–10], the expanding fireball is modulated by *in medio* collisions [11–13] and radiation [14, 15]. Also, shadowing effects [16] are accounted for. In addition; hydro dynamical processes are parameterized on freeze-out hyper surfaces of the expanding fireball [17, 18]; and forms in superposition with hard particles the simulated distributions. In order to simulate the azimuthal anisotropy; the fireball transverse radius R

$$R(b,\phi) = R_f(b)\frac{\sqrt{1-\epsilon^2(b)}}{\sqrt{1+\epsilon(b)\cos 2\phi}}, \tag{3}$$

is modulated in terms of azimuth $\phi$ and the impact-parameter dependent model-eccentricity $\epsilon(b)$; with $R_f$ as the fireball RMS radius. Also, the linear transverse rapidity profile, and thus the particle velocity are modulated in terms of a second model parameter $\delta(b)$. The modulation of the transverse velocity $u^T$ is

$$u^x = \sqrt{1+\delta(b)}\sinh\tilde{\rho_u}\cos\phi, \quad u^y = \sqrt{1-\delta(b)}\sinh\tilde{\rho_u}\sin\phi, \tag{4}$$

where the transverse rapidity is further modulated by event plane angles, e.g. $\tilde{\rho_u} = \tilde{\rho_u}(\Psi_3)$.

## 3. Results

The investigation of evolutionary processes is done by relating different projections of the azimuthal anisotropy and particle spectra, thus investigating in-collisional processes. The simulations are array run on the UiO cluster ABEL [1] with an ATLAS setting and compared to available experimental projections. First; the transverse momentum spectra are simulated for unidentified particles and displayed with experimental data in addition to a hydro dynamical projection (Fig.1). The simulated particle spectra are seen to agree with experimental data [19]

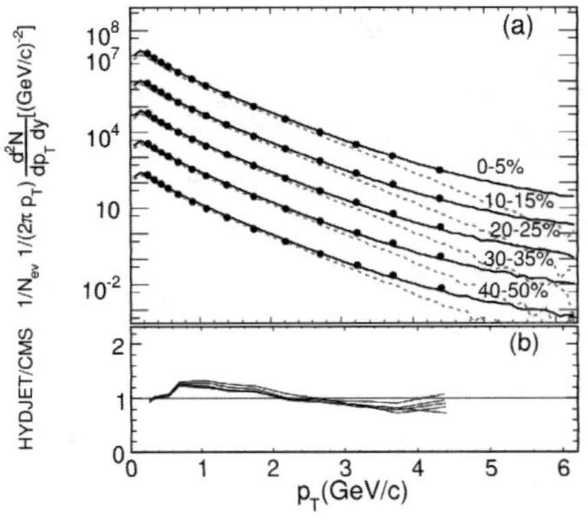


**Figure 1.** Transverse momentum spectra for unidentified particles (——) and hydro dynamical particles (- - - -). Particle spectra are simulated for:$\sigma/\sigma_{geo} = 0 - 5\%$; $\sigma/\sigma_{geo} = 10 - 15\%$; $\sigma/\sigma_{geo} = 20 - 25\%$; $\sigma/\sigma_{geo} = 30 - 35\%$; and $\sigma/\sigma_{geo} = 40 - 50\%$. Experimental data [19] are denoted by ● . Comparison between experimental data and simulations in the lower pane.

[1] The Abel cluster is owned by the University of Oslo and the Norwegian metacenter for High Performance Computing (NOTUR); and operated by the Department for Research Computing at USIT, the Oslo University IT-Department.

Journal of Physics: Conference Series **668** (2016) 012046 doi:10.1088/1742-6596/668/1/012046

for all simulated centralities. The hydro dynamical projection displays thermal characteristics for the soft transverse momentum regime. In contrast; the power law characterized fragmentation regime is observed for the transverse momentum regime 2 GeV/c $\leq p_T$.

Further, the ratios of the sixth order triangular- and elliptic factorizations are investigated. Agreement with experimental data is seen for the regime $p_T \leq 3.5$ GeV/c in Fig.2 (a). Deviations are seen for central collisions; although qualitative behaviour is seen to be reproduced. The magnitude of the ratio is here seen as strongly centrality dependent due to the rising ellipticity (coherency), which provides for a suppressed ratio. The flow is more prominent in the triangular plane for the soft transverse momentum regime; and for higher transverse momentum ($\sim$ 2 GeV/c), the elliptic plane provides coherency, thus depressing the ratio. This depression is seen to depend on centrality and thus the level of fragmentation. The higher transverse momentum regime (2.5 GeV/c$\leq p_T$ GeV/c) displays a rising trend; thus moving toward a triangular-like flow. The hydro dynamical projection of the $v_3^2/v_2^3$ ratio of Fig.2 (b) displays a ratio which is

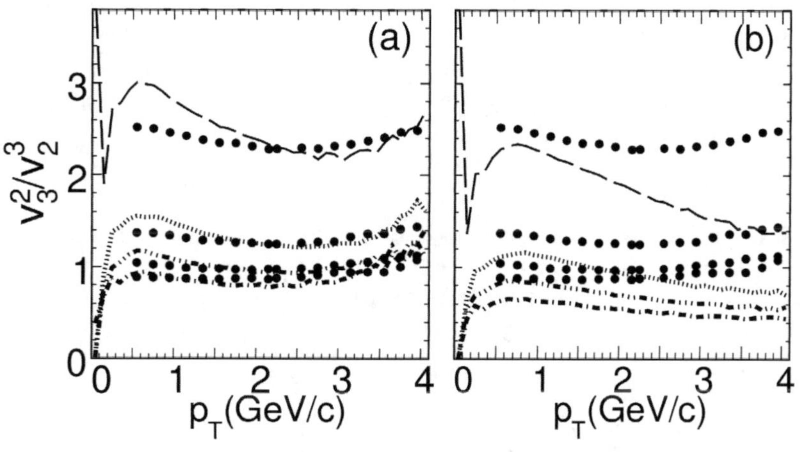


**Figure 2.** Ratio $v_3^2/v_2^3$ for unidentified- (a); and *hydro dynamical* particles (b). Experimental data for inclusive particles are included for comparison [20] and are denoted by ●. Simulations are made for: $\sigma/\sigma_{geo}$ = 10 – 20% (— — —); $\sigma/\sigma_{geo}$ = 20 – 30% (······); $\sigma/\sigma_{geo}$ = 30 – 40% (— · · —); and $\sigma/\sigma_{geo}$ = 40 – 50% (— · —).

differentiated in transverse momentum; thus not displaying scaling. From the hydro dynamical projection; hard processes are now seen to contribute to the approximate scaling; thus, the elliptic plane is seen to provide coherency for the hydro dynamical evolution. Also from the plots; the ratio is seen to be less dependent on hydrodynamics for more peripheral hits, thus deviating from scaling to a lesser degree.

## 4. Discussion and Conclusion

The finer coherence dependent structure observed in flow, makes flow a main observable of heavy ion collisions in combination with the particle spectra. The azimuthal anisotropy provides information regarding initial conditions, *in medio* processes; and hadronization processes. The agreement with experimental spectra for the soft regime, using a parameterized Cooper-Frye freeze-out [17, 18], is high (Fig.1). The postulated extensive hydrodynamics provides an exponential hydrodynamic projection; which is calculated on the fireball freeze-out hyper surfaces. The agreement between simulations and experimental data for the high transverse momentum regime works in favour of the present model which implements jet production by hard vertex production in the PYTHIA model [21]; which is subsequently quenched by the PYQUEN [8, 9] part of HYDJET++. The flow ratio under investigation, in Fig.2, displays momentum space characteristics, thus carrying information regarding particle production mode. From the plots, the jets are seen to provide for the fluctuating triangular plane; thus, the ratio displays sensitivity to fluctuations and jets. Also, the ratios dependence on production mode is seen to be connected to initial geometry, thus suppressed by the coherent elliptic plane. Relatively, the ratio is seen to be more dependent on fragmentation for peripheral hits. In conclusion: through the simulated azimuthal anisotropy and particle spectra; *in medio* scattering is seen to provide greater medium effects in central collisions in contrast to the more fragmentation characterized peripheral collisions. Thus, media properties and processes are manifesting themselves in the simulations which are related to the laboratory observables.

## Acknowledgments

The author wish to thank his supervisors L. Bravina, E. Zabrodin; and the associated research group. In addition, the author wish to thank the SQN and DENSE MATTER 2015 organizing committees; and also the city of Dubna for an exceptional event and a lovely stay.

## References

[1] Arsene I *et al.* 2005 *Nucl. Phys.* A **757** 1
[2] Back B B *et al.* 2005 *Nucl. Phys.* A **757** 28
[3] Adams J *et al.* 2005 *Nucl. Phys.* A **757** 102
[4] Adcox K *et al.* 2005 *Nucl. Phys.* A **757** 184
[5] Voloshin S and Zhang Y 1996 *Z. Phys.* C **70** 665
[6] Poskanzer A M and Voloshin S A 1998 *Phys. Rev.* C **58** 1671
[7] Bravina L, Johansson B H B, Zabrodin E E, Eyyubova G, Korotkikh V L, Lokhtin I P, Malinina L V, Petrushanko S V and Snigirev A M 2014 *Physical Review* C **89** 024909
[8] Lokhtin I P and Snigirev A M 2006 *Eur. Phys. J.* C **45** 211
[9] Lokhtin I P and Snigirev A M 2010 *www.lokhtin.web.cern.ch/lokhtin/pyquen/*
[10] Lokhtin I P, Malinina L V, Petrushanko S V, Snigirev A M, Arsene I and Tywoniuk K 2009 *Comp. Phys. Comm.* **180** 779
[11] Bjorken J D 1982 *Fermilab publication FERMILAB-PUB-82/59-THY*
[12] Braaten E and Thoma M 1991 *Phys. Rev.* D **44** 1298
[13] Lokhtin I P and Snigirev A M 2000 *Eur. Phys. J.* C **16** 527
[14] Baier R, Dokshitzer Y L, Mueller A H and Shiff D 1999 *Phys. Rev.* C **60** 064902
[15] Baier R, Dokshitzer Y L, Mueller A H and Shiff D 2001 *Phys. Rev.* C **64** 057902
[16] Tywoniuk K, Arsene I, Bravina L, Kaidalov A B and Zabrodin E 2007 *Phys. Lett.* B **657** 170
[17] Amelin N S, Lednicky R, Lokhtin I P, Malinina L V, Snigirev A M, Karpenko I A, Sinyukov Y M, Arsene I and Bravina L 2006 *Phys. Rev.* C **74** 064901
[18] Amelin N S, Lednicky R, Pocheptsov T A, Lokhtin I P, Malinina L V, Snigirev A M, Karpenko I A and Sinyukov Y M 2008 *Phys. Rev.* C **77** 014903
[19] Chatrchyan S *et al.* 2013 *Phys. Rev.* C **87** 014902
[20] Aad G *et al.* 2012 *Phys. Rev.* C **86** 014907
[21] Sjostrand T, Mrenna S and Skands P 2006 *JHEP* **05** 26

Journal of Physics: Conference Series **668** (2016) 012047 doi:10.1088/1742-6596/668/1/012047

# Chiral symmetry aspects in the open charm sector

**T Buchheim[1,2], T Hilger[3] and B Kämpfer[1,2]**

[1] Helmholtz-Zentrum Dresden-Rossendorf, PF 510119, D-01314 Dresden, Germany

[2] Technische Universität Dresden, Institut für Theoretische Physik, D-01062 Dresden, Germany

[3] University of Graz, Institute of Physics, NAWI Graz, A-8010 Graz, Austria

E-mail: t.buchheim@hzdr.de, thomas.hilger@uni-graz.at, kaempfer@hzdr.de

**Abstract.** QCD sum rules serve as tools to investigate changing hadronic properties in a hot and/or dense nuclear medium. The role of chiral symmetry breaking and restoration effects in a medium can be addressed also in the heavy-light meson sector. Thus, we consider Weinberg sum rules which refer to chiral partner mesons composed of a light and a heavy quark.

## 1. Introduction

Within the standard model of particle physics, the current quark masses are generated by the Higgs mechanism. Due to strong interaction process, the hadrons acquire masses by far exceeding these current masses in the $u$, $d$ sector. Still, the masses of hadrons, in particular mesons, composed of a $u$ (or $d$) and a heavy quark (e. g. $c$) exceed the quark masses too. While ab initio QCD evaluations allow for an access to the hadron spectrum [1] somewhat indirect methods also relate hadron properties, encoded in spectral functions for instance, with QCD quantities most notable the QCD condensates characterizing the vacuum state. In a strongly interacting environment, e. g. within nuclear matter, the latter ones are supposed to be modified [2]. As a consequence, the hadron spectral functions can experience induced modifications, too. The celebrated QCD sum rules [3] represent a formalized framework wherein this connection of hadron properties and QCD condensates can be quantified.

Among the infinite tower of condensates as expectation values of QCD operators, the chiral condensate $\langle\bar{q}q\rangle$ is a particularly important quantity, signaling dynamical chiral symmetry breaking (DCSB). Here, we highlight some chiral symmetry aspects in the open charm meson sector. It is the beam energy range of forthcoming experiments at FAIR, NICA and J-Park where the onset of charm production is expected. Thus, charm degrees of freedom continue the previous focus on strangeness as probes of compressed nuclear matter and medium modifications of hadrons and as a tool for controlling our understanding of hadrons in themselves.

## 2. QCD sum rules to study in-medium chiral effects

A starting point of sum rule evaluations is the causal current-current correlator which reads in a medium $\Pi(q) = \int d^4x \, e^{iqx} \langle \, \mathrm{T}\left[j(x)j^\dagger(0)\right]\rangle_{T,n}$ [4], where the current $j$ refers to a construction representing the quantum numbers and essential structure features of the hadron species under consideration. The symbol $\langle\cdots\rangle_{T,n}$ denotes Gibbs averaging at finite temperature ($T$) and/or baryon density ($n$); at zero temperature and density, it reduces to the vacuum expectation value. From analyticity of the correlator and from Cauchy's formula one can derive an in-medium

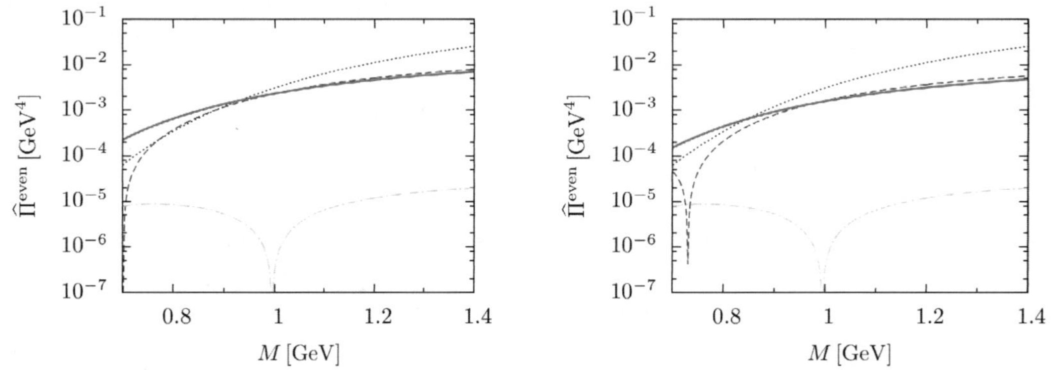


**Figure 1.** Modulus of the contributions to the Borel transformed $D$ meson sum rule in vacuum (left panel) and at nuclear saturation density (right panel) [15]. The chiral condensate (red solid) dominates at small Borel mass $M$ the non-perturbative contributions (blue dashed); the perturbative contribution (black dotted) and the four-quark condensate contribution (green dot-dashed) represent the largest and smallest contributions to the OPE, respectively.

dispersion relation by relating the phenomenological spectral density $\text{Im}\Pi(s)/\pi = \rho_{T,n}(s)$ at low energies with the correlator evaluated by Wilson's operator product expansion (OPE) in the deep Euclidean region:

$$\int_{-\infty}^{+\infty} d\omega \frac{\rho_{T,n}(\omega,\vec{q})}{\omega - q_0} \tanh\left(\frac{\omega}{2T}\right) = \sum_O C_O(q)\langle O\rangle_{T,n} \tag{1}$$

with the asymptotic expansion in QCD condensates $\langle O\rangle_{T,n}$ related to the non-perturbative, long-range effects parameterizing the complex QCD ground state. The Wilson coefficients $C_O$ are complex valued functions computable in perturbation theory. Equation (1) connects hadronic spectral properties encoded in $\rho_{T,n}$ to the QCD ground state via condensates. Hence, changes of hadron properties (masses, mass splittings of (anti-)particles, widths) carry signals of the way the vacuum changes in a nuclear environment [5] quantified by condensates, such as the chiral condensate which would vanish in a chiral restoration scenario.

The chiral condensate can also be probed fairly well by mesons composed of a heavy ($Q$) and a light ($q$) quark. For instance, in the $\rho$ and $\omega$ meson sum rules the chiral condensate contribution is suppressed numerically due to the light-quark mass, i. e. the renormalization group invariant combination $m_q\langle\bar{q}q\rangle$, whereas gluon and four-quark condensates are important instead [6, 7]. Such four-quark condensates have also turned out to be significant for nucleons [8, 9]. As a consequence, numerically relevant chiral symmetry effects enter the sum rule via chirally odd four-quark condensates [10]. However, the chiral condensate has a sizable impact on spectral properties in the heavy-light meson sector, since the heavy quark acts as an amplification factor, i. e. the scale dependent combination $m_Q\langle\bar{q}q\rangle$ enters the sum rule [11–13]. The chiral condensate dominates in fact the contributions to the pseudo-scalar $D$ meson OPE in the region of interest (Borel window) in vacuum as well as at nuclear saturation density, see Fig. 1 [14, 15]. Thus, the $D$ meson sum rule is highly sensitive to chiral symmetry effects mediated by $\langle\bar{q}q\rangle$.

## 3. Chiral symmetry transformations in the open charm sector

Chiral symmetry transformations act as flavour rotation on the left (L) and right (R) handed quark spinors separately: $\phi_{\text{L,R}} \longrightarrow e^{-i\Theta^a_{\text{L,R}}t^a}\phi_{\text{L,R}}$. The quantities $\Theta^a_{\text{L,R}}$ are the rotation parameters and $t^a$ denote the generators of the symmetry groups $SU(N_{\text{f}})_{\text{L,R}}$, where $N_{\text{f}}$ is the number of involved quark flavours. The mass term of the QCD Lagrangian breaks the symmetry.

However, if only light quark flavours are under consideration the approximation by massless quarks is justified and the QCD Lagrangian is invariant under such transformations. Adding a heavy quark with mass $m_Q$ leaves the mass term of the QCD Lagrangian not invariant under chiral transformations, but one may restrict solely the flavour rotations on the light part of the flavour vector $\phi = (u, d, Q)^{\mathrm{T}} = \phi_{\mathrm{L}} + \phi_{\mathrm{R}}$ with $\phi_{\mathrm{L,R}} = \frac{1}{2}(1 \mp \gamma_5)\phi$ by choosing those Gell-Mann matrices as generators which merely reduce to Pauli matrices forming the subgroup $SU(2)_{\mathrm{L}} \times SU(2)_{\mathrm{R}}$ of the full chiral symmetry group $SU(3)_{\mathrm{L}} \times SU(3)_{\mathrm{R}}$. These restricted transformations leave invariant the mass term $\bar{\phi}M\phi$ with $M = \mathrm{diag}(0, 0, m_Q)$, but alter the chiral condensate, which can be transformed into its negative by a suitable choice of rotation parameters. Thus, condensates that change under chiral transformations are dubbed chirally odd and serve as elements of order parameters of chiral symmetry.

One can show that chiral partner meson currents, such as pseudo-scalar and scalar $D$ meson currents, can be transformed into each other by an appropriate finite chiral transformation suggesting degenerate spectral properties of chiral partner mesons. However, the experimentally observed heavy-light meson masses deviate strongly from one another. This discrepancy signals DCSB, which is driven by order parameters, such as the chiral condensate. The vacuum mass splitting of chiral partners (e.g. $P(0^-)$ states $\pi_{140}$, $D^0_{1865}$, $D^{\pm}_{s\,1986}$ vs. $S(0^+)$ states $f_{0\,400-550}$, $D^0_{2318}$, $D^{\pm}_{s\,2318}$ and $V(1^-)$ states $\rho_{775}$, $D^0_{2007}$, $D^{\pm}_{s\,2112}$ vs. $A(1^+)$ states $a_{1\,1230}$, $D^0_{2421}$, $D^{\pm}_{s\,2460}$) appears to be approximately 400 MeV for light unflavoured as well as heavy-light mesons. However, as the breaking pattern of light mesons is driven by four-quark condensates, whereas the one of open flavour mesons is driven by the chiral condensate, the modifications of spectra in both sectors might differ from each other. Therefore, also the impact of four-quark condensates on open flavour mesons is investigated as a first step towards a comprehensive understanding of DCSB phenomenology [16–19].

## 4. Weinberg sum rules

Weinberg sum rules (WSR), known from current algebra, link moments of the difference of chiral partner spectra, $\Delta\rho = \rho^{(P)}_{T,n} - \rho^{(S)}_{T,n}$ or $\rho^{(V)}_{T,n} - \rho^{(A)}_{T,n}$, to order parameters of chiral symmetry, such as the pion decay constant, the chiral condensate and chirally odd four-quark condensates. In this spirit, chirally odd condensates quantify the difference of chiral partner spectra. Such difference sum rules attracted much attention as they can be utilized to investigate the medium behavior of vector and axial-vector spectral functions in the light meson sector. The WSR of chiral partner mesons $\rho$ and $a_1$ have been studied extensively, e.g. to study chiral restoration driven by the order parameters of chiral symmetry [20], to further restrict spectral functions containing more than one low lying resonance [21] as well as to check their compatibility with chiral mixing [22] and hadronic models [23].

According to the arguments of Sec. 3, one can set up Weinberg type sum rules as differences of chiral partner sum rules of heavy-light mesons in the $P-S$ as well as $V-A$ channels [24–26]. Evaluation of in-medium sum rules distinguishes even and odd parts of the OPE with respect to the meson energy. We are able to extend the previous findings in the $P-S$ channel contributing solely to the even OPE $\int_{-\infty}^{\infty} d\omega\, \omega\, e^{-\omega^2/M^2} \tanh\left[\omega/(2T)\right] \Delta\rho(\omega)$ by (cf. [16, 24] for details)

$$e^{-m_Q^2/M^2}\left[ -2m_Q\langle\bar{q}q\rangle + \left(\frac{m_Q^3}{2M^4} - \frac{m_Q}{M^2}\right)\langle\bar{q}g\sigma Gq\rangle - \left(\frac{m_Q^3}{2M^4} - \frac{m_Q}{M^2}\right)\langle\Delta\rangle\right] \tag{2}$$

to mass dimension 6 by adding the chirally odd four-quark condensate contribution occurring in the odd OPE $\int_{-\infty}^{\infty} d\omega\, e^{-\omega^2/M^2} \tanh\left[\omega/(2T)\right] \Delta\rho(\omega)$ by

$$\frac{2e^{-m_Q^2/M^2} m_Q}{M^4} g^2 \langle \bar{q}t^a q \sum_f q_f^\dagger t^a q_f\rangle\,. \tag{3}$$

The expectation value $\langle \bar{q} g \sigma G q \rangle$ is the mixed quark-gluon condensate of mass dimension 5 and $\langle \Delta \rangle = \langle \bar{q} g \sigma G q \rangle - 8 \langle \bar{q} D_0^2 q \rangle$ [24] is a medium-specific condensate [18] of mass dimension 5.

Pseudo-scalar $D$ and $B$ meson sum rule analyses in the nuclear medium [14] show that the odd OPE drives the mass splitting of anti-$D(B)$ and $D(B)$ mesons. However, at zero nuclear density, particles and anti-particles are degenerate even at finite temperature involving a vanishing odd OPE (3). Although, this requires a zero chirally odd four-quark condensate, chiral symmetry is not restored since the chirally odd condensates entering the even OPE (2) do not vanish per se. Therefore, chiral symmetry restoration does not emerge by vanishing of a single chirally odd condensate, but requires vanishing of all chirally odd condensates.

## 5. Summary

In summary, the role of chiral symmetry breaking and restoration in the open charm meson sector is addressed. We consider Weinberg sum rules which refer to chiral partners. Specific models, e. g. [27], should fulfil them, also evidencing that many chirally odd condensates beyond the chiral condensate $\langle \bar{q} q \rangle$ appear parameterizing the non-degeneracy of chiral partners prior to restoration. Individual $D$ meson QCD sum rules for pseudo-scalar, scalar, vector and axial-vector channels depend on four-quark condensates, however, with overwhelming impact of the chiral condensate.

## Acknowledgments

This work is supported by BMBF 05P12CRGHE and FWF P25121-N27.

## References

[1] Wittig H 2008 *Theory and Experiments* (*Landolt-Börnstein* vol I/21A) ed Schopper H (Springer-Verlag)
[2] Thomas R, Zschocke S, Hilger T and Kämpfer B 2006 *Proceedings, 44th International Winter Meeting on Nuclear Physics (Bormio 2006)* (*Preprint* hep-ph/0602163)
[3] Shifman M A, Vainshtein A I and Zakharov V I 1979 *Nucl. Phys. B* **147** 385
[4] Hatsuda T, Koike Y and Lee S H 1993 *Nucl. Phys. B* **394** 221
[5] Weise W 1994 *Nucl. Phys. A* **574** 347c
[6] Rapp R, Wambach J and van Hees H 2010 *Relativistic Heavy Ion Physics* (*Landolt-Börnstein* vol I/23) ed Stock R (Springer-Verlag)
[7] Thomas R, Hilger T and Kämpfer B 2008 *Prog. Part. Nucl. Phys.* **61** 297
[8] Thomas R, Hilger T, Zschocke S and Kämpfer B 2007 *AIP Conf. Proc.* **892** 274
[9] Thomas R, Hilger T and Kämpfer B 2007 *Nucl. Phys. A* **795** 19
[10] Hilger T, Thomas R, Kämpfer B and Leupold S 2012 *Phys. Lett. B* **709** 200
[11] Rapp R *et al.* 2011 *Lect. Notes Phys.* **814** 335
[12] Zschocke S, Hilger T and Kämpfer B 2011 *Eur. Phys. J. A* **47** 151
[13] Hilger T and Kämpfer B 2010 *Nucl. Phys. Proc. Suppl.* **207-208** 277
[14] Hilger T, Thomas R and Kämpfer B 2009 *Phys. Rev. C* **79** 025202
[15] Buchheim T, Hilger T and Kämpfer B 2015 *Phys. Rev. C* **91** 015205
[16] Hilger T, Buchheim T, Kämpfer B and Leupold S 2012 *Prog. Part. Nucl. Phys.* **67** 188
[17] Buchheim T, Hilger T and Kämpfer B 2014 *J. Phys. Conf. Ser.* **503** 012006
[18] Buchheim T, Hilger T and Kämpfer B 2015 *Nucl. Part. Phys. Proc.* **258** 2013
[19] Buchheim T, Hilger T and Kämpfer B 2014 *EPJ Web Conf.* **81** 05007
[20] Hohler P M and Rapp R 2014 *Phys. Lett. B* **731** 103
[21] Hohler P M and Rapp R 2012 *Nucl. Phys. A* **892** 58
[22] Holt N P M, Hohler P M and Rapp R 2013 *Phys. Rev. D* **87** 076010
[23] Kwon Y, Sasaki C and Weise W 2010 *Phys. Rev. C* **81** 065203
[24] Hilger T, Kämpfer B and Leupold S 2011 *Phys. Rev. C* **84** 045202
[25] Hilger T and Kämpfer B 2009 *Proceedings, 47th International Winter Meeting on Nuclear Physics (Bormio 2009)* (*Preprint* 0904.3491)
[26] Hilger T, Schulze R and Kämpfer B 2010 *J. Phys.* **G37** 094054
[27] Sasaki C 2014 *Phys. Rev. D* **90** 114007

# Strange meson-baryon interaction in hot and dense medium: recent progress for a road to GSI/FAIR

**D Cabrera**[1,2], **L Tolos**[1,3], **J Aichelin**[4] **and E Bratkovskaya**[1,2]

[1] Frankfurt Institute for Avanced Studies, J. W. Goethe-Universität, Frankfurt am Main, Germany
[2] Institut für Theoretische Physik, J. W. Goethe-Universität, Frankfurt am Main, Germany
[3] Institut de Ciències de l'Espai (IEEC/CSIC), Universitat Autònoma de Barcelona, Cerdanyola del Vallès, Spain
[4] Subatech, UMR 6457, IN2P3/CNRS, École de Mines de Nantes, Université de Nantes, France

E-mail: cabrera@fias.uni-frankfurt.de

**Abstract.** We report recent results on the dynamics of strange hadrons in two-body reactions relevant for near-threshold production in heavy-ion collisions at GSI/FAIR and NICA-Dubna. In particular, $\bar{K}N$ scattering in hot and dense nuclear matter is studied within a chiral unitary framework in coupled channels, setting up the starting point for implementations in microscopic off-shell transport approaches. We focus on the calculation of transition rates with special attention to the excitation of hyperon resonances and isospin effects. Additionally, we explore "unconventional" strangeness generation by meson-meson and meson-baryon interactions in connection with recent HADES observations of deep sub-threshold $\phi$ and $\Xi$ production.

Hadrons with strangeness, embedded in strongly interacting matter at finite temperature and density, have been matter of intense investigation in the last decades in connection with the study of exotic atoms [1], the analysis of strangeness production in heavy-ion collisions (HICs) [2, 3, 4] and the microscopic dynamics ruling the composition of neutron stars [5]. In particular, understanding the dynamics of light strange mesons in a nuclear environment has posed a challenge to theoretical models. The $\Lambda(1405)$ resonance, sitting below the $\bar{K}N$ threshold in vacuum, is excited at a higher energy in nuclear matter due to Pauli-blocking on the nucleons. This leads to a rapid change of the $\bar{K}N$ scattering amplitude from repulsive in vacuum to attractive in the medium already at small densities, as required by kaonic-atom phenomenology [1], quickly deviating from the low-density approximation. The production of $K$ and $\bar{K}$ close to threshold energy has also been thoroughly investigated in HICs at SIS energies [2, 3, 4, 6, 7]. The analysis of experimental data in conjunction with microscopic transport approaches has allowed to draw solid conclusions regarding the production mechanisms and the freeze-out conditions of strange mesons. Still, a simultaneous description of all observables involving antikaon production is missing [4]. For example, recent experimental data on the $v_1$, $v_2$ flow coefficients of strange mesons show sensitivity to the details of the in-medium meson-baryon interaction, leaving room for improvements within hadronic models accounting for medium effects.

The properties of strange mesons in nuclear matter at finite temperature were studied within a self-consistent coupled-channel approach based on the SU(3) meson-baryon chiral Lagrangian (see [8, 9] and references therein). Such kind of approaches successfully describes the $\bar{K}$ meson interactions in nuclear matter, accounting for a shallow attractive $\bar{K}$ potential at nuclear matter

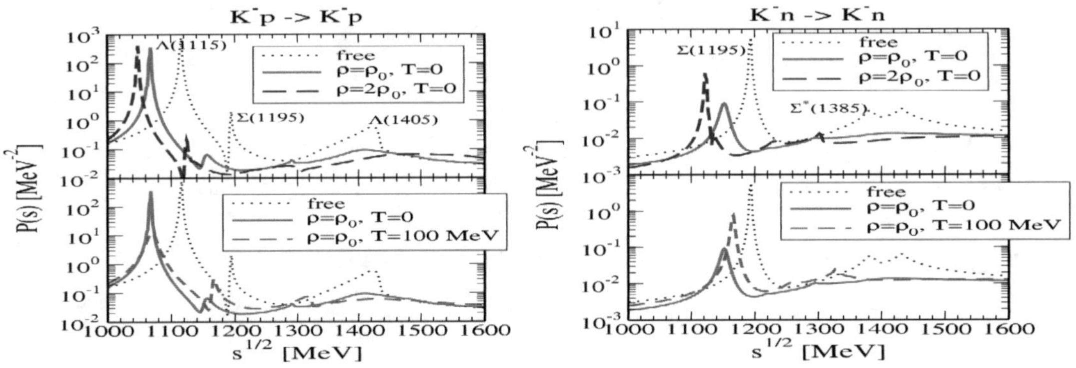


**Figure 1.** In-medium transition probability $\mathcal{P}$ at zero total three-momentum of the meson-baryon pair for the elastic $K^-p$ (left) and $K^-n$ (right) reactions.

density (40-60 MeV) and a broad spectral function. Our present results improve on this model by implementing the unitarization of $\bar{K}N$ scattering amplitudes in both $s$ and $p$ waves at finite density and temperature [10]. This formulation gives access to full off-shell in-medium scattering amplitudes in the SU(3) set of coupled channels. For $K^-p$ scattering these include: $K^-p$, $\bar{K}^0n$, $\pi^0\Lambda$, $\pi^0\Sigma^0$, $\eta\Lambda$, $\eta\Sigma^0$, $\pi^+\Sigma^-$, $\pi^-\Sigma^+$, $K^+\Xi^-$ and $K^0\Xi^0$; whereas for $K^-n$ scattering we have $K^-n$, $\pi^0\Sigma^-$, $\pi^-\Sigma^0$, $\pi^-\Lambda$, $\eta\Sigma^-$ and $K^0\Xi^-$ [compare to, e.g., the previous analysis [6] within the Hadron-String-Dynamics (HSD) transport model]. Another important asset of our approach is that, whereas it achieves a good description of $K^-p$ vacuum low-energy scattering observables with a minimal number of parameters, the model is highly predictive also in the $K^-n$ sector. The relevance of isospin effects in strangeness production near threshold was pointed out in [4].

The calculation of dynamical quantities in transport theory requires that the pertinent reaction rates or transition probabilities are folded with the spectral functions of the particles in the initial and final states [6]. The transition probability for a given reaction, $\mathcal{P}$, is defined fully off-shell as the angular integrated modulus-squared scattering amplitude (including all partial waves, $L = 0, 1$), averaged over the total angular momentum ($J = 1/2, 3/2$). The cross section for the process $i \to j$ then follows as $\sigma_{ij} = \frac{1}{8\pi} \frac{M_i M_j}{s} \frac{\tilde{q}_j}{\tilde{q}_i} \mathcal{P}_{ij}$, with $\tilde{q}_i$ the c.m. three-momentum and $M_i$ the baryon mass in channel $i$. The total cross section for a given reaction (e.g. $K^-p$) is then given by the sum over the partial cross sections for all the permitted coupled channels, $\sigma^i_{\rm tot} = \sum_j \sigma_{ij}$. We emphasize here that in our model we also have access to the angular dependence of the scattering amplitudes, which is non-trivial due to the contribution of the (deeply) sub-threshold hyperon excitations in the $p$-wave ($\Lambda$, $\Sigma$ and $\Sigma^*$). The angular information is a novel aspect of our study and allows for a full determination of the kinematics of the outgoing particles.

In Fig. 1 the transition probability for the $K^-p$ and $K^-n$ elastic reactions is depicted as a function of the meson-baryon center-of-mass energy at total vanishing three-momentum, evidencing the importance of isospin effects. The $K^-p$ state is an admixture of isospin $I = 0, 1$ and therefore the two isoscalar $\Lambda$ resonances and the isovector $\Sigma(1195)$ show up in the spectrum. The $\Sigma^*(1385)$ couples weakly to the $\bar{K}N$ system and cannot be resolved in the $K^-p$ elastic reaction. The $K^-n$ reaction is pure $I = 1$ and consequently only the isovector hyperon excitations are present in the right panel of Fig. 1. This explains the dramatic difference between the $K^-p$ and $K^-n$ cross sections in vacuum, the former being dominated by a quick fall due to a strong Flatté effect on the resonance shape of the $\Lambda(1405)$. In the medium, the structure of the $\Lambda(1405)$ is practically washed out at normal matter density. The $p$-wave ground states experience moderately attractive mass shifts as density is increased, whereas the $\Sigma^*(1385)$ is largely broadened due to the opening of in-medium decay channels such as $\Sigma^* \to \Lambda N N^{-1}$, $\Sigma N N^{-1}$ (cf. Fig. 5 in [10]). The effect of temperature is particularly appreciable as a broadening of the $p$-wave resonances as compared to the vacuum case. Our results are complementary to the ones obtained in our previous work [9], where the $\bar{K}$ spectral function and nuclear optical potential were provided. Altogether, they permit a systematic accounting of medium effects in the $S = -1$ sector within transport approaches such as PHSD and IQMD.

The implementation of suitable hadronic models in PHSD offers a unique opportunity to address some of the standing puzzles regarding the production of strange mesons and baryons, such as the $\phi$ and the $\Xi$. The observation of deep sub-threshold $\phi$ and $\Xi$ production by the HADES Collaboration in 1.23 AGeV Au+Au collisions [11] calls for a review of mechanisms that may have been missed in the analysis performed by different transport simulation groups, involving the production from resonance decay [12] as well as from two-body hadronic processes [13]. We briefly report here on recent progress on the theoretical assessment of pertinent meson-meson and meson-baryon reactions in the light of chiral approaches.

A suitable candidate is the $\eta\pi \to \phi\pi$ reaction as the advocated hidden strangeness content of the $\eta$ could help circumvent OZI suppression in meson-meson interactions as compared to $\phi$ production in $NN$ collisions. Moreover, the $\eta\pi$ system couples strongly to the isovector $a_0(980)$, which may enhance the cross section as it sits not far below the $\phi\pi$ threshold with a peculiar cusp-like energy dependence. Unfortunately, the mismatch between initial and final states regarding $C$-parity ($G$-parity) in the case of $\eta\pi^0 \to \phi\pi^0$ ($\eta\pi^\pm \to \phi\pi^\pm$) suppresses this reaction and the relevant chiral amplitudes vanish at leading order (LO)[1].

Next we have explored reactions involving the interaction of the $\eta$ with nucleons and deltas to produce a $\phi$ in the final state, e.g. $\eta N(\Delta) \to \phi N(\Delta)$. Here the challenge dwells on the theoretical side, since a model for this process has to couple meson-baryon states with pseudoscalar and vector mesons from the SU(3) octet and baryons from the $1/2^+$ octet and the $3/2^+$ decuplet. The low- and intermediate-energy interactions of pseudoscalar and vector mesons with baryons have been described successfully in coupled-channel unitarized approaches based on chiral symmetry (see, e.g., [14] and references therein). However, vector and pseudoscalar mesons are decoupled in these approaches whereas there is no reason, based on symmetry or dynamics, to expect such phenomenon. With our goal in mind we recourse to the approach in Ref. [15], which implements an extension of the Weinberg-Tomozawa interaction of standard SU(3) meson-baryon ChPT by means of SU(6) spin-flavor symmetry. The model comprehends several sectors of total spin, isospin and strangeness (with negative parity), where a number of baryonic resonances are dynamically generated in the energy range from the $N^*(1535)$ up to ~2.2 GeV from the unitarization of the LO scattering amplitudes. In particular we note that two resonances slightly above the $\phi N$ threshold emerge in this approach, with $J^P = 1/2^-, 3/2^-$. Such states, which also couple strongly to $K^*\Lambda$ and $K^*\Sigma$, have been found crucial to correctly interpret the data shape and energy dependence of the $\gamma p \to K^0\Sigma^+$ reaction by the CBELSA/TAPS Collaboration [16]. Claims for the need of a $N^*$ resonant state in that energy region have also been made from the analysis of $\Lambda(1520)$ photoproduction data by the LEPS Collaboration.

We focus on the relevant channels producing a $\phi$ meson in the final state with total strangeness $S = 0$ whereas other sectors (involving multi-strange hadrons, such as the $\Xi$) will be studied in the future. In Fig. 2 we depict the cross section, as a function of the center-of-mass energy, for the $\eta N \to \phi N$ reaction (total isospin $I = 1/2$). In spite of the resonant states present above 2 GeV the cross section remains below the mbarn scale. The reason for this is twofold: First, the LO chiral amplitude for this process is vanishing, the coupling proceeding via an intermediate $KY$ state. Second, whereas the states above 2 GeV have a large coupling to the final state, they are practically decoupled from $\eta N$, which lies far below in energy. Motivated by the

[1] The small branching ratio of $\phi \to \pi^+\pi^-\eta$ as compared to the dominant hadronic decay of the $\phi$ is a clear indication of this suppression.

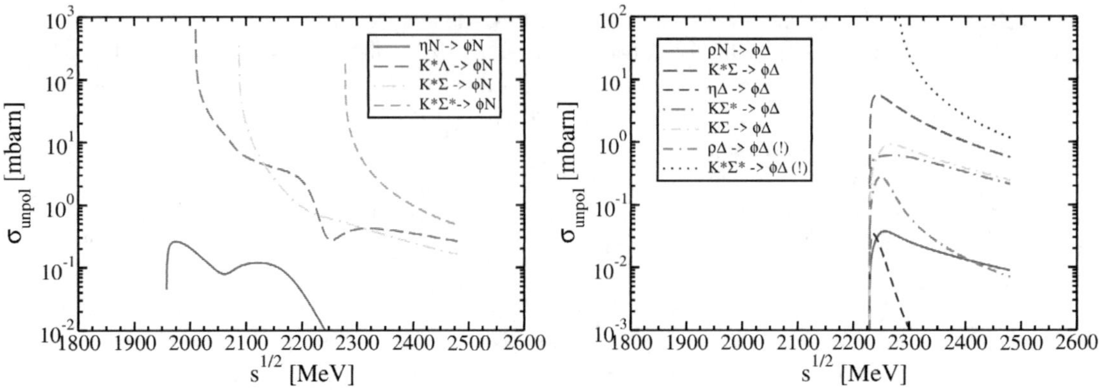


**Figure 2.** Unpolarized cross section for $\phi$ production from several isospin 1/2 (left) and 3/2 (right) meson-baryon reactions. The (!) sign indicates processes involving up to $J = 5/2$.

recent experimental findings above, we have also calculated the cross sections for $K^*Y \to \phi N$, with $Y = \Lambda, \Sigma, \Sigma^*$. Although the $K^*(892)$ is produced as a secondary particle, preliminary simulations in PHSD indicate that it could be sufficiently abundant in the hadronic phase as to influence $\phi$ production at least at SPS energies and beyond. Our results show sizable cross sections in the range of 10-100 mbarn close to threshold energies, exhibiting the typical rapid drop of strangeness-exchange reactions and signalling the presence of nearby resonant states. Other reactions without net strangeness content have also been evaluated resulting in smaller cross sections, typically below 5 mbarn [e.g. $\rho N(\Delta) \to \phi N$]. Production in the $I = 3/2$ sector (i.e. with $\phi\Delta$ as final state) is also found to be relevant, as is shown if the right panel of Fig. 2. Further conclusions on present and future measurements of $\phi$ and $\Xi$ production can be drawn from detailed studies within PHSD. Such developments are in progress.

## Acknowledgments
We acknowledge stimulating discussions with W. Cassing, E. Oset and H. Stroebele, and support from HIC for FAIR, from BMBF under Project No. 05P12RFFCQ, from Grants No. FPA2010-16963 (Ministerio de Ciencia e Innovación), No. FP7-PEOPLE-2011-CIG under Contract No. PCIG09-GA-2011-291679, No. FPA2013-43425-P, and the EC-RII HadronPhysics3 (Grant Agreement No. 283286). L.T. acknowledges support from the Ramón y Cajal Research Programme (Ministerio de Ciencia e Innovación).

## References
[1] Friedman E Gal A 2007 *Phys. Rept.* **452** 89
[2] Fuchs C 2006 *Prog. Part. Nucl. Phys.* **56** 1
[3] Forster A *et al.* 2007 *Phys. Rev.* C **75** 024906
[4] Hartnack C Oeschler H Leifels Y Bratkovskaya E L Aichelin J 2012 *Phys. Rept.* **510** 119
[5] Kaplan D B Nelson A E 1986 *Phys. Lett.* B **175** 57
[6] Cassing W Tolos L Bratkovskaya E L Ramos A 2003 *Nucl. Phys.* A **727** 59
[7] Tolos L Polls A Ramos A Schaffner-Bielich J 2003 *Phys. Rev.* C **68** 024903
[8] Tolos L Ramos A Oset E 2006 *Phys. Rev.* C **74** 015203
[9] Tolos L Cabrera D Ramos A 2008 *Phys. Rev.* C **78** 045205
[10] Cabrera D Tolos L Aichelin J Bratkovskaya E 2014 *Phys. Rev.* C **90** 055207
[11] Lorenz M [HADES Collaboration] 2014 *Nucl. Phys.* A **931** 785
[12] Steinheimer J Bleicher M 2015 *arXiv:1503.07305*
[13] Moreau P Li F Ko C M Cassing W Bratkovskaya E L 2015 *these proceedings*
[14] Oset E Ramos A Garzon E J Molina R Tolos L Xiao *et al.* 2012 *Int. J. Mod. Phys.* E **21** 1230011
[15] Gamermann D Garcia-Recio C Nieves J Salcedo L L 2011 *Phys. Rev.* D **84** 056017
[16] Ramos A Oset E 2013 *Phys. Lett.* B **727** 287

# Charmonium production in pp, p–Pb and Pb–Pb collisions at forward rapidity with ALICE at the LHC

**Arianna Batista Camejo for the ALICE Collaboration**
Université Blaise Pascal and Laboratoire de Physique Corpusculaire, Clermont-Ferrand, France
E-mail: abatista@cern.ch

**Abstract.** This contribution presents recent ALICE measurements on charmonium production at forward rapidity in pp collisions at $\sqrt{s} = 7$ TeV, p–Pb at $\sqrt{s_{\rm NN}} = 5.02$ TeV and Pb–Pb at $\sqrt{s_{\rm NN}} = 2.76$ TeV. The charmonium production cross sections in pp collisions are in good agreement with calculations within the NRQCD approach. The comparison between $J/\psi$ and $\psi$(2S) yields in p–Pb collisions indicates the presence of Cold Nuclear Matter effects on charmomium production in the final state. In Pb–Pb collisions the measured $J/\psi$ suppression is in agreement with models that introduce $J/\psi$ regeneration mechanisms through the recombination of $c\bar{c}$ pairs produced in the medium.

## 1. Introduction

Charmonia have been an intense subject of study since their discovery 40 years ago [1, 2]. Their production involves different energy scales allowing to test different regimes of quantum chromodynamics (QCD). Experimental and theoretical efforts have since then addressed the question of charmonium production mechanisms in nucleon–nucleon collisions [3]. Also interesting, charmonium states have been proposed as a probe of the Quark–Gluon Plasma (QGP) formation in high-energy heavy-ion collisions [4]. In the presence of the hot medium their yields are expected to be modified with respect to those in nucleon–nucleon collisions due to color screening effects on the $c\bar{c}$ binding potential. Charmonium suppression in nucleus–nucleus collisions was observed by a number of experiments at SPS, RHIC and LHC [5–8]. At LHC energies, however, a smaller suppression compared to that measured at RHIC has been observed at low transverse momentum, $p_{\rm T}$, suggesting the presence of regeneration mechanisms in the medium [9–11]. Those effects are usually quantified by measuring the nuclear modification factor $R_{\rm AA}$, defined as the ratio of charmonium production yields in nucleus–nucleus and nucleon–nucleon collisions scaled by the average number of binary collisions. Furthermore, Cold Nuclear Matter (CNM) effects like the modification of the nucleon Parton Distribution Functions in the nucleus (nPDF), multiple scattering of partons in the nucleus, nuclear absorption or dissociation by comoving particles can also affect charmonium production yields. Charmonium suppression in nucleon–nucleus collisions was reported as well by experiments at SPS, RHIC and LHC [12–15]. The ALICE Collaboration studied $J/\psi$ and $\psi$(2S) production in pp, p–Pb and Pb–Pb collisions. The data were collected during LHC Run 1 at mid- and forward rapidity. At forward rapidity the study is performed in the dimuon decay channel with muon tracks reconstructed in the forward muon spectrometer. Details about the ALICE detector and the analysis procedures can be found in any of the cited ALICE publications (e.g. [11, 14, 16]).

## 2. Results from pp collisions at $\sqrt{s} = 7$ TeV

The results presented in this section were obtained from the pp at $\sqrt{s} = 7$ TeV data collected during 2011 [16]. Figure 1 shows the $p_{\mathrm{T}}$ dependence of the differential production cross section for

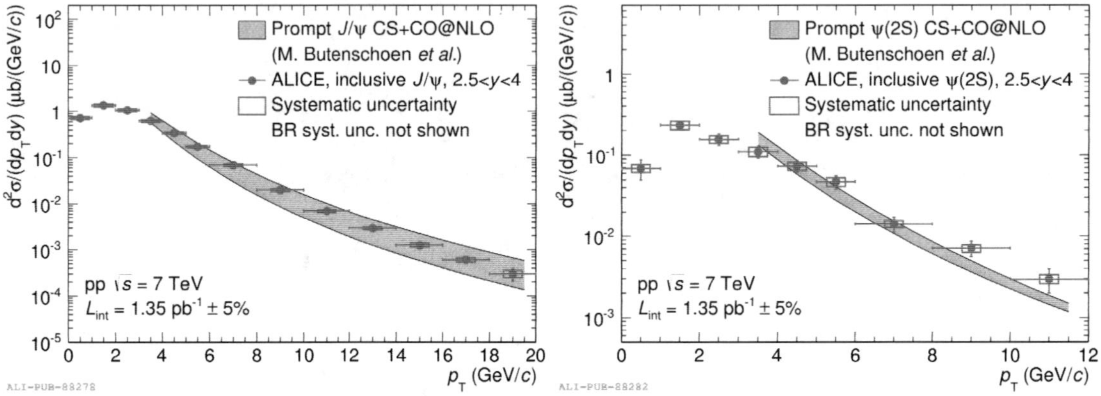


**Figure 1.** Inclusive $J/\psi$ (*left*) and $\psi(2S)$ (*right*) double differential production cross section as a function of $p_{\mathrm{T}}$. The data are compared with NLO NRQCD calculations [17].

inclusive $J/\psi$ (left panel) and $\psi$(2S) (right panel) in the rapidity range $2.5 < y < 4$. The data are compared with results from theoretical predictions for prompt $J/\psi$ and $\psi$(2S) production in the NRQCD framework, including NLO Color Singlet (CS) and Color Octet (CO) contributions [17]. A constant scaling factor was applied to the calculations to account for the prompt-to-inclusive comparison; good agreement between the data and the theoretical calculations is observed.

## 3. Results from p–Pb collisions at $\sqrt{s_{\mathrm{NN}}} = 5.02$ TeV

The measurements in this section were performed on the p–Pb data sample recorded in 2013, where two center-of-mass (cms) rapidity domains were explored: backward rapidity ($-4.46 < y_{\mathrm{cms}} < -2.96$) in the configuration in which the lead beam goes towards the muon spectrometer and forward rapidity ($2.03 < y_{\mathrm{cms}} < 3.53$) in the opposite case. Figure 2 (left

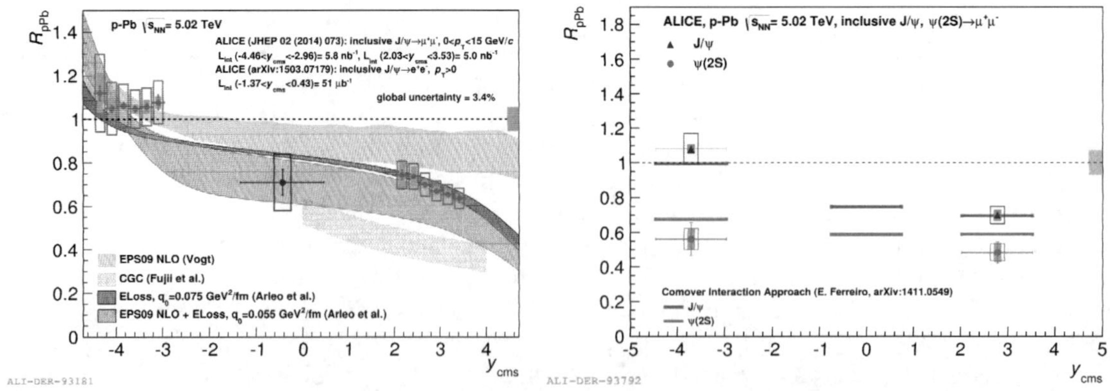


**Figure 2.** *Left:* inclusive $J/\psi$ $R_{\mathrm{pPb}}$ as a function of rapidity. *Right:* inclusive $J/\psi$ and $\psi(2S)$ $R_{\mathrm{pPb}}$ vs rapidity. The different theoretical calculations are described in the text.

panel) shows the nuclear modification factor $R_{\mathrm{pPb}}$ for inclusive J/$\psi$ in the transverse momentum interval $0 < p_{\rm T} < 15$ GeV/$c$ as a function of rapidity [14]. $J/\psi$ suppression is observed at mid- and forward rapidity, while at backward rapidity $R_{\rm pPb}$ is compatible with unity within uncertainties. The data are compared with several models: a NLO CEM calculation that uses the EPS09 shadowing parameterization [18]; a prediction based on the Color Glass Condensate framework [19] and a calculation including coherent parton energy loss processes with and without shadowing effects [20]. Models including coherent energy loss processes in the nuclear matter are able to describe the experimental results.Figure 2 (right panel) shows the results for $\psi$(2S) and $J/\psi$ in two rapidity bins for $p_{\rm T} > 0$ [15]. A stronger suppression is observed in the case of $\psi$(2S) compared to $J/\psi$ in particular in the backward rapidity region. Effects like shadowing and coherent energy loss are not expected to be sensitive to the final charmonium state, thus the theoretical predictions shown in the left panel of Figure 2 are essentially the same for both resonances. Nuclear absorption is also not expected to affect differently the $J/\psi$ and $\psi$(2S) yields at the LHC (see for instance discussions in [15]). However, other final state mechanisms may be at place: predictions within the comover interaction approach are shown in the right panel of Figure 2 [21]. The dissociation by comovers is expected to have a stronger impact on the $\psi$(2S) yields due to its larger size compared to the $J/\psi$ meson; additionally, the effect should be stronger in the Pb-going configuration (backward rapidity) due to the higher comover density. The inclusion of this effect results in a good qualitative agreement with the experimental measurements for both resonances.

## 4. Results from Pb–Pb collisions at $\sqrt{s_{\rm NN}} = 2.76$ TeV

Figure 3 shows the inclusive $J/\psi$ nuclear modification factor $R_{\rm PbPb}$ as a function of $p_{\rm T}$ in the centrality range 0–90%. The Pb–Pb data sample used for these results was collected at the end of 2011 [11]. In the left panel, we can see the comparison with the product $R_{\rm pPb}^{\rm forw} \times R_{\rm pPb}^{\rm backw}$, which can be interpreted as an extrapolation of CNM effects to $R_{\rm PbPb}$ [22]. The extrapolation is valid if one assumes that shadowing is the main CNM effect involved in $J/\psi$ production and given that the Bjorken-$x$ ranges probed by the $J/\psi$ production process in Pb–Pb collisions at $\sqrt{s_{\rm NN}} = 2.76$ TeV and at forward and backward rapidity in p–Pb collisions at $\sqrt{s_{\rm NN}} = 5.02$ TeV, are approximately the same. From the comparison, it can be inferred that CNM effects cannot explain the $J/\psi$ production yields in Pb–Pb collisions. At low-$p_{\rm T}$, one would expect a stronger $J/\psi$ suppression just as a result of nucleus related effects; $R_{\rm PbPb}$ being greater than $R_{\rm pPb}^{\rm forw} \times R_{\rm pPb}^{\rm backw}$ constitutes a hint of $J/\psi$ production enhancement in Pb–Pb collisions. The results in the high-$p_{\rm T}$ region suggests that the strong suppression observed in Pb–Pb collisions is associated to hot nuclear matter effects. In the right panel of Figure 3 data are compared with two transport models [23, 24], primordial and regenerated $J/\psi$ contributions are shown separately as well. The calculations of the two models differ essentially in the rate equation that controls $J/\psi$ dissociation and regeneration. The uncertainty bands are related to the inclusion of CNM effects and the $c\bar{c}$ cross section. Both predictions show a fair agreement with the data. The $J/\psi$ $R_{\rm AA}$ centrality dependence [11] and elliptic flow measurements [25], not shown here, confirms the regeneration model ansatz.

## 5. Conclusions

ALICE studied charmonia production at forward rapidity in all collision systems available during the LHC Run 1. A selection of results was presented here. $J/\psi$ and $\psi$(2S) production cross sections in pp collisions at 7 TeV are in agreement with calculations including NLO CS+CO contributions within the NRQCD framework. The $J/\psi$ nuclear modification factor in p–Pb collisions at 5.02 TeV indicates a higher impact of CNM effects at forward than at backward rapidities. The rapidity dependence can be explained by models including nPDF related effects and coherent energy loss processes. The $\psi$(2S) nuclear modification factor shows no significant dependence on rapidity, a stronger suppression compared to that of $J/\psi$ was observed. The

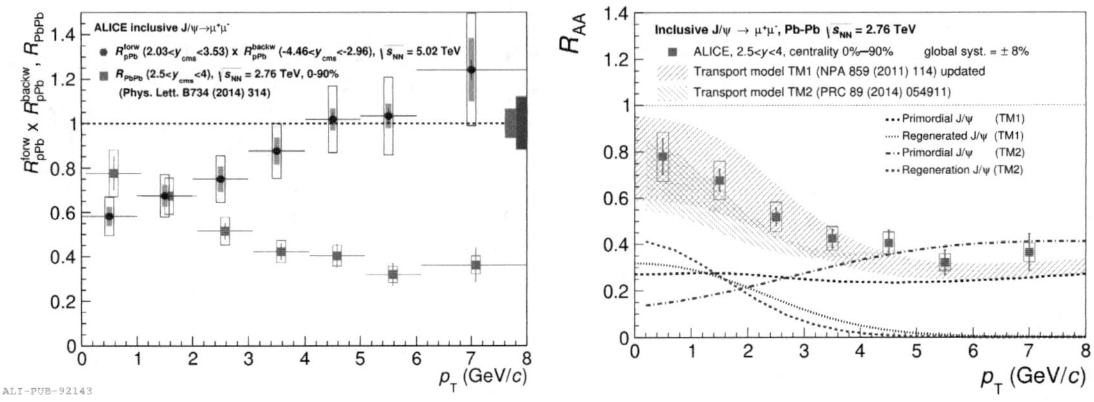


**Figure 3.** $J/\psi$ $R_{\rm PbPb}$ as a function of $p_{\rm T}$ in the centrality range 0–90%. In the left panel a result from a CNM extrapolation is also shown. In the right panel the data are compared with two transport models [23, 24].

results could be interpreted in terms of a strong dissociation effect at backward rapidity due to the interaction with comovers, while the results at forward rapidity are dominated by shadowing effects. The $p_{\rm T}$ dependence of the $J/\psi$ nuclear modification factor in Pb–Pb collisions at 2.76 TeV is in good agreement with transport models, thus favoring the interpretation of $c\bar{c}$ recombination mechanisms at the LHC energies.

## 6. Acknowledgement

The author would like to thank the French institutions 'Région Auvergne' and CNRS-IN2P3 for their financial support.

## References

[1] Auber J J *et al* 1974 *Phys. Rev. Lett.* **33** 1404
[2] Augustin J E *et al* 1974 *Phys. Rev. Lett.* **33** 1406
[3] Brambilla N *et al* 2011 *Eur. Phys. J.* C **71** 1534
[4] Matsui T and Satz H 1986 *Phys. Lett.* B **178** 416
[5] Arnaldi R *et al* (NA60 Collaboration) 2007 *Phys. Rev. Lett.* **99** 132302
[6] Adare A *et al* (PHENIX Collaboration) 2011 *Phys. Rev.* C **84** 054912
[7] Aad G *et al* (ATLAS Collaboration) 2011 *Phys. Lett.* B **697** 294
[8] Chatrchyan S *et al* (CMS Collaboration) 2012 *JHEP* **05** 063
[9] Braun-Munzinger P and Stachel J 2000 *Phys. Lett.* B **490** 196
[10] Thews R L *et al* 2001 *Phys. Rev.* C **63** 054905
[11] Adam J *et al* (ALICE Collaboration) 2015 arXiv:1506.08804
[12] Arnaldi R *et al* (NA60 Collaboration) 2012 *Phys. Lett.* B **706** 263
[13] Adare A *et al* (PHENIX Collaboration) 2013 *Phys. Rev.* C **87** 034904
[14] Abelev B *et al* (ALICE Collaboration) 2014 *JHEP* **02** 073
[15] Abelev B *et al* (ALICE Collaboration) 2014 *JHEP* **12** 073
[16] Abelev B *et al* (ALICE Collaboration) 2014 *Eur. Phys. J.* C **74** 2974
[17] Butenschoen M and Kniehl A 2011 *Phys. Rev.* D **84** 051501
[18] Albacete J L *et al* 2013 *Int. J. Mod. Phys.* E **22** 1330007
[19] Fujii H and Watanbe K 2013 *Nucl. Phys.* A **915** 1
[20] Arleo F *et al* 2013 *JHEP* **05** 155
[21] Ferreiro E G 2015 *Phys. Lett.* B **749** 98
[22] Adam J *et al* (ALICE Collaboration) 2015 *JHEP* **06** 055
[23] Zhao X and Rapp R 2011 *Nucl. Phys.* A **859** 114
[24] Zhou K *et al* 2014 *Phys. Rev.* C **89** 054911
[25] Abbas E *et al* (ALICE Collaboration) 2013 *Phys. Rev. Lett.* **111** 162301

# Large Transverse Momenta and Tsallis Thermodynamics

**J. Cleymans**
UCT-CERN Research Centre and Physics Department, University of Cape Town, South Africa
E-mail: jean.cleymans@uct.ac.za

**M. D. Azmi**
HEP Lab, Physics Department, Aligarh Muslim University, Aligarh - 202002, India
E-mail: danish.hep@gmail.com

**Abstract.** The charged particle transverse momentum ($p_T$) spectra measured by the ATLAS and CMS collaborations in proton - proton collisions at $\sqrt{s} = 0.9$ and 7 TeV have been studied using Tsallis thermodynamics. A thermodynamically consistent form of the Tsallis distribution is used for fitting the transverse momentum spectra at mid-rapidity. It is found that the fits based on the proposed distribution provide an excellent description over 14 orders of magnitude with $p_T$ values up to 200 GeV/c.

## 1. Introduction

It is by now well-known that the Tsallis distribution gives excellent fits to the transverse momentum distributions observed at the Relativistic Heavy Ion Collider (RHIC) [1, 2, 3] and at the Large Hadron Collider (LHC) [4, 5, 6, 7, 8] with only three parameters, $q$, $T$ and $dN/dy$ (or, alternatively, a volume $V$ [9, 10, 11]). The parameter $q$ is referred to as the Tsallis parameter and discussed elsewhere [12] in detail.

It was recently shown that these fits extend to values [14] of $p_T$ up to 200 GeV/c [15, 16]. This is unexpected because in this kinematic range hard scattering processes become important [17]. A description of the high $p_T$ results has been discussed in [18] where a model using a combination of Tsallis at low $p_T$ and QCD hard scattering at high $p_T$ was considered. The present analysis shows that the Tsallis distribution describes measurements up to the highest $p_T$ using the same parameters as those obtained at low $p_T$.

A power law based on the Tsallis distribution [13] is used to fit the $p_T$ spectra of charged particles measured by the ATLAS and CMS collaborations. The ATLAS collaboration has reported the transverse momentum in an inclusive phase space region taking into account at least two charged particles in the kinematic range $|\eta| < 2.5$ and $p_T > 100$ MeV [7]. The CMS collaboration has presented the differential transverse momentum distribution covering a $p_T$ range up to 200 GeV/c, the largest range ever measured in a colliding beam experiment [14].

The results can be compared to those obtained in [15, 16, 17, 19, 20] where very good fits to transverse momentum distributions were presented. We confirm the quality of the fits but obtain different values of the parameters albeit using a different version of the Tsallis model.

Journal of Physics: Conference Series **668** (2016) 012050 doi:10.1088/1742-6596/668/1/012050

## 2. Tsallis Distribution

The Tsallis distribution is defined as

$$f(E) \equiv \left[1 + (q-1)\frac{E-\mu}{T}\right]^{-\frac{1}{q-1}}, \tag{1}$$

which at large energies behaves as

$$\lim_{E\to\infty} f(E) = \left(\frac{E}{T}\right)^{-\frac{1}{q-1}}, \tag{2}$$

so that the scale is set by $T$ and the asymptotic behaviour is set by $q$. For high energy physics a consistent form of Tsallis thermodynamics for the particle number density $n$, energy density, $\epsilon$, and pressure $P$ are given by [22, 23]

$$n = g\int \frac{d^3p}{(2\pi)^3}\left[1 + (q-1)\frac{E-\mu}{T}\right]^{-\frac{q}{q-1}}, \tag{3}$$

$$\epsilon = g\int \frac{d^3p}{(2\pi)^3} E\left[1 + (q-1)\frac{E-\mu}{T}\right]^{-\frac{q}{q-1}}, \tag{4}$$

$$P = g\int \frac{d^3p}{(2\pi)^3}\frac{p^2}{3E}\left[1 + (q-1)\frac{E-\mu}{T}\right]^{-\frac{q}{q-1}}. \tag{5}$$

where $T$ and $\mu$ are the temperature and the chemical potential, $V$ is the volume and $g$ is the degeneracy factor. This introduces only one new parameter $q$ which for transverse momentum spectra is always close to 1. The consistency conditions

$$d\epsilon = Tds + \mu dn, \qquad dP = nd\mu + sdT, \tag{6}$$

are satisfied. This is shown explicitly for one of the relation $n = \partial P/\partial\mu$.

$$\begin{aligned}
\frac{\partial P}{\partial \mu} &= g\int \frac{d^3p}{(2\pi^3)}\frac{p^2}{3E}\frac{\partial}{\partial\mu}f^q \\
&= -g\int \frac{d^3p}{(2\pi^3)}\frac{p^2}{3E}\frac{d}{dE}f^q \\
&= -g\frac{4\pi}{(2\pi^3)}\int_0^\infty dp\,\frac{p^4}{3E}\frac{d}{dE}f^q \\
&= -g\frac{4\pi}{(2\pi^3)}\int_0^\infty dp\,\frac{p^3}{3}\frac{d}{dp}f^q \qquad \text{using} \quad EdE = pdp \\
&= g\frac{4\pi}{(2\pi^3)}\int_0^\infty dp\,p^2 f^q \\
&= n
\end{aligned}$$

## 3. Transverse Momentum Distributions

From the expression for the total number of particles:

$$N = gV\int \frac{d^3p}{(2\pi)^3}\left[1 + (q-1)\frac{E-\mu}{T}\right]^{-\frac{q}{q-1}}, \tag{7}$$

we obtain the corresponding momentum distribution

$$E\frac{dN}{d^3p} = gVE\frac{1}{(2\pi)^3}\left[1+(q-1)\frac{E-\mu}{T}\right]^{-\frac{q}{q-1}}. \tag{8}$$

In terms of the rapidity and transverse mass variables, $E = m_T \cosh y$, this becomes (at mid-rapidity $y = 0$ and for $\mu = 0$)

$$\left.\frac{d^2N}{dp_T\,dy}\right|_{y=0} = gV\frac{p_T m_T}{(2\pi)^2}\left[1+(q-1)\frac{m_T}{T}\right]^{-\frac{q}{q-1}}, \tag{9}$$

and, for charged particles it is given by a sum over the most abundant ones, $\pi^\pm, K^\pm, p, \bar{p}$. The results are shown in figures (1) and (2) and have been discussed in more detail in [24].

$$\left.\frac{d^2N(\text{charged})}{dp_T\,dy}\right|_{y=0} = \sum_{i=\pi,K,p,\ldots} g_iV\frac{p_T m_T}{(2\pi)^2}\left[1+(q-1)\frac{m_T}{T}\right]^{-\frac{q}{q-1}}, \tag{10}$$

The resulting parameters are listed in table 1.

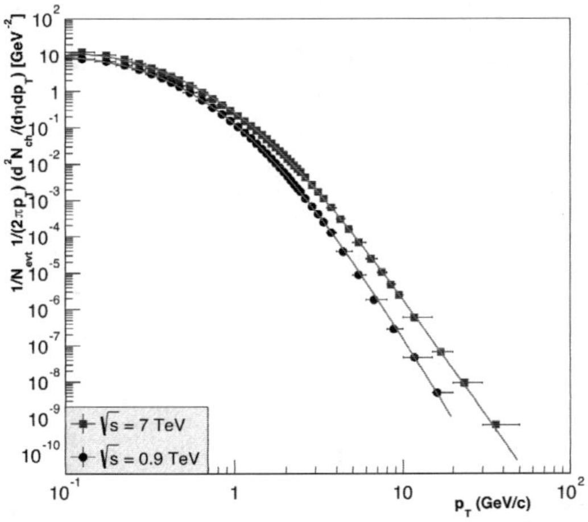


**Figure 1.** Charged particle multiplicities as a function of the transverse momentum measured by the ATLAS detector for events with $n_{ch} \geq 2$, $p_T$ >100 MeV and $|\eta| < 2.5$ at $\sqrt{s} = 0.9$ and 7 TeV in proton - proton collisions [7] fitted with Tsallis distribution.

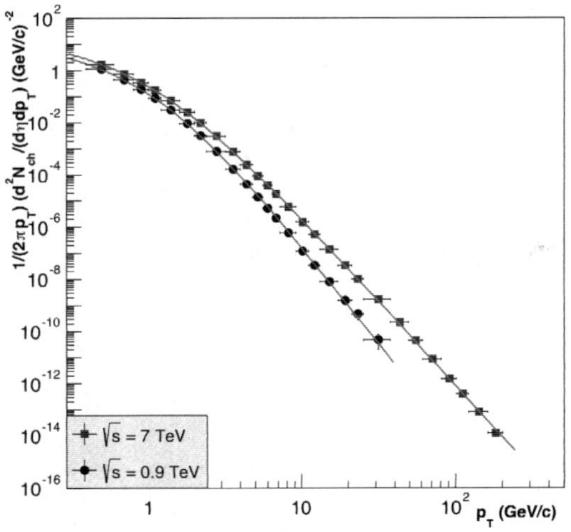


**Figure 2.** Charged particle differential transverse momentum yields measured within $|\eta| < 2.4$ by the CMS detector in proton - proton collisions at $\sqrt{s} = 0.9$ and 7 TeV [14] fitted with Tsallis distribution.

## 4. Conclusion

It is quite remarkable that the transverse momentum distributions measured up to 200 GeV/c in $p_T$ can be described consistently over 14 orders of magnitude by a straightforward Tsallis distribution. The advantages are that the thermodynamic conistency conditions are satisfied:

$$n = \frac{\partial P}{\partial \mu} \qquad \text{etc...,}$$

and the parameter $T$ truly deserves its name since $T = \partial E/\partial S$.

**Table 1.** Values of the $q$, $T$ and $R$ parameters and $\chi^2/NDF$ obtained from fits to the $p_T$ spectra measured by the ATLAS [7] and CMS [14] detectors.

| **Experiment** | $\sqrt{s}$ **(TeV)** | $q$ | $T$ **(MeV)** | $R$ **(fm)** | $\chi^2/NDF$ |
|---|---|---|---|---|---|
| ATLAS | 0.9 | 1.129 ± 0.005 | 74.21 ± 3.55 | 4.62 ± 0.29 | 0.657503/36 |
| ATLAS | 7 | 1.150 ± 0.002 | 75.00 ± 3.21 | 5.05 ± 0.07 | 4.35145/41 |
| CMS | 0.9 | 1.129 ± 0.003 | 76.00 ± 0.17 | 4.32 ± 0.29 | 0.648806/17 |
| CMS | 7 | 1.153 ± 0.002 | 73.00 ± 1.42 | 5.04 ± 0.27 | 0.521746/24 |

## References

[1] B. I. Abelev et al. (STAR collaboration), Phys. Rev. **C 75**, 064901 (2007).
[2] A. Adare et al. (PHENIX collaboration), Phys. Rev. **D 83**, 052004, (2011);
[3] A. Adare et al. (PHENIX collaboration), Phys. Rev. **C 83**, 064903 (2011).
[4] K. Aamodt, et al. (ALICE collaboration), Phys. Lett. **B 693**, 53 (2010).
[5] V. Khachatryan, et al. (CMS collaboration), JHEP **02**, 041 (2010);
[6] V. Khachatryan, et al. (CMS collaboration), Phys. Rev. Lett. **105**, 022002 (2010).
[7] G. Aad, et al. (ATLAS collaboration), New J. Phys. **13**, 053033 (2011).
[8] K. Aamodt, et al. (ALICE collaboration), Eur. Phys. J. **C 73**, 2662 (2013).
[9] M. Rybczynski, Z. Wlodarczyk, Eur. Phys. J. **C 74**, 2785 (2014).
[10] J. Cleymans, G. I. Lykasov, A. S. Parvan, A. S. Sorin, O. V. Teryaev, D. Worku, Phys. Lett. **B 723** 351, 2013.
[11] M. D. Azmi, J. Cleymans, J. Phys. **G 41**, 065001, 2014.
[12] G. Wilk, Z. Wlodarczyk, Phys. Rev. Lett. **84**, 2770, 2000.
[13] C. Tsallis, J. Stat. Phys. **52**, 479 (1988).
[14] S. Chatrchyan, et al. (CMS collaboration), JHEP **1108**, 086 (2011).
[15] C. Y. Wong and G. Wilk, Phys. Rev. **D 87**, 114007 (2013).
[16] L. J. L. Cirto, C. Tsallis, C. Y. Wong, G. Wilk, arXiv:1409.3278 [hep-ph].
[17] C. Y. Wong, G. Wilk, L. J. L. Cirto, C. Tsallis, Phys. Rev. D **91** (2015) 114027.
[18] K. Urmossy, G. G. Barnaföldi, S. Harangozó, T. S. Biró and Z. Xu, arXiv:1501.02352 [hep-ph].
[19] G. Wilk, Z. Wlodarczyk, Physics **A 413**, 53 (2014).
[20] G. Wilk, Z. Wlodarczyk, arXiv: 1403.3508[hep-ph].
[21] J. M. Conroy, H. G. Miller, A. R. Plastino, Phys. Lett. **A 374**, 4581, 2010.
[22] J. Cleymans and D. Worku, J. Phys. **G 39**, 025006 (2012).
[23] J. Cleymans and D. Worku, Eur. Phys. J. **A 48**, 160 (2012).
[24] M.D. Azmi, J. Cleymans arXiv: 1501.07127[hep-ph].

Journal of Physics: Conference Series **668** (2016) 012051 doi:10.1088/1742-6596/668/1/012051

# Toward a simultaneous description of $R_{AA}$ and $v_2$ for heavy quarks

**S. K. Das[1,2], F. Scardina [1,2], S. Plumari[1,2] and V. Greco[1,2]**

[1] Department of Physics and Astronomy, University of Catania, Via S. Sofia 64, I- 95125 Catania, Italy

[2] Laboratori Nazionali del Sud, INFN-LNS, Via S. Sofia 62, I-95123 Catania, Italy

E-mail: santosh@lns.infn.it

**Abstract.** The two key observables related to heavy quarks that have been measured in RHIC and LHC energies are the nuclear suppression factor $R_{AA}$ and the elliptic flow $v_2$. The Simultaneous description of these two observables is a top challenge for all the existing models. We highlight how a consistent combination of four ingredients i.e., the temperature dependence of the energy loss, the full solution of the Boltzmann collision integral for the momentum evolution of heavy quark, the hadronization by coalescence, and the hadronic rescattering, are responsible to address a large part of such a puzzle. We consider four different models to evaluate the temperature dependence of drag coefficients of the heavy quark in the QGP. All these four different models are set to reproduce the same $R_{AA}$ as of the experiments. We show that for the same $R_{AA}$, the $v_2$ could be quite different depending on the interaction dynamics as well as other ingredients.

## 1. Introduction

Theoretical calculations predict that hadronic matter at high temperatures and densities dissolves into a deconfined state of quarks and gluons - called Quark Gluon Plasma (QGP).The experimental efforts at Relativistic Heavy Ion Collider (RHIC) and Large Hadron Collider (LHC) energies is aimed at to create and characterized the properties of QGP. The heavy hadrons (hadrons which contain at least one heavy quark, mainly c and b ) constitute a unique probe of the QGP properties, because they are produced in the early stage of the collisions and they are the witness to the entire space-time evolution of the system.

The two key observables related to heavy quarks that have been measured in RHIC and LHC are the nuclear suppression factor $R_{AA}$ [1, 2, 3], which is the ratio between the the $p_T$ spectra of heavy flavored hadrons (D and B) produced in Au+Au collisions with respect to those produced in p+p collisions, and the elliptic flow $v_2$ [2, 4], which is a measure of the anisotropy in the angular distribution. Several theoretical efforts have been made to study the $R_{AA}$ and the $v_2$ measured in experiments within the Fokker Planck (FP) approach [5, 6, 7, 8, 9, 10] and the relativistic Boltzmann approach [11, 12, 13, 14, 17, 16]. However all the approaches shown some difficulties to describe both the $R_{AA}$ and $v_2$ simultaneously.

## 2. Results

To address the simultaneous description of $R_{AA}$ and $v_2$, we start with the time evolution of $R_{AA}$ and $v_2$ to know how they develop during the expansion of the QGP. To study the time evolution of $R_{AA}$ and $v_2$, we have solved the Fokker Planck (FP) equation stochastically in terms of the Langevin equation, for detail we refer to our earlier work [10]. The elastic collisions of heavy quarks with the bulk has been considered within the framework of pQCD. The divergence associated with the t-channel diagrams due to massless intermediate particle exchange has been regularized introducing the Debye screening mass $m_D = 4\pi\alpha_s T$ with the running coupling $\alpha_s$. For the bulk evolution we are using a transport bulk which can reproduce some of the gross features of the bulk [19, 18]. At RHIC energy, $Au + Au$ at $\sqrt{s} = 200$, we simulate the fireball with initial temperature in the center is $T_i = 340$ MeV and the initial time is $\tau_i = 0.6$ fm/c. for the detail of the initialization, we refer to our early work. Initially the charm quark are distributed according to the charm production in pp collisions [20].

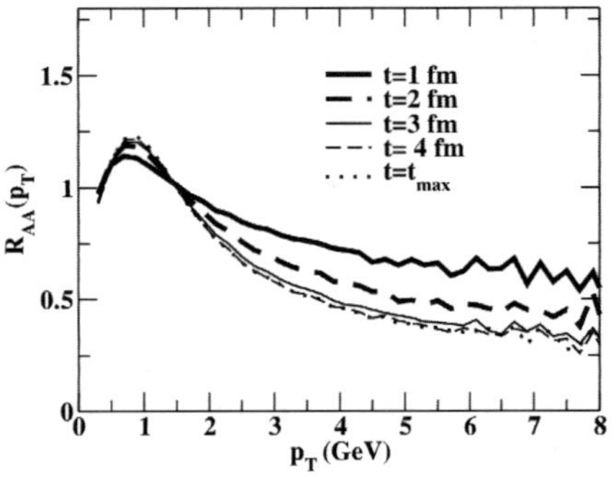


**Figure 1.** Time evolution of $R_{AA}$.

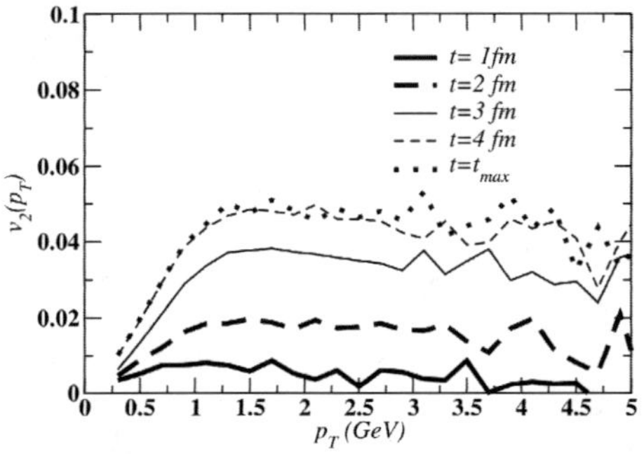


**Figure 2.** Time evolution of $v_2$.

As shown in Fig 1, the $R_{AA}$ is develop at the very early stage of the evolution and get saturated within 3-4 fm where as the $v_2$ (in Fig2) is develop at the latter stage of the evolution. Initially the bulk have zero $v_2$. First the bulk will develop its own $v_2$ and then it will transfer it to the heave flavor. Hence the $R_{AA}$ is sensitive to the magnitude of the drag coefficient at the early stage of the evolution i.e at $T_i$ where as the $v_2$ is very sensitive to the magnitude of the drag coefficient at the latter stage of the evolution i.e at $T_c$ . This highlights the T-dependence of the drag coefficient and may play a significance role for a simultaneous description of $R_{AA}$ and $v_2$ as they are sensitive to different range of T. To study the impact of T-dependence of drag coefficients on heavy quark observable, we have consider four different model to calculate the drag and diffusion coefficients of heavy quark in QGP.

Model-I (pQCD): The elastic collisions of heavy quarks with the bulk has been considered within the framework of pQCD.
Model-II (AdS/CFT): In this case we have considered the drag force from the gauge/string duality i.e. AdS/CFT [21] which reads $\Gamma_{conf} = C\frac{T^2_{QCD}}{M_{HQ}}$, where $C = 2.1 \pm 0.5$.(also see [22])
Model-III: In the third case, we have employed a quasi particle model (QPM) [23, 24, 25] with T-dependent quasi-particle masses, $m_q = 1/3g^2T^2$, $m_g = 3/4g^2T^2$, plus a T-dependent background field known as bag constant, tuned to the thermodynamics of the lattice QCD. Such a fit lead to the coupling, $g^2(T) = \frac{48\pi^2}{[(11N_c-2N_f)ln[\lambda(\frac{T}{T_c}-\frac{T_s}{T_c})]^2}$, where $\lambda$=2.6 and $T/T_s$=0.57.
Model-IV ($\alpha_{QPM}(T), m_q = m_g = 0$): In this case, we are considering a model where the light quarks and gluons are massless but the coupling is taken from the QPM model discussed in Model III.

This fourth case has been considered to have a drag which is decreasing with $T$ as obtained in the T-matrix approach [6, 26].The T-dependence of the drag coefficients have been shown in Fig 3 obtained within the four model discussed above at $p_T = 100$ MeV. These are the rescaled drag coefficients which can reproduced the same $R_{AA}$ as of the experiment at RHIC energy.

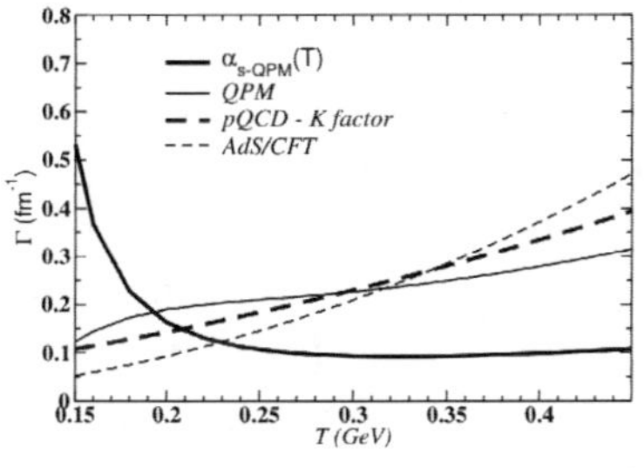


**Figure 3.** Drag coefficients as a function of temperature.

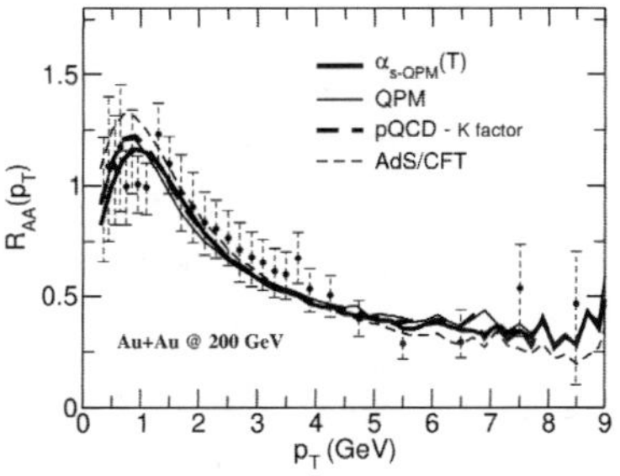


**Figure 4.** $R_{AA}$ as a function of $p_T$.

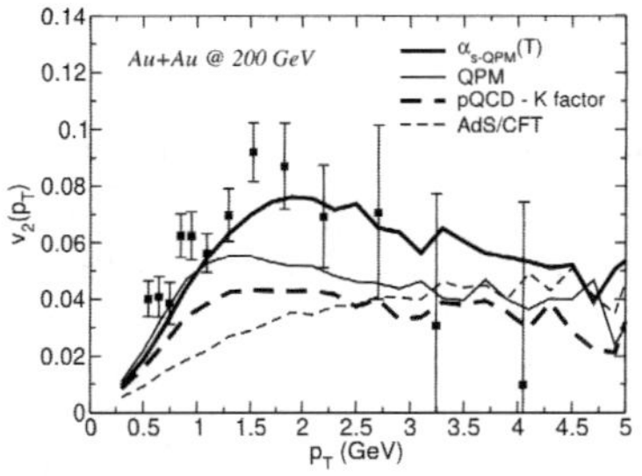


**Figure 5.** $v_2$ as a function of $p_T$.

In Fig 4 we have shown the $R_{AA}$ as a function of $p_T$ for the four different models obtained within the Langevin dynamics at RHIC energy. The $v_2$ for the same $R_{AA}$ has been plotted in Fig 5 for all models as a function of $p_T$. We found that for the same $R_{AA}$, the $v_2$ build-up can be quite different depending on the T-dependence of the interaction. Larger the interaction at $T_c$ larger is the $v_2$ [10]. Similar effect has also been found in the light quark sector [27, 28]. This suggests, the correct temperature dependence of drag coefficient has a significant role for a simultaneous reproduction of $R_{AA}$ and $v_2$.

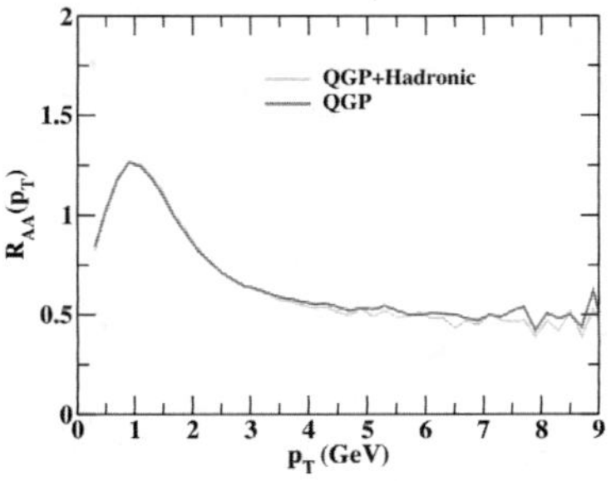


**Figure 6.** $R_{AA}$ as a function of $p_T$ at RHIC energy for QGP and QGP+hadronic phase.

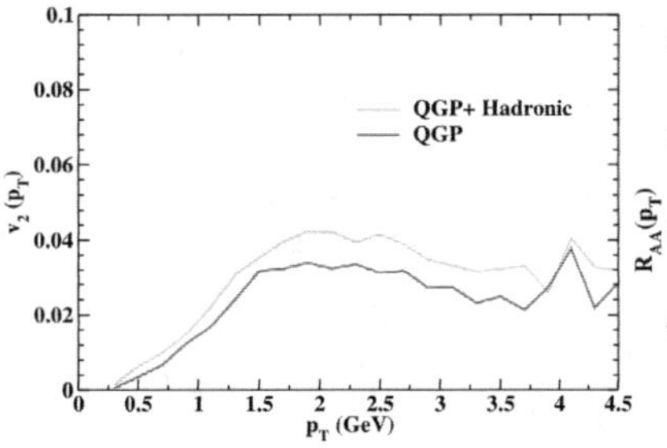


**Figure 7.** $v_2$ as a function of $p_T$ at RHIC energy for QGP and QGP+hadronic phase.

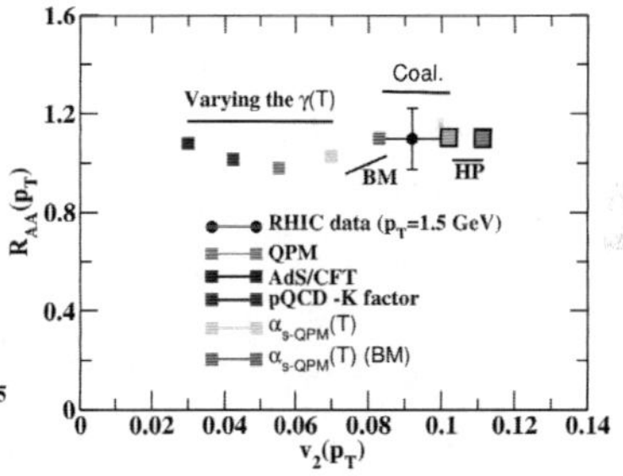


**Figure 8.** $R_{AA}$ vs $v_2$ at $p_T = 1.5$ GeV.

Recently it has been shown [13] that the full solution of the Boltzmann (BM) integral i.e. without the assumption of small collisional exchanged momenta, leads in general to a larger $v_2$ than that of the Fokker Plank case which is a approximation of the Boltzmann equation under the assumption of small exchanged momenta. Also heavy quark hadronization by coalescence further enhances the $v_2$ [6, 29].

To study the role of the hadronic phase (HP) on the heavy quark observables, we study the propagation of D mesons in the hadronic medium consist of pions, kaons and eta. The elastic interaction between the D meson with the hadronic matter has been treated within the framework of chiral perturbation theory [30]. In Fig 6 we have shown the impact of the hadronic medium on $R_{AA}$ which is almost unnoticeable. This is because the $R_{AA}$ develops at the early stage of the evolution and get saturated due to radial flow within 3-4 fm. The impact of the hadronic medium on $v_2$ has been shown in Fig 7. The hadronic medium further enhance the $v_2$ around 20%. In Fig 8 we have shown how the $v_2$ build-up depends on the T-dependence of the drag coefficients for the same $R_{AA}$. Then the $v_2$ gets a boost from the Boltzmann equation

(BM) in terms of evolution over the Fokker Planck equation. The $v_2$ is further enhanced due to hadronization by coalescence. Finally the $v_2$ gets the hadronic phase boost where the $R_{AA}$ remain the same.

## 3. Summary and outlook

In summary, we have shown how the $v_2$ build-up for the same $R_{AA}$ depends on the T-dependence of the drag coefficients which is the key for a simultaneous description of $R_{AA}$ and $v_2$. We have also highlighted how the $v_2$ gets a boost from the Boltzmann dynamics to study the momentum evolution of heavy quark and from hadronization by coalescence. Then we have shown the effect of the hadronic medium on $R_{AA}$ and $v_2$. The impact of radiative processes on heavy quark observables will also be investigated in an upcoming study along with all these four ingredients within a single framework . It will also be interesting to study the role of the pre-equilibrium phase [31] on $R_{AA}$ and $v_2$ relation.

## Acknowledgements

We acknowledge the support by the ERC StG under the QGPDyn Grant n. 259684.

## References

[1] Abelev B I *et al.* (STAR Collaboration) 2007 *Phys. Rev.* Lett. **98**, 192301.
[2] Adare A *et al.* (PHENIX Collaboration) 2007 *Phys. Rev.* Lett. **98**, 172301.
[3] Abelev B *et al.* (ALICE Collaboration) 2012 *J. High Ener. Phys.* **09**, 112.
[4] B. Abelev *et al.* (ALICE Collaboration) 2013 *Phys. Rev.* Lett. **111** 102301.
[5] Hees H van, Greco V and Rapp R 2006 *Phys. Rev.* C **73**, 034913
[6] Hees H van, Mannearelli M, Greco V and Rapp R 2008 *Phys. Rev. Lett.* **100** 192301
[7] Das S K, Alam J, Mohanty 2010 *Phys. Rev.* C **82** 014908 ; Majumdar S, Bhattacharyya T, Alam J, Das S K 2011 *Phys. Rev.* C **84** 044901
[8] Alberico W M *et al.* 2011 *Eur. Phys. J.* C **71** 1666
[9] Cao S., Qin G.-Y., Bass S A 2013 *Phys. Rev.* C **88**,4, 044907: 2015 *Phys.Rev.* C **92** 2, 024907
[10] Das S K, Scardina F, Plumari S and Greco V (2015) *Phys. Lett.* B **747**, 260
[11] Gossiaux P B, Aichelin J 2008 *Phys. Rev.* C **78** 014904
[12] Uphoff J, Fochler O, Xu Z and Greiner C 2011 *Phys. Rev.* C **84** 024908; Phys. Lett. B **717** (2012) 430
[13] Das S K, Scadina F, Plumari S, Greco V 2014 *Phys. Rev.* C **90** 044901.
[14] Song T *et al.* 2015 *Phys Rev.* C **92** 014910
[15] Scadina S, Das S K, Plumari P, Greco V 2015 J.Phys.Conf.Ser. *636* 1, 012017; 2014 J.Phys.Conf.Ser. *535* 012019
[16] Younus M, Coleman-Smith C E, Bass S A and Srivastava D K, arXiv:1309.1276 [nucl-th].
[17] Scadina S, Das S K, Plumari P, Greco V 2015 J.Phys.Conf.Ser. *636* 1, 012017; 2014 J.Phys.Conf.Ser. *535* 012019
[18] Ruggieri M, Scardina F, Plumari S, Greco V 2013 *Phys. Lett.* B **727** 177
[19] Scardina F, Colonna M, Plumari S and Greco V 2013 *Phys. Lett. B* **724** 296
[20] Cacciari M,Nason P, Vogt R 2005 *Phys. Rev. Lett.* **95**, 122001
[21] Maldacena J M 1998 Adv. Theor. Math. Phys. **2**, 231
[22] Das S K, Davody A 2014 *Phys. Rev.* C **89**, 054912.
[23] Plumari S, Alberico W M, Greco V, Ratti C 2011 *Phys. Rev.* D **84**, 094004
[24] Das S K, Chandra V, Alam J 2014 *J. Phys.* G **41** 015102.
[25] H. Berrehrah *et al.* (2014) *Phys.Rev.* C **89**, 054901
[26] Riek F, Rapp R 2010 *Phys. Rev.* C **82**, 035201
[27] Liao J, Shuryak E 2009 *Phys. Rev. Lett.* **102** 202302
[28] Scardina F, Toro M Di, Greco V 2010 *Phys. Rev.* C **82** 054901.
[29] Greco V, Ko C M, Rapp R 2004 *Phys. Lett.* B **595** 202
[30] Das S K, Ghosh S, Sarkar S, Alam J 2012 *Phys Rev* D **85** 074017; 2014 *Phys Rev* D **88**, 017501
[31] Das S K, Ruggieri M, Mazumder S, Greco V, Alam J 2015 *J. Phys.* G **42** 9, 095108

# Pion and kaon in the Beth-Uhlenbeck approach

**A. Dubinin**[1], **D. Blaschke**[1,2,3], **A. Radzhabov**[4]

[1] Instytut Fizyki Teoretycznej, Uniwersytet Wrocławski, 50-204 Wrocław, Poland
[2] Bogoliubov Laboratory for Theoretical Physics, JINR Dubna, 141980 Dubna, Russia
[3] National Research Nuclear University (MEPhI), 115409 Moscow, Russia
[4] Matrosov Institute for System Dynamics and Control Theory, Irkutsk 664033, Russia

E-mail: aleksandr.dubinin@ift.uni.wroc.pl

**Abstract.** In the present work the Mott effect for pions and kaons is described within a Beth-Uhlenbeck approach on the basis of the PNJL model. The contribution of these degrees of freedom to the thermodynamics is encoded in the temperature dependence of their phase shifts. A comparison with results from $N_f = 2 + 1$ lattice QCD thermodynamics is performed.

## 1. Introduction

One of the central problems in the investigation of the transition from hadronic to quark matter is a microphysical description of the dissociation of hadrons into their quark constituents. This Mott transition occurs under extreme conditions of high temperatures and densities as they are provided, e.g., in ultrarelativistic heavy-ion collisions or in the interiors of compact stars. Since an ab-initio description of QCD thermodynamics within simulations of lattice QCD (LQCD) is yet limited to finite temperatures and low chemical potentials only, the development of effective model descriptions is of importance. Here, we develop a relativistic Beth-Uhlenbeck approach to the description of mesonic bound and scattering states in a quark plasma [1, 2] further by including the strange sector. To this end we employ the PNJL model which is particularly suitable for addressing the appearance of pions and kaons as both, quasi Goldstone bosons of the broken chiral symmetry and pseudoscalar meson bound states. Within this framework the confinement of colored quark states is effectively taken into account by coupling the chiral quark dynamics to the Polyakov loop and its effective potential. The model is widely used to describe quark-gluon thermodynamics in the meanfield approximation [3, 4], but has also been developed to address mesonic correlations [5, 6, 7, 8, 9, 10, 11, 12]. The relativistic Beth-Uhlenbeck approach is the appropriate tool to develop a unified description of quark-gluon and hadron thermodynamics including the transition between both asymptotic regimes of QCD.

In the next section the basic formulae for the thermodynamic potential, the phase shifts and the pressure in the Beth-Uhlenbeck approach are presented, followed by a discussion of numerical results compared to LQCD data in Sect. 3 and conclusions in Sect. 4.

## 2. Relativistic Beth-Uhlenbeck approach for the PNJL model

The Lagrangian for the 3 flavor PNJL model is given by

$$\mathscr{L} = \bar{q}[\mathrm{i}\partial\!\!\!/ - m_0 + \gamma_0(\mu - \mathrm{i}A_4)]q + G_\mathrm{S}\sum_{a=0}^{8}\left[(\bar{q}\lambda^a q)^2 + (\bar{q}\mathrm{i}\gamma_5\lambda^a q)^2\right] - \mathscr{U}(\Phi,\bar{\Phi};T)\ .$$

Here, $q$ denote the quark fields with 3 colors and flavors, $\lambda^a$ are the Gell-Mann matrices in flavor space ($a = 0, 1, 2, 3 \ldots 8$), $G_S$ is dimensionful coupling constant. The Polyakov-loop potential $\mathscr{U}(\Phi, \bar{\Phi}; T)$ is chosen in the polynomial form [4], with the parameters taken from that reference. The gluon background field in the Polyakov gauge is a diagonal matrix in color space $A_4 = \mathrm{diag}(\phi_3 + \phi_8, -\phi_3 + \phi_8, -2\phi_8)$. The Polyakov loop field $\Phi$ is defined via the color trace over the gauge-invariant average of the Polyakov line $L(\vec{x})$ [4].

The thermodynamic potential in Gaussian approximation has the form [11]

$$\Omega_{\text{Gauß}} = \mathscr{U}(\Phi, \bar{\Phi}; T) + \frac{\sigma_{\text{MF}}^2}{4G_{\text{S}}} + \Omega_{\text{Q}} + \Omega_{\text{M}} , \tag{1}$$

where in the quark ($\Omega_{\text{Q}}$) and meson ($\Omega_{\text{M}}$) contributions the zero-point energy terms are removed ("no sea" approximation). The quark contribution is given by

$$\Omega_{\text{Q}} = -\frac{2N_c N_f}{3} \int \frac{dp}{2\pi^2} \frac{p^4}{E_p} \left[ f_\Phi^+(E_p) + f_\Phi^-(E_p) \right] , \tag{2}$$

where $f_\Phi^\pm(E_p)$ are the generalized Fermi distribution functions

$$f_\Phi^+(E_p) = \frac{(\bar{\Phi} + 2\Phi Y)Y + Y^3}{1 + 3(\bar{\Phi} + \Phi Y)Y + Y^3} , \; f_\Phi^-(E_p) = \frac{(\Phi + 2\bar{\Phi}\bar{Y})\bar{Y} + \bar{Y}^3}{1 + 3(\Phi + \bar{\Phi}\bar{Y})\bar{Y} + \bar{Y}^3} . \tag{3}$$

with the abbreviation $Y = \mathrm{e}^{-(E_p - \mu)/T}$ and $\bar{Y} = \mathrm{e}^{-(E_p + \mu)/T}$ (cf. Ref. [5]). The meson contribution reads [2]

$$\Omega_{\text{M}} \doteq -d_M \int \frac{d^3q}{(2\pi)^3} \int \frac{d\omega}{2\pi} \left[ g^+(\omega + \mu_M) + g^-(\omega - \mu_M) \right] \delta_M(\omega, \mathbf{q}) .$$

Here $d_M$ are the meson degeneracy factors: $d_\pi = 3$ for pions and $d_K = 4$ for kaons; $g(E) = (\mathrm{e}^{\beta E} - 1)^{-1}$ is the Bose function, $\mu_M = \mu_i - \mu_j$ is the chemical potential of a meson M composed of quark $i$ with chemical potential $\mu_i$ and antiquark $j$ with chemical potential $-\mu_j$. The phase shift at rest $\delta_{\text{M}}(\omega, \mathbf{q} = \mathbf{0}) = \delta_{\text{M}}(\omega)$ is determined by

$$\delta_{\text{M}}(\omega) = -\,\mathrm{Im} \ln \left[ \beta^2 S_{\text{M}}^{-1}(\omega - \mu_M + \mathrm{i}\eta) \right] . \tag{4}$$

The propagator and the polarization function in the present work should be understood as to be evaluated at the shifted energy $z = \omega - \mu_M + \mathrm{i}\eta$

$$\begin{aligned} S_{\text{M}}^{-1}(z) &= G_{\text{S}}{}^{-1} - \Pi_{\text{M}}(z) , \\ \Pi_{\text{M}}(z) &= 4\{ \left( I_1^i + I_1^j \right) - [(\omega - \mu_M)^2 - (M_i - M_j)^2] I_2^{ij}(\omega - \mu_M)] , \end{aligned} \tag{5}$$

where the 1-loop integrals are given by

$$I_1^i = -N_c \int \frac{dp}{4\pi^2} \frac{p^2}{E_i} \left[ f_\Phi^+(E_i) + f_\Phi^-(E_i) \right] , \tag{6}$$

$$\begin{aligned} I_2^{ij} = -N_c \int \frac{d^3p}{(2\pi)^3} \Big[ &\frac{1}{2E_i} \frac{1}{(E_i + \omega)^2 - E_j^2} f_\Phi^+(E_i) - \frac{1}{2E_i} \frac{1}{(E_i - \omega)^2 - E_j^2} f_\Phi^-(E_i) \\ &+ \frac{1}{2E_j} \frac{1}{(E_j - \omega)^2 - E_i^2} f_\Phi^+(E_j) - \frac{1}{2E_j} \frac{1}{(E_j + \omega)^2 - E_i^2} f_\Phi^-(E_j) \Big] . \end{aligned} \tag{7}$$

## 3. Results

The parameters used for the numerical studies in the present work are the bare quark masses $m_{u,d} = 5.5$ MeV and $m_s = 138.6$ MeV, the three-momentum cutoff $\Lambda = 602$ MeV and the scalar coupling constant $G_S\Lambda^2 = 2.317$. With these parameters one finds in vacuum a constituent quark mass of 367 MeV, a pion mass of 135 MeV and pion decay constant $f_\pi = 92.4$ MeV. We report here results for the case of vanishing chemical potentials.

The solution for the pion phase shift is shown in Figure 1 as function of the energy variable $\omega$. The jump of the phase shift from 0 to $\pi$ indicates the position of a bound state in the spectrum below the threshold of the continuum states which is located where the phase shift starts decreasing towards zero. With increasing temperature the threshold moves to lower $\omega$-values and the phase shift jumps from $\pi$ to 0 when the Mott temperature is reached where the bound state gets dissociated. This behavior is in accordance with the Levinson theorem, for details see [11]. In the kaon phase shift arises one more threshold at low energy because

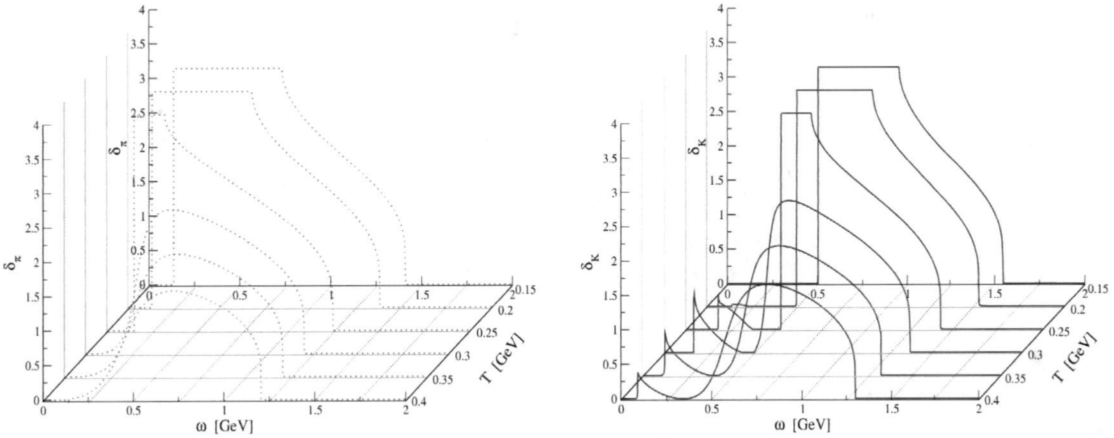

**Figure 1.** Phase shift of pions (left panel) and kaons (right panel) as function of the energy for different temperatures from $T = 150$ MeV to 400 MeV.

the dynamical masses of the quarks composing the kaon are different, for more information see Ref.[9, 10]. In the left panel of Figure 2 the pressure of pions and kaons is presented. It shows a typical behavior: first increasing with temperature towards the Stefan-Boltzmann limit which isn't reached since due to the chiral phase transition the continuum threshold lowers and this induces a reduction of the meson gas pressure already before the Mott temperature is reached. Above the Mott temperature, the growing meson width leads to a stronger reduction of the pressure with a rather sharp onset of this effect. In the right panel of Figure 2 we compare the total pressure of the $N_f = 2+1$ flavor PNJL model with recent LQCD data from [13, 14]. While our results agree with LQCD below $T \simeq 150$ MeV, a discrepancy opens up beyond this temperature which is mainly due the fact that the chiral transition temperature in the PNJL model is too high ($\sim 250$ MeV) when compared to the pseudocritical temperature of $T_c = 154 \pm 9$ MeV from LQCD [15]. Above a temperature of $T \simeq 300$ MeV the model pressure reaches towards the Stefan-Boltzmann limit while the LQCD pressure stays below it.

## 4. Conclusions

The main result this work is the addition of the strange sector to the relativistic Beth-Uhlenbeck approach which allows to compare with $N_f = 2+1$ LQCD data. At the present level of description, there is a discrepancy between our results and the LQCD ones for temperatures

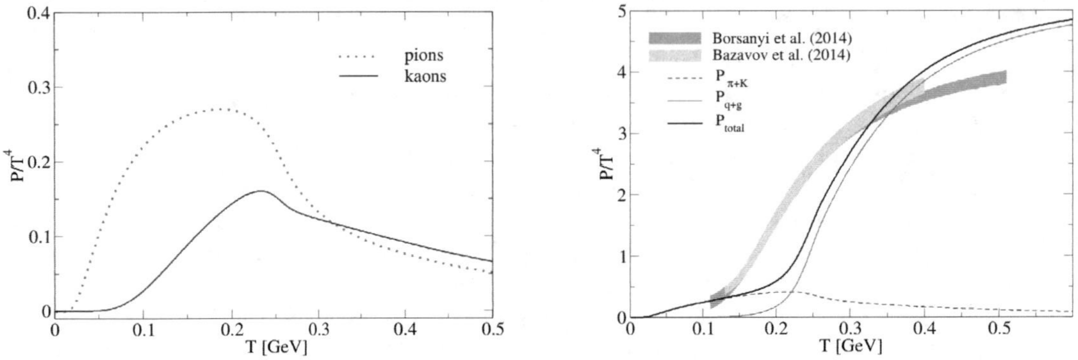


**Figure 2.** Left panel: pressure of pions (red dotted line) and kaons (blue solid line). Right panel: total pressure of $N_f = 2+1$ flavor PNJL model with pseudoscalar meson correlations as a function of temperature (black solid line) and LQCD data (colored bands [13, 14]).

$T \gtrsim 150$ MeV which can be reduced as follows. First, by including higher lying hadronic states into the calculation of the total pressure. Second, by lowering the Mott temperature. Third, by accounting for virial corrections (perturbative) to the quark-gluon pressure corresponding to residual interactions of quarks and gluons in the plasma. These improvements are subject of our current work.

**Acknowledgement**
We gratefully acknowledge discussions about lattice QCD data with O. Kaczmarek. The work of A.D. was supported by the Polish National Science Centre (NCN) under grant number UMO - 2013/09/B/ST2/01560 and by the Institute for Theoretical Physics of the University of Wroclaw under internal grant number 1439/M/IFT/15. D.B. and A.R. received support by NCN under grant number UMO-2011/02/A/ST2/00306.

## 5. References

[1] Hüfner J, Klevansky S P, Zhuang P and Voss H 1994 Annals Phys. **234** 225
[2] Blaschke D, Buballa M, Dubinin A, Röpke G and Zablocki D 2014 Annals Phys. **348** 228
[3] Fukushima K 2004 Phys. Lett. B **591**, 277
[4] Ratti C, Thaler M A and Weise W 2006 Phys. Rev. D **73**, 014019
[5] Hansen H, Alberico W M, Beraudo A, Molinari A, Nardi M and Ratti C 2007 Phys. Rev. D **75**, 065004
[6] Radzhabov A, Blaschke D, Buballa M and Volkov M K 2008 Phys. Atom. Nucl. **71**, 1981
[7] Radzhabov A, Blaschke D, Buballa M and Volkov M K 2011 Phys. Rev. D **83**, 116004
[8] Wergieluk A, Blaschke D, Kalinovsky Y L and Friesen A 2013 Phys. Part. Nucl. Lett. **10**, 660
[9] Yamazaki K and Matsui T 2013 Nucl. Phys. A **913**, 19
[10] Yamazaki K and Matsui T 2014 Nucl. Phys. A **922**, 237
[11] Blaschke D, Dubinin A and Buballa M 2015 Phys. Rev. D **91**, 125040
[12] Blaschke D, Dubinin A and Turko L 2015 Phys. Part. Nucl. **46**, 732
[13] Borsanyi A, Fodor Z, Hoelbling C, Katz S D, Krieg S and Szabo K K 2014 Phys. Lett. B **730**, 99
[14] Bazavov A, *et al.* [HotQCD Collaboration] 2014 Phys. Rev. D **90**, 094503
[15] Kaczmarek O, *et al.* 2011 Phys. Rev. D **83**, 014504

# Hydrodynamic modeling of QGP expansion using an exact solution of Riemann problem

**Zuzana Fecková[1,2] and Boris Tomášik[2,3]**

[1] University of Pavol Jozef Šafárik, Šrobárova 2, 04001 Košice, Slovakia
[2] Matej Bel University, Tajovského 40, 97401 Banská Bystrica, Slovakia
[3] FNSPE, Czech Technical University in Prague, Břehová 7, 11519 Prague 1, Czech Republic

E-mail: zuzana.feckova1@student.upjs.sk, boris.tomasik@umb.sk

**Abstract.** Hydrodynamic modelling of quark-gluon plasma requires sophisticated numerical schemes that have low numerical viscosity and are able to cope with high gradients of energy density that may appear in initial conditions. We propose to use the Godunov method with an exact Riemann solver for ideal hydrodynamic modelling to meet these conditions. We present the results of numerical tests of the method, such as the sound wave propagation and the shock tube problem, which show both high precision and low numerical viscosity.

## 1. Introduction

Relativistic hydrodynamics has proved to be a useful tool for studying the expansion of the quark-gluon plasma created in heavy ion collisions, e.g. [1, 2]. In the present paper we will introduce a new numerical scheme for such modelling based on the exact solution of the Riemann problem with an arbitrary equation of state and a possible source term. It aims for high precision and the suitability to work with complicated initial conditions and to capture shocks that might arise from jets propagating in the medium [3, 4, 5].

## 2. Numerical method

Ideal relativistic hydrodynamic equations express the conservation of energy and momentum as well as conserved charges, e.g. the baryon number in heavy-ion collisions. In collisions at highest energies the net baryon density is practically zero, thus we only consider energy and momentum:

$$\partial_\mu T^{\mu\nu}_{(0)} = 0, \qquad (1)$$
$$T^{\mu\nu}_{(0)} = (\epsilon + p)u^\mu u^\nu - p g^{\mu\nu},$$

and the equation of state which completes the system of equations relates only the pressure and the energy density $p = p(\epsilon)$.

The presented equations are solved numerically using a Godunov scheme with slope-limited linear reconstruction of states. The flows of conserved quantities are calculated at every cell boundary [6, 7] by exactly solving the Riemann problem for an arbitrary equation of state with the presence of tangential velocities [8, 9]. The Riemann problem is solved by reconstructing the corresponding wave pattern at the boundary considering possible shock and rarefaction waves and a contact discontinuity.

## 3. Numerical tests

*3.1. Sound wave problem*

We simulate a sound wave over one wavelength in the numerical grid with these initial conditions:

$$p_{init}(x) = p_0 + \delta p \sin\frac{2\pi x}{\lambda}, \tag{2}$$
$$v_{init}(x) = \frac{\delta p}{c_s(\epsilon_0 + p_0)} \sin\frac{2\pi x}{\lambda},$$

with parameters $p_0 = 10^3$ fm$^{-4}$, $\delta p = 10^{-1}$ fm$^{-4}$. The variation of pressure is sufficiently small $\delta p \ll p_0$, we can thus consider the linearized analytic solution, which is a sound wave of velocity $c_s$. To evaluate the precision of the scheme the numerical solution $p$ is compared to the analytic one denoted $p_s$ using $L1$ norm [10]:

$$L(p(N_{cell}), p_s) = \sum_{i=1}^{N_{cell}} |p(x_i; N_{cell}) - p_s(x_i)| \frac{\lambda}{N_{cell}}, \tag{3}$$

We have studied the dependence of $L1$ norm on the number of cells in the numerical grid, which is shown in the left panel of Fig. 1. As expected, the precision improves with finer discretization and is comparable to other numerical schemes presented in [10].

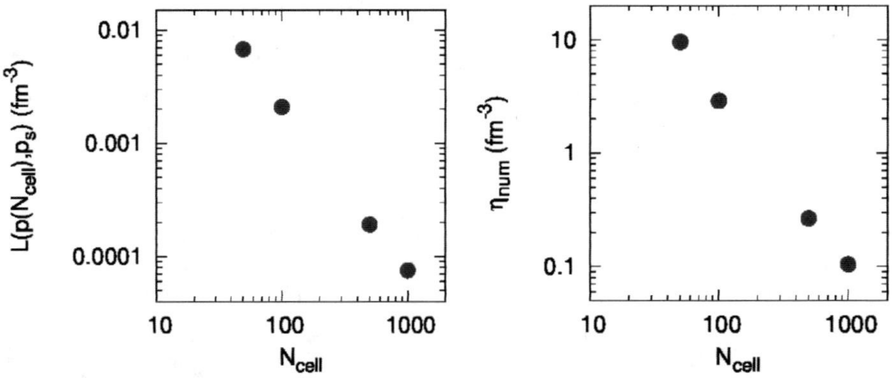


**Figure 1.** Dependence of $L1$ norm (left) and numerical viscosity $\eta_{num}$ (right) on the number of cells in the numerical grid.

The $L1$ norm can also be used to assess the numerical dissipation that is due to limited precision. The quark-gluon plasma is expected to have very low viscosity, therefore it is very important to control the numerical viscosity, evaluated as

$$\eta_{num} = -\frac{3\lambda}{8\pi^2} c_s(\epsilon_0 + p_0) \ln\left[1 - \frac{\pi}{2\lambda\delta p} L(p(N_{cell}), p_s)\right]. \tag{4}$$

We show the dependence of numerical viscosity $\eta_{num}$ on the number of cells in the grid in the right panel of Fig. 1. Its values are adequately small, comparable again to other numerical schemes used in heavy ion collision modelling [10].

To compare the numerical dissipation with the possible values of physical dissipation a more suitable parameter is the ratio of viscosity to entropy density $\eta/s$. We have estimated the value of this ratio for the numerical viscosity in our scheme and in Fig. 2 we are comparing its values with the estimated values for pion gas [11] and the well-known quantum limit $\eta/s = 1/4\pi$ from AdS/CFT calculations [12].

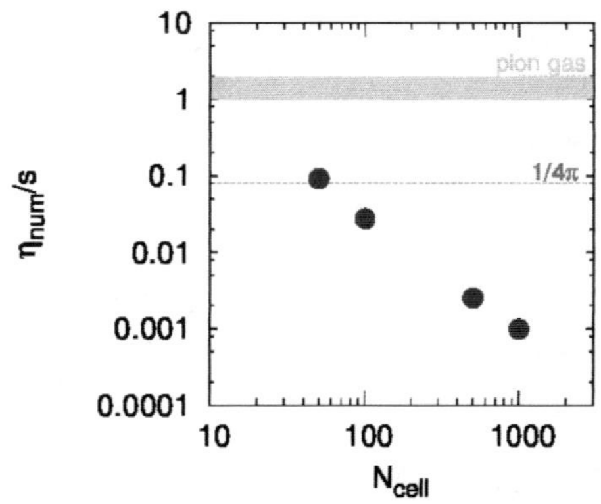


**Figure 2.** Dependence of the numerical viscosity to entropy density ratio $\eta_{num}/s$ (blue points) on the number of the cells compared to the limiting value $\eta/s = 1/4\pi$ (in red) and $\eta/s$ of pion gas (green band).

### *3.2. Shock tube problem*

To test the scheme's capability to cope with large discontinuities in energy density and velocity as well as its ability to resolve the shock and rarefaction wave as precisely as possible we impose special initial conditions in the numerical grid. They consist of two constant states separated by a discontinuity. The imposed energy density and tangential velocity vary in the grid: in the left (right) half they are: $\epsilon_L = 1$ GeV($\epsilon_R = 20$ GeV), $v_L^t = 1/3c$ ($v_R^t = 1/2c$). The initial normal velocity is $v^x = c/2$ over the whole grid. The analytic solution is such that with time the discontinuity is dissolved via waves: a rarefaction wave propagates into the region of higher energy density, a shock wave into the region of lower energy density and the two are connected by a contact discontinuity. In Fig. 3 we compare the analytic solution to the numerical solution with profiles of the energy density and the normal velocity in the grid after 100 time-steps. We conclude that the wave patterns are very well reproduced by the numerical solution.

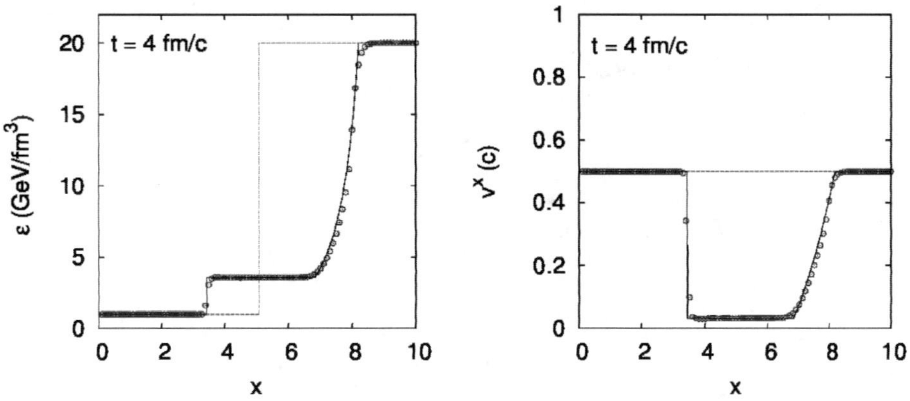


**Figure 3.** Profile of energy density $\epsilon$ (left) and normal velocity $v$ (right) in the numerical grid after 100 time-steps (our scheme in blue circles, analytic solution in black solid line, initial conditions in red dashed line).

Finally, in Fig. 4 we show the importance of linear reconstruction of states in the numerical grid. In the left panel, the profile of the tangential velocity after 100 time-steps when using a

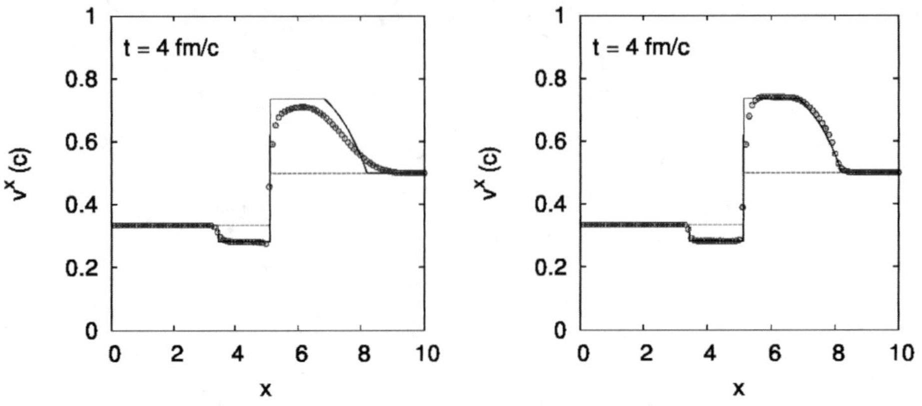


**Figure 4.** Profile of tangential velocity using a piecewise constant distribution of variables (left) and using a linear reconstruction of states (right) in the numerical grid after 100 time-steps (our scheme in blue circles, analytic solution in black solid line, initial conditions in red dashed line).

piecewise constant distribution of quantities for the computation is shown, while in the right panel we show the same profile when using a slope-limited linear reconstruction of the states [7]. The linear reconstruction is clearly necessary to obtain good results for the tangential velocity.

## 4. Summary

We have built and tested an ideal relativistic hydrodynamic scheme based on exact solution of the Riemann problem in one spatial dimension with presence of tangential velocity. The sound wave propagation test shows a good precision and low numerical viscosity which will become important when introducing dissipation into the model. The shock tube problem reveals that the scheme is able to capture shock and rarefaction waves very well. We will extend this scheme to three dimensions and then apply it in the description of the flow in ultrarelativistic nuclear collisions.

*Acknowledgements*
Supported in parts by APVV-0050-11, VEGA 1/0469/15 (Slovakia) and MŠMT grant LG13031 (Czech Republic).

## References

[1] Romatschke P 2010 *Int. J. Mod. Phys. E* **19** 1
[2] Gale C, Jeon S and Schenke B 2013 *Int. J. Mod. Phys. A* **28** 1340011
[3] Satarov L M, Stoecker H and Mishustin I N 2005 *Phys.Lett. B* **627** 64
[4] Casaderrey-Solana J, Shuryak E V, Teaney D 2006 *Nucl.Phys. A* **774** 577
[5] Betz B, Noronha J et al. 2009 *Phys.Rev. C* **79** 034902
[6] Godunov S K 1959 *Mat. Sb.* **47** 357
[7] E. F. Toro: Riemann Solvers and Numerical Methods for Fluid Dynamics, Springer, Berlin, 1999
[8] Pons J A 2000 *J. Fluid Mech.* **422** 152
[9] Mach P, Pietka M 2010 *Phys. Rev. E* **81** 046313
[10] Akamatsu Y, Inutsuka S, Nonaka C and Takamoto M 2014 *J. Comput. Phys.* **256** 34
[11] Mitra S, Sarkar S 2013 *Phys. Rev. D* **87** 094026
[12] Policastro G, Son D T and Starinets A O 2001 *Phys. Rev. Lett.* **87** 081601

# Measurement of D-meson production in Pb–Pb collisions at the LHC with ALICE

**Andrea Festanti for the ALICE Collaboration**

Università degli Studi di Padova and INFN Sezione di Padova, via Marzolo 8, 35131 Padova, Italy

E-mail: andrea.festanti@pd.infn.it

**Abstract.** The measurement of D-meson production in heavy-ion collisions at LHC energy provides insights into the mechanisms of interaction of charm quarks in the hot and dense medium formed in these collisions. ALICE results on the D-meson nuclear modification factor and azimuthal anisotropy in Pb–Pb collisions at $\sqrt{s_{\rm NN}} = 2.76$ TeV are presented.

## 1. Introduction

Heavy-flavour hadrons are suitable to probe the conditions of the high-energy-density Quark-Gluon Plasma (QGP) medium formed in ultra-relativistic heavy-ion collisions. Heavy quarks are mainly produced in hard scattering processes in the early stage of Pb–Pb collisions. The time-scale of their production ($\Delta\tau < 1/2m_{c,b} \sim 0.07$ fm/$c$ for charm and $\sim 0.02$ fm/$c$ for beauty) is, in general, shorter than the formation time of the QGP, $\tau_0 \sim 0.1 - 1$ fm/$c$. During their propagation through the medium, heavy quarks interact with its constituents and lose energy. QCD energy loss is expected to occur via both inelastic (radiative energy loss via medium-induced gluon radiation) [1] and elastic (collisional energy loss) processes [2]. The energy loss for quarks is expected to be smaller than for gluons, due to the smaller colour coupling factor of quarks with respect to gluons. In addition, the "dead-cone effect" should reduce small-angle gluon radiation for heavy quarks with moderate energy compared to their mass, thus further attenuating the effect of the medium [3].

The nuclear modification factor $R_{\rm AA}(p_{\rm T}) = ({\rm d}N_{\rm AA}/{\rm d}p_{\rm T})/(\langle T_{\rm AA}\rangle \cdot {\rm d}\sigma_{\rm pp}/{\rm d}p_{\rm T})$, where $\langle T_{\rm AA}\rangle$ is the average nuclear overlap function for a given centrality class, is sensitive to the interaction of hard partons with the medium. At large $p_{\rm T}$, $R_{\rm AA}$ is expected to be mostly sensitive to the average energy loss of heavy-quarks in the hot and dense medium. The questions whether low-momentum heavy quarks can reach thermal equilibrium with the medium constituents and participate in the collective expansion of the system [4, 5], and whether heavy quarks can hadronise also via recombination with other quarks from the medium [5, 6] are still open. These questions are addressed by studying $R_{\rm AA}$ at low and intermediate $p_{\rm T}$ and measuring the azimuthal anisotropy of heavy-flavour hadron production with respect to the reaction plane, defined by the beam axis and the impact parameter of the collision. The hadronisation mechanisms of the c quark are also investigated through the measurement of $\rm D_s^+$ production in nucleus–nucleus collisions compared to that in pp collisions [7].

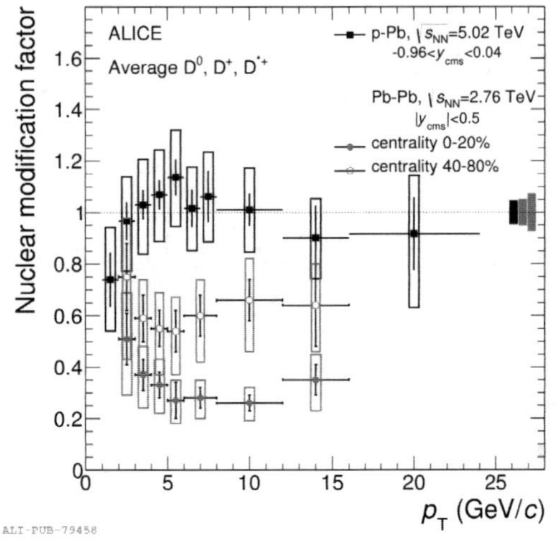


**Figure 1.** Average $R_{\mathrm{pPb}}$ of prompt $\mathrm{D}^0$, $\mathrm{D}^+$ and $\mathrm{D}^{*+}$ mesons [10] compared to the prompt D-meson $R_{\mathrm{AA}}$ in Pb–Pb collisions in the 0–20% and 40–80% centrality classes [11].

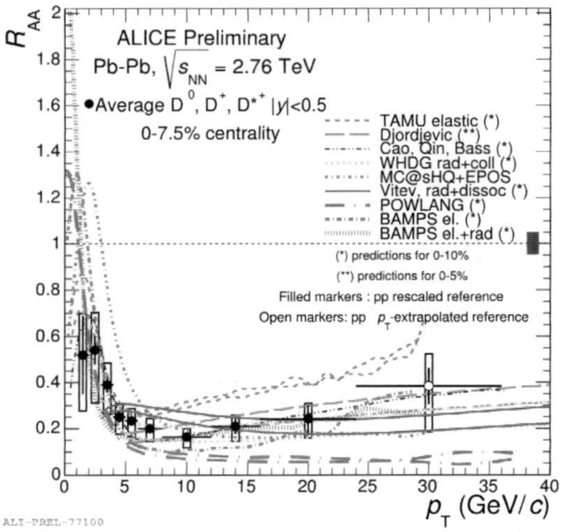


**Figure 2.** Average prompt D-meson $R_{\mathrm{AA}}$ in Pb–Pb collisions in the 0–7.5% centrality class compared to theoretical models including parton energy loss.

## 2. D-meson reconstruction

The decays $\mathrm{D}^0 \to \mathrm{K}^-\pi^+$, $\mathrm{D}^+ \to \mathrm{K}^-\pi^+\pi^+$ and $\mathrm{D}^{*+} \to \mathrm{D}^0\pi^+$, and their charge conjugates, were reconstructed at mid-rapidity ($|y| < 0.8$) in minimum-bias Pb–Pb collisions using the ALICE central barrel detectors. The D-meson selection was based on the precise reconstruction of the primary and secondary (decay) vertices, which is provided by the Inner Tracking System (ITS). Charged pions and kaons were identified using the information provided by the Time Projection Chamber (TPC) and the Time Of Flight (TOF) detectors [8]. The reference proton–proton cross section at $\sqrt{s_{\mathrm{NN}}} = 2.76$ TeV, needed to compute $R_{\mathrm{AA}}$, was obtained by a pQCD-based energy scaling of the $p_{\mathrm{T}}$-differential cross section measured at $\sqrt{s} = 7$ TeV [9].

## 3. D-meson nuclear modification factor

A large suppression of the D-meson $R_{\mathrm{AA}}$ (factor 3-5) was observed for $p_{\mathrm{T}} > 5$ GeV/$c$ in central Pb–Pb collisions at $\sqrt{s_{\mathrm{NN}}} = 2.76$ TeV (Figure 1) [11]. The comparison of $R_{\mathrm{AA}}$ with the D-meson nuclear modification factor measured in p–Pb collisions at $\sqrt{s_{\mathrm{NN}}} = 5.02$ TeV [10] shows that the expected cold nuclear matter effects are smaller than the uncertainties on $R_{\mathrm{AA}}$ for $p_{\mathrm{T}} > 3$ GeV/$c$. Therefore, the suppression observed in central Pb–Pb collisions is predominantly induced by final-state effects due to the presence of a hot and dense partonic medium. Figure 2 shows the D-meson $R_{\mathrm{AA}}$ measured in Pb–Pb collisions in the centrality class 0–7.5%, compared to theoretical models including charm interactions with the medium constituents. The large suppression observed, e.g. of a factor 6 at $p_{\mathrm{T}} = 10$ GeV/$c$, is described by the models that include radiative and collisional heavy-quark energy loss.

Figures 3 and 4 show the D-meson $R_{\mathrm{AA}}$ as a function of centrality (quantified in terms of the average number of participant nucleons in the Pb–Pb collision) [12] along with the $R_{\mathrm{AA}}$ of charged pions [13] and non-prompt J/$\psi$ mesons measured by the CMS Collaboration [15], respectively. The focus here is on the study of the parton energy loss, thus, the results are presented for the high-$p_{\mathrm{T}}$ interval $8 - 16$ GeV/$c$ for the pions and D mesons and for $6.5 < p_{\mathrm{T}} < 30$ GeV/$c$ for J/$\psi$ from B-meson decays. The $R_{\mathrm{AA}}$ of charged pions and D mesons are compatible within uncertainties in all centrality classes. The average difference of the $R_{\mathrm{AA}}$ values

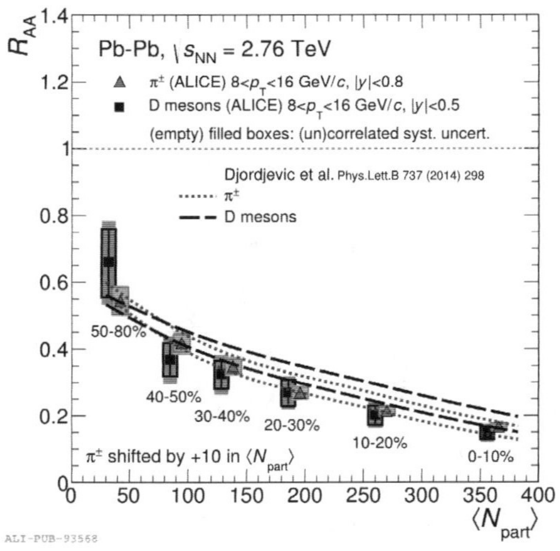


**Figure 3.** $R_{\mathrm{AA}}$ of D mesons [12] and charged pions [13] as a function of centrality compared to a pQCD model including mass dependent radiative and collisional energy loss [14].

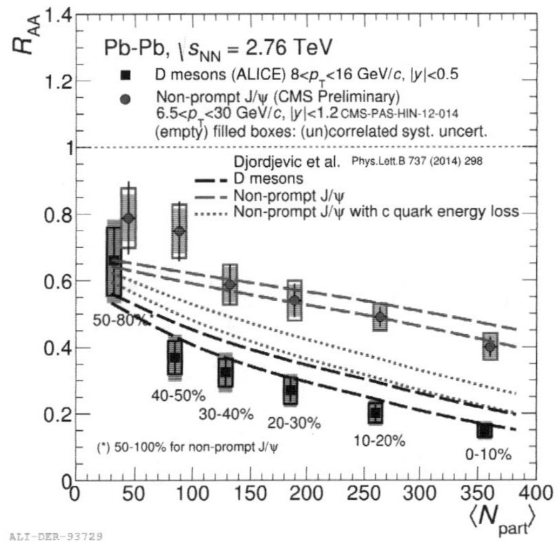


**Figure 4.** D [12] and non-prompt $J/\psi$ meson [15] $R_{\mathrm{AA}}$ vs. centrality compared to a pQCD model including mass dependent radiative and collisional energy loss [14].

of D mesons and non-prompt $J/\psi$ in the 0–10% and 10–20% centrality classes is larger than zero with a significance of 3.5 $\sigma$. A pQCD model including mass-dependent radiative and collisional energy loss [14] describes well the similarity of the D-meson and charged-pion $R_{\mathrm{AA}}$. In this model, the colour-charge dependence of energy loss is compensated by the softer fragmentation and $p_{\mathrm{T}}$ spectrum of gluons with respect to those of c quarks leading to a very similar $R_{\mathrm{AA}}$ for D mesons and pions. This calculation results in a larger suppression of D mesons with respect to non-prompt $J/\psi$, in agreement with the data. The large difference in the two $R_{\mathrm{AA}}$ derives predominantly from the quark-mass dependence of parton energy loss, as demonstrated by the non-prompt $J/\psi$ $R_{\mathrm{AA}}$ calculated considering the c-quark mass in the energy loss of b quarks (dotted lines in Figure 4) [14, 16, 17].

## 4. D-meson azimuthal anisotropy

$v_2$ is defined as the second coefficient of the Fourier expansion of the $p_{\mathrm{T}}$-dependent particle distribution $\mathrm{d}^2N/\mathrm{d}p_{\mathrm{T}}\mathrm{d}\varphi$ in azimuthal angle $\varphi$ relative to the reaction plane. It quantifies the momentum azimuthal anisotropy of final-state D mesons that might result from the spatial asymmetry of the initial parton distribution in non-central collisions, as a consequence of the collective expansion of the QGP. Figure 5 shows the D-meson $v_2$ in Pb–Pb collisions in the 30–50% centrality class [18, 19] compared to theoretical models including charm-quark interactions with the medium constituents. The value of $v_2$ is larger than zero in the interval $2 < p_{\mathrm{T}} < 6$ GeV/$c$ with a significance of about 5 $\sigma$ and it is best described by models including mechanisms that transfer to the charm quarks the elliptic flow of the medium during the system expansion (namely collisional processes and hadronisation by recombination with light quarks). A smaller suppression is observed in the in-plane direction (Figure 6) [19]; the ordering $R_{\mathrm{AA}}^{\mathrm{out-of-plane}} < R_{\mathrm{AA}}^{\mathrm{in-plane}}$ is equivalent to the observation of $v_2 > 0$ and, thus, consistent with the expectations from collective flow.

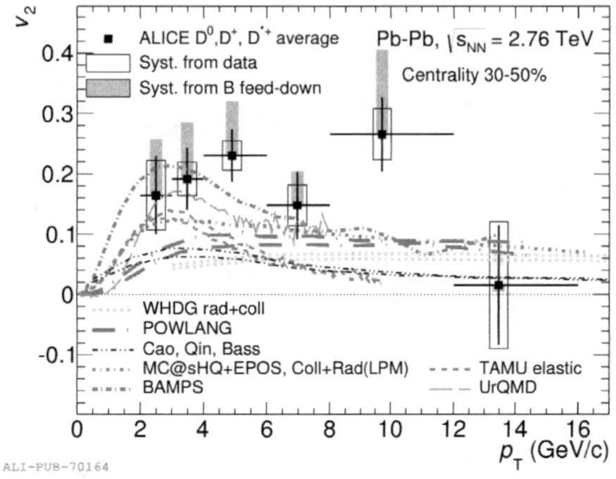


**Figure 5.** Model comparison for the average D-meson $v_2$ in the 30–50% centrality class [18, 19].

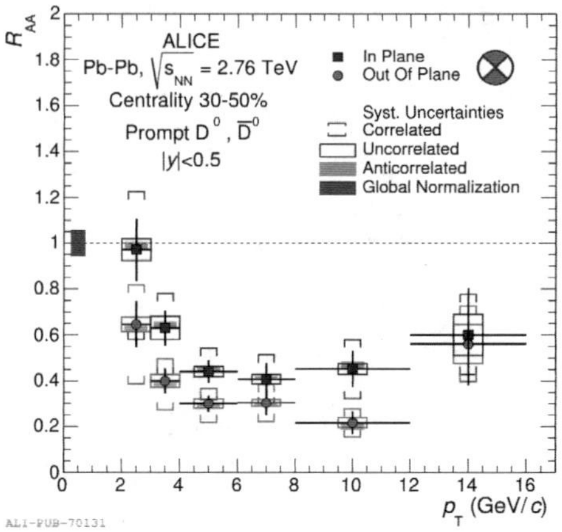


**Figure 6.** $D^0$ $R_{AA}$ in the 30–50% centrality class in two 90°-wide orthogonal azimuthal intervals [19].

## 5. Conclusions

The results obtained by ALICE using the data from the LHC Run-1 (2010–2013) indicate a strong suppression of the D-meson production in central Pb–Pb collisions for $p_T > 3$ GeV/$c$, which is mainly due the interactions of heavy quarks with the hot and dense medium. The smaller $R_{AA}$ observed for D mesons with respect to non-prompt J/$\psi$ confirms the mass-dependent nature of the energy-loss mechanisms. The non-zero $v_2$ of D mesons and the azimuthal dependence of the $D^0$ $R_{AA}$ indicate that, during the collective expansion of the medium, the interactions between its constituents and the charm quarks transfer to the latter information on the azimuthal anisotropy of the system. During the LHC Run-2 we expect to collect a data sample larger by a factor 5-10 with respect to Run-1, depending on collision centrality. It will be, thus, possible to measure the D-meson $R_{AA}$ and $v_2$ with a better precision and in an extended $p_T$ range.

## References

[1] Baier R, Dokshitzer Y, Mueller A, Peigné S and Schiff D 1997 *Nuclear Physics B* **484** 265–282 ISSN 0550-3213
[2] Braaten E and Thoma M H 1991 *Phys. Rev. D* **44**(4) 1298–1310
[3] Armesto N, Salgado C A and Wiedemann U A 2004 *Phys. Rev. D* **69**(11) 114003
[4] Batsouli S, Kelly S, Gyulassy M and Nagle J 2003 *Physics Letters B* **557** 26–32 ISSN 0370-2693
[5] Greco V, Ko C and Rapp R 2004 *Physics Letters B* **595** 202–208 ISSN 0370-2693
[6] Andronic A, Braun-Munzinger P, Redlich K and Stachel J 2003 *Physics Letters B* **571** 36–44 ISSN 0370-2693
[7] Barbano A (for the ALICE Collaboration) 2015 *These proceedings*
[8] Abelev B B *et al.* (ALICE) 2014 *Int. J. Mod. Phys.* **A29** 1430044 (*Preprint* 1402.4476)
[9] Abelev B B *et al.* (ALICE) 2012 *JHEP* **01** 128 (*Preprint* 1111.1553)
[10] Abelev B B *et al.* (ALICE) 2014 *Phys. Rev. Lett.* **113** 232301 (*Preprint* 1405.3452)
[11] Abelev B B *et al.* (ALICE) 2012 *JHEP* **09** 112 (*Preprint* 1203.2160)
[12] Adam J *et al.* (ALICE) 2015 (*Preprint* 1506.06604)
[13] Abelev B B *et al.* (ALICE) 2014 *Physics Letters B* **736** 196–207 ISSN 0370-2693
[14] Djordjevic M, Djordjevic M and Blagojevic B 2014 *Physics Letters B* **737** 298–302 ISSN 0370-2693
[15] CMS Collaboration (CMS) 2012 *CMS-PAS-HIN-12-014*
[16] Nahrgang M, Aichelin J, Gossiaux P B and Werner K 2014 *Phys. Rev. C* **89**(1) 014905
[17] He M, Fries R J and Rapp R 2013 *Nuclear Physics A* **910911** 409–412 ISSN 0375-9474
[18] Abelev B B *et al.* (ALICE) 2013 *Phys.Rev.Lett.* **111** 102301 (*Preprint* 1305.2707)
[19] Abelev B B *et al.* (ALICE) 2014 *Phys. Rev.* **C90** 034904 (*Preprint* 1405.2001)

# Strong Decay of $\phi(1s)$, $\Upsilon(4s)$ and $\Sigma^{\pm *}$ in Magnetic Field

**Peter Filip**

Institute of Physics, Slovak Academy of Sciences, Dúbravská cesta 9, Bratislava 84511, Slovakia

E-mail: `peter.filip@savba.sk`

**Abstract.** The influence of static magnetic field on decays $\phi(1020) \to K\bar{K}$, $\Upsilon(4s) \to B\bar{B}$, and $\Sigma^{\pm *} \to \Lambda\pi^{\pm}$ is discussed. In particular, we estimate how much the ratio of decay widths $\Gamma(K^+K^-)/\Gamma(K^0\bar{K}^0)$ and $\Gamma(B^+B^-)/\Gamma(B^0\bar{B}^0)$ can be modified by magnetic fields of strength $B \approx 10^{14}$ T created in relativistic collisions of heavy nuclei at LHC and RHIC colliders. Due to interaction of charged $K$ and $B$ mesons with the magnetic field, branching ratios for $K^+K^-$ and $B^+B^-$ decays of $\phi(1020)$ and $\Upsilon(4s)$ mesons may become reduced by 50% or completely closed. We study also the influence of magnetic field on strong decay $\Sigma^{\pm *}(1385) \to \Lambda + \pi^{\pm}$.

## 1. Introduction

It has been suggested recently [1], that static magnetic fields of strength $B \approx 10^{15}$ T can prolong the lifetime of $\rho(770)$ mesons via suppression of $\rho^0 \to \pi^+\pi^-$ decay channel, as a consequence of increasing energy of the lowest Landau level of charged pions in the magnetic field. Magnetic fields created in heavy ion collisions [2] at LHC and RHIC accelerators may reach extremal values $B \approx 10^{14} - 10^{15}$ T for a very short time, and the presence of expanding partonic plasma (QGP) can slow down [3] rapid decrease of the magnetic field intensity.

We have investigated the response of $s-$quark containing hadronic resonances $\Lambda^*(1520)$, $\Xi^*(1535)$ and $K^*(892)$ to such extremal magnetic fields [4], and it was found that particular strong decay channels (containing charged daughter particles) are affected much more than neutral channels. In this contribution we investigate the behavior of baryonic resonance $\Sigma^{+*}$ in very strong static magnetic fields.

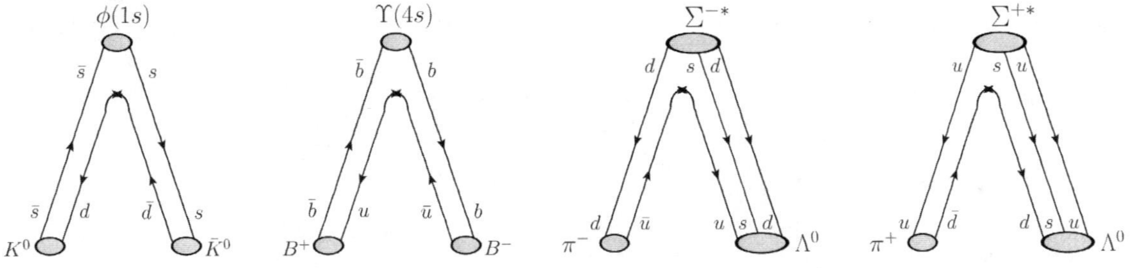


**Figure 1.** Strong decays of $\phi(1020)$, $\Upsilon(4s)$ mesons and $\Sigma^{\pm *}$ baryons by $(u\bar{u})$, $(d\bar{d})$ pair creation.

Additionally, we estimate how much decay widths of $\phi(1020) \to K^+K^-$ and $\Upsilon(4s) \to B^+B^-$ channels are suppressed in the magnetic field, due to the reduction of phase-space occurring in the magnetic field. The effects we discuss here can be of experimental significance, only if lifetimes of particles considered are comparable to the duration of extremal magnetic field created in heavy ion collisions. The lifetime $\tau_{\Sigma^*} \approx 5$fm/c of $\Sigma^*(1385)$ resonance is reasonably short, while lifetimes of $\Upsilon(4s)$ and $\phi(1020)$ mesons are considerably longer: 10fm/c and 46fm/c.

## 2. Energy of charged baryons in the magnetic field

Dispersion relation of proton [5] in fields $B < B_p^c = 1.5 \cdot 10^{16}$ T can be expressed as

$$E[B] = \left[ p_z^2 + \left( \sqrt{M_p^2 + (1-2s_z)\frac{B}{B_p^c} M_p^2} - (2s_z)\kappa_p \frac{B}{2B_p^c} M_p \right)^2 \right]^{1/2} \tag{1}$$

where critical field $B_p^c = M_p/2\mu_N = M_p^2 c^2/e\hbar$, anomalous part of proton magnetic moment $\kappa_p = 1.79 = \mu_p/\mu_N - 1$, and $\mu_N = 3.152 \cdot 10^{-14}$ T. For weak fields $B \ll B_c^p$ and $p_z \to 0$ one has

$$E[B, s_z] \approx m_p + eB/2m_p - (2s_z)[1+\kappa_p]\mu_N B \ . \tag{2}$$

Term $E = eB/2m_p$ corresponds to ($n = 0$) Landau level energy of proton, and interaction of proton magnetic moment $\mu_p = (1+\kappa_p)\mu_N$ with external field is simply $E = -\vec{\mu}_p \vec{B}$. One can rewrite Eq. (1) also for $\Sigma$ hyperons. Anomalous magnetic moment is $\kappa_\Sigma = (M_\Sigma/M_p)\mu_\Sigma/\mu_N - 1$, and small field approximation $B \ll B_\Sigma^c = B_p^c(M_\Sigma/M_p)^2$ gives formula identical to Equation (2) with replacements $m_p \to m_\Sigma$ and $\kappa_p \to \kappa_\Sigma$.

For $J = 3/2$ baryons ($\Sigma^{\pm *}$, $\Xi^{-*}$ and $\Omega^-$), Equation (2) can still be used, if term $(2s_z)$ is substituted by $(2J_z/3)$. For neutral baryons ($\Lambda^0$ for example) Landau energy term $eB/2m$ vanishes and term $[1+\kappa_p]\mu_N \longrightarrow \mu_{\Lambda^0}$, while magnetic moment interaction of spinless hadrons (pseudoscalars $\pi, K, \eta$) is zero $\mu = 0$ (if their magnetic polarizability [6] is neglected [7]).

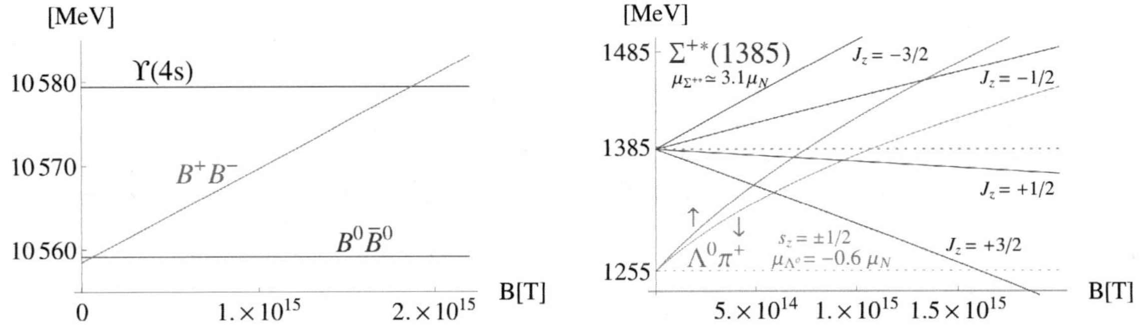


**Figure 2.** Energy plots for $\Upsilon(4s) \to B\bar{B}$ and for $\Sigma^{+*} \to \Lambda^0 + \pi^+$ decays in the magnetic field.

In Figure 2, we show the energy plot for $\Upsilon(4s) \to B\bar{B}$ and $\Sigma^{+*} \to \Lambda^0 + \pi^+$ strong decays (using $p_z = 0$). Large $\Sigma_{3/2}^{+*}$ quadruplet energy splitting comes from $\mu_{\Sigma^{+*}} \approx 3.1\mu_N = 2\mu_u + \mu_s$ magnetic moment interaction. Behavior of $\Sigma^{-*}$ is similar ($\mu_{\Sigma^{-*}} \approx -2.6\mu_N = 2\mu_d + \mu_s$) to that shown for $\Sigma^{+*}$, while quadruplet splitting of $\Sigma^{0*}$ is much smaller. We observe that 25% of $\Sigma^{+*}$ decays are energetically forbidden at $B = 5 \cdot 10^{14}$ T, and only 50% are possible at $B = 10^{15}$ T. Allowed $\Sigma^{+*}$ decays at $B > 10^{15}$ T would behave as originating from a polarized $J$=3/2 baryon.

Decay $\Upsilon(4s) \to B^+B^-$ is kinematically closed for $B \geq 2 \cdot 10^{15}$ T, while $\Upsilon(4s) \to B^0\bar{B}^0$ channel remains unaffected in our approximation. Landau level energy $E_L = |Q|B/2m$ is always increasing, while interaction of particle magnetic moment $E = -\vec{\mu}\vec{B}$ splits the energy of particles with non-zero magnetic moments. As energy difference between decaying hadron and its decay products decreases, phase space of that particular decay channel gets reduced (see next section).

Journal of Physics: Conference Series **668** (2016) 012055 doi:10.1088/1742-6596/668/1/012055

### 3. Magnetic field dependence of decay widths

Isospin conservation for strong decays of $\phi(1020)$ and $\Upsilon(4s)$ mesons into $K^+K^-$, $K^0\bar{K}^0$ and $B^+B^-$, $B^0\bar{B}^0$ pairs predicts equal branching ratios for neutral and charged decay channel (Clebsch-Gordan coefficients are $\pm\sqrt{1/2}$). This agrees quite well with $\mathrm{BR}(\Upsilon \to B^+B^-) = 51.6\%$ and $\mathrm{BR}(\Upsilon \to B^0\bar{B}^0) = 48.4\%$ observed experimentally (a small deviation from equality is expected due to Coulomb effects [8]). However, $\mathrm{BR}(\phi(1020) \to K^+K^-) = 49.2\%$ and $\mathrm{BR}(\phi(1020) \to K^0\bar{K}^0) = 34.0\%$ which gives experimental ratio $R^{\phi}_{\pm 0} = \Gamma^{\phi}_{\pm}/\Gamma^{\phi}_{00} = 1.45 \pm 0.04$. Such a deviation from the expectation may originate from phase-space effects.

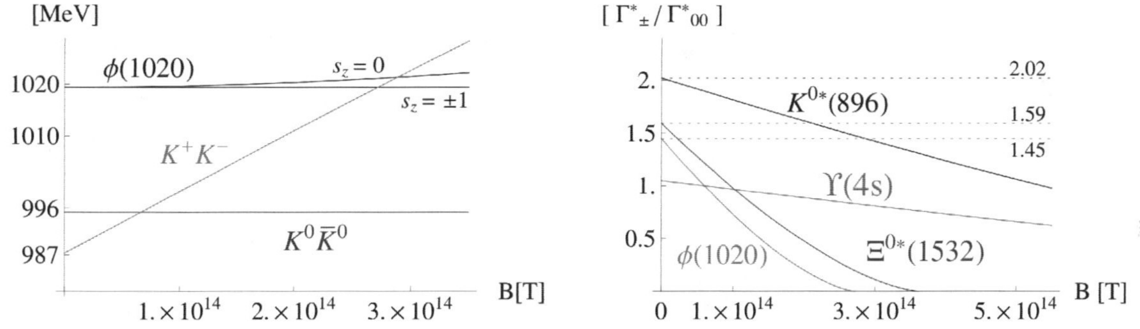


**Figure 3.** (**Left**) Energy plot for $\phi(1020) \to K\bar{K}$ decays in magnetic field, with Zeeman splitting of $\phi(s_z = 0, \pm 1)$ substates explicitly shown. (**Right**) Ratio of charged $\Gamma_{\pm}$ to neutral $\Gamma_{00}$ partial decay widths for strong decays $\Upsilon(4s) \to B\bar{B}$, $\phi(1020) \to K\bar{K}$, $\Xi^{0*} \to \Xi\pi$, and $K^{0*} \to K\pi$.

Spin $J$=1 of hadrons $\phi(1020)$ and $\Upsilon(4s)$, decaying into the pair of spinless particles, is transformed into $P$-wave angular momentum $L$=1 of pseudoscalar meson pair. In such case, phase-space is proportional to the third power of momentum [8]. Decay width is expressed as

$$\Gamma[B] \approx |\mathcal{M}|^2 \, p^3[B] \qquad \text{where} \qquad p[B] = \frac{1}{2}\sqrt{(M_R^2 - 4m^2[B])}\,. \tag{3}$$

$M_R$ is mass of decaying resonant state and $m = m_1 = m_2$ are equal masses of two pseudoscalar decay products, which can depend on external magnetic field. Indeed, using masses $K^{\pm} = 493.7\mathrm{MeV/c^2}$ and $K^0 = 497.6\mathrm{MeV/c^2}$ gives different decay momenta $p_{K^{\pm}} = 127\mathrm{MeV/c}$ and $p_{K^{\pm}} = 110\mathrm{MeV/c}$ in vacuum ($B$=0), which translates into expected ratio $R^{\phi}_{\pm 0} = 1.52$ of decay widths (while we still neglect Coulomb effects [8]). In the magnetic field, one has

$$R^{\Upsilon,\phi}_{\pm 0} = \Gamma_{\pm}[B]/\Gamma_{00} = p^3_{\pm}[B]/p^3_{00} \quad \text{and} \quad \Gamma^*_{tot}[B] = \Gamma^*_{\pm}[B] + \Gamma^*_{00} + \Gamma^*_{rest}\,, \tag{4}$$

if dependence [10] of strong decay amplitude $|\mathcal{M}|^2$ in Equation (3) on external fields is neglected. We assume $\Gamma^{*[B]}_{rest} \approx 0$ for $\Upsilon(4s)$, while $\Gamma^*_{rest} = 0.75\mathrm{MeV} = \Gamma_{\rho\pi} + \Gamma_{\eta\gamma}$ for $\phi(1020)$ meson [9].

In Figure 3 we show magnetic field dependence of ratios $R^* = \Gamma_{\pm}[B]/\Gamma_{00}$ of charged and neutral decay widths for strong decays of $\phi(1020)$, $\Upsilon(4s)$, $\Xi^{0*}$, and $K^{0*}$ hadrons estimated according to Equation (4). For $\phi(1020)$ decay, Coulomb correction [8] factor is used to meet the experimental data [9] in Figure 3 with value $R^{\phi}_{\pm 0} = \Gamma^{\phi}_{\pm}/\Gamma^{\phi}_{00} = 1.45$.

Charged decay widths $\Gamma^*_{\pm}[B]$ can be reduced even more than we suggest here: only $p_z$ momentum component is continuous in $B_z$ magnetic field, and $\int d^3\vec{p}/(2\pi\hbar)^3$ has to be replaced by the sum [11] of Landau levels $(eB/c)\sum_{n=0}^{n_{max}}[2 - \delta^0_n] \int dp_z/(2\pi\hbar)^2$ for phase-space evaluation. Figure 4 shows our estimate of lifetime $\tau^* = 1/\Gamma^*_{tot}$ for $\phi(1020)$, $\Upsilon(4s)$, $\Xi^{0*}$, and $K^{0*}$ in $B$ field.

## 4. Summary and Conclusions

We have studied the influence of static magnetic field on strong decay properties of $\phi(1020)$, $\Upsilon(4s)$ mesons and $\Sigma^*(1385)$ baryonic resonance. It is found, that charged decay channels $\phi \to K^+K^-$ and $\Upsilon(4s) \to B^+B^-$ can become completely closed in magnetic fields $B_\phi > 3 \cdot 10^{14}$ T and $B_\Upsilon > 2 \cdot 10^{15}$ T. For strong decays of $\Sigma^{\pm *}$ resonance we find no critical field strength. However, decay $\Sigma^{+*} \to \Lambda + \pi^+$ of $J_z = 3/2$ and $J_z = 1/2$ substates becomes forbidden in field $B > 10^{15}$ T (see Fig. 2), which results in the effective polarization of decaying $\Sigma^{+*}$ baryons in magnetic field. Our estimates are based on the reduction of phase-space, occurring due to increasing minimal energy ($n = 0$ Landau level) of charged decay products in the magnetic field.

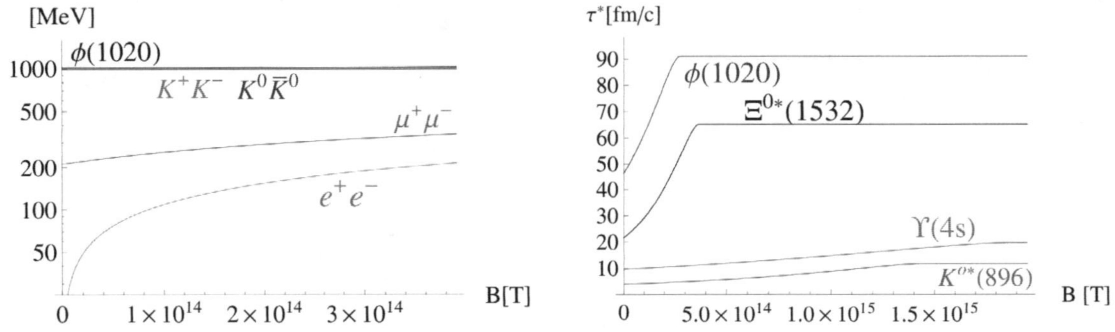


**Figure 4.** Energy plot for $\phi(1020) \to l^+l^-$ decays in magnetic field, and estimated lifetimes of $\Upsilon(4s), \phi(1020), \Xi^{0*}$, and $K^{0*}$ obtained from magnetic field dependences of their decay channels.

We also find that lifetimes of $\phi(1020)$, $\Upsilon(4s)$ mesons and other (e.g. $K^{0*}$ and $\Xi^{0*}$) hadrons increase in the magnetic field. Since duration of extremal magnetic fields in relativistic collisions of nuclei is rather short [2], effects studied here might be observable only for short-living resonant states, $K^*$ for example. In Figure 4 we suggest that phase-space for $\phi \to e^+e^-$ decay changes more than $\phi \to \mu^+\mu^-$, but separate study is needed to understand the magnetic field influence on dilepton decays. Behavior of $\Sigma^{\pm *}$ is also non-trivial and more detailed studies are needed.

**Acknowledgments**

This work was supported by Slovak Grant Agency VEGA under contract No. 2/0197/14. The author is indebted to the organizers of Strange Quark Matter 2015 conference for the kind invitation to present a talk, and also for their hospitality at JINR.

## References

[1] Chernodub M N 2011 *Phys. Rev. B* **82** 085011
[2] Deng W T and Huang C G 2012 *Phys. Rev. C* **85** 044907
[3] Tuchin K 2010 *Phys. Rev. C* **82** 034904
[4] Filip P 2015 *Journal of Physics: Conf. Series* **636** 012013
[5] Suh I S and Mathews G J 2001 *The Astrophysical Journal* **546** 1126
[6] Filip P 2013 *Nuclear Physics B (Proc. Suppl.)* **245** 251
[7] Filip P 2014 *Physics of Particles and Nuclei* **45** 7
[8] Lepage G P 1990 *Phys. Rev. D* **42** 3251
[9] Amsler C et al. (Particle Data Group) 2008 *Phys. Lett. B* **667** 1
[10] Chernodub M N 2014 *Mod. Phys. Lett. A* **29** 1450162 (*Preprint* arXiv:1001.0570).
[11] Landau L D and Lifshitz E M 1938 *Statistical Mechanics* (Clarendon: Oxford)

Journal of Physics: Conference Series **668** (2016) 012056 doi:10.1088/1742-6596/668/1/012056

# Prospects for heavy-flavour measurements with the ALICE inner and forward tracker upgrade

**F. Fionda, on behalf of the ALICE Collaboration**
University and INFN - Cagliari, Italy
E-mail: fiorella.fionda@cern.ch

**Abstract.** During the second long shutdown (LS2) of the LHC the ALICE detector will be improved with the installation of an upgraded Inner Tracking System (ITS) and a new Muon Forward Tracker (MFT). These detectors will crucially contribute to the precise characterization of the high-temperature, strongly-interacting medium created in ultra-relativistic Pb–Pb collisions at $\sqrt{s_{\rm NN}} = 5.5$ TeV. In the central barrel, the upgraded ITS will consist of seven cylindrical layers of silicon pixel detectors, starting at a radial distance of 22.4 mm from the beam axis. At forward rapidity, the MFT will be composed of five silicon pixel planes added in the acceptance of the existing Muon Spectrometer ($-4 < \eta < -2.5$), upstream to the hadron absorber. Detailed results on the expected performances for heavy-flavour (HF) measurements down to low transverse momentum, with the upgraded ITS and MFT, will be given for central Pb–Pb collisions for various benchmark analyses, assuming an integrated luminosity of 10 nb$^{-1}$, as foreseen for the ALICE upgrade programme.

**Introduction.** The primary goal of ALICE (A Large Ion Collider Experiment) is to explore the properties of the strongly-interacting matter at high temperature and energy density, where the existence of the Quark–Gluon Plasma (QGP) is expected. Heavy quarks, i.e. charm and beauty, are primarily created in initial hard scatterings, therefore they may carry information from the system at its early stage. The heavy-flavour production in heavy-ion collisions is related to the following physics topics.

- The in-medium parton energy-loss mechanism, studied through measurements of the nuclear modification factor ($R_{\rm AA}$). In particular, it is crucial to characterize the dependencies of energy loss on the parton colour-charge, mass, and energy, as well as on the properties of the medium itself.
- The thermalization of charm and beauty quarks in the QGP through measurements of the azimuthal-flow anisotropy ($v_2$). This is especially sensitive to the partonic equation of state.
- The in-medium hadronization of heavy quarks, studied by measuring the baryon/meson ratio for charm-hadrons ($\Lambda_{\rm c}$/D) and for beauty-hadrons ($\Lambda_{\rm b}$/B), as well as the $\rm D_s$ production w.r.t. the non-strange D mesons.

Covering a wide rapidity range is crucial for a comprehensive understanding of the observed phenomena and to constrain theoretical models. In particular measurements at large rapidities will give access to the study of the gluon dynamics at small Bjorken-$x$ values, where phenomena related to the saturation of low-momentum gluons can be best investigated.

**Goals of the future ALICE heavy-flavour programme.** The ALICE [1] detector consists of a central barrel ($|\eta| < 0.9$) placed inside a large solenoidal magnet providing a 0.5 T field, and a Muon Spectrometer located at forward rapidity ($-4 < \eta < -2.5$).
At central rapidity the $R_{AA}$ and the $v_2$ of D mesons in Pb–Pb collisions were measured from LHC Run 1 data (2009-2013), down to $p_T = 1$ and 2 GeV/$c$, respectively. Very precise measurements down to $p_T = 0$ for D mesons, including also $D_s$, represent one of the primary goals of the future ALICE HF programme. Another important target is the exclusive $\Lambda_c$ baryon reconstruction, which is the most challenging HF measurement, because the small mean proper decay length ($c\tau = 59\,\mu$m) prevents an effective background rejection based on the decay topology selection. A sizable part of the HF programme at central rapidity includes the reconstruction of beauty-hadrons, both via inclusive and exclusive decay channels. Another ambitious goal is the full reconstruction of the $\Lambda_b$ baryon that will open the possibility to investigate the beauty baryon sector in Pb–Pb collisions.
At forward rapidity the existing Muon Spectrometer provided interesting results from Run 1 data. However it has an important limitation since no track constraints in the region of the primary vertex are available. This is mainly due to the large distance of the tracking stations from the primary vertex in combination with the multiple scattering induced on the muon tracks by the frontal absorber. Therefore the details of the vertex region are completely smeared out. The installation of a high-position-resolution tracker would improve the resolution of muon tracks in the primary vertex region, allowing to disentangle open charm and beauty contribution to muon production and enabling the separation between prompt and displaced $J/\psi$ mesons. Furthermore it would help in rejecting background muons coming from semi-muonic decays of pions and kaons, which represent a large background source both in single muon and dimuon analyses.

**The ALICE Upgrade Strategy.** The ALICE upgrade plans foreseen for the LS2 (2018-2019) are discussed in the ALICE Upgrade Letter of Intent [4]. The primary goal of ALICE is to perform high precision measurements of rare probes down to low transverse momentum. These probes, characterized by very low signal-to-background ratios, cannot be selected by a dedicated trigger. Improved performance for such measurements can be obtained with the following strategy.
• A large increase of the luminosity integrated for minimum-bias collisions: the upgrade of the readout capability of several detectors, combined with the new Offline-Online computing system (the "$O^2$" project [3]), will allow us to record Pb–Pb collisions at an interaction rate of about 50 kHz. In particular, the goal is to collect a Pb–Pb sample of at least 10 $nb^{-1}$ (corresponding to about $10^{11}$ interactions). The largest project within this context concerns the upgrade of the Time Projection Chamber [2] consisting of the replacement of the existing Multi-Wire Proportional Chambers (MWPC) redout by the Gas Electron Multipliers (GEM) detectors[1].
• Improvement of the tracking and vertexing resolution: at central rapidity an upgraded Inner Tracking System (ITS) will significantly improve the tracking performance, in particular the installation of a thinner beam pipe with a smaller radius (from the present value of 29.4 mm down to 18.6 mm) will allow us to position the first detection layer closer to the interaction point. At forward rapidity a new telescope of silicon pixel detectors, the Muon Forward Tracker (MFT), will be added to the Muon Spectrometer upstream of the hadron absorber.
The installation of the upgraded ITS and the MFT, combined with the high then available statistics, will enable several precise HF measurements in a wide rapidity window and down to zero transverse momentum. In this respect ALICE will be unique at the LHC.

[1] The GEM detectors, unlike the MWPC, provide intrinsic ion blocking capabilities preventing back-drifting of ions from the amplification region to the drift volume and therefore enabling the TPC to operate in a continuous, "ungated" readout mode at high rate.

**The upgrade of the Inner Tracking system at central rapidity.** The new ITS, described in the ITS Technical Design Report (TDR) [6, 5], will improve the impact parameter resolution (by a factor 3 and 6 in $r\varphi$ and $z$, direction respectively), the tracking efficiency, the transverse momentum resolution and the readout capabilities. The improvement of the impact parameter resolution will be realized by placing the first ITS layer at 22.4 mm from the beam line (present value: 39 mm) as well as by reducing the total material budget per layer: the usage of Monolithic Active Pixel Sensors (MAPS) and the optimization of front-end electronics can reduce it from the present value of 1.14% $X_0$ to 0.3% $X_0$ for the three innermost layers. Both the impact parameter at high $p_{\rm T}$ and the transverse momentum resolution will be improved by increasing the detector segmentation: all layers will be segmented in pixels with dimensions $30 \times 30\ \mu m^2$ (to be compared with $50 \times 425\ \mu m^2$ for the present ITS). Finally the tracking efficiency and the $p_{\rm T}$ resolution will be further improved by increasing the number of detector layers to seven (present ITS: six layers).

**The new Muon Forward Tracker.** The MFT detector layout and the expected physics performance are described in the MFT Technical Design Report [7]. The MFT detector will surround the vacuum beam-pipe and will be positioned inside the ITS outer barrel (four outermost ITS layers), in front of the hadron absorber of the Muon Spectrometer. It will consist of 5 layers (each one made by two half-disks) of silicon pixel sensors sharing the same pixel technology as the new ITS [6]. Each layer will be equipped by two detection planes and the corresponding total material budget is expected to be 0.6% $X_0$. The muon tracks reconstructed in the muon spectrometer will be extrapolated through the absorber and matched with track segments reconstructed in the MFT planes, enabling the precise determination of the muon production vertex. Detailed performance studies show that the offset resolution is better than 100 μm above 1 GeV/$c$, which is enough to perform a reliable separation of muons from charm-hadron ($c\tau \sim 100\,\mu$m) and beauty-hadron ($c\tau \sim 500\,\mu$m) decays.

**Expected performance for Heavy-Flavour measurements.** A few selected highlights of the HF physics performance expected for the LHC Runs 3 and 4 (foreseen from 2020 onwards after the LS2), assuming $L_{\rm int} = 10$ nb$^{-1}$, are shown in the following.

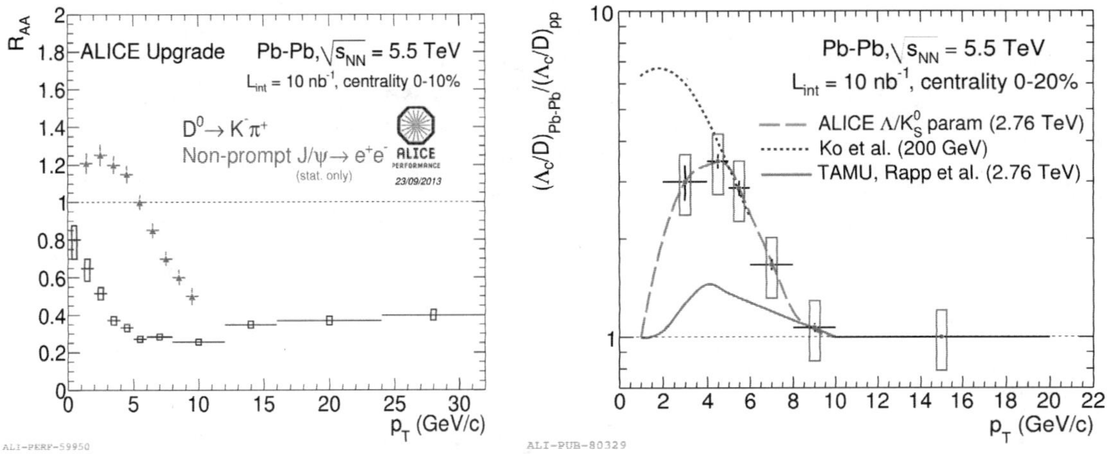


**Figure 1.** Left: Nuclear modification factor of prompt D$^0$ and J/$\psi$ from beauty-hadron decays [6]. Right: Double ratio ($\Lambda_c$/D) in most central Pb–Pb (0-10%) w.r.t. pp collisions [6]. Both figures refer to the central rapidity region.

The left-hand panel of Fig. 1 shows the $R_{\rm AA}$ of prompt D$^0$ and the one of J/$\psi$ from beauty-

hadron decays. The precision of the measurements will give the possibility to study the mass-dependence of the charm and beauty quarks energy loss in the QGP in a wide transverse momentum range. In the right-hand panel of the same figure the expected performance for the double ratio $\Lambda_c/D$ is shown and it is compared to theoretical predictions. The achievable precision will allow us to distinguish among several models providing a better understanding of the hadronization mechanisms.
The left panel of Fig. 2 shows the expected performance for the measurement of muons from charm-hadron decays down to $p_T^{\mu} = 1$ GeV/$c$ (similar results, not shown here, will be available for beauty decay muons down to $p_T^{\mu} = 3$ GeV/$c$). The small statistical uncertainties suggest that precise measurements of charm and beauty production cross sections will be available at forward rapidity down to very low $p_T$. In the right-hand side of Fig. 2 the non-prompt $D^0$ $R_{AA}$ at central rapidity is shown with the expected performance for the $R_{AA}$ of non-prompt J/$\psi$ at forward rapidity. Both measurements represent a key tool for accessing beauty production in Pb–Pb collisions down to zero $p_T$ and covering a large rapidity window.

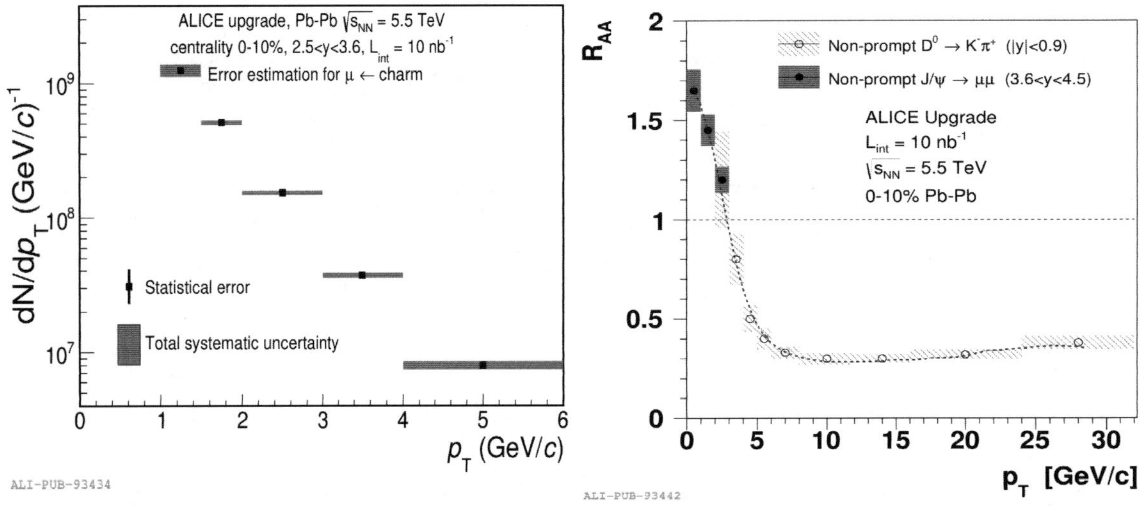


**Figure 2.** Left: Transverse momentum distribution of single muons from charm-hadrons at forward rapidity [7]. Right: $R_{AA}$ of beauty mesons via displaced J/$\psi$ at forward rapidity, shown together with the expected performance for the displaced $D^0$ channel at central rapidity [6, 7].

**Conclusions.** During the LS2 the ALICE detector will undergo a major upgrade. The installation of the upgraded ITS and the MFT within this programme will significantly extend the ALICE physics reach in the HF sector in a large rapidity window. In particular the precision for existing measurements in the charm and beauty sectors will significantly be improved down to $p_T = 0$ and new measurements will be available for the first time in Pb–Pb collisions.

## References.

[1] ALICE Coll. 2008 *JINST* **3** S088002
[2] ALICE Coll. 2013 *Upgrade of the ALICE Time Projection Chamber* CERN-LHCC-2013-020
[3] ALICE Coll. 2015 *Technical Design Report of the ALICE Online-Offline Computing System* CERN-LHCC-2015-006
[4] ALICE Coll. 2012 *J. Phys. G* **41** 087002
[5] ALICE Coll. 2013 *Addendum to the ALICE Letter of Intent* CERN-LHCC-2013-014
[6] ALICE Coll. 2013 *J. Phys. G* **41** 087001
[7] ALICE Coll. 2015 *Technical Design Report of the ALICE Muon Forward Tracker* CERN-LHCC-2015-001

# Equilibration of hadrons in HICs via Hagedorn States

**M Beitel, K Gallmeister and C Greiner**
Institut für Theoretische Physik, Goethe-Universität Frankfurt am Main,
Max-von-Laue-Str. 1, 60438 Frankfurt am Main, Germany

E-mail: gallmeister@th.physik.uni-frankfurt.de

**Abstract.** Hagedorn states (HS) are a tool to model the hadronization process which occurs in the phase transition region between the quark gluon plasma (QGP) and the hadron resonance gas (HRG). These states are believed to appear near the Hagedorn temperature $T_H$ which in our understanding equals the critical temperature $T_c$. A covariantly formulated bootstrap equation is solved to generate the zoo of these particles characterized baryon number $B$, strangeness $S$ and electric charge $Q$. These hadron-like resonances are characterized by being very massive and by not being limited to quantum numbers of known hadrons. All hadronic properties like masses, spectral functions etc. are taken from the hadronic transport model Ultra Relativistic Quantum Molecular Dynamics (UrQMD). Decay chains of single Hagedorn states provide a well description of experimentally observed multiplicity ratios of strange and multi-strange particles. In addition, the final energy spectra of resulting hadrons show a thermal-like distribution with the characteristic Hagedorn temperature $T_H$. Box calculations including these Hagedorn states are performed. Indeed, the time scales leading to equilibration of the system are drastically reduced down to $2 \ldots 5\,\mathrm{fm}/c$.

## 1. Introduction

Back in the 60's of the last century and before the advent of quantum chromodynamics (QCD) as the theory of strong interactions, R. Hagedorn [1] proposed the existence of a whole zoo of massive, unobserved hadronic resonance states. The spectrum of these particles, known as Hagedorn spectrum, exhibits the specific feature of being exponential in the infinite mass limit with the slope given by the so called Hagedorn temperature $T_H$. This temperature denotes the limiting temperature for hadronic matter since any partition function of a HRG with Hagedorn-like mass spectrum diverges as long as $T > T_H$. Above the Hagedorn temperature a new state of matter, namely the QGP, shall be realized. Thus, Hagedorn states provide a tool to understand the phase transition from HRG to QGP and back. The Hagedorn states are created in multi-particle collisions most abundantly near $T_H$ which in our understanding equals to the critical temperature $T_c$. Hagedorn states are color neutral objects which are allowed to have any quantum numbers as long as they are compatible with its mass. The appearance of Hagedorn states in multi-particle collisions and their role was already discussed in [2, 3, 4, 5, 6]. Their appearance near $T_c$ can explain, as shown in [4, 5, 6], the fast chemical equilibration of (multi-) strange baryons $B$ and their anti-particles $\bar{B}$ at Relativistic Heavy Ion Collider (RHIC) energies. The inclusion of Hagedorn states in a hadron resonance gas model provides also a lowering of the speed of sound, $c_s$ and of the shear viscosity over entropy density ratio $\eta/s$ at the phase transition and being in good agreement with lattice calculations [7, 8, 9, 10].

Journal of Physics: Conference Series **668** (2016) 012057 doi:10.1088/1742-6596/668/1/012057

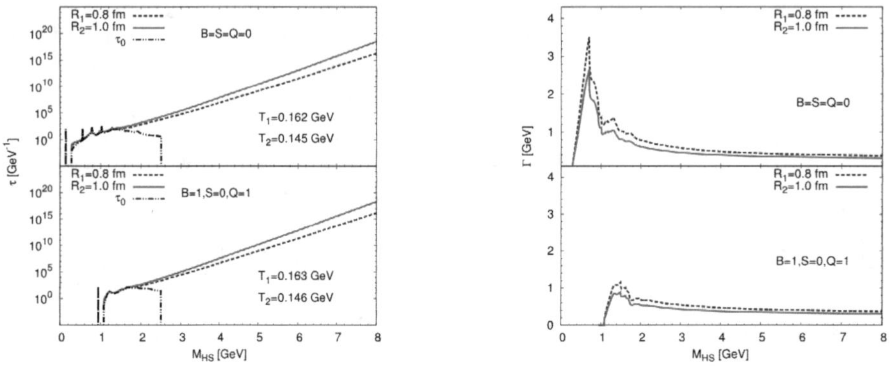


**Figure 1.** Hagedorn spectrum (left) and total decay width (right) for meson-like ($B = S = Q = 0$) (upper panel) and baryon-like ($B = 1, S = 0, Q = 1$) (lower panel) Hagedorn states.

The starting point of all calculations provided here is the postulate of the statistical bootstrap model (SBM) stating that fireballs consist of fireballs which in turn consist of fireballs etc. . As shown in [11] (cf. also [12]) the mathematical formulation of this postulate in the formulation of [13, 14] leads to the evolving bootstrap equation for the density of states,

$$\tau_{\vec{C}}(m) = \frac{R^3}{3\pi\, m} \sum_{\vec{C}_1,\vec{C}_2} \iint \mathrm{d}m_1 \mathrm{d}m_2\, m_1 \tau_{\vec{C}_1}(m_1)\, m_2 \tau_{\vec{C}_2}(m_2) \times\, p_{\mathrm{cm}}(m, m_1, m_2)\, \delta(\vec{C} - \vec{C}_1 - \vec{C}_2)\ , \tag{1}$$

where the conservation of the quantum numbers baryon number, strangeness and electrical charge $\vec{C} = (B, S, Q)$ is guaranteed. A major aspect was to restrict on Hagedorn states, which just consist of two constituents in order to allow for generation processes as $2 \to 1$ and decay as $1 \to 2$. This is some necessary constraint in order to allow for a microscopical calculation in a transport model simulation based on cross sections. The above set of integral equations of Volterra type is solved numerically by disretizing the mass range and consecutively stepping to higher mass bins. The zeroth order input are the known hadrons according their spectral functions as implemented in the transport model UrQMD [15]. The radius $R$ is the only free parameter of the model. Choosing reasonable values as $R = 0.8\,\mathrm{fm}$ ($1.0\,\mathrm{fm}$) yield Hagedorn temperatures, i.e. the slopes of the exponentially rising Hagedorn spectra $\tau_{\vec{C}}(m) \sim m^a \exp(m/T_H)$ as $T_H = 162\,\mathrm{MeV}$ ($145\,\mathrm{MeV}$) nearly independent of the quantum numbers, as indicated in fig. 1.

In order to treat the decay and the production of Hagedorn states, the decay width $\Gamma$ and the production cross section $\sigma = \pi R^2$ are connected via detailed balance as

$$\Gamma_{\vec{C}}(m) = \frac{\sigma}{2\pi^2\, \tau_{\vec{C}}(m)} \sum_{\vec{C}_1,\vec{C}_2} \iint \mathrm{d}m_1 \mathrm{d}m_2\, \tau_{\vec{C}_1}(m_1)\, \tau_{\vec{C}_2}(m_2) \times\, p^2_{\mathrm{cm}}(m, m_1, m_2)\, \delta(\vec{C} - \vec{C}_1 - \vec{C}_2)\ , \tag{2}$$

which represents the second major equation of the model. Corresponding Hagedorn spectra and their decay width are shown in fig. 1. One observes, that the slope decreases with increasing volume of the Hagedorn state. Correspondingly also the width decreases. Nevertheless, for larger masses, the width is nearly constant in the range of some hundreds of MeV, and also stays finite in the infinite mass limit.

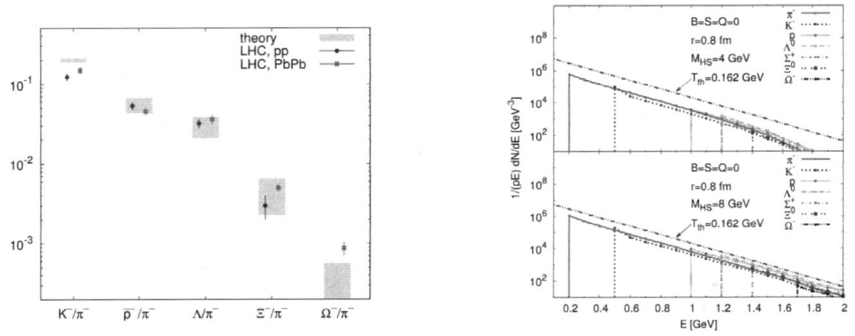


**Figure 2.** Left: Comparison of particle multiplicity ratios from theory vs. p-p at $\sqrt{s_{NN}} = 0.9\,\mathrm{TeV}$ [16] and Pb-Pb at $\sqrt{s_{NN}} = 2.76\,\mathrm{TeV}$ [17, 18, 19], both from ALICE at LHC. The bands show the range of calculations from cascade decay of a single charge neutral Hagedorn state with radius $R = 0.8\,\mathrm{fm}$ and initial mass $M_{HS} = 4\,\mathrm{GeV}$ and $M_{HS} = 8\,\mathrm{GeV}$. Right: Energy spectra of hadrons stemming from cascade decay of charge neutral Hagedorn state with radius $R = 0.8\,\mathrm{fm}$ and initial mass $M_{HS} = 4\,\mathrm{GeV}$ and $M_{HS} = 8\,\mathrm{GeV}$.

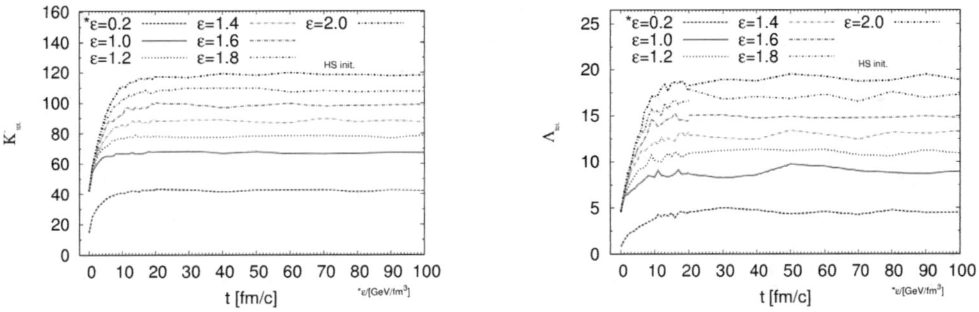


**Figure 3.** The time dependence of $K^-$ (left) and $\Lambda$ hadrons in a box calculation, initialized with Hagedorn states, for different initial energy densities as indicated in the plot.

Having the total decay width and the corresponding composition, one is able to calculate the branching ratios for the decay into light Hagedorn states and/or hadrons. Performing such cascading decay simulations, the resulting hadron multiplicity coincide very well with experimentally observed values[11]. This is visualized in fig. 2, where the multiplicity ratios for strange and multi-strange particles are compared with experimental values measured at LHC. The most striking result is the fact, that the energy spectra of these decay hadrons *look* thermal, i.e. the functional behavior is exponential, as shown in fig. 2. The slope parameter coincides with the underlying Hagedorn temperature.

All these ingredients were implemented into UrQMD [15], which can handle the creation and the decay of Hagedorn states and their component dynamically. This enables us to perform box calculations and study the time evolution and the equilibration of the system. The model is able to handle different initialization scenarios, as e.g. the initialization with only nucleons (or pions) with a given particle and energy density. While this may stand for some 'bottom-up' scenario, alternatively one may start with a given Hagedorn state distribution and look into the resulting hadron distributions. This indicates a 'top-down' scenario, where the Hagedorn states may stem from some preceding deconfined phase.

Fig. 3 shows the time dependence of kaons and lambdas for box calculations, where only Hagedorn states according to different energy densities were initialized. The given multiplcities

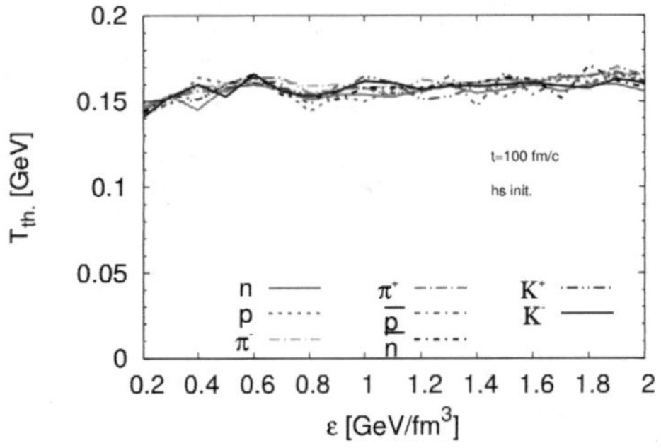


**Figure 4.** The slope parameter of the energy distribution of different particle species for different initialization energy densities.

at some time are the resulting muliplicities, if all particles would decay down to the stable remnants. One indeed observes very short time scales for the equilibration: While the kaons are almost equilibrated after $2\,\mathrm{fm}/c$, the number of lambdas reaches values close to the maximum value already after $5\,\mathrm{fm}/c$. Not shown here, but the last statement holds also for nucleons, while pions are even faster equilibrated than the kaons. The multiplicity ratios are almost stable for all energy densities, except for the smallest considered one, i.e. for $\varepsilon = 0.2\,\mathrm{GeV/fm}^3$. In addition we find, that the energy spectra of the hadrons again follow an exponential shape, with slope parameters very close to the Hagedorn temperature $T_H = 162\,\mathrm{MeV}$, see fig. 4.

We conclude, that Hagedorn states are a valuable tool for the understanding of the phase transition between a deconfined and a hadronic phase. While already the decay particles of one single Hagedorn state *look* thermal with the Hagedorn temperature as the only scale, a collective system of Hagedorn states leads to small equilibration times of hadrons in the order of few $\mathrm{fm}/c$. These remarkable features may also open the door for alternative descriptions of heavy ion collisions, where directly from a pure gluonic phase without quarks a transition to the confined phase is possible [20].

[1] Hagedorn R 1965 *Nuovo Cim. Suppl.* **3** 147–186
[2] Greiner C and Leupold S 2001 *J. Phys. G* **27** L95–L102
[3] Greiner C, Koch-Steinheimer P, Liu F M, Shovkovy I A and Stoecker H 2005 *J. Phys. G* **31** S725–S732
[4] Noronha-Hostler J, Greiner C and Shovkovy I 2008 *Phys. Rev. Lett.* **100** 252301
[5] Noronha-Hostler J, Greiner C and Shovkovy I 2010 *J. Phys. G* **37** 094017
[6] Noronha-Hostler J, Beitel M, Greiner C and Shovkovy I 2010 *Phys. Rev. C* **81** 054909
[7] Noronha-Hostler J, Noronha J and Greiner C 2009 *Phys. Rev. Lett.* **103** 172302
[8] Majumder A and Muller B 2010 *Phys. Rev. Lett.* **105** 252002
[9] Noronha-Hostler J, Noronha J and Greiner C 2012 *Phys. Rev. C* **86** 024913
[10] Jakovac A 2013 *Phys. Rev. D* **88** 065012
[11] Beitel M, Gallmeister K and Greiner C 2014 *Phys. Rev. C* **90** 045203
[12] Yellin J 1973 *Nucl. Phys. B* **52** 583–594
[13] Frautschi S C 1971 *Phys. Rev. D* **3** 2821–2834
[14] Hamer C and Frautschi S C 1971 *Phys. Rev. D* **4** 2125–2137
[15] Bass S A *et al.* 1998 *Prog. Part. Nucl. Phys.* **41** 255–369
[16] Aamodt K *et al.* (ALICE) 2011 *Eur. Phys. J. C* **71** 1594
[17] Abelev B *et al.* (ALICE Collaboration) 2013 *Phys. Rev. C* **88** 044910
[18] Abelev B B *et al.* (ALICE Collaboration) 2013 *Phys. Rev. Lett.* **111** 222301
[19] Abelev B B *et al.* (ALICE Collaboration) 2014 *Phys. Lett. B* **728** 216–227
[20] Stoecker H *et al.* 2015 (*Preprint* 1509.00160) and these proceedings

Journal of Physics: Conference Series **668** (2016) 012058 doi:10.1088/1742-6596/668/1/012058

# Soft particle production and study of collective phenomena with the ALICE detector at the LHC

**B Guerzoni for the ALICE Collaboration**

INFN and University of Bologna, Italy

E-mail: barbara.guerzoni@bo.infn.it

**Abstract.** After the successful completion of the first LHC run, a comprehensive set of light flavour hadron measurements at midrapidity in pp, p-Pb and Pb–Pb collisions is available for detailed comparisons with soft particle-production models. The observation of a medium in local thermal equilibrium in nucleus-nucleus collisions is supported by the success of thermal and hydrodynamic models in the description of hadron yields and spectral shapes, respectively. However, the presence of collective phenomena in small systems created in pp and p-Pb collisions is still debated. In this paper, similarities and differences between large and small collision systems with respect to a mass dependent evolution of the spectral shapes will be highlighted. In addition, the experimental significance of the deviations of particle yields observed between data and thermal model expectations will be discussed.

## 1. Introduction

The ALICE Collaboration has shown that the transverse momentum ($p_{\mathrm{T}}$) spectra of pions, kaons, and protons measured in Pb–Pb collisions at LHC energy are harder than the equivalent spectra measured in Au-Au collisions at RHIC energy [1, 2, 3]. The spectra get progressively harder for more central events and are also observed to be harder for more massive particles. This effect can be interpreted in terms of hydrodynamic evolution of the medium produced in the collision that induces a stronger radial flow when the center of mass energy and/or the centrality of the collision increase which pushes heavier particles to higher $p_{\mathrm{T}}$. A similar evolution of the spectra with multiplicity is also observed in p-Pb collisions at $\sqrt{s}$= 5.02 TeV and pp collisions at $\sqrt{s}$= 7 TeV. In the next section, the pion, kaon and proton spectra produced in pp and p–Pb collisions will be compared with the same results measured in Pb–Pb collisions. The similarities that are suggestive of collective behaviour also in small colliding systems will be analyzed. Afterwards, the comparison of the integrated yields of identified particles with predictions from thermal models, fundamental to characterize the medium at the chemical freeze-out, will be discussed.

## 2. Collective phenomena in pp, p-Pb and Pb–Pb collisions

In Fig. 1 the $p_{\mathrm{T}}$ spectra of pions, kaon and protons produced in Pb–Pb collisions at $\sqrt{s}$= 2.76 TeV in 10 centrality classes are reported [1]. It is evident that the shape of the spectra evolves both with the centrality of the collision (becoming harder with increasing centrality) and with the particle mass (the heavier the particle mass, the flatter the spectrum). This effect can be interpreted in terms of hydrodynamic models in which the collective radial expansion becomes stronger with the centrality of the collisions due to the higher particle density. As a consequence, the hardening of the spectra is more pronounced for heavier particles. A similar evolution of the

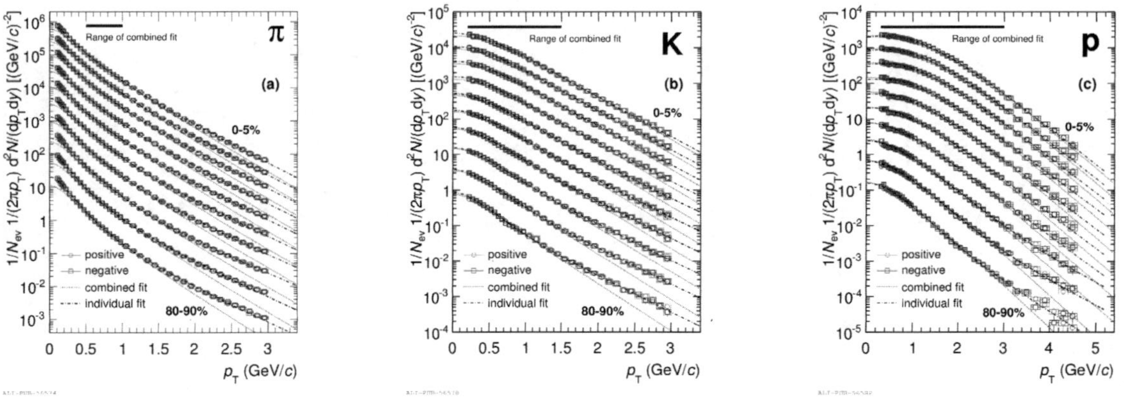


**Figure 1.** $p_T$ spectra of positive (circles) and negative (squares) pions, kaons and protons as a function of centrality [1]. Dashed curves: blast wave fits to individual particles. Dotted curves: combined blast wave fits.

spectra with the multiplicity of the particles produced in the collision and with the particle mass has been observed also in p-Pb collisions at $\sqrt{s}$= 5.02 TeV [4] and in pp collisions at $\sqrt{s}$= 7 TeV [5]. In Fig. 2 it can be observed that the $(p+\overline{p})/(\pi^+ + \pi^-)$ ratio as a function of $p_T$ in high (red) and low (purple) multiplicity events shows a similar behaviour in the three colliding systems even if the absolute values of the ratios are different. As multiplicity increases, this ratio decreases for low transverse momenta and increases at intermediate transverse momentum. Since in Pb–Pb collisions this behaviour is explained in terms of hydrodynamic expansion of the medium (radial flow), this similarity could suggest that also in pp and p-Pb collisions a collective behaviour may be present. To have a more quantitative comparison of the evolution of the spectra with

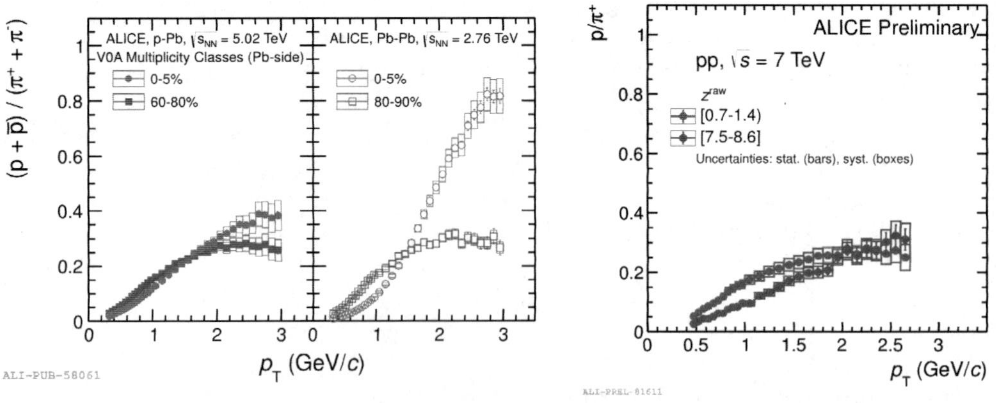


**Figure 2.** $p_T$ dependence of $(p+\overline{p})/(\pi^+ + \pi^-)$ ratio in high (red) and low (purple) multiplicity events in p-Pb (left), Pb–Pb (center) and pp (right) collisions.

the particle multiplicity in the three colliding systems, a simultaneous fit of pions, kaons and proton spectra with a blast wave function [6] in each multiplicity class has been performed. The blast wave function is a simplified hydrodynamic model which assumes a locally thermalized medium, expanding collectively with a common velocity field and undergoing an instantaneous common freeze-out. It can be used to systematically compare the spectral shapes in different colliding systems. It has three free parameters: the kinetic freeze-out temperature $T_{kin}$, the mean radial expansion velocity $\langle \beta_T \rangle$ and the velocity profile $n$. In the left panel of Fig. 3 the results of the simultaneous fits to pions, kaons and protons spectra in pp, p-Pb, Pb–Pb collisions and PYTHIA8 simulations of pp collisions for different multiplicity classes are reported. It can be seen that the evolution with the multiplicity of the pp and p-Pb parameters is similar to the one in Pb–Pb collisions with $\langle \beta_T \rangle$ increasing and $T_{kin}$ decreasing with multiplicity (from left to right). This observation seems to be consistent with the presence of radial flow also in pp and p-Pb collisions. In Fig. 3 (left) it is also noteworthy that the same evolution of the parameters can be obtained fitting the $\sqrt{s}$= 7 TeV pp spectra simulated with PYTHIA8, a model not including any collective system expansion, once the Color Reconnection (CR) mechanism is activated. It seems hence that other processes not related to hydrodynamic collectivity, like the Color Reconnection, could be responsible for the observed results in small colliding system and mimic the effects of radial flow [7]. Also the evolution of the $n$-$\langle \beta_T \rangle$ correlation with the multiplicity is similar in the three colliding systems with the increse of $\langle \beta_T \rangle$ and the decrease of $n$ with the multiplicity.

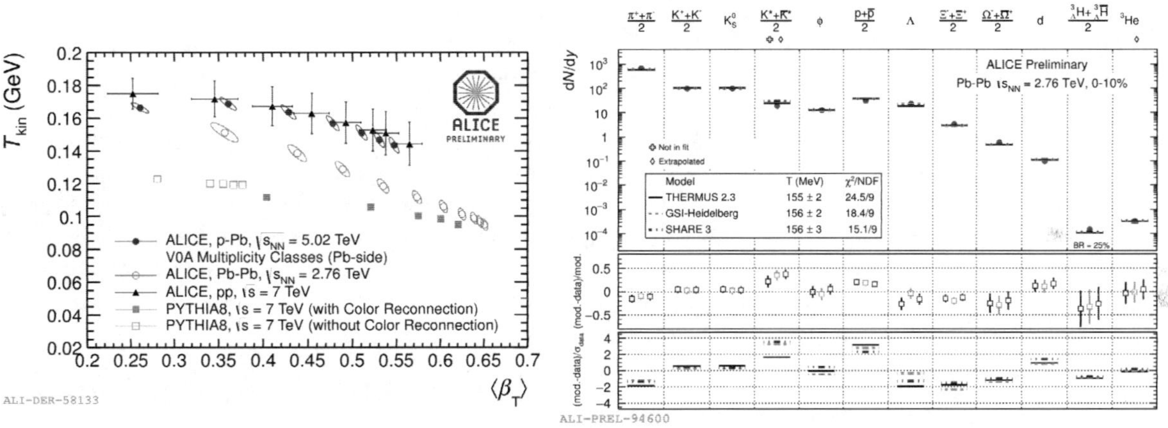


**Figure 3.** (left) Results of blast-wave fits of pp, p-Pb and Pb–Pb spectra compared to MC simulations from PYTHIA8 with and without Color Reconnection for increasing centrality classes from left to right. (right) Particle yields measured in central Pb–Pb collisions compared to results of fit with THERMUS, GSI and SHARE thermal models.

## 3. Thermal models in Pb–Pb collisions

In the right panel of Fig. 3, the integrated yields ($\mathrm{d}N/\mathrm{d}y$) of identified particles measured in central Pb–Pb collisions are reported. They are interpreted in terms of three thermal models which describe the properties of the system at chemical freeze-out. The thermal models have three main parameters: the baryochemical potential $\mu_{\mathrm{B}}$ that at the LHC energies is fixed to zero, the chemical freeze-out temperature $T_{\mathrm{ch}}$ and the volume $V$. Some models introduce three additional parameters $\gamma_{s,c,q}$ that take into account deviation from equilibrium production for strange, charm or light quarks. Equilibrium thermal models ($\gamma_s = \gamma_c = \gamma_q = 1$) were able to reproduce the particle yields measured at RHIC energies but they were found to overestimate the proton production by a factor 1.5 once their parameters were extrapolated to LHC energy. The results of the fit to the data of three different thermal models implementing full equilibrium, Thermus 2.3 [8], GSI-Heidelberg [9] and SHARE 3 [10], are reported in Fig. 3 (the K* resonances are not included in the fit). The disagreement of the protons with respect to the equilibrium model expectations is still present and is the main contributor to the $\chi^2/ndf \sim 2$. Furthermore, the value of the temperature obtained from the fit by the three models are compatible ($\sim$ 156 MeV) and lower than the value predicted from the extrapolation of RHIC data ($\sim$ 164 MeV).

Many hypotheses have been formulated to explain the deviations of particle yields from the thermal models. At lower energies the particle abundances were described by thermal models assuming thermal and chemical equilibrium, hence assuming the modification of particle ratios in the hadronic phase to be negligible. Deviations from thermal predictions can be explained assuming that final state interactions in the hadronic phase are not negligible and affects baryons and mesons differently [11, 12]. One process that might be important is baryon annihilation, in line with the observation of the lower value of the protons compared to the model predictions. A non-equilibrium configuration and a quark-gluon plasma undergoing a sudden hadronization without rescattering can be represented by the SHARE model [10] (in Fig. 3 SHARE was implemented in its equilibrium configuration with $\gamma_s = \gamma_c = \gamma_q = 1$). In such a configuration it is possible to fit all hadrons except nuclei with $\chi^2/ndf \sim 1$. Another possible explanation is that, since a single freeze-out temperature cannot describe the data, the transition temperature could be flavour-dependent [13].

## 4. Conclusions

The ALICE Collaboration has measured the $p_{\mathrm{T}}$ spectra of identified particles systematically in different colliding systems and for different centralities. In the low $p_{\mathrm{T}}$ region the spectra get harder with the increase of the particle multiplicity and with the particle mass, not only in Pb–Pb, but also in p-Pb and pp collisions. This common behaviour seems to indicate that also in pp and p-Pb collisions a collective radial expansion is present. A similar spectra evolution can be observed also in events generated with PYTHIA8 when the Color Reconnection is included. Therefore, further studies are needed to determine the nature of collective behaviour, if indeed present, in these smaller colliding systems. Finally, it was shown that equilibrium thermal models successfully describe the integrated yields of the measured particle species, though they everestimate proton production in central Pb–Pb collisions.

## References

[1] Abelev B et al (ALICE Collaboration) 2013 *Phys. Rev.* B **88** 044910
[2] Abelev B I et al (STAR Collaboration) 2009 *Phys. Rev.* C **79** 034909
[3] Adler S S et al (PHENIX Collaboration) 2004 *Phys. Rev.* C **69** 034909
[4] Abelev B et al (ALICE Collaboration) 2014 *Phys. Lett.* B **728** 25-38
[5] Andrei C 2014 *Nucl. Phys.* A **931** 888-892
[6] Schnedermann et al 1993 *Phys. Rev.* C **48** 2462
[7] Ortiz A et al 2013 *Phys. Rev. Lett.* **111** 042001
[8] Wheaton S et al 2009 *Comp. Phys. Commun.* **180** 84
[9] Andronic A et al 2009 *Phys. Lett.* B **673** 142
[10] Petran M et al 2014 *Comput. Phys. Commun.* **185** 2056-2079
[11] Becattini F et al 2014 *Phys. Rev.* C **90** 054907
[12] Becattini F et al 2013 *Phys. Rev. Lett.* **111** 082302
[13] Bellwied R et al 2013 *Phys. Rev. Lett.* **111** 202302

Journal of Physics: Conference Series **668** (2016) 012059 doi:10.1088/1742-6596/668/1/012059

# Dileptons in a coarse-grained transport approach

**Hendrik van Hees[1,2], Stephan Endres[1,2], Janus Weil[2] and Marcus Bleicher[1,2]**

[1]Institut für Theoretische Physik, Universität Frankfurt, Max-von-Laue-Straße 1, 60438 Frankfurt, Germany

[2]Frankfurt Institute for Advanced Studies, Ruth-Moufang-Straße 1, 60438 Frankfurt, Germany

E-mail: endres@th.physik.uni-frankfurt.de

**Abstract.** We calculate dilepton spectra in heavy-ion collisions using a coarse-graining approach to the simulation of the created medium with the UrQMD transport model. This enables the use of dilepton-production rates evaluated in equilibrium quantum-field theory at finite temperatures and chemical potentials.

Due to their penetrating nature, the invariant-mass and transverse-momentum spectra of dileptons are considered unique probes for the in-medium properties of hadrons in strongly interacting hot and dense matter, created in heavy-ion collisions [1, 2]. The dilepton-production rates are proportional to the electromagnetic current-correlation function of the source [3] and thus provide insight into the corresponding excitations carrying the quantum numbers of the electromagnetic current. Particularly from the restoration of chiral symmetry one expects substantial modifications of the spectral shape of this current-correlation function. Particularly one hopes to gain insight into the microscopic realization of the chiral-symmetry restoration, e.g., whether it is due to dropping in-medium hadron masses as predicted by the Brown-Rho-scaling conjecture or models implementing the vector manifestation of chiral symmetry or substantial broadening of the pertinent light-vector-meson spectral shapes due to hadronic interactions within the medium ("melting-resonance scenario") as predicted by quantum-field-theoretical effective hadronic many-body theory (for a review see [1]).

Since the dileptons are produced and emitted undisturbed by final-state interactions during the entire evolution of the hot and dense medium, first a good understanding of all relevant microscopic sources of electromagnetic radiation in this medium is important.

For heavy-ion collisions at ultrarelativistic energies, in the early hot stage the relevant degrees of the medium are quarks and gluons forming a quark-gluon plasma (QGP) close to local thermal equilibrium, as is inferred from the successful description of the hadronic bulk observables like $p_T$ spectra and angular anisotropies $v_n$ by hydrodynamical models for the evolving medium. Here the main source of dileptons is the quark-antiquark annihilation $q + \overline{q} \to \ell^+ + \ell^-$ (with $\ell = \mathrm{e}, \mu$). In the later stages of the fireball evolution (or, in the case of heavy-ion collisions at lower energies, the entire evolution) the medium is described as a hot hadronic gas. In the low invariant-mass range, $2m_\ell \leq M \lesssim 1\text{GeV}$ the dilepton sources are thus hadronic and phenomenologically well described within a vector-meson dominance (VMD) model [4], i.e., the assumption that the electromagnetic current of the hadrons is proportional to the light-vector meson field. This also implies that the electromagnetic transition form factors for the various Dalitz decays of hadron resonances contributing substantially to the dilepton yield in the low-mass tail are describable

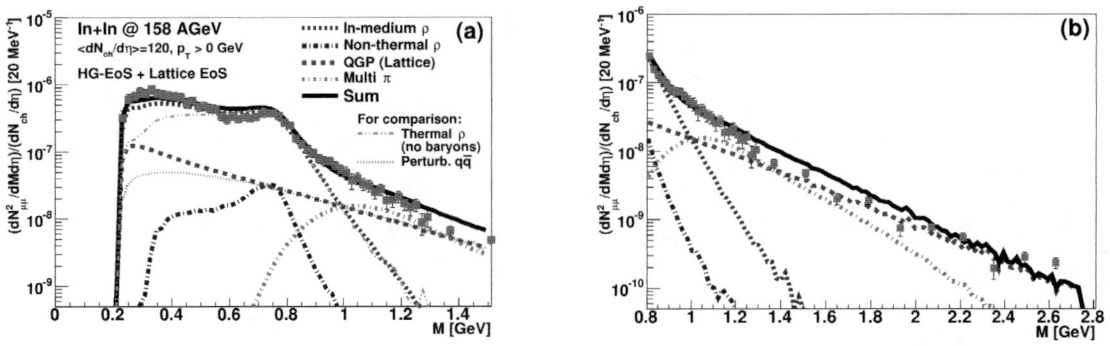


**Figure 1.** The dimuon spectra in $158A$ GeV In+In collisions at the CERN SPS in (a) the low-mass (LMR) and (b) the intermediate-mass region (IMR) in comparison to the data from the NA60 Collaboration [5, 6, 7]. In the LMR the dilepton yield is dominated by the decays of the in-medium $\rho$ mesons (dotted blue line) and in the IMR by QGP and multi-pion hadronic processes. The comparison with a dilepton rate based on a thermal-$\rho$ contribution, neglecting the effect of baryon couplings, shows the importance of the these contributions for the tremendous broadening of the $\rho$-spectral function as well as the yield in the very-low-mass range close to the two-muon threshold. Microscopically these contributions originate from Dalitz decays of various baryon resonances.

within the VMD model. In the intermediate-mass range also "multi-pion processes" become relevant.

All these microscopic processes have to be evaluated in a hot and dense medium, which is at present available in detailed many-body quantum-field theoretical calculations only for a medium in thermal equilibrium, with some chemical off-equilibrium taken into account in terms of effective baryon densities and pion- as well as kaon-chemical potentials [8, 9]. It has been quite well established that the observed excess of low-mass dilepton production in heavy-ion collisions as compared to the expectations from the "hadronic cocktail" (i.e., from a simple scaling of the findings in pp collisions by the number of nucleon collisions in the nuclear reaction) can be understood by effective hadronic models that predict a tremendous broadening of the light-vector mesons' spectral shape with small mass shifts [10, 11, 12, 13].

On the other hand, one also has to describe the evolution of the hot and dense medium in a reliable way. This is usually done either by employing macroscopic descriptions, reaching from quite simple thermal-fireball parametrizations ("blast-wave fits") [14] to detailed ideal or viscous hydrodynamics simulations [15, 16, 17, 18, 12, 19], or microscopic transport simulations [20, 21, 22, 23, 24, 25, 26, 27, 28, 29].

In the latter approach it is challenging to fully implement the medium modifications of the partonic and hadronic dilepton sources. In our approach this problem is solved by mapping the kinetic description of the fireball evolution via Monte-Carlo transport simulations to a state close to local thermal equilibrium. In this work, an ensemble of collision events simulated by the UrQMD model is used to calculate a coarse-grained energy-momentum tensor and net-baryon-number current on a space-time grid. Then, using the Eckart definition of the local rest frame, a temperature and baryo-chemical potential can be extracted via an equation of state (EoS), making use of a description of the energy-momentum tensor in terms of an anisotropic-fluid ansatz [30], where the pressure anisotropy, particularly relevant in the early stages of the fireball evolution, is taken into account. For the EoS we match a parametrization of lattice-QCD results [31] with a hadron-resonance gas EoS [32], which match very well in the region of the

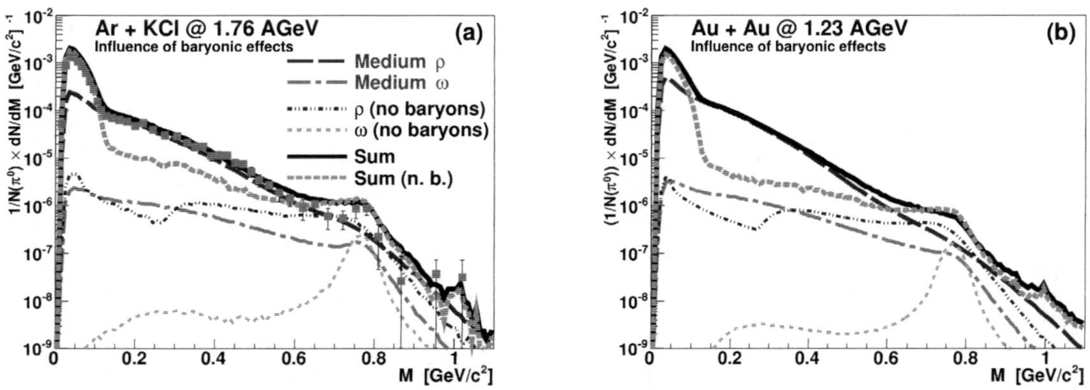


**Figure 2.** (Color online) (a) Dielectron invariant-mass spectrum for Ar+KCl collisions at $E_{\text{lab}} = 1.76A\,\text{GeV}$ and (b) for Au+Au at $E_{\text{lab}} = 1.23A\,\text{GeV}$. The results are normalized to the average total number of $\pi^0$ per event and shown within the HADES acceptance. The results for Ar+KCl are compared to the data from the HADES Collaboration [35]. The comparison of the dilepton spectra between using the full model and omitting the contributions of direct baryon couplings to the $\rho$ and $\omega$ self-energies underlines the prevalence of the corresponding baryon-resonance Dalitz-decay processes processes for the in-medium modifications of these vector mesons, shifting substantial spectral strength to the low-invariant-mass region.

parton-hadron transition around the critical temperature of $T_c \simeq 170$ MeV.

In recent work [33, 34], we have combined this coarse-graining method for the bulk-medium evolution of the strongly interacting fluid via the UrQMD transport model with state-of-the art in-medium dilepton-production rates [8, 9] in order to evaluate the dilepton invariant-mass and $p_T$ spectra in $158A$ GeV In+In collisions as measured by the NA60 Collaboration at the CERN SPS and in $1.76A$ GeV Ar+KCl and $1.23A$ GeV Au+Au collisions as measured by the HADES collaboration at the GSI SIS.

As exemplified by the comparison of the so calculated dimuon invariant-mass spectra in $158A$ GeV In+In Collisions with the NA60 data (Fig. 1) and corresponding dielectron invariant-mass spectra for $1.76A$ GeV Ar+KCl collisions with data from the HADES collaboration (Fig. 2, left panel), the model successfully describes the data. In the right panel of Fig. 2 we additionally provide our prediction for the $1.23A$ GeV Au+Au collisions as also measured by the HADES collaboration.

Both figures underline the importance of the interaction of the light vector mesons with baryons in the medium [in the present model [1] N(938), Δ(1232), N(1440), N(1520), Δ(1620), Δ(1700), N(1720), Δ(1900), N(2000)]. As expected, these processes are prevalent at the low GSI-SIS energy with high net-baryon densities but also relevant at CERN-SPS energies and even at higher RHIC energies. This is due to the fact that the strong interaction is CP invariant and thus not the net-baryon density, $N_B - N_{\overline{B}}$, but the total baryon-antibaryon density, $N_B + N_{\overline{B}}$, is relevant for the in-medium modifications of the light-vector mesons' spectral distributions.

In the near future we also plan to apply the model to dilepton production at RHIC energies as well as photon production in heavy-ion collisions.

## Acknowledgments

We thank Ralf Rapp for providing the parameterization of the spectral function and many fruitful discussions. This work was supported by BMBF, HIC for FAIR and H-QM.

Journal of Physics: Conference Series **668** (2016) 012059 doi:10.1088/1742-6596/668/1/012059

## References


[1] Rapp R and Wambach J 2000 *Adv. Nucl. Phys.* **25** 1
[2] Rapp R, Wambach J and van Hees H 2009 *Landolt-Börnstein* **I/23** 4–1
[3] McLerran L D and Toimela T 1985 *Phys. Rev. D* **31** 545
[4] Sakurai J J 1960 *Ann. Phys. (NY)* **11** 1–48
[5] Arnaldi R *et al.* (NA60 Collaboration) 2009 *Eur. Phys. J. C* **59** 607–623
[6] Specht H J (NA60 Collaboration) 2010 *AIP Conf.Proc.* **1322** 1–10 URL `http://dx.doi.org/10.1063/1.3541982`
[7] Arnaldi R *et al.* (NA60 Collaboration) 2009 *Eur. Phys. J. C* **61** 711–720
[8] Rapp R and Wambach J 1999 *Eur. Phys. J. A* **6** 415–420
[9] Gale C and Rapp R 1999 *Phys. Rev. C* **60** 024903
[10] van Hees H and Rapp R 2006 *Phys. Rev. Lett.* **97** 102301
[11] van Hees H and Rapp R 2008 *Nucl. Phys. A* **806** 339
[12] Dusling K, Teaney D and Zahed I 2007 *Phys. Rev. C* **75** 024908
[13] Ruppert J, Gale C, Renk T, Lichard P and Kapusta J I 2008 *Phys. Rev. Lett.* **100** 162301
[14] Rapp R 2013 *Adv. High Energy Phys.* **2013** 148253
[15] Teaney D, Lauret J and Shuryak E 2001 (*Preprint* `arXiv: nucl-th/0110037`)
[16] Hirano T and Tsuda K 2002 *Phys. Rev. C* **66** 054905
[17] Kolb P F and Heinz U 2003 Published in R. C. Hwa, X.-N. Wang (Ed.), Quark Gluon Plasma 3, World Scientific
[18] Nonaka C and Bass S A 2007 *Phys. Rev. C* **75** 014902
[19] Vujanovic G, Young C, Schenke B, Rapp R, Jeon S *et al.* 2014 *Phys. Rev. C* **89** 034904
[20] Cassing W, Bratkovskaya E, Rapp R and Wambach J 1998 *Phys. Rev. C* **57** 916–921
[21] Schenke B and Greiner C 2006 *Phys. Rev. C* **73** 034909
[22] Schenke B and Greiner C 2007 *Phys. Rev. Lett.* **98** 022301
[23] Schenke B and Greiner C 2007 *Nucl. Phys. A* **785** 170–173
[24] Bratkovskaya E, Cassing W and Linnyk O 2009 *Phys. Lett. B* **670** 428–433
[25] Cassing W and Bratkovskaya E 2009 *Nucl. Phys. A* **831** 215–242
[26] Barz H, Kampfer B, Wolf G and Zetenyi M 2010 *Open Nucl. Part. Phys. J.* **3** 1–15
[27] Linnyk O, Bratkovskaya E, Ozvenchuk V, Cassing W and Ko C 2011 *Phys. Rev. C* **84** 054917
[28] Weil J, van Hees H and Mosel U 2012 *Eur. Phys. J. A* **48** 111
[29] Weil J, Mosel U and Metag V 2013 *Phys. Lett. B* **723** 120–125
[30] Florkowski W and Ryblewski R 2011 *Phys. Rev. C* **83** 034907
[31] He M, Fries R J and Rapp R 2012 *Phys. Rev. C* **85** 044911
[32] Zschiesche D, Schramm S, Schaffner-Bielich J, Stoecker H and Greiner W 2002 *Phys. Lett. B* **547** 7–14
[33] Endres S, van Hees H, Weil J and Bleicher M 2015 *Phys. Rev. C* **91** 054911
[34] Endres S, van Hees H, Weil J and Bleicher M 2015 *Phys. Rev. C* **92** 014911
[35] Agakishiev G *et al.* (HADES Collaboration) 2011 *Phys. Rev. C* **84** 014902

# Fluctuations in Strong-Coupling Heavy Quark Energy Loss

**W. A. Horowitz**

Department of Physics, University of Cape Town, Private Bag X3, Rondebosch 7701, South Africa

E-mail: wa.horowitz@uct.ac.za

**Abstract.** We present recent and new results for observables related to heavy quarks from a strong-coupling energy loss model that includes fluctuations in the energy loss. We find that the inclusion of these next-to-leading order fluctuations yields results consistent with all measured data. Correlation measurements and bottom decays to electrons are potential future measurements that can distinguish between weak- and strong-coupling energy loss models.

## 1. Introduction

In high-energy nuclear physics, we seek a self-consistent theoretical framework that provides quantitative, falsifiable predictions that describe experimental data in order to extract quantitatively the properties of the region of the phase diagram of quantum chromodynamics (QCD) accessible to modern collider experiments. The mathematical tools used to understand this quark-gluon plasma (QGP) have broad applications in condensed matter physics at applied mathematics; the properties of the QGP may have important implications for the large scale structure of the universe.

Two obvious extreme assumptions about the properties of the QGP, for which it seems one has analytic theoretical tools to analyze, are that its constituents are slightly thermally modified versions of the quarks and gluons in the QCD Lagrangian, which can be probed using perturbative QCD (pQCD), or that the plasma is strongly-coupled, which perhaps can be understood with the methods of AdS/CFT.

AdS/CFT has had a number of successes in describing several aspects of the QGP: 1) the lattice calculation of the energy density as a function of temperature [1, 2], 2) the rapid hydrodynamization of the QGP [3, 4], and 3) the extremely small viscosity to entropy density ratio [4, 5, 6].

We aim in this proceedings to extend the application of AdS/CFT to heavy quark observables.

## 2. Heavy Quark Energy Loss in AdS/CFT

It was shown in [7] that the three-momentum $p^i$ of an on-shell heavy quark moving at constant velocity in a strongly-coupled thermal bath of $\mathcal{N} = 4$ SYM plasma evolves as:

$$\frac{dp_i}{dt} = -\mu\, p_i + F_i^L + F_i^T, \tag{1}$$

Journal of Physics: Conference Series **668** (2016) 012060 doi:10.1088/1742-6596/668/1/012060

where the drag coefficient [8, 9, 10] of a heavy quark of mass $M_Q$ in a plasma of temperature $T$ with 't Hooft coupling $\lambda$ is $\mu = \pi\sqrt{\lambda}T^2/2M_Q$. and the fluctuating momentum kicks are correlated as

$$\langle F_i^L(t_1)F_j^L(t_1)\rangle = \kappa_L \hat{p}_i\hat{p}_j g(t_2 - t_1) \quad (2)$$

$$\langle F_i^T(t_1)F_j^T(t_1)\rangle = \kappa_T(\delta_{ij} - \hat{p}_i\hat{p}_j) g(t_2 - t_1), \quad (3)$$

where $\hat{p}_i = p_i/|\vec{p}|$, $\kappa_T = \pi\sqrt{\lambda}T^3\gamma^{1/2}$, $\kappa_L = \gamma^2\kappa_T$, and $g$ is a function known only numerically.

One may make a back of the envelope estimate for the momentum at which the fluctuations above become important [11], finding $\gamma \lesssim \gamma_{crit}^{fluc} = \frac{M_Q^2}{4T^2}$. There is a well-known speed limit for the heavy quark drag setup in which the derivation of Eq. (1) was performed [7, 12]: $\gamma < \gamma_{crit}^{sl} = \left(1 + \frac{2M_Q}{\sqrt{\lambda}T}\right)^2 \sim \frac{4M_Q^2}{\lambda T^2}$. While the second speed limit is parametrically smaller than the first, the finite size of the coupling and mass lead one to conclude that the fluctuations must be taken into account, even at very small momenta (and that the latter speed limit is at extremely large momenta $\mathcal{O}(100$ GeV$)$ for bottom quarks).

The implementation of the fluctuations in Eq. (1) are nontrivial for two reasons: first, the noise is *colored* and second, the fluctuations are *multiplicative*. For small enough momenta, both these issues can be overcome; see [11].

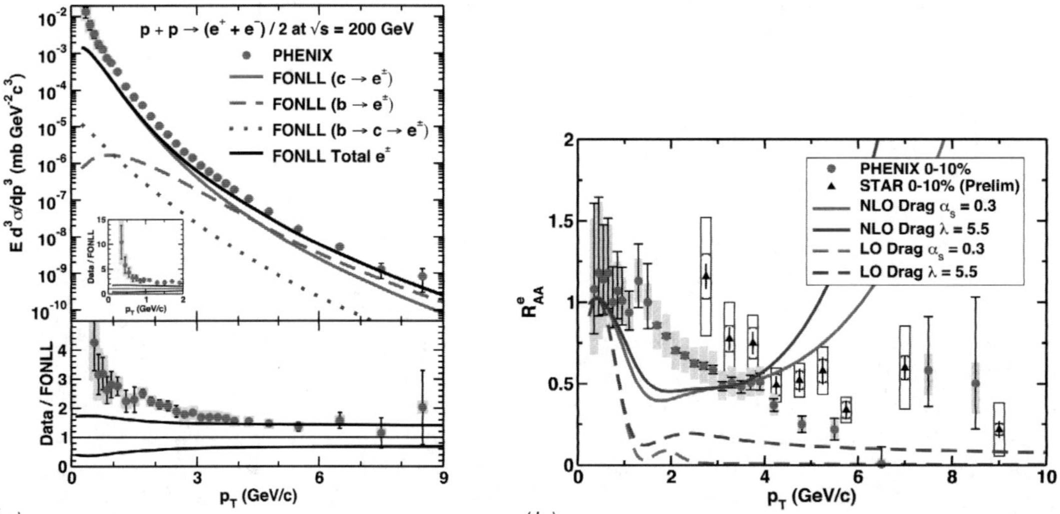


(a) (b)

Figure 1: (a) FONLL predictions [13, 14, 15] for the spectrum of electrons from heavy flavor decay at RHIC compared to data from PHENIX [16]. (b) Strong-coupling energy loss predictions at leading (dashed) and next-to-leading order (solid) for two schemes (red, blue) of translating QCD parameters to $\mathcal{N} = 4$ SYM parameters [11] compared to data from PHENIX [16] and preliminary data from STAR [17].

## 3. Strong-coupling Energy Loss Model Results

In order to compare with data from $A + A$ collisions, we must first demonstrate that we have the production mechanism in $p + p$ collisions under control. Fig. 1 (a) compares the spectrum of electrons from the decay of heavy flavor mesons from RHIC and from FONLL [13, 14, 15]. The spectra of open heavy quarks is readily available from an online generator [18], and the FONLL $D$ and $B$ meson fragmentation functions are well documented [15]. The decay of the mesons to electrons in FONLL is not as well documented, and the results of this work [11] as shown in Fig. 1 (a) differ significantly from those shown in [16], with the results presented here diverging much more from the experimental results at low, $p_T \lesssim 3$ GeV/c. Thus the comparison of electrons from $A + A$ is restricted from above by the latter speed limit above and from below by the currently inaccurate production/fragmentation function mechanism.

Nevertheless, we show in Fig. 1 (b) the comparison of the strong-coupling energy loss model predictions for electrons from heavy flavor decay to RHIC data [16, 17], and then $D$ and $B$ meson suppression to LHC data [19, 20] in Fig. 2. The plots include both the leading and next-to-leading order predictions from AdS/CFT [11]. One of the subtleties involved in comparing the AdS/CFT calculations to data is the map between the parameters ($\alpha_s$, $N_c$, and $T$) in QCD and the parameters in $\mathcal{N} = 4$ SYM. Two schemes for translating between the theories were proposed in [7]; we report the results of both schemes here as an attempt to provide a reasonable exploration of the parameter space. (One would hope that a theory dual to QCD would produce results somewhere between the predictions presented here.) In the range of applicability, 3 GeV/c $\lesssim p_T^e \lesssim$ 4 GeV/c, the theory and data agree for electron suppression. Similarly, in the range of $p_T^D \lesssim$ 10 GeV/c and for all current $B$ meson measurements, the theory and data agree. In fact, the fast rise in $R_{AA}$ for the observables for momenta above the scale set by the speed limit is precisely the sign of the calculation breaking down. A more correct treatment of the above equations would include autocorrelations, which would mitigate some of that spurious rapid rise. Additionally, a derivation of the energy loss that allowed the heavy quark to dynamically slow down (a highly non-trivial calculation) would include the Larmor-like energy loss experienced by accelerating probes in AdS/CFT [21, 22]. It is not yet known how much of a quantitative difference including this additional energy loss channel would make as a function of momentum, but an implementation of this improved energy loss theory would almost certainly compare more favorably with data.

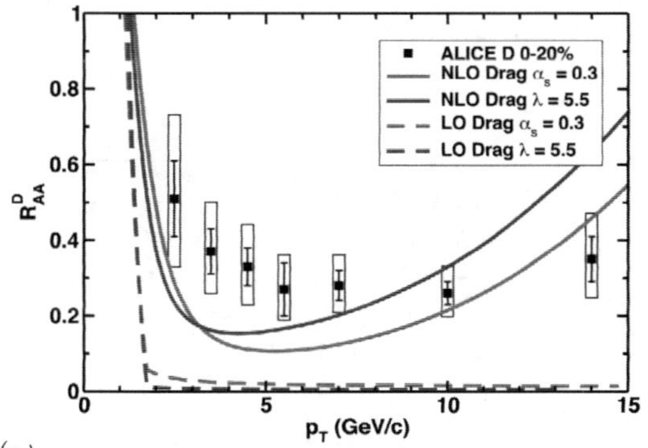


(a)

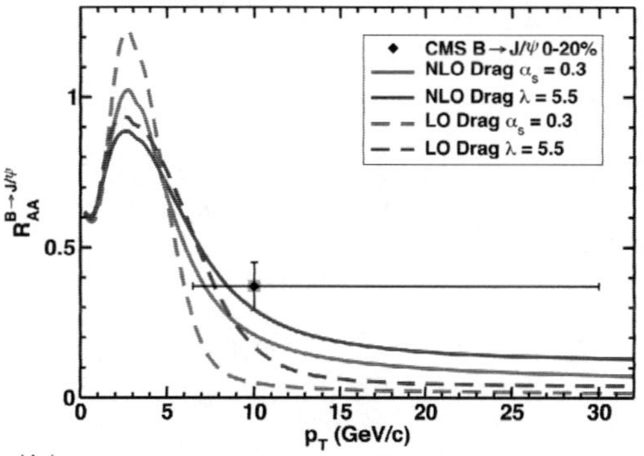


(b)

Figure 2: $R_{AA}(p_T)$ predictions from AdS/CFT at leading (dashed) and next-to-leading (solid) order for two schemes (red, blue) of translating QCD parameters to $\mathcal{N} = 4$ SYM parameters [11] compared to (a) $D$ meson data from ALICE [19] and (b) non-prompt $J/\psi$ mesons compared to data from CMS [20].

In Fig. 3 we show a novel result: the very large difference in predictions for bottom quark to electron $R_{AA}(p_T)$ from the strong-coupling energy loss model presented here and the WHDG weak-coupling energy loss model [23, 24]. One can readily see that, even with the inclusion of the fluctuations in the strong-coupling energy loss, the AdS/CFT predictions are significantly more suppressed than those from pQCD.

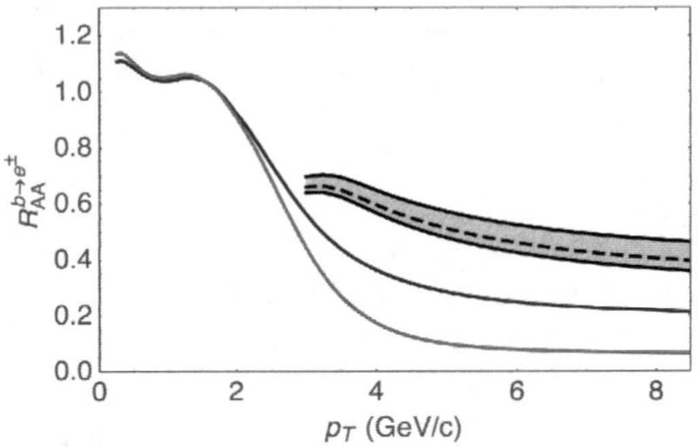


Figure 3: $R_{AA}^{b\to e^\pm}(p_T)$ predictions from AdS/CFT at next-to-leading order for two schemes (red, blue) of translating QCD parameters to $\mathcal{N}=4$ SYM parameters [11] compared to predictions (black) from the WHDG energy loss model [23, 24] in 0-20% central PbPb collisions at $\sqrt{s}=2.76$ ATeV.

## 4. Conclusions

We showed in these proceedings that a strong-coupling energy loss model including fluctuations can qualitatively describe the heavy flavor data measured at RHIC and LHC. Additionally, we provided a novel, falsifiable prediction for the suppression of electrons from bottom decay at LHC from both the AdS/CFT energy loss model presented here and from the pQCD-based WHDG energy loss model. Future correlations measurements of heavy flavor, such as the decays from $b\bar{b}$ events, should provide even more insight into the energy loss mechanism at work in QGP.

## 5. Acknowledgments

The author wishes to thank the South African National Research Foundation and SA-CERN for support.

## 6. References

[1] Gubser S S, Klebanov I R and Tseytlin A A 1998 *Nucl. Phys.* **B534** 202–222 (*Preprint* hep-th/9805156)
[2] Borsanyi S, Endrodi G, Fodor Z, Katz S D, Krieg S, Ratti C and Szabo K K 2012 *JHEP* **08** 053
[3] Chesler P M and Yaffe L G 2011 *Phys. Rev. Lett.* **106** 021601 (*Preprint* 1011.3562)
[4] Gale C, Jeon S, Schenke B, Tribedy P and Venugopalan R 2013 *Phys. Rev. Lett.* **110** 012302
[5] Gubser S S, Klebanov I R and Peet A W 1996 *Phys. Rev.* **D54** 3915–3919 (*Preprint* hep-th/9602135)
[6] Kovtun P, Son D T and Starinets A O 2005 *Phys. Rev. Lett.* **94** 111601 (*Preprint* hep-th/0405231)
[7] Gubser S S 2008 *Nucl. Phys.* **B790** 175–199 (*Preprint* hep-th/0612143)
[8] Gubser S S 2006 *Phys. Rev.* **D74** 126005 (*Preprint* hep-th/0605182)
[9] Herzog C P, Karch A, Kovtun P, Kozcaz C and Yaffe L G 2006 *JHEP* **07** 013 (*Preprint* hep-th/0605158)
[10] Casalderrey-Solana J and Teaney D 2006 *Phys. Rev.* **D74** 085012 (*Preprint* hep-ph/0605199)
[11] Horowitz W A 2015 *Phys. Rev.* **D91** 085019 (*Preprint* 1501.04693)
[12] Casalderrey-Solana J and Teaney D 2007 *JHEP* **04** 039 (*Preprint* hep-th/0701123)
[13] Cacciari M, Greco M and Nason P 1998 *JHEP* **05** 007 (*Preprint* hep-ph/9803400)
[14] Cacciari M, Frixione S and Nason P 2001 *JHEP* **03** 006 (*Preprint* hep-ph/0102134)
[15] Cacciari M, Frixione S, Houdeau N, Mangano M L, Nason P and Ridolfi G 2012 *JHEP* **10** 137
[16] Adare A *et al.* (PHENIX) 2011 *Phys. Rev.* **C84** 044905 (*Preprint* 1005.1627)
[17] Mustafa M (STAR) 2013 *Nucl. Phys.* **A904-905** 665c–668c (*Preprint* 1210.5199)
[18] http://www.lpthe.jussieu.fr/~cacciari/fonll/fonllform.html
[19] Abelev B *et al.* (ALICE) 2012 *JHEP* **09** 112 (*Preprint* 1203.2160)
[20] Chatrchyan S *et al.* (CMS) 2012 *JHEP* **05** 063 (*Preprint* 1201.5069)
[21] Mikhailov A 2003 (*Preprint* hep-th/0305196)
[22] Fadafan K B, Liu H, Rajagopal K and Wiedemann U A 2009 *Eur. Phys. J.* **C61** 553–567 (*Preprint* 0809.2869)
[23] Wicks S, Horowitz W, Djordjevic M and Gyulassy M 2007 *Nucl. Phys.* **A784** 426–442
[24] Horowitz W A and Gyulassy M 2011 *Nucl. Phys.* **A872** 265–285 (*Preprint* 1104.4958)

# Baryon stopping signal for mixed phase formation in HIC

**Yu. B. Ivanov**

National Research Centre "Kurchatov Institute", 123182 Moscow, Russia
National Research Nuclear University "MEPhI" (Moscow Engineering Physics Institute), 115409 Moscow, Russia

E-mail: y.b.ivanov@yandex.ru

**Abstract.** It is argued that an irregularity in the baryon stopping is a natural consequence of onset of deconfinement occurring in the compression stage of a nuclear collision. It is an effect of the softest point inherent in an equation of state (EoS) with a deconfinement transition. In order to illustrate this effect, calculations within the three-fluid model were performed with three different EoS's: a purely hadronic EoS, an EoS with a first-order phase transition and that with a smooth crossover transition.

## 1. Introduction

One of the main goals of the current experiments at RHIC and SPS and forthcoming experiments at FAIR and NICA facilities is to determine a kind of the deconfinement transition in dense baryonic matter and to find the collision energy (thereby the baryonic density) at which this transition starts. In this paper it is argued that the baryon stopping in nuclear collisions can be a sensitive probe of the deconfinement onset. Let us start with discussion in terms of the conventional (i.e. one-fluid) hydrodynamics.

The form of the resulting rapidity distribution of net-baryons depends on the spatial form of the produced fireball. If the fireball is almost spherical, the expansion of the fireball is essentially 3-dimensional which results in a peak at the midrapidity in the rapidity distribution. This statement is a theorem that can be proved in few lines. If a the fireball is strongly deformed (compressed) in the beam direction, i.e. has a form of a disk, its expansion is approximately 1-dimensional that produces a dip at the midrapidity, which is confirmed by numerous simulations, see e.g. [1]. In terms of the fluid mechanics this is a consequence of interaction of two rarefaction waves propagating from opposite peripheral sides of decaying disk toward its center [2].

The formation of this fireball is already a matter of dynamics at the early compression stage of the nuclear collision. A softest point [3] characteristic of EoS's with a phase transition plays an important role in this compression dynamics. At the softest point the system exhibits the weakest resistance to its compression as compared with that in adjacent regions of the EoS. At low collision energies the softest point is not reached in the collision process, the system remains stiff and therefore the produced fireball is almost spherical. As a result, the baryon rapidity distribution is peaked at the midrapidity. When the incident energy gets high enough, the softest-point region of the EoS starts to dominate during the compression stage, the system weaker resists to the compression and hence the resulting fireball becomes more deformed, i.e.

more of the disk shape. Then its expansion is close to the 1-dimensional pattern and, as a result, we have a dip at the midrapidity. With energy rise, the stiffness of the EoS (in the range relevant to compression stage) grows, the system stats to be more resistant to the compression and hence the produced fireball becomes less deformed. The expansion of this fireball results in a peak or, at least, to a weaker dip at the midrapidity as compared to that at the "softest-point" incident energy. With further energy rise, the initial kinetic pressure overcomes the stiffness of the EoS and makes the produced fireball strongly deformed again, which in its turn again results in a dip at the midrapidity.

Thus, even without any nonequilibrium, we can expect a kind of a "peak-dip-peak-dip" irregularity in the incident energy dependence of the form of the net-proton rapidity distributions. Nonequilibrium also contributes to this irregularity. At a phase transformation the hadronic degrees of freedom are changed to partonic ones. In particular, the dip at the midrapidity in ultrarelativistic nuclear collisions occurs because the baryon charges of colliding nuclei traverse through each other rather than results from the 1-dimensional expansion of a disk-like fireball.

It is important to emphasize that the "peak-dip-peak-dip" irregularity is a signal from the hot and dense stage of the nuclear collision.

In the present paper this qualitative pattern is illustrated by calculations within a model of the three-fluid dynamics (3FD) [4] employing three different equations of state (EoS): a purely hadronic EoS [5] (hadr. EoS) and two versions of EoS involving deconfinement [6]: an EoS with the first-order phase transition (2-phase EoS) and that with a smooth crossover transition (crossover EoS). The softest points in these EoS's are illustrated in Ref. [7]. The hadronic EOS [5] possesses no softest point, i.e. stiffness of the EoS changes monotonously. Results on the stopping power were reported in Refs. [8] more in detail.

## 2. Results of simulations

A direct measure of the baryon stopping is the net-baryon (i.e. baryons-minus-antibaryons) rapidity distribution. However, since experimental information on neutrons is unavailable, we have to rely on net-proton (i.e. proton-minus-antiproton) data. Presently there exist experimental data on proton (or net-proton) rapidity spectra at AGS [9] and SPS [10] energies.

Figure 1 presents calculated rapidity distributions of net-protons in central collisions at AGS and SPS energies and their comparison with available data. Difference between protons and net-protons is negligible at the AGS energies. As seen from Fig. 1, the distributions within the first-order-transition scenario indeed exhibit the above-discussed "peak-dip-peak-dip" irregularity in contrast to results obtained within the purely hadronic and crossover scenarios. The experimental distributions exhibit a qualitatively similar behavior as that in the 2-phase-EoS scenario. However, quantitatively the 2-phase-EoS results certainly disagree with data in the energy region $8A$ GeV $\leq E_{lab} \leq 40A$ GeV.

In order to quantify the above-discussed "peak-dip-peak-dip" irregularity, it is useful to make use of the method proposed in Ref. [8]. For this purpose the data on the net-proton rapidity distributions are fitted by the following simple formula

$$\frac{dN}{dy} = a\left(\exp\{-(1/w_s)\cosh(y-y_s)\} + \exp\{-(1/w_s)\cosh(y+y_s)\}\right) , \qquad (1)$$

where $a$, $y_s$ and $w_s$ are parameters of the fit. The form (1) is a sum of two thermal sources shifted by $\pm y_s$ from the midrapidity which is put to be $y_{\rm mid} = 0$ as it is in the collider mode. The parameters of the two sources are identical (up to the sign of $y_s$) because only collisions of identical nuclei are considered. The above fit has been done by the least-squares method and applied to both available data and results of calculations.

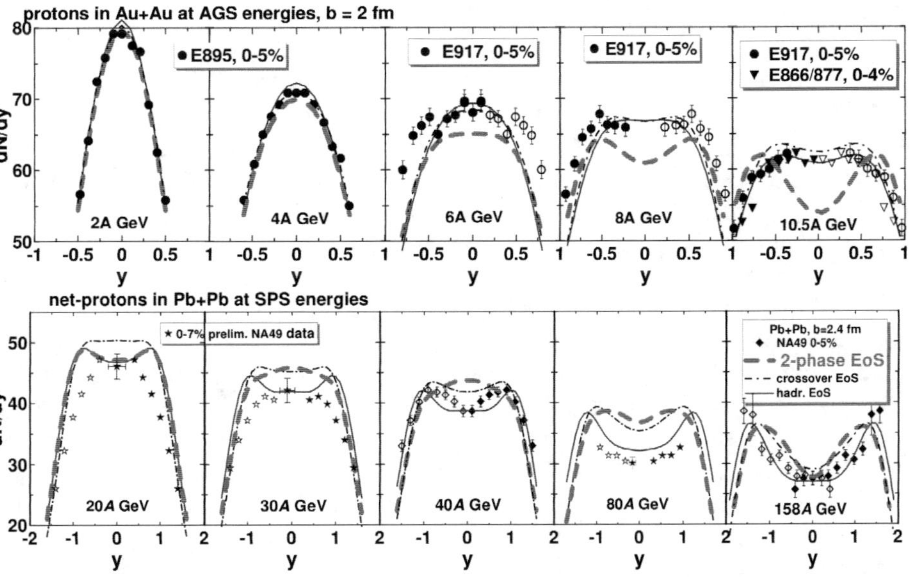


**Figure 1.** Rapidity spectra of protons for AGS energies (upper raw of panel) and net-protons SPS energies (lower raw of panels) from central collisions of Au+Au (AGS) and Pb+Pb (SPS). Experimental data are from [9, 10]. The percentage shows the fraction of the total reaction cross section, corresponding to experimental selection of central events.

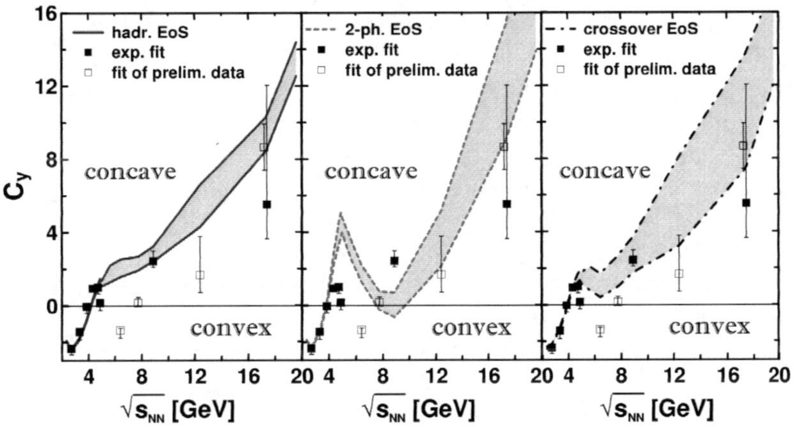


**Figure 2.** Midrapidity reduced curvature [see. Eq. (2)] of the (net)proton rapidity spectrum as a function of the center-of-mass energy of colliding nuclei as deduced from experimental data and predicted by 3FD calculations with different EoS's: the hadronic EoS (hadr. EoS) [5] (left panel), the EoS involving a first-order phase transition (2-ph. EoS, middle panel) and the EoS with a crossover transition (crossover EoS, right panel) into the quark-gluon phase [6]. Upper bounds of the shaded areas correspond to fits confined in the region of $|y|/y_{\rm beam} < 0.7$, lower bounds, $|y|/y_{\rm beam} < 0.5$.

A useful quantity, which characterizes the shape of the rapidity distribution, is a reduced curvature of the spectrum at midrapidity defined as follows

$$C_y = \left( y_{\rm beam}^3 \frac{d^3N}{dy^3} \right)_{y=0} / \left( y_{\rm beam} \frac{dN}{dy} \right)_{y=0} = (y_{\rm beam}/w_s)^2 \left( \sinh^2 y_s - w_s \cosh y_s \right), \qquad (2)$$

where $y_{\rm beam}$ is the beam rapidity in the collider mode. The second part of Eq. (2) presents this curvature in terms of parameters of fit (1). Excitation functions of $C_y$ deduced both from experimental data and from results of the 3FD calculations with different EoS's are displayed in Fig. 2. To evaluate errors of the $C_y$ values deduced from data, errors produced by the least-squares method were estimated. The uncertainty associated with the choice of the rapidity range turned out to be the dominant one for the $C_y$ quantities deduced from simulation results. Therefore, in Fig. 2 results for the curvature $C_y$ are presented by shaded areas with borders corresponding to the fit ranges $|y| < 0.7\, y_{\rm beam}$ and $|y| < 0.5\, y_{\rm beam}$.

The irregularity in data is distinctly seen here as a wiggle irregularity in the energy dependence of $C_y$. Of course, this is only a hint to irregularity since this wiggle is formed only due to preliminary data of the NA49 collaboration. A remarkable observation is that the $C_y$ excitation function in the first-order-transition scenario manifests qualitatively the same wiggle irregularity (middle panel of Fig. 2) as that in the data fit, while the hadronic scenario produces purely monotonous behavior. The crossover EoS represents a very smooth transition, therefore, it is not surprising that it produces only a weak wiggle in $C_y$.

## 3. Conclusions

In conclusion, the irregularity in the baryon stopping is a natural consequence of deconfinement occurring at the compression stage of a nuclear collision and thus is a signal from the hot and dense stage of the nuclear collision. It is an effect of the softest point of a EoS. As it was demonstrated in Ref. [11], this irregularity is a very robust signal of a first-order phase transition that survives even under conditions of a very limited acceptance. Updated experimental results are badly needed to analyze a trend of the "peak-dip-peak-dip" irregularity. It would be highly desirable if new data are taken within the same experimental setup.

Fruitful collaboration with D. Blaschke is gratefully acknowledged. I am grateful to A.S. Khvorostukhin, V.V. Skokov, and V.D. Toneev for providing me with the tabulated 2-phase and crossover EoS's. The calculations were performed at the computer cluster of GSI (Darmstadt). This work was partially supported by grant NS-215.2012.2.

## References

[1] Russkikh V N, Ivanov Y B, *Phys. Rev.* C **76** (2007) 054907.
[2] Landau L D and Lifshitz E M, *"Fluid Mechanics"* (Pergamon Press, Oxford, 1979).
[3] Hung C M and Shuryak E. V., *Phys. Rev. Lett.* **75** (1995) 4003.
[4] Ivanov Yu B, Russkikh V N, and Toneev V D, *Phys. Rev.* C **73** (2006) 044904.
[5] Galitsky V M and Mishustin I N, *Sov. J. Nucl. Phys.* **29**, 181 (1979).
[6] Khvorostukhin A S, Skokov V V, Redlich K, and Toneev V D, *Eur. Phys. J.* C **48** (2006) 531.
[7] Nikonov E G, Shanenko A A and Toneev V D, *Heavy Ion Phys.* **8** (1998) 89.
[8] Ivanov Y B, *Phys. Lett.* B **690** (2010) 358; *Phys. At. Nucl.* **75** (2012) 621; *Phys. Lett.* B **721** (2013) 123; *Phys. Rev.* C **87** (2013) 064904.
[9] Klay J L *et al.* [E-0895 Collaboration], *Phys. Rev.* C **68** (2003) 054905; Barrette J *et al.* (E877 Collaboration), *Phys. Rev.* C **62** (2000) 024901; Back B B *et al.*, (E917 Collaboration), *Phys. Rev. Lett.* **86** (2001) 1970; Stachel J, *Nucl. Phys.* **A654** (1999) 119c.
[10] Appelshäuser H *et al.* (NA49 Collaboration), *Phys. Rev. Lett.* **82** (1999) 2471; Anticic T *et al.* (NA49 Collaboration), *Phys. Rev.* C **69** (2004) 024902; Alt C *et al.* (NA49 Collaboration), *Phys. Rev.* C **73** (2006) 044910; Blume C (NA49 Collaboration), *J. Phys.* **G34** (2007) S951; Anticic T. *et al.* [NA49 Collaboration], *Phys. Rev.* C **83** (2011) 014901; Alt C *et al.* [NA49 Collaboration], *Phys. Rev.* C **68** (2003) 034903.
[11] Ivanov Y B and Blaschke D, *Phys. Rev.* C **92** (2015) 024916.

Journal of Physics: Conference Series **668** (2016) 012062 doi:10.1088/1742-6596/668/1/012062

# Optical conductivity and string theory

**Jakub Jankowski**

Institute of Physics, Jagiellonian University, ul. Łojasiewicza 11, 30-348 Kraków, Poland

E-mail: jakubj@th.if.uj.edu.pl

**Abstract.** In this note we summarize some facts about optical conductivity and the corresponding holographic models attempting to describe it.

## 1. Optical conductivity

The optical properties of solids provide an important tool to study their microscopic dynamics. In certain materials with strong electron correlations a quantum critical state of matter can occur. In this situation the system looks on average the same, regardless of the time- and length scale on which it is observed. This implies, that scale invariant correlations should drive the response functions, revealing some universal patterns determined by the quantum mechanical nature of the fluctuations. Such regularities were reported for the mid-infrared frequency range of the optical conductivity of optimally doped copper-oxide superconductors [1, 2]. In more detail in ref. [1] two interesting regions of frequencies were identified from the data.

The low frequency region ($\omega < T$), where measurement time was large compared to the time scale set by inverse temperature. In this region system follows classical relaxation dynamics with a relaxation time $\tau$. Conductivity in this case is given by the usual Drude formula

$$\sigma(\omega) = \frac{K\tau}{1 - i\omega\tau} \,. \tag{1}$$

The second region ($T < \omega < \Omega$) directly probes the scale invariance of the quantum critical state. A non-universal cut-off $\Omega$ is called the optical gap and is required by the f-sum rule [1]. In this region the optical conductivity $\sigma(\omega) = \sigma_1(\omega) + i\sigma_2(\omega)$ follows the universal scaling law

$$|\sigma(\omega)| \sim \omega^{2/3} \,, \tag{2}$$

with a frequency independent phase angle $\arctan(\sigma_2(\omega)/\sigma_1(\omega)) \approx 60^o$. The experimental data are shown in figure 1. A remarkable thing is the quite robust temperature independence of this scaling relation.

## 2. Holographic models

Up to date we have no microscopic understanding of how quantum criticality emerges from the strong correlations of the normal state near optimal doping. An emergent scale invariance implies that the low energy effective theory takes a form of a strongly coupled conformal field theory (CFT) [3]. A widely used tool to analyze certain strongly coupled CFT systems is the anti-de Sitter conformal field theory (AdS/CFT) correspondence, a setup in which a

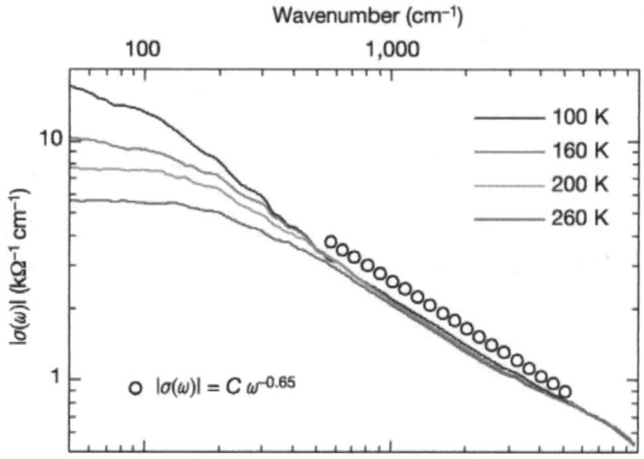


**Figure 1.** Universal power law of the optical conductivity from ref. [1].

nongravitational system is mapped to a theory with gravity in a higher dimensional spacetime. Over the past few years, it has been applied to model many interesting phenomena, including superconductivity and superfluidity, Fermi surfaces and non-Fermi liquids (see [4] for a review). The main advantage of this method is that it is based on conserved currents, rather than on a quasi-particle description, which makes it more suitable for understanding the relation between the infrared and the ultraviolet degrees of freedom. A natural question to ask is whether any form of power-like scaling can occur in the optical conductivity computed in a holographic model?

The simplest bulk description of a translationally invariant system at finite temperature and charge density is given by a charged black hole solution [4]. This was then improved by the coupling to a bulk scalar field, which could provide a model for a solid state lattice and in turn account for a realistic description of the charge transport [5]. Because the relevant physics is in two dimensional layers, the CFT in question lives in $(2+1)$ spacetime dimensions. Thus, the gravitational dual is a theory in four dimensional spacetime.

In more detail, the action used in [5] in $d = 3+1$ dimensions reads

$$S = \frac{1}{16\pi G_N}\int d^4x\sqrt{-g}\left[R - \frac{1}{2}\nabla_a\phi\nabla^a\phi + 6 + \phi^2\right] . \tag{3}$$

States in a CFT are mapped to solutions of the equations of motion following from the action (3), which have some specified boundary conditions. In Poincare coordinates $\{\mathbf{x}, z\} = \{t, x, y, z\}$ any solution of the Klein-Gordon equation for chosen parameters has the near boundary asymptote

$$\phi(\mathbf{x}, z) \sim \phi_0(\mathbf{x})z + \phi_1(\mathbf{x})z^2 . \tag{4}$$

In particular, in the boundary theory this corresponds to the operator $\hat{O}$ of conformal dimension $\Delta = 2$ that has a source $\phi_0(\mathbf{x})$ and corresponding expectation value $\langle\hat{O}(\mathbf{x})\rangle = \phi_1(\mathbf{x})$. The simplest model of a lattice was to impose a one dimensional source of the form of a single Fourier mode, i.e.

$$\phi_0(x) = A_0\cos(kx) . \tag{5}$$

Such a choice causes explicit translational symmetry breaking in a chosen direction. The claim of ref. [5] is that with this assumption both frequency regions described in the previous section are reproduced. This has motivated further research and explorations of more realistic and transparent models of holographic lattices [6, 7, 8].

To analyze the data, rather than using log-log plots another quantity was chosen for the study:

$$\alpha = 1 + \omega \frac{|\sigma|''}{|\sigma|'} \ . \tag{6}$$

A value of $\alpha = -2/3$ would correspond to the mid-infrared conductivity of the cuprates. In contrast to the claim of [5] no constant $\alpha$ as a function of $\omega$ was found. Sample results, taken from ref. [8], are shown in figure 2. This leads to the conclusion, that holographic lattices do not encode the power-law optical conductivity.

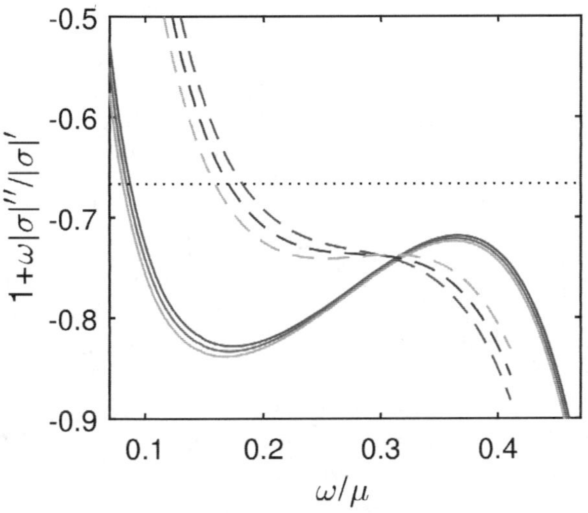


**Figure 2.** The exponent as a function of frequency. Figure taken from ref. [8]. Dashed and solid lines correspond to two slightly different lattice models.

However, one point which could be improved, is to replace an unrealistic, spatially delocalized lattice of a single Fourier mode with a lattice of localized atoms in the form of (possibly regularized) Dirac delta functions. A case where the source was set to be

$$\phi_0(t, x, y) = A_0\, \delta(x) \ , \tag{7}$$

i.e., the Dirac delta supported on a $x = 0$ line was considered in ref. [9]. The solution turned out to impose strong constraints on the choice of matter interactions of the dual gravitational theory. Only the potential

$$V(\phi) = -6 \cosh\left(\frac{\phi}{\sqrt{3}}\right) \ , \tag{8}$$

for the scalar filed allows solutions with such boundary conditions. This form of interactions is one of possibilities which come from consistent compactifications of ten dimensional string theories [10]. This shows, that seemingly innocent modifications of the source causes highly non-trivial consequences. This might be an indication that similar impact would be imprinted on a conductivity computed in such a background.

Having a theory in which one can consistently study localized atoms we can go a step ahead and consider a regularized lattice of atoms, e.g., in a form motivated by linearized solution of a regularized Dirac delta source

$$\phi_0(x) = \frac{A_0 \sinh(z_0)}{\cosh(z_0) - \cos(x)} \ . \tag{9}$$

The parameter $z_0$ regularizes the configuration and the limit $z_0 = 0$ gives exact delta functions placed at positions $x = 2\pi n, (n \in \mathbb{Z})$. Sample configurations are shown in fig. 3.

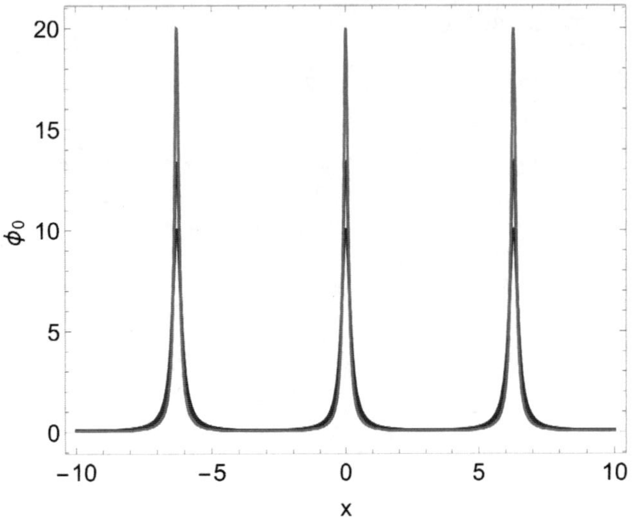


**Figure 3.** Sample lattices which can be considered in the holographic models. Parameters: $A_0 = 1$, $z_0 = 0.1,\ 0.15,\ 0.2$.

Another possible direction is to formulate a holographic dual of a system in which translational symmetry is broken spontaneously, not explicitly. This is closer to the cases of actual systems which exhibit this form of symmetry breaking.

Up to date the only setup which successfully reproduced the $\omega^{-2/3}$ law is based on unparticles, i.e., scale-invariant excitations [11]. Given the fact that the radial direction in the $AdS$ space describes dual RG flow and contains excitations at all energy scales, in principle it contains the correct ingredients to capture the unparticle dynamics. This suggests that in some construction of holographic models, perhaps different that the one discussed above, should be able to reproduce the power law.

## Acknowledgments

This research was supported by the NCN post-doctoral internship grant DEC-2013/08/S/ST2/00547.

## References

[1] van de Marel D, *et al.*, 2003 *Nature* **425** 271
[2] Hwang J, Timusk T and Gu G D 2007 *Journal of Physics Condensed Matter* **19** 125208
[3] Sachdev S 2008 *Nature Physics* **4** 173
[4] Hartnoll S A 2009 *Class. Quant. Grav.* **26** 224002
[5] Horowitz G T, Santos J E and Tong, D 2012 *JHEP* **1207**, 168
[6] Rangamani M, Rozali M and Smyth D 2015 *JHEP* **1507** 024
[7] Donos A and Gauntlett J P 2014 *JHEP* **1404** 040
[8] Langley B W, Vanacore G and Phillips P W 2015 *JHEP* **1510** 163
[9] Janik R A, Jankowski J and Witkowski P 2015 *JHEP* **1507** 050
[10] Duff M J and Liu J T 1999 *Nucl. Phys. B* **554** 237
[11] Limtragool K and Phillips P 2015 *Phys. Rev. B* **92** 155128

# Estimating $\eta/s$ of QCD matter at high baryon densities

**Iu. Karpenko[1,2], M. Bleicher[1,3], P. Huovinen[1,3] and H. Petersen[1,3,4]**

[1] Frankfurt Institute for Advanced Studies, Ruth-Moufang-Straße 1, 60438 Frankfurt am Main, Germany
[2] Bogolyubov Institute for Theoretical Physics, 14-b, Metrolohichna st., 03680 Kiev, Ukraine
[3] Institute for Theoretical Physics, Johann Wolfgang Goethe Universität, Max-von-Laue-Str. 1, 60438 Frankfurt am Main, Germany
[4] GSI Helmholtzzentrum für Schwerionenforschung GmbH, Planckstr. 1, 64291 Darmstadt, Germany

E-mail: karpenko@fias.uni-frankfurt.de

**Abstract.** We report on the application of a cascade + viscous hydro + cascade model for heavy ion collisions in the RHIC Beam Energy Scan range, $\sqrt{s_{\rm NN}} = 6.3\ldots200$ GeV. By constraining model parameters to reproduce the data we find that the effective (average) value of the shear viscosity over entropy density ratio $\eta/s$ decreases from 0.2 to 0.08 when collision energy grows from $\sqrt{s_{\rm NN}} \approx 7$ to 39 GeV.

## 1. Introduction

The principal goal of the Beam Energy Scan program at Relativistic Heavy Ion Collider (RHIC) is to study the transition from baryon-free to baryon-rich regime of strongly interacting QCD matter by decreasing the collision energy. One of questions is what the transport properties of the medium with high baryon density, created at lower beam energies, are.

To address this question we use a state-of-the-art event-by-event viscous hydrodynamic + cascade (or hybrid) model to describe the evolution of the collision system.

## 2. The model, its parameter space and results

The hybrid model consists of the pre-equilibrium stage, the fluid dynamical stage and the final hadronic cascade. The pre-thermal stage of evolution is described by UrQMD cascade [1, 2]. The interactions are calculated until a hypersurface of a constant Bjorken proper time $\tau = \sqrt{t^2 - z^2} = \tau_0$, which is a parameter of the model. At $\tau = \tau_0$ the energy, momentum and charges of particles are smoothly distributed to fluid cells according to a Gaussian profile. The contribution of each particle to a fluid cell $\{i, j, k\}$ is given by:

$$\left\{\Delta P^{\mu}_{ijk}, \Delta N^{0}_{ijk}\right\} = \left\{P^{\mu}, N^{0}\right\} \cdot C \cdot \exp\left(-\frac{\Delta x_i^2 + \Delta y_j^2}{R_\perp^2} - \frac{\Delta \eta_k^2}{R_\eta^2}\gamma_\eta^2 \tau_0^2\right). \quad (1)$$

where $\Delta x_i$, $\Delta y_j$, $\Delta \eta_k$ are the differences between particle's position and the coordinates of the hydrodynamic cell $\{i, j, k\}$. Due to the lack of knowledge about thermalization process, $R_\perp$ and

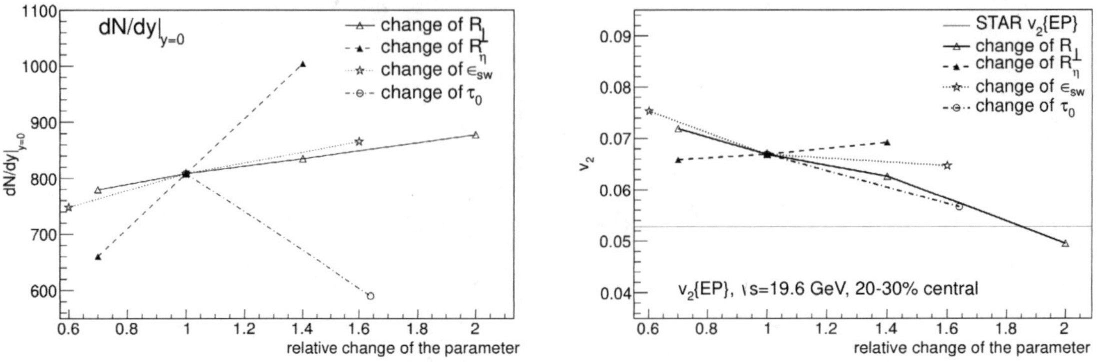


**Figure 1.** Parameter dependence of the total yield at midrapidity (left) and $p_T$ integrated elliptic flow (right) in 20-30% central Au-Au collisions at $\sqrt{s_{\rm NN}} = 19.6$ GeV. Figures are from Ref. [6].

$R_\eta$ are another free parameters of the model. These distributions are used as an initial state for 3-dimensional hydrodynamic expansion, described by relativistic viscous hydrodynamic code `vHLLE` [3]. We use the equation of state based on Chiral model [4] in the fluid stage. We set bulk viscosity and baryon charge diffusion to zero in the fluid phase, and study the effects of shear viscosity only. When the energy density reaches a certain value $\epsilon = \epsilon_{\rm sw}$, the hydrodynamic medium is converted to particles (particlized) according to Cooper-Frye prescription [5]. The last stage of evolution - hadron rescatterings and decays - is described again with UrQMD cascade. For more details see Ref. [6].

As a result, the model has three parameters related to the initial state (starting time $\tau_0$ and Gaussian smearing radii $R_\perp$, $R_\eta$), one parameter of the hydrodynamic medium (shear viscosity over entropy ratio $\eta/s$) and one parameter related to switching from hydrodynamics to cascade (switching energy density $\epsilon_{\rm sw}$). The main goal is to extract the value of the *physical* parameter $\eta/s$ from comparison with available experimental data. This goal is, however, complicated by the fact that the observables in the model depend significantly on all the mentioned parameters, not only $\eta/s$. A somewhat stronger sensitivity to the initial state parameters compared to hydrodynamic studies at full RHIC or LHC energies [7] is explained by shorter duration of hydrodynamic phase at lower beam energies.

To learn how exactly final observables depend on the model parameters, we consider the setup with $R_\perp = R_\eta = 1.0$ fm, $\tau_0 = 2R/(\gamma v_z)$, $\eta/s = 0$ and $\epsilon_{\rm sw} = 0.5$ GeV/fm$^3$ as a "default" scenario. On Fig. 1 we show how deviations of parameter values from the "default" scenario affect the multiplicity of all hadrons at midrapidity and $p_T$ integrated elliptic flow of all charged hadrons at one particular collision energy $\sqrt{s_{\rm NN}} = 39$ GeV. One can summarize the dependencies seen in Fig. 1 in a following way:

- Increased $R_\perp$ smoothens the initial energy density profile in the transverse plane, which leads to smaller gradients and less explosive transverse expansion. Increased $R_\perp$ also results in decreased ellipticity an initial energy density profile, which is hydrodynamically translated into smaller final elliptic flow.
- In a similar manner, the increase of $R_\eta$ leads to shallower longitudinal gradients and weaker longitudinal expansion. Thus more energy remains at midrapidity to form stronger transverse expansion, which increases $v_2$. On the other hand, larger $R_\eta$ also results in larger initial entropy of the fluid, which considerably increases the final particle multiplicity.
- Increased $\tau_0$ leads to a shorter lifetime of the hydrodynamic phase, as a result of longer

| $\sqrt{s_{\rm NN}}$ [GeV] | $\tau_0$ [fm/c] | $R_\perp$ [fm] | $R_\eta$ [fm] | $\eta/s$ |
|---|---|---|---|---|
| 7.7 | 3.2 | 1.4 | 0.5 | 0.2 |
| 8.8 (SPS) | 2.83 | 1.4 | 0.5 | 0.2 |
| 11.5 | 2.1 | 1.4 | 0.5 | 0.2 |
| 17.3 (SPS) | 1.42 | 1.4 | 0.5 | 0.15 |
| 19.6 | 1.22 | 1.4 | 0.5 | 0.15 |
| 27 | 1.0 | 1.2 | 0.5 | 0.12 |
| 39 | 0.9* | 1.0 | 0.7 | 0.08 |
| 62.4 | 0.7* | 1.0 | 0.7 | 0.08 |
| 200 | 0.4* | 1.0 | 1.0 | 0.08 |

**Table 1.** Collision energy dependence of the model parameters, which allows the model to reproduce the experimental data in BES region and higher RHIC energies.

pre-thermal phase. At the same time $\tau_0$ enters the Gaussian energy/momentum smearing profile. Thus its increase acts opposite to the increase of $R_\eta$. The results show that the second effect is stronger.

- Increased $\epsilon_{\rm sw}$ shortens the effective lifetime of the hydrodynamic phase. The shorter time to develop radial and elliptic flows is not fully compensated by the longer cascade phase, which results in smaller final $v_2$. Since the total entropy is conserved in the ideal hydrodynamic expansion, but increases in the cascade stage, the final particle multiplicity increases with the increase of $\epsilon_{\rm sw}$.

The found dependencies allow us to adjust the values of parameters individually for every collision energy to approximately reproduce the available experimental data points in the BES region. To simplify the analysis, we keep the same value of switching energy density $\epsilon_{\rm sw}$ at all collision energies. The adjusted values of the parameters are shown in Table 1. Some of the results are shown in Fig. 2.

The shear viscosity over entropy density ratio $\eta/s$ in fluid phase is taken to be independent of temperature and chemical potentials, and changes only with collision energy. Therefore the value of $\eta/s$ can be thought of as effective one for given computational setup. The parameter adjustment procedure yields $\eta/s$ values which decrease with increasing collision energy. At lower collision energies the fluid phase probes thermodynamic region of lower temperature, limited by the particlization temperature from below, and higher chemical potential. One expects that physical $\eta/s$ depends on temperature and chemical potential, and not on collision energy. Therefore we conclude that the $\sqrt{s_{\rm NN}}$ dependence of the effective $\eta/s$ we found hints that in more realistic simulations $(\eta/s)(T, \mu_B)$ should increase with increasing baryon chemical potential $\mu_B$.

## 3. Summary

To summarize, the 3 dimensional cascade + viscous hydro + cascade model is applied for the description of heavy ion collisions in the collision energy range of RHIC Beam Energy Scan program, $\sqrt{s_{\rm NN}} = 6.3\ldots200$ GeV. The parameter space of the model is explored. The values of model parameters are adjusted for different collision energies in order to approach the existing experimental data for rapidity and $p_T$ distributions, and elliptic flow coefficients of hadrons. As a result it is found that the value of effective shear viscosity over entropy density ratio $\eta/s$ decreases from 0.2 to 0.08 when the center of mass collision energy increases from 7.7 to 39 GeV, and stays at 0.08 for $\sqrt{s_{\rm NN}} > 39$ GeV.

## Acknowledgements

The simulations have been performed at the Center for Scientific Computing (CSC) at the Goethe-University Frankfurt. The authors acknowledge the financial support by the Helmholtz International Center for FAIR and Hessian LOEWE initiative. HP acknowledges funding by the

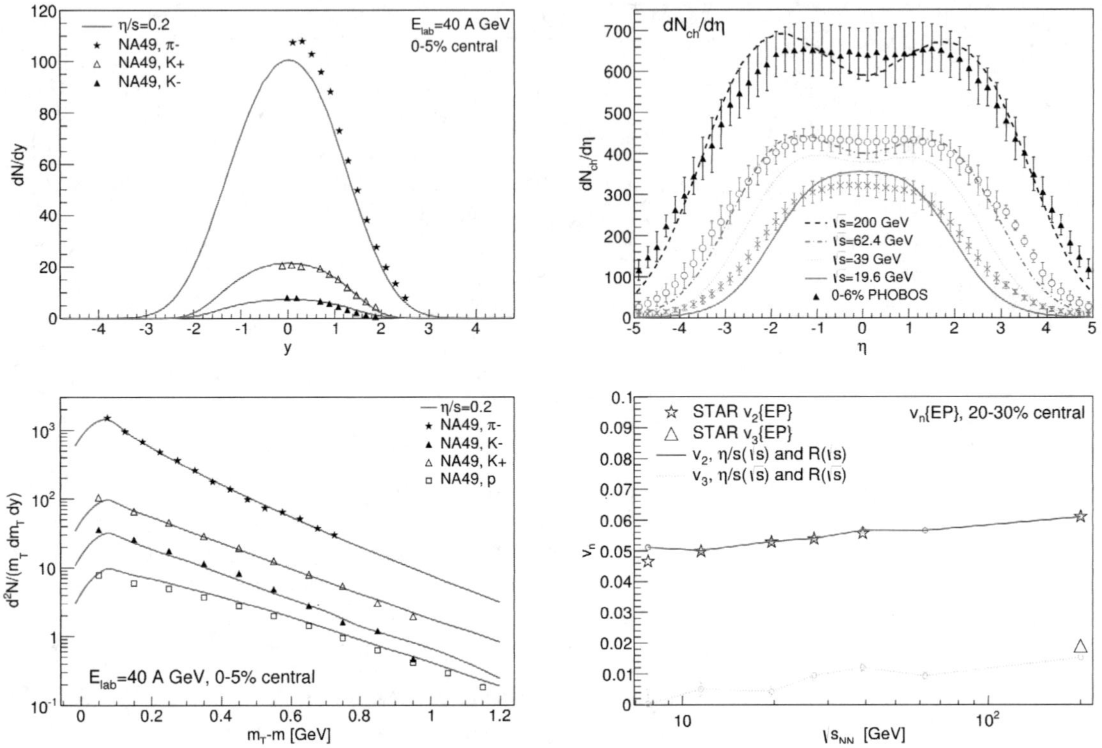


**Figure 2.** Rapidity (top left) and transverse momentum (bottom left) distributions of pions and kaons at $E_{\mathrm{lab}} = 40$ A GeV Pb-Pb collisions, pseudorapidity distributions of all charged hadrons at $\sqrt{s} = 19.6, 62.4$ and 200 GeV Au-Au collisions (top right), and $p_T$ integrated flow coefficients for different collision energies [10] (bottom right), all calculated in the model and compared to the experimental data from NA49 [8] and PHOBOS [9]. Figures are from Ref. [6].

Helmholtz Young Investigator Group VH-NG-822. The work of PH was supported by BMBF under contract no. 06FY9092.

## References

[1] S. A. Bass *et al.*, Prog. Part. Nucl. Phys. **41** (1998) 255 [nucl-th/9803035].
[2] M. Bleicher *et al.*, J. Phys. G **25** (1999) 1859 [hep-ph/9909407].
[3] I. Karpenko, P. Huovinen and M. Bleicher, Comput. Phys. Commun. **185** (2014) 3016
[4] J. Steinheimer, S. Schramm and H. Stocker, J. Phys. G **38** (2011) 035001 [arXiv:1009.5239 [hep-ph]].
[5] F. Cooper and G. Frye, Phys. Rev. D **10** (1974) 186.
[6] I. A. Karpenko, P. Huovinen, H. Petersen and M. Bleicher, Phys. Rev. C **91** (2015) 6, 064901 [arXiv:1502.01978 [nucl-th]].
[7] K. Werner, I. Karpenko, T. Pierog, M. Bleicher and K. Mikhailov, Phys. Rev. C **82** (2010) 044904 [arXiv:1004.0805 [nucl-th]].
[8] S. V. Afanasiev *et al.* [NA49 Collaboration], Phys. Rev. C **66** (2002) 054902 [nucl-ex/0205002].
[9] B. Alver *et al.* [PHOBOS Collaboration], Phys. Rev. C **83** (2011) 024913 [arXiv:1011.1940 [nucl-ex]].
[10] L. Adamczyk *et al.* [STAR Collaboration], Phys. Rev. C **86** (2012) 054908 [arXiv:1206.5528 [nucl-ex]].

Journal of Physics: Conference Series **668** (2016) 012064 doi:10.1088/1742-6596/668/1/012064

# Hyperon puzzle and the RMF model with scaled hadron masses and coupling constants

**E E Kolomeitsev[1], K A Maslov[2], D N Voskresensky[2]**

[1]Matej Bel University, SK-97401 Banska Bystrica, Slovakia
[2]National Research Nuclear University (MEPhI), 115409 Moscow, Russia

E-mail: Evgeni.Kolomeitsev@umb.sk

**Abstract.** The equation of state of cold baryonic matter is studied within a relativistic mean-field model with hadron masses and coupling constants depending on a scalar field. We demonstrate that if the effective nucleon mass stops to decrease with a density increase at densities $n > n_* > n_0$, where $n_0$ is the nuclear saturation density, the equation of state stiffens for these densities and the limiting neutron star mass increases. The stabilization of the nucleon mass can be realised if in the equation of motion for the scalar mean-field there appear a term sharply varying in a narrow vicinity of the field value corresponding to the density $n_*$. We show several possible realizations of this mechanism getting sufficiently stiff equations of state. The appearance of hyperons in dense neutron star interiors is accounted for. The obtained equations of state remain sufficiently stiff if the reduction of the $\phi$ meson mass is incorporated. Thereby, the hyperon puzzle can be resolved.

## 1. Introduction

In this contribution we present the results of our studies of the equation of state (EoS) for the cold nuclear matter. We work on the extension of a relativistic mean-field (RMF) model [1] including a possibility for scaling of hadron masses and meson-baryon coupling constants. As in [1], all hadron masses are scaled in the same way, whereas the coupling constants are scaled differently.

The current challenge is to construct such an EoS that can describe a neutron star (NS) with the mass of $(2.01 \pm 0.04)M_\odot$, where $M_\odot$ is the mass of the Sun, which was recently observed, cf. [2, 3]. For that the EoS should be rather stiff. Simultaneously the EoS has to respect the constraint derived from the flow of particles produced in heavy-ion collisions [4]. That oppositely requires the EoS to be rather soft. The additional complication in the description of heavy NSs arises if hyperons are included into consideration, since their presence softens the EoS, cf. [5].

## 2. The RMF model with scaling

The nuclear energy-density functional of the RMF model including members of the baryon SU(3) octet ($B = N, \Lambda, \Sigma, \Xi$) the interactions of which are mediated by the scalar meson ($\sigma$) and non-strange members of the vector-meson SU(3) nonet $V = \omega, \rho, \phi$, is

$$E[f] = \sum_B \int_0^{p_{\mathrm{F},B}} \frac{p^2 dp}{\pi^2} \sqrt{p^2 + m_B^{*2}(f)} + \frac{m_N^4 f^2}{2C_\sigma^2} + U(f) + \sum_V \frac{C_V^2 \tilde{n}_V^2}{2m_N^2 \eta_V(f)}. \quad (1)$$

Here $p_{\mathrm{F},B} = (3\pi^2 n_B)^{1/3}$ is the baryon Fermi momentum with $n_B$ standing for the density of baryon $B$. The effective baryon density $\tilde{n}_V$, being a source for the mean-field of meson $V$, is defined as $\widetilde{n}_V = \sum_B x_{VB} q_V n_B$, where $q_V$ stands for the baryon charge for $V = \omega$, the baryon third isospin projection for $V = \rho$, and the baryon strangeness for $V = \phi$. The effective baryon mass $m_B^*$ differs from the vacuum mass $m_B$ because of the coupling to the scalar field, $m_B^*(f) = m_B - x_{\sigma B} m_N\, f$, where $f = g_{\sigma N}\sigma/m_N$ is the dimensionless scalar field, and $g_{\sigma N}$ is the $\sigma N$ coupling constant. The functions $\eta_V(f)$ are the key ingredients, which differ our approach from other RMF models. These functions stem from the ratios of scalings of vector-meson masses to scalings of hadron-nucleon coupling constants and are tuned in our approach to render the best agreement of the resulting EoS with empirical constraints. The scalar field is found from the minimization of the functional (1) with respect to $f$: $\delta E[f]/\delta f = 0$. The coupling constants $C_{\sigma,\omega\,\rho}$ are to be determined from the saturation properties of the nuclear matter, therefore we have $x_{\sigma N} = x_{\omega N} = x_{\rho N} = 1$. Since the $\phi$ meson does not couple to a nucleon, if the strict Okubo-Zweig-Iizuka selection rule is applied, i.e. $x_{\phi N} = 0$, the coupling constant $C_\phi$ is related to the coupling of the $\omega$ meson as $C_\phi = C_\omega m_\omega/m_\phi$, where $m_\omega = 783\,\mathrm{MeV}$, $m_\phi = 1020\,\mathrm{MeV}$. The coupling constants of vector mesons to hyperons are chosen traditionally according to the quark SU(6) symmetry

$$x_{\omega\Lambda} = x_{\omega\Sigma} = 2x_{\omega\Xi} = \tfrac{2}{3}\,,\ \ x_{\rho\Sigma} = 2x_{\rho\Xi} = 2\,,\ \ x_{\phi\Lambda} = x_{\phi\Sigma} = x_{\phi\Xi} = -\tfrac{\sqrt{2}}{3}\,,\ \ x_{\rho\Lambda} = x_{\phi N} = 0. \tag{2}$$

The coupling constants of hyperons with the scalar mean field are derived from the hyperon binding energies $\mathcal{E}^H_{\mathrm{bind}}$ in isospin-symmetric matter (ISM) at the saturation density $n = n_0$ given by [1] $\mathcal{E}^H_{\mathrm{bind}}(n_0) = C_\omega^2 m_N^{-2} x_{\omega H} n_0 - m_N + m_N^*(n_0)$, and the empirical values $\mathcal{E}^\Lambda_{\mathrm{bind}}(n_0) = -28\,\mathrm{MeV}$, $\mathcal{E}^\Sigma_{\mathrm{bind}}(n_0) = 30\,\mathrm{MeV}$ and $\mathcal{E}^\Xi_{\mathrm{bind}}(n_0) = -15\,\mathrm{MeV}$. The repulsive $\Sigma$ potential prevents the appearance of $\Sigma$ hyperons in all models considered below.

For the description of the NS one should add to (1) the energies of leptons. The NS composition is governed by the weak processes bringing the matter in the $\beta$-equilibrium and relating chemical potential of all species to the electron and neutron chemical potentials. Together with the electroneutrality condition these relations allow to express the densities of all baryonic and leptonic species as functions of the total baryon density $n = \sum_B n_B$. The pressure of the NS matter follows then as $P(n) = n\partial E/\partial n - E$. With $P = P(E)$ at hand, the NS masses and radii are calculated with the help of the Tolman-Oppenheimer-Volkoff (TOV) equation. In the crust region we match smoothly our EoS with the BPS EoS [6].

## 3. A novel mechanism how one can stiffen the EoS in an RMF model

With appropriate choices of the scaling functions $\eta_V(f)$, and $U(f)$ and, perhaps, also with a modification of $m_B^*(f)$, our model can be reduced to various traditional RMF models. For example, putting

$$\eta_\omega(f) = \eta_\rho(f) = 1, \quad U(f) = m_N^4(bf^3/3 + cf^4/4) \tag{3}$$

we recover the extension of the original Walecka model proposed by Boguta and Bodmer [7], the so-called non-linear Walecka (NLW) model, which allows to fit satisfactory the nuclear saturation properties, which we chose here for an illustration as follows: the saturation density $n_0 = 0.16\,\mathrm{fm}^{-3}$, the binding energy per nucleon $\mathcal{E}_0 = -16\,\mathrm{MeV}$, the compressibility modulus $K = 250\,\mathrm{MeV}$, the effective nucleon mass $m_N^*(n_0)/m_N = 1 - f_0 = 0.8$, and the symmetry energy $E_{\mathrm{sym}} = 30\,\mathrm{MeV}$. For this choice of parameters and scalings (3) the limiting mass of a NS is 1.92 $M_\odot$ that is below the experimental limit.

In [8] we proposed a simple method how one can stiffen the EoS in an RMF model. The strategy is to reduce the growth of the scalar field above a value $f_{\mathrm{cut}}$. As argued in [8], this can be achieved by adding to the original energy-density functional (1) a function of the scalar field,

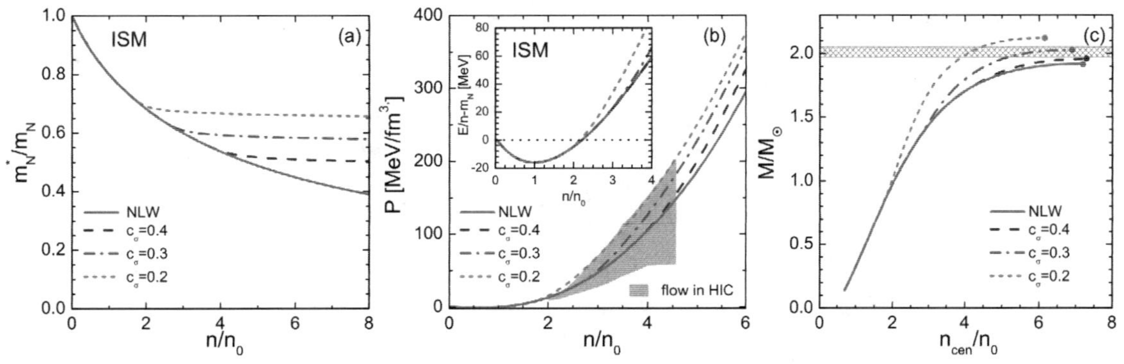


**Figure 1.** Results for the NLW and NLWcut models. Panel (a): the nucleon effective mass in the ISM as a function of the nucleon density. Panel (b): Pressure $P$ as a function of the nucleon density in the ISM for various values $c_\sigma$; the insertion shows the binding energy per nucleon. The shadowed area is the constraint on the pressure from the particle flow in heavy-ion collisions (HIC) obtained in [4]. Panel (c): The NS mass as a function of the central density. The hatched band denotes the uncertainty in the value of the measured heaviest NS mass [3].

which is vanishingly small for $f \sim f(n_0) < f_{\rm cut}$ and is sharply increased for $f \sim f_{\rm cut}$, so that its second derivative becomes su ciently large. Then, the self-consistent solution of the equation for $f$ does not let the function $f(n)$ to grow significantly above $f_{\rm cut}$ within a broad interval of densities for $n > n_*$, where $n_*$ is determined by $f_{\rm cut}$. We will illustrate this mechanism at hand of the following example: we replace $U$ in (3) by $U(f) + \Delta U(f)$, where

$$\Delta U(f) = \alpha \ln[1 + \exp(\beta(f - f_{\rm cut}))] , \quad \alpha = m_\pi^4 , \quad \beta = 120. \tag{4}$$

We will refer to such an extension as the $\sigma$-cut scheme and call the model as the NLWcut model. Applying the method to models with different values of $f_0$ it is convenient to parameterise $f_{\rm cut} = f_0 + c_\sigma(1 - f_0)$, and allow different values of $c_\sigma$ to vary. The NLW model is recovered from the NLWcut model for $c_\sigma \gtrsim 1$ corresponding to $f_{\rm cut} \gtrsim 1$.

On panel (a) in Fig. 1 we show the effective nucleon mass $m_N^*$ as a function of the nucleon density in ISM. We see that the function $m_N^*(n)$ flattens off for densities $n > n_* \simeq 1.9\, n_0$, $3.0\, n_0$, and $4.0\, n_0$ for $c_\sigma = 0.2$, 0.3, and 0.4, respectively. The influence of the quenching of the nucleon-mass decrease on the EoS is illustrated on panel (b) in Fig. 1, where we show the total pressure $P(n)$ and the binding energy per nucleon as functions of $n$ in the ISM. We observe that the replacement (4) leads to a sizable increase of the pressure for $n > n_*$. The shadow region shows the constraint on the pressure extracted from the particle flow in HIC [4]. We see that the EoSs in the NLWcut models satisfy this constraint for $c_\sigma \gtrsim 0.2$. In difference from the pressure, the energy density and corresponding binding energy increase much weaker after the inclusion of the new term (4). The NS masses as functions of the central density are shown on panel (c) in Fig. 1. The stiffening of the EoS reflects in an increase of the NS mass at a given central density. The proposed $\sigma$-cut scheme allows us to shift the maximum NS mass from $1.92\, M_\odot$ for the original NLW model to $1.96\, M_\odot$, $2.03\, M_\odot$, and $2.12\, M_\odot$ for the NLWcut models with $c_\sigma = 0.4$, 0.3, and 0.2, respectively.

## 4. KVOR, KVORcut and MKVOR models

In [1] the model KVOR was proposed with scaling functions $\eta_\rho > 1$ and $\eta_\omega < 1$ for $n > n_0$, which were chosen to shift the threshold of direct Urca (DU) process, $n \to p + e + \nu$, to higher densities (via the $\eta_\rho$ tuning) and to increase the maximum neutron star mass (via the $\eta_\omega$ tuning).

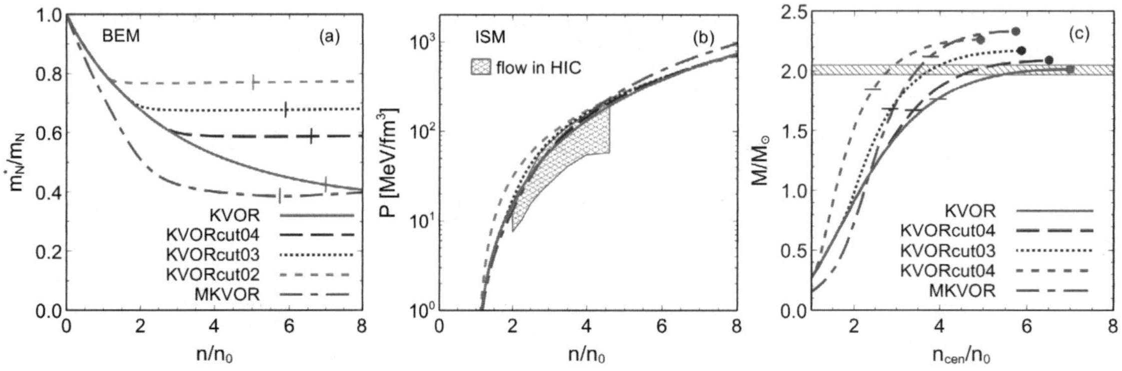


**Figure 2.** Panel (a): The nucleon mass as a function of the density in KVOR, KVORcut and MKVOR models in BEM. Vertical bars indicate central densities in the stars with the maximum masses for the corresponding model. Panel (b): Pressure as a function of the nucleon density for ISM. Double-hatched area is the constraint from the particle flow in heavy-ion collisions [4]. Panel (c): The NS mass versus the central density. Bold dots show maximum available masses. The band shows uncertainty range of the measured mass of PSR J0348+0432 [3]. Horizontal dashes show the DU thresholds.

The KVOR model exploits the saturation parameters $m_N^*(n_0)/m_N = 0.805$, $K = 275\,\mathrm{MeV}$, $J_0 = 32\,\mathrm{MeV}$, and other parameters being the same as for the NLW model discussed above. The KVOR EoS was shown in [9] to satisfy appropriately the majority of experimental constraints known by that time. Hyperons were not included in the original KVOR model. However even without hyperons the KVOR EoS yields the maximum mass $2.01M_\odot$ that fits the constraint [3] only marginally.

We discuss now several extensions of the KVOR model [1], which can better fulfill the modern constraints on the nuclear EoS. To stiffen the EoS we use the method described in the previous section but now will change not the potential $U(f)$ but the $\eta_\omega(f)$ function as

$$\eta_\omega^{\mathrm{KVOR}}(f) \to \eta_\omega^{\mathrm{KVOR}}(f) + \frac{a_\omega}{2}\left[1 + \tanh(b_\omega(f - f_{\mathrm{cut},\omega}))\right]. \tag{5}$$

We will call such extensions the KVORcut models. We consider three choices of parameters $a_\omega$, $b_\omega$ and $f_{\mathrm{cut},\omega}$: KVORcut04 with $a_\omega = -0.5$, $b_\omega = 55.76$, $f_{\mathrm{cut},\omega} = 0.454$; KVORcut03 with $a_\omega = -0.5$, $b_\omega = 46.78$, $f_{\mathrm{cut},\omega} = 0.365$; KVORcut02 with $a_\omega = -0.2$, $b_\omega = 74.55$, $f_{\mathrm{cut},\omega} = 0.249$. In [10] we introduced another RMF model with scaling — the MKVOR model — where the mechanism of the $f$ stabilization is implemented not in the $\eta_\omega(f)$ function but in the $\eta_\rho(f)$ function, which rises for $f < 0.5$ similarly to the KVOR model but drops fast for $f > 0.52$. Therefore in the MKVOR model the nucleon mass decreases with the density increase in the ISM matter stronger than in the $\beta$-equilibrium matter (BEM). All other scaling functions we have chosen anew in order to satisfy better various constraints on the EoS form microscopic calculations, see [10] for details.

The nucleon effective masses for KVORcut and MKVOR models are shown in Fig. 2, panel (a), as functions of the baryon density for BEM. In all models the effective hadron masses first decrease with the density and then saturate at some approximately constant values. The pressure as a function of the nucleon density in the ISM is shown on panel (b) in Fig. 2. We see that pressure for the KVORcut04 model is only slightly higher than that of the KVOR one. The curve for the KVORcut03 model goes on top of the upper boundary of the experimentally allowed region. The MKVOR-model curve goes through the region for $n < 4n_0$ but escapes it at higher density. Oppositely, the curve for the KVORcut02 model lies above the region for

$n < 3.5n_0$ but enters it for higher $n$. Panel (c) in Fig. 2 demonstrates the NS mass as a function of the central density for our EoSs. The band shows the uncertainty range of the masses for PSR J0348+0432 [3] which is fulfilled by all models. We see that for the KVORcut models the lesser the value $f_\omega$ is chosen, the larger the value of the maximum mass is and the smaller the central density corresponding to $M_{\rm max}$ becomes. The MKVOR EoS is the stiffest among the EoSs considered here. Dashes on the right panel indicate the DU thresholds. The MKVOR model yields the highest DU threshold density. Thus, the KVORcut03 and MKVOR models are the most promising models for the simultaneous fulfillment of the particle-flow and NS-mass constraints.

## 5. Inclusion of hyperons

We discuss now the results of the hyperonization phase transition on the EoS within our RMF models. We will consider three schemes for the inclusion of hyperons:
(i) The hyperons are included with the vector-meson coupling parameters $x_{\omega H}$ and $x_{\rho H}$ according to (2) and the scalar coupling parameters $x_{\sigma H}$ constrained by the hyperon binding energies as outlined in section 2. We switch off the hyperon coupling to $\phi$ mesons, putting $x_{\phi H} = 0$. The models with hyperons obtained within such a scheme are labeled as the KVORH, KVORHcut03 and MKVORH models.
(ii) Next we incorporate the $\phi$-meson mean field and assume the very same scaling of its mass as for all other hadrons, i.e., $m^*_\phi(f) = m_\phi(1-f)$, but with the unscaled hyperon-$\phi$ coupling constants, which gives the scaling function $\eta_\phi(f) = (1-f)^2$. The resulting models are denoted as KVORH$\phi$, KVORHcut03$\phi$ and MKVORH$\phi$.
(iii) In addition to the $\phi$-meson mass scaling we take into account a possible change of the $x_{\sigma H}$ parameters with an increase of density. We assume these parameters being unchanged for densities $n \lesssim n_0$ and decreasing for higher densities, so that they reach and stay zero for the baryon density $n > n_{cH}$, where $n_{cH}$ are critical densities for hyperonization. With such a parametrization we will exploit vacuum masses of the hyperon $H$ for $n > n_{cH}$. Such extensions of the models are call KVORHcut03$\phi\sigma$ and MKVORH$\phi\sigma$ models.

As known, the inclusion of hyperons results usually in a substantial decrease of the maximum NS mass. The difference between NS masses with and without hyperons proves to be so large for reasonable hyperon fractions in the standard RMF approach that in order to solve the puzzle one needs to start with a very stiff EoS without hyperons that would be in odds with microscopical calculations and with the constraints from heavy-ion collisions.

The NS masses as functions of the central baryon density for various models with hyperons are shown in Fig. 3. Comparing with Fig. 2 we see that the appearance of hyperons reduces the maximum NS mass for the KVORH and KVORHcut03 models by $0.35\,M_\odot$ and $0.33\,M_\odot$, respectively, bringing them below the empirically known pulsar mass. We find also that hyperons cannot be incorporated in the MKVOR model, since the solution for $f(n)$ does not exist for all densities. Including the $\phi$-meson with an $f$-dependent mass and scaling of $\sigma H$ coupling constants (schemes (ii) and (iii) above) we obtain maximum NS mass of $1.97\,M_\odot$ for KVORHcut03$\phi$ that marginally agrees with the observational constraint and $2.07\,M_\odot$ for KVORHcut03$\phi\sigma$ that fully fulfils the maximum NS mass constraint. For the extensions of the MKVOR models we find the maximum mass of $2.22\,M_\odot$ for MKVORH$\phi$ and $2.29\,M_\odot$ for MKVORH$\phi\sigma$; both values are well above the observational maximum NS mass constraint.

Mass-radius relations for KVORHcut03$\phi$, KVORHcut03$\phi\sigma$, MKVORH$\phi$ and MKVORH$\phi\sigma$ EoSs are shown on panel (c) in Fig. 3. We see that the MKVORH$\phi$ and MKVORH$\phi\sigma$ EoSs satisfy all presented constraints, whereas the KVORHcut03$\phi$ and KVORHcut03$\phi\sigma$ EoSs render star radii slightly larger than the corridor of allowed values (shown by black thin lines) extracted in [13] from the analysis of empirical data.

Journal of Physics: Conference Series **668** (2016) 012064 doi:10.1088/1742-6596/668/1/012064

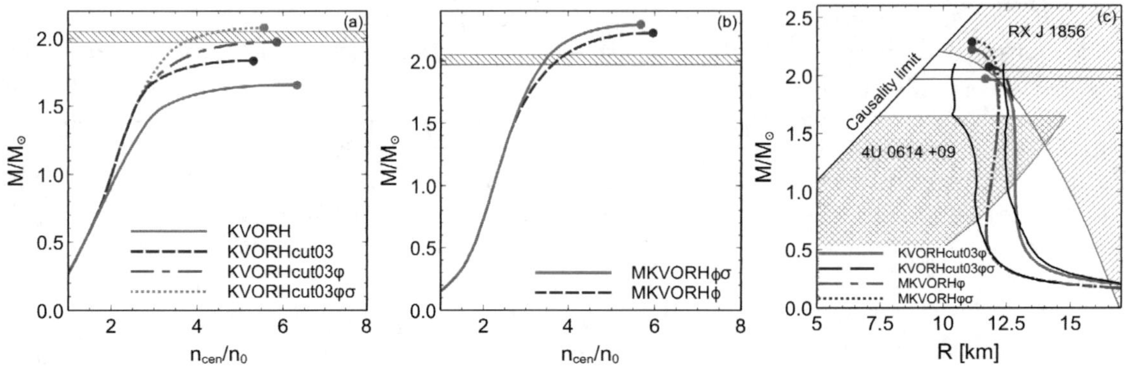


**Figure 3.** Panels (a) and (b): NS mass versus the central baryon density for various models with hyperons. Panel (c): Mass-radius relation for KVORcut03 and MKVOR models including hyperons and constraints from thermal radiation of the isolated NS RX J1856 [11] and from quasi-periodic oscillations in the LMXBs 4U 0614+09 [12]. The black thin lines show the radius constraint from the analysis [13]. The horizontal band shows the uncertainty range of masses for PSR J0348+0432 [3].

## 6. Conclusion

We presented a novel mechanism of stiffening the EoS within an RMF model with hadron masses and coupling constants depending on the scalar field. The key idea is to quench the growth of the scalar field and correspondingly a decrease of the nucleon mass with a density increase. Three different approaches are outlined where the quenching is achieved by the modification of the $\sigma$ mean-field self-interaction or by changing in the vector ($\omega$) and isovector ($\rho$) sectors of the energy density functional. Within these schemes we discuss the $\sigma$-cut versions of the traditional non-linear Walecka model, the $\omega$-cut versions of the KVOR model proposed in [1] (KVORcut models) and a new model MKVOR combining various schemes. The hyperons are included and it is demonstrated that KVORcut and MKVOR models can support su ciently heavy neutron stars if a decrease of the $\phi$-meson mass or the hyperon-$\sigma$ coupling constants are incorporated.

This work was supported by the Ministry of Education and Science of the Russian Federation (Basic part), by the Slovak Grants No. APVV-0050-11 and VEGA-1/0469/15, and by NewCompStar , COST Action MP1304. E.E.K and K.A.M. acknowledge the hospitality of Joint Institute of Nuclear Research in Dubna (Russia), where part of this work was done. E.E.K acknowledges the support by the Plenipotentiary of the Slovak Government to JINR Dubna.

## References

[1] Kolomeitsev E E and Voskresensky D N 2005 *Nucl. Phys. A* **759** 373.
[2] Demorest P *et al* 2010 *Nature* **467** 1081.
[3] Antoniadis J *et al* 2013 *Science* **340** 6131.
[4] Danielewicz P, Lacey R and Lynch W G 2002 *Science* **298** 1592.
[5] Schaffner-Bielich J 2010 *Nucl. Phys. A* **835** 279.
[6] Baym G, Pethick C and Sutherland P 1971 *Astrophys. J.* **170** 317.
[7] Boguta J and Bodmer A R 1977 *Nucl. Phys. A* **292** 413.
[8] Maslov K A, Kolomeitsev E E and Voskresensky D N 2015 arXiv:1508.03771
[9] Klähn T *et al* 2006 *Phys. Rev. C* **74** 035802.
[10] Maslov K A, Kolomeitsev E E and Voskresensky D N 2015 *Phys. Lett. B* **748** 369
[11] Trümper J E *et al* 2004 *Nucl. Phys. B (Proc. Suppl.)* **132** 560.
[12] van Straaten S et al. 2000 *Astrophys. J.* **540** 1049.
[13] Lattimer J M 2012 *Ann. Rev. Nucl. Part. Sci.* **62** 485.

Journal of Physics: Conference Series **668** (2016) 012065 doi:10.1088/1742-6596/668/1/012065

# Strangeness production and long-range correlations in pp collisions in string fusion approach

**Vladimir Kovalenko, Vladimir Vechernin**
Saint Petersburg State University, Russia
E-mail: vladimir.kovalenko@cern.ch

**Abstract.** The effects of string fusion on the correlations in strange particles production in proton-proton collisions at high energy are studied in the framework of a Monte Carlo string-parton model. The model is based on the strings formation in elementary dipole-dipole collisions. The hardness of the elementary interaction is defined by a transverse size of the colliding dipoles. The interaction between strings is realized in the accordance with the string fusion model prescriptions by the introduction of the lattice in the impact parameter plane and taking into account the finite rapidity length of strings. The particles species differentiation is implemented according to Schwinger mechanism. The parameters of the model are fixed with the experimental data on total inelastic cross section and charged multiplicity. In the framework of the model the long-range correlation functions with an accounting of strangeness have been studied. A new intensive event-by-event observable has been proposed, which characterizes the fraction of strange particles in the event. The predictions on the correlations between strangeness, multiplicity and mean transverse momentum are obtained for pp collisions at 7 TeV.

## 1. Introduction

Long-range (forward-backward) correlation studies between observables in two separated rapidity windows are considered [1] as a tool for investigation of the initial stages of the hadronic and nuclear collisions. Due to the non-perturbative nature of multiparticle production in a soft region, one has to apply the semiphenomenological approaches, such as the model of quark-gluon string formation. At high energies, because of multiple partonic interactions, the formation of several pairs of strings becomes possible. The interaction between the strings could be observed as collective phenomena in pp collisions. Particularly, according to the string fusion model [2, 3, 4], the multiplicity, transverse momentum of produced particles, as well as the yield of strangeness are modified compared to the ones of independent strings. This results in a non-trivial correlation between strangeness, multiplicity and transverse momentum.

Numerically, the long-range correlations are studied in terms of correlation functions and correlation coefficients. A correlation function by definition is the mean value $\langle B\rangle_F$ of variable B in the backward window as a function of another variable F in the forward rapidity window: $f_{B-F}(F) = \langle B\rangle_F$. The correlation coefficient is a slope of correlation function: $b_{B-F} = \frac{df_{B-F}(F)}{dF}|_{F=<F>}$. The role of the observables, $B$ and $F$, in backward and forward rapidity windows can play multiplicity ($n$) or mean event transverse momentum ($p_t = \frac{1}{n}\sum_{i=1}^{n} p_{ti}$) [4, 5, 6].

In the present paper, we introduce an additional variable for forward-backward correlation studies, which is defined as an event fraction of strange particles in a given rapidity interval –

$S$. We study correlation functions in the framework of the Monte Carlo model [7, 8], that incorporates string collectivity effects in the form of string fusion [2, 3, 4], and provide the predictions for $pp$ collisions at 7 TeV.

## 2. Monte Carlo model

The Monte Carlo model [7, 8] is based on the partonic picture of nucleon interaction. It preserves the energy and angular momentum conservation in the nucleon initial state and uses the dipole approach [9] for description of elementary partonic collisions. Multiplicity and transverse momentum are obtained in the approach of color strings, stretched between projectile and target partons. The interaction of strings is realized in accordance with the string fusion model prescriptions [2, 3, 4]. Namely, the mean multiplicity $\mu$ and the mean transverse momentum $p_T$ of the particles produced from a cluster of $k$ overlapping strings are related to those $(\mu_1, p_{T1})$ from a single string, as follows: $\mu = \sqrt{k}\mu_1, p_T = \sqrt[4]{k}p_{T1}$. For the multiplicity from one string (or a cluster of fused strings) we used Poisson distribution, with Gaussian transverse momentum spectra of produced particles.

In order to provide the possibility of a direct comparison with experimental data, including the correct description of the transverse momentum spectra, the MC model [7, 8] has been extended [9] by taking into account the hardness of elementary collision. For this purpose the mechanism similar to the one in DIPSY event generator has been incorporated in our model with the string fusion. It was assumed that the hardness an elementary collision is inversely proportional to the transverse size of the interacting dipoles: $d_i = |\vec{r_1} - \vec{r_2}|, d_i' = |\vec{r_1}' - \vec{r_2}'|$. Therefore, the mean transverse momentum of particles produced by a single string has the contributions from both edges of the string plus the additional constant term $p_0$, corresponding to the intrinsic string transverse momentum: ${p_T}_1^{\,2} = \frac{1}{d_i^2} + \frac{1}{d_i'^2} + p_0^2$. Accordingly, in the version with string fusion, the transverse momentum of a cluster of strings: $p_T^4 = \sum_{i=1}^{k} {p_T}_{1_i}^{\,4}$, where ${p_T}_{1_i}^{\,2} = \frac{1}{d_i^2} + \frac{1}{d_i'^2} + p_0^2$.

Parameters of the model are constrained from the data on total inelastic cross-section and multiplicity [10]. We used the string radius $r_{\rm str} = 0.2$ fm, and the intrinsic string transverse momentum $p_0 = 0.2$ GeV/$c$, which provides a reasonable description of the transverse momentum distribution in $pp$ collisions at the LHC energies.

The particles species differentiation in this model is implemented in the framework of Schwinger mechanism [11, 12], according to which the probability distribution of the emission of charged particles with mass $m_\nu$ and transverse momentum $p_{t_\nu}$ is proportional to $g_\nu \exp\left(-\frac{\pi(p_{t_\nu}^2 + m_\nu^2)}{t}\right)$. Here $t$ is a string tension, and $g_\nu$ is a weighting factor, responsible for cascade fragmentation like the $\rho$ meson decay into several pions. In the present model, the string tension is modified according to usual string fusion prescription: $t = t_0\sqrt{\eta}$, where $\eta$ is the number of overlapped strings. The values of $p_{t_\nu}$ and $g_\nu$ are listed in Table 1.

**Table 1.** Parameters of particles differentiation.

| $\nu$ | Particle species | $m_\nu$, GeV | $g_\nu$ |
|---|---|---|---|
| 0 | Pions from $\rho$ decays | 0.775 | 3 |
| 1 | Direct pions | 0.135 | 1 |
| 2 | Kaons | 0.494 | 1 |
| 3 | Protons | 0.938 | 1 |

## 3. Results

In Figure 1 the $n-n$, $p_t-n$ and $p_t-p_t$ correlation functions for kaons are shown. All three correlation functions have non-linear behavior with a tendency of saturation at high multiplicities and transverse momenta. These features are similar to those of correlation functions of non-identified hadrons [8]; however, they differ quantitatively.

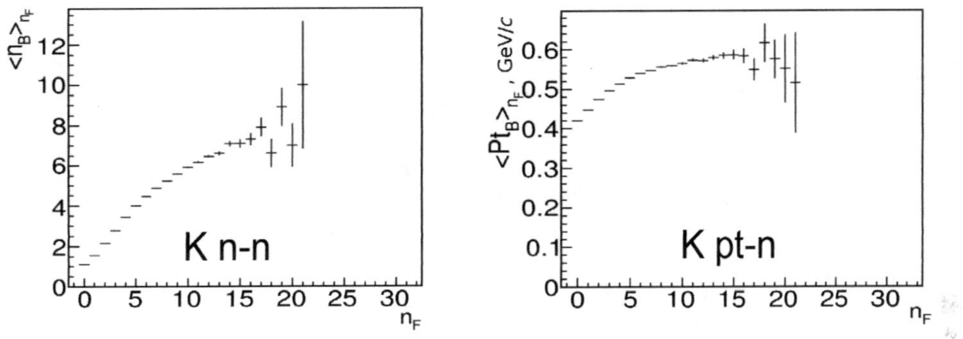


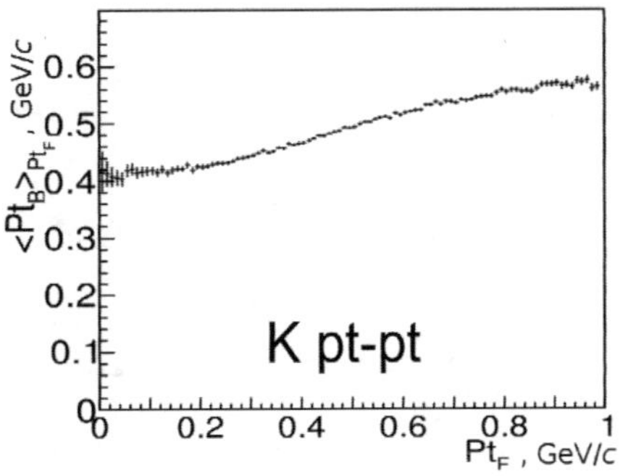


**Figure 1.** Correlation functions for kaons, produced in $pp$ collisions at $\sqrt{s}$ = 7 TeV in rapidity windows (-2.5, -0.5), (0.5, 2.5): $n-n$ (left), $p_t-n$ (right), $p_t-p_t$ (bottom), calculated in Monte Carlo model with string fusion.

In Figure 2 the results for new types of correlation functions are shown. They could be calculated only by taking into account the particle differentiation. One can see that the fraction of the strange particles ($S$) in the backward window behaves non-monotonously with charged multiplicity in the forward window: it increases at small multiplicities, then saturates and even decreases at high multiplicities. Similar behaviour is obtained for strangeness – transverse momentum ($S-p_t$) correlation function. This could be related to the growth of the non-strange resonance fraction at high string density, which after a strong decay produce more pions than strange particles. Comparison of $p_t-S$ and $S-p_t$ correlations demonstrates that the use of the transverse momentum in the forward window for event classification provides more pronounced correlation than the strangeness fraction.

The results also show non-zero strangeness-strangeness correlation, although the slope is rather smaller in $pp$ collisions. We should note that these dependencies could be obtained only when the model includes string fusion, without the fusion the correlations would be zero. Therefore, these correlations could serve as an indicator of string fusion when comparing to experimental data.

It is important, that the strangeness fraction, as well as mean event transverse momentum is an intensive variable. Hence, the $S-S$, $p_t-S$ correlation coefficients are expected to be robust against the peculiarities of the event selection in the collisions with nuclei, as it was established for $p_t-p_t$ correlations [13].

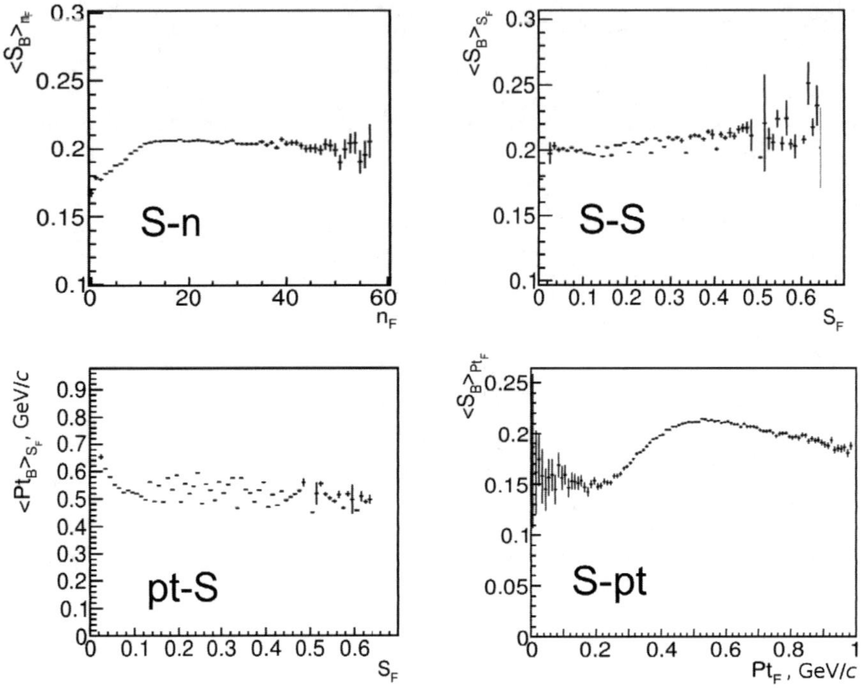


**Figure 2.** Correlation functions in $pp$ collisions at $\sqrt{s} = 7$ TeV in rapidity windows (-2.5, -0.5), (0.5, 2.5): $S - n$ (top left), $S - S$ (top right), $p_t - S$ (bottom left), $S - p_t$ (bottom right), calculated in Monte Carlo model with string fusion.

## 4. Conclusions

Long-range correlations between observables in separated rapidity window are calculated in the framework of the string fusion model, taking into account finite rapidity length of the strings, hardness of the process and particle differentiation according to Schwinger mechanism. The $n - n$, $p_t - n$ and $p_t - p_t$ correlation functions for strange particles have been studied. Four new types of correlation functions are introduced by an accounting of the mean event strangeness. The predictions for $pp$ interactions at 7 TeV are obtained.

## Acknowledgments

The authors acknowledge Saint-Petersburg State University for a research grant 11.38.242.2015. V.K. acknowledges the support of Special SPbSU Rector's Scholarship, G-RISC stipend and Dynasty Foundation Scholarship.

## References

[1] Alessandro B *et al.* (ALICE Collaboration) 2006 *J. Phys.* G **32** pp 1295–2040 (Section 6.5.15, pp 1749–1752)
[2] Amelin N S, Armesto N, Braun M A, Ferreiro E G and Pajares C 1994 *Phys. Rev. Lett.* **73** pp 2813–2816
[3] Braun M A, Pajares C and Vechernin V V 2000 *Phys. Let. B* **493** pp 54–64
[4] Braun M A, Kolevatov R S, Pajares C and Vechernin V V 2004 *Eur. Phys. J. C* **32** pp 535–546
[5] Vechernin V V and Kolevatov R S 2007 *Phys. Atom. Nucl.* **70** pp 1797–1808
[6] Vechernin V V and Kolevatov R S 2007 *Phys. Atom. Nucl.* **70** pp 1809–1818
[7] Kovalenko V N 2013 *Phys. Atom. Nucl.* **76** pp 1189–1195
[8] Kovalenko V and Vechernin V 2012 *PoS* (Baldin ISHEPP XXI) 077, arXiv:1212.2590 [nucl-th]
[9] Kovalenko V and Vechernin V 2015 *DESY Conf. Proc.* **2014-04**, 82 (pp 691–694), DOI:10.3204/DESY-PROC-2014-04/82, arXiv:1410.3884 [hep-ph]
[10] Kovalenko V 2013 *PoS* (QFTHEP 2013) 052
[11] Schwinger J 1951 *Phys. Rev.* **82** pp 664–679
[12] Biro T S, Nielsen H B and Knoll J 1984 *Nucl. Phys.* B. **245** pp 449–468
[13] Kovalenko V and Vechernin V 2014 *EPJ Web of Conferences* **66**, 04015

# Search of strangelets and "forward" physics on the collider

**A.B.Kurepin**

Institute for Nuclear Research, Russian Academy of Sciences, 117312, 60-th October Anniversary prospect, 7A, Moscow, Russia

E-mail: kurepin@inr.ru

**Abstract**. A new stage of the collider experiments at the maximum energy of protons and nuclei at the LHC may lead to the discovery of new phenomena, as well as to confirm the effects previously observed only at very high energies in cosmic rays. A specific program of the experiments is so-called "forward" physics, i.e. the study of low-angle processes. Of the most interesting phenomena can be noted the detection in cosmic rays events called Centauro, which could be explained as the strangelets production. Centauro represent events with small multiplicity and with a strong suppression of electromagnetic component. Since the energy of the beams at the collider and kinematic parameters of the forward detectors CASTOR (CMS), TOTEM, LHCf and the ADA and ADC (ALICE) are close to the parameters and energies of abnormal events in cosmic rays, it is possible to reproduce and investigate in details these events in the laboratory.

## 1. Introduction

The "forward" physics will be one of the most important part of the experimental program at the LHC during the Run 2 in 2015 – 2018 years for proton-proton and lead-lead collisions with the ever highest energy and increased luminosity. Besides of the forward detectors existing at each of the main four installations the specialized systems CASTOR, TOTEM, LHCf will be in operation. For the ALICE the new ADA and ADC detectors are installed at very large rapidity. The main physics potential for these detectors is the measurement of the total elastic and inelastic cross sections and the study of the diffraction. The diffraction processes contribute about 25 % of all inelastic proton interactions at high energies. Because of the central barrel structure and perfect particle identification of ALICE the decay of the exited by diffraction of one or two protons could be investigated. Single proton diffraction at high energy on colliders was measured by ISR [1] and Tevatron [2]. First data on single and double proton diffraction at ALICE were obtained at three energies [3]. The advantage of the AD detectors of ALICE is the increasing of the detectors efficiency to smaller diffraction masses and the extension of the rapidity gaps for the study of central diffraction. The new experimental data on the diffraction will provide the important information on the non perturbation QCD.

However at the same time the forward detectors could be used for the search of the rare exotic phenomena observed in some cosmic rays experiments. Indeed, the energy of the LHC beams in Run 2 at $\sqrt{s}$ = 13 TeV for proton-proton collisions corresponds to $10^{17}$ eV energy of cosmic rays. For lead-lead collisions $\sqrt{s}$ = 5.27 TeV per nucleon correspond to 1.5 x $10^{16}$ eV per nucleon or 3x$10^{18}$ eV total heavy nucleus energy.

The most interesting unusual cosmic rays events are called Centauros, observed in the emulsion chamber experiments at Mt. Chacaltaya and in the Pamir at energies above $10^{15}$ eV. These events have abnormal ratio of hadrons to photons yield. Centauros exhibit relatively small multiplicity and strong suppression of the electromagnetic component [4]. Also some hadron rich events are accompanied by a strongly penetrating component [5,6] or long living cascades with many maxima in the transition curve [7,8]. Some new results by re measurement of Centauros were published [9]. Several models have been proposed by assumption that the Centauros events are the results of nucleus-nucleus collisions with formation of a quark-gluon plasma.

Due to high baryochemical potential the creation of $u$-$\bar{u}$ and $d$-$\bar{d}$ quarks are blocked by Pauli principle and the fragmentation of gluons is going to $s$-$\bar{s}$ pairs. Therefore the production of pions is suppressed as well as gamma from neutral pions decay [10, 11]. Then strangelets are formed via a mechanism of strangeness distillation and could be identified as the strongly penetrating particles in forward direction with large rapidity [12]. Simulations based on this model predicted the mass of strangelets up to several hundreds of GeV and large transverse momentum due to the large mass [13]. The signature for observation of Centauro events or strangelets in heavy ion collisions at LHC with the forward detectors could be small multiplicities, predominance of detected hadrons compared to gamma and large transverse momenta. At high energy, compared to the energy of cosmic rays, there was only one attempt for the search of strangelets on the collider RHIC at $\sqrt{s}$ = 200 GeV in gold-gold collisions by the STAR experiment [14]. Assuming that the strangelets have nearly straight trajectory in the magnetic field due to a small charge to mass ratio, the search was made to find a large energy deposition with a narrow transverse shower profile in the Zero degree calorimeter. No candidates were observed and an upper limit was obtained of the order $10^{-5}$ to $10^{-6}$, depending on the possible strangelets mass. In the next chapters two proposed experiments for the search of strangelets at LHC are considered.

## 2. Search of strangelets by the CASTOR calorimeter.

The CASTOR calorimeter was constructed especially for the search of Centauros like events or strangelets in the heavy ion collisions at the LHC. Firstly it was proposed to be installed at the Point 2 of the ALICE [12], but afterwards it was positioned about 14 m from the Point 3 of the CMS [15]. The CASTOR calorimeter consists of 80 tungsten and quartz layers placed symmetric around the beam pipe and tilted at 45 degrees to the beam axes ( Fig.1 ). CASTOR is azimuthally segmented into 16 semi octants and divided longitudinally into 18 sections. It covers the pseudorapidity interval $5.2 < \eta < 6.6$. The photo multipliers with the air light guides were calibrated by electron, pion and muon beams. First 8 sections (~14.7 $X_0$) comprise the electromagnetic calorimeter and the remaining 72 (~ 9.47 $\lambda_i$) the hadronic calorimeter. The structure of the CASTOR calorimeter permits to search for the events with unusual ratio of electromagnetic to hadron emission. The longitudinal transition curve can display the long living cascade with many maxima structure like it was observed in some of Centauro cosmic ray events. The alignment effects seen in some events could be detected due to the azimuthally segmentation. The additional multiplicity information could be obtained from the TOTEM – T2 tracker placed in front of CASTOR.

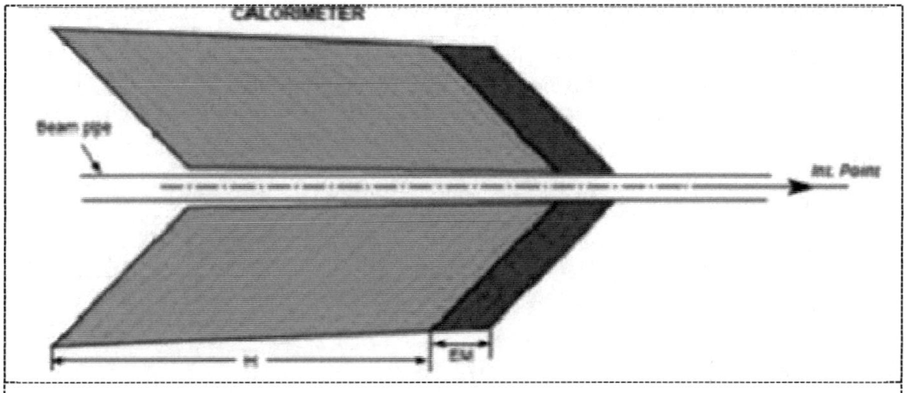


**Figure 1.** Construction of the CASTOR calorimeter.

## 3. The use of the ADA –ADC detectors of ALICE for the search of exotic events

The important contribution for the "forward" physics at the LHC could be obtained by the new detector system ADA - ADC of ALICE. The two scintillator hodoscopes are installed at the distance 17 m from IP 2 at the A – side and 19.5 m at the C – side of ALICE ( Fig.2 ). They cover the pseudorapidity intervals $4.8 < \eta < 6.3$ and $-7.0 < \eta < 5$ at A and C sides correspondently. The kinematics of these detectors is close to that of the Centauro events in pseudorapidity and in transverse momentum. Each of the AD detectors consists of 4 scintillators arranged around the beam pipe. Wave shifter bars are glued on the side of each scintillator. Optical fibers bundles collect the light from the wave shifters and transmit to the PMTs. The centrality information will be provided by the Central Barrel detectors. Strangelets with large rapidity and high transverse momentum will give a large amplitude signal in one of the scintillatotrs of the AD detector and small multiplicity measured in other AD scintillators and in other ALICE forward detectors FMD, T0 and V0.

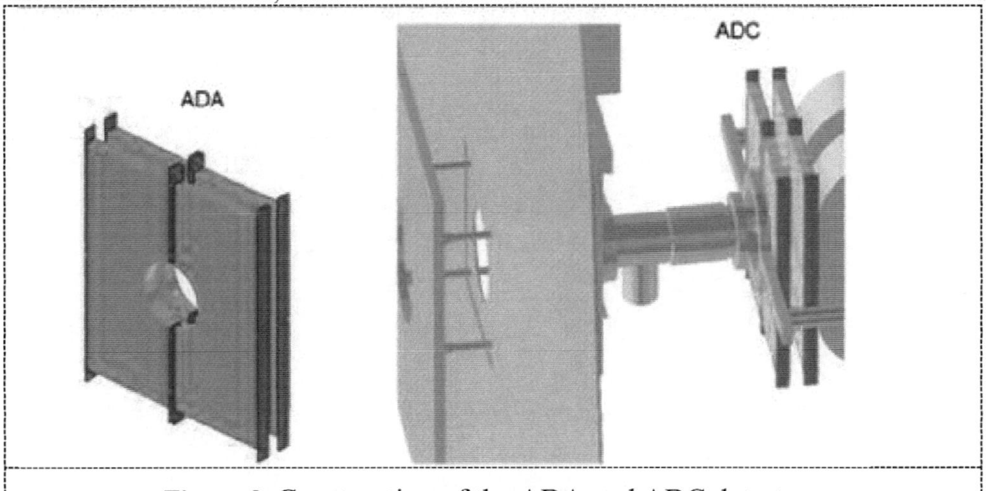


**Figure 2.** Construction of the ADA and ADC detectors.

## 4. Conclusion

The forward detectors CASTOR and AD will provide a possibility for the search of exotic events during Run 2 at the LHC. This will complement the diffraction physics program and could solve the nature of Centauro events observed in the cosmic rays experiments.

## References

[1] Armitage J C M et al. 1982 Nucl. Phys. B194 365

[2] Affolder T et al. 2001 Phys. Rev. Lett. 87 141802
[3] ALICE Collaboration 2013 Eur. Phys. J. C73 2456
[4] Lattes C M G, Fugimoto Y and Hasegawa S 1980 Phys. Rep. 65 151
[5] Hasegawa S and Tamada M 1996 Nucl. Phys. B474 225
[6] Barsadzei L T et al. 1992 Nucl. Phys. B370 365
[7] Arisawa T et al. 1994 Nucl. Phys. B424 241
[8] Buja Z et al. 1981 Proc. 17th ICRC, Paris Vol. 11 104
[9] Ohsawa A, Shibuya E and Tamada M 2006 J. Phys. G: Nucl. Part. Phys. 32 2333
[10] Panagiotou A D et al. 1992 Phys. Rev. D45 3134
[11] Gladysz-Dziadus E and Wlodarczyk Z 1997 J. Phys. G: Nucl. Part. Phys. 23 2057
[12] Angelis A L S et al. 1999 arXiv:hep-ph/99082210
[13] Angelis A L S et al. 2004 Phys. At. Nucl. 67 396
[14] Abelev B I et al. 2007 arXiv:nucl-ex/0511047
[15] Norbeck E et al. 2007 CMS CR 2007/013

Journal of Physics: Conference Series **668** (2016) 012067 doi:10.1088/1742-6596/668/1/012067

# Study on initial geometry fluctuations via correlation of finite distributions of secondary particles in nucleus-nucleus interactions

**A I Fedosimova, A Sh Gaitinov, I A Lebedev and A T Temiraliev**

Institute of Physics and Technology, Almaty 050032, Kazakhstan

E-mail: lebedev@sci.kz

**Abstract.** The study of the peculiarities of the distribution of secondary particles depending on the degree of centrality and the degree of asymmetry of the interacting nuclei, is performed. The number of multicharged fragments of the projectile nucleus ($N_f$) in interactions of sharply asymmetric nuclei depends on the centrality degree of interaction. In the majority of peripheral interactions $N_f$ is equal 1. In the most of the central interactions there are $N_f > 1$ or $N_f$=0. At that, the events with $N_f$ =1 are separated clearly in the distribution of the total charge of the fragments of a projectile nucleus ($Q$) depending on the nature of the correlation of the number of fragments of the target nucleus ($N_h$) and the multiplicity of secondary particles from interaction area ($n_s$). The peak in the field of the maximum values of $Q$ characterizes peripheral events with various degree of the periphery on a growing branch of average $N_h$-$n_s$ curve. And the peak in the field of small $Q$ characterizes the central events with various degree of centrality on the decreasing branch of average $N_h$-$n_s$ curve.

## 1. Introduction

According to the present-day conceptions, the hadron substance transits within the interactions of the nuclei, at the high energies, into a state of the quark-gluon plasma (QGP), in which the quarks and gluons stay in a quasi-free state [1]. The expansion of secondary particles, formed from the QGP cluster of nuclear matter, leads to the collective nature of the formation of secondary particles. It can be detected on the basis of the correlation analysis of finite distributions of secondary particles. For the research of the multiparticle correlations different approaches and methods are used [2]-[5].

On the other hand, the dynamics of the interaction of nuclei is determined not only by the energy and mass of the interacting nuclei, but also the geometry of the nucleus-nucleus collisions. Accounting for the influence of fluctuations of the initial states in individual interactions allows to explore the true dynamic correlation of the final state of interactions of excited hadronic systems [6]-[7].

In central collisions the number of generated secondary particles is maximal. If the collision is peripheral, then the overlap of the interacting nuclei is incomplete and resulting fireball expands asymmetrically. Thus, the degree of centrality of interaction significantly affects the parameters of the spatial (angular) distribution of secondary particles [8]-[9].

The initial state, about which there is usually very little direct experimental information, leads to significant peculiarities in the distribution of fragments and the multiplicity of secondary particles. The study of the peculiarities allows to give a physical interpretation of the results on the base of the differences in the initial states of collisions [10]-[11].

In the present work we analyze the peculiarities of the distribution of secondary particles depending on the degree of centrality and the degree of asymmetry of the interacting nuclei on the base of experimental data of interactions of gold nuclei $^{197}Au$ at 10.7 AGeV with photoemulsion nuclei [12].

The nuclear emulsion represents a certain combination of light and heavy nuclei. On the one hand, it allows to analyze various types of nuclear interactions obtained in exactly the same experimental conditions. On the other hand, it introduces additional problems with the identification of the target nucleus.

Peculiarities of distribution of secondary particles and fragments in depending on the initial conditions of collisions, we analyzed in two complementary directions.

The first direction is the study of the parameters of the fragmentation of the projectile nucleus for separation of the peripheral interactions with a small number of interacting nucleons and with a large value of the total charge of the fragments of a projectile nucleus.

The second direction was based on the study of the dependence of the number of fragments of the target nucleus and the multiplicity of secondary particles to estimate the centrality degree of the interaction and to separate interactions with light and heavy emulsion nuclei.

## 2. The fragmentation of projectile nucleus

In interaction of huge gold nucleus with smaller nucleus of the photoemulsion the quantity of multi-charged fragments depends on centrality degree of the interaction.

Schematic visualization of events of varying degrees of centrality is presented in Figure 1. In peripheral interactions the appearance of events with one multi-charged fragment are more probably in comparison with deeply central interactions. The total remainder charge of the fragments of the projectile nucleus is less for events with larger overlap of the interacting nuclei.

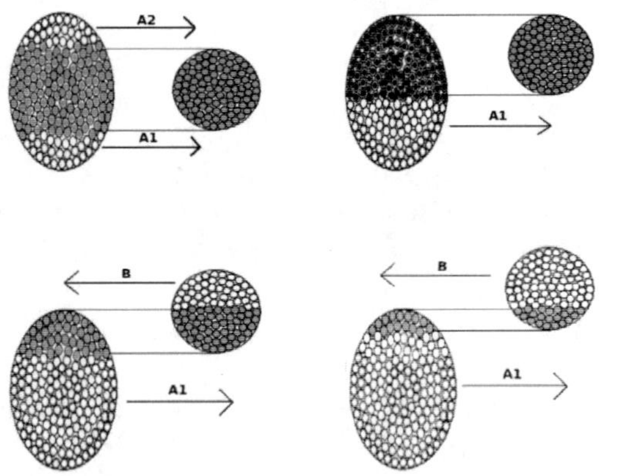


**Figure 1.** Schematic visualization of events of varying degrees of centrality and periphery.

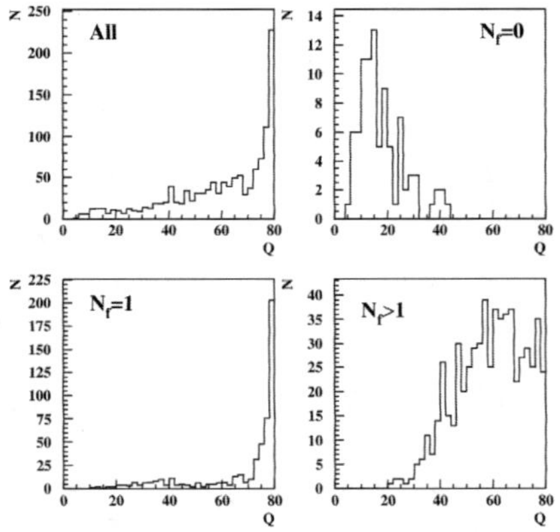


**Figure 2.** The distribution of the total charge $Q$ of the fragments of a projectile nucleus for events with different numbers of multi-charged fragments $N_f$ for interactions of 10.7 AGeV $^{197}Au$ with emulsion nuclei.

Journal of Physics: Conference Series **668** (2016) 012067 doi:10.1088/1742-6596/668/1/012067

In Figure 2 the distributions of total charge $Q$ of projectile nucleus fragments for events with various number of multi-charged fragments $N_f$ are presented.

From Figure 2 it follows that in interactions with one multi-charged fragment, in the most of the events large total charge of the projectile nucleus fragments, is discovered.

The region of $Q > 70$ is separated visually on a low background of other events. Thus, events with one multi-charged fragment and $Q > 70$, probably, characterize the peripheral interaction.

Events with several multi-charged fragments mainly are characterized by lower values of $Q$ and likely characterize more central collisions.

Special attention should be paid to the fact that there is a significant number of events in which multi-charged fragments are absent. That is, under certain conditions, the complete destruction of the huge gold nucleus at interaction with significantly less emulsion nuclei, is discovered.

## 3. The dependence of the number of fragments and the multiplicity of secondary particles

One of the most optimal parameters for estimating the centrality degree of interaction and separation of events with light and heavy emulsion nuclei, is the dependence of the number of fragments of the target nucleus and the multiplicity of secondary particles $n_s$. This correlation dependence is presented in Figure 3.

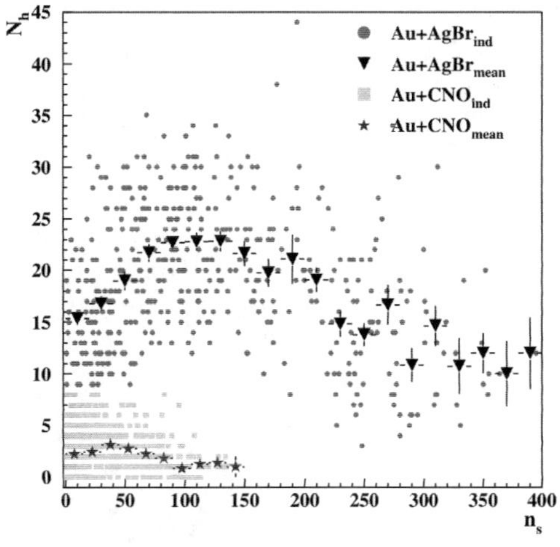


**Figure 3.** Correlation of the total number of fragments of the target nucleus $N_h$ and the multiplicity of secondary particles $n_s$ in interactions of 10.7 AGeV $^{197}Au$ with heavy $(AgBr)$ and light $(CNO)$ emulsion nuclei.

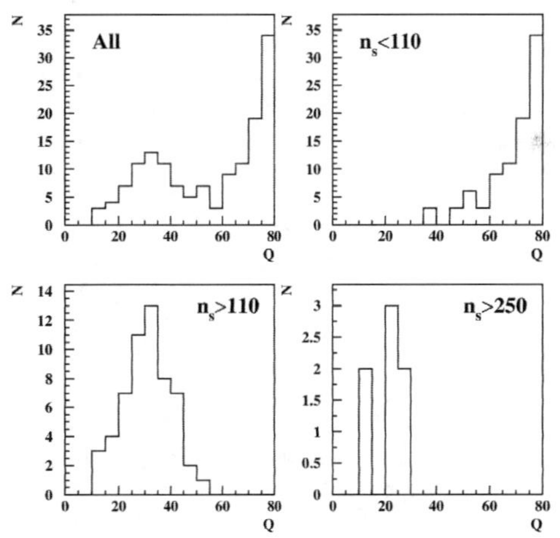


**Figure 4.** The distribution of the total charge of the projectile fragments, depending on the multiplicity $n_s$ for interaction $Au + AgBr$ with one multi-charged fragment $N_f = 1$.

As it can be seen from Figure 3, events of interaction with light $(CNO)$ and heavy $(AgBr)$ emulsion nuclei can be separated quite well. Selection of interactions with light nuclei is limited to two requirements.

First, the maximum number of fragments of the target nucleus may not exceed 8, which corresponds to the charge of the largest of light emulsion nuclei - nuclei of oxygen.

Secondly, the maximal multiplicity $n_s$ in interactions with light nuclei emulsion is significantly lower compared to the interactions with heavy emulsion nuclei. Using of this fact allows to separate $Au + AgBr$ events with large multiplicity, in which the number of fragments of the target nucleus is less than 8, from $Au + CNO$ events.

For analysis of peripheral collisions in Figure 3 the mean values of $N_h$ for groups of interactions with heavy and light emulsion nuclei, by described above criteria, are presented separately.

In $Au+AgBr$ interactions the average dependence shows the steady growth in the field before $n_s$=110. Then it decreases and at large $n_s$ >250 it goes on the plateaus. Similar behavior shows and $Au + CNO$ dependence, but at a lower multiplicity: $n_s$=40 and $n_s$ >100, respectively. This behavior reflects the degree of peripheral interaction. This assumption is confirmed by the difference in the total charge of the fragments of a projectile nucleus for the respective areas of $n_s$. This distribution of $Q$ is presented in Figure 4.

As it can be seen from Figure 4 there is a clear two-peak distribution. The peak in the region of maximal values of $Q$ characterizes peripheral events with varying degrees of periphery on the rising branch of the average $N_h$-$n_s$ curve. The peak at low values of $Q$ characterizes the central event with varying degrees of centrality on the decreasing branch of the average $N_h$-$n_s$ curve.

## 4. Conclusion

In summary, we formulate the following main conclusions of the study of the peculiarities of the distribution of secondary particles depending on the degree of centrality and the degree of asymmetry of the interacting nuclei.

The quantity of multi-charge fragments of projectile nucleus $N_f$ in interactions of sharply asymmetric nuclei essentially depends on the centrality degree. In the majority of peripheral interactions $N_f$ is equal 1. In the most of the central interactions there are $N_f > 1$ or $N_f$ =0.

$N_f$ =1 area is rather well divided in distribution of a total charge of projectile nucleus fragments $Q$. The peak in the field of the maximum values $Q$ characterizes peripheral events with various degree of the periphery on a growing branch of average $N_h$-$n_s$ curve, which is characterizing dependence of number of fragments of the target nucleus and multiplicity of secondary particles from interaction area. And the peak in the field of small values $Q$ characterizes the central events with various degree of centrality on the decreasing branch of average $N_h$-$n_s$ curve.

## Acknowledgments

This work was supported by grant N4824/GF4 of Ministry of Education and Science of Kazakhstan Republic.

## References

[1] Shuryak E 2009 Physics of strongly coupled quark-gluon plasma *Prog. in Part. and Nucl. Phys.* **62** 48
[2] Adamovich M I *et al* 2000 Azimutal correlations of secondary particles in $^{32}S$ induced interactions with Ag(Br) nuclei at 4.5 GeV/c/nucleon *Part. and Nucl. Lett.* **101** 76
[3] Adamovich M I *et al* 2001 Factorial Moments of $^{28}Si$ Induced Interactions with Ag(Br) Nuclei *APH N.S. Heavy Ion Phys.* **113** 213
[4] Lebedev I A and Shaikhatdenov B G 1997 *J.Phys.* G **23** 637
[5] Kvochkina T N *et al* 2000 *J.Phys.* G **26** 35
[6] Luzum M and Petersen H 2014 *J. Phys.* G **41** 063102
[7] Adare A, Luzum M and Petersen H 2013 *Physica Scripta* **87** 048001
[8] Jia J and Mohapatra S 2013 *Eur.Phys.J.* C **73** 1
[9] Schenke B, Tribedy P and Venugopalan R 2014 *Phys.Rev.* C **89** 064908
[10] Voloshin S A 2012 *Prog. in Part. and Nucl. Phys.* **67** 541
[11] Vovchenko V, Anchishkin D and Csernai L P 2013 *Phys. Rev.* C. **88** 014901
[12] Adamovich M I *et al* 1999 *Eur.Phys.J.* A **5** 429

Journal of Physics: Conference Series **668** (2016) 012068 doi:10.1088/1742-6596/668/1/012068

# Thermal and non-thermal charmed meson production in heavy ions collisions at the LHC

**I P Lokhtin**[1,*], **A V Belyaev**[1], **G Kh Eyyubova**[1,2], **G Ponimatkin**[3] and **E Yu Pronina**[1]

[1] Skobeltsyn Institute of Nuclear Physics, Lomonosov Moscow State University, Moscow, Russia

[2] Faculty of Nuclear Sciences and Physical Engineering, Czech Technical University in Prague, Prague, Czech Republic

[3] Ostrov Industrial High School, Ostrov, Karlovy Vary District, Czech Republic

E-mail: Igor.Lokhtin@cern.ch

**Abstract.** The phenomenological analysis of the LHC data on $p_T$-spectrum and elliptic flow of $J/\psi$ and D mesons in PbPb collisions at $\sqrt{s_{NN}} = 2.76$ TeV is presented. The charmed meson production pattern in PbPb collisions may be reproduced by two-component model HYDJET++ including thermal and non-thermal components. The significant part of D-mesons is found to be in a kinetic equilibrium with the created medium, while $J/\psi$-mesons are not.

## 1. Introduction

Studying the charmed hadron production is a particularly useful tool to probe properties of hot and dense matter created in relativistic heavy ion collisions. The modern pattern of multi-particle production in (most central) heavy ion collisions at RHIC and LHC agrees with the formation of hot strongly-interacting matter with hydrodynamical properties ("quark-gluon fluid"), which absorbs energetic quarks and gluons due to their multiple scattering and medium-induced energy loss (see, e.g., [1]). Within such paradigm, a number of questions on heavy flavor production arises. Are heavy quarks thermalized in quark-gluon plasma? What is the mass dependence of in-medium quark energy loss? Are charmed hadrons in a kinetic equilibrium with the medium? How the specific pattern of quarkonium suppression is related to the interplay between thermal and non-thermal mechanisms of $J/\psi$-meson production? In this paper, some results of the phenomenological analysis of LHC data on momentum spectra and elliptic flow of charmed hadrons (D, $J/\psi$) in PbPb collisions at $\sqrt{s_{NN}} = 2.76$ TeV within two-component model HYDJET++ are presented. The comparison with RHIC results is also discussed.

## 2. HYDJET++ model

HYDJET++ is the model of relativistic heavy ion collisions, which includes two independent components: the soft hydro-type state ("thermal" component) and the hard state resulting from the medium-modified multi-parton fragmentation ("non-thermal" component) [2].

The soft component represents the hadronic state generated on the chemical and thermal freeze-out hypersurfaces obtained from the parametrization of relativistic hydrodynamics with preset freeze-out conditions (the adapted event generator FAST MC [3, 4]). Hadron multiplicities are calculated using the effective thermal volume approximation and Poisson multiplicity distribution around its mean value, which is supposed to be proportional to a number of participating nucleons for a given impact parameter of an AA collision. To simulate the elliptic flow effect, the hydro-inspired parametrization is implemented for the momentum and spatial anisotropy of a thermal hadron emission source.

The approach used for the hard component is based on the PYQUEN partonic energy loss model [5]. The simulation of single hard NN sub-collision by PYQUEN is constructed as a modification of the jet event obtained with PYTHIA_6.4 generator [6]. It takes into account medium-induced rescattering, collisional and radiative energy loss of hard partons in the expanding quark-gluon fluid, realistic nuclear geometry and nuclear shadowing effect. The mean number of jets produced in an AA event is calculated as a product of the number of binary NN sub-collisions at a given impact parameter per the integral cross section of the hard process with the minimum transverse momentum transfer $p_{\rm T}^{\rm min}$. The partons produced in hard processes with the momentum transfer lower than $p_{\rm T}^{\rm min}$, are considered as being "thermalized". So, their hadronization products are included "automatically" in the soft component of the event.

The input parameters of the model for soft and hard components have been tuned from fitting to heavy ion data on various observables for inclusive hadrons at RHIC [2] and LHC [7].

Charmed meson production in HYDJET++ includes soft and hard components as well. Thermal production of $D$, $J/\psi$ and $\Lambda_c$ hadrons is treated within the statistical hadronization approach [8, 9]. Momentum spectra of charm hadrons are computed according to the thermal distribution, and the multiplicities $N_c$ ($C = D, J/\psi, \Lambda_c$) are calculated through the corresponding thermal numbers $N_c^{\rm th}$ as $N_c = \gamma_c^{n_c} N_c^{\rm th}$, where $\gamma_c$ is the charm enhancement factor (or charm fugacity), and $n_c$ is the number of charm quarks in a hadron $C$. The fugacity $\gamma_c$ can be treated as a free parameter of the model, or calculated through the number of charm quark pairs obtained from PYTHIA and multiplied by the number of NN sub-collisions. Non-thermal charmed hadrons are generated by PYQUEN taking into account in-medium energy loss of heavy (b, c) quarks within "dead-cone" generalization [10] of BDMPS model [11, 12] for radiative loss and high-momentum transfer limit [13, 14, 15] for collisional loss.

## 3. $J/\psi$-meson production in PbPb collisions at the LHC

Some time ago it was demonstrated in [16] that if $J/\psi$-mesons are produced in HYDJET++ at the same freeze-out parameters as for inclusive (light) hadrons, then simulated $p_{\rm T}$- and $y$-spectra in 20% of most central AuAu collisions at RHIC energy $\sqrt{s_{\rm NN}} = 200$ GeV are much wider than ones measured by PHENIX [17]. However the assumption that the thermal freeze-out for $J/\psi$-mesons happens at the same temperature as chemical freeze-out (with reduced collective velocities) allows HYDJET++ to reproduce the data. It turned out that the similar situation holds at the LHC. Figure 1 shows the comparison of HYDJET++ simulations with the ALICE data [18] for $p_{\rm T}$-spectrum of $J/\psi$-mesons in 20% of most central PbPb collisions at $\sqrt{s_{\rm NN}} = 2.76$ TeV. One can see that if thermal freeze-out for $J/\psi$'s happens at the same temperature as chemical freeze-out (with reduced collective velocities), then simulated spectra match the data up to $p_{\rm T} \sim 3$ GeV/$c$. The discrepancy at high $p_{\rm T}$ may indicate on the necessity to tune PYTHIA for charmonium production in this kinematic domain. The $p_{\rm T}$-dependence of the elliptic flow coefficient $v_2$ measured by ALICE [19] is also reproduced by HYDJET++ simulations in such a case (figure 2). The contribution of non-thermal component is important at high $p_{\rm T}$.

Thus we conclude that $J/\psi$-meson production in heavy ion collisions at RHIC and LHC may be reproduced by HYDJET++ model under the assumption that thermal freeze-out of charmonia happens appreciably before thermal freeze-out of light hadrons [20], presumably at chemical freeze-out (with reduced radial and longitudinal collective velocities). Our interpretation of this is that the significant part of $J/\psi$-mesons (up to $p_{\rm T} \sim 3$ GeV) is thermalized, but without being in a kinetic equilibrium with the medium.

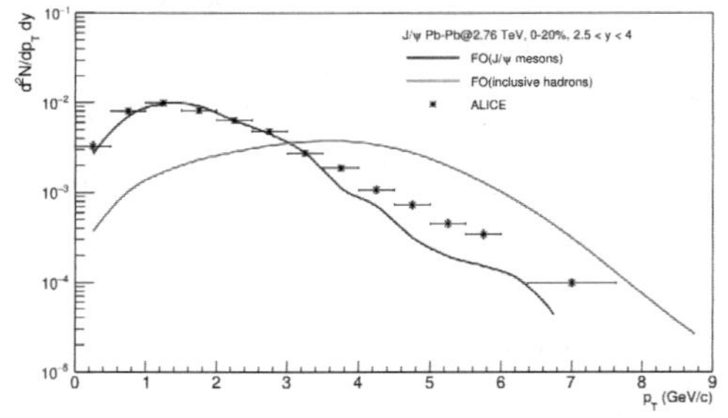


**Figure 1.** Transverse momentum spectrum of $J/\psi$-mesons at rapidity $2.5 < y < 4$ in 20% of most central PbPb collisions at $\sqrt{s_{\rm NN}} = 2.76$ TeV. The points are ALICE data [18], histograms are simulated HYDJET++ events (red – freeze-out parameters as for inclusive hadrons, blue – "early" thermal freeze-out).

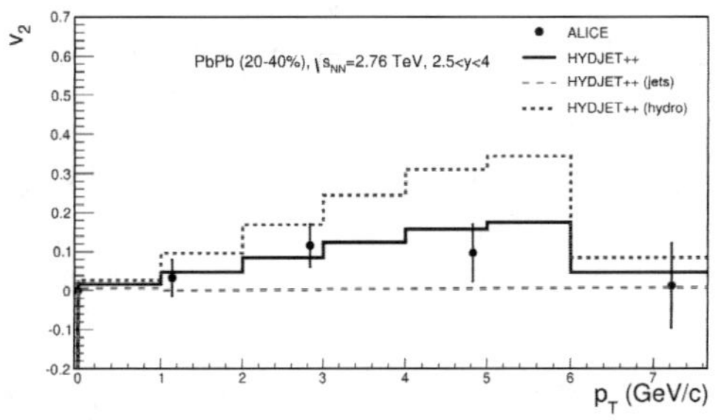


**Figure 2.** Elliptic flow coefficient $v_2(p_{\rm T})$ of $J/\psi$-mesons at rapidity $2.5 < y < 4$ for 20–40% centrality of PbPb collisions at $\sqrt{s}_{\rm NN} = 2.76$ TeV. The points are ALICE data [19], histograms are simulated HYDJET++ events (blue – soft component, red – hard component, black - both components).

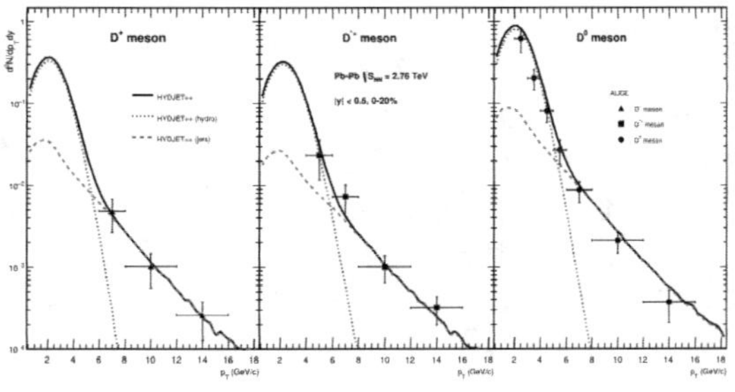


**Figure 3.** Transverse momentum spectra of D-mesons at rapidity $\mid y \mid < 0.5$ in 20% of most central PbPb collisions at $\sqrt{s_{\rm NN}} = 2.76$ TeV. The points are ALICE data [22], histograms are simulated HYDJET++ events (blue – soft component, red – hard component, black - both components).

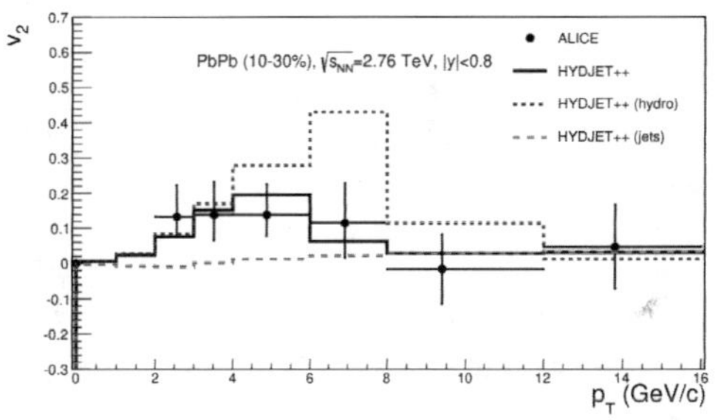


**Figure 4.** Elliptic flow coefficient $v_2(p_{\rm T})$ of $D^0$-mesons at rapidity $\mid y \mid < 0.8$ for 10–30% centrality of PbPb collisions at $\sqrt{s}_{\rm NN} = 2.76$ TeV. The points are ALICE data [23], histograms are simulated HYDJET++ events (blue – soft component, red – hard component, black - both components).

## 4. D-meson production in PbPb collisions at the LHC

At first we have checked that $p_{\rm T}$-spectrum of D-mesons measured by STAR [21] in 10% of most central AuAu collisions at $\sqrt{s_{\rm NN}} = 200$ GeV is reproduced by HYDJET++ simulation with fugacity $\gamma_c = 7$ and the same freeze-out parameters as for $J/\psi$-mesons (but not for inclusive hadrons). It means that D-mesons like $J/\psi$-mesons are not in a kinetic equilibrium with the medium at RHIC. However the situation gets changed dramatically at the LHC. Figure 3 shows the comparison of HYDJET++ simulations with the ALICE data [22] for $p_{\rm T}$-spectra of $D^0$, $D^{\pm}$ and $D^{*\pm}$ mesons in 20% of most central PbPb collisions at $\sqrt{s_{\rm NN}} = 2.76$ TeV. The freeze-out parameters as for inclusive hadrons have been used, the fugacity value $\gamma_c = 11.5$ being fixed from $J/\psi$ yield. The model results are close to the data within the experimental uncertainties. Note that $p_{\rm T}$-spectra and nuclear modification factors $R_{\rm AA}(p_{\rm T})$ (the ratio of particle yields in AA and pp collisions normalized on the mean number of NN sub-collisions) of D-mesons are reproduced by HYDJET++ up to very high $p_T$. So the treatment of heavy quark energy loss in hard component of the model (PYQUEN) seems quite successful. In addition, HYDJET++ is able to reproduce the ALICE data [23] on $p_T$-dependence of the elliptic flow coefficient $v_2$ (figure 4).

Thus in contrast to RHIC, thermal freeze-out of D-mesons at the LHC happens simultaneously with thermal freeze-out of light hadrons. Therefore the significant part of D-mesons (up to $p_T \sim 4$ GeV/$c$) seems to be in a kinetic equilibrium with the medium.

## 5. Summary

Momentum spectra and elliptic flow of D and $J/\psi$ mesons in PbPb collisions at the LHC are reproduced by two-component model HYDJET++ including thermal (soft) and non-thermal (hard) components. Thermal freeze-out of D-mesons happens simultaneously with thermal freeze-out of light hadrons. Thermal freeze-out of $J/\psi$-mesons happens appreciably before freeze-out of light hadrons, presumably at chemical freeze-out (with reduced radial and longitudinal collective velocities). Thus the significant part of D-mesons (up to $p_T \sim 4$ GeV/c) seems to be in a kinetic equilibrium with the medium, while $J/\psi$ mesons are not. The production of charmed mesons at high transverse momenta is determined by non-thermal charm production mechanism and medium-induced heavy quark energy loss.

## 6. Acknowledgments

Discussions with L.V. Bravina, V.L. Korotkikh, L.V. Malinina, A.M. Snigirev, E.E. Zabrodin and S.V. Petrushanko are gratefully acknowledged. We thank our colleagues from ALICE and CMS collaborations for fruitful cooperation. This work was supported by the Russian Scientific Fund under Grant No. 14-12-00110 in a part of computer simulation of $p_T$-spectrum and elliptic flow of charmed mesons in PbPb collisions. G.E. acknowledges the European Social Fund within the framework of realizing the project Support of Inter-sectoral Mobility and Quality Enhancement of Research Teams at Czech Technical University in Prague, CZ.1.07/2.3.00/30.0034.

## References

[1] *Proc. of 24th Int. Conf. on Ultra-Relativistic Nucleus-Nucleus Collisions (Quark Matter 2014)*, Darmstadt, Germany, 19-24 May, 2014 *Nucl. Phys. A* **931**
[2] Lokhtin I P et al. 2009 *Comput. Phys. Commun.* **180** 779
[3] Amelin N S et al. 2006 *Phys. Rev.* C **74** 064901
[4] Amelin N S et al. 2008 *Phys. Rev.* C **77** 014903
[5] Lokhtin I P and Snigirev A M 2006 *Eur. Phys. J.* C **45** 211
[6] Sjostrand T, Mrenna S and Skands P 2006 *J. High Energy Phys.* JHEP05(2006)026
[7] Lokhtin I P et al. 2012 *Eur. Phys. J.* C **72** 2045
[8] Andronic A, Braun-Munzinger P, Redlich K and Stachel J 2003 *Phys. Lett.* B **571** 36
[9] Andronic A, Braun-Munzinger P, Redlich K and Stachel J 2007 *Nucl. Phys.* A **789** 334
[10] Dokshitzer Yu L and Kharzeev D 2001 *Phys. Lett.* B **519** 199
[11] Baier R, Dokshitzer Yu L, Mueller A H, Peigne S and Schiff D 1997 *Nucl. Phys.* B **483** 291
[12] Baier R, Dokshitzer Yu L, Mueller A H and Schiff D 1999 *Phys. Rev.* C **60** 064902
[13] Bjorken J D 1982 *Preprint* Fermilab Pub-82/29-THY
[14] Braaten E and Thoma M 1991 *Phys. Rev.* D **44** 1298
[15] Lokhtin I P and Snigirev A M 2000 *Eur. Phys. J.* C **16** 527
[16] Lokhtin I P et al. 2010 *J. Phys.: Conf. Series* **270** 012060
[17] Adare A et al. (PHENIX Collaboration) 2007 *Phys. Rev. Let.* **98** 232301
[18] Adam J et al. (ALICE Collaboration) 2015 *Preprint* arXiv:1506.08804
[19] Abbas E et al. (ALICE Collaboration) 2013 *Phys. Rev. Let.* **111** 162301
[20] Bugaev K A, Gazdzicki M and Gorenstein M I 2001 *Phys. Lett.* B **523** 255
[21] Adamczyk L et al. (STAR Collaboration) 2014 *Phys. Rev. Let.* **113** 142301
[22] Abelev B et al. (ALICE Collaboration) 2012 *J. High Energy Phys.* JHEP09(2012)112
[23] Abelev B et al. (ALICE Collaboration) 2014 *Phys. Rev.* C **90** 034904

Journal of Physics: Conference Series **668** (2016) 012069 doi:10.1088/1742-6596/668/1/012069

# Additional strange resonances from Lattice QCD

**Michał Marczenko**

Institute of Theoretical Physics, University of Wrocław, pl. Maksa Borna 9, PL-50204 Wrocław, Poland

E-mail: michal.marczenko@ift.uni.wroc.pl

**Abstract.** Recent Lattice QCD (LQCD) studies suggest that there are missing resonances in the strange sector of the Hadron Resonance Gas (HRG) model. By adopting the continuous Hagedorn mass spectrum, we present how different medium compositions influence the HRG predictions of conserved charge fluctuations. It is shown that missing strange resonances may be partially accounted for by applying the Hagedorn mass spectrum extracted from experimentally established hadrons. On the other hand, the strange-baryonic spectra, extracted from LQCD results for fluctuations, are found to be consistent with the unconfirmed states in the Particle Data Group (PDG) database, whilst the strange-mesonic spectrum points towards yet-undiscovered states in the intermediate mass region.

## 1. Introduction

The confined phase of QCD, the theory of strong interactions, is commonly modeled with the Hadron Resonance Gas (HRG) [1]. Its predictions of the QCD equation of state at finite temperature were confirmed by Lattice QCD (LQCD) findings [2, 3, 4]. Recent LQCD results for fluctuations of conserved charges, however, have pointed to limitations in the HRG description [5]. In particular, HRG results for the baryon-strangeness correlations and strangeness fluctuations are below those from LQCD (Fig. 1), whilst the results for the equation of state and baryon number fluctuations remain in agreement. The discrepancy in the strange-hadronic sector of HRG was attributed to missing resonances in the Particle Data Group (PDG) database [6].

To quantify the LQCD equation of state in the confined phase, many extensions of HRG were proposed. One such extension, which gained a lot of attention over the years due to its simplicity and pertinence in phenomenology of particle production, assumes an asymptotic continuous exponential mass spectrum. It was introduced by Rolf Hagedorn in the context of the Statistical Bootstrap Model [7, 8] and then explored within many others.

The experimental PDG spectrum, unlike the Hagedorn one, saturates at $\sim 2.5$ GeV. This may be due to the complicated decay properties and large widths of the heavy resonances, which make their final assessment and experimental confirmation challenging. Hence, the Hagedorn mass spectrum, because of its asymptotic behavior, accounts for possible undiscovered heavy resonances.

In this contribution, we study whether the inclusion of the heavy resonances, by adopting the Hagedorn mass spectrum, can reduce or resolve the disparities between the HRG and LQCD results on baryon-strangeness and strangeness fluctuations. In order to identify the possible origins of the missing resonances, we study the influence of different medium compositions on

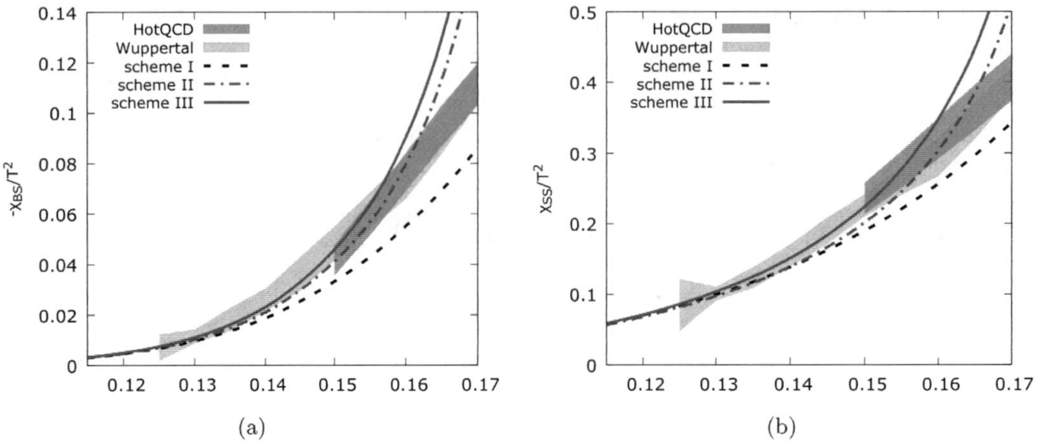


**Figure 1.** Lattice QCD results of HotQCD [3] and Budapest-Wuppertal Collaboration [4] for: (a) baryon-strangeness correlations, (b) strangeness fluctuations. Also shown are the HRG results for the discrete mass spectrum (dashed line), continuous mass spectrum obtained by fitting to the PDG (broken-dashed line) and LQCD results (continuous line).

fluctuations of conserved charges. In particular, we compare the predictions of the HRG model and its extended version with the continuous Hagedorn mass spectrum, extracted from the experimentally established discrete PDG spectra and LQCD results on the baryon-strangeness correlations and strangeness fluctuations.

## 2. Hadron Resonance Gas and Hagedorn mass spectrum

In the confined phase of QCD, at finite temperature and density, the effective degrees of freedom are hadrons and their resonances. In the simplest form of the HRG model, the interaction effects and resulting widths of hadrons and resonances are neglected, and the medium constituents are treated effectively as point-like and independent [1]. Therefore, all the information about the medium composition enter through the mass spectrum. In this contribution, we consider the continuum-extended Hagedorn mass spectrum [9], namely

$$\rho(m) = \sum_i d_i \delta(m - m_i) + \rho^H(m)\theta(m - m_x), \quad \text{where} \quad \rho^H(m) = \frac{A\, e^{m/T_H}}{(m^2 + m_0^2)^{5/4}}. \tag{1}$$

The factor $d_i = (2J_i + 1)$ is the spin degeneracy of particle $i$ with mass $m_i$, less than the mass of the first resonance $m_x$. $T_H$ is the Hagedorn limiting temperature, $A$ and $m_0$ are additional free model parameters. Noting the mass spectrum in Eq. (1) is additive, it can be decomposed into a sum of mesonic and baryonic contributions, as well as a sum of particles with definite quantum numbers. Here we assume that the Hagedorn temperature is the same for all different quantum number sectors [9].

In the grand canonical ensemble, the total thermodynamic pressure of an uncorrelated gas of particles (and antiparticles) is given by

$$\hat{P} \equiv \frac{P}{T^4} = \pm \int \mathrm{d}m\, \rho(m) \int \frac{\mathrm{d}\hat{p}}{2\pi^2}\, \hat{p}^2 \left[\ln\left(1 \pm \lambda e^{-\hat{\epsilon}}\right) + \ln\left(1 \pm \lambda^{-1} e^{-\hat{\epsilon}}\right)\right], \tag{2}$$

where $\hat{p} = p/T$, $\hat{m} = m/T$, $\hat{\epsilon} = \sqrt{\hat{p}^2 + \hat{m}^2}$ and $\pm$ sign refers to baryons and mesons respectively. The fugacity term reads $\lambda = \exp(B\hat{\mu}_B + S\hat{\mu}_S + Q\hat{\mu}_Q)$, where $\hat{\mu} = \mu/T$. For particles with $B = S = Q = 0$, the antiparticle term must be neglected to avoid double counting. Note that

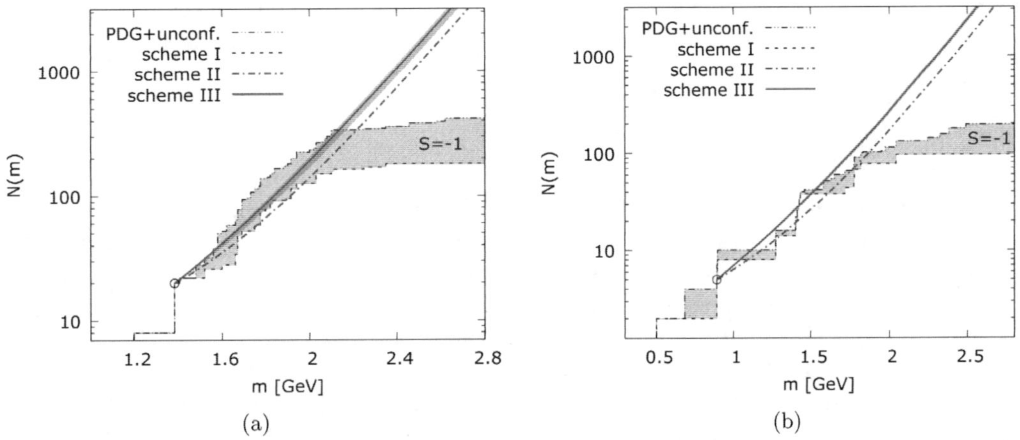


**Figure 2.** Cumulative PDG mass spectra [9] of strangeness −1 (dashed line) for baryons (a) and mesons (b). Also shown are the cumulative spectra that include the unconfirmed states (doubly broken-dashed line). The broken-dashed lines and the continuous lines are obtained by fitting the continuous spectra to the PDG and to LQCD results respectively. The errors in the continuous lines come from the fitting procedure in the latter case [9].

the mass spectrum $\rho(m)$ in Eq. (2) is a model dependent quantity. This allows us to study the effects of different medium compositions.

In the HRG model, the second order fluctuations of conserved charges are probed by general susceptibilities, i.e., by taking the second derivatives of Eq.(2), with respect to the chemical potential, namely

$$\hat{\chi}_{\rm xy} = \left. \frac{\partial^2 \hat{P}}{\partial \hat{\mu}_{\rm x} \partial \hat{\mu}_{\rm y}} \right|_{\hat{\mu}_x = \hat{\mu}_y = 0}, \tag{3}$$

where $(\rm x, y)$ are conserved charges, which, for our purpose, we restrict to the net-baryon number $B$ and net-strangeness $S$.

To study the implications of different medium compositions on the observables, we consider a comparison of the predictions of the discrete mass spectrum, obtained by neglecting the continuous part of Eq.(1) (scheme I), with the predictions of the Hagedorn mass spectrum, extracted from the PDG (scheme II) and from the lattice results on $\hat{\chi}_{\rm BS}$ and $\hat{\chi}_{\rm SS}$ (scheme III). For scheme I, we take all the established hadrons from the PDG database. Matching the Hagedorn spectrum to the LQCD results in scheme III was performed with the additional assumption that all the missing strange contribution in the baryonic sector comes solely from the $|S| = 1$ sector [9].

In Fig. 1, we compare obtained results for the baryon-strangeness correlations $\hat{\chi}_{\rm BS}$ and strangeness fluctuations $\hat{\chi}_{\rm SS}$. The corresponding cumulative spectra are shown in Fig. 2. The discrepancies between the HRG and LQCD results for the baryon-strangeness correlations and strangeness fluctuations are clearly visible in scheme I (Fig. 1). The spectra extracted in scheme II show that they can be, partially, resolved only at sufficiently high temperatures. This finding already indicates the existence of additional resonances, in the intermediate mass region of the PDG data, that could contribute substantially, due to their lower masses. The matching performed in scheme III shows that the disparities can be completely resolved. Moreover, the corresponding $|S| = 1$ baryonic spectrum, shown in Fig. 2(a), follows the trend of the unconfirmed states in the PDG. However, the $|S| = 1$ mesonic spectrum obtained in scheme

III, shown in Fig. 2(b), exceeds that of the PDG, even after the inclusion of the unconfirmed states. This may suggest the existence of still unobserved resonances in the strange-mesonic sector below 2 GeV.

## 3. Conclusions

Motivated by the recent Lattice QCD (LQCD) results on the baryon-strangeness correlations and strangeness fluctuations, which suggest the existence of unconfirmed resonances in the strange sector of the Hadron Resonance Gas (HRG), we compared the influence of different medium compositions on the HRG results. In particular, we compared the predictions of the experimentally established discrete mass spectrum with the continuous Hagedorn mass spectra, obtained from the fit to the PDG and to the LQCD observables.

We have shown that the continuous Hagedorn mass spectra, extracted from the PDG data, partially remove the disparities with the lattice results. On the other hand, the Hagedorn mass spectra, matched to LQCD results for fluctuations, not only remove those differences but also point to existence of extra states in the strange-hadronic sector. As a result, the extracted strange-baryonic spectrum is consistent with the trend dictated by the unconfirmed states in the PDG. For the strange-mesonic spectrum, however, it suggests the existence of yet-undiscovered states in the intermediate mass region.

Nonetheless, one has to remember that, in the HRG model, the contribution to $\hat{\chi}_{BS}$ and $\hat{\chi}_{SS}$ comes only from the hadronic sector with open strangeness, whilst, in general, the contribution from the non-strange sector is also possible.

Morevoer, the point-like approximation for broad resonances is known to overestimate their contribution to the observables. Hence, they have to be treated with great care when implemented in effective models [10, 11, 12]. Further investigation and new LQCD results are needed to clarify these issues.

## Acknowledgments

The author acknowledges the financial support from the organizers of 15th International Conference on Strangeness in Quark Matter and the Helmholtz International Summer School "Dense Matter 2015", fruitful discussions with collaborators P. M. Lo, K. Redlich and C. Sasaki. The author would also like to thank M. A. R. Kaltenborn for proofreading the manuscript. This work was partly supported by the Polish National Science Center (NCN), under Maestro Grant No. DEC-2013/10/A/ST2/00106.

## References

[1] Braun-Munzinger P, Redlich K and Stachel J 2004 *Quark Gluon Plasma 3* Hwa R C *et al* (Editors) World Scientific Publishing 491
[2] Borsányi S, Fodor Z, Hoelbling C, Katz S D, Krieg S and Szabó K K 2014 *Phys. Lett.* B **730** 99
[3] Bazavov A *et al* [HotQCD Collaboration] 2012 *Phys. Rev.* D **86** 034509
[4] Borsányi S, Fodor Z, Katz S D, Krieg S, Ratti C and Szabó K K 2012 *JHEP* **1201** 138
[5] Bazavov A *et al* 2014 *Phys. Rev. Lett.* **113** 072001
[6] Olive K A *et al* [Particle Data Group Collaboration] 2014 *Chin. Phys.* C **38** 090001
[7] Hagedorn R 1965 *Nuovo Cim. Suppl.* **3** 147
[8] Hagedorn R 1971 *CERN yellow report* 71-12
[9] Lo P M, Marczenko M, Redlich K and Sasaki C 2015 *Phys. Rev* C **92** 055206
[10] Friman B, Lo P M, Marczenko M, Redlich K and Sasaki C 2015 *Phys. Rev.* D **92** 074003
[11] Broniowski W, Giacosa F and Begun V 2015 *Phys. Rev.* C **92** 034905
[12] Blaschke D, Dubinin A and Turko L 2015 *Phys. Part. Nucl.* **46** 732

Journal of Physics: Conference Series **668** (2016) 012070 doi:10.1088/1742-6596/668/1/012070

# Blast wave fits with resonances to $p_t$ spectra from nuclear collisions at the LHC

**Ivan Melo**[a,b] **and Boris Tomášik**[b,c]

[a] Žilinská univerzita, Univerzitná 1, 01026 Žilina, Slovakia
[b] Univerzita Mateja Bela, FPV, Tajovského 40, 97401 Banská Bystrica, Slovakia
[c] FNSPE, Czech Technical University in Prague, Břehová 7, 11519 Prague 1, Czech Republic

E-mail: boris.tomasik@cern.ch

**Abstract.** We report our results for the freeze-out temperature and transverse flow profile obtained from fits to hadronic spectra measured by the ALICE collaboration. The influence of resonance decays is important and cannot be simply accounted for without the inclusion of their decays into the fits.

## 1. Introduction

Hadron spectra produced in ultrarelativistic nuclear collisions are shaped by the dynamical state of the fireball at the moment of its breakup. The azimuthally integrated spectra are particularly characterised by the temperature and the transverse expansion velocity.

We have analysed data by ALICE collaboration on transverse momentum spectra of $\pi$, $K$, $p$ [1], $K_0$ and $\Lambda$ [2], $\Xi$ and $\Omega$ [3], $K^*$ and $\Phi$ [4]. For a single species there is an ambiguity for the slope of the spectrum which goes like

$$T^* = T + m\langle v_t\rangle^2 . \tag{1}$$

This is resolved by common fitting of different species.

Many hadrons originate from decays of resonances. They crucially influence the shape of the spectrum. We will show below that this cannot be accounted for simply by shifting the extracted temperature. Also, it cannot be eliminated by choosing a fiducial $p_t$ interval for data fitting.

The notorious problem of calculations with resonances is that they cannot be done reasonably analytically and one has to Monte-Carlo-simulate spectra which are then compared with real data. This is costly in terms of both CPU time and disk capacity. We have used the Monde Carlo event generator DRAGON [5] for the generation of theoretical spectra. It is a MC realisation of the blast wave model which was adapted to be identical to the one used by ALICE collaboration for data fiting [1].

## 2. The model

The blast wave model used in our fitting is characterised by the theoretical prescription

$$\frac{dN}{dy\, d^2p_t} = \int d\Sigma_\mu p^\mu\, f(p_\mu u^\mu, x) = \int d^4x\, S(x,p)\, , \tag{2}$$

Journal of Physics: Conference Series **668** (2016) 012070 doi:10.1088/1742-6596/668/1/012070

**Table 1.** Selected fit results using the fiducial intervals as determined by the ALICE collaboration [1]. Compared are sets of results with and without inclusion of resonance decays.

| | no resonances | | | with resonances | | |
|---|---|---|---|---|---|---|
| centrality | $T$ (MeV) | $\langle v_t \rangle$ | $n$ | $T$ (MeV) | $\langle v_t \rangle$ | $n$ |
| 0–5% | 98 | 0.645 | 0.73 | 82 | 0.662 | 0.69 |
| 10–20% | 102 | 0.637 | 0.73 | 90 | 0.649 | 0.71 |
| 30–40% | 110 | 0.605 | 0.81 | 102 | 0.616 | 0.79 |
| 50–60% | 122 | 0.527 | 1.15 | 126 | 0.541 | 1.03 |
| 70–80% | 142 | 0.439 | 1.51 | 170 | 0.423 | 1.55 |

where $d\Sigma_\mu p^\mu$ indicate integration of the flux of particles through the three-dimensional freeze-out hypersurface and $f(p_\mu u^\mu, x)$ is the statistical distribution. The emission is conveniently expressed by the emission function

$$S(x,p)\,d^4x = g_i\,\delta(\tau-\tau_{fo})\,m_t\cosh(y-\eta_s)\Theta(R-r)\frac{1}{\exp\left(\frac{p_\mu u^\mu(x)-\mu(x)}{T}\right)\pm 1}\,\tau d\tau\,d\eta_s\,r\,dr\,d\varphi\,, \quad (3)$$

where we use $\tau = \sqrt{t^2-z^2}$, $\eta_s = \frac{1}{2}\ln\frac{t+z}{t-z}$, polar coordinates $r$, $\varphi$, and $g_i$ is spin degeneracy. The source is characterised by freeze-out Bjorken time $\tau_{fo}$ and transverse radius $R$. Statistical distributions are taken with the argument $p_\mu u^\mu$ which gives the energy in the local rest frame of the fluid. Here, $u_\mu$ is the local expansion four-velocity with longitudinal and transverse expansion $u^\mu = \gamma(1,\vec{v}) = (\gamma, \gamma v_t\cos\varphi, \gamma v_t \sin\varphi, \gamma v_l)$, where $v_l = \tanh\eta_s$ and the transverse velocity grows with $r$ like $v_t(r) = \eta_f\,(r/R)^n$ with model parameters $\eta_f$ and $n$. The mean transverse velocity is $\langle v_t\rangle = 2\eta_f/(1+n)$.

Resonances are also produced according to the emission function (3). Then they are let to decay with the assumption of constant matrix elements and uniform filling of the phase space.

## 3. The influence of resonances

Chemical composition of the fireball plays a crucial role as it determines which part of the final state hadrons is due to resonance decays. We apply here the scenario of two freeze-outs: chemical freeze-out at a higher temperature is followed by further cooling and expansion which ends up in the thermal freeze-out. The abundances of all species—including resonances—are determined by chemical equilibrium at chemical freeze-out temperature of 152 MeV and baryochemical potential of 1 MeV [6]. This assumption is rather extreme and actually means that the kinetic freeze-out follows the chemical so quickly that no resonances have enough time to decay.

Often, resonance contributions are left out in the fits to transverse momentum spectra. It is argued then that this simplification is justified if the fit is performed in a fiducial region where $p_t$ is high enough so that resonance contributions are not important and still low enough so that hard processes play no role. Unfortunately, the common lore that resonance contribution is concentrated only at low $p_t$ is wrong! We have checked this by performing fits to fiducial $p_t$-intervals used in [1].

Some of our results are summarised in Table 1. Our fit results with no resonances included agree with those by ALICE [1]. The inclusion of resonances, however, *does* change the results. Moreover, the change depends on centrality. While in central collisions the temperature *drops* by 10 MeV when resonances are included, in peripheral collisions it *rises* by 10 MeV.

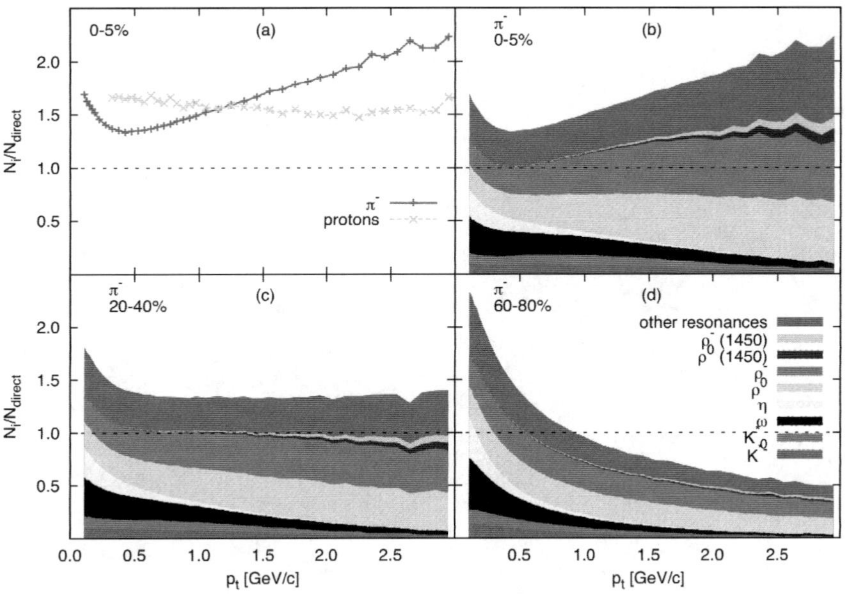


**Figure 1.** The ratio of hadrons from resonance decays to those produced directly as a function of $p_t$. (a) pions and protons in 0–5% central collisions; (b)-(d) the ratio is broken up into contributions from individual resonance species, different panels show different centralities. Simulations correspond to our best fits to the indicated centralities. The used values of $T$ and $\langle v_t \rangle$ are indicated in Fig. 2.

To see the cause of such shifts, we show in Fig. 1 how the resulting spectra are build up from contributions of individual resonance species. This was calculated with our best fits to the given centrality classes, as reported below. In peripheral collisons we indeed observe that resonance production populates mainly the low $p_t$ region. This feature still survives in the 20–40% centrality class. In the most central collisions, however, the relative contribution from resonance decays grows with increasing $p_t$.

The figure also demostrates the reason for such a behaviour. In central collisons temperature is low and transverse flow strong. Low $p_t$ region is populated by decays of $\eta$ and $\omega$. The former is light and the latter decays into three pions. Hence, there is not much energy left for the momentum of the pions. The higher-$p_t$ tail is mainly populated by decays of $\rho$ and heavier resonances. When they decay, much energy is available for the kinetic energy of the pions. Pions get a kick from decays of such resonances. Also, due to low temperature not many direct pions have high $p_t$ and so the share of pions from heavy resonance decays grows. Strong transverse flow boosts also the heavy resonances and helps to pronounce the kick to high $p_t$. This changes when the temperature grows and the transverse expansion is weaker (peripheral collisions). There, directly produced pions can better populate also higher $p_t$ and the share of resonance production at high $p_t$ does not grow anymore.

Qualitatively similar results for resonance production, although not so much pronounced as for pions, are obtained for protons, as well. (Shown in the supplementary online material.)

Thus there is i) no fiducial interval where resonance decay contribution can be ignored; ii) no simple recipe stating that the inclusion of resonances will just shift the fit results by any fixed value, since even qualitatively the influence of resonance decays depends on the state of the fireball.

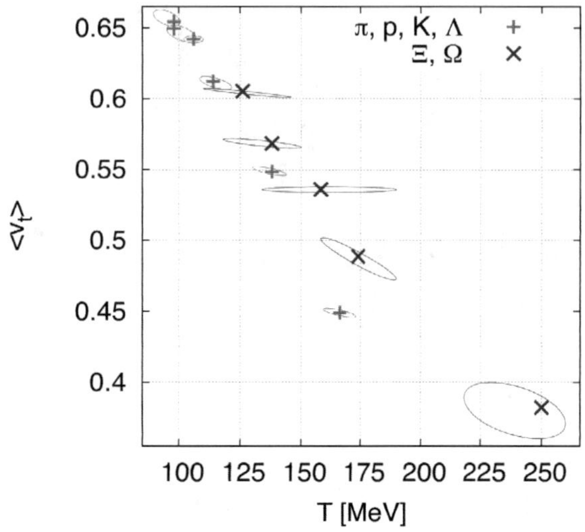


**Figure 2.** Summary of the fitted temperatures and transverse flows. Different values show results for different centralities with top left being the most central and bottom right the most peripheral. The centralities for common fits to $\pi$, $K$, $p$, and $\Lambda$ spectra (symbol +) are: 0–5%, 5–10%, 10–20%, 20–40%, 40–60%, 60–80%. Centralities for $\Xi$ and $\Omega$ spectra (symbol $\times$) are: 0–10%, 10–20%, 20–40%, 40–60%, 60–80%. Elipses indicate our estimates of 99% confidence levels (for $\pi$, $K$, $p$, and $\Lambda$) and $1\sigma$ intervals (for $\Xi$ and $\Omega$).

## 4. Fit results

The obtained freeze-out temperatures grow and the transverse expansion velocity decreases as we move from central collisions to more peripheral. The overview of results can be found in Figure 2. With the model fitted to pions, kaons, nucleons, and $\Lambda$'s we were able to reproduce also the $K^*$ and $\phi$ spectra. However, the spectra of multistrange hadrons, $\Xi$ and $\Omega$ did not fall into this systematics. When we fitted them separately, the obtained temperatures were generally higher and transverse flows generally weaker than those of more abundant species.

More detailed description of the procedure and results can be found in [7].

## 5. Conclusions

Our results indicate that between the chemical and thermal freeze-out the temperature drops considerably in central collisions. It opens the question whether assuming abundances of all resonances given by chemical freeze-out temperature is justified. We shall reconsider it soon.

It should also be noted that a successful fit on $p_t$ spectra including resonances and assuming common chemical and thermal freeze-out was recently reported in [8].

*Acknowledgements*

Supported in parts by APVV-0050-11, VEGA 1/0469/15 (Slovakia) and MŠMT grant LG13031 (Czech Republic). Computing performed in HPCC of Matej Bel University supported by grants ITMS 26230120002 and 26210120002. BT acknowledges the support of attendance to this conference by the Plenipotentiary of the Slovak Government to JINR Dubna.

## References

[1] Abelev B B *et al.* [ALICE collaboration] 2013, *Phys. Rev. C* **88** 044910.
[2] Abelev B B *et al.* [ALICE collaboration] 2014 *Phys. Rev. Lett.* **111** 222301.
[3] Abelev B B *et al.* [ALICE collaboration] 2014 *Phys. Lett. B* **728** 216.
[4] Abelev B B *et al.* [ALICE Collaboration] 2015 *Phys. Rev. C* **91** 024609.
[5] Tomášik B 2009 *Comput. Phys. Commun.* **180** 1642.
[6] Milano L [ALICE Collaboration] 2013 *Nucl. Phys. A* **904-905** 531c.
[7] Melo I and Tomášik B, arXiv:1502.01247 [nucl-th].
[8] Begun V, Florkowski W and Rybczynski M 2014 *Phys. Rev. C* **90** 054912; and *Phys. Rev. C* **90** 014906.

Journal of Physics: Conference Series **668** (2016) 012071 doi:10.1088/1742-6596/668/1/012071

# Bose-Einstein correlations of charged and neutral kaons in pp and Pb–Pb collisions at the LHC with the ALICE experiment

**Konstantin Mikhaylov (for the ALICE collaboration)**
Institute for Theoretical and Experimental Physics, Moscow, Russia
Joint Institute for Nuclear Research (JINR), Dubna, Russia
E-mail: Konstantin.Mikhaylov@cern.ch

**Abstract.** Due to the effects of quantum statistics and final state interactions, momentum correlations of two or more particles at small relative velocities, i.e. at small relative momenta in their center-of-mass system, are sensitive to the space-time characteristics of the production processes at the level of fm ($10^{-15}$ m). Kaons are the perfect tool to study Bose-Einstein correlations due to the fact that they are less influenced by resonance decays and therefore probe more effectively directly produced particles. In conjunction with femtoscopic measurements of pions and protons, they can also reveal properties of collective dynamics in heavy-ion collisions. We report on the results of Bose-Einstein correlations of charged and neutral kaons in pp collisions at $\sqrt{s} = 7$ TeV and in Pb–Pb collisions at $\sqrt{s_{\rm NN}} = 2.76$ TeV by the ALICE experiment at the LHC. The results are compared with existing data from Bose-Einstein correlations of identical pions at LHC energies, and of kaons in pp collisions. A comparison of experimental data with theoretical expectations is also carried out.

## 1. Introduction

Femtoscopic correlations are the tool to measure directly the spatial and temporal scales of the extremely small and short-lived systems created in particle or nuclear collisions with accuracy of 1 fm. The method is based on the Bose-Einstein or Fermi-Dirac symmetric properties of quantum states and uses quantum correlations between identical particles due to symmetrization or antisymmetrization of the wave functions [1]. The source radii extracted from two-particle correlations at low relative momenta describe the system at kinetic freeze-out, i.e. the last stage of particle interactions. Pion femtoscopy, which is the most common femtoscopic analysis, has shown signatures of hydrodynamic flow in heavy-ion collisions, manifesting as a decrease in the source radii with increasing transverse mass $m_{\rm T} = \sqrt{k_{\rm T}^2 + m^2}$ [2], where $k_{\rm T} = |\mathbf{p}_{\rm T,1} + \mathbf{p}_{\rm T,2}|/2$ is the average transverse momentum of the pair. This behavior can be interpreted as one of the signatures of the formation of a deconfined quark matter in these collisions [3]. However, a necessary condition for collective behavior is for *all* particles created in the collision, not just pions, to experience hydrodynamic flow. We report on the results of femtoscopic studies of $\pi^{\pm}\pi^{\pm}$, $\rm K^{\pm}K^{\pm}$, $\rm K^0_S K^0_S$, pp, and $\overline{\rm p}\overline{\rm p}$ correlations from Pb–Pb collisions at $\sqrt{s_{\rm NN}} = 2.76$ TeV by the ALICE experiment at the LHC [4]. The femtoscopic radii are extracted from one-dimensional correlation functions in terms of the invariant momentum difference for a range of collision centralities and $m_{\rm T}$ values.

## 2. Data analysis

The dataset analyzed in this contribution is from Pb–Pb collisions at $\sqrt{s_{\mathrm{NN}}} = 2.76$ TeV at the LHC measured by the ALICE detector [5]. About 8 million events from 2010 and about 40 million events from 2011 were used. Events were classified according to their centrality determined using the measured amplitudes in the V0 detectors. Charged particle tracking is performed using the Time Projection Chamber (TPC) and the Inner Tracking System (ITS). The ITS allows for high spatial resolution in determining the primary (collision) vertex. The determination of the momenta of the particles was performed using tracks reconstructed with the TPC and constrained to the primary vertex. In order to reduce the number of secondaries, primary tracks were selected based on the distance of closest approach (DCA) to the primary vertex. Additional track selections based on the quality of the track momentum fit and the number of detected space points in the TPC were used. Pairs formed by primary particles sharing more than 5% of TPC clusters were rejected. In the neutral kaon analysis, the secondary daughter tracks used global (TPC and ITS) track reconstruction. Particle identification (PID) was carried out using both the TPC and the Time-of-Flight (TOF) detector in the pseudorapidity range $|\eta| < 0.8$. For the TPC PID, a parametrized Bethe-Bloch formula was used to calculate the specific energy loss ($<\mathrm{d}E/\mathrm{d}x>$) in the detector expected for a particle with a given mass and momentum. For PID with the TOF, the particle mass hypothesis was used to calculate the expected time-of-flight as a function of track length and momentum.

## 3. Results

The experimental two-particle correlation function is defined as $C(\mathbf{q}) = A(\mathbf{q})/B(\mathbf{q})$, where $A(\mathbf{q})$ is the measured distribution of same-event pair momentum difference, $\mathbf{q} = \mathbf{p}_1 - \mathbf{p}_2$, and $B(\mathbf{q})$ is the reference distribution of pairs from mixed events. The charged kaon correlation functions were fitted using the Bowler-Sinyukov formula $C(q) = N\left[1 - \lambda + \lambda K(q)\left(1 + \exp\left(-R^2q^2\right)\right)\right]$, where $K(q)$ factor approximates the influence of the Coulomb interaction on the correlation function. The neutral kaon correlation functions were fitted by $C(q) = [1 - \lambda + \lambda C'(q)]$, where $C'(q) = 1 + e^{-q^2R^2} + C_{\mathrm{strongFSI}}(q, R)$ and $C_{\mathrm{strongFSI}}$ is the strong final state interaction (FSI) contribution to the $\mathrm{K^0_SK^0_S}$ correlation function due to the near-threshold resonances, $\mathrm{f_0}(980)$ and $\mathrm{a_0}(980)$.

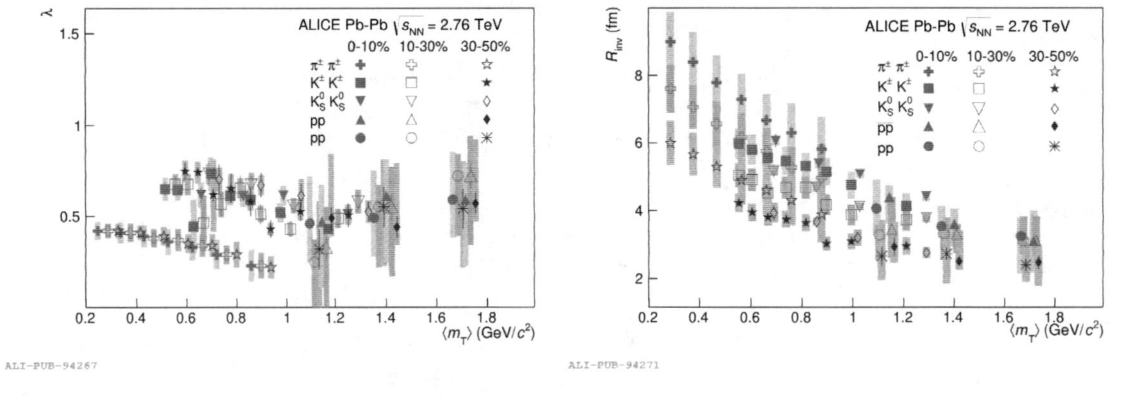


**Figure 1.** Left: $\lambda$ parameters ($\lambda_{\mathrm{pp}} + \lambda_{\mathrm{p}\Lambda}$ in case of (anti)proton pairs) vs. $m_{\mathrm{T}}$ for the three centralities considered for $\pi^{\pm}\pi^{\pm}$, $\mathrm{K^{\pm}K^{\pm}}$, $\mathrm{K^0_SK^0_S}$, pp, and $\overline{\mathrm{p}}\overline{\mathrm{p}}$ [4]. Right: $R_{\mathrm{inv}}$ parameters versus $m_{\mathrm{T}}$ for the three centralities considered for $\pi^{\pm}\pi^{\pm}$, $\mathrm{K^{\pm}K^{\pm}}$, $\mathrm{K^0_SK^0_S}$, pp, and $\overline{\mathrm{p}}\overline{\mathrm{p}}$ [4]. Left and right: Statistical (thin lines) and systematic (boxes) uncertainties are shown.

Fig. 1(left panel) shows the extracted $\lambda$ parameters as a function of $m_{\mathrm{T}}$ for several centralities.

The values of $\lambda$ are less than unity due to long-lived resonances which dilute the correlation functions and also lead to non-Gaussian shapes of the correlation functions, especially in the one-dimensional case. Results for kaons and protons are consistent with each other at similar $m_{\mathrm{T}}$. The values of $\lambda$ for pions are lower than for kaons due to the stronger influence of resonances. An additional cause could be a partial coherence of pions. Fig. 1(right panel) shows the extracted $R_{\mathrm{inv}}$ parameters as a function of $m_{\mathrm{T}}$ for several centralities. For overlapping $m_{\mathrm{T}}$, the radius parameters are mostly consistent with each other within uncertainties, though the pion radii are generally larger than the kaon radii. The radius parameters show increasing size with increasing event multiplicity as would be expected from a simple geometric picture of the collisions. They also show a decreasing size with increasing $m_{\mathrm{T}}$ as would be expected in the presence of collective radial flow [3]. Both of these dependencies can be seen in previous $\pi^{\pm}\pi^{\pm}$ femtoscopic measurements [2] and also reinforce the interpretation that collective flow is present in these collisions for pions, kaons (neutral and charged), and protons alike. Deviations from exact $m_{\mathrm{T}}$-scaling of $R_{\mathrm{inv}}$ can be explained as a consequence of the increase of the Lorentz factor with decreasing particle mass. In a hydrodynamic model [6], scaling is observed for the three-dimensional radii measured in the Longitudinally Co-Moving System (LCMS). The transformation from LCMS to the Pair Rest Frame (PRF) involves a boost along the outward direction only, where the boost value is proportional to the transverse velocity of the pair and inversely proportional to the particle mass (for similar $m_{\mathrm{T}}$). Thus, a smaller mass leads to an increase in the boosted $R_{\mathrm{out}}$ and, subsequently, $R_{\mathrm{inv}}$ in the PRF. Indeed, we observe such an effect in the data, as the pion radii are systematically higher than the kaon radii at the same $m_{\mathrm{T}}$.

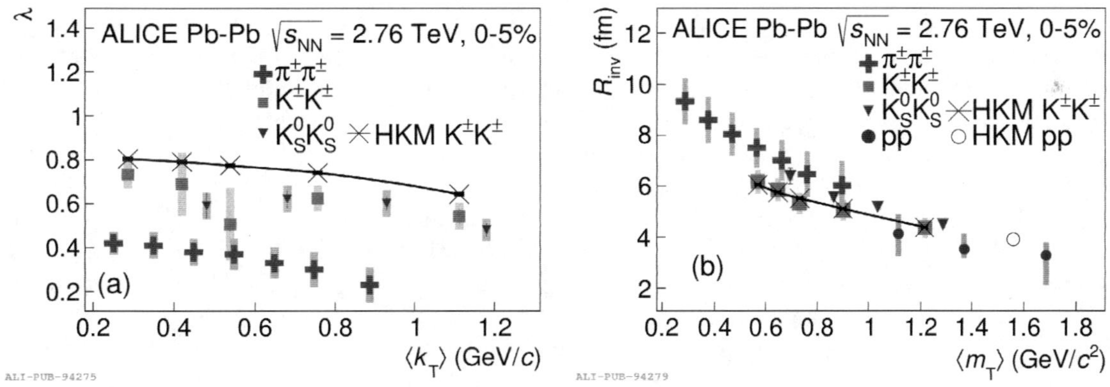


**Figure 2.** Comparison of the HKM model (see text) with measured kaon $\lambda$ (a) and $R_{\mathrm{inv}}$ (b) parameters for 0-5% centrality [4]. Statistical (thin lines) and systematic (boxes) uncertainties are shown.

A comparison of a hydrodynamic flow and kinetics model, HKM [7], with the measured $R_{\mathrm{inv}}$ and $\lambda$ parameters for 0-5% centrality is shown in Fig. 2. The HKM values in Fig. 2 are extracted from $\mathrm{K}^{\pm}\mathrm{K}^{\pm}$, but the predictions for $\mathrm{K}^0_{\mathrm{S}}\mathrm{K}^0_{\mathrm{S}}$ and $\mathrm{K}^{\pm}\mathrm{K}^{\pm}$ are consistent with each other. The $R_{\mathrm{inv}}$ predicted by the HKM are in a very good agreement with the experimental kaon data. The experimental data for the neutral kaons are again slightly higher than for the charged kaons, but this difference is still within systematic uncertainties. The HKM predicts a decrease of $\lambda$ with increasing $k_{\mathrm{T}}$, which is in accordance with both sets of kaon data, however, the model slightly overpredicts the experimental data. It is shown in [7] that the most important resonances for KK pairs, K*(890) and $\phi$(1020), do not significantly influence the $\lambda$ parameter (due to their low contribution), and the discrepancy between the

Journal of Physics: Conference Series **668** (2016) 012071 doi:10.1088/1742-6596/668/1/012071

model and the experimental data can be explained by the lower experimental kaon purity and deviations of the experimental correlation function shape from a Gaussian distribution. The $K^{\pm}K^{\pm}$ correlation radii in Fig. 3 show an increase with multiplicity in agreement with the $\pi\pi$ radii, at 900 GeV and 7 TeV, and the $K^0_S K^0_S$ radii, as was observed for $\pi\pi$ correlations in heavy-ion collisions. These radii also decrease with increasing $m_T$ for the large multiplicity bins $11 < N_{ch} < 22$ and $N_{ch} > 22$. Such tendency was found for pions and neutral kaons $K^0_S K^0_S$ in pp collisions and pions in heavy-ion collisions at LHC energies. In the low multiplicity bin $1 < N_{ch} < 11$ charged kaons show a completely different $m_T$ dependence of the radii: these radii increase with $m_T$. This effect is qualitatively similar to that of pions. The $m_T$ dependence at high multiplicity show a decreasing size with increasing $m_T$ as would be expected by models incorporating some collective expansion even in small systems [9].

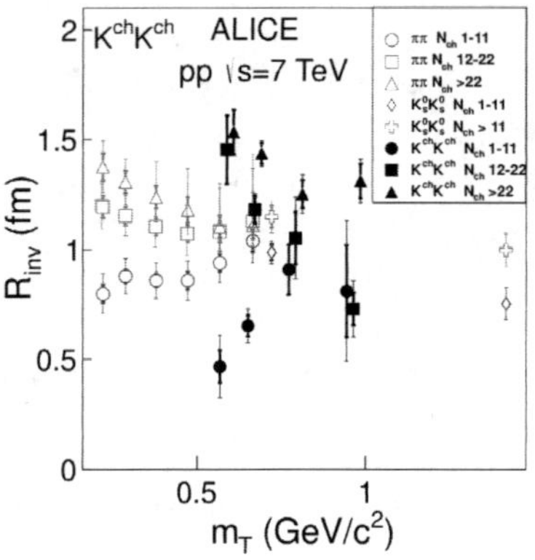


**Figure 3.** One-dimensional charged kaon radii versus $m_T$ from [8].

## 4. Summary

Results from femtoscopic studies of $\pi^{\pm}\pi^{\pm}$, $K^{\pm}K^{\pm}$, $K^0_S K^0_S$, pp, and $\overline{\text{p}}\overline{\text{p}}$ correlations from Pb–Pb collisions at $\sqrt{s_{\text{NN}}} = 2.76$ TeV with ALICE at the LHC have been presented. The femtoscopic radii and $\lambda$ parameters were extracted from one-dimensional correlation functions in terms of the invariant momentum difference. It was found that the emission source sizes of kaons and protons measured in these collisions exhibit transverse mass scaling within uncertainties, which is consistent with hydrodynamic model predictions assuming collective flow. The deviation from the scaling for pions can be explained as a consequence of the increase of the Lorentz factor with decreasing particle mass during the transformation from LCMS to PRF systems [6]. The extracted $\lambda$ parameters are found to be less than unity, as is expected due to long-lived resonances and non-Gaussian correlation functions. The predictions of the hydrokinetic model (HKM) for the one-dimensional femtoscopic radii for charged and neutral kaons and protons coincide well with the observations. The $m_T$ dependence in pp collisions at high multiplicity is expected by models incorporating some collective expansion even in small systems. K.M. acknowledges partial support by the RFBR-CNRS grants No 14-02-93107 and 14-02-93108.

## References

[1] G. Goldhaber, S. Goldhaber, W. Lee, A. Pais, *Phys. Rev.* 120 (1960) 325.
[2] M. A. Lisa, S. Pratt, R. Soltz and U. Wiedemann, *Ann. Rev. Nucl. Part. Sci.* 55, 357 (2005).
[3] S. V. Akkelin, Yu. M. Sinyukov, *Phys. Lett. B* 356, 525-530 (1995).
[4] J. Adam *et al.* [ALICE Collaboration], *arXiv:1506.07884 [nucl-ex]*.
[5] K. Aamodt *et al.* [ALICE Collaboration], *JINST* 3, S08002 (2008).
[6] A. Kisiel, M. Galazyn, and P. Bozek, *arXiv:409.4571 [nucl-th]*.
[7] V.M. Shapoval, P. Braun-Munzinger, Iu.A. Karpenko, Yu.M. Sinyukov, *arXiv:1404.4501 [hep-ph]*.
[8] B. Abelev *et al.* [ALICE Collaboration], *Phys. Rev. D* 87, 052016 (2013)
[9] Yu.M. Sinyukov and V.M. Shapoval, *Phys. Rev. D* 87 094024 (2013); K. Werner *et al.* *arXiv:1104.2405 [hep-ph]*

# (Anti-)strangeness production in heavy-ion collisions

**P. Moreau**[1,2], **F. Li**[3], **C.-M. Ko**[3], **W. Cassing**[4], **E. L. Bratkovskaya**[1,2]

[1] Frankfurt Institute for Advanced Studies, Johann Wolfgang Goethe Universität, Frankfurt am Main, Germany

[2] Institut für Theoretische Physik, Johann Wolfgang Goethe Universität, Frankfurt am Main, Germany

[3] Cyclotron Institute and Department of Physics and Astronomy, Texas A&M University, College Station, USA

[4] Institut für Theoretische Physik, Universität Gießen, Germany

E-mail: moreau@fias.uni-frankfurt.de

**Abstract.** The production and dynamics of strange and antistrange hadrons in heavy-ion reactions from $\sqrt{s_{NN}} \approx 3$ GeV to 200 GeV is analyzed within the Parton-Hadron-String-Dynamics (PHSD) transport model. The PHSD results for strange baryon and antibaryon production are roughly consistent with the experimental data starting from upper SPS energies. Nevertheless, hadronic final state flavor-exchange reactions are important for the actual abundances, in particular at large rapidities where hadronic dynamics, parton fragmentation and string decay dominate. A striking disagreement between the PHSD results and the available data persists, however, for bombarding energies below $\sqrt{s_{NN}} \approx 8$ GeV where the strangeness production is significantly underestimated as in earlier HSD studies. This finding implies that the strangeness enhancement seen experimentally at FAIR/NICA energies cannot be attributed to a deconfinement phase transition or crossover but probably involves the approximate restoration of chiral symmetry in the hadronic phase.

The production of strange hadrons in heavy-ion collisions has long been suggested to provide a sensible probe for the heavy-ion dynamics and its degrees-of-freedom at relativistic energies [1, 2]. However, while the production of kaons and antikaons in Pb+Pb collisions (158 A GeV or $\sqrt{s_{NN}} = 17.3$ GeV) at top SPS energies turned out to be well described by hadron/string models such as the Hadron-String-Dynamics (HSD) or URQMD [3], the abundances of antistrange baryons were clearly underestimated. Furthermore, the kaon and hyperon abundances have been underestimated by up to a factor of two at the lower bombarding energies of 4 to 40 A GeV [3]. This surprising result has led to the suggestion that a quark-gluon plasma (QGP) might have already been produced at lower energies [4]. But this raises the question why we did not see a similar strangeness enhancement at the top super-proton-synchrotron (SPS) or even the Relativistic Heavy Ion Collider (RHIC) energies where a QGP is more likely produced. The answer to this problem is not straightforward since to differentiate between the hadronic and partonic degrees-of-freedom and their impacts on the various hadronic final-state interactions requires a fully microscopic description of these reactions. To this aim, an off-shell covariant transport approach – denoted as the Parton-Hadron-String Dynamics (PHSD) – has been developed to incorporate all necessary ingredients. This model has proven in the past to provide a good description of light hadron abundances, collective flow and electromagnetic radiation from nucleus-nucleus collisions [5, 6, 7, 8, 9, 10] up to the Large-Hadron-Collider (LHC) energies.

We recall that the PHSD model is a covariant dynamical approach for strongly interacting systems formulated on the basis of Kadanoff-Baym equations [11, 12] or off-shell transport equations in the phase-space representation. In the Kadanoff-Baym theory, the field quanta are described in terms of dressed propagators with complex selfenergies. Whereas the real part of the selfenergies can be related to mean-field potentials of Lorentz scalar or vector type, the imaginary parts provide information about the lifetime and/or reaction rates of time-like particles [12]. Once the proper complex selfenergies of the degrees-of-freedom are known, the time evolution of the system is fully governed by off-shell transport equations for quarks and hadrons (as described in Refs. [11, 12]). The PHSD model includes the creation of massive quarks via hadronic string decay to partons above the critical energy density $\sim 0.5$ GeV/fm$^3$ and quark fusion forming a hadron in the hadronization process. With some caution, the latter process can be considered as a simulation of a crossover transition since the underlying EoS in PHSD is a crossover [12]. At energy densities close to the critical energy density, the PHSD describes a coexistence of the quark-hadron mixture. This approach allows for a simple and transparent interpretation of lattice QCD results for thermodynamic quantities as well as correlators and leads to effective strongly interacting partonic quasiparticles with broad spectral functions. For a review on off-shell transport theory we refer the reader to Ref. [12]. Results from the PHSD model and their comparison with experimental observables for heavy-ion collisions from the lower SPS to RHIC energies can be found in Refs. [5, 8, 9, 10]. In the hadronic phase, i.e. for energies densities below the critical energy density, the PHSD approach is identical to the HSD model [13, 14], which has been successfully employed for p-p, p-A and A-A reactions up to top SPS energies.

We here present an overview of (anti-) strangeness production in central Au+Au (Pb+Pb) collisions from $\sqrt{s_{NN}} \approx 3$ GeV to 200 GeV within the PHSD with a particular focus on the multi-antistrange baryon production. Compared to the HSD study in Ref. [14], we now have incorporated in the hadronic sector all strangeness exchange reactions $\pi+N \leftrightarrow K+Y, N+N \leftrightarrow N+K+Y, K+N \leftrightarrow \pi+Y$, $\pi+\Xi \leftrightarrow K+Y, Y+Y \leftrightarrow N+\Xi$, and $K+\Xi \leftrightarrow \Omega+\pi, \Xi+\Xi \leftrightarrow \Omega+Y$ in meson-baryon and baryon-baryon collisions following Refs. [15, 16, 17]. It is essential to recall that in PHSD (as in HSD) the dominant three-body channels $B + B \leftrightarrow$3 mesons are included on the basis of 'detailed-balance' [18], where the dominant mesons involve the $\pi, \rho, \omega, K, K^*$ degrees-of-freedom.

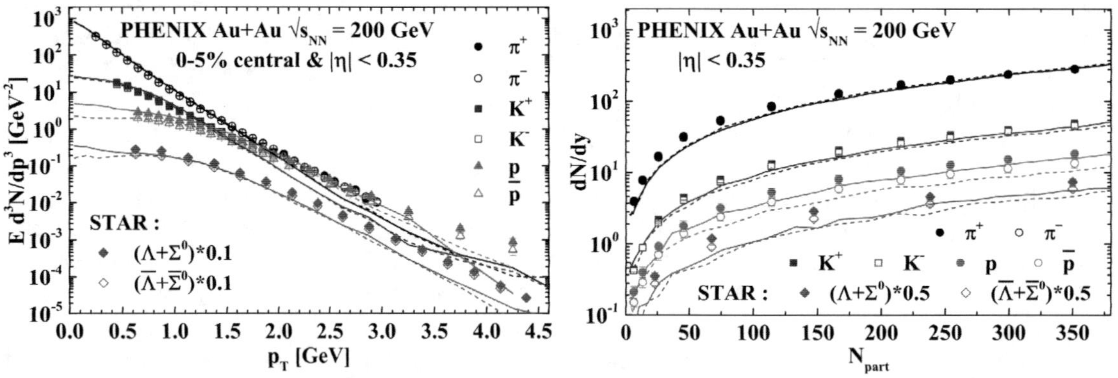


**Figure 1.** The invariant yield $Ed^3N/dp^3$ as a function of the transverse momentum $p_T$ (left window) and the yield $dN/dy$ as a function of the number of participating nucleons $N_{part}$ (right window) at midrapidity for pions, kaons, protons, antiprotons, $\Lambda$ and $\Lambda$ for Au+Au collisions at $\sqrt{s_{NN}} = 200$ GeV in comparison to the experimental data from the PHENIX [19] and STAR collaboration [20].

Journal of Physics: Conference Series **668** (2016) 012072 doi:10.1088/1742-6596/668/1/012072

We start with PHSD results for the transverse momentum distributions of particles (solid lines) and antiparticles (dashed lines), shown in the left window of Fig. 1 for 5% central Au+Au collisions at $\sqrt{s_{NN}} = 200$ GeV in comparison to the data from PHENIX and STAR [19, 20]. Apart from a reasonable description of the data, we find baryons and antibaryons to differ only at low $p_T < 0.5$ GeV/c where some net absorptions of antibaryons becomes visible. The centrality dependence of the particle yield (at midrapidity) is displayed in the right window of Fig. 1 as a function of the number of participants $N_{part}$ in comparison to the PHENIX and STAR data [19, 20]. Here we find only a small difference between protons and antiprotons, whereas hyperons and antihyperons are about the same roughly in line with the data. We mention that the differences between baryons and antibaryons increase drastically when going down in bombarding energy, where nucleons from the target and projectile are shifted to the midrapidity region and dominate over antibaryons, which are preferentially produced in the hadronization process from the QGP at midrapidity.

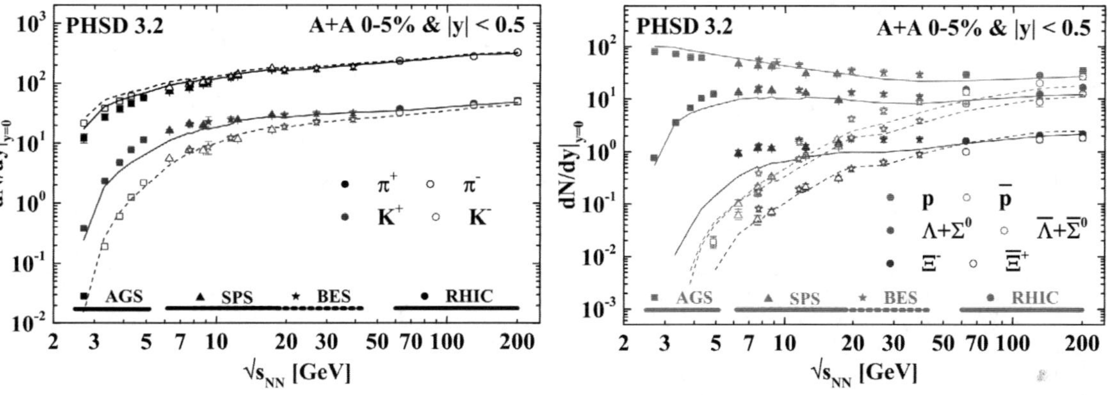


**Figure 2.** (Left window) The yield at midrapidity for pions and kaons as a function of $\sqrt{s_{NN}}$ for central Au+Au and Pb+Pb collisions in comparison to the experimental data from [21] for AGS energies, from [22] for SPS energies, from [23] for the preliminary results from the BES program at RHIC, and from [24] for RHIC energies. (Right window) The yield at midrapidity for protons, antiprotons, $\Lambda$, $\Lambda$, $\Xi^-$ and $\Xi^+$ as a function of $\sqrt{s_{NN}}$ for central Au+Au and Pb+Pb collisions in comparison to the experimental data from [25] for AGS energies, from [26] for SPS energies, from [23] for the preliminary results from the BES program at RHIC, and from [24, 27] for RHIC energies.

We further display in the left window of Fig. 2 the excitation function for $\pi^\pm$ and $K^\pm$ mesons from 5% central Au+Au collisions (at midrapidity) as a function of $\sqrt{s_{NN}}$ in comparison to the available data. We observe an overestimation of pion production by PHSD at low invariant energy and an underestimation of $K^+$ mesons as in the HSD model [3]. An underestimation of strangeness at low invariant energy is also seen in the baryon-antibaryon sector (cf. Fig. 2 (right window)) where the hyperons $(\Lambda + \Sigma^0)$ and cascade baryons $(\Xi^-)$ are missed substantially. On the other hand, the antihyperons $(\Lambda + \Sigma^0)$ and $\Xi^+$ are reasonably described, and this can be traced back to the dominant production in the hadronization process from the QGP.

A closer analysis of the explicit rapidity distribution of baryons and antibaryons shows that antibaryon production close to midrapidity can essentially be attributed to hadronization from the QGP, whereas at forward or backward rapidities (closer to the target and projectile rapidities) hadrons are mainly produced from fragmentation or hadronic string decay in the PHSD. This allows for a scan in the partonic fraction of the dynamics when analyzing separately baryons and antibaryons as a function of rapidity.

In summary: we have extended the PHSD transport approach to include all strangeness exchange reactions $\pi + N \leftrightarrow K + Y, N + N \leftrightarrow N + K + Y, K + N \leftrightarrow \pi + Y$, $\pi + \Xi \leftrightarrow K + Y, Y + Y \leftrightarrow N + \Xi$, and $K + \Xi \leftrightarrow \Omega + \pi, \Xi + \Xi \leftrightarrow \Omega + Y$ in meson-baryon and baryon-baryon collisions, thus allowing for a detailed study of multistrange baryon production in the hadronic phase and to compare their production in the hadronization process from the QGP. We find that an underestimation of strangeness production at low invariant energy is seen in the meson as well as baryon sector (cf. Fig. 2), whereas the multistrange baryon/antibaryon dynamics at the top SPS and RHIC energy are rather well described by hadronization from the QGP at midrapidity. This finding implies that the strangeness enhancement seen experimentally at FAIR/NICA (AGS and lower SPS) energies cannot be attributed to a deconfinement phase transition (or crossover) but probably involves the approximate restoration of chiral symmetry in the hadronic phase, which would lead to a change of hadron properties and thus an enhanced production of multistrange baryons.

**Acknowledgements**
The authors acknowledge the support by BMBF, HIC for FAIR and the HGS-HIRe for FAIR. The computational resources were provided by the LOEWE-CSC.

## References

[1] Koch P Müller B Rafelski J 1986 *Phys. Rep.* **142** 167
[2] Voloshin S A Poskanzer A M Snellings R 2010 *Landolt-Boernstein New Series* I/**23** 5-54
[3] Bratkovskaya E L *et al.* 2004 *Phys. Rev.* C **69** 054907
Bratkovskaya E L *et al.* 2004 *Phys. Rev. Lett.* **92** 032302
[4] Gazdzicki M Gorenstein M I 1999 *Acta Phys. Polon.* B **30** 2705
[5] Konchakovski V Cassing W Toneev V D 2015 *J. Phys.* G **42** 055106
[6] Cassing W Bratkovskaya E L 2009 *Nucl. Phys.* A **831** 215
[7] Bratkovskaya E L Cassing W Konchakovski V P Linnyk O 2011 *Nucl. Phys.* A **856** 162
[8] Toneev V D *et al.* 2012 *Phys. Rev.* C **85** 034910
[9] Konchakovski V P *et al.* 2012 *Phys. Rev.* C **85** 011902
[10] Linnyk O *et al.* 2011 *Phys. Rev.* C **84** 054917
Linnyk O *et al.* 2012 *Phys. Rev.* C **85** 024910
Linnyk O *et al.* 2013 *Phys. Rev.* C **87** 014905
[11] Juchem S Cassing W Greiner C 2004 *Phys. Rev.* D **69** 025006
Juchem S Cassing W Greiner C 2004 *Nucl. Phys.* A **743** 92
[12] Cassing W 2009 *E. Phys. J.* ST **168** 3
[13] Cassing W Bratkovskaya E L 1999 *Phys. Rep.* **308** 65
[14] Cassing W Bratkovskaya E L Juchem S 2000 *Nucl. Phys.* A **674** 249
[15] Li C H Ko C M 2002 *Nucl. Phys.* A **712** 110
[16] Li F Chen L W Ko C M Lee S H 2012 *Phys. Rev.* C **85** 064902
[17] Flaminio V Moorhead W G Morrison D R O Rivoire N 1983 *CERN-HERA-83-02*
[18] Cassing W 2002 *Nucl. Phys. A* **700** 618
[19] Adler S S *et al.* [PHENIX Collaboration] 2004 *Phys. Rev.* C **69** 034909
[20] Agakishiev G *et al.* [STAR Collaboration] 2012 *Phys. Rev. Lett.* **108** 072301
[21] Klay J L *et al.* [E-0895 Collaboration] 2003 *Phys. Rev.* C **68** 054905
Ahle L *et al.* [E866 and E917 Collaborations] 2000 *Phys. Lett.* B **476** 1
Ahle L *et al.* [E866 and E917 Collaborations] 2000 *Phys. Lett.* B **490** 53
[22] Alt C *et al.* [NA49 Collaboration] 2008 *Phys. Rev.* C **77** 024903
Afanasiev S V *et al.* [NA49 Collaboration] 2002 *Phys. Rev.* C **66** 054902
[23] Das S [STAR Collaboration] 2014 *arXiv:1412.0499* [nucl-ex]
Zhu X [STAR Collaboration] 2014 *J. Phys. Conf. Ser.* **509** 012004
[24] Abelev B I *et al.* [STAR Collaboration] 2009 *Phys. Rev.* C **79** 034909
[25] Klay J L *et al.* [E895 Collaboration] 2002 *Phys. Rev. Lett.* **88** 102301
Pinkenburg C *et al.* [E895 Collaboration] 2002 *Nucl. Phys.* A **698** 495
[26] Alt C *et al.* [NA49 Collaboration] 2006 *Phys. Rev.* C **73** 044910
Alt C *et al.* [NA49 Collaboration] 2008 *Phys. Rev.* C **78** 034918
Alt C *et al.* [NA49 Collaboration] 2005 *Phys. Rev. Lett.* **94** 192301
[27] Aggarwal M M *et al.* [STAR Collaboration] 2011 *Phys. Rev.* C **83** 024901

# Bose–Einstein correlations of charged kaons in $p + p$ collisions with the STAR detector

**Grigory Nigmatkulov (for the STAR Collaboration)**

National Research Nuclear University MEPhI (Moscow Engineering Physics Institute), 31 Kashirskoe shosse, Moscow, 115409, Russia

E-mail: ganigmatkulov@mephi.ru / nigmatkulov@gmail.com

**Abstract.** We report the results of charged kaon Bose–Einstein Correlations (BEC) measured in proton-proton collisions at $\sqrt{s}$ =200 and 510 GeV with the STAR detector at Relativistic Heavy Ion Collider (RHIC). The one-dimensional correlation functions are studied as a function of the charged particle multiplicity. The femtoscopic radii, $R$, and the correlation strength, $\lambda$, are extracted. The dependence of the source radii and the correlation strengths on the particle multiplicity are investigated.

## 1. Introduction

Particle correlations carry information about the multiparticle production process. Two-particle Bose–Einstein Correlations (BEC) at low relative momenta (also known as *correlation femtoscopy* or *HBT intensity interferometry*) are sensitive to the quantum statistics and allow one to measure the space–time extent of the particle emitting source [1, 2]. These correlations originate from the symmetrization of the two–particle wave function of identical particles and lead to an enhancement of pair particle production with small relative momenta. The first BEC were observed in the $p\bar{p}$ annihilation as an enhancement of the identical–pion production with small opening angles [3, 4]. In the 1970s they were refined by Kopylov and Podgoretsky [5, 6] and shown to be useful for the study of the emitting source parameters. Since then the femtoscopic analysis was extensively used in many experiments [7, 8]. In "elementary" leptonic and hadronic collisions the BEC measurements provide a test of different models [9, 10]. Usually the femtoscopic analysis is applied for pions however using kaons may provide additional information about the particle emitting source.

## 2. Charged kaon femtoscopic analysis

The results presented in this paper are based on the analysis of the $p + p$ collisions at $\sqrt{s}$=200 and 510 GeV measured by Solinoidal Tracker At RHIC (STAR) in the 2012 and 2011 runs, respectively. The STAR Time Projection Chamber (TPC) was used for charged particle tracking, reconstruction of the particle momentum, and particle identification by measuring the ionization energy loss ($dE/dx$) [11]. The TPC has a full $2\pi$ azimuthal acceptance and covers the pseudorapidity range $|\eta| < 1.3$. The multiplicity of each collision was determined via the measured number of charged particles within the pseudorapidity range $|\eta| < 0.5$ and the transverse momentum $p_T > 0.15$ GeV/$c$. To ensure a more uniform detector acceptance and reduce pile-up interactions, only events with the position of the reconstructed primary vertex within ±50 cm of the center of STAR along the beam direction were used in the analysis. All tracks in the TPC were required to have more than 15 measured space points along the trajectory. The tracks with pseudorapidity $|\eta| > 1$ were excluded from the analysis. In order to avoid track splitting, the ratio of the number of reconstructed space points to possible space points along the track was required to be greater than 0.52.

The charged kaon identification in the momentum range $0.2 < p < 0.55$ GeV/$c$ was performed by using the information from TPC. It was required that the measured energy loss of a track must be less than two standard deviations, $\sigma$, from the predicted energy loss of a kaon and more than $2\sigma$ from the energy loss prediction of any other hypothesis. In the momentum range $0.55 < p < 1.55$ GeV/$c$, charged particle identification was performed by measuring the flight time ($t_{TOF}$) in the Time-Of-Flight (TOF) system [12, 13]. The particles with $0.18 < m^2 < 0.35$ (GeV/$c^2$)$^2$ were selected as kaons. The $m^2$ is calculated by $m^2 = p^2\left((t_{TOF} \cdot c/l)^2 - 1\right)$, where $l$ is the flight path length of the particle and $c$ is the speed of light.

Usually the femtoscopic correlation functions are sensitive to the two-track reconstruction effects (track splitting and track merging) due to the fact that the particles at low relative momenta have close trajectories. In hadronic collisions the track splitting (one single particle is reconstructed as two tracks with close momenta) and track merging (two particles are reconstructed as one track) effects have a small influence on the correlation functions. However the same cuts as were used for removing these effects in STAR analysis of $Au+Au$ collisions [14] were applied.

The one-dimensional correlation function $C(Q)$ of two particles with the three-momenta $\mathbf{p_1}$ and $\mathbf{p_2}$ is constructed as a ratio:

$$C(Q) = A(Q)/B(Q), \tag{1}$$

where $A(Q)$ is the measured two-particle distribution in the given event. The quantity $B(Q)$ is the reference distribution, where the quantum statistical correlations are absent, and is constructed by mixing particles from different events. The quantity $Q = \sqrt{(\mathbf{p_1} - \mathbf{p_2})^2 - (E_1 - E_2)^2}$ is the relative four-momentum of the particles. In order to extract the emitting source radius, $R$, the correlation functions were fit by the Bowler–Sinyukov formula:

$$C(Q) = N\left(1 - \lambda + \lambda K(Q)(1 + e^{-R^2Q^2})\right) D(Q), \tag{2}$$

where $N$ is a normalization factor, $\lambda$ is the strength of the correlations, and $K(Q)$ is the Coulomb function integrated over a spherical source of 1 fm. The quantity $D(Q)$ is the "baseline" term that takes into account all non-femtoscopic correlations which arising, for instance, due to the energy and momentum conservation [15]. In the current analysis, a quadratic polynomial was used for the $D(Q)$.

The two–kaon correlation functions were studied for the three multiplicity ranges: (1–6), (7–12) and (>12). In order to correct for the non-femtoscopic effects, the Monte Carlo event generator PYTHIA-6.4.28 [16] with Perugia 0 Tune [17] was used. Since the PYTHIA event generator does not contain Bose–Einstein correlations and final state interactions (Coulomb and strong), the deviations of the simulated correlation functions from unity will correspond to the non-femtoscopic correlations. The comparison of the simulated charged kaon correlation functions, where PYTHIA events were filtered through the analysis cuts, to the experimentally measured distributions for each collision energy and multiplicity bin is shown in Fig. 1. The experimentally measured (solid circles) and Monte Carlo simulated (empty circles) correlation functions were normalized in the region $0.5 < Q < 0.8$ GeV/$c$ such that the correlation functions $C(Q) \to 1$. One may see that PYTHIA gives a good description of experimental correlation functions in the region $Q > 0.5$ GeV/$c$. The decrease of the measured experimental correlation functions at low momenta, $Q < 0.1$ GeV/$c$, is due to the Coulomb repulsion between the kaons.

In order to correct for the non-femtoscopic effects, one may construct the double ratio which is defined as a ratio of the experimentally measured correlation function to the simulated one. Fig. 2 shows these double ratios for $p + p$ collisions at $\sqrt{s}$ =200 (top panels) and 510 GeV (bottom panels) and the three multiplicity ranges.

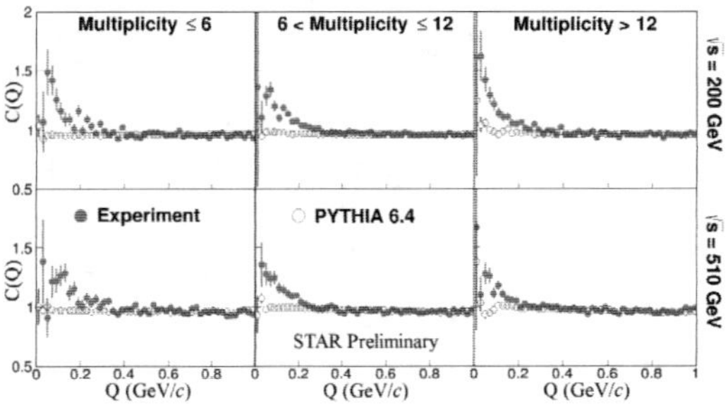


**Figure 1.** Correlation functions for charged kaon pairs from $p + p$ collisions at $\sqrt{s}$ = 200 (top panels) and 500 GeV (bottom panels) measured by STAR (filled circles) and simulated with the PYTHIA event generator (open circles). Columns from left to right show the three multiplicity ranges.

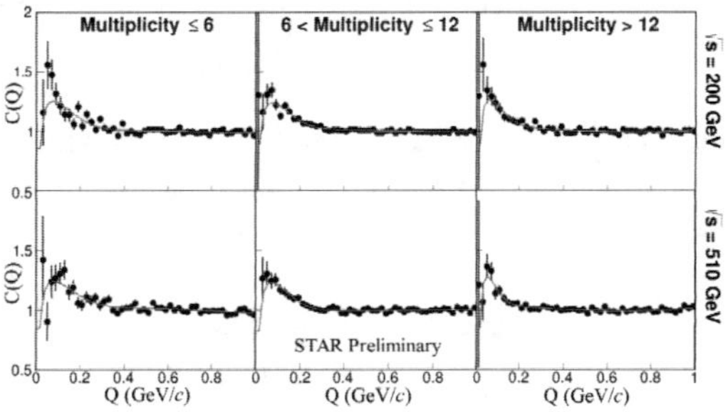


**Figure 2.** Charged kaon correlation functions after the correction on the PYTHIA simulated distributions obtained for $p+p$ collisions at $\sqrt{s} = 200$ (top panels) and 500 GeV (bottom panels) and the three event multiplicity ranges. The lines show the fits to the data by using the Eq. 2.

The correlation functions were fitted with the function shown as Eq. 2. Due to the imperfections of the simulation in the $Q > 0.8$ GeV/$c$ region, the non-femtoscopic term $D(Q) = 1 + aQ + bQ^2$ was used. Fig. 3 shows the extracted source radii (a) and correlation strengths (b) parameters measured at $\sqrt{s}$ =200 (stars) and 510 (squares) GeV as a function of the particle multiplicity. Only statistical uncertainties are shown.

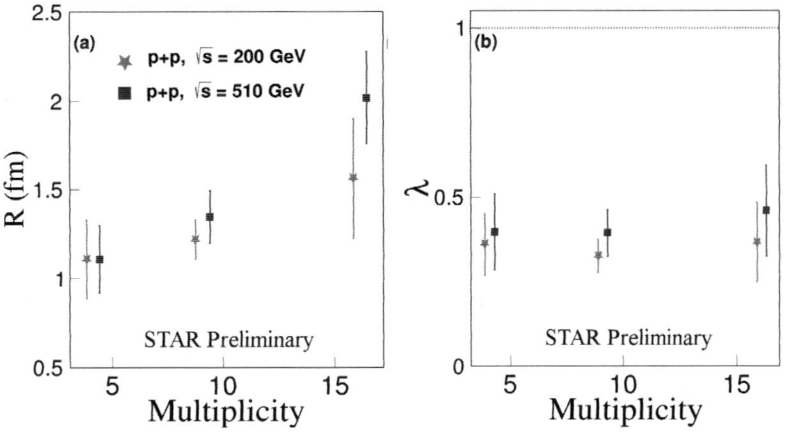


**Figure 3.** The extracted emitting source radii (a) and correlation strength (b) as a function of particle multiplicity measured at $\sqrt{s} = 200$ (stars) and $\sqrt{s} = 510$ (squares) GeV. Error bars are statistical only.

Journal of Physics: Conference Series **668** (2016) 012073 doi:10.1088/1742-6596/668/1/012073

As one can see the source radii slightly increase with the particle multiplicity for both collision energies. The observed multiplicity scaling may be understood as the increase of the initial geometrical region of overlap of the colliding objects [18]. The measured femtoscopic parameters, $R$ and $\lambda$, are consistent for both energies within the statistical errors.

## 3. Summary

The systematic femtoscopic analysis of two–kaon correlation functions from $p+p$ collisions at $\sqrt{s}$ =200 and 510 GeV measured by STAR has been presented. The one–dimensional correlation functions were studied for the three event multiplicity ranges: (1–6), (7–12) and (>12). The measured emitting source parameters, $\lambda$ and $R$, are generally independent of the collision energy. A slight increase of the source radii with event multiplicity for both $\sqrt{s}$ =200 and 510 GeV collision energies has been observed.

## References

[1] Lednicky R 2004 *Phys. Atom. Nucl.* **1** 73
[2] Alexander G 2003 *Rep. Prog. Phys.* **66** 481
[3] Goldhaber G *et al.* 1959 *Phys. Rev. Lett.* **3** 191
[4] Goldhaber G *et al.* 1960 *Phys. Rev.* **120** 300
[5] Kopylov G I and Podgoretsky M I 1972 *Sov. J. Nucl. Part.* **15** 219
[6] Kopylov G I 1974 *Phys. Lett.* B **50** 472
[7] Lisa M A *et al.* 2005 *Ann. Rev. Part. Sci.* **55** 357
[8] Chajecki Z 2009 *Acta Phys. Polon.* B **40** 1119
[9] Sinyukov Y M and Shapoval V M 2013 *Phys. Rev.* D **87** 094024
[10] Nilsson M S, Bravina L V, Zabrodin E E, Malinina L V and Bleibel J 2011 *Phys. Rev.* D **84** 054006
[11] Anderson M *et al.* 2003 *Nucl. Instrum. Meth.* A **499** 659
[12] Bonner B *et al.* 2003 *Nucl. Instrum. Meth.* A **503** 181
[13] Sun Y J *et al.* 2008 *Nucl. Instrum. Meth.* A **593** 307
[14] Adams J *et al.* 2005 *Phys. Rev.* D **71** 044906
[15] Chajecki Z and Lisa M 2008 *Phys. Rev.* C **78** 064903
[16] Sjöstrand T, Mrenna S and Skands P 2006 *JHEP* **05** 026
[17] Skands P 2010 *Phys. Rev.* D **82** 074018
[18] Barshay S 1983 *Phys. Lett.* B **130** 220

15th International Conference on Strangeness in Quark Matter (SQM2015) IOP Publishing
Journal of Physics: Conference Series **668** (2016) 012074 doi:10.1088/1742-6596/668/1/012074

# Recent results on two-particle correlations in ALICE

**Alice Ohlson for the ALICE Collaboration**
CERN, Geneva, Switzerland
E-mail: alice.ohlson@cern.ch

**Abstract.** The discovery of correlations between particles separated by several units of pseudorapidity in high-multiplicity pp and p–Pb collisions, reminiscent of structures observed in Pb–Pb collisions, was a challenge to traditional ideas about collectivity in heavy ion collisions. In order to further explore long-range correlations and provide information to theoretical models, correlations between forward trigger muons and mid-rapidity associated hadrons were measured in p–Pb collisions at $\sqrt{s_{NN}} = 5.02$ TeV. The results demonstrate that the nearside and awayside ridges extend to $\Delta\eta \sim \pm 5$ and that the $v_2$ of muons, obtained from subtracting the correlation functions in high- and low-multiplicity events, is $(16 \pm 6)\%$ higher in the Pb-going than in the p-going direction. The results are compared with AMPT simulations.

## 1. Introduction

Two-particle angular correlations are used to study many aspects of the physics of heavy ion collisions, in particular jet fragmentation and collective effects. The distributions in relative azimuthal angle ($\Delta\varphi = \varphi_{trig} - \varphi_{assoc}$) and relative pseudorapidity ($\Delta\eta = \eta_{trig} - \eta_{assoc}$) between trigger and associated particles are constructed to obtain the per-trigger yield, $\frac{1}{N_{trig}} \frac{d^2N}{d\Delta\varphi d\Delta\eta}$.

Correlation functions in collisions such as pp show characteristic features attributed to jet production: a nearside peak localized around $(\Delta\varphi, \Delta\eta) = (0, 0)$, representing pairs where the trigger and associated particles are fragments of the same jet, and the awayside peak localized around $\Delta\varphi = \pi$ but extended in $\Delta\eta$, representing pairs in which the trigger and associated particles are in back-to-back jets. In heavy ion collisions, the same jet features are observed, in addition to structures around $\Delta\varphi = 0$ and $\Delta\varphi = \pi$ extended in $\Delta\eta$. These 'ridges' are often attributed to hydrodynamic flow behavior in the quark-gluon plasma (QGP) and quantified by the Fourier coefficient, $v_2$.

It was therefore surprising when a nearside ridge was observed in high multiplicity collisions of small systems, pp [1] and p–Pb [2]. Subtraction of the correlations in low-multiplicity events from the high-multiplicity correlation functions revealed a symmetric ridge on the awayside [3, 4], which is reminiscent of the ridges attributed to flow in heavy ion collisions. This 'double ridge' structure can be quantified by the parameter $v_2$, and further studies demonstrated that the $v_2$ in p–Pb shows similar mass ordering as was observed in Pb–Pb collisions [5, 6]. It is important to note that the physical mechanism leading to a non-zero $v_2$ is still under theoretical debate, and the presence of $v_2$ does *not* necessarily imply the existence of hydrodynamics or a QGP in small systems.

Several theoretical explanations have been proposed to explain the ridge in pp and p–Pb collisions, such as higher-order glasma graphs within a Color Glass Condensate (CGC) picture, hydrodynamics in small systems, and others (see for example [7, 8]). In order to differentiate

Journal of Physics: Conference Series **668** (2016) 012074 doi:10.1088/1742-6596/668/1/012074

between the models and constrain future theoretical calculations, it would be important to measure the ridge to larger $\Delta\eta$, as well as to measure the dependence of $v_2$ on the pseudorapidity of the trigger particle. Both of these points are addressed in the muon-hadron analysis performed in ALICE [9], which will be discussed in detail in the following.

## 2. Muon-hadron correlations – Analysis details

In the analysis described below, the correlation functions between muons at forward rapidities and charged hadrons at mid-rapidity are constructed in order to investigate the long-range behavior of the double ridge structure for $-5 < \Delta\eta < -1.5$. Two beam configurations made it possible to measure the $v_2$ coefficients for muons going in the direction of the proton beam and in the direction of the lead ion beam.

### *2.1. ALICE detector*

Proton-lead collisions were recorded with the ALICE detector [10]. The main subsystems used in the muon-hadron analysis were: the Forward Muon Spectrometer (FMS), the Inner Tracking System (ITS), the Time Projection Chamber (TPC), and the V0 system.

Trigger muons were detected within the FMS, which has a pseudorapidity coverage of $-4 < \eta_{lab} < -2.5$. The composition of parent particles of the detected muons depends on transverse momentum ($p_{\rm T}$): at low $p_{\rm T}$ the muons predominantly come from weak decays of pions and kaons, while at high $p_{\rm T}$ the muons are largely the result of heavy flavor decays.

Two sets of associated hadrons are used. (1) "Tracks:" charged tracks are fully reconstructed in the ITS and TPC, and have a transverse momentum $0.5 < p_{\rm T} < 4$ GeV/$c$ and uniform coverage in $0 < \varphi < 2\pi$ and $|\eta| < 1$. (2) "Tracklets:" short track segments are reconstructed in the Silicon Pixel Detector (SPD), the two innermost layers of the ITS. The tracklet momentum is correlated with the differences of the azimuthal ($\Delta\varphi_h$) and polar angles ($\Delta\theta_h$) of hits in the SPD layers, and so a tracklet sample with $\langle p_{\rm T}\rangle \approx 0.75$ GeV/$c$ is obtained by requiring that $\Delta\varphi_h < 5$ mrad.

The multiplicity of charged particles incident on the V0 detectors is used to classify the overall event activity. Only two of the four rings in each V0 detector are used in order to achieve symmetric coverage: the innermost two rings of the V0C ($-3.7 < \eta_{lab} < -2.7$) and the outermost two rings of the V0A ($2.8 < \eta_{lab} < 3.9$).

### *2.2. Collected data sets*

During 2013 the LHC provided proton and lead beams at beam energies of 4 TeV and 1.58 ATeV, respectively, for a center-of-mass collision energy of $\sqrt{s_{\rm NN}} = 5.02$ TeV. Because of the unequal beam energies, the center-of-mass rapidity is shifted by 0.465 units in the direction of the proton beam (in the following all pseudorapidities will be quoted in the laboratory frame, $\eta_{lab}$). The beams were provided in both configurations: p–Pb, with the proton beam going in the direction of the muon arm, and Pb–p, with the lead beam facing into the muon arm.

Muon-tracklet correlations were investigated in both p–Pb and Pb–p collisions. Because of the triggering scheme chosen during the Pb–p data-taking period, the TPC was not read out in the trigger cluster for the majority of the data set, and therefore muon-track correlations were studied only in p–Pb data.

### *2.3. Analysis methodology*

The muon-track and muon-tracklet correlation functions were measured in high-multiplicity (the top 20% of the analyzed event sample) and in low-multiplicity (60-100%) events. As in [3], the low-multiplicity correlations are subtracted from the high-multiplicity correlation functions to

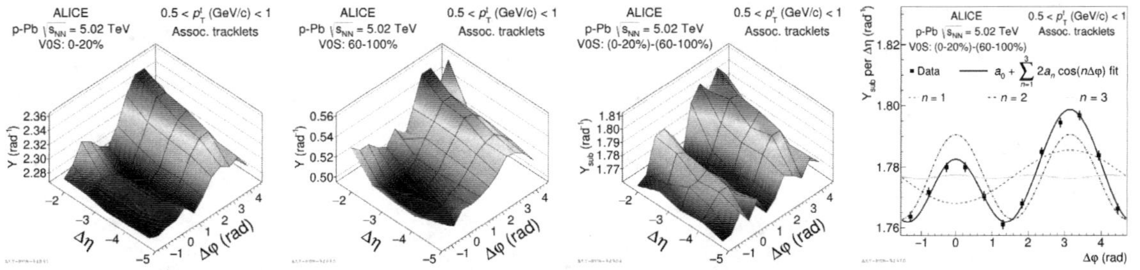


**Figure 1.** Muon-tracklet correlations in p–Pb collisions. From left to right: the correlation function in high-multiplicity events in which the nearside ridge is clearly visible, the correlation function in low-multiplicity events where there is no obvious ridge, the subtracted correlations which show the double ridge structure, and the $\Delta\varphi$ projection with a Fourier cosine series fit.

remove structures associated with jet fragmentation. After subtraction, the correlation functions are projected onto $\Delta\varphi$, and then fit with a Fourier cosine series,

$$a_0 + \sum_{n=1}^{3} 2a_n \cos(n\Delta\varphi). \quad (1)$$

From the coefficients of the Fourier series, the pair azimuthal anisotropy $V_{2\Delta}\{\text{2PC,sub}\}$ can be obtained:

$$V_{2\Delta}\{\text{2PC,sub}\} = \frac{a_2}{a_0 + b}, \quad (2)$$

where $b$ is the baseline in the low-multiplicity event class. Assuming factorization of the $v_2$ values for trigger and associated particles, the $v_2$ of trigger muons can be obtained by dividing $V_{2\Delta}\{\text{2PC,sub}\}$ by the $v_2$ of associated tracks or tracklets, which is obtained from track-track and tracklet-tracklet correlations, respectively:

$$v_2^{\mu}\{\text{2PC,sub}\} = \frac{V_{2\Delta}^{\mu-h}\{\text{2PC,sub}\}}{v_2^h} = \frac{V_{2\Delta}^{\mu-h}\{\text{2PC,sub}\}}{\sqrt{V_{2\Delta}^{h-h}\{\text{2PC,sub}\}}}. \quad (3)$$

The agreement between $v_2^{\mu}\{\text{2PC,sub}\}$ measured in muon-track and muon-tracklet correlations (which have very different $p_{\mathrm{T}}$-distributions for the associated hadrons) demonstrates that the factorization assumption is valid. An example set of correlation functions from the muon-tracklet correlations analysis in p–Pb collisions is shown in Fig. 1.

## 3. Muon-hadron correlations – Results and model comparisons

The $v_2^{\mu}\{\text{2PC,sub}\}$ of muons has been measured as a function of $p_{\mathrm{T}}$ in the p- and Pb-going directions, and the results are shown in Figs. 2 and 3. It is observed that $v_2^{\mu}\{\text{2PC,sub}\}$ in the Pb-going direction is consistently higher than $v_2^{\mu}\{\text{2PC,sub}\}$ in the p-going direction. The ratio, shown in Fig. 3, is roughly independent of $p_{\mathrm{T}}$ within statistical and systematic uncertainties. A constant fit to the ratio yields $1.16 \pm 0.06$.

The data are compared with an AMPT [11] simulation in which the muon decay products are scaled to account for the efficiency of the absorber in the ALICE FMS. In Fig. 2 it is seen that while AMPT qualitatively describes the $p_{\mathrm{T}}$-dependence at low $p_{\mathrm{T}}$, there are significant quantitative differences in the $p_{\mathrm{T}}$-dependence and $\eta$-dependence between data and the model. At high $p_{\mathrm{T}}$ (above $p_{\mathrm{T}} \sim 2$ GeV/$c$), where muon production is dominated by heavy flavor decays, AMPT does not describe the data well. This could be because heavy flavor muons have a non-zero $v_2$, or the parent particle composition or $v_2$ values in data and AMPT are different.

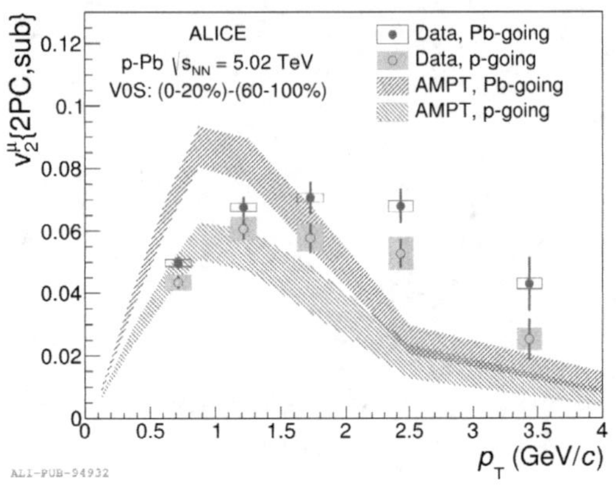


**Figure 2.** $v_2^{\mu}\{2\text{PC,sub}\}$ in the Pb-going and p-going directions as a function of $p_{\mathrm{T}}$ compared to AMPT.

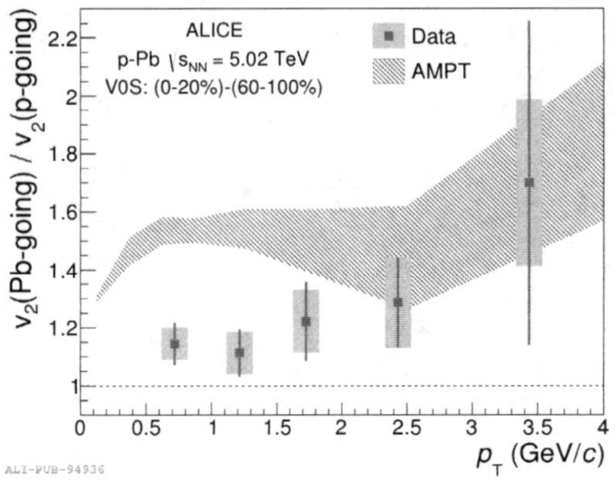


**Figure 3.** Ratio of $v_2^{\mu}\{2\text{PC,sub}\}$ in the Pb-going and p-going directions as a function of $p_{\mathrm{T}}$ compared to AMPT.

Theoretical treatments of forward-central muon-hadron correlations indicate that the observables measured here are sensitive to the dynamics and possible collective effects in small systems [12]. However, current theoretical calculations cannot be directly compared with experimental results, because the effects of the absorber are included in the experimental data (unfolding such effects could not be done in a model-independent way). Future model calculations should use the efficiencies provided in [9] in order to compare directly to the experimental results.

## 4. Conclusions

Correlations between forward trigger muons and mid-rapidity associated hadrons can provide insight into the $\eta$-dependence of long-range correlations in p–Pb collisions. The analysis is performed using muons detected in the FMS and charged hadrons reconstructed in the central barrel in ALICE in p–Pb and Pb–p collisions at $\sqrt{s_{\mathrm{NN}}} = 5.02$ TeV. The results demonstrate that the nearside and awayside ridges extend to $\Delta\eta \sim \pm 5$ with trigger particles out to $|\eta| \sim 4$. Furthermore, the $v_2^{\mu}\{2\text{PC,sub}\}$ of muons is $(16\pm6)\%$ higher in the Pb-going direction than in the p-going direction, and this ratio is roughly independent of momentum. While there is qualitative agreement between data and AMPT at low $p_{\mathrm{T}}$, the model does not yet quantitatively describe the data. Future measurements and model comparisons will allow us to gain understanding of the production mechanisms of long-range correlations in small systems.

## 5. References

[1] Khachatryan V *et al.* (CMS) 2010 *JHEP* **09** 091
[2] Chatrchyan S *et al.* (CMS) 2013 *Phys. Lett.* **B718** 795–814
[3] Abelev B *et al.* (ALICE) 2013 *Phys. Lett.* **B719** 29–41
[4] Aad G *et al.* (ATLAS) 2013 *Phys. Rev. Lett.* **110** 182302
[5] Abelev B B *et al.* (ALICE) 2013 *Phys. Lett.* **B726** 164–177
[6] Khachatryan V *et al.* (CMS) 2015 *Phys. Lett.* **B742** 200–224
[7] Bzdak A, Schenke B, Tribedy P and Venugopalan R 2013 *Phys. Rev.* **C87** 064906
[8] Dusling K and Venugopalan R 2013 *Phys. Rev.* **D87** 094034
[9] Adam J *et al.* (ALICE) 2015 *submitted to Phys. Lett. B* (*Preprint* 1506.08032)
[10] Aamodt K *et al.* (ALICE) 2008 *JINST* **3** S08002
[11] Lin Z W, Ko C M, Li B A, Zhang B and Pal S 2005 *Phys. Rev.* **C72** 064901
[12] Bozek P, Bzdak A and Ma G L 2015 *Phys. Lett.* **B748** 301–305

# Directed flow in heavy-ion collisions from the PHSD transport approach

**A. Palmese and W. Cassing**
Institut für Theoretische Physik, Universität Gießen, Germany
E-mail: Alessia.Palmese@theo.physik.uni-giessen.de

**Abstract.** Recent STAR data for the directed flow of protons, antiprotons, charged pions and kaons obtained within the beam energy scan (BES) program are analyzed within the Parton-Hadron-String-Dynamics (PHSD/HSD) transport models. Both versions of the kinetic approach are used to clarify the role of partonic degrees of freedom. The PHSD results, simulating a partonic phase and its coexistence with a hadronic one, are roughly consistent with the STAR data. Generally, the semi-qualitative agreement between the measured data and model results supports the idea of a crossover type of quark-hadron transition which softens the nuclear EoS but shows no indication of a first-order phase transition. Furthermore, the directed flow of kaons and antikaons is evaluated in the PHSD approach from $\sqrt{s_{NN}} \approx 5$ - 12 GeV which shows a high sensitivity to the hadronic potentials in the FAIR/NICA energy regime $\sqrt{s_{NN}} \leq 8$ GeV.

High energy heavy-ion collisions are the unique experimental way to study the Equation of State (EoS) of strongly interacting matter. New experiments on this topic are currently in preparation such as the CBM at FAIR and the MPD at NICA. In order to extract information about the behaviour of matter, one has to analyse the data with respect to the production of various hadronic particles, both mesons and baryons. This analysis is highly non-trivial, since at these energy regimes, the baryon chemical potential cannot be neglected and, unfortunately, the EoS of the matter at finite baryon chemical potential is still quite uncertain. The possibility to investigate the deconfinement phase transition at high baryon density and moderate temperature is one of the main goals of the scientific community in the recent years. Since the QGP is created only for a short time (of a couple of fm/c) it is quite challenging to study its properties and to find the most sensible probes. The study of the particle azimuthal angular distribution in momentum space with respect to the reaction plane is an important tool to probe the hot, dense matter created in heavy-ion collisions [1]. The *directed flow* refers to a collective sidewards deflection of particles and is characterized by the first-order harmonic $v_1$ of the Fourier expansion of the particle azimuthal angular distribution with respect to the reaction plane [2]. Apart from first measurements in the early nineties and till recent times, the directed flow was studied mainly theoretically although some experimental information from the Schwerionen-Synchrotron (SIS) to Super-Proton-Synchrotron (SPS) energies is available [3]. It is generally assumed that the directed flow is generated during the nuclear passage time [4]. The directed transverse flow therefore probes the onset of bulk collective dynamics during thermalization, thus providing valuable information also on the pre-equilibrium stage.

The interest in the directed flow $v_1(y)$ has recently been enhanced considerably due to new STAR data obtained in the framework of the beam-energy-scan (BES) program [5]. The directed flow of identified hadrons – protons, antiprotons, positive and negative pions and more recently also kaons and antikaons [6] – has been measured with high precision for semi-central Au+Au collisions in the energy range $\sqrt{s_{NN}}$ =(7.7-39) GeV. These data provide a promising basis for studying direct-flow issues as discussed above and have been addressed already by the Frankfurt group [7] limiting themselves to the energy $\sqrt{s_{NN}}$ <20 GeV where hadronic processes are expected to be dominant. However, the authors of Ref. [7] did not succeed to describe the data and to obtain conclusive results which led to the notion of the 'directed flow puzzle'. Our study aims to analyze these STAR results in the whole available energy range including in particular antiproton and $K^{\pm}$ data [8, 9].

The description of the PHSD model employed here is described elsewhere [10, 11]. It involves dynamical partons with a scalar selfenergy, a covariant hadronization scheme as well as hadronic rescattering with (optionally) mean-field potentials. In the hadronic phase, i.e. for energies densities below the critical energy density $\epsilon_c \approx 0.5$ GeV/fm$^3$, the PHSD approach is identical to the Hadron-String-Dynamics (HSD) model [12, 13].

The directed flow

$$v_1(y) = \langle \cos(\psi - \Psi_{RP}) \rangle = \left\langle \frac{p_x}{\sqrt{p_x^2 + p_y^2}} \right\rangle , \tag{1}$$

is computed for all hadrons separately as a function of rapidity $y$ within the experimental momentum acceptance. The characteristic slope $F$ of the $v_1(y)$ distributions at midrapidity, $F := dv_1/dy|_{y=0}$, is presented in Fig. 1 (l.h.s.) for all cases considered. In fact, in a first approximation the $v_1$ flow in the center-of-mass system may be well fitted by a linear function $v_1(y) = F\ y$ within the rapidity interval $-0.5 < y < 0.5$. For protons there is a qualitative agreement of the HSD/PHSD results with the experiment measurements: the slope $F > 0$ at low energies, however, exceeding the experimental values by up a factor of about two; the slope crosses the line $F$ =0 at $\sqrt{s_{NN}}$ ~20 GeV, which is twice larger than the experimental crossing point, and then stays negative and almost constant with further energy increase. However, the absolute values of the calculated proton slopes in this high energy range are on the level of -(0.010-0.015), while the measured ones are about -0.005. The standard UrQMD model results, as cited in the experimental paper [5] and in the more recent theoretical work [7], are displayed in Fig. 1 (l.h.s.) by the wide and narrow shaded areas, respectively. These results for protons are close to those from the HSD and essentially overestimate the slope for energies below ~30 GeV but at higher energy become negative and relatively close to the experiment. The predictions for the pure hadronic version of the transport model HSD (dotted lines in Fig. 1 (l.h.s.)) slightly differ from the PHSD results which overpredict the negative proton slope at higher RHIC energies.

For the antiproton slopes we again observe an almost quantitative agreement with the BES experiments [5]: with increasing collision energy the HSD and PHSD slopes grow and then flatten above 20-30 GeV. The HSD results saturate at $v_1(0) = 0$, while the PHSD predictions stay negative and in good agreement with experiment (see Fig. 1 (l.h.s.)). It is noteworthy to point out that these PHSD predictions strongly differ from the UrQMD results which no longer describe the data for $\sqrt{s_{NN}}$ ≤20 GeV but are in agreement with the measurements for higher energies.

The differences between the calculations and experimental data become apparent for the charged pion slopes at $\sqrt{s_{NN}}$ ≤11 GeV: the negative minimum of the charged pion slope is deeper than the measured one. The HSD and PHSD results practically coincide at low energy (due to a minor impact of partonic degrees-of-freedom) but dramatically differ from those of the UrQMD model for $\sqrt{s_{NN}}$ ≤20 GeV (see Fig.1 (l.h.s.)).

Thus, in agreement with the STAR experimental data, in the considered energy range the PHSD model predicts for protons a smooth $F(\sqrt{s_{NN}})$ function which is flattening at $\sqrt{s_{NN}} \geq 20$ GeV and reveals no signatures of a possible first-order phase transition as expected

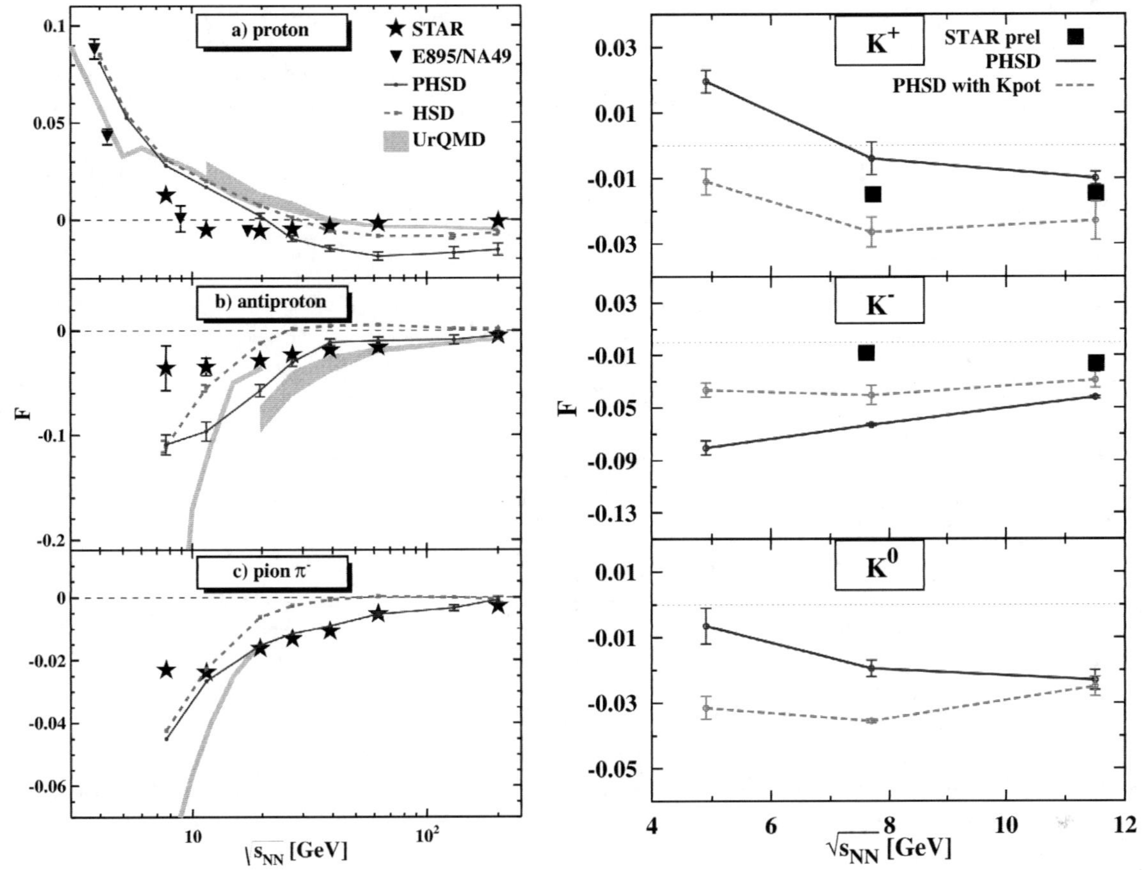


**Figure 1.** (l.h.s.) The beam energy dependence of the directed flow slope $F$ at midrapidity for protons, antiproton and charged pions from semicentral Au+Au collisions [8]. The shaded band corresponds to the UrQMD results as cited in [5]. The experimental data are from the STAR collaboration [5] along with results of prior experiments using comparable cuts. (r.h.s.) The directed flow slope $F$ for $K^+, K^-, K^0$ as a function of the invariant energy $\sqrt{s_{NN}}$ with (dashed purple lines) and without kaon potentials (solid blue lines) in the PHSD calculations.

in Refs. [14, 15]. For antiprotons the slope at midrapidity manifests a wide but shallow negative minimum for $\sqrt{s_{NN}}$ ≈30 GeV while the measured slope is a monotonically increasing function. It is noteworthy that the new STAR data are consistent with the PHSD results which include a crossover transition by default due to a matching of the partonic EoS to lattice QCD results.

Strange hadrons and in particular kaons and antikaons provide additional information on the reaction dynamics. In relativistic mean-field models the dispersion relation for kaons and antikaons in the nuclear medium can be written as [16, 17, 18]

$$\omega^2_{K^\pm}(\rho_N, \mathbf{p}) = \pm\frac{3}{4}\frac{\omega}{f_K^2}\rho_N + m_K^2 + \mathbf{p}^2 - \frac{\Sigma_{KN}}{f_K^2}\rho_s. \qquad (2)$$

In Eq. (2) $\Sigma_{KN}$ is the kaon-nucleon sigma term ($\approx$ 400 MeV), $m_K$ denotes the bare kaon mass, $f_K \approx$ 100 MeV is the kaon decay constant, while $\rho_s$ and $\rho_N$ stand for the scalar and vector nucleon densities, respectively. This leads to repulsive kaon mean fields $U_K > 0$ and attractive mean fields for the antikaons $U_{\bar{K}} < 0$ at finite baryon density [16].

The microscopic calculation of kaon and antikaon potentials (or mean fields) is more involved and can e.g. be worked out within G-matrix theory [17, 18]. These calculations show that the potentials also explicitly depend on the $K^\pm$ momentum $p$ with respect to the local rest frame of the system. Such a momentum-dependence can be incorporated by introducing momentum dependent formfactors in the scalar and vector potentials and fitting the parameters to the results from G-matrix theory up to twice nuclear matter density. In order to shed some light on the possible effects of $K^\pm$ potentials on their directed flow - especially at FAIR/NICA energies - we have incorporated the momentum-dependent potentials for kaons and antikaons as displayed in Fig. 5 of Ref. [9].

We directly step on with the preliminary results for Au+Au collisions at impact parameters 6 to 9 fm for invariant energies of $\sqrt{s_{NN}}$ from 5 to 12 GeV, i.e. at the FAIR/NICA energies and the low RHIC energy. The actual results are shown in Fig. 1 (r.h.s.) for $K^+, K^-$ and $K^0$ mesons as a function of the invariant energy $\sqrt{s_{NN}}$ with (dashed purple lines) and without kaon potentials (solid blue lines) in the PHSD calculations. These directed flows are comparable in size with those from the pions. The focus in this case is on the role of the kaon/antikaon potentials. We find a remarkable sensitivity to the kaon/antikaon potentials in this low energy (FAIR/NICA) domain; the positive flow of kaons even changes sign when including the repulsive potentials, whereas the antikaon flow is decreased in size substantially by the attractive mean fields. The $K^0$ flow presents features similar to $K^+$, since they are associated with the same repulsive potential (except for Coulomb forces). Above $\sqrt{s_{NN}} \sim 12$ GeV there is no longer a sizeable sensitivity to the kaon/antikaon potentials. Comparing our results for $K^+, K^-$ with the data, we infer that the presently adopted kaon potential is too strong, while the antikaon potential is too soft.

We speculate that the inclusion of a restoration of chiral symmetry at high baryon density and/or temperature might lead to a solution of the currently remaining problems. Moreover, precise experimental studies at FAIR, NICA or within the BES II program at RHIC will provide more information on the kaon/antikaon potentials.

## Acknowledgements

The authors acknowledge inspiring discussions with E. L. Bratkovskaya, V. P. Konchakovski , O. Linnyk and V. D. Toneev.

## References

[1] Voloshin S A Poskanzer A M Snellings R 2010 *Landolt-Boernstein New Series* I/**23** 5-54
[2] Poskanzer A M Voloshin S A 1998 *Phys. Rev.* C **58** 1671
[3] Senger P *et al.* 2011 *Lect. Notes Phys.* **814** 681
[4] Sorge H 1997 *Phys. Rev. Lett.* **78** 2309
[5] Adamczyk L *et al.* 2014 *Phys. Rev. Lett.* **112** 162301
[6] Pandit Y 2015 *J. Phys. Conf. Ser.* **636** 012001
[7] Steinheimer J *et al.* 2014 *Phys. Rev.* C **89** 054913
[8] Konchakovski V Cassing W Toneev V D 2014 *Phys. Rev.* C **90** 014903
[9] Cassing W *et al.* 2015 *EPJ Web Conf.* **95** 01004
[10] Cassing W Bratkovskaya E L 2009 *Nucl. Phys.* A **831** 215
[11] Bratkovskaya E L Cassing W Konchakovski V P Linnyk O 2011 *Nucl. Phys.* A **856** 162
[12] Cassing W Bratkovskaya E L 1999 *Phys. Rep.* **308** 65
[13] Cassing W Bratkovskaya E L Juchem S 2000 *Nucl. Phys.* A **674** 249
[14] Rischke D H *et al.* 1995 *Heavy Ion Phys.* **1** 309
[15] Rischke D H 1996 *Nuclear Physics* A **610** 88c
[16] Oeschler H Leifels Y Bratkovskaya E L Aichelin J 2012 *Phys. Rept.* **510** 119
[17] Tolós L Ramos A Polls A 2002 *Phys. Rev.* C **65** 054907
[18] Cassing W Tolós L Bratkovskaya E L Ramos A 2003 *Nucl. Phys.* A **727** 59

# The cool potential of gluons

**André Peshier**[⋆]
Department of Physics, University of Cape Town, Rondebosch 7700, South Africa
E-mail: Andre.Peshier@uct.ac.za

**Dino Giovannoni**
Department of Physics and Electronics, Rhodes University, Grahamstown 6140, South Africa
E-mail: D.Giovannoni@ru.ac.za

**Abstract.** We put forward the idea that the quark-gluon plasma might exist way below the usual confinement temperature $T_c$. Our argument rests on the possibility that the plasma produced in heavy-ion collisions could reach a transient quasi-equilibrium with 'over-occupied' gluon density, as advocated by Blaizot et al. Taking further into account that gluons acquire an effective mass by interaction effects, they can have a positive chemical potential and therefore behave similarly to non-relativistic bosons. Relevant properties of this dense state of interacting gluons, which we dub serried glue, can then be inferred on rather general grounds from Maxwell's relation.

## 1. Introduction

Experiments at RHIC and LHC have firmly established that the quark-gluon plasma produced in heavy ion collisions approaches local equilibrium rapidly, at a rate which allows for successful hydrodynamical descriptions of the parton phase despite its short lifetime of some fm/c. It seems reasonable to consider that this approach to equilibrium is initially dominated by binary collisions of partons [1]. If indeed gluon number changing processes cannot keep up with the fast kinetic equilibration, the quark-gluon plasma might exit sufficiently long in a *transient* quasi-equilibrium state that is characterized not only by local temperature, collective flow velocity and quark chemical potential(s), but also by a non-zero *gluon* chemical potential $\mu \neq 0$, quantifying that the gluon phase space density has not yet reached the asymptotic Bose distribution with $\mu_\infty = 0$. As a consequence thereof, it has been argued that the high gluon occupation in the initial state, which can be described as a color glass condensate, may lead to the formation of a gluonic Bose-Einstein condensate in heavy ion collisions [2].

Here we will put forward another noteworthy implication of this *transient* quasi-equilibrium phase with gluon overpopulation. In a thermal medium massless partons acquire a dynamical mass $m \sim gT$, which is a key mechanism to understand thermodynamic properties of the strongly interacting quark-gluon plasma, see e. g. [3]. In the present context it implies that the chemical potential $\mu$, which must be smaller than the minimal energy of the dressed gluons $\omega(k) = (k^2 + m^2)^{1/2}$, can be positive (opposed to $\mu \leq 0$ assumed in [2], and in general in perturbative QCD). This *dense and interacting* phase at $0 < \mu < m$ we dub **serried glue**.

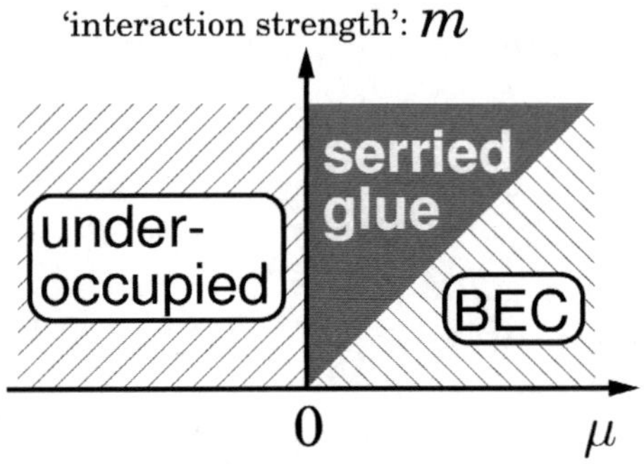


**Figure 1.** Schematic 'phase diagram' of the gluon plasma in *transient* equilibrium with $\mu \neq 0$. Only with the dynamical gluon mass $m$ taken into account, the window for the serried glue phase opens between the under-occupied phase at $\mu < 0$ and the Bose-Einstein condensate at $\mu_{\rm BEC} = \min[\omega(k)] = m$.

## 2. QCD quasiparticles and Maxwell flow

In order to study the thermodynamic properties of the serried glue phase, we generalize a QCD quasiparticle model [3]: In the *fully* equilibrated system, i. e. at $\mu_\infty = 0$, dressed gluons can be approximated as quasiparticles with a temperature-dependent mass $m^2(T) = \frac{1}{2}\, G^2(T)T^2$ where

$$G^2(T) = b/\ln\Big((\lambda T + T_s)/T_c\Big)^2 \tag{1}$$

is an *effective* coupling, with $b = 16\pi^2/(11 - \frac{2}{3}\,n_f)$ for $n_f$ flavors. Thermodynamic properties follow from the thermodynamic potential $-p(T)V$. For simplicity we focus here on the quenched limit, $n_f = 0$; then the quasiparticle pressure is that of an ideal gas of bosons with mass $m(T)$ and degeneracy $d_g = 16$, less a residual quasiparticle interaction contribution,

$$p(T) = p_{\rm id}(T; m(T)) - B(T)\,. \tag{2}$$

By thermodynamic consistency, $B(T)$ is related to (and can be obtained, up to an integration constant, from) the quasiparticle mass $m(T)$ by

$$\frac{dB(T)}{dT} = \frac{\partial p_{\rm id}(T;m)}{\partial m}\bigg|_T \frac{dm(T)}{dT}\,.$$

As consequence thereof, the entropy density takes a particularly simple form (without a residual interaction term like in the pressure), viz.

$$s(T) = \partial p(T)/\partial T = s_{\rm id}(T; m(T))\,. \tag{3}$$

After fixing the parameters $\lambda$ and $T_s$ by matching to lattice QCD results [4], the scheme is fully specified and will provide the 'initial' condition for a flow equation derived now.

For $\mu \neq 0$, both the quasiparticle mass and the pressure will depend on the chemical potential. Similar to the entropy density ((3) is now complemented by a $\mu$-dependence), we find the gluon particle density equal to the density of an ideal boson gas with $T, \mu$-dependent mass,

$$n(T,\mu) = \partial p(T,\mu)/\partial\mu = n_{\rm id}(T,\mu; m(T,\mu))\,. \tag{4}$$

Obviously, as derivatives of $p(T,\mu)$, the entropy and particle density must satisfy Maxwell's relation $\partial s/\partial\mu = \partial n/\partial T$, which translates into a condition for the quasigluon mass $m(T,\mu)$,

$$\frac{\partial m}{\partial \mu} + A\Big(T,\mu; m(T,\mu)\Big)\,\frac{\partial m}{\partial T} = 0\,. \tag{5}$$

This flow equation is similar to a relation derived in [5] for the effective coupling $G$ in the quark-gluon plasma at finite *quark* chemical potential $\mu_Q$, which led to an early successful prediction of the curvature of the critical line to the hadronic phase in the $T\mu_Q$ phase diagram.

We only mention (without detailing it here) that the expression $A$ in (5) is a non-linear function of $T, \mu, m$ involving integrals of the gluon distribution $f(k) = (\exp[(\omega(k) - \mu)/T] - 1)^{-1}$. Note that nonetheless (5) is a *linear* partial differential equation which can be readily solved by the *method of characteristics* (with the 'initial' condition $m(T, 0)$ obtained from fitting the usual $\mu = 0$ lattice QCD results, as announced). We then notice immediately that with only derivatives of $m$ in the flow equation (5), i. e. no summand term $\tilde{A}(T, \mu; m)$, the mass $m(T, \mu)$ is *constant* along the characteristics. Therefore, the characteristic line $\mu^c(T, T_i)$ from a given 'initial' point $(T_i, 0)$ into the $\mu > 0$ region is limited by $\mu \leq \mu^{\max}(T_i) = m(T_i, 0)$. This upper bound $\mu^{\max}$ is in fact reached (with vanishing slope $\partial\mu^c(T, T_i)/\partial T$) and earmarks the onset of Bose-Einstein condensation. The corresponding temperatures at the end points of the characteristic lines, which are needed to fully specify the condensate phase boundary $\mu^\star_{\text{BEC}}(T)$, can be be calculated analytically in the weak-coupling limit or numerically from the flow equation (5). As a qualitative feature we point out that the inverse relation $T^\star_{\text{BEC}}(\mu)$ has two branches, see Fig. 2 – an obvious result of the continuous mapping by the flow (5) of the 'initial' condition $m(T, 0) \propto G(T)T$ which is *non-monotonous* (being the product of a decreasing and an increasing function).

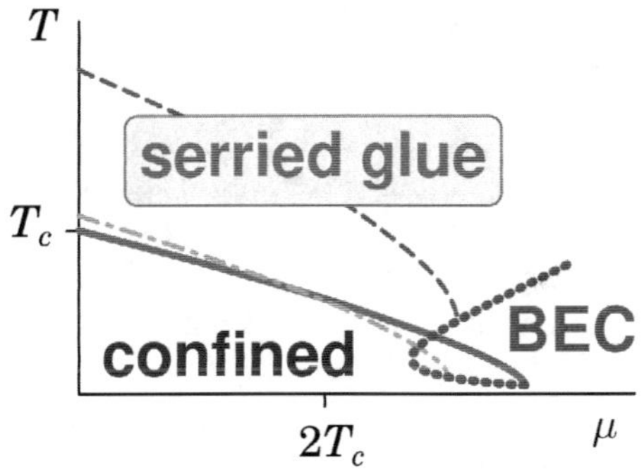


**Figure 2.** Illustrating the Maxwell flow (5): the end points of the characteristics specify the boundary to the Bose-Einstein condensate (dotted green line). The characteristic line $\mu^c(T, T_c)$ from the 'critical point' $(T_c, 0)$ can be roughly related to the confinement phase boundary. Note that the characteristics from a vicinity of $T_c$ intersect (see dash-dotted orange line) since the solution of (5) is locally non-unique: $m(T, \mu)$ develops a fold (as explained in the text, this has no physics relevance).

## 3. Results and conclusion

Although Bose-Einstein condensation in heavy-ion collisions is certainly interesting, we focus here on another aspect. Sufficient for our argument, and easy to show analytically from (5), is that the slope of the characteristics is negative. Following [5] and relating the characteristic line $\mu^c(T, T_c)$ from the 'critical point' $(T_c, 0)$ to the phase boundary with the confined phase, we find that the confinement temperature $T_c^\star(\mu)$ decreases markedly with $\mu$ in the serried glue phase, see full red line in Fig. 2.

In this context we remark that for large effective coupling the solution of (5) turns out to be multi-valued in a small region, indicated by characteristics from a small vicinity of $T_c$ intersecting each other, cf. Fig. 2. However, this circumstance has no physical relevance since the pressure, when calculated 'along' these descending characteristics, becomes negative 'before' the solution of (5) becomes non-unique. This is illustrated in Fig. 3 where $p(T, \mu) = 0$ is indicated by the dashed purple line – a 'conservative bound' for the confinement transition occurring in fact at non-zero pressure (i. e. somewhat above the dashed line), which thus excludes seemingly ambiguous solutions of the flow equation.We recall a corresponding observation in case of the quark chemical potential [5].

Of interest for phenomenology are quantitative estimates of the magnitude of the expected effects. Keeping in mind that for simplicity we are discussing here the quenched limit where $T_c \simeq 0.26\,\text{GeV}$, we may expect a pure gluon plasma in the serried phase to exist down to

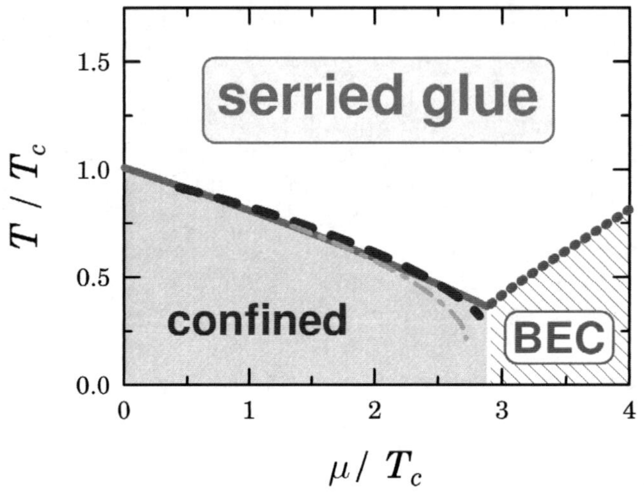


**Figure 3.** Proposed phase diagram of (the gluon sector) of QCD matter. As discussed in the text, the system will undergo a transition to the confined phase 'before' it reaches on the descending characteristics $p(T, \mu) = 0$ (indicated by the dashed purple line), hence the multi-valued solution of (5) in a region below the dash-dotted orange line is irrelevant.

temperatures of some 0.1 GeV, if the gluon chemical potential reaches $\mu^\star \sim 3T_c$ where Bose-Einstein condensation would set in. We remark that such a *'massive'* effect[1] is not a surprise: massive bosons with chemical potential close to $\mu^{\max} = m$ behave like non-relativistic particles, $\omega(k) - \mu \to (k^2 + m^2)^{1/2} - m = k^2/2m + \ldots$ Thus the 'Bose divergence', $(e^x - 1)^{-1} \approx x^{-1}$ for small $x$, is enhanced compared to perturbatively massless gluons ($\mu^{\max} \to 0$), which can over-compensate the suppression effects due to larger kinetic energy with a dynamical mass.

In conclusion, we propose a new phase 'serried glue' in the QCD phase diagram. While we discussed here only the quenched limit, we expect also for the physical case the possibility of reaching temperatures significantly below the commonly assumed lowest value $T_c \approx 0.16$ GeV, catalyzed by interacting (i. e. dressed) gluons in over-occupied states [6]. We consider this idea as reasonably robust insofar as it relies on the indubitable emergence of dynamical mass scales in the medium and the fundamental Maxwell relation (which governs not only the quasiparticle model used here, but also more sophisticated resummation approaches like [7]). The only remaining assumption, the existence of a sufficiently long-lived transient equilibrium with approximately conserved gluon number, remains to be substantiated. If confirmed, the implications of a 'super-cool serried glue' phase would be far-reaching for hydrodynamic simulations and thus for heavy-ion physics in general.

## References

[1] J. P. Blaizot, F. Gelis, J. F. Liao, L. McLerran and R. Venugopalan, Nucl. Phys. A **873**, 68 (2012).
[2] J. P. Blaizot, J. Liao and L. McLerran, Nucl. Phys. A **920**, 58 (2013).
[3] A. Peshier, B. Kampfer, O. P. Pavlenko and G. Soff, Phys. Rev. D **54**, 2399 (1996).
[4] S. Borsanyi, G. Endrodi, Z. Fodor, S. D. Katz and K. K. Szabo, JHEP **1207**, 056 (2012).
[5] A. Peshier, B. Kampfer and G. Soff, Phys. Rev. C **61**, 045203 (2000).
[6] D. Giovannoni and A. Peshier, to be published.
[7] A. Peshier, Phys. Rev. D **63**, 105004 (2001).

[1] Even though it might be slightly misleading, we are tempted to call it gluon *'super-cooling'*, in particular because it is also a transient phenomenon like the usual super-cooling.

# Energy scan of correlations in p+p and Be+Be from NA61/SHINE

**A. Seryakov for NA61/SHINE collaboration**
St. Petersburg State University, Russia

E-mail: seryakov@yahoo.com

**Abstract.** The existence of the critical point (CP) of strongly interacting matter is still an open problem. An extensive strong interactions program including a search of the CP and the study of the onset of deconfinement was started by the NA61/SHINE experiment at the CERN SPS. A two dimensional scan of the phase diagram is performed to search for the CP and to shed light on the phase transition region. This program includes studies of hadron production in proton-proton, proton-nucleus and nucleus-nucleus interactions measured in a wide range of colliding energy and system size. Correlations between various observables measured at midrapidity as well as in separated rapidity intervals are considered as additional and sensitive tools of this phase diagram scan. We present NA61/SHINE results of studies of energy dependence of two-particle correlations of pseudo-rapidity and azimuthal angle in p+p collisions at the SPS and the first results on correlations between multiplicity and mean transverse momentum in $^{7}Be + {}^{9}Be$ collisions at 150A GeV/c obtained for separated pseudo-rapidity intervals (so called long-range correlations). Comparison with data calculations using the EPOS 1.99 model are also discussed.

## 1. Introduction

We present preliminary results of two-particles correlation analysis in p+p collisions (previously shown in [1], except figure.3) and pseudo-rapidity correlations in $^{7}Be + {}^{9}Be$ interactions at 150A GeV/c. Two-particles correlations are an unique tool to disentangle different sources of correlations. Pseudo-rapidity correlations are a powerful method for probing the initial conditions for the formation of the QGP [2][3]. A detailed description of the NA61/SHINE setup can be found in [4].

### *1.1. Two particles correlations*

Such correlations were studied as a function of the difference in pseudo-rapidity $\Delta\eta = |\eta 1 - \eta 2|$ and azimuthal angle $\Delta\varphi = |\varphi 1 - \varphi 2|$ between two particles in the same event. The correlation function is calculated as:

$$C(\Delta\eta,\Delta\varphi) = \frac{N^{pairs}_{mixed}}{N^{pairs}_{data}}\frac{S(\Delta\eta,\Delta\varphi)}{M(\Delta\eta,\Delta\varphi)};\;\; S(\Delta\eta,\Delta\varphi) = \frac{d^2N^{signal}}{d\Delta\eta d\Delta\varphi};\;\; M(\Delta\eta,\Delta\varphi) = \frac{d^2N^{mixed}}{d\Delta\eta d\Delta\varphi};$$

where $S(\Delta\eta,\Delta\varphi)$ and $M(\Delta\eta,\Delta\varphi)$ are the distributions for pairs from data and mixed events, respectively. The $\Delta\varphi$ range is folded. Pseudo-rapidity is calculated in the center of mass system with the pion mass assumption for all charged particles, which allows to compare our results with the RHIC and LHC results.

*1.2. Pseudo-rapidity correlations*

Pseudo-rapidity correlations are defined as correlations between observables $B$ and $F$ in different pseudo-rapidity windows. The correlation parameter is computed as:

$$\boldsymbol{b_{rel}[B,F]} = \frac{\langle \boldsymbol{F}\rangle}{\langle \boldsymbol{B}\rangle}\frac{\langle \boldsymbol{BF}\rangle - \langle \boldsymbol{B}\rangle\langle \boldsymbol{F}\rangle}{\langle \boldsymbol{F}^2\rangle - \langle \boldsymbol{F}\rangle^2}$$

where $<B>$ and $<F>$ are the event averages in the «backward» and «forward» pseudo-rapidity windows respectively. In this work we present the following types of pseudo-rapidity correlations:

- Backward multiplicity $N_B$ and forward multiplicity $N_F$: $b_{rel}[N_B, N_F]$
- Backward mean transverse momentum $Pt_B$ and $N_F$: $b_{rel}[Pt_B, N_F]$
- Backward and forward mean transverse momenta $Pt_B$ and $Pt_F$: $b_{rel}[Pt_B, Pt_F]$

## 2. Analysis procedure

All results were corrected for biases due to trigger, off-line selection, track and event selection etc. by using events obtained from EPOS 1.99 model event generator before and after GEANT detector simulation and reconstruction. Electrons and positrons were removed by a dE/dx cut. Possible effects from hard scattering were reduced by a cut on particle transverse momentum $p_T$< 1,5 GeV/c. All results were calculated inside NA61/SHINE acceptance, the influence of which will be shown in the example of the EPOS 1.99 model in the next chapter. Only statistical errors are shown for pseudo-rapidity correlations.

## 3. Results

3.1. *Two particles correlations*

The correlation function $C(\Delta\eta, \Delta\varphi)$ for p+p collisions of different energies and different charged pair combinations is shown in figure. 1. There are several notable structures: a maximum at (Δη,Δφ)= (0,π), likely a result of resonance decays and momentum conservation. It is stronger for unlike-sign pairs and significantly weaker for same charge pairs; an enhancement at (Δη,Δφ) = (0,0), probably due to Coulomb interactions (unlike-sign pairs) and quantum statistics (same charge pairs). Results of calculations with the EPOS 1.99 model are also shown in figure. 3. The model calculations are consistent with the enhancement seen at (0,π) (compare figure. 3 left and right). Moreover, the model results in the NA61/SHINE acceptance and 4π acceptance are very similar (compare figure.3 center and right).

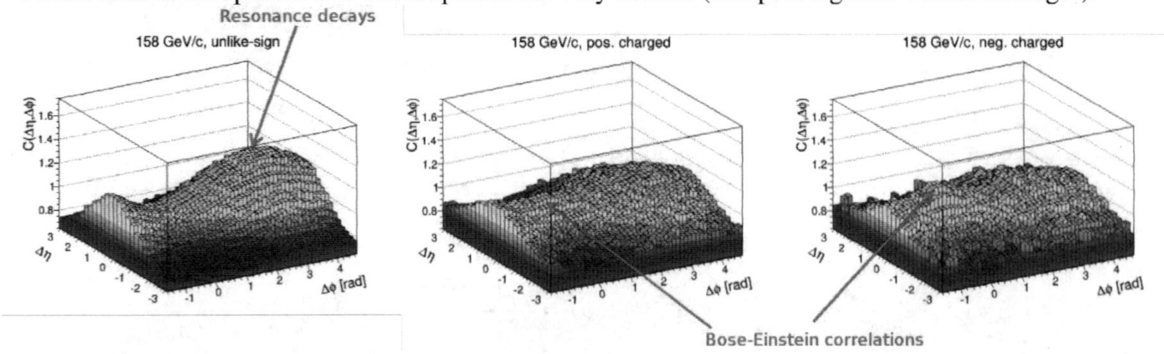


**Figure 1.** Results on ΔηΔφ correlations for inelastic p+p interactions at 158 GeV/c. Results for unlike-sign (left), pos. charged (middle) and neg. charged (right) pairs are shown. The maximum at (Δη,Δφ) = (0,π) is probably a result of resonance decays and momentum conservation. The enhancement at (Δη,Δφ) = (0,0) is likely due to Coulomb interactions and quantum statistics.

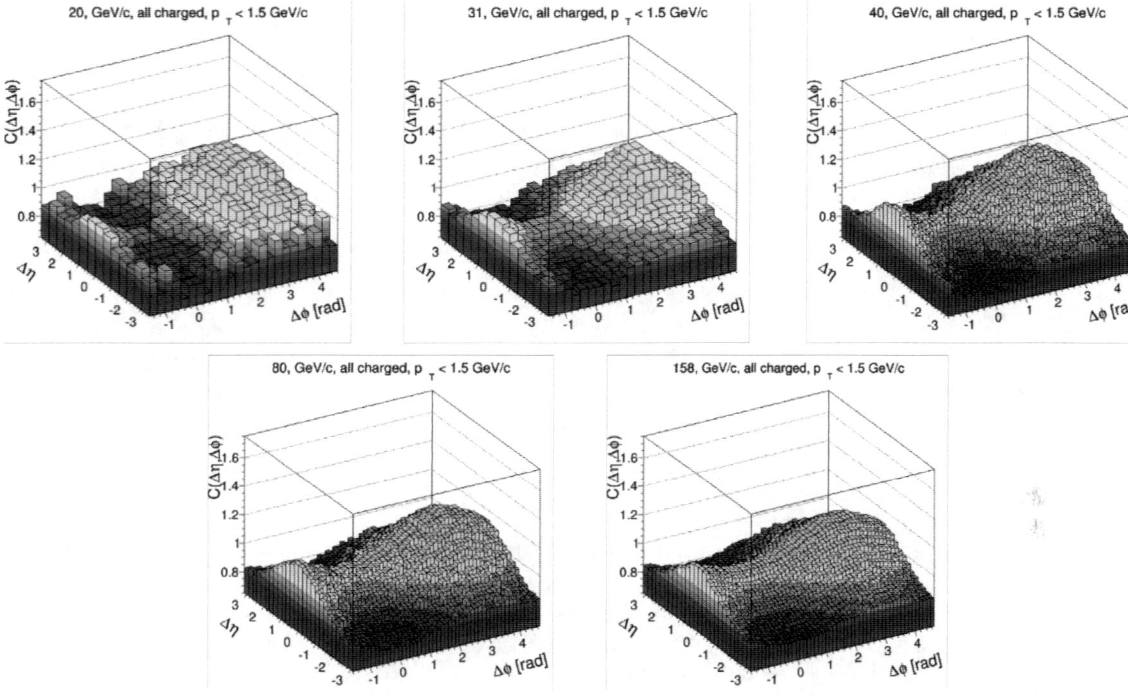


**Figure 2.** Energy dependence of $\Delta\eta\Delta\varphi$ correlations for inelastic p+p interactions.

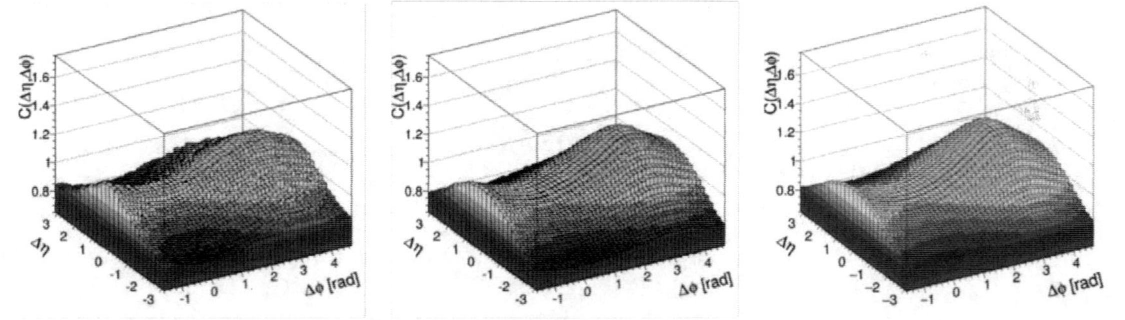


**Figure 3.** $\Delta\eta\Delta\varphi$ correlations for all charged particles in inelastic p+p interactions at 158 GeV/c. NA61/SHINE (left), EPOS 1.99 inside the NA61/SHINE acceptance (middle) and EPOS 1.99 in full $4\pi$ acceptance (right).

*3.2. Pseudo-rapidity correlations*

We used two sets of windows for the study of pseudo-rapidity correlations in $^{7}Be+^{9}Be$ collisions at 150A GeV/c: short-range (figure 4) and with fixed forward window (figure 6). A strong dependence of the correlations on the window position is observed. EPOS 1.99 describes the data only qualitatively.

## 4. Conclusions

Measurements in p+p collisions show structures in two-particle correlations of pseudo-rapidity and azimuthal angle coming mainly from resonance decays, conservation laws, quantum statistics and Coulomb interactions. The enhancement at (0,0) increases, the hill at $(0,\pi)$ decreases with energy.

Correlations in pseudo-rapidity of multiplicity and mean transverse momentum in $^{7}Be+^{9}Be$ collisions at 150A GeV/c strongly depend on the position of the chosen pseudo-rapidity windows. Significant multiplicity–mean transverse momentum long-range correlations were observed (figure 6) that might be a signal of collective effects.

**Acknowledgments**
This work was supported by the Polish National Center for Science (grants 2011/03/N/ST2/03691, 2012/04/M/ST2/00816 and 2013/11/N/ST2/03879), the Federal Agency of Education of the Ministry of Education and Science of the Russian Federation and SPbSU research grant 11.38.193.2014.

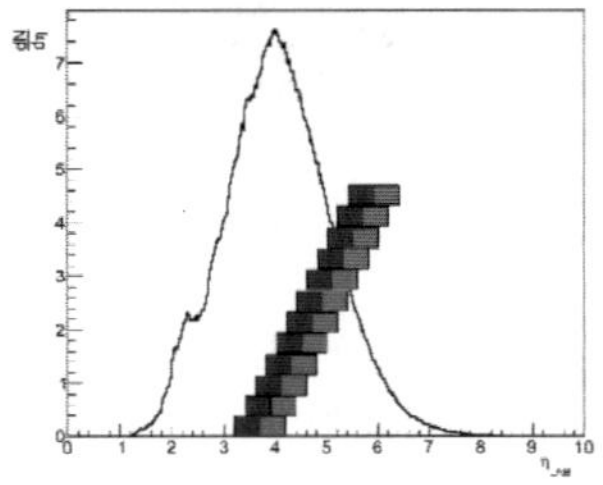

**Figure 4.** Illustration of connected η-windows configurations.

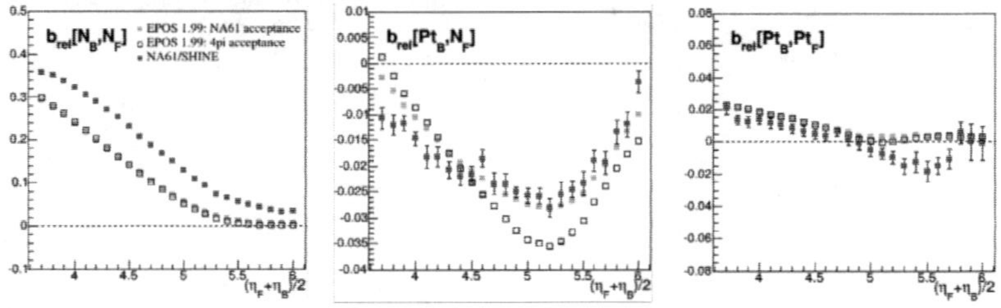


**Figure 5.** Pseudo-rapidity multiplicity and event mean transverse momenta correlation coefficient as a function of windows connection point (see figure 4) for Be7+Be9 150A GeV/c. EPOS 1.99 inside NA61/SHINE acceptance (grey squares) and $4\pi$-acceptance (open squares) are shown for comparison.

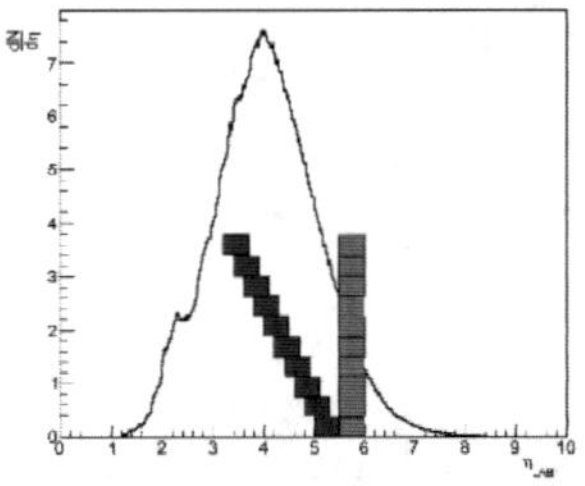

**Figure 6.** Illustration of disconnected η-windows configurations.

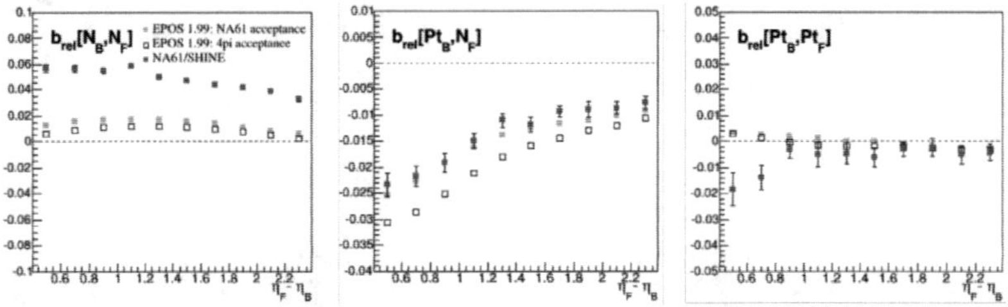


**Figure 7.** Pseudo-rapidity multiplicity and event mean transverse momenta correlation coefficient as a function of distance between windows (see figure 6) for Be7+Be9 150A GeV/c. EPOS 1.99 inside NA61/SHINE acceptance (grey squares) and $4\pi$-acceptance (open squares) are shown for comparison.

**References**

[1] Maksiak B for the NA61/SHINE Coll. 2014 Two particle correlations from the energy scan with p+p interactions *Proceedings of Science PoS(CPOD2014)055*
[2] Braun M and Pajares C. 1992 *Phys. Lett.* B **287** 154
[3] McLerran L 2002 *Nucl.Phys.* A **699** 73c
[4] Abgrall N *et al.* (NA61 Coll.) 2014 *JINST* **9** P06005

Journal of Physics: Conference Series **668** (2016) 012078 doi:10.1088/1742-6596/668/1/012078

# Charged particle production in Pb-Pb and p-Pb collisions measured by the ATLAS detector

**Evgeny Shulga for the ATLAS collaboration**

National Research Nuclear University "MEPhI", Kashirskoe shosse 31, Moscow, Russian Federation

E-mail: eshulga@cern.ch

**Abstract.** The ATLAS experiment at the Large Hadron Collider measures charged hadron spectra in Pb+Pb and p+Pb collisions. The results are compared to the pp spectra of charged hadrons at the same centre-of-mass energy. Charged hadron distributions from Pb+Pb are compared to charged particle cross-sections in pp collisions at $\sqrt{s} = 2.76$ TeV, reference cross-section for p+Pb at $\sqrt{s_{NN}} = 5.02$ TeV is reconstructed using $\sqrt{s} = 2.76$ TeV and 7 TeV pp results. These allow for a detailed comparison of the collision systems in a wide transverse momentum and rapidity ranges in different centrality intervals. The nuclear modification factors $R_{\mathrm{AA}}$ and $R_{\mathrm{pPb}}$ are presented as a function of centrality, $p_{\mathrm{T}}$, $\eta$. The charged particle $R_{\mathrm{AA}}$ is found to vary significantly as a function of transverse momentum, and shows a pronounced minimum at about 7 GeV. Above 60 GeV, $R_{\mathrm{AA}}$ is consistent with a flat, centrality-dependent, value within the uncertainties. $R_{\mathrm{pPb}}$ results show strong rapidity dependence in the region of so-called Cronin peak at about 2 GeV and above $p_{\mathrm{T}}$ of 10 GeV show anomalous enhancement.

## 1. Introduction

Charged hadrons of high transverse momentum ($p_{\mathrm{T}}$) are a tool that can be used to study strongly interacting matter [1, 2] produced in high-energy heavy-ion collisions (Pb+Pb). It is expected from first results of the LHC experiments [3–5] that the resulting spectrum of hadrons, which originate from parton fragmentation [6, 7], is modified. Additionally proton-nucleus (p+Pb) collisions probe the physics of the initial state of ultra-relativistic heavy ion collisions without the effects of thermalization, playing an important role in Pb+Pb collisions evolution [8]. p+Pb spectra can provide insight on the effect of an extended nuclear target on the dynamics of soft and hard scattering processes and subsequent particle production. For an extensive review of predictions see [9].

## 2. Analysis

For Pb+Pb and p+Pb collisions the expected particle production rate can be quantified using the "nuclear modification factor" which is calculated as the ratio of differential yield of charged particles in Pb+Pb or p+Pb collisions divided by the average value of the nuclear thickness function $\langle T_{\mathrm{Pb}} \rangle$ (in units of inverse barns) [10] and differential charged-particle production cross section measured in pp collisions. Differential yield of charged particles can be measured as a function of centrality, eta, y (or $y^{\star}$ in case of p+Pb, which due to different beam energies has its centre-of-mass system boosted by $-0.465$ [11]).

This proceedings presents a high-statistics measurement of charged-hadron spectra, $R_{AA}$ and $R_{pPb}$ as a function of $p_T$ and intervals of collision centrality for $0.5 < p_T < 150$ GeV. The collision centrality is estimated using the sum of transverse energy measured in the ATLAS [12] forward calorimeter, $3.1 < |\eta| < 4.9$, in both parts of the detector for Pb+Pb [13] and only in the outgoing Pb beam direction for p+Pb [10], the dependence of p+Pb collisions on the number of participating nucleons ($N_{part}$) was estimated using two models: the Glauber model [14] and the Glauber-Gribov colour-fluctuation approach [15], with the relative fluctuations determined by the parameter $\Omega$. Pb+Pb data recorded by the ATLAS experiment in the 2010 and 2011 HI physics runs of the LHC at $\sqrt{s} = 2.76$ TeV, and the pp data at the same value of $\sqrt{s}$ recorded in the 2011 and 2013 physics runs. The total integrated luminosity of the combined Pb+Pb sample is $0.15\,\text{nb}^{-1}$, and that of the combined pp sample is $4.2\,\text{pb}^{-1}$. The proton-proton data sets used to obtain reference measurement for p+Pb results at $\sqrt{s} = 7$ TeV with total integrated luminosity 130 $\mu\text{b}^{-1}$ were obtained in April 2010. For p+Pb studies events were preselected within the vertex range $|z_{vtx}| < 150$ mm, and at least one hit in each of the MBTS detectors. In addition, a time difference of less than 10 ns was required between the two MBTS sides.

The events used in this analysis are selected from several triggered samples. A minimum bias (MB) samples were obtained requiring signal in the Minimum Bias Trigger Scintillator. For jet triggered samples, jets were reconstructed in events that passed the MB requirement using the anti-kt algorithm [16] with the distance parameter R=0.4. Events were selected by the jet trigger if they contained jets with $E_T$ above a certain threshold. Multiple thresholds were defined. Only a fraction of all events which fired a trigger was randomly selected to be recorded for further analysis. This fraction was set differently for each trigger. Total spectra were obtained by merging spectra from MB and jet triggers, all of them scaled by reciprocal of the fraction to obtain MB spectra up to high $p_T$ [17].

Charge particle tracks are reconstructed in the ATLAS Inner Detector [13]. Tracks are measured using a combination of silicon pixel detector (Pixel), silicon microstrip detector (SCT), and a straw tube transition radiation tracker (TRT), all immersed in a 2 T axial magnetic field. Charged particles typically traverse 3 layers of silicon pixel detectors, 4 layers of double sided microstrip detector, and 36 straws. Each track is required to have at least 1 hit in the Pixel detector, a hit in the innermost layer if such hit is expected by the tracking model, and at least 6 hits in the SCT. Additionally, tracks with $p_T > 10$ GeV are required to produce at least 8 hits in the TRT. These requirements select tracks with good $p_T$ resolution and suppress the contribution of poorly reconstructed tracks. Yet they limit the analysis to the coverage of the TRT detector, thus all tracks are selected with $|\eta| < 2$. To ensure that the tracks originate from the event vertex of Pb+Pb(p+Pb) collisions, the transverse impact parameter, $d_0$, and $z_0 sin\theta$ ($z_0$ is the longitudinal impact parameter) are required to be less than 3(1.5) mm. Tracks selected for the analysis are also required to satisfy the conditions on the impact parameters significances: $|d0/\sigma d_0| < 3$ and $|z_0 sin\theta/\sigma z_0 sin\theta| < 3$. The $d_0$ and $z_0 sin\theta$ parameters and their uncertainties are estimated by a vertex finding algorithm. The jet is reconstructed with a radius parameter of $\Delta R = 0.2$ in Pb+Pb and $\Delta R = 0.4$ in p+Pb and pp collisions. The track-to-jet matching requirement is applied to all tracks with $p_T > 15$ GeV in p+Pb and pp collisions and $p_T > 20$ GeV in Pb+Pb collisions both in the MB and jet-triggered samples, to reduce the amount of mis-measured tracks at high $p_T$. Tracks associated with electrons and muons with $p_T$ above 10 GeV coming predominately from electroweak decays bosons are subtracted from the measured spectra. Reconstructed charged particle spectra in different collision systems are corrected for mis-measured tracks and secondary particles, for limited momentum resolution, and for tracking inefficiency. The resulting spectra are unfolded using the iterative Bayesian unfolding [18] to correct for the finite detector $p_T$ resolution.

Once the differential pp cross section at $\sqrt{s}$=2.76 and 7 TeV have been measured, the differential pp cross section at $\sqrt{s_{NN}}$=5.02 TeV is obtained by interpolation. The interpolation

Journal of Physics: Conference Series **668** (2016) 012078 doi:10.1088/1742-6596/668/1/012078

is proportional to $\ln(\sqrt{s})$ and it is performed for every $p_{\rm T}$ bin in each rapidity interval used in the p+Pb analysis.

## 3. Results

Figure 1(left) shows the nuclear modification factor $R_{\rm AA}$ as a function of $p_{\rm T}$ in five centrality intervals. The $R_{\rm AA}$ measured as a function of $p_{\rm T}$ shows a characteristic non-flat $p_{\rm T}$ shape which becomes more pronounced for more central collisions. In all centrality and rapidity intervals studied the $R_{\rm AA}$shows a characteristic shape with a local maximum at $p_{\rm T} \approx 2$ GeV decreasing to a minimum at $p_{\rm T} \approx 7$ GeV. It increases again towards higher $p_{\rm T}$ with a change in the slope at $p_{\rm T} \approx 60$ GeV, where the rise becomes much less steep. Above this $p_{\rm T}$, the $p_{\rm T}$ dependence is consistent with a flat behaviour within statistical and systematic uncertainties. This shape is present in all measured centrality and pseudorapidity intervals, but it has the largest maximum-to-minimum ratio in the most central collisions. Figure 1(right) shows $R_{\rm AA}$ as a function of the mean number of participating nucleons, $\langle N_{\rm part}\rangle$, in the four momentum intervals corresponding to the local maximum, local minimum, plateau and intermediate region of $R_{\rm AA}$ values. Overall decrease of $R_{\rm AA}$ with $\langle N_{\rm part}\rangle$ is pronounced, the dependence is strongest around the minimum of $R_{\rm AA}$ ($p_{\rm T} \approx 7$ GeV) and weakest in the $p_{\rm T}$ region above 60 GeV [19].

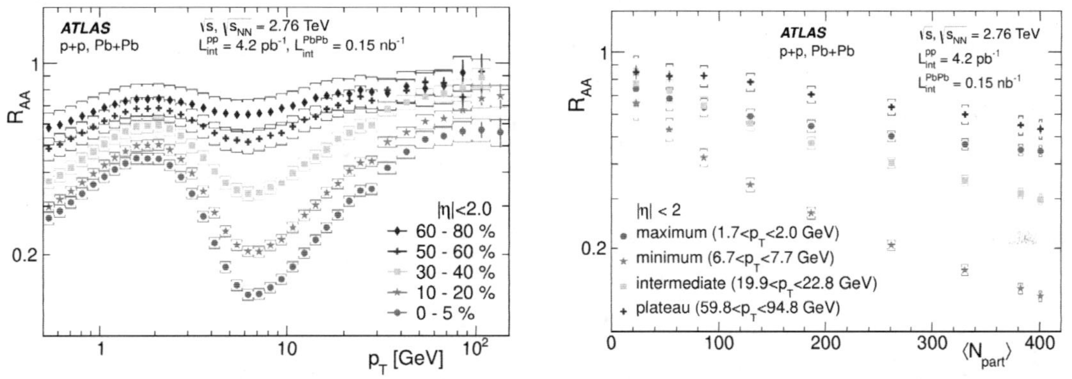


Figure 1: (left) The $R_{\rm AA}$ as a function of $p_{\rm T}$ measured in five centrality intervals $0-5\%$, $10-20\%$, $30-40\%$, $50-60\%$ and $60-80\%$. (right) The $R_{\rm AA}$ as a function of $\langle N_{\rm part}\rangle$ for four $p_{\rm T}$ intervals corresponding to the maximum $R_{\rm AA}$ $(1.7 < p_{\rm T} < 2.0GeV)$, minimum $R_{\rm AA}(6.7 < p_{\rm T} < 7.7GeV)$, intermediate $R_{\rm AA}$ $(19.9 < p_{\rm T} < 22.8GeV)$ and plateau of $R_{\rm AA}$ $(59.8 < p_{\rm T} < 94.8GeV)$. The statistical uncertainties, which are smaller than the size of markers in all but a few cases, are shown with vertical bars and the systematic uncertainties with brackets.

Figure 2 shows the $R_{\rm pPb}$ distributions for 2 centrality classes $(0-10\%$ and $60-90\%)$ and for the rapidity interval $-1.5 < y^{\star} < 2$.The nuclear modification factors increase with momentum in the region $0.1 < p_{\rm T} < 2$ GeV, then reach a maximum and decrease up to $p_{\rm T} \approx 8$ GeV. Above that momentum the nuclear modification factor is constant within the experimental uncertainties. The magnitude of the peak strongly depends both on rapidity and centrality. It increases from peripheral to the central collisions and from the proton beam direction to the Pb beam direction [11]. The choice of the different geometrical model strongly affects the level of the plateau in the central events and its magnitude with respect to peripheral. Higher $p_{\rm T}$ ranges have been also measured, showing a clear trend of increasing enhancement towards higher $p_{\rm T}$. This trend does not have a strong rapidity dependence but is more marked in peripheral events [17]. The results for high $p_{\rm T}$ measurement seem to show the same trend at high $p_{\rm T}$ as was observed by the CMS collaboration [20] but not the ALICE collaboration [21].

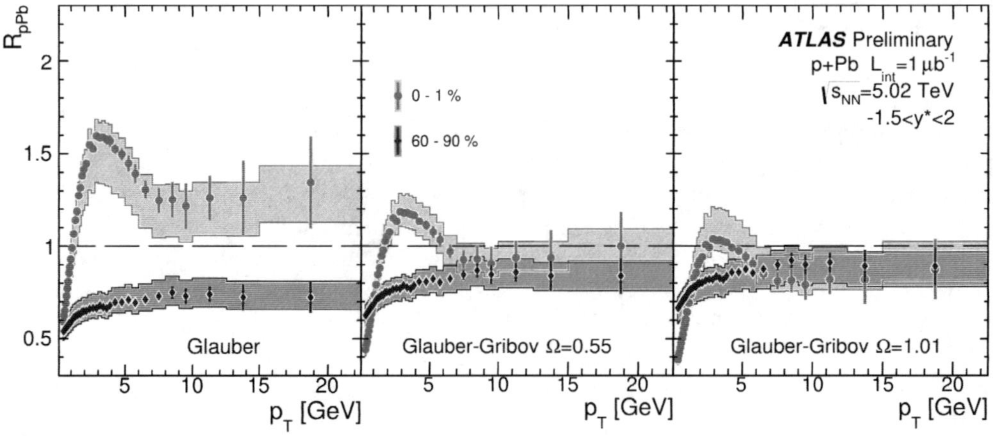


Figure 2: $R_{\mathrm{pPb}}$ as a function of $p_{\mathrm{T}}$ extracted from the invariant yields integrated over $-1.5 < y^{\star} < 2$ for the $0-1\%$ and $60-90\%$ centrality intervals, and for the three different geometrical models from left to right (Glauber, Glauber-Gribov $\Omega = 0.55$ and Glauber-Gribov $\Omega = 1.01$). Statistical errors are indicated with vertical lines and the systematic uncertainties on the invariant yields are indicated by a shaded area. The total systematic uncertainty, which includes the uncertainty on $\langle T_{\mathrm{pA}} \rangle$ or its ratios are indicated by lines of the same colour.

## 4. Acknowledgements

This research is supported by The Ministry of education and science of Russian Federation, project 3.472.2014/K.

## References


[1] D. G. d'Enterria, Eur. Phys. J. **C43** (2005) 295–302, arXiv:nucl-ex/0504001 [nucl-ex].
[2] J. Casalderrey-Solana and C. A. Salgado, Acta Phys. Polon. **B38** (2007) 3731–3794, arXiv:0712.3443 [hep-ph].
[3] ATLAS Collaboration, Phys. Rev. Lett. **105** (2010) 252303, arXiv:1011.6182 [hep-ex].
[4] CMS Collaboration, Phys. Rev. **C84** (2011) 024906, arXiv:1102.1957 [nucl-ex].
[5] ALICE Collaboration, J. Adam et al., Phys. Lett. **B746** (2015) 1–14, arXiv:1502.01689 [nucl-ex].
[6] ATLAS Collaboration, Phys. Lett. **B739** (2014) 320–342, arXiv:1406.2979 [hep-ex].
[7] CMS Collaboration, Phys. Rev. **C90** no. 2, (2014) 024908, arXiv:1406.0932 [nucl-ex].
[8] L. P. Csernai, J. I. Kapusta, and L. D. McLerran, Phys. Rev. Lett. **97** (2006) 152303, arXiv:nucl-th/0604032.
[9] J. Albacete, N. Armesto, R. Baier, G. Barnafoldi, J. Barrette, et al., Int.J.Mod.Phys. **E22** (2013) 1330007, arXiv:1301.3395.
[10] ATLAS Collaboration, arXiv:1508.00848 [hep-ex].
[11] Tech. Rep. ATLAS-CONF-2013-107, CERN, Geneva, Nov, 2013.
[12] ATLAS Collaboration, JINST **3** (2008) S08003.
[13] ATLAS Collaboration, Phys. Lett. **B707** (2012) 330–348, arXiv:1108.6018 [hep-ex].
[14] M. L. Miller, K. Reygers, S. J. Sanders, and P. Steinberg, Ann. Rev. Nucl. Part. Sci. **57** (2007) 205–243, arXiv:nucl-ex/0701025 [nucl-ex].
[15] M. Alvioli and M. Strikman, Phys. Lett. **B722** (2013) 347–354, arXiv:1301.0728.
[16] M. Cacciari, G. P. Salam, and G. Soyez, JHEP **04** (2008) 063, arXiv:0802.1189 [hep-ph].
[17] Tech. Rep. ATLAS-CONF-2014-029, CERN, Geneva, May, 2014.
[18] ATLAS Collaboration, arXiv:1104.2962 [hep-ex].
[19] ATLAS Collaboration, arXiv:1504.04337 [hep-ex].
[20] CMS Collaboration, Eur. Phys. J. **C75** no. 5, (2015) 237, arXiv:1502.05387 [nucl-ex].
[21] ALICE Collaboration, B. B. Abelev et al., Eur. Phys. J. **C74** no. 9, (2014) 3054, arXiv:1405.2737 [nucl-ex].

Journal of Physics: Conference Series **668** (2016) 012079 doi:10.1088/1742-6596/668/1/012079

# Centrality dependence of elliptic flow of multi-strange hadrons in Au+Au collisions at $\sqrt{s_{NN}} = 200$ GeV

**Shusu Shi (for the STAR collaboration)**

Key Laboratory of Quarks and Lepton Physics (MOE) and Institute of Particle Physics, Central China Normal University, Wuhan, 430079, China

E-mail: shishusu@gmail.com

**Abstract.** We present recent results of the mid-rapidity elliptic flow ($v_2$) for multi-strange hadrons and the $\phi$ meson as a function of centrality in Au + Au collisions at the center of mass energy $\sqrt{s_{NN}} = 200$ GeV. The transverse momentum dependence of $\phi$ and $\Omega$ $v_2$ is similar to that of pion and proton, indicating that the heavier strange ($s$) quark flows as strongly as the lighter up ($u$) and down ($d$) quarks. These observations constitute a clear piece of evidence for the development of partonic collectivity in heavy-ion collisions at the top RHIC energy. In addition, the mass ordering of $v_2$ breaks between the $\phi$ and proton at low transverse momenta in the 0-30% centrality bin, possibly due to the effect of late hadronic interactions on the proton $v_2$.

## 1. Introduction

In high energy nucleus nucleus collisions, the produced particles are anisotropic in momentum space. The elliptic flow, $v_2$, which is the second Fourier coe cient of the azimuthal distribution of produced particles with respect to the reaction plane, is defined as $v_2 = \langle \cos 2(\varphi - \Psi) \rangle$, where $\varphi$ is the azimuthal angle of produced particle and $\Psi$ is the azimuthal angle of the reaction plane. It has been shown that elliptic flow is sensitive to the early stage of heavy-ion collisions and equation of state of the formed system [1]. However, early dynamic information might be obscured by later hadronic rescatterings [2]. Multi-strange hadrons and the $\phi$ meson are believed to be less sensitive to hadronic rescatterings in the late stage of collisions, as their freeze-out temperatures are close to the phase transition temperature and their hadronic interaction cross sections are expected to be small [3, 4, 5]. Previous measurements of $\phi$ and $\Omega$ $v_2$ from STAR were limited by statistics [6, 7], thus the $p_T$ and centrality dependence of $\phi$ and $\Omega$ $v_2$ are not clear. Furthermore, with high precision measurements of $\phi$ meson $v_2$, one can compare it with proton $v_2$ in the low $p_T$ region. It may provide information on the effect of hadronic rescatterings in the late stage of the collision.

## 2. Results and Discussions

In these proceedings, $v_2$ measurements of multi-strange hadrons from the STAR experiment at $\sqrt{s_{NN}} = 200$ GeV Au + Au collisions are presented. About 730 million minimum bias events recorded by STAR in 2010 and 2011 were used in the analysis. The Time Projection Chamber (TPC) was used for centrality definition and event plane determination. The centrality was determined by the number of tracks from the pseudorapidity region $|\eta| \leq 0.5$. The particle identification (PID) was achieved via energy loss in TPC and flight time in the Time of Flight (TOF). The multi-strange hadrons and the $\phi$ meson were reconstructed though the following decay channels: $\phi \to K^+ + K^-$, $\Xi^- \to \Lambda + \pi^-$ ($\overline{\Xi}^+ \to \overline{\Lambda} + \pi^+$) and $\Omega^- \to \Lambda + K^-$ ($\overline{\Omega}^+ \to \overline{\Lambda} + K^+$). The $\eta$ sub-event plane method is used for the $v_2$ measurement [8] . In this method, one defines the event flow vector for each particle based on particles measured in the opposite hemisphere in pseudorapidity. An $\eta$ gap of $|\Delta\eta| > 0.1$ between positive and negative pseudorapidity sub-events is introduced to suppress non-flow effects.

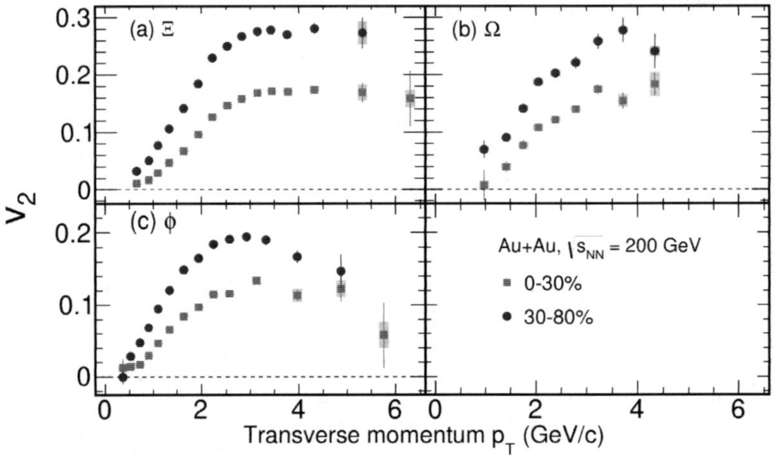


**Figure 1.** The $v_2$ as a function of $p_T$ at midrapidity ($|y| < 1.0$) for (a) $\Xi^- + \overline{\Xi}^+$ (b) $\Omega^- + \overline{\Omega}^+$ and (c) $\phi$ in Au + Au collisions at $\sqrt{s_{NN}} = 200$ GeV for 0-30% and 30-80% centrality. The systematic uncertainties are shown by shaded boxes and the statistical uncertainties by vertical lines.

Figure 1 [9] shows the elliptic flow $v_2$ as a function of transverse momentum $p_T$ at midrapidity ($|y| < 1.0$) for (a) $\Xi^- + \overline{\Xi}^+$, (b) $\Omega^- + \overline{\Omega}^+$ and (c) $\phi$ in Au + Au collisions at $\sqrt{s_{NN}} = 200$ GeV for 0-30% and 30-80% centrality. A clear centrality dependence of $v_2(p_T)$ is observed for multi-strange hadrons and $\phi$ meson, which is similar to that of light and strange hadrons previously measured by the RHIC experiments [10, 11]. The $v_2$ values are larger in peripheral collisions (30-80% centrality) than those in central collisions (0-30% centrality). This is consistent with an interpretation in which the anisotropy in final momentum space is driven by the anisotropy in initial position space.

Figure 2 [9] shows the $v_2$ as a function of $p_T$ for $\pi$, $p$ (panel (a)) and $\phi$, $\Omega$ (panel (b)) in Au + Au collisions at $\sqrt{s_{NN}} = 200$ GeV for 0-80% centrality. A comparison between $v_2$ of $\pi$ and $p$, consisting of up ($u$) and down ($d$) light constituent quarks is shown in panel (a). Correspondingly, panel (b) shows a comparison of $v_2$ of $\phi$ and $\Omega$ containing $s$ constituent quarks. This is the first measurement of $\Omega$ baryon $v_2$ up to 4.5 GeV/$c$ with high precision. In the low $p_T$ region ( $p_T$<2.0 GeV/$c$), the $v_2$ of $\phi$ and $\Omega$ follows mass ordering. At intermediate $p_T$ ( 2.0<$p_T$<5.0 GeV/$c$), a baryon-meson separation is observed. It is evident that the $v_2(p_T)$ of hadrons consisting only of strange constituent quarks ($\phi$ and $\Omega$) is similar to that of light hadrons, $\pi$ and $p$. However the $\phi$ and $\Omega$ do not participate strongly in the hadronic interactions, because of the smaller hadronic cross sections compared to $\pi$ and $p$. It suggests that most of the collectivity is developed during the partonic phase in Au + Au collisions at $\sqrt{s_{NN}} = 200$ GeV.

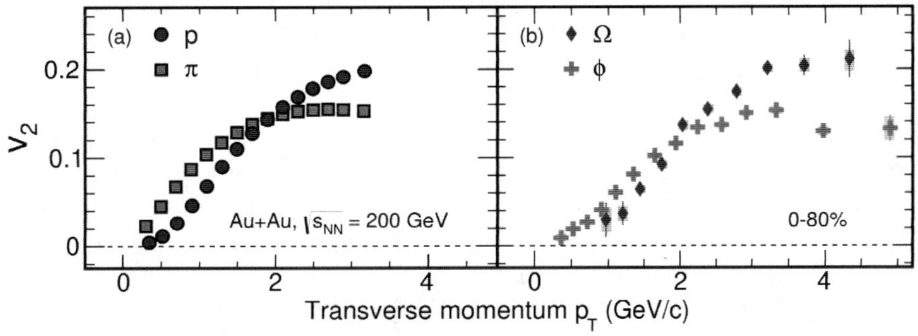


**Figure 2.** The $v_2$ as function of $p_T$ for $\pi$, $p$ (a) and $\phi$, $\Omega$ (b) in Au+Au collisions at $\sqrt{s_{NN}}$ = 200 GeV for 0-80% centrality. The systematic uncertainties are shown by the shaded boxes while vertical lines represent the statistical uncertainties.

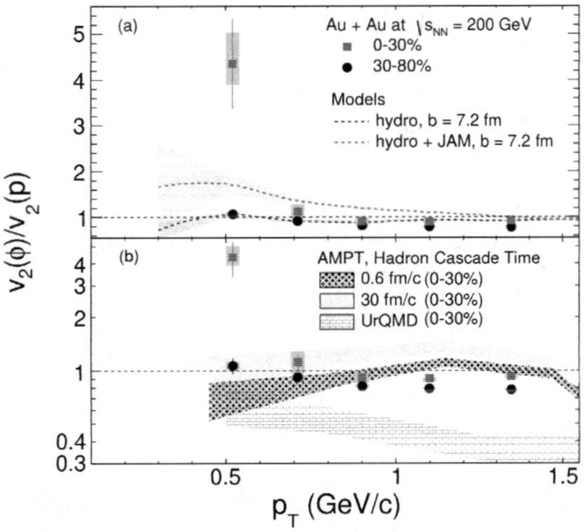


**Figure 3.** The ratio of $v_2(\phi)$ to $v_2(p)$ as function of $p_T$ in Au + Au collisions at $\sqrt{s_{NN}}$ = 200 GeV for 0-30% and 30-80% centrality. Shaded bands are the systematic uncertainties and vertical lines are the statistical uncertainties. The bands in panel (a) and (b) represent the hydro and transport model calculations for $v_2(\phi)/v_2(p)$, respectively.

Hydrodynamical model calculations predict that $v_2$ as a function of $p_T$ for different particle species follows mass ordering, where the $v_2$ of heavier hadrons is lower than that of lighter hadrons [12]. The identified hadron $v_2$ measured in experiment indeed proves the mass ordering in the low $p_T$ region ($p_T$<1.5 GeV/$c$) [10]. Recently, Hirano *et al.* predict the mass ordering of $v_2$ could be broken between $\phi$ mesons and protons at low $p_T$ ($p_T$<1.5 GeV/$c$) based on a model with ideal hydrodynamics plus hadron cascade process [2]. As the model calculations assign a smaller hadronic cross section for $\phi$ mesons compared to protons, the broken mass ordering is regarded as the different hadronic rescattering contributions on the $\phi$ meson and proton $v_2$. Figure 3 [9] shows the ratios of $\phi$ $v_2$ to proton $v_2$ from model calculations and experimental data. This ratio is larger than unity at $p_T \sim$ 0.5 GeV/$c$ for 0-30% centrality. It indicates breakdown of the expected mass ordering in that momentum range. This could be due to a large effect of hadronic rescattering on the proton $v_2$, which qualitatively agrees with hydro + hadron cascade calculations indicated by the shaded red band in panel (a) of Fig. 3. A centrality dependence of $v_2(\phi)$ to $v_2(p)$ ratio is observed in the experimental data. Namely, the breakdown of mass ordering of $v_2$ is more pronounced in 0-30% central collisions than in 30-80% peripheral collisions. In the central events, both hadronic and partonic interactions are stronger than in peripheral events. Therefore, the larger effect of late stage hadronic interactions relative to the partonic collectivity produces a greater breakdown of mass ordering in the 0-30% centrality data than in the 30-80%. This observation indirectly supports the idea that the $\phi$ meson has a smaller hadronic interaction cross section. The ratio of $\phi$ $v_2$ to proton $v_2$ was also studied by using the transport models AMPT [13] and UrQMD [14]. The panel (b) of Fig. 3 shows the $v_2(\phi)$ to $v_2(p)$ ratio for 0-30% centrality from AMPT and UrQMD models. The black shaded band is from AMPT with a hadronic cascade time of 0.6 fm/$c$ while the yellow band is for a hadronic cascade time of 30 fm/$c$. Larger hadronic cascade time is equivalent to stronger hadronic interactions. It is clear that the $v_2(\phi)/v_2(p)$ ratio increases with increasing hadronic cascade time. This is attributed to a decrease in the proton $v_2$ due to an increase in hadronic re-scattering while the $\phi$ meson $v_2$ is less affected. The ratios from the UrQMD model are much smaller than unity (shown as a blue shaded band in the panel (b) of Fig. 3). The UrQMD model lacks partonic collectivity, thus the $\phi$ meson $v_2$ is not fully developed.

## 3. Summary

In summary, high-statistics elliptic flow measurements for multi-strange hadrons ($\Xi$ and $\Omega$) and $\phi$ meson as a function of centrality and $p_T$ were reported in Au + Au collisions at $\sqrt{s_{NN}} = 200$ GeV. The $p_T$ dependence of $\phi$ and $\Omega$ $v_2$ is observed to be similar to that of light hadrons, $\pi$ and $p$, indicating that most of the collectivity is developed in the initial partonic phase for light and strange hadrons. The comparison between the $\phi$ and $p$ $v_2$ at low $p_T$ shows that there is a possible violation of hydro-inspired mass ordering between $\phi$ and $p$. Model calculations suggest that the $p_T$ dependence of $v_2(\phi)$ to $v_2(p)$ ratios can be qualitatively explained by the different effects of late-stage hadronic interactions on the $\phi$ and proton $v_2$.

## 4. Acknowledgments

This work was supported in part by the National Natural Science Foundation of China under grant No. 11475070, National Basic Research Program of China (973 program) under grand No. 2015CB8569 and self-determined research funds of CCNU from the colleges' basic research and operation of MOE under grand No. CCNU15A02039.

[1] S. A. Voloshin, A. M. Poskanzer and R. Snellings, in Landolt-Boernstein, Relativistic Heavy Ion Physics, Vol. 1/23, p. 5-54 (Springer-Verlag, 2010). arXiv:0809.2949.
[2] T. Hirano *et al.*, Phys. Rev. **C 77**, 044909 (2008).
[3] J. Adams *et al.*, Nucl. Phys. **A 757** 102 (2005).
[4] A. Shor, Phys. Rev. Lett. **54**, 1122 (1985).
[5] H. van Hecke, H. Sorge and N. Xu, Phys. Rev. Lett. **81**, 5764 (1998).
[6] J. Adams *et al.*, Phys. Rev. Lett. **95**, 122301 (2005).
[7] S. S. Shi (for the STAR collaboration), Nucl. Phys. A **830**, 187c (2009); Nucl. Phys. A **862-863**, 263 (2011).
[8] A. M. Poskanzer and S. A. Voloshin, Phys. Rev. **C 58**, 1671 (1998).
[9] L. Adamczyk *et al.*, arXiv:1507.05247 (2015).
[10] B. I. Abelev *et al.*, Phys. Rev. **C 77**, 054901 (2008).
[11] A. Adare *et al.*, Phys. Rev. **C 85**, 064914 (2012).
[12] P. Huovinen *et al.*, Phys. Lett. **B 503**, 58 (2001).
[13] Z.-W. Lin *et al.*, Phys. Rev. **C 72**, 064901 (2005).
[14] S. A. Bass *et al.*, Prog. Part. Nucl. Phys. **41**, 255 (1998).

Journal of Physics: Conference Series **668** (2016) 012080 doi:10.1088/1742-6596/668/1/012080

# Spectra and elliptic flow for $\Lambda$, $\Xi$, and $\Omega$ in 200 A GeV Au+Au collisions

**Xiangrong Zhu**[1,2] **and Huichao Song**[1,2,3]

[1]Department of Physics and State Key Laboratory of Nuclear Physics and Technology, Peking University, Beijing 100871, China
[2]Collaborative Innovation Center of Quantum Matter, Beijing 100871, China
[3]Center for High Energy Physics, Peking University, Beijing 100871, China

**Abstract.** Using VISHNU hybrid model, we calculate the $p_T$-spectra and elliptic flow of $\Lambda$, $\Xi$, and $\Omega$ in 200 A GeV Au+Au collisions. Comparisons with the STAR measurements show that the model generally describes these soft hadron data. We also briefly study and discuss the mass ordering of elliptic flow among $\pi$, $K$, $p$, $\Lambda$, $\Xi$, and $\Omega$ in minimum bias Au+Au collisions.

## 1. Introduction

The main goal of relativistic heavy-ion collisions is to create the Quark-gluon Plasma (QGP) and study its properties. Due to their small hadronic cross sections, multi-strange hadrons, such as $\Xi$ and $\Omega$, experience early chemical and thermal freeze-out after hadronization [1, 2]. Their anisotropy flow are mainly developed in the QGP phase and less contaminated by the hadronic evolution, which are expected to provide valuable information for the QGP. In our early paper [2], the spectra and elliptic flow of strange and multi-strange hadrons at the LHC have been studied with VISHNU hybrid model. This proceeding will extend the calculations to top RHIC energies and compare the results with recent STAR data.

## 2. Setup of the calculations

VISHNU hybrid model [3] combines (2+1)-d relativistic viscous hydrodynamics (VISH2+1) [4] for the QGP fluid expansion with a microscopic hadronic transport model(UrQMD) [5] for the hadron resonance gas evolution. The hydrodynamic simulations input an equation of state (EoS) s95p-PCE constructed from recent lattice QCD data [6]. The transition from hydrodynamics to the hadron cascade is controlled by a switching temperature which is set at 165 MeV. We input smooth initial conditions that are generated from MC-Glauber model through averaging a large number of fluctuating initial entropy density profiles (Here, each initial density distribution is recenterd and rotated to align the major and minor axes before averaging, which is called as *initialization in the participant plane*) [7]. Following [8, 9], we set the initial time $\tau_0 = 0.6$ fm/$c$, the specific shear viscosity $(\eta/s)_{QGP} = 0.08$ and specific bulk viscosity $(\zeta/s)_{QGP} = 0$. The normalization factor of the the initial entropy density profiles is tuned to reproduce the final charged hadron multiplicity in the most central Au+Au collisions at $\sqrt{s_{NN}}$=200 GeV [10]. These inputs and settings once nicely described the $p_T$ spectra and differential elliptic flow for all charged hadron and for pions and protons in 200 A GeV Au+Au collisions.

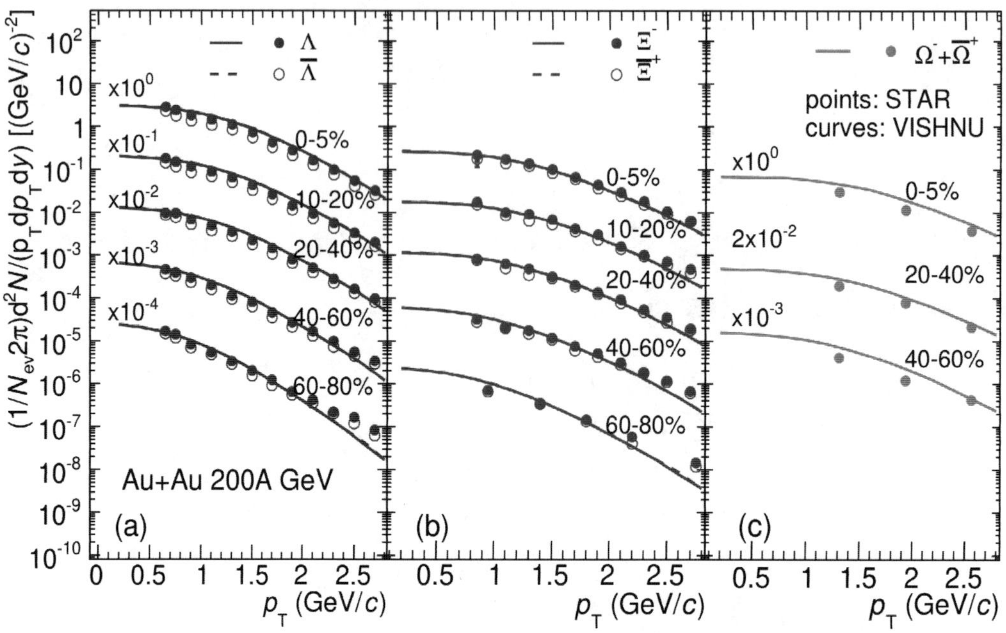


**Figure 1.** (Color online) Transverse momentum spectra of Λ, Ξ and Ω at various centralities in 200 A Au+Au collisions. Experimental data are from the STAR measurements [11]. Theoretical curves are calculated from VISHNU with MC-Glauber initial conditions, $\tau_0 = 0.6$ fm/c and $\eta/s = 0.08$.

## 3. Results

The $p_T$-spectra and elliptic flow of pions and protons in 200 A Au+Au collisions have been studied with VISHNU hybrid model in our early paper [3]. We showed that, with MC-Glauber initial conditions, $\tau_0 = 0.6$ fm/c and $\eta/s = 0.08$, VISHNU nicely describes these low $p_T$ data of pions and protons at various centrality bins.

Figure 1 presents transverse momentum spectra of Λ, Ξ and Ω in 200 A GeV Au+Au collisions [1]. In general, VISHNU describes the $p_T$-spectra of Λ and Ξ, but slightly over-predicts the production of Ω for all centrality bins. In spite of the normalization issues, VISHNU nicely fits the slope of the $p_T$-spectra for Λ, Ξ and Ω at various centralities. Together with the early nice descriptions of the $p_T$-spectra for pions, kaons and protons [8], it indicates that VISHNU generates a proper amount of radial flow during the QGP and hadronic evolution at the top RHIC energy.

Figure 2 shows the differential elliptic flow of Λ, Ξ and Ω in 200 A GeV Au+Au collisions. The theoretical curves are calculated from VISHNU using smooth initial conditions from MC-Glauber (with an average in participant plane). The STAR data are measured by the event plane method [11, 12], which contain the contributions from event-by-event fluctuations and the non-flow effects. To account the fluctuation contributions, we divide the flow data by $\langle \varepsilon_{\rm part}^{\alpha} \rangle^{\alpha}$ and the theoretical results by $\bar{\varepsilon}_{\rm part}$, where the exponent $\alpha$ depends on the event-plane resolution $R$ and details of the $v_2$ extraction method [8, 13]. The detail values of $\langle \varepsilon_{\rm part}^{\alpha} \rangle^{\alpha}$ and $\bar{\varepsilon}_{\rm part}$ can be found in [8]. Our past research has shown that VISHNU nicely fits the elliptic flow for all charged

[1] For the STAR measurement, the Λ spectra has been corrected with the feed-down contributions from weak decays. Λ, Ξ and Ω from VISHNU are final hadrons after collisions and strong resonance decays since the UrQMD hadronic evolution does not contain any weak decay channels.

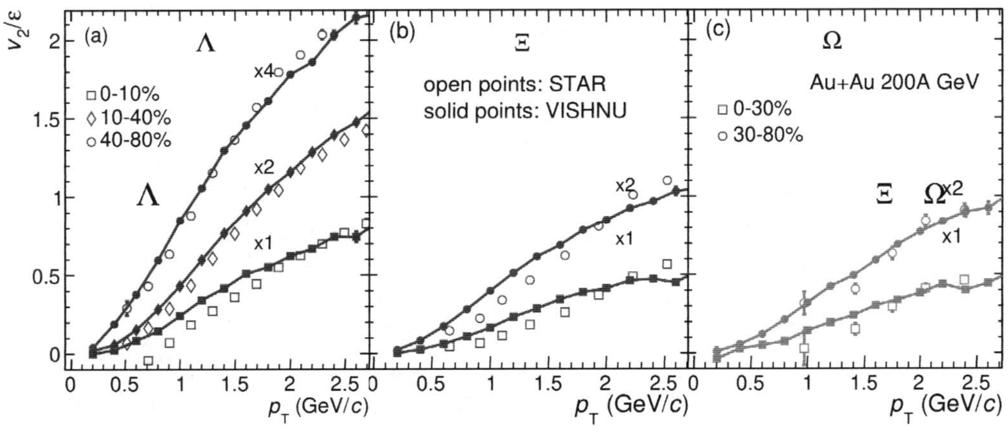


**Figure 2.** (Color online) Differential elliptic flow of Λ, Ξ and Ω in 200 A GeV Au+Au collisions. Experimental data are from STAR [11, 12], theoretical curves are calculated from VISHNU with the same inputs as for Fig. 1.

hadrons and for identified hadrons $\pi$ and $p$ in 200 A GeV Au+Au collisions with MC-Glauber, $\tau_0 = 0.6$ fm/c and $\eta/s = 0.08$ [8]. With the same inputs, we extend the VISHNU calculations for the strange and multi-strange hadrons with high statistical simulations. Fig. 2 shows that, VISHNU generally fits the elliptic flow of Λ, Ξ and Ω in 200 A GeV Au+Au collisions. For 0-10% centrality, the VISHNU results are slightly above the data for all three hadrons. It is worthwhile to mention that, for 2.76 A TeV Pb+Pb collisions, early VISHNU calculations also slightly over-predict the elliptic flow of Λ, Ξ and Ω at 10-20% and 30-40% centralities. In fact, the UrQMD simulations in VISHNU assigns small cross-sections to Ξ and Ω, leading to an early decoupling of these two hadrons. Compared with pure hadronic simulations [15], this leads to an insufficient development of elliptic flow for these two multi-strange hadrons in central and semi-central collisions at top RHIC and the LHC energies.

It is generally believed that the mass ordering of elliptic flow among various identified hadrons reflects the interplay between radial and elliptic flows, providing more information on the QGP fireball evolution. Figure 3 presents the mass ordering of differential elliptic flow among $\pi$, $K$, $p$, Λ, Ξ, and Ω in minimum bias Au+Au collisions. Compared with the STAR measurements, VISHNU correctly describes the $v_2$ mass ordering among $\pi$, $K$, $p$, and Ω, but fails to reproduce the mass ordering of Λ and Ξ. In our early investigations [2, 16], a similar situation was found in 2.76 A TeV Pb+Pb collisions at 10-20% and 40-50% centralities. Although VISHNU correctly generates the $v_2$ mass ordering among $\pi$, $K$, $p$, it slightly under-predicts the proton $v_2$ at lower transverse momentum. Together with the slightly over-perditions of the Λ and Ξ data, this leads to an inverse $v_2$ mass-ordering among $p$, Λ and Ξ. The effects from the initial flow or improved UrQMD hadronic cross-sections may solve this issue within the framework of VISHNU, which should be investigated in the near future.

## 4. Summary

In this proceeding, we present VISHNU calculations on the soft hadron data of strange and multi-strange hadrons in 200 A GeV Au+Au collisions. With the parameter settings used in our early calculations, VISHNU generally describes the $p_T$-spectra and elliptic flow of Λ, Ξ and Ω for various centrality bins. We also compare the $v_2$ mass ordering among $\pi$, $K$, $p$, Λ, Ξ, and Ω with the STAR measurements and find that VISHNU correctly describes the $v_2$ mass ordering among $\pi$, $K$, $p$, and Ω, but fails to reproduce the mass ordering among $p$, Λ, and Ξ.

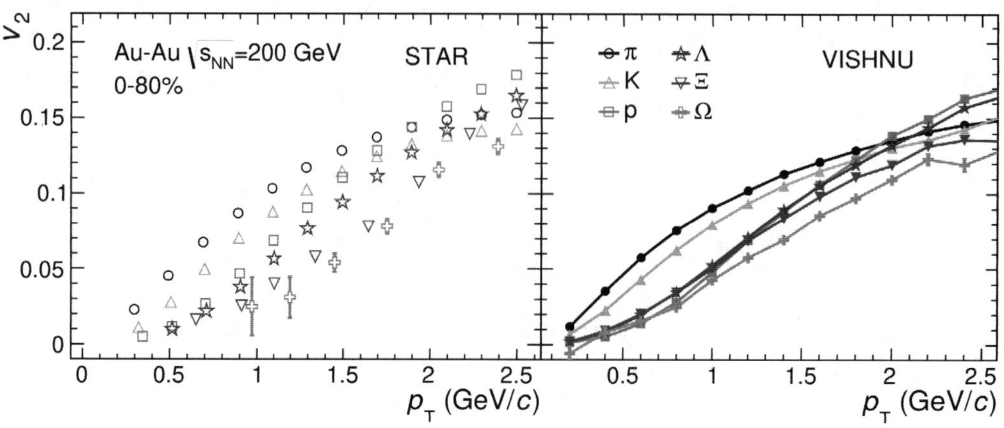


**Figure 3.** (Color online) Differential elliptic flow of $\pi$, $K$, $p$, $\Lambda$, $\Xi$, and $\Omega$ in minimal bias Au+Au collisions. Left panel: STAR measurements [12, 14], right panel: `VISHNU` calculations.

## Acknowledgments

We thank X. Zhu and S. Shi for helpful discussion. This work was supported by the NSFC and the MOST under Grants No. 11435001 and No. 2015CB856900. X.Z. was also partially supported by the China Postdoctoral Science Foundation under Grant No. 2015M570879. We gratefully acknowledge extensive computing resources provided to us on Tianhe-1A by the National Supercomputing Center in Tianjin, China.

## References

[1] van Hecke H, Sorge H and Xu N 1998 *Phys. Rev. Lett.* **81** 5764
[2] Zhu X, Meng F, Song H, and Liu Y X 2015 *Phys. Rev.* C **91** 034904 ; Song H, Meng F, Xin X and Liu Y X 2014 *J. Phys. Conf. Ser.* **509** 012089
[3] Song H, Bass S A and Heinz U 2011 *Phys. Rev.* C **83** 024912; Song H 2015 *Pramana* **84** 5 703-15, [arXiv:1401.0079 [nucl-th]]
[4] Song H and Heinz U 2008 *Phys. Lett.* B **658** 279; 2008 *Phys. Rev.* C **77** 064901; 2008 *Phys. Rev.* C **78** 024902; Song H 2009 *Ph.D thesis* (*Preprint* arXiv:0908.3656 [nucl-th])
[5] Bass S A *et al.* 1998 *Prog. Part. Nucl. Phys.* **41** 255; Bleicher M *et al.* 1999 *J. Phys.* G **25** 1859
[6] Huovinen P and Petreczky P 2010 *Nucl. Phys.* A **837** 26-53
[7] Hirano T and Nara Y 2009 *Phys. Rev.* C **79** 064904
[8] Song H, Bass S A, Heinz U, Hirano T and Shen C 2011 *Phys. Rev.* C **83** 054910; 2012 *Phys. Rev.* C **86** 059903
[9] Song H 2013 *Nucl. Phys.* A **904-905** 114c-21c [arXiv:1210.5778 [nucl-th]].
[10] Abelev B I *et al.* (STAR Collaboration) 2009 *Phys. Rev.* C **79** 034909
[11] Abelev B I *et al.* (STAR Collaboration) 2008 *Phys. Rev.* C **77** 054901
[12] Adamczyk L *et al.* (STAR Collaboration) 2015 (*Preprint* arXiv:1507.05247 [nucl-ex])
[13] Ollitrault J Y, Poskanzer A M and Voloshin S A 2009 *Phys. Rev.* C **80** 014904
[14] Adams J *et al.* (STAR Collaboration) 2007 *Phys. Rev. Lett.* **98** 062301
[15] Shen C, Heinz U, Huovinen P and Song H 2011 *Phys. Rev.* C **84** 044903
[16] Song H, Bass S and Heinz U W 2014 *Phys. Rev.* C **89** 034919

Journal of Physics: Conference Series **668** (2016) 012081 doi:10.1088/1742-6596/668/1/012081

# Are flow measurements at RHIC reliable?

**Arkadiy Taranenko**[a] **and Vladislav Vishnyakov**[a]

[a] National Research Nuclear University MEPhI (Moscow Engineering Physics Institute), Kashirskoe highway 31, Moscow, 115409, Russia

E-mail: AVTaranenko@mephi.ru

**Abstract.** The measurements of collective flow effects in particle production have provided invaluable insights on the transport properties of the strongly interacting matter produced in relativistic heavy-ion collisions at RHIC. The detailed comparison of flow measurements from PHENIX and STAR experiments at RHIC have been presented and discussed. For elliptic flow $v_2$ of charged hadrons from Au+Au collisions at 200 GeV the two data sets overlap excellently for centralities > 20%, they increasingly diverge at small centralities, with a 30% difference between STAR an PHENIX in the 0-5% centrality bin. For $v_3$ values the agreement is much worse and coming from the difference in STAR measurements. More investigations are needed to understand the reason for such differences.

## 1. Introduction

Ultra-relativistic heavy ion collisions offer the opportunity to study the strongly interacting matter under extreme conditions. In nuclear collisions at the Relativistic Heavy Ion Collider (RHIC) with energies up to $\sqrt{s_{NN}} = 200$ GeV and at the Large Hadron Collider (LHC) at energies of $\sqrt{s_{NN}} = 2.76$ TeV, nuclear matter is heated to temperatures up to 200-500 MeV. At these high energies quantum chromodynamics predict a cross-over transition to a new state of matter, so called quark-gluon plasma (QGP) [1]. The most important observables supporting the discovery of the QGP are the high values of elliptic flow and the suppression of high $p_T$ hadrons that can be explained within assuming a strongly coupled quark-gluon liquid to be formed [2].

The collective expansion, also called flow, originates from the initial pressure gradients in the created hot and dense matter. These pressure gradients transform the initial spatial deformations and inhomogeneities of the created matter into momentum anisotropies of the final state particle production, which are experimentally characterized by so-called flow harmonics $v_n$ [3]. Anisotropic flow can probe the properties of the system created in heavy-ion interactions, such as the equation of state (EoS) and transport coefficients like the shear viscosity to the entropy density ratio ($\eta/s$).

Every physical quantity which can be experimentally measured is important for a verification of a theory and therefore should be obtained properly. A good measurement should be reproducible; in particular, it should be done in such a way that one can easily compare results from different experiments, using different detectors.

## 2. Flow analysis methods

There are several methods to measure the magnitude of the flow signal $v_n$. The various methods have different sensitivities to non-flow effects and $v_n$ fluctuations. Non-flow refers to correlations between particles which are independent of the initial geometry of the collision system, such as resonance decays, HBT correlations, final state interactions and jets.

- *Event Plane method*
  The Event Plane (EP) method [5] uses the anisotropic flow itself to determine the event plane. Within this method, one first identifies the event plane angle ($\Psi_n$) in each event using a specific detector at forward rapidity and then calculates the correlation of particles near mid rapidity as $v_n = \langle cos[n(\phi - \Psi_n)]\rangle / Res\{\Psi_n\}$, where $Res\{\Psi_n\}$ is the event plane resolution factor. Note that detectors evaluating $\Psi_n$ and detectors measuring the azimuthal angle of the particle $\phi$ should have the rapidity gap to suppress the auto- and non-flow correlations. The two-particle cumulant method is based on a similar idea.
- *The 4-particle cumulant method*
  The advantage of the cumulant method is that the multi-particle cumulant removes the contribution of non-flow correlations from lower-order correlations [6]. The flow contribution can be obtained by subtracting the 2-particle correlation from the 4-particle correlation.
- *Lee-Yang Zero (LYZ) method*
  As opposed to the four-particle cumulant method which is sensitive to the correlations of four particles, LYZ method [7] is sensitive to the correlations of all the particles. Thus it is supposed to remove non-flow correlations to all orders.

Event-by-event fluctuations of $v_n$ affect each method of measurement differently. The event plane and two-particle cumulant methods have a positive contribution while the multi-particle methods (4-particle cumulant and LYZ) have a negative contribution. As a result of fluctuations, an event-plane measurement of $v_n$ yields an ambiguous measure lying somewhere between the event-averaged mean value $\langle v_n\rangle$ and the root-mean-square value $\langle v_n^2\rangle^{1/2}$. Where exactly depends on the event plane resolution $Res\{\Psi_n\}$, which strongly depends on the experimental setup.

## 3. Comparison of PHENIX and STAR flow results

An important step toward reducing the systematic uncertainty associated with flow measurements at RHIC, is a detailed comparison of differential flow results $v_n(p_T$, centrality). The flow measurements at RHIC where performed by four collaborations: BRAHMS, PHENIX, PHOBOS and STAR. However, the published results for differential flow measurments can be found only for STAR and PHENIX collaborations.

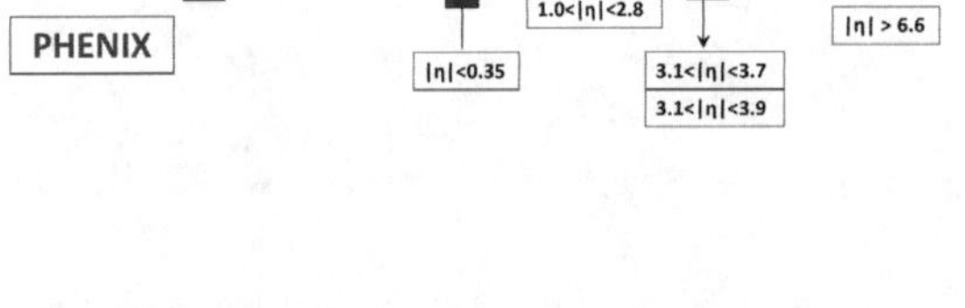


**Figure 1.** The pseudo-rapidity $\eta$ acceptance of the PHENIX and STAR detectors used for event-plane reconstruction, along with the mid-rapidity detectors used for charged hadron measurements.

The flow measurements in the PHENIX collaboration were performed using the EP method, where one correlates the azimuthal angles $\phi$ of the charged tracks in the PHENIX central arms ($|\eta| \leq 0.35$) with the azimuth of the estimated n-order event plane $\Psi_n$, determined via hits in the two beam-beam counters (BBCs), reaction plane hodoscopes (RXNs) and zero-degree calorimeters (ZDCs). The respective $\eta$ coverage for these event-plane detector pairs are: $1.0 < |\eta_{\rm RXN}| < 2.8$, $3.1 < |\eta_{\rm BBC}| < 3.9$ and $|\eta_{\rm ZDC}| > 6.5$, see figure 1. The event centrality is determined by using the total charge deposited in the BBCs.

The STAR collaboration uses both: multi-particle methods (4-particle cumulant and LYZ method) and EP method. In EP method, one correlates the azimuthal angles $\phi$ of the charged tracks reconstructed in the STAR Time Projection Chamber (TPC) ($|\eta| < 1.0$) with with the azimuth of the estimated n-order event plane $\Psi_n$, determined via tracks in TPC ($|\eta_{TPC}| < 1.0$ ) and FTPC ($2.5 < |\eta_{FTPC}| < 4.0$), hits in ZDCs ($|\eta_{ZDC}| > 6.5$), see figure 1. The centrality determination is based on the number of charged tracks in the TPC with track quality cuts. To suppress non-flow effects for EP measurements using TPC one applies $\Delta\eta$ gap between tracks used for EP construction and tracks used for the measurements of $v_n$.

In order to make proper comparison for EP method we should use the event plane detectors with the same $\eta$ coverage and similar event plane resolution. This is the case for STAR/PHENIX ZDC/SMD and for STAR FTPC/PHENIX BBC. Unfortunately, we have only one colliding system and one beam energy for the proper comparison of the published elliptic flow ($v_2$) results between STAR and PHENIX: the $v_2$ values of charged hadrons from Cu+Cu collisions at $\sqrt{S_{NN}} = 200$ GeV obtained using the EP from STAR FTPC [9] and PHENIX BBC [10]. Figure 2 compares the centrality dependence of the ratio of $v_2$ obtained using EP from STAR FTPC and PHENIX BBC for four bins in $p_T$. The difference between two data sets are less than 5-10% within a typical systematic uncertainty of the measurements of 5%. So even with the similar $\Delta\eta$-gap and event plane resolution there are some differences between two experiments.

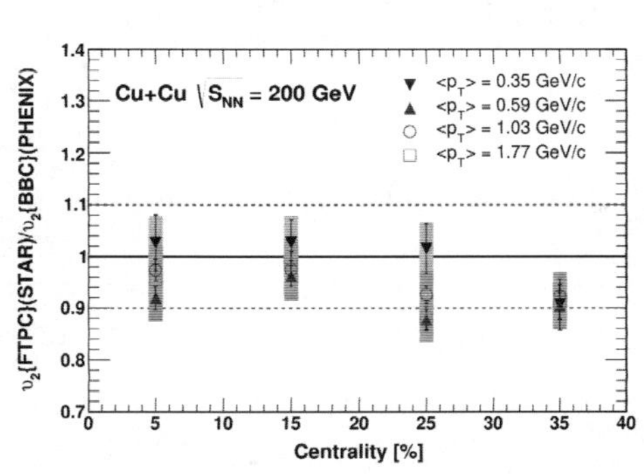


**Figure 2.** Centrality dependence of the ratio of STAR $v_2(p_T)$ values [9] obtained using EP from STAR FTPC to the $v_2$ values [10] obtained using EP from PHENIX BBC. The data are for charged hadrons from Cu+Cu collisions at $\sqrt{S_{NN}} = 200$ GeV. The 5% systematic errors are shown.

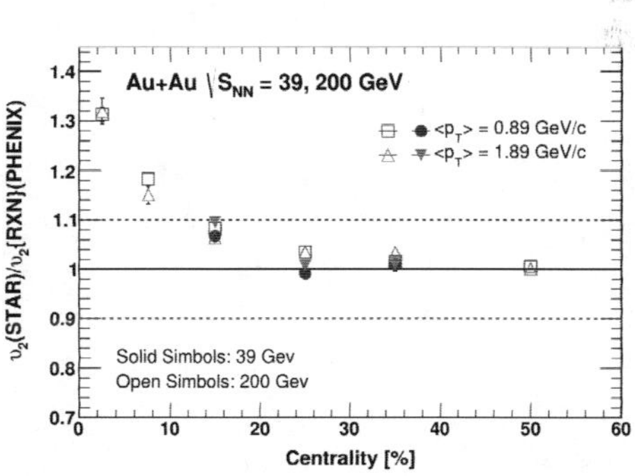


**Figure 3.** Centrality dependence of the ratio of STAR $v_2(p_T)$ values obtained using EP from STAR TPC to the $v_2$ values obtained using EP from PHENIX RXN. The data are for charged hadrons from Au+Au collisions at $\sqrt{S_{NN}} = 39$ and 200 GeV.

However, it is hard to make proper comparison for the $v_2$ measurements of charged hadrons emerged from Au+Au collisions at $\sqrt{S_{NN}} = 200$ GeV. This is the reference point for the Beam Energy Scan program at RHIC and for comparison with LHC results. Here we can make the comparison between $v_2$ results obtained using the PHENIX EP RXN ($1.0 < |\eta_{RXN}| < 2.8$) and

STAR TPC EP with $\Delta\eta > 0.5$. The open symbols from figure 3 show the centrality dependence of the ratio of STAR $v_2(p_T)$ values to the $v_2$ values obtained from PHENIX for charged hadrons at $\sqrt{S_{NN}}$ = 200 GeV. While the two data sets overlap excellently for centralities > 20%, they increasingly diverge at small centralities, with a 30% difference between STAR an PHENIX in the 0-5% centrality bin. The PHENIX EP measurements are less affected by non-flow effects due to larger $\Delta\eta$ gap. However, it is hard to explain the difference by non-flow effects only as we see the same difference between STAR and PHENIX results for Au+Au collisions at $\sqrt{S_{NN}}$ = 39 GeV. The excess of the STAR over the PHENIX data is almost uniform in $p_T$ and could be explained by a small shift in the centrality definitions between the experiments.

Figure 4 compares $v_3$ values obtained with STAR TPC and FTPC event plane detectors to PHENIX RXN EP $v_3$ measurements. One can see a huge difference ( 40-50%) between TPC and FTPC data and < 10% difference between STAR TPC to PHENIX RXN results on the other hand. To understand this enormous difference between STAR results more comparisons should be made, e.g. for the elliptic flow between FTPC and TPC EP.

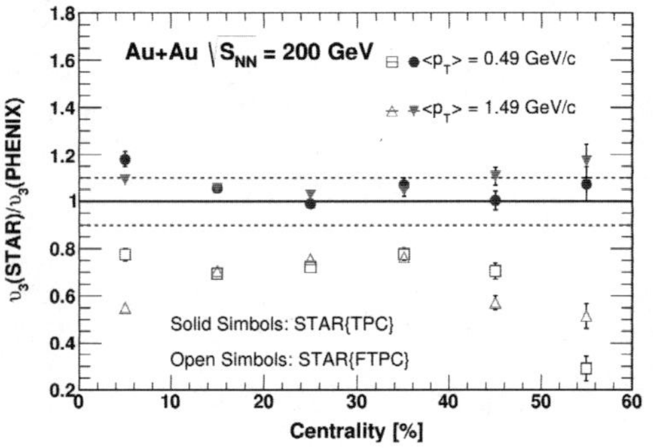


**Figure 4.** The centrality dependence of the ratio of the $v_3$ values of STAR{TPC}(solid) and STAR{FTPC}(open) to the PHENIX $v_3(p_T)$ points fit for the two $p_T$ bins at $\sqrt{S_{NN}}$ = 200 GeV Au+Au collisions. The points are extracted from [11].

## 4. Conclusion

The comparison of the differential flow measurements $v_n(p_T$, centrality) from PHENIX and STAR experiments was presented and discussed. The significant differences are found for both $v_2$ and $v_3$ values for charged hadrons from Au+Au collisions at 200 GeV. The understanding of the origin of this difference will be an important step toward reducing the systematic uncertainty associated with flow measurements at RHIC.

## References

[1] Aoki Y *et al.* 2006 *Nature* **443** 675.
[2] Adams J *et al.* [STAR Collaboration] , Nucl. Phys. **A757** (2005) 102,
[3] Voloshin S and Zhang Y 1996 *Z. Phys.* C **70** 665.
[4] Bhalerao R S and Ollitrault J Y 2006*Phys. Lett.* B **641** 260.
[5] Poskanzer A M and Voloshin S A 1998 *Phys. Rev.* C **58** 1671.
[6] Borghini N, Dinh P M and Ollitrault J Y 2001 *Phys. Rev.* C **63** 054906.
[7] Bhalerao R S, Borghini N and Ollitrault J Y 2004 *Phys. Lett.* B **580** 157.
[8] Abelev B I *et al.* [STAR Collaboration] 2008 *Phys. Rev.* C **77** 054901.
[9] Abelev B I *et al.* [STAR Collaboration] 2010 *Phys. Rev.* C **81** 44902.
[10] Adare A *et al.* [PHENIX Collaboration] 2007 *Phys. Rev. Lett.* **98** 162301.
[11] Adamczyk L *et al.* [STAR Collaboration] 2013 *Phys. Rev.* C **88** 14904.

# Transport coefficients and quark-hadron phase transition(s) from PLSM in vanishing and finite magnetic field

**Abdel Nasser Tawfik**

Egyptian Center for Theoretical Physics (ECTP), Modern University for Technology and Information (MTI), 11571 Cairo, Egypt
World Laboratory for Cosmology And Particle Physics (WLCAPP), 11677 Cairo, Egypt

E-mail: a.tawfik@eng.mti.edu.eg

**Abstract.**
In peripheral heavy-ion collisions, localized, short-lived and extremely huge magnetic field can be generated. Its possible influences on the quark-hadron phase transition(s) and the transport properties of the hadronic and partonic matter shall be analysed from the Polyakov linear-sigma model. Our calculations are compared with recent lattice QCD calculations.

## 1. Introduction

The heavy-ion collision can be divided into several successive processes including formation of quark-gluon plasma (QGP) and hadronization, etc. Some of such processes are believed to exist shortly after the big bang and explain the interior of compact stellar objects. On the other hand, most of experimental devices are designed to detect hadrons, leptons and electromagnetically interacting particles. Thus, quarks and gluons can't directly be detected and therefore properties such as viscosity of QGP still represent a great challenge for the particle scientists. The nonnegligible elliptic flow measured at the Relativistic Heavy-Ion Collider (RHIC) and recently at the Large Hadron Collider (LHC) is a striking observation referring to the fact that the viscosity property turns to be measurable in the heavy-ion collision experiments, on one hand and the hydrodynamic response of produced matter to the initial geometry leads to such particle production asymmetry, on the other hand.

Most of the experimental signals revealing the QGP formation and the elliptic flow, for instance, strengthen with increasing the collision centrality. Furthermore, the peripheral collisions introduce remarkable effects, as well. The local imbalance in the momentum that carried by the colliding nucleons generates a nonvanishing local angular-momentum [1, 2] and the relativistic motion of the spectator nucleons generates currents of net positive charge moving off-center (opposite to each other). Both types of nuclear collisions generate a huge magnetic field [3], $\mathcal{O}(m_\pi^2)$.

In quantum electrodynamics (QED), phenomena such as magnetic catalysis in which the magnetic field dynamically generates masses, and Meissner effect where the magnetic field changes the order of the phase transition in type-I superconductor from second to first are examples on the possible influence of such large magnetic field on the phase-transitions in quantum choromodynamics (QCD) and/or the properties of the hadronic and partonic matter.

Influence of this magnetic field on the chiral phase-transition has been examined by using the linear-sigma model [4]. In recent works, we have implemented the Polyakov linear-sigma model (PLSM) in characterizing various QCD phenomena and quantities. Thermodynamic quantities including normalized and non-normalized higher-order moments of the particle multiplicity have been investigated [5, 6]. The chiral phase-structure of (pseudo)-scalar and (axial)-vector meson masses in thermal and dense medium has been determined [7]. Also, it was reported on some properties of QGP in nonzero magnetic field [8].

## 2. Polyakov linear-sigma model in nonzero magnetic filed

Let us assume that the generated magnetic field ($eB$) is directed along $z$-axis. From magnetic catalysis [9] and by using Landau quantization, the magnetic field modifies the quarks dispersion relations,

$E_{B,f} = [p_z^2 + m_f^2 + |q_f|(2n+1-\sigma)B]^{1/2}$, where $f$ runs over all quarks flavors, $n$ is the quantized Landau number, $\sigma$ is related to the spin quantum number and $S$; $\sigma = \pm S/2$. From dimensional reduction of the magnetic catalysis property [9], $\int \frac{d^3p}{(2\pi)^3} \longrightarrow |q_f|B/(2\pi)\sum_{\nu=0}^{\infty}\int dp_z(2-\delta_{0\nu})/(2\pi)$, where $2-\delta_{0\nu}$ stands for degenerate Landau levels and $2n+1-\sigma$ is replaced by the quantum number $\nu$

In mean field approximation, the PLSM grand potential is given as [5], $\Omega(T,\mu) = U(\sigma_x,\sigma_y) + \mathcal{U}(\phi,\phi^*,T)+\Omega_{\bar{\psi}\psi}(T;\phi,\phi^*,B)$. The mesonic potential $\mathcal{U}(\phi,\phi^*,T)$ is assumed to vanish at high temperatures $\gtrsim \Lambda_{QCD}$. The purely mesonic potential is given in Ref. [11]. Assuming Landau quantization and from magnetic catalysis, the quarks and antiquark contribution to PLSM potential becomes [8]

$$\Omega_{\bar{q}q}(T,\mu_f,B) = -2\sum_{f=l,s}\frac{|q_f|BT}{(2\pi)^2}\sum_{\nu=0}^{\nu_{max\,f}}(2-\delta_{0\nu})\int_0^{\infty} dp_z \ln\left[1+3\left(\phi+\phi^*\mathcal{E}\right)\mathcal{E}+\mathcal{E}^3\right], \tag{1}$$

where $\mathcal{E} = \exp[-(E_{B,f}-\mu_f)/T]$, $E_{B,f}$ is the modified dispersion relation and $\mu_f$ refers to the chemical potential of $f$-quark. In vanishing magnetic field, the quarks potential reads

$$\Omega_{\bar{q}q}(T,\mu_f) = -2T\sum_{f=l,s}\int_0^{\infty}\frac{d^3\vec{p}}{(2\pi)^3}\ln\left[1+3\left(\phi+\phi^* e^{-\frac{E_f-\mu_f}{T}}\right)\times e^{-\frac{E_f-\mu_f}{T}}+e^{-3\frac{E_f-\mu_f}{T}}\right] \tag{2}$$

## 3. Results and discussion

From Green-Kubo formula, the bulk viscosity can be related to the correlation functions of trace of the energy-momentum tensor. From the relaxation time approximation, the deviation of energy-momentum tensor from its local equilibrium which corresponds to the difference between the distribution function near and at equilibrium, $\delta n = n - n_0$, can be used to determine bulk and shear viscosities, [12]

$$\zeta(T,\mu,eB) = \frac{3}{2T}\sum_f\frac{|q_f|B}{2\pi}\sum_\nu\int\frac{dp_z}{2\pi}(2-\delta_{0\nu})\frac{\tau_f}{E_{B,f}^2}\left[\frac{|\vec{p}|^2}{3}-c_s^2E_{B,f}^2\right]^2 f_f(T,\mu)\left[1+f_f(T,\mu)\right], \tag{3}$$

$$\eta(T,\mu,eB) = \frac{1}{5T}\sum_f\frac{|q_f|B}{2\pi}\sum_\nu\int\frac{dp_z}{2\pi}(2-\delta_{0\nu})\frac{|\vec{p}|^4\tau_f}{E_{B,f}^2} f_f(T,\mu)\left[1+f_f(T,\mu)\right]. \tag{4}$$

The relaxation time ($\tau$) can be related to the quark decay constant ($\Gamma$). In doing this, two models can be implemented:

- The dynamic quasi-particle model (DQPM) which describes well the phenomenology of interacting massless quarks and gluons as non-interacting massive quasi-particles [13]. The flavor-blind reaction rates for quarks and gluons are inversely dependent on the particle width, $\tau_q \sim 1/\Gamma_q$ and $\tau_g \sim 1/\Gamma_g$, respectively.
- The transport theory with Boltzmann master equation which can be used to calculate the momentum loss and the relaxation time, as well [14]. To this end, screening of long-range quark-quark interactions is assumed for the process of interpenetrating quark plasmas. For a spatially uniform quark plasma flowing with respect to $N_f - 1$ plasmas with another type of flavors, the relative flow velocity is assumed to relax in time due to the collisions [15]. In partonic and hadronic phases, the relaxation time, respectively, scales with the temperature $\tau \sim \exp[-T]$ for $T > T_c$ and $\tau \sim 1/T$ for $T < T_c$.

In vanishing magnetic field and at zero chemical potential, the temperature dependence of bulk viscosity ($\xi$) normalized to thermal entropy ($s$) is presented in left-hand panel of Fig. 1. The ratio of shear viscosity to thermal entropy ($\eta/s$) is given in right-hand panel. At temperatures $\gtrsim T_c$, $\xi/s$ has a good agreement with the lattice calculations [16, 17, 18, 19]. LSM [20], DQPM and NJL models [?] are also good with the lattice. The sharp raise of bulk viscosity slightly above the phase transition (as the entropy is assumed to remain constant) indicates instability in the hydrodynamic flow of the plasma as confirmed in RHIC observables [22]. The agreement between PLSM and lattice QCD calculations for $\eta/s$ is excellent above $T_c$.

The top panel of Fig. 2 shows the results of $\zeta/s$ vs. temperature in finite magnetic field. Again the results at $eB = 0$ GeV$^2$ are depicted as reference (solid curve). The results at $eB = 0.0$ GeV$^2$ (solid), $eB = 0.2$ GeV$^2$ (dotted) and $eB = 0.4$ GeV$^2$ (dot-dashed) show a remarkable dependence on temperature. For instance, the finite magnetic field seems to enhance the appearance of the characterizing peak at $T_c$. The peaks are related to minima. At very low temperature, all curves seem to bundle together to large $\zeta/s$-values.

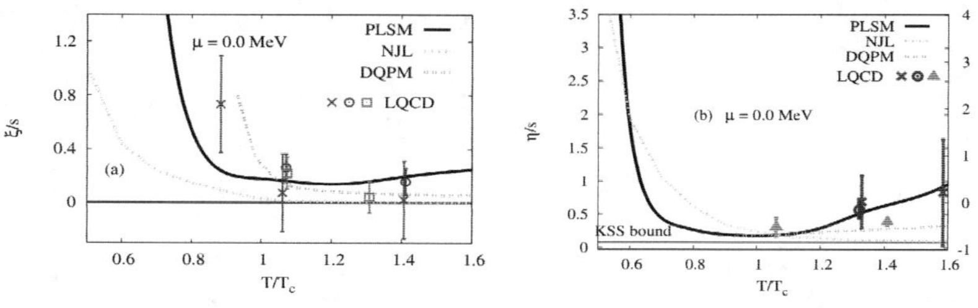


**Figure 1.** Left-hand panel: the ratio of bulk viscosity and thermal entropy ($\xi/s$) calculated from PLSM (solid curve) and compared with lattice QCD simulations [16] (cross points) and [17] (circle points) and (square points) is given as function of temperature at vanishing chemical potential and magnetic field. Right-hand panel shows the ratio of shear viscosity to thermal entropy ($\eta/s$) calculated from PLSM (solid curve) and compared with the available lattice QCD simulations [16] (cross points), [17] (circle points) and (square points) and [18] (triangle points). The Kovtun-Son-Starinets (KSS) bound is also depicted. Both panels are compared with NJL (dotted dash) and DQPM (double dotted) [?].

In right-hand panel of Fig. 2, the ratio $\eta/s$ is depicted as function of temperature at different values of $eB$. The solid curve represents the results at a vanishing magnetic field, while $eB = 0.2$, $eB = 0.15$ and $0.4$ GeV$^2$ are given as dotted, dot-dashed curves, receptively. KSS limit is represented by dashed line. At low temperatures, $\eta/s$ becomes very large. A peak positioned at $T_c$ starts to appear when the magnetic field increases while the peak's height is not sensitive to $eB$. The remarkable effect around $T_c$ refers to instability in the hydrodynamic flow of quarks and hadrons. At extremely high temperatures, the coupling becomes weak and the hadrons are entirely liberated into quarks and gluons. Thus, it is likely that the deconfinement phase-transition is completed. At temperatures $\gtrsim T_c$ the values of $\eta/s$ are not sensitive to the change in $e\,B$.

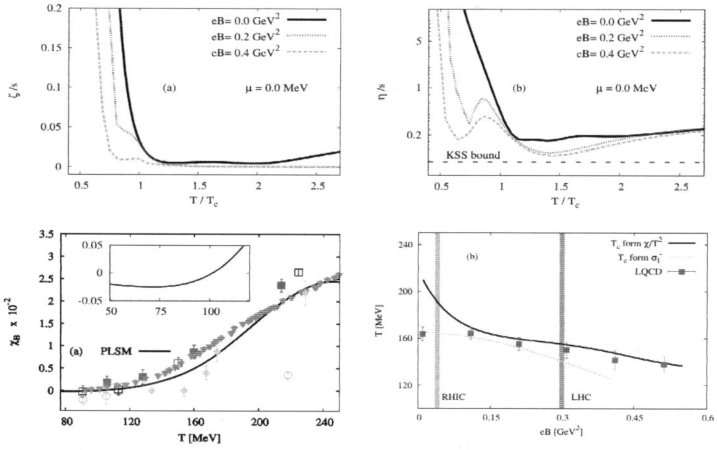


**Figure 2.** Top panel: the influence of magnetic fields, $eB = 0.0$ GeV$^2$ (solid), $eB = 0.2$ GeV$^2$ (dotted) and $eB = 0.4$ GeV$^2$ (dot-dashed) on $\zeta/s$ is given as function of temperature at vanishing chemical potential. In right-hand panel, the same as in left-hand panel but for $\eta/s$. A very small increase can be observed at high temperature. Bottom panel: the temperature dependence of magnetic susceptibility $\chi_B$ is given at $eB = 0.2$ GeV$^2$ and vanishing baryon chemical potential. The results are compared with different lattice simulations (symbols) [23]. Right-hand panel: the chiral phase-diagram relates the critical temperature to $eB$ through $\chi/T^2$ (solid curve) and quark condensate $\sigma_l$ (dashed curve). The vertical bands refer to magnetic fields at RHIC and LHC energies.

The bottom panel of Fig. 2 shows the magnetic susceptibility ($\chi_B$), which is determined from the second derivative of PLSM free energy with respect to $eB$, as function of temperature at $eB = 0.2$ GeV$^2$ and a vanishing baryon chemical potential. The magnetic susceptibility is a dimensionless quantity referring to the degree of magnetization as a response to the non-zero magnetic field. The results from PLSM are compared with various lattice simulations (symbols) and with the HRG model. It is worthwhile to notice the negative slope of $\chi_B$ (inside-box in middle panel). This signals that the QCD matter is diamagnetic. At high temperatures, i.e. restoring broken chiral symmetry, we observe a transition between dia- and para-magnetism. It is likely that the QCD matter is para-magnetic is very high temperatures.

In right-hand panel, influences of the magnetic field strength on the chiral phase-diagram are analysed and compared with recent lattice QCD results (circles with errorbars) [23]. The vertical bands indicate the value of the magnetic fields at RHIC ($\sim m_\pi^2$) and LHC energies ($\sim 10 - 15\, m_\pi^2$). There is a small suppression in the chiral condensates, i.e. magnetic catalysis. In PLSM we implement two methods in order to determine the quasi-critical temperature (dotted curve). The first one implements the higher-order moments of quark multiplicity, i.e. normalized quark susceptibility $\chi_q/T^2$ (solid curve). The second one uses the intersection of light-quark chiral condensate $\sigma_l$with the deconfinement order-parameters (dashed curve). We find that the critical temperature decreases with increasing $eB$, i.e. inverse magnetic catalysis. An excellent agreement is apparently achieved between our calculations and the lattice QCD, especially at small magnetic field $0 \leq eB \leq 0.2$ GeV$^2$ (second method). The first method agrees well with the lattice data at a wider range of magnetic fields $0.13 \leq eB \leq 0.55$ GeV$^2$. Again, the first method apparently overestimates lattice data at low temperature, while the second method slightly underestimates these at high temperature.

## 4. Conclusion

The magnetic field which is likely generated in peripheral heavy-ion collisions due to local imbalance in momentum carried out by the participants and off-center relativistic motion of spectator's electric charges seems to influence the transport properties (bulk and shear viscosity) in both hadron and quarks phases. Also, the quark-hadron phase transitions are remarkable influenced by such large magnetic field.

In vanishing magnetic field, we find that increasing $T$ decreases $\zeta/s$. But at high temperatures, a small increase can be observed. In finite magnetic filed, there is a remarkable dependence on temperature; appearance of characterizing peaks at $T_c$ (enhanced with increasing $eB$) and very large values of $\zeta/s$ observed at low temperatures. For $\eta/s$, the peaks around $T_c$ gradually appear with increasing $eB$ but their heights seem not being sensitive to $eB$. At temperatures $\gtrsim T_c$, $\eta/s$ is also not sensitive to the change in $e\,B$.

The magnetic susceptibility $\chi_B$ which measures the degree of magnetization as a response of finite magnetic field seems to raise with the temperature. Both PLSM and lattice QCD calculations are in good agreement with each other. It is worthwhile to highlight the slightly negative $\chi_B$ takes place at low temperature. This refers to diamagnetic properties of the QCD matter. Switching to high temperatures, i.e. restoring broken chiral symmetry, a transition between dia- and para-magnetic properties is observed. At very high temperatures, the QCD matter is likely para-magnetic.

In nonzero magnetic field, we have implemented two methods in order to determine the critical temperatures with changing $eB$. Magnetic catalysis is clearly observed in QCD matter. The agreement with the lattice calculations is fairly good.

## References

[1] Z.-T. Liang and X.-N Wang, Phys. Rev. Lett. **94**, 102301 (2005).
[2] F. Becattini, F. Piccinini and J. Rizzo, Phys. Rev. C **77**, 024906 (2008).
[3] D. E. Kharzeev, L. D. McLerran and H. J. Warringa, Nucl. Phys. A **803**, 227 (2008).
[4] E. S. Fraga and A. J. Mizher, Phys. Rev. D **78**, 025016 (2008).
[5] A. Tawfik, N. Magdy and A. Diab, Phys. Rev. C **89**, 055210 (2014).
[6] A. Tawfik and N. Magdy, J. Phys. G **42**, 015004 (2015) .
[7] A. Tawfik and A. Diab, Phys. Rev. C **91**, 015204 (2015).
[8] A. Tawfik and N. Magdy, Phys. Rev. C **90**, 015204 (2014).
[9] I. A. Shovkovy, Lect. Notes Phys. **871**, 13 (2013).
[10] S. Ghosh, S. Mandal and S. Chakrabarty, Phys. Rev. C **75**, 015805 (2007).
[11] B. J. Schaefer and M. Wagner, Phys. Rev. D **79**, 014018 (2009).
[12] A. Tawfik, and M. Wahba, Annalen Phys. **522**, 849 (2010).
[13] A. Peshier, Phys. Rev. D **70**, 034016 (2004).
[14] P. Zhuang, J. Hufner, S.P. Klevansky, and L. Neise, Phys. Rev. D **51**, 3728 (1995).
[15] R. Marty, E. Bratkovskaya, W. Cassing, J. Aichelin and H. Berrehrah, Phys. Rev. C **88**, 045204 (2013).
[16] H. B. Meyer, Phys. Rev. D **76**, 101701 (2007); H. B. Meyer, Phys. Rev. Lett. **100**, 162001 (2008).
[17] S. Sakai and A. Nakamura, PoS **LAT2007**, 221 (2007).
[18] A. Nakamura and S. Sakai, Phys. Rev. Lett. **94**, 072305 (2005).
[19] F. Karsch, D. Kharzeev and K. Tuchin, Phys. Lett. B **663**, 217 (2008).
[20] H. Mao, J. Jin and M. Huang, J. Phys. G **37**, 035001 (2010).
[21] P. K. Kovtun, D. T. Son and A. O. Starinets, Phys. Rev. Lett. **94**, 111601 (2005).
[22] G. Torrieri and I. Mishustin, Phys. Rev. C **78**, 021901 (2008).
[23] G. Bali, F. Bruckmann, G. Endrodi, S. Katz, and A. Schaefer, JHEP **1408**, 177 (2014).

# D-meson production in pp and p-Pb collisions measured with ALICE at the LHC

**Cristina Terrevoli, for the ALICE Collaboration**
Universita' degli Studi di Padova and INFN, via Marzolo 9, 35131, Padova, Italy
E-mail: cristina.terrevoli@pd.infn.it

**Abstract.** The open heavy-flavour production studied with ALICE at the LHC in pp collisions at $\sqrt{s}$ = 7 TeV and p–Pb collisions at $\sqrt{s_{\rm NN}}$ = 5.02 TeV is presented. Focus is given to the dependence of the production on the multiplicity of charged particles produced in the collision.

## 1. Introduction

Heavy quarks (charm and beauty) can be used to study several aspects of QCD in high-energy hadronic collisions. They are produced in hard partonic scattering processes occurring in the initial stage of the collisions with large momentum transfer. Therefore, the study of their production in pp collisions provides a precision test of perturbative QCD (pQCD) calculations. In proton-nucleus collisions, heavy-flavour production could be modified by the presence of a nucleus in the colliding system. The cross section and $p_{\rm T}$ distributions of heavy-flavour hadrons could be influenced by the modification of the parton distributions functions in the nucleus (nPDF), especially at small Bjorken $x$ (see e.g. [1]), by the $k_{\rm T}$-broadening caused by soft scatterings of the partons in the incoming nucleus [2], and by possible energy loss of the produced heavy quarks in the nucleus [3]. These effects are commonly called Cold Nuclear Matter (CNM) effects. They are explored through the measurement of the nuclear modification factor $R_{\rm pPb}$, i.e. the comparison of p–Pb measurements with pp results, to test whether the yield and the transverse momentum distributions follow or not a scaling with the number of binary nucleon-nucleon collisions in p–Pb collisions.

The study of the nuclear modification factor in different event multiplicity classes is also interesting to assess the possible modification of the $p_{\rm T}$ spectrum in high multiplicity events. Moreover, the measurement of heavy-flavour production, both in pp and p–Pb collisions, as a function of the multiplicity of charged particles produced in hadronic collisions can improve our understanding of their production mechanisms and could give insight into the role of Multi-Parton Interactions (MPI), i.e. several partonic interactions occurring in a single collision between two nucleons. If the production of heavy quarks is affected by the presence of additional hard scatterings between incoming partons, their yield should increase with multiplicity. The shape of the increasing trend depends on how the MPI contribute to the production of particles: if they equally contribute to soft and hard production, from small to large multiplicities, the increase would be linear. Indications of the relevance of MPI in heavy-quark production have been observed by NA27 [4] and recently by ALICE in the $J/\psi$ and D-meson production measurement as a function of charged-particle multiplicity [5, 6].

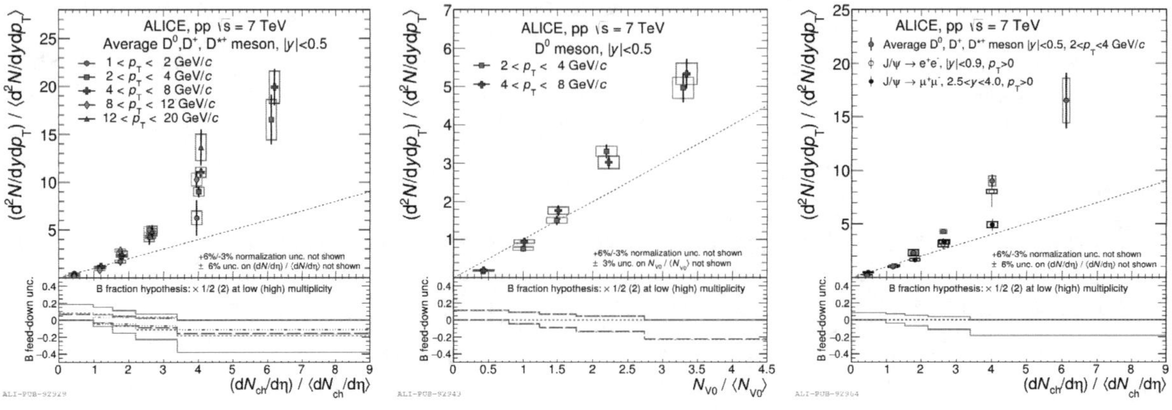


**Figure 1.** Self-normalized average D-meson yields as a function of multiplicity in pp collisions for different $p_{\mathrm{T}}$ ranges [6]. Left: multiplicity estimates at mid rapidity. Middle: multiplicity estimate at forward rapidity. Right: D-meson (red) and $J/\psi$ (black) yields as a function of the relative charged-particle multiplicity at mid rapidity.

## 2. D-meson reconstruction in ALICE

pp collisions at $\sqrt{s}$ =7 TeV ($3 \times 10^8$ events) and p–Pb collisions at $\sqrt{s_{\mathrm{NN}}}$ =5.02 TeV ($10^8$ events) collected by the ALICE detector [7], with a minimum-bias trigger, in 2010 and 2013, respectively, were analyzed. Prompt $\mathrm{D}^0$, $\mathrm{D}^+$, $\mathrm{D}^{*+}$ were reconstructed via their hadronic decay channels and selected via the reconstruction of decay vertices displaced by a few hundred $\mu$m from the interaction vertex. Events were divided in classes using the charged-particle multiplicity measured at mid rapidity through the number of tracklets with $|\eta| < 1$ reconstructed by associating pairs of hits in the Silicon Pixel Detector (SPD) or at forward and backward rapidity using the charge collected by the V0 scintillator array.

## 3. D-meson production in pp collisions

D-meson $p_{\mathrm{T}}$-differential cross sections were measured in pp collisions at $\sqrt{s}$ = 7 TeV [8] and $\sqrt{s}$ = 2.76 TeV [9]. They are well described by pQCD calculations [10, 11]. A pQCD-based energy scaling of the 7 TeV cross section to 2.76 TeV is in good agreement with the measured cross section at the same energy, confirming the robustness of the $\sqrt{s}$ extrapolation needed to define the references for Pb–Pb and p–Pb measurements.

The study of D-meson production, evaluated for various multiplicity and $p_{\mathrm{T}}$ intervals, is presented via the D-meson self-normalized yield, i.e. the corrected per-event yield normalized to the multiplicity-integrated value:

$$\frac{(\mathrm{d}^2N^{\mathrm{D}}/\mathrm{d}y\mathrm{d}p_{\mathrm{T}})^j}{\langle \mathrm{d}^2N^{\mathrm{D}}/\mathrm{d}y\mathrm{d}p_{\mathrm{T}}\rangle} = \left(\frac{1}{N^j_{\mathrm{events}}}\frac{N^j_{\mathrm{raw\ D}}}{\epsilon^j_{\mathrm{prompt\ D}}}\right) \Big/ \left(\frac{1}{N_{\mathrm{MB\ trigger}}/\epsilon_{\mathrm{MB\ trigger}}}\frac{\langle N_{\mathrm{raw\ D}}\rangle}{\langle \epsilon_{\mathrm{prompt\ D}}\rangle}\right)$$

where the index $j$ identifies the multiplicity interval, $N^j_{\mathrm{raw\ D}}$ is the raw yield, $\epsilon^j_{\mathrm{prompt\ D}}$ represents the reconstruction and selection efficiencies for prompt D mesons and $N^j_{\mathrm{events}}$ the number of events analysed in each multiplicity interval. The results for $\mathrm{D}^0$, $\mathrm{D}^+$, $\mathrm{D}^{*+}$ are compatible with each other. The average self-normalized yield of the three species increases with increasing multiplicity and does not depend on $p_{\mathrm{T}}$ within uncertainties in the interval 1-20 GeV/$c$, as shown in Fig 1. A similar trend is observed when the multiplicity is estimated at mid rapidity (Fig 1 left) or at forward rapidity, (middle), demonstrating that the trend is not connected

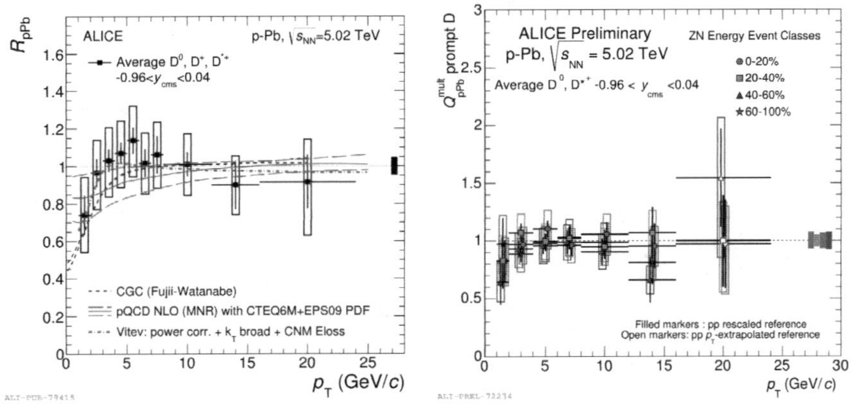


**Figure 2.** Left: Average $R_{\rm pPb}$ of D mesons as a function of $p_{\rm T}$ compared to models [15]. Right: Average $p_{\rm T}$-differential D-meson $Q_{\rm pPb}$ in four ZNA event activity classes.

to a possible bias due to the pseudorapidity region of the multiplicity measurements [6]. The self-normalized yield of D mesons is comparable with that of inclusive J/$\psi$ measured at mid rapidity, and the increasing trend seems to be stronger at mid rapidity than at forward rapidity, as shown in Fig 1 (right). Furthermore, an analogous trend is observed comparing D mesons and non-prompt J/$\psi$. The similar trends of charm (open and hidden) and beauty suggest that their multiplicity dependence is not, or only slightly, affected by the hadronisation stage. The yield increase can be described by calculations taking into account the influence of the interactions between colour sources in the percolation model [12], and the contribution of MPI, through the PYTHIA-8 [13] and the EPOS 3 event generators [14], that also include the effect of the initial conditions of the collision followed by a hydrodynamic evolution. However more precise measurements are needed for a conclusive discrimination of the possible origins of the effect.

## 4. D-meson production in p-Pb collisions

Figure 2 (left) shows the D-meson $R_{\rm pPb}$ (average of $\rm D^0$, $\rm D^+$, $\rm D^{*+}$) as a function of $p_{\rm T}$ in minimum-bias events, compared with theoretical calculations [15]. $R_{\rm pPb}$ is compatible with unity within uncertainties in the measured $p_{\rm T}$ range. Predictions based either on Color Glass Condensate calculations [16], or next-to-leading order (NLO) pQCD calculations of D-meson production, including EPS09 nPDFs [17], or $k_{\rm T}$-broadening [18] with CNM energy loss [19] can describe the measurement by considering only initial-state effects, but the comparison is not conclusive, due to the sizeable uncertainties at low $p_{\rm T}$. The analysis of D-meson production was also carried out in four event activity classes and the nuclear modification factor was calculated as: $Q_{\rm pPb} = \frac{({\rm d}N^{\rm D}/{\rm d}p_{\rm T})^{\rm cent}_{\rm pPb}}{\langle T^{\rm cent}_{\rm pPb}\rangle \times ({\rm d}\sigma^{\rm D}/{\rm d}p_{\rm T})_{\rm pp}}$, where $({\rm dN^D}/{\rm d}p_{\rm T})^{\rm cent}_{\rm pPb}$ is the yield of D mesons in p–Pb collisions in a given event activity class, $({\rm d}\sigma^{\rm D}/{\rm d}p_{\rm T})_{\rm pp}$ is the cross section of D mesons in pp collisions at the same $\sqrt{s}$, and $\langle T^{\rm cent}_{\rm pPb}\rangle$ is the average nuclear overlap function in a given event activity class. In contrast to the multiplicity-integrated $R_{\rm pPb}$, $Q_{\rm pPb}$ is influenced by potential biases in the centrality estimation that are not related to nuclear effects [20]. The least biased estimator is based on the energy deposited by nuclear fragments in the Pb-going side (ZNA), measured with the Zero Degree Neutron Calorimeter. The D-meson $Q_{\rm pPb}$ estimated with the ZNA is consistent within uncertainties with binary collision scaling of the yield in pp collisions, it is independent of the event activity, as it is shown in Fig 2 (right). D-meson per-event yields were also studied in p–Pb collisions as a function of the relative charged-particle multiplicity, in a similar way as for the pp data. The D-meson self-normalized yields increase with the relative charged-particle

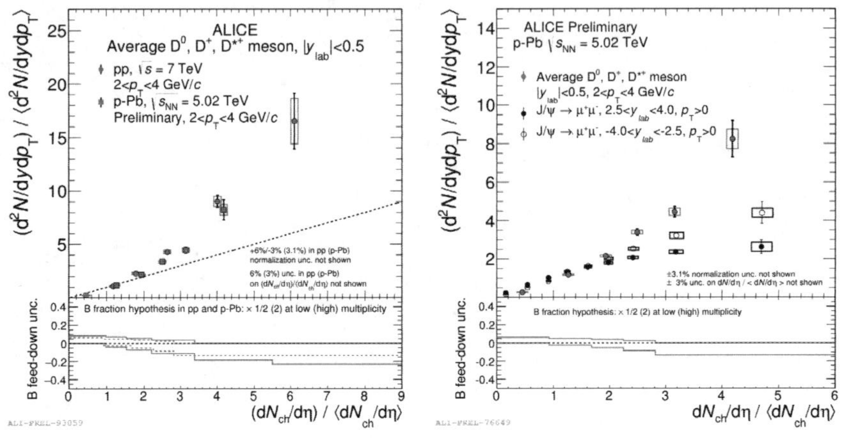


**Figure 3.** Left: Self-normalized D-meson yield as a function of multiplicity in pp (red) and p–Pb (green) collisions. Right: Self-normalized D-meson and $J/\psi$ yields in p–Pb collisions.

multiplicity, and the increase is independent of $p_{\mathrm{T}}$ within the measurement uncertainties, Fig 3 (left). This behavior is similar to that of the corresponding measurements in pp collisions. The increasing trend is present also in the $J/\psi$ measured in p–Pb collisions at backward and forward rapidity, Fig 3 (right). In p–Pb collisions, in addition to the MPI, the contribution from multiple binary nucleon–nucleon collisions should also be considered.

## 5. Conclusions

D-meson self-normalized yields show an increasing trend with increasing multiplicity, and the trend is similar for pp and p–Pb collisions. The $Q_{\mathrm{pPb}}$ measured in event-activity classes defined with the least-biased estimator (ZNA) are consistent with the multiplicity-integrated $R_{\mathrm{pPb}}$ and with unity. This suggests that there is no evidence of a deviation of D-meson production in p-Pb from binary scaling, even in high event-activity collisions.

## References

[1] Eskola K J, Paukkunen H, Salgado C A, 2009 *JHEP* **0904** 065
[2] Lev M and Petersson B, 1983 *Z.Phys.* **C21** 155
[3] Vitev I, 2007 *Phys.Rev.* **C75** 064906
[4] Aguilar-Benitez A *et. al.*, [NA27 Coll.] 1988 *Z.Phys.* **C41** 191
[5] Abelev B *et al.* [ALICE Coll.] 2012 *Phys.Lett.* **B712** 165-175
[6] Abelev B *et al.* [ALICE Coll.] 2015 arXiv:1505.00664
[7] Abelev B *et al.* [ALICE Coll.] 2014 *Int. J. Mod. Phys.* **A 29**, 1430044
[8] Abelev B *et al.* [ALICE Coll] 2012 *JHEP* **1201** 128
[9] Abelev B *et al.* [ALICE Coll] 2012 *JHEP* **1207** 191
[10] Cacciari M, Frixione S, Houdeau N, Mangano M L, Nason P, Ridolfi G, 2012 *JHEP* **1210** 137
[11] Kniehl B A, Kramer G, Spiesberger H, 2012 *EPJ* **C72** 2082
[12] Ferreiro E G, Pajares C, 2012 *PRC* **86** 034903
[13] Sjostrand T, Mrenna S, Skands P, 2008 *Comp. Phys.* **C178** arXiv:0710.3820
[14] Werner K, Guiot B, Karpenko I, Pierog T, 2014 *PRC* **89** 064903
[15] Abelev B *et al.* [ALICE Coll.] 2014 *Phys.Rev.Lett.* **113** 232301
[16] Fuji H, Watanabe K, 2013 *Nucl. Phys.* **A 920** 78
[17] Mangano L, Paolo N, Ridolfi G, 1992 *Phys. B* **373** 295
[18] Sharma R, Vitev I, Zhang B, 2009 *Phys. Rev.* **C 80** 054902
[19] Abelev B *et al.* [ALICE Coll.] 2012 *JHEP* **09** 112
[20] Adam J *et al.* [ALICE Coll.] 2015, *Phys. Rev.* **C 91** 064905

Journal of Physics: Conference Series **668** (2016) 012084 doi:10.1088/1742-6596/668/1/012084

# Hydrodynamic helicity and strange hyperon polarization in heavy-ion collisions

**M. Baznat**[1,2], **K. Gudima**[1,2], **A. Sorin**[1,3], **O. Teryaev**[1,3]

[1]Joint Institute for Nuclear Research, 141980 Dubna, Russia
[2]Institute of Applied Physics, Academy of Sciences of Moldova, MD-2028 Kishinev, Moldova
[3]National Research Nuclear University (MEPhI), 115409 Moscow, Russia

E-mail: `teryaev@theor.jinr.ru`

**Abstract.**
We study the P-odd effects related to the vorticity of the medium formed in noncentral heavy ion collisions. Using the kinetic Quark-Gluon String Model, we perform numerical simulations of the vorticity and hydrodynamical helicity for various atomic numbers, energies and centralities. We observe vortical structures typically occupying a relatively small fraction of the fireball volume. In the course of numerical simulations a noticeable hydrodynamic helicity was observed manifesting a specific mirror behaviour with respect to the reaction plane. The effect is maximal at the NICA and FAIR energy range.

## 1. Introduction

The local violation [1] of discrete symmetries in strongly interacting QCD matter is now under intensive theoretical and experimental investigation. The Chiral Magnetic Effect (CME) uses the (C)P-violating (electro)magnetic field emerging in heavy-ion collisions in order to probe the (C)P-odd effects in QCD matter.

There is an even more interesting counterpart of this effect, the Chiral Vortical Effect (CVE)[2] due to coupling to P-odd medium vorticity leading to the induced electromagnetic and [3] all conserved-charge currents, in particular the baryonic ones.

The zeroth component of axial current and the corresponding axial charge are related to the hydrodynamic helicity

$$H \equiv \int dV(\vec{v} \cdot \vec{w}),$$

being the projection of velocity $\vec{v}$ to vorticity $\vec{w} = curl\ \vec{v}$. This quantity is manifesting the recently discovered [4] and confirmed [5] phenomenon of separation, i.e. mirror behavior (same magnitude but different sign) in the half-spaces separated by the reaction plane.

Another important P-odd observable is the baryon polarization. The mechanisms analogous to the CVE (known as axial vortical effect, see [6] and references therein) lead to an induced axial current of strange quarks which may be converted to polarization of $\Lambda$-hyperons [3, 7, 4]. Another mechanism of this polarization is provided [8] by so-called thermal vorticity in the hydrodynamic approach.

Noncentral heavy-ion collisions could naturally generate a rotation (global or local, both related to vorticity) with an angular velocity normal to the reaction plane, which is their generic qualitative feature.

We paid special attention to the pseudoscalar characteristics of the vorticity, that is the hydrodynamic helicity $H \equiv \int dV(\vec{v} \cdot \vec{w})$ which is related to a number of interesting phenomena in hydrodynamics and plasma physics, such as the turbulent dynamo (providing possibly an additional mechanism of magnetic field generation in the later stages of heavy-ion collisions) and Lagrangian chaos. It might be compared to the analog of topological charge $Q = \int d^3x J^0(x)$ where the current $J^\mu = \epsilon^{\mu\nu\rho\gamma} u_\nu \partial_\rho u_\gamma$ (as usual, the four-velocity $u_\nu \equiv \gamma\,(1, \vec{v}\,))$ contributes to the hydrodynamic anomaly [9] and the polarization of hyperons [3, 7]. The calculation of the topological charge which is the correct relativistic generalization of the hydrodynamic helicity leads to the extra factor $\gamma^2$ in the integrand. Still as the helicity itself is a more traditional quantity, we use it for the numerical calculations.

## 2. Angular momentum conservation in the kinetic model

The natural source of the P-odd observables in heavy-ion collisions is the pseudovector of angular momentum. The question immediately emerges whether it is conserved in the course of evolution governed by Quark-Gluon String Model (QGSM)[10, 11]. To check this we calculated the angular momentum at various moments of collision taking into account both the contributions of participants and spectators.

We found that the participants carry about 20% of angular momentum and that the total angular momentum of participants and spectators is conserved with a rather good accuracy. We also observe clear correlation between the fireball angular momentum and hydrodynamic helicity, pointing to the global rotation as a source of vorticity.

## 3. Large-scale structures of vorticity fields

We study the qualitative structure of velocity and vorticity fields. Our observation is that while velocity fields represent the "small bang" picture, vorticity fields form relatively thin toroidal tire-like structures.

## 4. Hyperon polarization

We compare two rather distinct methods of determining the hyperon polarization. The first one corresponds to the earlier suggested [3] and explored [4] relation to the induced axial current while the second one follows the procedure [8] based on the thermal vorticity.

The hydrodynamical helicity should give rise to the polarization of $\Lambda-$hyperons with the sign differing for the particles with "up" and "down" $y-$components of their momenta, so that the hyperons acquire the helicity in the course of their motion transverse to the reaction plane. As we already suggested earlier [3], the effect is pronounced at moderate (NICA) energies due to a large (strange) chemical potential. The current investigation shows that, luckily, the helicity at these energies is also noticeable.

For a semi-quantitative estimate of this effect one may use the average strange chiral charge produced by the zeroth component of the respective current

$$Q_5^s = \frac{N_c}{2\pi^2} \int d^3x \mu^2(x) \epsilon^{ijk} u_i \partial_j u_k = \frac{<\mu^2> N_c H}{2\pi^2},$$

where we use the mean value theorem to extract the value of the square of strange chemical potential at some point inside the integration region and get the helicity from the remaining integral. Assuming that the strange chirality is carried by the $\Lambda-$ hyperons whose average

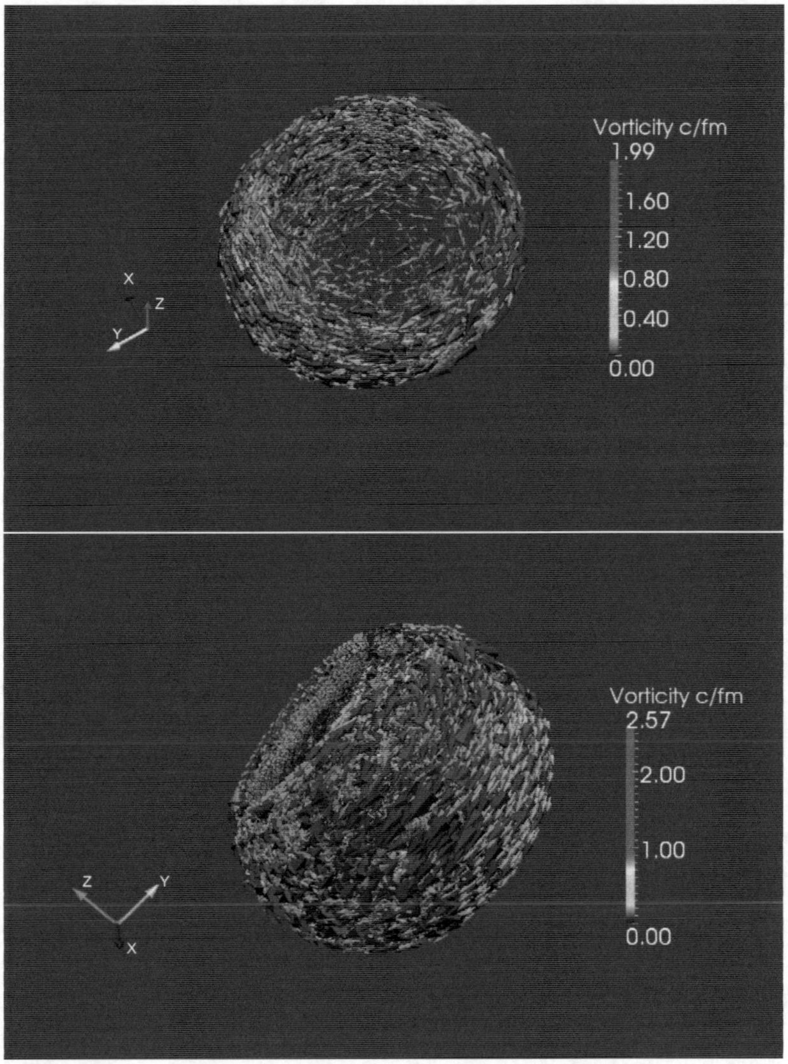


**Figure 1.** Upper: Tire-like structure of vorticity field for fixed orientation of reaction plane. Lower: Tire-like structure of vorticity field for orientations of reactions plane averaged in the whole $2\pi$ region of azimuthal angles.

number in each event is $< N_\Lambda >$ one gets the estimate for its average polarization as

$$< P_\Lambda >\sim \frac{< \mu^2 > N_c H}{2\pi^2 < N_\Lambda >}.$$

For numerical estimates at NICA energies, we take $H = 30$ fm$^2$ $(c = 1)$ and, as typical values, $< \mu^2 >= 900$ MeV$^2$ and $< N_\Lambda >= 15$ to get $< P_\Lambda >\sim 0.8\%$. This value is not large, but does not exclude the opportunity to measure the effect. Note that it is indirectly supporting the actual calculations of helicity as the obtained expression respects the density matrix positivity [15] limit $P_\Lambda \leq 1$. Should the helicity be much larger, a much larger number of hyperon production (and/or $K^*-$mesons) would be required to preserve the positivity of the density matrix. This is an example of the situation when the spin-dependent effects may be used [15] to bound the spin-averaged cross-sections from below.

Of course, more detailed calculations of polarization taking into account the spatial distribution of the chemical potential and the kinematics of produced hyperons will be required

## 5. Conclusions and Outlook

We have investigated vorticity and hydrodynamic helicity in noncentral heavy-ion collisions in the framework of the kinetic quark-gluon string model. We have observed that the vorticity is predominantly localized in a relatively thin layer ($2 \div 3\ fm$) on the boundary between the participant and spectator nucleons. This might be qualitatively understood in the spirit of the core-corona type models [13, 14].

Thus, the gradients of the velocities in the region occupied by the participants are small due to the compensation of momenta between the target and projectile particles in the c.m. frame. As a result the vorticity is substantial only in the thin transition layer between the participant (i.e., core) and the spectator (i.e., corona) regions. We found the novel effect of the helicity separation in heavy-ion collisions when it has the different signs below and above of the reaction plane. We have investigated its dependence on the type of nuclei and collision energy and observed that it is maximal in the NICA energy range. We have also calculated the degree of alignment of the velocity and vorticity which is maximal for the Beltrami flows whose relativistic generalization is currently under investigation.

This pattern may be compared with the distribution of P-odd correlation of particle momenta, so-called handedness [5] which manifests a similar separation phenomenon. The handedness averages to zero when averaged over the whole momentum space and shows mirror pattern in its separate octants.

We used the obtained values of helicity for estimates of $\Lambda$ hyperons polarization in heavy-ion collisions at NICA energy range due to the earlier suggested [3] mechanism. The resulting polarization is about 1% and may be studied experimentally. Of course, more detailed theoretical investigations are required.

In particular, the discovery of an extra $T^2$ term [16] raises again the question why polarization was not observed at RHIC. Here one may refer to the (exponential) dilution of polarization by temperature effects (similar to what happens at much larger scale in polarized targets) although this problem certainly requires further investigation.

## Acknowledgements

We acknowledge useful discussions with F. Becattini, L. Csernai and R. Usubov.

## References

[1] K. Fukushima, D.E. Kharzeev and H.J. Warringa, Phys. Rev. D **78**, 074033 (2008)
[2] D. Kharzeev and A. Zhitnitsky, Nucl. Phys. A **797**, 67 (2007)
[3] O. Rogachevsky, A. Sorin and O. Teryaev, Phys. Rev. C **82**, 054910 (2010).
[4] M. Baznat, K. Gudima, A. Sorin and O. Teryaev, Phys. Rev. C **88**, 061901 (2013).
[5] O. Teryaev and R. Usubov, arXiv:1406.4451 [nucl-th].
[6] T. Kalaydzhyan, Phys. Rev. D **89**, no. 10, 105012 (2014)?
[7] J. -H. Gao, Z. -T. Liang, S. Pu, Q. Wang and X. -N. Wang, Phys. Rev. Lett. **109**, 232301 (2012)
[8] F. Becattini, L. Csernai and D. J. Wang, Phys. Rev. C **88**, no. 3, 034905 (2013) [arXiv:1304.4427 [nucl-th]].
[9] D.T. Son and P. Surowka, Phys. Rev. Lett. **103**, 191601 (2009)
[10] V.D. Toneev, N.S. Amelin, K.K. Gudima, S.Yu. Sivoklokov, Nucl. Phys. A **519**, 463c (1990).
[11] N.S. Amelin, E.F. Staubo, L.S. Csernai et al., Phys.Rev. C **44**, 1541 (1991).
[12] L. P. Csernai, D. D. Strottman and C. Anderlik, Phys. Rev. C **85** (2012) 054901 [arXiv:1112.4287 [nucl-th]].
[13] J. Aichelin and K. Werner, Phys. Rev. C **79**, 064907 (2009); Erratum - inid. C **81**,0299902 (2010).
[14] J. Steinheimer and M. Bleicher, Phys. Rev. C **84**, 024905 (2011).
[15] X. Artru, M. Elchikh, J. -M. Richard, J. Soffer and O. V. Teryaev, Phys. Rept. **470**, 1 (2009)
[16] K. Landsteiner, E. Megias, L. Melgar and F. Pena-Benitez, JHEP **1109** (2011) 121 [arXiv:1107.0368 [hep-th]].

# Quarkonium dissociation in an anisotropic QGP

**Lata Thakur and Binoy Krishna Patra**

Department of Physics, Indian Institute of Technology Roorkee, Roorkee -247667, India

E-mail: thakurphyom@gmail.com, binoyfph@iitr.ac.in

**Abstract.** We have studied the properties of quarkonium states in an anisotropic quark-gluon plasma (QGP). We address the effects of non-perturbative confining force and momentum anisotropy on the heavy quark potential at finite temperature, which have been resolved by correcting both perturbative and non-perturbative terms of the Cornell potential through the dielectric function in terms of hard-loop resumed gluon propagator in real-time formalism. We have found that the real-part of the potential becomes stronger and thus makes the quarkonia more bound whereas the (magnitude) imaginary-part too becomes larger and thus contribute more to the thermal width, compared to the medium-contribution of the Coulomb term alone. In addition, the presence of anisotropy makes the real-part of the potential stronger but the imaginary-part is weakened slightly. The potential obtained in anisotropic medium is nonspherical. Therefore, we have used the first-order perturbation theory to estimate the shift in energy eigenvalues due to the small anisotropic correction to the energy eigenvalues from the spherically-symmetric part in isotropic medium and determine their dissociation temperatures. The presence of non-perturbative term in the potential results the quarkonia to dissociate at higher temperatures.

## 1. Introduction

At the very early time of collision, asymptotic weak-coupling enhances the longitudinal expansion substantially than the radial expansion hence the system becomes colder in the longitudinal direction than in the transverse direction. As a result, an anisotropy in the momentum space sets in and causes the parton system unstable with respect to the chromomagnetic plasma modes [1] and facilitates to isotropize the system [2,3]. The QGP with an anisotropic momentum distribution has been subject to several recent investigations as a simple model for the early stages of heavy ion collisions [1,4,5]. Anisotropy leads to an angular dependence of the $Q\overline{Q}$ potential [6]. The effect of anisotropy on the the heavy-quark potential and subsequently on the dissociation of quarkonia states have already been investigated in an anisotropic medium [6,7,8,9,10] extensively, but in these works only the Coulombic term was considered and the string term was assumed to vanish. We have investigated the effect of anisotropy on the real as well as imaginary part of the medium modified heavy quark potential obtained by correcting both the perturbative and non-perturbative terms of the $Q\bar{Q}$ potential at T=0 through the dielectric function and also studied the properties of quarkonium states. The effect of string term on the heavy quarkonium bound states was studied in [11] through a potential by correcting the full Cornell potential not its Coulomb part alone, with a dielectric function encoding the effects of the deconfined medium. However, their work was limited to an isotropic medium, whereas we have extended our work to an anisotropic medium. The general approach in our work was to consider small anisotropies $\xi \ll 1$.

## 2. Medium modification to heavy quark potential in an anisotropic medium

Anisotropic QGP system is obtained by replacing the isotropic momentum space distribution functions by [1]

$$f_{\rm aniso}(\mathbf{k}) = f_{\rm iso}\left(\sqrt{\mathbf{k}^2+\xi(\mathbf{k}.\mathbf{n})^2}\right) \quad , \tag{1}$$

where $f_{\rm iso}(k)$ is either a Bose or a Fermi distribution and $f_{\rm aniso}(\mathbf{k})$ is anisotropic distribution, which is obtained from an isotropic distribution $f_{iso}(|\mathbf{k}|)$ by removing particles with a large momentum component along the direction of anisotropy, $\mathbf{n}$ [1]. The anisotropy parameter ($\xi$) is defined by

$$\xi = \frac{\langle \mathbf{k}_T^2\rangle}{2\langle k_L^2\rangle} - 1 \ , \tag{2}$$

where $k_L = \mathbf{k}.\mathbf{n}$ and $\mathbf{k}_T = \mathbf{k} - \mathbf{n}(\mathbf{k}.\mathbf{n})$ are the components of momentum parallel and perpendicular to the direction of anisotropy, $\mathbf{n}$, respectively. In order to study the effect of anisotropy on the in-medium potential, one need to calculate first the self-energy in an anisotropic medium. With the specified anisotropic distribution function, one can compute the gluon self-energy analytically [12], which can be used to calculate the gluon propagator. By using the gluon propagator in an anisotropic medium, we have calculated the dielectric permittivity for small anisotropic limit ($\xi \ll 1$) in the temporal axial gauge from the definition [13]

$$\epsilon^{-1}(k) = -\lim_{\omega\to 0}\omega^2\frac{k_ik_j}{k^2}\Delta_{ij}(\omega,k) = \frac{k^2}{k^2+m_D^2\left(1+\frac{\xi}{6}(3\cos 2\beta_n - 1)\right)} \ . \tag{3}$$

Once we have obtained the dielectric permittivity in anisotropic medium (3), we have substituted it in the Fourier transform of the medium modified potential [11] and then evaluate its inverse Fourier transform to obtain the medium modified potential in weak anisotropic medium ($\xi \ll 1$):

$$\begin{aligned} ReV(r,\theta_n,\xi,T) &= \left(\frac{2\sigma}{m_D} - \alpha m_D\right)\frac{e^{-\hat{r}}}{\hat{r}} - \frac{2\sigma}{m_D\hat{r}} + \frac{2\sigma}{m_D} - \alpha m_D \\ &+ \xi\left(\frac{2\sigma}{m_D}\frac{e^{-\hat{r}}}{\hat{r}}\left[\frac{e^{\hat{r}}-1}{\hat{r}^2} - \frac{5e^{\hat{r}}}{12} - \frac{1}{\hat{r}} + \frac{\hat{r}}{12} - \frac{1}{12}\right]\right. \\ &- \frac{\alpha m_D}{2}\frac{e^{-\hat{r}}}{\hat{r}}\left[\frac{e^{\hat{r}}-1}{\hat{r}^2} - \frac{1}{\hat{r}} + \frac{\hat{r}}{6} - \frac{1}{2}\right] \\ &+ \left(\frac{2\sigma}{m_D}\frac{e^{-\hat{r}}}{\hat{r}}\left[3\frac{e^{\hat{r}}-1}{\hat{r}^2} - \frac{e^{\hat{r}}}{4} - \frac{3}{\hat{r}} - \frac{\hat{r}}{4} - \frac{5}{4}\right]\right. \\ &\left.\left. - \frac{\alpha m_D}{2}\frac{e^{-\hat{r}}}{\hat{r}}\left[3\frac{e^{\hat{r}}-1}{\hat{r}^2} - \frac{3}{\hat{r}} - \frac{\hat{r}}{2} - \frac{3}{2}\right]\right)\cos\,2\theta_n\right) . \end{aligned} \tag{4}$$

## 3. Imaginary part of the medium modified potential: Thermal width, $\Gamma_{\rm aniso}$

The imaginary part in anisotropic medium and its effects on the thermal widths of the resonance states have been studied [10,14,15] with the medium-modification to the perturbative (Coulomb) term only. Besides the binding energy as the dissociation criteria decay width is the another criteria to calculate the dissociation temperature. The imaginary part of the potential arises due to the singlet-to-octet transitions induced by the dipole vertex as well as due to the Landau damping in the plasma, *i.e.* scattering of the gluons with space-like momentum off the thermal excitations in the medium. We follow their work by including the medium corrections to both perturbative (Coulombic) and non-perturbative (string) terms in a weakly anisotropic medium. We have obtained the imaginary part of the potential by the leading anisotropic correction to

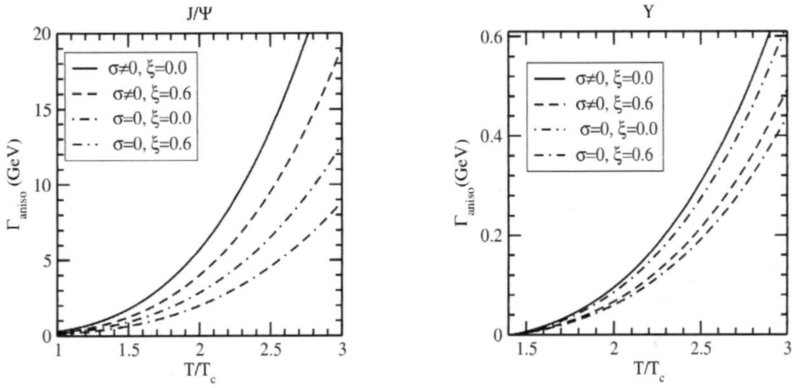


**Figure 1.** The thermal width for the $J/\psi$ and $\Upsilon$ states in the anisotropic medium.

the imaginary part of the (temporal component) symmetric propagator in real time formalism as:

$$\begin{aligned} ImV_{(\mathrm{aniso})}(r,\theta_r,T) &= -T\left(\frac{\alpha\hat{r}^2}{3}+\frac{\sigma\hat{r}^4}{30m_D^2}\right)\log(\frac{1}{\hat{r}}) \\ &+\xi T\left[\left(\frac{\alpha\hat{r}^2}{5}+\frac{3\sigma\hat{r}^4}{140m_D^2}\right)-\cos^2\theta_r\left(\frac{\alpha\hat{r}^2}{10}+\frac{\sigma\hat{r}^4}{70m_D^2}\right)\right]\log(\frac{1}{\hat{r}})\ . \end{aligned} \tag{5}$$

For weakly anisotropic medium, the imaginary part is a small perturbation to vacuum potential and thus provides an estimate for the (thermal) width:

$$\begin{aligned} \Gamma_{(\mathrm{aniso})} &= \int d^3\mathbf{r}|\Psi(r)|^2\left[\alpha T\hat{r}^2\log(\frac{1}{\hat{r}})\left(\frac{1}{3}-\xi\frac{3-\cos 2\theta_r}{20}\right)\right. \\ &\left.+\frac{2\sigma T}{m_D^2}\hat{r}^4\log(\frac{1}{\hat{r}})\frac{1}{20}\left(\frac{1}{3}-\xi\frac{2-\cos 2\theta_r}{14}\right)\right] \\ &= T\left(\frac{4}{\alpha m_Q^2}+\frac{12\sigma}{\alpha^2 m_Q^4}\right)\left(1-\frac{\xi}{2}\right)m_D^2\log\frac{\alpha m_Q}{2m_D}\ . \end{aligned} \tag{6}$$

## 4. Properties of quarkonium states in an Anisotropic Medium

We choose the matrix method [16] to solve the stationary Schrödinger equation with the isotropic part of the potential in an anisotropic medium in order to understand the in-medium properties of the quarkonium states. The corrected energy eigen value comes from the solution of Schrödinger equation of the isotropic component $V_{\mathrm{iso}}(r,T)$, using the matrix method plus the first-order perturbation due to the anisotropic component $V_{\mathrm{aniso}}(r,\theta;\xi,T)$ (4) through the quantum mechanical perturbation theory. The real and imaginary part of the binding energies for the $J/\psi$ and $\Upsilon$ states are computed numerically and Figure 2 shows their variation with different values of anisotropies. The study of the temperature dependence of the binding energies are poised to provide a wealth of information about the dissociation pattern of quarkonium states in an anisotropic thermal medium that can be used to determine the dissociation temperatures of different quarkonium states.

## Acknowledgments

We would like to thank U. S. Kakade. LT would also like to thank MHRD, India for financial assistance.

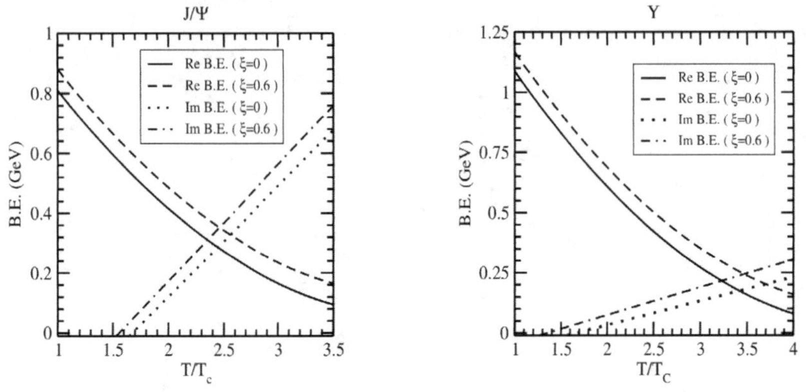


**Figure 2.** Variation of the real and imaginary part of the binding energies for $J/\psi$ and $\Upsilon$ states for different anisotropies.

| Method | State | $\xi = 0.0$ | $\xi = 0.3$ | $\xi = 0.6$ |
|---|---|---|---|---|
| Re B.E.=Im B.E. | $J/\psi$ | 1.33 | 1.34 | 1.35 |
| | $\Upsilon$ | 1.91 | 1.93 | 1.94 |
| $\Gamma$=2B.E. | $J/\psi$ | 1.02 | 1.06 | 1.12 |
| | $\Upsilon$ | 1.88 | 1.92 | 2.02 |

**Table 1.** Dissociation temperatures of $J/\psi$ and $\Upsilon$ states for different value of anisotropies with Debye mass ($m_D = 1.4\ m_D^{\rm LO}$)

[1] Romatschke P and Strickland M 2003 *Phys. Rev.* D **68** 036004.
[2] Arnold P, Lenaghan J, Moore G D, and Yaffe L G 2005 *Phys. Rev. Lett.* **94** 072302.
[3] Mrowczynski S 1993 *Phys. Lett.* B **314** 118.
[4] Romatschke P and Strickland M 2004 *Phys. Rev.* D **70** 116006.
[5] Mrowczynski S, Rebhan A and Strickland M 2004 *Phys. Rev.* D **70** 025004.
[6] Dumitru A, Guo Y and Strickland M 2008 *Phys. Lett* B **662** 37.
[7] Dumitru A, Guo Y, Mocsy A and Strickland M 2009 *Phys.Rev.* D **79** 054019.
[8] Burnier Y, Laine M and Vepsalainen M 2009 *Phys.Lett.* B **678** 86.
[9] Beraudo A, Blaizot J P, Ratti C 2008 *Nucl. Phys.* A **806** 312.
[10] Dumitru A, Guo Y and Strickland M 2009 *Phys.Rev.* D **79** 114003.
[11] Agotiya V, Chandra V and Patra B K 2009 *Phys. Rev.* C **80** 025210.
[12] Mrowczynski S and Thoma M H 2000 *Phys. Rev.* D **62** 036011.
[13] Kapusta J I and Gale C 1996 *Finite Temperature Field Theory Principle and Applications* (Cambridge University Press, Cambridge, 2nd ed.
[14] Dumitru A 2011 *Prog. Theor. Phys. Suppl.* **187** 87.
[15] Margotta M, McCarty K, McGahan C, Strickland M, Elorriaga D Y 2011 *Phys.Rev.* D **83** 105019.
[16] Ghatak A 2002 *Basic Quantum Mechanics* (Macmillan Publishers India Ltd, New Delhi) p 235.

Journal of Physics: Conference Series **668** (2016) 012086 doi:10.1088/1742-6596/668/1/012086

# Centrality Dependence of Particle Production in p–A collisions measured by ALICE

**Alberica Toia (for the ALICE Collaboration)**
Goethe University Frankfurt, GSI Darmstadt
E-mail: alberica.toia@cern.ch

**Abstract.** We present the centrality dependence of particle production in p–Pb collisions at $\sqrt{s_{NN}}$ = 5.02 TeV measured by the ALICE experiment, including the pseudo-rapidity and transverse momentum spectra, with a special emphasis on the event classification in centrality classes and its implications for the interpretation of the nuclear effects.

## 1. Introduction

Nuclear modification factors measured by ALICE in minimum bias (MB) p–Pb collisions for charged particles [1], heavy flavor and jets show no deviations from unity at high-$p_{\mathrm{T}}$, demonstrating that the observed strong suppression in Pb–Pb collisions is due to final state effects. However several measurements [2, 3, 4, 5] of particle production in the low and intermediate $p_{\mathrm{T}}$ region can not be explained by an incoherent superposition of pp collisions, but rather call for coherent and collective effects. As their strength increases with multiplicity, a more detailed characterisation of the collision geometry is needed. Moreover, a knowledge of the geometry dependence is necessary to interpret the suppression pattern of J/$\psi$ in Pb–Pb collisions relative to the effects observed in the nuclear medium produced in p–Pb collisions.

ALICE has carried out detailed studies of the centrality determination in p–Pb collisions, and the possible biases induced by the event selection on the scaling of hard processes in a selected event sample. The centrality determination consists in relating a Glauber model, which calculates the geometric properties of the event ($N_{\mathrm{coll}}$), to a measured observable related to the event activity in a specific rapidity region [6], via the conditional probability to observe a certain activity for a given $N_{\mathrm{coll}}$. Specifically, particle production measured by detectors at mid-rapidity can be modeled with a negative binomial distribution (NBD). The zero-degree energy is related to the number of the slow nucleons emitted in the nucleus fragmentation process, which we model with a Slow Nucleon Model (SNM) [7]. However, the connection of the measurement to the collision geometry has to be validated, eg by correlating observables from kinematic regions causally disconnected after the collision, or by comparing a Glauber MC with data for a known process, as eg the deuteron dissociation probability at RHIC [8].

In addition the consistency of the approach must be demonstrated. As in general the selection in a system with large relative fluctuations can induce a bias, one needs to identify the physics origin of the bias in order to correct centrality dependent measurements. In p–Pb collisions, the relative large size of the multiplicity fluctuations has the consequence that a centrality selection based on multiplicity may select a biased sample of nucleon-nucleon collisions. In essence,

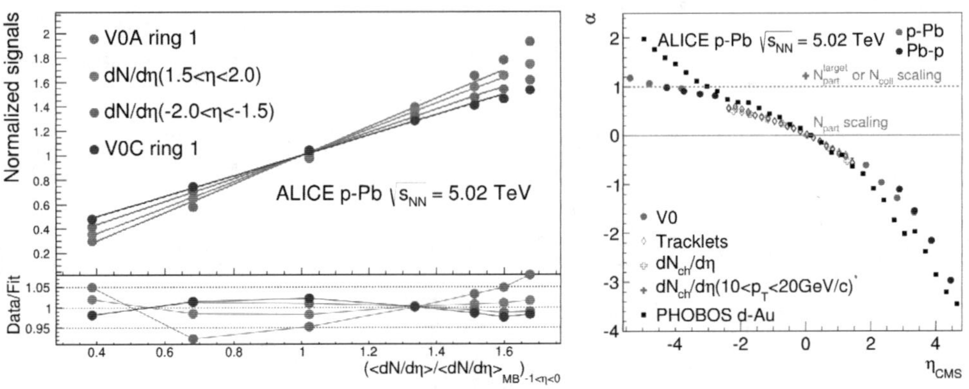

**Figure 1.** Left: Normalized signal from various observables versus the normalized charged-particle density, fit with a linear function of $N_{\text{part}}$. Right: Results from the fits as a function of the pseudorapidity covered by the observables. The red horizontal lines indicate the ideal geometrical scalings. The most central point is excluded from the fit, to avoid pile-up effects.

by selecting high (low) multiplicity one chooses not only large (small) average $N_{\text{part}}$ but also positive (negative) multiplicity fluctuations per nucleon-nucleon collision. These fluctuations are partly related to qualitatively different types of collisions, described in all recent Monte Carlo generators by impact parameter dependence of the number of particle sources via multi-parton interaction. However also other types of biases have an influence on the nuclear modification factor: the jet-veto effect, due to the trivial correlation between the centrality estimator and the presence of a high-$p_{\text{T}}$ particles in the event; the geometric bias, resulting from the mean impact parameter between nucleons rising for the most peripheral events.

## 2. The ALICE approach

The ALICE approach aims to a centrality selection with minimal bias and, therefore, uses the signal in the Zero Degree Calorimeter (ZNA). In this case we cannot establish a direct connection to the collision geometry but we can study the correlation of two or more observables that are causally disconnected after the collision, e.g. because they are separated in rapidity. In centrality classes selected by ZNA, we study the dependence of various observables in different $\eta$ and $p_{\text{T}}$ regions on the charged particle density at mid-rapidity. Fig.1 indicate a monotonic change of the scaling with rapidity. The correlation between the ZDC energy and any variable in the central part shows unambiguously the connection of these observables to geometry. Exploiting the findings from the correlation analysis described, we make use of observables that are expected to scale as a linear function of $N_{\text{coll}}$ or $N_{\text{part}}$, to calculate $N_{\text{coll}}$:

- $N_{\text{coll}}^{\text{mult}}$: the multiplicity at mid-rapidity proportional to the $N_{\text{part}}$;
- $N_{\text{coll}}^{\text{Pb-side}}$: the target-going multiplicity proportional to $N_{\text{part}}^{\text{target}}$;
- $N_{\text{coll}}^{\text{high-}p_{\text{T}}}$: the yield of high-$p_{\text{T}}$ particles at mid-rapidity is proportional to $N_{\text{coll}}$.

These scalings are used as an ansatz, in the ALICE socalled *hybrid method* to calculate $N_{\text{coll}}$, rescaling the MB value $N_{\text{coll}}^{\text{MB}}$ by the ratio of the normalized signals to the MB one. We therefore obtain 3 sets of values of $N_{\text{coll}}$. The relative difference does not exceed 10%. We have performed a consistency check, correlating ZNA and V0A centrality measurements, to establish their relation to the centrality. The $N_{\text{coll}}$ distributions for centrality classes selected with ZNA, obtained from the SNM-Glauber fit, are convolved with the NBD from the fit to the V0A distribution. These

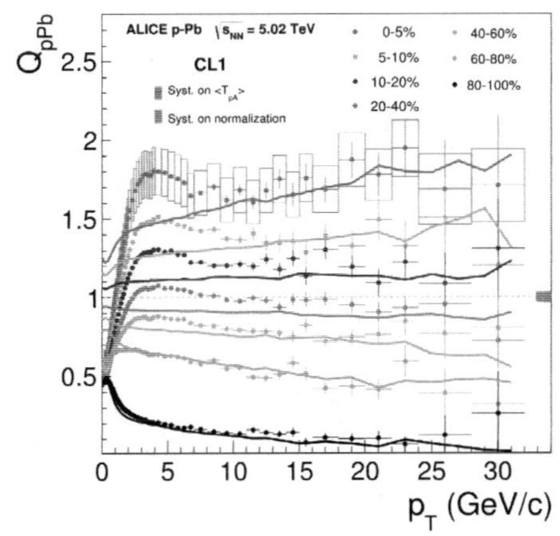


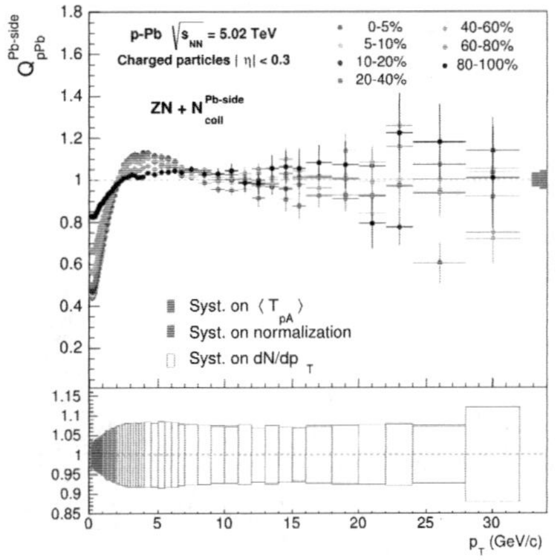


**Figure 2.** $Q_{\rm pPb}$ calculated with CL1 estimator (left), the lines are the G-PYTHIA calculations; with the hybrid method (right), spectra are measured in ZNA-classes and $N_{\rm coll}$ are obtained with the assumption that forward multiplicity is proportional to $N_{\rm part}^{\rm target}$.

are compared to the data and a good agreement is found. Moreover, a good agreement is also achieved with an unfolding procedure, where the $N_{\rm coll}$ distributions have been fitted to the data using the parameters from the NBD-Glauber fit. For each V0A distribution selected by ZNA, we find the $N_{\rm coll}$ distribution that, convolved with the $\rm NBD_{MB}$, fits the data, i.e. the parameters of the fit are the relative contributions of each $N_{\rm coll}$ bin. The existence of $N_{\rm coll}$ distributions that folded with NBD agree with measured signal distributions is a necessary condition for ZNA to behave as an unbiased centrality selection This procedure, which actually does not work for a biased centrality selection (eg CL1) shows that the energy measured by ZN is connected to the collision geometry.

## 3. Physics Results

### *3.1. Nuclear Modification Factors*

The nuclear modification factors $Q_{\rm pPb}$ calculated with multiplicity based estimator (shown in Fig.2 for CL1, where centrality is based on the tracklets measured in $|\eta| < 1.4$) widely spreads between centrality classes. They also exhibit a negative slope in $p_{\rm T}$, mostly in peripheral events, due to the jet veto bias, as jet contribution increases with $p_{\rm T}$. The $Q_{\rm pPb}$ compared to G-PYTHIA, a toy MC which couples Pythia to a p-Pb Glauber MC, show a good agreement, everywhere in 80-100%, and in general at high-$p_{\rm T}$, demonstrating that the proper scaling for high-$p_{\rm T}$ particle production is an incoherent superposition of pp collisions, provided that the bias introduced by the centrality selection is properly taken into account, as eg in G-PYTHIA. With the hybrid method, using either the assumption on mid-rapidity multiplicity proportional to $N_{\rm part}$, or forward multiplicity proportional to $N_{\rm part}^{\rm target}$, the $Q_{\rm pPb}$ shown in Fig. 2, are consistent with each other, and also consistent with unity for all centrality classes, as observed for MB collisions, indicating the absence of initial state effects. The observed Cronin enhancement is stronger in central collisions and nearly absent in peripheral collisions. The enhancement is also weaker at LHC energies compared to RHIC energies.

### *3.2. Charged particle density*

Charged particle density is also studied as a function of $\eta$, for different centrality classes, with different estimators. In peripheral collisions the shape of the distribution is almost fully

Journal of Physics: Conference Series **668** (2016) 012086 doi:10.1088/1742-6596/668/1/012086

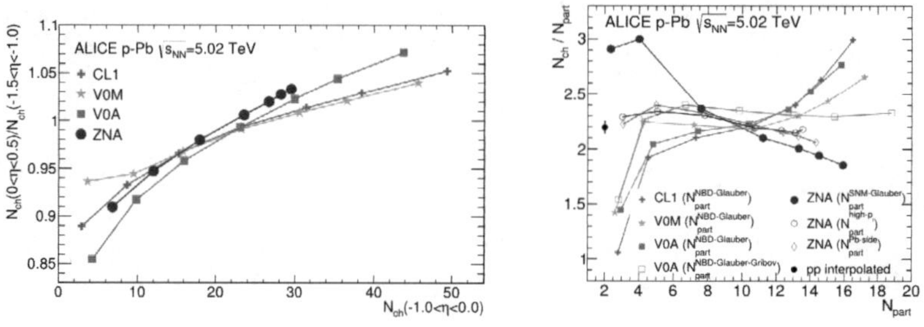


**Figure 3.** Left: Asymmetry of particle yield, as a function of the pseudorapidity density at mid-rapidity. Right: Pseudorapidity density of charged particles at mid-rapidity per participant as a function of $N_{\text{part}}$ for various centrality estimators.

symmetric and resembles what is seen in proton-proton collisions, while in central collisions it becomes progressively more asymmetric, with an increasing excess of particles produced in the direction of the Pb beam. We have quantified the evolution plotting the asymmetry between the proton and lead peak regions, as a function of the yield around the midrapidity (see Fig. 3, left): the increase of the asymmetry is different for the different estimators. Fig. 3 right shows $N_{\text{ch}}$ at mid-rapidity divided by $N_{\text{part}}$ as a function of $N_{\text{part}}$ for various centrality estimators. For Multiplicity-based estimators (CL1, V0M, V0A) the charged particle density at mid rapidity increases more than linearly, as a consequence of the strong multiplicity bias. This trend is absent when $N_{\text{part}}$ is calculated with the Glauber-Gribov model, which shows a relatively constant behavior, with the exception of the most peripheral point. For ZNA, there is a clear sign of saturation above $N_{\text{part}} = 10$, due to the saturation of forward neutron emission. None of these curves points towards the pp data point. In contrast, the results obtained with the hybrid method, using either $N_{\text{part}}^{\text{target}}$-scaling at forward rapidity or $N_{\text{coll}}$-scaling for high-$p_{\text{T}}$ particles give very similar trends, and show a nearly perfect scaling with $N_{\text{part}}$, which naturally reaches the pp point. This indicates the sensitivity of the $N_{\text{part}}$-scaling behavior to the Glauber modeling, and the importance of the fluctuations of the nucleon-nucleon collisions themselves.

## 4. Conclusions

Multiplicity Estimators induce a bias on the hardness of the pN collisions. When using them to calculate centrality-dependent $Q_{\text{pPb}}$, one must include the full dynamical bias. However, using the centrality from the ZNA estimator and our assumptions on particle scaling, an approximate independence of the mid-rapitity multiplicity on the number of participants is observed. Furthermore, at high-$p_{\text{T}}$ the p–Pb spectra are found to be consistent with the pp spectra scaled by the number of binary collisions for all centrality classes. Our findings put strong constraints on the description of particle production in high-energy nuclear collisions.

## References

[1] B. Abelev et al. (ALICE Collaboration), arXiv:1405.2737, 2014.
[2] B. Abelev et al. (ALICE Collaboration), Phys.Lett. B719, 29-41, 2013.
[3] B. Abelev et al. (ALICE Collaboration), Phys.Lett. B727, 371-380, 2013.
[4] B. Abelev et al. (ALICE Collaboration), Phys.Lett. B728, 25-38, 2014.
[5] B. Abelev et al. (ALICE Collaboration), Phys.Lett. B726, 164-177, 2013.
[6] B. Abelev et al. (ALICE Collaboration), Phys. Rev. C 88 (2013) 044909.
[7] J. Adam et al. (ALICE Collaboration), Phys. Rev. C 91 (2015) 064905
[8] A. Adare et al. (PHENIX Collaboration), Phys.Rev. C90 (2014) 3, 034902
[9] B.B. Back et al. (PHOBOS Collaboration), Phys.Rev. C72, 031901, 2005.

# Self-similarity of strangeness production in $pp$ collisions at RHIC

**M Tokarev**[1] **and I Zborovský**[2]

[1] Joint Institute for Nuclear Research, Dubna, Moscow region, Russia
[2] Nuclear Physics Institute, Academy of Sciences of the Czech Republic, Řež, Czech Republic

E-mail: tokarev@jinr.ru

E-mail: zborov@ujf.cas.cz

**Abstract.** New experimental data on transverse momentum spectra of strange particles ($K^0_S, K^-, K^*, \phi, ...$) produced in $pp$ collisions at $\sqrt{s}$ = 200 GeV obtained by the STAR and PHENIX collaborations at RHIC are analysed in the framework of $z$-scaling approach. Scaling properties of the data $z$-presentation are illustrated. Self-similarity of strange particle production is discussed. A microscopic scenario of constituent interactions developed within the $z$-scaling approach is used to study constituent energy loss, proton momentum fraction and recoil mass in dependence on the transverse momentum, strangeness, and mass of the inclusive particle. The obtained results can be useful for understanding strangeness origin, for searching for new physics with strange probes and can serve as a benchmark for complex analyses of self-similar features of strange production in heavy ion collisions.

## 1. Introduction

The production of particles with high transverse momenta from the collisions of hadrons and nuclei at high energies has relevance to constituent interactions at small scales. One of the leading principles governing these interactions is self-similarity. This principle reflects hadron structure at different scales, interaction of hadron constituents and processes of fragmentation into colorless particles. Strange particles as well as other probes (direct photons, jets, lepton pairs) play an important role to study fundamental features of the produced matter in hadron and nuclear interactions and provide information about its transition into observed particles.

The scaling behavior of particle production related to self-similarity of hadron interactions at constituent level is manifested by the $z$-scaling [1, 2]. The concept of this scaling was used for analysis of inclusive spectra obtained at the accelerators U70, SppS, SPS, ISR, Tevatron and RHIC [3]-[9]. The experimental spectra reveal striking similarity over a wide range of kinematic variables when expressed by the variable $z$. The $z$-scaling is treated as a manifestation of the self-similarity of the structure of the colliding objects, interaction mechanism of their constituents, and processes of fragmentation into real hadrons. The approach can be a suitable tool to search for phase transitions and the critical point in hadron and nuclear matter. The parameters of the $z$-scaling, $c, \delta$ and $\epsilon$, have physical interpretation as the heat capacity of the produced matter, the fractal dimension of the structure of hadrons or nuclei and the fractal dimension of the fragmentation process, respectively. Universality of the scaling is given by its flavour independence. It means that spectra of particles with different flavour content can be described by a universal function $\Psi(z)$ when using the scale transformation $z \to \alpha_F z$, $\Psi \to \alpha_F^{-1}\Psi$.

In this contribution we illustrate properties of the $z$-presentation of the transverse momentum distributions of strange particles produced in $pp$ collisions. We present results of new analysis of data on inclusive cross sections obtained by the STAR and PHENIX collaborations at RHIC. The dependence of the momentum fractions and recoil mass on the collision energy, transverse momentum and mass of the inclusive particle is used to estimate the constituent energy loss.

## 2. $z$-Scaling

The $z$-scaling has been suggested to express general regularities found in the inclusive hadron production in high energy proton-(anti)proton and nucleus-nucleus collisions. It manifests itself in the fact that the inclusive spectra of various types of particles can be described with a universal scaling function. The function $\Psi(z)$ depends on a single variable $z$ in a wide range of the transverse momentum, registration angles, collision energies and centralities. The scaling variable is expressed by the formula:

$$z = z_0 \cdot \Omega^{-1}. \tag{1}$$

Here $z_0$ and $\Omega$ are functions of kinematic variables:

$$z_0 = \frac{\sqrt{s_\perp}}{(dN_{ch}/d\eta|_0)^c m_N} \tag{2}$$

$$\Omega = (1-x_1)^{\delta_1}(1-x_2)^{\delta_2}(1-y_a)^{\epsilon_a}(1-y_b)^{\epsilon_b}. \tag{3}$$

The variable $z$ is proportional to the transverse kinetic energy of the selected binary constituent sub-process required for production of the inclusive particle with mass $m$ and its partner (antiparticle). The multiplicity density $dN_{ch}/d\eta|_0$ of charged particles in the central region, the nucleon mass $m_N$ and the parameter $c$ completely determine the dimensionless quantity $z_0$. The parameter $c$ has meaning of the specific heat of the medium produced in the collisions.

The quantity $\Omega$ is proportional to the relative number of the constituent configurations which include the binary sub-processes corresponding to the momentum fractions $x_1$ and $x_2$ of colliding hadrons (nuclei) and to the momentum fractions $y_a$ and $y_b$ of the secondary objects produced in these sub-processes. The parameters $\delta_1$ and $\delta_2$ are fractal dimensions of the colliding objects, and $\epsilon_a$ and $\epsilon_b$ stand for the fractal dimensions of the fragmentation process in the scattered and recoil direction, respectively. For unpolarized processes we assume the later to have the same value $\epsilon_a = \epsilon_b = \epsilon_F$ which depends on the type of the inclusive particle. The selected binary sub-process, which results in production of the inclusive particle and its recoil partner (antiparticle), is defined by the maximum of $\Omega(x_1, x_2, y_a, y_b)$ with the kinematic constraint

$$(x_1P_1 + x_2P_2 - p/y_a)^2 = M_X^2. \tag{4}$$

Here $M_X = x_1M_1 + x_2M_2 + m/y_b$ is the mass of the recoil system in the sub-process. The 4-momenta of the colliding objects and the inclusive particle are $P_1, P_2$ and $p$, respectively. The microscopic scenario of constituent interactions developed in the framework of $z$-scaling is based on dependencies of the momentum fractions on the collision energy, transverse momentum and centrality. The scaling variable $z$ has property of a fractal measure. It grows in a power-like manner with the increasing resolution $\Omega^{-1}$ defined with respect to the constituent sub-processes which satisfy (4). The scaling function $\Psi(z)$ is expressed in terms of the inclusive cross section $Ed^3\sigma/dp^3$, multiplicity density $dN/d\eta$, and total inelastic cross section $\sigma_{in}$. All these quantities are measurable for the inclusive reaction $P_1 + P_2 \to p + X$. The function $\Psi(z)$ is determined by the expression:

$$\Psi(z) = -\frac{\pi}{(dN/d\eta)\sigma_{in}} J^{-1} E\frac{d^3\sigma}{dp^3}. \tag{5}$$

Here $J$ is Jacobian of the transformation from the variables $\{p_T^2, y\}$ to $\{z, \eta\}$. The function $\Psi(z)$ is normalized as follows:

$$\int_0^\infty \Psi(z)dz = 1. \tag{6}$$

Equation (6) allows us to interpret $\Psi(z)$ as probability density of the production of the inclusive particle with the corresponding value of the variable $z$.

## 3. Scaling properties of strangeness production

The self-similarity of production of different hadron species in $pp$ collisions provides the basis for analysing scaling properties of the hadron spectra for more complicated systems. The hadrons with strange quarks constitute a special interest as they are the lightest particles with quantum numbers which are absent in the net amount in the initial state.

Figure 1(a) shows $z$-presentation of the transverse momentum spectra of strange mesons and baryons measured in $pp$ collisions in the central rapidity region at RHIC. The symbols representing data on differential cross sections include baryons which consist of one, two and three strange valence quarks. The open circles and diamonds correspond to the respective spectra of $K_s^0$ and $K^-$ mesons measured by the STAR collaboration [10, 11]. The data on the neutral short-lived $K^*(892)$ resonance obtained by the STAR [12] and PHENIX [13] collaborations are depicted by the triangles down and triangles up, respectively. The open squares correspond to the PHENIX data on $\phi$-meson production [14] detected in the $K^+K^-$ and $e^+e^-$ decay channels. The $z$-presentation of spectra of strange baryons shown in the figure is based on data collected by the STAR collaboration [10]. The distributions of $\Lambda$, $\Xi^-$ and $\Omega$ baryons are depicted by stars, black circles and black squares, respectively. The spectra of strange baryon resonances $\Sigma^*(1385)$ (full triangles up) and $\Lambda^*(1520)$ (full triangles down) were taken from [15].

The solid line represents the scaling curve obtained from analysis of pion production in $pp/pp$ collisions ($\epsilon_\pi = 0.2$, $\alpha_\pi = 1$) over a wide range of kinematic variables. The curve is consistent with the energy, angular and multiplicity independence of the scaling function for different hadrons. One can see that the corresponding fragmentation dimension $\epsilon_F$ for strange mesons is larger than for pions, suggesting larger energy loss by production of mesons with strangeness content. The fragmentation dimension for strange baryons grows with the number of the strange valence quarks. This is connected with the increase of energy losses.

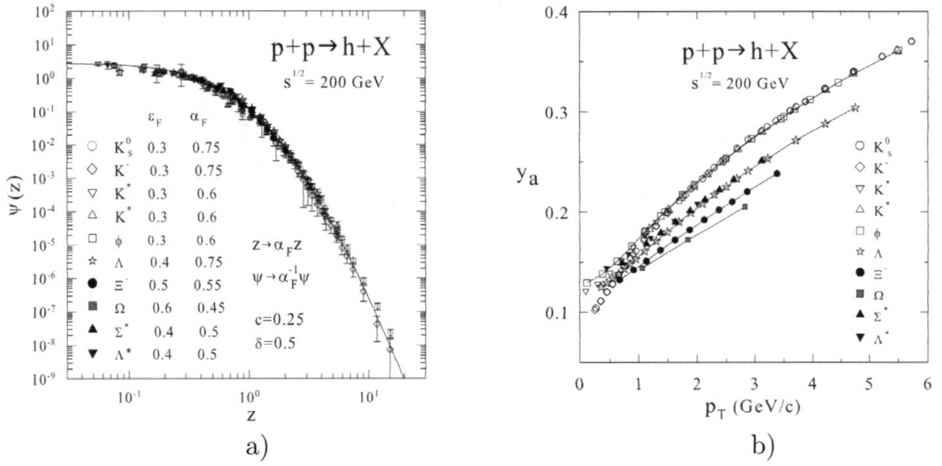


**Figure 1.** (a) Inclusive transverse momentum spectra of strange particle production in $pp$ collisions in $z$-presentation. (b) Relative energy loss $\Delta E/E = (1 - y_a)$ in dependence on $p_T$.

The $p_T$-dependence of the energy loss by production of strange hadrons is shown in Fig.1(b). The relative energy loss $\Delta E/E = (1 - y_a)$ depends on value of the fragmentation dimension $\epsilon_F$. As one can see, the relative energy loss decreases with increasing transverse momentum for all particles. For particles with a given $p_T > 1$ GeV/c, the energy loss is larger for strange baryons than for strange mesons. The growth indicates increasing tendency with larger number of strange valence quarks inside the strange baryon. Such a tendency corresponds to the $z$-scaling universality (Fig.1(a)) for hadron production including strange mesons and baryons.

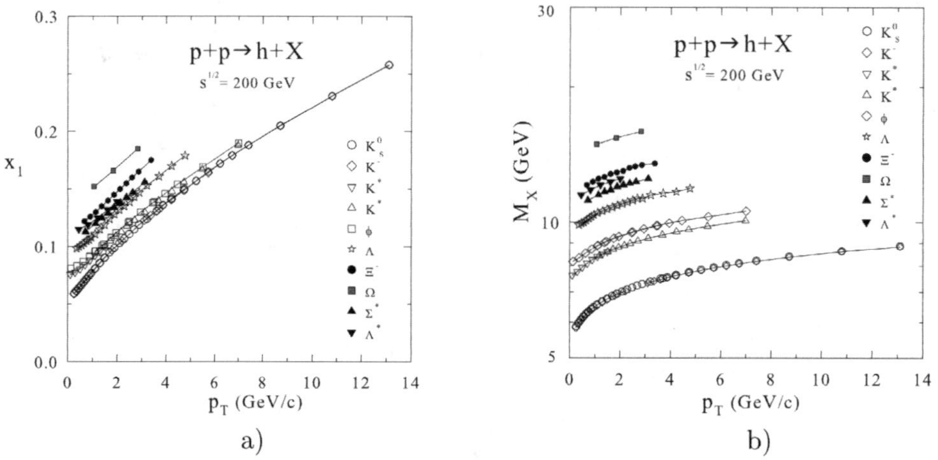


**Figure 2.** The $p_T$-dependence of the momentum fraction $x_1$ (a) of the colliding proton and the recoil mass $M_X$ (b) for production of strange hadrons.

Figure 2(a) shows $p_T$-dependence of the momentum fraction $x_1$ for production of different hadrons. The fraction $x_1$ characterizes amount of energy (momentum) of the incoming proton carried by the interacting constituent which underlies production of the inclusive particle with momentum $p_T$. One can see that, for given strangeness and $p_T$, the value of $x_1$ increases with the mass of the strange meson. The same tendency is observed for the recoil mass $M_X$ (Fig.2(b)).

In summary, we have analysed new transverse momentum spectra of strange particles measured by the STAR and PHENIX collaborations in $pp$ collisions at $\sqrt{s}$=200 GeV using data $z$-representation. The analysis confirmed flavor independence of the scaling function $\Psi(z)$. The universality of description of the spectra is consistent with a hierarchy of the energy loss in dependence on hadron strangeness and for a given strangeness with the increasing tendency of the recoil mass $M_X$ and the constituent energy with the mass of the strange hadron.

[1] Zborovský I and Tokarev M V 2007 *Phys. Rev.* D **75** 094008
[2] Zborovský I and Tokarev M V 2009 *Int. J. Mod. Phys.* A **24** 1417
[3] Tokarev M V 2006 *Phys. Part. Nucl. Lett.* **3** 7
[4] Tokarev M V 2007 *Phys. Part. Nucl. Lett.* **4** 676
[5] Tokarev M V and Zborovský I 2012 *Phys. At. Nucl.* **7** 700
[6] Tokarev M V and Zborovský I 2010 *Phys. Part. Nucl. Lett.* **7** 171
[7] Tokarev M V (for the STAR Collab.) 2011 *Phys. At. Nucl.* **74** 799
[8] Tokarev M V and Zborovský I 2012 *J. Mod. Phys.* **3** 815
[9] Tokarev M V and Zborovský I 2013 *Nucl. Phys. Proc. Suppl.* B **245** 231
[10] Abelev B I *et al* 2007 *Phys. Rev.* C **75** 064901
[11] Agakishiev G *et al* 2012 *Phys. Rev. Lett.* **108** 072302
[12] Adams J *et al* 2005 *Phys. Rev.* C **71** 064902
[13] Adare A *et al* 2014 *Phys. Rev.* C **90** 054905
[14] Adare A *et al* 2014 *Phys. Rev.* D **83** 052004
[15] Abelev B I *et al* 2006 *Phys. Rev. Lett.* **97** 132301

15th International Conference on Strangeness in Quark Matter (SQM2015) IOP Publishing
Journal of Physics: Conference Series **668** (2016) 012088 doi:10.1088/1742-6596/668/1/012088

# Heavy Hadrons in Dense Matter

**Laura Tolos[1,2], Carmen Garcia-Recio[3], Carlos Hidalgo-Duque[4], Juan Nieves[4], Olena Romanets[5], Lorenzo Luis Salcedo[3] and Juan M. Torres-Rincon[6]**

[1]Instituto de Ciencias del Espacio (IEEC/CSIC), Campus UAB, Carrer de Can Magrans s/n, 08193 Cerdanyola del Valles, Spain
[2]Frankfurt Institute for Advanced Studies, Johann Wolfgang Goethe University, Ruth-Moufang-Str. 1, 60438 Frankfurt am Main, Germany
[3]Departamento de Física Atómica, Molecular y Nuclear, and Instituto Carlos I de Física Teórica y Computacional, Universidad de Granada, E-18071 Granada, Spain
[4]Instituto de Física Corpuscular (centro mixto CSIC-UV), Institutos de Investigación de Paterna, Aptdo. 22085, 46071, Valencia, Spain
[5]KVI, University of Groningen, Zernikelaan 25, 9747AA Groningen, The Netherlands
[6]Subatech, UMR 6457, IN2P3/CNRS, Université de Nantes, École de Mines de Nantes, 4 rue Alfred Kastler 44307, Nantes, France

E-mail: `tolos@ice.csic.es`

**Abstract.** We study the behavior of dynamically-generated baryon resonances with heavy-quark content within a unitarized coupled-channel theory in matter that fulfills heavy-quark spin symmetry constraints. We analyze the implications for the formation of charmed mesic nuclei and the propagation of heavy mesons in heavy-ion collisions from RHIC to FAIR.

## 1. Introduction

The study of the properties of hadrons under extreme conditions is one of the main research activities of several experimental programs and, in particular, of the forthcoming FAIR (Germany) project. The aim is to move from the light-quark sector to the heavy-quark domain and to face new challenges where charm and new symmetries, such as heavy-quark symmetries, will play a significant role.

One of the primary goals is to understand the nature of newly discovered states and, in particular, baryonic states with charm and bottom degrees of freedom. In that respect, approaches based on unitarized coupled-channel dynamics have shown a tremendous success in the past. Recently, a unitarized coupled-channel scheme that incorporates heavy-quark spin symmetry (HQSS) [1] constraints has been developed [2, 3, 4, 5, 6, 7, 8]. HQSS is a proper symmetry of the strong interaction that appears when the quark masses become larger than the typical confinement scale. Furthermore, nuclear medium corrections have been implemented [9, 10, 11] to study the properties of the newly discovered heavy baryonic states in dense matter and their influence on heavy mesons in nuclear matter and nuclei.

In this work we study the properties of heavy hadrons in dense matter. We aim at investigating nuclear medium effects on dynamically-generated heavy baryonic resonances and the consequences for the formation of charmed mesic nuclei as well as the propagation of heavy mesons under extreme conditions.

**Table 1.** Compositeness of $\Lambda$ states in the charm (left table) and bottom (right table) sectors (taken from Ref. [8]). The masses ($M_R$) and widths ($\Gamma_R$) are given in MeV.

| State | $J^P$ | $M_R$ | $\Gamma_R$ | $1-Z$ | Channel | $\|g_i\|$ | $X_i$ |
|---|---|---|---|---|---|---|---|
| $\mathbf{\Lambda_c(2595)}$ | $\frac{1}{2}^-$ | 2619 | 1.2 | 0.878 | $\pi\Sigma_c$ | 0.31 | −0.012 |
| | | | | | $DN$ | 3.49 | 0.275 |
| | | | | | $D^*N$ | 5.64 | 0.465 |
| $\mathbf{\Lambda_c(2595)}$ | $\frac{1}{2}^-$ | 2617 | 90 | 0.401 | $\pi\Sigma_c$ | 2.36 | 0.325 |
| | | | | | $DN$ | 1.64 | 0.027 |
| | | | | | $D^*N$ | 1.43 | 0.024 |
| $\mathbf{\Lambda_c(2625)}$ | $\frac{3}{2}^-$ | 2667 | 55 | 0.365 | $\pi\Sigma_c^*$ | 2.19 | 0.268 |
| | | | | | $D^*N$ | 2.03 | 0.057 |

| State | $J^P$ | $M_R$ | $\Gamma_R$ | $1-Z$ | Channel | $g_i$ | $X_i$ |
|---|---|---|---|---|---|---|---|
| $\mathbf{\Lambda_b(5912)}$ | $\frac{1}{2}^-$ | 5878 | 0 | 0.956 | $\pi\Sigma_b$ | 0.04 | 0.000 |
| | | | | | $\bar{B}N$ | −4.55 | 0.205 |
| | | | | | $\bar{B}^*N$ | −7.70 | 0.539 |
| $\mathbf{\Lambda_b(5912)}$ | $\frac{1}{2}^-$ | 5949 | 0 | 0.865 | $\pi\Sigma_b$ | 1.31 | 0.698 |
| | | | | | $\bar{B}N$ | −2.90 | 0.096 |
| | | | | | $\bar{B}^*N$ | 1.91 | 0.038 |
| $\mathbf{\Lambda_b(5920)}$ | $\frac{3}{2}^-$ | 5963 | 0 | 0.818 | $\pi\Sigma_b^*$ | 1.54 | 0.581 |
| | | | | | $\bar{B}^*N$ | 4.16 | 0.185 |

## 2. Compositeness of dynamically generated heavy $\Lambda$ states

Heavy baryonic states are dynamically generated by the scattering of mesons and baryons within a unitarized coupled-channel approach. In this work we employ a model that explicitly incorporates HQSS [1]. HQSS predicts that all types of spin interactions involving heavy quarks vanish for infinitely massive quarks, thus, connecting vector and pseudoscalar mesons containing heavy quarks. Furthermore, chiral symmetry fixes the lowest order interaction between Goldstone bosons and other hadrons by means of the Weinberg-Tomozawa (WT) term. This predictive model includes all basic hadrons (pseudoscalar and vector mesons, and $1/2^+$ and $3/2^+$ baryons) and it reduces to the WT interaction in the sector where Goldstone bosons are involved while incorporating HQSS in the sector where heavy quarks participate. This scheme is justified in view of the reasonable outcome of the SU(6) extension in the three-flavor sector [12] and on a formal plausibleness in the vector-meson exchange picture of the interaction in the heavy pseudoscalar meson-baryon sector.

The extended WT model with HQSS constraints is used as the kernel of the on-shell Bethe-Salpeter equation in coupled channels so as to calculate the scattering amplitudes. The poles of the scattering amplitudes are the dynamically-generated heavy baryonic resonances. In this work we present results in the sector with heavy (charm/bottom) (H), strange (S) and isospin (I) content such as $H = 1, S = 0, I = 0$, where the $\Lambda$ states are found [8].

We study the generalized Weinberg's sum rule [13] to estimate the importance of the different meson-baryon channels for the generation of the $\Lambda$ states. In Table 1 we show the spin-parity ($J^P$), mass ($M_R$) and width ($\Gamma_R$) of the different charmed and bottomed $\Lambda$ states, together with the absolute value of the coupling to the dominant meson-baryon channels ($|g_i|$). The quantity $\sum_i X_i = 1 - Z$ represents the *compositeness* of the hadronic state in terms of all the considered channels, and $Z$ is referred to as its *elementariness*. A small value of $Z$ indicates that the state is well described by the contributions explicitly considered, namely, $s$-wave meson-baryon channels, while a larger value of $Z$ indicates that, for that state, significant pieces of information are missing in the model.

We obtain two $J^P = 1/2^-$ and one $J^P = 3/2^-$ $\Lambda$ states. We find that the $\Lambda$ states which are bound states (the three $\Lambda_b$) or narrow resonances (one $\Lambda_c(2595)$) are well described as molecular states composed of $s$-wave meson-baryon pairs. The $1/2^-$ wide $\Lambda_c(2595)$ as well as the $3/2^-$ $\Lambda_c(2625)$ states display smaller compositeness. With respect to the detailed composition of the states, we find that the first $\Lambda(1/2^-)$ states of each flavor couple strongly to pseudoscalar-$N$ and vector-$N$ channels. For the second $\Lambda(1/2^-)$ states, the main observation is its sizable coupling to the lightest channels $\pi\Sigma_c(\Sigma_b)$. Another observation is the similar structure of the second $\Lambda(1/2^-)$ and $\Lambda(3/2^-)$ states, which appear as HQSS or spin-flavor partners.

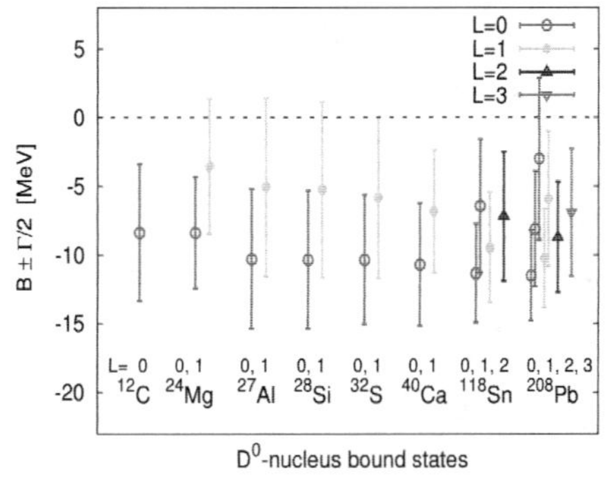


**Figure 1.** $D^0$-nucleus bound states (taken from Ref. [10]).

## 3. Charmed mesons in nuclei

The in-medium modifications of the dynamically-generated $\Lambda$ states (and others) have important consequences on the properties of heavy mesons in matter. In particular, the properties of open-charm mesons in matter have been object of recent analysis due to the consequences for charmonium suppression. Moreover, the existence of charmed meson bound states in nuclei have been predicted in $^{208}$Pb [14], using an attractive $D$ and $\bar{D}$ -meson potential in the nuclear medium within a quark-meson coupling (QMC) model. The experimental observation of these bound states, though, can be problematic since their widths could be very large compared to the separation of the levels.

Within our model, we obtain that the $D^0$-nucleus states are weakly bound (see Fig. 1), in contrast to previous results using the QMC model. Moreover, those states have significant widths [10], in particular, for $^{208}$Pb. Only $D^0$-nucleus bound states are possible since the Coulomb interaction forbids the formation of bound states for $D^+$ mesons. As for $\bar{D}$ mesons in nuclei, not only $D^-$ but also $\bar{D}^0$ bind in nuclei [11] .

The information on bound states is very valuable for gaining insight into the meson-nucleus interaction at the PANDA experiment at FAIR. Nevertheless, the experimental observation of $D$ and $\bar{D}$-meson bound states is a difficult task. Open-charm mesons with high momenta would be produced in antiproton-nucleus collisions at PANDA and it is a challenge to bind them in nuclei [10].

## 4. Heavy-meson propagation in hot dense matter

Information on the properties of heavy mesons in matter can be also achieved by analyzing the heavy-meson propagation in the hot and dense medium created in heavy-ion collisions [15, 16]. The heavy-meson propagation can be studied by means of solving the corresponding Fokker-Planck equation. The two relevant quantities to be determined are the drag ($F_i$) and diffusion coefficients ($\Gamma_{ij}$) of heavy mesons in hot dense matter. These are obtained from an effective field theory that incorporates both the chiral and HQSS in the meson [17] and baryon sectors [18, 19].

One interesting observable is the behaviour of the spatial diffusion coefficient $D_x$ that appears in Fick's diffusion law in medium, given in terms of the scalar $F(p)$ and $\Gamma(p)$ coefficients [18, 19]. In Fig. 2 we show $2\pi T D_x$ for $D$ and $\bar{B}$ mesons following isentropic trajectories ($s/n_B$=ct) from RHIC to FAIR energies. For the $D$ mesons, we observe that the dependence of the $2\pi T D_x$ on the entropy per baryon is similar in the hadronic and quark phase (below and above the transition temperature of $T_c \sim 150$MeV, respectively). The possible matching between curves in both phases for a given $s/n_B$ seems to indicate the possible existence of a minimum in the

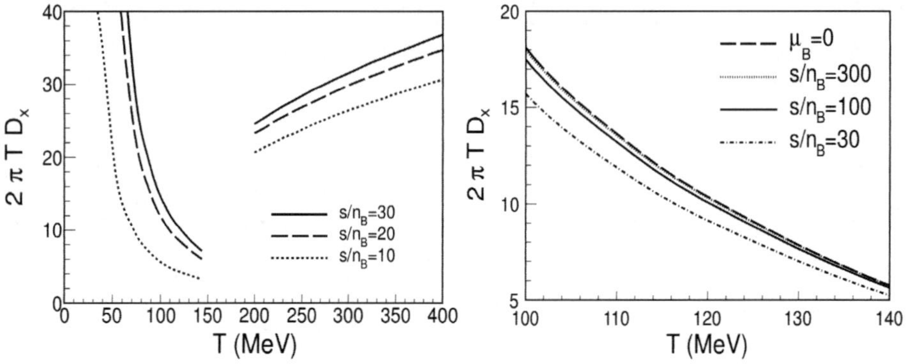


**Figure 2.** The coefficient $2\pi T D_x$ for $D$ meson (left plot) and $\bar{B}$ meson (right plot) (taken from Ref. [18, 19]). For recent updates in the high-temperature phase for $D$ meson, see Refs. [15, 20].

$2\pi T D_x$ at the phase transition [15, 20].

## Acknowledgments

C. H.-D. thanks the support of the JAE-CSIC Program. L.T. acknowledges support from the Ramón y Cajal Research Programme from Ministerio de Ciencia e Innovación and from FP7-PEOPLE-2011-CIG under Contract No. PCIG09-GA-2011-291679. This research was supported by Spanish Ministerio de Economía y Competitividad and European FEDER funds under contracts FPA2010-16963, FIS2011-28853-C02-02, FPA2013-43425-P, FIS2014-59386-P, FIS2014-51948-C2-1-P and FIS2014-57026-REDT, and Junta de Andalucía grant FQM-225.

## References

[1] Isgur N and Wise M B 1989 *Phys. Lett. B* **232** 113; Neubert M 1994 *Phys. Rept.* **245** 259; Manohar A V and Wise M B 2000 *Camb. Monogr. Part. Phys. Nucl. Phys. Cosmol.* **10** 1
[2] Garcia-Recio C, Magas V K, Mizutani T, Nieves J, Ramos A, Salcedo L L and Tolos L 2009 *Phys. Rev. D* **79** 054004
[3] Gamermann D, Garcia-Recio C, Nieves J, Salcedo L L and Tolos L 2010 *Phys. Rev. D* **81** 094016
[4] Romanets O, Tolos L, Garcia-Recio C, Nieves J, Salcedo L L and Timmermans R G E 2012 *Phys. Rev. D* **85** 114032
[5] Garcia-Recio C, Nieves J, Romanets O, Salcedo L L and Tolos L 2013 *Phys. Rev. D* **87** 034032
[6] Garcia-Recio C, Nieves J, Romanets O, Salcedo L L and Tolos L 2013 *Phys. Rev. D* **87** 074034
[7] Tolos L 2013 *Int. J. Mod. Phys. E* **22** 1330027
[8] Garcia-Recio C, Hidalgo-Duque C, Nieves J, Salcedo L L and Tolos L 2015 *Phys. Rev. D* **92** 034011
[9] Tolos L, Garcia-Recio C and Nieves J 2009 *Phys. Rev. C* **80** 065202
[10] Garcia-Recio C, Nieves J and Tolos L 2010 *Phys. Lett. B* **690** 369
[11] Garcia-Recio C, Nieves J, Salcedo L L and Tolos L 2012 *Phys. Rev. C* **85** 025203
[12] Gamermann G, Garcia-Recio C, Nieves J and Salcedo L L 2011 *Phys. Rev. D* **84** 056017
[13] Weinberg S 1963 *Phys. Rev.* **130** 776
[14] Tsushima K, Lu D H, Thomas A W, Saito K and Landau R H 1999 *Phys. Rev. C* **59** 2824
[15] Ozvenchuk V, Torres-Rincon J M, Gossiaux P B, Tolos L and Aichelin J 2014 *Phys. Rev. C* **90** 054909
[16] Song T, Berrehrah H, Cabrera D, Torres-Rincon J M, Tolos L, Cassing W and Bratkovskaya E 2015 *Phys. Rev. C* **92** 014910
[17] Abreu L M, Cabrera D, Llanes-Estrada F J and Torres-Rincon J M 2011 *Annals Phys.* **326** 2737
[18] Tolos L and Torres-Rincon J M 2013 *Phys. Rev. D* **88** 074019
[19] Torres-Rincon J M, Tolos L and Romanets O 2014 *Phys. Rev. D* **89** 074042
[20] Berrehrah H, Gossiaux P B, Aichelin J, Cassing W, Torres-Rincon J M and Bratkovskaya E 2014 *Phys. Rev. C* **90** 051901

# Reference calculations for subthreshold $\Xi$ production

**Boris Tomášik**[a,b] **and Evgeni E Kolomeitsev**[a]

[a] Univerzita Mateja Bela, FPV, Tajovského 40, 97401 Banská Bystrica, Slovakia

[b] Czech Technical University in Prague, FNSPE, Břehová 7, 11519 Prague 1, Czech Republic

E-mail: boris.tomasik@cern.ch

**Abstract.** We present a minimal statistical model designed for the description of rare-hadron multiplicities in nucleus-nucleus collisions at energies below the threshold of the particle production in binary elementary collisions. Differences to more conventional canonical statistical model are explained. The minimal statistical model is applied to the description of multiplicity ratios involving $\Xi$ hyperons, which are measured by the HADES collaboration at GSI-SIS. It is argued that the HADES data cannot be reproduced by the model based on the statistical equilibrium and the strangeness conservation. The data remain underpredicted even when in-medium potentials acting on hadrons are taken into account. This hints to non-equilibrium production of the $\Xi$ hyperons and their continuous freeze-out.

## 1. Introduction

Data obtained by HADES collaboration [1] show that doubly strange $\Xi$ hyperon can be produced in nuclear collisions at energies well below the production threshold for nucleon-nucleon collisions. In Ar+KCl collisions at $E_{\rm beam} = 1.76$ $A$GeV the observed ratio of $\Xi$ to $(\Lambda + \Sigma^0)$ multiplicities is $\mathcal{M}_\Xi/\mathcal{M}_{\Lambda+\Sigma^0} = (5.6 \pm 1.2^{+1.8}_{-1.7}) \times 10^{-3}$, and some $\Xi$'s are also seen in Au+Au collisions at 1.23 $A$GeV. Statistical model with canonical suppression of strangeness production predicts a value of the order $(10^{-4})$ for the ratio.

## 2. Isospin symmetry and total strangeness content

The colliding nuclei with the atomic number $A$ have usually different number of protons $(Z)$ and neutrons $(A - Z)$ and thereby the matter created in their collisions has a non-zero total isospin projections quantified by the asymmetry coefficient $\eta = (A - Z)/Z$ being different from one. In our calculations we assume that this asymmetry is reflected in the observed multiplicities of hadrons belonging to one isospin multiplet, e.g,

$$\frac{\mathcal{M}_{K^0}}{\mathcal{M}_{K^+}} = \frac{\mathcal{M}_{K^-}}{\mathcal{M}_{\bar{K}^0}} = \frac{\mathcal{M}_{\Xi^-}}{\mathcal{M}_{\Xi^0}} = \frac{\mathcal{M}_n}{\mathcal{M}_p} = \eta. \tag{1}$$

For the Ar+KCl collisions we have $\eta \approx 1.14$.

We can use the relations (1) to determine the total amount of produced $s\bar{s}$ quark pairs (distributed among hadrons). In a baryon-rich environment, created in collisions at SIS energies, the $\bar{s}$ antiquarks can only be accommodated in kaons. Furthermore, under such conditions kaons do not find partners for inelastic scattering and once produced they are guaranteed to leave the system. Thus the number of $s\bar{s}$ pairs can be reconstructed as

$$\mathcal{M}_{s\bar{s}} = \mathcal{M}_{K^+} + \mathcal{M}_{K^0} = (1 + \eta)\mathcal{M}_{K^+} \,. \tag{2}$$

Once the numbers of produced $s$ and $\bar{s}$ quarks are known we assume in our minimal statistical model [2] that the $s$ quarks are distributed among strange hadrons (anti-kaons and hyperons) with relative abundances determined by thermal and chemical equilibria and by the strangeness conservation. For our analysis we use the ratios which formally do not depend on the produced strangeness [2]

$$\frac{\mathcal{M}_{K^-}}{\mathcal{M}_{K^+}} = 2.54^{+1.21}_{-0.91} \times 10^{-2}, \quad \frac{\mathcal{M}_{\Lambda+\Sigma^0}}{\mathcal{M}_{K^+}} = 1.46^{+0.49}_{-0.37}$$
$$\frac{\mathcal{M}_{\Sigma^++\Sigma^-}}{2\,\mathcal{M}_{K^+}} = 0.30^{+0.23}_{-0.17}, \quad \frac{\mathcal{M}_{\Xi^-}}{\mathcal{M}_{K^+}\mathcal{M}_{\Lambda+\Sigma^0}} = 0.20^{+0.16}_{-0.12}. \tag{3}$$

These ratios have the same total number of strange an anti-strange quarks in both numerator and denominator.

## 3. The minimal statistical model

At subthreshold energies no strange particles are created in incident nucleon-nucleon collisions and all strangeness must be produced *during* the evolution of the hot fireball created in the collision. Creation of strangeness is a very rare event, with $K^+$ multiplicity

$$\mathcal{M}_{K^+} = (2.8 \pm 0.4) \times 10^{-2}. \tag{4}$$

The amount of $s\bar{s}$ pairs is thus far below the saturation and can be treated perturbatively. It will not only scale with the volume, but also we can assume that it will steadily grow with time. If — for simplicity — we assume ideal hydrodynamics, then the only length scale is given by $V^{1/3}$, where $V$ is the fireball volume, and it must determine also the lifetime [3]. At fixed impact parameter, the probability of creating one $s\bar{s}$ pair is given as

$$W = \lambda V^{4/3}. \tag{5}$$

For a fixed volume $V$, the probability to have $n$ of the $s\bar{s}$ pairs is Poissonian

$$P^{(n)}_{s\bar{s}} = e^{-W}\frac{W^n}{n!}. \tag{6}$$

The HADES data have been obtained after averaging over collisions with different centralities. The dependence on $V^{4/3}$ in (5) makes the averaging non-trivial. If we denote the centrality averaging by angle brackets, $\langle\cdots\rangle$, we can relate the parameter $\lambda$ to the measured kaon multiplicity as follows

$$\lambda = \frac{\langle W\rangle}{\langle V^{4/3}\rangle} = \frac{\mathcal{M}_{s\bar{s}}}{\langle V^{4/3}\rangle} = \frac{\mathcal{M}_{K^+}(1+\eta)}{\langle V^{4/3}\rangle}. \tag{7}$$

In the minimal statistical model we assume that in an event with volume $V$ and $n$ pairs of $s\bar{s}$ quarks, the $s$ quarks are distributed into hadrons according to the probability

$$P^{(n)}_a = (z^{(n)}_s)^{s_a} V\, p_a = (z^{(n)}_s)^{s_a} V\, e^{B_a\mu_B/T}\frac{m_a^2 T}{2\pi}K_2\left(\frac{m_a}{T}\right). \tag{8}$$

This is the probability to find hadron $a$ with the mass $m_a$, baryon number $B_a$ and strangeness content $s_a$. The chemical potential is determined by the fireball freeze-out density taken $0.6\rho_0$, where $\rho_0 = 0.16\,\mathrm{fm}^{-3}$ is the nuclear saturation density. The probability (8) is normalised — for each $n$ separately — by the factor $z^{(n)}_s$.

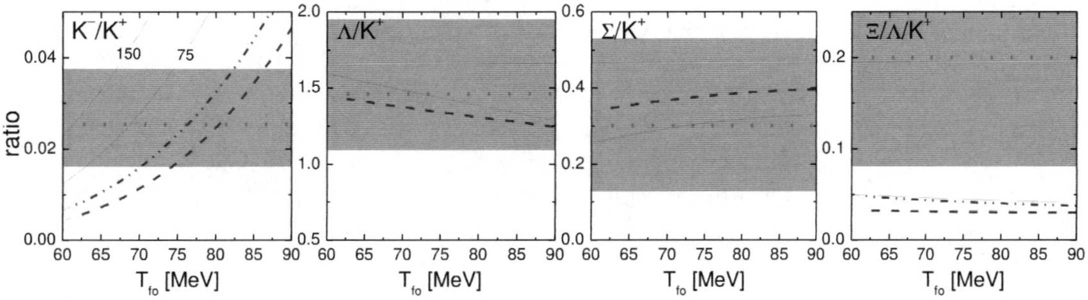


**Figure 1.** Multiplicity ratios as functions of the temperature. Red dotted lines with purple bands show the measured data values (3). Dashed lines are minimal statistical model predictions according to (10). Solid lines are predictions with in-medium potentials included. For the $\mathcal{M}_{K^-}/\mathcal{M}_{K^+}$ ratio we show results with two values of $\bar{K}$ optical potential acting on $\bar{U}$: $-75$ and $-150$ MeV (other curves are calculated with $-75$ MeV). Dash-doubly-dotted curves show ratios with only nucleon potentials included.

In total, average multiplicity of hadron species $a$ is then given by the sum

$$\mathcal{M}_a = \langle M_a^{(1)} \rangle + \langle M_a^{(2)} \rangle + \langle M_a^{(3)} \rangle + \dots \tag{9}$$

where the r.h.s. lists all contributions to the average multiplicity from events with one, two, three, etc. $s\bar{s}$ pairs. Note that $M_a^{(n)}$ is understood as the number of hadrons $a$ produced in events with $n$ $s\bar{s}$ pairs, divided by the number of all events. These multiplicities are determined with the help of $P_a^{(n)}$ and $P_{s\bar{s}}^{(n)}$. The $\Xi$ hyperons, caring two $s$ quarks, are produced only in events with two and more $s\bar{s}$ pairs.

## 4. The results

Our result for the $\Xi$ multiplicity ratio is

$$\frac{\mathcal{M}_\Xi}{\mathcal{M}_{\Lambda+\Sigma^0}\mathcal{M}_{K^+}} = \frac{\langle V^{5/3}\rangle\langle V\rangle}{2\langle V^{4/3}\rangle^2}\eta\frac{p_\Xi/(p_{\bar{K}}+p_\Lambda+p_\Sigma)}{\langle V\rangle(p_\Lambda+\frac{\eta p_\Sigma}{\eta^2+\eta+1})}\,. \tag{10}$$

In the pre-factor we collected the modifications with respect to the "usual" statistical model with the canonical suppression. This and all other calculated ratios are displayed in Fig. 1 by dashed lines as functions of temperature. We see that the calculated $\mathcal{M}_\Xi/\mathcal{M}_{\Lambda+\Sigma^0}\mathcal{M}_{K^+}$ is clearly below the measured error band. In Fig. 1 we also revised the data on $\Sigma/K^+$ ratio based on the actual isospin ratio from (1), see [2].

By trying to improve the agreement of our calculations with the data we have introduced density-dependent scalar and vector potentials $S_a$ and $V_a$ [2]. They modify the statistical distribution function by replacing the dispersion relation to $E_a(p) = \sqrt{(m_a+S_a)^2+p^2}+V_a$. The chosen values were $V_N = V_\Delta = \frac{3}{2}V_\Lambda = \frac{3}{2}V_\Sigma = 3V_\Xi = 130\,\mathrm{MeV}\rho_B/\rho_0$, and $S_N = -190\,\mathrm{MeV}\rho_B/\rho_0$. The scalar potentials for strange particles follow from the relation $S_a = (U_a - V_a(\rho_0))\rho_B/\rho_0$, where we use $U_\Lambda = -27\,\mathrm{MeV}$, $U_\Sigma = 24\,\mathrm{MeV}$, $U_\Xi = -14\,\mathrm{MeV}$, $U_{\bar{K}} = -75$ or $-150\,\mathrm{MeV}$ and $V_{\bar{K}} = 0$. This leads to changes in our theoretical results which allow to explain $K^-/K^+$, $\Lambda/K^+$ and $\Sigma/K^+$ ratios by one freeze-out temperature $\sim 70\,\mathrm{MeV}$. However, the changes are not large enough to bring the $\Xi/\Lambda/K^+$ ratio into agreement with the measured data.

It is interesting that accounting for the trigger on central collisions actually makes the disagreement even bigger. The culprit is in the volume factors which *reduce* the overall result when selection on larger volumes is made.

## 5. A prediction for Au+Au collisions at 1.23 GeV

Although the model disagrees with the Ar+KCl data, for curiosity we can make a prediction also for the Au+Au collisions recently recorded by HADES. As seen from Fig. 1, the temperature dependence of $\mathcal{M}_{\Xi}/\mathcal{M}_{\Lambda+\Sigma^0}\mathcal{M}_{K^+}$ is very weak. The crucial difference to the smaller system is in the volume which is almost five times bigger for Au+Au than for Ar+KCl. As the ratio (10) is proportional to $V^{-1}$, we expect that it will be five-times smaller in gold-induced reactions than in Ar+KCl. Below the strangeness production threshold this statement is independent of the collision energy.

## 6. Conclusions

In comparison to the canonical statistical model, the minimal statistical model assumes that strangeness must be produced during the fireball evolution and takes into account the proper scaling of rare species multiplicity with the volume. This leads to non-trivial differences between the two models and the result of the minimal statistical model on centrality-averaged multiplicities depends on a set of averaged volume factors.

In any case, the conclusion we could draw from our study is that cascade production is non-statistical. We speculate that it might be caused by immediate decoupling of the produced $\Xi$'s from the bulk matter. At an early hot stage some $\Xi$'s are produced which are later not annihilated at lower temperature. Those particles chemically decouple from the system.

We speculate that the most viable for such a production of $\Xi$ would be the strangeness recombination reactions, either involving $\bar{K}$ and a hyperon, or two hyperons. Note that we have provided parametrised cross-sections for such reactions in [4]. An interesting option appears in the $\bar{K}\Lambda$ channel, which in vacuum has a high reaction threshold of 1609 MeV. Due to effective lowering of kaon mass in the baryon-rich environment it can decrease so that eventually it hits the $\Xi^*$ resonance around 1530 MeV. This could increase the cross section for $\Xi$ production considerably.

Kinetic calculations aiming the the validation of these mechanisms are planned for the near future.

*Acknowledgements*
This work was partially supported by APVV-0050-11, VEGA 1/0469/15 (Slovakia) and MŠMT grant LG13031 (Czech Republic). We also want to acknowledge the support towards attendance of this conference by the Plenipotentiary of the Slovak Government to JINR Dubna.

## References

[1] Agakishiev G et al 2009 *Phys. Rev. Lett.* **103** 132301.
[2] Kolomeitsev E E, Tomášik B and Voskresensky D N 2012 *Phys. Rev. C* **86** 054909.
[3] Russkikh V N and Ivanov Yu B 1992 *Nucl. Phys. A* **543** 751.
[4] Tomášik B and Kolomeitsev E E 2012 *Acta Phys. Polonica B Proc. Suppl.* **5** 201.

# The ATLAS experience and its relevance to the data acquisition of the BM@N experiment at the NICA complex

**K G Tomiwa[1*], I Slepnev[2], S Bazylev[2] and B Mellado[1*] on behalf of Wits HEP group**

[1] School of Physics, University of the Witwatersrand, Johannesburg 2050, South Africa
[2] Joint Institute of Nuclear Research, Dubna, Russia

E-mail: kehinde.gbenga.tomiwa@cern.ch

**Abstract.** The quest to understand the world around us has increased the size of high energy physics experiments and the processing rate of the data output from high energy experiments. The Large Hadron Collider is the largest experimental set-up known, with ATLAS detector as one of the detectors built to record proton-proton collision at about 10 PB/s (Petabit/s) around the LHC interaction point. With the Phase-II upgrade in 2022 this data output will increase by at least 10 times higher than those of today due to luminosity increase, this poses a serious challenge on processing and storage of the data. Also the BM@N fixed target experiment is expected to have event size of about 80,000 bytes/Event, leading to huge amount of data output to be processed in real time. Experimentalists handle these challenges by developing High-throughput electronic, with the capability of processing and reducing big data to scientific data in real time. One of these high-throughput electronics is the Super Readout Driver (sROD) and ARM-based processing unit (PU) developed for ATLAS TileCal detector by the University of the Witwatersrand. The sROD is designed to process data from Tile Calorimeter at 40 MHz. This work takes a look at the architecture of the data acquisition (DAQ) system of the BM@N detectors and the adaptation of the high-throughput systems to last stage of the BM@N DAQ system.

## 1. Introduction

The field of High Energy Physics (HEP) has contributed to our understanding of the world around us. These contributions are results of various experiments, ranging from fixed target experiment to collision experiments. The size of these experimental setups has grown from table top experiments like the Rutherford experiment to giant collision experiments like the CERN Large Hadron Collider (LHC). These experiments increase our understanding of matter by probing into the constituent of matter. In collision and fixed target experiments particles interactions are observed by detectors. Active materials in detectors interact with particle to produce signals which are converted to electrical signals. These signals are shaped, amplified and processed to observable by data acquisition electronics. Physics experiments rely on these electronics, therefore they are important to our experiments. The usage of electronics poses challenge of big data. Example is the ATLAS detector records proton-proton collisions provided by the Large Hadron Collider (LHC) at CERN produces 10 PB/s data every every 50 ns, with

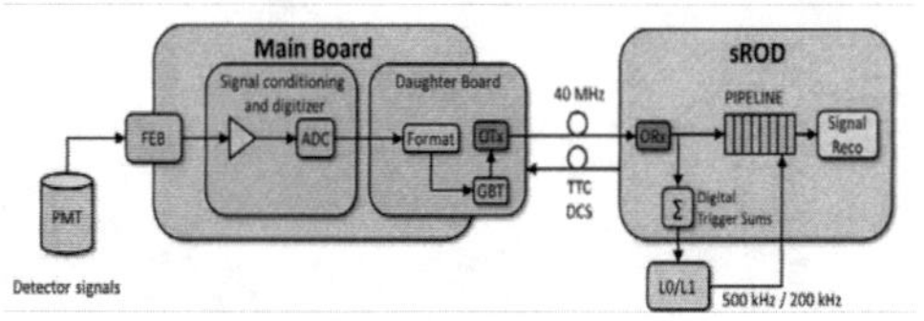


**Figure 1.** ATLAS TileCal future front-end electronics

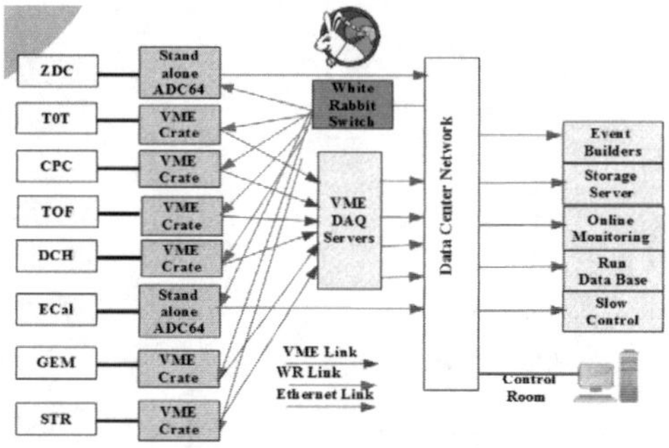


**Figure 2.** BM@N DAQ System.

the plans to upgrade the detector, the output will be about ten times its previous recording energy which is about 7 TeV (14 TeV collision energy for Phase-II) [1].

## 2. ATLAS Tile Calorimeter Detector and Upgrade

The ATLAS Tile Calorimeter (TileCal) is a hadronic calorimeter for the ATLAS experiment at Large Hadron Collider (LHC)[1], that provides precise measurement of energy and position of electrons, photons, isolated hadrons, taus and jets and measures missing transverse energy. The TileCal detector covers pseudorapidity region of $-1.7 < \eta < 1.7$ which makes it symmetric with respect to the interaction point [2].

Hadronic showers created at the interaction point of TileCal create light in the scintillating tiles, which is transmitted by wavelength shifting fibers and readout by the photomultiplier tube (PMT). The PMT acts as a transducer that converts light to electric signals, these signals are collected and digitized by the front-end electrons. In the TileCal readout architecture Fig. 1, the signals from PMT are shaped and seperated to high (HG) and low (LG) gain branches in the 3-in-1 card located on the main board. The HG and LG signal are sampled with the LHC bunch-crossing frequency of 40MHz using a 10-bit ADC in the TileCal Data Management Unit (DMU) chip which is located in the digitizer board [3]. The back-end electronics can infer the time and energy of channel's signal from the digitized samples. The L1 trigger is formed by analogue sum of signals (Signature from particles) which represent trigger tower of cell $\Delta\eta \times \Delta\phi = 0.1 \times 0.1$.

After Phase-II upgrades in 2022, the data output from the ATLAS Tile Calorimeter will increase significantly. The University of the Witwatersrand has contributed in the following ways towards the Phase-II upgrade in terms of instrumentation.

## 3. Portable Readout Module for TileCal electronics

The plan to upgrade ATLAS TileCal imposes several challenges on the TileCal's readout electronics. The Prometeo system is a test-beach system developed for full certification of the readout electronics by performing multiple tests. Prometeo has been designed to be self-contained and portable, with the ability to readout all channels at the LHC bunch crossing frequency. The Prometeo consist of several parts like the Main board (FPGA based board), Dual QSFP FMC, ADC FMC card and host to test electronics for ATLAS upgrade. The Prometeo's prototypes have been installed at CERN by the University of the Witwatersrand for further development and testing.

### *3.1. Super Readout (sROD)*

The sROD (Fig. 1) is an off-detector Electronics at the back-end of the TileCal readout chain, the module uses one Xilinx Virtex 7 and one Xilinx Kintex 7 FPGAs as the processing core. The

Journal of Physics: Conference Series **668** (2016) 012090 doi:10.1088/1742-6596/668/1/012090

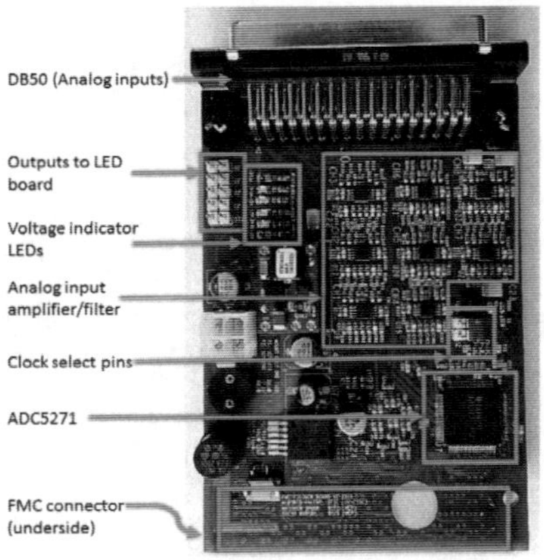


**Figure 3.** FMC ADC Trigger board [5]

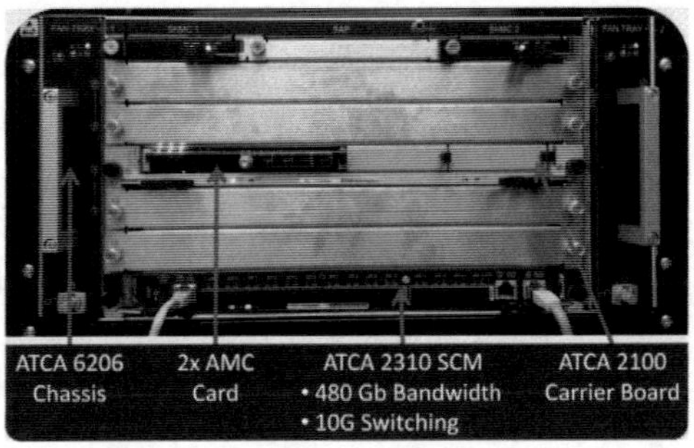


**Figure 4.** ATCA Chassis [6]

board is a double mid-size Advanced Mazzanine Card (AMC) designed to be compatible with the ATLAS Advance telecommunications computing architecture (ATCA) carrier or the micro Telecommunications computing Architecture (uTCA) system (the new replacement for current VME crates of the DAQ system). The sROD performs several functions like data processing and data reception from ADC boards; Timing, Trigger and control (TTC), Detector Control System (DCS), data reconstruction and transmission to the Read-Out Subsystem (ROS); it also performs trigger data preprocessing and transmission to the Level-1 Calorimeter trigger system (L1Calo)[4].

### *3.2. ARM-based Processing Unit (PU)*

The ARM-based Processing unit tries to address the problem of High Volume throughput computing, by arranging ARM system on chips (SoCs) in cluster configuration. This configuration enhances the performance of the PU, it can process at least 40 GB/s of raw data fed through the I/O interface (PCI-Express , XAUI or bonded SFP+Connectors) [5]

### *3.3. ADC Trigger Board*

The ADC Trigger board is designed to condition and digitize analog trigger signals coming from the TileCal front-end adder cards. The board uses 12-bit ADC5271 8-channel ADC chips to sample data at 40 Mega Sample Per Second. Serialized data is sent at the rate of 480 Mbps per channel resulting in a total data flow of 7.68 Gbps. The board is powered by a 12V power supply, from which different voltages are created (+D3.3V, +-A3.3V and +-5V) to power differential amplifiers and ADC chips [6].

### *3.4. ATCA Framework*

Because of the ATLAS Calorimeter Phase II upgrade, the frontend and backend electronics will be optimized to allow full data rate of 40 MHz to exit the detector. This optimization will include changing the current trigger system to digital signal that will be processed by digital electronics like the sROD. This calls for replacement for module manager (VME crates) by the Advanced Telecommunication Computing Architecture chassis. ATCA chassis is an intelligent shelf with high speed backend connectivity. It provides functionality which enable the integration of CERN's detector control system [7].

## 4. BM@N DAQ system and Wits Proposed contrition

The Baryonic matter at Nuclotron (BM@N) project in the Nuclotron-Based Ion collider Facility (NICA) at the Joint Institute of Nuclear Research (JINR) is an experiment proposed to the study of AU+AU collision, with beam energy of $\sqrt{S_{NN}} = 11GeV/n$ [8]. The proposed particle detectors will be used to record particle produced by the fixed target experiment. The HEP group of the university of Witwatersrand has proposed sROD as a solution to the Trigger and Data acquisition system. The DAQ system will be responsible for transporting and assembling of event data all the way from the front-end buffer to the logging disk. Because of the large size of event and limited available storage size, trigger systems are employed for event data selection. The usual way of doing this is to move event data from the on-detector electronics into the front-end buffer at the bunch crossing rate of collision (20 to 100 kHZ for BM@N). While the event data are in buffer, trigger signals are used to move data to Read-out drivers where event building is done, the ROD also records data to storage disk.

Trigger systems are usually in hierarchies depending on the size of data output, the first level trigger (L1 trigger) is clock-driven, it reads data from on-detector electronics at some time interval. While the higher level trigger (second and third level trigger) are event driven, they are usually implemented in software by executing chains of reconstruction and signature algorithms to analyse the presence of interesting event.

For BM@N experiment, signal from central event would be read at 20 kHz, the approximate event size would be in order of 60Mbps. Minimum bias event size would be 600 Mbps at 100 kHz. The Level-1 trigger acceptance of the BM@N detector will be formed by signals from Zero Degree Calorimeter (ZDC) based on centrality of event. L1 trigger are produced withing 700 to 1000 nano seconds after interaction to initiate readout of all sub component of the BM@N.

## References

[1] ATLAS Collaboration 2011 Eur. Phys. J. C721849. 63

[2] ATLAS tile calorimeter interface card, K. Anderson, A. Sen-Gupta, J. E. Pilcher, H. Sanders, F. Tang, 256 R. Teuscher, and H. Wu, in 8th Workshop on Electronics for LHC Experiments.

[3] K. Anderson, A. Gupta, F. Merritt, M. Oreglia, J. Pilcher, H. Sanders, M. Shochet, F. Tang, R. Teuscher, H. Wu, G. Blanchot, M. Cavalli-Sforza, I. Korolkov, Design of the front-end analog electronics for the ATLAS tile calorimeter, Nuclear Instruments and Methods in Physics Research, Section A: Accelerators, Spectrometers, Detectors and Associated Equipment, Volume 551, Pages 469-473.

[4] Carrio F, Castillo V, Ferrer A, Fiorini L, Hernandez Y, Higon E, Mellado B, March L, Moreno P, Reed R, Solans C, Valero A and Valla J 2013 *ATLAS Collaboration JINST 9 (2014) C02019* **2013-020**

[5] Cox M, Reed R, and Mellado B, 2015 The development of a general purpose ARM-based processing unit for the ATLAS TileCal sROD *JINST C01007* **10**

[6] A high data throughput ADC board for the PROMETEO test-bench of the ATLAS Tile Calorimeter, Matthew Spoor, Oscar Kureba, Charles Sandrock, to be published in the book of proceeding of HEPP 2015 workshop, iThemba Labs, February 2015.

[7] Robert Reed, An ATCA framework for the upgrade ATLAS readout electronics at LHC, SA Institute of Physics, ISBN: 978-0-620-653916.

[8] The data acquisition system for a fixed target experiment at NICA complex at JINR and its connection to the ATLAS TileCAl readout electronics. K. Tomiwa, I Slepnev, S Bazylev, to be published in the book of proceeding of HEPP 2015 workshop, iThemba Labs, February 2015.

# Heavy mesons in a hadronic medium: interaction and transport coefficients

**J M Torres-Rincon[1], L M Abreu[2], D Cabrera[3], O Romanets[4] and L Tolos[3,5]**

[1] Subatech, UMR 6457, IN2P3/CNRS, Université de Nantes, École de Mines de Nantes, 4 rue Alfred Kastler 44307, Nantes, France
[2] Instituto de Física, Universidade Federal da Bahia, 40210-340, Salvador, BA, Brazil
[3] Frankfurt Institute for Advanced Studies. Johann Wolfgang Goethe University, Ruth-Moufang-Strasse 1, 60438 Frankfurt am Main, Germany
[4] KVI, University of Groningen, Zernikelaan 25, 9747AA Groningen, The Netherlands
[5] Instituto de Ciencias del Espacio (IEEC/CSIC), Campus UAB Carrer de Can Magrans s/n, 08193 Cerdanyola del Valles, Spain

E-mail: `torresri@subatech.in2p3.fr`

**Abstract.** We review the recent results of heavy meson diffusion in thermal hadronic matter. The interactions of $D$ and $\bar{B}$ mesons with other hadrons (light mesons and baryons) are extracted from effective field theories based on chiral and heavy-quark symmetries. When these guiding principles are combined with exact unitarity, physical values of the cross sections are obtained. These cross sections (which contain resonant contributions) are used to calculate the drag and diffusion coefficients of heavy mesons immersed in a thermal and dense medium. The transport coefficients are computed using a Fokker-Planck reduction of the Boltzmann equation.

## 1. Introduction

In relativistic heavy-ion collisions (HICs) heavy-flavor dynamics is considered one of the cleanest processes carrying direct information from the quark-gluon plasma (QGP) phase. During the last years —and thanks to the experimental heavy-flavor results from the Relativistic Heavy-Ion Collider (RHIC) and the Large Hadron Collider (LHC)— a lot of effort has been done in the theoretical side to understand a series of observables and infer the properties of the deconfined medium. Of special interest are the elliptic flow $v_2$ and the nuclear modification factor $R_{AA}$ of $D$ mesons, and of electrons coming from the heavy hadron decays. These observables are sensitive to the microscopic details of hot matter and the interaction among their components. In particular, the transport coefficients of heavy particles affect these quantities (in a similar fashion in which $v_2$ of light systems depends on the value of $\eta/s$ and $\zeta/s$). For these reasons, interaction and transport coefficients of heavy particles are fundamental ingredients to interpret the experimental results of heavy particles and get knowledge of matter at extreme conditions.

The last stage of a HIC is described by the expansion and final decoupling of the hadronic phase. In this medium, the heavy-flavor dynamics is dominated by the propagation of charmed and bottomed mesons and baryons. Therefore, a complete description of a HIC should contain the evolution of these states in an expanding medium, and consequently, information about the slowing down of the momentum carried by them. The relative amount of momentum stopping

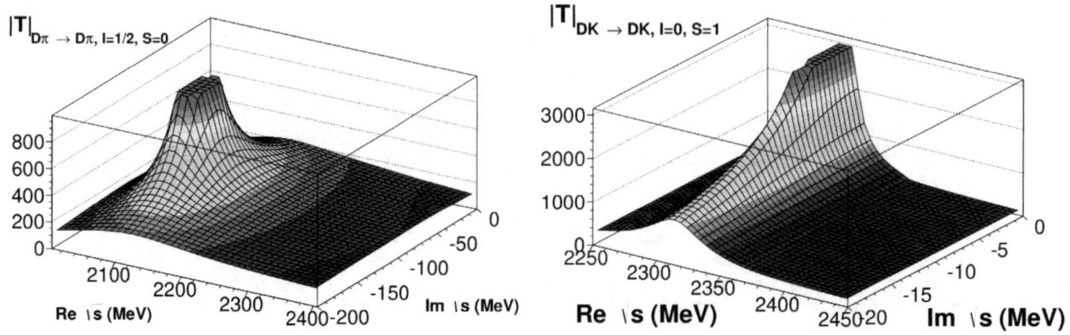


**Figure 1.** Many resonant states are generated when exact unitarity is imposed to the EFT scattering amplitudes. Left panel: Resonance in the $(I,S) = (1/2,0)$, $D\pi \to D\pi$ channel associated to the $D_0(2400)$ state. Right panel: Bound-state in the $(I,S) = (0,1)$, $DK \to DK$ channel associated to the $D^*_{s0}(2317)$ state.

and diffusion is measured by the drag force and diffusion coefficients, respectively. To compute these transport coefficients from a microscopical framework, it is necessary to define a sensible effective theory for the interaction between heavy hadrons and the particles of the thermal bath.

In this work we review our recent results on the microscopical calculation of the heavy-meson transport coefficients in a hot and dense medium using an up-to-date hadronic interaction based on effective field theories (EFTs) and demanding exact unitarity. In particular, we will present our description for the interaction of $D$ and $\bar{B}$ mesons with lighter hadronic states, as well as the transport coefficients as functions of temperature and baryochemical potential.

## 2. Heavy-meson interaction

The EFT for the $D$ and $\bar{B}$ mesons follow quite in parallel because the meson mass (hard scale) dominates over all other scales [1, 2]. For this reason the heavy-quark spin symmetry (HQSS) is one the key ingredients in the construction of the effective Lagrangian. At leading order, the interactions of $D$, $D^*$, $\bar{B}$ and $\bar{B}^*$ mesons are formally the same. This fact is reminiscent of the spin-flavor symmetry of the heavy-quark effective theory [1]. However physical masses explicitly break this symmetry, leading to different interactions. Due to the presence of the pseudo-Goldstone bosons $(\pi, K, \bar{K}, \eta)$ the chiral symmetry is also exploited in the EFT.

We construct the effective Lagrangians for both meson-meson and meson-baryon interactions. The former can be arranged in a power-counting scheme giving a systematic expansion for the effective Lagrangian [3, 2, 4]. We consider the chiral expansion up to next-to-leading order, whereas the scattering amplitudes are expanded at lowest order in the (inverse) heavy-meson-mass expansion. In the meson-baryon sector the Lagrangian is based on a $SU(6)\times$HQSS symmetry, kept at leading order. The resulting Lagrangian coincides with a Weinberg-Tomozawa contact vertex [5, 6, 7]. In both sectors, we account for all possible elastic and inelastic channels involving a heavy-meson and a lighter hadron.

We apply a unitarization scheme to ensure that the scattering-matrix elements satisfy the unitarity condition exactly. This is done by solving a Bethe-Salpeter equation for the scattering amplitude in a coupled-channel basis [8, 9]. This method provides reliable interaction rates for $D, D^*, \bar{B}, \bar{B}^*$ mesons scattering off light hadrons: $\pi, K, \bar{K}, \eta, N, \Delta$ (many other intermediate channels are also included e.g. $D_s\pi, \Lambda_c\eta$... but they are not considered here as asymptotic states).

In Fig. 1 we show two channels in which resonant states are found in the scattering amplitude

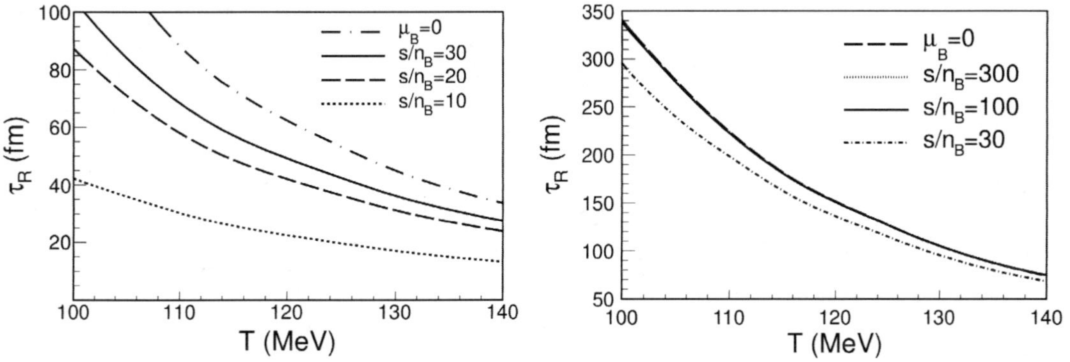


**Figure 2.** Relaxation time for $D$ mesons (left panel) and $\bar{B}$ mesons (right panel) as a function of temperature for several isentropic trajectories.

(note that the physical scattering amplitude is taken along the real-energy axis). A typical resonance —with a finite decay width— and a bound state —without decay width— are clearly seen in the left and right panels, respectively. It is important to notice that the EFT alone cannot give physical interactions without the application of a unitarization method, because the unitarization might give rise to resonances or bound states which produce strong modifications to the cross sections.

## 3. Transport coefficients and application

The scattering amplitudes are now used to compute the transport coefficients of heavy mesons. The momentum distribution (initially out of equilibrium) evolves in time because of the rearrangement of momentum due to collisions. If we focus on the first two moments of the distribution —i.e. average momentum and the distribution width—, rather than the distribution itself, their time evolution is controlled by the drag force $F$ (that stops the heavy meson) and two diffusion coefficients $\Gamma_0, \Gamma_1$ (measuring the momentum broadening along the parallel and transverse directions of motion). These coefficients determine the Fokker-Planck equation governing the time evolution of the distribution function. Derived quantities are the relaxation time of the average momentum $\tau_R = 1/F$ and the spatial diffusion coefficient $D_x$ (which measures the spreading of heavy particles in the medium). We plot these coefficients for a heavy-meson momentum of 100 MeV as functions of temperature and entropy per baryon in Figs. 2 and 3.

Based on estimates of non-relativistic kinetic theory and assuming a comparable interaction strength, the drag force should roughly scale with the inverse of the heavy-meson mass. For this reason the relaxation time of the bottomed mesons is approximately $m_B/m_D \sim 3$ times larger than the one for charmed mesons (see Fig. 2). On the other hand, the diffusion coefficient $D_x$ does not carry an explicit dependence on the heavy-meson mass. Therefore it should be approximately equal for $D$ and $\bar{B}$ mesons, as seen in Fig. 3 (different cross sections and subleading effects in the mass break this simple picture). Many other results can be found in our recent works [8, 9].

It is interesting to mention that the diffusion coefficient has also been computed in the dynamical quasiparticle model for a quark-gluon plasma at finite temperature and chemical potential in Ref. [10]. The results —with a sizable reduction with respect to perturbative quantum chromodynamics (QCD) calculations— lie very close to our hadronic results around the transition temperature. This provides a nice agreement of the different models at both sides of the crossover region (also with lattice-QCD calculation around the transition temperature).

The results summarized here have been applied to a 2+1D MonteCarlo simulation of an

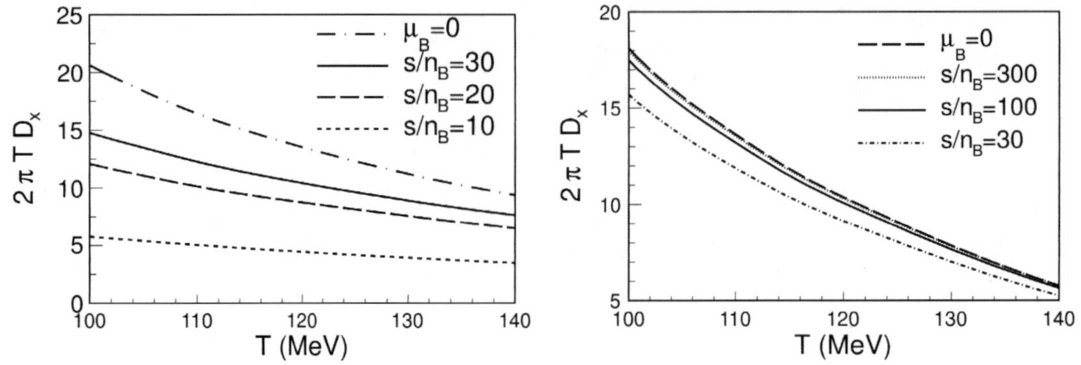


**Figure 3.** Spatial diffusion coefficient for $D$ mesons (left panel) and $\bar{B}$ mesons (right panel) as a function of temperature and entropy per baryon.

ideal expanding plasma, showing that the hadronic interaction does not modify appreciably the $R_{AA}$ but contributes to the increase of the $v_2$ [11]. Finally, in a dynamical quasiparticle model simulation the effect of hadronic interaction is a bit stronger, shifting the peak of $R_{AA}$ at higher momenta and increasing $v_2$ up to a factor of 2 [12]. These examples show that the hadronic interaction presents a nonzero effect on the final observables, bringing the $R_{AA}$ and $v_2$ closer to the experimental results of RHIC experiments. We expect that the hadronic contribution will be more important for low-energy collisions like those at future Facility for Antiproton and Ion Research (FAIR) or the Nuclotron based Ion Collider fAcility (NICA).

## Acknowledgments

This work has been financed by grants FPA2010-16963, FPA2013-43425-P (Spain), BMBF (Germany) under project no. 05P12RFFCQ, EU Integrated Infrastructure Initiative Hadron Physics Project under Grant Agreement n. 227431 and Grant No. FP7-PEOPLE-2011-CIG under Contract No. PCIG09-GA-2011-291679.

JMTR acknowledges financial support from programme TOGETHER from Région Pays de la Loire and the European I3-Hadron Physics programme. LMA thanks the Brazilian agencies CNPq (Project no. 308890/2014-0) and CAPES for financial support. DC acknowledges support from Centro Nacional de Fisica de Partículas, Astropartículas y Nuclear (Consolider - Ingenio 2010) and LT acknowledges support from the Ramón y Cajal research programme.

## References

[1] Manohar A V and Wise M B 2000 *Heavy quark physics* (Cambridge, UK: Cambridge University Press)
[2] Abreu L M, Cabrera D, Llanes-Estrada F J and Torres-Rincon J M 2011 *Annals Phys.* **326** 2737
[3] Guo F K, Hanhart C, Meissner U G 2009 *Eur. Phys. J. A* **40** 171
[4] Abreu L M, Cabrera D and Torres-Rincon J M 2013 *Phys. Rev. D* **87** 034019
[5] Romanets O, Tolos L, Garcia-Recio C, Nieves J, Salcedo L L and Timmermans R G E 2012 *Phys. Rev. D* **85** 114032
[6] Garcia-Recio C, Nieves J, Romanets O, Salcedo L L and Tolos L 2013 *Phys. Rev. D* **87** 034032
[7] Garcia-Recio C, Nieves J, Romanets O, Salcedo L L and Tolos L 2013 *Phys. Rev. D* **87** 074034
[8] Tolos L and Torres-Rincon J M 2013 *Phys. Rev. D* **88** 074019
[9] Torres-Rincon J M, Tolos L and Romanets O 2014 *Phys. Rev. D* **89** 074042
[10] Berrehrah H, Gossiaux P B, Aichelin J, Cassing W, Torres-Rincon J M and Bratkovskaya E 2014 *Phys. Rev. C* **90** 051901
[11] Ozvenchuk V, Torres-Rincon J M, Gossiaux P B, Tolos L and Aichelin J 2014 *Phys. Rev. C* **90** 054909
[12] Song T, Berrehrah H, Cabrera D, Torres-Rincon J M, Tolos L, Cassing W and Bratkovskaya E 2015 *Phys. Rev. C* **92** 014910

15th International Conference on Strangeness in Quark Matter (SQM2015) IOP Publishing
Journal of Physics: Conference Series **668** (2016) 012092 doi:10.1088/1742-6596/668/1/012092

# Towards the quark–gluon plasma Equation of State with dynamical strange and charm quarks

**Florian Burger[1], Ernst–Michael Ilgenfritz[2], Maria Paola Lombardo[4], Michael Müller–Preussker[1] and Anton Trunin[2,3]***

[1] Physics Department, Humboldt University Berlin, Newtonstr. 15, 12489 Berlin, Germany
[2] Bogoliubov Laboratory of Theoretical Physics, Joint Institute for Nuclear Research, Joliot–Curie Str. 6, 141980 Dubna, Russia
[3] Samara State Aerospace University, Moskovskoye Shosse 34, 443086 Samara, Russia
[4] Frascati National Laboratory, National Institute for Nuclear Physics, Via Enrico Fermi 40, 00044 Frascati (Rome), Italy

E-mail: amtrnn@gmail.com

**Abstract.** We present an ongoing project aimed at determining the thermodynamic Equation of State (EoS) of quark–gluon matter from lattice QCD with two generations of dynamical quarks. We employ the Wilson twisted mass implementation for the fermionic fields and the improved Iwasaki gauge action. Relying on $T = 0$ data obtained by the ETM Collaboration the strange and charm quark masses are fixed at their physical values, while the pion mass takes four values in the range from 470 MeV down to 210 MeV. The temperature is varied within a fixed–lattice scale approach. The values for the pseudocritical temperature are obtained from various observables. For the EoS we show preliminary results for the pure gluonic contribution obtained at the pion mass value 370 MeV, where we can compare with previously obtained results with $N_f = 2$ degenerate light flavours.

The Equation of State (EoS) plays a key role in the description of the quark–gluon plasma, serving as important and necessary input for hydrodynamical models describing the space–time evolution of hot QCD matter in relativistic heavy ion collisions. At the moment, lattice QCD is the only known way to non–perturbatively calculate the EoS from first principles and without application of phenomenological models, although this calculation is currently possible only in the limit of an (almost) vanishing baryon chemical potential [1]. Detailed EoS results were presented recently by the Budapest–Wuppertal [2] and HotQCD [3] collaborations, both with $N_f = 2+1$ flavours of staggered fermions. This type of fermions on the lattice is attractive due to the relatively low computational costs, but is known to have certain conceptual problems from the theoretical point of view ("rooting trick" problem). As an alternative, one can consider a Wilson–type fermionic action, which is theoretically safe but computationally demanding. In this talk we report on the current status of the "twisted mass at finite temperature" (tmfT) project [4] devoted in particular to the calculation of the EoS with $N_f = 2+1+1$ flavours of twisted mass Wilson fermions. To our knowledge, this project represents a first study of the Equation of State with almost realistic masses of strange and charm quarks implemented with a Wilson–type fermion discretization. In the works [5] and [6], where the $c$ quark was also taken into account, the staggered approach was used, and in Ref. [7] charm effects were estimated on

* Presenter

the basis of a $N_f = 2+1$ analysis. The recent progress for the EoS with Wilson fermions includes $N_f = 2$ twisted mass results [8] and $N_f = 2+1$ studies with various action improvements [9, 10].

The Wilson twisted mass fermionic action for light and heavy quark doublets has the following form [11]:

$$
\begin{aligned}
S_f^{\rm light}[U,\chi_l,\overline{\chi}_l] &= \sum_{x,y} \overline{\chi}_l(x)\left[\delta_{x,y} - \kappa D_{\rm W}(x,y)[U] + 2i\kappa a\mu_l\gamma_5\delta_{x,y}\tau^3\right]\chi_l(y)\,, \qquad (1)\\
S_f^{\rm heavy}[U,\chi_h,\overline{\chi}_h] &= \sum_{x,y} \overline{\chi}_h(x)\left[\delta_{x,y} - \kappa D_W(x,y)[U] + 2i\kappa a\mu_\sigma\gamma_5\delta_{x,y}\tau^1 + 2\kappa a\mu_\delta\delta_{x,y}\tau^3\right]\chi_h(y),
\end{aligned}
$$

where $D_{\rm W}[U]$ is the usual Wilson operator, $a$ is the lattice spacing, and $\chi_{l,h}$ are quark spinors in the twisted basis. The hopping parameter $\kappa$ is set to its coupling dependent critical value $\kappa_c(\beta)$ leading to the so-called "maximal twist" of the action (1) with the property of automatic $\mathcal{O}(a)$ improvement for expectation values of any operator [12]. The parameter $\mu_l$ describes the mass of the degenerate light quark doublet, which is still unphysically large in our study: the charged pion mass values $m_{\pi^\pm}$ considered at present are 210, 260, 370 and 470 MeV. The heavy twisted mass parameters $\mu_\sigma$ and $\mu_\delta$ have been tuned in the unitary approach to reproduce approximately the physical $K$ and $D$ meson mass values within the accuracy of 10%, thus allowing for a realistic treatment of $s$ and $c$ quarks. Three lattice spacing values $a$ are available for the bare gauge coupling parameter $\beta = 6/g^2 = 1.90$, 1.95 and 2.10 corresponding to $a = 0.094$, 0.082 and 0.065 fm, respectively [13].

The temperature $T = 1/(aN_\tau)$ is varied by changing $N_\tau = 3, 4, \ldots, 24$, the latter being the lattice extent in the fourth Euclidean direction (*fixed–scale approach* [9]). The linear spatial extent varies between $N_\sigma = 24$ and 48, such that the bound of the aspect ratio $N_\sigma/N_\tau \geq 2$ is guaranteed as the worst case. In comparison with the more often used fixed–$N_\tau$ approach, such a choice allows to reduce the amount of required $T = 0$ computations, necessary to carry out the obligate subtractions in computing the EoS. Fortunately, we can rely on the mass parameters and on the scale setting by the European Twisted Mass (ETM) Collaboration (see [13] and references therein).

For determining the pseudocritical temperature(s) of the crossover region we have computed the renormalized Polyakov loop

$$\langle \text{Re L}\rangle_R = \langle \text{Re L}\rangle \exp[V(r_0)/2T]\,, \qquad (2)$$

where $V(r_0)$ is static quark potential at the Sommer scale $r_0 \simeq 0.5$ fm and the renormalized subtracted chiral condensate

$$\Delta_{l,s} = \frac{\langle\bar\psi\psi\rangle_l - \frac{\mu_l}{\mu_s}\langle\bar\psi\psi\rangle_s}{\langle\bar\psi\psi\rangle_l^{T=0} - \frac{\mu_l}{\mu_s}\langle\bar\psi\psi\rangle_s^{T=0}}\,, \qquad (3)$$

where $\langle\bar\psi\psi\rangle_l$ and $\langle\bar\psi\psi\rangle_s$ are the light and strange quark condensates, respectively. The latter has been obtained in the Osterwalder–Seiler setup [14, 12], which avoids mixing in the heavy quark sector. The mass $\mu_s$ has been determined as to reproduce the physical $\bar s\gamma_\mu s$–mass. Moreover, we have determined the unrenormalized disconnected chiral susceptibility

$$\sigma^2_{\bar\psi\psi} = \frac{V}{T}\left(\langle(\bar\psi\psi)^2\rangle_l - \langle\bar\psi\psi\rangle_l^2\right), \qquad V = a^4N_\sigma^3N_\tau. \qquad (4)$$

The inflexion points of $\langle\text{Re L}\rangle_R$ and $\Delta_{l,s}$ as well as the maxima of $\sigma^2_{\bar\psi\psi}$ versus $T$ determine the crossover temperatures $T_L, T_\Delta$ and $T_\chi$, respectively, as shown in Table 1. In Figure 1, as an

**Table 1.** Crossover temperatures for different pion mass values $m_{\pi^\pm}$. $N_f = 2$ results are taken from Ref. [8] (all numbers in units of MeV).

| Number of flavours | $m_{\pi^\pm}$ | $T_\chi$ | $T_\Delta$ | $T_L$ |
|---|---|---|---|---|
| $N_f = 2+1+1$ | 210 | 152(5) | 164(3) | – |
| | 260 | 170(5) | – | – |
| | 370 | 184(4) | 192(2) | 201(3) |
| | 470 | 199(6) | – | – |
| $N_f = 2$ | 360 | 193(13) | – | 219(3)(14) |
| | 430 | 208(14) | – | 225(3)(14) |

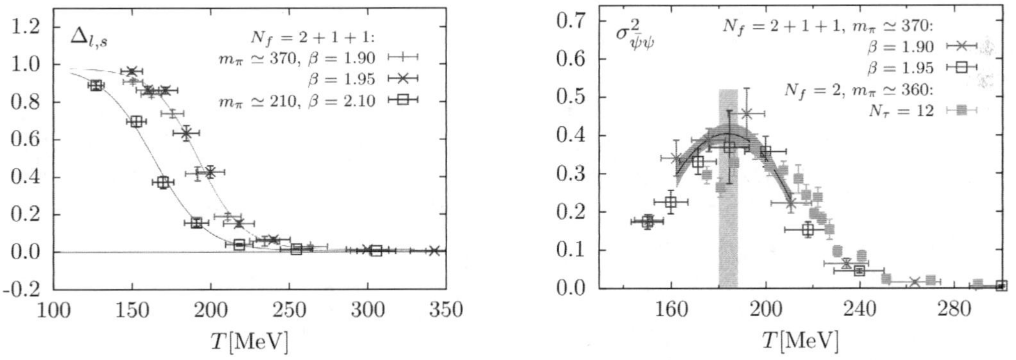


**Figure 1.** $\Delta_{l,s}$ (left) and $\sigma^2_{\bar\psi\psi}$ (right) versus $T$.

example, we show the behaviour of $\Delta_{l,s}$ for the pion mass values $m_{\pi^\pm} = 370$ and 210 MeV as well as of $\sigma^2_{\bar\psi\psi}$ for $m_{\pi^\pm} = 370$ MeV (the latter in comparison with the case $N_f = 2$ [8]).

The trace anomaly or interaction measure, from which the pressure $p$ as well as the energy density $\epsilon$ can be finally extracted as a function of $T$, is given by the logarithmic derivative of the lattice partition function $\mathcal{Z}$

$$I(T) = \epsilon - 3p, \qquad \frac{I(T)}{T^4} = -\frac{1}{T^3 V}\left(\frac{d\ln\mathcal{Z}}{d\ln a}\right)_{\rm sub}, \tag{5}$$

where the subtracted expectation values are defined as $\langle\ldots\rangle_{\rm sub} \equiv \langle\ldots\rangle_T - \langle\ldots\rangle_{T=0}$. This leads to the expression

$$\begin{aligned}\frac{I(T)}{T^4} &= \frac{1}{T^3V}\sum_i \frac{db_i}{da}\langle\frac{\partial S}{\partial b_i}\rangle_{\rm sub} = N_\tau^4\Big\{\Big(-a\frac{d\beta}{da}\Big)\Big(\frac{c_0}{3}\langle\sum_P \mathrm{Re\,Tr}\,U_P\rangle_{\rm sub} + \frac{c_1}{3}\langle\sum_R \mathrm{Re\,Tr}\,U_R\rangle_{\rm sub} \\ &+ \frac{\partial\kappa_c}{\partial\beta}\langle\bar\chi D_{\rm W}[U]\chi\rangle_{\rm sub} - 2a\mu_l\frac{\partial\kappa_c}{\partial\beta}\langle\bar\chi i\gamma_5\tau^3\chi\rangle_{\rm sub}\Big) + 2\kappa_c\Big(a\frac{d(a\mu_l)}{da}\Big)\langle\bar\chi i\gamma_5\tau^3\chi\rangle_{\rm sub} + \text{heavy terms}\Big\}, \\ I(T) &= I_{\rm gauge} + I_{\rm light} + I_{\rm heavy}\end{aligned} \tag{6}$$

with the renormalization functions $db_i/da$ to be fixed from ETM Collaboration $T = 0$ data. Preliminary results for $I/T^4$ are presented in Figure 2, where the pure gauge field (tree–level corrected) contribution as the most dominant one is shown for the case $m_{\pi^\pm} = 370$ MeV and

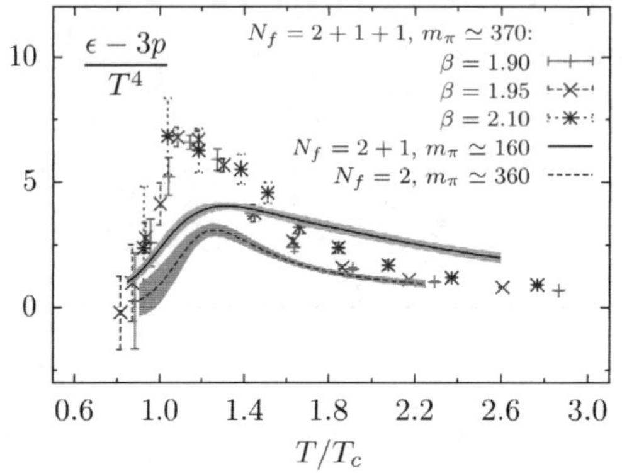


**Figure 2.** Tree–level corrected gauge field contribution to the trace anomaly for $m_{\pi^\pm} \simeq 370$ MeV in comparison with the fitted full $N_f = 2$ result [8] as well as with the continuum extrapolated $N_f = 2+1$ case [3].

the three lattice spacings. We compare with the full $N_f = 2$ result at approximately the same pion mass [8] as well as with the fitted results for $N_f = 2+1$ dynamical staggered fermion flavour degrees of freedom close to the physical point [3]. The comparison demonstrates that our $N_f = 2+1+1$ results turned out to be in the right ball park.

The computation of the light and heavy quark contributions — being most interesting at higher temperatures — is under way. Their magnitude turns out to be rather sensitive with respect to the mass renormalization functions to be determined from the ETM Collaboration analysis of the hadronic spectrum at $T = 0$.

## Acknowledgments

We acknowledge generous support by the HLRN supercomputer centers Berlin and Hannover, the Supercomputing Center of Lomonosov Moscow State University, as well as by the HybriLIT group of JINR. The work was partly financed by the Heisenberg–Landau program of BLTP JINR and BMBF Germany, by the "Dynasty" foundation, and by the Ministry of Education and Science of Russia under SSAU Competitiveness Enhancement Program 2013–2020.

## References

[1] Philipsen O 2012 *Prog. Part. Nucl. Phys.* **70** 55
[2] Borsanyi S *et al* 2014 *Phys. Lett.* B **730** 99
[3] Bazavov A *et al* 2014 *Phys. Rev.* D **90** 094503
[4] Burger F *et al* 2013 *Proceedings of Science* PoS(LATTICE2013)153
[5] Bazavov A *et al* 2013 *Proceedings of Science* PoS(LATTICE2013)154
[6] Ratti C *et al* 2013 *Nucl. Phys.* A **904–905** 869c
[7] Borsanyi S *et al* 2010 *J. High Energy Phys.* JHEP11(2010)077
[8] Burger F *et al* 2015 *Phys. Rev.* D **91** 074504
[9] Umeda T *et al* 2012 *Phys. Rev.* D **85** 094508
[10] Borsanyi S *et al* 2012 *J. High Energy Phys.* JHEP08(2012)126
[11] Chiarappa T *et al* 2007 *Eur. Phys. J.* C **50** 373
[12] Frezzotti R and Rossi G C 2004 *J. High Energy Phys.* JHEP 04(2004)007
[13] Alexandrou C *et al* 2014 *Phys. Rev.* D **90** 074501
[14] Osterwalder K and Seiler E 1978 *Ann. Phys.* **110** 440

Journal of Physics: Conference Series **668** (2016) 012093 doi:10.1088/1742-6596/668/1/012093

# J/$\psi$ and $\psi(2S)$ measurement in $p$+$p$ collisions at $\sqrt{s}$ = 200 and 500 GeV with the STAR experiment

**Barbara Trzeciak**[1] **for the STAR Collaboration**

[1]Faculty of Nuclear Sciences and Physical Engineering, Czech Technical University in Prague, Brehova 7, 115 19 Praha 1, Czech Republic

E-mail: `trzecbar@fjfi.cvut.cz`

**Abstract.** In this paper, results on the J/$\psi$ cross section and polarization measured via the dielectron decay channel at mid-rapidity in $p+p$ collisions at $\sqrt{s}$ = 200 and 500 GeV in the STAR experiment are discussed. Also, first measurements of the J/$\psi$ production as a function of the charged-particle multiplicity density and of $\psi(2S)$ to J/$\psi$ ratio at $\sqrt{s}$ = 500 GeV are reported.

## 1. Introduction

J/$\psi$ and $\psi(2S)$ are bound states of charm ($c$) and anti-charm ($\bar{c}$) quarks. Final color-singlet charmonium physical state can be produced via either color-singlet (CS) or color-octet (CO) intermediate $c\bar{c}$ state. Based on this, there are different models that try to describe $c\bar{c}$ pair production and its hadronization to the physical charmonium state, such as Color Singlet Model (CSM), Color Evaporation Model (CEM), or more sophisticated Non-Relativistic QCD (NRQCD) calculations. However, the charmonium production mechanism in elementary particle collisions is not yet exactly known. For many years measurements of the charmonium cross sections have been used to test different production models. While many models can describe relatively well the experimental data on the J/$\psi$ cross section in $p+p$ collisions [1–8], they have different predictions for the J/$\psi$ polarization. Therefore, measurements of the J/$\psi$ polarization may allow to discriminate among different models and provide new insight into the J/$\psi$ production mechanism. Further constraints for models can be obtained by looking at the J/$\psi$ production vs relative charged-particle event multiplicity.

## 2. J/$\psi$ cross-section measurements at $\sqrt{s}$ = 200 and 500 GeV

STAR has measured inclusive J/$\psi$ $p_T$ spectra in $p$+$p$ collisions at $\sqrt{s}$ = 200 and 500 GeV via the dielectron decay channel ($B_{ee}$ = 5.9%) at mid-rapidity ($|y| < 1$). These results are compared to different model predictions in order to understand the J/$\psi$ production mechanism in elementary collisions. The left panel of Fig. 1 shows STAR low and high-$p_T$ measurements of J/$\psi$ $p_T$ spectra [2, 9] at $\sqrt{s}$ = 200 GeV compared to different model predictions: CEM [10], NNLO* CS [12] and NLO NRQCD calculations with both color-singlet and color-octet contributions (NLO CS+CO) [11]. The NNLO* CS model prediction is for the direct J/$\psi$ production only, so it does not include contributions from $\psi(2S)$, $\chi_C$ and $B$-meson decays to J/$\psi$, which may account for as much as 40% of the inclusive J/$\psi$ production. The right panel of Fig. 1 shows high-$p_T$ results at $\sqrt{s}$ = 500 GeV compared to NLO NRQCD prediction for the prompt J/$\psi$

production [11, 13, 14]. In both cases NLO NRQCD calculations describe the data well at a higher $p_T$ range, $p_T \gtrsim 4$ GeV/$c$. The result at $\sqrt{s} = 200$ GeV is also in a good agreement with the CEM prediction.

Also, $x_T$ scaling of J/$\psi$ cross section is observed: $\frac{d^2\sigma}{2\pi p_T dp_T dy} = g(x_T)/(\sqrt{s})^n$, where $x_T = 2p_T/\sqrt{s}$, with $n = 5.6 \pm 0.2$ at mid-rapidity and $p_T > 5$ GeV/$c$ for a wide range of colliding energies [1]. At $\sqrt{s} = 500$ GeV the same $x_T$ scaling of high-$p_T$ J/$\psi$ production is seen [15].

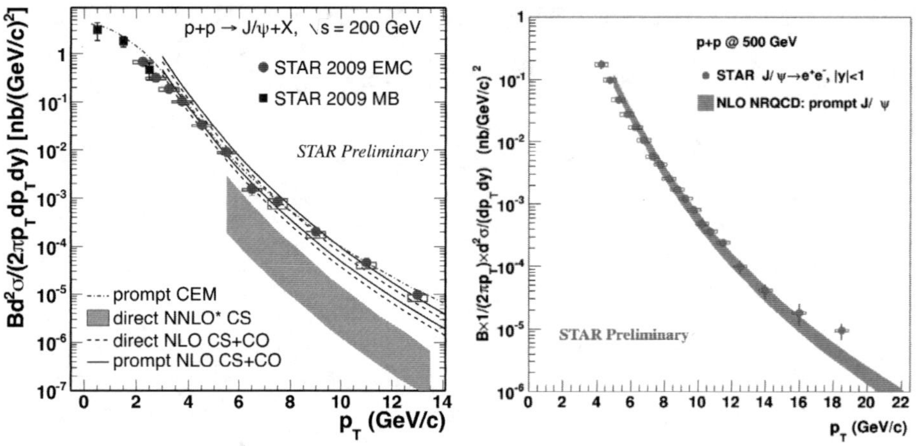


**Figure 1.** J/$\psi$ invariant cross section vs $p_T$ in $p$+$p$ collisions at mid-rapidity, left: at $\sqrt{s} = 200$ GeV, low [9] and high $p_T$ [2] results are shown as blue squares and red circles, respectively, and are compared to different model predictions [10–12], right: at $\sqrt{s} = 500$ GeV, compared to NRQCD calculations [11, 13, 14].

## 3. J/$\psi$ and $\psi(2S)$ measurements at $\sqrt{s} = 500$ GeV

The left panel of Fig. 2 shows $\psi(2S)/J/\psi$ ratio from STAR (red full circle) compared to measurements of other experiments at different colliding energies, in $p + p$ and $p$+A collisions. The STAR data point is consistent with the observed trend, and no collision energy dependence of the $\psi(2S)$ to J/$\psi$ ratio is seen with the current precision.

STAR has also measured J/$\psi$ production vs event activity. The right panel of Fig. 2 shows relative J/$\psi$ yield as a function of relative charged particle event multiplicity. The measurement is done for integrated $p_T$ range in the dimuon decay channel and for $p_T > 4$ GeV/$c$ in the dielectron decay channel, shown as red and blue points, respectively. We observe a strong correlation between the J/$\psi$ yield and event activity, and for $p_T > 4$ GeV/$c$ there is a stronger than linear growth for higher multiplicities. The same behaviour has been observed for J/$\psi$ and open charm by the LHC experiments. There is also a hint of $p_T$ dependence of the correlation, but further analysis, especially in the dimuon decay channel for higher event activity bins is required in order to draw strong conclusions. Results are also compared to the default PYTHIA 8 tune that includes multiple parton-parton interactions, for both analysed $p_T$ ranges, and to the percolation model [16] which is based on a string screening assumption. Both models can describe the observed increase of the J/$\psi$ relative yield.

## 4. J/$\psi$ polarization measurements at $\sqrt{s} = 200$ and 500 GeV

$J/\psi$ polarization is analyzed via the angular distribution of the decay electrons that is described by: $\frac{d^2N}{d(\cos\theta)d\phi} \propto 1 + \lambda_\theta \cos^2\theta + \lambda_\phi \sin^2\theta \cos 2\phi + \lambda_{\theta\phi} \sin 2\theta \cos\phi$, where $\theta$ and $\phi$ are polar and azimuthal angles, respectively; $\lambda_\theta$, $\lambda_\phi$ and $\lambda_{\theta\phi}$ are the angular decay coefficients. The first STAR

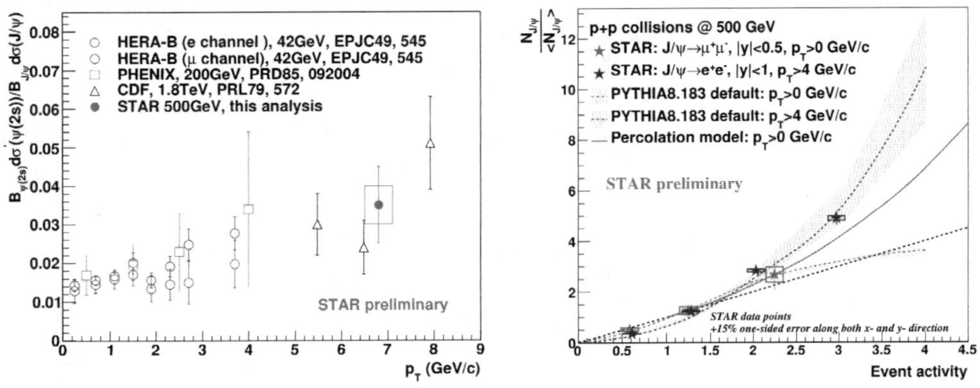


**Figure 2.** Left: ratio of $\psi(2S)$ to J/$\psi$ in $p$+$p$ collisions at $\sqrt{s} = 500$ GeV from STAR (red circle) compared to results from other experiments at different energies. Right: Relative J/$\psi$ yield vs event activity compared to the default PYTHIA 8 tune and to the percolation model [16].

measurement of the $J/\psi$ polarization was performed at $\sqrt{s} = 200$ GeV at mid-rapidity and 2 $< p_T < 6$ GeV/$c$, in the helicity frame (HX) [17]. Due to limited statistics only $\lambda_\theta$ polarization parameter, related to the polar anisotropy was extracted as a function of $p_T$. The result was compared to NRQCD calculations [18] and the NLO$^+$ CSM prediction [19]. A trend observed in the RHIC data is towards longitudinal polarization as $p_T$ increases and, within experimental and theoretical uncertainties, the result is consistent with the NLO$^+$ CSM prediction.

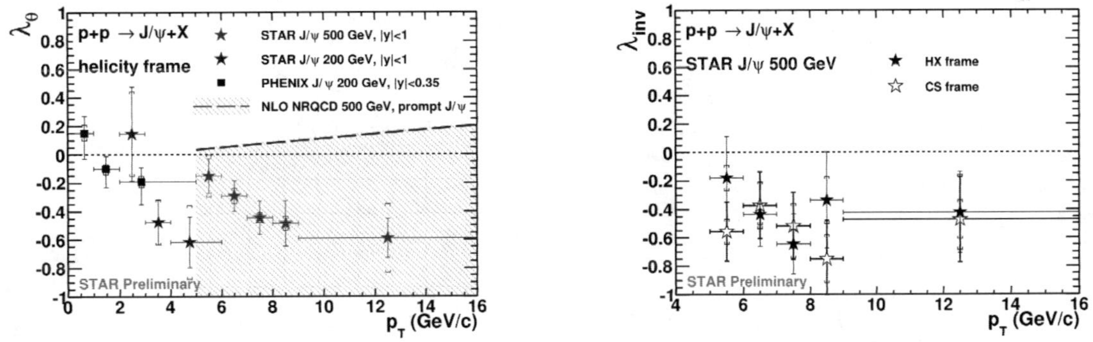


**Figure 3.** Left: J/$\psi$ polarization parameter $\lambda_\theta$ vs $p_T$ at $\sqrt{s} = 500$ GeV at mid-rapidity compared to results at $\sqrt{s} = 200$ GeV and to the NLO NRQCD calculations [14, 20–22]. Right: $\lambda_{inv}$ parameters vs $p_T$ in the HX (full symbols) and CS frames (open symbols).

The J/$\psi$ polarization measurements have been extended to a wider $p_T$ range of 5-16 GeV/$c$ using $p$+$p$ dataset at $\sqrt{s} = 500$ GeV. Both $\lambda_\theta$ and $\lambda_\phi$ parameters have been extracted in the HX and the Collins-Soper (CS) frames. This has also allowed to obtain the frame invariant parameter $\lambda_{inv} = (\lambda_\theta + 3\lambda_\phi)/(1 - \lambda_\phi)$ in both frames which is a good cross-check between measurements since any arbitrary choice of a reference frame should give the same value of the $\lambda_{inv}$ [23]. The left panel of Fig. 3 shows $\lambda_\theta$ in the HX frame as a function of $p_T$ at $\sqrt{s} = 500$ GeV compared to results at $\sqrt{s} = 200$ GeV and to a prediction of the NLO NRQCD [14, 20–22]. We observe similar trend of $\lambda_\theta$ going towards negative values with increasing $p_T$ in both analysed colliding energies. And at $\sqrt{s} = 500$ GeV we have also checked that the $\lambda_\phi$ parameter related to the azimuthal anisotropy is consistent with 0. Since the lower limit in the NRQCD calculations is unconstrained we do not draw conclusions from the comparison to data, and the new data points should help to constrain color-octet Long-Distance Matrix Elements for the model. The right panel of Fig. 3 shows $\lambda_{inv}$ parameters as a function of $p_T$ extracted in the HX and CS frames, shown as full and open symbols, respectively. $\lambda_{inv}$ has negative values and the results are in agreement between the frames.

## 5. Summary

In summary, STAR has measured the inclusive J/$\psi$ cross section and polarization in $p+p$ collisions at $\sqrt{s}$ = 200 and 500 GeV as a function of $p_T$. The measurements are compared to different model predictions of the J/$\psi$ production. The $p_T$ spectra are described well by the NRQCD calculations. The polarization parameter $\lambda_\theta$ at $\sqrt{s} = 200$ GeV is consistent with the NLO$^+$ CSM. At both colliding energies $\lambda_\theta$ has a trend towards negative values with increasing $p_T$ in the HX frame. And at $\sqrt{s} = 500$ GeV in the HX frame the $\lambda_\phi$ parameter is consistent with 0. Also, frame invariant parameters, $\lambda_{inv}$, agree in the HX and CS frames. The first measurement of $\psi(2S)/J/\psi$ ratio in $p+p$ collisions at $\sqrt{s} = 500$ GeV is reported and compared with results from other experiments. No collision energy dependence is observed. A strong correlation is observed between relative J/$\psi$ yield and the event activity which is stronger than linear at higher multiplicities at $p_T > 4$ GeV/$c$.

## Acknowledgements

This publication was supported by the European social fund within the framework of realizing the project „Support of inter-sectoral mobility and quality enhancement of research teams at Czech Technical University in Prague", CZ.1.07/2.3.00/30.0034.

## References

[1] Abelev B I *et al.* (STAR) 2009 *Phys. Rev.* **C80** 041902 (*Preprint* 0904.0439)
[2] Adamczyk L *et al.* (STAR) 2013 *Phys. Lett.* **B722** 55–62 (*Preprint* 1208.2736)
[3] Adare A *et al.* (PHENIX) 2012 *Phys. Rev.* **D85** 092004 (*Preprint* 1105.1966)
[4] Abe F *et al.* (CDF) 1997 *Phys. Rev. Lett.* **79** 572–577
[5] Acosta D *et al.* (CDF) 2005 *Phys. Rev.* **D71** 032001 (*Preprint* hep-ex/0412071)
[6] Aad G *et al.* (ATLAS) 2011 *Nucl. Phys.* **B850** 387–444 (*Preprint* 1104.3038)
[7] Khachatryan V *et al.* (CMS) 2011 *Eur. Phys. J.* **C71** 1575 (*Preprint* 1011.4193)
[8] Aaij R *et al.* (LHCb) 2011 *Eur. Phys. J.* **C71** 1645 (*Preprint* 1103.0423)
[9] Kosarzewski L (STAR) 2012 *Acta Phys. Polon. Supp.* **5** 543–548
[10] Frawley A D, Ullrich T and Vogt R 2008 *Phys. Rept.* **462** 125–175 (*Preprint* 0806.1013)
[11] Ma Y Q, Wang K and Chao K T 2011 *Phys. Rev.* **D84** 114001 (*Preprint* 1012.1030)
[12] Artoisenet P, Campbell J M, Lansberg J P, Maltoni F and Tramontano F 2008 *Phys. Rev. Lett.* **101** 152001 (*Preprint* 0806.3282)
[13] Ma Y Q, Wang K and Chao K T 2011 *Phys. Rev. Lett.* **106** 042002 (*Preprint* 1009.3655)
[14] Shao H S, Han H, Ma Y Q, Meng C, Zhang Y J and Chao K T 2015 *JHEP* **05** 103, and private communication (*Preprint* 1411.3300)
[15] Trzeciak B (STAR) 2015 *J. Phys. Conf. Ser.* **612** 012038 (*Preprint* 1412.7341)
[16] Ferreiro E G and Pajares C 2012 *Phys. Rev.* **C86** 034903, and private communication (*Preprint* 1203.5936)
[17] Adamczyk L *et al.* (STAR) 2014 *Phys. Lett.* **B739** 180–188 (*Preprint* 1311.1621)
[18] Chung H S, Yu C, Kim S and Lee J 2010 *Phys. Rev.* **D81** 014020 (*Preprint* 0911.2113)
[19] Lansberg J P 2011 *Phys. Lett.* **B695** 149–156 (*Preprint* 1003.4319)
[20] Chao K T, Ma Y Q, Shao H S, Wang K and Zhang Y J 2012 *Phys. Rev. Lett.* **108** 242004 (*Preprint* 1201.2675)
[21] Shao H S and Chao K T 2014 *Phys. Rev.* **D90** 014002 (*Preprint* 1209.4610)
[22] Shao H S, Ma Y Q, Wang K and Chao K T 2014 *Phys. Rev. Lett.* **112** 182003, and private communication (*Preprint* 1402.2913)
[23] Faccioli P, Lourenco C, Seixas J and Wohri H K 2010 *Eur. Phys. J.* **C69** 657–673 (*Preprint* 1006.2738)

# Low-mass dimuon measurements in pp, p-Pb and Pb-Pb collisions with ALICE at the LHC

**Antonio Uras**

IPNL, Université Claude Bernard Lyon-I and CNRS-IN2P3, Villeurbanne, France

E-mail: antonio.uras@cern.ch

**Abstract.**
Low-mass dimuon production, including light vector mesons $\rho$, $\omega$, $\phi$, provides key information on the hot and dense state of strongly interacting matter produced in high-energy heavy-ion collisions. In particular, strangeness production can be studied via $\phi$ meson measurements, while the detailed description of the full dimuon mass spectrum down to the kinematic threshold can be used to reveal in-medium modifications of hadron properties and the thermal emission arising from the medium. Measurements in pp and p-A systems, in absence of hot nuclear matter effects, must be used as a reference to test our knowledge of the processes expected to contribute to dilepton production. Dimuon production is studied with the ALICE apparatus at the LHC at forward rapidity ($2.5 < y < 4$) with the Muon Spectrometer. In this contribution, results on low-mass dimuon production are shown, for various center-of-mass energies per nucleon pair, in pp, p-Pb, and Pb-Pb collisions.

## 1. Introduction

Low-mass dimuon production at the LHC energies is dominated by the 2-body and Dalitz decays of the light neutral mesons $\eta, \rho, \omega, \eta'$ and $\phi$ — the so-called "hadronic cocktail" — with dimuons from open charm and open beauty decay processes contributing to the invariant mass continuum. An increased production cross section for the open charm and open beauty processes, relative to the hadronic cocktail, marks the main evolution with respect to the low-mass dilepton measurements at the SPS [1] and RHIC [2, 3] energies.

Describing the observed low-mass dimuon production in pp collisions allows for the study of particle production mechanisms in elementary hadronic interactions in the light quark sector, while observations in p-A collisions are sensitive to initial state effects affecting particle production in cold nuclear matter. In heavy-ion collisions, dimuons deploy their full potential as penetrating probes: insensitive to the strong color field dominating inside the QCD matter, leptons arrive to the detectors almost unaffected, providing an ideal tool to probe the whole evolution of the system. Low-mass dimuon measurements in heavy-ion collisions, in particular, allow the identification of hot nuclear matter effects related to the creation of an extended volume of quark-gluon plasma (QGP), to its hydrodynamical evolution, and to the chiral symmetry restoration expected to occur within the deconfined system. In this case, key-observables are strangeness production via $\phi$ meson measurement, the spectral function modification of the short-lived $\rho$ meson, and the observation of direct thermal radiation from the QGP phase. The possibility to observe the $\eta$ meson via its prominent Dalitz decay as well as the $\omega$ meson via its 2-body decay completes the rich spectrum of information which can be accessed by the analysis of the low-mass dimuon spectrum. In this contribution, the main results from the available ALICE low-mass dimuon measurements performed during the LHC Run 1 in pp, p-Pb and Pb-Pb collisions are reviewed.

The ALICE experiment at the CERN LHC measures low-mass dimuon production at forward rapidity, complementing the low-mass dielectron observations performed at mid-rapidity. Identification and measurement of muons are performed in the Muon Spectrometer [4], covering the pseudo-rapidity region $2.5 < \eta_{\mathrm{lab}} < 4$. Starting from the interaction point (IP), the Muon Spectrometer is composed of the following elements: a hadron absorber which filters out hadrons; a 3 T·m integrated field dipole magnet; a set of five tracking stations, each one composed of two cathode pad chambers with a spatial resolution of about 100 $\mu$m in the bending direction; an iron wall, which absorbs the residual secondary hadrons emerging from the absorber; the muon trigger system, consisting of four planes of resistive plate chambers with a time resolution of about 2 ns. Uncorrelated muon pairs coming from semi-muonic decays of $\pi$ and $K$ mesons represent the main background source affecting measurements in the dimuon channel, especially at low invariant masses. In all the measurements presented in the following, the combinatorial background is evaluated using an event mixing technique [5] and subtracted from the raw dimuon invariant mass spectrum.

## 2. Results in pp Collisions

Low-mass dimuon production in pp collisions were measured by ALICE at $\sqrt{s} = 2.76$ and 7 TeV [6, 5]. These results allowed to exclude, for the first time in elementary hadronic reactions at the LHC energies, any significant contribution to the low-mass dimuon spectrum beyond the already cited hadronic cocktail and the open charm and open beauty processes, see left panel of Figure 1. The $\phi$ signal could be extracted and the differential cross-sections was measured as a function the transverse momentum ($p_{\mathrm{T}}$) for $1 < p_{\mathrm{T}} < 5$ GeV/$c$ (see right panel of Figure 1) and compared to the predictions from `PHOJET` [7] and the `ATLAS-CSC`, `D6T` [8], the `Perugia0` and `Perugia11` [9] tunes of `PYTHIA` [10]. At both energies, an overall good agreement is found between predictions and data, with the exception of the `Perugia0` and `Perugia11` tunes of `PYTHIA` which underestimate the measured cross section by a factor between 1.5 and 2. The integrated cross section over the accessible $p_{\mathrm{T}}$ range $1 < p_{\mathrm{T}} < 5$ GeV/$c$ and $2.5 < y < 4$ was measured to be $\sigma_\phi = 0.940 \pm 0.055$ (stat.) $\pm\, 0.044$ (syst.) mb and $\sigma_\phi = 0.566 \pm 0.084$ (stat.) $\pm\, 0.076$ (syst.) mb at $\sqrt{s} = 2.76$ and 7 TeV, respectively.

## 3. Results in p-Pb Collisions

Data in p-Pb collisions were taken by ALICE at $\sqrt{s_{\mathrm{NN}}} = 5.02$ TeV [6], with the Muon Arm accessing two different rapidity regions[1]: $2.03 < y < 3.53$ (p-going) and $-4.46 < y < -2.96$ (Pb-going). In the following, these two rapidity ranges will be referred to as "forward" and "backward", respectively.

The analysis of the p-Pb data shares the same strategy of the pp data: after subtraction of the combinatorial background, estimated by means of an event mixing technique, the invariant mass spectrum of the signal was described in terms of the expected sources. The analysis focused on the $\phi$ meson measurement: the $\phi$ cross section, integrated over the accessible $p_{\mathrm{T}}$ range, $1 < p_{\mathrm{T}} < 7$ GeV/$c$, is shown as a function of rapidity in the left panel of Figure 2, exhibiting a significant asymmetry between the forward and backward rapidity regions. The comparison between the data and the predictions by `HIJING` [11] and `DPMJET` [12] clearly shows how the models — which successfully described charged particle production at mid-rapidity in the same collision system [13] — fail to properly reproduce the shape and the normalisation of the observed rapidity dependence of the $\phi$ cross section. Still, `HIJING` prediction qualitatively

[1] The sign of $y_{\mathrm{cms}}$ is defined by assuming the proton beam to have positive rapidity.

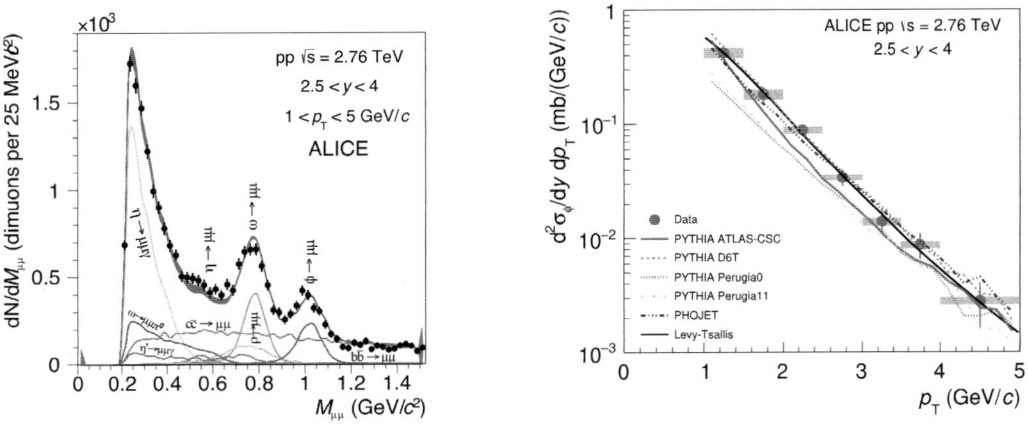


**Figure 1.** Left: Fit to the low-mass dimuon spectrum in pp collisions at $\sqrt{s} = 2.76$ TeV; cyan boxes: systematic uncertainty from background subtraction; red band: uncertainty in the relative normalization of the sources. Right: $\phi$ cross section as a function of $p_T$ compared with predictions from PHOJET and several tunes of PYTHIA.

reproduces the forward-backward asymmetry observed in the data, as well as — ignoring the normalisation — the shape of the $y$-dependence in the backward region. DPMJET, on the contrary, fails to reproduce even qualitatively the observed forward-backward asymmetry. Averaging over the available $p_T$ range, the discrepancy between the data and the predictions from HIJING and DPMJET amounts to $\sim 18\,\%$ and $\sim 57\,\%$, respectively, at backward rapidity and $\sim 5\,\%$ and $\sim 9.5\,\%$, respectively, at forward rapidity.

The $\phi$ meson nuclear modification factor $R_{\mathrm{pPb}}$ is defined as the ratio between the production cross section $\sigma_{\mathrm{pPb}}(p_T)$ in p-Pb collisions and the cross section $\sigma_{\mathrm{pp}}(p_T)$ in pp collisions evaluated at $\sqrt{s} = 5.02$ TeV, scaled by $A_{\mathrm{Pb}}$: $R_{\mathrm{pPb}}(p_T) = \sigma_\phi^{\mathrm{pPb}}(p_T)/(\sigma_\phi^{\mathrm{pp}}(p_T) \cdot A_{\mathrm{Pb}})$ , where $A_{\mathrm{Pb}}$ is the nuclear mass number for the Pb nucleus. Since for the pp cross section $\sigma_\phi^{\mathrm{pp}}$ at $\sqrt{s} = 5.02$ TeV no direct measurement is currently available, it was evaluated by interpolating the already cited measurements in the rapidity interval $2.5 < y < 4$ at $\sqrt{s} = 2.76$ and 7 TeV. The nuclear modification factor $R_{\mathrm{pPb}}$ as a function of $p_T$ is shown in the center and right panels of Figure 2 for the backward and forward rapidity regions. In the p-going direction a rising trend of $R_{\mathrm{pPb}}$ is observed from $\sim 0.5$ to $\sim 1$, when going from $p_T = 1$ GeV/$c$ to $p_T = 4$ GeV/$c$. This observation is compatible with the behaviour of charged particles at forward rapidity at RHIC energies [14, 15], and at mid-rapidity at LHC energies [16]. In the Pb-going direction, on the other hand, an enhancement is observed for $R_{\mathrm{pPb}}$, reaching values as large as $\sim 1.6$ around $p_T = 3$-4 GeV/$c$. In both rapidity regions, observations are found to be in agreement with a recent measurement of $\phi$ meson production at the RHIC [17]. An interpretation of these results, either in terms of an initial-state (Cronin-like) effect or a final-state effect related to radial flow in p-Pb, is not possible yet, due to a general lack of theoretical predictions for soft particle production at forward rapidity in p-A collisions at the LHC energies.

## 4. Results in Pb-Pb Collisions

ALICE collected data in Pb-Pb collisions at $\sqrt{s_{\mathrm{NN}}} = 2.76$ TeV. Due to the hardware trigger threshold at $p_T \sim 1$ GeV/$c$ for the single muons involved in this data taking, the acceptance for low-mass dimuons is negligible below $p_T = 2$ GeV/$c$. As for the pp and p-Pb analyses, the low-mass dimuon spectrum is described, after subtraction of the combinatorial background, in terms of the superposition of the hadronic cocktail and the open charm and open beauty

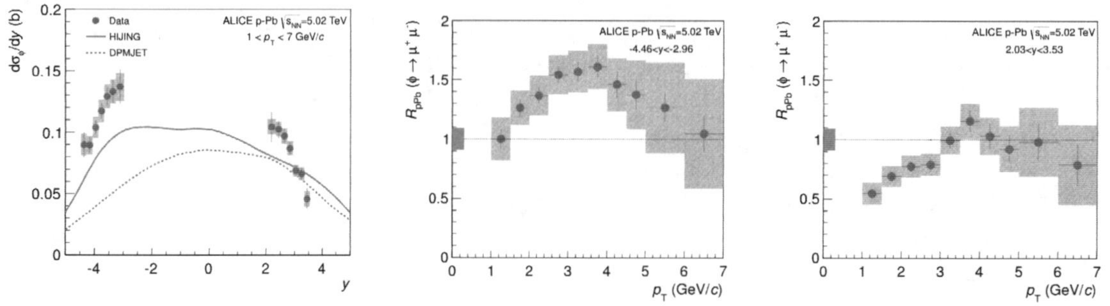

**Figure 2.** Left: $\phi$ cross section as a function of $p_T$ in p-Pb collisions at $\sqrt{s_{NN}} = 5.02$ TeV compared with predictions from DPMJET and HIJING. Center and right: nuclear modification factor $R_{pPb}$ as a function of $p_T$ in the backward and forward rapidity regions.

decay processes. In this way, the signal for the $(\rho + \omega)$ and $\phi$ mesons could be extracted in the $p_T$ interval $2 < p_T < 5$ GeV/$c$ and for $2.5 < y < 4$. The signal-over-background ratio at the $\phi$ peak ranges from $\sim 0.1$ for the most central collisions (0-20 % centrality), to $\sim 3$ for peripheral collisions (60-90 % centrality). Strangeness enhancement in Pb-Pb collisions could thus be studied considering the evolution of the $\phi/(\rho + \omega)$ ratio as a function of the centrality. This ratio, shown as $BR_{\phi\to\mu\mu}\sigma_\phi/(BR_{\rho\to\mu\mu}\sigma_\rho + BR_{\omega\to\mu\mu}\sigma_\omega)$, is plotted in the left panel of Figure 3. The centrality of the collision is expressed in terms of the number of participant nucleons $N_{part}$ — i.e. the number of nucleons that undergo at least one inelastic collision with another nucleon. The pp value ($N_{part} = 2$) is also shown. The ratio increases from pp to Pb-Pb and tends to saturate already in semi-peripheral collisions in the 40-60 % centrality class. The value for 0-20 % most central Pb-Pb collisions is about two times the one measured in pp collisions. It should be noted that, due to the limited $p_T$ range accessible and the different masses of the considered particles, this measurement may also be sensitive to the radial flow mechanism, preventing firm conclusions to be established in terms of genuine strangeness enhancement.

The nuclear modification factor for the $\phi$ meson is defined as $R_{AA} = \langle\phi\rangle/(\sigma_{pp}\langle T_{AA}\rangle)$, where $\langle\phi\rangle$ is the $\phi$ yield in a given kinematic range, $\sigma_{pp}$ is the $\phi$-production cross section in pp collisions in the same kinematic range, and $\langle T_{AA}\rangle$ is the average nuclear overlap function of the considered centrality class. The $R_{AA}$ was extracted for the $\phi$ meson as a function of the centrality, using as a reference the pp results obtained at the same nucleon-nucleon center-of-mass energy as for the Pb-Pb collisions. The centrality dependence of $R_{AA}(\phi)$, again expressed as a function of $N_{part}$, is shown in the right panel of Figure 3. In the figure, the results from the $\phi$ measurement at mid-rapidity in the $KK$ channel [18], in the same $p_T$ range as the dimuon channel, are also reported. The comparison between the two channels suggests a systematic difference between the centrality dependence of the $R_{AA}$ measured at mid and forward rapidities, related to a different centrality dependence of the $\phi$ yield per participating nucleon in the two rapidity ranges (not shown in this contribution). Further investigations are currently ongoing on this point, unfortunately limited by the narrow $p_T$ range accessible with the dimuon channel.

## References

[1] Uras A (NA60 Collaboration) 2012 *Acta Phys. Polon. Supp.* **5** 465–470 (*Preprint* 1201.0270)
[2] Adamczyk L *et al.* (STAR Collaboration) 2012 *Phys. Rev.* **C86** 024906 (*Preprint* 1204.1890)
[3] Tserruya I (PHENIX Collaboration) 2013 *Nucl. Phys.* **A904-905** 225c–232c (*Preprint* 1211.6002)
[4] Aamodt K *et al.* (ALICE Collaboration) 2008 *JINST* **3** S08002
[5] Abelev B *et al.* (ALICE Collaboration) 2012 *Phys.Lett.* **B710** 557–568 (*Preprint* 1112.2222)
[6] Adam J *et al.* (ALICE Collaboration) 2015 (*Preprint* 1506.09206)
[7] Engel R and Ranft J 1996 *Phys.Rev.* **D54** 4244–4262 (*Preprint* hep-ph/9509373)

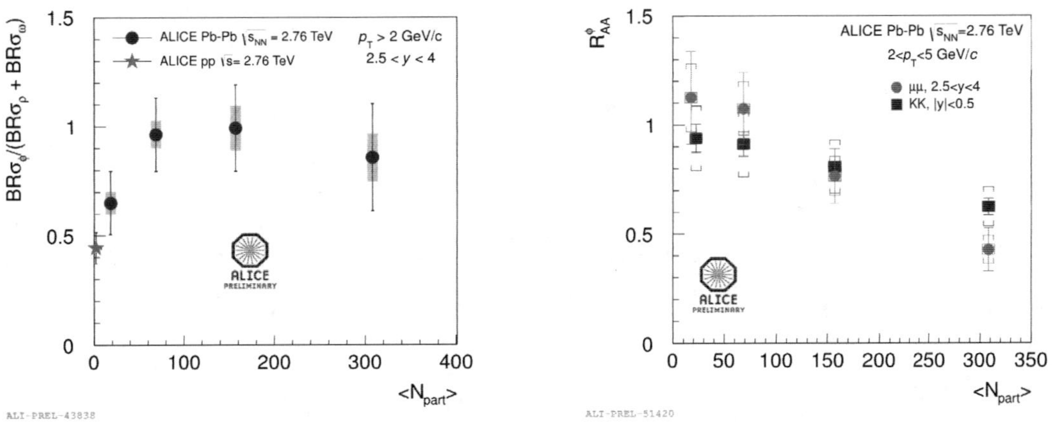


**Figure 3.** Left: ratio $BR_{\phi\to\mu\mu}\sigma_\phi/(BR_{\rho\to\mu\mu}\sigma_\rho + BR_{\omega\to\mu\mu}\sigma_\omega)$ in Pb-Pb collisions at $\sqrt{s_{\rm NN}} = 2.76$ TeV, as a function of $N_{\rm part}$. Right panel: nuclear modification factor $R_{\rm AA}$ for the $\phi$ meson, as a function of $N_{\rm part}$; forward rapidity measurement in the $\mu\mu$ channel is compared with the mid-rapidity one in the $KK$ channel.

[8] Field R 2008 *Acta Phys.Polon.* **B39** 2611–2672
[9] Skands P Z 2010 *Phys.Rev.* **D82** 074018 (*Preprint* 1005.3457)
[10] Sjostrand T, Mrenna S and Skands P Z 2006 *JHEP* **0605** 026 (*Preprint* hep-ph/0603175)
[11] Wang X N and Gyulassy M 1991 *Phys.Rev.* **D44** 3501–3516
[12] Roesler S, Engel R and Ranft J 2000 1033–1038 (*Preprint* hep-ph/0012252)
[13] Abelev B *et al.* (ALICE Collaboration) 2013 *Phys.Rev.Lett.* **110** 032301 (*Preprint* 1210.3615)
[14] Arsene I *et al.* (BRAHMS Collaboration) 2004 *Phys.Rev.Lett.* **93** 242303 (*Preprint* nucl-ex/0403005)
[15] Back B *et al.* (PHOBOS Collaboration) 2004 *Phys.Rev.Lett.* **93** 082301 (*Preprint* nucl-ex/0311009)
[16] Abelev B *et al.* (ALICE Collaboration) 2013 *Phys.Rev.Lett.* **110** 082302 (*Preprint* 1210.4520)
[17] Adare A *et al.* (PHENIX Collaboration) 2015 (*Preprint* 1506.08181)
[18] Abelev B B *et al.* (ALICE Collaboration) 2014 (*Preprint* 1404.0495)

# Measurements of strange hadrons $K_S^0$, $\Lambda$ and $\Xi$ from Au+Au collisions at $\sqrt{s_{NN}}$ = 14.5 GeV in STAR

**Muhammad Usman Ashraf (for the STAR Collaboration)**
Department of Engineering Physics, Tsinghua University, Beijing 100084, China
E-mail: musman_mughal10@yahoo.com

**Abstract.**
We report the measurements of the production of strange hadrons $K_S^0$, $\Lambda$ and $\Xi$ at mid rapidity in Au+Au collisions at $\sqrt{s_{NN}}$ = 14.5 GeV from the Beam Energy Scan (BES) program at the BNL Relativistic Heavy Ion Collider (RHIC). The collision energy dependence of strange hadron yields are also presented. To understand the recombination and parton energy loss mechanisms, nuclear modification factors and baryon to meson ratios are measured and also compared with the statistical and thermal model.

## 1. Introduction
The Quantum Chromodynamics (QCD) phase diagram lies at the heart of what the heavy-ion physics program is all about [1, 2]. The main goals of Beam Energy Scan program at RHIC are to search for the QCD critical point and the phase boundary between partonic and hadronic phases [3, 4]. To carry out this program in detail, the STAR detector is uniquely positioned due to its large uniform acceptance and excellent particle identification capabilities. In the QCD phase diagram, the wide region of temperature T and baryon chemical potential $\mu_B$ can be accessed by systematically studying the Au+Au collisions at $\sqrt{s_{NN}}$ = 7.7 – 39 GeV.

To identify the phase boundary and onset of deconfinement, strange hadrons are proposed as excellent probes. The enhanced production of strange hadrons in nucleus-nucleus collisions with respect to proton-proton collisions at the same center of mass energy has been suggested as a signature of Quark-Gluon Plasma (QGP) in these collisions [5]. Till now, strange hadron yields have been comprehensively measured in many experiments at different accelerator facilities [6–13]. In these experiments, substantial strangeness enhancement especially for multi-strange hyperons has indeed been observed. Generally, the strange hadron yields in nuclear collisions are close to those expected within the statistical hadron gas models [14–17]. In heavy-ion collisions in BES, the precise measurements of strange hadron yields will certainly lead to a better understanding of strangeness production mechanism in nuclear collisions.

The nuclear modification factor, $R_{CP}$, is the $N_{bin}$ scaled ratio of the central to peripheral $p_T$ distributions. If the hard processes are not affected by the presence of the medium, and scale with the number of binary collisions, $R_{CP}$ would be equal to unity. At RHIC, it has been observed that at high $p_T$, the $R_{CP}$ of different particles is much less than unity for Au+Au collisions at $\sqrt{s_{NN}}$ = 200 GeV [18], presumably due to the partonic energy loss in the medium. It has also been observed that the $R_{CP}$ at intermediate $p_T$, of baryons and mesons follow different trends. The baryon-to-meson ratios show a strong centrality dependence at intermediate $p_T$ range, in more central Au+Au collisions, are much higher as compared to those in elementary collisions [3,5,18]. These kinds of phenomena may be explained by the recombination/coalescence model, which is an indication of formation of strongly coupled quark-gluon plasma (sQGP) [18]. In order to locate the phase transition boundary, it will be quite interesting to measure the energy evolution of baryon-to-meson ratios and nuclear modification factor, and investigate the collision energy at which these phenomena are prevalent [3].

In this paper, we present the new experimental results on strangeness, obtained from Au+Au collisions at $\sqrt{s_{NN}}$ = 14.5 GeV. Previous measurement on strange hadron production in $\sqrt{s_{NN}}$ = 7.7, 11.5, 19.6, 27 and 39 GeV has been reported in Refs. [19–22].

## 2. Experimental data analysis

In this analysis, we have used the Au+Au collision events which are collected by the minimum bias trigger. To ensure nearly uniform acceptance of the detector, the events are required to have a primary Z vertex along the beam direction within $\pm 70$ cm from the center of the Time Projection Chamber (TPC) [23] for Au+Au collisions at $\sqrt{s_{NN}}$ = 14.5 GeV. We obtain nearly 18 million Au+Au minimum bias triggered events at $\sqrt{s_{NN}}$ = 14.5 GeV after selecting the events. From TPC, within the pseudorapidity window $|\eta| < 0.5$, the collision centrality is determined by measuring the raw charged hadron multiplicity.

The strange hadrons ($K^0_S$, $\Lambda$ and $\Xi$) in STAR have been reconstructed using the topology of their weak decay channels [24, 25], $K^0_S \to \pi^+\pi^-$ with branching ratio 69.2%, $\Lambda \to p\pi$ with branching ratio of 63.9% and $\Xi \to \Lambda\pi$ with branching ratio of 99.9%. The decay daughters $\pi$ and $p$ are identified by the TPC of the STAR detector.

## 3. Results

After making corrections in the raw spectra for reconstruction efficiency and geometrical acceptance, we obtain the $p_T$ spectra and yield ($dN/dy$) of the strange hadrons at mid rapidity in different centralities for Au+Au collisions at $\sqrt{s_{NN}}$ = 14.5 GeV. The $\Lambda$ ($\overline{\Lambda}$) spectra have been corrected for the feed down contribution from $\Xi^-$($\overline{\Xi}^+$) weak decays.

Figure 1 shows the collision energy dependence of the particle yield ($dN/dy$) at mid rapidity for $K^0_S$, $\Lambda$ and $\Xi$ for the most central (0 – 5%) Au+Au collisions STAR data at $\sqrt{s_{NN}}$ = 14.5 GeV and higher energies compared to the corresponding data from NA49, NA57 and CERES. The data from NA57 and NA49 are from the most central Pb+Pb collisions. The STAR BES data at $\sqrt{s_{NN}}$ = 14.5 GeV lie on a trend established by the previous STAR data and corresponding data from NA49 and NA57. The $\Lambda$ dN/dy seems to show non-monotonic $\sqrt{s_{NN}}$ dependence, which possibly originates from the interplay of slow rise of $\Lambda$ total multiplicity and the change of shape in rapidity distribution in the same energy range [9]. The third point from lower to higher energy in the Fig.1 is STAR BES data at $\sqrt{s_{NN}}$ = 14.5 GeV, which is in general consistent with the NA49. The data of $K^0_S$ from CERES seems to be in line with STAR BES data as well.

Figure 2 (left) shows the nuclear modification factor $R_{CP}$ of $K^0_S$, $\Lambda$ and $\Xi$ in Au+Au collisions at $\sqrt{s_{NN}}$ = 14.5 GeV at mid rapidity from STAR BES. The $R_{CP}$ data at 7.7, 11.5, 19.6, 27 and 39 GeV are from reference [19,20]. If the nucleus-nucleus collisions are just simple superpositions of nucleon-nucleon collisions the $R_{CP}$ will be unity. If these ratios deviate from unity it means there are some contributions from nuclear or medium effects. At $\sqrt{s_{NN}} \geq$ 19.6 GeV, the $K^0_S$ $R_{CP}$ $\leq 1$ at $p_T >$ 1.5 GeV/c and much less than those of baryons. The high $p_T$ suppression of $K^0_S$ and baryon/meson separation is qualitatively consistent with results from RHIC higher energies [17]. However, for $\sqrt{s_{NN}} \leq$ 14.5 GeV, the $R_{CP}$ data seem to be qualitatively different from that at $\sqrt{s_{NN}} \geq$ 19.6 GeV, although the maximum reachable $p_T$ is smaller at these energies. There is no suppression for $K^0_S$ at $p_T >$ 1.5 GeV/c; and at intermediate $p_T$ the baryon/meson separation becomes less significant. Figure 2 (right) shows the $\overline{\Lambda}/\mathrm{K}^0_\mathrm{S}$ ratio from Au+Au collisions in STAR

t

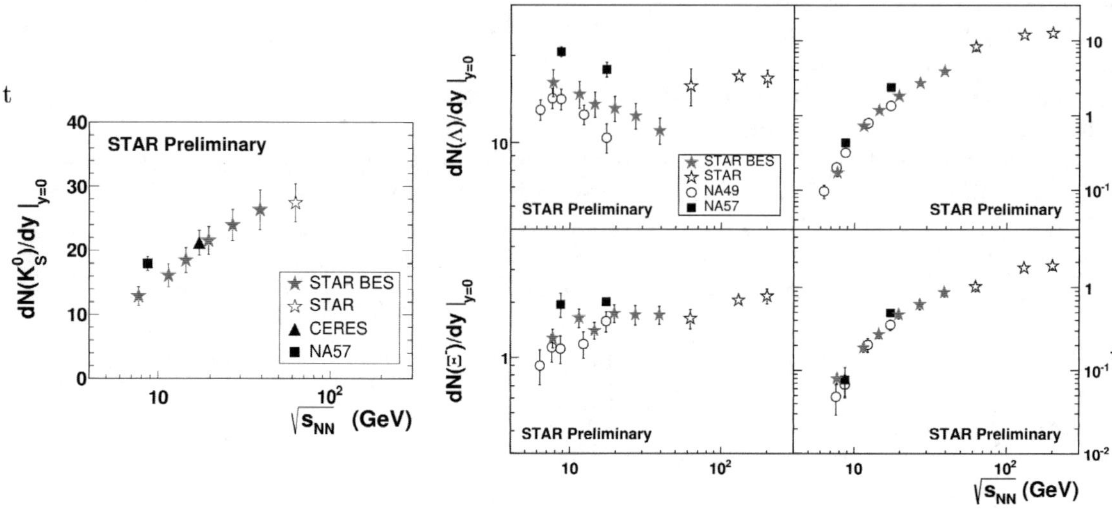


**Figure 1.** The strange hadrons $(dN/dy)$ at mid rapidity in the most central Au+Au collisions as a function of $\sqrt{s_{NN}}$ . The results at higher energies from the STAR experiment are also shown [24,25], NA57 [10,26], NA49 [9] and AGS [27]. Errors are the quadratic sum of statistical and systematic errors. The rapidity ranges are $|y| < 0.5$ for STAR and NA57, $|y| < 0.4$ for NA49 $\Lambda$ $(\overline{\Lambda})$ and $|y| < 0.5$ for NA49 $\Xi^-(\overline{\Xi}^+)$. CERES data are the extrapolated values based on the measurements at backward rapidity.

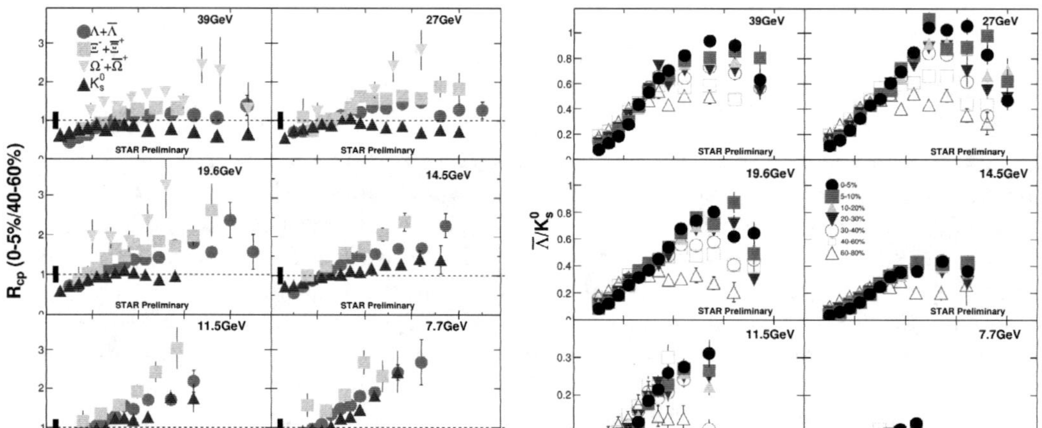


**Figure 2.** (left) The nuclear modification factor $R_{CP}$, of $K^0_S$, $\Lambda$ and $\Xi$ as a function of $p_T$ in Au+Au collisions at $\sqrt{s_{NN}} = 7.7 - 39$ GeV including 14.5 GeV at mid rapidity $|y| < 0.5$. Errors are statistical only. The left black band is the normalization error from $N_{bin}$. (Right) The ratio of $\overline{\Lambda}/K^0_S$ as a function of $p_T$at $|y| < 0.5$ from Au+Au collisions at $\sqrt{s_{NN}} = 7.7 - 39$ GeV including 14.5 GeV. Errors are statistical only.

BES as a function of $p_T$ in different centralities. At $\sqrt{s_{NN}} \geq 19.6$ GeV, in the most central collisions the $\overline{\Lambda}/K^0_S$ can reach the maximum value of unity at $p_T$ nearly 2.5 GeV/c, while the

Journal of Physics: Conference Series **668** (2016) 012095 doi:10.1088/1742-6596/668/1/012095

maximum value attained in the peripheral collisions is only about 0.3–0.5. This means that there is baryon enhancement at intermediate $p_T$ for $\sqrt{s_{NN}} \geq 19.6$ GeV, which is similar to that observed at RHIC higher energies. However, at $\sqrt{s_{NN}} \leq 14.5$ GeV for Au+Au collisions, the difference in $\overline{\Lambda}/K^0_S$ in the measured $p_T$ range is less significant between 0–5% and 40–60%. This shows much less baryon to meson enhancement at intermediate $p_T$ in Au+Au collisions at $\sqrt{s_{NN}}$ $\leq$ 14.5 GeV.

## 4. Summary

We presented recent measurements on strange hadron productions in the STAR beam energy scan. The yields of strange hadron from STAR BES program seem to follow a trend with published data from RHIC, SPS and AGS. For $\sqrt{s_{NN}} \leq 14.5$ GeV, we have also observed: $K^0_S$ $R_{CP}$ are larger than unity for $p_T > 1.5$ GeV/$c$; the difference in $\overline{\Lambda}/K^0_S$ ratio between central and peripheral collisions gets reduced.

## Acknowledgments

Muhammad Usman Ashraf thanks the support by the Major State Basic Research Development Program in China (No. 2014CB845400), the National Natural Science Foundation of China (Grant No. 11335005), and the Tsinghua University Initiative Scientific Research Program.

## References

[1] Stephanov M A *Prog. Theor. Phys. Suppl.* **153**, 139 (2004), hep-ph/0402115.
[2] Mohanty B *Nucl. Phys. A* **830**, 899c (2009), 0907.4476.
[3] Aggarwal M M *et al* 2011 *Phys. Rev. C* **83** 024901
[4] Aggarwal M M *et al.* (STAR Collaboration.), *Phys. Rev. Lett.* **105** (2010) 022302.
[5] Rafelski J and Müller B 1982 *Phys. Rev. Lett.* **48** 1066
[6] Chung P *et al.* (E895 Collaboration), *Phys. Rev. Lett.* **91** (2003) 202301.
[7] Albergo S *et al.*, *Phys. Rev. Lett.* **88** (2002) 062301.
[8] Back B B *et al.* (E917 Collaboration), *Phys. Rev. Lett.* **87** (2001) 242301.
[9] Alt C *et al.* (NA49 Collaboration), *Phys. Rev. C* **78** (2008) 034918.
[10] Antinori F *et al.* (NA57 Collaboration), *J. Phys. G* **37** (2010) 045105.
[11] Abelev B I *et al.* (STAR Collaboration), *Phys. Rev. C* **77**(2008), 044908.
[12] Adcox K *et al.* (PHENIX Collaboration), *Phys. Rev. Lett.* **89** (2002) 092302.
[13] Abelev B B *et al.* (ALICE Collaboration) 2013 arXiv:1307.5530; arXiv:1307.5543
[14] Becattini F *et al.* 2001 *Phys. Rev. C* **64** 024901
[15] Braun-Munzinger P *et al.* 2002 *Nucl. Phys. A* **697** 902
[16] Andronic A *et al.* 2006 *Nucl. Phys. A* **772** 167
[17] Redlich K and Tounsi A 2002 *Eur. Phys. J. C* **24** 589
[18] Lamont M A C (for the STAR Collaboration) 2006 *J. Phys. Conf. Ser.* **50** 192
[19] F. Zhao (STAR Collaboration), *AIP Conf. Proc.* 1441 (2012) 797.
[20] X. Zhu (STAR Collaboration), *Acta Phys. Pol. B Proc. Suppl.* 5 (2012) 213, arXiv:1203.5183v1 [nucl-ex].
[21] X. Zhu (STAR Collaboration), arXiv:1203.5182v1 [nucl-ex].
[22] X. Zhang (STAR Collaboration),*Acta Phys. Pol. B Proc. Suppl.* 5 (2012) 509, arXiv:1203.5193v1 [nucl-ex].
[23] Ackermann K H, *et al.* (STAR Collaboration), *Nucl. Instrum. Meth. A* **499** (2003) 624.
[24] Adler C *et al.* (STAR Collaboration) 2002 *Phys. Rev. Lett.* **89** 092301
[25] Adams J *et al.* (STAR Collaboration) 2004 *Phys. Rev. Lett.* **92** 182301
[26] Anticic T *et al.* [ NA49 Collaboration ], Phys. Rev. Lett. **93**, 022302 (2004). [arXiv:nucl-ex/0311024 [nucl-ex]].
[27] Ahle L *et al* (E866 and E917 Collaborations) 2000 *Phys. Lett. B* **490** 53;

Journal of Physics: Conference Series **668** (2016) 012096 doi:10.1088/1742-6596/668/1/012096

# Charmonium production at mid-rapidity in Pb-Pb and p-Pb collisions with ALICE

**Steffen Georg Weber on behalf of the ALICE Collaboration**
Research Division, GSI Helmholtzzentrum für Schwerionenforschung, Darmstadt, Germany
E-mail: S.Weber@gsi.de

**Abstract.** We present an overview of the ALICE measurements on the production of J/$\psi$ in Pb-Pb collisions at a centre-of-mass energy per nucleon pair of $\sqrt{s_{\mathrm{NN}}} = 2.76\,\mathrm{TeV}$ and p-Pb collisions at $\sqrt{s_{\mathrm{NN}}} = 5.02\,\mathrm{TeV}$ at mid-rapidity ($|y_{\mathrm{lab}}| < 0.8$) down to zero transverse momentum. The cold nuclear matter effects estimated from the p-Pb measurements and their impact on the interpretation of Pb-Pb results are discussed, based on comparison of data to model calculations.

## 1. Introduction

Charmonium production has been proposed as a sensitive probe of the hot and dense medium created in ultra-relativistic heavy-ion collisions. The original idea [1] was that in nucleus-nucleus collisions the binding potential between the charm and anticharm quarks is screened due to the high density of colour charges via a mechanism similar to the Debye screening. Results at SPS [2] and RHIC [3] energies indeed showed a suppression of the J/$\psi$ yield in most central collisions, beyond what was expected from cold nuclear matter (CNM) effects.

At LHC energies, charm-anticharm quark pairs are abundantly produced in initial hard partonic processes. For this reason, it has been suggested that a new production mechanism based on the (re)combination of independently produced charm and anticharm quarks to charmonium pairs becomes relevant. These models assume either a creation of charmonium states at the hadronization stage, like the Statistical Hadronization Model (SHM) [4] or a continuous dissociation and regeneration throughout the lifetime of the hot medium, addressed in transport models [5].

The experimental observable used to quantify modifications of charmonium production in nucleus-nucleus collisions compared to proton-proton collisions is the nuclear modification factor, $R_{\mathrm{AA}}$, defined as

$$R_{\mathrm{AA}} = \frac{1}{\langle T_{\mathrm{AA}} \rangle} \frac{\mathrm{d}^2 N_{\mathrm{AA}}^{\mathrm{J}/\psi}/\mathrm{d}p_{\mathrm{T}}\mathrm{d}y}{\mathrm{d}^2 \sigma_{\mathrm{pp}}^{\mathrm{J}/\psi}/\mathrm{d}p_{\mathrm{T}}\mathrm{d}y},$$

where $T_{\mathrm{AA}}$ is the nuclear overlap function determined by Glauber model calculations.

In addition to hot medium effects, also cold nuclear matter effects can influence the production of charmonia, the most relevant ones at LHC energies are listed below. The parton distribution functions (PDF) in a nucleus are modified compared to the one in a free nucleon, this effect is known as shadowing [6]. There is a limited knowledge on gluon PDFs in heavy ion collisions at the moment, new data on charmonium production can provide further constraints.

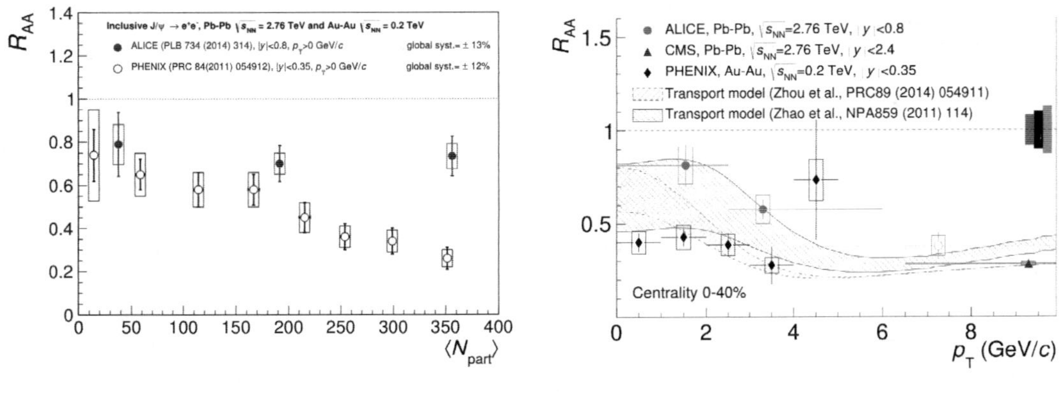


**Figure 1.** Left: The nuclear modification factor $R_{\mathrm{AA}}$ measured by ALICE in Pb-Pb collisions at $\sqrt{s_{\mathrm{NN}}} = 2.76\,\mathrm{TeV}$ as a function of the mean number of participants $\langle N_{\mathrm{part}}\rangle$[7], compared with PHENIX [8] results in Au-Au collisions at $\sqrt{s_{\mathrm{NN}}} = 200\,\mathrm{GeV}$. Right: $R_{\mathrm{AA}}$ as function of transverse momentum [9], compared with measurements from CMS [10] and PHENIX [3] and with theoretical models [11, 12].

Another way to describe the depletion of the nuclear gluon PDFs in the kinematical domain relevant for the charmonium production is via gluon saturation, which is addressed in the Color Glass Condensate (CGC) model [13]. Other theoretical models take into account energy loss mechanisms on the partonic level, before and after the hard collision [14].

The cold nuclear matter effects are experimentally accessible via proton-nucleus collisions, where the creation of a hot medium is not expected. These measurements are thus crucial for the interpretation of data from nucleus-nucleus collisions.

## 2. Experimental Results

ALICE has measured inclusive $\mathrm{J}/\psi$ production at mid-rapidity in the dielectron decay channel $\mathrm{J}/\psi \rightarrow \mathrm{e}^+\mathrm{e}^-$ and in forward rapidity in the dimuon decay channel $\mathrm{J}/\psi \rightarrow \mu^+\mu^-$. In both cases, $\mathrm{J}/\psi$ are reconstructed down to $p_{\mathrm{T}} = 0$. Here, measurements at mid-rapidity will be discussed. The pseudorapidity coverage for measurements in Pb-Pb collisions is $|\eta| < 0.8$, for p-Pb collisions, the coverage is shifted to $-1.37 < \eta_{\mathrm{cms}} < 0.43$ in the centre-of-mass-frame due to the different per nucleon energy of the two collision partners. The main detectors which have been used for the presented results are the Inner Tracking System (ITS) for collision vertex determination and tracking, the Time Projection Chamber (TPC), for tracking and particle identification, and the V0 and Zero Degree Calorimeter (ZDC) detectors for centrality determination and event selection.

The raw $\mathrm{J}/\psi$ yield has been obtained using the bin counting method from the invariant mass distribution of dielectron pairs in the window $2.92 < m_{\mathrm{ee}} < 3.16\,\mathrm{GeV}/c^2$ after combinatorial background subtraction using the mixed-event technique. For the calculation of the nuclear modification factor, the pp reference was obtained with an interpolation procedure including data at mid-rapidity from ALICE [15], PHENIX [16] and CDF [17].

The presented results from Pb-Pb collisions are based on data taken in 2010 and 2011 at a collision energy of $\sqrt{s_{\mathrm{NN}}} = 2.76\,\mathrm{TeV}$ with an integrated luminosity $\mathcal{L}_{\mathrm{int}} = 27.7\,\mu\mathrm{b}^{-1}$, the results from p-Pb collisions are based on data from 2013 at a collision energy of $\sqrt{s_{\mathrm{NN}}} = 5.02\,\mathrm{TeV}$ with $\mathcal{L}_{\mathrm{int}} = 51.4\,\mu\mathrm{b}^{-1}$. More details of the analysis can be found in [7] for the Pb-Pb analysis and in [18] for the p-Pb analysis.

The left panel of Fig. 1 shows the $p_{\mathrm{T}}$-integrated nuclear modification factor $R_{\mathrm{AA}}$ from Pb-

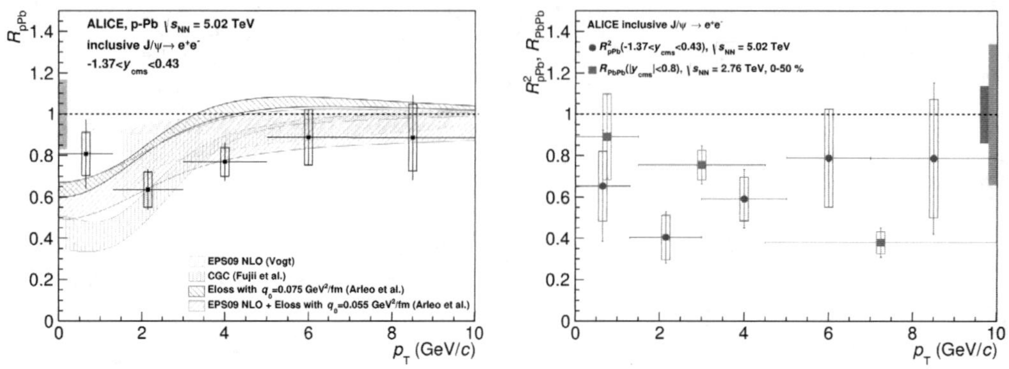


**Figure 2.** Left: The nuclear modification factor $R_{\mathrm{pPb}}$ measured by ALICE in p-Pb collisions at $\sqrt{s_{\mathrm{NN}}} = 5.02\,\mathrm{TeV}$ as a function of $p_{\mathrm{T}}$ [18], together with model calculations [19, 20, 21]. Right: Squared nuclear modification factor $R^2_{\mathrm{pPb}}$ as an estimate for CNM effects in Pb-Pb collisions and $R_{\mathrm{AA}}$from Pb-Pb collisions at $\sqrt{s_{\mathrm{NN}}} = 2.76\,\mathrm{TeV}$.

Pb collisions at mid-rapidity as a function of the average number of participating nucleons $\langle N_{\mathrm{part}}\rangle$. The ALICE results [7] are compared to measurements by PHENIX in Au-Au collisions at $\sqrt{s_{\mathrm{NN}}} = 200\,\mathrm{GeV}$ [8]. The results for central collisions are significantly different, with the ALICE data showing a factor of three less suppression than the PHENIX results. Also the PHENIX results indicate a strong centrality dependence, with a suppression increasing with centrality, whereas the ALICE results are compatible with no centrality dependence. The same behaviour is seen at forward rapidity [22]. The pt dependence of the J/$\psi$ $R_{\mathrm{AA}}$ for the 40% most central collisions is shown in the right panel of Fig. 1. Our data is compared to PHENIX results in Au-Au collisions at $\sqrt{s_{\mathrm{NN}}} = 200\,\mathrm{GeV}$ [3], CMS results at higher $p_{\mathrm{T}}$ [10] and model calculations [11, 12]. The ALICE data suggests a much lower degree of suppression at low $p_{\mathrm{T}}$ compared to lower energy measurements. Our data is well described by transport model predictions by Zhao et al. [12]. Newer predictions by Zhou et al. [11] are systematically below the measurement and exhibit a $p_{\mathrm{T}}$ dependence similar to the one in the data. The model calculations show substantial theoretical uncertainties due to the limited knowledge of charm cross section and cold nuclear matter effects at LHC energies.

Another experimental observable adopted to gain further insight on charmonium production mechanisms is the mean transverse momentum $\langle p_{\mathrm{T}}\rangle$ of J/$\psi$ produced in different collisions systems. ALICE measurements [9] have shown that the $\langle p_{\mathrm{T}}\rangle$ for Pb-Pb collisions is significantly smaller than that for pp collisions. Such a behaviour is not observed at smaller centre-of-mass energies from meausrements by PHENIX and NA50, for which no significant system size dependence of $\langle p_{\mathrm{T}}\rangle$ is seen. This might indicate the onset of processes which either deplete the high $p_{\mathrm{T}}$ region or enhance the J/$\psi$ production at low $p_{\mathrm{T}}$ in heavy-ion collisions at the LHC. The latter effect would be expected as a consequence of a significant contribution from charmonium recombination.

On the left panel of Fig. 2 the nuclear modification factor in p-Pb collisions $R_{\mathrm{pPb}}$ is shown as a function of $p_{\mathrm{T}}$, together with model calculations. These consider either pure gluon shadowing effects [19], gluon saturation [20] or partonic energy loss, with or without inclusion of nuclear shadowing contributions [21]. All considered models describe the data reasonably well.

In order to compare the results obtained in Pb-Pb and in p-Pb collisions, it is important to notice that for the investigated energies and rapidity regions, the probed Bjorken-$x$ ranges are comparable. Using $2 \to 1$ ($gg \to \mathrm{J}/\psi$) kinematics, in Pb-Pb collisions the probed range is $7.0 \times 10^{-4} < x < 3.5 \times 10^{-3}$ and in p-Pb collisions $6.1 \times 10^{-4} < x < 3.0 \times 10^{-3}$. Under the assumption that shadowing is the main CNM-related mechanism that plays a role in the charmonium production and that its effect on the two colliding nuclei in Pb-Pb collisions can be factorized, the squared nuclear modification factor $R^2_{\mathrm{pPb}}$ can be considered as an estimate of CNM effects in Pb-Pb collisions [23, 24].

The right panel of Fig. 2 shows $R^2_{\mathrm{pPb}}$ together with $R_{\mathrm{AA}}$ as a function of $p_{\mathrm{T}}$. The $R^2_{\mathrm{pPb}}$ shows a clear $p_{\mathrm{T}}$ dependence with a stronger effect at low $p_{\mathrm{T}}$, whereas the $R_{\mathrm{AA}}$ shows the largest suppression at high $p_{\mathrm{T}}$. These differences suggest that the J/$\psi$ suppression in Pb-Pb collisions cannot be explained by CNM effects alone. The data at low $p_{\mathrm{T}}$ hint at the presence of charmonium production in Pb-Pb consistent with (re)combination scenarios, whereas at high $p_{\mathrm{T}}$ suppression effects seem to be present in the hot medium.

## 3. Conclusions

Measurements by ALICE of inclusive J/$\psi$ production at mid-rapidity as a function of centrality and transverse momentum in Pb-Pb collisions at $\sqrt{s_{\mathrm{NN}}} = 2.76\,\mathrm{TeV}$ and in p-Pb collisions at $\sqrt{s_{\mathrm{NN}}} = 5.02\,\mathrm{TeV}$ have been presented. Uniquely at the LHC, these measurements reach down to $p_{\mathrm{T}} = 0$. In Pb-Pb collisions, a suppression of J/$\psi$ production is observed, with no strong centrality dependence and a weak suppression at low transverse momentum. The results are in contrast to measurements at RHIC energies, which show an increase of the suppression with centrality, especially at low $p_{\mathrm{T}}$. Our data can be well described by theoretical models which take into account a (re)combination component in the J/$\psi$ production.

In p-Pb collisions, a suppression of J/$\psi$ production at low transverse momenta was observed, which can be well described by models taking into account either shadowing effects, gluon saturation, or partonic energy loss. An estimate of CNM effects in Pb-Pb via the squared nuclear modification factor $R^2_{\mathrm{pPb}}$ in p-Pb collisions underlines the importance of hot nuclear matter effects for the interpretation of the results in Pb-Pb collisions and supports the idea of charmonium production in the hot medium or at hadronization.

## References

[1] Matsui T and Satz H 1986 *Phys. Lett.* B **178** 416
[2] Alessandro B *et al.* (NA50) 2005 *Eur. Phys. J.* C **39** 335–345
[3] Adare A *et al.* (PHENIX) 2007 *Phys. Rev. Lett.* **98** 232301
[4] Braun-Munzinger P and Stachel J 2000 *Phys. Lett.* B **490** 196–202
[5] Thews R L, Schroedter M and Rafelski J 2001 *Phys. Rev.* C **63**(5) 054905
[6] Eskola K J, Paukkunen H and Salgado C A 2009 *JHEP* **04** 065
[7] Abelev B *et al.* (ALICE) 2014 *Phys. Lett.* B **734** 314–327
[8] Adare A *et al.* (PHENIX) 2011 *Phys. Rev.* C **84** 054912
[9] Adam J *et al.* (ALICE) 2015 *JHEP* **07** 051
[10] Chatrchyan S *et al.* (CMS) 2012 *JHEP* **05** 063
[11] Zhou K, Xu N, Xu Z and Zhuang P 2014 *Phys. Rev.* C **89** 054911
[12] Zhao X and Rapp R 2011 *Nucl. Phys.* A **859** 114–125
[13] Gelis F, Iancu E, Jalilian-Marian J and Venugopalan R 2010 *Ann. Rev. Nucl. Part. Sci.* **60** 463–489
[14] Sharma R and Vitev I 2013 *Phys. Rev.* C **87** 044905
[15] Aamodt K *et al.* (ALICE) 2011 *Phys. Lett.* B **704** 442–455 [Erratum: Phys. Lett.B718,692(2012)]
[16] Adare A *et al.* (PHENIX) 2012 *Phys. Rev.* D **85** 092004
[17] Acosta D *et al.* (CDF) 2005 *Phys. Rev.* D **71** 032001
[18] Adam J *et al.* (ALICE) 2015 *JHEP* **06** 055
[19] Albacete J, Armesto N, Baier R, Barnafoldi G, Barrette J *et al.* 2013 *Int.J.Mod.Phys.* E **22** 1330007
[20] Fujii H and Watanabe K 2013 *Nucl.Phys.* A **915** 1–23
[21] Arleo F and Peigne S 2013 *JHEP* **1303** 122
[22] Abelev B *et al.* (ALICE) 2012 *Phys. Rev. Lett.* **109** 072301
[23] Vogt R 2010 *Phys. Rev.* C **81**(4) 044903
[24] Ferreiro E, Fleuret F, Lansberg J and Rakotozafindrabe A 2009 *Phys. Lett.* B **680** 50 – 55

Journal of Physics: Conference Series **668** (2016) 012097 doi:10.1088/1742-6596/668/1/012097

# Analytical Expressions for the Hard-Scattering Production of Massive Partons

**Cheuk-Yin Wong**

Physics Division, Oak Ridge National Laboratory, Oak Ridge, Tennessee 37831, USA

**Abstract.** We obtain explicit expressions for the two-particle differential cross section $E_c E_\kappa d\sigma(AB{\to}c\kappa X)/d\boldsymbol{c}\, d\boldsymbol{\kappa}$ and the two-particle angular correlation function $d\sigma(AB{\to}c\kappa X)/d\Delta\phi\, d\Delta y$ in the hard-scattering production of massive partons in order to exhibit the "ridge" structure on the away side in the hard-scattering process. The single-particle production cross section $d\sigma(AB{\to}cX)/dy_c c_T dc_T$ is also obtained and compared with the ALICE experimental data for charm production in $pp$ collisions at 7 TeV at LHC.

## 1. Introduction

Knowledge of massive quark production processes in $pp$ collisions provides useful insight to guide our intuition in heavy quark production in nucleus-nucleus collisions. Analytical expressions for these processes summarize important features and essential dependencies so as to facilitate the uncovering of dynamical effects wherever they may occur. Similar analyses in massless quark production have led to new insights in the dominance of the hard-scattering process over a large $p_T$ domain and have paved the way for locating the boundary between the hard-scattering process and the flux-tube fragmentation process in high-energy $pp$ collisions [1, 2].

Accordingly, we would like to obtain $E_c E_\kappa d\sigma(AB{\to}c\kappa X)/d\boldsymbol{c}\, d\boldsymbol{\kappa}$ for the two-particle differential cross section in the production of massive partons $c$ and $\kappa$. From such a general result, we integrate out the transverse momenta and obtain the two-particle angular correlation function $d\sigma/d\Delta\phi d\Delta y$ where $\Delta\phi{=}\phi_\kappa{-}\phi_c$ and $\Delta y{=}y_\kappa{-}y_c$, exhibiting analytically the "ridge" structure on the away side at $\Delta\phi{\sim}{\pm}\pi$ in the hard-scattering process. We subsequently examine $d\sigma(AB{\to}cX)/dy_c c_T dc_T$ for the single-particle spectrum and compare with ALICE experimental data for charm production in $pp$ collisions at 7 TeV at LHC [3].

## 2. Hard Scattering Integral for $E_c E_\kappa d\sigma(AB \to c\kappa X)/d\boldsymbol{c}\, d\boldsymbol{\kappa}$

In the parton model, the hard-scattering cross section for $AB \to c\kappa X$ is given by [4]

$$d\sigma(AB \to c\kappa X) = \sum_{ab} \int K_{ab} dx_a d\boldsymbol{a}_T dx_b d\boldsymbol{b}_T G_{a/A}(x_a, \boldsymbol{a}_T) G_{b/B}(x_b, \boldsymbol{b}_T) d\sigma(ab \to c\kappa), \qquad (1)$$

where $(x_a, \boldsymbol{a}_T)$ and $(x_b, \boldsymbol{b}_T)$ represent the momenta and $G_{a/A}$ and $G_{b/B}$ the structure functions of the incident partons $a$ and $b$ respectively, and $K_{ab}$ is the correction factor which can be obtained perturbatively [5] or it can also be approximated nonperturbatively [6]. The quantity $d\sigma(ab \to c\kappa)$ is the cross section element for the process $ab \to c\kappa$,

$$d\sigma(ab \to c\kappa) = \frac{1}{4[(a\cdot b)^2 - m_a^2 m_b^2]^{1/2}} |T_{fi}|^2 \frac{d^3c}{(2\pi)^3 2E_c} \frac{d^3\boldsymbol{\kappa}}{(2\pi)^3 2E_\kappa} (2\pi)^4 \delta^4(a+b-c-\kappa). \qquad (2)$$

Here, we normalize the Dirac fields by $\bar{u}u = 2m$. The quantity $|T_{fi}|^2$ is related to $d\sigma/dt$ by

$$|T_{fi}|^2 = 16\pi[\hat{s} - (m_a + m_b)^2][\hat{s} - (m_a - m_b)^2]\frac{d\sigma(ab \to c\kappa)}{dt}. \tag{3}$$

We consider the simplified case with $m_a = m_b = 0$ and treat $a_T, b_T$ as small perturbations. The cross section element is then

$$d\sigma(ab \to c\kappa) = \frac{s_{ab}}{2\pi}\frac{d\sigma(ab \to c\kappa)}{dt}\frac{d^3c}{E_c}\frac{d^3\kappa}{E_\kappa}\delta^4(a + b - c - \kappa), \tag{4}$$

where $\hat{s} = s_{ab} = (a + b)^2$ that is different from $s = s_{AB} = (A + B)^2$. We get

$$\frac{E_c E_\kappa d\sigma(AB \to c\kappa X)}{d^3c\, d^3\kappa} = \sum_{ab}\int K_{ab} dx_a d\boldsymbol{a}_T dx_b d\boldsymbol{b}_T G_{a/A}(x_a, \boldsymbol{a}_T) G_{b/B}(x_b, \boldsymbol{b}_T)\frac{\hat{s}}{2\pi}\frac{d\sigma(ab \to c\kappa)}{dt}\delta^4(a + b - c - \kappa). \tag{5}$$

We consider a fractorizable structure function with a Gaussian intrinsic transverse momentum distribution,

$$G_{a/A}(x_a, \boldsymbol{a}_T) = G_{a/A}(x_a)\frac{1}{2\pi\sigma^2}e^{-\boldsymbol{a}_T^2/2\sigma^2}. \tag{6}$$

Upon integrating over the transverse momenta $\boldsymbol{a}_T$ and $\boldsymbol{b}_T$, we obtain

$$\frac{d\sigma(AB \to c\kappa X)}{dy_c c_T dc_T d\phi_c\, dy_\kappa \kappa_T d\kappa_T d\phi_\kappa} = \sum_{ab}\int K_{ab} dx_a dx_b G_{a/A}(x_a) G_{b/B}(x_b)\frac{e^{-\frac{(\boldsymbol{c}_T + \boldsymbol{\kappa}_T)^2}{4\sigma^2}}}{2(2\pi\sigma^2)}\frac{\hat{s}}{2\pi}\frac{d\sigma(ab \to c\kappa)}{dt}$$
$$\times\delta(a_0 + b_0 - (c_0 + \kappa_0))\delta(a_z + b_z - (c_z + \kappa_z)). \tag{7}$$

To carry out the integration over $x_a$ and $x_b$, we write out the momenta in the infinite momentum frame,

$$a = (x_a\frac{\sqrt{s}}{2} + \frac{a^2 + a_T^2}{2x_a\sqrt{s}},\ \boldsymbol{a}_T,\ x_a\frac{\sqrt{s}}{2} - \frac{a^2 + a_T^2}{2x_a\sqrt{s}}), \tag{8}$$

$$b = (x_b\frac{\sqrt{s}}{2} + \frac{b^2 + b_T^2}{2x_b\sqrt{s}},\ \boldsymbol{b}_T, -x_b\frac{\sqrt{s}}{2} + \frac{b^2 + b_T^2}{2x_b\sqrt{s}}), \tag{9}$$

$$c = (x_c\frac{\sqrt{s}}{2} + \frac{c^2 + c_T^2}{2x_c\sqrt{s}},\ \boldsymbol{c}_T,\ x_c\frac{\sqrt{s}}{2} - \frac{c^2 + c_T^2}{2x_c\sqrt{s}}), \tag{10}$$

$$\kappa = (x_\kappa\frac{\sqrt{s}}{2} + \frac{\kappa^2 + \kappa_T^2}{2x_\kappa\sqrt{s}},\ \boldsymbol{\kappa}_T, -x_\kappa\frac{\sqrt{s}}{2} + \frac{\kappa^2 + \kappa_T^2}{2x_\kappa\sqrt{s}}), \tag{11}$$

where $x_c$ and $x_\kappa$ can be represented by $y_c$ and $y_\kappa$

$$x_c = \frac{m_{cT}e^{y_c}}{\sqrt{s}}, \quad x_\kappa = \frac{m_{\kappa T}e^{y_\kappa}}{\sqrt{s}}. \tag{12}$$

The two delta functions in Eq. (7) can be integrated to yield

$$\frac{d\sigma(AB \to cdX)}{dy_c c_T dc_T d\phi_c\, dy_\kappa \kappa_T d\kappa_T d\phi_\kappa} = \sum_{ab} K_{ab} x_a G_{a/A}(x_a) x_b G_{b/B}(x_b)\frac{e^{-\frac{(\boldsymbol{c}_T + \boldsymbol{d}_T)^2}{4\sigma^2}}}{2\pi(4\pi\sigma^2)}\frac{d\sigma(ab \to c\kappa)}{dt}, \tag{13}$$

$$\text{where} \quad x_a = x_c + \frac{\kappa^2 + \kappa_T^2}{x_\kappa s} - \frac{b^2 + b_T^2}{x_b s} = \frac{m_{cT}e^{y_c}}{\sqrt{s}} + \frac{m_{\kappa T}e^{-y_\kappa}}{\sqrt{s}} - \frac{b^2 + b_T^2}{x_b s},$$
$$x_b = x_\kappa + \frac{c^2 + c_T^2}{x_c s} - \frac{a^2 + a_T^2}{x_a s} = \frac{m_{\kappa T}e^{y_\kappa}}{\sqrt{s}} + \frac{m_{cT}e^{-y_c}}{\sqrt{s}} - \frac{a^2 + T^2}{x_a s}. \tag{14}$$

The above explicit formula gives the cross section for the production of $c$ and $\kappa$, when the elementary cross section $d\sigma(ab \to c\kappa)/dt$ is given explicitly in terms of its depending variables.

Journal of Physics: Conference Series **668** (2016) 012097 doi:10.1088/1742-6596/668/1/012097

## 3. $c\kappa$ angular correlation $d\sigma(AB \to c\kappa X)/d\Delta\phi\, d\Delta y$

We can represent $\boldsymbol{c}$ and $\boldsymbol{\kappa}$ by $(y_c, \phi_c)$ and $(y_c + \Delta y, \phi_c + \Delta\phi)$, respectively. After averaging over $y_c$ and $\phi_c$, and integrating over $c_T, \kappa_T$, the correlation function (13) from the process $ab \to c\kappa$ is

$$\frac{d\sigma(AB \to c\kappa X)}{d\Delta\phi\, d\Delta y} = K_{ab} x_a G_{a/A}(x_a) x_b G_{b/B}(x_b) \delta_\sigma(\Delta\phi), \tag{15}$$

$$\text{where } \delta_\sigma(\Delta\phi) = \frac{1}{(4\pi\sigma^2)} \int_0^\infty c_T dc_T \int_0^\infty \kappa_T d\kappa_T \exp\{-\frac{c_T^2 + 2c_T\kappa_T \cos\Delta\phi + \kappa^2}{4\sigma^2}\} \frac{d\sigma(ab \to c\kappa)}{dt}. \tag{16}$$

The above analytical expression assumes a simple form for $a$=$b$ and $c$=$\kappa$. The structure function can be represented in the form $x_a G_{a/A}(x_a) \propto (1 - x_a)^{g_a}$ for which the two-particle angular correlation function becomes

$$\frac{d\sigma(AB \to c\kappa X)}{d\Delta\phi\, d\Delta y} \sim A\Big[1 - \frac{2m_{cT}}{\sqrt{s}}[\cosh y_c + \cosh(y_c + \Delta y)] + 2(\frac{m_{cT}}{\sqrt{s}})^2 [1 + \cosh(2y_c + \Delta y)]\Big]^{g_a} \delta_\sigma(\Delta\phi). \tag{17}$$

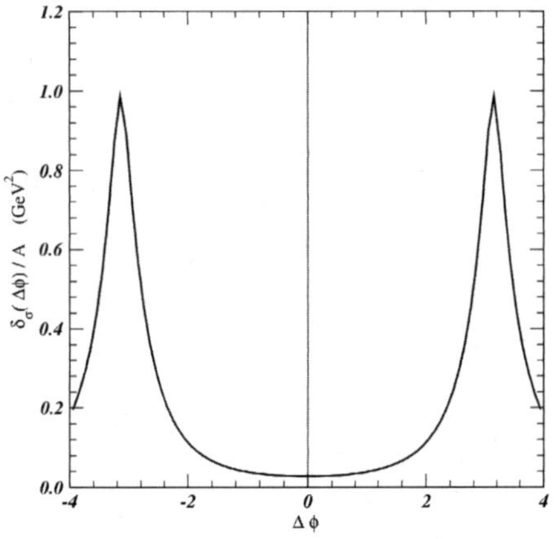


**Fig. 1** The correlation function $\delta_\sigma(\Delta\phi)/A$

If we consider $d\sigma(ab \to c\kappa)/dt$ to be approximately of the form

$$\frac{d\sigma(ab \to c\kappa)}{dt} = \frac{A}{(1 + c_T^2/m_c^2)^{n/2}}, \tag{18}$$

where $n$=4 from pQCD, then the integration over $c_T$ and $\kappa_T$ in Eq. (16) gives $\delta_\sigma(\Delta\phi)/A$ as shown in Fig. 1. The correlation function has maxima at $\Delta\phi \sim \pm\pi$ and a minimum at $\Delta\phi$=0. It is relatively flat in $\Delta y$ because $m_{cT}/\sqrt{s} \ll 1$. This gives the distribution in the form of a ridge structure on the away side at $\Delta\phi \sim \pm\pi$.

## 4. Production of massive quarks by gluons

We consider the process $gg \to c\bar{c}$, where analytical expressions $d\sigma/dt$ have been obtained earlier by Cambridge [7] and Glück, Owen, and Reya [8]. In the notation of [8], the cross section is

$$\frac{d\sigma(gg \to c\bar{c})}{dt} = \frac{\pi\alpha_s^2}{64\hat{s}^2}\Big[12M_{ss} + \frac{16}{3}M_{tt} + \frac{16}{3}M_{uu} + 6M_{st} + 6M_{su} - \frac{2}{3}M_{tu}\Big]. \tag{19}$$

Upon writing out the above quantity as a function of $c_T, \bar{c}_T, \bar{y} = (y_c + y_{\bar{c}})/2$ and $\Delta y = y_{\bar{c}} - y_c$, we get the heavy-quark pair production cross section

$$\frac{d\sigma(AB \to c\bar{c}X)}{dy_c c_T dc_T d\phi_c\, dy_{\bar{c}} \bar{c}_T d\bar{c}_T d\phi_\kappa} \sim AK_{ab}(1 - x_a)^{g_a}(1 - x_b)^{g_b} \frac{e^{-\frac{(\boldsymbol{c}_T + \bar{\boldsymbol{c}}_T)^2}{4\sigma^2}}}{2\pi(4\pi\sigma^2)} \frac{d\sigma(gg \to c\bar{c})}{dt}, \tag{20}$$

$$\text{where } \frac{d\sigma(gg \to c\bar{c})}{dt} = \frac{\pi\alpha_s^2}{4^5 m_{cT}^4 \cosh^4\bar{y}}\Bigg\{\Big[\frac{12}{\cosh^2\bar{y}} + \frac{64}{3}\cosh 2\bar{y} - 24\Big] + (\frac{m_c}{m_{cT}})^2\Big[\frac{64}{3} + \frac{24\sinh^2\bar{y}}{\cosh^2\bar{y}} - \frac{8}{3}\Big] + \frac{m_c^4}{m_{cT}^4}\Big[-\frac{64}{3}\frac{\cosh 2\bar{y}}{\cosh^2\bar{y}} + \frac{8}{3}\frac{1}{\cosh^2\bar{y}}\Big]\Bigg\},$$

$x_a = 2m_{cT}\cosh\bar{y}e^{\Delta y/2}$, and $x_b = 2m_{cT}\cosh\bar{y}e^{-\Delta y/2}$. This shows the back-to-back correlation of $\boldsymbol{c}_T$ and $\bar{\boldsymbol{c}}_T$ and the relatively flat distribution as a function of $\Delta y$.

## 5. Single-particle charm production

We need to integrate over $\boldsymbol{\kappa}$ in Eq. (5) to get the single-particle distribution of $\boldsymbol{c}$. Neglecting intrinsic $p_T$ and integrating over $\boldsymbol{\kappa}$, we get

$$\frac{E_c d\sigma(AB\to cX)}{d^3c} = \sum_{ab}\int dx_a dx_b G_{a/A}(x_a) G_{b/B}(x_b) \frac{\hat s}{\pi}\frac{d\sigma(ab\to cd)}{dt}\delta(\hat s+\hat t+\hat u - m_c^2 - m_\kappa^2). \quad (21)$$

The integral over $x_a$ and $x_b$ can be evaluated by the saddle point method [1], and we get for $\bar y = y_c \approx 0$,

$$\frac{E_c d\sigma(AB\to cX)}{d^3c} \propto K_{ab}(1-x_a)^{g_a+1/2}(1-x_b)^{g_b+1/2}\frac{1}{\sqrt{x_c}}\frac{d\sigma(gg\to c\bar c)}{dt} \quad (22)$$

$$= \frac{AK_{ab}\alpha_s^2}{m_{cT}^{4.5}\cosh^4\bar y}\left\{\left[\frac{12}{\cosh^2\bar y}+\frac{64}{3}\cosh 2\bar y - 24\right]+\frac{m_c^2}{m_{cT}^2}\left[\frac{64}{3}+\frac{24\sinh^2\bar y}{\cosh^2\bar y}-\frac{8}{3}\right]+\frac{m_c^4}{m_{cT}^4}\left[-\frac{64}{3}\frac{\cosh 2\bar y}{\cosh^2\bar y}+\frac{8}{3}\frac{1}{\cosh^2\bar y}\right]\right\},$$

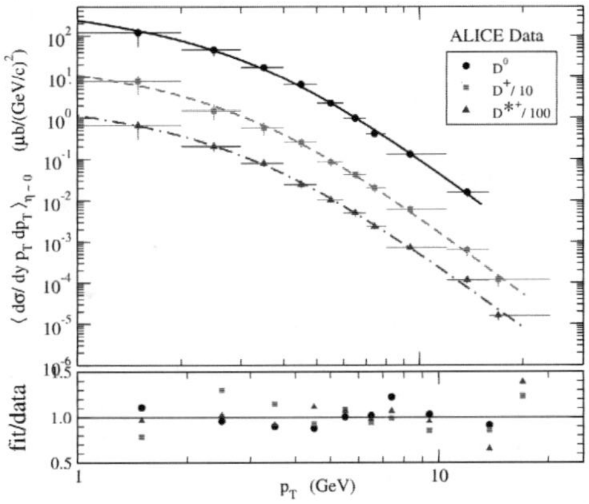


**Fig. 2** . ALICE data for charm production.

which contains $1/m_{cT}^4$, $m_c^2/m_{cT}^6$, and $m_c^4/m_{cT}^8$. We examine the ALICE charm production cross section data in *pp* collisions at 7 TeV [3] shown in Fig. 2. If we parametrize the data at mid-rapidity as $d\sigma/dy_c c_T dc_T \sim a/(1+p_T^2/m_0^2)^{n/2}$, we find that the ALICE data can be fitted by a set of parameters given by

| $d\sigma/dyp_Tdp_T\|_{\eta\sim 0} = a/(1+p_T^2/m_0^2)^{n/2}$ | | | |
|---|---|---|---|
| Data | $a$ [$\mu b/(\mathrm{GeV/c})^{-2}$] | $n$ | $m_0$ (GeV) |
| $D^0$ | 1600 | 5.8 | 3.5 |
| $D^+$ | 780 | 5.9 | 3.5 |
| $D^{*+}$ | 808 | 5.7 | 3.5 |

The extracted values of $n$ and $m_0$ are greater than those from the expected lowest-order results of $n = 4.5$ and $m_c \sim 1.5$ GeV in Eq. (22). This may arise from the final-state interactions correction factor $K_{ab}$ [6] for production of the $c$-$\bar c$ pair in the color singlet state. The attractive color-singlet interaction between $c$ and $\bar c$ enhances the production of the pair at lower $p_T$ and increases the value of the effective mass of the produced charm meson.

## 6. Conclusion

We present analytical expressions for the hard-scattering production of massive quarks in order to guide our intuition, point out essential dependencies, and summarize important features. They will facilitate future comparisons with experimental data and pave the way for a better understanding of particle production processes.

The research was supported in part by the Division of Nuclear Physics, U.S. Department of Energy under Contract DE-AC05-00OR22725.

## References

[1] C.Y.Wong and G.Wilk, Acta Phys. Pol. B **43**, 2047 (2012); C.Y.Wong and G.Wilk, Phys. Rev. D**87**, 114007 (2013); C.Y.Wong, G.Wilk, L.J.L.Cirto and C.Tsallis, EPJ Web Conf. **90**, 04002 (2015); C.Y.Wong,G.Wilk, L.J L.Cirto and C.Tsallis, Phys.Rev **D91**,114027 (2015).
[2] C.Y.Wong, arxiv:1505.07817, in press in Physical Review D.
[3] B. Abelev *et al.*, (ALICE Collaboration), JHEP 01, 128 (2012).
[4] See for example the review by J. F. Owens, Rev. Mod. Phys. **59**,465 (1987).
[5] See for example, T. Sjöstrand *et al.*, Comput. Phys. Commun. **191**,159 (2015), arXiv:1410.3012 .
[6] L. Chatterjee and C.Y. Wong, Phys. Rev. C **51**, 2125 (1995); C. Y. Wong and L. Chatterjee, Heavy Ion Phys. **4**, 201 (1996).
[7] B. L. Combridge, Nucl. Phys. **B151**, 429 (1979).
[8] M. Glück, J. F. Owens, and E. Reya, Phys. Rev. **D17**, 2324 (1978).

# Chemical and thermal freeze-out of identified hadrons at the LHC

**Xiangrong Zhu**[1], **Huichao Song**[1,2,3]

[1]Department of Physics and State Key Laboratory of Nuclear Physics and Technology, Peking University, Beijing 100871, China
[2]Collaborative Innovation Center of Quantum Matter, Beijing 100871, China
[3]Center for High Energy Physics, Peking University, Beijing 100871, China

**Abstract.** This proceedings contribution briefly summarizes our recent `VISHNU` hybrid model investigations on the chemical and thermal freeze-out of various hadrons species in 2.76 A TeV Pb+Pb collisions. Detailed analysis on the evolution of particle yields and the last elastic collisions distributions during the hadronic evolution reveals that the two multi-strange hadrons, $\Xi$ and $\Omega$, experience early chemical and thermal freeze-out when compared with other hadron species.

## 1. Introduction

In relativistic heavy-ion collisions at top RHIC and the LHC energies, a hot and dense matter – the Quark-Gluon Plasma (QGP) has been created. During its expansion, the QGP fireball quickly cools down, which then undergoes the phase transition and produces a large amount of hadrons. These hadrons subsequently experience frequent inelastic and elastic collisions during the hadronic evolution. With the termination of inelastic and elastic collisions, the evolving system reaches the chemical and thermal freeze-out, respectively.

Traditionally, the chemical freeze-out temperature $T_{ch}$ and the baryon chemical potential $\mu_b$ are extracted from the particle yields of various hadrons using the statistical model [1, 2, 3], which gives a uniform $T_{ch}$ and $\mu_b$ for all hadrons species. Similarly, pure hydrodynamic simulations implement a uniform thermal freeze-out temperature to construct the freeze-out hyper-surface, which decouples various hadrons from the bulk matter [4].

Microscopically, the related inelastic and elastic scattering channels are different for each hadron species. As a result, the chemical and thermal freeze-out procedures are very possibly hadron-species dependent. In this proceedings contribution, we will briefly summarize our recent investigations on the chemical and thermal freeze-out of various hadrons in 2.76 A TeV Pb+Pb collisions [5], based on dynamical simulations from `VISHNU` hybrid model.

## 2. Setup of the calculations

Our calculations implement `VISHNU` hybrid model [6] that combines (2+1)-d relativistic viscous hydrodynamics (`VISH2+1`) [7] for the QGP fluid expansion with a microscopic hadronic transport model (`UrQMD`) [8] for the hadron resonance gas evolution. The hydrodynamic calculations input the equation of state (EoS) s95p-PCE [9] with the transition to `UrQMD` at a constant temperature

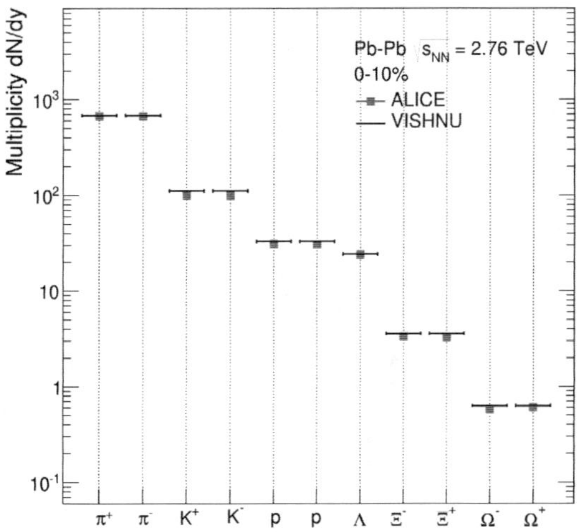


**Figure 1.** (Color online)Particle yields of $\pi$, $K$, $p$, $\Lambda$, $\Xi$, and $\Omega$ in the most central Pb+Pb collisions at $\sqrt{s_{NN}}$=2.76 TeV. The experimental data are from the ALICE Collaboration [16, 17, 18]. Theoretical curves are from `VISHNU` hybrid model calculations, using `MC-KLN` initial conditions, $\eta/s = 0.16$ and $T_{sw} =$ 165 MeV.

of 165 MeV. Following Ref. [10, 11], we input smooth initial conditions generated from the MC-KLN model [12, 13], where the normalization factor of the initial entropy density profiles is tuned to reproduce the multiplicity of all charged hadrons in the most central Pb+Pb collisions [14]. The hydrodynamic simulations start at $\tau_0 = 0.9$ fm/$c$ with the specific shear viscosity $(\eta/s)_{QGP}$ set to 0.16, and the bulk viscosity set to zero [10, 15].

## 3. Results

The chemical and thermal freeze-out of the evolving system are respectively related to the terminations of inelastic and elastic collisions. After the chemical freeze-out, the hadron yields no longer change. After the thermal freeze-out, the momentum distributions of various hadrons no longer change. In this section, we will briefly summarize our recent investigations on the chemical and thermal freeze-out of various hadron species at the LHC, based on dynamical simulations from `VISHNU` hybrid model.

Figure 1 presents the `VISHNU` calculations of the particle yields for $\pi$, $K$, $p$, $\Lambda$, $\Xi$, and $\Omega$ in the most central Pb+Pb collisions, which nicely fit the corresponding experimental data measured by ALICE. It was found that the baryon and anti-baryon ($B$-$\bar{B}$) annihilations during the hadronic evolution reduce the proton and anti-proton ($p/\bar{p}$) yields by $\sim$ 30%, largely improving the description of the $p/\bar{p}$ data [11]. Ref. [11] and [5] also showed that the `VISHNU` hybrid model also nicely fits the multiplicities, $p_T$ spectra and differential elliptic flow of $\pi$, $K$, $p$, $\Lambda$, $\Xi$, and $\Omega$ at various centrality bins. With the good descriptions of these related soft hadron data, it is meaningful to further investigate the chemical and thermal freeze-out procedures of various hadrons at the LHC.

Figure 2 shows the time evolution for the particle yields of $\pi$, $K$, $p$, $\Lambda$, $\Xi$ and $\Omega$ during the hadronic evolution, simulated from `VISHNU` hybrid model with and without $B$-$\bar{B}$ annihilations. For the simulations without $B$-$\bar{B}$ annihilations, the yields of these two multi-strange baryons $\Xi$ and $\Omega$ almost do not change. We also notice that their yields quickly reach saturation for the case with $B$-$\bar{B}$ annihilations. This reveals that $\Xi$ and $\Omega$ experience early chemical freeze-out after the hadronazition. In contrast, the yields of $p$ slowly decreases during the hadronic evolution especially for the case with $B$-$\bar{B}$ annihilations, indicating later chemical freeze-out of protons. On the production of $\Lambda$ and $K$, the $B$-$\bar{B}$ annihilations almost balance with other inelastic collision channels. As a result, the yields of $\Lambda$ and $K$ only slightly change during the hadronic evolution with $B$-$\bar{B}$ annihilations [5].

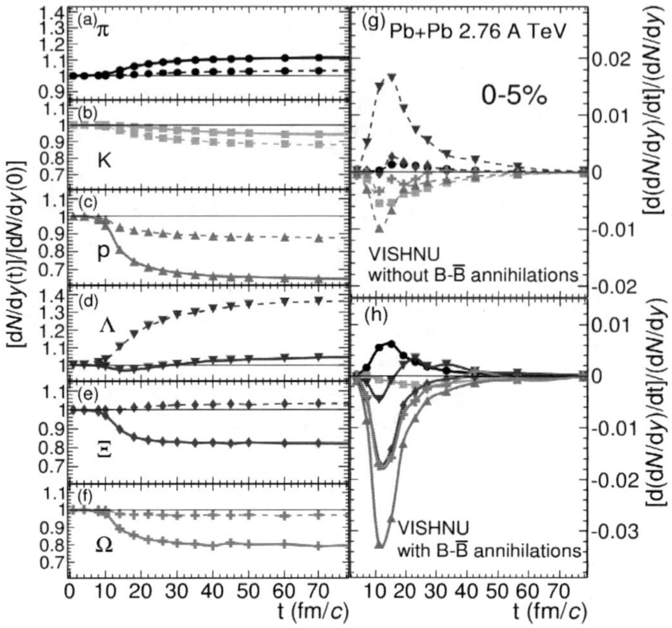


**Figure 2.** (Color online) Left panels (a)-(f): time evolution of relative particle yield density $\frac{dN}{dy}(t)/\frac{dN}{dy}(0)$ for $\pi$, $K$, $p$, $\Lambda$, $\Xi$, and $\Omega$ during the UrQMD expansion of VISHNU. Here, $\frac{dN}{dy}(t)$ and $\frac{dN}{dy}(0)$ denote the particle yield density at mid-rapidity at later evolution time and at the starting time, respectively. Right panels (g) and (h): time evolution of the changing rates for the corresponding particle yield densities. Solid/dashed lines denote the VISHNU simulations with/without $B$-$\bar{B}$ annihilations.

The right panels of Fig. 2 presents the changing rates of the particle yields of these identified hadrons. For the simulations without $B$-$\bar{B}$ annihilations, the changing rates for $K$, $p$, and $\Lambda$ show wide peaks along the time axis, illustrating that the associated inelastic collisions are still frequent after 10-20 fm/$c$. Due to small hadronic cross sections, multi-strange hadrons $\Xi$ and $\Omega$ rarely interact with other hadrons. Therefore, the annihilations with their own anti-particles lead to wide peaks of the changing rates for $\Xi$ and $\Omega$ in panel (h), which delay their chemical freeze-out, when compared with the case without $B$-$\bar{B}$ annihilations. However, these multi-strange hadrons still experience earlier chemical freeze-out in the full VISHNU simulations compared with other identified hadrons [5].

Figure 3 presents the thermal freeze-out time distributions of $\pi$, $K$, $p$, $\Lambda$, $\Xi$ and $\Omega$ in the most central Pb+Pb collisions. Here, the thermal freeze-out time distributions are extracted through analyzing the space-time information of the last collisions for various hadron species. For $\Xi$ and $\Omega$, the peaks of their thermal freeze-out time distributions are located around 10 fm/$c$, while the peaks of $p$ and $\Lambda$ distributions are around 20-30 fm/$c$. Due to a large amount of hadronic scatterings during the late UrQMD evolution, the thermal freeze-out time distributions of $\pi$ and $K$ widely spread along the time axis. Fig. 3 thus illustrates that thermal freeze-out procedures are hadron species dependent. Compared with other hadrons, the two multi-strange hadrons $\Xi$ and $\Omega$ experience early thermal freeze-out due to their much smaller hadronic cross sections [5].

## 4. Summary

In this proceedings contribution we have briefly reviewed our recent investigations on the chemical and thermal freeze-out of various hadron species in 2.76 A TeV Pb+Pb collisions within the framework of VISHNU hybrid model. We found that the chemical and thermal freeze-out procedures are hadron-species dependent. Due to their much smaller hadronic cross sections, the two multi-strange hadrons $\Xi$ and $\Omega$ experience earlier chemical and thermal freeze-out than other hadrons. To extract the possible effective chemical freeze-out temperatures of various hadron species requires detailed analysis of the space-time distributions of the last inelastic collisions, which we would like to leave it to future study.

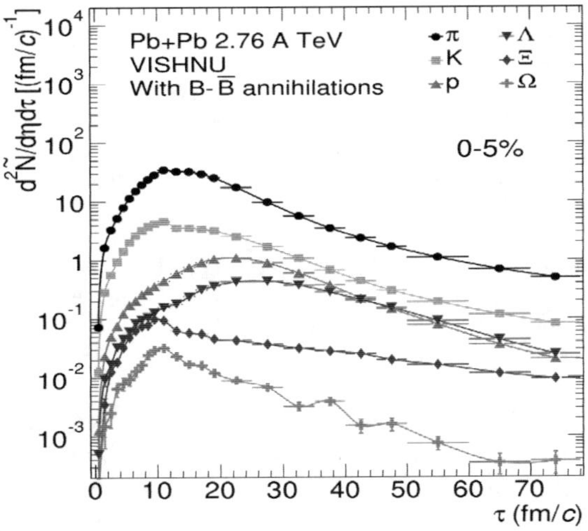


**Figure 3.** (Color online) Thermal freeze-out time distributions for $\pi$, K, p, $\Lambda$, $\Xi$ and $\Omega$ in the most central Pb+Pb collisions, calculated from `VISHNU` with $B$-$\bar{B}$ annihilations.

**Acknowledgments**

This work was supported by the NSFC and the MOST under Grants No. 11435001 and No. 2015CB856900 and the China Postdoctoral Science Foundation under Grant No. 2015M570879. We gratefully acknowledge extensive computing resources provided to us on Tianhe-1A by the National Supercomputing Center in Tianjin, China.

**References**

[1] Braun-Munzinger P, Magestro D, Redlich K, and Stachel J 2001 *Phys. Lett.* B **518** 41
[2] Andronic A, Braun-Munzinger P and Stachel J 2006 *Nucl. Phys.* A **772** 167-99
[3] Becattini F, Manninen J and Gazdzicki M 2006 *Phys. Rev.* C **73** 044905
[4] Kolb P F, Sollfrank J and Heinz U W 2000 *Phys. Rev.* C **62** 054909
[5] Zhu X, Meng F, Song H and Liu Y X 2015 *Phys. Rev.* C **91** 034904; Song H, Meng F, Xin X and Liu Y X 2014 *J. Phys. Conf. Ser.* **509** 012089
[6] Song H, Bass S A and Heinz U 2011 *Phys. Rev.* C **83** 024912; Song H 2015 *Pramana* **84** 5 703-15
[7] Song H and Heinz U 2008 *Phys. Lett.* B **658** 279; 2008 *Phys. Rev.* C **77** 064901; 2008 *Phys. Rev.* C **78** 024902; Song H 2009 *thesis proceeding* (*Preprint* arXiv:0908.3656 [nucl-th])
[8] Bass S A *et al.* 1998 *Prog. Part. Nucl. Phys.* **41** 255; Bleicher M *et al.* 1999 *J. Phys.* G **25** 1859
[9] Huovinen P and Petreczky P 2010 *Nucl. Phys.* A **837** 26-53
[10] Song H, Bass S A and Heinz U 2011 *Phys. Rev.* C **83** 054912
[11] Song H, Bass S and Heinz U W 2014 *Phys. Rev.* C **89** 034919
[12] Drescher H J and Nara Y 2007 *Phys. Rev.* C **75** 034905
[13] Hirano T and Nara Y 2009 *Phys. Rev.* C **79** 064904
[14] Aamodt K *et al.* (ALICE Collaboration) 2011 *Phys. Rev. Lett.* **106** 032301
[15] Song H 2013 *Nucl. Phys.* A **904-905** 114c-21c
[16] Abelev B *et al.* (ALICE Collaboration) 2013 *Phys. Rev.* C **88** 044910
[17] Abelev B B *et al.* (ALICE Collaboration) 2013 *Phys. Rev. Lett.* **111** 222301
[18] Abelev B B *et al.* (ALICE Collaboration) 2014 *Phys. Lett.* B **728** 216

# Features of triangular flow of strange and non-strange hadrons at LHC

**E E Zabrodin[1,2], L V Bravina[1], B H Brusheim Johansson[1], J Crkovska[3], G Kh Eyyubova[1,3], V L Korotkikh[2], I P Lokhtin[2], L V Malinina[2,4], S V Petrushanko[2] and A M Snigirev[2]**

[1] Department of Physics, University of Oslo, Norway
[2] Skobeltsyn Institute of Nuclear Physics, Moscow State University, Russian Federation
[3] Faculty of Nuclear Sciences and Physical Engineering, Czech Technical University in Prague, Czech Republic
[4] Joint Institute for Nuclear Researches, Dubna, Russian Federation

E-mail: eugen.zabrodin@fys.uio.no

**Abstract.** Triangular flow of strange and non-strange hadrons in Pb+Pb collisions at LHC energies is studied within the HYDJET++ model, which combines hard processes in a hot and dense partonic medium with parametrized hydrodynamics. Together with an extensive table of resonances, this circumstance enables us to investigate (i) the interplay between hard and soft processes and (ii) between jets and final-state interactions. Jets are found to be the main reason for violation of the NCQ scaling for $v_3$ at LHC, whereas decays of resonances improve the scaling fulfillment. Comparison with the experimental data is performed.

## 1. Introduction

Anisotropic flow of secondary particles and quenching of jets, traversing the hot and dense medium, are among the few signals especially sensitive to creation of quark-gluon plasma (QGP) in ultra-relativistic heavy-ion collisions. These two phenomena are usually considered separately as a soft and a hard signal, respectively. In previous studies [1, 2, 3] we pointed out that the interplay between the soft hydro-like processes and the jets might have important consequences for the particle anisotropic flow at transverse momentum $p_T \geq 1$ GeV/$c$. For our analysis the HYDJET++ model [4, 5] was employed. The model relies on parametrized ideal hydrodynamics with more than 360 possible hadronic states and resonances, including charmed hadrons, after the chemical freeze-out accompanied by treatment of hard jets propagating in the QGP. In the last case both collisional and radiative energy losses are taken into account [6]. Below we focus on implementation of the anisotropic flow components in HYDJET++.

The angular distribution of hadrons in the azimuthal plane is decomposed in the Fourier series as [7]

$$E\frac{d^3N}{d^3p} = \frac{1}{2\pi}\frac{d^2N}{p_t dp_t dy}\left\{1 + 2\sum_{n=1}^{\infty} v_n \cos\left[n(\phi - \Psi_n)\right]\right\} , \quad (1)$$

$$v_n = \langle \cos\left[n(\phi - \Psi_n)\right]\rangle , \quad (2)$$

with $p_t$, $y$ and $\phi$ being the transverse momentum, the rapidity and the azimuthal angle of a particle, respectively. $\Psi_n$ is the angle of the reaction plane of $n$-th harmonic, and the coefficients $v_n$ are dubbed *directed* flow ($v_1$), *elliptic* flow ($v_2$), *triangular* flow ($v_3$), and so forth. Experiments show that $v_1$ is extremely weak in heavy-ion collisions at LHC, therefore, we concentrate on $v_2$ and $v_3$. To take into account both ellipticity and triangularity of the initial state, resulting into appearance of corresponding flow harmonics in the particle momentum distribution, HYDJET++ utilizes the following typical parametrizations of the freeze-out hypersurface of the fireball produced in a collision of two similar nuclei with the impact parameter $b$

$$R(b,\phi) = R_{ell}(b,\phi)\{1+\epsilon_3(b)\cos\left[3(\phi-\Psi_3)\right]\}\,, \tag{3}$$

$$R_{\rm ell}(b,\phi) = R_{\rm fo}(b)\frac{\sqrt{1-\epsilon^2(b)}}{\sqrt{1+\epsilon(b)\cos 2\phi}}\,. \tag{4}$$

Here $R_{\rm fo}(0)$ is the model parameter that determines the scale of the fireball transverse size at freeze-out, and two spatial anisotropy parameters, $\epsilon(b)$ and $\epsilon_3(b)$, are responsible for emergence of elliptical and triangular anisotropy, respectively. Our aim is to investigate the influence of hard processes and final-state interactions on the triangular flow of both strange and non-strange hadrons.

## 2. Results

*Interplay of soft processes and jets.* Lead-lead collisions at $\sqrt{s_{\rm NN}} = 2.76$ TeV with centrality $0 \leq \sigma/\sigma_{geo} \leq 50\%$ are considered. As shown in [8, 9], HYDJET++ provides a good quantitative

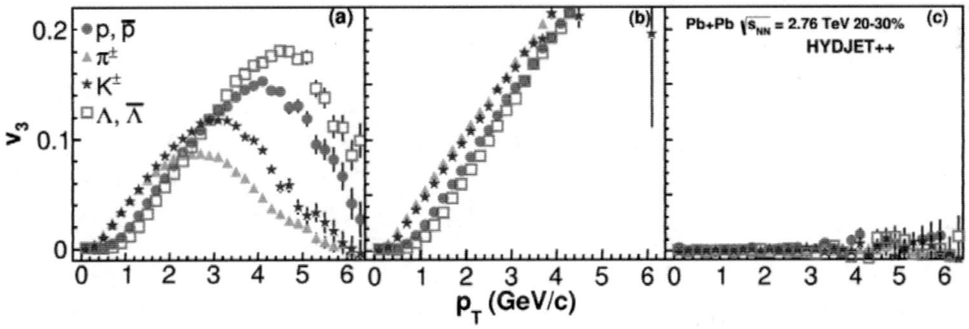


**Figure 1.** $v_3(p_T)$ of $\pi^{\pm}$, $K^{\pm}$, $p+\bar{p}$, $\Lambda+\bar{\Lambda}$. Particles considered: (a) all hadrons, (b) particles produced in hydro processes only and (c) hadrons from hard processes only.

description of the triangular flow of charged particles within the indicated centrality range, but $v_3$ of hadronic species was not studied yet. Partial triangular flows of $p+\bar{p}$, $\pi^{\pm}$, $K^{\pm}$ and $\Lambda+\bar{\Lambda}$ in collisions with centrality $20-30\%$ are shown in Fig. 1. Here the plot (a) displays the resulting $v_3$ of particle species as a function of transverse momentum $p_T$, whereas the plots (b) and (c) show, respectively, the triangular flow generated separately in hydro-module and jet-module of the program. The combination of these two processes helps one to understand the origin of the hump-like structure of total $v_3(p_T)$ distributions and also breaking of the mass ordering. Indeed, in pure hydro calculations presented in the middle window $v_3(p_T)$ grows almost linearly at $0.5 \leq p_T \leq 4$ GeV/$c$, and meson triangular flow is stronger than the baryon one. Although hadrons produced from the fragmentation of jets should carry zero anisotropic flow, they can develop a weak $v_3$ at $p_T \geq 3$ GeV/$c$ because of the well-known jet quenching process. The reason of the flow falloff at $p_T \geq 2$ GeV/$c$ and simultaneous breaking of the mass ordering is as follows. At small and intermediate transverse momenta spectra of particles are dominated by hadrons produced in soft processes. At higher $p_T$ the spectra are dominated by hadrons from jets. The latter carry weak $v_3$, therefore the total signal drops. The transverse momentum, at which the crossing between soft and hard parts of particle spectrum takes place, increases with rising mass of the particles, thus resulting in crossing of meson and baryon branches of $v_3(p_T)$ at $p_T \geq 2.5$ GeV/$c$.

Figure 2 presents the ratio $v_3^{1/3}(p_T)/v_2^{1/2}(p_T)$ calculated in four centrality bins in comparison with the ATLAS data. The model results and the data agree within the error bars. The ratio seems to be insensitive to transverse momentum in the interval $1 \leq p_T \leq 6$ GeV/$c$, however, the separate ratios of individual contributions of directly produced particles, decays of resonances and jets, such as $v_{3(hydro)}^{1/3}(p_T)/v_{2(hydro)}^{1/2}(p_T)$, reveal the dependence on $p_T$, see [10].

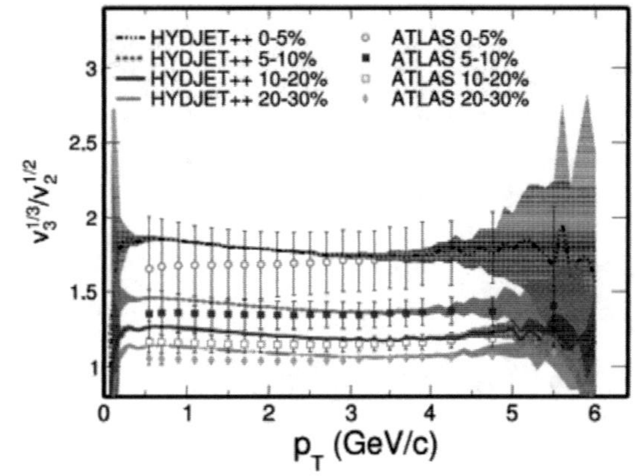


**Figure 2.** Ratio $v_3^{1/3}/v_2^{1/2}$ for different centralities in comparison with ATLAS data.

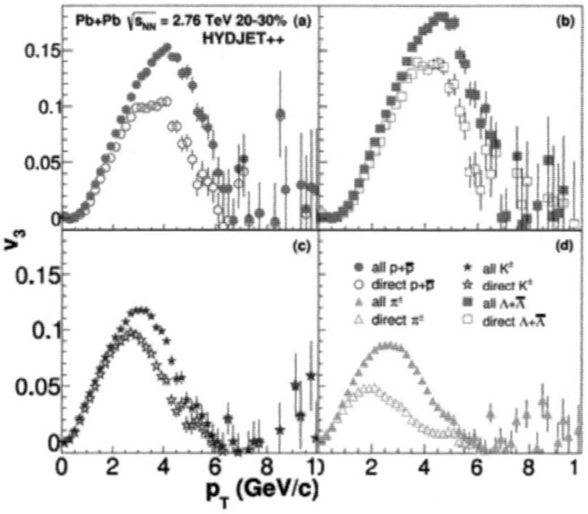


**Figure 3.** $v_3(p_T)$ of directly produced hadrons (open symbols) and the same spectra after the feeddown from resonances (full symbols).

*Final-state interactions: decays of resonances.* Partial $v_3(p_T)$ spectra of only directly produced particles, which are frozen at the freeze-out hypersurface or decoupled from jets, are shown in Fig. 3 together with the spectra obtained after the decays of resonances. The reaction centrality is $20-30\%$. One may note that decays of resonances modify the distributions of all hadrons in a similar way - the maximum of the spectrum increases, and its position is shifted to a bit higher transverse momenta. The most increase is observed for charged pions. In contrast, charged kaons and $\Lambda + \bar{\Lambda}$ demonstrate just moderate rise of their $v_3(p_T)$-maxima of about 15%.

*Number-of-constituent-quark (NCQ) scaling.* The idea behind is to use the $kE_T = m_T - m_0$ dependence of $v_3$ rather than the $p_T$ one, and to divide both $v_3$ and $kE_T$ of a hadron on its number of valence quarks. The distribution functions of elliptic flow of identified hadrons were found nearly coinciding at $kE_T \leq 1$ GeV at RHIC [11, 12]. Since then this phenomenon is called NCQ scaling. Worsening of NCQ scaling at LHC was predicted in [1] and observed later in [13]. Let us analyse the $v_3/n_q$ distributions in HYDJET++ presented in Fig. 4. Top plots show the $v_3/n_q$ of charged pions, charged kaons, (anti)protons and (anti)Lambdas separately for (a) directly produced hadrons, (b) hadrons produced in soft processes and in decays of resonances, and (c) final distributions where all processes are taken into account. The bottom row displays the ratios of distributions to that of (anti)protons. The NCQ scaling of triangular flow certainly

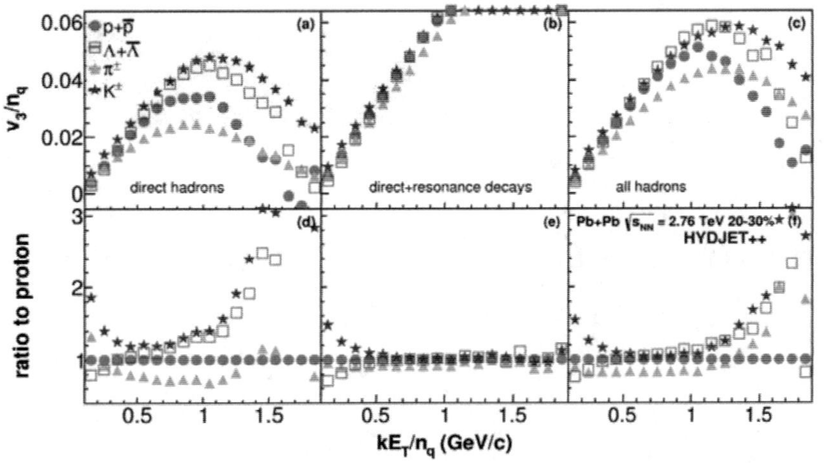


**Figure 4.** $v_3/n_q(kE_T/n_q)$ distributions for (a) directly produced particles, (b) particles produced in hydro-like processes with resonance decays, and (c) all particles. Bottom windows present the ratios to the $p+\bar{p}$ distribution.

holds for hydro-produced hadrons, and decays of resonances make the agreement between the spectra even better by increasing the $p_T$-yields of mesons at $p_T \geq 2.5$ GeV/$c$. Hadrons from the jets, in contrast, lead to worsening of the scaling.

## 3. Conclusions

The influence of the interplay between soft and hard processes and final-state interactions on the triangular flow of identified hadrons is studied in Pb+Pb collisions at $\sqrt{s_{NN}} = 2.76$ TeV within the HYDJET++ model. Hadrons emitted by jets dominate the hard part of particle spectrum, thus causing the drop of resulting triangular flow at high transverse momenta, because these hadrons develop almost zero flow. Decays of resonances increase the maxima of $v_3(p_T)$ distributions and shift it to higher $p_T$ values. The number-of-constituent-quark scaling for particle $v_3$ is observed for hadrons produced in soft processes. Hadrons originated from the jet fragmentation carry almost no flow, therefore, the NCQ scaling of $v_3/n_q(kE_T/n_q)$ is worsening.

*Acknowledgments.* G.E. acknowledges the European Social Fund within the framework of realizing the project Support of Inter-sectoral Mobility and Quality Enhancement of Research Teams at Czech Technical University in Prague, CZ.1.07/2.3.00/30.0034.

## References

[1] Eyyubova G et al. 2009 *Phys. Rev. C* **80** 064907
[2] Zabrodin E et al. 2010 *J. Phys. G* **37** 094060
[3] Bravina L, Brusheim Johansson B H, Eyyubova G and Zabrodin E, 2013 *Phys. Rev. C* **87** 034901
[4] Lokhtin I et al. 2009 *Comput. Phys. Commun.* **180** 779
[5] Lokhtin I et al. 2012 *Eur. Phys. J. C* **72** 2045
[6] Lokhtin and Snigirev A 2006 *Eur. Phys. J. C* **46** 211
[7] Voloshin S and Zhang Y 1996 *Z. Phys. C* **70** 665
[8] Bravina L et al. 2014 *Eur. Phys. J. C* **74** 2807
[9] Bravina L et al. 2014 *Phys. Rev. C* **89** 024909
[10] Brusheim Johansson B H et al. *these Proceedings*
[11] Adler S S et al. (PHENIX Collaboration) 2003 *Phys. Rev. Lett.* **91** 182301
[12] Adams J et al. (STAR Collaboration) 2004 *Phys. Rev. Lett.* **92** 052302
[13] Noferini F et al. (ALICE Collaboration) 2013 *Nucl. Phys. A* **904-905** 483c

# Excitation function Analysis for Charmonium Production in Heavy-Ion Collisions

**Kai Zhou**

Institute for Theoretical Physics, Johann Wolfgang Goethe-University, Max-von-Laue-Str. 1, 60438 Frankfurt am Main, Germany

E-mail: zhou@th.physik.uni-frankfurt.de

**Abstract.** Both color screening and regeneration are hot medium effects on charmonium production in heavy ion collisions. While they affect in an opposite way the charmonium yield, their competition in transverse dynamics bring sensitivity to the ratio of averaged transverse momentum square for charmonium, which thus can reveal more nature of the QCD medium created from the collisions. We make an excitation analysis based on transport approach to illustrate such a picuture.

Since 1986 and firstly in light of color screening argument for the initial proposing [1], $J/\Psi$ has long been considered as an effective probe to the hot and dense QCD matter created from heavy ion collisions. A suppression for their production in A+A compared with in p+p collisions is motivated and extensively studied both in experiment and theory. Along with more released experimental data, it's realized that apart from color screening, many other medium effects come into play during the collision evolution, which make the picture for charmonium production complicated. Before any hot medium effects acting, the cold nuclear matter also impacts the $J/\Psi$ production through nuclear absorption, Cronin effect and nuclear shadowing effect. After entering the created partonic medium, the color screening will alter the binding inside the bound state. Different suppression mechanism availably works here at different temperature domain. When the temperature is higher than the charmonium dissociation temperature, no bound states survive based on spectral analysis where no peak signal is left due to color screening, thus they will melt into the medium. Below this melting threshold, the partons inside the medium can break-up the quarkonium bound states through collisions, like the gluon dissociation and quasi-free dissociation, which respectively correspond to Singlet-to-Octet transition and Landau damping in terms of EFT (effective field theory). Another important concept that arises here is regeneration [2], thus the recombination to bound state from those uncorrelated charm and anti-charm quark pairs being in favor of their more abundant hard production along with increasing collision energy. Through regeneration, the hot medium effects on heavy quarks would also leave a mark on heavy quarkonium production.

Actually from the theoretical point of view it's still not clear enough about the quarkonium evolution inside QGP. Like, which suppression mechnism works dominantly? Which is the proper potential, free energy F or internal energy U, for quarkonium in medium? Also, in a rapidly cooling QGP medium, a potential analysis might not possess enough validity since a bound state would hardly always lie in a fixed eigenstate of the evolving Hamiltonian describing in medium quarkonium. Even further, if a seperate treatment for heavy quarkonia evolution taking the

QGP medium only as background really works is somehow questionable, as recently pointed out in Ref. [3]: according to the associated large entropy for heavy quarkonium in medium observed from Lattice QCD study (especially near $T_c$) [4], the coupling between medium constituent and quarkonium interquarks is stronger even than the binding between the interquarks inside the bound state. This makes it unjustifiable to treat the bound state as a color neutral entity to be influenced by the surrounding medium, since actually each of their interquarks are strongly entangled with the medium to be consistent with the Lattice results [4].

Nevertheless, in the present proceeding we provide a transport model based analysis for the medium effects on charmonium production, and we especially focus on their excitation function to illustrate the picture given by kinetic approach. We now start with a brief introduction to the transport model we adopted in the analysis. Since a charmonia is heavy and can hardly reach equilibrium with the surronding medium, its phase space distribution function, $f_\Psi(\mathbf{x}, \mathbf{p}, t)$, is governed by the following Boltzmann-type transport equation [5],

$$\frac{\partial f_\Psi}{\partial t} + \mathbf{v}_\Psi \cdot \nabla f_\Psi = -\alpha_\Psi f_\Psi + \beta_\Psi, \tag{1}$$

where the loss and gain terms $\alpha$ and $\beta$ are the main hot medium effects describing dissociation and regeneration, while the cold matter effects are reflected in the initial condition. Taking gluon dissociation as the dominant suppression mechanism, $\alpha$ is the momentum integration of the dissociation cross section multiplied by the thermal gluon distribution. The color screening induced melting is also considered, where the melting temperature $T_d$ is evaluated from potential approach. Using detailed balance the regeneration cross section can be got from the gluon dissociation cross section. By convoluting it with the in-medium charm quarks's distribution we obtain the continuous regeneration rate $\beta(\mathbf{x}, \mathbf{p}_\Psi, t)$. Considering the experimentally observed large quench factor [6] and anisotropic flow [7] of open charm mesons, we here approximate the charm quarks' distribution to be kinetically thermalizaed which is a strong interaction limit. On the level of kinetic approach, the medium information is then encoded in the loss and gain term through the distribution function, thus the appeared local temperature $T(\mathbf{x}, t)$ and fluid velocity $u_\mu(\mathbf{x}, t)$ those being controled by ideal hydrodynamics in the present model analysis. Using the above transport model, we have studied the nuclear modification factor $R_{AA}$ and transverse momentum distribution and also the anisotropic flow for charmonium production [5]. Along with increasing collision energy, the medium becomes hotter and we naturally expect stronger $J/\Psi$ suppression to be observed. However, it's found that in mid-rapidity the suppression magnitude of $J/\Psi$ at RHIC is similar to that at SPS and smaller than that at LHC. To make the advantages protruded in explaining these phenomenon from transport model, here we turn to analyze their collision energy dependence thus the excitation function.

As the collision becomes more violent, more charm quark pairs are produced through hard process, the regeneration component gradually dominate the charmonium production. The fraction of the regenerated $J/\Psi$s, $g_{AA} = N_{AA}^{reg}/N_{AA}$, calculated from our transport model, is shown in the left of Fig. 1 as a function of collision energy, where $N_{AA}^{reg}$ is $J/\Psi$ yield from regeneration. At lower energy collisions like at SPS, only very few charm quarks are produced initially, then almost no regeneration happens. At RHIC, the initial production is still dominant but the regeneration becomes equally important especially in mid-rapidity. At LHC, the final charmonium yield is dominated by regeneration in both mid and forward rapidiy. The lower fraction in forward rapidity merely reflects the rapidity distribution of charm quarks.

Since the transverse motion in heavy ion collisions which is developed during the dynamical evolution of the system can directly reflect the multiple scattering and also parton density in the medium, as has been well documented in light quarks sector [8], for charmonium production, the transverse momentum distribution also can reveal more information about their evolution and be more sensitive to the medium effects. We propose the following new defined nuclear

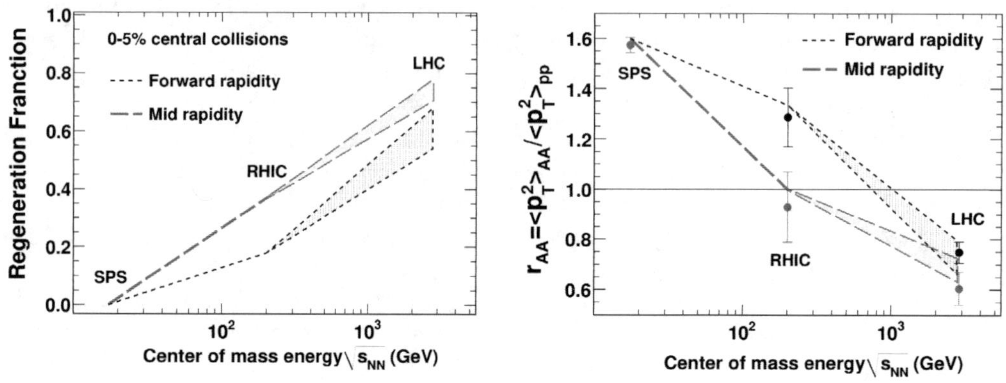


**Figure 1.** (Color online) (left) $J/\psi$ regeneration fraction in central Au+Au collisions at RHIC and Pb+Pb collisions at SPS and LHC. (right) $r_{AA} = \langle p_t^2 \rangle_{AA} / \langle p_t^2 \rangle_{pp}$ in central Au+Au collisions at RHIC and Pb+Pb collisions at SPS and LHC. The band at LHC comes from charm quark cross section's uncertainty. Experimental data are also shown for SPS [9], RHIC [10] and LHC [11].

modification factor for transverse momentum square

$$r_{AA} = \frac{\langle p_t^2 \rangle_{AA}}{\langle p_t^2 \rangle_{pp}}. \tag{2}$$

We show in the right of Fig. 1 the energy dependence of $r_{AA}$ in central collisions. As been shown in left of Fig. 1, at lower energy collisions almost all of the finally observed charmonia are from survived previously produced ones, then the initial state gluon multiple scattering, i.e. the Cronin effect, is dominant and together with leakage effect and medium suppression would increase their transverse momentum leading to $r_{AA} > 1$. On the contrary, at very high energy collisions, LHC, due to the copiously production of charm quarks, the recombination of c and $\bar{c}$ into charmonium, thus the regeneration becomes voilently significant and dominate the whole charmonium production finally. Since the heavy quark would be quenched when passing through the medium, the regenerated $J/\Psi$s sourced from the in-medium charm quarks thus possess lower transverse momentum compared to initial production. In between the above two limits, at RHIC, in mid-rapidity the competition between initial production which controls high $p_T$ charmonium production and the regeneration which inherits the low momentum of kinetically thermalized charm quarks leads to a cancelled influence on transver momentum square, while in forward rapidity $r_{AA}$ is around 1.3 due to the still overwhelming initial production. Our prediction is also confirmed by the experimental measurements as shown in the plot. Being unlike the yield $R_{AA}$ which appears cancelling out each other's increasing between regeneration and suppression for initial production, the newly defined $r_{AA}$ lends additional sensitivity to the transverse dynamics and presents a robust excitation feature: as a consequence of the increasing regeneration with collision energy, the interplay between regeneration and initial contribution just leads to a decreasing trend for $r_{AA}$ from SPS to LHC, manifesting a hotter medium creating stronger hot medium effects impacting in higher energy collisions.

From another aspect, due to the strong interaction with the medium, the charm quarks' distribution becomes steeper although they are initially hard, consequently the regeneration for charmonium decreases with increasing momentum and hardly contribute to high $p_T$ $J/\Psi$s. Therefore, the high $p_T$ $J/\Psi$s are characterized solely by Debye screening effect and also medium induced suppression on the initially produced charmonia. Fig. 2 shows the excitation function

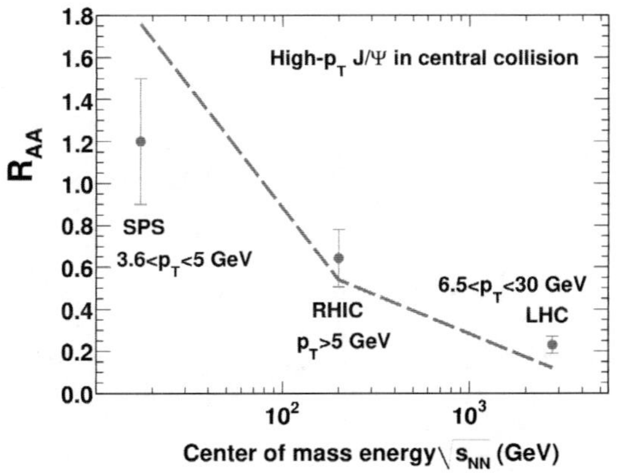


**Figure 2.** High $p_T$ $J/\psi$s' $R_{AA}$ in mid-rapidity in central Au+Au collisions at RHIC and Pb+Pb collisions at SPS and LHC. Experimental data are shown for SPS [12], RHIC [13] and LHC [14].

for high $p_T$ $J/\Psi$s' $R_{AA}$ in central collisions. The decreasing trend obviously re-authenticate the natural expectation : stronger hot medium effects are active in higher energy collisions.

To summarize, we analyze the excitation behavior of charmonium production in heavy ion collisions based on transport model. The color screening and regeneration are both medium induced effects on charmonium production, they affect the production yield in an opposite way. At the same time, they drastically influence the transverse momentum dynamics for charmonium. We promote the ratio $r_{AA} = \langle p_t^2 \rangle_{AA} / \langle p_t^2 \rangle_{pp}$ for charmonia as a much more sensitive probe to the hot and dense QCD medium, its decresing trend along with collision energy robustly indicate a hotter QCD matter created from higher energy collsions. The high $p_T$ $J/\Psi$ from another angle also helps to distringuish the medium among different energy collisions.

## Acknowledgments

I'd like to acknowledge financial support from Frankfurt Institute for Advanced Studies.

## References

[1] T. Matsui and H. Satz, Phys. Lett. **B178**, 417(1986).
[2] P. Braun-Munzinger and J. Stachel, Phys. Lett. **B490**, 196(2000); M. Gorenstein, A. Kostyuk, H. Stöcker and W. Greiner, Phys. Lett. **B509**, 277(2001); L. Grandchamp and R. Rapp, Nucl. Phys. **A709**, 415(2002).
[3] D.E. Kharzeev, Phys. Rev. **D90**, 074007(2014).
[4] O. Kaczmarek, F. Karsch, P. Petreczky and F. Zantow, Phys. Lett. **B43**, 41-47(2002); P. Petreczky and K. Petrov, Phys. Rev. **D70**, 054503(2004).
[5] Y. Liu, Z. Qu, N. Xu and P. Zhuang, Phys. Lett. **B678**, 72(2009); K. Zhou, N. Xu and P. Zhuang, Nucl. Phys. **A834**, 249c-252c(2010); K. Zhou, N. Xu, Z. Xu and P. Zhuang, Phys. Rev. **C89**, 054911(2014); Z. Tang, N. Xu, K. Zhou and P. Zhuang, J. Phys. **G41**, 124006(2014).
[6] B.I. Abelev *et al.* [STAR Collaboration], Phys. Rev. Lett. **98**, 192301(2007); A. Adare *et al.*, [PHENIX Collaboration], Phys. Rev. Lett. **98**, 172301(2007); B. Abelev *et al.*, [ALICE Collaboration], JHEP **09**, 112(2012).
[7] A. Adare, et al., [PHENIX Collaboration], Phys. Rev. Lett. **98**, 172301(2007); Z.C. Valle, [ALICE Collaboration], Nucl. Phys. **A**904-905, 178c(2013).
[8] I. Arsene *et al.*, [BRAHMS Collaboration], Nucl. Phys. **A757**, 1(2005); J. Adams *et al.*, [STAR Collaboration], Nucl. Phys. **A757**, 102(2005); K. Adcox *et al.*, [PHENIX Collaboration], Nucl. Phys. **A757**, 184(2005).
[9] N.S. Topilskaya, et al., [NA50 Collaboration], Nucl. Phys. **A715**, 675(2003).
[10] A. Adare, et al., [PHENIX Collaboration], Phys. Rev. Lett. **98**, 232301(2007).
[11] J. Adam, et al., [ALICE Collaboration], arXiv:1506.08804; J. Adam, et al., [ALICE Collaboration], JHEP **1507**, 051(2015).
[12] M.C. Abreu, et al., Phys. Lett. **B499**, 85-96(2001).
[13] Z. Tang, et al., [STAR Collaboration], J. Phys. **G38**, 124107(2011).
[14] C. Miranov, [CMS Collaboration], Nucl. Phys. **A904**, 194(2013).

Journal of Physics: Conference Series **668** (2016) 012101 doi:10.1088/1742-6596/668/1/012101

# Phase diagram of the three-flavor color superconducting PNJL model

**Alexander Ayriyan**[a], **David Blaschke**[b,c,d], **Rafal Lastowiecki**[b]

[a]Laboratory for Information Technologies, JINR Dubna, Russia
[b]Institute of Theoretical Physics, University of Wroclaw, Poland
[c]Bogoliubov Laboratory for Theoretical Physics, JINR Dubna, Russia
[d]National Research Nuclear University (MEPhI), Moscow, Russia

E-mail: ayriyan@jinr.ru

**Abstract.** The phase diagram of a three-flavor Polyakov-loop Nambu-Jona-Lasinio model is analyzed for symmetric matter with a parametrization consistent with the 2 $M_\odot$ mass constraint from the pulsars PSR J1614-2230 and PSR J0348+0432. A coexistence of partial chiral symmetry restoration, diquark condensation (2SC phase) and the hadronic (confined) phase is conjectured that entails the existence of a quadruple point and is accessible by trajectories of constant entropy per baryon for heavy-ion collisions in the NICA/FAIR energy range.

## 1. Three-flavor color superconducting PNJL model and phase diagram

The phase diagram (PD) of quantum chromodynamics (QCD) is at the center of attention in modern physics. It is an object of intensive study at already existing heavy-ion collision facilities (LHC, RHIC, SPS) and will also be investigated at the planned ones (FAIR, NICA). In light of the difficulties of lattice simulations at a finite chemical potential, the class of Nambu–Jona-Lasinio (NJL) models has been applied in order to extract information and cast predictions on the phase structure of the QCD at high densities. Models extended by the Polyakov-loop (PNJL) have been shown to reproduce lattice results at $\mu = 0$ [1] and have been used to predict the PD [2]. In this contribution we report on the PhD for three-flavor quark matter described with the Polyakov-loop extension of Ref. [3]. By solving the gap equations for the order parameters related to symmetries and their breaking in the color superconducting three-flavor PNJL model [3] as functions of temperature $T$ and baryochemical potential $\mu$ one arrives at a PhD with three order parameters characterizing the phase structure:

(i) chiral condensates (quark masses) - chiral symmetry breaking in light and strange sectors;
(ii) diquark gaps (color superconductivity) - 2SC and CFL phases;
(iii) Polyakov loop (confinement) - Z(3) symmetry breaking.

## 2. Results and Discussion

We consider the isospin symmetric case. The model parameters were constrained such that in applications for hybrid compact stars with quark matter core [4] the condition from the measurement of high masses $\sim 2\ M_\odot$ for the two pulsars PSR J1614-2230 and PSR J0348+0432 [5] would be fulfilled. The results are shown in the left panel of Fig. 1. The colored lines with

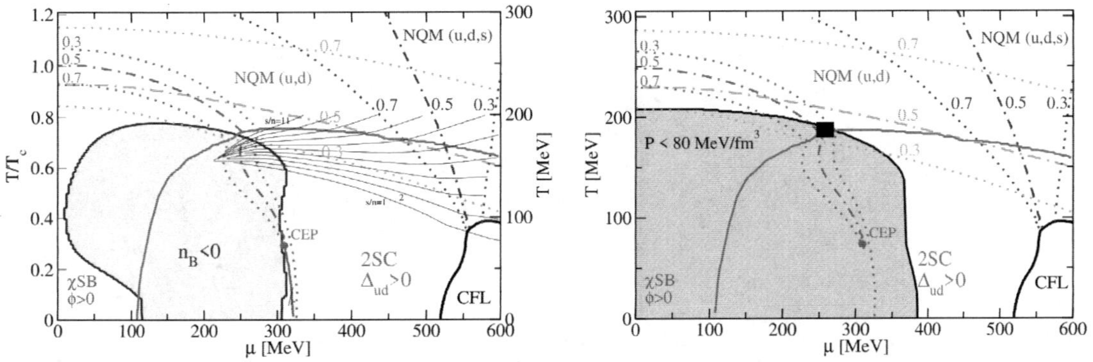


**Figure 1.** (Color online) Phase diagram of the 3FCS PNJL model for symmetric quark matter without (left) and with (right) hadron confinement. For details see text.

label "0.5" show the 50% reduction of the maximum value of the order parameter and serve as indicators of the crossover transition for light (green) and strange (blue) quark condensate as well as Polyakov loop (orange). The red (black) bold solid line denotes the second order transition to the 2SC (CFL) diquark condensate. Lines of constant entropy per baryon indicate paths for the dynamical evolution of a fireball in a heavy-ion collisions at NICA/FAIR energies which traverse the region of the color superconducting 2SC phase.

The occurrence of negative baryon densities and pressures (yellow region) in the phase diagram due to adjustment of color neutrality [6] is a feature this model has in common with other quark matter approaches which attempt to model confinement, e.g., bag models [7] and nonlocal covariant models with complex-conjugated mass poles [8]. Both, the pressure instability and color neutrality problem, can be removed simultaneously when color neutral hadronic bound states of quarks are introduced. Their dominance with positive pressure in the confined, chiral symmetry broken phase cures these problems. We define the phase border of confined hadronic matter (grey region in the right panel of Fig. 1) by $P_{\rm crit} \sim 80 \pm 3$ MeV/fm$^3$, the conjectured universal pressure at chemical freezeout [9]. We emphasize the accessibility at NICA/FAIR energies of a "quarkyonic matter" region with partial chiral symmetry restoration inside the hadronic world, as well as a quadruple point (black square) where the four phases meet.

## Acknowledgments

We acknowledge collaboration with D. Alvarez-Castillo, F. Sandin and H. Grigorian. A.A. was supported by JINR grant No. 1460201 and by the Bogoliubov-Infeld program for visiting Wroclaw; D.B. received funding from NCN under grant UMO-2011/02/A/ST2/00306.

## References

[1] C. Ratti, M. A. Thaler and W. Weise, Phys. Rev. D **73**, 014019 (2006).
[2] K. Fukushima, Phys. Rev. D **77**, 114028 (2008) [Phys. Rev. D **78**, 039902 (2008)].
[3] D. Blaschke, S. Fredriksson, H. Grigorian, A. M. Oztas and F. Sandin, Phys. Rev. D **72**, 065020 (2005)
[4] T. Klähn, R. Lastowiecki and D. B. Blaschke, Phys. Rev. D **88**, no. 8, 085001 (2013)
[5] P. Demorest *et al.*, Nature **467**, 1081 (2010); J. Antoniadis *et al.*, Science **340**, 6131 (2013).
[6] D. Gomez Dumm, D. B. Blaschke, A. G. Grunfeld and N. N. Scoccola, Phys. Rev. D **78**, 114021 (2008)
[7] T. Klähn and T. Fischer, Astrophys. J. **810**, no. 2, 134 (2015).
[8] S. Benic, D. Blaschke and M. Buballa, Phys. Rev. D **86**, 074002 (2012).
[9] J. Rafelski and M. Petran, Acta Phys. Polon. Supp. **7**, 35 (2014)

# Equation of State in Non-Zero Magnetic Field

**Nada Ezzelarab, Abdel Magied Diab and Abdel Nasser Tawfik**
Egyptian Center for Theoretical Physics (ECTP), Modern University for Technology and Information (MTI), 11571 Cairo, Egypt
World Laboratory for Cosmology And Particle Physics (WLCAPP), Cairo, Egypt

E-mail: nada.ezzelarab@eng.mti.edu.eg

**Abstract.** The Polyakov linear-sigma model (PLSM) and Hadron Resonance Gas (HRG) model are considered to study the hadronic and partonic equation(s) of state, the pressure, and response to finite magnetic field, the magnetization. The results are compared to recent lattice QCD calculations. Both models are in fairly good agreement with the lattice.

In heavy-ion collisions, especially the peripheral ones, the magnetic field likely has significant effects, which can be coupled to experimental observables, for instance, in the STAR experiment at the relativistic heavy-ion collider (RHIC) and ALICE experiment at the large hadron collider (LHC). Influences of such huge magnetic fields, $\mathcal{O}(m_\pi^2)$, on the hadronic and partonic matter and on the structure of the quark-gluon phase was analyzed in various models, such as the hadron resonance gas (HRG) [1] and also estimated in lattice QCD simulations [2, 3]. Coupling Polyakov loops to the linear-sigma model (PLSM) introduces color charge interactions to the pure gauge field. PLSM has revealed interesting features from studying the responses of QCD equation(s) of state to non-zero magnetic field [4]. In the present work, we compare our calculations to pressure and magnetization from HRG and PLSM at vanishing and finite magnetic field with recent lattice QCD calculations [3].

The influence of finite magmatic field to HRG and PLSM can be divided into two parts :

- The first one, the dispersion relation, becomes subject of modification [5, 6]. Assuming that the magnetic field is directed along $z$-axis, then in HRG, $E_{\ell s_z}(B) = \left[m^2 + P_{||}^2 + 2g_s(\ell + s_z + 1/2)B\right]^{1/2}$, and in PLSM, $E_{B,f}(B) = \left[p_z^2 + m_f^2 + |q_f|(2n + 1 - \sigma)B\right]^{1/2}$, where $n$ is Landau quantum number, and $\sigma = \pm S/2$ is related to the spin quantum number.
- The second part is the phase space, which gets modifications because of the effects of the magnetic catalysis,

$$\int \frac{d^3p}{(2\pi)^3} \longrightarrow \frac{|q_f|B}{2\pi} \sum_{\nu=0}^{\infty} \int \frac{dp}{2\pi}(2 - 1\delta_{0\nu}), \tag{1}$$

  where $2 - 1\delta_{0\nu}$ stands for degenerate Landau levels.

Treating hadron resonances as a collision-free gas [7, 8, 9, 10] allows us to estimate various thermodynamic quantities from the grand canonical partition function [7, 8]. In non-zero magnetic fields, One has

$$\ln Z(T, V, \mu) = \frac{V g_i}{(2\pi)^3} |Q_i| e B_z \sum_n \sum_{s_z} \int_0^\infty \pm dp_z \ln\left\{1 \pm \exp[(\mu_i - E_i^{B_z})/T]\right\}, \tag{2}$$

where $\pm$ stands fermions and bosons, respectively. $V$ is the volume and $g_i$ is the degeneracy for $i^{th}$ hadron.

In the mean field approximation of PLSM, the partition function expresses the exchange of energy between particle and antiparticle at temperature ($T$) and baryon chemical potential ($\mu_f$). At finite magnetic field and assuming Landau quantization [12], the quark-antiquark potential is given by

$$\begin{aligned}\Omega_{\bar{\psi}\psi}(T,\mu_f,B) &= -2\sum_f \frac{|q_f|BT}{(2\pi)^2}\sum_{\nu=0}^{\infty}(2-\delta_{0\nu})\int_0^{\infty} dp_z \\ &\left\{\ln\left[1+3\left(\phi+\phi^* e^{-\frac{E_{B,f}-\mu_f}{T}}\right)e^{-\frac{E_{B,f}-\mu_f}{T}}+e^{-3\frac{E_{B,f}-\mu_f}{T}}\right]\right. \\ &\left.\ln\left[1+3\left(\phi^*+\phi e^{-\frac{E_{B,f}+\mu_f}{T}}\right)e^{-\frac{E_{B,f}+\mu_f}{T}}+e^{-3\frac{E_{B,f}+\mu_f}{T}}\right]\right\}, \qquad (3)\end{aligned}$$

where antiquarks can be taken into consideration by adding a second integral with $\phi$ and $-\mu$.

The magnetization gives one of the most fundamental properties of the QCD matter $\mathcal{M} = (T/V)\partial \ln \mathcal{Z}/\partial eB$, where $e \neq 0$ is the elementary electric charge. Positive magnetization refers to para-, while negative $\mathcal{M}$ refer to diamagnetism. Left two panels (a) and (b) of Fig. 1 present the magnetization calculated form HRG model and PLSM, respectively, at different values of magnetic fields as function of temperature (curves). The results are compared to recent lattice QCD simulations (symbols) [3]. The agreement is fairly good. We conclude that the QCD matter possesses paramagnetic property for a wide range of temperatures. The right two panels, (c) and (d), illustrate the thermodynamic pressure. Again, the agreement between HRG and PLSM calculations, and lattice QCD simulations is good.

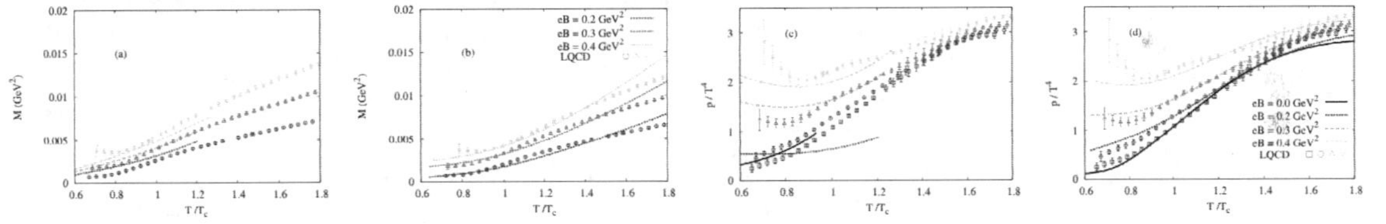


**Figure 1.** The temperature dependence of the magnetization (left two panels) and pressure (right two panels) is calculated from HRG and from PLSM at different magnetic fields, $eB = 0.0 - 0.4$ GeV$^2$ (curves) and compared with recent lattice QCD calculations (symbols) [13].

## References

[1] G. Endrodi, JHEP **1304**, 023 (2013).
[2] G. S. Bali, F. Bruckmann, G. Endrodi, and A. Schafer, PoS **LATTICE2013**, 182 (2014).
[3] G. Bali, F. Bruckmann, G. Endrodi, S. Katz, and A. Schaefer, JHEP **1408**, 177 (2014).
[4] A. Tawfik, and N. Magdy, Phys. Rev. C **90**, 015204 (2014).
[5] J. F. Letessier, and J. Rafelski, *"Hadrons and Quark-Gluon Plasma"*, (Cambridge university press, Cambridge, 2004).
[6] L. D. Landau, and E. M. Lifshitz, *Quantum Mechanics non-relativistic theory*, (Butterwoth Heinemann, Haidelberg, 1965).
[7] F. Karsch, K. Redlich, and A. Tawfik, Eur. Phys. J. **C29**, 549 (2003).
[8] K. Redlich, F. Karsch, and A. Tawfik, J. Phys. **G30**, S1271 (2004).
[9] A. Tawfik, Phys. Rev. **D71**, 054502 (2005).
[10] A. Tawfik, J. Phys. **G31**, S1105 (2005).
[11] R. Venugopalan, and M. Prakash, Nucl. Phys. **A546**, 718 (1992).
[12] A. Tawfik, A. Diab, and M. T. Hussein, "*Chiral Magnetic Effects form Extended SU(3) Linear-sigma Model*", submitted to Phys. Rev. C (2015).
[13] G. S. Bali, F. Bruckmann, G. Endrodi, S. D. Katz, A. Schafer, JHEP **177** (2014).

# Prospects of Elliptic Flow Studies at NICA/MPD

**Nikolay Geraksiev for the MPD Collaboration**
FPET, Plovdiv University Paisii Hilendarski, Plovdiv, Bulgaria
VBLHEP, Joint Institute for Nuclear Research, Dubna, Russia

E-mail: nikolay.geraksiev@gmail.com

**Abstract.** As a key observable, anisotropic flow presents a unique insight into heavy ion collision physics. The presented poster reveals the prospects of studying elliptic flow at the NICA/MPD facility through the UrQMD model. Here, results for the elliptic flow of simulated and reconstructed hadrons at the planned NICA energy range are presented.

## 1. Introduction

The NICA/MPD project is designed to provide a broad range of heavy ion experimental research opportunities [1, 2, 3]. Measurements of collective flow are used to study the dense barionic matter created in relativistic heavy ion collisions. The anisotropic flow is characterized through a Fourier expansion of the invariant triple differential distributions:

$$E\frac{\mathrm{d}^3N}{\mathrm{d}^3\mathbf{p}} = \frac{1}{2\pi}\frac{\mathrm{d}N}{p_T\mathrm{d}p_T\mathrm{d}y}\left(1 + 2\sum_{n=1}^{\infty}\nu_n\cos(n(\phi - \Psi_{RP}))\right), \tag{1}$$

where $E$ is the energy of the particle, p the momentum, $p_T$ the transverse momentum, the azimuth angle, $y$ the rapidity, and $\Psi_{RP}$ the reaction plane angle. The Fourier coefficients are $p_T$ and $y$ dependent and are given by

$$\nu_n(p_T, y) = \langle\cos(n(\phi - \Psi_{RP}))\rangle, \tag{2}$$

where the angular brackets denote an average over the particles summed over all events in the ($p_T$, $y$) bin under study. Special attention is paid to the elliptic flow $\nu_2(p_T)$. At relatively low transverse momenta ($p_T$ < 2 GeV/c), the internal pressure gradients take dominance in the fireball expansion and can be related to the degree of thermalization, viscosity, and Equation of State of the produced matter. Hyperon flow is of particular interest. The wide range of collision systems at NICA will provide a valuable insight into the reaction dynamics on hyperon production and flow.

## 2. Results

The analysis procedure was carried out using the MPDroot software. A total of $3 \cdot 10^5$ $Au^{79+}$ collisions with impact parameter in the range of 0 - 9 fm and energy $\sqrt{s_{NN}}$ = 11A GeV were generated with the UrQMD 3.3 model [4]. The produced particles were transported and reconstructed through the MPD software chain using only TPC and TOF detectors. A track selection criteria of minimum 10 TPC points was required. For this study, the Bayesian approach based on energy loss $dE/dx$ from TPC and mass measurements $M^2$ from TOF were used. Particle identification quality was estimated (see Fig. 2 top) as a function of momentum. Reconstruction of $\Lambda(\bar{\Lambda})$ hyperons was performed using the decay mode $\Lambda \rightarrow p + \pi^-$. The secondary vertex reconstruction utilizes a Kalman filtering approach, and a set of topological cuts was applied to the daughter particle candidates [5]. The achieved invariant mass distribution for $\Lambda$-hyperons is shown in Fig. 2 bottom. For the flow analysis, the well-established event-plane method was used [6]. The event-plane angle was reconstructed by all identified charged TPC tracks using two sub-events in negative and positive rapidity. Resolution of the event-plane angle and integrated flow are then estimated (see Fig. 1 ). Particles are then correlated to the event-plane revealing the observed differential distribution $\nu_n^{obs}(p_T)$; however they need to be corrected by the event-plane resolution to obtain the final results given by $\nu_n(p_T) = \nu_n^{obs}(p_T)/R_n$ (see Fig. 3 ).

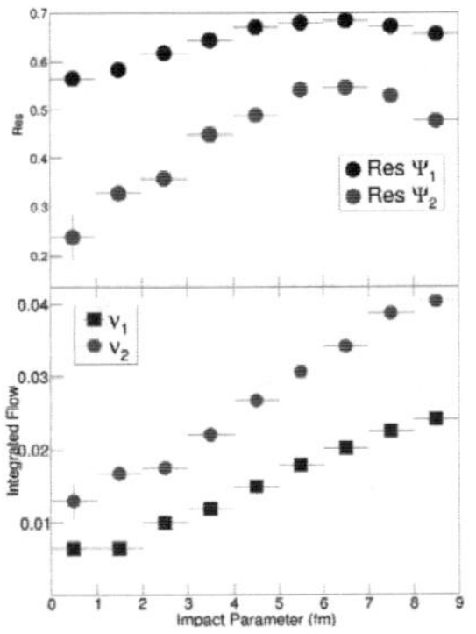


**Figure 1.** Event-plane angle resolution (*top*) and integrated flow (*bottom*) for first and second harmonics versus the impact parameter.

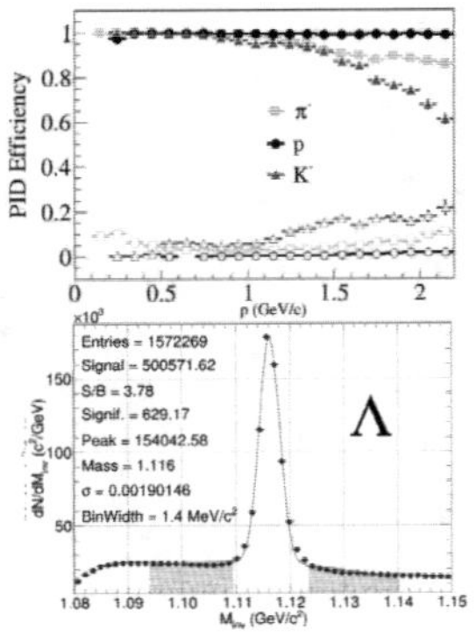


**Figure 2.** Particle identification efficiency/contamination versus particle momentum (*top*). Reconstructed $\Lambda$ invariant mass (*bottom*).

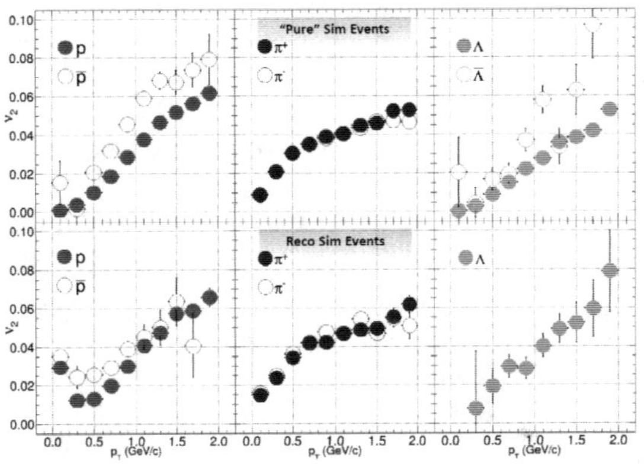


**Figure 3.** Differential elliptic flow versus transverse momentum for protons, pions and lambda particles: model results (*top*), identified and reconstructed particles (*bottom*).

## 3. Conclusion

Several points have been made on the prospects of hadron flow studies at NICA/MPD, and the important role hyperons may provide for the better understanding of the initial fireball conditions. Although the data set is limited, the results are promising.

## References

[1] NICA Conceptual Design Report. 2008. http://nica.jinr.ru/files/NICA CDR.pdf.
[2] NICA White Paper. 2011. http://nica.jinr.ru/files/WhitePaper.pdf.
[3] MPD Conceptual Design Report. 2011. http://nica.jinr.ru/files/MPD CDR en.pdf.
[4] Bass S.A. et al. // Prog. Part. Nucl. Phys. 1998. V. 41. P. 225.
[5] Fruehwirth R. //Nucl. Instr. and Meth. 1987. V. 262. P. 444.
[6] Voloshin S., Zhang Y. // Z. Phys. C. 1996. V. 70. P. 665.

# A feasibility study of hypernuclei reconstruction at NICA/MPD

**M. Ilieva*, V. Kolesnikov, D. Suvarieva, V. Vasendina, A. Zinchenko**
Joint Institute for Nuclear Research, 141980 Dubna, Moscow region, Russia
E-mail: maiailieva@mail.bg

**Abstract.** One of the main tasks of the NICA/MPD physics program is a study of the strangeness production in nuclear collisions. Heavy strange objects (including hypernuclei) could provide essential signatures of the excited and compressed baryonic matter.

Study of hypernuclei is important for: understanding the strangeness degrees of freedom in hadronic systems; study of all populated regions in the three-dimensional chart of the nuclides; study of hyperon-nucleus and hyperon-hyperon interactions.

The Monte Carlo simulation results presented show that the start version of the MPD detector will provide good conditions for hypernuclei reconstruction in Au+Au collisions at NICA.

## 1. Detector performance

The main goal of MPD detector is to explore the properties of nuclear matter under extreme density and temperature conditions. To reach this goal, the detector is designed as a $4\pi$ spectrometer capable of detecting charged hadrons, electrons and photons in heavy-ion collisions in the energy range of the NICA collider. The detailed description of the MPD geometry can be found in Ref. [1].

The present analysis is based on the detectors covering the mid-rapidity region ($|\eta|$ <1.3): the main tracker Time Projection Chamber (TPC) and barrel Time-Of-Flight system (TOF).

The track reconstruction method is based on the Kalman filtering technique and the number of TPC points per track is required to be greater than 10 to ensure a good precision of momentum and $dE/dx$ measurements.

Particle identification (PID) in the MPD experiment will be achieved by combining specific energy loss ($dE/dx$) and time-of-flight measurements. The basic detector parameters, namely, $dE/dx$ and TOF resolutions $\sigma_{dE/dx} \approx 6\%$ and $\sigma_{TOF} \approx 100$ ps provide a high degree of selectivity for hadrons at momenta below 2 GeV/$c$.

## 2. Event generator and data sets

The MPD software framework MpdRoot[2] is based on FairRoot and provides a powerful tool for detector performance studies, development of algorithms for reconstruction and physics analysis of the data.

The event samples for the present studies have been produced with the DCM-QGSM generator[3] at $\sqrt{s}$ = 5A GeV. The analyzed statistics was $5 \cdot 10^5$ and $6 \cdot 10^7$ central events (0 − 3 fm) for studies of ${}^3_\Lambda$H and ${}^4_\Lambda$He, respectively, corresponding to about 30 minutes and 60 hours of running time at the NICA collision rate of 6 kHz [1]. Produced by the event generators particles have been transported through the detector using the GEANT3 transport package (describing particle decays, secondary interactions, etc.).

## 3. Results

${}^{3}_{\Lambda}$H and ${}^{4}_{\Lambda}$He hypernuclei were reconstructed using their decay modes into two (2-prong ${}^{3}_{\Lambda}$H) or three (3-prong ${}^{3}_{\Lambda}$H and ${}^{4}_{\Lambda}$He) charged tracks. The signal event topology (Fig. 1) defines the selection criteria: relatively large distance of the closest approach DCA to the primary vertex of decay products ($dca_{1,2}$), small track-to-track separation in the decay vertex ($dca_{12}$), relatively large decay length of the mother particle ($path$). Both the DCA and two-track separation cuts are more efficient if applied in $\chi^2$ - space, i.e if normalized to their respective errors. The reconstruction results are presented in Figs. 2 – 4, where the invariant mass distributions are

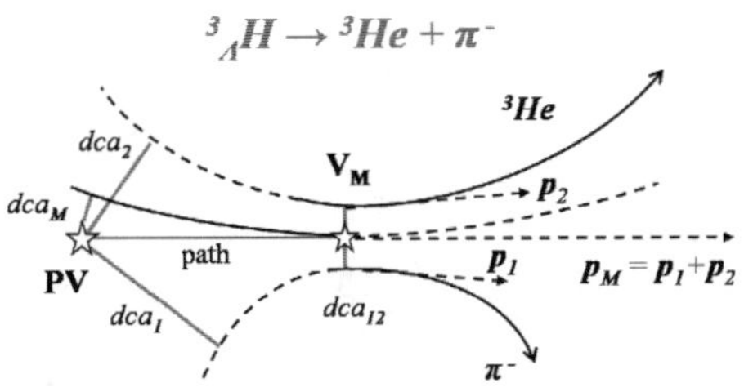


**Figure 1.** Event topology of two-particle decay of a charged particle (e.g. ${}^{3}_{\Lambda}$H$\rightarrow$${}^{3}$He+$\pi^-$).

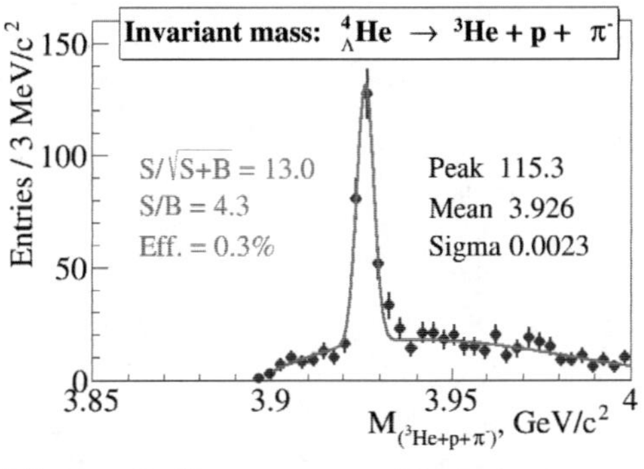


**Figure 2.** Reconstructed invariant mass of ${}^{3}$He, proton and $\pi^-$.

shown for 2-prong (${}^{3}_{\Lambda}$H $\rightarrow$${}^{3}$He+$\pi^-$) and 3-prong (${}^{3}_{\Lambda}$H $\rightarrow p + d + \pi^-$, ${}^{4}_{\Lambda}$H $\rightarrow$${}^{3}$He+$p + \pi^-$) decay modes. One can see narrow peaks ($\sigma$ of the Gaussian fit $\approx 2$ MeV/$c^2$) with quite high significance and signal-to-background ratio.

From the results presented (see also [4]) one can conclude that MPD offers good opportunities for studies of heavy strange probes at the NICA collider.

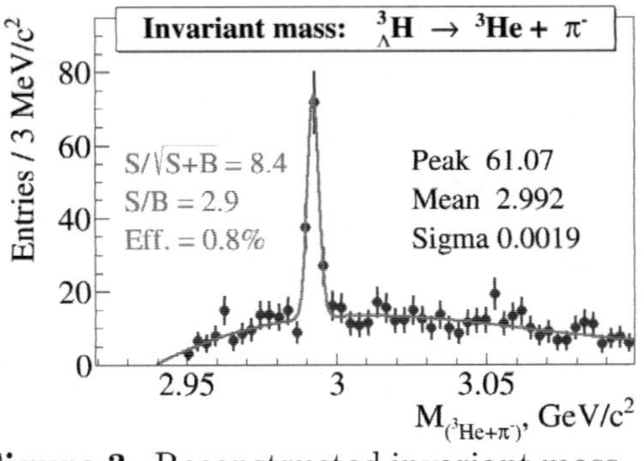


**Figure 3.** Reconstructed invariant mass of ${}^{3}$He and $\pi^-$.

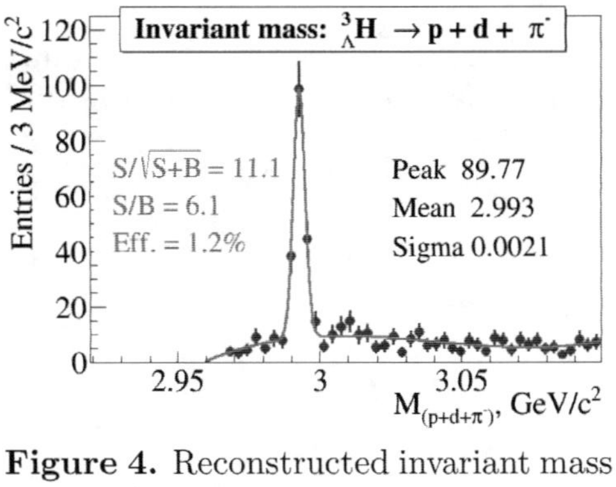


**Figure 4.** Reconstructed invariant mass of proton, deuteron and $\pi^-$.

## References

[1] K.U. Abraamyan (MPD Collaboration), Nucl. Instrum. Meth. **A628**, 99 (2011).
[2] http://mpd.jinr.ru/.
[3] A. S. Botvina *et al.*, Rhys. Rev. C **84**, 064904 (2011).
[4] M. Ilieva *et al*, Phys. Part. Nucl. Lett. **12**, no. 4, 618 (2015).

Journal of Physics: Conference Series **668** (2016) 012105 doi:10.1088/1742-6596/668/1/012105

# Conformal atoms in Supergravity

**Jakub Jankowski**

Institute of Physics, Jagiellonian University, ul. Łojasiewicza 11, 30-348 Kraków, Poland

E-mail: jakubj@th.if.uj.edu.pl

**Abstract.** In this note, we summarize recent attempts to construct holographic models of atoms and lattices in the context of applications to solid state physics. The simplest setup turned out to impose strong constraints on the choice of matter interactions of dual gravitational theory.

## 1. Introduction

In recent years, there has been a strong interest in applications of gauge/gravity duality to solid state systems [1]. One of the main directions in the current research is to incorporate a key element of condensed matter: *the lattice.* The aim of this research is to develop a simple, albeit fairly realistic model, for holographic lattices and defects and, in order to get a better understanding of the following physics, in particular, striking regularities in the electric transport. A quintessential model of a solid state lattice is the venerable Kronig-Penney model, within which a periodic lattice of Dirac delta functions is imposed in the potential of the Schrödinger equation. This allows for quasi-analytical calculations of key physical quantities. Recently, a similar setup was considered on the dual gravity side [2], and below we shortly review the main result.

## 2. A holographic model and its vacuum solution

Let us start with the action in $d = 3 + 1$ dimensions

$$S = \frac{1}{16\pi G_N} \int d^4x\sqrt{-g}\left[R - \frac{1}{2}\nabla_a\phi\nabla^a\phi - V(\phi)\right] , \tag{1}$$

where we choose

$$V(\phi) = -6 - \phi^2 - \sum_{k=1}^{\infty} c_k\phi^{2k} , \tag{2}$$

with the coefficients $c_k$ being arbitrary for the moment. With such a definition, the leading order terms give the right cosmological constant and determine the mass of the scalar to be $m^2 = -2$. The vacuum solution of this theory is empty $AdS_4$ space with a vanishing scalar, while the equilibrium configuration of finite charge is mapped to a charged black hole solution [1]. In Poincare coordinates $\{\mathbf{x}, z\} = \{t, x, y, z\}$ the near-boundary asymptotic is

$$\phi(\mathbf{x}, z) \sim \phi_0(\mathbf{x})z + \phi_1(\mathbf{x})z^2 . \tag{3}$$

In the boundary theory, this corresponds to the operator $\hat{O}$ of conformal dimension $\Delta = 2$ that has a source $\phi_0(\mathbf{x})$ and corresponding expectation value $\langle\hat{O}(\mathbf{x})\rangle = \phi_1(\mathbf{x})$. We set $\phi_0(t,x,y) = A_0\,\delta(x)$, i.e., the Dirac delta supported on a $x = 0$ line.

Due to the symmetry of the problem, the full set of Einstein-scalar equations reduces to an ODE system, with variables $x = rf(\alpha)$, $z = rg(\alpha)$ where $0 \le \alpha \le \pi/2$. The ansatz for the metric reads

$$ds^2 = \frac{1}{A(\alpha)^2}\left(\frac{d\alpha^2}{p^2} + \frac{dr^2 - dt^2 + dy^2}{r^2}\right), \tag{4}$$

where $r > 0$ and $p > 0$ is a constant determined from equations of motion. The conformal boundary consists of two parts, $\alpha_0 = \pm\pi/2$, joined together along the defect. The solution can be found in a systematic expansion:

$$A(\alpha) = \sum_{n=0}^{\infty} A_n(\alpha)\epsilon^{2n}, \qquad \phi(\alpha) = \sum_{n=0}^{\infty} f_n(\alpha)\epsilon^{2n+1}, \qquad p = \sum_{n=0}^{\infty} p_n\epsilon^{2n}, \tag{5}$$

where the lowest order is the $AdS$ solution. Boundary conditions are $\partial_\alpha A_n(0) = 0$ , $\partial_\alpha f_n(0) = 0$ , $f_n(0) = 0$ . The purely localized Dirac delta source imposes the additional condition

$$f_n'\left(\frac{\pi}{2}\right) = 0 , \tag{6}$$

as an equation for the coefficients of the scalar potential (2), which turn out to be uniquely fixed order by order: $c_1 = 1/36$, $c_2 = 1/3240$, $c_3 = 1/544320$, $c_4 = 1/146966400$, $c_5 = 1/58198694400$. Those are exactly the same as the first terms in the Taylor series expansion of

$$V(\phi) = -6\cosh\left(\frac{\phi}{\sqrt{3}}\right) . \tag{7}$$

On the other hand, in the full numerical simulation, only the potential (7) gives a solution with the desired boundary conditions [2]. This leads to the conclusion that a backreacted conformal defect with a Dirac delta function source on a line only exists for this potential. The above potential comes from a certain Kaluza-Klein compactification of $D = 11$ supergravity after truncation of the equations of motion to $\mathcal{N} = 2$ supersymmetry [3].

## 3. Outlook

It is natural to suspect that the above simple conclusion should have a simple, purely field theoretical argument, for example in analogy to deforming a conformal field theory by marginal or exactly marginal operators. This, however, remains an open question.

## Acknowledgments

This research was supported by the NCN post-doctoral internship grant DEC-2013/08/S/ST2/00547.

## References

[1] Hartnoll S A 2009 *Class. Quant. Grav.* **26** 224002
[2] Janik R A, Jankowski J and Witkowski P 2015 *JHEP* **1507** 050
[3] Duff M J and Liu J T 1999 *Nucl. Phys. B* **554** 237

Journal of Physics: Conference Series **668** (2016) 012106 doi:10.1088/1742-6596/668/1/012106

# Study of clusters and hypernuclei production with the NICA/MPD experiment

**V. Kireyeu**

The Joint Institute for Nuclear Research, Dubna, Russia
E-mail: skvitek@gmail.com

**Abstract**. We report on the first results on the dynamical modeling of cluster formation with the combined PHSD+SACA model at Nuclotron and NICA energies. Based on present predictions of the combined model we study the possibility to detect such clusters and hypernuclei in the BM@N and MPD/NICA detectors.

## PHSD+SACA model

The Parton-Hadron-String Dynamics (PHSD) [1] is a microscopic off-shell transport approach that describes the full evolution of heavy-ion collision at relativistic energies. The Simulated Annealing Clusterisation Algorithm (SACA) [2] is based on the search for nucleon configurations with a minimal binding energy; SACA takes randomly one nucleon and adds it to another fragment until the most bound configuration is found. For our studies, we combine the PHSD and SACA approaches; PHSD produces hadrons and stops at a certain time, and then SACA takes over and forms clusters and hypernuclei. It is very important to choose an appropriate starting time for the clusterisation algorithm.

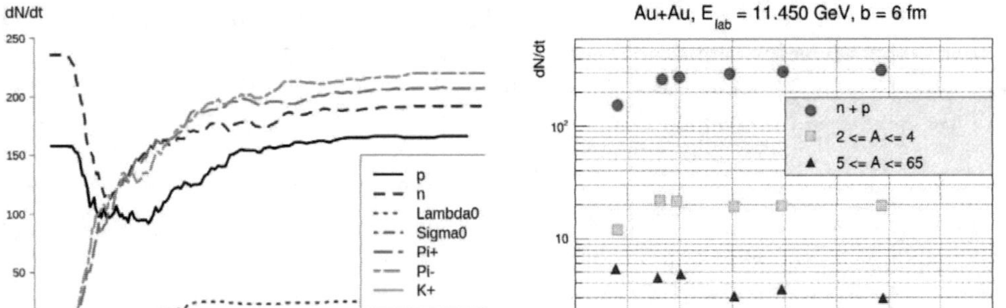


Fig.1. (left) Hadrons multiplicity over PHSD evolution time; (right) multiplicity of fragments over selected starting times.

In the left panel of Fig.1, particles multiplicity over PHSD evolution time is shown. For our studies we chose starting time t = 4, 8, 10, 15, 20, and 30 fm/c. In the right panel, the multiplicity of different kind of particles and fragments for selected times is shown. After 15 fm/c, multiplicity is almost stable, so we decided to choose t = 15 fm/c as a starting time for SACA simulations at 11.45 GeV. Fig. 2 shows the comparison of PHSD+SACA predictions with dN/dy distributions - for protons (left panel) and deuterons (right panel) - measured at the AGS by the E-802 experiment [3] for central Au+Au collisions at 11.45 AGeV. As one can see, the combined model reproduces the experimental data well.

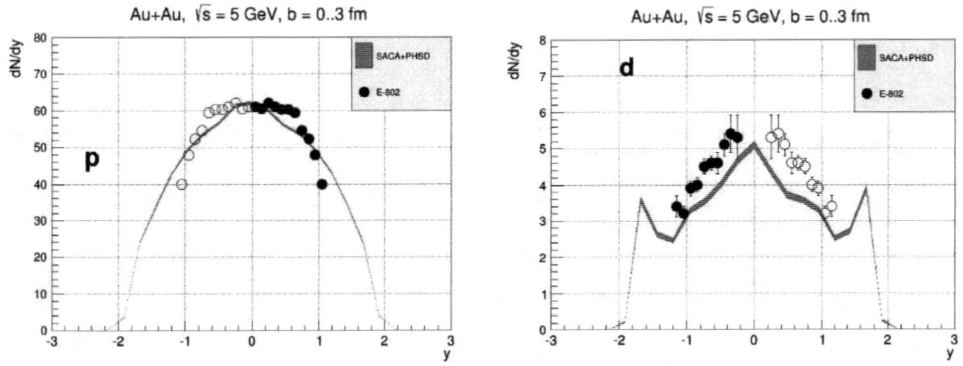


Fig.2. dN/dy distributions for protons (left panel) and deuterons (right panel) in the central Au+Au collisions at 11.45. Lines are PHSD+SACA predictions; black filled circles are the E-802 experimental data; black open circles are the mirrored data.

### Multi-Purpose Detector feasibility study

One of the tasks of the Multi-Purpose Detector [4] is to study the strangeness production. The left panel of the Fig. 3 shows the yields of hypernuclei and hyperons of PHSD+SACA model for Au+Au collisions at 11.45 GeV. Preliminary results of the MPD feasibility study for hypertrinon are shown in the right panel. Hypertriton signal is clearly viewed, with a Signal-to-Background ratio of 3.3, and an efficiency of about 0.4%.

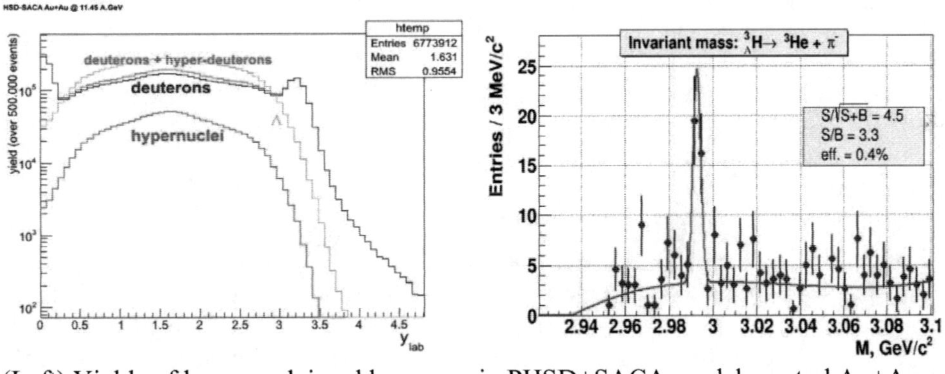


Fig.3. (Left) Yields of hypernuclei and hyperons in PHSD+SACA model, central Au+Au collisions at 11.45 GeV; (right) Invariant mass distribution for He3 and п-.

### Conclusion

PHSD+SACA model can produce clusters and hypernuclei and reproduce experimental data for 11.45 GeV; these predictions have already been used for MPD performance studies. PHSD+SACA model is actively developing, and there is some polishing work to do.

### Acknowledgments

I am very thankful to A. Le Fevre, E. Bratkovskaya, J. Aichelin and Y. Leifels for their assistance and comments.

### References

[1] Bratkovskaya E.L., Cassing W., Nucl.Phys. A856 (2011) 162-182.
[2] Puri R.K., Aichelin J., J.Comput.Phys. 162 (2000) 245-266.
[3] Ahle L. et al, Phys. Rev. C, 60 064901
[4] Abraamyan Kh.U et al. (MPD Collaboration), NIM A 628 (2011) 99-102.

# Short path length pQCD corrections to energy loss in the quark gluon plasma.

**Isobel Kolbe**[1], **W. A. Horowitz**[1]

[1]Department of Physics, University of Cape Town, Rondebosch 7701, South Africa

E-mail: `isobel.kolbe@gmail.com`

**Abstract.**
The twin identifications of high-$p_T$ enhancement and low-$p_T$ collective behaviour in the shockingly small systems of interacting particles created in $pA$ collisions calls for a detailed theoretical energy loss analysis. We study the way in which energy is dissipated in the QGP created in $pA$ collisions by calculating the short path length corrections to the DGLV energy loss formulae that have produced excellent predictions for AA collisions. We find that, shockingly, because of the large formation time assumption ( used in the DGLV calculation), a highly non-trivial cancellation of correction terms results in a null short path length correction to the DGLV energy loss formula. We investigate the effect of relaxing the large formation time assumption in the final stages of the calculation and find, because of the separation distance between production and scattering centre is integrated over from 0 to $\infty$, $\gtrsim 100\%$ corrections, even in the large path length approximation employed by DGLV.

The presence of what appears to be collective behaviour in seemingly impossibly small systems of QGP created in $pPb$ or $dAu$ experiments demands a detailed analysis of energy loss in the QGP. We must attempt to discover whether or not high momentum observables can be understood in pA experiments within the framework of perturbative QCD (pQCD), as they have been in AA collisions [1]. If a QGP does exist, the experiments must show energy loss.

While not monopolizing QGP observables, $v_2$ and $R_{AA}$ offer important insight into collective behaviour and jet quenching, the careful study of which is the key to gleaning information about the QGP. Jet quenching has been exceptionally well described by a number of energy loss formalisms, but in this work the GLV (Gyulassy, Levai, Vitev) setup [2–4] was followed as it is pedagogically insightful. The GLV description of radiative energy loss has been immensely successful in describing the energy loss of partons moving through a QGP that is created in very central $AA$ collisions.

However, the DGLV (Djordjevic massive GLV generalization) derivation relies explicitly on the assumption that the separation distance between the production point of the parton and the interaction with a scattering centre is large as compared to the Debye screening mass. We relaxed the assumption of large separation distance in order to account for small systems in which one might expect the parton to interact with the medium shortly after production. That is, we changed the length scale of the problem by taking $\frac{1}{\mu} \ll \lambda_{mfp} \ll L \Rightarrow \frac{1}{\mu} \ll L$, which means that we no longer assume that the system size $L$ is larger than the mean free path $\lambda_{mfp}$, but only that it is much larger than the Debye screened length $\frac{1}{\mu}$. Now, a major simplification arises in the DGLV calculation from the the large formation time assumption, $\mu \gg \omega_i \equiv \frac{\mathbf{k}^2}{2\omega} = \frac{1}{\tau}$, and this assumption unexpectedly played a crucial simplifying role in the short separation distance generalization as well. In fact, although the large formation time assumption reduces the number of correction terms at the amplitude level of the calculation, the assumption also causes a highly non-trivial cancellation to occur at the square-sum level,

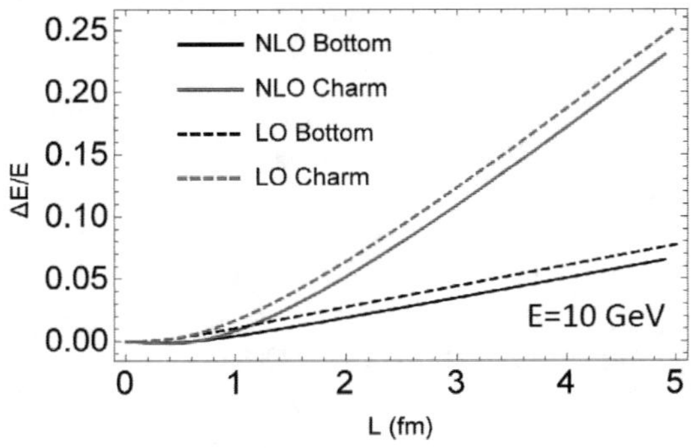


(a)

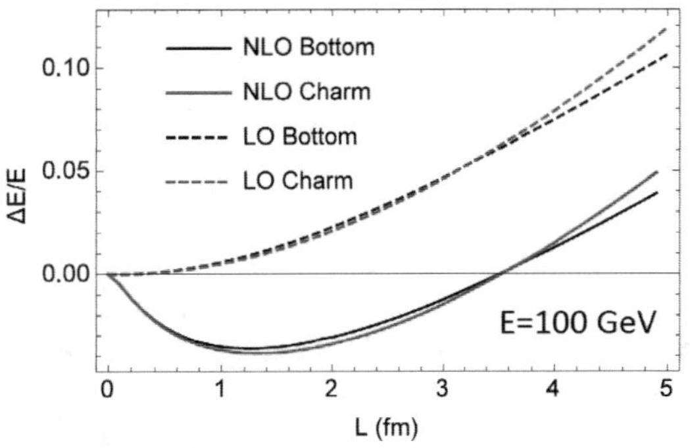


(b)

**Figure 1.** Energy loss per unit energy as a function of the static thickness of the medium for Leading Order (dashed) and Leading Order with NLO corrections (solid) for bottom and charm quarks at (a)10 GeV and (b) 100 GeV. ($\mu = 0.5$ GeV, $E = 10$ GeV, $\lambda_{mfp} = 1$ fm, $C_R = 4/3$, $C_A = 3$, $\alpha_s = 0.3$, charm mass 1.3 GeV, bottom mass 4.75 GeV Ter-Mikayelian mass of the gluon as $\mu/\sqrt{2}$)

resulting in a null correction. However, if one relaxes the large formation time assumption at the square-sum stage of the calculation, the DGLV energy loss formula, generalised for short separation distances, becomes

$$\Delta E^{(1)}_{ind} = \frac{C_R\alpha_s LE}{\pi\lambda_g}\int\frac{d^2\mathbf{q}_1}{\pi}\frac{\mu^2}{(\mu^2+\mathbf{q}_1^2)^2}\frac{d^2\mathbf{k}}{4\pi}\int d\Delta z\bar{\rho}(\Delta z)\times\left[-\frac{2(1-\cos\{(\omega_1+\tilde{\omega}_m)\Delta z\})}{(\mathbf{k}-\mathbf{q}_1)^2+M^2x^2+m_g^2}\times\right.$$
$$\times\left(\frac{(\mathbf{k}-\mathbf{q}_1)\cdot\mathbf{k}}{m_g^2+\mathbf{k}^2+x^2M^2}-\frac{(\mathbf{k}-\mathbf{q}_1)^2}{(\mathbf{k}-\mathbf{q}_1)^2+M^2x^2+m_g^2}\right)$$
$$+\frac{1}{2}e^{-\mu\Delta z)}\left\{\left(\frac{\mathbf{k}}{m_g^2+\mathbf{k}^2+x^2M^2}\right)^2\times\left(1-\frac{2C_R}{C_A}\right)\left(1-\cos\{(\omega_0-\tilde{\omega}_m)\Delta z\}\right)\right.$$
$$\left.\left.+\frac{\mathbf{k}\cdot(\mathbf{k}-\mathbf{q}_1)}{(\mathbf{k}^2+m_g^2+x^2M^2)((\mathbf{k}-\mathbf{q}_1)^2+M^2x^2+m_g^2)}\left(\cos\{(\omega_0-\tilde{\omega}_m)\Delta z\}-\cos\{(\omega_0-\omega_1)\Delta z\}\right)\right\}\right]. \tag{1}$$

In equation (1), $\bar{\rho}(\Delta z)$ is the distribution of scattering centres which can be chosen freely. We used an exponential distribution to account for the rapidly expansion of the medium such that $\bar{\rho}(\Delta z) = \frac{2}{L}\exp\{-\frac{2\Delta z}{L}\}$.

Figure 1 shows the energy loss per unit energy as a function of the size of the system at (a)10 GeV and (b)100 GeV. We see that the fact that the distribution of scattering centres is (even in DGLV) integrated from 0 to $\infty$, meaning that the short separation distance corrections are relevant even for large systems. Also, particularly at high energies, we see 'corrections' of $\sim 100\%$. We conclude that the large formation time assumption is problematic and that future energy loss calculations will have to take small formation times into account.

## 1. References

[1] Djordjevic M and Gyulassy M 2004 *J. Phys. G Nucl. Part. Phys.* **30** S1183–S1187
[2] Gyulassy M, Levai P and Vitev I 1999 *Nucl. Phys. B* **571** 197–233 (*Preprint* 9907461)
[3] Gyulassy M, Levai P and Vitev I 1999 *Nucl. Phys. A* **661** 637c
[4] Gyulassy M, Levai P and Vitev I 2001 *Nucl. Phys. B* **594** 371–419 ISSN 05503213 (*Preprint* 0006010)

Journal of Physics: Conference Series **668** (2016) 012108 doi:10.1088/1742-6596/668/1/012108

# Investigation of hypernuclei with Nuclotron beams

**A. Korotkova for HyperNIS collaboration**
Joint Institute for Nuclear Research, 141980 Dubna, Moscow region, Russia

E-mail: Annega@jinr.ru

**Abstract.** At the Laboratory of High Energy Physics (JINR, Dubna) a unique approach was elaborated with hypernuclei produced by the excitation of the accelerator beam nuclei and decay observed at a distance of tens of cm behind the production target. While the very first experiments were carried out in the Synchrophasotron beams, a more extensive hypernuclei research programme is planned for the Nuclotron accelerator and the new spectrometer created by the authors (Hypernuclei and NIS experiments). An investigation of hydrogen hypernuclei isotopes is in the first line of the experimental research plans. Namely, lifetimes and cross sections of ${}_{\Lambda}^{4}H$ and ${}_{\Lambda}^{3}H$ together with the search for ${}_{\Lambda}^{6}H$ are among the tasks. The next experiment can be the investigation of ${}_{\Lambda}^{6}He$.

## 1. Introduction

The experimental program of the HyperNIS project initially was aimed to investigate the role which strangeness plays in nuclei. Hypernuclear research programme was proposed for the Nuclotron accelerator with a new HyperNIS spectrometer [1].

## 2. Physics motivation

Recently, evidence from Frascati for three ${}_{\Lambda}^{6}H$ hypernuclei was reported [2, 3]. In the concluding remarks at the closing of the 11th International Conference on Hypernuclei and Strange Particle Physics (held in 2012 in Barcelona), the first observation of ${}_{\Lambda}^{6}H$ was mentioned by T.Nagae [4] as one of the four main achievements in hypernuclear physics reached during the last years. On the other hand, the E10 collaboration at J-PARC experiment in 2013 did not observe the missing mass peak corresponding to ${}_{\Lambda}^{6}H$ production [5]. However, A.Gal at the Barcelona conference noted, that there will be a low probability to see signal from ${}_{\Lambda}^{6}H$ at J-PARC experiment due to high transferred momentum and, consequently, high momentum of produced hypernuclei. Theoretical predictions are strongly model dependent and are controversial as well. For example, E.Hiyama and others have calculated [6] that ${}_{\Lambda}^{6}H$ is not a stable nucleus and should decay into ${}_{\Lambda}^{4}H$+n+n, if one takes into account the parameters of ${}^{5}H$ resonance measured up to now. At the same time, there are estimates [7, 8] showing that the binding energy for ${}_{\Lambda}^{6}H$ should be about a few MeV. So, it is necessary to carry out an experiment, which can test ${}_{\Lambda}^{6}H$ hypernucleus without doubt.

## 3. Method to produce relativistic hypernuclei

A unique approach suggested in the Laboratory of High Energy Physics (JINR, Dubna) will be used. In this approach the hypernuclei under study are being produced by the excitation of the beam nuclei, and the hypernuclei decay is being observed at a distance of tens of centimetres behind the production target. Thus, passage of the hypernuclei "beam" through vacuum or a matter can be investigated in order to obtain experimental estimation of the lifetimes, production cross sections and binding energy of the lightest neutron-rich hypernuclei. It should be noted that the approach is adequate for hypernuclei lifetime measurements because of minimal background in case of hypernuclei decay in the vacuum.

## 4. Present status and results

An extracted deuteron beam with typical kinetic energy of 3 GeV/nucleon and intensity of ($10^4 \div 10^5$) $s^{-1}$ was used in the test runs. In the Nuclotron run (June 2014), a $^7$Li beam was used as well. Data for proportional chambers alignment were taken in test runs and trigger tuned. A few alignment program codes were developed and investigated, all necessary calculations and fits were carried out. The trigger of the HyperNIS spectrometer works perfectly. Now, installing and commissioning of the new FEE are carried out.

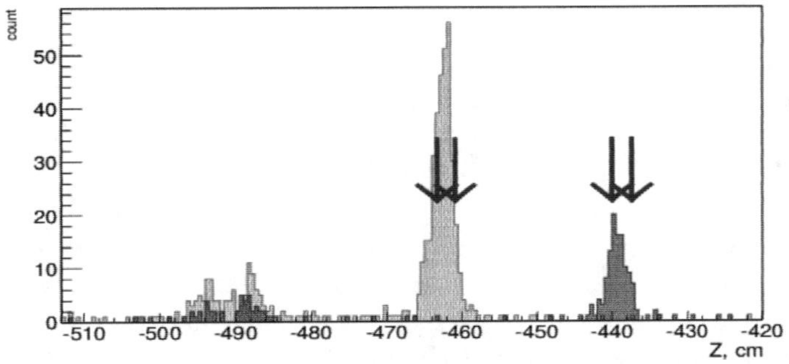


**Figure 1.** Positions of reconstructed vertices - results of a test of alignment codes and vertex reconstruction procedure. The target was placed at two positions (red and blue histograms). Arrows show edges of target. Positions of two trigger monitor counters are also seen. Vertex position - point of minimal distance d between two tracks (d<2mm).

## References

[1] A.V.Averyanov et al, Phys. At. Nucl. **71**, (2008), 2101.
[2] M.Agnelo et al.,Phys. Rev. Lett. **108** (2012) 042501.
[3] E.Botta, Nucl. Phys. **A942** 2013, 119.
[4] T.Nagae, Nucl. Phys.**A942** 2013, 559
[5] H. Sugimura et al., Phys. Lett. **B 729**, 39 (2014).
[6] E.Hiyama et al. arXiv:1304.0317v1 1 Apr 2013, Nucl. Phys. **A908**, (2013) 29.
[7] A.Gal, D.J.Millener, arXiv:1305.6716v4 28 Aug 2013.
[8] B.F.Gibson, I.R.Afnan, Nucl. Phys. **A(2014)**, in press.

Journal of Physics: Conference Series **668** (2016) 012109 doi:10.1088/1742-6596/668/1/012109

# From lepton protoplasm to the genesis of hadrons

**S. M. Eliseev**[1] **and O. S. Kosmachev**[2]

Joint Institute for Nuclear Research, 141980, Moscow region, Dubna, Russia

E-mail: [1]`selis@theor.jinr.ru`, [2]`kos@theor.jinr.ru`

**Abstract.** Theory of matter under extreme conditions opens a new stage in particle physics. It is necessary here to combine Dirac's elementary particle physics with Prigogine's dynamics of nonequilibrium systems. In the article we discuss the problem of the hierarchy of complexity. What can be considered as the lowest level of the organization of extreme matter on the basis of which the self-organization of the complex form occur?

**New physics** appears if and only if we change the assumptions or if they are complemented by original assumptions of related disciplines under the condition of the synthesis of these disciplines. In other cases we have **news** of the known physics.

**The theory of matter under extreme conditions** is the beginning of the formation kinetics of elementary particles. In this theory merging of **Dirac's physics** of elementary particles with **Prigogine's dynamics** of nonequilibrium systems is necessary. Full wording and subsequent design of this conceptual program represent extraordinary difficulties as they affect the foundations of modern physics of the microworld. One of the initial problems of the program is the problem of **the hierarchy of complexity**. What can be considered simple on the basis of which the self-organization of complex occurs?

The exclusive role of leptons is associated with their participation in most cases of nuclear and subnuclear processes. In addition, they are the final products of the decays of all hadrons. In recent years we have achieved a holistic description of the **lepton sector** [1-2]. Based on five assumptions a complete and closed set of lepton equations was obtained. They describe the massive and massless, charged and neutral, and stable and unstable leptons in a free state. The presence of individual structure of each lepton equation is established. Assumptions are necessary and sufficient conditions based on which Dirac received his fundamental equation (1928). Next, Pauli got the equation for the two-component neutrino (1932) using the Dirac's algorithm, and Majorana derived the equation for massive neutrinos (1937). Each of these lepton equations is determined by its own group. Groups are not isomorphic and differ in their structure. Just the structure is the carrier of individual characteristics of each lepton equation and, as a consequence, the properties of leptons. So the structure is an invariant feature that distinguishes one lepton equation from another. Given these circumstances, one can speak about the differences in the structures of the leptons themselves or their characteristics and thereby about guarantees of existence of their own quantum numbers. The presence of these structures is simply a mathematical fact. It was found that in the lepton sector relativism (or rather all four connected components of the Lorentz group) is the original structure-forming factor.

The decays of **tau leptons** are of particular interest for the studied range of problems [3-5]. They can decay directly into leptons and hadrons with further disintegration of the latter again into leptons. Therefore, stability of the end products of decay of hadrons to leptons proves that the structure lepton is simpler than the structure of hadrons. On the other hand, it is easy to verify that stable leptons do not contain substructures of the type of quarks. Therefore, the fact of the decay of Tau leptons directly into hadrons is obvious evidence of the possibility to describe **hadrons and quarks in terms of the same components like leptons**. In general, the decay of unstable particles from a free state (i.e. in a vacuum) with the release of energy can be considered as the process opposite to the synthesis (self-organization) not in a vacuum but in an environment where processes of energy dissipation are possible. The environment, which is equally suitable for the genesis of leptons and hadrons, is the **lepton protoplasm** [1-2].

We have the answer to the question of the total lepton number of equations that are valid in the framework of five assumptions (see [1-2]). From here, together with a detailed description of the structure of each lepton equation, it was found that in the lepton sector there is an **analogue of confinement**. From this it follows that in the structures of the unstable leptons there appear maximal invariant substructures that do not exist in a free state. They satisfy the requirement to be wave processes in free states, but they do not contain necessary elements for the formation of the current probability conservation. This situation may be a prelude to a non-standard description of confinement of quarks without unnecessary suppositions.

**Theory of matter under extreme conditions** should be a testing ground for building and validating new physics. Here physics faces the most important task: how, starting from something simple, to recreate all that was accumulated by particle physics during the time of its existence. The amazing diversity of the final particles in a single act of collision of heavy ions does not give reasons to assume that private methods to explain the birth of those or other particles will create understanding of the problem. It is a natural situation when a realistic unification of strong and electroweak interactions will require the description of quarks and leptons in a unified and strictly relativistic basis, because there is no alternative for the description of mutual transitions binding energy and matter. In this case, the lepton sector is the most adequate starting point for advancing in this direction.

From the standpoint of extreme matter one can take a fresh look at famous models or phenomenona. Here is the blurring of the notion of the on-mass surface as well as of the concept of structure that separates particles into the fermions with their spin structure and bosons with their vector structure. Therefore, the achievements of supersymmetry can be claimed in the analysis of numerous transitions during highly non-equilibrium phase in heavy ion collisions.

## References

[1] Kosmachev O S 2015 Quartet of tau-leptos and possibility of their oscillations *Proc. of Science* PoS ISHEPPXXII (2015) 030

[2] Kosmachev O S 2012 The simplest and universal constituents for description of leptons and hadrons *Phys. Atom. Nucl.* **75** 885

[3] Sudhir Kumar Gupta et al. 2012 Tau lepton charge asymmetry and new physics at the LHC Phys.Rev. D **86** 119901

[4] Aaltonen T et al. 2013 Search for Supersymmetry with Like-Sign Lepton-Tau Events at CDF *Phys. Rev. Lett.* **110** 201802

[5] Chatrchyan C et al. (CMS Collaboration) 2013 Search for physics beyond the standard model in events with tau leptons in pp collisions at sqrt(s) = 7 TeV *Eur. Phys. J.* C **73** 2493

# Stable droplet formation of quark-gluon plasma in quark-hadron phase transition

**Yogesh Kumar**

Department of Physics, Deshbandhu College, University of Delhi, New Delhi - 110019, India

E-mail: yogesh.du81@gmail.com

**Abstract.** In this contribution, a model is used to describe the droplet formation of quark-gluon plasma in quark-hadron phase transition with the effect of quark mass. The total free energy including curvature term with the effect of quark mass indicates the propensity of the system for the formation of stable droplet of quark-gluon plasma. It is found that there is a large positive barrier height enhancement in the free energy evolution from earlier results. Also, the results are in good agreement to predict the order of phase transition.

## 1. Introduction

The Lattice Quantum Chromo Dynamics (QCD) predicts the phase transition at the critical temperature around 170 MeV [1]. Since both theoretical and experimental studies are not yet settled in predicting the order of phase transition between the two phases, phenomenological studies are appropriate as presented in this work to form a stable droplet of QGP and predict the order of phase transition.

## 2. Model description and free energy droplet formation

In the earlier work, we have computed free energy with the effect of curvature using dynamical quark mass [2]. Here, I extend the work by using finite quark mass instead of dynamical quark mass, to calculate free energy in order to demonstrate the robustness of our numerical results. This finite value of quark mass is strongly dependent on temperature and is well behaved near or above the critical temperature. The quark mass is defined as [3 – 5]

$$m_q(T) = \sqrt{\gamma_q}\; g(k)\; T. \tag{1}$$

All parameters are suitably defined in Ref. [2, 3]. The authors [2] studied free energy with the inclusion of a curvature term. Using Ref. [2], I reconstruct the free energy, $F_q$ for finite quark mass having three flavors with inclusion of a curvature term. The other free energy term for gluons, pions and interface are taken from Ref. [2]. The modified total free energy is expressed as

$$F_{Total} = F_q + F_{gluon} + F_{\pi} + F_{surface}\ . \tag{2}$$

## 3. Results and conclusion

The figures 1 and 2 describe the formation of stable QGP droplet. With large enough barrier height, the QGP droplet is more stable and large in fig. 1 as compared to fig. 2 in the presence of quark mass with suitable choice of factors. The smooth cut at the phase boundary shows clear cut information about the stability of the QGP droplet with various values of temperature. In fig. 1, bunching of curves ranging $T = 150 - 230$ MeV provide a more realistic picture to describe the stability of QGP droplet and a stable QGP droplet occurs at $R \approx 10$ fm. It indicates that the use of finite quark mass with factor $\gamma_q = 1/6$ and $\gamma_g = 6\gamma_q$ including a curvature term enhances the droplet size and produces more stability in the plasma evolution from earlier work [2].

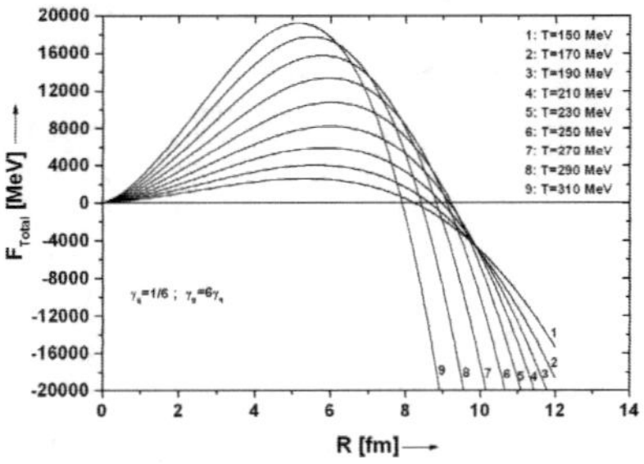


**Figure 1.** Modified free energy $F_{Total}$ with droplet size for various temp. T.

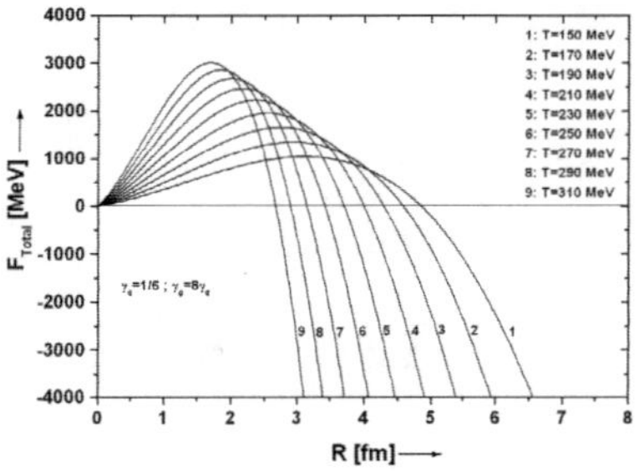


**Figure 2.** Modified free energy $F_{Total}$ with droplet size for various temp. T.

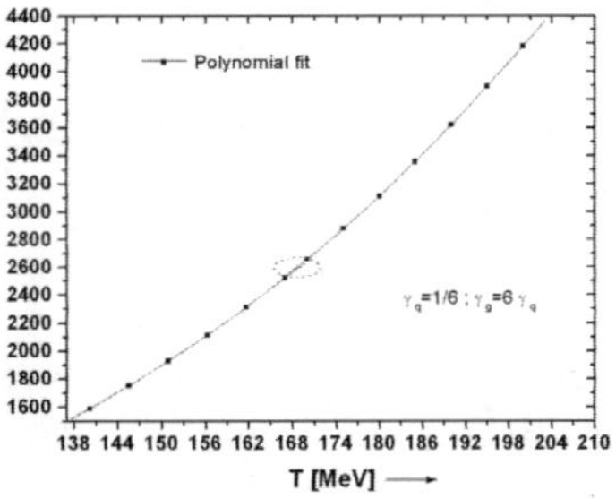


**Figure 3.** Variation of entropy S with temperature T.

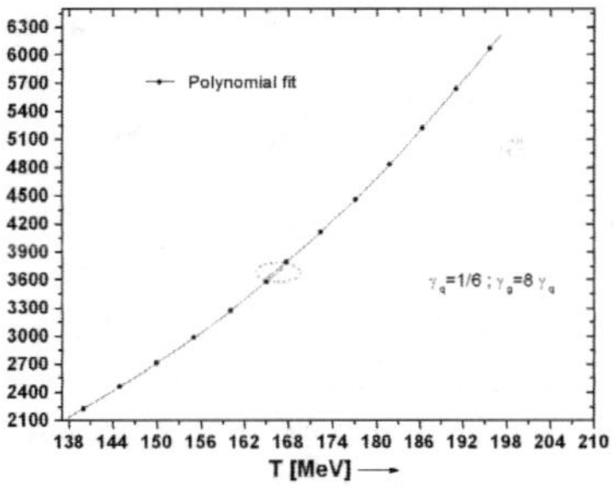


**Figure 4.** Variation of entropy S with temperature T.

In figures 3 and 4, there exists a weak discontinuity around temperature 168 MeV. It indicates a phase transition of first order that is in good agreement with Ref. [2]. Above all, the results are interesting, while enhanced to form a large stable QGP droplet, and predicts a first order phase transition with the effect of quark mass. The results are compared with earlier work [2, 5].

## Acknowledgments
I am very grateful to R. Baier for useful discussions.

## References
[1] Borsanyi S *et al* [Wuppertal-Budapest Collaboration] 2010 *JHEP* **1009** 073
[2] Singh S S, Gosain D S, Kumar Y and Jha A K 2010 *Pram. J. Phys.* **74** 27
[3] Kumar Y and Singh S S 2012 *Can. J. Phys.* **90** 955
[4] Singh S S and Kumar Y 2014 *Int. J. Mod. Phys. A* **29** 1450110
[5] Ramanathan R *et al* 2007 *Pram. J. Phys.* **68** 757, Ramanathan R *et al* 2004 *Phys. Rev.* C **70** 027903

# Limits on the variability of the Fine-Structure Constant over cosmological space-time

**T D Le**
Department of Physics, Center for Natural Science, Vietnam

E-mail: ldthong@hotmail.com

**Abstract.** The fine-structure constant plays an important role in the measurement of the strength of the electromagnetic interaction and it quantifies the strength with electrons bound within atoms and molecules. A new method provides complementary constraints on its value with the most precise estimations made so far.

## 1. Introduction

Many modern theories predict that the values of fundamental constants vary over space and time such as the fine-structure constant, $\alpha = \frac{e^2}{4\pi\varepsilon_0 \hbar c}$, which is motivated by theories unifying the fundamental interactions [1]. The search for variations of these constants can be detected by comparing transitions between levels in atoms and molecules from astronomical observed data that have a different functional dependence on these constants. Up to very recently, tests of $\alpha$-variation over cosmological time scales were based on the comparisons of atomic clocks and observations of high redshift objects with their corresponding laboratory values at z = 0 [2-5]. The most sensitive current limits on $\alpha$ comes from measurements using microwave and mm-molecular transitions such as hydrogen give null results for $\Delta\alpha/\alpha$ at the level of a few parts per million (ppm) [6-8]. Here, we propose to use the absorption lines identified as Fe V and Ni V transitions toward the white dwarf G191-B2B [9] with the aim of constraining past spatial and temporal variations in $\alpha$. The current analysis is based on absorption lines, the constraints on $\Delta\alpha/\alpha$ are stronger than those derived by other methods.

## 2. Data analysis

To extract whether α changed over the cosmological time and space scales, we have applied the method first suggested by Le [10] to the fine splitting of the transition lines of Fe V, Ni V, observed in the spectra of white dwarf G191-B2B with high resolution far-UV spectrum. According to the this method, the relationship between the laboratory wavelengths and those observed near the white dwarf is approximately written as $\frac{\Delta\alpha}{\alpha} \approx \frac{1}{2}\left(\frac{\frac{1}{2}\left(\frac{\lambda_2(t)}{\lambda_1(t)}\right)-1}{\frac{1}{2}\left(\frac{\lambda_2(0)}{\lambda_1(0)}\right)-1} - 1\right)$, provided that $\Delta\alpha/\alpha$ is very small. Thus, any variation of α could be revealed by comparing different transitions in different atoms in cosmic and laboratory, one can directly infer the possible variation of $\alpha$ at different epochs, space-time points, regions of the Universe and the present value. With those measurements and existing of Fe V and Ni V in the white dwarf star G191-B2B recorded by the Hubble Space Telescope Imaging Spectrograph, we constrain variation of $\alpha$ and open a new way for future radio astronomy measurement of higher accuracy to test more stringently whether the $\alpha$ varies. This uncertainty can be used to estimate systematic error in the determination of $\Delta\alpha/\alpha$. The discussions of possible sources of systematical and statistical errors of the method are as described in [9-10]. Combining these results with other stringent limits on $\Delta\alpha/\alpha$ obtained from comparisons of transition wavelengths in some molecules such as $H_2$, $NH_3$, CH and CD constrain $\Delta\mu/\mu$ in the future [11-13]. Fig. 1 and 2 show the limit obtained from the analysis the Fe V and Ni V absorption lines from the white dwarf star G191-B2B spectra to constrain past variations in $\alpha$.

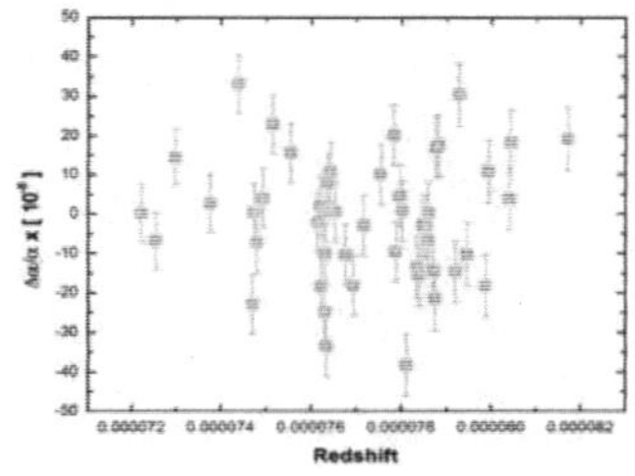


**Figure 1.** Plot for the components of Fe V.

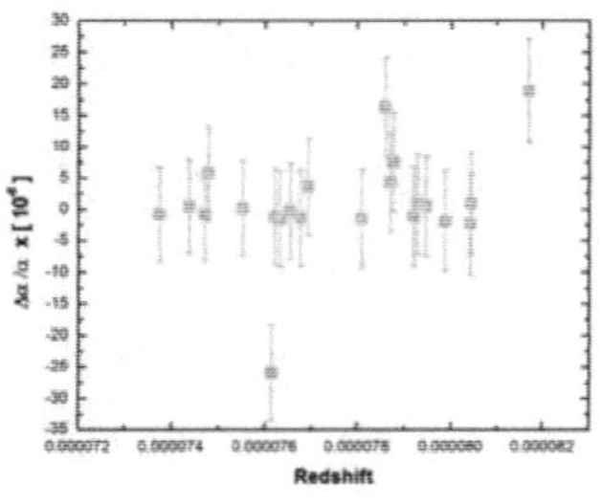


**Figure 2.** Plot for the components of Ni V.

## 3. Conclusions

The most stringent limit derived here is statically more constraining than that derived in the past test by the methods used in previous studies [13, 15, 18-20]. This remarkably consistent indication for a possible variation of $\alpha$ certainly deserves further investigation on a large number of systems, aimed at reducing the final error bar. It suggests how dedicated astronomical measurements can improve these constrain in the future with very high resolution spectroscopic studies are needed to probe the variations in $\alpha$ with much better accuracy. With improvement from astronomical measurements and in laboratory wavelengths, the study of the variation of the fundamental constants such as the fine structure constant should also be of utmost for a complete understanding of fundamental physics.

## References


[1] Uzan J P 2011 *Living. Rev. Relativ.***14** 2.
[2] Leefer N, C, Weber T M *et al.* 2013 *Phys. Rev. Lett.* **111** 060801.
[3] Lentati L, Carilli C, Alexander P, *et al* 2013 *Mon. Not. R. Astron. Soc.* 430 2454.
[4] Rahmani H, Maheshwari N & Srianand R 2014 *Mon. Not. R. Astron. Soc.* **439** 70.
[5] Webb J K, King J A, Murphy M T *et al.*2011 *Phys. Rev. Lett.***107** 191101.
[6] Weiss A, Walter F, Downes D *et al.* 2012 *Astrophys. J.* **102** 753.
[7] Levshakov S A, *et al.* 2012 *Astron. Astrophys.* **L9** 540.
[8] Weiβ A, Walter F, Downes D *et al.*2012 *Astrophys. J.***753** 102.
[9] Berengut J C, Flambaum V V, Ong A *et al.* 2013 *Phys. Rev. Lett.* **111** 010801.
[10] Le T D 2014 *AIP Conf.Proc.* **1594** 23.
[11] Murphy M T, Flambaum V V, Muller S & Henkel C 2008 *Science.***320** 1611.
[12] Truppe S, Hendricks R J, Tokunaga S K *et al.* 2013 *Nature Communs.* 4 2600.
[13] de Nijs A J, Ubachs W & Bethlem H L 2012 *Phys. Rev. A.* **86** 032501.
[14] Bagdonaite J, Jansen P *et al.* 2012 *Science.* **4** 46.
[15] Levshakov S A, Kozlov G & Reimers D 2011 *Astrophys. J.***738** 26.

# On the pair correlations of neutral $K, D, B$ and $B_s$ mesons with close momenta produced in inclusive multiparticle processes

**Valery V Lyuboshitz**[1] and **Vladimir L Lyuboshitz**

Joint Institute for Nuclear Research, 141980 Dubna, Moscow Region, Russia

E-mail: `Valery.Lyuboshitz@jinr.ru`

**Abstract.** The phenomenological structure of inclusive cross-sections of the production of two neutral $K$ mesons in collisions of hadrons and nuclei is studied taking into account the strangeness conservation in strong and electromagnetic interactions. Expressions for the correlation functions are obtained. It is shown that the correlation functions of the pairs $K_S^0 K_S^0$ and $K_L^0 K_L^0$, produced in the same inclusive process, coincide, and the difference between those for the pairs $K_S^0 K_S^0$ and $K_S^0 K_L^0$ is conditioned by the generation of pairs of non-identical kaons $K^0 \bar{K}^0$. Analogous pair correlations for neutral heavy mesons $D^0$, $B^0$ and $B_s^0$ are also analyzed.

## 1. Consequences of the strangeness conservation for neutral kaons

In the work [1] the properties of the density matrix of two neutral $K$ mesons, following from the strangeness conservation in strong and electromagnetic interactions, have been investigated.

Due to the strangeness conservation, all the non-diagonal elements of the two-kaon density matrix are equal to zero except those between the states $|K^0\rangle^{(\mathbf{p}_1)}|\bar{K}^0\rangle^{(\mathbf{p}_2)}$ and $|\bar{K}^0\rangle^{(\mathbf{p}_1)}|K^0\rangle^{(\mathbf{p}_2)}$ with the zero strangeness ( $\mathbf{p}_1$ and $\mathbf{p}_2$ are the kaon momenta ). The internal states of $K^0$ and $\bar{K}^0$ mesons are $|K^0\rangle = \frac{1}{\sqrt{2}}\,(|K_S^0\rangle + |K_L^0\rangle)$ and $|\bar{K}^0\rangle = \frac{1}{\sqrt{2}}\,(|K_S^0\rangle - |K_L^0\rangle)$, where $K_S^0$ is the $CP$-even short-lived neutral kaon and $K_L^0$ is the $CP$-odd long-lived one ( neglecting the weak effect of $CP$ non-invariance ). The strangeness conservation leads to the fact that all the double inclusive cross-sections of production of pairs $K_S^0 K_S^0$, $K_L^0 K_L^0$ and $K_S^0 K_L^0$ prove to be invariant with respect to the replacement of $K_S^0$ by $K_L^0$ and *vice versa* [1]: $f_{SS}(\mathbf{p}_1, \mathbf{p}_2) = f_{LL}(\mathbf{p}_1, \mathbf{p}_2)$, $f_{SL}(\mathbf{p}_1, \mathbf{p}_2) = f_{LS}(\mathbf{p}_1, \mathbf{p}_2)$. In doing so, we have $f_{SS}(\mathbf{p}_1, \mathbf{p}_2) - f_{SL}(\mathbf{p}_1, \mathbf{p}_2) = \mathrm{Re}\,\rho_{K^0\bar{K}^0 \to \bar{K}^0 K^0}(\mathbf{p}_1, \mathbf{p}_2)$, where $\rho_{K^0\bar{K}^0 \to \bar{K}^0 K^0}(\mathbf{p}_1, \mathbf{p}_2)$ is the nonzero non-diagonal element of the two-kaon density matrix.

## 2. Pair correlations of neutral kaons with close momenta

Within the model of one-particle sources [2-7], in the case of identical states $K_S^0 K_S^0$ and $K_L^0 K_L^0$ we obtain the following expressions for the correlation functions $R_{SS}$, $R_{LL}$ ( see also, e.g., [8] ):

$$R_{SS}(\mathbf{k}) = R_{LL}(\mathbf{k}) = \lambda_{K^0 K^0}\left[1 + F_{K^0}(2\mathbf{k}) + 2\, b_{\mathrm{int}}(\mathbf{k})\right] +$$

$$+\lambda_{\bar{K}^0 \bar{K}^0}\left[1 + F_{\bar{K}^0}(2\mathbf{k}) + 2\,\tilde{b}_{\mathrm{int}}(\mathbf{k})\right] + \lambda_{K^0 \bar{K}^0}\left[1 + F_{K^0 \bar{K}^0}(2\mathbf{k}) + 2\, B_{\mathrm{int}}(\mathbf{k})\right]. \qquad (1)$$

[1] Presenter at SQM 2015

Here $\mathbf{k}$ is the momentum of one of the kaons in the c.m. frame of the pair, and the quantities $\lambda_{K^0K^0}$, $\lambda_{\bar{K}^0\bar{K}^0}$, and $\lambda_{K^0\bar{K}^0}$ are the respective relative fractions of the average numbers of produced pairs $K^0K^0$, $\bar{K}^0\bar{K}^0$, and $K^0\bar{K}^0$. The "formfactors" $F_{K^0}$, $F_{\bar{K}^0}$, and $F_{K^0\bar{K}^0}$ appear due to the contribution of Bose statistics, and the quantities $b_{\text{int}}$, $\tilde{b}_{\text{int}}$, and $B_{\text{int}}$ describe the contributions of the $S$-wave final-state interaction of two $K^0$ mesons, two $\bar{K}^0$ mesons, and of the $K^0$ and $\bar{K}^0$ mesons (for more details, see [8]). The function $R_{SS} = R_{LL}$ is normalized to unity at large relative momenta.

Meantime, for the non-identical states $K_S^0K_L^0$ the correlation functions have the form [8]

$$R_{SL}(\mathbf{k}) = R_{LS}(\mathbf{k}) = \lambda_{K^0K^0}\left[1 + F_{K^0}(2\mathbf{k}) + 2\,b_{\text{int}}(\mathbf{k})\right] +$$

$$+\lambda_{\bar{K}^0\bar{K}^0}\left[1 + F_{\bar{K}^0}(2\mathbf{k}) + 2\,\tilde{b}_{\text{int}}(\mathbf{k})\right] + \lambda_{K^0\bar{K}^0}\left[1 - F_{K^0\bar{K}^0}(2\mathbf{k})\right]. \tag{2}$$

In accordance with Eqs. (1) and (2), the following relation holds:

$$R_{SS}(\mathbf{k}) - R_{SL}(\mathbf{k}) = 2\lambda_{K^0\bar{K}^0}\left[F_{K^0\bar{K}^0}(2\mathbf{k}) + B_{\text{int}}(\mathbf{k})\right]. \tag{3}$$

So, the difference between the correlation functions of the pairs of identical and non-identical neutral kaons $K_S^0K_S^0$, $K_S^0K_L^0$ is conditioned exclusively by the generation of $K^0\bar{K}^0$ -pairs.

## 3. Correlations of neutral heavy mesons

Formally, analogous relations are valid also for the neutral heavy mesons $D^0$, $B^0$ and $B_s^0$, generated in inclusive multiple processes with charm (beauty) conservation. Here, neglecting the weak effects of $CP$ violation, the quasistationary states are also states with definite $CP$ parity – e.g., $|B_S^0\rangle = \frac{1}{\sqrt{2}}\left(|B^0\rangle + |\bar{B}^0\rangle\right)$ ( $CP$-even ) and $|B_L^0\rangle = \frac{1}{\sqrt{2}}\left(|B^0\rangle - |\bar{B}^0\rangle\right)$ ( $CP$-odd ).

However, for $D^0$, $B^0$, and $B_s^0$ mesons, it is practically impossible to distinguish the $CP$-even and $CP$-odd states by the difference in their lifetimes – since, quite contrary to the case of $K^0$ mesons, the respective lifetimes are almost the same. These states, in principle, can be identified through the purely $CP$-even and $CP$-odd decay channels; however, in fact the branching ratios for such decays are very small – for example, $Br\,(D^0 \to \pi^+\pi^-) = 1.62 \cdot 10^{-3}$ ( $CP = +1$ ), $Br\,(D^0 \to K^+K^-) = 4.25 \cdot 10^{-3}$ ( $CP = +1$ ), $Br\,(B_s^0 \to J/\Psi\ \pi^0) < 1.2 \cdot 10^{-3}$ ( $CP = +1$ ), and $Br\,(B^0 \to J/\Psi\ K_S^0) = 9 \cdot 10^{-4}$ ( $CP = -1$ ) .

Just as for neutral $K$ mesons, the correlation functions for the pairs of states of neutral $D$, $B$, and $B_s$ mesons with the same $CP$ parity ($R_{SS} = R_{LL}$) and for the pairs of states with different $CP$ parity ($R_{SL}$) do not coincide, and the difference between them is conditioned exclusively by the production of pairs $D^0\bar{D}^0$, $B^0\bar{B}^0$ and $B_s^0\bar{B}_s^0$, respectively; e.g., for $B_s^0$ mesons we have

$$R_{SS}(\mathbf{k}) - R_{SL}(\mathbf{k}) = 2\lambda_{B_s^0\bar{B}_s^0}\left[F_{B_s^0\bar{B}_s^0}(2\mathbf{k}) + B_{\text{int}}(\mathbf{k})\right]; \tag{4}$$

here $\lambda_{B_s^0\bar{B}_s^0}$ is the relative fraction of generated pairs $B_s^0\bar{B}_s^0$, and the contributions of Bose statistics ( $F_{B_s^0\bar{B}_s^0}$ ) and the $S$-wave $B_s^0\bar{B}_s^0$ -interaction ( $B_{\text{int}}$ ) are determined analogously to the case of $K^0\bar{K}^0$-pairs ( see [8] ).

## References


[1] Lyuboshitz V L 1976 *Yad. Fiz.* **23** 1266 [ *Sov. J. Nucl. Phys.* **23**, 673 ]
[2] Lyuboshitz V L and Podgoretsky M I 1979 *Yad. Fiz.* **30** 789 [ *Sov. J. Nucl. Phys.* **30** 407 ]
[3] Podgoretsky M I 1989 *Fiz. Elem. Chast. At. Yadra* **20** 628 [ *Sov. J. Part. Nucl.* **20** 266 ]
[4] Lednicky R and Lyuboshitz V L 1982 *Yad. Fiz.* **35** 1316 [ *Sov. J. Nucl. Phys.* **35** 770 ]
[5] Lyuboshitz V L 1985 *Yad. Fiz.* **41** 820 [ *Sov. J. Nucl. Phys.* **41** 523 ]
[6] Lyuboshitz V L 1988 *Yad. Fiz.* **48** 1501 [ *Sov. J. Nucl. Phys.* **48** 956 ]
[7] Lednicky R, Lyuboshitz V V and Lyuboshitz V L 1998 *Yad. Fiz.* **61** 2161 [ *Phys. At. Nucl.* **61** 2050 ]
[8] Lyuboshitz V L and Lyuboshitz V V 2007 *Pis'ma v EchaYa* **4** 654 [ *Phys. Part. Nucl. Lett.* **4** 388 ]

Journal of Physics: Conference Series **668** (2016) 012113 doi:10.1088/1742-6596/668/1/012113

# Our Understanding on Landau-Raychaudhuri Cosmology

**Abdel Magied Diab and Abdel Nasser Tawfik**

Egyptian Center for Theoretical Physics (ECTP), Modern University for Technology and Information (MTI), 11571 Cairo, Egypt

World Laboratory for Cosmology And Particle Physics (WLCAPP), Cairo, Egypt

E-mail: a.diab@eng.mti.edu.eg

**Abstract.** Recent developments to the generalized uncertainty principles (GUPs) and the modified dispersion relations (MDRs) play an important role in estimating the quantum corrections to the cosmic line element. We first apply both approaches to de Sitter-Schwarzschild and Reissner-Nördstrom black holes. Then from the fixed point method, the modified Landau-Raychaudhuri equations are derived in emergent cosmic space. We conclude that (non-)singular big bang solutions are obviously model dependent.

The predictions about minimum length uncertainty and maximum measurable momentum modify Heisenberg uncertainty principle (HUP), the so-called Generalized Uncertainty Principle (GUP), in a straight forward manner. The GUP has been confirmed by various observations such as and quantum optics [1] and seems to provide evidence for the existence of minimal length uncertainty [2]. Also, one of the most puzzling paradoxes in experimental high-energy physics is the time delay observed in the ultra high-energy cosmic rays (UHECRs). When assuming Lorentz invariance violation, this can be interpreted, and the dispersion relation becomes subject to modification (MDR) [3, 4], accordingly. Both approaches are applied to de Sitter-Schwarzschild and Reissner-Nördstrom black holes.

The unique spherically symmetric vacuum solution of General Relativity is the Schwarzschild metric [5]. Details on how the modified Landau-Raychaudhuri equations in emergent cosmic space of de Sitter-Schwarzschild and Reissner-Nördstrom black holes vary with GUP and MDR and depend on the equations of state ($p = \omega\rho$) that characterize the cosmic geometry can be found in [6, 7]. The modified Landau-Raychaudhuri equations in de Sitter-Schwarzschild black holes–due to the minimal length and higher-order GUP and MDR approaches, respectively–can be determined as

$$\dot{H}\left[1+\frac{\alpha^2}{4}H^2+\cdots\right] = -3H^2\left[2(1+\omega)+\frac{1}{4}\alpha^2 H^2+\cdots\right], \tag{1}$$

$$\dot{H}\left[1+\alpha^2 H^2-\alpha H+\cdots\right] = -3H^2\left[6(1+\omega)+\frac{3\alpha^2}{8}H^2-4\alpha H+\cdots\right], \tag{2}$$

$$\dot{H}[1-\frac{15}{32}(\alpha_1^2-4\alpha_2)H^2-\frac{3}{8}\alpha_1 H^2] = -3H^2\left[96(1+\omega)+18\alpha_1 H^2+\frac{45}{8}(\alpha_1^2-4\alpha_2)H^4\right]. \tag{3}$$

The first equation is related to the linear GUP. From the quadratic GUP, the second equation is obtained. The last equation gives the influence of MDR. From the line element in Reissner-Nordström geometry which has static electrical charged $Q$ [8], the modified Landau-Raychaudhuri equations for linear and quadratic GUP and MDR approaches, respectively, read

$$\dot{H}\left[1+\frac{\alpha^2}{4}QH^4+\alpha^2\frac{4Q^4}{3\pi}H^6+\cdots\right]=-3H^2\left[\frac{2}{3}(1+\omega)+\frac{2\alpha^2}{3}(QH^2+Q^2H^4)H^4+\cdots\right], \quad (4)$$

$$\dot{H}\left[1-\alpha H+\alpha^2(1+Q)H^2+\cdots\right]=-3H^2\left[2(1+\omega)-\frac{4}{3}\alpha H+\frac{\alpha^2}{4}H^2+\cdots\right], \quad (5)$$

$$\begin{aligned}\dot{H}\left[1-\frac{5}{128}(\alpha_1^2-4\alpha_2)H^4-\alpha_1\left(1+8Q\right)H^4+\cdots\right] &= -3H^2\left[2(1+\omega)+\frac{3\alpha_1}{8}(1+3Q)H^2\right.\\ &\left.- \frac{9\alpha_1 Q^2}{64}H^4+\frac{5}{6}(\alpha_1^2-4\alpha_2)H^4+\cdots\right]. \quad (6)\end{aligned}$$

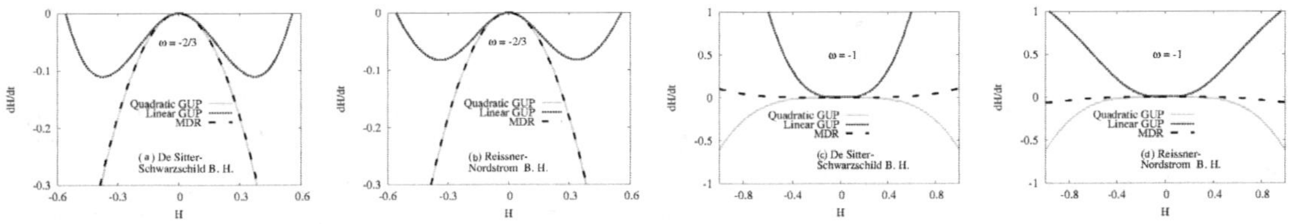


**Figure 1.** A depiction of $dH/dt$ as function of $H$ for de Sitter-Schwarzschild ((a) and (c)) and Reissner-Nördstrom ((b) and (d)) black holes at $\omega = -2/3$ and $-1$, with the minimal length GUP (Linear GUP) (solid curve), higher order GUP (Quadratic GUP) (dotted curve) and modified dispersion relation (MDR) (dashed curve).

The results are given in Fig. 1. The solid curve stands for linear GUP at $\omega = -2/3$ and $-1$. We find two fixed points, $H_1 = 0$ and $H_2 \sim 1/\gamma\ell_p = H_P$, which point to non-singular big bang. Accordingly, the Universe age can be determined from $\tau = \int_{H_1}^{H_2} dH/\dot{H} \longrightarrow \infty$, resulting in an infinite time-interval to reach the fixed points. It is obvious that these results are not observed from the other approaches and for other equations of state. Thus, we merely conclude that (non-)singular solutions depend on the metric types, the equations of states, and the quantum gravity approaches. These results address the need for further systematic studies and highlight the marginal impacts of the conclusion claimed in Ref. [9]. Last but not least, the absence of a good theory for quantum gravity should not permit advocating conclusions about singular or nonsingular big bang based on model-dependent approaches.

## References

[1] I. Pikovski, M. R. Vanner, M. Aspelmeyer, M. S. Kim and C. Brukner, Nature Phys. **8**, 393 (2012).
[2] F. Scardigli and R. Casadio, Int. J. Mod. Phys. D **18**, 319 (2009).
[3] A. Tawfik and A. Diab, Int. J. Mod. Phys. D **23**, 1430025 (2014).
[4] A. Tawfik and A. Diab, Rept. Prog. Phys. **78**, 126001 (2015).
[5] S. M. Carroll, "Spacetime and Geometry: An Introduction to General Relativity" (University of Chicago, New York, 2004).
[6] A. Tawfik and A. Diab, Int. J. Mod. Phys. A **30**, 1550059 (2015).
[7] A. Tawfik and E. Abou El Dahab, Int. J. Mod. Phys. A **30**, 1550030 (2015).
[8] H. Reissner, *Annalen der Physik* **50**, 106 (1916)
[9] A. F. Ali, *Phys. Lett. B* **732**, 335 (2014)

# Renormdynamics and Hadronization

**Nugzar Makhaldiani**
Joint Institute for Nuclear Research
Dubna, Moscow Region, Russia
E-mail: mnv@jinr.ru

**Abstract.** Independently radiating valence quarks and corresponding negative binomial distribution presents phenomenologically preferable mechanism of hadronization in multiparticle production processes. Main properties of the renormdynamics, corresponding motion equations and their solutions are considered.

## 1. Renormdynamics

We say that we find New Physics (NP) when either we find a phenomenon which is forbidden by SM in principal - this is the qualitative level of NP - or we find significant deviation between precision calculations in SM of an observable quantity and corresponding experimental value. Renordynamics (RD) unifies different renormgroups in one society. Quantum field theory (QFT) and Fractal calculus (FC) provide Universal language of fundamental physics (see e.g. [1]). In QFT existence of a given theory means, that we can control its behavior at some scales (short or large distances) by renormalization theory [2]. If the theory exists, than we want to solve it, which means to determine what happens on other (large or short) scales. This is the problem (and content) of Renormdynamics. The result of the Renormdynamics, the solution of its discrete or continual motion equations, is the effective QFT on a given scale (different from the initial one). We will call RDF functions $g_n = f_n(t)$, which are solutions of the RD motion equations

$$\dot{g}_n = \beta_n(g), 1 \leq n \leq N. \tag{1}$$

In the simplest case of one coupling constant, the function $g = f(t)$, is constant $g = g_c$ when $\beta(g_c) = 0$, or is invertible (monotone). Indeed,

$$\dot{g} = f'(t) = f'(f^{-1}(g)) = \beta(g). \tag{2}$$

Each monotone interval ends by UV and IR fixed points and describes corresponding phase of the system. Note that, the simplest case of the classical dynamics, the Hamiltonian system with one degree of freedom, is already two dimensional, so we have not an analog of one charge renormdynamics. Then the regular Hamiltonian systems of the classical mechanics are defined on the even dimensional phase space, so there is not an analog of the three dimensional renormdynamics for the coupling constants of the SM. The fixed points of renormdynamics belong to the set of zeros of the polynomial system of equations $\beta_n(g) = 0, 1 \leq n \leq N$, in the perturbative renormdynamics. Describing qualitative and numerical properties of the set, finding the number and multiplicity, evaluation and finding the numerical values of the zeros is the task of contemporary algebraic and computational geometry. QCD is the theory of the strong interactions with, as only inputs, one mass parameter for each quark species and the value of the QCD coupling constant at some energy or momentum scale in some renormalization scheme. The quantitative values and qualitative content of the given field theory depends on the scale. In QCD e.g. the effective action have the following form

$$S(\mu) = \frac{1}{g^2(\mu)} \int d^D x \mathcal{L}(A(\mu)), \tag{3}$$

variation with respect the change of scale gives

$$\delta S = -2\frac{\beta(g)}{g^3}\delta\mu S + \frac{1}{g^2}\int d^D x \frac{\delta \mathcal{L}}{\delta A}\delta A \tag{4}$$

and the following two statements are equivalent,

$$\delta S = 0, \ \beta(g) = 0 \Leftrightarrow \delta S = 0, \ \frac{\delta \mathcal{L}}{\delta A} = 0 \tag{5}$$

So, from renorminvariance of the effective action follows that, at the conformal symmetric points, the motion equations for fields are satisfied. Generalization for the several coupling constants and other models is obvious. In string theory, the connection between conformal invariance of the effective theory on the parametric world sheet and the motion equations of the fields on the embedding space is well known [3]. More recent topic in this direction is AdS/CFT Duality [4].

## 2. Hadronization

From dimensional considerations [5], the following combination of cross sections [6] must be universal function

$$< n > \frac{\sigma_n}{\sigma} = \Psi(\frac{n}{< n >}) \tag{6}$$

Corresponding relation for the inclusive cross sections is [7]

$$< n(p) > \frac{d\sigma_n}{\bar{dp}}/\frac{d\sigma}{\bar{dp}} = \Psi(\frac{n}{< n(p) >}) \tag{7}$$

Generating function (GF) for NBD can be presented as

$$(1 + \frac{< n >}{k}(1-h))^{-k} = \int_0^\infty e^{-<n>z(1-h)}\Psi(z)dz \tag{8}$$

which means that Poisson GF weighted by KNO distribution gives NBD GF. This is the exact and universal picture of hadronization in multiparticle production processes.

## References

[1] Makhaldiani N 2011 *Physics of Particles and Nuclei Letters* **8** 325
[2] Collins J C 1984 *Renormalization* (London: Cambrige University Press)
[3] Green M B, Schwarz J H and Witten E 1987 *Super String Theory* (London: Cambrige University Press)
[4] Maldacena J M 1998 *Adv. Theor. Math. Phys.* **2** 231
[5] Makhaldiani N V 2013 *Physics of Atomic Nuclei* **76** 1169
[6] Koba Z, Nielsen H B and Olesen P 1972 *Nucl. Phys.* B **40** 317
[7] Matveev V A, Sisakian A N and Slepchenko L A 1976 *Nucl. Phys.* **23** 432

Journal of Physics: Conference Series **668** (2016) 012115 doi:10.1088/1742-6596/668/1/012115

# Jet–induced modifications of the characteristic of the bulk nuclear matter

**P Marcinkowski[1], M Słodkowski[1], D Kikoła[1], J Sikorski[3], J Porter-Sobieraj[2], P Gawryszewski[1] and B Zygmunt[2]**

[1] Faculty of Physics, Warsaw University of Technology, Koszykowa 75, 00-662 Warsaw, PL
[2] Faculty of Mathematics and Information Science, Warsaw University of Technology, Koszykowa 75, 00-662 Warsaw, PL
[3] Faculty of Physics, University of Warsaw, Pasteura 5, 02-093 Warsaw, PL

E-mail: `patryk.marcinkowski@protonmail.ch`

**Abstract.** We present our studies on jet–induced modifications of the characteristics of bulk nuclear matter. To describe such matter, we use efficient relativistic hydrodynamic simulations in (3+1)-dimension, employing the Graphics Processing Unit (GPU) in the parallel programming framework. We use Cartesian coordinates in the calculations to ensure a high spatial resolution that is constant throughout the evolution of the system. We show our results on how jets modify the hydrodynamics fields and discuss the implications.

## 1. Simulation setup

The dynamics of the bulk matter is simulated using the ideal relativistic hydrodynamic equations with a source term [1]

$$\partial_\mu T^{\mu\nu} = \partial_\mu \left((e+p)u^\mu u^\nu - pg^{\mu\nu}\right) = S^\nu, \tag{1}$$

where $u = u(x)$ represents a conserved variable, which includes energy, momentum, and charge densities. The source term represents energy deposited by jets in the system, and it is defined as

$$S^\nu(\vec{x}) = \sum_{i=1}^{n_{jet}} \left(-\frac{\mathrm{d}E}{\mathrm{d}t}, -\frac{\mathrm{d}\vec{M}}{\mathrm{d}t}\right) \delta^{(3)}(\vec{x} - \vec{x_i}(t)), \tag{2}$$

where $n_{jet}$ is the number of jets. We implemented several algorithms for solving the problem, including WENO [2] ($5^{th}$ and $7^{th}$ order) and Musta-Force [3]. We use $7^{th}$ order WENO, which has proven to have high performance and good quality in test problems with steep gradients. The numerical scheme is integrated over time using the $3^{rd}$ order Runge-Kutta algorithm.

The energy loss of jets propagating through the bulk nuclear matter is described by using two mechanisms, radiation of gluons and collisions of partons in a dense medium, and is given by

$$\left(-\frac{\mathrm{d}E}{\mathrm{d}x}\right) = \kappa_{rad}\frac{C_R}{C_F}T^3 x + \kappa_{coll}\frac{C_R}{C_F}T^2, \tag{3}$$

where $T$ is local temperature and $\kappa_{rad}, \kappa_{coll}, C_R, C_F$ coefficients depend on jet flavor (quark or gluon) and its energy [4]. In the simulations, we assumed $\kappa_{rad} = 4, \kappa_{coll} = 2.5, C_R/C_F = 1$. We assume that $\frac{\mathrm{d}E}{\mathrm{d}x}$ is small compared to the jet energy, and the jet is not modified by the medium.

The code is written using NVIDIA CUDA programming framework purposed for parallel computing on graphical processing units (GPU). Our implementation [5, 6, 7, 8] speeds up more than two orders of magnitude in comparison to the equivalent CPU execution.

## 2. Results

We simulated propagation and energy loss $\left(\frac{\mathrm{d}E}{\mathrm{d}x}\right)$ of a high-energy parton through the medium and analyzed the evolution of the system. The initial condidtions were based on ellipsoidal flow [9] with initial simulation time $t_0 = 1$ fm/c.

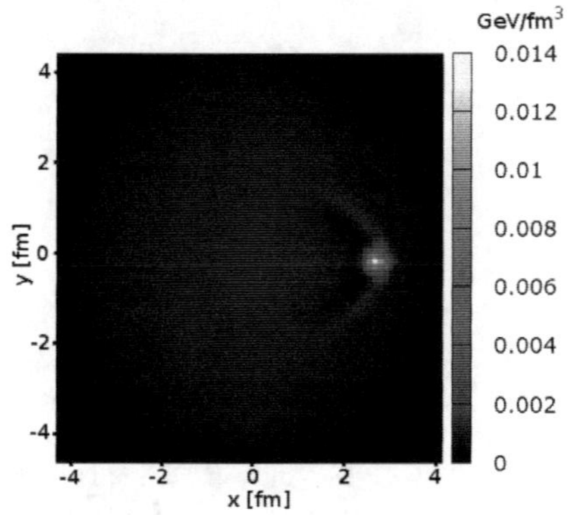


**Figure 1.** Energy density in plane $xy$ at $z = 0$ after $t = 2.4$ fm/c.

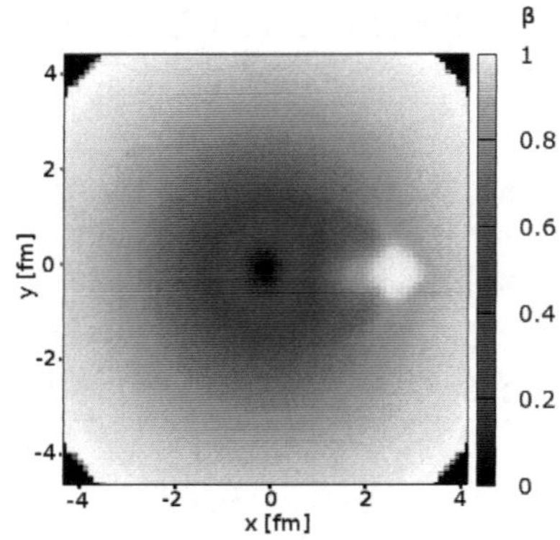


**Figure 2.** Velocity profile in plane $xy$ at $z = 0$ after $t = 2.4$ fm/c.

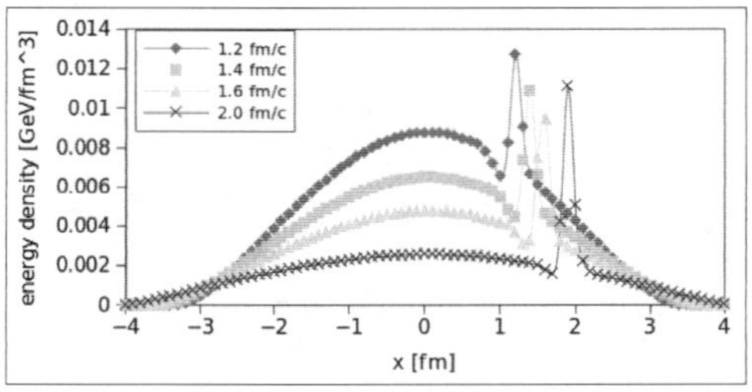


**Figure 3.** Evolution of the system: energy density 1D sections at $y = z = 0$ at different points of time (from 1.2 fm/c to 2 fm/c after simulation start). Simulation parameters: $\mathrm{d}x = 0.1$ fm, $\mathrm{d}t = 0.02$ fm/c, grid size $256 \times 256 \times 256$, EOS $p = e/3$.

In figures 1 and 2, there is a clean signal of jet propagation in the medium, a Mach cone is visible both in energy and velocity distributions. Figure 3 shows the evolution of energy density with time. There is a clear indication that jet energy loss has a significant impact on the system. The area where the jet interacted with the system has much larger energy density compared to the surrounding matter and the difference increases with time.

In the future, we plan to employ more realistic jet energy loss algorithm and add freeze–out to study the effect on the experimental observables: elliptic flow and higher flow harmonics.

## References

[1] Tachibana Y and Hirano T 2014 *Phys. Rev.* C **90** 021902
[2] Zhang X and Shu C W 2012 *J. Comput. Phys.* **231**(5) 2245-58
[3] Toro E F 2006 *Appl. Numer. Math.* **56**(10-11) 1464-79
[4] Casalderrey-Solana J, Gulhan D C, Milhano J G, Pablos D and Rajagopal K 2014 *J. High Energy Phys.* JHEP10(2014)019
[5] Porter-Sobieraj J, Cygert S, Kikoła D, Sikorski J and Słodkowski M 2015 *Concurrency and Computation: Practice and Experience* **27**(6) 1591-602
[6] Cygert S, Kikoła D, Porter-Sobieraj J, Sikorski J and Słodkowski M 2014 *Lect. Notes Comput. Sc.* **8384** 500-9
[7] Sikorski J, Cygert S, Porter-Sobieraj J, Słodkowski M, Krzyżanowski P, Książek N and Duda P 2014 . *J. Phys. Conf. Ser.* **509** 012059
[8] Cygert S, Porter-Sobieraj J, Kikoła D, Sikorski J and Słodkowski M 2013 *Fed. Conf. on Computer Science and Information Systems (Kraków)* pp 441-6
[9] Sinyukov Y M and Karpenko I A 2005 *Preprint* nucl-th/0505041

Journal of Physics: Conference Series **668** (2016) 012116 doi:10.1088/1742-6596/668/1/012116

# Late time behaviour for Hard Parton Evolution using the Schwinger-Keldysh Formalism

**Ben Meiring**[1] **and W A Horowitz**[2]

Department of Physics, University of Cape Town, Private Bag X3, Rondebosch 7701, South Africa

E-mail: [1]`mrnben002@myuct.ac.za` and [2]`wa.horowitz@uct.ac.za`

**Abstract.** AdS/CFT computations have been used to describe the energy loss of QCD-like particles moving through a strongly coupled plasma, but little is understood regarding the initial conditions of these jets. We are investigating an analytic description of jet initial conditions through the time evolution of hard partons by means of the Schwinger-Keldysh Formalism.

We have derived a general formula for the late time expectation value of the total 4-momentum in terms of the differential cross-section and the initial 4-momenta of the system. As a consistancy check, we have found the same expression for $\lim_{x^0\to\infty}\langle\int d^3x\, T_{\mu\nu}(x)\rangle$ for an interacting scalar field theory to leading order.

## 1. Introduction

A good description of the collision process involving heavy nuclei (such as Pb or Au) is given in [1, 2]. In such a system we take interactions to begin at $\tau = 0$, when the nuclei collide. The 'hard' processes (those that involve a transfer energy of $\gtrsim 10$ GeV) occur at a time scale $\tau \sim Q^{-1}$ by the uncertainty principle. Interactions that are able to result in high transverse momentum (such as hard parton production) occur very early in the collision before the 'softer' processes that result in thermalization, which begin at $\tau \sim 0.2$ fm/$c$ at an energy scale of $\sim 1$ GeV. Typically an expansion of QCD phenomena at weak coupling can be used to describe collisions with high momentum transfer [3].

### *1.1. The Expectation Value of the Energy Momentum Tensor*

In the earliest moments of the collision ($\tau < 0.2$ fm/$c$), before the equilibration of a Quark Gluon Plasma (QGP) we assume that the the evolution of hard partons will be qualitatively similar to the evolution of partons that result from a proton-proton (p-p) collision. We aim to understand the $T_{\mu\nu} = \sum\limits_i \partial_\mu\psi_i\partial_\nu\psi_i - g_{\mu\nu}\mathcal{L}$ (as defined in [4]) of hard partons resulting from collisions in a vacuum (similar to the evolution present after a proton-proton interaction). We will use the Schwinger-Keldysh real time formalism to perform a perturbative expansion at weak coupling for the expectation value of $T_{\mu\nu}$ due to the interaction of colliding particles. $\langle T_{\mu\nu}\rangle(x)$ in a theory with interaction Hamiltonian given by $H$, due to an initial state given by $|\text{in}\rangle$ is given by the expansion

$$\langle T_{\mu\nu}\rangle(x) = \langle\text{in}|\, \underset{\rightarrow}{T}\exp\left(i\int_{-\infty}^{x^0} dz_1\,\hat{H}_{\text{I}}(z_1)\right) T_{I\mu\nu}(x)\underset{\leftarrow}{T}\exp\left(-i\int_{-\infty}^{x^0} dz_1\,\hat{H}_{\text{I}}(z_1)\right)|\text{in}\rangle\,, \quad (1)$$

where the $I$ subscripts indicate quantities in the interaction picture.

## 2. Results

We found a general result for the late time expectation value of the total momentum $\int d^3x\, \hat{T}_{\mu 0} = \hat{P}_\mu$ for a system two colliding particles. For clarity we will only state the result of Equation 2 for the spatial momentum in the center of mass frame.

$$\lim_{t\to\infty} \int d^2b\, \langle \vec{\hat{P}} \rangle = \vec{0} = \int d\Pi_{p_f} \left( \sum_f \vec{p}_f \right) \frac{d\sigma}{d\Pi_{p_f}}^{(\text{in}\to\{\vec{p}_f\})}, \tag{2}$$

where $\langle \vec{P} \rangle$ is the expectation value of the canonical momentum, $\vec{p}_f$ are the eigenvalues of the final state particles, $\int d\Pi_{p_f}$ is an integral over the final state momentum phase space and

$$\frac{d\sigma}{d\Pi_{p_f}}^{(\text{in}\to\{\vec{p}_f\})} = \int d^2b \left( \prod_{i=A,B} \int \frac{d^3k_i}{(2\pi)^3} \frac{\phi_i(\vec{k}_i)}{\sqrt{2E_i}} \int \frac{d^3\bar{k}_i}{(2\pi)^3} \frac{\bar{\phi}_i(\vec{\bar{k}}_i)}{\sqrt{2\bar{E}_i}} \right)$$
$$\times\, e^{i(\vec{b}^{\perp}\cdot(\vec{\bar{k}}_B^{\perp} - \vec{k}_B^{\perp}))} \left( {}_{\text{in}}\langle \{\bar{\mathbf{k}}_i\} | T^\dagger | \{\mathbf{p}_f\} \rangle_{\text{out}} \right) \left( {}_{\text{out}}\langle \{\mathbf{p}_f\} | T | \{\mathbf{k}_i\} \rangle_{\text{in}} \right). \tag{3}$$

Here $\left( {}_{\text{out}}\langle \{\mathbf{p}_f\} | T | \{\mathbf{k}_i\} \rangle_{\text{in}} \right) = i\mathcal{M}(\{k_i\} \to \{p_f\})(2\pi)^4 \delta^{(4)}(\sum_i k_i - \sum_f p_f)$ gives the transition amplitude for a set of momentum eigenstates $|\{k_i\}\rangle$ at $t = -\infty$ to evolve to momentum eigenstates $|\{p_f\}\rangle$ at $t = \infty$, $\phi_i$ and $\bar{\phi}_i$ are wavepackets, and $\vec{b}^{\perp}$ gives the impact parameter. The definition of 3 is motivated by equation (4.76) in [4]. The interpretation of Equation 2 is that given an initial state $|\text{in}\rangle$ with total spatial momentum in the centre of mass frame $\sum_i \vec{p}_i = 0$, the final state particles must distribute themselves in momentum space so as to keep the total momentum conserved.

### *2.1. The Elastic Collisional $\lambda\psi_1^2\psi_2^2$ Model*

We checked Equation 2 for a theory of two interacting scalar fields, $\mathcal{L} = \frac{1}{2}(\partial\psi_1)^2 - \frac{1}{2}m^2\psi_1^2 + \frac{1}{2}(\partial\psi_2)^2 - \frac{1}{2}m^2\psi_2^2 - \lambda\psi_1^2\psi_2^2$, with initial state given by two particle wave packets of species $\psi_1$ and $\psi_2$. We find that in the center of mass frame, to leading order in the coupling $\lambda$,

$$\lim_{t\to\infty} \int d^2b\, \langle \vec{P} \rangle = \vec{0} = \underbrace{\int \frac{d^3p d^3q}{(2\pi)^6 2E_p 2E_q}}_{\int d\Pi} \underbrace{(\vec{p} + \vec{q})}_{\sum_f p^f} \underbrace{\lambda^2 \delta^{(4)}(p+q)}_{\frac{d\sigma}{d\Pi}^{(\text{in}\to\{\vec{p},\vec{q}\})}}. \tag{4}$$

As we would expect, the late time total momentum is given by a distribution of final state particles in the rest frame weighted by the differential cross-section.

## 3. Conclusion

We have found a consistent and general result for the late time behaviour of the total 4-momentum for any given quantum field theory in terms of the initial 4-momentum and the differential cross-section. We have verified the expression explicitly for the $\lambda\psi_1^2\psi_2^2$ model. Going forward, we will attempt to find finite time results for the expectation value of the energy momentum tensor associated with the evolution of hard partons.

## References

[1] Gelis F 2001 *The Initial Stages of High Energy Heavy Ion Collisions* Ph.D. thesis The Institut de Physique Thorique (IPhT) habilitation thesis URL `http://ipht.cea.fr/Pisp/francois.gelis/Physics/habilitation-thesis.pdf`

[2] Iancu E 2014 QCD in heavy ion collisions *Proceedings, 2011 European School of High-Energy Physics (ESHEP 2011)* (*Preprint* 1205.0579)

[3] Gross D J and Wilczek F 1973 *Phys. Rev. Lett.* **30**(26) 1343–1346

[4] Peskin M E and Schroeder D V

# Radiation tolerance studies of silicon microstrip sensors for the CBM Silicon Tracking System

**I. Momot[1,2], M. Singla[3], M. Teklishyn[2,3], V. Pugatch[2], and J. Heuser[3]**

[1] Goethe University Frankfurt, Frankfurt am Main, Germany
[2] Kiev Institute for Nuclear Research (KINR), Kiev, Ukraine
[3] GSI Helmholtzzentrum fr Schwerionenforschung GmbH, Darmstadt, Germany

E-mail: i.momot@gsi.de

**Abstract.** Double-sided silicon microstrip sensors will be used in the Silicon Tracking System of the CBM experiment. During experimental run they will be exposed to a radiation field of up to $1{\times}10^{14}$ 1 MeV $n_{eq}$ cm$^{-2}$. Radiation tolerance studies were made on prototypes from two different vendors. Results from these prototype detectors before and after irradiation to twice that neutron fluence are discussed.

## 1. Introduction

The Silicon Tracking System (STS) [1] is the core detector of the Compressed Barionic Matter (CBM) experiment at the future Facility for Antiproton and Ion Research (FAIR) in Darmstadt, Germany. The STS is located in the dipole magnet to provide track reconstruction and momentum determination of charged particles from beam-target interactions. According to the CBM operation parts of the tracking stations will be exposed to a total particle fluence up to $1{\times}10^{14}$ 1 MeV $n_{eq}$ cm$^{-2}$ afterwards they will be replaced. Thus radiation tolerance is an important requirement for the STS detectors.

## 2. Detectors under test

The STS will employ double-sided silicon microstrip sensors mounted onto a low-mass carbon fibre support structure. The strips on the p side of the sensors are tilted with respect to the n side by 7.5$^{\circ}$ as displayed in Fig.1(a). To read out the sensor only from one edge, as required by the detector ladder structure, the end strips from one edge were connected to the end strips on the other edge as shown in Fig.1(b). This interconnection can be provided via double metallization (DM) or using external interstrip cables (SMwC). The central strips were the full-length strips without any interconnections.

The sensors were tested for electrical parameters (leakage current and bulk capacitance) and also for the charge collection both before and after irradiation (performed at KIT, Germany) using a $\beta$-source ($^{90}$Sr) to mimic minimum ionizing particles (MIP). All the measurements were made in a light-tight set-up with temperature control. The signals were readout with two laboratory read-out systems (nXYTER [2] and Alibava [3]) for the cross check collected results. In case of nXYTER, in order to collect charge, produced by MIPs only, a 2 mm Aluminum absorber was placed below the sensor to absorb low-energy electrons, particles were selected by requiring a coincidence with a signal in the Si-pad detector. In case of Alibava read-out,

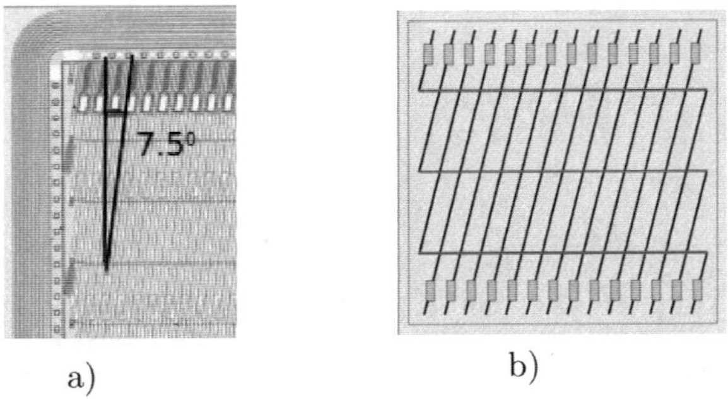


**Figure 1.** (a) Microscopic view on the p-side of a prototype detector with strips tilted by 7.5 °, (b) Detector topology to read out inclined strips. The interconnections of the corner strips are visible as horizontal lines.

scintillator and amplifier behind the sensor were used. The spectrum from the $\beta$-source was fitted with a Landau-Gauss convolution and the most probable value is interpreted as collected charge. The results for charge collection efficiency, based on an ADC-to-charge calibration and the expected charge in the detector according to the wafer thickness, are shown in Fig. 2 (nXYTER set-up) and on the Fig.3 (Alibava set-up). The sensor naming encodes the prototype generation (5 or 6), the manufacturer (H = Hamamatsu, C = CiS), the detector height/strip length in cm (4 or 6), and the wafer number.

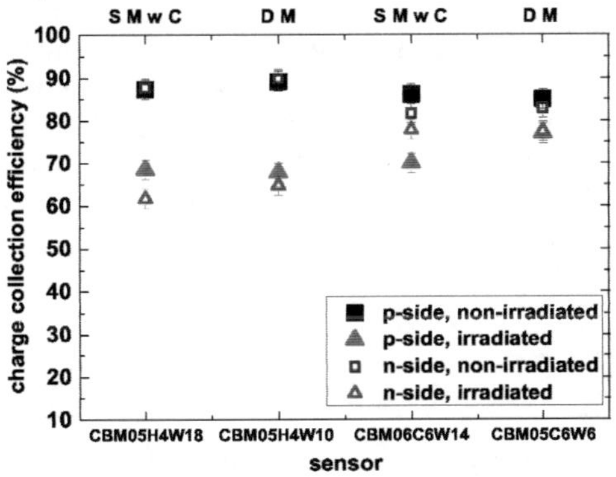


**Figure 2.** Charge collection efficiency results with $^{90}$Sr, before and after irradiation. Measurements were performed at 470 V sensor bias and $-8\,^{\circ}$C environment.

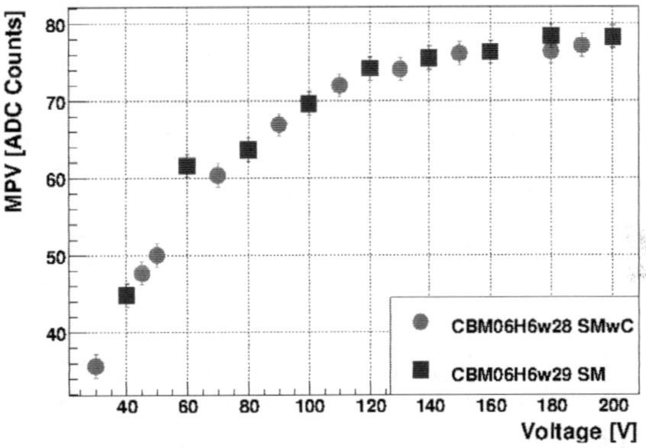


**Figure 3.** Dependence charge collection of applied voltage in terms of Most Probable Value.

## 3. Results

The prototype sensors from two vendors, in two technological configurations, show a reduction of charge collection by 15% to 25% after irradiation to twice the maximum neutron fluence expected in the CBM experiment. Measurements with the latest prototype sensors are forthcoming.

## References

[1] J. Heuser et al., Technical Design Report for the CBM Silicon Tracking System, GSI Report 2013-4, http://repository.gsi.de/record/54798
[2] A.S. Brogna et al., N-XYTER, a CMOS read-out ASIC for high resolution time and amplitude measurements on high rate multi-channel counting mode neutron detectors, Nucl. Instrum. Methods A 568 (2006) 301-308
[3] http://www.alibavasystems.com/

# Beam energy and centrality dependence of multiplicity fluctuations in heavy ion collisions

**Maitreyee Mukherjee, Sumit Basu, Subikash Choudhury and T K Nayak**

Variable Energy Cyclotron Centre, 1/AF, Bidhan Nagar, Kol-700064, India

E-mail: maitreyee.mukherjee@gmail.com

**Abstract**. Multiplicity distributions of produced particles and their event-by-event fluctuations are studied using the AMPT model in the Default and String melting modes. In addition to being sensitive to the QCD phase transition, these fluctuations provide baselines for other event-by-event measurements. The collision energy and centrality dependence of fluctuations are estimated for heavy-ion collisions from $\sqrt{s_{NN}}$=7.7 GeV to 2.76 TeV. The multiplicity fluctuations expressed in terms of scaled variance decreases from peripheral to central collisions for all energies, except for that of the Pb+Pb collisions at $\sqrt{s_{NN}}$= 2.76 TeV. The beam energy dependence shows an increase of multiplicity fluctuations with increasing beam energy.

## 1. Introduction

Event-by-event fluctuations of thermodynamic quantities have been proposed as the basic tools for understanding the particle production mechanisms and to probe the QCD phase transition. The fluctuations of experimentally accessible quantities, such as particle multiplicities, mean transverse momenta, temperature, particle ratios, and other global observables are related to the thermodynamic properties of the system, such as the entropy, specific heat, chemical potential and matter compressibility [1]. A non-monotonic behaviour of multiplicity fluctuations may signal the onset of deconfinement and can be effectively used to probe the critical point in the QCD phase diagram [2]. Measurements at the vanishing $\mu_B$ at LHC energies set the scale of all theoretical calculations, and one can accurately calculate several quantities and their fluctuations. Multiplicity fluctuations are normally characterized by the scaled variances of the multiplicity distributions, defined as,

$$\omega_{ch} = \frac{\sigma_n^2}{<n>} \tag{1}$$

where <n> and $\sigma_n^2$ are the mean and variance of the multiplicity distribution, respectively [3].

## 2. Analysis Strategy

Narrower centrality bins are important to choose to get rid of the inherent fluctuations within a centrality bin. For this analysis, centrality is selected from the minimum-bias distribution for the number of participants. It is desirable to choose wider centrality bins, such as 5% or 10% cross sections, because of lack of statistics. In these cases, we have centrality bin width effect (i.e, effect caused by non-uniformity of charged particle distributions), which needs to be corrected. These are done by dividing one centrality bin into smaller bins and weight the moments. Thus, it is possible to minimize the statistical fluctuations sufficiently and to evaluate the dynamical fluctuations.

## 3. Results from the multiplicity fluctuations

The physics origin of the fluctuations in the multiplicity distributions are inherent in the width of the distributions. The analysis is performed with the generated events from AMPT-model in the default and the string-melting (SM) modes. For the analysis, transverse momentum considered is $p_T < 2.0$ GeV/c and the pseudorapidity range is $|\eta| < 0.5$.

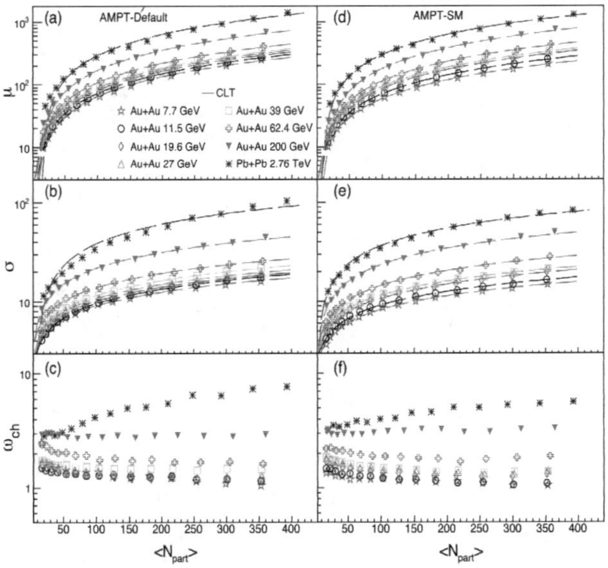


**Figure 1.** Results for mean, sigma and scaled variances from lower to higher energies. The lines denote CLT-fits.

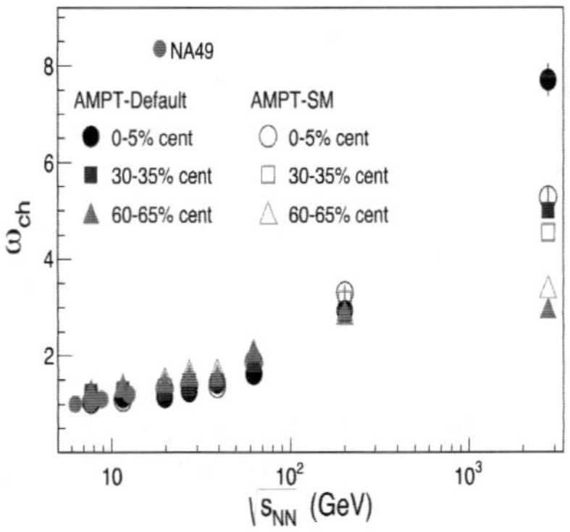


**Figure 2.** Beam-energy dependence of the multiplicity fluctuations

The results from the multiplicity fluctuations are shown in figure 1. It has been observed that, Central Limit Theorem (CLT)-fits are better for $\mu$ and $\sigma$ for lower energies. For energies from 7.7 GeV to 200 GeV, $\omega_{ch}$ shows increasing trend for central to peripheral collisions. For 2.76 TeV, the trend for scaled variance is opposite.

Beam-energy dependence of the multiplicity fluctuations is shown in figure 2. The fluctuations show an increasing trend from lower to higher energies. The fluctuations at low energies are close to each other for all centralities. Fluctuations increase for increasing beam energies for all RHIC energies. At 2.76 TeV and central collisions the fluctuations are larger for AMPT-default compared to the string melting mode. The results from AMPT matches with the results from NA49-experiment.

## References

[1] Adare A *et al*, *Phys. Rev.* **C 78**, 044902 (2008).
[2] Stephanov M A, Rajagopal K, and Shuryak E V, **Phys. Rev. Lett. 81**, 4816 (1998).
[3] Aggarwal M M et al, *Phys. Rev.* **C 65,** 054912 (2002).

Journal of Physics: Conference Series **668** (2016) 012119 doi:10.1088/1742-6596/668/1/012119

# Nucleus probing for short-range correlations using neutrino-nucleon interactions

**Kajetan Niewczas**

Institute of Theoretical Physics, University of Wrocław,
pl. M. Borna 9, 50-204 Wrocław, PL

E-mail: kajetan.niewczas@gmail.com

**Abstract.** Short-range nucleon-nucleon correlations (NN SRC) contain important information about nuclear structure and dynamics within the nuclei. It has been extensively studied through two-nucleon knockout reactions, in both pion and electron scattering experiments. A very recent approach is to probe NN SRC using neutrino scattering, however, with limited results due to the intrinsic uncertainty of the four-momentum transfer. Here, we present one of the latest attempts to reconstruct this information.

## 1. Introduction

Short-range correlations (SRC) are baryonic configurations in nuclei, which are defined by local properties of nucleon-nucleon (NN) interactions rather than by mean field [1]. Contributions from NN SRC is a significant factor in understanding of the dynamics within nuclei. So far, various experiments successfully probed the nuclei for SRC pairs with both hadron and electron scattering. In the case of neutrino scattering, there are vast limitations, namely the intrinsic uncertainty of the incident neutrino energy and the four-momentum transfer. There are also various distortions coming from the final state interactions. These limitations make NN SRC probing via neutrinos experimentally challenging.

In general, the most promising venue to probe NN correlations are two-proton knockout processes. The experiment conducted by ArgoNeuT Collaboration provides precise calorimetric data and full 3d reconstruction of the charged current (CC) neutrino events. During the operation time, ArgoNeuT detected only 30 such knockout events [2]. Despite the low statistics, a comparison with Monte Carlo simulations is essential in order to set a theoretical background for future experiments.

## 2. NuWro analisys

NuWro is a Monte Carlo event generator for simulation of neutrino-nucleus scattering, developed by Wroclaw Neutrino Group [3]. Using NuWro we performed a simulation of the ArgoNeuT experiment. Namely, we took neutrino beams, provided by ArgoNeuT Collaboration, and set up an argon target. We also implemented the finite size of the detector and performed a proper event selection [4].

For a specific subsample, ArgoNeuT performed a reconstruction of the essential quantities, such as incident neutrino energy and a momentum transfer [2]. NuWro provides us with complete

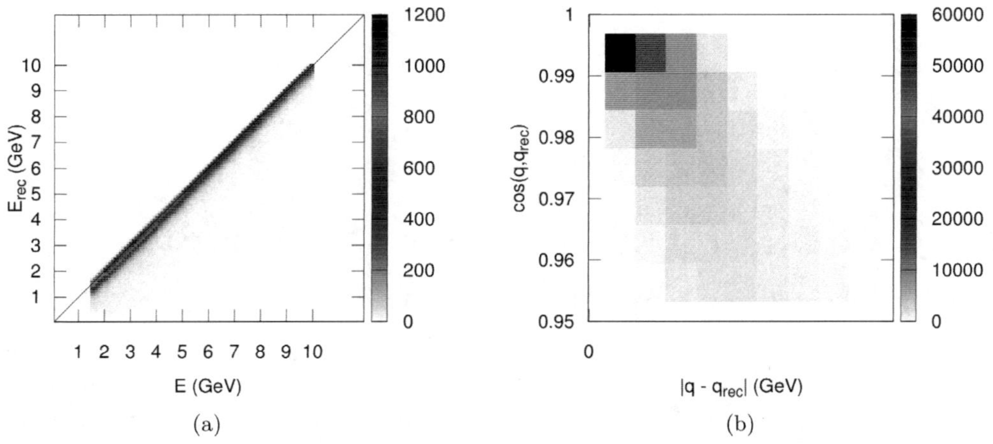


**Figure 1.** (Color online) The reconstruction accuracy analysis using data from NuWro simulations for the Fermi gas model [4]: (a) the reconstructed against the actual value of the incident neutrino energy, (b) consine of the reconstructed and the real momentum trasfer against the modulus of their difference. Colors represent number of events in each bin.

knowledge of the each event. We tested how accurately the reconstruction formula matches the actual simulation data.

In our simulations we used two different models of momentum distribution within argon target. Namely, the Fermi gas and the spectral function approach [4]. In this analysis the results for two models do not severely differ, hence only the former is presented.

The reconstruction consists of two steps. Namely, the reconstruction of the incident neutrino energy and the momentum transfer [2]. The analysis of each step is shown in Fig. 1.

As seen in Fig. 1(a), the neutrino energy is reconstructed very well. The vast majority of the events are disctributed around the diagonal. The procedure is, however, more likely to underestimate incident neutrino energy.

Turning to the momentum transfer, as seen in Fig. 1(b), the momentum transfer direction is preserved, as the cosine of the angle between the real and the reconstructed one tends to be 1. The length of the momentum transfer is also reconstructed very well. Note that the momentum transfer reconstruction depends on the first step, however it includes additional assumptions [2].

Although, the experimental event statistics is very limited, the reconstruction formula gives promising results. New experimental data and further studies are needed to investigate the subject.

## Acknowledgments
The author acknowledges the financial support from the organizers of 15th International Conference on Strangeness in Quark Matter and the Helmholtz International Summer School "Dense Matter 2015", and fruitful discussions with professor J. T. Sobczyk. The author would also like to thank M. Marczenko for helpful comments and M. A. R. Kaltenborn for proofreading the manuscript.

## References
[1] Arrington J, Higinbotham D W, Rosner G, and Sargsian M 2012 *Prog. Part. Nucl. Phys.* **67** 898
[2] Acciarri R *et al.* (ArgoNeuT Collaboration) 2014 *Phys. Rev.* D **90** 012008
[3] NuWro http://borg.ift.uni.wroc.pl/nuwro
[4] Niewczas K, Sobczyk J T (paper in preparation)

# Lattice simulation of two-color QCD with $N_f = 2$ at non-zero baryon density

**V V Braguta[1,2], A Yu Kotov[3], A A Nikolaev[2], S N Valgushev[3]**

[1] Institute for High Energy Physics NRC "Kurchatov Institute", Protvino, 142281 Russia
[2] Far Eastern Federal University, School of Biomedicine, Vladivostok, 690950 Russia
[3] Institute for Theoretical and Experimental Physics, Moscow, 117218 Russia

E-mail: nikolaev.aa@dvfu.ru

**Abstract.** We present the results of lattice simulations of two-color QCD with two fermion flavors and non-zero chemical potential. Dependencies of the chiral condensate and baryon number density on the quark chemical potential were studied. We found that an increase of the quark chemical potential leads to chiral symmetry restoration. At small values of chemical potential, our results for the baryon density agree with ChPT predictions.

## 1. Introduction

One of the possible approaches to the study of the phase diagram of QCD is lattice simulations, where the action is discretized on the $R^4$ space, and functional integrals are numerically calculated using the Monte Carlo method [1] [2]. However, at the moment, there is no way, in lattice QCD, to perform direct simulations at finite chemical potential, due to the sign problem of the fermionic determinant [1]. Therefore, we shall study a simpler theory: QFT with $SU(2)$ gauge group and two degenerate flavors, where no sign problem arises [3], in order to better understand the qualitative features of the QCD phase diagram and to analyze the effect of a nonzero chemical potential on the properties of QGP.

Previous investigations of the $QC_2D$ phase structure were performed in [4, 5, 6] with Wilson fermions and in [7][8] with $N_f = 4$ and 8 flavors of staggered fermions respectively. However, Kogut-Susskind fermions are more suitable to study the $QC_2D$ phase diagram, because they preserve an analog of the chiral symmetry in the lattice formulation [9], and Wilson fermions explicitly break it [1]. In this work we considered two flavors of staggered fermions.

## 2. Lattice formulation

LQCD partition function for $N_f = 2$ staggered fermions has the following form:

$$Z = \int DU \, det\Big[M^\dagger(\mu_q)M(\mu_q)\Big]^{\frac{1}{4}} e^{-S_G[U]}, \quad (1)$$

where the functional integration is performed over the $SU(2)$ group manifold, $\mu_q$ denotes quark chemical potential, $M(\mu_q)$ is the Dirac operator for the Kogut-Susskind fermions, and $S_G[U]$ is the Wilson action of gauge fields. Root of the power 4 from the fermionic deteminant in (1) is introduced to provide $N_f = 2$ [9]. For the details of the formulation and simulation parameters we refer the reader to [10].

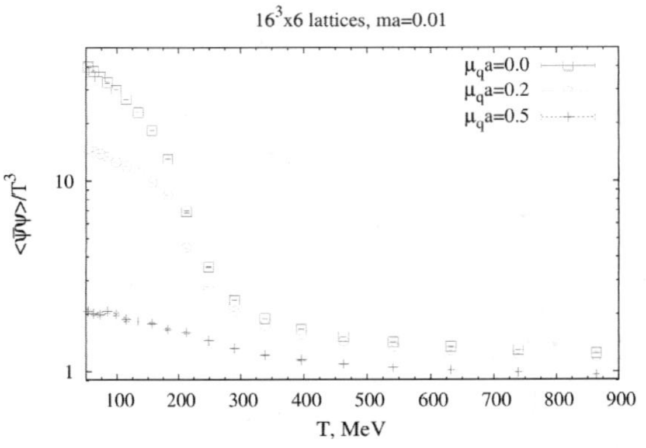


**Figure 1.** Chiral condensate as a function of $T$ at three different values of the quark chemical potential. Logarithmic scale is employed for the ordinate axis.

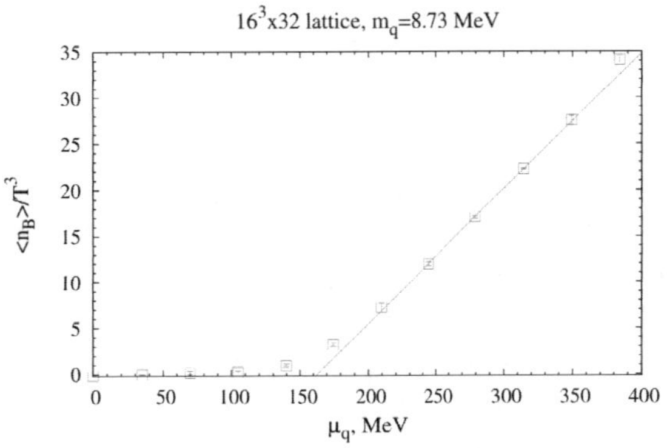


**Figure 2.** Baryon number density as a function of the quark chemical potential (the temperature is fixed at 54.6 MeV).

## 3. Numerical results and conclusions

Figure 1 shows the dependences of the chiral condensate on temperature for various $\mu_q$ values ($a$ in the titles inside the figure denotes lattice spacing). It can be seen, that an increase in the quark chemical potential leads to a significant decrease in the chiral condensate, i.e. to the recovery of the chiral symmetry. These results are in agreement with the previous findings obtained for two flavors of Wilson fermions [5].

The dependence of the baryon number density on the quark chemical potential at low temperature is shown on the figure 2. One can see, that at small values of $\mu_q$ baryon number density stays approximately zero till the value of $\mu_q^c \approx 160$ MeV, then it begins to rise almost linearly. It is important to note, that this $\mu_q^c$ value equals to nearly the half of the pion mass, $m_\pi = 350(5)$ MeV on the $16^3 \times 32$ lattice at $\mu_q = 0$. Obtained behaviour of $\langle n_B \rangle$ agrees well with ChPT predictions [8, 11].

## Acknowledgments

We are grateful to M. Müller-Preussker and V.G. Bornyakov for stimulating discussions and to A.V. Barilov, A.A. Golubev, V.A. Kolosov, I.E. Korolko, and M.M. Sokolov for assistance in the organization of calculations at the ITEP supercomputer.

## References

[1] Gattringer C and Lang C B 2010 *Quantum chromodynamics on the lattice* (Springer)
[2] Bazavov A *et al* 2012 *Phys. Rev.* D **85** 054503
[3] Kogut J B, Stephanov M A and Toublan D 1999 *Phys. Lett.* B **464** 183-191
[4] Makiyama T *et al* 2015 Phase structure of two-color QCD at real and imaginary chemical potentials; lattice simulations and model analyses *Preprint* arXiv:1502.06191
[5] Cotter S, Giudice P, Hands S and Skullerud J I 2013 *Phys. Rev.* D **87** 034507
[6] Skullerud J I, Cotter S, Giudice P, Hands S, Kim S and Mehta D B, 2012 *Proc. XXX Int. Symp. on Lattice Field Theory (Cairns)* PoS(Lattice 2012)091
[7] Kogut J B, Toublan D and Sinclair D K 2002 *Nucl. Phys.* B **642** 181-209
[8] Hands S, Kogut J B, Lombardo M P and Morrison S E 1999 *Nucl. Phys.* B **558** 327-346
[9] Rothe H J 2012 *Lattice gauge theories: an introduction* (World Scientific)
[10] Braguta V V, Kotov A Yu, Nikolaev A A and Valgushev S N 2015 *JETP Letters* **101** 732-734
[11] Splittorff K, Son D T and Stephanov M A 2001 *Phys. Rev.* D **64** 016003

Journal of Physics: Conference Series **668** (2016) 012121 doi:10.1088/1742-6596/668/1/012121

# A Feasibility Study of Hyperon and Hypernuclei Reconstruction at NICA with BM@N Detector

**D. Suvarieva*, M. Ilieva, M. Kapishin, V. Kolesnikov, V. Vasendina, A. Zinchenko**

Joint Institute for Nuclear Research, 141980 Dubna, Moscow region, Russia

E-mail: DilyanaSuvarieva@mail.bg

**Abstract.** Heavy strange objects (hyperons and hypernuclei) could provide essential signatures of the excited and compressed baryonic matter. Their reconstruction and identification should be one of the most important tasks of any experiment with heavy ions. At NICA, it is planned to study hyperons both in the collider mode (MPD detector) and the fixed-target one (BM@N setup). The results on $\Lambda$, $\Xi^-$ hyperon and ${}^3_\Lambda$H hypernuclei reconstruction in Monte Carlo simulated event samples of gold-gold collisions with the BM@N detector are presented.

## 1. Introduction

At sufficiently high temperature and/or baryon density, which could be realized in heavy ion collisions, the existence of a new phase of matter, the Quark-Gluon Plasma (QGP) is predicted. At these temperatures and densities the excitation of the QCD vacuum can lead to the restoration of chiral symmetry and deconfinement of color charges. These phenomena are the goals for the project Nuclotron-based Ion Collider fAcility (NICA) [1] which is being developed at JINR, Dubna. The project includes design and construction of a heavy ion collider for the energy range $\sqrt{s}$ = 4-11 GeV.

## 2. BM@N detector

A successful operation of the NICA complex will require the existing machine Nuclotron to be upgraded in order to accelerate Au nuclei. After that, the improved Nuclotron beams will also be used to run a fixed target experiment BM@N [2]. The detector will allow to study A+A collisions by measuring a variety of observables. Particle yields, ratios, transverse momentum spectra, rapidity and angular distributions, as well as fluctuations and correlations of hadrons will be studied as a function of the collision energy and centrality.

A sketch of the proposed experimental set-up is shown in Fig. 1. It combines high precision track measurements with time-of-flight information for particle identification and total energy measurements for event characterization. The charged track momenta and multiplicity will be measured with the set of GEM (Gas Electron Multipliers) detectors located downstream of the target inside the analyzing magnet of 0.8 T and drift chambers (Straw, DCH) situated outside the magnetic field. Design parameters of the time-of-flight detectors based on multigap Resistive Plate Chambers (mRPC-1, -2) with a strip read-out allow efficient discrimination between particle species with momentum up to a few GeV/c. The Zero Degree Calorimeter

(ZDC) is designed for the collision centrality analysis by measuring the energy of forward going particles.

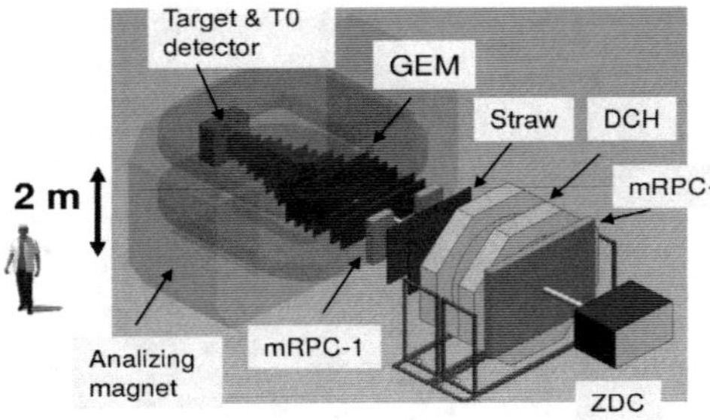


**Figure 1.** BM@N detector.

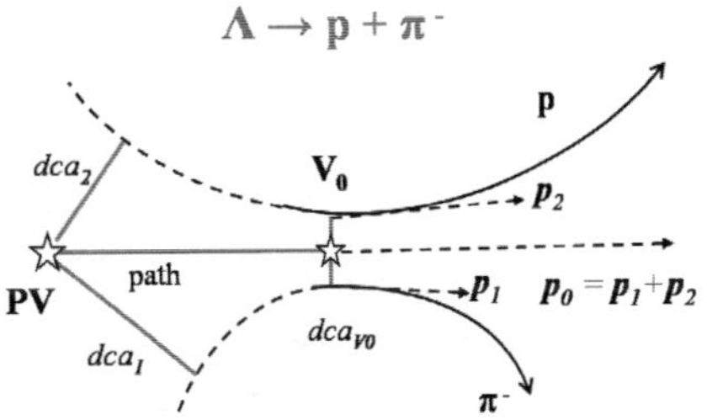


**Figure 2.** Event topology of a neutral particle two-prong decay.

## 3. Event samples and results

The study was done for central Au-Au collisions at beam kinetic energy $T_0 = 4.5$ GeV/nucleon ($\sqrt{s} = 3.5$A GeV). The event samples were produced with the UrQMD [3] and DCM-QGSM (Dubna Cascade Model - Quark-Gluon String Model) [4] generators for hyperon and hypernuclei studies, respectively. $\Lambda$ hyperons and ${}^3_\Lambda$H hypernuclei were reconstructed using their decay modes into two charged tracks, while for $\Xi^-$-hyperons their cascade decay chain into $\Lambda+\pi^-(\Lambda \to p+\pi^-)$ was used. The signal event topology (decay of a relatively long-lived particle into two or more tracks - Fig. 2) defines the selection criteria: relatively large distance of the closest approach ($dca_{1,2}$) to the primary vertex of decay products, small track-to-track separation in the decay vertex ($dca_{V0}$) and relatively large decay length of the mother particle ($path$).

Figures 3 – 5 illustrate the quality of hyperon and hypernuclei reconstruction in the BM@N detector with the GEM tracker. The results on $\Lambda$, $\Xi^-$ and ${}^3_\Lambda$H reconstruction have been obtained for 10 thousand, 900 thousand and 2.6 million central events, respectively.

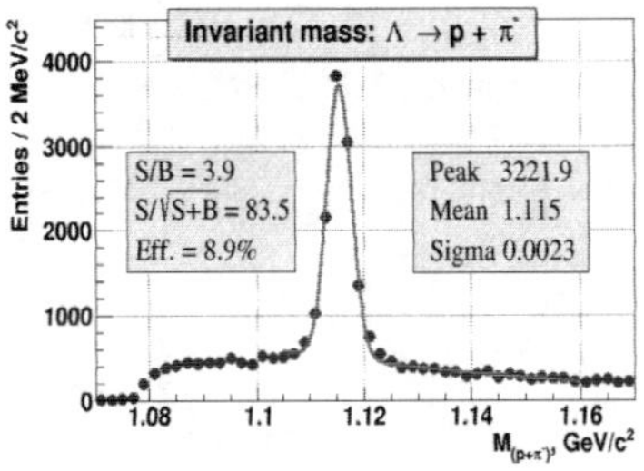


**Figure 3.** Reconstructed invariant mass of proton and $\pi^-$.

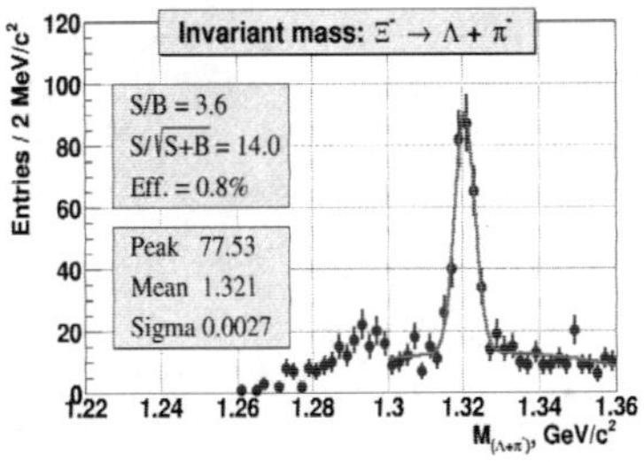


**Figure 4.** Reconstructed invariant mass of $\Lambda$ candidate and $\pi^-$.

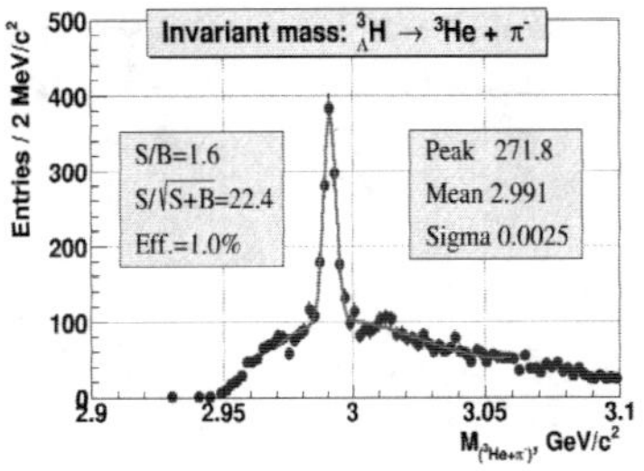


**Figure 5.** Reconstructed invariant mass of $^3$He and $\pi^-$.

## 4. Conclusions

It has been demonstrated that the BM@N start version will provide a good opportunity to reconstruct hyperons and hyperniclei with high enough yields.

## References

[1] http://nica.jinr.ru
[2] http://nica.jinr.ru/files/BM@N/BMN_CDR.pdf
[3] http://urqmd.org/
[4] A. S. Botvina *et al.*, Rhys. Rev. C **84**, 064904 (2011).

Journal of Physics: Conference Series **668** (2016) 012122 doi:10.1088/1742-6596/668/1/012122

# Electrical conductivity of (an-)isotropic quark gluon plasma

**Binoy Krishna Patra, Lata Thakur and P. K. Srivastava**
Department of Physics, Indian Institute of Technology Roorkee, Roorkee -247667, India
E-mail: binoyfph@iitr.ac.in, thakurphyom@gmail.com, prasu111@gmail.com

**Abstract.** In this article, we have calculated one of the transport coefficient viz. electrical conductivity of the quark gluon plasma (QGP) phase which exhibits an anisotropy in the momentum space. Relativistic Boltzmann's kinetic equation has been solved in the relaxation-time approximation to obtain the electrical conductivity. We have used the quasiparticle description to define the basic properties of QGP through the distribution functions of partons.

## 1. Introduction
A recent observation at heavy ion collisions is that a strong electromagnetic field is created in the peripheral collisions. Thus a lot of interest arises in the study of electrical conductivity ($\sigma_{\rm el}$) which is a transport coefficient and quantifies the effect of electric field on the medium produced in these experiments [1-3]. With the discovery of "most perfect fluid ever generated" in heavy ion collision experiments, another important observation has been made that the fluid possesses momentum-space anisotropies in the local rest frame (LRF) [4]. This has important implications for both dynamics and signatures of the QGP. In this work our main motivation is to calculate the electrical conductivity of an anisotropic QGP phase using the Relativistic Boltzmann's kinetic equation. We have used the quasiparticle description to define the basic properties of QGP as earlier we have shown that quasiparticle description successfully provides the proper and realistic thermodynamical and transport behaviour of QGP phase [5].

## 2. Model Description
One can obtain the expression for $\sigma_{\rm el}$ by solving relativistic Boltzmann's kinetic equation under an external electromagnetic field using relaxation time approximation as follows [6]:

$$\sigma_{\rm el}^{\rm iso} = \frac{2}{T}\sum_f g_f q_f^2 \int \frac{d^3p}{(2\pi)^3}\frac{p^2}{E_f^2}\tau_f \times f_f^0(1-f_f^0), \quad (1)$$

where the subscript $f$ implies summation over the flavors. Here we have taken up, down and strange flavors only.
The anisotropic distribution relevant for uRHICs can be approximated by removing particles with the large momentum component along the direction of anisotropy, $\mathbf{n}$ as [6]:

$$f_{\rm aniso}(\mathbf{p}) = f_{iso}\left(\sqrt{p^2+\xi(\mathbf{p}\cdot\mathbf{n})^2}\right), \quad (2)$$

where $f_{iso}$ is an arbitrary isotropic distribution function and $\xi$ is the anisotropic parameter. For weakly anisotropic systems ($\xi \ll 1$), one can expand the quark distribution function as follows :

$$f_{\text{aniso}}(x, \mathbf{p}; T) = \frac{1}{e^{E_f/T}+1} - \frac{\xi}{2E_f T}(\mathbf{p}\cdot\mathbf{n})^2 \frac{e^{E_f/T}}{(e^{E_f/T}+1)^2}$$
$$= f^0 - \frac{\xi}{2E_f T}(\mathbf{p}\cdot\mathbf{n})^2 f^{02} e^{E_f/T}, \quad (3)$$

where $\mathbf{p} \equiv (p\sin\theta\cos\phi, p\sin\theta\sin\phi, p\cos\theta)$ and $\mathbf{n} \equiv (\sin\alpha, 0, \cos\alpha)$. $\alpha$ is the angle between $\mathbf{p}$ and $\mathbf{n}$. After substituting the anisotropic distribution function in Eq. (1), the expression of electrical conductivity in anisotropic medium is modified as [6]:

$$\sigma_{\text{el}}^{\text{aniso}}(\mu_q = 0) = \frac{1}{\pi^2 T}\sum_f g_f q_f^2 \int dp \frac{p^4}{E_f^2}\tau_f f_f^0(1-f_f^0) - \xi\frac{1}{6\pi^2 T}\sum_f g_f q_f^2 \int dp \frac{p^6}{E_f^3}\tau_f (f_f^0)^2 e^{E_f/T}$$
$$= \sigma_{\text{el}}^{\text{iso}} - \xi A, \quad (4)$$

We also generalize the electrical conductivity in an anisotropic medium for $\mu_q \neq 0$ [6]:

$$\sigma_{\text{el}}^{\text{aniso}}(\mu_q \neq 0) = \frac{1}{2\pi^2 T}\sum_f g_f q_f^2 \int dp \frac{p^4}{E_f^2} \times \left[\tau_f f_f^0(1-f_f^0) + \tau_{\bar{f}} \bar{f}_f^0(1-\bar{f}_f^0)\right] \quad (5)$$
$$- \xi\frac{1}{12\pi^2 T^2}\sum_f g_f q_f^2 \int dp \frac{p^6}{E_f^3} \times \left[e^{(E_f-\mu_f^q)/T}\tau_f (f_f^0)^2 + e^{(E_f+\mu_f^q)/T}\tau_{\bar{f}}(\bar{f}_f^0)^2\right],$$

where $f_f^0$ and $\bar{f}_f^0$ are the equilibrium distribution function for a quark and anti-quark of flavour $f$, respectively.

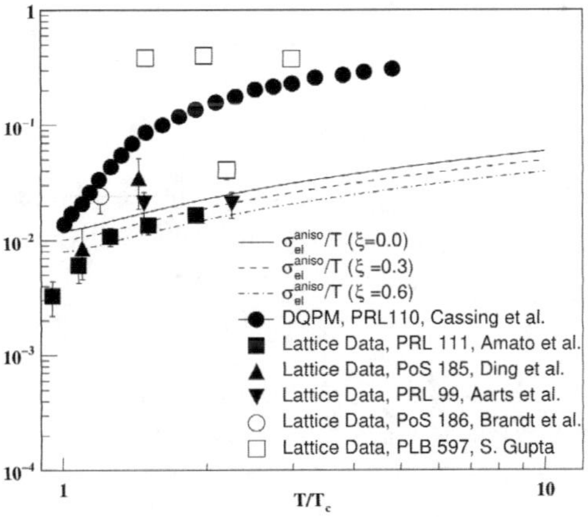


**Figure 1.** Variation of $\sigma_{el}^{aniso}$ with respect to temperature for different values of $\xi$.

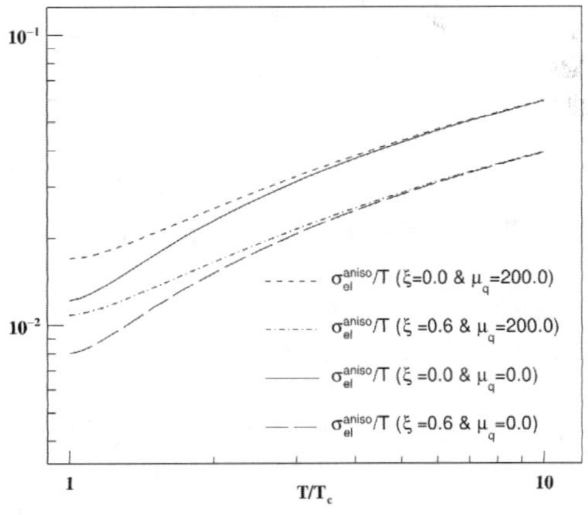


**Figure 2.** Variation of $\sigma_{el}^{aniso}$ with respect to temperature at zero as well as finite $\mu_q$.

[1] Puglisi A, Plumari S and Greco V 2014 *Phys. Rev.* D **90**, 114009.
[2] Fernandez-Fraile D and Nicola A G 2006 *Phys. Rev.* D **73**, 045025.
[3] Greif M, Bouras I, Greiner C, Xu Z 2014 *Phys. Rev.* D **90**, 094014.
[4] Strickland M 2014 *Nucl. Phys.* A **926**, 92.
[5] Srivastava P K, Tiwari S K and Singh C P 2010 *Phys. Rev.* D **82**, 014023.
[6] Srivastava P K, Thakur L, Patra B K 2014 *Phys. Rev.* C **91** 044903.

15th International Conference on Strangeness in Quark Matter (SQM2015) IOP Publishing
Journal of Physics: Conference Series **668** (2016) 012123 doi:10.1088/1742-6596/668/1/012123

# Topological susceptibility from $N_f = 2+1+1$ lattice QCD at nonzero temperature

**Anton Trunin[1,2]*, Florian Burger[3], Ernst–Michael Ilgenfritz[1], Maria Paola Lombardo[4] and Michael Müller–Preussker[3]**

[1] Bogoliubov Laboratory of Theoretical Physics, Joint Institute for Nuclear Research, Joliot–Curie Str. 6, 141980 Dubna, Russia
[2] Samara State Aerospace University, Moskovskoye Shosse 34, 443086 Samara, Russia
[3] Physics Department, Humboldt University Berlin, Newtonstr. 15, 12489 Berlin, Germany
[4] Frascati National Laboratory, National Institute for Nuclear Physics, Via Enrico Fermi 40, 00044 Frascati (Rome), Italy

E-mail: amtrnn@gmail.com

**Abstract.** We present results for the topological susceptibility at nonzero temperature obtained from lattice QCD with four dynamical quark flavours. We apply different smoothing methods, including gradient Wilson flow and over–improved cooling, before calculating the susceptibility. It is shown that the considered smoothing techniques basically agree among each other, and that there are simple scaling relations between flow time and the number of cooling/smearing steps. The topological susceptibility exhibits a surprisingly slow decrease at high temperature.

The non–trivial topological structure of gauge fields and the computation of the topological susceptibility $\chi_{top}$ is discussed in lattice QCD since long time. Recent considerations (see [1] and references therein) ranging from the restoration of the $U_A(1)$ symmetry at high temperature (or density) to the abundance of cosmic axions [2] are calling for a better knowledge of $\chi_{top}$ as a function of the temperature and quark masses.

In this study we calculate $\chi_{top}$ in the temperature range $150 < T < 500$ MeV using lattice configurations generated with $N_f = 2+1+1$ dynamical Wilson twisted mass fermions at finite temperature [3]. The heavy doublet of $s$ and $c$ quarks has its mass parameters matching the physical $K$ and $D$ meson masses, while the HMC simulations still require light quarks being unnaturally heavy. The configurations are generated at three coupling values $\beta = 1.90$, 1.95, and 2.10 (for later continuum extrapolations) with the Wilson twisted mass fermionic action tuned to maximal twist, taking benefit from automatic $\mathcal{O}(a)$ improvement. Several charged pion masses are considered, but here we restrict ourselves to one value $m_{\pi^\pm} \simeq 370$ MeV.

We focus on the comparison between different smoothing methods for lattice gauge fields necessary to get $\chi_{top}$ by the gluonic method. The smoothing techniques under consideration are the Wilson flow [4], Wilson and over–improved cooling [5], and stout–link smearing [6]. We use the Wilson flow to set two stopping scales determined by

$$t^2\langle E\rangle\big|_{t=t_0} = 0.3 \quad \text{or} \quad t^2\langle E\rangle\big|_{t=t_1} = 0.66, \qquad E = \frac{1}{2N_\tau N_\sigma^3}\sum_x \mathrm{Tr}[F_{\mu\nu}F^{\mu\nu}(x)], \tag{1}$$

* Presenter

Journal of Physics: Conference Series **668** (2016) 012123 doi:10.1088/1742-6596/668/1/012123

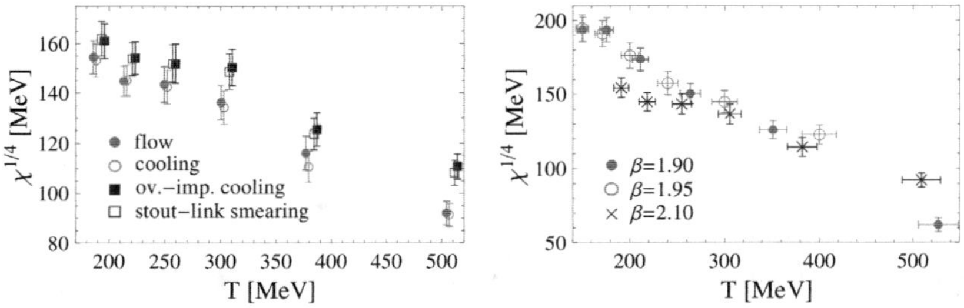


**Figure 1.** $T$ dependence of $\chi_{top}^{1/4}$ from Wilson flow. Left: comparison with all other smoothing methods for $\beta = 2.10$ and $t_0$. Right: at stopping time $t_0$ for three different $\beta$ values.

where $F_{\mu\nu}(x)$ is the field strength tensor on the lattice. We match ensemble averages $\langle E \rangle$ obtained in other methods to the values measured with the Wilson flow at $t_0$ and $t_1$ in order to relate these to corresponding numbers of cooling steps. Empirically we find in a good agreement for both the stopping criteria $t_0$ and $t_1$ that $N_{\text{cool}}^{\text{Wilson}} \simeq 3\tau$ [7], $N_{\text{cool}}^{\text{ov.-imp.}} \simeq 5\tau$ and $N_{\text{smear}}^{\text{stout}} \simeq 12\tau$, the latter for our choice $\rho_{\text{smear}} = 0.06$ ($\tau$ defined by $t = a^2\tau$, $a$ is the lattice spacing).

If the diffusion radius $R = \sqrt{8t}$ at $t_0$ or $t_1$ satisfies the condition $R \ll 1/T$ the topological susceptibility $\chi_{top}(T)$ can be safely determined as

$$\chi_{top} = \frac{\langle Q_{top}^2 \rangle}{V}, \qquad Q_{top} = \frac{a^4}{32\pi^2} e^{\mu\nu\rho\sigma} \sum_x \text{Tr}[F_{\mu\nu} F_{\rho\sigma}(x)], \qquad V = a^4 N_\tau N_\sigma^3. \tag{2}$$

The spatial and temporal lattice sizes are $N_\sigma = 24$, 32 and $N_\tau = 4 \ldots 16$, respectively, and the statistics is varying in the range $200 \ldots 1000$ configurations, depending on the respective temperature. One can see from the left panel of Figure 1, where the methods are compared for the finest lattice spacing (at $\beta = 2.10$), that the pairs of Wilson flow and Wilson cooling, as well as over–improved cooling and stout–link smearing, give almost indistinguishable $\chi_{top}(T)$ values throughout the considered temperature range. In the same manner, the agreement between the two stopping criteria $t_0$ and $t_1$ can be checked independently for each algorithm. The results for $\chi_{top}$ from Wilson and over–improved cooling agree (within errors) for low temperatures. Starting from approximately $T \gtrsim 300$ MeV, the over–improved result for $\chi_{top}$ turns out somewhat larger than the results for Wilson flow and Wilson cooling as one would expect.

In the right panel of Figure 1, $\chi_{top}^{1/4}(T)$ is shown as calculated at three different $\beta$ values (different lattice spacings $a$). The curves beyond $\sim 200$ MeV can be reasonably approximated with linear fits, with slopes visibly flattening with the growth of $\beta$ (towards $a \to 0$). Note that the value of crossover temperature obtained from the chiral susceptibility is $T_c = 184(4)$ MeV. Thus, we come to the preliminary conclusion that for $N_f = 2 + 1 + 1$ and $m_{\pi^\pm} \simeq 370$ MeV the topological susceptibility decreases very slowly beyond $T_c$ in the continuum limit, in contrast to the rapid fall–off observed in the quenched approximation ($N_f = 0$), and to the gradual descent taking place in the $N_f = 2$ case [8].

We are grateful to the HLRN supercomputer centers Berlin and Hannover, the Supercomputing Center of Lomonosov Moscow State University, and to the HybriLIT group of JINR for computational resources. The work was supported by the Heisenberg–Landau program of BLTP JINR and BMBF of Germany, by the "Dynasty" foundation, and by the Ministry of Education and Science of Russia under Competitiveness Enhancement Program 2013–2020.

## References

[1] Müller–Preussker M 2014 *Proceedings of Science* PoS(LATTICE2014)003
[2] Berkowitz E, Buchoff M I and Rinaldi E 2015 *Phys. Rev.* D **92** 034507
[3] Burger F *et al* 2013 *Proceedings of Science* PoS(LATTICE2013)153
[4] Lüscher M 2010 *J. High Energy Phys.* JHEP08(2010)071
[5] Perez M G, Gonzalez–Arroyo A, Snippe J and van Baal P 1994 *Nucl. Phys.* B **413** 535
[6] Moran P J and Leinweber D B 2008 *Phys. Rev.* D **77** 094501
[7] Bonati C and D'Elia M 2014 *Phys. Rev.* D **89** 105005
[8] Bornyakov V G *et al* 2013 *Phys. Rev.* D **87** 114508

15th International Conference on Strangeness in Quark Matter (SQM2015) IOP Publishing
Journal of Physics: Conference Series **668** (2016) 012124 doi:10.1088/1742-6596/668/1/012124

# Feasibility Study of $\phi(1020)$ Production at NICA

**Lyubka Yordanova for the MPD Collaboration**

Veksler and Baldin Laboratory of High Energy Physics, Joint Institute for Nuclear Research, 141980 Dubna, Moscow region, Russia

E-mail: lyubka.s.yordanova@gmail.com

**Abstract.** The goal of this article is to provide overview of the new accelerator complex NICA at JINR and to present the current results of the Multi-Purpose Detector (MPD) performance for $\phi$-meson production. In our study we use the channel decay $\Phi \rightarrow K^+K^-$ to detect the formation of the $\phi$-meson. UrQMD event generator is used and central events at $\sqrt{s} = 11$ GeV are analyzed. The obtained values of the parameters from the invariant mass distribution are consistent with the PDG values. This study shows that the measurement of $\phi$-meson is feasible at NICA/MPD.

## 1. Introduction

The Nuclotron-based Ion Collider fAcility (NICA) is a new accelerator complex being constructed at JINR. The global scientific goal of the NICA/MPD project is to explore the phase diagram of strongly interacting matter in the region of highly compressed baryonic matter. NICA's aim is to provide collisions of heavy ions over a wide range of atomic masses, from Au+Au collisions at $\sqrt{s_{NN}} = 11$ GeV (for $Au^{79+}$) and an average luminosity of $L = 10^{27}$ cm$^{-2}$ s$^{-1}$ to proton-proton collisions with $\sqrt{s_{pp}} = 20$ GeV and $L \sim 10^{32}$ cm$^{-2}$ s$^{-1}$. [1]

## 2. Overview of the MPD detector

The MPD experimental program includes simultaneous measurements of observables that are presumably sensitive to high nuclear density effects and phase transitions. MPD covers a large phase space, comprises high efficiency and excellent particle identification capabilities. The main tracker is the time projection chamber (TPC) supplemented by the inner tracker (IT). IT and TPC provide precise tracking, momentum determination and vertex reconstruction. The high performance time of flight (TOF) system should identify charged hadrons and nuclear clusters in a broad pseudorapidity range. The electromagnetic calorimeter (ECAL) should identify electrons, photons and measure their energy. The zero degree calorimeter (ZDC) should provide event centrality and event plane determination. [2]

## 3. Motivation for study of $\phi$-meson

The $\phi$ vector meson is the lightest bound state of hidden strangeness. It is expected to have a very small cross-section for interactions with non-strange hadrons. Its observables should remain largely undisturbed by the hadronic rescattering phase of the system's evolution. Previous experimental measurements of the $\phi$/K- ratio as a function of centrality showed the possibility of $\phi$ production via $K^+$+ $K^-$ coalescence in the hadronic stage. The observed ratios are flat as a function of centrality. These properties make the $\phi$-meson an excellent probe of the hot and dense medium created in nucleus-nucleus collisions. [3, 4, 5, 6]

## 4. Data set
In this study the channel decay $\Phi \rightarrow K^+K^-$ is chosen to detect the formation of the $\phi$-meson. UrQMD event generator is used and Au-Au collisions are generated. The number of the analyzed central events is 300 000 at $\sqrt{s} = 11$ GeV. The detector configuration used in the study includes TPC and TOF for track reconstruction and particle identification.

## 5. Reconstruction of $\phi$-meson
The kaon daughter particles are identified through their ionisation energy loss (dE/dx ) in TPC and also measurements in TOF. A selection of kaon pairs by track quality cuts and particle identification (PID) is performed. The invariant mass is calculated by combination of all $K^+$ and $K^-$ pairs of the same event. Since not all charged kaons in each event originate from $\phi$-meson decays, the $\phi$-meson extracted signal is above a large combinatorial background of uncorrelated pairs. The mixed-event background distribution is estimated, scaled and subtracted from the signal distribution. The obtained peak from the invariant mass distribution is fitted by a Breit-Wigner function and the characteristics of the $\phi$-meson such as its mass and its width are found.

## 6. Results
The recent results of the $\phi$-meson study are shown in Figure 1.The values of the parameters obtained by the Breit-Wigner fit (Width = $4.956 \pm 0.245$ MeV/$c^2$ and $M_{inv} = 1019.030 \pm 0.118$ MeV/$c^2$) are consistent with the values given in literature. All these preliminary results show that the measurement of $\phi$-meson is feasible at NICA/MPD.

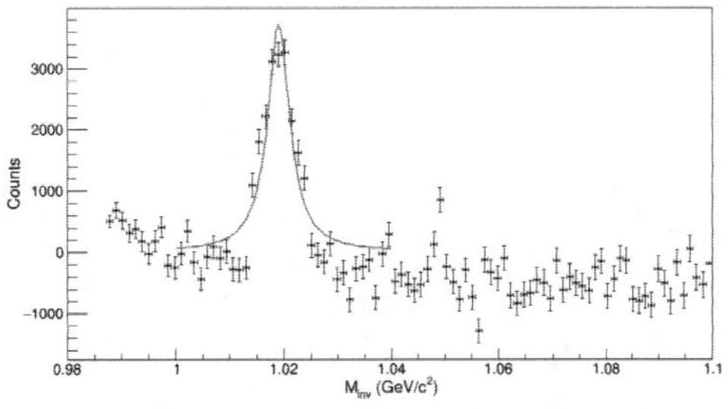


**Figure 1.** $\phi$-meson invariant mass

## 7. Summary
The current results show that the study of the $\phi$-meson is feasible and it can be used to probe the properties of the medium created at NICA/MPD. NICA facilities provide unique capabilities for studying fundamental properties of the theory of strong interactions (QCD).

## References
[1] MPD Collaboration 2014 *NICA White Paper*
[2] MPD Collaboration 2011 *The Multi-Purpose Detector-MPD*
[3] STAR Collaboration 2009 *Phys.Rev.* C **79** 064903
[4] PHENIX Collaboration 2005 *Phys. Rev.* C **72** 014903
[5] S. Pal, C.M. Ko, and Z. Lin 2002 *Nucl.Phys.* A **707** 525-539
[6] STAR Collaboration 2004 *J. Phys.* G **30** S543-S548

Journal of Physics: Conference Series **668** (2016) 012125 doi:10.1088/1742-6596/668/1/012125

# Simulation of electromagnetic and strange probes of dense nuclear matter at NICA/MPD

**A. Zinchenko*, V. Kolesnikov, V. Vasendina**

Joint Institute for Nuclear Research, 141980 Dubna, Moscow region, Russia

E-mail: `Alexander.Zinchenko@jinr.ru`

**Abstract.** The main task of the NICA/MPD physics program is a study of the properties of nuclear matter under extreme conditions achieved in collisions of heavy ions. These properties can reveal themselves through different probes, the most promising among those being the lepton-antilepton pairs and strange hadrons. In this paper the MPD performance for measuring the electron-positron pairs and strange hyperons in central Au+Au collisions at NICA energies is presented.

## 1. MPD detector performance

The main goal of studying heavy-ion collisions is to explore the properties of nuclear matter under extreme density and temperature conditions in order to resolve the most fundamental problems of the underlying theory - confinement and chiral symmetry breaking. To reach this goal, the MPD detector is designed as a $4\pi$ spectrometer capable of detecting charged hadrons, electrons and photons in heavy-ion collisions in the energy range of the NICA collider. The detailed description of the MPD geometry can be found in Ref. [1].

The present analysis is based on the detectors covering the mid-rapidity region ($|\eta| <1.3$): the main tracker Time Projection Chamber (TPC) and barrel Time-Of-Flight system (TOF). For dilepton simulations the barrel ElectroMagnetic Calorimeter (EMC) is also used.

The track reconstruction method is based on the Kalman filtering technique and the number of TPC points per track is required to be greater than 10 to ensure a good precision of momentum and $dE/dx$ measurements. In addition, we have restricted our study to the mid-rapidity region with $|\eta| < 1.3$. The track finding efficiency in TPC for primary and secondary tracks is above 95% for momentum $p > 100$ MeV/$c$. The momentum resolution $\Delta p/p < 2\%$ at $p_T < 1.5$ GeV/$c$. The primary vertex resolution in transverse and longitudinal directions $\sigma_x$ and $\sigma_z < 0.15$ mm in central collisions at track multiplicity in TPC $> 500$.

Particle identification (PID) in the MPD experiment will be achieved by combining specific energy loss ($dE/dx$) and time-of-flight measurements. The basic detector parameters, namely, $dE/dx$ and TOF resolutions $\sigma_{dE/dx} \approx 6\%$ and $\sigma_{TOF} \approx 100$ ps will provide a high degree of selectivity for hadrons at momenta below 2 GeV/$c$.

An identified hadron candidate is assumed to lie within the boundaries of the PID ellipse ($3\sigma$ around the nominal position for a given particle specie) in the $dE/dx - M^2$ space, where $M^2$ is the reconstructed particle mass squared. In addition, the probability for a given particle to belong to each of the species can be calculated knowing the widths of the corresponding distributions (along the $dE/dx$ and $M^2$ axes) and the difference from the predicted position for the specie. It was found that by requiring this probability to be greater than 0.75 one can obtain high PID efficiency and low contamination.

## 2. Simulations: event generators, data sets and results

The software framework for the MPD experiment MpdRoot[2] is based on FairRoot and provides a powerful tool for detector performance studies, development of algorithms for reconstruction and physics analysis of the data.

The event sample used for the study of hyperons was produced with the UrQMD generator at $\sqrt{s} = 9A$ GeV. The analyzed statistics of $4 \cdot 10^4$ central events $(0-3$ fm) corresponds to about 2 minutes of running time at the NICA collision rate of 6 kHz [1]. For dilepton simulations the Pluto generator was used at $\sqrt{s} = 7A$ GeV with accumulated statistics of $2 \cdot 10^7$ central events, corresponding to about 17 hours of NICA running time. Produced by the event generators particles have been transported through the detector using the GEANT3 transport package (describing particle decays, secondary interactions, etc.).

Multistrange hyperons were reconstructed using their decay modes into a charged particle and a $\Lambda$ hyperon followed by $\Lambda$ decay into a proton and a pion. The result for $\Xi^-$ hyperon simulations is presented in Fig. 1.

The measurement of the dilepton yield is rather complicated. Huge combinatorial background is the experimental challenge. Already the start version of MPD (TPC, TOF and EMC) allows to select electrons with a hadron supression factor up to $10^5$. Since the particle identification suppresses the hadron contamination, the main source of the remaining background is photon conversion in the detector material. The conversion pairs are rejected by a special topological cut. If only one lepton from a conversion pair is reconstructed, it can survive the cut. Such single tracks are rejected by a low-momentum cut $p_T > 200$ MeV/$c$ and the requirement to satisfy the primary track selection criteria.

Figure 2 shows the invariant mass spectrum of reconstructed low-mass dielectrons after background subtraction (dots) and the one for true dielectrons from the Pluto generator (line). The obtained $S/B$ ratio in the mass region $0.2 < M_{e^+e^-} < 1.2$ GeV/$c^2$ is about 10%.

From the results presented (see also [3, 4]) one can conclude that MPD offers good opportunities for studies of strange and electromagnetic probes at the NICA collider.

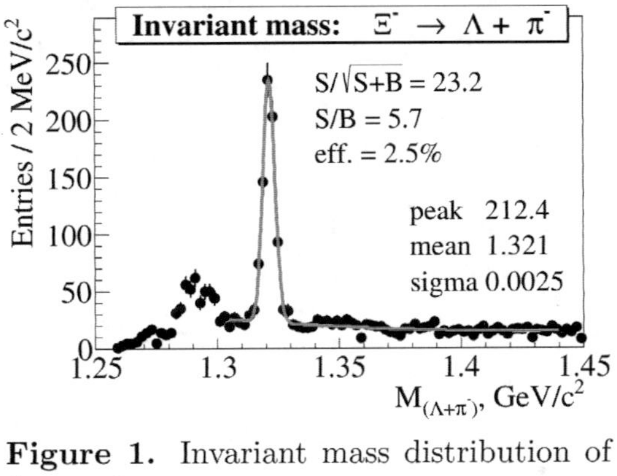


**Figure 1.** Invariant mass distribution of $\Lambda\pi^-$ pairs.

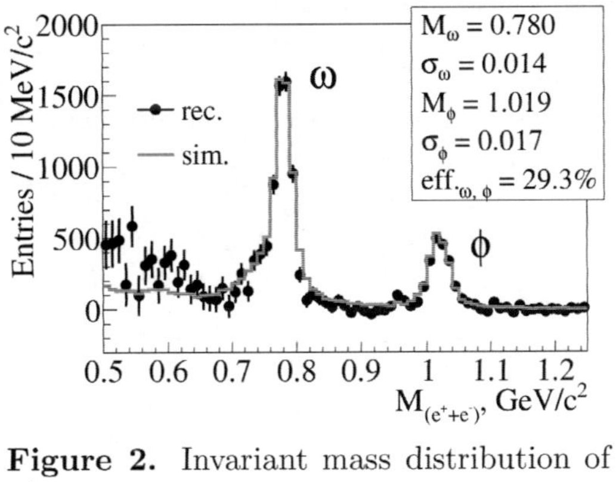


**Figure 2.** Invariant mass distribution of $e^+e^-$ pairs.

## References

[1] K.U. Abraamyan (MPD Collaboration), Nucl. Instrum. Meth. **A628**, 99 (2011).
[2] http://mpd.jinr.ru/.
[3] V. Vasendina *et al*, Phys. Part. Nucl. Lett. **10**, no. 7, 769 (2013).
[4] M. Ilieva *et al*, Phys. Part. Nucl. Lett. **12**, no. 4, 618 (2015).

# SQM2015: Summary of Theory Talks

**J. Cleymans**
UCT-CERN Research Centre and Physics Department, University of Cape Town, South Africa
E-mail: `jean.cleymans@uct.ac.za`

**Abstract.** A summary of plenary talks presented at the 15th Strangeness in Quark Matter 2015 conference held at the JINR, Dubna July 6-11, 2015.

## 1. Overview

During the "*Strangeness in Quark Matter 2015*" (SQM2015) conference, held at the JINR in Dubna from July 6-11 a total of about 150 talks were presented of which approximately half were theory talks. In preparation for this overview generous use was made of the material submitted to these proceedings. Due to the large number of talks this overview is limited to the submitted plenary talks.

## 2. Chemical Equilibrium.

One of the important lessons learned from the study of heavy ion collisions in the past few years is how well chemical equilibrium works at all beam energies. At high energies this was shown by results from the ALICE collaboration [1], at low beam energies it was shown by the HADES collaboration [2]. In both cases the agreement with chemical equilibrium is remarkable. At the same time there are also clear deviations, as seen in the yields of $\Xi$'s at low energy and of protons at high energies. The agreement is made more striking by plotting the results in Fig. (1) in a form first suggested about two decades ago by Becattini [3]. It is worth emphasizing that all yields of hadrons are established at a single freeze-out temperature to a very good approximation.

The deviations at lower energies were addressed in the talk of M. Bleicher [4] where it was shown that by adding new decay channels of heavier resonances one can explain the additional yield of $\Xi$'s and $\phi$'s.

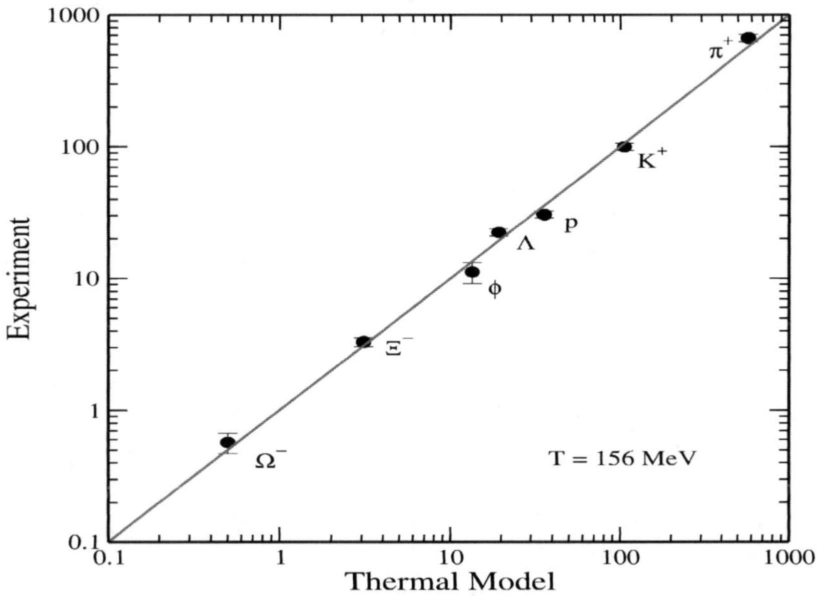


## 3. Heavy Quarks and Transport Models.

The agreement is generally very good and predictions for higher beam energies were presented. The production of heavy flavors was considered by several speakers [6, 7, 8, 9]. Due to the heavy masses these are mainly produced in the early stages of the collisions when the initial energy has not been degraded and, due to the smaller cross sections, they provide information on the early stages of the collisions.

The behavior of jets moving through the hot and dense medium created in heavy-ion collisions was discussed by Djordjevic [5]. A formalism was presented for calculating the energy loss and the suppression of jets taking into account radiative and collisional energy loss. The calculations were applied to angular averaged values of $R_{AA}$ which are less sensitive to the evolution of the medium. The results were compared to $R_{AA}$ for various particles, including light and heavy hadrons, for a wide range of probes, centralities and beam energies. It was noted [9] that the values of $R_{AA}$ and $v_2$ are described well by many models and it is necessary to go beyond these observables in heavy-ion collisions. More differential observables like correlations and higher-order flow coefficients of D mesons as well as the elliptic flow of B mesons indicate to have a

larger discriminating power and are good candidates for reaching another level of understanding of the heavy-flavor dynamics and its relation to the bulk medium evolution. As a result they can help us to identify the most dominant features of the heavy-quark-medium interaction in order to connect experimental data to fundamental properties of QCD.
Detailed results were presented based on the Parton-Hadron-String Dynamics approach [6] where it was concluded that collisional energy loss was dominant up to $p_T$ values of 15 GeV/c in relativistic heavy-ion collisions.

The present theoretical understanding of quarkonium production was presented in great detail by Lansberg [8]. It was strongly argued that the study of quarkonium production in proton-nucleus collisions is of fundamental importance if one wishes to not only properly interpret quantitatively but also qualitatively interpret the physics underlying the quarkonium suppression in nucleus-nucleus collisions. The case of the sequential suppression of $\Upsilon$ measured by CMS is a prime example. Just measuring quarkonium production in nucleus-nucleus collisions is not enough. One of the mechanisms which can explain the relative suppression is the scattering with comoving hadrons. This is included in the comover interaction model which was introduced long ago to explain part of the anomalous quarkonium suppression in AA collisions.

The talk of Beraudo [7] considered transport approaches based on the Boltzmann and Langevin equations. The importance of future beauty measurements as well as heavy flavor studies in small systems like p-Pb was emphasized. It was shown that correlations would be helpful to resolve some of the outstanding issues.

A new approach to identify isotopes and hypernuclei in transport models was presented by Le Fèvre [10]. The new algorithm includes pairing and asymmetry energies as well as other effects is able to describe the nuclear binding energy and allow for the calculation of isotope yields. It was shown that the asymmetry and pairing potentials have a strong influence on the yields and on the momentum anisotropies for the (hyper-)isotopes. The fragment formation is sensitive to the density dependence of the asymmetry energy and the pairing energy. However, fragments test this dependence only for densities below the saturation density.

## 4. Lattice QCD

An investigation was presented by Wittig [12] about the existence of the H-dibaryon using lattice QCD. These are very demanding calculations and a bound H-dibaryon was only found at unphysically large pion masses. There is a need for more and better lattice data.

An important ingredient in understanding heavy-ion experiments is the production of hadronic resonances and their subsequent decays. Information about this is far from complete [14]. Related to this, interesting results about strange resonances were presented by Kaczmarek [11] as obtained from Lattice QCD. A large list of strange baryons and mesons is found in lattice QCD which are not listed by the Particle Data Group [14] . These directly affect calculations done using the thermal model. It would be very helpful to also have information about non-strange baryons and mesons as this has implications for the calculation of the $K^+/\pi^+$ ratio using the thermal model. Furthermore, there is no information about decay channels from lattice QCD. Finally and most importantly, one needs experimental confirmation about all these new hadrons!

A report about heavy quarkonia from Lattice QCD was presented by Rothkopf [13]. Combining an effective description for heavy quarks with finite temperature lattice QCD we compute in-medium spectra providing insight in heavy quarkonium melting in a static thermal medium. On the other hand spectral information was used to extract the complex in-medium potential which provides the basis for a dynamical description of quarkonium real-time evolution via the Schrödinger equation. It was pointed out that bottomonium S-wave and P-wave will survive up to very high temperatures of at least T = 250 MeV.

## 5. Conclusions

In conclusion, an impressive amount of progress has been made in the theoretical understanding of relativistic heavy ion collisions. Many suggestions were made for future measurements that will be improve our present models. The next series of experiments at various laboratories and different beam energies will contribute greatly to the better understanding.

## References

[1] M. Floris, Journal of Physics: Conference Series (these proceedings).
[2] M. Lorenz, Journal of Physics: Conference Series (these proceedings).
[3] F. Becattini, Z. Phys. C69 (1996) 485.
[4] M. Bleicher, Journal of Physics: Conference Series (these proceedings).
[5] M. Djordjevic, Journal of Physics: Conference Series (these proceedings).
[6] E. Bratkovskaya, Journal of Physics: Conference Series (these proceedings).
[7] A. Beraudo, Journal of Physics: Conference Series (these proceedings).
[8] J.P. Lansberg, Journal of Physics: Conference Series (these proceedings).
[9] M. Nahrgang, Journal of Physics: Conference Series (these proceedings).
[10] A. Le Fèvre, Journal of Physics: Conference Series (these proceedings).
[11] A. Bazavov et al. Phys. Rev. Lett. 111 (2013) 082301.
[12] J. Green et al. PoS LATTICE2014 (2014) 107.
[13] A. Rothkopf, arXiv:1509.03715.
[14] K.A. Olive et al. (Particle Data Group), Chin. Phys. **C38**, 090001 (2014).

Journal of Physics: Conference Series **668** (2016) 012127 doi:10.1088/1742-6596/668/1/012127

# Summary talk: Experiments at low energies

**Yvonne Leifels**

GSI Helmholtzzentrum für Schwerionenforschung, Darmstadt,
Planckstr. 1, 64291 Darmstadt, Germany

E-mail: y.leifels@gsi.de

**Abstract.**

In heavy-ion collisions at beam energies $\sqrt{s_{NN}}$ between 1 and 150A GeV highest baryonic densities are reached at rather moderate temperatures. By varying the beam energy and the system size a broad range of the QCD phase diagram is scanned where several interesting phenomena are predicted by theoretical models. Apart from possible phase transitions and existence of a critical point in this regime, the production of strangeness and the interaction of strange particles with the surrounding hot and dense nuclear medium constitutes a prominent probe not only to address the underlying reaction mechanisms and production processes but in particular to constrain densities and temperatures reached in the course of the collision.

Recent results on heavy-ion collisions in this beam energy regime obtained by various experimental collaborations are summarized, with special emphasis on strangeness production, rare probes, and critical phenomena. The importance of data on elementary reactions (i.e., pp, p+nucleus, and $\pi$+nucleus) as a bench mark for theoretical models and their relevance for understanding the underlying mechanisms of heavy-ion collisions are being discussed. Several interesting observables have been presented in various contributions, which give further motivation for the construction of high-rate experiments at new accelerator facilities.

## 1. Introduction

The conjectured phase diagram of QCD, usually presented in terms of temperature $T$ and baryonic chemical potential $\mu_B$, is of fundamental interest. At small baryonic chemical potential, the hadronic phase is assumed to be separated from the quark-gluon phase by a cross-over transition. At the critical point the cross over transition is expected to evolve with rising baryonic chemical potential into a supposedly first-order phase transition. Super-conducting phases might appear at even higher baryonic potential and low temperature [1, 2]. Theoretical predictions of the properties of both the critical point and the phase boundary at high baryon chemical potential are still lacking experimental verification and/or constraints.

The relevant regime in the QCD phase diagram of high baryonic potential and moderate temperature can be accessed experimentally with heavy-ion collisions in the beam energy regime $\sqrt{s_{NN}} < (20 - 100)$ GeV. During the course of these reactions densities up to several times normal nuclear matter density $\rho_0$ are reached. Such collisions of heavy-ions have been studied at SPS/CERN, AGS/BNL and in the low-energy scan of RHIC/BNL, and–in the lowest energy regime–at the accelerators of GSI and formerly at Berkeley. Several observables have been suggested to give access to the properties of the phase boundary and the critical point (i.e., the K/$\pi$ ratio as a function of beam energy, fluctuations as a measure of critical phenomena etc.). Experimental findings, like the enhancement in the $K^+/\pi^+$ ratio at $\sqrt{s_{NN}} \approx 10$ GeV, have motivated various efforts to understand the origin of these observations and to characterize heavy-ion collisions in this energy regime with a very high degree of precision.

In addition, the study of heavy-ion collisions at low energies offers unique opportunities to investigate particle production, strangeness production in particular, in a high-density and moderate temperature environment. The high density phase is baryon dominated and rather long lived. Most of the particles are produced below or close to their production thresholds in NN collisions. Hence, the production rates are very sensitive to density fluctuations and in-medium effects. Despite low production yields modern detector set-ups capable to sustain high event rates are able to accumulate enough statistics even on very rare probes. Relying now on a broad range of particle species, the chemical freeze-out can be more reliably characterized in the framework of thermal models.

Another interesting aspect is the formation of hyper-nuclei in heavy-ion collisions by coalescence of nucleons with hyperons, which is–according to thermal model predictions–highest at these moderate beam energies [3]. Measuring yields and phase-space distributions will give hints on the underlying production mechanisms and the interaction of hyperons with nucleons in a dense medium. From the astrophysical side, the precise knowledge of the hyperon-nucleon interaction is of crucial importance, since $\Lambda$-hyperons are supposed to be stable in neutron stars and might form an important constituent of neutron star matter.

## 2. Strangeness production at SIS energies

The HADES experiment at GSI has measured strangeness production in Ar+KCl reactions at 1.8A GeV [4]. It was possible not only to reconstruct $\Phi$-mesons–their production in elementary processes is inhibited by the OZI rule–via their decay into charged kaons but also double strange $\Xi^-$-hyperons ($\Xi(1315)^-$) by measuring their decay into $\Lambda + \pi^-$. Rather high production yields have been deduced for both particles. A $\Phi/K^-$ ratio of 0.37±0.13 was measured [5]. Such a high ratio means that the $K^-$ spectra might be affected by the decay of $\Phi$-mesons substantially. After an upgrade of the HADES detector, it was possible to investigate Au+Au collisions at $E_{beam}$ = 1.25A GeV with an event rate of 50 kHz, thus enabling to collect enough statistics to measure not only kaon production yields and spectra but also to access rare probes like the $\Phi$-meson with a production threshold in $E_{NN} = 2.59$ GeV in NN collisions. The $\Phi/K^-$ ratio is even higher than in the 2A GeV region. Interestingly, this experimental data can be described within the framework of a thermal model [6], when the $\Phi$-Meson is included without strangeness suppression; strangeness is conserved only in a radius smaller $R_C$ than the system radius $R_V$. In a first attempt describing Au+Au results (not including systematical errors), $T_{chem} = 47 \pm 5$ MeV, $\mu_B = 799 \pm 34$ MeV, and $R_c/R_v = 0.3 \pm 0.2$ have been extracted.

Generally, all particle yields in heavy-ion reactions at SIS energies can be described within thermal models with the exception of the $\Xi^-$ data. However, it was found that the chemical freeze-out temperature $T_{chem}$ is equal or slightly smaller than the kinetic temperature $T_{kin}$, which is extracted from the particle spectra at mid-rapidity [5, 7]. This is unexpected, because the kinetic freeze-out should occur after the chemical one at a lower temperature. Variant findings were reported for p+Nb reactions at 3.5 GeV [5]. Thermal model fits agree with the experimental particle yields, except–again–for the $\Xi^-$ production. Particle yields for p+Nb and Ar+KCl collisions including the thermal model fits are presented in 1. Such a result is rather remarkable. since thermal equilibration is not readily expected for proton induced reactions at such low energies. Of particular interest is, however, that for p-induced reactions chemical temperatures $T_{chem}$ are *larger* than the kinetic temperatures $T_{kin}$. One might speculate whether the larger flow contribution in collisions of heavy-ions is responsible for the high kinetic temperatures observed in this data. But similar findings have also been reported for a rather small heavy-ion system (Al+Al), where essentially no flow exists [7]. These results certainly call for more systematic measurements varying energy and system size.

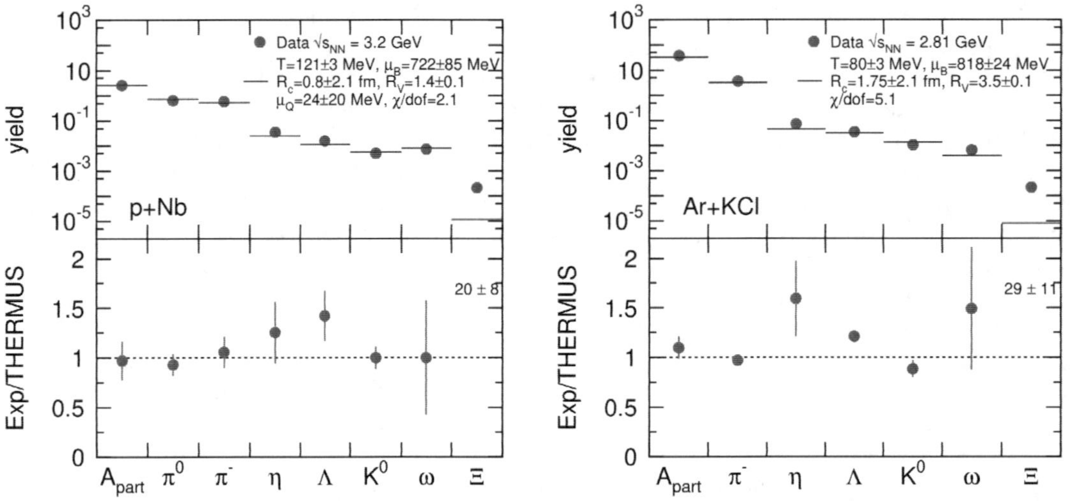


**Figure 1.** Preliminary HADES data on particle production in p+Nb at 3.5 GeV (left) and Ar+KCl collisions at 1.85A GeV (right) in comparison to THERMUS fits [6], as presented by [5].

It has been shown that the particle ratios measured in heavy-ion collisions can be reproduced by models like UrQMD [8]. If microscopic transport models also reproduce the difference between kinetic and chemical temperatures (including the gross properties of the reactions like stopping and flow), one might investigate the mechanisms leading to the apparent thermal equilibration in heavy-ion reactions within those models.

## 3. Strangeness production at SIS energies and the role of heavy resonances

Several theoretical attempts to describe strangeness production at SIS energies based on microscopic transport models have been presented [8, 9] during this conference. In one of them [8], high mass resonances are contributing to the strangeness production, in particular to the production of Φ-mesons and the double strange $\Xi^-$. It is possible to reproduce not only the Φ yields with such an assumption but also the production rates of $\Xi^-$-hyperons. Such conjectures have to be verified by experimental data. Proton-induced reactions deliver information on elementary production mechanisms of strange particles, form a normalization for heavy-ion reaction data, and yield bench mark information for microscopic transport models.

Data on proton-induced reaction in the SIS energy regime measured by the HADES collaboration have been presented [10]. Data of p+p at $E_{beam} = 3.5$ GeV are used to extract a complete cross section data base of channels relevant for $K^0$ production and now can serve as reference measurement for the $K^0$ in p+A data, where the interaction between Kaons and nucleons is studied.

It was possible to study the role of heavy resonances in the production of strange particles with the reaction channel p+p → $\Sigma(1382)^+$+ $K^+$+ n. Angular distributions of the $\Sigma(1385)^+$ in different reference frames are found to be compatible with the hypothesis that 33% of the $\Sigma(1385)^+$ result from the decay of an intermediate broad $\Delta^{++}$ excitation at about 2000 MeV/c$^2$.

Those two examples corroborate the necessity of measuring elementary reactions in conjunction with heavy-ion reactions. At low energies strangeness production is a complex process, which is not completely understood, and additional information on elementary processes is definitely needed.

Journal of Physics: Conference Series **668** (2016) 012127 doi:10.1088/1742-6596/668/1/012127

## 4. Production of nuclei and hyper matter

Nuclei are formed in heavy-ion collisions essentially at all energies. At SIS energies a substantial part of the colliding nuclei matter is emitted from the hot and dense zone is bound into fragments even in the most central collisions. The number of formed nuclei is dramatically decreased at LHC energies. But on the other hand, the energy is high enough that nuclei and anti-nuclei are formed with essentially the same rates, with a penalty factor of 300 for each additional nucleon [11].

One or more of the nucleons are replaced by a hyperon in hyper-nuclei. Study of hyper-nuclei in heavy-ion collisions has several advantages over production in elementary reactions: the momentum transfer is large, which allows determining lifetimes with a high-degree of precision; one could produce hyper-nuclei with an exotic n/p ratio by employing radioactive nuclei; or one might observe multi-strange objects at higher incident energies and study the properties of the EOS of hyper matter. Hyper-nuclei are supposedly formed by coalescence. At least at higher energies their production yields are compatible with thermal-model predictions [11]. At lower incident energies other processes are discussed. Hyper-nuclei might be formed by re-scattering hyperons into the spectator matter and subsequent capture within the residues. Such a mechanism requires that the spectator matter is moving not too fast to ensure that the phase space distribution of the hyperons, which are produced in the hot and dense collision zone, overlaps with that of the spectator matter. Scattering of pions into the spectator matter followed by strangeness production via the reaction $\pi$+ N $\rightarrow$ $K^+$+Y is an alternative mechanism to form hyper-nuclei in the spectator residues. Heavier (and more exotic) hyper-nuclei could be formed much more easily by such processes than by coalescence [12].

The HYPHI collaboration has investigated hyper-nuclei production at SIS energies in the spectator region. In $^6$Li + $^{12}$C collisions at $E_{beam}$ = 2A GeV $^3_\Lambda$H and $^4_\Lambda$H have been reconstructed by their weak decays into $\pi^-$ and $^3$He or $\alpha$, respectively. An interesting finding, which is also reported by other experiments (STAR and ALICE [11]), is that the $^3_\Lambda$H lifetime is much shorter than previously measured [13] and significantly shorter than the lifetime of the $\Lambda$-hyperon (263 ps). How the $\Lambda$N interaction and the hyper-triton lifetime are connected is still unresolved.

Cross sections as well as phase space distributions of $^3_\Lambda$H and $^4_\Lambda$H hyper-nuclei have been reconstructed by the HYPHI collaboration for the first time at such low incident energies [13]. These results have been compared to predictions of a microscopic transport code in combination with a specific clusterization algorithm [14]. Since usually only nucleons are transported in microscopic models of heavy-ion reactions, additional codes are needed to form clusters. The FRIGA code calculates binding energies (including a $\Lambda$N potential and quantum effects like shell and pairing energies) and searches the most probable clusters with a simulated annealing mechanism. The model results describe the experimental data of the HYPHI reasonably well.

## 5. Search for the critical point

Theoretical attempts to predict the existence and location of the critical point in the phase diagram of $T$ versus $\mu_B$ lead to conflicting results. A scan of the phase diagram by varying the sizes of colliding nuclei (change of rescattering probability) and energies of the collisions (change of $\mu_B$) is a promising experimental search strategy, which is being pursued at different accelerators and setups. A study of hadron production in central Pb+Pb collisions by NA49 at SPS, revealed structures in the energy dependence of the $<K^+> / <\pi^+>$ ratio at mid-rapidity (see Fig. 2), a sudden change in the number of pions per participant, and a plateau in the energy dependence of the inverse slope parameter. Such structures have been predicted to be an onset of the phase transition to the QGP. Non-monotonic behaviour of the energy dependence of certain observables and a coinciding maximum of several fluctuation measures, like particle multiplicities or average transverse momenta of produced particles, would indicate the existence and the location of the critical point. Amongst others, such a program was started by the NA49

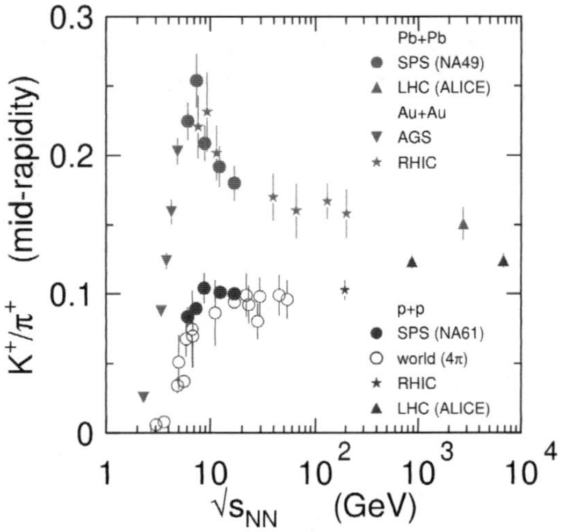


**Figure 2.** Data sets from different experiments for $K^+/\pi^+$ ratio of Au+Au/Pb+Pb and p+p collisions as a function of $\sqrt{s_{NN}}$.

collaboration [16] and is now followed systematically by the NA61/SHINE experiment and the low energy scan program pursued by the STAR collaboration at RHIC.

New high precision p+p and Be+Be data was presented by the SHINE collaboration [15], which serves as bench mark and normalization of the heavy-ion data. There are indications that the structures observed in Pb+Pb are already present in p+p but on a much smaller level as can be seen in Fig. 2. Theoretical models do not describe all features of the p+p data. The search for maxima in fluctuation observables has been deemed to not be conclusive enough. In medium size systems there is an indication of a maximum observed, but, for heavier systems, high statistical data must still be obtained in future experimental campaigns.

## 6. Outlook

High rate experiments to investigate heavy-ion collisions are undertaken at SIS18 energies employing detectors, which are able to withstand high intensity beams. The large statistics allows investigating strangeness production and the search for rare probes like $\Phi$-mesons, double strange $\Xi^-$ [5], as well as hyper-nuclei [13]. Parallel to the measurements of heavy-ion reactions, proton induced reactions are studied to understand reactions channels for strangeness production at threshold energies. Partial wave analysis (PWA) was discussed [10] as another important tool to unravel the role of high mass resonances in the strangeness production. But still more data is needed to disentangle the possible processes.

The search for the critical point was continued with measurements in p+p and Be+Be systems at SPS energies. Hints were found in the energy and system size dependence fluctuation observables, which are sensitive to the critical point and the experiments will be continued with heavier systems to complete these systematic efforts.

The STAR detector will extend the available data set to the lowest possible energies available at RHIC in the near future by making the beam collide with a fixed target. This method will enhance the luminosity at those beam energies and reduce the available energy in the center-of-mass system (compare Fig. 3).

The renewed interest in the high density, moderate temperature region of the QCD phase diagram is one of the important research lines at the two accelerator projects currently pursued in Europe: FAIR and NICA. The first stage of the FAIR accelerator will come into operation in 2021 with the SIS100. SIS100 will deliver high intensity beam upto 11A GeV for Au ions and 29

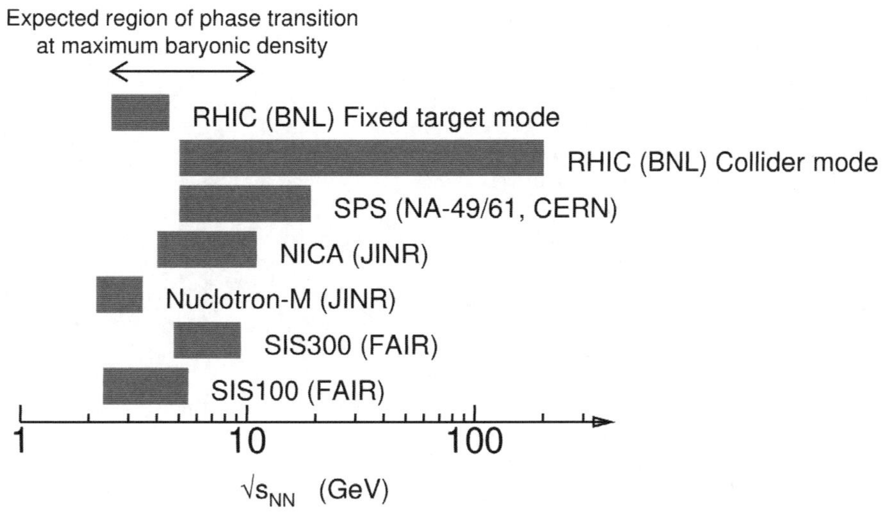


**Figure 3.** Energy ranges in terms of $\sqrt{s_{NN}}$ accessible by existing and future accelerators in the region of the QCD phase diagram of high $\mu_B$ and moderate $T$ after [17].

GeV for protons; NICA will offer experiments at a collider with heavy-ions at $\sqrt{s_{NN}} = 3.5$ GeV with average luminosity of $10^{27}$ starting in 2019. An overview on the future accelerator ranges in $\sqrt{s_{NN}}$ is shown in Fig. 3. Plans for new experiments aiming on the investigation of the QCD phase diagram at high baryonic densities have been presented during this conference; CBM at FAIR [18], and BM@N and MPD at the accelerators in Dubna [17]. BM@N (Baryonic Matter at Nuclotron) is a fixed target experiment at the Nuclotron (JINR) to study A+A collisions by measuring a variety of observables, i.e., particle yields, ratios, transverse momentum spectra, rapidity, and angular distributions, as well as fluctuations and correlations of hadrons, whereas the Multi-Purpose Detector MPD will be an experiment at the NICA collider at the JINR. The Compressed Baryonic Matter experiment (CBM) is laid out as a fixed-target experiment at the heavy-ion synchrotrons SIS-100/300 at the FAIR facility. The detector will record both pA and AA collisions at beam energies between 10 and 45 AGeV with interaction rates of 10 MHz. Hence, it will be possible to study extremely rare probes, which are not accessible by any other accelerator and/or detector setup operating in this energy regime.

Many measurements have been undertaken at SPS, AGS, and lately at RHIC investigating $J/\Psi$ suppression, strangeness production, elliptic and directed flow, HBT (femtoscopy), event-by-event fluctuations, and many more to understand the QCD phase diagram in the region of high baryonic densities and moderate temperatures. Several interesting non-monotonic behaviours of observables have been found, which motivated the development of the new detector and accelerator concepts. During the next years data will become available, which will shed new light on this highly interesting and exciting region of the QCD phase diagram.

Finally, I want to express my gratitude to the organizers and all participants for the interesting and fruitful conference, the stimulating discussions and the inspiring atmosphere.

## 7. References

[1] Weise W 2012 *Progress in Nuclear and Particle Physics* **67** 299-311
[2] Fukushima K and Hatsuda T 2010 *Reports on Progress in Physics* **67** 014001
[3] Andronic A et al 2011 *Phys. Lett.* B **697** 203-207
[4] Schmah A 2010 *J. Phys.* G **37** 094004

[5] Lorenz M *contribution to this conference*
[6] Wheaton S, Cleymans J, Hauer M 2009 *Comp. Phys. Commun.* **180** 84
[7] Lopez X et al. 2008 *J. Phys.* **G** 044020
[8] Bleicher M 2015 *contribution to this conference*
[9] Cabrera D 2015 *contribution to this conference*
[10] Fabbietti L 2015 *contribution to this conference*
[11] Chinellato D 2015 *contribution to this conference*
[12] Botvina A 2015 *contribution to this conference*
[13] Rappold C 2015 *contribution to this conference*
[14] Le Fèvre A 2015 *contribution to this conference*
[15] Larsen D 2015 *contribution to this conference*
[16] Seyboth P 2015 *contribution to this conference*
[17] Kekelidze V 2015 *contribution to this conference*
[18] Friese V 2015 *contribution to this conference*

Journal of Physics: Conference Series **668** (2016) 012128 doi:10.1088/1742-6596/668/1/012128

# Phase diagram in the entanglement PNJL model

**Alexandra Friesen**[a], **Yuri Kalinovsky**[b], **Vyacheslav Toneev**[a]
[a]Bogoliubov Laboratory for Theoretical Physics, JINR Dubna, Russia
[b]Laboratory for Information Technologies, JINR Dubna, Russia

E-mail: avfriesen@theor.jinr.ru

**Abstract.** Effects of the vector interaction in the Nambu-Jona-Lasinio model with Polyakov loop are studied in combination with the entanglement interaction between the quark and pure gauge sectors. We investigate the QCD phase diagram and find that the first order chiral phase transition at finite baryon chemical potentials and its critical endpoint disappear for sufficiently large values of the vector interaction constant $G_v$. The presence of an entanglement interaction between quark and pure gauge sectors leads to an increase of the value $G_v$ for which the first order transition disappears. The influence of a nonzero $G_v$ on the curvature of the crossover boundary in the $T-\mu$ plane nearby $\mu$= 0 is also examined for both cases.

## 1. Vector interaction and entanglement PNJL model

To introduce the couplings between quarks and pure gauge sector in NJL model, in the work [1] it was suggested to include of an effective potential into model. The inclusion of the vector interaction in NJL or PNJL models, discussed for example in [2–4], leads to the dependence of equation of state on the vector coupling constant $G_{\rm v}$. It was suggested that the first order transition area becomes smaller and the critical point appears at higher chemical potential and at lower temperature when constant $G_{\rm v}$ reaches critical value. The phase diagram in the $(T, \mu)$

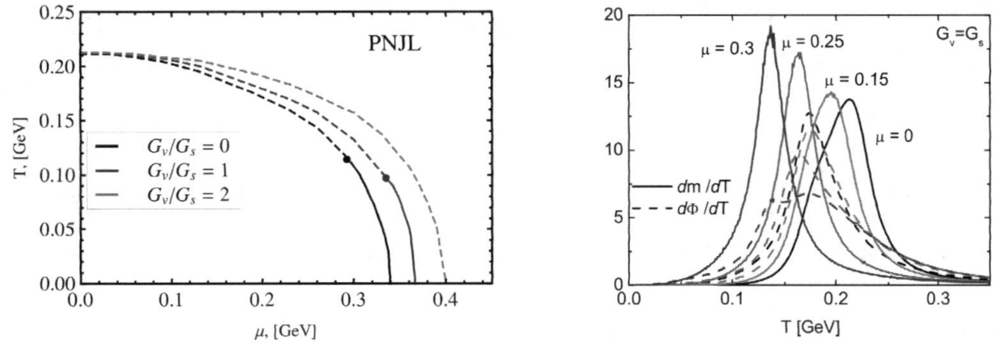


**Figure 1.** Left panel: Phase diagram of the PNJL model with vector interaction. Right panel: temperature dependence of the order parameters for various chemical potential in the PNJL model with $\tilde{G}_{\rm v} = \tilde{G}_{\rm s}$.

plane is presented in Fig. 1 (left panel). The phase diagram of the PNJL model at $G_{\rm v}/G_s = 0$

represents the modern view on the QCD phase diagram structure: there is a crossover transition at high temperature (low density) and a first order transition at low temperature (high density), which join in a critical endpoint. It was obtained that for the vector coupling $G_\mathrm{v}/G_s > 1.5$ the first order transition region disappears completely and there exists a crossover transition only.

Lattice calculations found that the temperature of the chiral transition for $N_f = 2$ at low $\mu$ is $T_c^{\mathrm{lat}} = 0.173 \pm 8$ GeV. However, in the PNJL calculation $T_c$ is higher, and $T_c$ depends on the parameter of the effective potential $T_0 = 0.27$ GeV (the temperature of the deconfinement transition in the pure gauge sector). When $T_0$ is renormalized to reach agreement with the lattice results, the chiral phase transition (the order parameter is the mass $m$) and the deconfinement transition (the order parameter is the Polyakov loop field $\Phi$) take place at different temperatures (see Fig. 1, right panel), while the lattice results show their coincidence. It was suggested [5] that there exists an entanglement between the chiral and deconfinement transitions. Therefore, the renormalized scalar and vector four-quark couplings can be introduced as functions of the Polyakov loop field [4, 6]. The phase diagram for the EPNJL model is shown in Fig. 2. In the

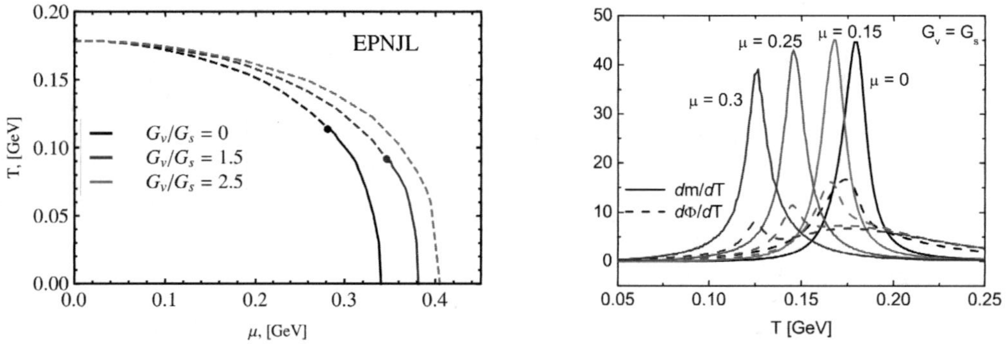


**Figure 2.** Left panel: Phase diagram of the EPNJL model with vector interaction. Right panel: temperature dependence of the order parameters for various chemical potentials in the EPNJL model with $\tilde{G}_\mathrm{v} = \tilde{G}_\mathrm{s}$.

case of the EPNJL model the critical temperature at $\mu = 0$ turns out to be the same as in the lattice calculations and the appropriate critical end-points are situated at higher temperature than in PNJL case. As in the PNJL model at $G_v \neq 0$ the first order transition region becomes smaller and then disappears. Moreover, as can be seen in the right panel of Fig 2, in the EPNJL model the temperatures of the chiral phase transition and the deconfinement transition coincide.

## Acknowledgments

We are thankful to P. Costa and M. Ilgenfritz for useful discussions. This work has been supported by the RFBR under grant No. 13-01-00060a (Yu.L.K) and by the Heisenberg-Landau and Bogoliubov-Infeld grants of JINR.

## References

[1] C. Ratti, M. A. Thaler, and W. Weise, Phys. Rev. **D 73**, 014019 (2006).
[2] K. Kashiwa, H. Kuono, M. Matsuzaki and M. Yahiro, Phys. Lett. **B 662**, 26 (2008).
[3] Y. Sakai *et al.*, Phys. Rev. **D 78**, 076007 (2008).
[4] A. V. Friesen, Yu. L. Kalinovsky, V. D. Toneev, Int. J. Mod. Phys. **A 30**, 1550089 (2015).
[5] Y. Sakai, T. Sasaki, H. Kouno, M. Yahiro, Phys. Rev. **D 82**, 076003 (2010).
[6] J. Sugano *et al.*, arXiv:1405.0103.

Journal of Physics: Conference Series **668** (2016) 012129 doi:10.1088/1742-6596/668/1/012129

# Particle correlators and possible local parity violation in nuclear collisions

**V Okorokov and P Parfenov**

National Research Nuclear University "MEPhI" (Moscow Engineering Physics Institute), Kashirskoe shosse, 31 Moscow, 115409, Russia

E-mail: VAOkorokov@mephi.ru

**Abstract.** The complex topology structure of the vacuum of quantum chromodynamics (QCD) can lead to local violation of the $P/CP$invariance of strong interactions. The preliminary estimations of magnetic field and correlators within the model of chiral magnetic effect are presented for types of nuclei and collision energies corresponded to RHIC and the LHC beams. These correlators characterized the asymmetry of the emission of charged particles with respect to the reaction plane in non-central nucleus-nucleus collisions are calculated for two various nuclear densities, namely, for approach of rigid sphere and for two-component Fermi model. The centrality dependence is under consideration for correlators integrated over other kinematic parameters. The correlator magnitudes depend on collision energy weakly for $Cu + Cu$ and $Au + Au$ collisions. The same-charge correlator shows noticeable increasing with growth of atomic number for symmetric collisions. Also magnetic field and centrality dependence of correlator are obtained for asymmetric $Cu + Au$ collisions.

## 1. Introduction

The Lagrangian of QCD contains natural terms that can break the $P/CP$ symmetry [1]. But the experimental data show us the global $P/CP$ symmetry conservation in the strong interactions that means the $P/CP$ breaking terms should be close to zero (fine-tuned problem). There is a fundamental connection between geometry and fundamental properties of QCD Lagrangian. The vacuum of QCD is a very complicated matter with rich structure, which can corresponds to the fractal-like geometry [2]. At zero temperature, transition between metastable domains of QCD vacuum which possess of various properties with respect to the $P/CP$ symmetries is the tunneling through a potential barrier [3] but at finite temperatures such transitions can be classical called also sphalerons [4]. In the last case, the temperature of the physical system should be larger than the typical energy of sphaleron in QCD $\Lambda_{\text{QCD}} \approx 0.2$ GeV [5]. In the presence of background Abelian electromagnetic field the sphalerons and / or decays of metastable domains of QCD vacuum can lead to the separation of secondary charged particles toward the magnetic field. This phenomenon called also chiral magnetic effect (CME) [6] is the experimental manifestation of the local topologically induced $P/CP$ parity violation in the strong interactions.

## 2. Magnetic field in heavy-ion collisions

The magnetic field for nucleus-nucleus collisions can be obtained, for instance, using the Lienard–Wiechert potentials. This approach is described in detail elsewhere [6]. In the paper the

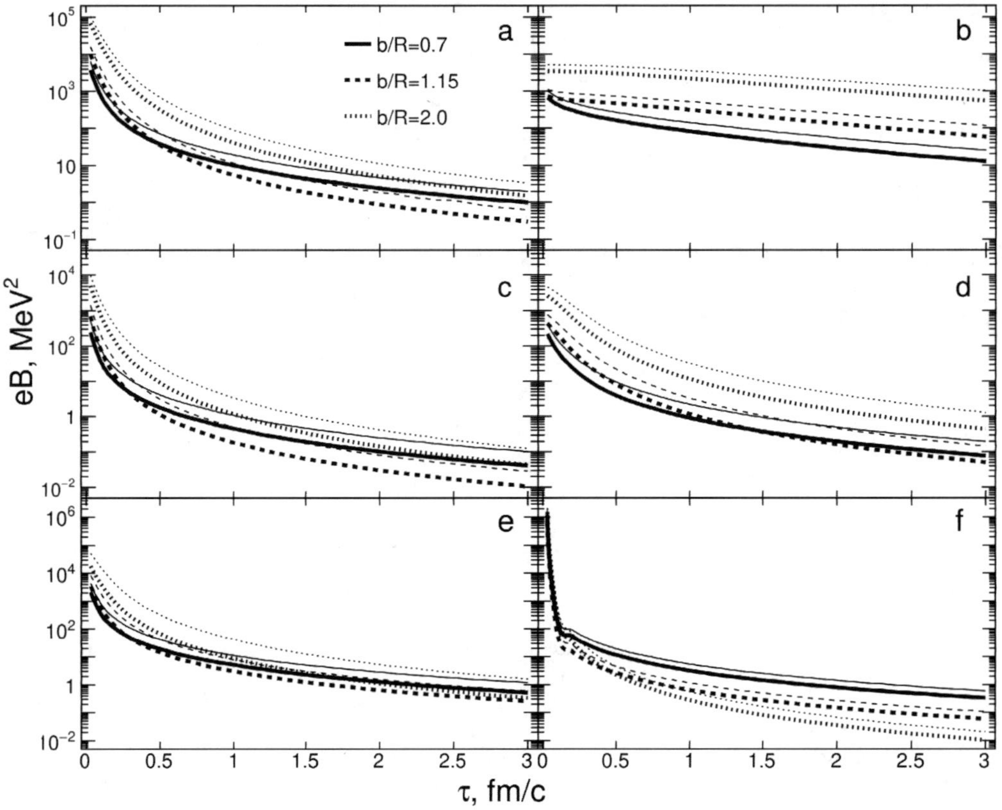


**Figure 1.** Dependence of the strength of magnetic field on proper time in nucleus-nucleus collisions at the origin ($x'_\perp = 0$) and central pseudorapidity ($\eta = 0$): $Au+Au$ at $\sqrt{s_{NN}} = 200$(a) and 7.7 GeV (b); $Cu+Cu$ at $\sqrt{s_{NN}} = 200$(c) and 62.4 GeV (d); $Cu+Au$ at $\sqrt{s_{NN}} = 200$ GeV (e), $Pb+Pb$ at $\sqrt{s_{NN}} = 2.76$ TeV. The solid lines show $eB(\tau)$ for central collisions, dashed lines – for mid-central, dotted lines – for peripheral. Thin lines correspond to the limit case of the rigid sphere, thick lines – to the Fermi parametrization.

strength of the magnetic field ($eB$) in collisions of various ion beams are compared for two parameterizations of nuclear density, namely, the sphere with sharp boundary, i.e. rigid sphere [6] and the following equation within the framework of two-component Fermi model

$$\rho_i^{\mathrm{F}}(\mathbf{x}'_\perp) \propto \varepsilon[1+\exp(-\varepsilon/a)]^{-1}\theta_i(\varepsilon^2), \quad \varepsilon(\mathbf{x}'_\perp) \equiv \sqrt{R_i^2 - [\mathbf{x}'_\perp \pm \mathbf{b}/2 \mp (R_2^2 - R_1^2)\mathbf{e}_\perp/2b]^2}, \quad (1)$$

Here $b$ is the impact parameter, $R_i$ – the radius of the nucleus moved along of the $z$-axis in the positive ($i = 1$) and negative ($i = 2$) directions, $a$ is the cutoff factor, $\kappa_i \equiv R_i/a$, the nuclear densities are normalized to unit as well in [6]. It would be stressed that the Eq. (1) is valid for asymmetric, in general case, nucleus-nucleus collisions. The strength of Abelian magnetic field is studied for two parameterizations of nuclear density for wide set of nucleus-nucleus collisions at RHIC and the LHC energies ($\sqrt{s_{NN}}$), namely, for $Cu+Cu$ at $\sqrt{s_{NN}} = 22.6, 62.4$ and 200 GeV, $Cu+Au$ at $\sqrt{s_{NN}} = 200$ GeV, $Au+Au$ at $\sqrt{s_{NN}} = 7.7-200$ GeV, $Pb+Pb$ at $\sqrt{s_{NN}} = 2.76$ TeV and $U+U$ at $\sqrt{s_{NN}} = 193$ GeV. The preliminary $eB$ dependence on proper time ($\tau$) obtained for several collision types and $\sqrt{s_{NN}}$ is shown in Fig. 1 for central ($b/R = 0.7$), mid-central ($b/R = 1.15$) and peripheral ($b/R = 2.0$) interactions. At large $\tau$ as expected in accordance with

Journal of Physics: Conference Series **668** (2016) 012129 doi:10.1088/1742-6596/668/1/012129

[7] the $eB$ increases significantly with the decrease of the collision energy in both the $Au + Au$ (Fig. 1a, b) and the $Cu + Cu$ (Fig. 1c, d) with opposite behaviour of peak $eB$ values at very small $\tau$. The $eB$ shows the much more sharper decrease for $Pb + Pb$ collisions at the LHC energy for small $\tau$ (Fig. 1f) than that for highest RHIC energy $\sqrt{s_{NN}} = 200$ GeV. As seen the transition from the simplest case of the rigid sphere to the more realistic Fermi distribution for nuclear density leads to some decrease of the strength of magnetic field especially for RHIC energies.

## 3. Correlators of the charged particles

In the framework of CME model the theoretical correlator is given by [7, 8]

$$\langle \mathbf{K}^{\mathrm{T}}_{\alpha\beta} \rangle = \frac{\pi^2}{16} \frac{\langle \Delta_\alpha \Delta_\beta \rangle}{\langle N_\alpha N_\beta \rangle}, \quad \langle \Delta_\alpha \Delta_\beta \rangle = \kappa \alpha_S \left( \frac{\pi R}{2} \sum_f q_f^2 \right)^2 \varepsilon_{\alpha\beta} \int_{S_\perp} d^2 x_\perp \xi_{\alpha\beta}(x_\perp) \int_{\tau_i}^{\tau_f} d\tau \tau \left[ eB(\eta, \tau) \right]^2 . \quad (2)$$

Here $\varepsilon_{\pm\pm} = -0.5\varepsilon_{\pm\mp} = 0.5$, $\xi_{\pm\pm}(x_\perp) \equiv \sum_{i=+,-} \xi_i^2(x_\perp, \lambda)$, $\xi_{\pm\mp}(x_\perp, \lambda) \equiv \prod_{i=+,-} \xi_i(x_\perp, \lambda)$, $\alpha_{\mathrm{S}}$ – renormalized strong coupling constant, $N_\alpha$ – multiplicity of particles with electric charge sign $\alpha$ in event, $q_f$ – electric charge (in units of $e$) of quark with flavor $f$, $\Delta_\alpha$ is the difference between total charges with fixed sign $\alpha$ on each side of the reaction plane, numerical coe cient $\kappa \sim 1$, $x_\perp$ – position of nucleon from one of the colliding nuclei in the plane perpendicular to the beam axis, $S_\perp$ – area of overlap region of colliding nuclei in the transverse plane, functions $\xi_\pm(x_\perp, \lambda)$ take into account the suppression effect via screening length $\lambda$ [6]. The correlators $\langle \Delta_\alpha \Delta_\beta \rangle$ are calculated for set of collision types and energies indicated above. The variation of the type of nucleus and $\sqrt{s_{NN}}$ allow the investigation of the energy and atomic number dependencies of $\langle \Delta_\alpha \Delta_\beta \rangle$ as well as the influence of both the parametrization of nuclear density and the $\lambda$ on correlator values. The preliminary results are shown in Figs. 2 and 3.

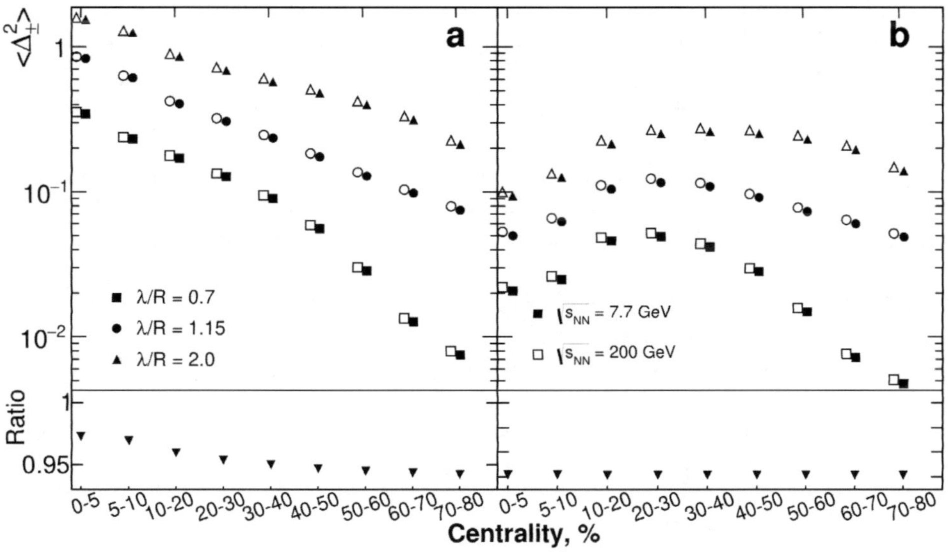


**Figure 2.** Centrality dependence of correlators for same-charged particle pairs in $Au + Au$ collisions for rigid sphere (a) and Fermi parametrization (b) for nuclear density at different $\lambda/R$ and initial energies. The low panels show the ratio of correlator values at $\sqrt{s_{NN}} = 200$ GeV to the $\langle \Delta_\alpha \Delta_\beta \rangle$ at $\sqrt{s_{NN}} = 7.7$ GeV. Due to the structure of correlators this ratio is independent on $\lambda$.

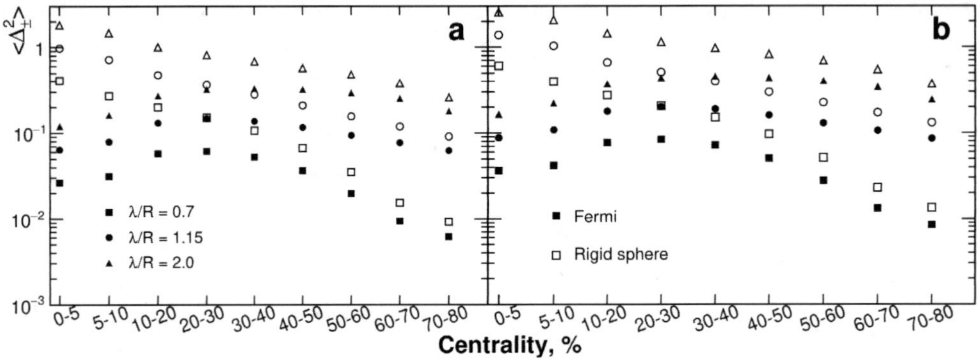


**Figure 3.** Centrality dependence of the correlator $\langle\Delta^2_\pm\rangle$ for $Pb + Pb$ collisions at $\sqrt{s_{NN}}$ = 2.76 TeV (a) and for $U + U$ at $\sqrt{s_{NN}}$ = 193 GeV (b) for various screening lengths and parameterizations for nuclear density.

The larger $\lambda$ leads to noticeable increase of same-charge correlator $\langle\Delta^2_\pm\rangle$ in $Au+Au$ collisions at some fixed $\sqrt{s_{NN}}$ (Fig. 2). On the other hand as seen in Fig. 2 the values of $\langle\Delta^2_\pm\rangle$ depend weakly on collision energy for wide range of $\sqrt{s_{NN}} = 7.7-200$ GeV. It should be noted that the growth of multiplicity of secondary particles with increase of $\sqrt{s_{NN}}$ leads to the some decrease of the magnitudes of $\langle\mathbf{K}^{\mathrm{T}}_{\alpha\beta}\rangle$ as well as of corresponding experimental observables for higher energy nucleus-nucleus collisions. Therefore the energy dependence of $\langle\Delta^2_\pm\rangle$ obtained here agrees with expectations from [6, 7, 8]. In general the correlator values for same-charged particle pairs are smaller for Fermi parametrization than that for approach of rigid sphere at corresponding centralities. But the type of parametrization for nuclear density influences on the $\langle\Delta^2_\pm\rangle$ mostly for central collisions and this influence decreases for more peripheral collisions. The correlator values change weakly with increasing of atomic number (Fig. 3) at least for same-charge pair in heavy ion ($Pb + Pb$ and $U + U$) collisions.

## 4. Conclusion

The strength of magnetic field in nucleus-nucleus collisions as well as multiplicity-independent correlators within the framework of the chiral magnetic effect are calculated for two parameterizations of nuclear density, namely, for rigid sphere and for two-component Fermi model. The calculations are made for wide set of beam types and collision energies. The more realistic Fermi distribution leads to some decrease of the strength of magnetic field especially for RHIC energies. For the first time the estimations of the magnetic field are calculated in asymmetric $Cu + Au$ collisions. The correlator magnitudes depend on collision energy weakly for $Cu+Cu$ and $Au+Au$. Comparison of correlator values with experimental observables is in the progress.

## References

[1] Olive K A et al. 2014 *Chin. Phys.* C **38** 090001
[2] Okorokov V A and Sandrakova E V 2013 *Univ. J. Phys. Appl.* **1** 196
[3] t Hooft G 1976 *Phys. Rev. Lett.* **37** 8; *Phys. Rev.* D **14** 3432 [Erratum. 1978 *ibid* **18** 2199]
[4] Lee T D 1973 *Phys. Rev.* D **8** 1226
[5] Adler S L 1969 *Phys. Rev.* **177** 2246
[6] Kharzeev D E, McLerran L D and Warringa H J 2008 *Nucl. Phys.* A **803** 228
[7] Okorokov V A 2009 *arXiv: 0908.2522 [nucl-th]*; 2013 *Yad. Fiz. and Engin.* **4** 805
[8] Okorokov V A 2013 *Int. J. Mod. Phys.* E **22** 1350041

15th International Conference on Strangeness in Quark Matter (SQM2015) IOP Publishing
Journal of Physics: Conference Series **668** (2016) 012130 doi:10.1088/1742-6596/668/1/012130

# Possible effect of mixed phase and deconfinement upon spin correlations in the $\Lambda\bar{\Lambda}$ pairs generated in relativistic heavy-ion collisions

**Valery V Lyuboshitz**[1] **and Vladimir L Lyuboshitz**

Joint Institute for Nuclear Research, 141980 Dubna, Moscow Region, Russia

E-mail: `Valery.Lyuboshitz@jinr.ru`

**Abstract.** Spin correlations for the $\Lambda\Lambda$ and $\Lambda\bar{\Lambda}$ pairs, generated in relativistic heavy-ion collisions, and related angular correlations at the joint registration of hadronic decays of two hyperons, in which space parity is not conserved, are analyzed. The correlation tensor components can be derived from the double angular distribution of products of two decays by the method of "moments" . The properties of the "trace" of the correlation tensor ( a sum of three diagonal components ), determining the relative fractions of the triplet states and singlet state of respective pairs, are discussed. Spin correlations for two identical particles ($\Lambda\Lambda$) and two non-identical particles ($\Lambda\bar{\Lambda}$) are considered from the viewpoint of the conventional model of one-particle sources. In the framework of this model, correlations vanish at sufficiently large relative momenta. However, under these conditions, in the case of two non-identical particles ($\Lambda\bar{\Lambda}$) a noticeable role is played by two-particle annihilation ( two-quark, two-gluon ) sources, which lead to the difference of the correlation tensor from zero. In particular, such a situation may arise when the system passes through the "mixed phase".

## 1. General structure of the spin density matrix of the pairs $\Lambda\Lambda$ and $\Lambda\bar{\Lambda}$

Spin correlations for $\Lambda\Lambda$ and $\Lambda\bar{\Lambda}$ pairs, generated in relativistic heavy-ion collisions, and respective angular correlations at joint registration of space-parity nonconserving hadronic decays of two hyperons give important information on the character of multiple processes .

The spin density matrix of the $\Lambda\Lambda$ and $\Lambda\bar{\Lambda}$ pairs, just as the spin density matrix of two spin-1/2 particles in general, can be presented in the following form [1,2,3]:

$$\hat{\rho}^{(1,2)} = \frac{1}{4}\left[\hat{I}^{(1)} \otimes \hat{I}^{(2)} + (\hat{\boldsymbol{\sigma}}^{(1)}\mathbf{P}_1) \otimes \hat{I}^{(2)} + \hat{I}^{(1)} \otimes (\hat{\boldsymbol{\sigma}}^{(2)}\mathbf{P}_2) + \sum_{i=1}^{3}\sum_{k=1}^{3} T_{ik}\hat{\sigma}_i^{(1)} \otimes \hat{\sigma}_k^{(2)}\right]; \qquad (1)$$

in doing so, $tr_{(1,2)}\hat{\rho}^{(1,2)} = 1$.

Here $\hat{I}$ is the two-row unit matrix, $\hat{\boldsymbol{\sigma}} = (\ \hat{\sigma}_x,\ \hat{\sigma}_y,\ \hat{\sigma}_z\ )$ is the vector Pauli operator ( $x, y, z \to 1, 2, 3$ ), $\mathbf{P}_1$ and $\mathbf{P}_2$ are the polarization vectors of first and second particle ( $\mathbf{P}_1 = \langle\hat{\boldsymbol{\sigma}}^{(1)}\rangle$, $\mathbf{P}_2 = \langle\hat{\boldsymbol{\sigma}}^{(2)}\rangle$ ), $T_{ik} = \langle\hat{\sigma}_i^{(1)} \otimes \hat{\sigma}_k^{(2)}\rangle$ are the correlation tensor components. In the general case $T_{ik} \neq P_{1i}P_{2k}$. The tensor with components $C_{ik} = T_{ik} - P_{1i}P_{2k}$ describes the spin correlations of two particles .

[1] Speaker at SQM 2015

## 2. Spin correlations and angular correlations at joint registration of decays of two $\Lambda$ particles into the channel $\Lambda \to p + \pi^-$

It is essential that any decay of an unstable particle may serve as an analyzer of its spin state .

The normalized angular distribution at the decay $\Lambda \to p + \pi^-$ takes the form:

$$\frac{d\,w(\mathbf{n})}{d\,\Omega_{\mathbf{n}}} = \frac{1}{4\,\pi}(1 + \alpha_\Lambda \mathbf{P}_\Lambda\,\mathbf{n}). \quad (2)$$

Here $\mathbf{P}_\Lambda$ is the polarization vector of the $\Lambda$ particle, $\mathbf{n}$ is the unit vector along the direction of proton momentum in the rest frame of the $\Lambda$ particle, $\alpha_\Lambda$ is the coefficient of $P$-odd angular asymmetry ( $\alpha_\Lambda = 0.642$ ). The decay $\Lambda \to p + \pi^-$ selects the projections of spin of the $\Lambda$ particle onto the direction of proton momentum; the analyzing power equals $\boldsymbol{\xi} = \alpha_\Lambda \mathbf{n}$ .

Now let us consider the double angular distribution of flight directions for protons formed in the decays of two $\Lambda$ particles into the channel $\Lambda \to p + \pi^-$, normalized by unity ( the analyzing powers are $\boldsymbol{\xi}_1 = \alpha_\Lambda \mathbf{n}_1$, $\boldsymbol{\xi}_2 = \alpha_\Lambda \mathbf{n}_2$ ). It is described by the following formula [2,3]:

$$\frac{d^2\,w(\mathbf{n}_1,\mathbf{n}_2)}{d\,\Omega_{\mathbf{n}_1}\,d\,\Omega_{\mathbf{n}_2}} = \frac{1}{16\,\pi^2}\left[1 + \alpha_\Lambda \mathbf{P}_1\,\mathbf{n}_1 + \alpha_\Lambda \mathbf{P}_2\,\mathbf{n}_2 + \alpha_\Lambda^2 \sum_{i=1}^{3}\sum_{k=1}^{3} T_{ik} n_{1i} n_{2k}\right], \quad (3)$$

where $\mathbf{P}_1$ and $\mathbf{P}_2$ are polarization vectors of the first and second $\Lambda$ particle, $T_{ik}$ are the correlation tensor components, $\mathbf{n}_1$ and $\mathbf{n}_2$ are unit vectors in the respective rest frames of the first and second $\Lambda$ particle, defined in the common ( unified ) coordinate axes of the c.m. frame of the pair ( $i, k = \{1,2,3\} = \{x,y,z\}$ ) .

The polarization parameters can be determined from the angular distribution of decay products by the method of moments [2,3] .

The angular correlation, integrated over all angles except the angle between the vectors $\mathbf{n}_1$ and $\mathbf{n}_2$ and described by the formula [2,3,4,5]

$$d\,w(\cos\theta) = \frac{1}{2}\,(\,1 + \frac{1}{3}\,\alpha_\Lambda^2 T\,\cos\theta)\sin\theta d\,\theta = \frac{1}{2}\,[\,1 - \alpha_\Lambda^2\,(W_s - \frac{W_t}{3})\cos\theta]\sin\theta d\,\theta, \quad (4)$$

is determined only by the "trace" of the correlation tensor $T = W_t - 3W_s$ ( $W_s$ and $W_t$ are relative fractions of the singlet state and triplet states, respectively ), and it does not depend on the polarization vectors ( single-particle states may be unpolarized ).

## 3. Correlations at the joint registration of the decays $\Lambda \to p + \pi^-$ and $\bar{\Lambda} \to \bar{p} + \pi^+$

Due to CP invariance, the coefficients of $P$-odd angular asymmetry for the decays $\Lambda \to p + \pi^-$ and $\bar{\Lambda} \to \bar{p} + \pi^+$ have equal absolute values and opposite signs: $\alpha_{\bar{\Lambda}} = -\alpha_\Lambda = -0.642$ . The double angular distribution for this case is as follows [2,3]:

$$\frac{d^2\,w(\mathbf{n}_1,\mathbf{n}_2)}{d\,\Omega_{\mathbf{n}_1}\,d\,\Omega_{\mathbf{n}_2}} = \frac{1}{16\,\pi^2}\left[1 + \alpha_\Lambda \mathbf{P}_\Lambda\,\mathbf{n}_1 - \alpha_\Lambda \mathbf{P}_{\bar{\Lambda}}\,\mathbf{n}_2 - \alpha_\Lambda^2 \sum_{i=1}^{3}\sum_{k=1}^{3} T_{ik} n_{1i} n_{2k}\right] \quad (5)$$

(here $-\alpha_\Lambda = +\alpha_{\bar{\Lambda}}$ and $-\alpha_\Lambda^2 = +\alpha_\Lambda \alpha_{\bar{\Lambda}}$ ) .

Thus, the angular correlation between the proton and antiproton momenta in the rest frames of the $\Lambda$ and $\bar{\Lambda}$ particles is described by the expression:

$$d\,w(\cos\theta) = \frac{1}{2}\,(\,1 - \frac{1}{3}\,\alpha_\Lambda^2 T\,\cos\theta)\sin\theta d\,\theta = \frac{1}{2}\,[\,1 + \alpha_\Lambda^2\,(W_s - \frac{W_t}{3})\cos\theta]\sin\theta d\,\theta, \quad (6)$$

where $\theta$ is the angle between the proton and antiproton momenta .

## 4. Spin correlations at the generation of $\Lambda\bar{\Lambda}$ pairs in multiple processes

Further we will use the model of one-particle sources [6], which is the most adequate one in the case of collisions of relativistic ions .

Two $\Lambda$ particles are identical particles. Spin and angular correlations at their decays, taking into account Fermi statistics and final-state interaction, were considered previously in the works [2,7].

We will be interested in spin correlations at the decays of $\Lambda\bar{\Lambda}$ pairs . In the framework of the model of independent one-particle sources, spin correlations in the $\Lambda\bar{\Lambda}$ system arise only on account of the difference between the interaction in the final triplet state ( $S = 1$ ) and the interaction in the final singlet state . At small relative momenta, the $s$-wave interaction plays the dominant role as before, but, contrary to the case of identical particles ( $\Lambda\Lambda$ ), in the case of non-identical particles ( $\Lambda\bar{\Lambda}$ ) the total spin may take both the values $S = 1$ and $S = 0$ at the orbital momentum $L = 0$ . In doing so, the interference effect, connected with quantum statistics, is absent .

If the sources emit unpolarized particles, then, in the case under consideration, the correlation function describing momentum-energy correlations has the following structure ( in the c.m. frame of the $\Lambda\bar{\Lambda}$ pair ):

$$R(\mathbf{k},\mathbf{v}) = 1 + \frac{3}{4}\, B_t^{(\Lambda\bar{\Lambda})}(\mathbf{k},\mathbf{v}) + \frac{1}{4}\, B_s^{(\Lambda\bar{\Lambda})}(\mathbf{k},\mathbf{v}). \tag{7}$$

The spin density matrix of the $\Lambda\bar{\Lambda}$ pair is given by the formula :

$$\hat{\rho}^{(\Lambda\bar{\Lambda})} = \hat{I}^{(1)} \otimes \hat{I}^{(2)} + \frac{B_t^{(\Lambda\bar{\Lambda})}(\mathbf{k},\mathbf{v}) - B_s^{(\Lambda\bar{\Lambda})}(\mathbf{k},\mathbf{v})}{4\, R(\mathbf{k},\mathbf{v})}\, \hat{\boldsymbol{\sigma}}^{(1)} \otimes \hat{\boldsymbol{\sigma}}^{(2)}, \tag{8}$$

and the components of the correlation tensor are as follows:

$$T_{ik} = \frac{B_t^{(\Lambda\bar{\Lambda})}(\mathbf{k},\mathbf{v}) - B_s^{(\Lambda\bar{\Lambda})}(\mathbf{k},\mathbf{v})}{4 + 3\, B_t^{(\Lambda\bar{\Lambda})}(\mathbf{k},\mathbf{v}) + B_s^{(\Lambda\bar{\Lambda})}(\mathbf{k},\mathbf{v})}\, \delta_{ik}; \tag{9}$$

here the contributions of final-state triplet and singlet $\Lambda\bar{\Lambda}$ interaction are determined by the expression obtained in the works [2,7] .

At sufficiently large values of $k$, one should expect that [7]:

$$B_s^{(\Lambda\bar{\Lambda})}(\mathbf{k},\mathbf{v}) = 0, \qquad B_t^{(\Lambda\bar{\Lambda})}(\mathbf{k},\mathbf{v}) = 0.$$

In this case the angular correlations in the decays $\Lambda \to p + \pi^-$, $\bar{\Lambda} \to \bar{p} + \pi^+$, connected with the final-state interaction, are absent :

$$T_{ik} = 0, \quad T = 0.$$

## 5. Angular correlations in the decays $\Lambda \to p + \pi^-$ and $\bar{\Lambda} \to \bar{p} + \pi^+$ and the "mixed phase"

Thus, at sufficiently large relative momenta (for example, $k \gtrsim m_\pi$ ) one should expect that the angular correlations in the decays $\Lambda \to p + \pi^-$ and $\bar{\Lambda} \to \bar{p} + \pi^+$ , connected with the interaction of the $\Lambda$ and $\bar{\Lambda}$ hyperons in the final state ( i.e. with one-particle sources ) are absent . But, if at the considered energy the dynamical trajectory of the system passes through the so-called "mixed phase", then the two-particle sources, consisting of the free quark and antiquark , start playing a noticeable role . For example, the process $s\,\bar{s} \to \Lambda\bar{\Lambda}$ may be discussed .

In this process, the charge parity of the pairs $s\bar{s}$ and $\Lambda\bar{\Lambda}$ is equal to $C = (-1)^{L+S}$, where $L$ is the orbital momentum and $S$ is the total spin of the fermion and antifermion. Meantime, the $CP$ parity of the fermion-antifermion pair is $CP = (-1)^{S+1}$ .

In the case of one-gluon exchange, $CP = 1$, and then $S = 1$, i.e. the $\Lambda\bar{\Lambda}$ pair is generated in the triplet state; in doing so, the "trace" of the correlation tensor $T = 1$ .

Even if the frames of one-gluon exchange are overstepped, the quarks $s$ and $\bar{s}$, being ultrarelativistic, interact in the triplet state ( $S = 1$ ) . In so doing, the primary $CP$ parity $CP = 1$, and, due to the $CP$ parity conservation, the $\Lambda\bar{\Lambda}$ pair is also produced in the triplet state. Let us denote the contribution of two-quark sources by $x$ . Then at large relative momenta

$$T = x > 0.$$

Apart from the two-quark sources, there are also two-gluon sources being able to play a comparable role. Analogously with the annihilation process $\gamma\gamma \to e^+e^-$, in this case the "trace" of the correlation tensor is described by the formula ( the process $g\,g \to \Lambda\bar{\Lambda}$ is implied ):

$$T = 1 - \frac{4\,(1-\beta^2)}{1 + 2\,\beta^2\,\sin^2\theta - \beta^4 - \beta^4\,\sin^4\theta}, \tag{10}$$

where $\beta$ is the velocity of $\Lambda$ ( and $\bar{\Lambda}$) in the c.m. frame of the $\Lambda\bar{\Lambda}$ pair, $\theta$ is the angle between the momenta of one of the gluons and $\Lambda$ in the c.m. frame (see [8]). At small $\beta$ ( $\beta \ll 1$ ) the $\Lambda\bar{\Lambda}$ pair is produced in the singlet state ( total spin $S = 0$, $T = -3$ ), whereas at $\beta \approx 1$ – in the triplet state ( $S = 1$, $T = 1$ ). Let us remark that at ultrarelativistic velocities $\beta$ ( i.e. at extremely large relative momenta of $\Lambda$ and $\bar{\Lambda}$ ) both the two-quark and two-gluon mechanisms lead to the triplet state of the $\Lambda\bar{\Lambda}$ pair ( $T = 1$ ).

## 6. Summary

So, it is surely advisable to investigate the spin correlations of $\Lambda\Lambda$ and $\Lambda\bar{\Lambda}$ pairs produced in relativistic heavy ion collisions ( see also, e.g., [9] ) .

The spin correlations, as well as the momentum–energy ones, make it possible to determine the space–time characteristics of the multiple particle generation region . In doing so, the best way of studying the spin correlations is the method of angular correlations – method of moments.

Finally, it should be emphasized that, in the general case, the appearance of angular correlations in the decays $\Lambda \to p + \pi^-$ and $\bar{\Lambda} \to \bar{p} + \pi^+$ with the nonzero values of the "trace" of the correlation tensor $T$ at large relative momenta of the $\Lambda$ and $\bar{\Lambda}$ particles may testify to the passage of the system through the "mixed phase" [9] .

## References

[1] Lyuboshitz V L and Podgoretsky M I 1997 *Yad. Fiz.* **60** 45
[2] Lyuboshitz V L 2000 *Proc. of the XXXIV PNPI Winter School "Physics of Atomic Nuclei and Elementary Particles"* ( Saint-Petersburg ) p 402
[3] Lednicky R and Lyuboshitz V L 2001 *Phys. Lett.* B **508** 146
[4] Alexander G and Lipkin H J 1995 *Phys. Lett.* B **352** 162
[5] Lednicky R, Lyuboshitz V V and Lyuboshitz V L 2003 *Yad. Fiz.* **66** 1007
[6] Podgoretsky M I 1989 *Fiz. Elem. Chast. At. Yadra* **20** 628
[7] Lednicky R and Lyuboshitz V L 1982 *Yad. Fiz.* **35** 1316
[8] McMaster W H 1961 *Rev. Mod. Phys.* **33** 8
[9] Lyuboshitz V L and Lyuboshitz V V 2010 *Yad. Fiz.* **73** 836 [ *Phys. At. Nucl.* **73** 805 ]

**Institute of Physics**
Dirac House, Temple Back
Bristol BS1 6BE UK

ISSN: 1742-6588
ISBN 978-1-5108-1959-7

9 781510 819597

CURRAN ASSOCIATES INC.
proceedings.com